AF412967

CE and CEC Reviews 2001

Edited by Ziad El Rassi

CE and CEC Reviews 2001

Edited by Ziad El Rassi

Weinheim · New York · Chichester · Brisbane · Singapore · Toronto

Editor: Prof. Ziad El Rassi
Senior Deputy Editor of ELECTROPHORESIS
Department of Chemistry
Oklahoma State University
Stillwater, OK 74078-3071
USA

Library of Congress Card No.: applied for

British Library Cataloguing-in-Publication Data:
A catalogue record for this book
is available from the British Library

Die Deutsche Bibliothek – CIP-cataloguing-in-Publication Data
A catalogue record for this book is available from Die Deutsche Bibliothek
ISBN 3-527-30255-7

© WILEY-VCH Verlag GmbH, D-69469 Weinheim (Federal Republic of Germany), 2001

Printed on acid-free paper

Composition and Printing: Rheinhessische Druckwerkstätte, E. Dietl GmbH & Co.KG, Alzey.
Bookbinding: J. Schäffer GmbH & Co.KG, Grünstadt.
Printed in the Federal Republic of Germany.

Preface

The present book is the start of a book series that will be published on yearly basis. The aim of the book is to compile all review articles on capillary electrophoresis (CE), capillary electrochromatography (CEC) and related topics published in the journal ELECTROPHORESIS in a calendar year. The compilation of comprehensive and critical review articles in a book format is expected to provide the interested reader a handy reference book for advances and perspectives in CE, CEC and related topics.

This first book of the book series "CE and CEC Reviews" describes in depth various timely topics in CE and CEC by experts in the field of electrically-driven microcolumn separations. The book contains 38 review articles arranged in six major parts. Part I groups 4 reviews which (i) survey past and recent events in microslab gel electrophoresis, CE, CEC and auxiliary micromethods and (ii) describe some fundamental aspects concerning DNA electrophoresis and dispersive phenomena in electromigration separation methods. Part II consists of 6 review articles describing recent progress in electromigration techniques and their methodologies and applications. These include capillary isotachophoresis and its application to food analysis, CEC, capillary affinity electrophoresis, CE-based immunoassays and CE of metal ions. Microfabrication, miniaturization, capillary array electrophoresis and ultra-thin-layer gel electrophoresis form the subject of Part III of this book, which has six different review articles. Part IV with its 7 consecutive articles is dedicated to (i) advances in stationary phases and pseudo-stationary phases in CEC and electrokinetic capillary chromatography, respectively, (ii) progress in polymeric media for separation of rigid, submicron-sized particles and for DNA sequencing, (iii) isoelectric buffers for CE of peptides and proteins and (iv) nonaqueous CE. Chiral separation principles in CE and CEC and their applications are the topics of 5 different review articles, which make up the bulk of Part V of this book. Finally, Part VI groups 10 different reviews on sample derivatization and detection issues, which are the most challenging aspects ofcapillary electroseparation methods. The relatively high theoretical plate numbers that are readily generated by CE and CEC are meaningless unless accompanied by high detection sensitivity systems and sensitivity enhancement schemes specially when dealing with analytes at trace levels. In this regard, efficient derivatization schemes (review article No. 29) and on-line pre-concentration approaches (Reviews Nos. 30–32) are important sample manipulation steps for sensitivity enhancement by CE and CEC separations. Review No. 33 discusses all kinds of detection systems suitable after CE and CEC separations, while Reviews 34–38 describe in details advances made in specific detection systems and applications.

I would like to thank all the contributors for their fine articles which will certainly impact the area of electrically-driven microcolumn separations by providing a better understanding of the underlying phenomena and by enlarging the scope of their applications. Also, I wish to acknowledge the financial support of my research program at Oklahoma State University by the United States Department of Agriculture/National Research Initiative Competitive Grants Program.

Stillwater, December 2000

Ziad el Rassi
Editor

Notice to Readers

Whenever citing an article from this book please refer to the original journal's citation which is given for each article in the table of Contents. Further to this and for the same reason the original journal's pagination as given in the column titles of each page has also remained unchanged. A consecutive pagination of the book pages is to be found at the bottom of each page; author's and subject index are referring to those page numbers.

Thank you!

Contents

Overviews and fundamental aspects

Techniques, methodologies and applications

Microfabrication, miniaturization and capillary array electrophoresis

Electrophoresis 2000, *21*, 3–11

3

Retrospect

Volker Neuhoff

Max-Planck-Institut für
experimentelle Medizin,
Göttingen, Germany

Microelectrophoresis and auxilary micromethods

In this retrospect of approximately 30 years of work with micromethods, some of them developed in our own laboratory, their principles and application to different separation problems are described, such as one- and two-dimensional microelectrophoresis in capillaries and microslab gels, isoelectric focusing in capillaries or microslab gels, microchromatography, microphotometry, and microfluorometry for qualitative and quantitative evaluation of separation patterns. In addition, some useful auxiliary methods are also described, *e.g.*, a method for quantitative protein determination in a microliter volume when neither the volume nor the protein content in that volume are known, and methods for the determination of glycoproteins, amino acids, and sugars in the picomole range.

Keywords: Microelectrophoresis / Micromethods / Microchromatography / Microphotometry / Microfluorometry / Retrospect   EL 3538

Contents

1 Introduction

The development of microelectrophoresis as well as of some auxiliary micromethods was dictated by the need to

analyze samples of biological origin, available in only minute quantities. Microscale polyacrylamide gel electrophoresis was first described in 1964 by Pun and Lombrozo [1] for the fractionation of brain proteins. In 1965, Grossbach [2] used 5 µL Drummond microcaps for this technique. The separation volume was further reduced in 1966 by Hydén *et al.* [3], and McEwen and Hydén [4], who used 2 µL capillaries for the fractionation of brain proteins. In 1968 Neuhoff [5] introduced polyacrylamide gel optimized for the microfractionation of water-soluble brain proteins. Such gels were later found to be suitable for a variety of separations. Hydén and Lange [6] used microdisc electrophoresis for the analysis of protein changes in different brain areas as a function of intermittent training. Griffith and LaVelle [7] analyzed the developmental protein changes in facial nerve nuclear regions by this method. Ansorg, Dames and Neuhoff [8] used microdisc electrophoresis to study the effect of different extraction procedures on the pattern of brain proteins, and Althaus *et al.* [9] used the method for the analysis of water-soluble proteins following post-tetanic potentiation of the monosynaptic reflexes in cat spinal cords. Grossbach [10] used microdisc electrophoresis for the analysis of chromosomal activity in the salivary glands of *Camtochirononmus*. These are a few examples of early applications of microelectrophoresis of proteins. Later this micromethod proved most suitable for a plethora of applications in the wide field of biochemistry.

With a little experience, disc electrophoresis on the microscale is hardly more difficult than the conventional technique. It requires only a small amount of equipment, well corrected vision, and steady hands. In addition to reduced sample equipment, it has the advantage of giving results

---

Correspondence: Prof. (em.) Dr. Volker Neuhoff, Wilhelmstr. 11, D-31582 Nienburg, Germany

                    1   0173-0835/00/0101-0003 $17.50+.50/0

in good agreement with those obtained by the macro-method, however, in appreciably shorter time. The lower detection limit for a single protein band in a 5 μL gel with 450 μM ID is 10^{-9} g of albumin, following amido black 10B staining, 0.1–0.2 ng with silver staining [11], and 0.1 ng using the background-free colloidal staining with Coomassie brilliant blue according to Neuhoff *et al.* [12, 13]. Microinterferometric protein determination in separated fractions in the nanogram range was described by Hydén and Lange [14]. For 0.1–0.5 μg total protein in a mixture, fractionated by electrophoresis in 5 μL capillaries, this allows 5000–10000 estimations to be performed with 1 mL of a solution containing 1 mg protein.

2 Microdisc electrophoresis

2.1 Electrophoresis in homogeneous capillary gels

The performance of these micromethods has been described in full detail previously [15] and therefore only the main features of the technique will be outlined here. Generally, 5 or 10 μL capillaries (Drummond Microcaps, Brand Intraend) were used with 0.42 and 0.60 mm ID allowing easy sample application with the aid of capillary pipettes. Capillaries of 2 μL or even 0.5 μL could also be used freehand, without special equipment, but for sample application (0.1 or 0.05 mm ID [2, 16]), a micromanipulator as well as a stereomicroscope were required. Batch-wise cleaning of the capillaries prior to use [15] was necessary to facilitate capillary attraction for filling, simply by dipping into a suitable polymerization mixture. Homogeneous gels were prepared by filling to approximately two-thirds of their total volume, pressing into a plasticine cushion covered with Parafilm, and careful overlayering with water. In a moist chamber polymerization normally required 1 h. As in the standard technique, this was followed by rapidly polymerizing a few mm of a 5% stacking gel. Gel composition as well as discontinuous or continuous buffer systems in the gel and at the electrodes could be optimized for any separation problem, *e.g.*, depending on protein size. Additives such as SDS, Triton X-100, urea, etc. could be used. Separations were run at 60–120 V and 60–120 μA per gel, under these conditions cooling was normally not necessary, owing to the advantageous surface-to-volume ratio of the capillary gels; therefore no problems due to Joule heating were created during the 20–60 min required for electrophoresis. A power supply allowing independent control of the current for each capillary [15] proved useful. After electrophoresis the gels with a low acrylamide concentration could be extracted from the capillaries with water pressure [15]. Staining was initially performed in 1% amido black 10 B in 7.5% acetic acid or Coomassie brilliant blue for only 10 min, followed

by destaining in several baths of acetic acid in approximately 30 min. Utilizing the colloidal properties of Coomassie brilliant blue G-250, Neuhoff *et al.* [12, 13] have described a protein staining with clear background at nanogram sensitivity, without destaining and stable protein fixation. This colloidal staining procedure allows for quantitative determination of stained proteins following polyacrylamide gel electrophoresis [13], and can of course also be used for gels of standard size.

2.2 Microgradient gels

Protein mixtures are generally separated with a higher resolution in gradient gels. Gradient capillary gels could be more easily prepared than homogeneous gels (with a stacking gel), since the top of the gradient gel can be adjusted to only 1 or 2% acrylamide, which is adequate for protein stacking. Microgradient gels could be prepared in capillaries of different diameter or length in batches using a special apparatus [15] or, according to Rüchel *et al.* [17], freehand with single capillaries, by using capillary attraction. For this purpose, the polymerization mixture is divided in two parts, one solution contains only ammonium peroxodisulfate as catalyst, in a suitable concentration and buffer. The second solution contains the acrylamide, *N,N'*-methylenebisacrylamide (Bis) and TEMED, again in suitable concentration and buffer. The latter solution finally determines the slope of the acrylamide concentration along the linear gradient, depending on the separation problem. The maximum concentration of acrylamide can be as high as 53%. The slope can additionally be modified by adding sucrose to the mixture [15, 17].

The acrylamide gradient was prepared by first filling (by capillary attraction with the ammonium peroxodisulfate solution) half of the capillary, followed by filling to the rim (by dipping in the correct acrylamide mixture), this polymerization mixture thereby being linearly diluted in the opposite direction. With this procedure, at the top, where the acrylamide concentration has its minimum, the catalyst starting the polymerization has its maximum. This gradient formation could easily be followed by dissolving a dye in the acrylamide solution. The linearity of the gradient could be densitometrically checked by incorporation of albumin into the acrylamide solution and staining. After filling with the gradient mixture, the capillary was immediately transferred into a small beaker with a solution of 50% sucrose and some ammonium peroxodisulfate to ensure sealing of the capillary and complete polymerization of the highly concentrated acrylamide at the lower end of the gradient [15]. The polymerized gels could be stored for prolonged periods at 4°C. Gradient gels are excellently suited for all types of SDS electrophoresis and can especially be used to demonstrate the presence of artifacts [18] in micro- as well as in macrogels if the conditions of SDS electrophoresis are not carefully controlled. The high resolving power of these microgradient gels is shown by the fact that the difference of only one nucleotide is enough for separation of tRNA species in a 1–40% microgradient gel [19].

2.3 Isoelectric focusing in capillary gels

With the development of carrier ampholytes [20, 21] it became possible to fractionate proteins according to their isoelectric point. Isoelectric focusing (IEF) was initially carried out in saccharose gradients [22], in column with a volume of 190 or 440 µL, and later in polyacrylamide gels [23, 24]. Haglund [25] has described the historical development, the theoretical foundation, and early application of this important and powerful method. Riley and Colemann [26] and Catsimpoolas [27] described a microversion of IEF in cylindrical polyacrylamide gels, requiring only 10 µg of a single protein or 200–400 µg of a protein mixture for a single fractionation. Quentin and Neuhoff [28], Grossbach [16], and Gainer [29], working independently, refined the IEF methods further for routine separations in 5 or 10 µL capillaries, with the same equipment as used for homogeneous or gradient microgels. Although only microgram amounts were required for one-dimensional IEF, for two-dimensional separations, up to 20 µg per 10 µL gel of a protein mixture could be applied without overloading effects. Micro-IEF proved excellently suited for fractionation of isoenzymes followed by enzyme activity determinations [15, 28, 30–33]. All technical details of IEF in capillary gels have been extensively described [15, 28, 30–33]. A major advantage of microelectrophoresis was the short separation time, *e.g.*, approximately 10 min for IEF. By trial and error experiments the separation conditions could easily be optimized, including current and voltage (normally 200 V). Excessive voltage was found to produce artifacts, as found, *e.g.* on fractionation of LDH isoenzymes [30, 31] (see Fig. 1).

3 One- and two-dimensional microslab gel electrophoresis

All three types of microelectrophoresis in capillaries, described in Section 2, are well suited for reproducible fractionation of complex protein mixtures. However, they have the disadvantage that a direct comparison of different samples is not possible in the same way as in slab gels. Poehling and Neuhoff [35] have adapted the widely used one-dimensional [36] and two-dimensional [37, 38] slab gel technique to a flexible and sensitive microsystem.

3.1 Preparation of microslab gels

Microgels were cast in chambers prepared from microscope slides trimmed to a size of 2.5 × 3 cm^2 or 3 × 3.5 cm^2. Approximately 1–2 mm wide plastic strips of suitable thickness were used as spacers. Gel chambers were laterally sealed by Teflon clamps, dental wax, or electrical tape. Sealing of the lower end of the chamber prior to filling with the gel solution was not necessary with spacers less than 0.25 mm as capillary forces are sufficient to keep the solution between the plates. Chambers wider than 0.25 mm were sealed by pressing them into a cushion of plasticine coated with Parafilm. For fixation of the polyacrylamide microslab gels to the glass plates (a procedure that is optimal for photometric or autoradiographic evaluation after staining and drying), the glass plates were first silanized according to Radola [39].

3.2 Homogeneous microslab gels

Homogeneous separating gels were prepared with 10, 15, or 20% acrylamide and a buffer of 375 mM Tris-HCl,

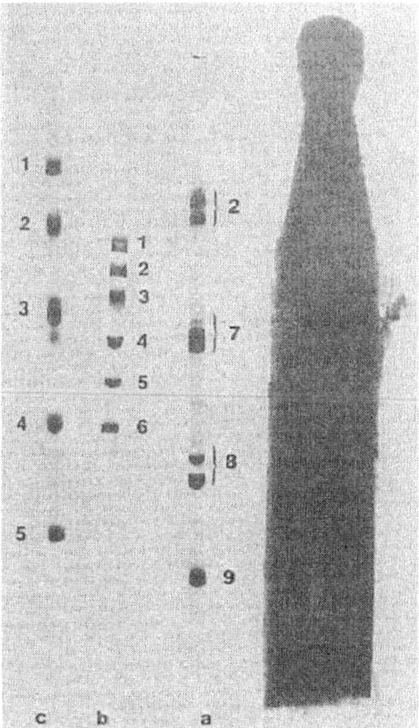

Figure 1. 10 µL gels prepared in capillaries beside a normal matchstick. (a) Isoelectric focusing of a mixture of marker proteins (2 µg total protein) in a 5% polyacrylamide gel containing 9 M urea and 2% carrier ampholytes (pH 2–11). (b) 1–30% acrylamide gradient gel with 0.6 µg of marker proteins. (c) 15% homogeneous polyacrylamide gel with 0.6 µg of marker proteins. Marker proteins: 1, phosphorylase *b*; 2, bovine serum albumin; 3, ovalbumin; 4, carbonic anhydrase; 5, soybeam trypsin inhibitor; 6, α-lactalbumin; 7, β-lactoglobulin, 8, myoglobin (horse); 9, cytochrome *c*.

pH 8.8. The stacking gels were made of 3% or 6% acrylamide and 125 mM Tris-HCl, pH 6.8. Chambers with spacers of 0.3–1 mm thickness were filled in a vertical position using fine glass pipettes while chambers with less than 0.3 mm could be filled by capillary attraction. A Teflon comb was used in the stacking gel to form the sample wells of 1 mm width and 3 mm depth (see Fig. 2). The sample, containing 15–20% glycerol and 0.001% bromophenol blue, was carefully underlayered in the sample wells filled with a suitable buffer. Sample loading was especially important in 1 mm wide wells because uneven distribution of the sample solution in the wells reduced resolution. Lateral streaking or bending in vertical one-dimensional slab gel electrophoresis could be completely avoided if a comb-gel made of electrophoresis buffer was placed on top of the stacking gel or sample gel [40]. An example of the resolution that could be achieved with a 15% homogeneous SDS microslab gel (3 × 3.5 × 0.03 cm^3) is shown in Fig. 2. High molecular weight proteins were well separated into sharp bands in this system, while lower molecular weight proteins appear more diffuse.

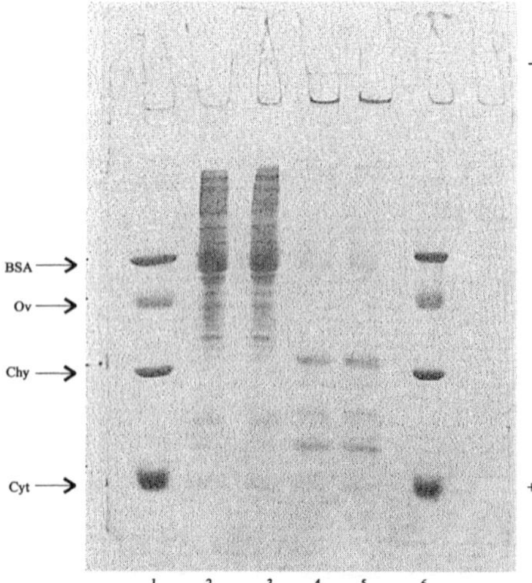

Figure 2. SDS-electrophoresis in a homogeneous polyacrylamide gel prepared in a 3 × 3.5 cm^2 chamber and swollen after staining to 3.5 × 4 cm^2. Stacking gel: 6% acrylamide, 125 mM Tris-HCl, pH 6.8, and 0.1% SDS; thickness of the gel, 0.35 mm; electrode buffer, 50 mM Tris/glycine, pH 8.4, and 0.1% SDS; electrophoresis: 60 V, 15 min, and 120 V, 50 min. Sample wells (1) and (6) are loaded with a mixture containing 0.1 µg each of bovine serum albumun (BSA), ovalbumin (Ov), chymotrypsinogen A (Chy), and cytochrome *c* (Cyt). Wells (2) and (3) are each loaded with 0.6 µg soluble rat brain proteins, and (4) and (5) with 0.3 µg rat myelin. All samples contained 1% SDS, 15% glycerol, and 0.001% bromophenol blue.

A novel multiphasic buffer system for high resolution SDS polyacrylamide gel electrophoresis in microslab gels of dansylated and nondansylated proteins/peptides in the relative molecular mass range (M_r) of 100000–1000 was described by Wiltfang *et al.* [41]. The system, allowing complete stacking and destacking of proteins/peptides, utilizes Bicine and sulfate as trailing and leading ion, respectively, and Bistris and Tris as counterions in the stacking and separating phase, respectively. Through selection of two different counterions, the stacking limits of a multiphasic buffer system can be further widened, thus making it applicable to gel electrophoresis of a larger spectrum of rapidly migrating species, such as SDS proteins/peptides and nucleic acids. Dansylated proteins/peptides were detected owing to fluorescence either directly in the gel or following electroblotting onto anion-exchange or polyvinylidene difluoride membranes, with detection sensitivity of approximately 1 ng for the latter. Nondansylated proteins/peptides were either detected within the gels by colloidal Coomassie staining [11, 12] or by electroblotting onto polyvinylidene difluoride membranes, followed by colloidal gold staining. Prior to both staining procedures the proteins/peptides were pretreated with glutardialdehyde in the presence of borate at near-neutral pH values to generate protein/peptide polymers of poor solubility. In contrast to conventional fixation procedures even small polypeptides (M_r 1000) were immobilized and approximately 15 ng and 0.75 ng could be detected after colloidal Coomassie or gold staining, respectively.

3.3 Gradient microslab gels

Gradient microslab gels can also be prepared freehand, as are the capillary gels, by using capillary forces to form convex as well as linear gradients [35]. It is, however, easier and more reproducible to use a Plexiglas chamber, prefilled to the rim with water or 0.1% SDS, for simultaneous preparation of multiple polyacrylamide gradient gels [35], which were displaced by a preformed gradient polymerization mixture slowly pumped through a central hole in the bottom of the chamber. Preparation of stacking gels, sample wells, and sample application were as described above. The reproducibility of fractionation was strongly dependent on the reproducibility of the gradient, which is difficult to achieve and requires more experience with "handmade" gradient gels. For routine use, microslab gels, 0.3–0.5 mm thick, with sample wells 0.7–1 mm wide and 3 mm deep, are optimal. A suitable sample volume is approximately 1 µL of a solution containing 1 mg/mL protein. An increase in sensitivity is achieved by using both thinner slabs and smaller sample wells to achieve concentration of the proteins into sharp band. However, the absolute detection limit is dependent on the stainability of a protein and the staining protocol. Ultrathin slab gels for macroscale separations by IEF [39, 42] have been described. These could be easily converted to the microscale. Kinzkofer and Radola [43] have described miniaturized ultrathin-layer isoelectric focusing in 20–50 μm polyacrylamide gels on a polyester backing. By the criterion of coalescence of proteins migrating from different application sites, a steady state was reached, in pH 4–9 carrier ampholyte gradients, on 1 cm (separation distance) gels after 2 min (15 Vh), on 2 cm gels after 4 min (40 Vh) and on 3 cm gels after 10 min (110–130 Vh) total focusing time, with field strengths up to 800–1000 V/cm. (Fig. 3). The miniaturized systems were found to combine short focusing time with high resolution and operational simplicity.

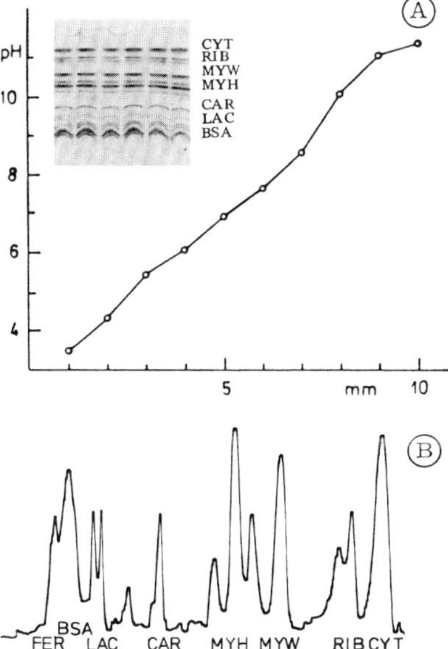

Figure 3. Miniaturized ultrathin-layer isoelectric focusing of pH marker proteins on a 1 cm separation distance. Gel layer: 20 μm, pH 4–9, Servalyt T carrier ampholytes. Prefocusing: 150 V/cm, 1 min; focusing: 800–1000 V/cm, 1 min. (A) pH Gradient. Inset: Detail of a 1 × 3 cm gel stained with Serva Violet 49. Sample application: 0.3 cm from the anode. (B) Densitometric scan of the pattern in (A), 1:20 expansion. Marker proteins: FER – ferritin (p*I* 4.4), BSA – bovine serum albumin (p*I* 4.9), LAC – β-lactoglobulin (p*I* 5.15), CAR – carbonic anhydrase (p*I* 5.9), MYH –horse myoglobin (p*I* 7.1 and 7.4), MYW – sperm whale myoglobin (p*I* 7.7 and 8.3), RIB – ribonuclease (p*I* 8.3), CYT –cytochrome c (p*I* 10.1). Reproduced from [43], with permission.

3.4 Two-dimensional microslab gel electrophoresis

For the two-dimensional separation of protein mixtures in microslab gels, IEF in microliter capillaries in the first dimension is combined with electrophoresis in microslab gels for the second dimension, the most suitable size being 3×3.5 cm^2, 0.25–0.1 mm thick. A crucial point in a two-dimensional separation is the embedding of the first-dimensional IEF on top of the slab gel. If not done properly, gel electrophoresis transfer of the proteins from the first-dimensional gel will be incomplete. When a fast-polymerizing 6% acrylamide gel, with 1% SDS in order to avoid diffusion of the separated proteins in the first-dimensional gel, is used as embedding medium, up to 33% of the total protein could be retained by the first-dimensional gel [35]. When 0.3% agarose with 1% SDS is used as the embedding medium these artifacts could be avoided.

The two-dimensional microgel system had a good resolving power, as was demonstrated, *e.g.*, for total spinal cord proteins from a 30-day-old control rat and a rat with experimental phenylketonuria [44–46]. Other examples include separations of synovial tissue, synovial fluid, and serum in patients with rheumatoid arthritis and related diseases [47]. One- and two-dimensional microelectrophoresis were successfully applied to differential diagnosis of proteinuric diseases [48]. Photometric evaluation of two-dimensional electrophoretograms has a high sensitivity of 5×10^{-4} OD, because it takes advantage of the noise characteristics of photometric signals [49]. Digital image analysis was used to locate and evaluate the individual spots on the electropherogram. This procedure of segmentation detects the faintest protein spots visible by eye and separates even confluent spots reliably and automatically. The combination of scanning photometry and image analysis is a sensitive, versatile, and accurate method for the evaluation of slab gels [49]. The process of densitometric evaluation of gels can be described as a series of message channels connected one behind the other, allowing for the basic principles of information theory to be applied [50].

4 Auxiliary micromethods

4.1 Protein determination

The quantitative determination of protein is a common prerequisite for many biochemical analyses. The different methods currently available often have the drawback that reagents used for preparation of biological samples, *e.g.*, SDS, Triton X-100, NP-40, mercaptoethanol, and urea, interfere with the analysis. In such cases, a protein precipitation step is often required, which is not only time-consuming but may also affect quantification. The volume routinely used for protein determination is often excessively large and, for microdeterminations, special and often cumbersome modifications are necessary. Neuhoff *et al.* [51] have described a method of overcoming these disadvantages and furthermore allowing quantitative determination of small amounts of protein, even if the volume used (in the range between 1–5 µL) is unknown. The method is based on spot analysis first introduced in 1859 by H. Schiff, and frequently applied 50–60 years ago but nowadays almost forgotten. By spotting sample solutions onto a suitable support, the contents will be concentrated over a very small area. During fixation, staining, and washing out of excess stain, all interfering components not firmly bound to the sample being analyzed are almost completely washed out. With this protein determination procedure [51] it is possible to determine the protein concentration in a microliter volume when neither the volume nor the protein content in that volume are known.

For the calculation of protein concentration, the analyzed volume is usually needed, and the amount of protein in that volume is then determined. Both values can be obtained by a single determination since there is a linear relationship between spot size and the applied volume [1–5 µL]. The spot size is further determined by the protein concentration in that volume, as is shown [51] for protein concentrations of 0.1–2.0 mg/mL. The densitometric evaluation of the stained spot gives both values needed for the determination of a protein concentration in an unknown volume between 1–5 µL: absorption as a measure of the amount of protein present in the spot and the diameter of the spot as a measure of the volume needed. With a set of calibration curves using, *e.g.*, albumin as standard, a simple iterative procedure may by applied [51] to determine the protein concentration. The whole protein determination procedure with unknown volumes has a standard deviation of 4.1%. Gutenberg *et al.* [53] have shown that a dot-blot assay for quantitation of nanogram amounts of protein is also possible in the presence of carrier ampholytes and that the procedure is applicable to large numbers of samples. The same principle can be used for the characterization of glycoproteins in the nanogram range [54] and, in cases of limited amounts of sample, the determination of protein and glycoprotein can be performed with a single spot. First the protein is determined with the chromophore Hoechst 2495 which, after elution, is measured fluorometrically. In the second step, the spots are washed first with PBS buffer and then incubated with FITC-lectin. Control experiments have shown that the values obtained by this double determination are the same as those found using two different spots.

4.2 Determination of carbohydrates in the picomole range

For a detailed analysis of pure glycoproteins, often available only in minute amounts from biological material, it may be useful to have a highly sensitive method for the resolution and quantitation of carbohydrate content. A microprocedure to determine carbohydrates in the picomole range has been described by Seiler *et al.* [55]. Carbohydrates are reacted with dansyl hydrazine and amino sugars with dansyl chloride, respectively. Two-dimensional microchromatography [15] on 3 × 4 cm^2 micropolyamide sheets are used to resolve the carbohydrate derivatives. With a carbohydrate mixture, containing 10 mM each of fucose, ribose, arabinose, xylose, galatose, glucose, mannose, 2-deoxy-D-ribose, N-acetylgalactosamine, *N*-acetylglucosamine, glucuronic acid, and galacturonic acid, only 0.01–0.02 µL is needed for a microchromatogram, with sample loading under a stereomicroscope. The reaction of the carbohydrates and uronic acids is stoichiometric, thereby allowing quantitative determination, either by elution [15] with ethanol, followed by fluorometric measurements, or by automatic scanning fluorescence microscopy [56–58]. Glucosamine and galactosamine do not react with dansyl hydrazine under acidic reaction condition but easily react with dansyl chloride to form a dansyl compound which can be separated chromatographically in the presence of dansyl amino acids. *N*-Acetylamino sugars react readily with dansyl hydrazine but not with dansyl chloride, because the amide nitrogen is not protonated under acidic conditions.

4.3 Determination of amino acids in the picomole range

Dansyl chloride (dans-Cl, 1-dimethylamino-napthalene-5-chloride) was first used by Weber for the preparation of fluorescent conjugates of albumin. This reagent has subsequently found almost as wide an application as Fischer's naphthalene sulfonyl chloride or Sanger's 2,4-dinitrofluorobenzene (for review see [59–61]). Its usefulness stems from the fact that its reaction products with amino acids, amines, peptides, proteins, phenols, imidazoles, and sulfhydryl groups have an intense yellow-orange fluorescence and can be separated easily with suitable chromatographic systems [60–63]. In 1969 Neuhoff [15] successfully adapted this technique to the microscale on 3 × 3 cm^2 polyamide sheets with sample application, under a stereomicroscope. As little as 2×10^{-13} mol of the dansylated amino acid could be detected. However, in order to obtain consistent data, even when only semiquantitative results are required, a number of factors have to be taken into account, particularly in the microscale (for details see [15]).

Microdetermination of dansylated amino acids from small biological samples, *e.g.*, isolated nerve cells, small cores of material from particular regions of the brain, biopsy material, the usual agents for extraction, such as picric acid, ethanol, sulfosalicylic acid, acetone/hydrochloric acid, ethanol/HCl, or perchloric acid, are unsuitable because they interfere with microchromatography, with resultant poor separation. After a single microhomogenization [15] of weighed pieces of tissue [64] in 0.05 M NaHCO$_3$, 85% of the free amino acids were found in the supernatant, the remaining 15% were extracted almost completely in a second homogenization step. The proteins, interfering with the dansyl reaction [65], were precipitated by addition of 90% acetone and kept at –20°C to complete precipitation. A mixture of 30–40 dansyl derivatives can be separated on 3 × 3 cm^2 or 3 × 4 cm^2 polyamide microlayers after two-dimensional chromatography [15, 66]. Compounds having two reacting groups in the molecule, *e.g.*, serotonin (-NH$_2$ and -OH), can easily form double dansylated products in addition to the two monodansyl derivatives [67], depending mostly on the pH. Each bisdansyl compound will show specific migration during two-dimensional chromatography. Identification of compounds with more than two reactive groups (many of the biologically active amines) as dansyl products [15] is difficult because more than three different dansyl products are formed. Microchromatograms of dansyl compounds could most accurately be evaluated by fluorescence scanning microscopy. Alternatively, they were evaluated fluorometrically after elution of the isolated dansyl spot with ethanol. With ^{14}C-labeled dansyl chloride autoradiography could be performed [15], which also proved well suited for turnover studies if radioactively labeled amino acids are reacted with unlabeled dansyl chloride. Lowest accuracy with variations up to ±10% are obtained after cumbersome scraping off the spot area with a microknife [15], followed by scintillation counting. Many applications of the dansyl microprocedure for the analysis of free amino acids from different biological materials have been described [15, 68–87]. The method was also suitable for determining *C*- and *N*-terminal amino acids from minute amounts of purified protein [15].

4.4 Photometry and fluorometry of microgels and microchromatograms

Any densitometer or photometer fitted with a suitable device for microgels [15, 56, 57] could be used, also for the evaluation of gels stained for enzyme activity measurements [15]. A single scan was not sufficient and the only correct procedure was the complete scanning of both one-dimensional and two-dimensional microgel slabs [49, 88–95]. Automatic evaluation of two-dimensional thin-layer chromatograms with scanning fluorescence microscopy [49, 88–91, 95] clearly demonstrated the applicability of signal processing, image analysis, and pattern recognition to microphotometry by an enlarged dynamic range of the photometer, improved reproducibility through noise reduction, discrimination between objects and background, feature extraction, and, in addition, automatic classification [95] of the results.

4.5 Auxiliary equipment for microanalysis

Ideally, anyone working with microscale methods should have available a small workshop (and a practical machinist), since much of the equipment can be easily made or adapted from normal equipment. For example, capillary centrifugation can be performed in a special capillary rotor, but a normal test tube rotor can also be adapted [15]. Analytical high-speed centrifugation for determination of sedimentation coefficients with volumes less than 1 µL can be achieved with any analytical centrifuge if the usual analytical cell is adapted for capillaries [15, 96]. Microdialysis chambers for variable volume between 15 and 500 µL [15] are easily prepared in a workshop or are otherwise commercially available (*e.g.*, from E. S. Schütt, Jr., Göttingen, Germany). Homogenization is an important step in microprocedures. The use of a steel wire loop for the homogenization of isolated cells in a capillary was first described by Eichner [97]. Easier to handle are the nerve-canal drills driven by a dental drilling machine [15] in opposite direction in order not to pull out the content of the capillary. For larger volumes than 5 or 10 µL, suitable homogenizers can easily be made in the laboratory [15]. The production of capillary pipettes, repeatedly used in microelectrophoresis, was easy, provided a suitable burner was available [15, 98]. Two-dimensional microimmunodiffusion [15, 99, 100] proved suitable for immunoprecipitation with volumes of less than 1 µL of antigen and antibody solutions, with the additional advantage of results that were already visible within minutes or hours, in contrast to days with the classical Ouchterlony test. Using the same approach, complex formation could be studied in other systems [15, 101, 102], *e.g.*, for the characterization of a polymerase-template complex [15, 103].

5 Concluding remarks

The experience of the author in more than 30 years has shown that the use of micromethods requires no more technical expertise and critical understanding than other methods. The same critical appraisal of the results is needed whether they are obtained with microscale or macroscale methods. However, microscale methods afforded the obvious advantage of faster results in addition to much lower sample demands. Therefore, the author is somewhat surprised that micromethods, as de-

Electrophoresis 2000, *21*, 3–11

scribed in this retrospect, are used rather seldom. One reason may be that the manual steps required a higher level of precision than in the corresponding macroscale approaches. This potential shortcoming is overcome by the new generation of miniaturized systems with increasing degrees of automation boosting their spreading application.

Received April 27, 1999

6 References

[1] Pun, J. Y., Lombrozo, K., *Analyt. Biochem.* 1964 *9*, 9–20.

[2] Grossbach, U., *Biochim. Biophys. Acta* 1965, *107*, 180–182.

[3] Hydén, H., Bjurstam, K., McEwen, B., *Anal. Biochem.* 1966, *17*, 1–15.

[4] McEwen, B., Hydén, H., *J. Neurochem.* 1966, *13*, 823–833.

[5] Neuhoff, V., *Arzneimittel-Forsch.* 1968, *18*, 35–39.

[6] Hydén, H., Lange, P. W., *Proc. Nat. Acad. Sci. USA* 1972, *69*, 1980–1984.

[7] Griffith, A., LaVelle, A., *Exp. Neurol.* 1971, *33*, 360–371.

[8] Ansorg, R., Dames, W., Neuhoff, V., *Arzneimittel-Forsch.* 1971, *21*, 699–710.

[9] Althaus, H. H., Briel, G., Dames, W., Neuhoff, V., *Sonderforschungsbereich 33*, Göttingen, 1969–1972, pp. 107–121.

[10] Grossbach, U., *Chromosoma* 1969, *28*, 136–187.

[11] Poehling, H.-M., Neuhoff, V., *Electrophoresis* 1981, *2*, 141–147.

[12] Neuhoff, V., Arold, N., Taube, D., Ehrhardt, W., *Electrophoresis* 1988, *9*, 255–262.

[13] Neuhoff, V., Stamm, R., Pardowitz, I., Arold, N., Ehrhardt, W., Taube, D., *Electrophoresis* 1990, *11*, 101–117.

[14] Hydén, H., Lange, P. W., *J. Chromatogr.* 1968, *35*, 336–351.

[15] Neuhoff, V., *Micromethods in Molecular Biology*, Springer-Verlag, Berlin 1973.

[16] Grossbach, U., *Biochem. Biophys. Res. Commun.* 1972, *49*, 667–672.

[17] Rüchel, R., Mesecke, S., Wolfrum, D. I., Neuhoff, V., *Hoppe Seylers Z. Physiol. Chem.* 1973, *354*, 1351–1368.

[18] Rüchel, R., Mesecke, S., Wolfrum, D. I., Neuhoff, V., *Hoppe Seylers Z. Physiol. Chem.* 1975, *356*, 1283–1288.

[19] Sprinzel, M., Wolfrum, D. I., Neuhoff, V., *FEBS Lett.* 1975, *50*, 54–56.

[20] Kolin, A., *J. Chem. Phys.* 1954, *22*, 1628–1629.

[21] Svensson, H., *Acta Chem. Scand.* 1961, *15*, 325–341.

[22] Vesterberg, O., Svensson, H., *Acta. Chem. Scand.* 1966, *20*, 820–834.

[23] Dale, G., Latner, A., *Lancet* 1968, *36*, 847–848.

[24] Wrigley, C. W., *J. Chromatogr.* 1968, *36*, 326–365.

[25] Haglund, H., *Methods Biochem. Anal.* 1971, *19*, 1–104.

[26] Riley, R. F., Coleman, M. K., *J. Lab. Clin. Med.* 1968, *72*, 714–720.

[27] Catsimpoolas, N., *Anal. Biochem.* 1968, *26*, 480–482.

[28] Quentin, C.-D., Neuhoff, V., *Int. J. Neurosci.* 1972, *4*, 17–24.

[29] Gainer, H., *Anal. Biochem.* 1973, *51*, 646–650.

[30] Bispink, G., Neuhoff, V., in: Radola, B. J. and Graesslin, D. (Eds.), *Electrofocusing and Isotachophoresis*, Walter de Gruyter, Berlin 1977, pp. 135–146.

[31] Gustke, H. H., Neuhoff, V., *Hoppe Seylers Z. Physiol. Chem.* 1978, *359*, 1481–1489.

[32] Gustke, H. H., Neuhoff, V., *Hoppe Seylers Z. Physiol. Chem.* 1979, *360*, 605–608.

[33] Huether, G., Neuhoff, V., *Histochem. J.* 1981, *13*, 207–225.

[34] Baumann, J., Chrambach, A., *Anal. Biochem.* 1976, *77*, 216–225.

[35] Poehling, H.-M., Neuhoff, V., *Electrophoresis* 1980, *1*, 90–102.

[36] Laemmli, U. K., *Nature* 1970, *227*, 680–685.

[37] O'Farrel, P. H., *J. Biol. Chem.* 1975, *250*, 4007–4021.

[38] Klose, J., Blohm, M., Gerner, L., in: Neubert, D., Merker, H. J., Kwasigroch, T., (Eds.), Georg Thieme, Stuttgart 1977, pp. 303–313.

[39] Radola, B. J., *Electrophoresis* 1980, *1*, 43–56.

[40] Neuhoff, V., Cheong-Kim, K.-S., Altland, K., *Electrophoresis* 1986, *7*, 56–57.

[41] Wildfang, J., Arold, N., Neuhoff, V., *Electrophoresis* 1991, *12*, 352–366.

[42] Görg, A., Postel, W., Westermeier, R., *Anal. Biochem.* 1978, *89*, 60–70.

[43] Kinzkofer, A., Radola, B. J., *Electrophoresis* 1981, *2*, 174–183.

[44] Lane, J. D., Neuhoff, V., *Naturwissenschaften* 1980, *67*, 227–233.

[45] Lane, J. D., Schöne, B., Langenbeck, U., Neuhoff, V., *Biochim. Biophys. Acta* 1980, *627*, 144–156.

[46] Huether, G., Neuhoff, V., *J. Int. Metab. Dis.* 1981, *4*, 67–68.

[47] Fritz, P., Arold, N., Mischlinski, A., Wisser, H., Oeffinger, B., Neuhoff, V., Lascner, W., Koenig, G., *Rheumatol. Int.* 1990, *10*, 177–183.

[48] Weber, M. H., Cheong-Kim, K.-S., Schott, K.-J., Neuhoff, V., *Electrophoresis* 1986, *7*, 134–141.

[49] Kronenberg, H., Zimmer, H. G., Neuhoff, V., *Electrophoresis* 1980, *1*, 27–32.

[50] Pardowitz, I., Ehrhardt, W., Neuhoff, V., *Electrophoresis* 1990, *11*, 400–406.

[51] Neuhoff, V., Philipp, K., Zimmer, H.-G., Mesecke, S., *Hoppe Seylers Z. Physiol. Chem.* 1979, *369*, 1657–1670.

[52] Feigl, F., *Tüpfelanalyse. II. Organischer Teil*, Akademische Verlagsgesellschaft MBH, Frankfurt/Main 1960.

[53] Gutenberger, M., Neuhoff, V., Hamp, R., *Anal. Biochem.* 1991, *196*, 99–103.

[54] Neuhoff, V., Ewers, E., Huether, G., *Hoppe Seylers Z. Physiol. Chem.* 1981, *362*, 1427–1434.

[55] Seiler, P., Thorn, W., Neuhoff, V., *Funkt. Biol. Med.* 1984, *3*, 85–94.

[56] Zimmer, H.-G., Neuhoff, V., *GIT Fachz. Lab.* 1975, *19*, 481–484.

[57] Lane, J. D., Zimmer, H.-G., Neuhoff, V., *Hoppe Sylers Z. Physiol. Chem.* 1979, *360*, 1405–1408.

[58] Taube, D., Neuhoff, V., *J. Chromatogr.* 1988, *437*, 411–421.

[59] Seiler, N., *Methods Biochem. Anal.* 1970, *18*, 259–337.

[60] Seiler, N., Wiechmann, M., in: Niederwieser, A., Pataki, G., (Eds.), *Progress in Thin-Layer Chromatography and Related Methods*, Ann Arbor Science Publications, Ann Arbor 1970, *Vol. 3*, pp. 95–144.

[61] Woods, K. R., Wang, K. T., *Biochim. Biophys. Acta* 1967, *133*, 369–370.

[62] Wang, K. T., Weinstein, B., in: Niederwieser, A., Pataki, G., (Eds.) *Progress in Thin-Layer Chromatography and Related Methods*, Ann Arbor Science Publications, Ann Arbor 1972, *Vol. 3*, pp. 177–231.

[63] Gray, W. R., Hartley, B. S., *Biochem. J.* 1963, *89*, 59.

[64] Neuhoff, V., *Anal. Biochem.* 1971, *41*, 270–271.

[65] Neuhoff, V., Poehling, H.-M., *Hoppe Seylers Z. Physiol. Chem.* 1980, *361*, 77–78.

[66] Zimmer, H.-G., Neuhoff, V., Schulze, E., *J. Chromatogr.* 1976, *124*, 120–122.

[67] Neuhoff, V., Weise, M., *Arzneim. Forsch.* 1970, *20*, 368–372.

[68] Leonard, B. E., Neuhoff, V., Tonge, S. R., *Z. Naturforsch.* 1974, *29c*, 184–186.

[69] Osborne, N. N., Wu, P. H., Neuhoff, V., *Brain. Res.* 1974, *74*, 175–181.

[70] Osborne, N. N., Neuhoff, V., *Brain. Res.* 1974, *74*, 366–369.

[71] Quentin, C.-D., Behbehani, A. W., Schulte, F. J., Neuhoff, V., *Neuropediatrie* 1974, *5*, 138–145.

[72] Behbehani, A. W., Quentin, C.-D., Schulte, F. J., Neuhoff, V., *Neuropediatrie* 1974, *5*, 258–270.

[73] Quentin, C.-D., Behbehani, A. W., Schulte, F. J., Neuhoff, V., *Neuropediatrie* 1974, *5*, 271–278.

[74] Neuhoff, V., Behbehani, A. W., Quentin, C.-D., Prinz, H., *Hoppe Seylers Z. Physiol. Chem.* 1974, *355*, 891–894.

[75] Leonhard, B. E., Neuhoff, V., Tonge, S. R., *Z. Naturforsch.* 1974, *29c*, 767–772.

[76] Osborne, N. N., Neuhoff, V., *Brain Res.* 1974, *80*, 251–264.

[77] Behbehani, A. W., Quentin, C.-D., Neuhoff, V., *Neurobiology* 1975, *5*, 52–59.

[78] Neuhoff, V., Behbehani, A. W., Quentin, C.-D., Briel, C., *Neurobiology* 1975, *5*, 254–261.

[79] Richter-Landsberg, C., Neuhoff, V., *Naturwissenschaften* 1975, *62*, 491.

[80] Leonard, B. E., Neuhoff, V., Tonge, S. R., *J. Neurosci. Res.* 1975, *1*, 83–92.

[81] Schulze, E., Neuhoff, V., *Hoppe Seylers Z. Physiol. Chem.* 1976, *357*, 593–600.

[82] Stenzel, K., Neuhoff, V., *J. Neurosci. Res.* 1976, *2*, 1–9.

[83] Osborne, N. N., Stahl, W. L., Neuhoff, V., *J. Chromatogr.* 1976, *123*, 212–215.

[84] Osborne, N. N., Neuhoff, V., *J. Chromatogr.* 1977, *134*, 489–496.

[85] Poehling, H.-M., Wys, U., Neuhoff, V., *Physiol. Plant Pathol.* 1980, *16*, 59–61.

[86] Ulmar, G. Neuhoff, V., *Exp. Neurol.* 1980, *69*, 99–109.

[87] Meyer, W., Poehling, H.-M., Neuhoff, V., *Comp. Biochem. Physiol.* 1980, *67c*, 83–86.

[88] Zimmer, H.-G., Neuhoff, V., in: Brauer W., (Ed.) *Informatik-Fachberichte*, Springer Verlag, Berlin 1977, pp. 12–20.

[89] Kronberg, G., Zimmer, H.-G., Neuhoff, V., *Fresenius Z. Anal. Chem.* 1978, *290*, 133–134.

[90] Kronberg, G., Zimmer, H.-G., Neuhoff, V, *Microscop. Acta* 1979, *82*, 223–228.

[91] Zimmer, H.-G., *J. Microscop.* 1979, *116*, 365–372.

[92] Zimmer, H.-G., Neuhoff, V., *Naturwissenschaften* 1981, *68*, 464–470.

[93] Zimmer, H.-G., Kronberg, H., Berstein, R., Neuhoff, V., *Pattern Recogn.* 1981, *13*, 79–82.

[94] Kronberg, G., Zimmer, H.-G., Neuhoff, V., *Microsc. Acta (Suppl.)* 1980, *4*, 217–221.

[95] Kronberg, G., Neuhoff, V., in: Triendle, E. (Ed.) *Informatik-Fachberichte, Volume 17*, Springer Verlag, Berlin 1978, pp. 334–337.

[96] Neuhoff, V., Rödel, E., *Hoppe Seylers Z. Physiol. Chem.* 1973, *354*, 1541–1549.

[97] Eichner, D., *Experientia* 1966, *22*, 620.

[98] Edstrøm, J.-E., Neuhoff, V., in: Neuhoff, V. (Ed.), *Micromethods in Molecular Biology*, Springer-Verlag, Berlin 1973, pp. 215–256.

[99] Neuhoff, V., Schill, W.-B., *Hoppe Seylers Z. Physiol. Chem.* 1968, *349*, 795–800.

[100] Neuhoff, V., Mesecke, S., *Hoppe Seylers Z. Physiol. Chem.* 1977, *358*, 1623–1637.

[101] Neuhoff, V, Schill, W.-B., Sternbach, H., *Biochem. J.* 1970, *117*, 623–631.

[102] Hydén, H., Rönnbäck, L., *J. Neurol. Sci.* 1978, *39*, 241–246.

[103] Neuhoff, V., Schill, W.-B., Sternbach, H., *Hoppe Seylers Z. Physiol. Chem.* 1969, *350*, 335–340.

Review

Haleem J. Issaq

Separation Technology
Group, Chemical Synthesis
and Analysis Laboratory,
SAIC Frederick,
NCI-FCRDC,
Frederick, MD, USA

A decade of capillary electrophoresis

Since the introduction of the first commercial capillary electrophoresis (CE) instrument a decade ago, CE applications have become widespread. Today, CE is a versatile analytical technique which is successfully used for the separation of small ions, neutral molecules, and large biomolecules and for the study of physicochemical parameters. It is being utilized in widely different fields, such as analytical chemistry, forensic chemistry, clinical chemistry, organic chemistry, natural products, pharamaceutical industry, chiral separations, molecular biology, and others. It is not only used as a separation technique but to answer physicochemical questions. In this review, we will discuss different modes of CE such as capillary zone electrophoresis, micellar electrokinetic chromatography, capillary gel electrophoresis, capillary isoelectric focusing, and capillary electrochromatography, and will comment on the future direction of CE, including array capillary electrophoresis and array microchip separations.

Keywords: Capillary electrophoresis / Micellar electrokinetic chromatography / Capillary electrochromatography / Biomolecules / DNA fragments / Proteins / Natural products / Review   EL 3842

Contents

Correspondence: Dr. Haleem J. Issaq, Separation Technology Group, Chemical Synthesis and Analysis Laboratory, SAIC Frederick, NCI-FCRDC, P.O. Box B, Frederick, MD 21702, USA
E-mail: issaqh@mail.ncifcrf.gov
Fax: +301-846-1438

Abbreviations: CGE, capillary gel electrophoresis; **DMMAPS**, 3-(*N,N*-dimethylmyristylammonio)propanesulfonate; **FLEC**, 1-(9-fluorenyl)ethyl chloroformate; **FMOC**, 9-fluorenylmethyl chloroformate; **RF-MEKC**, reversed flow-MEKC; **SGE**, slab gel electrophoresis

1 Introduction

Electrophoresis as an analytical technique was first introduced by Tiselius in 1930 [1]. In his thesis he described the separation of blood plasma proteins, namely albumin from α-, β- and γ-globulin, using electrophoresis. For this pioneering work Tiselius was awarded the Nobel Prize in 1948. Since then, many advancements in electrophoresis, such as paper and gel electrophoresis, have been introduced. Filter paper has been widely used as a supporting medium since the late 1940s for the separation of ionized compounds such as amino acids, lipids, nucleotides, and charged sugars [2]. Two-dimensional paper electrophoresis was carried out to enhance the resolution by changing the buffer's pH, *i.e.*, first-dimensional electrophoresis at a particular pH and the second-dimensional run at a different pH.

Gelatin and agar gels as supports in electrophoresis have been used for at least a hundred years [2]. A major advance in gel electrophoresis took place when it was realized that polyacrylamide (PAA) gels under denaturing conditions using urea or sodium dodecyl sulfate (SDS) can be employed for the resolution of proteins [3]. SDS-polyacrylamide gel was not only used to separate proteins but also to estimate their molecular weight. Two-dimensional gel electrophoresis is used to separate cellular proteins having relatively large differences in molecular weight and charge density in two sequential steps: first by their charge and then by their mass [4]. Other historical advances of gel electrophoresis are discussed in [2].

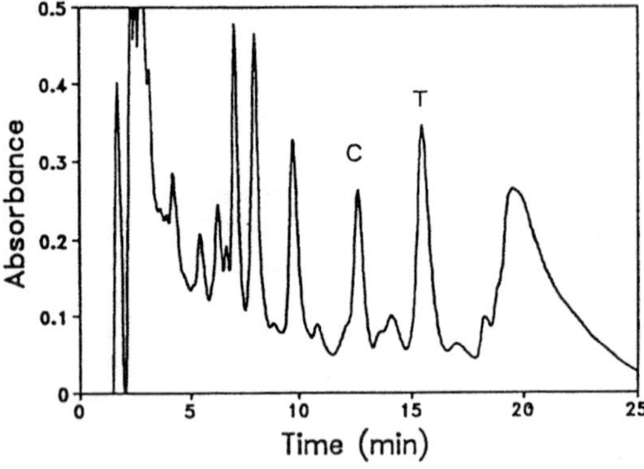

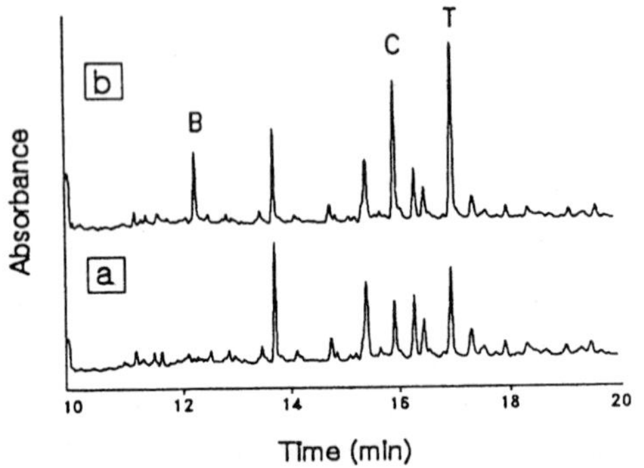

Figure 1. Comparative separation of a Western Yew needle extract by HPLC (upper) and CE (lower). (A) Unspiked and (B) spiked needle extract. T, taxol; C, cephalomannine; B; baccatin III. Reprinted from [63], with permission.

In 1967, Hjertén [5] showed that it was possible to carry out electrophoretic separations in a 300 μm glass tube and to detect the separated compounds by ultraviolet absorption (UV), and called it free zone electrophoresis. Although other researchers used electrophoresis in tubes, glass, and Teflon [6–8], electrophoresis in a tube did not become popular until 1981 when Jorgenson and Lukacs [9] published their work in which they demonstrated the high resolving power of capillary zone electrophoresis (CZE). The simple and efficient instrumental setup, which is the basis of most commercial instruments on the market today, used a narrow ID fused-silica capillary, less than 100 μm, high voltage, 30 kV, and on-column UV detection for the separation of ionic species. In 1984, Terabe *et al.* [10] introduced micellar electrokinetic chromatography (MEKC) for the separation of neutral compounds by adding a micelle, SDS, to the buffer solution. In 1988, the first commercial instrument was marketed. Since then, many advances and applications have taken place [11–119]. A review of the application of CE in bio-

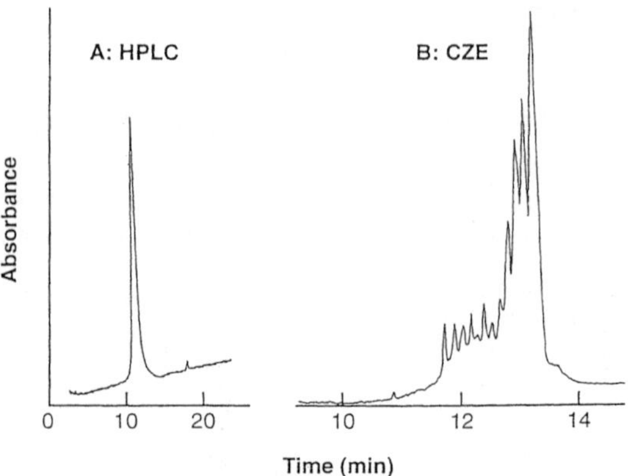

Figure 2. Comparative separation of a 66 residue peptide by (A) HPLC and (B) CEZ. Reprinted from [63], with permission.

logical sciences was published by Karger *et al.* [11] in 1989. For a more recent historical review of advancement in CE instrumentation, the reader is referred to the excellent write-up by Camilleri [2].

This review will present examples of the application of CE in different fields. In the last ten years our laboratory has been involved in the development and application of CE for the separation of small as well as large biomolecules; therefore, where applicable, examples of our work will be presented. Otherwise, examples from the literature will be used to illustrate the utility and application of CE. This review is not meant to be a comprehensive one; rather, it is presented here to show the speed of advancement and applicability of this microanalytical technique in many different chemical, biological, and analytical laboratories. For recent reviews see [12, 13]. The review is divided into the following sections: why CE?; inorganic ions and organic acids; amino acids, peptides, and proteins; natural products; nucleic acids; clinical applications; illicit drugs; chiral separations; miscellaneous separations; physicochemical studies; capillary electrochromatography; and future directions.

2 Why CE?

The first question we have been asked on numerous occasions is why use CE when other separation techniques are available? What is so special about CE and what are the advantages of using it considering that it is only a microanalytical technique? The advantage of CE over gas chromatography (GC), high performance liquid chromatography (HPLC), thin-layer chromatography (TLC), and slab gel electrophoresis (SGE) is its applicability for the separation of widely different compounds, inor-

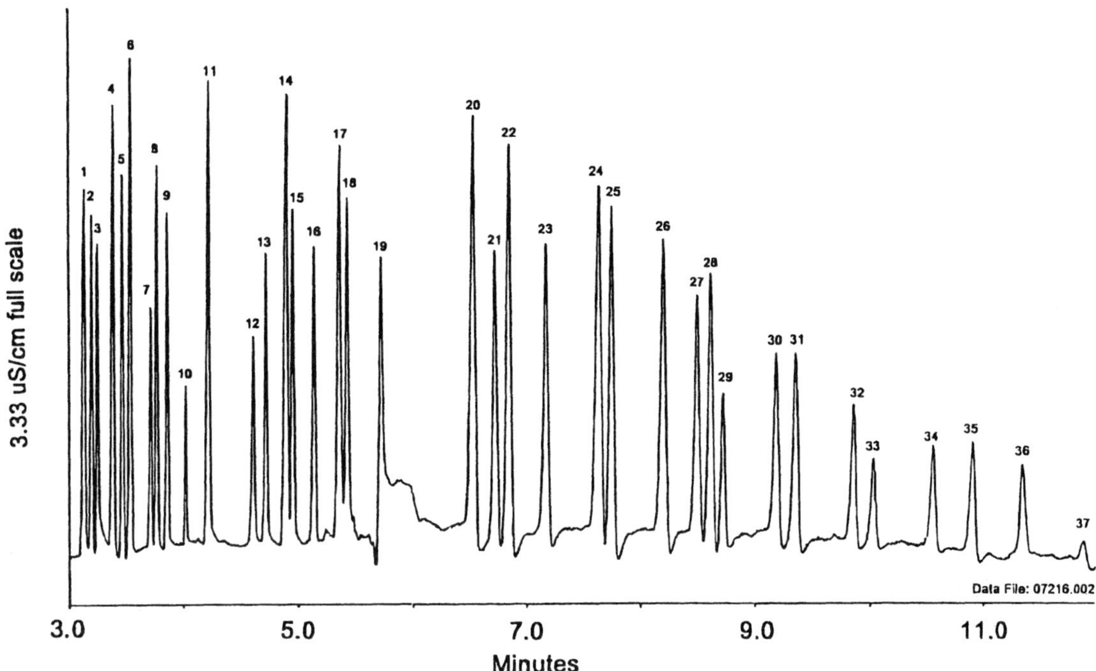

Figure 3. An electropherogram of the CZE separation of 37 anions in a 60 cm × 50 µm using a 50 mM CHES, 20 mM lithium hydroxide monohydrate, 0.03% Triton X-100, pH 9.2, at an applied potential of 25 kV (negative polarity). Pressure injection at 25 mbar for 12 s. Analysis was carried out at room temperature. For peak identification and concentration see [15]. Reprinted from [15], with permission.

ganic ions, organic molecules and large biomolecules, using the same instrument and in most cases the same column while changing only the composition of the running buffer. This can not be said about any of the other separation techniques. In addition, CE possesses the highest resolving power of any liquid separation technique, due to its plug flow and minimal diffusion. The CE resolving power is illustrated in Fig. 1, which compares the separation of a needle extract of the Western Yew, *Taxus brevifolia* (Taxacaea) by HPLC and CE, and Fig. 2 which compares the separation of a 66-residue peptide by both CE and HPLC. Compared to SGE, CE is faster, easier, simpler, can be automated, and gives quantitative data. Unlike SGE and TLC, where two-dimensional separations are achieved easily, in CE, as in HPLC and GC, such separations can not be performed. However, CE can be used on-line with other instrumental techniques such as HPLC, mass spectrometry (MS) and nuclear magnetic resonance (NMR). The amount of material needed for a CE experiment is minute, nanoliters of sample and microliters of buffer, compared to the other liquid separation techniques which require microliters of sample and milliliters of solvent. The resulting CE waste, mostly

an aqueous buffer, but sometimes with a small percentage of an organic modifier, is environmentally safe and can be discarded without any danger to the environment. This is not true when TLC or HPLC is used, where large amounts of organic solvent waste are produced, or GC, where volatile compounds, which are sometimes toxic, escape into the environment.

3 Inorganic ions and organic acids

CE is well suited for the separation of inorganic ions using indirect UV detection. Hjertén [5] in his initial publication presented data showing the separation of Bi from Cu ions. Since then, many studies have been published dealing with the separation of inorganic ions, acids, and bases. A notable one is the separation of 36 anions in 3 min [14]. In another study Jones *et al.* [15] resolved 37 anions in 12 min (Fig. 3). The separation of metal ions was first reported by Hjertén [5]. Weston *et al.* [16] resolved 11 metal ions from each other in under 8 min. In our laboratory we developed a method for the separation of nitrate from nitrite at pH 3, where they are widely separated and detected at 214 nm where they absorb strongly, while most other negative ions have negligible absorbance. The method was applied for the analysis of nitrate and nitrite in urine and water (Fig. 4) [17, 18].

4 Amino acids, peptides, and proteins

The separation of amino acids, peptides, and proteins by CE has been a popular one. Many studies have been published in the last decade dealing with this subject.

4.1 Amino acids

The 20 amino acids possess different chemical properties – basic, acidic, polar, hydrophilic and hydrophobic – which complicates their separation by CZE according to only their charge density using a simple buffer. Therefore, MEKC is the method of choice for the separation of a mixture of amino acids where SDS is added to the buffer to effect the separation by introducing partitioning as a mechanism of separation in addition to charge density. The sensitive detection of amino acids is achieved by

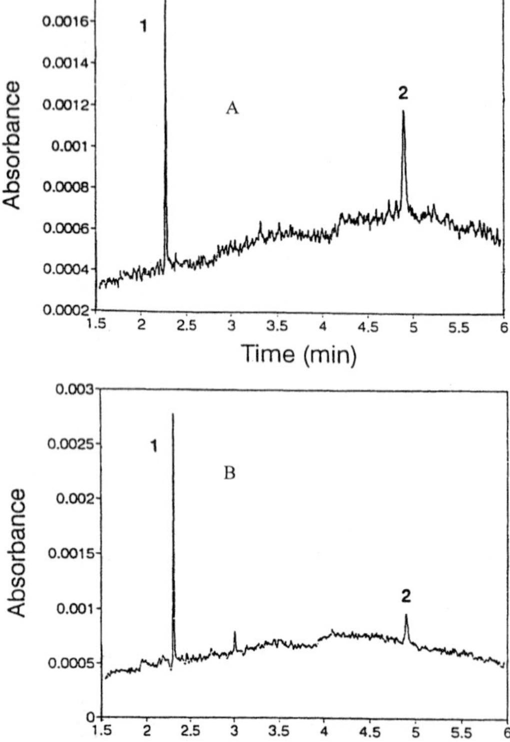

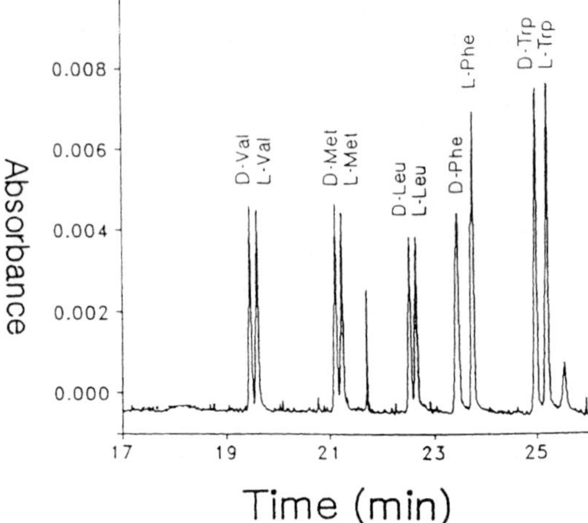

Figure 5. MEKC separation of a racemic amino acid mixture derivatized with FMOC. Buffers: 5 mM sodium borate (pH 9.2), 150 mM SDS, and 40 mM γ-CD. Capillary, 50 μm × 67 cm; voltage, 10 kV; UV detection, at 200 nm. Reprinted from [19], with permission.

Figure 4. Electropherogram of the separation of nitrate and nitrite at pH 3 (A) and (B) urine. Buffer, 25 mM phosphate containing 0.5% 3-(*N,N*-dimethylmyristylammonio) propanesulfonate (DMMAPS) and 1.0% Brij-35; applied voltage, −15 kV; column, 10%T, polyacrylamide-coated fused silica, column dimensions: L_{total} = 57 cm, $L_{detection}$ = 50 cm; 75 μm ID; instrument, Beckman Model P/ACE System 2000; detection, 214 nm; solutes: 1, nitrate (2.5 μg/mL); 2, nitrite (2.5 μg/mL). Reprinted from [18], with permission.

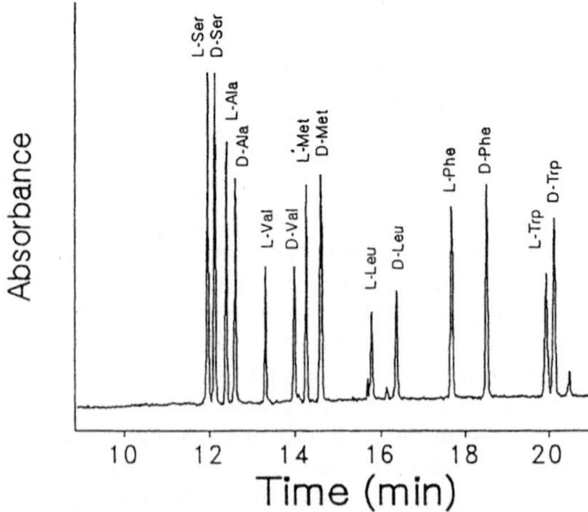

Figure 6. MEKC separation of a racemic amino acid mixture derivatized with (+)-FLEC. Buffers: 10 mM sodium phosphate, pH 6.8, 25 mM SDS, and 15% ACN. Reprinted from [21] with permission.

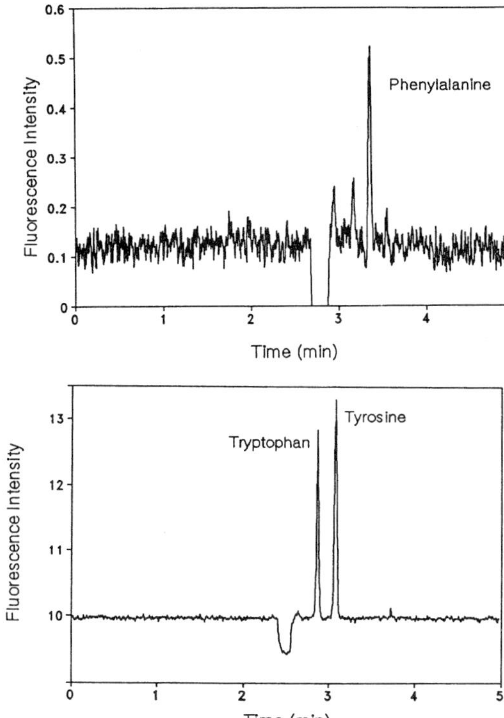

Figure 7. Electropherogram of phenylalanine, tryptophan and tyrosine with UV-LIF detection. Buffer: 5 mM sodium borate, pH 9.1. Capillary, 75 µm ID × 70 cm total length, 60 cm effective length; applied voltage, 25 kV; samples: 1×10^{-4} M phenylalanine, 5×10^{-8} M tryptophan and 5×10^{-6} M tryosine; injection, 15 s, gravity. Reprinted from [22], with permission.

laser-induced fluorescence (LIF). In our laboratory, amino acids were resolved by MEKC and detected with pulsed UV-LIF after derivatization with 9-fluorenylmethyl chloroformate [19]. A recent review by Issaq and Chan [20] discussed the separation and detection of amino acids and their enantiomers by CE. The separation of the D- and L-amino acid enantiomers can be accomplished in three ways: (i) addition of a chiral compound, such as cyclodextrin (CD) to the running buffer (Fig. 5); (ii) by reacting the amino acids with a chiral reagent, such as 1-(9-fluorenyl)-ethyl chloroformate (FLEC), to form two stereoisomers which can be easily resolved (Fig. 6) [21]; or (iii) by using a chiral column. The three aromatic amino acids, tryptophan, phenylalanine, and tyrosine can be detected with UV-LIF without derivatization with a fluorescent tag using the KrF laser detection system which was built in our laboratory (Fig. 7) [20, 22]. The separation of tryptophan and its metabolites [23] and hydroxy proline and other secondary amino acids in biological samples [24] was reported.

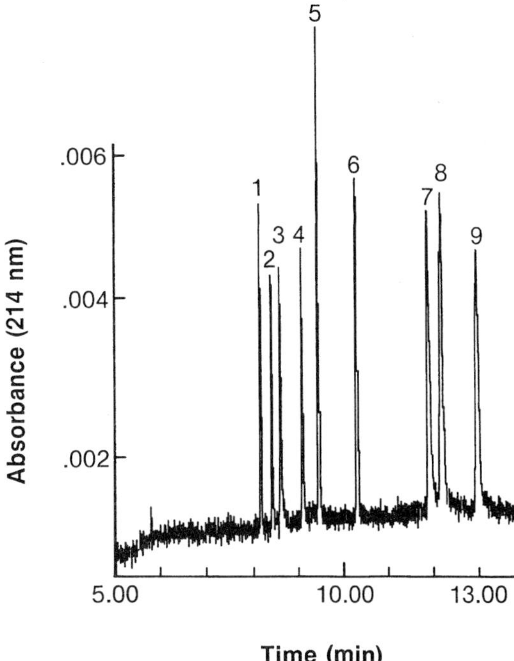

Figure 8. Electropherogram of bioactive peptides. Peak 1, bradykinin; 2, bradykinin fragment 1–5; 3, substance P; 4, [Arg^8]vasopressin; 5, luteinizing hormone releasing hormone; 6, bombesin; 7, leucine enkephalin; 8, methionine enkephalin; 9, oxytocin. Reprinted from [32], with permission.

4.2 Peptides

Many studies have dealt with the separation of peptides. In our laboratory, CE was used to separate a group of bioactive peptides (Fig. 8) and is used routinely to check the purity of synthetic peptides (see, for example, Fig. 2). Peptides are detected after separation by UV absorption or by LIF after derivatization with a fluorescent reagent [24]. Pulsed UV-LIF was used for the detection of peptides without derivatization if the peptides contained one of the three aromatic amino acids (Fig. 9) [22]. We also used CE to predict the mobility of peptides [26, 27] and the relation between mobility of dipeptides and charge density (Fig. 10) [28]. CE was used for peptide mapping of proteins after enzymatic digestion [29]. For a comprehensive review of CE of peptides, see [30].

4.3 Proteins

The electrophoretic separation of proteins in untreated fused silica capillaries has not been very successful due to the interaction of the protein with the silanol groups on

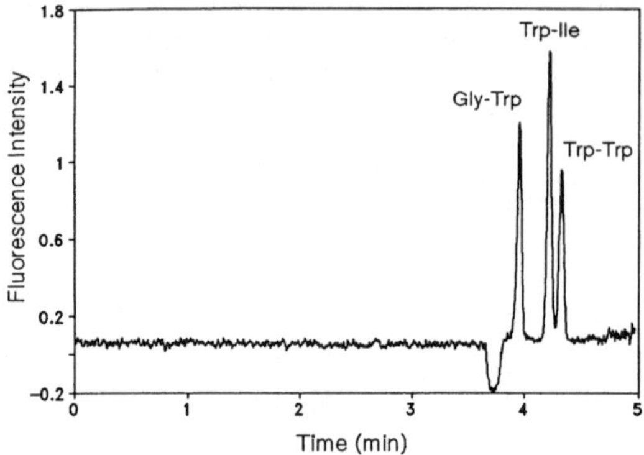

Figure 9. Electropherogram of dipeptides with LIF detection. Buffer, 10 mM sodium phosphate, pH 7.0; capillary, 75 µm ID × 65 cm total length, 55 cm effective length; applied voltage, 23 kV; samples: 2.1 × 10⁻⁷ M Gly-Trp, 8.1 × 10⁻⁸ M Trp-Ile, and 6.9 × 10⁻⁸ M Trp-Trp; injection, 15 s gravity. Reprinted from [22] with permission.

the inner capillary wall. Therefore, many studies used buffer modifiers or neutrally coated capillaries to eliminate or suppress the protein/wall interaction. A solution to the wall interaction (adsorption) of proteins was suggested by Lauer and McManigill [31] who stated that CZE of proteins in untreated fused silica capillaries is possible if coulombic repulsion between protein and the capillary can overcome adsorption tendencies. This coulombic repulsion can be achieved either by raising the pH of the buffer solution above the isoelectric point of the protein or by dynamically modifying the interfacial double layer between the wall and the bulk solution with selected ions. Adsorption of proteins to the capillary wall is eliminated by physically coating the capillary wall with methyl cellulose [6], polyacrylamide [6, 32, 33], polyethyleneimine [34], polyethylene glycol [35], nonionic surfactant coatings [36], or *via* silane derivatization [6, 37, 38]. Another procedure for minimizing the adsorption of proteins is through the use of additives to the buffer. For example, one could add a high concentration of alkali salts [39], 1,3-diaminopropane [40], and cationic surfactant FC 135 [41] to the running buffer. The separation of a mixture of glycoproteins was resolved using a borate buffer containing 1 mM putrescine [42]. Capillary isoelectric focusing (CIEF) is used in our laboratory for two purposes, to check the purity of the protein and to determine its isoelectric point. Also, CIEF is a powerful technique with which the analyst can determine the protein isoforms. We recently determindthe isoforms of antibodies by CIEF (work in progress). Proteins are detected by their UV absorption or by LIF with or without derivatization with a fluorescent tag. In our laboratory we used UV-LIF, employing a KrF pulsed

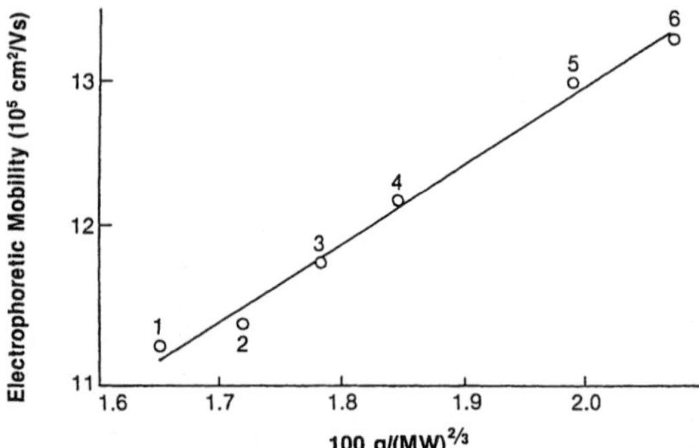

Figure 10. Electrophoretic mobility for dipeptides *versus* charge-to-size parameter. Solutes: 1, phenylalanine-phenylalamine (FF); 2, phenylalamine-aspartic acid (FD); 3, phenylalamine-leucine (FL); 4, phenylalamine-valine (FV); 5, phenylalanine-alanine (FA); 6, phenylalanine-glycine (FA). Reprinted from [28] with permission.

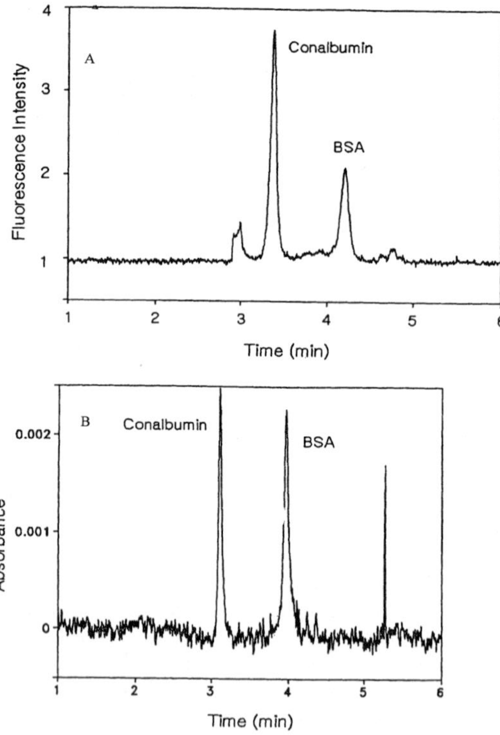

Figure 11. Electropherogram of proteins with (A) KrF-LIF detection and (B) UV absorption detection at 214 nm. Capillary, 75 µm ID × 57 cm total length; applied voltage, 22 kV; samples: 1.3 × 10⁻⁶ M conalbumin, 1.5 × 10⁻⁶ M BSA; injection, 3 s pressure for UV detection and 2.5 × 10⁻⁸ M conalbumin and 3 × 10⁻⁸ M BSA for LIF detection. Reprinted from [22] with permission.

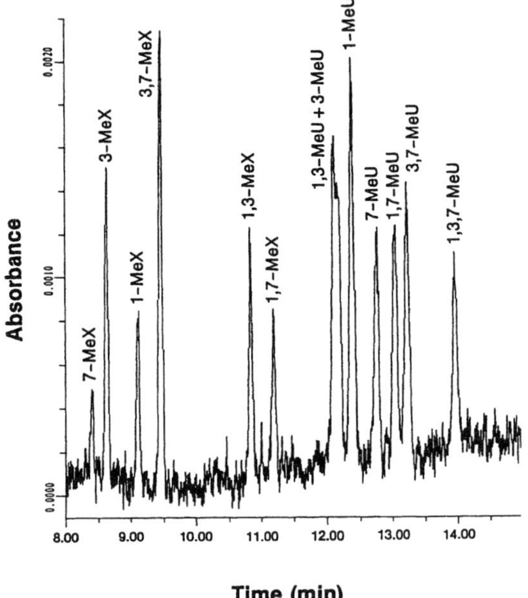

Figure 12. Electropherogram of a mixture of seven methyl-substituted xanthines and six methyl-substituted uric acids. Reprinted from [48] with permission.

UV laser for the detection of native proteins that contained aromatic amino acid residues. Figure 11 compares the detection of two proteins in their native form by UV absorption and UF-LIF [22]. In Section 11 we will discuss the application of CE for the determination of protein/drug and protein/DNA interaction. For a comprehensive review of protein separations, the reader is referred to Chapter 9 in [30].

5 Natural products

Natural products are the organic and inorganic compounds found in nature: in plants (leaves, needles, bark, roots, flowers, and seeds); in marine organisms (plants, animals, and microbes); in the microbial fauna found in highly diverse and sometimes extreme environments, and in the soil. The pharmaceutical industry in its drug discovery effort has relied heavily on natural products, if not as source of drugs, then as sources of novel bioactive chemotypes that can be developed into new drugs; an excellent example is penicillin.

CZE and MEKC were used for the separation of widely different compounds from natural sources, including antibiotics, flavonoids, isoflavonoids, steroids, coumarins, alkaloids, illicit drugs, nicotine, caffeine and related compounds, toxins such as aflatoxins [43], mycotoxins [43], mycotoxins, heptapeptide toxins, Chinese herbal prepara-

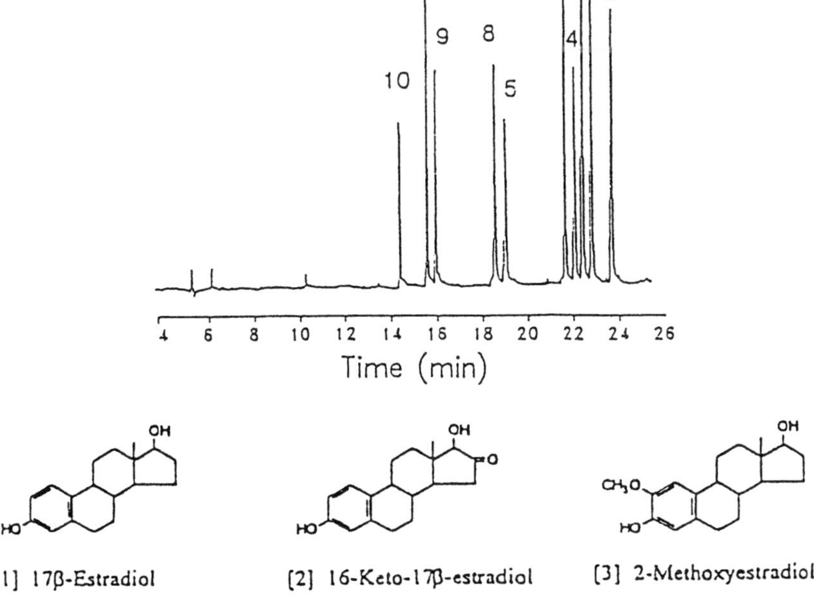

Figure 13. MEKC of ten estrogens. Buffers: 10 mM sodium phosphate, pH 7.0, containing 50 mM SDS and 20% methanol; capillary, 50 µm × 57 cm; voltage, 20 kV; pressure injection, 2 s; detection, absorption at 200 nm. Reprinted from [49] with permission.

tions, mineral elements, humic substances, and many other compounds. Two reviews of the subject have been published [44, 45]. The advantage of CZE and MEKC for the analysis of natural products is their high resolving power and applicability to a versatile group of compounds. In our laboratory, MEKC was used for the sepa-

ration of a bark and needle extract for the assay of taxol, an anticancer agent, from needle and bark extracts of the Western Yew [46]. Also, CE with LIF detection was used for the quantification of Michellamine, which has shown activity against the human immune deficiency virus (HIV) *in vitro*, from plant material [47]. The separation of

DNA Fragments

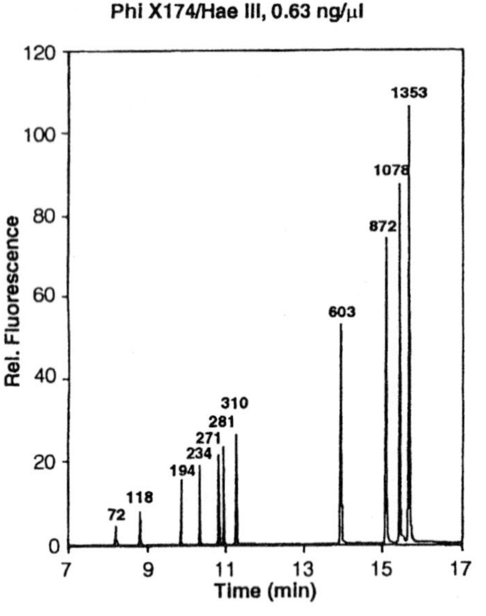

Single Stranded DNA

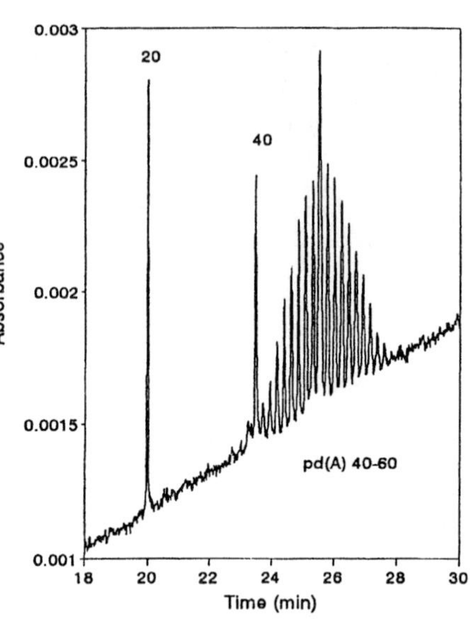

PCR Products

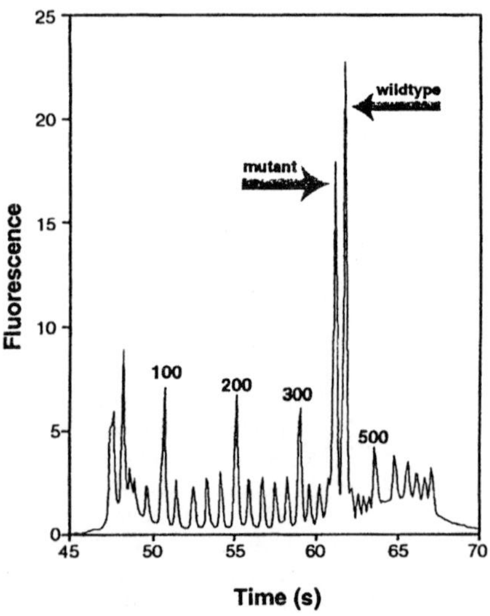

Figure 14. Electropherograms of the separation of DNA fragments (left), single-stranded DNA (middle) and PCR products (right). Reprinted from [51], with permission.

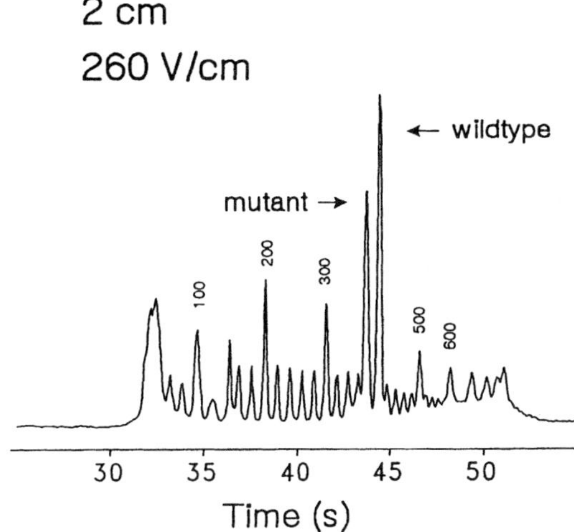

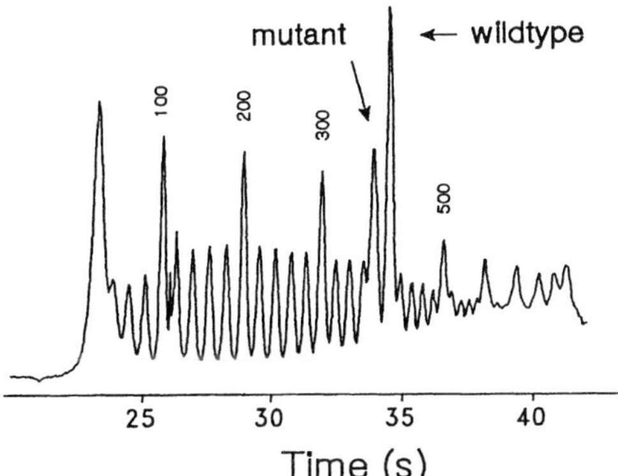

Figure 15. High-speed separation of PCR products using capillaries with effective lengths of 1 cm and 2 cm. Reprinted from [53], with permission.

methyl-substituted xanthines and uric acids (caffeine metabolites; Fig. 12) [48] and estrogen and nine related compounds (Fig. 13) was reported by our group [49]. Other examples are given in [44, 45].

6 Nucleic acids

Capillary gel electrophoresis (CGE) is a powerful analytical technique for the separation of double-stranded DNA fragments, single-stranded DNA, and polymerase chain reaction (PCR) products (Fig. 14). The success of CE as an analytical technique for the separation of nucleic acids is based on gel-filled capillaries. The capillary is filled with

a replaceable liquid gel, also known as entangled polymers, which separates the fragments by a sieving mechanism and can be replaced after each analysis. Hydroxyethylcellulose, hydroxymethylcellulose, methylcellulose, polyvinyl alcohol, liquid polyacrylamide, and others have been used as liquid polymers. For a comprehensive review of the subject, history, theory, and applications of CGE, consult [50]. Issaq *et al.* [51] examined the effect of different parameters on migration time, resolution, and speed of analysis of DNA fragments and PCR products. These parameters include column length, applied voltage, gel type and concentration, and buffer ionic strength. The results show that a 1 cm capillary at an applied voltage of 185 V/cm filled with commercial gel was adequate for the

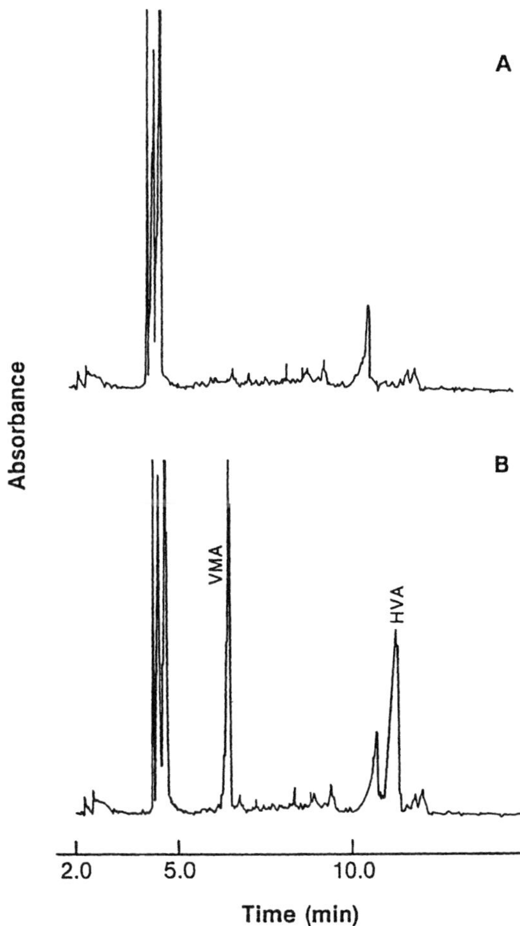

Figure 16. Electropherograms of normal infant urine. (A) Untreated normal infant urine; (B) normal infant urine spiked with vanillylmandelic acid (VMA) and homovanillic acid (HVA). Reprinted from [59], with permission.

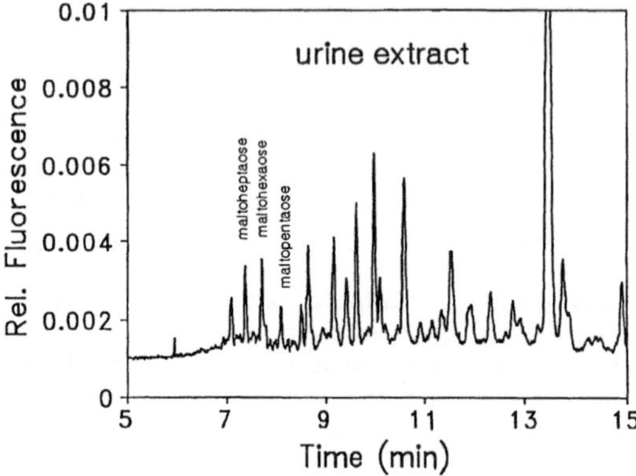

Figure 17. Electropherogram of the separation of three polysaccharides in urine. Reprinted from [61], with permission.

separation of small DNA fragments and PCR products. Resolution of large fragments is better at lower field strength at constant column length, and is directly proportional to column length at the same field strength. In a previous work from this laboratory we were able to resolve the wild-type (400 bp) and mutant (375 bp) PCR products in under 60 s using 1–2 cm effective length capillaries (Fig. 15) [52, 53]. In a recent CE study [54] we compared the effect of experimental temperature on the separation of DNA digest. The results showed that better resolution of the DNA fragments was obtained at 25°C than at 40 or 50°C. Array CE with gel-filled capillaries has been used for DNA sequencing [55], and this technique recently gained acceptance as the method of choice for DNA sequencing of the human genome. Also, microchip array technology is being used for DNA sequencing [56].

7 Clinical applications

Applications of CE in the clinical laboratory have been reported [57]. The determination of different compounds in biological fluids may at times not be a simple matter, due to the fact that biological fluids often have a complex matrix. For example, while a compound of interest may be present in blood or serum at the low mg/L, sodium and protein are present at 10 000 and 100 000 mg/L, respectively [58]. Also, the matrix may affect reproducibility, peak shape, and quantitation. However, selective detection may overcome some of the matrix effects. In our laboratory CE has been used for the separation of homovanillic and vanillylmandelic acids in baby urine (Fig. 16) [59], hydroxyproline in human urine and serum [60], and polysaccharides from human urine (Fig. 17) [61]. Also, we have reported the separation of retinoic acid isomers [62], caffeine metabolites [48], separation of nicotine and three of its metabolites in human urine (Fig. 18) [63] and sepa-

ration of ten estrogens and detection by UV (Fig. 13) [63], and the determination of estrogens by CE with electrochemical detection and microdialysis sampling [64]. Those interested in the CE analysis of drugs should consult [65].

8 Illicit drugs

The earliest work on the use of CE for the separation of illicit drugs was reported by Weinberger and Lurie [66] for the separation of heroin impurities employing a micellar buffer and UV detection. Later, Lurie *et al.* [67, 68] used MEKC with UV-LIF detection for the separation and detection of impurities in heroin and cocaine (Fig. 19). An improvement of 1000-fold in sensitivity over UV detection

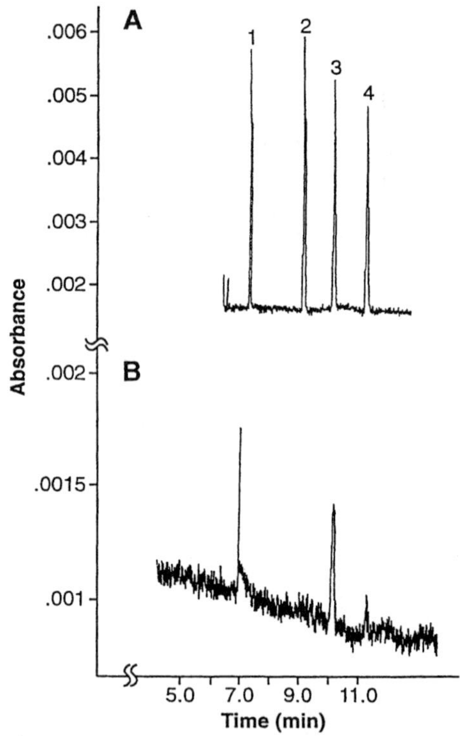

Figure 18. Separation of nicotine and three of its main metabolites. (A) Standard mixture; (B) smoker's urine sample. Buffer, 50 mM sodium acetate, pH 5.7; voltage, 20 kV; pressure injection, 3 s at 0.5 psi; detection, absorption of 260 nm; instrument, Beckman Model P/ACE System 5510; column, 10%T polyacrylamide-coated fused silica; column dimensions, L_{total} = 47 cm, $L_{detection}$ = 40 cm; 75 µm ID; solute concentration, 2–5 µg/mL in water. Solutes: 1, nicotine; 2, demethylcotinine; 3, cotinine; 4, *trans*-3-hydroxycotinine. Reprinted from [63], with permission.

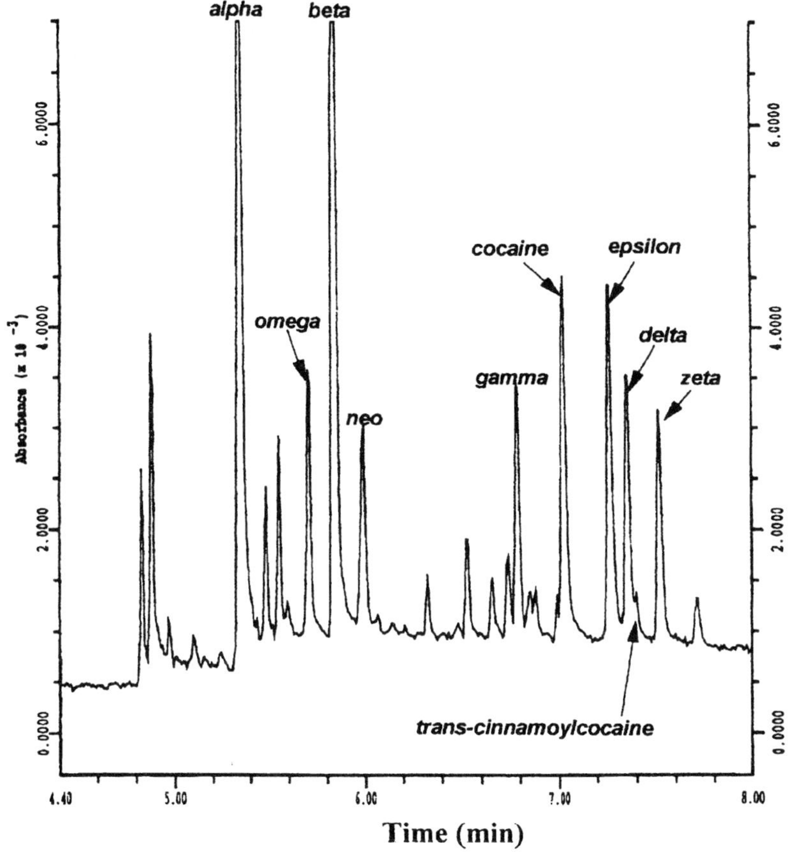

Figure   19. Electropherogram of an illicit cocaine HCL HPLC size exclusion extract. The run buffer contained 10% methanol and 90% of 10 mM β-CD-(SBE)-(IV), 7 mM phosphate (dibasic), pH 8.6. Reprinted from [68], with permission.

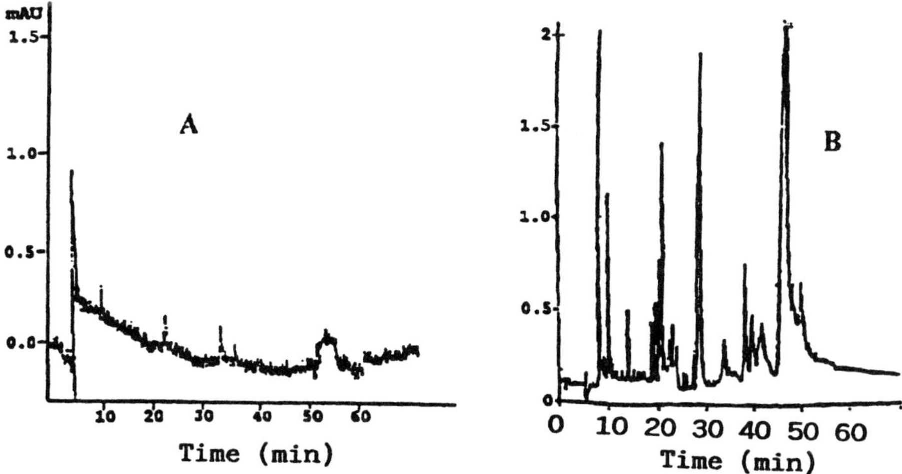

Figure 20. Comparison of (A) UV detection at 260 nm and (B) LIF detection with krypton-fluoride laser for the CE analysis of a refined Southwest Asian heroin hydrochloride exhibit. Reprinted from [67], with permission.

was achieved (Fig. 20). In the meantime, few studies have been published dealing with the separation of illicit drugs [69].

9 Chiral separations

The first to report on chiral separations by CE were Gassmann *et al.* [70]. They resolved a mixture of dansylated D- and L-amino acids by adding Cu(II)-L-histidine to the running buffer. Since then many studies have been published dealing with chiral separations, employing different complexing agents and materials. The most widely used compounds for chiral separations are the CDs and their modified derivatives. The first use of CDs in CE was reported by Terabe [71] and Terabe *et al.* [72] for the separation of positional isomers. Later, Guttman *et al.* [73] obtained chiral separations by incorporating CD within a polyacrylamide gel-filled capillary, and Fanali [74] reported on chiral CE separations by adding CDs to the running buffer. Issaq [75] compared the separation of chiral compounds by CE with other methods. For a detailed discussion of chiral separations by CE and MEKC, consult [76].

In our laboratory few procedures were used for the separation of D- and L-amino acids. 9-Fluorenylmethyl chloroformate (FMOC)-derivatized amino acids were resolved by MEKC in an SDS-CD phosphate buffer (Fig. 5) [19, 20]. The second approach involved reacting the chiral amino acid mixture with the chiral fluorescent agent FLEC to produce two diastereoisomers which can be resolved without the addition of a chiral reagent to the running buffer in a bare fused silica capillary (Fig. 6) [20, 21]. In both cases the resolved compounds were detected by UV-LIF. In another study, a polyacrylamide-coated capillary was used in a reversed-flow MEKC mode (RF-MEKC) to resolve a racemic mixture of danylated amino acids, and dipeptides using 1% beta CD-sulfobutyl ether in 10 mM phosphate buffer at pH 3.1 [43].

10 Miscellaneous separations

CE in all its formats, CZE, CGE, CIEF, MEKC, and RF-MEKC, have been applied to the separation of hundreds of widely different compounds. A multitude of books and journal articles dealing with CE separations have been published, and therefore it is difficult to cover such a wide topic in a short review. This review is thus limited to research topics that we investigated and published. RF-MEKC is a procedure which was developed in our laboratory for the separation of hydrophobic compounds. The principle is simple: electrokinetic chromatography with negatively charged CD or SDS added to the buffer was conducted in polayacrylamide-coated fused silica capillaries under suppression of electroosmotic flow with reversed polarity; *i.e.*, the flow to the detector is from the negative electrode to the positive electrode. In this proce-

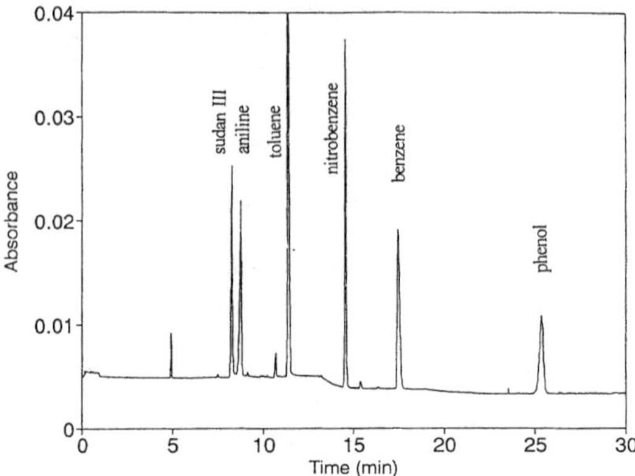

Figure 21. Order of migration *versus* solute hydrophobicity in RF-MEKC mode. Column, 10% linear polyacrylamide-coated fused silica; column dimensions, L_{total} = 57 cm, $L_{detection}$ = 50; 75 µm ID; buffer, 10 mM acetate and 50 mM SDS, pH 4.2; applied voltage, –20 kV; current, 30 µA; detection, 214 nm. Reprinted from [79], with permission.

dure the most hydrophobic solutes are eluted first, unlike MEKC where the most hydrophobic solutes elute last. This procedure was used for the separation of a mixture of polycyclic aromatic hydrocarbons (Fig. 21) [79], and a mixture of steroids (Fig. 22) [79] in addition to other hydrophobic compounds [77–80], and a racemic mixture of amino acids, dipeptides, aflatoxins, and chlorophenols [43]. Our group demonstrated the separation of pyridinecarboxylic acid isomers and related compounds [81], the separation of a mixture of catecholamine metabolites (Fig. 23) [59] the advantages of subambient temperature, 5°C, for the separation of heterocyclic nitrosoamino acid conformers (Fig. 24) [82], the separation of pyridinecarboxylic acid isomers and related compounds [83], and the separation of a retinoic acid mixture (Fig. 25).

11 Physicochemical studies

In this section we try to answer the following question: Is CE a separation technique only or can it be used to determine physicochemical parameters? To date, a search of the literature reveals that CE can indeed be used to determine physicochemical parameters. CE has been used for the determination [84] of: (i) pK values of weak electrolytes [85] and amphoteric compounds; (ii) pI values of proteins by CIEF; (ii) the mobility and its relation to charge/mass ratio in peptides and proteins, (iv) mobility of nucleic acids; (v) viscosity; (vi) binding and dissociation constants [86, 87]; (vii) the approximate molecular weight of proteins; (viii) the approximate number of base pairs of DNA fragments and PCR products; (ix) detection of DNA

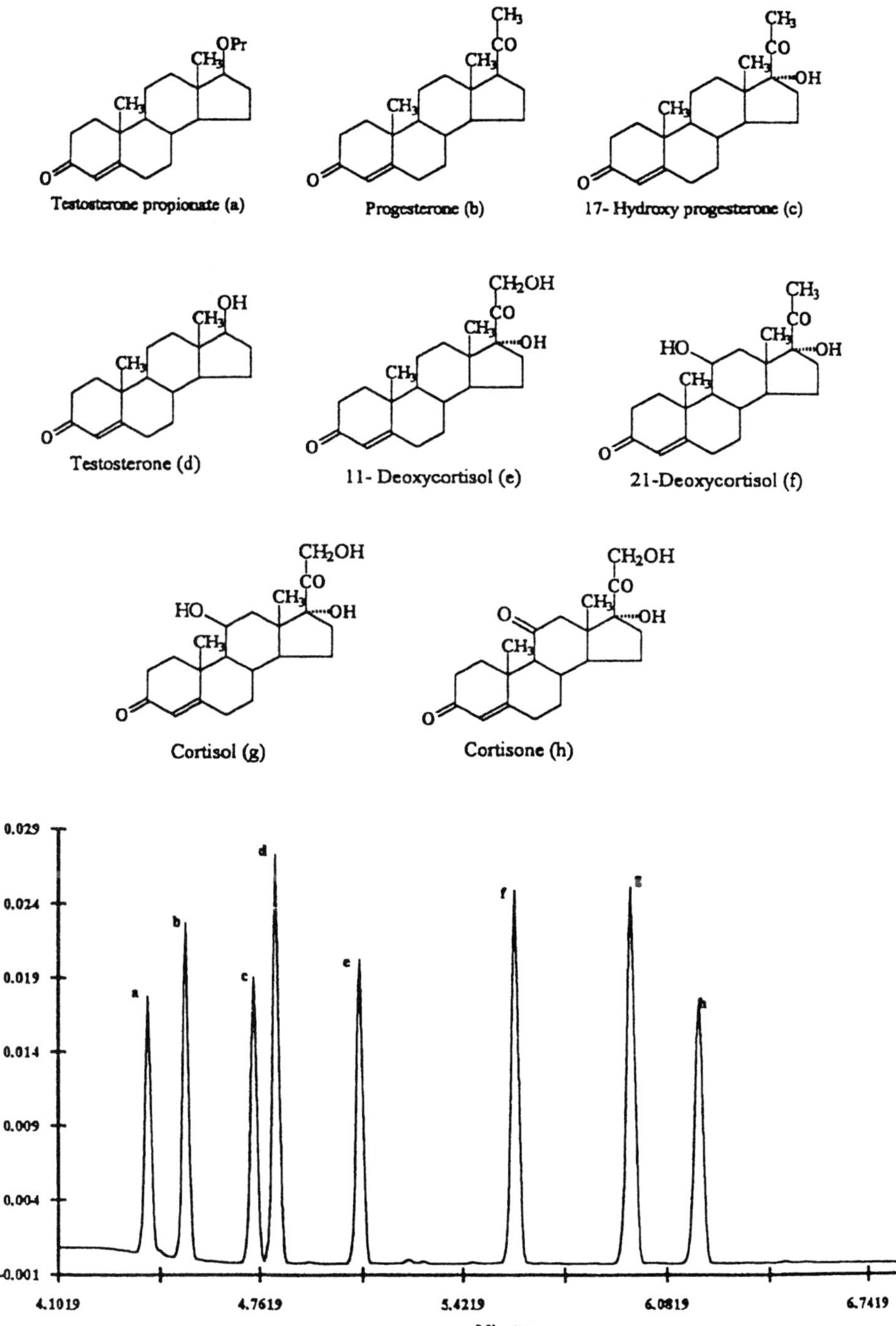

Figure 22. Separation of steroids on neutral (eCAP) capillary. Buffer, 100 mmol/L SDS, 20% v/v acetonitrile and 20 mmol/L ME, pH 6; voltage applied, 15 kV; temperature, 16°C. Reprinted from [79], with permission.

Electrophoresis 2000, *21*, 1921–1939

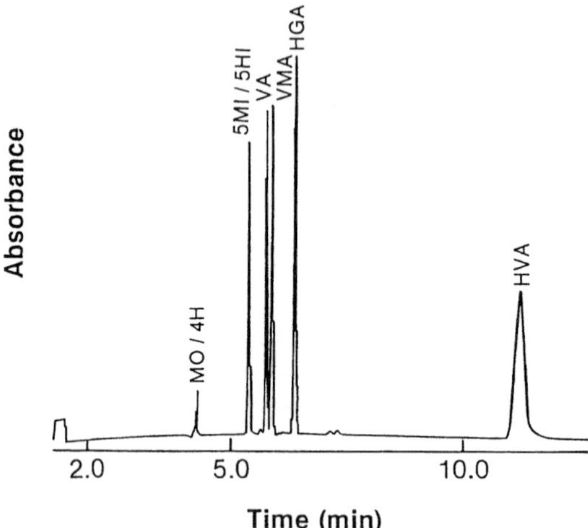

Figure 23. Electropherogram of a standard mixture of catecholamine metabolites. Buffer, 200 mM acetate, pH –4.10; applied voltage, 25 kV; current, 115 µA. Solutes: MO, mesityl oxide; 4 H, 4-hydroxy-3-methoxyphenol-glycol; 5 MI, 5-methoxyindole-3-acetic acid; 5 HI, 5-hydroxyindole-3-acetic acid; VA, 4-hydroxy-3-methoxy benzoic acid; VMA, vanillylmandelic acid; HGA, homogentisic acid; HVA, homovanillic acid. Reprinted from [59], with permission.

mutants [93–95]; (x) insulin content and excretion from single islets of Langerhans [88]; (xi) enzyme activity and quantity in single human erythrocyte [89]; and (xii) enzyme assay and activity [90, 92], *etc.* In our laboratory, CE was used to predict the mobility of peptides [26–28] and to determine protein/drug interaction and protein/DNA interaction (Fig. 26) [32]; RF-MEKC was used to determine the partition coefficients and hydrophobicity of organic compounds [78], CGE to estimate the size and number of base pairs in PCR products and DNA fragments [96], and CIEF to determine the purity and p*I* of proteins [40].

12 Capillary electrochromatography (CEC)

The column type (bare or coated), along with its filling, determined the mode of CE separation. A bare fused silica capillary and a simple buffer are used for CZE. The addition of a micelle to the buffer to resolve mainly neutral compounds is known as MEKC, while RF-MEKC uses a neutrally coated fused silica capillary and a micelle buffer for the separation of hydrophobic compounds. The addition of a chiral reagent to the buffer, simple or micelle, is used for the separation of racemic mixtures. The filling of a neutrally coated capillary with a carrier ampholyte is used for the determination of the p*I* of proteins by IEF. Filling the neutrally coated capillary with a liquid gel, entangled polymers, is used in CGE for the separation of DNA fragments and PCR products. In 1974, Pretorius and co-workers [97] showed that electroosmotic flow, not a pump, can be used to drive the mobile phase through a 1 mm glass LC column packed with 75–175 µm particles. Jorgenson and Lukacs [98] demonstrated the feasibility of using electroosmotic flow and 170 µm ID Pyrex glass tube, 68 cm long, packed with 10 µm octadecyl silica (ODS) particles, for the CEC separation of a mixture made of 9-methylanthracene and perylene. When CEC is compared to HPLC, better overall resolution, higher efficiency and zero back pressure, which will allow the use of up to 1 µm silica particles, are obtained with CEC. CEC has been used for the separation of different groups of compounds. Lurie *et al.* [99] applied CEC for the separation of drugs of forensic interest, while Yang and El Rassi [100] applied CEC for the separation of a mixture of urea herbicides (Fig. 27). The CEC separation of nucleosides and bases [101], small and large nucleic acids [102], carbohydrates [103] and polycyclic aromatic hydrocarbons were reported [104]. The application of CEC has been reviewed very recently by Dermaux and Sandra [105]. A comparison of CEC with CE reveals that CE is simpler, easier to perform, gives higher efficiency, and has wider application. To date, although CEC has been used for the separation of different groups of compounds, it still has not been fully utilized due to the problems of generating reproducible column inlet and outlet frits and the packing of the capillary column. The introduction of open tubular columns by Pesek *et al.* [106] may solve some of the above problems; however, the column capacity will be lower than that using a packed column. Another alternative is to mimic the packed bed by etching an array of support particles into a quartz substrate to produce an array of collocate monolith support structures. Such a microfabricated system was used for the CEC separation of peptides [107].

13 Future directions

The future of CE looks very promising. In the last ten years CE was evaluated and many theoretical studies have been carried out. Our group studied the effect of buffer and buffer modifiers on resolution, mobility and Joule heating [108–119, 28] and, as mentioned above, applied CE not only as a separation technique but to answer physiocochemical questions. Today CE is an established and useful microanalytical technique with widespread applications. As a matter of fact, CE has grown phenomenally. At least 15 books and hundreds of articles have been published; two specific conferences,

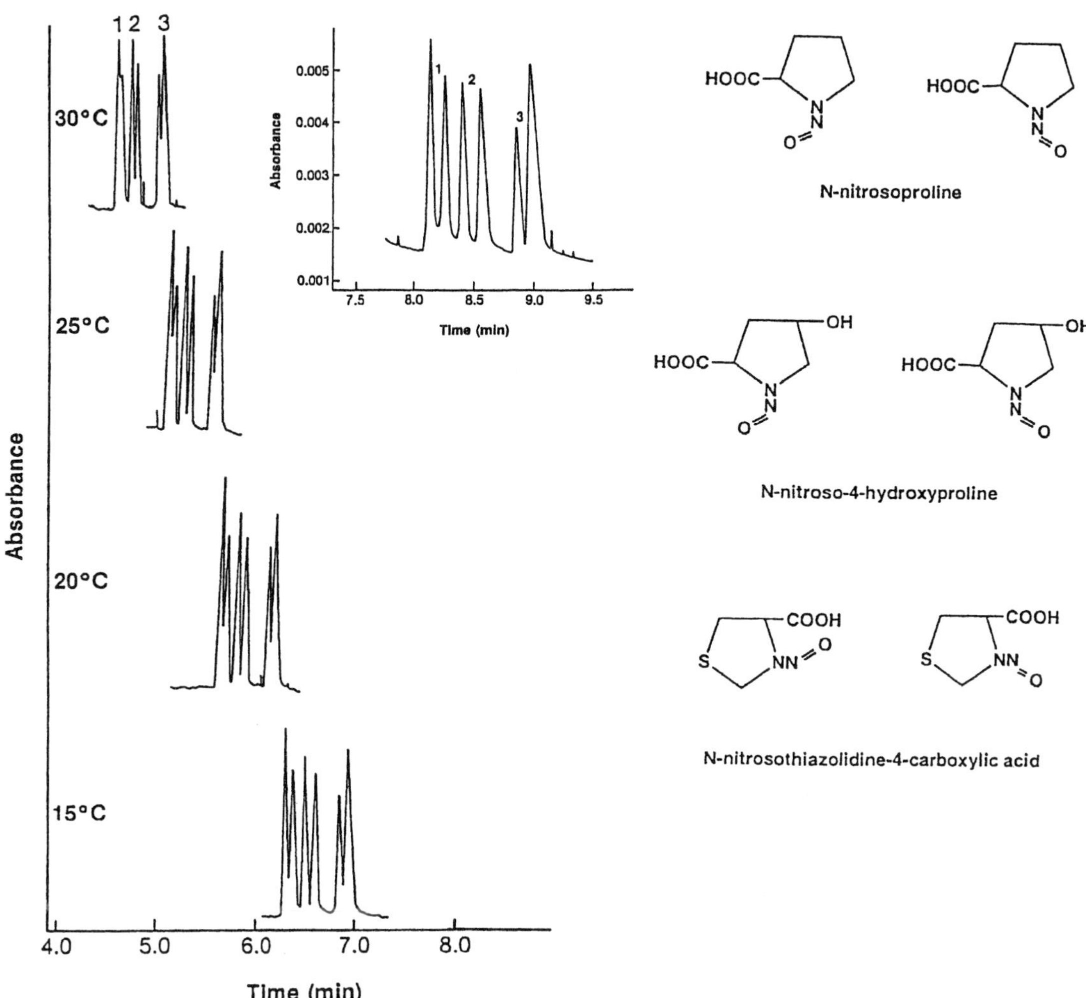

Figure 24. Electropherograms showing the separation of syn- and anti- conformers of selected nitrosoamine acids as a function of temperature. Solutes: 1, *N*-nitrosothiazolidine-4-carboxylic acid; 2, *N*-nitrosoproline; 3, *N*-nitroso-4-hydroxyproline. Instrument, Beckman Model P/ACE System5510; detection, 235 nm; 10%T polyacrylamide-coated fused silica; column dimensions L_{total} = 57 cm, $L_{detection}$ = 50 cm, 75 µm ID; buffer, 10 µm phosphate containing 2 mM DMMAPS and 0.1% Tween-20; pH 7.2, applied voltage, -25 kV; solute concentration, 2-5 µg/mL. Reprinted from [82], with permission.

HPCE and the Frederick CE Conference, were held in the USA; and many sessions in major meetings were devoted to CE theory, instrumentation, and applications. There is no doubt that CE is an acceptable analytical technique. CE has found a strong footing in molecular biology; determination of DNA fragments, PCR products and the sequencing of the human genome, in addition to studies of single cells, protein separations and peptide mapping all use CE. The future of CE is bright as the number of studies, development, and application continue to multiply.

In the next decade, CE will move into ultrafast separations – in seconds not minutes – using microchip technology; and high-efficiency, high-throughput analyses employing narrower and shorter capillaries (5–10 µm ID and 1–5 cm long) will be used in array format to achieve high throughput analyses by CE. Instruments with 96 and

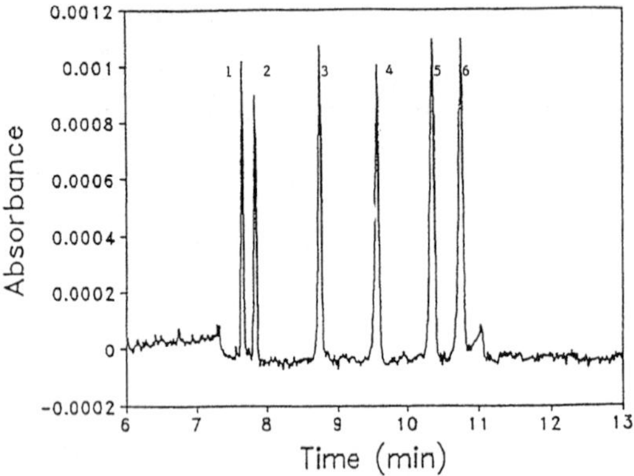

Figure 25. Separation of retinoic acid (RA) mixture from spiked rat plasma extract. Capillary, 75 µm × 47 cm; buffer, 20 mM Tris-borate, pH 8.5; 25 mM SDS, 20% acetonitrile; applied voltage, 13 kV; injection, 4 s, pressure. Peaks: 1, 13-*cis*-4-oxo-RA; 2, 4-oxo-RA; 3, 13-*cis*-acitretin; 4, 13-*cis*-RA; 5, 9-*cis*-RA, 6, all-*trans*-RA. Reprinted from [62], with permission.

384 capillaries will be produced, not only to sequence DNA and the human genome, but to be used for efficient and simultaneous multisample analysis. The era of analyzing one sample at a time will be out of fashion. We will witness fundamental changes for laboratory instrumentation. For example, market studies already project high growth for "lab-on-a-chip" technologies. A key component of such technology involves forcing fluids to move through a chip's channels; the principles involved are borrowed from CE. As an analytical technique, CE will become an established method in the pharmaceutical industry, the clinical laboratory, and the forensic laboratory, where high speed and accurate results are required. In the forensic laboratory, CE will be used for DNA typing, which at the present is being done by SGE, and for gun powder analysis. CE is well suited for the forensic laboratory because of its versatility, small sample requirements and speed of analysis. Apart from its use as an analytical tool, we envision a continued and expanded role for CE in the determination of physicochemical parameters. Multidimensional separations employing separation-separation and separation-spectroscopic techniques will be routinely used for

Ncp7/Drug Screen

Protein/DNA Binding

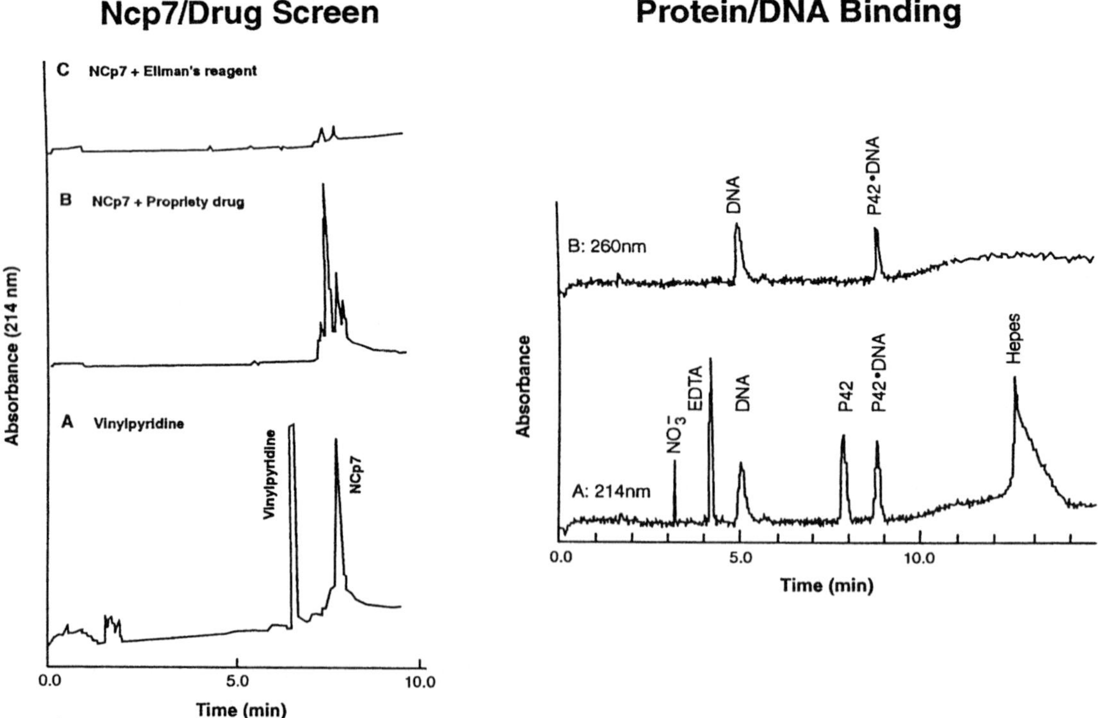

Figure 26. Electropherograms of Ncp7/drug (left) and protein/DNA (right) interaction. Reprinted from [32], with permission.

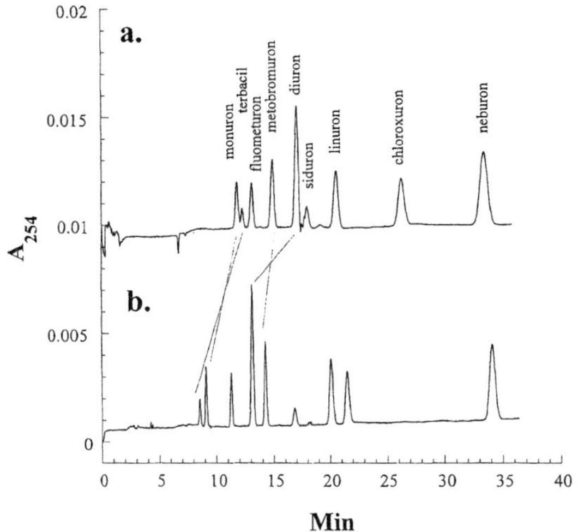

Figure 27. Electrochromatograms obtained with (a) methanol and (b) acetonitrile-containing mobile phases. Mobile phases: (a) 30% v/v 5 mM NaH₂PO₄, pH 6.0, mixed with 70% v/v methanol; (b) 50% v/v 5 mM NaH₂PO₄, pH 6.0, mixed with 50% v/v acetonitrile; column, ODS-Zorbax; length, 27 cm; detection window at 20 cm; running voltage, 20 kV; pressure injection, 20 × at 20 psi; temperature, 15°C; UV detection, 254 nm. Elution order in (a): 1, monurone; 2, terbacil; 3, fluometuron; 4, metobromuron; 5, diuron; 6, siduron; 7, linuron; 8, chloroxuron; 9, neburon; in (b): 1, terbacil; 2, monuron; 3, fluometuron; 4, diuron; 5, metobromuron; 6, siduron; 7, linuron; 8, chloroxuron; 9, neburon. Reprinted from [100], with permission.

the separation and identification of complex mixtures, biological and otherwise.

14 Conclusions

It is abundantly clear that CE has gained a respectable place in different fields dealing with analysis. In this review, mainly about our CE research and application in the last decade, we also, where appropriate, presented work that was done in other laboratories. Because we could not possibly include the applications using CE in this short review, the reader should refer to the more than 15 books and reviews, *e.g.*, the biannual reviews in *Analytical Chemistry*, that have been published on the applications of CE.

The author would like to thank Drs. George Janini and King Chean of SAIC Frederick and Professor Ziad El Rassi, Oklahoma State University, for constructive and helpful comments. This project was funded in whole or in part with Federal funds from the National Cancer Institute, National Institutes of Health, under Contract No. NO1-CO-56000. By acceptance of this article, the publisher or recipient acknowledges the right of the U.S. Government to retain a nonexclusive, royalty-free license and to any copyright covering the article. The content of this publication does not necessarily reflect the views or policies of the Department of Health and Human Services, nor does mention of trade names, commercial products, or organizations imply endorsement by the U.S. Government.

Received November 29, 1999

15 References

[1] Tiselius, A., Thesis, *Nova Acta Regiae Sociates Scientiarum Upsaliensis* 1930, Ser. IV, *7*, Number 4.

[2] Camilleri, P., in: Camilleri, P. (Ed.), *Capillary Electrophoresis, Theory and Practice*, CRC Press, Boca Raton, FL 1997, pp. 1–22.

[3] Swank, R. T., Munkers, K. D., *Anal. Biochem.* 1971, *39*, 462–466.

[4] O'Farrell, P. H., *Cell* 1978, *14*, 545–557.

[5] Hjertén, S., *Chromatogr. Rev.* 1967, *9*, 122–219.

[6] Neuhoff, V., Schill, W. B., Sternbach, H., *Biochem. J.* 1970, *117*, 623–627.

[7] Virtanen, R., *Acta Polytech. Scand.* 1974, *123*, 1.

[8] Mikkers, F. E., Everaerts, F. M., Verhegen, T. P. E. M., *J. Chromatogr.* 1979, *169*, 11–20.

[9] Jorgenson, J., Lukacs, K. D., *Anal. Chem.* 1981, *53*, 1298–1302.

[10] Terabe, S., Otsuka, K., Ichikawa, K., Tsuchuya, A., Ando. T., *Anal. Chem.* 1984, *56*, 111–113.

[11] Karger, B. L., Cohen, A. S., Gutman, A., *J. Chromatogr.* 1989, *492*, 585–614 and references therein.

[12] El Rassi, Z., *Electrophoresis* 1999, *20*, 2987–3330.

[13] El Rassi, Z., *Electrophoresis* 1997, *18*, 2121–2504.

[14] Jones, W. R., Jandik, P., *J. Chromatogr.* 1992, *608*, 385–393.

[15] Jones, W. R., Soglia, J., McGlynn, M., Haber, C., Reineck, J. Krstanovic, C., *Am. Lab.* 1996, *28*, 25–29.

[16] Weston, A., Brown, P., Jandik, P., Jones, W., *J. Chromatogr.* 1992, *608*, 395–402.

[17] Janini, G. M., Muschik, G. M., Issaq, H. J., *J. Capil. Electrophor.* 1994, *1*, 116–120.

[18] Janini, G. M., Chan, K. C., Muschik, G. M., Issaq, H. J., *J. Chromatogr. B* 1994, *657*, 419–423.

[19] Chan, K. C., Janini, G. M., Muschik, G. M., Issaq, H. J., *J. Chromatogr.* 1993, *653*, 93–97.

[20] Issaq, H. J., Chan, K. C., *Electrophoresis* 1995, *16*, 467–480.

[21] Chan, K. C., Muschik, G. M., Issaq, H. J., *Electrophoresis* 1995, *16*, 504–509.

[22] Chan, K. C., Janini, G. M., Muschik, G. M., Issaq, H. J., *J. Liq. Chromatogr.* 1993, *16*, 1877–1890.

[23] Chan, K. C., Muschik, G. M., Issaq, H. J., *J. Chromatogr. A* 1995, *718*, 203–210.

[24] Chan, K. C., Janini, G. M., Muschik, G. M., Issaq, H. J., *J. Chromatogr.* 1993, *622*, 269–273.

[25] Janini, G. M., Lukso, J., Issaq, H. J., *J. High Resolut. Chromatogr.* 1994, *17*, 102–103.

[26] Metral, C. J., Janini, G. M., Muschik, G. M., Issaq, H. J., *J. High Resolut. Chromatogr.* 1999, *22*, 373–378.

[27] Janini, G. M., Metral, C. J., Issaq, H. J., Muschik, G. M., *J. Chromatogr. A* 1999, *848*, 417–433.

[28] Issaq, H. J., Janini, G. M., Atamna, I. Z., Muschik, G. M., Lukszo, J., *J. Liq. Chromatogr.* 1992, *15*, 1129–1142.

[29] Issaq, H. J., Chan, K. C., Janini, G. M., Muschik, G. M., *Electrophoresis* 1999, *20*, 1533–1537.

[30] deGoor, T. V., Apffel, A., Chakel, T., Hancock, W., in: Landers, J. P. (Ed.), *Handbook of Capillary Electrophoresis*, CRC Press, Boca Raton, FL 1998, pp. 213–258.

[31] Lauer, H. H., McManigill, D., *Anal. Chem.* 1986, *58*, 166–170.

[32] Janini, G. M., Fisher, R. J., Henderson, L. E., Issaq, H. J., *J. Liq. Chromatogr.* 1995, *18*, 3617–3628.

[33] Hjertén, S., *J. Chromatogr.* 1985, *347*, 191–198.

[34] Towns, J. K., Regnier, F. E., *J. Chromatogr.* 1990, *516*, 69–78.

[35] Bruin, G. J. M., Chang, J. P., Kuhlman, R. H., Zegers, K., Kraak, J. C., Poppe, H., *J. Chromatogr.* 1989, *471*, 429–436.

[36] Towns, J. K., Regnier, F. E., *Anal. Chem.* 1991, *63*, 1126–1132.

[37] Jorgenson, J. W., Lukacs, K. D., *Science* 1983, *222*, 266–272.

[38] Gilges, M., Usmann, H., Kleemiss, H., Motsch, S. R., Schomburg, G., *J. High Resolut. Chromatogr.* 1992, *15*, 452–457.

[39] Green, J. S., Jorgenson, J. W., *J. Chromatogr.* 1989, *478*, 63–70.

[40] Bullock, J. A., Yuan, L.-C., *J. Microcol. Sep.* 1991, *3*, 241–248.

[41] Muijselaar, W. G. H., de Bruijn, C. H. M. M., Everaerts, F. M., *J. Chromatogr.* 1992, *605*, 115–123.

[42] Landers, J. P., Oda, R. P., Madden, B. J., Spelsberg, T. C., *Anal. Biochem.* 1992, *205*, 115–124.

[43] Janini, G. M., Muschik, G. M., Issaq, H. J., *Electrophoresis* 1996, *17*, 1575–1583.

[44] Issaq, H. J., *Electrophoresis* 1997, *18*, 2438–2452.

[45] Issaq, H. J., *Electrophoresis* 1999, *20*, 3190–3202.

[46] Chan, K. C., Alvarado, A. B., McGuire, M. T., Muschik, G. M., Issaq, H. J., Snader, K. M., *J. Chromatogr. B* 1994, *657*, 301–306.

[47] Chan, K. C., Majadly, F., McCloud, T. G., Muschik, G. M., Issaq, H. J., Snader, K. M., *Electrophoresis* 1994, *15*, 1310–1315.

[48] Atamna, I. Z., Janini, G. M., Muschik, G. M., Issaq, H. J., *J. Liq. Chromatogr.* 1991, *14*, 427–436.

[49] Chan, K. C., Muschik, G. M., Issaq, H. J., Siiteri, P. K., *J. Chromatogr. A* 1995, *690*, 149–154.

[50] Righetti, P. G., Gelfi, C., in: Righetti, P. G. (Ed.), *Capillary Electrophoresis in Analytical Biotechnology*, CRC Press, Boca Raton, FL 1996, pp. 431–476.

[51] Issaq, H. J., Chan, K. C., Muschik, G. M., *Electrophoresis* 1997, *18*, 1153–1158.

[52] Chan, K. C., Muschik, G. M., Issaq, H. J., Garvey, K. J., Generlette, P. L., *Anal. Biochem.* 1996, *243*, 133–139.

[53] Chan, K. C., Muschik, G. M., Issaq, H. J., *J. Chromatogr. B* 1997, *695*, 113–115.

[54] Issaq, H. J., Xu, H., Chan, K. C., Dean, M. C., *J. Chromatogr. B* 2000, *738*, 243–248.

[55] Dovichi, N. J., in: Landers, J. P. (Ed.), *Handbook of Capillary Electrophoresis* CRC Press, Boca Raton, FL 1996, *pp. 545–565*.

[56] Woolley, A. T., Sensabaugh, G. F., Mathies, R. A., *Anal. Chem.* 1997, *69*, 2182–2186.

[57] Oda, R. P., Bush, V. J., Landers, J. P., in: Landers, J. P. (Ed.), *Handbook of Capillary Electrophoresis*, CRC Press, Boca Raton, FL 1996, pp. 639–673.

[58] Shihabi, Z. K., in: Landers, J. P. (Ed.), *Handbook of Capillary Electrophoresis*, CRC Press, Boca Raton, FL 1996, pp. 457–477.

[59] Issaq, H. J., Delviks, K., Janini, G. M., Muschik, G. M., *J. Liq. Chromatogr.* 1992, *15*, 3193–3201.

[60] Chan, K. C., Janini, G. M., Muschik, G. M., Issaq, H. J., *J. Chromatogr.* 1993, *653*, 93–97.

[61] Chan, K. C., Issaq, H. J., unpublished results.

[62] Chan, K. C., Lewis, K. C., Phang, J. M., Issaq, H. J., *J. High Resolut. Chromatogr.* 1993, *16*, 558–562.

[63] Issaq, H. J., Chan, K. C., Muschik, G. M., Janini, G. M., *J. Liq. Chromatogr.* 1995, *18*, 1273–1288.

[64] Perkins, M., Zhong, M., Issaq, H. J., Davies, M., Lunte, S., *Amer. Assoc. Pharm. Scientists Meeting*, Boston, MA, Nov. 6, 1997.

[65] Cohen, A. S., Terabe, S., Deyl, Z. (Eds.), *Capillary Electrophoresis of Drugs*, Elsevier, Amsterdam 1996.

[66] Weinberger, R., Lurie, I. S., *Anal. Chem.* 1991, *63*, 823–827.

[67] Lurie, I. S., Chan, K. C., Spartley, T. K., Casale, J. F., Issaq, H. J., *J. Chromatogr. B* 1995, *669*, 3–13.

[68] Lurie, I. S., Hays, P. A., Casale, J. F., Moore, J. M., Castel, D. M., Chan, K. C., Issaq, H. J., *Electrophoresis* 1998, *19*, 51–56.

[69] Heeren, F. V., Thorman, W., *Electrophoresis* 1997, *18*, 2415–2426.

[70] Gassmann, E., Kuo, J. E., Zare, R. N., *Science* 1985, *230*, 813–815.

[71] Terabe, S., *Trends Anal. Chem.* 1989, *8*, 129–134.

[72] Terabe, S., Ozuaki, H., Otsuka, K., Ando, T., *J. Chromatogr.* 1985, *332*, 211–217.

[73] Gutman, A., Paulus, A., Cohen, A. S., Grinberg, N., Karger, B. L., *J. Chromatogr.* 1988, *448*, 41–53.

[74] Fanali, S., *J. Chromatogr.* 1989, *474*, 441–446.

[75] Issaq, H. J., *Instrum. Sci. Technol.* 1994, *22*, 119–149.

[76] Otsuka, K., Terabe, S., in: Guzman, N. A., (Ed.), *Capillary Electrophoresis Technology*, Marcel Dekker, New York, NY 1993, pp. 617–619.

[77] Janini, G. M., Muschik, G. M., Issaq, H. J., *J. Chromatogr. B* 1996, *683*, 29–35.

[78] Janini, G. M., Muschik, G. M., Issaq, H. J., *J. High Resolut. Chromatogr.* 1995, *18*, 171–174.

[79] Janini, G. M., Issaq, H. J., Muschik, G. M., *J. Chromatogr. A* 1997, *792*, 125–141.

[80] Issaq, H. J., Janini, G. M., Muschik, G. M., *HPCE '96*, Orlando, FL, USA, January 21–25, 1996.

[81] Janini, G. M., Chan, K. C., Barnes, J. A., Muschik, G. M., Issaq, H. J., *J. Chromatogr.* 1993, *653*, 321–327.

[82] Janini, G. M., Muschik, G. M., Issaq, H. J., *J. High Resolut. Chromatogr.* 1994, *17*, 753–755.

[83] Janini, G. M., Chan, K. C., Barnes, J. A., Muschik, G. M., Issaq, H. J., *J. Chromatogr.* 1993, *653*, 321–327.

[84] Issaq, H. J., *Eastern Analytical Symposium*, Somerset, NJ, USA, November 15–20, 1998.

[85] Cai, J., Smith, T. J., El Rassi, Z., *Electrophoresis* 1992, *13*, 30–32.

[86] Heegard, N. H. H., *Electrophoresis* 1998, *19*, 367–464.

[87] Robey, F. A., in: Landers, J. P. (Ed.), *Handbook of Capillary Electrophoresis*, CRC Press, Boca Raton, FL 1996, pp. 591–609.

[88] Schultz, N. M., Huang, L., Kennedy, R. T., *Anal. Chem.* 1995, *67*, 924–929.

[89] Tan W., Yeung, E. S., *Anal. Biochem.* 1995, *226*, 74–79.

[90] Wu, D., Regnier, F. E., *Anal. Chem.* 1993, *65*, 2029–2035.

[91] Xu, Q., Yeung, E. S., *Nature* 1995, *373*, 681–683.

[92] Schultz, N. M., Tao, L., Rose, D. J., Kennedy, R. T., in: Landers, J. P. (Ed.), *Handbook of Capillary Electrophoresis*, CRC Press, Boca Raton, FL 1996, pp. 611–637.

[93] Khrapko, K., Hanekamp, J. S., Thilly, W. G., Belenkii, A., Foret, F., Karger, B. L., *Nucleic Acids Res.* 1994, *22*, 364–369.

[94] Gelfi, C., Righetti, P. G., Trav, M., Father, S., *Electrophoresis* 1997, *18*, 724–731.

[95] Li-Sucholeiki, X.-C., Khrapko, K., Andre, P. C., Marcelino, L. A., Karger, B. L., Thilly, W. G., *Electrophoresis* 1999, *20*, 1224–1232.

[96] Issaq, H. J., Janini, G. M., Chan, K. C., Muschik, G. M., *Eastern Analytical Symposium*, Somerset, NJ, USA, November 16–21, 1997.

[97] Pretorius, V., Hopkins, B. J., Shieke, J. D., *J. Chromatogr.* 1974, *99*, 23–29.

[98] Jorgenson, J. E., Lukacs, K. D., *J. Chromatogr.* 1981, *218*, 209–216.

[99] Lurie, I. S., Conver, T. S., Meyers, R. P., Bailey, C. G., *10th Annual Frederick Conference on CE* Frederick, MD, USA, October 18–20, 1999

[100] Yang, C., El Rassi, Z., *Electrophoresis* 1999, *20*, 2337–2342.

[101] Zhang, M., El Rassi, Z., *Electrophoresis* 1999, *20*, 31–36.

[102] Zhang, M., Yang, C., El Rassi, Z., *Anal. Chem.* 1999, *71*, 3277–3282.

[103] Yang, C., El Rassi, Z., *Electrophoresis* 1998, *19*, 2061–2067.

[104] Yan, C., Dadoo, R., Shao, H., Zare, R. N., *Anal. Chem.* 1995, *67*, 2026–2029.

[105] Dermaux, A., Sandra, P., *Electrophoresis* 1999, *20*, 3027–3065.

[106] Pesek, J. J., Matyska, M., Menezes, S., *J. Chromatogr. A* 1999, *853*, 151–158.

[107] Bing, H., Junyan, J., Regnier, F. E., *J. Chromatogr. A* 1999, *853*, 257–262.

[108] Issaq, H. J., Atamna, I. Z., Metral, C. J., Muschik, G. M., *J. Liq. Chromatogr.* 1990, *13*, 1247–1259.

[109] Atamna, I. Z., Metral, C. J., Muschik, G. M., Issaq, H. J., *J. Liq. Chromatogr.* 1990, *13*, 2517–2527.

[110] Atamna, . Z., Metral, C. J., Muschik, G. M., Issaq, H. J., *J. Liq. Chromatogr.* 1990, *13*, 3201–3210.

[111] Issaq, H. J., Atamna, I. Z., Muschik, G. M., Janini, G. M., *Chromatographia* 1991, *32* 155–161.

[112] Atamna, I. Z., Issaq, H. J., Muschik, G. M., Janini, G. M., *J. Chromatogr.* 1991, *588*, 315–320.

[113] Janini, G. M., Issaq, H. J., *J. Liq. Chromatogr.* 1992, *15*, 927–960.

[114] Issaq, H. J., Janini, G. M., Atamna, I. Z., Muschik, G. M., Lukszo, J., *J. Liq. Chromatogr.* 1992, *15*, 1129–1136.

[115] Janini, G. M., Issaq, H. J., in: Guzman, N. A. (Ed.), *Capillary Electrophoresis: Theory, Methodology and Applications*, Marcel Dekker, New York 1993, *pp. 119–160*.

[116] Janini, G. M., Chan, K. C., Barnes, J. A., Muschik, G. M., Issaq, H. J., *Chromatographia* 1993, *35*, 497–492.

[117] Janini, G. M., Chan, K. C., Muschik, G. M., Issaq, H. J., *J. Liq. Chromatogr.* 1993, *16*, 3591–3607.

[118] Issaq, H. J., Janini, G. M., Chan, K. C., Rassi, Z. E., in: Brown, P. R., Grushka, E. (Eds.), *Advances in Chromatography*, Volume 35, Marcel Dekker, New York 1995, pp. 101–170.

[119] Issaq, H. J., Horng, P. L., Janini, G. M., Muschik, G. M., *J. Liq. Chromatogr. Rel. Technol.* 1997, *20*, 167–182

Electrophoresis 2000, *21*, 3873–3887

Review

Gary W. Slater
Claude Desruisseaux
Sylvain J. Hubert
Jean-François Mercier
Josée Labrie
Justin Boileau
Frédéric Tessier
Marc P. Pépin

Department of Physics,
University of Ottawa,
Ottawa, Canada

Theory of DNA electrophoresis: A look at some current challenges

Although electrophoresis is one of the basic methods of the modern molecular biology laboratory, new ideas are being suggested at an accelerated rate, in large part because of the pressing demands of the biomedical community. Although we now have, at least for some methods, a fairly good theoretical understanding of the physical mechanisms that lead to the observed peak spacings, widths and shapes, this knowledge is often too qualitative to be used to guide further technical developments and improvements. In this article, we review some selected elements of the current state of our theoretical ignorance, focusing mostly on DNA electrophoresis, and we offer several suggestions for further theoretical investigations.

Keywords: Gel electrophoresis theory / DNA separation and sequencing / Capillary electrophoresis / Free-flow electrophoresis / Ratchet separation systems / Simulations / Review
EL 4197

Contents

Correspondence: Dr. Gary W. Slater, Department of Physics, University of Ottawa, 150 Louis-Pasteur, Ottawa, Ontario, Canada K1N 6N5
E-mail: gslater@science.uottawa.ca
Fax: +613-562-5190

Abbreviations: BD, Brownian dynamics; **BRF**, biased reptation with fluctuation model; **BRM**, biased reptation model; **ELFSE**, end-labeled free-solution electrophoresis; **LPA**, linear polyacrylamide; **MC**, Monte-Carlo; **MD**, Molecular Dynamics; **PA**, polyacrylamide

1 Introduction

The development of automated capillary electrophoresis (CE) and capillary array electrophoresis (CAE) are the most recent technological advances that have made possible the completion of the sequencing of the human genome years ahead of the original estimates [1]. Besides sequencing further genomes for comparison and other applications (for example in the agricultural sciences), genetic testing and diagnostic will now represent the major pressure for the development of even higher throughput and more robust DNA analysis technologies. Electrophoretic technologies are likely to remain an important tool for years to come, although we are now seeing a clear evolution in the way electrophoresis is being used. For instance, the recent microchannel systems based on various entropic and ratchet concepts represent drastic departures from standard sieving-based electrophoretic processes. These novel ideas will be discussed in this review.

We have published a review article covering some of the new, innovative separation ideas in 1998 [2]. Obviously, much progress has been made in most of these directions since then. Instead of updating our review, however, we decided to choose a different path. Standard review articles typically discuss the progress made over the last several years, showing the successes and predicting the probable course of the research for the next little while. Werner von Braun once said: "Fundamental science is what I do when I don't know what I am doing". Based on this unique philosophy, our review will actually focus on the state of our ignorance. A review of some of the theoretical questions that remain to be answered, is a useful first step towards designing new fundamental research programs (combining both experimental and theoretical investigations). We hope that this original approach will be useful to experimentalists wishing to carry out fundamental studies of electrophoretic processes, and to theoreticians who are looking for new challenges.

The plan of the article is simple: each section covers a different technological or theoretical topic. The sections typically describe several key questions, why they remain unanswered, what (if anything) has been done recently about it, and in some cases which path research might take. We thus offer questions, not answers. These questions apply to a variety of separation modes (CE, slabs, ratchets, free-flow, microchannels, *etc.*; Fig. 1 presents a schematic picture of the different CE and microchip separation modes that will be covered in this review) and uses (mapping, sequencing, preparative, sizing). Similarly, they are related to a number of fundamental theoretical issues (ionic effects, CE and microchannel walls, sieving mechanisms, diffusion and band broadening, resolution, and read length, entropic effects) and they may require a range of theoretical methods (scaling analysis, simple analytical models, complex computer simulations). Obviously, it is impossible to cover all aspects of DNA electrophoresis in a single review article. As many of us know, it is even hard to keep track of the evolution of the discipline! This is why *Electrophoresis's* excellent annual review articles are so important. Our reference list is by no mean complete. Our choice of questions is also far from being complete: the choice we made is certainly biased by our very own interests, as one would expect. We can assure the reader, however, that the scientific community ignores a lot more than what we discuss in this review!

2 How can we design a sieving matrix for DNA CE sequencing?

Polyacrylamide (PA) gels are no longer the electrophoretic sieving matrices of choice for small DNA fragments, especially for sequencing applications, since CE methods

now dominate the technology. Although entangled solutions of very long polymers behave roughly like cross-linked gels (compare, *e.g.*, Figs. 1d and 1f), there are differences. Moreover, the larger number of parameters related to the polymer solution (the chemical composition, the molecular size and its polydispersity, the stiffness, the hydrophilicity or -phobicity, the polymer architecture, *etc.*) makes it more difficult to narrow the search for the optimal choice. Among the (numerous) factors that we must take into account, we find: (i) the possibility of using uncoated capillaries if the polymer self-coats the walls and kills the EOF [3]; (ii) the viscosity of the solution must be such that injection is mechanically doable; (iii) the entanglements must be long-lived compared to the relevant time scales [4]; (iv) the effective pore size of the solution must optimize the resolution [4]. In some cases, the polymer solution is self-organizing into microphases [5, 6]; the periodicity and geometry of these structures are then additional factors.

Perhaps the best studied sieving polymer is polydimethylacrylamide (pDMA) [3]. Heller [7–9] published a superb analysis of the performance of pDMA. His results show that the biased reptation model with fluctuations (BRF) [10–15], the best model that we currently have for understanding the electrophoresis of DNA fragments in dense porous systems, does not quite explain all the observations. Cottet *et al.* [4], on the other hand, described how theoretical concepts can be used to compare the performance of different polymers. Together, these studies provide a systematic methodology to characterize the sieving properties of polymer solutions. Unfortunately, most studies use less-than-optimal and hard-to-compare methodologies and data analysis approaches.

Perhaps the key questions where theory is lacking in predictive power are: (i) Is it preferable to have microstructures or homogeneous (nonassociating) polymer solutions? (ii) Is it preferable to have linear or branched polymers? (iii) How can we mix different polymers to optimize the resolution/read length? (iv) What length scales and time scales must be optimized when designing the polymer solution? (v) What happens when the concentration c of the solution is close to the overlap concentration c^*? (vi) what is the role of the polymer dynamics, especially on band broadening/diffusion processes? (vii) Can we use extremely high-fields in spite of the fact that the gel is elastic and compliant?

The best reported results appear to come from linear polyacrylamides (LPAs) [16], although there is no fundamental reason why PA should remain the leading polymer sieving matrix for DNA CE sequencing. There is certainly a lot of room for theoretical studies and computer modeling. Our knowledge of the physics of electrophoresis in polymer solutions is quite limited compared to what we previously achieved for gel electrophoresis. We believe that computer simulations will play an important role in the next few years since it is now possible to do realistic simulations of very detailed analyte-gel systems using a variety of computational tools.

3 Can end-labeled free-solution electrophoresis (ELFSE) be developed for sequencing applications?

Our group started developing a new method for DNA separation, which we called ELFSE, in 1994 [17]. ELFSE goes against intuition since it is usually taken for granted that separation of DNA fragments in the absence of a sieving matrix is impossible (indeed, the DNA free solution mobility μ_0 is normally molecular size-independent [15, 18]). ELFSE is based on the idea that one can create an "hydrodynamic asymmetry" by attaching a neutral object (a buoy!) at one end of the DNA molecules (see Fig. 1a). This object breaks the charge-to-friction symmetry and the mobility of a DNA molecule with M bases is then given by

$$\mu(M) = \mu_0 \frac{M}{M + \alpha} \tag{1}$$

where α is the effective friction coefficient in the "buoy" (measured in number of equivalent DNA bases). Recently, we reported the separation of dsDNA fragments [19] and the sequencing of about 100 bases in $\approx$ 18 min [20]. This method would be perfect for CE and microchip technologies since it does not require the injection of viscous polymer solutions in small ID channels. Equation (1) clearly indicates that the number of DNA bases that we can sequence with ELFSE, is directly proportional to the "size" α of the buoy. Larger buoys would lead to increased read lengths [20]. However, it is not trivial to find a large monodisperse object that could act as an efficient ELFSE label. Although larger labels would considerably slow down the separation, previous experimental work demonstrated that high fields can be used with ELFSE since biased reptation does not take place in free solution [19]. With a streptavidin buoy ($\alpha \approx 30$), about 100 bases could be sequenced. To achieve 500–600 bases, we probably need a value as large as $\alpha \approx 250$ [17, 20]. If you add to this that this label must be perfectly monodisperse, uncharged and hydrophilic, you have quite a challenge!

On the theoretical side, many fundamental issues need further clarification, including: (i) What is the diffusion coefficient (and the relevant band broadening processes) of the labeled DNA molecule during ELFSE? (ii) What kind of molecular architecture would optimize the performance of an ELFSE label? (iii) In particular, what would be the effect of a deformable label [21, 22]? (iv) Since one applies both electric and mechanical forces to the DNA molecules during ELFSE, the free-draining nature of the polyelectrolyte is modified in a fundamental way [22–26]. What is the impact of this breakdown of the free-draining concept on the performance of ELFSE? Our group is currently making progress in most of these directions, but the lack of experimental or numerical systems to test the predictions of the new theoretical framework that we are developing is a problem.

4 Are artificial electrophoresis "sieving" structures better?

The development of artificial electrophoresis sieving media is certainly a major step towards optimizing DNA separation methods and integrating them on chip within complete microanalysis environments. These novel sieving structures typically consist in arrays of obstacles etched on the surface of a silicon wafer using lithographic technologies. Precise control over the geometry of the sieving matrix is therefore possible, and this design flexibility probably constitutes a key advantage of artificial structures over traditional separation media. Examples of artificial separation microstructures include matrices of posts that mimic gel fibers (shown schematically in Fig. 1h) [27–30], channels with alternating deep and shallow regions that form entropic trap arrays (Fig. 1i) [31, 32], and asymmetric arrays of obstacles that act as Brownian motion rectifiers [28, 33–35] (see also the Brownian ratchet Section 1.1 herein).

The challenges brought about by these devices are at the outset practical ones. With resolution and operation speed already approaching those of leading separation technologies for DNAs in the kbp range [28, 34, 36], question concerning their scalability arise. One may thus ask: Will artificial structures ever reach sequencing resolution or fractionate multimegabase strands? Chou *et al.* [28] anticipate that sequencing will indeed be possible in arrays of sub-μm posts with ultrahigh (and pulsed) electric fields, but this has yet to be demonstrated. Channel structures are best suited for the separation of very long DNA molecules since they contain no small scale obstacle that inhibit chain relaxation, but Mbp molecules have not been tested so far. As for Brownian rectifiers, they have the advantages of sorting globular objects of practically any type (possibly even cells) and operating continuously [34]. It thus seems likely that various structures will eventually cooperate in a single chip environment to allow preparation and analysis of molecules over a wide range of sizes and types. We should also note that band-broadening remains an important practical limitation to high resolution and reliable detection in artificial sieving structures.

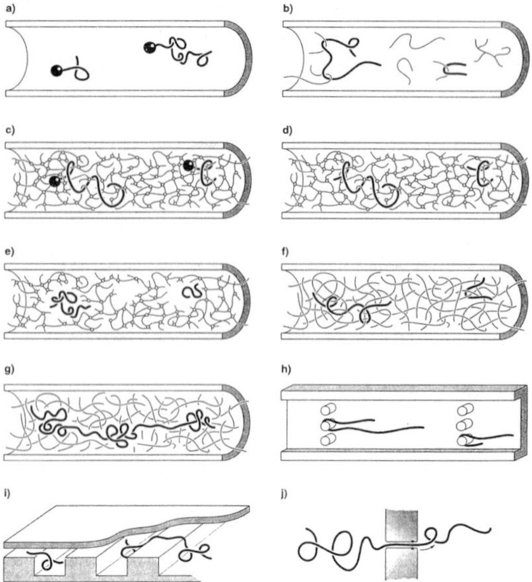

Figure 1. Schematic pucture showing some of the DNA separation mechanisms and ideas discussed in this article. (a) In ELFSE, a neutral buoy slows down the smaller DNA molecules during free-flow CE; (b) collisons between long DNA molecules and the polymer chains of an ultradilute (unentangled) polymer solution provide very fast CE separations; (c) in trapping electrophoresis, the large particle attached at the end of the DNA molecules gets stuck when the leading DNA strand chooses too narrow a passage; (d) gel electrophoretic separation of DNA molecules *via* a reptation process; (e) entropic trapping dominates the migration of smaller DNA molecules that jump between the larger pores inside a disordered gel; (f) the CE separation of DNA molecules in entangled polymer solutions resembles the situation observed in gels (see d), except for the fact that the polymer matrix is dynamic; (g) in very dense polymer solutions, megabase DNA molecules migrate in an elongated conformation that seems to fill large "lakes" in an almost periodic manner; (h) artificial arrays of post can be designed at will using modern techniques; here, two DNA molecules escape from the collision with apost; in the process, they go through a U→J→I sequence of conformations; (i) artificially constructed entropic trapping system; the DNA molecules must squeeze to go through the narrower parts of the channel, and this process leads to size separation; (j) in nanopore technologies, a ssDNA molecule is pulled through a nm-size hole, one monomer at a time, and information about the sequence can be inferred from various physical measurements.

The availability of simple and regular geometries also presents unique opportunities from the standpoint of electrophoresis theory and modeling, as it alleviates difficulties associated with the intricate pseudorandom and irreproducible architecture of polymer gels. Simple models already seem to explain the dynamics of molecules through microfabricated arrays quite well. Not to mention that direct observation of DNA strands in artificial structures can reveal some of their fundamental properties and how these are related to the local geometry. Theoretical investigations can now adopt a more quantitative point of view and revisit issues such as the impact of molecular size, type, and topology on separation efficiency. Detailed mechanisms of electrophoretic separation can be uncovered, the effect of order and disorder in the disposition of sieving elements can readily be studied, and new ideas can be tested in conclusive ways on new custom-made devices. Given the dimensions of microfabricated arrays and the processing involved in their fabrication, the dynamics of biomolecules and their interactions with surfaces also suggest new challenges on the road to integrated systems.

Lastly, while microfabricated sieving structures clearly offer many advantages over polymer gels, we should question whether sieving altogether is the optimal technique for macromolecular sizing. After all, on-chip arrays and channels often reproduce structural features found in gels (albeit in a controlled and orderly fashion) and size information is still extracted indirectly through mobility measurements. But given that we can design structures of molecular dimensions practically at will, perhaps we can tackle the problem in a more straightforward way. The single molecule DNA sizing device constructed by Chou *et al.* [36], which allows an absolute measure of strand length *via* fluorescence detection, stands as a superb example of a direct sizing approach. Advances in separation science based on refined lithographic methods are thus not limited to the design of more efficient sieving media, but also include the research for alternative ways to sort, identify and size biomolecules.

Some of the questions that require further theoretical investigation thus include: (i) Is it possible to design a sieving structure or imagine a mode of operation that limits dispersion? (ii) Can we take advantage of different sieve geometries, perhaps more elaborate ones, to optimize separation? There is currently no theory that can guide the development of such sieves. (iii) Can EOF be used, together with geometrical constraints, to separate biomolecules? Of course, the work on ratchet-like systems will remain heavily influenced by theory since their operation depends on very subtle thermodynamic effects that would be hard to optimize in a purely empirical way.

5 How can we optimize CE in ultradilute polymer solutions?

In 1993-4, Barron *et al.* ([37, 38] see also [39–41]) showed that dsDNA fragments can be separated in polymer solutions well below their entanglement concentration c^* (see Fig. 1b). Such a result was believed to be impossible within the framework of the existing theories (*i.e.*, the reptation [15] and Ogston [15, 42, 43] models) since the absence of a well-defined polymer structure seems to eliminate any possibility for a sieving effect. Not only is this new process capable of separating dsDNA fragments, but the elution times are up to 25 times shorter than normally achieved with slab gels. Soon afterwards a theoretical model was suggested [44] and video-microscopy experiments were carried out [45–49]. The concept of transient entanglement coupling between the polyelectrolyte and the polymer chains provides a decent qualitative explanation of the experimental results [38, 44]. The optimal characteristics of the polymer (size, chemical composition, concentration, *etc.*) needed to separate polyelectrolytes in this new regime are now known, at least empirically [37–41]. For instance, hydrophilic polymers provide better resolution since they offer larger collision cross-sections than flexible polymers. The molecular weight of the polymer impacts the DNA sizes that one can separate: the resolution of large DNA fragments improves in the presence of long polymers whereas small DNAs are best resolved with small polymers. But a microscopic model with predictive powers is still missing.

The separation of long DNAs in slab-gel electrophoresis (GE) is achieved using pulsed fields instead of a steady fields [50]. Morris *et al.* [45–49] followed the same principle for CE in ultradilute polymer solutions. They found that pulsed-field CE (PFCE) indeed increases the range of DNA sizes that can be separated and, with the help of video-microscopy, they were able to observe the DNA transitions between full extension and random coil conformations (hence the transient entanglement coupling) responsible for the separation. They also showed that the DNA maximum extension is dependent upon the frequency and strength of the electric field.

The theory developed by Hubert *et al.* [44] suggests that the average number of polymers in contact with a DNA fragment at any given time, n, must be larger than unity for optimal results. In the $n < 1$ limit, for instance, the total drag force exerted by the neutral polymers on the DNA is rather small, and the large fluctuations of the electrophoretic velocity are expected to lead to very broad elution peaks. As far as we can tell, available theoretical or numerical data, *e.g.*, from video-microscopy experiments, do not provide a precise quantitative estimate of n.

Startchev *et al.* [51] interpreted their birefringence data in terms of the existence of dynamic 1-1 complexes. A related question was examined by Braun et al. [52] who tested a star-(polyethylene oxide) with 64 arms. However, this high molecular mass polymer did not improve the resolution. Fewer arms and/or randomly cross-linked/ branched polymers could perhaps increase the entanglement level with the DNAs. It seems logical to assume that polymer architectures that would maximize the number and/or duration of the collisions between the polymers and the DNA fragments, would improve the performance.

Computer simulations have recently given us more information on the dynamics of the transient entanglements between the polyelectrolyte and the randomly distributed polymer molecules. Starkweather *et al.* [53, 54] used Monte-Carlo simulations to tackle the problem. They observed that the polyelectrolyte velocity goes to a minimum when the contour length of the polymer is varied. The polymer contour length at the minimum corresponds approximately to the contour length of the polyelectrolyte itself. Our theoretical model also predicted maximum effects for such conditions [44]. This minimum was observed experimentally by the same group [55]. This clearly indicates that both polymers are involved during the escape process [38, 44]. Since the disentanglement process is at the heart of this new separation mechanisms, a complete theoretical model of this process would provide us with better tools to achieve optimal separation.

Separation in ultradilute polymer solutions represents a potentially fertile ground for innovations and investigations. Resolution is not likely to become competitive with more standard methods, but the high mobilities and the low viscosity of the solution are advantages that might make this technique quite useful for some specialized applications, especially in microchip format. Since this is a rather recent technique which has yet to find a commercial application, there is much to be done in terms of fundamental theoretical science. Here are a few suggestions: (i) What conditions would reduce diffusional band broadening in this regime? (ii) What is the impact of the electric field intensity on the separation? In particular, what is the role of polymer deformation (when it is dragged along by the DNA fragment)? (iii) How can we optimize polymer mixtures to increase the range of DNA sizes to be separated? In other words, we need a theory that would include the effects of polymer polydispersity. (iv) What DNA and polymer dynamical modes should we excite to optimize PFCE results? (v) What are the differences between the $n < 1$, $n > 1$ and $n \approx 1$ regimes of separation? (vi) What is the optimal polymer shape (and stiffness) for polyelectrolytes separation? (vii) Given the recent theories of Long *et al.* [22–26] on electrophoresis in the presence of mechanical forces (which is the case here), it would seem important to develop a theory (or a computer simulation) that includes hydrodynamic interactions.

6 Is there new physics when we use ultrahigh fields?

Although using higher and higher field intensities is a widespread dream in our era of CE and microchip technologies, there are practical reasons why extremely high fields are not usually beneficial. Besides the standard technical reasons (such as Joule heating and electronics issues), the separation mechanism itself often imposes limits. The best example is that of DNA sequencing by gel electrophoresis where high fields actually reduce the read length because of the phenomenon of biased reptation [15, 43] (contrarily to what is often assumed, however, it is the increased diffusion coefficient and not the reduced band spacing that limits read lengths at high field intensities [56]). It is important to stress the fact that "high" is a subjective term. For instance, 10 V/cm is a high field intensity in agarose gels [15, 57], while 15 V/cm is extremely low in PA gels [58, 59]. A real measure of the field intensity is provided by the scaled field parameter used in the framework of the reptation models [15, 43]. The latter includes the pore size of the gel, hence the large differences between various types of gels, polymer matrices or microstructures. Because of the presence of biased reptation in gel electrophoresis, very high field intensities may require new separtion mechanisms that do not loose their resolution when the field is increased. These may include dilute polymer solutions, custom-made artificial matrices and ELFSE, but in all these cases theory is still in its infancy. Of course, pulsed fields are likely to be of some help when high fields are used since the very deformed DNA molecules are expected to respond in a resonant way when the frequency of the pulses matches that of their internal relaxation modes.

High field problems are often extremely difficult to investigate theoretically because nothing can be studied in terms of simple approximations, perturbations, series expansions, *etc.* On the other hand, high fields may make thermal effects (such as Brownian forces) irrelevant. This is the idea first studied by Lee *et al.* [60]. These authors used a deterministic model to explain the motion of megabase DNA molecules in gels in a limit where one expects geometration, a nonreptation mode of migration first proposed by Deutsch and Madden [61–63], to dominate because hernias form easily. In a 1993 paper, the same authors proposed that very large molecules may actually form self-similar trees of hernias in the presence of a "high" field [64]. Such models are conceptually much sim-

pler to define, and certainly easier to study using computational tools and even analytical calculations. In the case of periodic microsieves, these calculations are made even easier because of the absence of randomness in the distribution of obstacles. More recently, Popelka *et al.* [65] have used a similar deterministic approach to calculate the effective diffusion coefficient of long DNA molecules during gel electrophoresis. Their results are quite remarkable given the simplicity of the calculations based on the "falling rope model" (see Fig. 1h for an example).

We believe that theoreticians and experimentalists must work togehter to better understand the new physics that takes place at extreme field intensities. There is a clear tendency in the literature to use the BRF (or the biased reptation model (BRM)) to analyse the data obtained with very high field intensities although it is well-known that these models should not work in such cases (because the presence of hernias destroys the concept of a reptation tube) [15]. However, we must admit that the BRF is indeed useful, even in such situations!

Given that very high field intensities are not yet common in practice, it is not surprising that the related theoretical background is rather poor. Let us suggest some important problems: (i) Einstein once asked himself: "How can it be that mathematics, being after all a product of human thought independent of experience, is so admirably adapted to the objects of reality?" Given the success of the BRF at explaining data that it shouldn't be able to explain, one can similarly ask: "How can it be that the reptation model, being after all a product of the human thought, is so efficient in practice?" (ii) Can we design new "microsieving" structures using simple deterministic models? In passing, such models are essentially equivalent to understanding how a piece of rope might fall down a pinball machine under the action of gravity (see Fig. 1h)! (iii) What happens in polymer solutions, especially when their concentration is close to their overlap concentration c^*, if we use a very high field intensity? Since it is likely that the DNA molecule will then tear the weak polymer matrix apart, one may wonder whether this effect can be used in practice.

7 Why can we separate Mbp DNAs in ultradense polymer solutions?

It has recently been demonstrated that *Schizosaccharomyces pombe* chromosomes (3.5 Mbp, 4.6 Mbp and 5.5 Mbp) and T4 DNA (166 kbp) can be separated using steady-field CE in a dense LPA solution [66]. Until then, the most efficient method for separating chromosomal DNA molecules had been pulsed-field gel electrophoresis (PFGE) [50]. The experimental results reported by Ueda *et al.* [66] clearly show the existence of two distinct migration regimes: U-shaped motion (see also [44, 67]) and I-shaped motion (observed only in ultradense solutions). *S. pombe* chromosomes have been separated in both migration regimes, but the latter regime yielded far improved resolutions [66]. Ueda *et al.* [66] also demonstrated that higher electric fields improve the resolution. To our knowledge, no theoretical or numerical studies predicted the possibility of extra-long DNA separation under such conditions. The I-shaped migration regime is monotonous: the back and front ends (or head and tail) of the chain never interchange. A remarkable characteristic of this regime is the formation of stationary high-density regions covering a much larger area than the average estimated pore size of the polymer matrix, an unexplained feature. Figure 1g depicts a few high-density regions during the I-shaped displacement of the chain, as described by the authors.

This I-shaped regime (perhaps it should be called the ξ-shaped regime) of motion had never (to our knowledge) been mentioned in literature before and is certainly not trivial. The closest observation we are aware of, is that of a linear (I-shaped) conformation regime for DNA motion in regular microlithographic arrays under high fields [28, 68, 69]. Under such conditions, a somewhat high density region may appear at the leading end of the chain. The following chain segments then fold themselves in this region to form a spherical blob, which will then extend two arms in the direction of the field and gently escape (see Fig. 1h). On the other hand, our group has previously shown that the presence of small amounts of heterogeneities in a gel matrix is enough to change the diffusion mechanism from reptation-like behaviour to an alternative kind of diffusion mechanism [58, 70] called entropic trapping (see Section 10). Appel *et al.* [71] have shown that monomer displacement of approximately 100 nm can be observed in a chemically cross-linked polymer matrix, most likely due to cooperative flucturations of the whole gel. Intuitively, the formation of the DNA high-density regions in our case likely results from an entropic force exerted on the DNA by large polymer density fluctuations.

Our group has unsuccessfully tried to represent this I-shaped regime of migration by Monte Carlo (MC) simulations in two different two-dimensional systems, both of which used the ergodic bond fluctuation model (BFM) [70, 72]. The first system consisted of a periodic tunnel of alternating constrictions and pores. The constrictive part of the tunnel was kept small enough so that the polymer could not fold. Our second model was a two-dimensional array where the macropores were represented by unconnected periodic potential wells. This latter system minimized the excluded volume effects and offered the

chain the possibility to exit a pore in a more or less random direction. In both cases, polymers of different sizes N where investigated to cover both regimes, *i.e.*, a small molecule that fits entirely into a single pore and a large molecule covering many pores simultaneously. Indeed, the polymer conformations were very similar to those observed by Ueda *et al.* [66]. However, while Ueda *et al.* [66] clearly found a decrease of mobility with respect to the molecular size, both models showed a behaviour fundamentally different from experimental observations, *i.e.*, a mobility that increases with respect to molecular size (unpublished). Our models seemed to represent much more accurately the case of entropic trapping of long DNA molecules in sub-μm size constrictions (in artificial microdevices) [31, 32] (manuscript in preparation). Therefore, a realistic and tractable model is yet to be designed.

A better comprehension of the dynamics of extra-long genetic material in very dense systems of flexible linear polymer chains would allow us to optimize this novel separation process. Some specific question to be investigated thus include: (i) What leads to the spontaneous formation of the DNA high-density regions? What is the right modeling approach given that the formation of these regions is probably due to the dynamics of the polymer chains themselves? (ii) How does the I-regime separation mechanism work (our simple MC models failed miserably)?

8 Can we exploit DNA trapping electrophoresis?

Trapping electrophoresis (TE) is a gel separation method where protein-labelled DNA fragments are electrophoresed in a gel. The original report on TE [73] used a streptavidin label, but this is completely different from ELFSE (Section 3). DNA molecules that encounter a pore smaller than the size of the label will temporarily get stuck (or trapped) in the gel (Fig. 1c). Escape is thermally activated, which explains why trapping is (exponentially) stronger for larger molecules. Our group carried out a detailed theoretical and experimental study of TE, and our results clearly demonstrated that TE alone cannot help extend the read length for DNA sequencing [74–77]. Although it is possible to achieve slightly better DNA separation for large trapped molecules (but only over extremely narrow ranges of molecular sizes), the stochastic nature of the detrapping process increases the diffusion coefficient and reduces the overall resolution. Moreover, TE is extremely sensitive to system imperfections and gradients [76, 77].

Several fundamental questions remain, including: (i) Can we use pulsed fields to control the TE detrapping process and the diffusion mechanisms in order to achieve improved resolution [73, 77]? Can we use gel concentration gradients and/or field gradients for this purpose? (ii) The current theories [74–78], which are based on the BRM, are surprisingly good at explaining the available experimental data even though the BRM is known to have rather important weaknesses. This remains a mystery. A BRF theory of TE could clarify this issue. (iii) Does TE take place in concentrated sieving polymer solutions? We expect highly reduced trapping effects in such cases because of the finite lifetime of the steric traps. Therefore, one might wonder whether TE would be more efficient in polymer solutions where trapping would actually be replaced by "increased" friction and less diffusional band broadening.

Recently, however, Griess and Serwer [79, 80] proposed a clever new ratchet idea based on TE. These authors demonstrated that two DNA molecules having the same length, one of them being labelled with a bulky object and the other remaining unlabelled, may actually move in opposite directions when gel-electrophoresed in specially designed pulsed fields! It would be interesting to see if this can be made to be useful for broad ranges of applications. It is obvious that numerous purification schemes can be based on this idea as long as one can selectively label the fragment that must be isolated. Although we published a theoretical study of this process [81], surprisingly few theory examined this new separation process.

9 Can we exploit entropic trapping?

Since the pioneering work of Muthukumar and Baumgartner [82–84] in the late 1980's, entropic trapping (ET) has been recognized as the third mode of polymer migration in gels, the other two being Rouse and reptation dynamics [70, 85, 86]. Rouse (usually called the Ogston theory – or model – in the electrophoresis literature) and reptation theories explain most observed phenomena related to gel electrophoresis of polyelectrolytes [15, 43]. At least two experimental groups have now established that ET does indeed take place in some "real" DNA gel electrophoresis systems, including the low-field sequencing of ssDNA fragments in polyacrylamide gels [58, 87–90]. ET occurs when the radius-of-gyration R_g of the DNA molecule is comparable to the mean pore size (a) of the gel, *i.e.*, when it is entropically expensive to migrate from pore to pore because of the deformation that the chain must endure in the narrow passages that connect the larger voids of size > R_g (see Fig. 1e). The "entropic barriers" between the pores lead to a slow activated jump process which is obviously speeded up by the external electric forces [43]. Rousseau *et al.* [58], for example, studied a PA gel where ET was killed by fields as low as 30 V/cm. Obviously, the new separation system built and investigated by the Craighead group [31, 32] is based on custom-made entropic barriers (see Fig. 1i). Therefore, we can say that ET can indeed be exploited in practical microsystems. We believe that this is only the beginning and that other entropy-related systems will be designed and tested in the next few years. Microsystems seem to be quite exciting for this purpose since the low-field limitation observed for gels does not seem to be very severe if the system size is properly chosen.

Two related ideas, which have not attracted much attention in the electrophoresis community, have been recently described. Asher's group [91, 92] developed a remarkably elegant macroporous gel, using colloidal suspensions and PA, in which ET must be the most important mode of diffusion for selected molecular sizes. This is a system design that could very easily be adapted to electrophoretic problems. Elsewhere, a system of "molecular traffic" was suggested where separation between molecular species could be achieved without any external force! Due to steric and analyte-gel interactions, the diffusion coefficients of the two analytes become very anisotropic and the latter exit the gel through different faces [93]! Again, one can easily imagine novel DNA separation schemes based on this general idea.

Here are some interesting problems to be studied that include ET effects: (i) Does ET occur when DNA molecules are electrophoresed in uncross-linked polymer solutions? Intuition would suggest no, but there is no convincing theory or experimental study at this time. (ii) How can we improve the selectivity of ET-based separation systems? Because ET is based on molecular escape from local (entropic) traps, the motion is quite erratic, the bands rather broad, and the resulting separation is not very efficient. In other words, how can we narrow the distribution of escape times for ET processes? (iii) Can we build gel-free ET systems that could separate DNA molecules of the same molecular weight but with different configurations (*e.g.*, linear, branched, circular, knotted, *etc.*)? This would be a nice achievement of a diffusion regime that started out as an oddity less than 15 years ago! With the recent work of Craighead's group [31, 32], ET is just beginning to invade the world of the lab-on-a-chip. Still, very little is known about this process.

10 Is thermal diffusion affected during electrophoresis?

Although thermal diffusion is an important factor that limits the resolution and read length of all separation methods, it has not been studied as extensively as the mobility itself. Of course, thermal diffusion cannot be eliminated using simple technological methods (unlike most if not all other sources of band broadening) because it is ruled by Mother Nature Herself. Experimentally, measuring the diffusion coefficient D during electrophoresis is quite difficult. Theoretically, calculating D is also rather difficult because being a measure of the fluctuations in the system, it is very sensitive to the details of the model and approximations used.

However, theory [13–15, 94, 95] and experimental data [96–98] agree on two major facts regarding the diffusion coefficient of DNA molecules during gel electrophoresis: (i) the Nerst-Einstein relation

$$\frac{D}{\mu} = \frac{k_B T}{Q} \tag{2}$$

between the diffusion coefficient D, the mobility μ and the total DNA charge Q is not valid except for very small DNA sizes and very low field intensities; (ii) D increases with the field E and becomes a weak function of the molecular size $M \sim Q$ in most realistic cases. Beautiful experiments performed by Tinland and co-workers [96–98] have recently resolved these issues, and their results agree remarkably well with the predictions of the reptation model. The recent work by Gas *et al.* [65], which was discussed in Section 6, is also an important new theoretical result to mention here.

However, there are still important problems that remain unresolved, including: (i) What is the diffusion coefficient D when DNA electrophoresis is done in an entangled polymer solutions instead of cross-linked gels (the structure assumed by the reptation model)? The dynamics of the matrix is most likely adding to the band broadening processes, but this has yet to be included in a comprehensive theory. This missing piece of the puzzle might be crucial for the optimization of CE and microchip sequencing systems. (ii) What is the diffusion coefficient D when DNA electrophoresis is done in unentangled polymer solutions (see Section 5)? The statistical nature of the DNA-polymer collisons is an interesting problem in itself, and its impact on band broadening has yet to be studied. (iii) What is the diffusion coefficient D when using ELFSE (see Section 3)?

11 Can we build optimized ratchet separation systems?

The concept of Brownian ratchets where net motion can be achieved even when the net external force is zero, was initially discussed by Curie [99] and Feynman *et al.* [100]. Separation schemes based on ratchets can be constructed, *e.g.*, by providing a potential (steric or otherwise)

which is asymmetric in space (temporal asymmetry can also do the trick). Rousselet *et al.* [101] used an asymmetric microelectrode configuration which lead to the net displacement of charged particles in spite of the fact that the net field was zero. Another asymmetric electrode device, proposed and tested by Bader *et al.* [102, 103], makes the particles diffuse during the field-off periods and fall into a potential well during the field-on periods. There is a preferred direction due to the asymmetry in the electrode arrangement. In fact, the probability of moving forward in the preferred direction depends on the zero-field diffusion coefficient D of the particles. Strong band-broadening occurs in this system but the authors were able to predict the conditions required to maximize the resolution. In fact, large band widths are likely to be a fundamental weakness of ratchet systems because of the fact that thermally activated motion always plays a central role.

Our group [104] previously suggested the idea of using a narrow channel with asymmetric structures to produce entropic "ratchet" barriers for moving molecules. In an unbiased pulsed (AC) field, the molecules then have a net velocity in the direction which least affects the entropy of the molecule, and separation occurs because the larger molecules have more internal entropy and are more affected by the structures. We further showed that if a small DC field is added to the system, one can actually make the long and short DNA molecules migrate in opposite directions. This system has not yet been built and tested, although Craighead's system (Fig. 1i) is essentially a symmetric version of it which uses straight entropic trapping instead of ratchet mechanics [31, 32].

Austin's group [28, 33, 34] has demonstrated that µm-sized sieving structures can produce ratchet-like systems which can be used to separate charged analytes, including DNA. Other groups [35, 105] have proposed special microsieves with asymmetric obstacles. The molecules migrate in a direction which is not exactly parallel to the direction of the electric field. The deviation from the direction of the applied field is proportional to the transverse diffusion coefficient of the analyte and smaller molecules move further away from the direction of the applied field. These sieving systems have the remarkable property that molecules can be separated continuously. Even membrane proteins have been separated using such an idea [106]. Since ratchet systems require delicate geometrical properties, they are ideally suited for microchannel devices. Current systems exploit mostly diffusion properties as a "sieving" parameter. However, other properties (such as internal entropy) can also be used. This is a new field of investigation and there is no doubt that numerous new ideas will be proposed in the next years.

Among the interesting theoretical issues that recent work has suggested, we may mention: (i) How can we control/limit band broadening in a system that is fundamentally stochastic? (ii) How can we maximize the range of molecular sizes that a ratchet can work with? Indeed, most ratchets are rather restricted for this purpose because of the physical size of their features. (iii) What properties, other than the diffusion coefficient or the internal entropy, can we exploit to build a ratchet that would separate biomolecules on the basis of smaller/more subtle differences?

12 Can we design a useful replacement to the Ogston model?

The traditional model used to express the electrophoretic mobility μ of small macromolecules and particles in a gel matrix is the Ogston-Morris-Rodbard-Chrambach Model, or OMRC (it is more commonly called the Ogston model) [42, 107–119]. This model basically assumes that the ratio $\mu^* = \mu/\mu_0$ (which is called the scaled or reduced mobility) between the mobility in the gel (μ) and the free-solution mobility (μ_0) is equal to the fractional volume (f) of the gel available to the analyte:

$$\mu^* = \frac{\mu}{\mu_0} = f(C, M) \qquad (3)$$

where f is a function of both the gel concentration (C) and the analyte molecular size M. Historically, this theory comes from papers by Morris and Ogston. Using geometrical and statistical arguments, Ogston [110] first showed that the available volume in a random array of fibers is given by an exponential function $\sim \exp(-KC)$, where $K = K(M)$ is the retardation factor of the spherical analyte. Later, in a experimental study, Morris [111] showed that the scaled mobility μ^* of some proteins in PA gels was approximately equal to their available volumes f (he actually reported $1.1 \times f$). Rodbard and Chrambach [112] combined these two results and assumed that the relation between the mobility and the available volume was not only approximate but exact and general. The final result is thus

$$\mu^* = \frac{\mu}{\mu_0} = f = e^{-KC} \qquad (4)$$

Note that the last part of Eq. (4) applies only to "Ogston" gels made of random arrays of long fibers [110]. In such cases, and for spherical analytes, we have $K = b\,(r+R)^2$, where b is a constant and r and R are the fiber and analyte radii, respectively. This leads directly to the Ferguson [109] plot ($\ln(\mu^*)$ *vs.* C), which is the standard method for analyzing electrophoretic data [107, 108, 112–114]. Although the basic assumption ($\mu/\mu_0 = f$) does not have

solid theoretical or experimental foundations, the Ogston model is widely used and can explain semiquantitatively a fair amount of experimental data.

To verify the fundamental assumption defined by Eq. (3), we must rely on very accurate experimental measurements or computer simulations. However, direct measurements have failed to validate the OMRC model because μ^* and f (C) can generally not be measured independently and with enough accuracy. We must therefore rely on computer simulations. Recently our group has proposed a new lattice model of gel electrophoresis [107, 108, 115, 117, 119] that allows both the mobility μ^* and the obstructed volume $\phi = 1$-f to be calculated exactly, thus offering a way to test the OMRC hypothesis. The basic elements of the model are the same as for the Ogston model (low field, steric analyte-fiber interactions, rigid analytes). Furthermore, the field lines are assumed not to be affected by the obstacles. Many types of systems were considered starting from 2-D periodic gels to realistic 3-D random gels. The results disagree with the relation $\mu/\mu_0 = f$ for all system studied. For instance, when the simulation data are expressed as the series

$$\mu^*(f) = 1 - a_1\,\phi - a_2\,\phi^2 - a_3\,\phi^3 - \ldots \qquad \phi \equiv 1\text{-}f \qquad (5)$$

both the magnitude and sign of the a_i coefficients change with the gel structure. Of course, the fact that the free volume f is insufficient to describe the electrophoretic sieving of small molecules is a very important and favorable conclusion towards the design of custom-made sieving matrices. Clearly, the Ogston model needs to be replaced by a more microscopic version that will take into account the precise architecture of the sieving matrix as well as the properties of the analyte. This model will thus have to include: steric effects (analyte-gel fiber collisions), percolation constraints (migration pathway tortuosity), hydrodynamic interactions (including the electroosmotic flow), possible specific analyte-gel interactions (attractive or repulsive), *etc.*

Another element of a better model would be the precise electric field lines inside the gel [116]. When dealing with small electric fields, most models (including ours) hold that the field lines are parallel to one another, even near nonconducting gel fibers. A more accurate model would take into account the conductivity of the fibers. This would possibly lessen the obstacle's ability to impede the mobility of the analytes. Recently, Locke and Trinh [116] have studied the impact of the local variations of the electrical field around nonconducting obstacles. They conclude that the Ogston assumption is valid in the limit of very small electrical fields and very dilute media, as well as in sieving media without percolation limits. Local variations in the electric field are currently being investigated by our group. Preliminary results indicate that, in the low field limit, local variations in the electric field do not change the analyte mobility, thus contradicting the results of Locke *et al.* (manuscript in preparation).

A study of the mobility of small flexible oligomers using our new model is currently underway. Preliminary results suggest that the Ogston concept of free volume is insufficient to describe the separation of such molecules (submitted for publication). For example, we have recovered the phenomenon of entropic trapping (see Section 10) while the standard Ogston model does not account for such effects. Although the original Ogston model was not designed for DNA molecules, it is often used to analyze DNA electrophoretic data when the molecular sizes are below the reptation threshold. Attractive analyte-gel interactions have also been investigated in our group [117]. Our results indicate that such interactions may lead to inflection points and curvature changes in the Ferguson plots, offering a possible explanation for their presence in certain experimental electrophoretic data [42, 118]. Binding also had a beneficial effect on the size resolution for small analyte sizes in the systems studied, as well as when being used in conjunction with a molecular trap [117].

Because of their inherent limitations, lattice models cannot be used to build a complete theory. For example, hydrodynamic interactions cannot be treated directly using lattice methods although they could play a key role in some transport properties. Other methods, such as Molecular Dynamics (MD), must therefore be developed. However, MD (or even Brownian Dynamics, BD) methods take far more computing resources and only approximate results can be obtained. On the other hand, computational methods allow the users to perfectly control the parameters and the interactions that are used. Such studies may become useful in the near future, especially in relation with the design of ultrasmall microchip systems.

This short overview thus suggests a few directions for further research: (i) How important are the local variations of the electric field inside a gel? (ii) Does the apparent disagreement between our studies and those of Locke *et al.* tell us something fundamental about this problem? (iii) How can we include the effects of high field intensities in a Ogston-like model? (iv) In the case of CE, what is the impact of the matrix dynamics on the Ogston model (indeed, the "gel" is made of entangled polymers which have their own dynamics [120])? (v) How can we study more complex analyte-gel interactions as well as the hydrodynamic effects? (vi) What is the effect of the fractal nature of the gels [121, 122] (vii) Which off-lattice simulation models offer good potential for the study of the Ogston regime?

13 Will nanopores replace electrophoretic sequencing of ssDNA?

Besides mass spectroscopy, only nanopore sequencing seems to be an idea that could be threatening for CE and related methods. This very original idea was suggested a few years ago [123–125], but sequencing has yet to be achieved. Essentially, the process consists in forcing a ssDNA molecule to move through a very narrow opening in a wall (ID < 5 nm), base by base (Fig. 1j). An electric field is used to drive the molecule through (conceptually, mechanical forces could also be used). During the passage, a physical property should be measured that would indicate which base is currently entering or leaving the channel. Measuring the current, Branton *et al.* [123–125] were able to estimate molecular sizes and the nature of blocks of identical bases, but the resolution was not sufficient to identify single bases. Since the bases go through the channel at the rate of roughly one per microsecond, the human genome could be sequenced entirely by a single channel in a few hours at most!

The technical issues are not simple, however. Theoretical studies of the forced translocation of a polymer through a narrow opening indicate that this is a rather stochastic process because the loss of entropy of the polymer chain near the wall acts as an entropic barrier [126–130]. This being a very new field, numerous theoretical questions have yet to be examined. However, the main practical questions are simply: (i) How can we regularize the rate at which the bases go through the channel? (ii) How can we make sure that the DNA molecules find this incredibly small channel entrance? Of course, unlike mass spectroscopy, nanopore sequencing is intrinsically microscopic. Therefore, it would be an ideal approach for integrated microchip sequencing labs.

14 Can we find new ways to use pulsed fields?

As far as DNA sequencing is concerned, the answer seems to be: no. Early results appeared interesting, but it was soon realized that the effects that were observed were mostly or entirely due to the reduced electric field intensity [131]. This conclusion was also reached by Slater and Drouin [56]. More recently, Kim and Yeung [132] have reported some positive effects. The fact that pulsed fields have been essentially useless for DNA gel electrophoretic sequencing, is actually expected theoretically. Indeed, pulsed fields have been a revolution for the separation of chromosomal DNA fragments in agarose gels because it was field-induced molecular orientation that was reducing the ability of these gels to fractionate molecules larger than about 40 kbp [50]. Molecular orientation is known to play a minor role in DNA sequencing [56, 58, 59], hence the inefficiency of pulsed fields.

However, there are still a few interesting questions related to pulsed fields: (i) Can pulsed fields be useful if we separate DNA molecules in weakly entangled polymer solutions ($c \approx c^*$)? Since the polymer matrix is likely to be deformed by the migration DNA fragments, changing the direction of the field periodically may couple the motion of the DNA to the relaxation modes of the deformed polymer matrix. This has yet to be studied. (ii) Can pulsed fields be used together with microsystems that exploit it entropic effects (see Sections 9 and 11)? Such an idea was suggested by our group a few years ago [104], but the theory was rather rudimentary. The idea has yet to be tested experimentally. (iii) How can we model the reported effects of pulsed fields in unentangled (ultradilute) polymer solutions (see Section 5)?

15 Discussion

Essentially, this review proposes dozens of research projects for the theoretical scientist (or, if you prefer, it proposes dozens of topics for graduate students). We looked at a few standard methods, such as CE in polymer solutions, and a wide range of newer or understudied ideas. Most of these separation processes are based on CE and related systems. This is not a coincidence. The move towards miniature systems is clear. CE has basically conquered the human genome. Microchip technologies will likely conquer the next Everest's, including the diagnostic world. Among the great challenges, we consider that, ELFSE, ratchet and entropic trapping systems are the most promising novel ideas for the world of microchips, while the nanopore approach is clearly the most threatening. We have to move away from viscous sieving polymer solutions, and these ideas offer ways to achieve separations without the efforts involved in the loading of such matrices in narrower and narrower channels. For the theoreticians, perhaps the most important goal should be to come up with models to guide the design of new microstructures. Such models must be able to optimize the geometry of the system to both increase the separation and reduce the diffusion (band broadening). We think that computer models and simulations will play a larger role in the future. Problems like the sieving of small analytes can now be treated in a more fundamental way, thanks to the much increased computing power now widely available [107, 108, 115, 117, 119, 121]. There are four relevant computer simulation methods for our purposes: MD, BD, MC and deterministic approaches. Each of them have strengths and weaknesses. MD is the most general one since it consists in numerically integrating the equations of motion for thousands to millions of particles (which are either atoms or "united atoms" representing groups of atoms such as monomers or solvent molecules). With increases in computer speed, the availability of supercomputers and the improvement of parallel algorithms for MD, large-scale MD simulations are becoming more and more common. Unfortunately, problems spreading over a wide range of time scales remain a computational challenge. Improvement would require the use of different computational methods in the same simulation to treat phenomena happening at vastly different frequencies. In the case of separation methods, one usually cares only about long-time displacements, so simplified MD models can be used. An efficient MD algorithm for electrophoresis problems is yet to be developed. Our group is currently studying this issue.

BD does not implicitly simulate the solvent particles. Instead, it replaces the effects of the solvent by a random force in a Langevin equation. Such an approximation allows the study of longer timescale problems. To include electrostatic and hydrodynamic interactions, a Debye-Huckel approximation and the Rotne-Prager tensor are often used. For example, a recent model by Noguchi [133] incorporates BD and anisotropic friction to study gel electrophoresis. By varying the degree of the anisotropic friction, one can pass from a gel to a dilute or semidilute polymer solution. However, chain stiffness, excluded volume or hydrodynamic interactions are not taken into account. Nonetheless, the simulation results are impressive. MC is a faster method but it has two disadvantages: it does not have an intrinsic timescale (because it is based on statistical sampling) and it does not conserve momentum transfer. Only Rouse-type behaviour is accessible with MC, since no hydrodynamic interactions are included. However, various groups are studying new approaches to these limitations. The MC algorithm derived by Viovy [15] to study pulsed or steady-field gel electrophoresis of large DNA molecules is by far the best simulation method in the field of electrophoresis. It is loosely based on the reptation theory, but it admits the possibility that hernias might escape the constraint of the tube.

Finally, we have deterministic approaches as discussed in Section 6. These algorithms are usually developed to study specific problems. However, they are easy to use and appear to be well suited to the high-field situations that we find in microchip technologies. The next step might be to develop microscopic models that include hydrodynamic and electrostatic effects at the molecular level. In the mean time, there is plenty of work to be done if theoreticians want to continue contributing to the experimental developments that will follow the completion of the human genome. This theoretical work requires precise experimental data. The marriage between theory and experiments is thus here to stay.

This work was supported by a Research Grant of the Natural Science and Engineering Research Council (NSERC) of Canada to GWS, and by NSERC scholarships to JFM and FT.

Received August 7, 2000

16 References

[1] *Science* 2000, *288*, 2294–2295 and 2304–2307.

[2] Slater, G. W., Kist, T. B. L., Ren, H., Drouin, G., *Electrophoresis* 1988, *19*, 1525–1541.

[3] Madabhushi, R. S., *Electrophoresis* 1998, *19*, 224–230.

[4] Cottet, H., Gareil, P., Viovy, J.-L., *Electrophoresis* 1998, *19*, 2151–2162.

[5] Wu, C., Liu, T., Chu, B., *Electrophoresis* 1998, *19*, 231–241.

[6] Rill, R. L., Locke, B. R., Liu, Y., van Winkle, D. H., *Proc. Natl. Acad. Sci. USA* 1998, *95*, 1534–1539.

[7] Heller, C., *Electrophoresis* 1998, *19*, 3114–3127.

[8] Heller, C., *Electrophoresis* 1999, *20*, 1962–1976 and 1978–1986.

[9] Heller, C., *Electrophoresis* 2000, *21*, 593–602.

[10] Viovy, J. L., Heller, C., in: Righetti, P. G. (Ed.), *Capillary Electrophoresis in Analytical Biotechnology.* CRC Press, Boca Raton, FL 1996, pp. 447–508.

[11] Duke, T. A. J., Viovy, J. L., Semenov, A. N., *Biopolymers* 1994, *34*, 239–247.

[12] Semenov, A. N., Duke, T. A. J., Viovy, J. L., *Phys. Rev. E* 1995, *51*, 1520–1537.

[13] Duke, T. A. J., Viovy, J. L., Semenov, A. N., *Biopolymers* 1994, *34*, 239–248.

[14] Semenov, A. N., Joanny, J.-F., *Phys. Rev. E* 1997, *55*, 789–799.

[15] Viovy, J.-L., *Rev. Modern Phys.* 2000, *72*, 813–872.

[16] Zhou, H., Miller, A. W., Sosic, Z., Buchholz, B., Barron, A., Kotler, L., Karger, B. L., *Anal. Chem.* 2000, *72*, 1045–1052.

[17] Mayer, P., Slater, G. W., Drouin, G., *Anal. Chem.* 1994, *66*, 1777–1780.

[18] Olivera, B. M., Baine, P., Davidson, N., *Biopolymers* 1964, *2*, 245–257.

[19] Heller, C., Slater, G. M., Mayer, P., Dovichi, N. J., Pinto, D., Viovy, J. L., Drouin, G., *J. Chromatogr. A* 1998, *806*, 113–121.

[20] Ren, H., Karger, A. E., Oaks, F., Menchen, S., Slater, G. W., Drouin, G., *Electrophoresis* 1999, *20*, 2501–2509.

[21] Hubert, S. J., Slater, G. W., *Electrophoresis* 1995, *16*, 2137–2142.

[22] Long, D., Ajdari, A., *Electrophoresis* 1996, *17*, 1161–1166.

[23] Long, D., Viovy, J. L., Ajdari, A., *Phys. Rev. Lett.* 1996, *76*, 3858–3861.

[24] Long, D., Viovy, J. L., Ajdari, A., *Biopolymers* 1996, *39*, 755–759.

[25] Long, D., Viovy, J. L., Ajdari, A., *J. Phys.: Condens. Matter* 1996, *8*, 9471–9475.

[26] Long, D., Dobrynin, V., Rubinstein, M., Ajdari, A., *J. Chem. Phys.* 1998, *108*, 1234–1244.

[27] Turner, S. W., Perez, A. M., Lopez, A., Craighead, H. G., *J. Vac. Sci. Technol. B* 1998, *16*, 3835–3840.

[28] Chou, C.-F., Austin, R. H., Bakajin, O., Tegenfeldt, J. O., Castelino, J. A., Chan, S. S., Cox, E. C., Craighead, H., Darnton, N., Duke, T., Han, J., Turner, S., *Electrophoresis* 2000, *21*, 81–90.

[29] Volkmuth, W. D., Duke, T., Wu, M. C., Austin, R. H., Szabo, A., *Phys. Rev. Lett.* 1994, *72*, 2117–2120.

[30] Volkmuth, W. D., Austin, R. H., *Nature* 1992, *358*, 600–602.

[31] Han, J., Turner, S. W., Craighead, H. G., *Phys. Rev. Lett.* 1999, *83*, 1688–1691.

[32] Han, J., Craighead, H. G., *Science* 2000, *288*, 1026–1028.

[33] Chou, C.-F., Bakajin, O., Turner, S. W. P., Duke, T. A. J., Chan, S. S., Cox, E. C., Craighead, H. G., Austin, R. H., *Proc. Natl. Acad. Sci. USA* 1999, *96*, 13762–13765.

[34] Duke, T. A. J., Austin, R. H., *Phys. Rev. Lett.* 1998, *80*, 1552–1555.

[35] Ertaş, D., *Phys. Rev. Lett.* 1998, *80*, 1548–1551.

[36] Chou, H., Spence, C., Scherer, A., Quake, S., *Proc. Natl. Acad. Sci. USA* 1999, *96*, 11–14.

[37] Barron, A. E., Soane, D. S., Blanch, H. W., *J. Chromatogr.* 1993, *652*, 3–16.

[38] Barron, A. E., Blanch, H. W., Soane, D. S., *Electrophoresis* 1994, *15*, 597–615.

[39] Sassi, A. P., Barron, A., Alonso-Amigo, M. G., Hion, D. Y., Yu, J. S., Soane, D. S., Hooper, H. H., *Electrophoresis* 1996, *17*, 1460–1469.

[40] Barron, A. E., Sunada, W. M., Blanch, H. W., *Electrophoresis* 1995, *16*, 64–74.

[41] Barron, A. E., Sunada, W. M., Blanch, H. W., *Electrophoresis* 1996, *17*, 744–757.

[42] Tietz, D., in: Chrambach, A., Dunn, M. J., Radola, B. J. (Eds.), *Advances in Electrophoresis*, VCH Publishers, Weinheim 1988, pp. 109–169.

[43] Slater, G. W., in: Heller, C. (Ed.), *Analysis of Nucleic Acids by Capillary Electrophoresis*, Vieweg & Son, Wiesbaden 1997, pp. 24–66.

[44] Hubert, S. J., Slater, G. W., Viovy, J. L., *Macromolecules* 1996, *29*, 1006–1009.

[45] Kim, Y., Morris, M. D., *Anal. Chem.* 1995, *67*, 784–786.

[46] Shi, X., Hammond, R. W., Morris, M. D., *Anal. Chem.* 1995, *67*, 1132–1138.

[47] Shi, X., Hammond, R. W., Morris, M. D., *Anal. Chem.* 1995, *67*, 3219–3222.

[48] Kim Y., Morris, M. D., *Electrophoresis* 1996, *17*, 152–160.

[49] Schwinefus, J. J., Morris, M. D., *Macromolecules* 1999, *32*, 3678–3684.

[50] Burmeister, M., Ulanovsky, L. (Eds.), *Pulsed-Field Gel Electrophoresis: Protocols, Methods, and Theories*, Humana Press, Totowa, NJ 1992.

[51] Starchev, K., Sturm, J., Weill, G., *Macromolecules* 1999, *32*, 348–352.

[52] Braun, B., Blanch, H. W., Prausnitz, J. M., *Electrophoresis* 1997, *18*, 1994–1997.

[53] Starkweather, M. E., Muthukumar, M., Hoagland, D. A., *Macromolecules* 1998, *31*, 5495–5501.

[54] Starkweather, M. E., Muthukumar, M., Hoagland, D. A., *Macromolecules* 1999, *32*, 6837–6840.

[55] Starkweather, M. E., Hoagland, D. A., Muthukumar, M., *Macromolecules* 2000, *33*, 1245–1253.

[56] Slater, G. W., Drouin, G., *Electrophoresis* 1992, *13*, 574–582.

[57] Heller, C., Duke, T. A. J., Viovy, J. L., *Biopolymers* 1994, *249–259*.

[58] Rousseau, J., Drouin, G., Slater, G. W., *Phys. Rev. Lett.* 1997, *79*, 1945–1948.

[59] Rousseau, J., Drouin, G., Slater, G. W., *Electrophoresis* 2000, *21*, 1464–1470.

[60] Lee, N., Obukhov, S., Rubinstein, M., *Electrophoresis* 1996, *17*, 1011–1017.

[61] Deutsch, J. M., *J. Chem. Phys.* 1989, *90*, 7436–7441.

[62] Deutsch, J. M., *Science* 1988, *240*, 922–924.

[63] Deutsch, J. M., Madden, T. L., *J. Chem. Phys.* 1989, *90*, 2476–2485.

[64] Obukhov, S. P., Rubinstein, M., *J. Phys. (France) II* 1993, *3*, 1455–1465.

[65] Popelka, S., Kabatek, Z., Viovy, J.-L., Gas, B., *J. Chromatogr. A* 1999, *838*, 45–53.

[66] Ueda, M., Oana, H., Baba, Y., Doi, M., Yoshikawa, K., *Biophys. Chem.* 1998, *71*, 113–123.

[67] Carlsson, C., Larsson, A., Jonsson, M., Nordén, B., *J. Am. Chem. Soc.* 1995, *117*, 3871–3872.

[68] Duke, T., *Curr. Opin. Chem. Biol.* 1988, *2*, 592–596.

[69] Duke, T. A. J., Austin, R. H., Cox, E. C., Chan, S. S., *Electrophoresis* 1996, *17*, 1075–1079.

[70] Slater, G. W., Wu, S. Y., *Phys. Rev. Lett.* 1995, *75*, 164–167.

[71] Appel, M., Fleischer, G., Karger, J., Chang, I., Fujara, F., Schonhals, A., *Colloid Polym. Sci.* 1997, *275*, 187–199.

[72] Carmesin, I., Kremer, K., *Macromolecules* 1988, *21*, 2819–2823.

[73] Ulanovsky, L., Drouin, G., Gilbert, W., *Nature* 1990, *343*, 190–192.

[74] Slater, G. W., Villeneuve, C., *J. Polym. Sci. B* 1992, *30*, 1451–1457.

[75] Desruisseaux, C., Slater, G. W., *Phys. Rev. E* 1994, *49*, 5885–5888.

[76] Slater, G. W., Desruisseaux, C., Villeneuve, C., Guo, H. L., Drouin, G., *Electrophoresis* 1995, *16*, 704–712.

[77] Desruisseaux, C., Slater, G. W., Drouin, G., *Electrophoresis* 1998, *19*, 627–634.

[78] Défontaines, A.-D., Viovy, J. L., *Electrophoresis* 1993, *14*, 8–17.

[79] Griess, G. A., Serwer, P., *Biophys. J.* 1998, *74*, A71.

[80] Serwer, P., Griess, G. A., *J. Chromatogr. B* 1999, *722*, 179–190.

[81] Desruisseaux, C., Slater, G. W., Kist, T. B. L., *Biophys. J.* 1998, *75*, 1228–1236.

[82] Baumgartner, A., Muthukumar, M., *J. Chem. Phys.* 1987, *87*, 3082–3088.

[83] Muthukumar, M., Baumgartner, A., *Macromolecules* 1989, *22*, 1937–1940.

[84] Muthukumar, M., Baumgartner, A., *Macromolecules* 1989, *22*, 1941–1946.

[85] de Gennes, P. G., *Scaling Concepts in Polymer Physics*, Cornell University Press, New York 1979.

[86] Doi, M., Edwards, S. F., *The Theory of Polymer Dynamics*, Oxford University Press, New York 1986.

[87] Arvanitidou, E., Hoagland, D. A., *Phys. Rev. Lett.* 1991, *67*, 1464–1467.

[88] Smisek, D. L., Hoagland, D. A., *Science* 1990, *248*, 1221–1223.

[89] Hoagland, D. A., Muthukumar, M., *Macromolecules* 1992, *25*, 6696–6698.

[90] Mayer, P., Slater, G. W., Drouin, G., *Appl. Theor. Electrophor.* 1993, *3*, 147–155.

[91] Liu, L., Li, P., Asher, S. A., *Nature* 1998, *397*, 141–144.

[92] Liu, L., Li, P., Asher, S. A., *J. Am. Chem. Soc.* 1999, *121*, 4040–4046.

[93] Clark, L. A., Ye, G. T., Snurr, R. Q., *Phys. Rev. Lett.* 2000, *84*, 2893–2896.

[94] Slater, G. W., *Electrophoresis* 1993, *14*, 1–7.

[95] Slater, G. W., Mayer, P., Grossman, P. D., *Electrophoresis* 1995, *16*, 75–83.

[96] Tinland, B., *Electrophoresis* 1996, *17*, 1519–1523.

[97] Tinland, B., Meistermann, L., Weill, G., *Phys. Rev. E* 2000, *61*, 6993–6998.

[98] Pluen, A., Tinland, B., Sturm, J., Weill, G., *Electrophoresis* 1998, *19*, 1548–1559.

[99] Curie, P., *J. Phys. (Paris) III* 1994, *3*, 343.

[100] Feynman, R. P., Leighton, R. B., Sands, M., *The Feynman Lectures on Physics*, Addison-Wesley, Rading, MA 1966, Vol. I, Chapter 46.

[101] Rousselet, J., Salome, L., Ajdari, A., Prost, J., *Nature* 1994, *370*, 446–448.

[102] Bader, J. S., Hammond, R. W., Henck, S. A., Deem, M. W., McDermott, G. A., Bustillo, J. M., Simpson, J. W., Mulhern, G. T., Rothberg, J. M., *Proc. Natl. Acad. Sci. USA* 1999, *96*, 13165–13169.

[103] Hammond, R. W., Bader, J. S., Henck, S. A., Deem, M. W., McDermott, G. A., Bustillo, J. M., Rothberg, J. M., *Electrophoresis* 2000, *21*, 74–80.

[104] Slater, G. W., Guo, H. L., Nixon, G. I., *Phys. Rev. Lett.* 1997, *78*, 1170–1173.

[105] Derényi, I., Astumian, R. D., *Phys. ERev. E* 1998, *58*, 7781–7784.

[106] van Oudenaarden, A., Boxer, S. B., *Science* 1999, *285*, 1046–1048.

[107] Slater, G. W., Guo, H. L., *Electrophoresis* 1996, *17*, 977–988.

[108] Slater, G. W., Guo, H. L., *Electrophoresis* 1996, *17*, 1407–1415.

[109] Ferguson, K. A., *Metabolism* 1964, *13*, 985–1002.

[110] Ogston, A. G., *Trans. Faraday Soc.* 1958, *54*, 1754–1757.

[111] Morris, C. J. O. R., in: Peeters, H. (Eds.), *Protides of the Biological Fluids, 14th Colloquium*, Elsevier, New York 1967, pp. 543–551.

[112] Rodbard, D., Chrambach, A., *Proc. Natl. Acad. Sci. USA* 1970, *65*, 970–977.

[113] Serwer, P., *Electrophoresis* 1983, *4*, 375–382.

[114] Slater, G. W., Rousseau, J., Noolandi, J., Turmel, C., Lalande, M., *Biopolymers* 1988, *27*, 509–524.

[115] Treurniet, J. R., Slater, G. W., *J. Chromatogr. A* 1997, *772*, 39–48.

[116] Locke, B. R., Trinh. S. H., *Electrophoresis* 1999, *20*, 3331–3334.

[117] Labrie, J., Mercier, J.-F., Slater, G. W., *Electrophoresis* 2000, *21*, 823–833.

[118] Tietz, D., Gombocz, E., Chrambach, A., in: Dunn, M. J. (Ed.), *Electrophoresis '86*, VCH Publishers, Weinheim 1986, pp. 253–258.

[119] Mercier, J.-F., Slater, G. W., *Electrophoresis* 1998, *19*, 1560–1565.

[120] Chrambach, A., Radko, S. P., *Electrophoresis* 2000, *21*, 259–265.

[121] Mercier, J.-F., Slater, G. W., *J. Chem. Phys.* 1999, *110*, 6066–6076.

[122] Starchev, K., Sturm, J., Weill, G., Groger, C.-H., *J. Phys. Chem. B* 1997, *101*, 5659–5663.

[123] Kasianowicz, J. J., Brandin, E., Branton, D., Deamer, D. W., *Proc. Natl. Acad. Sci. USA* 1996, *93*, 13770–13773.

[124] Akeson, M., Branton, D., Kasianowicz, J. J. Brandin, E., Deamer, D. W., *Biophys. J.* 1999, *77*, 3227–3233.

[125] Meller, A., Nivon, L., Brandin, E., Golovchenko, J., Branton, D., *Proc. Natl. Acad. Sci. USA* 2000, *97*, 1079–1084.

[126] Lubensky, D. K., Nelson, D. R., *Biophys. J.* 1999, *77*, 1824–1838.

[127] Muthukumar, M., *J. Chem. Phys.* 1999, *111*, 10371–10374.

[128] Park, P. J., Sung, W., *Phys. Rev. E* 1998, *57*, 730–734.

[129] Sebastian, K. L., Paul, A. K. R., *Phys. Rev. E* 2000, *62*, 927–939.

[130] Boehm, R. E., *Macromolecules* 1999, *32*, 7645–7654.

[131] Mouradian, S., Brumley, R. L., Smith, L. M., *Electrophoresis* 1994, *15*, 1084–1090.

[132] Kim, Y., Yeung, E. S., *Electrophoresis* 1997, *18*, 2901–2908.

[133] Noguchi, H., *J. Chem. Phys.* 2000, *112*, 9671–9678.

Electrophoresis 2000, *21*, 3888–3897

Review

Bohuslav Gaš[1]
Ernst Kenndler[2]

[1]Faculty of Science,
 Charles University,
 Prague, Czech Republic
[2]Institute of Analytical
 Chemistry,
 University of Vienna,
 Vienna, Austria

Dispersive phenomena in electromigration separation methods

A review on dispersive effects and on peak broadening in electromigration separation methods (capillary electrophoresis and electrochromatography) is presented, mainly covering papers published between the beginning of 1997 and the beginning of 2000. Most attention is drawn to work dealing with nonlinear effects that cause anomalous electromigration dispersion in electrolyte systems with two or multiple coions. Further, topics cover the comparison of electroosmotic and pressure-driven modes in electrochromatography, dispersive effects due to nonhomogeneous velocity fields in packed electrochromatography columns, to nonuniform electroosmotic flow, to sorption of analytes (mainly proteins) at the column wall or the stationary phase, and due to the influence of the nonideal column geometry like coiling or irregularities in shape.

Keywords: Capillary electrophoresis / Electrochromatography / Electromigration dispersion / Review

EL 4199

Contents

1 Introduction

Electromigration separation methods are becoming routine techniques in analytical laboratories. They are believed to attain a very high separation efficiency. This, indeed, can be achieved when all the possible deteriorating phenomena are avoided or at least suppressed and the only significant dispersive phenomenon is the inevitable molecular diffusion. Such an ideal situation, however, is hardly possible and in most practical applications additional dispersive sources are present as well. In the linear separation model, the separation efficiency, expressed by the plate height, H, or the plate number, N, is related to the total dispersion of the analyte peak, σ_{tot}:

$$H = \frac{\sigma_{tot}^2}{L}, N = \frac{L}{H} \qquad (1)$$

L is the migration distance from the injection point to the position of an on-column detector. Most dispersion sources of the analyte peak diminishing the separation efficiency are fully identified. In addition to the above mentioned molecular diffusion these are: (i) extracolumn effects, *e.g.*, from analyte sampling or detection; (ii) thermal effects due to Joule heating; (iii) electromigration dispersion; (iv) interaction of analytes with the inner column wall; (v) influence of the nonideal column geometry like coiling or irregularities in shape; and (vi) radial nonuniformity of the bulk flow in the column raised either due to laminar flow or nonuniform electroosmotic flow.

To be able to utilize the most favorable feature of electromigration separation methods, its high efficiency, it is therefore important to find the causes of dispersion in order to determine the conditions in which they are eliminated. All these causes have been intensively investigated, as reviewed in our previous paper [1]. Nonetheless, work on the explanation and elimination of disturbing effects has continued. A concentrated effort is now directed to two main topics: (i) to explain nonlinear effects

Correspondence: Dr. Bohuslav Gaš, Faculty of Science, Charles University, Albertov 2030, CZ-128 40 Prague 2, Czech Republic
E-mail: gas@natur.cuni.cz
Fax: +420-2-2491-9752

causing anomalous electromigration dispersion in capillary zone electrophoresis, and (ii) to understand dispersive effects and improve separation efficiency in capillary electrochromatography. This review is a continuation of the previous review paper. It covers work published between the beginning of 1997 and the beginning of 2000. Earlier papers are mentioned only when it is useful to point out continuity with newer ones. Papers dealing with dispersive effects in the separation of DNA fragments in sieving matrices are not included because the mechanism of their electromigration is different.

2 Anomalous electromigration dispersion

Electromigration dispersion causes a characteristic triangular deformation of analyte peaks and has been described earlier by a number of authors. Recently, however, some other features due to electromigration dispersion have been noticed, which had remained overlooked by routine workers but had started to raise interest of theoreticians. One is the "anomalous" dispersion of analytes, which can sometimes disable the separation, or, even worse, some analytes can apparently be missing from the sample without being noticed by the analyst. Electromigration dispersion is the consequence of an inherent nonlinearity, which is present in the basic laws describing electromigration in the electric field. The continuity equations describing the migration of n strong ions in the electric field are given in Eq. (2).

$$\frac{\partial c_i}{\partial t} = \frac{\partial}{\partial x}\left(D_i \frac{\partial c_i}{\partial x} + \mathrm{sgn}(z_i) u_i j \frac{c_i}{F \sum_{i=1}^{n} c_i u_i |z_i|} \right) \quad i = 1, \ldots, n \tag{2}$$

Here, c_i, D_i, u_i, z_i are the concentration, diffusion coefficient, ionic mobility, and relative charge, respectively, of the i-th ion. F is the Faraday constant and j is the current density. The division in the second term on the right hand side of the equation is responsible for this nonlinearity and is present also even in the simplest system with exclusively strong electrolytes. The denominator of the fraction in the second term is the conductivity. If the analyte is present in a significant concentration, it influences the conductivity of the background electrolyte (BGE), which can no longer be regarded as constant. The analyte peak then undergoes dispersion and various deformations of its shape that significantly decrease the separation efficiency. This phenomenon is called electromigration dispersion and has been described many times. One of the most fundamental descriptions of electromigration dispersion can be found in papers by Gebauer *et al.* [2, 3].

It is remarkable that this inherent nonlinearity becomes much more noticeable if the BGE contains more than one coion. (The coion is that ion of the BGE that migrates in the same direction and has the same effective charge as the ion of the analyte of interest). BGEs with two coions are often used in capillary zone electrophoresis (CZE) with indirect photometric detection, where one of the coions is light-absorbing and the second one has more appropriate acid-base properties for the particular system. Such systems, however, may exhibit unexpected features.

The manifiesting phenomenon appearing in the systems with multiple coions is a generation of new migrating peaks often called system peaks. Unfortunately, the term system peak is vague and has so far been used for designating various phenomena. In the following text we will distinguish between two kinds of system peaks. The first one is the "normal water gap", which is caused by an initial disturbance introduced by injecting the sample into the originally uniform concentration of the BGE. This system peak is characterized by a jump in the Kohlrausch regulating function at the original site of injection. It is simply shifted in the column due to a bulk flow of the BGE, which is raised by the electroosmotic (or possibly pressure-driven) flow. This peak often serves as a marker for the determination of the EOF velocity, though some other phenomena causing its possible additional movement have to be taken into account [4, 5].

The second kind of system peak arises in systems with two (or multiple) coions. It has a certain electrophoretic velocity and develops due to the mutual interaction of the ions present in the electrolyte. This interaction is of electrostatic nature, maintaining a macroscopic electroneutrality of the solution. This system peak is in fact not a peak of a real analyte; rather, it is a moving mutual jump in concentration of both coions. As its movement is raised due to the electromigration flux of ions rather than from the bulk flow, the Kohlrausch regulating function is constant at the site of this peak and has the same value as the original BGE.

Analogous system peaks or disturbances were noticed earlier in chromatography by Crommen *et al.* [6] in connection with indirect detection and were analyzed by Poppe [7], who called them "eigenpeaks". Poppe further explained that the response in indirect detection is closely related to the appearance of the eigenpeaks and tends to reach ± infinity, when the retention of the analyte peak gradually moves across the position of the system peak. He also studied an analogous situation in electrophoretic migration and was able to simulate the shapes of these peaks [8]. Such peaks in electrophoresis were also observed by Beckers [9], who studied BGEs with two coions [9]. He found system peaks with a mobility between the mobility of the two coions. He also noticed that an analyte ion in the vicinity of the system peak "interacts" with the system peak and both the sample and system peaks are enlarged and dispersed due to this interaction.

For studying such phenomena, Beckers *et al.* [10] employed a multiple use of the ITP steady-state model, which was described much earlier. Beckers [11, 12] was able to calculate the electrolyte composition in the CZE zone in systems with two coions and to determine parameters describing the sensitivity of indirect UV detection like the transfer ratio *TR*, which can be defined as:

$$TR = \left(\frac{dc_B}{dc_x}\right)_{c_x \to 0} \tag{3}$$

where c_B is the concentration of the absorbing ion of the BGE and c_x is the concentration of the analyte. The system peaks in systems with multiple coions were also carefully studied by Gebauer and Boček [13], Desiderio *et al.* [14], and Macka *et al.* [15]. The authors used the concept of vacancy electrophoresis for their explanation, in analogy to previously described vacancy chromatography. The system peaks and response in indirect UV detection were described in the paper of Lu and Westerlund [16].

An additional phenomenon connected with the use of multiple coions is severe broadening of the analyte peaks. This takes place when the velocity of an analyte peak is very close or matches the velocity of the system eigenpeak. In this case, a response of indirect detection at the site of the analyte, which is based on properties of background constituents, has a tendency to reach both highly positive and highly negative values, as explained by Poppe [7, 8]. Furthermore, the analyte undergoes massive dispersive forces causing its extensive spatial dispersion and consequently its concentration becomes low. In spite of the gradually diminishing concentration of the analyte, the response of the indirect detector is still huge and attains its characteristic zigzag shape, which was observed by Beckers [9] and Bullock *et al.* [17]. When a specific detector sensitive to some property of the analyte was used, like the direct photometric detector, the heavy spatial dispersion would cause its signal to be low and the analyte could seemingly get lost. This behavior in systems with two coions was predicted and experimentally confirmed by Beckers [11, 12]. He found that the sensitivity parameters of indirect UV detection behave singularly under certain conditions, *e.g.*, when the mobility of an analyte matches the mobility of the system eigenpeak.

The same phenomena were also observed by Williams *et al.* [18, 19]. In attempts to develop new buffers with low electromigration dispersion, the authors used a system with two coions and found a heavy peak shape distortion under some conditions. Their electropherograms could be successfully simulated using a computer program that solves the dynamics of electromigration. The authors determined by their simulation tool [20] that the mobility difference for two comigrating ions migrating symmetrically in the vicinity of the analyte ions should be less than 1×10^{-9} cm^{2}V^{-1}s^{-1} in ordert to avoid anomalous dispersion of the analyte peak. They proposed and synthesized a buffer which contains a large number of components with mobilities closer than 1×10^{-9} cm^{2}V^{-1}s^{-1}. Such a buffer allowed mobility matching without excessive dispersion normally inherent for systems with multiple coions.

A fundamental description of the anomalous peak broadening in systems with two coions was given by Gebauer *et al.* [21, 22]. The authors applied the concept of the peak shape diagrams previously introduced [2], using a quantity called velocity slope S_x, the quantity characterizing electromigration dispersion of a given analyte in a given BGE. The velocity slope is defined as

$$S_x = \left(\frac{dv_x}{da_x}\right)_{a_x \to 0} \tag{4}$$

where a_x is the molar fraction of analyte X in the sample zone and v_x is the electrophoretic velocity of analyte X.

Figure 1 shows one result from their work: calculated values of S_x for picric acid and salicylic acid as samples X when migrating in the BGE containing 16 mM tris(hydroxymethyl)aminomethane and a varying concentration of α-hydroxyisobutyric acid and gallic acid as coions. It is obvious that at a certain composition of the BGE, there are singular points where the values of S_x either for picric acid or salicylic acid theoretically reach ± infinity. At these particular compositions the electromigration dispersion of the analytes attains a big value and the peaks are heavily dispersed. This effect was confirmed experimentally. The composition in the sample CZE zone was also calculated by Xiong and Li [23] who determined the analytical sensitivity of indirect UV detection both in the electrolyte systems with absorbing coions and counterions. A simple model for calculating the concentration distribution, allowing the use of spreadsheet programs for calculation, was proposed by Mikkers [24].

In some cases electrolytes with seemingly only one coion can also exhibit system peaks. Performing CZE with a single coion electrolyte at low pH, Beckers [25] noticed

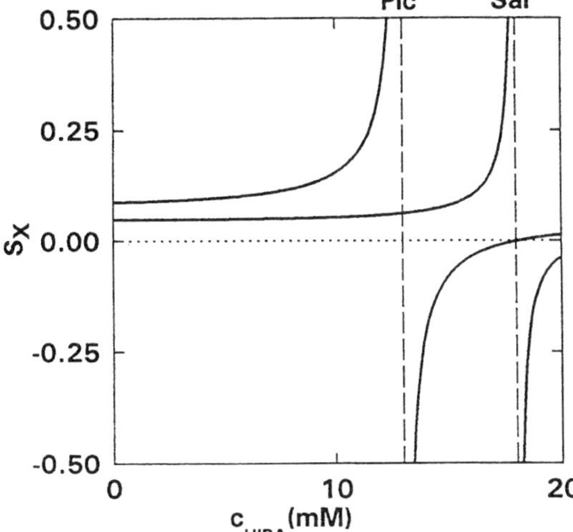

Figure 1. Calculated dependence of the velocity slope S_X (in 10^{-3} m s^{-1}) of picric acid (Pic) and salicylic acid (Sal) as analytes on the α-hydroxyisobutyric acid (HIBA) concentration in the BGE. The BGE consists of 16 mM tris(hydroxymethyl)aminomethane at various concentrations of HIBA and gallic acid (GA) as coions with c_{HIBA} + c_{GA} = 20 mM. Reprinted from [21], with permission.

that the system dip of the electroosmotic flow marker splits up and a new system peak migrates by its own electrophoretic velocity, which is not directly the velocity of the hydrogen ion. Depending on the concentration of ions in the sampled plug, either positive or negative peaks were observed. It was correctly explained later by Beckers [26] and Gebauer *et al.* [27] that these system peaks in the electrolyte with highly acidic and alkaline pH are a consequence of the fact that the hydrogen or hydroxide ions, respectively, have a significant concentration in comparison with the coion and act as a second coion with all the consequences as found in the two coion systems. The increase in sensitivity of indirect detection in electrolytes with two coions obtained at some conditions, on the other hand, can be used to increase the sensitivity of indirect absorbance detection as was proposed by Doble and Haddad [28].

Recently Boček *et al.* and Gaš *et al.* [29, 30] found that even simple multivalent buffers behave as systems with multiple coions when used in the pH region where two ionic forms are concurrently present. Gaš *et al.* [30] presented a model for optimization of the BGE systems for CZE, which allows taking into account trivalent ions at maximum. It also allows modeling highly acidic and alkaline electrolytes, where the presence of hydrogen and hydroxide ions, respectively, is significant. The model

enables predicting the parameters of the system that are experimentally available, like the transfer ratio *TR* or the molar conductivity detection response b_X, which expresses the sensitivity of conductivity detection:

$$b_X = \left(\frac{d\kappa}{dc_X}\right)_{c_X \to 0} \tag{5}$$

Here κ is the conductivity measured by the detector. The model also enables us to evaluate parameters of singular points leading to the appearance of eigenpeaks and anomalous dispersion. Figure 2 shows the molar conductivity response b_X, relative velocity slope S_X, and transfer ratio *TR* in their dependence on the ionic mobility of a strong anion when migrating in the popular and generally utilized phosphate buffer with pH about 7. The behavior of this buffer is unexpected and highly interesting.

In this buffer the acid-base equilibrium responsible for buffering ability in the buffer is between the doubly-charged and the singly-charged phosphates. It was revealed that such a system behaves as the electrolyte with two coions, which is proved by the singular point in Fig. 2 at a mobility of about 42×10^{-9} m^{2}V^{-1}s^{-1}. At this point all the differential quanties approach $\pm$ infinity. It means that a system peak having this mobility should be expected. Further, the signals of indirect detectors will be heavily distored and amplified if the mobility of any analyte of the sample comes close to this value. This is docu-

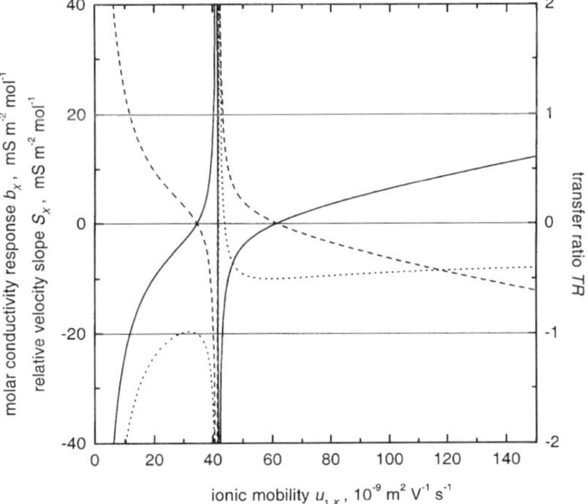

Figure 2. Dependence of differential quantities on the ionic mobility $u_{1,x}$ of a strong monovalent analyte anion. BGE: 5 mM phosphoric acid, 7.5 mM NaOH, 0.1 mM tetradecyltrimethylammonium hydroxide; calculated pH 7.24. Solid line: b_X, molar conductivity detection response. Dashed line: S_X, relative velocity slope. Dotted line: *TR*, transfer ratio. Reprinted from [30], with permission.

Electrophoresis 2000, *21*, 3888–3897

mented in Fig. 3 showing the experimental record of the conductivity detector, when the sample is composed of chloride, iodate and salicylate. As the mobility of iodate is 42.0×10^{-9} m²V⁻¹s⁻¹, which coincides with the mobility of the singularity, its conductivity signal will be expected to be extremely high and both positive and negative. The zigzag signal of iodate, which is much bigger than that of chloride and salicylate, confirms these expectations. On the other hand, the UV detector operating at 200 nm gives a different signal. As the UV absorbance of phosphates forming the BGE and of chloride from the sample is negligible, only direct signals for iodate and salicylate should be expected. In accordance with the theoretical prediction, the signal of iodate is almost lost due to its excessive dispersion, while the salicylic acid still gives a good and narrow peak.

3 Capillary electrochromatography and micellar systems

In terms of separation selectivity, liquid chromatography is superior over zone electrophoresis due to a wide display of partition mechanisms that can be tailored to fulfill specific demands of a particular class of analytic tasks. Therefore, capillary electrochromatography (CEC) tries to combine the benefits of both methods, *i.e.*, the electric driving of the liquid phase by means of the electroosmotic flow (EOF) with the partitioning principles that are at work in chromatography. Although the theory of separation in pressure-driven chromatography gives clear insight into the separation and dispersion mechansims, in CEC many questions are still unanswered. Therefore, a big effort to compare both modes of driving the mobile phase and to work out a more detailed theory of separation in CEC can be noticed.

Unlike the pressure-driven mode in HPLC, CEC utilizes the EOF to move the mobile phase through the packed bed. Theoretically, it should be supposed that the separation efficiency in CEC is generally better due to the plug-like radial velocity profile of the EOF (at least when the capillary or pore diameters are much greater than the Debye length) as the Taylor dispersion due to a parabolic velocity profile is absent. The real results in the packed columns and also in columns with continuous beds mostly confirm this straightforward assumption although the results reported by various authors are somewhat contradictory [31–41]. As the crucial phenomenon responsible for the solute transport in CEC is the EOF, a great deal of effort has been devoted to describe and understand complicated phenomena connected with its generation in packed columns. Especially Horváth's group contributed to this field. In the review on theoretical treatment of the EOF in packed-bed columns [42], Rathore and Horváth assessed Overbeek's *et al.* [43] and Dukhin's [44, 45] previous models and discussed a possible contribution of the electroosmosis of the 2ⁿᵈ type introduced by Dukhin. In experimental studies [46–48], they discussed various effects and parameters influencing separation. Among others they also mentioned the importance of the architecture of the column: if the column consists both of open and packed parts (see Fig. 4), differences in conductivities and other parameters in the two parts lead to a mis-

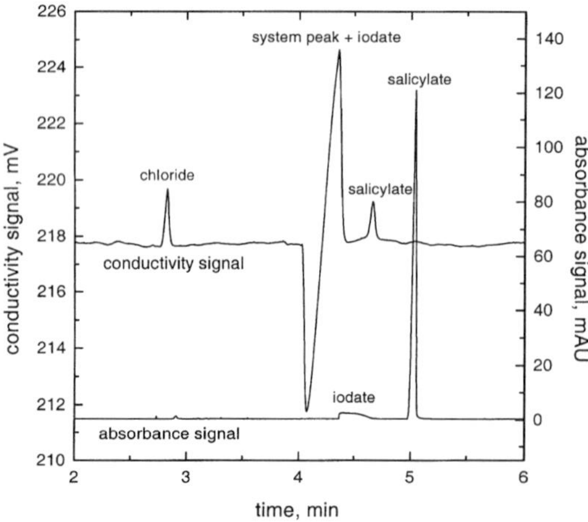

Figure 3. Experimental record of the electrophoretic separation with the conductivity and UV absorbance detector. Samples: chloride, iodate, salicylate, each 1 mM. BGE: 5 mM phosphoric acid, 7.5 mM NaOH, 0.1 mM tetradecyltrimethylammonium hydroxide; measured pH, 7.1. Conditions: uncoated fused-silica capillary, 75 μm ID, 360 μm OD, 80 cm total length, 66 cm length to the conductivity detector, 71.5 cm length to the UV detector; measuring wavelength, 200 nm; injection, 100 mbar·s; separation voltage, + 30 kV. Reprinted from [30], with permission.

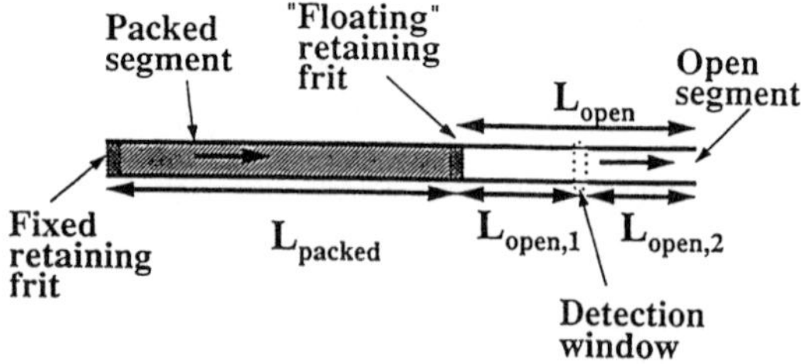

Figure 4. Schematic illustration of a duplex CEC column consisting of a packed and an open segment in series. The latter is divided into predetection and postdetection segments by the detector window. Reprinted from [48], with permission.

match in EOFs, which consequently cause local over-pressure, laminar flow, and an additional dispersion of peaks.

Further they found that some behavior of the packed columns could be ascribed to a generation of the intraparticle EOF inside the (porous) particles [49]. In almost all cases investigated by the authors, they reached better separation efficiency in the electroosmotically driven mode in comparison with the pressure-driven flow. The same was also found for the column with monolithic packing proposed by Gusev *et al.* [50] who compared the HPLC and CEC regimen. Jiskra *et al.* [51] examined the behavior of seven reversed-phase stationary phases both in pressure-driven and electroosmotically driven conditions. They found that chromatography parameters expressing selectivity of separation significantly differ in both eluent-driven modes, especially for porous packing. They explain it by different ligand orientations or possibly by an electroosmotic whirlwind effect in porous packing.

A novel insight into the dispersive effects due to the flow in various eluent-driven modes was recently brought by Tallarek *et al.* [52]. The authors used pulsed-field gradient NMR (PFG-NMR) spectroscopy to measure the axial displacement probability distribution of the solute molecules, getting direct insight into axial dispersion caused by the velocity field. They inspected the dispersion in open and packed columns, and in pressure- and electroosmotically driven modes. Using a unique experimental technique, the authors are able to "isolate" and determine the intrinsic dispersion arising due to the complex velocity field in the geometry of the column packing under conditions that do not include contributions from any other effects. The results of their study are convincing and can be summarized as follows: (i) As expected, even in rather thick open capillaries (150 µm ID), the EOF does not cause any additional contribution to axial dispersion, which could be addressed to longitudinal diffusion at the actual temperature. This was a striking contrast to the pressure-driven flow where the Taylor dispersion was fairly dominating. (ii) More interesting results were revealed in a packed capillary column (250 µm ID; diameter of packed beads $d_p = 40$ µm). The experimentally determined apparent axial dispersion enables us to calculate the reduced plate height h_a as

$$h_a = \frac{2D_{ap}}{vD_m} \tag{6}$$

where D_{ap} is the apparent axial dispersion coefficient determined from the axial displacement probability distributions, D_m is the molecular diffusion coefficient, $v = d_p u_{av}/D_m$ is the reduced flow velocity, and u_{av} is the aver-

age cross-sectional flow velocity. The authors analyzed the dependence of the reduced plate height on the reduced flow velocity by the Knox equation [53]

$$h_a = \frac{b}{v} + av^{1/3} + cv \tag{7}$$

Coefficient b accounts for obstructed longitudinal molecular diffusion. The second term expresses eddy dispersion so the coefficient a accounts for transverse diffusion and tortuosity of the flow. The third term with coefficient c includes mass-transfer resistance from possible stagnant fluid in the deep pools in the intraparticle network. The results of comparisons of the two driven modes of the flow through the same column are shown in Fig. 5. This figure shows that at the curves' minimum the reduced plate height is $h_{min} = 0.83$ for the pressure-driven mobile phase, while it is even lower, $h_{min} = 0.55$, for the electroosmotically driven flow. This confirms the superiority of the electroosmotically driven flow.

The nonlinear effects that play a basic role in the electromigration dispersive effects can be even more complicated in electrochromatography of charged analytes when they migrate in the column due to both an electrophoretic and a chromatographic transport mechanism. This was studied by Ståhlberg [54] who found that the mixed mechanism gives rise to strong nonlinear phenomena, causing either strong peak broadening or stabilized peaks that do not change their shape during migration. Moffatt *et al.* [55] were able to reach unusually high sepa-

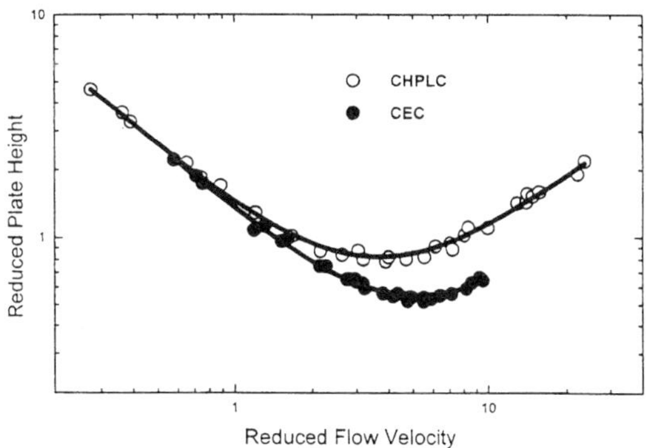

Figure 5. Reduced axial plate height ($h_a = H_a/d_p$) *versus* the reduced flow velocity ($v = d_p u_{av}/D_m$) for the capillary HPLC and CEC modes. Solid line: Eq. (7). For capillary HPLC: $a = 0.14$, $b = 1.25$, $c = 0.07$, $h_{min} = 0.83$. For CEC: $a = 0.06$, $b = 1.26$, $c = 0.04$, $h_{min} = 0.55$. Capillary column fused-silica 250 µm/360 µm ID/OD; packing 40 µm rehydroxylated spherically shaped particles; buffer: 10^{-3} M sodium tetraborate (pH 9.13); observation time, 120 ms. Reprinted from [52], with permission.

ration efficiencies during CEC of some compounds under reversed-phase conditions. They explained the sharpening of the analyte peaks by comigration of the analyte with the pulse of the solvent discontinuity that was injected together with the sample. An analogous sharpening of the analyte peaks was also observed by Pyell *et al.* [56] depending on the composition of solvent in the sampled zone. The authors, however, concentrated on another topic: the influence of the sample plug length on efficiency in a capillary electrochromatographic separation. The authors derived formulas giving maximum tolerable injection plug lengths (or other injection parameters like injection voltage or injection time) in order to reduce the deterioration of the separation efficiency.

The sources of peak dispersion in micellar electrokinetic chromatography (MEKC) are basically the same as in CZE: longitudinal diffusion, electromigration dispersion, radial temperature effects due to Joule heating sorption or any interaction with the inner wall, Taylor dispersion due to possible laminar flow profile, and various extracolumn effects. Additionally, some other effects specific for the presence of micelles interacting with analytes may be relevant: heterogeneity of micelles, intermicelle mass transfer and slow solute-micelle exchange kinetics. All these effects have been carefully analyzed previously (see, *e.g.*, [57–61]). Further work on this topic was also recently published by Thorsteindottir *et al.* [62, 63]. Much more obscure is the influence of an organic modifier present in the BGE, as it can influence the properties of both the micelles and the rest of the solution. This problem was studied by Seals *et al.* [64] to inspect the influence of 2-propanol as an organic modifer on the separation efficiency in MEKC with SDS micelles. They showed that the separation efficiency of weakly and strongly retained analytes varies in its dependence on the concentration of 2-propanol, which agreed well with the behavior predicted by the theory.

An analog to MEKC is suspension electrokinetic chromatography (SEKC) recently proposed by Göttlicher and Bächman [65]. Here separation is achieved by partitioning between chromatographic particles present in the suspension and the remaining solution. In their paper, the authors investigated the separation efficiency of polycyclic aromatic hydrocarbons by SEKC and found analogous sources of dispersion as in MEKC: longitudinal diffusion, thermal effects due to Joule heating, and nonuniform mobilities of the chromatographic particles.

Quirino and Terabe [66] proposed a novel method for on-line concentration (stacking) of neutral analytes in MEKC. They use an acidic BGE, which reduces the EOF directed to the cathode in bare capillaries, while the electropho-

retic mobility of the micelles is directed toward the detector at the anodic side. The electrophoretic mobility of micelles is higher in absolute value than the opposite EOF. This configuration gives rise to a stacking mechanism that is used for concentration of the samples originally diluted in a low-conductivity matrix. The authors noticed that the efficiency influenced by stacking is not only dependent on the field difference in the sampled plug and the BGE, but also on the presence of micelles in the sampled zone. They propose a mathematical model for describing this phenomenon, which was supported by experimental results.

4 Separation of proteins

The group of Chrambach [67–70] has been very active in finding sources of dispersion and peak asymmetry, especially in CZE of proteins, either in sieving media or in free solution. They considered, analyzed, and discussed many possible sources such as diffusion, Joule heat production, electromigration dispersion, initial zone length, analyte-wall interaction, microheterogeneity of the polymer, and finite rate of interaction with the polymer causing the interactive dispersion. Later, the same authors [71, 72] generalized their view and measured and discussed the dispersion of spherical particles of different sizes like globular proteins and various polystyrene carboxylates when moving by electromigration in buffered solution of uncrosslinked polyacrylamide and discussed extensively the mechanisms leading to the dispersion. Gysler *et al.* [73] measured the electromigration dispersion of protein zones in the CZE mode after previous stacking by an ITP step.

5 Geometry of the column, and plug length

Separation columns fabricated on a chip and microfabricated technology seem to be able to reach and exceed parameters attained in capillary columns. Due to their mostly more complicated geometry exhibiting bending, meandering or curvatures, the geometrical factors affecting the dispersion of peak become more important. Culbertson *et al.* [74] studied the influence of such geometrical factors of the separation channel in microchip on dispersion, namely lateral variations in both migration distance and field strength due to the curvature of the pathway. The results of their model were confirmed by experiments. The authors also proposed several methods to reduce geometrical dispersion. The group of Hjertén [75] polymerized the continuous beds in a chip format and compared the separation properties of such columns with the classical fused-silica capillary columns. They checked the electroosmotically driven electrochromatography and pressure-driven anion-exchange chromatography. They found basically the same van Deemter plots, both for chip and capillary format, which indicates that there are no additional dispersive sources in the chip format even when using continuous beds.

Bings *et al.* [75] showed that an appropriate connection of microfluidic devices to conventional 50 µm ID fused-silica capillaries is important to avoid deterioration of the separation efficiency. They pointed out that a conventional connection leads to a geometrical dead volume of 0.7 nL, which can cause significant peak broadening. The authors proposed a procedure to eliminate the dead volume and reach almost theoretical separation efficiency. On the other hand, some dead volume was not deteriorating in a new construction of the sleeve cell for optical detection in CE, which was proposed by Djordjevic *et al.* [77] to reach signal enhancement. They reported that an abrupt change of the capillary ID from 50 µm to 220 µm did not produce extensive band broadening. Peng and Chen [78] studied the contribution from the injection plug length to the total variance of the peak. They correctly pointed out that the front part of the plug inserted by the pressure injection does not have a rectangular axial profile but a rather sigmoidal profile due to the parabolic radial velocity profile during sampling, which should be considered in calculating the total injection length or variance.

6 EOF

It is well known that axially nonuniform electroosmotic velocity, either due to nonuniformity in zeta potential or due to different conductivities of solution, gives rise to local overpressures in the column and consequently to laminar Poiseuille flow. Andreev *et al.* [79] described these phenomena in a rectangular channel geometry by a mathematical model and showed that as low as 10% difference in zeta potentials at opposite walls of the channel can deteriorate resolution. Herr *et al.* [80] studied the EOF in capillaries with axially nonuniform zeta potential. For this purpose, they coupled capillaries with substantially different zeta potentials and imaged the resulting flow fields with a caged-fluorescence imaging technique. They also presented a simple analytical model for the velocity field and peak dispersion.

Ericson and Hjertén [81] used capillary columns filled with a newly synthesized polymerized continuous bed for the separation of proteins. The bed was derivatized with C18 groups to obtain reversed-phase separation and with ammonium groups for generation of the EOF. For the high resolution of proteins the authors used a gradient of acetonitrile concentration in the BGE. As the EOF is generally dependent on the composition of the electrolyte and especially the acetonitrile content, it should be supposed that it can be different in various parts of the column. This would lead again to the nonuniform EOF with possibly increased peak broadening. The authors, however, carefully verifield that the nonhomogeneity in this case did not contribute significantly to peak broadening. When controlling the EOF by an external radial field, Kašička *et al.* [82–84] paid attention to assure uniform intensity of the radial field along the capillary axis. Therefore, they developed several new low-conductive layers for the outer capillary surface and used a set of three high-voltage power supplies to form a desired uniform radial electric field.

An even more pronounced effect regarding possible dispersion due to EOF can take place in hydrodynamically closed capillaries. Here, the EOF inevitably gives rise to laminar viscous flow due to the hydrodynamically closed loop in spite of the fact that there is no bulk flow. This has been fully taken into account by Kaniansky *et al.* [85] when they described CZE in a closed separation system. They showed that the dispersion of peaks is huge when separation is run in a separation system without any additives suppressing the EOF. Therefore, they proposed several suitable electrolyte systems where the EOF is suppressed by additives like hydroxyethylcellulose derivatives or polyethylene glycol, enabling good efficiency of separation.

7 Adsorption

Sample sorption onto the inner capillary surface always reduces the separation efficiency due to its inherently nonequilibrium character, which has been known for a long time [86, 87]. Also recently, some work was devoted to the investigation of the peak broadening mechanism due to sorption. Zhukov *et al.* [88] introduced a simplified mathematical model describing the influence of totally irreversible sample adsorption on peak parameters. They showed that under these conditions the trailing part of the peak profile has the same form and velocity for any extent of adsorption while the velocity of the leading part decreases with increasing adsorption, which can cause the change in its shape.

Ward and Khaledi [89] studied the efficiency of electrophoretic separation of derivatized oligosaccharides in nonaqueous BGE with formamide as the organic solvent and compared the results with analogous aqueous BGEs. They found the same dispersion mechanisms as found in conventional aqueous systems, whereby wall adsorption was analyzed to be one of the most significant. However, at optimized conditions they reached a 1.5-fold enhancement in efficiency in the nonaqueous system.

Electrophoresis 2000, *21*, 3888–3897

8 Theory

A novel approach to the description of the separation process in CZE was presented by Liang *et al.* [90]. The authors utilized tools of irreversible thermodynamics and derived the relationship between the total entropy change of separation and conventional experimentally available parameters like plate height and resolution. They also expressed some irreversible processes taking place during separation in terms of corresponding forces and fluxes: temperature gradient, potential gradient, concentration gradient, electrophoretic dispersion, viscous flow, and wall adsorption.

Received August 31, 2000

9 References

[1] Gaš, B., Štědrý, M., Kenndler, E., *Electrophoresis* 1997, *18*, 2123–2133.

[2] Gebauer, P., Boček, P., *Anal. Chem.* 1997, *69*, 1557–1563.

[3] Gebauer, P., Čáslavská, J., Thormann, W., Boček, P., *J. Chromatogr. A* 1997, *772*, 63–71.

[4] Štědrý, M., Popelka, Š., Gaš, B., Kenndler, E., *Electrophoresis* 1996, *17*, 1121–1125.

[5] Kenndler-Blachkolm, K., Popelka, Š., Gaš, B., Kenndler, E., *J. Chromatogr. A* 1996, *734*, 351–356.

[6] Crommen, J., Schill, G., Westerlund, D., Hackzell, L., *Chromatographia* 1987, *24*, 252–260.

[7] Poppe, H., *J. Chromatogr.* 1990, *506*, 45–60.

[8] Poppe, H., *Anal. Chem.* 1992, *64*, 1908–1919.

[9] Beckers, J. L., *J. Chromatogr. A* 1994, *679*, 153–165.

[10] Everaerts, F. M., Beckers, J. L., Verheggen, T.P.E.M., *Isotachophoresis - Theory Instrumentation and Application*, Elsevier, Amstersam, Oxford, New York 1976.

[11] Beckers, J. L., *J. Chromatogr. A* 1996, *741*, 265–277.

[12] Beckers, J. L., *J. Chromatogr. A* 1997, *764*, 111–126.

[13] Gebauer, P., Boček, P., *J. Chromatogr. A* 1997, *772*, 73–79.

[14] Desiderio, C., Fanali, S., Gebauer, P., Boček, P., *J. Chromatogr. A* 1997, *772*, 81–89.

[15] Macka, M., Haddad, P. R., Gebauer, P., Boček, P., *Electrophoresis* 1997, *18*, 1998–2007.

[16] Lu, B., Westerlund, D., *Electrophoresis* 1998, *19*, 1683–1690.

[17] Bullock, J., Strasters, J., Snider, J., *Anal. Chem.* 1995, *67*, 3246–3252.

[18] Williams, R. L., Childs, B., Dose, E. V., Guiochon, G., Vigh, G., *Anal. Chem.* 1997, *69*, 1347–1354.

[19] Williams, R. L., Childs, B., Dose, E. V., Guiochon, G., Vigh, G., *J. Chromatogr. A* 1997, *781*, 107–112.

[20] Williams, R. L., Childs, B., Dose, E. V., Guiochon, G., Vigh, G., *J. Chromatogr. A* 1998, *814*, 199–204.

[21] Gebauer, P., Borecká, P., Boček, P., *Anal. Chem.* 1998, *70*, 3397–3406.

[22] Gebauer, P., Desiderio, C. Fanali, S., Boček, P., *Electrophoresis* 1998, *19*, 701–706.

[23] Xiong, X., Li, S.F.Y., *J. Chromatogr. A* 1999, *835*, 169–185.

[24] Mikkers, F.E.P., *Anal. Chem.* 1999, *71*, 522–533.

[25] Beckers, J. L., *J. Chromatogr. A* 1994, *662*, 153–166.

[26] Beckers, J. L., *J. Chromatogr. A* 1999, *844*, 321–331.

[27] Gebauer, P., Pantůčková, P., Boček, P., *Anal. Chem.* 1999, *71*, 3374–3381.

[28] Doble, P., Haddad, P. R., *Anal. Chem.* 1999, *71*, 15–22.

[29] Boček, P., Gebauer, P., Beckers, J. L., *Proceedings of the 13th International Symposium on High Performance Capillary Electrophoresis and Related Microscale Techniques*, Saarbrücken, Germany, February 20–24, 2000.

[30] Gaš, B., Coufal, P., Jaroš, M., Muzikář, J., Jelínek, I., *J. Chromatogr. A* 2000, in press.

[31] Robson, M. M., Cikalo, M. G., Myers, P., Euerby, M. R., Bartle, K. D., *J. Microcol. Sep.* 1997, *9*, 357–372.

[32] Ericson, C., Liao, J. L., Nakazato, K., Hjertén, S., *J. Chromatogr. A* 1997, *767*, 33–41.

[33] Van den Bosch, S. E., Heemstra, S., Kraak, J. C., Poppe, H., *J. Chromatogr. A* 1996, *755*, 165–177.

[34] Djordjevic, N. M., Fowler, P. W. J., Houdiere, F., Lerch, G., *J. Liq. Chromatogr. Rel. Techn.* 1998, *21*, 2219–2232.

[35] Xu, W. S. Regnier, F. E., *J. Chromatogr. A* 1999, *853*, 243–256

[36] Hilder, E. F., Klampfl, C. W., Macka, M., Haddad, P. R., Myers, P., *Analyst* 1999, *125*, 1–4.

[37] Tanaka, N., Nagayama, H., Kobayashi, H., Ikegami, T., Hosoya, K., Ishizuka, N., Minakuchi, H., Nakanishi, K., Cabrera, K., Lubda, D., *J. High Resol. Chromatogr.* 2000, *23*, 111–116.

[38] Banholczer, A., Pyell, U., *J. Chromatogr. A* 2000, *869*, 363–374.

[39] Crego, A. L., Martinez, J., Marina, M. L., *J. High Resol. Chromatogr.* 2000, *23*, 373–378.

[40] Ishizuka, N., Minakuchi, H., Nakanishi, K., Soga, N., Nagayama, H., Hosoya, K., Tanaka, N., *Anal. Chem.* 2000, *72*, 1275–1280.

[41] Hjertén, S., Vegvari, A., Srichaiyo, T., Zhang, H. X., Ericson, C., Eaker, D., *J. Capil. Electrophor.* 1998, *5*, 13–26.

[42] Rathore, A. S., Horváth, C., *J. Chromatogr. A* 1997, *781*, 185–195.

[43] Overbeek, J.T.G., Bungerberg de Jong, H. G. in: Kruyt, H. R., (Ed.), *Colloid Science II*, Elsevier, New York, Amsterdam, London, Brussels 1949, pp. 184–231.

[44] Dukhin, S. S., *Adv. Colloid Interface Sci.* 1991, *35*, 173–196.

[45] Dukhin, S. S., *Adv. Colloid Interface Sci.* 1991, *36*, 219–248.

[46] Choudhary G., Horváth, C., *J. Chromatogr. A* 1997, *781*, 161–183.

[47] Rathore, A. S., Horváth, C., *Anal. Chem.* 1998, *70*, 3069–3077.

[48] Rathore, A. S., Horváth, C., *Anal. Chem.* 1998, *70*, 3271–3274.

[49] Wen, E., Asiaie, R., Horváth, C., *J. Chromatogr. A* 1999, *855*, 349–366.

[50] Gusev, I., Huang, X., Horváth, C., *J. Chromatogr. A* 1999, *855*, 273–290.

[51] Jiskra, J., Cramers, C. A., Byelik, M., Claessens, H. A., *J. Chromatogr. A* 1999, *862*, 121–135.

[52] Tallarek, U., Rapp, E., Scheenen, T., Bayer, E., Van As, H., *Anal. Chem.* 2000, *72*, 2292–2301.

[53] Knox, J. H., *J. Chromatogr. Sci.* 1997, *15*, 352–354.

[54] Ståhlberg, J., *Anal. Chem.* 1997, *69*, 3812–3821.

[55] Moffatt, F., Cooper, P. A., Jessop, K. M., *Anal. Chem.* 1999, *71*, 1119–1124.

[56] Pyell, U. Rebscher, H., Banholczer, A., *J. Chromatogr. A* 1997, *779*, 155–163.

[57] Sepaniak, M. J., Cole, R. O., *Anal. Chem.* 1987, *59*, 472–476.

[58] Terabe, S., Otsuka, K., Ando, T., *Anal. Chem.* 1989, *61*, 251–260.

[59] Wallingford, R. A., Ewing, A. G., *Anal. Chem.* 1988, *60*, 258–263.

[60] Davis, J. M., *J. Microcol. Sep.* 1998, *10*, 479–489.

[61] Yu, L., Seals, T. H., Davis, J. M., *Anal. Chem.* 1996, *68*, 4270–4280.

[62] Thorsteinsdottir, M., Westerlund, D., Andersson, G., Kaufmann, P., *J. Chromatogr. A* 1998, *809*, 191–201.

[63] Thorsteinsdottir, M., Westerlund, D., Andersson, G., Kaufmann, P., *Chromatographia* 1998, *47*, 141–151.

[64] Seals, T. H., Davis, J. M., Murphy, M. R., Smith, K. W., Stevens, W. C., *Anal. Chem.* 1998, *70*, 4549–4562.

[65] Göttlicher, B., Bächman, K., *J. Chromatogr. A* 1997, *768*, 320–324.

[66] Quirino, J. P., Terabe, S., *Anal. Chem.* 1998, *70*, 149–157.

[67] Radko, S. P., Weiss, G. H., Chrambach, A., *J. Chromatogr. A* 1997, *781*, 277–286.

[68] Radko, S. P., Chrambach, A., Weiss, G. H., *J. Chromatogr. A* 1998, *781*, 253–262.

[69] Chrambach, A., Radko, S. P., *Electrophoresis* 1998, *19*, 1284–1287.

[70] Yarmola, E., Chrambach, A., *J. Phys. Chem. B* 1998, *102*, 4813–4818.

[71] Radko, S. P., Chrambach, A., *Electrophoresis* 1998, *19*, 1620–1624.

[72] Radko, S. P., Chrambach, A., *Electrophoresis* 1998, *19*, 2423–2431.

[73] Gysler, J., Jaehde, U., Schunack, W., *Fresenius J. Anal. Chem.* 1999, *365*, 398–403.

[74] Culbertson, C. T., Jacobson, S. C., Ramsey, J. M., *Anal. Chem.* 1998, *70*, 3781–3789.

[75] Ericson, C., Holm, J., Ericson, T., Hjertén, S., *Anal. Chem.* 2000, *72*, 81–87.

[76] Bings, N. H., Wang, C., Skinner, C. D., Colyer, C. L., Thibault, P., Harrison, D. J., *Anal. Chem.* 1999, *71*, 3292–3296.

[77] Djordjevic, N. M., Widder, M., Kuhn, R., *J. High Resol. Chromatogr.* 1997, *20*, 189–192.

[78] Peng, X., Chen, D. D. Y., *J. Chromatogr. A* 1997, *767*, 205–216.

[79] Andreev, V. P., Dubrovsky, S. G., Stepanov, V. Y., *J. Microcol. Sep.* 1997, *9*, 443–450.

[80] Herr, A. E., Molho, J. I., Santiago, J. G., Mungal, M. G., Kenny, T. W., *Anal. Chem.* 2000, *72*, 1053–1057.

[81] Ericson, C., Hjertén, S., *Anal. Chem.* 1999, *71*, 1621–1627.

[82] Kašička, V., Prusík, Z., Sázelová, P., Barth, T., Brynda, E., Machová, L., *J. Chromatogr. A* 1997, *772*, 221–230.

[83] Kašička, V., Prusík, Z., Sázelová, P., Brynda, E. Stejskal, J., *Electrophoresis* 1999, *20*, 2484–2492.

[84] Kašička, V., Prusík, Z., Sázelová, P., Chiari, M., Mikšík, I., Deyl, Z., *J. Chromatogr. B* 2000, *741*, 43–54.

[85] Kaniansky, D., Marák, J., Masár, M., Iványi, F., Madajová, E., Šimuničová, E., Zelenská, V., *J. Chromatogr. A* 1997, *772*, 103–114.

[86] Giddings, J. C., *Dynamics of Chromatography*, Marcel Dekker, New York 1965.

[87] Hjertén, S., *Chromatogr. Rev.* 1967, *9*, 122–219.

[88] Zhukov, M. Y., Ermakov, S. V., Righetti, P. G., *J. Chromatogr. A* 1997, *766*, 171–185.

[89] Ward, V. L., Khaledi, M. G., *J. Chromatogr. A* 1999, *859*, 203–219.

[90] Liang, H., Wang, Z. G., Lin, B. C., Xu, C. G., Fu, R. N., *J. Chromatogr. A* 1997, *763*, 237–251.

Review

Petr Gebauer
Petr Boček

Institute of Analytical
Chemistry,
Academy of Sciences of the
Czech Republic,
Brno, Czech Republic

Recent progress in capillary isotachophoresis

This article is a continuation of previous reviews and summarizes the progress of analytical capillary isotachophoresis in the years 1997–1999. Papers reviewed include theoretical and methodological aspects as well as analytical applications. Included are also papers using isotachophoresis and/or isotachophoretic principles as part of multidimensional separation schemes.

Keywords: Isotachophoresis / Capillary zone electrophoresis / Review   EL 4164

Contents

1 Introduction

This article is a continuation of our previous review [1] on the application and developments of capillary ITP. It is aimed to provide an overview on the work done in the field of analytical capillary ITP in the years 1997–1999. This time period has not shown any significant change in the trends which remain focused into three main directions: (i) theoretical and methodological research that includes also incorporation of ITP into novel multidimensional separation schemes: (ii) new analytical applications of ITP; (iii) ITP and transient ITP effects as an on-line preseparation technique for CZE. The mentioned time period has brought a rich production of new papers, first of all thanks to published proceedings of symposia of both the bian-

nual "ITP" series [2, 3] and the annual "HPCE" series [4–7]. To bring the review more up to date, we included also references from the recent paper symposium on ITP [8].

Published specialized reviews cover the application of ITP to the trace analysis of organic and inorganic pollutants [9], to the application of ITP in food analysis [10], and to analysis of inorganic ions in general [11]. In many cases ITP was reviewed together with CZE and other CE techniques. Published reviews comprise general theoretical principles [12], the role of complexation in the separation of inorganic ions [13, 14], sample stacking in CZE [15], methodological principles for inorganic analysis including ITP preconcentration techniques [16], and on-line preconcentration methods for CE in general [17]. Application reviews deal with analysis of inorganic species in environmental samples [18], food and feed samples [19], use of CE techniques in clinical toxicology [20], clinical and forensic analysis [21], pharmacy [22], pharmacokinetics [23], and for the separation of peptides and proteins [24–26]. Specialized reviews were published on the influence of organic solvents on separation selectivity [27] and on CE performed on microchips [28]. Papers on ITP were also included in the biannual fundamental review on CE [29].

2 Theory and methodology

Computer simulations remain a challenging alternative or complementary way to investigate the behavior of electrophoretic/ITP separation systems. Recent papers express the effort to bring the theoretical models as close to real systems as possible. This includes the development of new algorithms and models that bring both high numerical stability [30] and the possibility to work with realistic current densities [31] and electroosmotic flow [32, 33]. Computer simulations were used to investigate discontinuities of pH at zone boundaries in ITP systems with poorly buf-

Correspondence: Prof. Petr Boček, Institute of Analytical Chemistry, Academy of Sciences of the Czech Republic, Veveří 97, CZ-611 42 Brno, Czech Republic
E-mail: bocek@iach.cz
Fax: +4205-41212113

Abbreviations: BTP, bis[2-hydroxyethyl]iminotris[hydroxymethyl]methane; **EACA**, 6-aminocaproic acid; **HEC**, hydroxyethylcellulose; **HIBA**, α-hydroxyisobutyric acid; **HPMC**, hydroxypropylmethylcellulose; **LE**, leading electrolyte; **TE**, terminating electrolyte

fering leading electrolytes [34]. A paper on ITP at extreme pH values revealed that solvent ions, when present at non-negligible concentrations, may induce the formation of one additional moving boundary [35].

A new method for the precise estimation of ionic mobilities from the velocity of the moving boundary between the leading electrolyte (LE) and the terminating electrolyte (TE) in a geometrically and thermometrically well-defined channel was described [36]. The authors [37] published a new method to determine the stability constants of kinetically labile lanthanide-EDTA complexes from the reduction rate of their anionic ITP zones. ITP was used to determine unknown molecular masses of analytes, namely Jeffamines D230 and D400, by determining mobilities and concentrations from the measured step heights and specific zone resistance, and molar fractions and mass fractions from the measured zone lengths [38]. An interesting new approach to the problem of resolution and identification in ITP is based on the analyses of a sample at four different pH values [39]. A fingerprint of anions is suggested, depicted in a square diagram, two sides of which form the pK and absolute mobility axes and the other two the concentration axes in log scale, where an acid is represented by a pK-mobility point and by a dash to the concentration point.

A new theoretical approach to ITP with continuous pH and conductivity gradients was published [40]. It has extended the present models of electrofocusing to cover also conductivity changes and has shown that resolution and peak capacity in ITP with a continuous conductivity gradient is comparable to that of focusing in a pH gradient and to the separation power of CZE. The model was experimentally verified by separations in combined pH and conductivity gradients generated by synthetic polyampholytes [41]. The model was extended to a capillary with a nonconstant cross-section where a reasonable speed of separation and resolution could be achieved at a voltage drop across the separation capillary of only 200 V [42]. Methodological research has brought some new electrolyte components. Dithionate was suggested as a new LE for direct determination of chloride and other anions above pH 3 [43]. Polyethylene glycol was investigated as a complexing agent in acidic nonbuffered systems for the separation of mixtures of metal cations and a new approach for the estimation of polymer-metal stability constants from ITP data was proposed [44].

3 Analytical applications

Ammonium, calcium, magnesium, and potassium were determined in silage by using crown ether (18-crown-6) as a complexing agent (LE: 7.5 mM sulfuric acid, 7 mM 18-crown-6, 0,1% hydroxyproylmethylcellulose (HPMC); TE: 10 mM Bis-Tris-propane) [45]. The RSD of the method was between 4 and 10% and the LODs were below 200 ppm in the solid sample. ITP was used as a reference method to the CZE determination of free calcium in vegetables (LE: 5 mM HCl adjusted to pH 8.5 with Tris, 0.1% HPMC; TE: 10 mM sodium hexanoate, 0.5 mM EDTA) [46]. Yttrium and 14 lanthanide ions were separated simultaneously using α-hydroxyisobutyric acid (HIBA) and malonic acid as a second (helping) complexing agent (LE: 20 mM ammonia, 2 mM malonic acid, 7.5 mM HIBA, pH 4.8; TE: 20 mM carnitine hydrochloride) [47]. The superimposed complexing effects of both agents were investigated also theoretically and the reasons of the successful separation of yttrium from dysprosium without loss of the resolution of europium and gadollinium were explained [48]. Electrokinetic injections and transient ITP techniques were employed to increase sample loading and provided up to 700-fold on-column concentration for CZE of cations produced in nuclear fission [49]. Tributyltin and triphenyltin cations were determined simultaneously by ITP (LE: 10 mM KOH, adjusted to pH 5.0 with glutamic acid, 0.1% Triton X-100, 50% acetone; TE: 10 mM betaine hydrochloride) using a potential-gradient detector [50]. The RSDs were 1.5 and 1.9%, respectively, when 200 μL of a 10 mg/L concentration was injected; the LODs were 0.46 and 0.57 mg/L, respectively.

Gaseous sulfur dioxide was determined by ITP, after it had been adsorbed into a dilute solution of hydrogen peroxide and converted into sulfate, with a LOD of 2.3 ppm for sulfate concentrations between 5 and 100 ppm (LE: 8 mM HCl, 3 mM bis[2-hydroxyethyl]imino-tris[hydroxymethyl]methane (BTP), adjusted to pH 3.4 with β-alanine, 0.2% hydroxyethylcellulose (HEC); TE: 2 mM citric acid) [51]. The sulfur-oxidizing activity of *Thiobacillus ferrooxidans* was investigated by ITP monitoring of sulfate formation (LE: 1 mM HCl, 2 mM calcium chloride, adjusted to pH 4.0 with β-alanine; TE: 1 mM citric acid) and thiosulfate consumption (LE: 1 mM HCl + β-alanine, 4 mM calcium chloride; TE: 5 mM hexanoic acid) [52]. Aerosols of phosphorous and phosphoric acids were determined by ion chromatography and ITP [53]. The yields of the main ionogenic radiolytical degradation products of tributylphosphate (monobutylphosphate, dibutylphosphate, phosphoric acid, formic, acetic, propionic and butyric acids) as the result of gamma-irradiation of the water-tributylphosphate system were determined by ITP [54, 55]. The formation of acids in plaque after chewing various cereal-based foods and fruits was investigated by ITP, and formate, succinate, lactate, acetate and propionate were analyzed [56]. Produced water from oil production platforms were analyzed for organic acids by ITP [57]. The separation of hydroxycarboxylic acids by ITP in the presence of neutral CDs was investigated and complete separation was achieved using β-CD as additive to the LE [58].

ITP was used as a reference method in the study of the macrokinetics of methylmethacrylate degradation in a biotrickling filter by measuring the degradation product methacrylic acid in the drain of the filter (LE: 10 mM HCl adjusted to pH 4.1 with 6-aminocaproic acid (EACA); TE: 5 mM caprylic acid) [59]. 3-Methylhistidine was separated from histidine, 1-methylhistidine and other components of the acidic sample hydrolyzate of meat and meat products (LE: 5 mM ammonium hydroxide, 10 mM MES; TE: 10 mM EACA, 5 mM acetic acid) [60]. An ITP method for the determination of aldonic (mannonic, gluconic, arabinonic and glyceric) acids produced after alkaline oxidation of hexoses with palladium(II)-chloride was described (LE: 10 mM HCl adjusted to pH 3.0 with β-alanine, 0.2% Triton X-100; TE: 5.6 mM sodium acetate) [61]. ITP separations of the herbicides paraquat and diquat were performed in a glass microchip channel and monitored on-chip by normal Raman spectroscopy [62]. The 40 µm wide and 75 µm deep separation channels were chemically etched in a serpentine design to 21 cm total length. Raman isotachopherograms of the pesticides at starting concentrations as low as 2.3×10^{-7} M (60 ppb paraquat/80 ppb diquat) could be obtained. On-line preconcentration was achieved with the use of field-amplified injection into 30–100 mM sulfuric acid or sodium sulfate as the LE and ITP was run with 30–100 mM Tris + HCl as the TE. The analytes were concentrated to above 10^{-3} M at the detection window [63].

ITP with conductometric detection was used to separate and determine mannitol, sorbitol, dulcitol and xylitol in dosage forms, employing complex-formation equilibria with boric acid (LE: 10 mM HCl, 20 mM imidazole, 0.05% poly(vinyl alcohol), pH 7.1; TE: 20 mM boric acid, pH 8.1) [64]. The RSDs were 1.4–1.8% ($n = 6$) when determining 100–200 g/L of mannitol and/or 50–200 g/L of sorbitol. The same method was used to analyze sorbitol and xylitol in multicomponent pharmaceutical formulations [65]. Tramadol (2-dimethylaminomethyl-1-(3-methoxyphenyl)-cyclohexanol hydrochloride) was determined in various dosage forms (LE: 5 mM potassium picolinate, 5 mM picolinic acid, pH 5.25; TE: 10 mM formic acid) [66]. The psychoactive agent maprotiline was analyzed in human serum by cationic ITP (LE: 10 mM sodium acetate + acetic acid; TE: 10 mM β-alanine) after extraction with *n*-heptane [67]. Cationic ITP with conductivity detection was used to determine citalopram, fluoxetine, fluvoxamine and sertraline in antidepressant drugs [68]. The enantiomeric separation of several 2-arylpropionic acids has been studied by ITP using two chiral selectors added to the leading

electrolyte [69]. Enantiomeric resolution was recorded for fenoprofen, flurbiprofen, ibuprofen and ketoprofen with heptakis(2,3,6-tri-*O*-methyl)-β-CD used as an additive to the LE. Different conditions (LE: 10 mM hydrochlorid acid, adjusted to pH 4.8 with EACA, 0.1% polyvinyl pyrrolidone; TE: 5 mM MES) were used to determine fenoprofen [70] and ibuprofen [71] in serum. The analysis of fenoprofen, naproxen, ibuprofen and ketoprofen in human serum (LE: 10 mM HCl, adjusted with creatinine to pH 4.5, 0.1% methylhydroxyethylcellulose; TE: 10 mM MES, adjusted with Tris to pH 6.9) provided LODs in the 10^{-3} mM range [72]. Capillary ITP was used to asses the separation conditions of methadone enantiomers by recycling preparative ITP [73]. A 10 mM sodium acetate/acetic acid LE of pH 4.3, containing 5 mM (2-hydroxypropyl)-β-CD as the chiral selector, was found to provide best enantiomeric separation.

ITP with on-capillary Raman spectroscopy detection was used to separate the nucleosides adenosine, cytidine, guanosine and uridine as borate complexes (LE: 100 mM chloride, pH 7.5; TE: 100 mM borate, pH 10) [74]. The same technique using a fiberoptic Raman probe fitted with a microscope objective was used to obtain on-line spectra of adenosine 5′-monophosphate, cytidine 5′-monophosphate, guanosine 5′-monophosphate and uridine 5′monophosphate from their separated ITP zones (LE: 100 mM sodium sulfate; TE: 100 mM MOPS) [75]. 2-Chloropropionic, glyoxylic, and levulinic acids were evaluated as migration markers for ITP of ribonucleotides, separating them into three groups of triphosphate, diphosphate, and monophosphate (LE: HCl + EACA; TE: caproic acid) [76]. Ceramide-labeled serum lipoproteins were analyzed by ITP linked to laser-induced fluorescent detection and separated into a number of fractions and subfractions [77]. The ITP determination of splitting of phosphate from amifostine and *p*-nitrophenyl phosphate in serum and neuroblastoma cells was described (LE: 10 mM HCl; TE: 5 mM caproic acid, both adjusted to pH 6.0 with histidine) [78]. The dihydrofolate reductase reaction in presence of methotrexate and ascorbic acid was monitored by ITP analysis [79] of methotrexate polyglutamates (LE: 10 mM HCl; TE: 5 mM MES, both adjusted to pH 6.0 with histidine) and NADPH, NADP$^+$ and ascorbate (LE: 10 mM HCl, adjusted to pH 3.73 with β-alanine; TE: 10 mM caproic acid, pH 3.27).

4 Transient ITP and combined techniques

The use of stacking and combined techniques employing ITP effects and the on-line combination of ITP with other techniques, particularly with CZE, have experienced a growing number of published applications. The reason for this is the obvious need for increasing the sensitivity of the analyses that can be usually enhanced by 100–10 000 times. A recent paper compared the enhancement of the sensitivity of the following modes: transient ITP induced by the sample (sample self-stacking), transient ITP induced by the composition of the electrolyte system and the on-line combination of ITP and CZE [80]. The on-line combination was found to be by far superior. A study of this technique [81] has defined the potential interfering components of a sample for a given CZE analyte in both stages of the ITP-CZE combination and residual interfering components in the CZE stage. The major requirements for a successful application of this technique were found to be different separation mechanisms for the analyte in the ITP and CZE electrolyte systems and the minimum size of the sample fraction transferred into the CZE column. The investigation of sample self-stacking effects with complex samples containing more than one major stacking component revealed [82] that even in such a case it may considerably enhance the sensitivity of the analysis for minor analytes. It was shown that the conditions for their successful stacking depend on the mobilities of the species involved and on the concentration ratio of the stacking sample components. ITP superimposed on CZE in an ITP technique allowed to manipulate with the mobility window fo the samples to migrate in the ITP stack by adding some amount of the TE to the LE. A recent paper [83] has shown how to predict this mobility window by simple calculations.

The sample self-stacking principle was used to determine phosphite and phosphate impurities in foscarnet sodium (trisodium phosphonateformate hexahydrate) by CZE (BGE: 3.56 mM sulfanilic acid, 6.98 mM EACA) [84]. Foscarnet served as the leading-type stacker and glutamic acid was added to the sample to act as a second stacker of the terminating type. The sensitivity of CZE determination of nitrate and nitrite in seawater by using artificial seawater as the BGE was increased twice when adding chlorate (acting as a terminating-type stacker) to the sample [85]. Electrokinetic injection and transient ITP were used to preconcentrate on-line lanthanides for their CZE analysis with indirect LIF detection, resulting in LODs in the low-ppb range of 6–11 nM (LE: 10 mM sulfuric adic; TE: 80 mM Tris; BGE: 0.4 mM quinine sulfate, 8 mM HIBA) [86]. The same technique was used to stack large sample volumes in the cationic analysis of fission products – Eu(152), Cs(137) and Ba(137m) – by CZE with on-line indirect UV-absorbance and radioactivity detection (LE: 100 mM HCl; TE: 30 mM creatinine; BGE: 8 mM HIBA, pH 4.5) [87], to determine ultratraces of anions in silicon wafer surfaces [88], to determine adenosine deaminase activity in human erythrocytes (TE = BGE: boric acid + NaOH, pH 10.0; LE: phosphate from the sample) [89]. This combination was also used to increase the sensitivity of both cationic and anionic analy-

ses of various model peptides. For cations, the system used was LE = BGE: 10 mM ammonium acetate adjusted to pH 5.0 with acetic acid; TE: 10 mM acetic acid, both 50% v/v methanol [90]. For anions, two different systems were reported (LE = BGE: 10 mM HCl adjusted to pH 7.4 with ammonia; TE: 100 mM glycine adjusted to pH 8.6 with ammonia) [90], (LE: 0.1–5% ammonium hydroxide; TE = BGE: 2 mM ammonium acetate, 1% acetic acid, pH 2.9) [91]. Further, one can mention here peptides in plasma (LE = BGE: 10 mM ammonium acetate adjusted to pH 5.0 with acetic acid; TE: 10 mM acetic acid, both 50% v/v methanol) [92] and antisense oligonucleotides (LE: 50 mM HCl; TE: 50 mM butyric acid; BGE: 50 mM HCl, 6 M urea, 5% HEC, all adjusted to pH 7.5 with Tris) [93]. A study of the effect of electromigration dispersion in CZE with transient ITP of several basic model proteins and interleukin-6 (rhIL-6) has led to the introduction of an electromigration dispersion factor that measures the level of peak asymmetry, increasing with the analyte amount and mobility [94].

ITP preconcentration with hydrodynamic counterflow was used to enhance the detectability of amitriptyline and metoprolol in CZE (LE = BGE: 10 mM NaOH; TE: 6.1 mM tetrapentylammonium hydroxide, both adjusted to pH 2.75 with phosphoric acid) [95], to determine trace concentrations of recombinant human interleukin-3, recombinant human IL-6 and various basic model proteins (LE = BGE: 20 mM triethylamine (TEA) adjusted to pH 4.2 with acetic acid; TE: 20 mM β-alanine adjusted to pH 4.5 with acetic acid) [96], to monitor the formation of the interleukin-6 dimer (LE = BGE: 20 mM ammonium acetate adjusted to pH 4.2 with acetic acid; TE: 10 mM acetic acid) [97], to characterize recombinant cytokine fragments [98] and to preconcentrate a new cholinesterase inhibitor (NXX-066) prior to CZE analysis with LIF detection (LE = BGE: 10 mM NaOH adjusted to pH 2.72 with phosphoric acid; TE: 6.13 mM tetrahexylammonium chloride adjusted to pH 2.72 with phosphoric acid, 37,5% v/v methanol) [99].

ITP and CZE in directly coupled columns with a counterflow were used for the determination of neostigmine and propantheline (LE = BGE: 10 mM TEA; TE: 10 mM β-alanine, both adjusted to pH 5.0 with acetic acid) [100]. On-line combined ITP and CZE served for a highly sensitive determination of orotic acid in human urine (LE: 10 mM HCl adjusted to pH 6.8 with BTP, 0.02% hydroxypropylcellulose (HPC); TE: 10 mM glutamic acid; BGE: 30 mM glutamic acid, 7 mM spermine, pH 5.2; or LE: 10 mM HCl adjusted to pH 2.15 with glycine, 2% HPC; TE: 10 mM phosphoric acid; BGE: 30 mM phosphoric acid adjusted to pH 2.15 with glycine) [101]. Using a UV-scanning detector, a LOD of 3×10^{-7} M orotic acid was achieved. The same technique was applied to the analysis of L-ascorbic

Electrophoresis 2000, *21*, 3898–3904

acid in human body fluids with reaching a LOD of 0.09–0.15 mg/L (LE: 10 mM HCl adjusted to pH 3.3 with β-alanine, 0.01% HPC; TE: 50 mM propionic acid; BGE: 50 mM propionic acid adjusted to pH 3.8 with β-alanine) [102] and of hippurate in serum with a LOD 7×10^{-7} M (LE: 10 mM HCl adjusted to pH 5.5 with histidine; TE: 10 mM MES; BGE: 10 or 50 mM MES adjusted to pH 6.2 with histidine) [103]. A group of eight nitrophenols was separated by on-line ITP-CZE using host-guest complexation with β-CD and intermolecular interactions with polyvinyl pyrrolidone; concentration LODs at the low ppb level were reached with detection at 254 and 405 nm [104]. The same technique was used for the determination of μg/L levels of iron in water in form of the EDTA-Fe(III) complex with UV-absorbance detection (LE: 10 mM HCl, 20 mM histidine, 0.1% HPMC, pH 6.0; TE: 5 mM MES; BGE: 25 mM MES, 10 mM BTP, pH 6.6) [105] and for the separation of the enantiomers of tryptophan present in complex ionic matrices (LE: 10 mM HCl adjusted to pH 9.3 with BTP, 0.2% HEC; TE: 10 mM boric acid adjusted to pH 9.0 with BTP; BGE: 50 mM boric acid adjusted to pH 9.0 with BTP, 80 mM α-CD, 0.2% methylhydroxyethylcellulose) [106].

The column coupling technique is being used for a long time to enhance the selectivity and sensitivity of ITP by combining two different ITP electrolyte systems. Ammonium, sodium, potassium, magnesium, and calcium cations were determined in rain water by this technique [107]; complete resolution of all cations was reached at LODs in the range of 0.4 to 2.1×10^{-6} M using conductivity detection. Anionic trace impurities in glycerol (nitrate, sulfate, chlorate, nitrite, oxalate, fluoride, formate and phosphate) were determined up to an analyte-to-bulk ratio of $1:4 \times 10^{7}$ (preseparation LE: 10 mM HCl adjusted to pH 3.2 with β-alanine; analytical capillary LE: 5 mM HCl, BTP, adjusted to pH 3.6 with β-alanine; TE: citric acid) [108]. A similar approach was used to determine trace anionic impurities in acetic acid (nitrate, sulfate, nitrite, fluoride, formate, phosphate and oxalate) up to an analyte-to-excess ratio of $1:3 \times 10^{5}$ [109]. The analysis of creatinine in meat and meat products by cationic ITP (preseparation LE: 10 mM NH_4OH + 20 mM MES; analytical capillary LE: 5 mM NH_4OH + 10 mM MES; TE: 10 mM EACA + 5 mM acetic acid) after acidic extraction was described [110].

ITP served also as one of the steps in techniques that utilize the combination of different separation and detection principles. Transient ITP preconcentration was employed in testing microfabricated devices for CE combined with electrospray MS (BGE: 20 mM EACA adjusted to pH 4.4 with acetic acid; sample dissolved in 100 mM ammonium acetate) [111]. The same preconcentration principle was used to enhance the sensitivity of peptide analysis by CZE-MS with a sheath liquid electrospray interface [112].

Transient cationic ITP preconcentration was combined with the use of an on-capillary adsorptive phase to analyze pharmaceutically active peptides (BGE = TE: citric acid; LE: NH_4OH; elution solvent: 80% v/v acetonitrile/water) [113]. Interfacing of ITP with Fourier transform ion cyclotron resonance-MS was tested on analyses of proteins and oligonucleotides and has shown to be effective for separating minor components of protein mixtures for on-line mass spectral analysis (LE: 7 mM HCl, 13 mM β-alanine; TE: 10 mM caproic acid) [114]. The combination of ITP with electrothermal vaporization for inductively coupled plasma MS was used for identification of the organic ligands in selenium speciation in human milk [115]. The technique of ITP focusing was combined with the effect of micellar partition and successfully tested on the separation of dinitrophenyl and dansyl derivatives of amino acids and of position isomers of nitrophenols (LE: 10 mM HCl, 20 mM Tris, pH 8.0, 0.06% HPMC, 2.5 mg/L polyethyleneimide; TE: 50 mM CHES, 5 mM KOH; gradient formed by 8% Ampholines 3.5–10.0) [116]. Sequential injection followed by ITP focusing was used in an automated microanalysis technique where a small quantity of magnetic beads containing immobilized biomolecules was injected into a neutral hydrophilic-coated fused-silica capillary [117]. The method was employed to quantitate an antigen using antibodies immobilized on the beads.

5 Conclusions

Capillary ITP participates successfully in the progress of capillary electrophoresis. In particular it keeps its position in analyses of non-UV absorbing low molecular weight compounds due to the possibility to apply universal (*e.g.*, conductivity) detection principles to evaluate the zones. It is strongly involved in the development of ultrasensitive methods where it mostly operates as the first stage of the analysis since it offers efficient preconcentration and preseparation of diluted samples with trace analytes and bulk interfering components. Last but not least, the principles of ITP form the fundamental background of the theory of electrophoresis and their application and understanding elucidates a lot of phenomena observed in CZE. The theory of ITP and the related methodology are therefore always very attractive and new scientific results in these areas are strongly and positively reflected in both capillary electrophoresis and microchip separations.

Received September 4, 2000

6 References

[1] Gebauer, P., Boček, P., *Electrophoresis* 1997, *18*, 2154–2161.

[2] Gaš, B., Boček, P., Deyl, Z. (Eds.), *J. Chromatogr. A* 1997, *772*, (10th Internat. Symp. on CE and ITP, Prague, Czech Republic, 17–20 September 1996).

[3] Righetti, P. G. (Ed.), *J. Chromatogr. A* 1999, *838* (11th Internat. Symp. on Capillary Electroseparation Techniques, Venice, Italy, 4–7 October 1998).

[4] Yeung, E. S. (Ed.), *J. Chromatogr. A* 1999, *853* (12th Internat. Symp. on HPCE and Related Microscale Techniques, Palm Springs, CA, 23–28 January 1999).

[5] Fanali, S., Karger, B. L. (Eds.), *J. Chromatogr. A* 1998, *817* (11th Internat. Symp. on HPCE and Related Microscale Techniques, Orlando, FL, 1–5 February 1998).

[6] Terabe, S., Tanaka, N. (Eds.), . *Chromatogr. A* 1998, *802* (10th Internat. Symp. on HPCE and Related Microsclae Techniques, Kyoto, Japan, 8–11 July 1997).

[7] Hancock, W. S. (Ed.), *J. Chromatogr. A* 1997, *781* (HPCE '97, Orlando, FL, 1997).

[8] Křivánková, L., Boček, P. (Eds.), *Electrophoresis* 2000, *21*, 2747–2910.

[9] Onuska, F. I., Kaniansky, D., Onuska, K. D., Lee, M. L., *J. Microcol. Sep.* 1998, *10*, 567–579.

[10] Kvasnička, F., *Electrophoresis* 2000, *21*, 2780–2787.

[11] Valášková, I., Havránek, E., *J. Chromatogr. A* 1999, *836*, 201–208.

[12] Kašička, V., *Chem. Listy* 1997, *91*, 320–329.

[13] Liu, B. F., Liu, L. B., Cheng, J. K., *J. Chromatogr. A* 1999, *834*, 277–308.

[14] Janoš, P., *J. Chromatogr. A* 1999, *834*, 3–20.

[15] Beckers, J. L., Boček, P., *Electrophoresis* 2000, *21*, 2747–2767.

[16] Kaniansky, D., Masár, M., Marák, J., Bodor, R., *J. Chromatogr. A* 1999, *834*, 133–178.

[17] Osbourn, D. M., Weiss, D. J., Lunte, C. E., *Electrophoresis* 2000, *21*, 2768–2779.

[18] Valsecchi, S. M., Polesello, S., *J. Chromatogr. A* 1999, *834*, 363–385.

[19] Blatný, P., Kvasnička, F., *J. Chromatogr. A* 1999, *834*, 419–431.

[20] Thormann, W., Aebi, Y., Lanz, M., Caslavska, J., *Forens. Sci. Int.* 1998, *92*, 157–183.

[21] Von Heeren, F., Thormann, W., *Electrophoresis* 1997, *18*, 2415–2426.

[22] Holland, L. A., Chetwyn, N. P., Perkins, M. D., Lunte, S. M., *Pharmaceut. Res.* 1997, *14*, 372–387.

[23] Chen, S. H., Chen, Y. H., *Electrophoresis* 1999, *20*, 3259–3268.

[24] Kašička, V., *Electrophoresis* 1999, *20*, 3084–3105.

[25] Krull, I. S., Strong, R., Sosic, Z., Cho, B. Y., Beale, S. C., Wang, C. C., Cohen, S., *J. Chromatogr. B* 1997, *699*, 173–208.

[26] Szoko, E., *Electrophoresis* 1997, *18*, 74–81.

[27] Sarmini, K., Kenndler, E., *J. Chromatogr. A* 1997, *792*, 3–11.

[28] Dolník, V., Liu, S. R., Jovanovich, S., *Electrophoresis* 2000, *21*, 41–54.

[29] Beale, S. C., *Anal. Chem.* 1998, *70*, R279–R300.

[30] Ikuta, N., Hirokawa, T., *J. Chromatogr. A* 1998, *802*, 49–57.

[31] Martens, J. H. P. A., Reijenga, J. C., Boonkkamp, J. H. M. T., Mattheij, R. M. M., Everaerts, F. M., *J. Chromatogr. A* 1997, *772*, 49–62.

[32] Caslavska, J., Thormann, W., *J. Chromatogr. A* 1997, *772*, 3–17.

[33] Thormann, W., Zhang, C. X., Caslavska, J., Gebauer, P., Mosher, R. A., *Anal. Chem.* 1998, *70*, 549–562.

[34] Reijenga, J., Kašička, V., *Electrophoresis* 1998, *19*, 1601–1605.

[35] Ermakov, S. V., Zhukov, M. Y., Capelli, L., Righetti, P. G., *Electrophoresis* 1998, *19*, 192–205.

[36] Bednář, P., Stránský, Z., Ševčík, J., Dostál, V., *J. Chromatogr. A* 1999, *831*, 277–284.

[37] Hirokawa, T., Mao, Q. L., *J. Chromatogr. A* 1997, *786*, 377–381.

[38] Beckers, J. L., Verheggen, T. P. E. M., Triepels, H. J. P., Everaerts, F. M., *J. Chromatogr. A* 1999, *838*, 149–155.

[39] Wronski, M., *J. Chromatogr. A* 1997, *772*, 19–25.

[40] Šlais, K., *J. Chromatogr. A* 1997, *764*, 309–321.

[41] Rejtar, T., Šlais, K., *J. Chromatogr. A* 1999, *838*, 71–80.

[42] Šťastná, M., Šlais, K., *J. Chromatogr. A* 1997, *768*, 283–294.

[43] Meissner, T., Eisenbeiss, F., Jastorff, B., *J. Chromatogr. A* 1999, *838*, 81–88.

[44] Gogová, K., Zusková, I., Tesařová, E., Gaš, B., *J. Chromatogr. A* 1999, *838*, 101–109.

[45] Blatný, P., Kvasnička, F., Loučka, R., Šafářová, H., *J. Agric. Food Chem.* 1997, *45*, 3554–3558.

[46] Fukushi, K., Takeda, S., Wakida, S., Higashi, K., Hiiro, K., *J. Chromatogr. A* 1997, *759*, 211–216.

[47] Hirokawa, T., Hashimoto, Y., *J. Chromatogr. A* 1997, *772*, 357–367.

[48] Mao, Q., Hashimoto, Y., Manabe, Y., Ikuta, N., Nishiyama, F., Hirokawa, T., *J. Chromatogr. A* 1998, *802*, 203–210.

[49] Klunder, G. L., Andrews, J. E., Church, M. N., Spear, J. D., Russo, R. E., Grant, P. M., Andresen, B. D., *J. Radioanal. Nucl. Chem.* 1998, *236*, 149–153.

[50] Fukushi, K., Sagishima, K., Saito, K., Takeda, S., Wakida, S., Hiiro, K., *Anal. Chim. Acta* 1999, *383*, 205–211.

[51] Bektas, N., Brown, B. J. T., Fielden, P. R., *J. Chromatogr. A* 1999, *836*, 107–114.

[52] Janiczek, O., Mandl, M., Češková, P., *J. Biotechnol.* 1998, *61*, 225–229.

[53] Obrezkov, O. N., Krokhin, O. V., Makarov, A. Y., Guglya, E. B., Shpigun, O. A., *J. Anal. Chem. Engl. Tr.* 1998, *53*, 557–560.

[54] Kuruc, J., Petrů, A., Čech, R., Rajec, P., *J. Radioanal. Nucl. Chem. Art.* 1996, *208*, 351–368.

[55] Kuruc, J., Batista, R., Čech, R., Rajec, P., *J. Radioanal. Nucl. Chem. Art.* 1996, *208*, 369–391.

[56] Pollard, M. A., Higham, S. M., Curzon, M. E. J., Edgar, W. M., *Eur. J. Oral Sci.* 1996, *104*, 535–539.

[57] Utvik, T. I. R., *Chemosphere* 1999, *39*, 2593–2606.

[58] Sádecká, J., Hercegová, A., Polonská, J., Chilmonczyk, Z., *Analusis* 1999, *27*, 373–380.

[59] Deridder, R., Prickaerts, R. M. H., Reijenga, J. C., Verheggen, T. P. E. M., *J. Chromatogr. A* 1999, *862*, 237–242.

[60] Kvasnička, F., *J. Chromatogr. A* 1999, *838*, 191–195.

[61] Petrushevska-Tozi, L., Ristov, T., Kavrakovski, Z., *J. Chromatogr. A* 1997, *757*, 324–327.

[62] Walker, P. A., Morris, M. D., Burns, M. A., Johnson, B. N., *Anal. Chem.* 1998, *70*, 3766–3769.

[63] Walker, P. A., Shaver, J. M., Morris, M. D., *Appl. Spectrosc.* 1997, *51*, 1394–1399.

[64] Pospíšilová, M., Polášek, M., Procházka, J., *J. Chromatogr. A* 1997, *772*, 277–282.

[65] Pospíšilová, M., Polášek, M., Jokl, V., *J. Pharmaceut. Biomed. Anal.* 1998, *17*, 387–392.

[66] Pospíšilová, M., Polášek, M., Jokl, V., *J. Pharmaceut. Biomed. Anal.* 1998, *18*, 777–783.

[67] Čakrt, M., Buzinkaiová, T., Polonský, J., Kořínková, V., *Electrophoresis* 2000, *21*, 2834–2838.

[68] Buzinkaiová, T., Polonský, J., *Electrophoresis* 2000, *21*, 2839–2841.

[69] Sádecká, J., Skacani, I., Hercegová A., Polonský, J., *Chem. Anal.* 1998, *43*, 677–685.

[70] Sádecká, J., Hercegová, A., Polonský, J., *J. Chromatogr. B* 1999, *729*, 11–17.

[71] Sádecká, J., Polonský, J., *Pharmacie* 1999, *54*, 627–627.

[72] Hercegová, A., Sádecká, J., Polonský, J., *Electrophoresis* 2000, *21*, 2842–2847.

[73] Lanz, M., Caslavska, J., Thormann, W., *Electrophoresis* 1998, *19*, 1081–1090.

[74] Li, H. M., Walker, P. A., Morris, M. D., *J. Microcol. Sep.* 1998, *10*, 449–453.

[75] Walker, P. A., Morris, M. D., *J. Chromatogr. A* 1998, *805*, 269–275.

[76] Chen, S. J., Lee, M. L., *J. Microcol. Sep.* 1998, *10*, 423–430.

[77] Schlenck, A., Herbeth, B., Siest, G., Visvikis, S., *J. Lip. Res.* 1999, *40*, 2125–2133.

[78] Renner, S., Klingebiel, T., Niethammer, D., Bruchelt, G., Meissner, T., Eisenbeiss, F., *J. Chromatogr. A* 1999, *838*, 251–257.

[79] Renner, S., Kuçi, Z., d'Cruze, H., Niethammer, D., Bruchelt, G., *Electrophoresis* 2000, *21*, 2828–2833.

[80] Křivánková, L., Pantůčková, P., Boček, P., *J. Chromatogr. A* 1999, *838*, 55–70.

[81] Kaniansky, D., Marák, J., Laštinec, J., Reijenga, J. C., Onuska, F. I., *J. Microcol. Sep.* 1999, *11*, 141–153.

[82] Gebauer, P., Křivánková, L., Pantůčková, P., Boček, P., Thormann, W., *Electrophoresis* 2000, *21*, 2797–2808.

[83] Beckers, J. L., *Electrophoresis* 2000, *21*, 2788–2796.

[84] Andersson, E. K. M., *J. Chromatogr. A* 1999, *846*, 245–253.

[85] Fukushi, K., Ishio, N., Sumida, M., Takeda, S., Wakida, S., Hiiro, K., *Electrophoresis* 2000, *21*, 2866–2871.

[86] Church, M. N., Spear, J. D., Russo, R. E., Klunder, G. L., Grant, P. M., Andresen, B. D., *Anal. Chem.* 1998, *70*, 2475–2480.

[87] Klunder, G. L., Andrews, J. E., Grant, P. M., Andresen, B. D., Russo, R. E., *Anal. Chem.* 1997, *69*, 2988–2993.

[88] Ekmann, T., Bächmann, K., Fabry, L., Rufer, H., Pahlke, S., Kotz, L., *Chromatographia* 1997, *45*, 301–311.

[89] Adam, T., Ševčík, J., Švagera, Z., Fairbanks, L. D., Barták, P., *Electrophoresis* 1999, *20*, 564–568.

[90] Waterval, J. C. M., Laporte, C. J. L., Vanthof, R., Teeuwsen, J., Bult, A., Lingeman, H., Underberg, W. J. M., *Electrophoresis* 1998, *19*, 3171–3177.

[91] Tomlinson, A. J., Benson, L. M., Jameson, S., Naylor, S., *Electrophoresis* 1996, *17*, 1801–1807.

[92] Waterval, J. C. M., Krabbe, H., Teeuwsen, J., Bult, A., Lingeman, H., Underberg, W. J. M., *Electrophoresis* 1999, *20*, 2909–2916.

[93] Khan, K., Vanschepdael, A., Saisonbehmoaras, T., Vanaerschot, A., Hoogmartens, J., *Electrophoresis* 1998, *19*, 2163–2168.

[94] Gysler, J., Jaehde, U., Schunack, W., *Fresenius J. Anal. Chem.* 1999, *365*, 398–403.

[95] Enlund, A. M., Westerlund, D., *Chromatographia* 1997, *46*, 315–321.

[96] Bergmann, J., Jaehde, U., Schunack, W., *Electrophoresis* 1998, *19*, 305–310.

[97] Gysler, J., Mazereeuw, M., Helk, B., Heitzmann, M., Jaehde, U., Schunack, W., Tjaden, U. R., Vandergreef, J., *J. Chromatogr. A* 1999, *841*, 63–73.

[98] Gysler, J., Helk, B., Dambacher, S., Tjaden, U. R., Vandergreef, J., *Pharmaceut. Res.* 1999, *16*, 695–701.

[99] Enlund, A. M., Schmidt, S., Westerlund, D., *Electrophoresis* 1998, *19*, 707–711.

[100] Chen, S. J., Lee, M. L., *Anal. Chem.* 1998, *70*, 3777–3780.

[101] Procházková, A., Křivánková, L., Boček, P., *J. Chromatogr. A* 1999, *838*, 213–221.

[102] Procházková, A., Křivánková, L. Boček, P., *Electrophoresis* 1998, *19*, 300–304.

[103] Křivánková, L., Vraná, A., Gebauer, P., Boček, P., *J. Chromatogr. A* 1997, *772*, 283–295.

[104] Kaniansky, D., Masár, M., Marák, M. J., Madajová, V., Onuska, F. L., *J. Radioanal. Nucl. Chem. Art.* 1996, *208*, 331–350.

[105] Blatný P., Kvasnička, F., Kenndler, E., *J. Chromatogr. A* 1997, *757*, 297–302.

[106] Danková, M., Kaniansky, D., Fanali, S., Iványi, F., *J. Chromatogr. A* 1999, *838*, 31–43.

[107] Zelenský, I., Hybenová, A., Kaniansky, D., *Chem. Pap. Chem. Zvesti* 1997, *51*, 221–225.

[108] Meissner, T., Eisenbeiss, F., Jastorff, B., *J. Chromatogr. A* 1998, *810*, 201–208.

[109] Meissner, T., Eisenbeiss, F., Jastorff, B., *Fresenius J. Anal. Chem.* 1998, *361*, 459–464.

[110] Kvasnička, F., Voldřich, M., *Electrophoresis* 2000, *21*, 2848–2850.

[111] Zhang, B., Liu, H., Karger, B. L., Foret, F., *Anal. Chem.* 1999, *71*, 3258–3264.

[112] Larsson, M., Lutz, E. S. M., *Electrophoresis* 2000, *21*, 2859–2865.

[113] Waterval, J. C. M., Hommels, G., Teeuwsen, J., Bult, A., Lingeman, H., Underberg, W. J. M., *Electrophoresis* 2000, *21*, 2851–2858.

[114] Severs, J. C., Hofstadler, S. A., Zhao, Z., Senh, R. T., Smith, R. D., *Electrophoresis* 1996, *17*, 1808–1817.

[115] Michalke, B., Schramel, P., *Biol. Tr. Elem. Res.* 1997, *59*, 45–56.

[116] Šťastná, M., Šlais, K., *J. Chromatogr. A* 1999, *832*, 265–271.

[117] Rashkovetsky, L. G., Lyubarskaya, Y. V., Foret, F., Hughes, D. E., Karger, B. L., *J. Chromatogr. A* 1997, *781*, 197–204.

Review

František Kvasnička

Department of Food
Preservation and Meat
Technology,
Institute of Chemical
Technology, Technická 5,
Prague, Czech Republic

Application of isotachophoresis in food analysis

This review summarizes possibilities of capillary isotachophoresis for the determination of important analytes in food. The emphasis is on quantitative determinations in real food samples. This article covers papers published from 1980 to 1999.

Keywords: Capillary isotachophoresis / Food analysis / Review   EL 4090

Contents

1 Introduction

Laws governing isotachophoretic separation were laid down in 1897 by the German chemist Kohlrausch [1]. However, isotachophoresis (ITP) as a working analytical method can be considered to have started in the beginning of the 1970s, when the principles and instrumentation for analytical capillary ITP (CITP), including on-line detection, were developed [2]. In the next two decades (1970–1990) CITP underwent major development. Much work has been done on the theoretical description of the dynamics of the separation, calculating the composition of isotachophoretic zones. Furthermore, the problems concerning the selection of electrolyte systems, zone sta-

Correspondence: Dr. František Kvasnička, Department of Food Preservation and Meat Technology, Institute of Chemical Technology, Technická 5, 166 28 Prague 6, Czech Republic
E-mail: kvasnicf@vscht.cz
Fax: +420-2-311-6284

Abbreviations: 3-MeHis, 3-methylhistidine ([S]-1-methylimidazolyl-4-alanine or τ-methyl-L-histidine); **4-Mel**, 4-methylimidazole; **BTP**, 1, 3-bis- [tris-(hydroxymethyl)-methylamino]-propane, Bis-Tris-Propane; **CITP**, capillary isotachophoresis; **EACA**, ε-aminocaproic acid; **FEP**, fluorinated ethylene-propylene copolymer; **GlyGly**, glycylglycine; **HAc**, acetic acid; **His**, histidine; **His-HCl**, histidine hydrochloride; **HPMC**, hydroxypropylmethylcellulose; **IDS**, iminodisulphonate; **LE**, leading electrolyte; **NH₄Ac**, ammonium acetate; **PTFE**, polytetrafluoroethylene; **R**, resistance (conductivity detector signal); **TE**, terminating electrolyte; **α-HIBA**, α-hydroxyisobutyric acid

bility, separability of substances, temperature effects, and resolution have been solved. To date, three monographs on CITP have been published [3–5]. In the past few years, during which capillary zone electrophoresis (CZE) has occupied the major part of both research and applications of capillary electrophoresis, CITP has kept its position as a specialized technique with unique features that make it possible to reach much better results, in particular cases, than when using CZE. In the course of time, however, practical limits of the separation efficiency and sensitivity were reached, and a marked drawback of CZE was revealed, namely an unsatisfactory detection limit resulting from the fact that it is not possible to inject sufficiently large volumes of a sample directly into a capillary. This situation explains the comeback of ITP as a preconcentration and preseparation method for CZE. The interest in CITP is also documented by the continuation of the biannual ITP symposia (the last symposium was held in Venice, Italy) [6]. Several reviews dealing with the application of CITP in food analysis have been published [7–9], but all are rather old. The aim of this review is to provide a concise and critical view of the work done in the application of CITP in food analysis within the last 10 years.

2 Basic features of CITP

ITP is one of the basic electrophoretic techniques, and almost exclusively carried out in an instrument using a capillary tube as a separation compartment. The separation of the sample constituents as well as their detection is achieved in one experiment since on-column or post-column detection systems are integral parts of the separation unit. ITP separations are advantageously carried out in free solutions of appropriately chosen electrolytes. A high resolving power and low detection limits (subnanomole) are typical features of capillary ITP.

In CITP analysis the sample solution (up to 300 µL) is injected between leading and terminating solutions present in the separation compartment. The leading electrolyte solution contains an anion or cation having a higher effective mobility than those of the sample constituents to be analyzed. On the other hand the terminating electrolyte solution contains an anion or cation of a lower effective mobility. As in other electrophoretic separation methods the electric field is applied to separate the sample constituents. In dependence on the working conditions employed, the driving current is in the range of 1–500 µA. Since the current is constant, the voltage between the driving electrodes will change during the separation in the range of 1–15 kV.

The separation process in CITP is based on various migration velocities of the constituents in the mixed zone(s). The migration velocities depend on the field strength and the effective mobilities of the ionic constituents. The latter are influenced by pH, complex formation, solvent used, ion pairing, *etc.*. Therefore, the choice of separation conditions (the composition of the leading electrolyte) is of key importance in achieving the desired separation effects. When the separation is completed, the so-called "steady state" is achieved. The stack of sample zones migrates with an identical velocity (the name for ITP comes from the Greek language: ισο = equal, ταχσ = velocity, φορεεσθαι = to be dragged) through the detection sensors of the conductivity and UV-photometric detectors (some other detection principles such as fluorescence, radioactivity, and amperometry can be also employed). The zones are then detected as the well-known isotachophoretic steps consisting of step height (qualitative information) and step length (quantitative information). An important features of ITP is that the concentration of each zone is adjusted to a constant value, independent of the sample composition, and consequently the volume (length) of each zone relates directly to the absolute quantity of an analyte. The adjustment of the isotachophoretic zones is a well-known effect and is usually named regulating effect. The adjusted concentration of zone depends on the mobility of the analyte and concentration of the leading electrolyte. The regulating effect is highly advantageous: it may concentrate diluted samples by several orders of magnitude.

The isotachophoretic equipment consists of a narrow bore tube made of an insulating material (PTFE, fluorinated ethylene-propylene copolymer (FEP) with an ID < 1 mm and with a length ≈ 100–500 mm. In CITP analyzers the outlet side of the capillary is separated from the electrode chamber by a semipermeable membrane, which does not allow the electrolyte to flow out of the capillary due to different electrolyte levels in the electrode chambers. This arrangement with a closed capillary results in a reversed hydrodynamic flow in the center of the capillary with a negative effect of the zone boundaries. Particularly for anionic analysis a reduction of the EOF is required

(increasing viscosity, zeta potential reduction). Food samples represent a complex ionic mixture. There is only a limited possibility of determining ions, which occur at low concentration levels in complex ionic matrices. Rather laborious sample pretreatment is often necessary. The column-coupling configuration of the separation unit is very effective in solving problems encountered in this case because low detection limits can be achieved while the load capacity is very high. In addition, two-dimensional (2-D) ITP separations are thereby possible. In such a case, ionic macroconstituents can by determined in the preseparation capillary (wider diameter). With suitable timing of the column switching, only zones of interest are transferred into the analytical capillary (narrower diameter). A simultaneous determination of macro- and microconstituent with a molar ratio up to 10^4:1 is possible with appropriately chosen running conditions.

Column-coupling ITP is a method of wide applicability in the analysis of ionic constituents in various fields. This equipment enables an on-line combination of CITP-CITP (2-D ITP) [10] and/or a combination of CITP-CZE [11, 12]. Column-coupling ITP provides detection limits as low as 10^{-8}–10^{-9} mol/L (300 µL injection). Very often no sample pretreatment is necessary. In dependence on the required load capacity of the separation compartment, the applied driving currents, and the injected sample amount, the analysis times range between 5–25 min in general. A detailed description of the instrumentation can be found elsewhere [3, 5].

3 Application of CITP in food analysis

From the point of view of analytical chemists, food samples are very complex and heterogeneous, which causes difficulties in determining compounds of interest. Advantages of CITP applied to such analyses are: (i) Nonionic compounds (carbohydrates, starch, fat, *etc.*), which are frequently bulk components of the sample in question, do not move electrophoretically, and thus do not interfere with the analysis of ionic compounds. (ii) ITP separations are advantageously carried out in free solutions so that no interaction between analytes and separation media (*e.g.*, sorbent in HPLC column) occurs. (iii) Minimal sample treatment (usually extraction of solid sample followed by filtration and/or dilution in the case of liquid samples). (iv) Low running cost (two order of magnitude compared with HPLC). Disadvantages of CITP applied to food analysis are: (i) Only ionic compounds can be determined. (ii) Either cations or anions are separated and analyzed in one analytical run. Attempts were made to analyze both kinds of ions in one run (so-called bidirectional CITP), but it has more of a methodological asset than a practical one [13–16].

Compounds that are usually determined in foodstuff can be divided into three groups: (i) Naturally occurring food constituents (alkali and alkaline earth cations, inorganic anions, organic acids, amino acids, water-soluble vitamins, amines, proteins, antinutrients and toxins); (ii) food additives (preservatives, flavor enhancers, water-soluble artificial food dyes); (iii) food contaminants (pesticides, heavy metals, some inorganic anions).

3.1 Naturally occurring food constituents

With the help of the ITP technique, macroelements such as potassium, sodium, calcium and magnesium could be determined practically in all kinds of foodstuff. An extraction of solid samples by water or mineral acid (HNO_3, HCl) or a dilution of liquid samples is the only treatment of samples prior to ITP analysis. Several optimization approaches to the separation of these cations have been published. Authors used various counter- or co-counterions with complexing ability to achieve a separation of cations determined in well water [17], sugar, starch, and their products [18, 19], and drinking or mineral water [20]. These approaches are, however, not efficient for the separation of ammonium and potassium. Since the ammonium is a weaker base than potassium, this separation problem can be solved using a high pH leading electrolyte. If the buffer consisting of a mixture of potassium hydroxide and a suitable weak acid (boric acid [18] or histidine [21]) give a pH above 8, the effective mobility of ammonium is reduced and the separation can be achieved. Another possibility is the application of a suitable nonionic additive, *i.e.*, crown ether [22] or polyethylene glycol (PEG) [23]. Crown ether was applied for a simultaneous determination of ammonium, potassium, calcium, and magnesium in silage samples [24]. This electrolyte system is also applicable to various foodstuff. An example of such an analysis is the determination of elements mentioned above in orange juice (Fig. 1) carried out in cationic mode using 7.5 mM sulfuric acid + 7 mM 18-crown-6 as leading electrolyte and 5 mM Bis-Tris-propane (BTP) + 10 mM caproic acid as terminating electrolyte. Within 20 min information about the NH₄, K, Na, Ca, and Mg content in the sample is available [25].

Inorganic anions, such as halides, sulfate, sulfite, nitrate and phosphate are easily determined by ITP in tap water, mineral water, vegetables, fruits, and milk. From the separation point of view the critical group of anions are chloride, nitrate, and sulfate because of their similar ionic mobilities. As pH changes cannot affect the effective mobilities of these strong acids, the application of a suitable additive to the electrolyte is necessary. Boček *et al.* [26] used cadmium as counterion for the selectivity enhancement. A limited dynamic concentration range

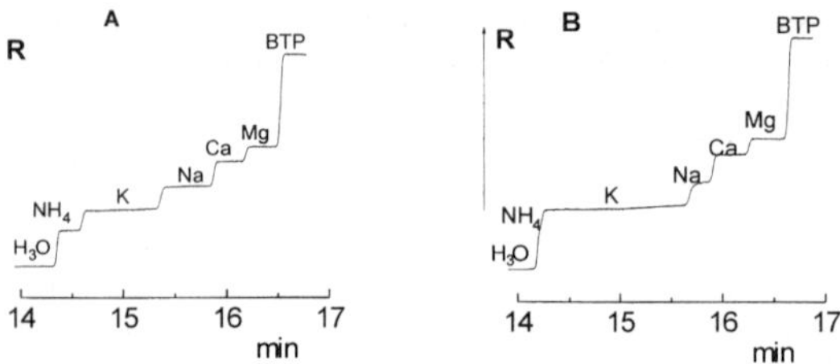

Figure 1. Separation of alkali and alkaline earth metals [25]. (A) Standard mixture: 3.6 mg/L ammonium, 16 mg/L potassium, 4.6 mg/L sodium, 4 mg/L calcium and 2.4 mg/L magnesium; (B) sample of a 50 times diluted orange juice. R, detector signal (increasing resistance). The analysis was performed on a single capillary (volume coupling system, 20 × 1.5 mm preseparation part, 100 × 0.5 mm separation capillary, and 70 × 0.4 mm detection capillary), isotachopherograph IONOSEP 900.1. The applied driving current was 80 µA, which was decreased to 30 µA during detection.

(molar ratio 1:100) in a single-capillary CITP instrument overcame the introduction of 2-D CITP in a column-coupling analyzer. Such an arrangement [27] used in water analysis enabled the determination of macroconstituents (nitrate, sulfate) in the preseparation and microconstituents (nitrite, fluoride, phosphate) in the analytical capillary. The simultaneous determination of chloride, nitrate and sulfate was reported by Vacík and Muselasová [17]. Unfortunately, the reaction of calcium hydroxide used as leading electrolyte with carbon dioxide causes rather low robustness of this system, which is probably unsuitable for routine analysis. Meissner *et al.* [28] used dithionate as leading anion for the simultaneous determination of chloride, nitrate and sulfate in drinking water. The authors have showed that the stability of dithionate is sufficient and such a system is suitable for routine analyses. Kvasnička *et al.* [18] described an electrolyte system for simultaneous determination of nitrate, sulfate, nitrite, sulfite, and phosphate in sugar juices, molasses, and white sugar. The presented method is an example of an application of CITP for the determination of ions in a noniono-genic sample matrix. Even 30% w/w water solution of sucrose could have been injected into the analyzer and detection limits at ppm levels were thus achieved. Similarly, Meissner *et al.* [29] used CITP (column-coupling instrument) for the determination of ionic impurities in glycerol and due to a large sample load they have reached an LOD of 10^{-5}% w/w.

The iminodisulfonate (IDS) is created within the sugar production by the reaction between sulfite and nitrite (especially when sulfite or bisulfite is added to the process). The IDS goes through the process as potassium salt resulting in higher ash content of white sugar in the final product. The CITP technique enables the determination of IDS in sugar juices and white sugar with sufficient sensitivity [30]. As an example, the determination of this compound in thick juice is given in Fig. 2. CITP was applied for the determination of fluoride [31], iodide [32], sulfite [33, 34], nitrite [35], chromate [36], phosphate [37], and cyclic trimetaphosphate [38] in various food samples. More CITP applications on the determination of inorganic ions can be found in a review published by Blatný and Krasnička [39].

Biogenic amines (histamine, cadaverine, putrescine, and other amines) are important constituents of some food products. These compounds are formed by microbial action during processing and storage of fish meat, cheese, fermented cabbage, and wine. The levels of these amines are one of the criterions of food quality. Rubach and co-workers [40] used an electrolyte system consisting of Ba^{++} as leading cation buffered with valine to pH 9.9 and Tris as terminating cation for the isotachophoretic determination of histamine in fish and fish products after water extraction of the sample. Ginzinger and Hohenauer [41] applied the same electrolyte system to the determination of histamine, tyramine and phenyl-ethylamine in cheese. Voldřich *et al.* [42] replaced Ba^{++} with K^+ as leading cation to speed up analysis. Using 2-D CITP (different concentration of leading electrolyte in preseparation and analytical capillary) the authors reached an LOD of 1 mg of histamine/kg of fish meat. In acidic (5% w/w TCA) extracts of meat (and meat products) creatinine could by easily determined in an electrolyte system consisting of NH_4OH buffered with MES to pH 6 as leading and ε-aminocaproic acid (EACA) as terminating electrolyte [43].

The determination of organic acids in various foodstuff is the main application field of CITP. The levels of organic acids are usually sufficiently high for direct determination without any treatment of sample. This is the great advantage of CITP in comparison with chromatographic techniques such as GLC and HPLC. The determination of organic acids in wine was one of the first applications of CITP in food analysis. Farkaš and Koval' [44, 45] determined tartaric, lactic, malic, succinic acetic and citric acids in different kinds of wine and monitored the content of these acids during the processing of wine. Screening of organic acids in 100 different kinds of wine from 13 European countries was done by Reijenga *et al.* [46].

A sugar juice containing ~95% of nonionogenic compounds and 5% of mainly ionogenic nonsugars is an almost ideal sample for ITP. CITP was used for organic acid monitoring during sugar processing [47–50]. Butyric acid could be a marker for an anaerobic microbial action that can take place in a liquid sugar during its storage. The analysis of the liquid sugar is given in Fig. 3. Isopropyl alcohol added to the leading electrolyte enabled the separation of isobutyric from butyric acid [51]. CITP proved to be a useful technique for the determination of volatile fatty acids in molasses [52, 53]. It was confirmed that formic acid influenced the yield of ethanol production from molasses.

The ratio of citric/isocitric acid is one of the recommended markers for the detection of citrus juice adulteration. CITP can serve as an alternative technique to enzymic methods or HPLC. Separation of these two acids is achieved

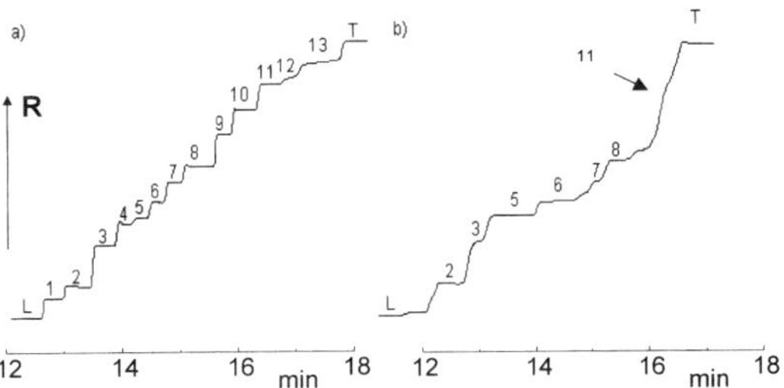

Figure 3. Isotachopherogram of (a) a standard mixture of 13 anions, each 0.1 m$_M$, and (b) liquid sugar diluted to a concentration of 5 g/100 mL. LE, 10 m$_M$ HCl + 22 m$_M$ EACA + 0.05% HPMC + 15% w/v isopropanol; TE, 5 m$_M$ caproic acid. L, chloride; 1, oxalate; 2, formate; 3, citrate; 4, phosphate; 5, lactate; 6, pyroglutamate; 7, aspartate; 8, acetate; 9, glutamate; 10, propionate; 11, butyrate; 12, isobutyrate; 13, valerate; T, caproate; R, detector signal (increasing resistance). The analysis was performed as in Fig. 1.

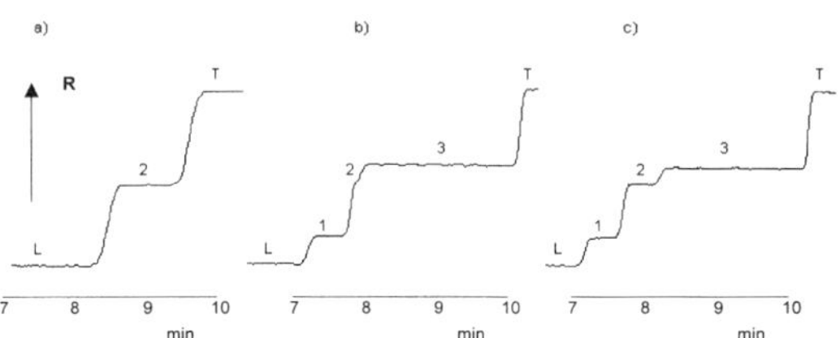

Figure 2. Determination of iminodisulfonate (IDS) in sugar solutions. (a) Standard of IDS, (b) thick juice and (c) thick juice with standard addition. L, chloride; 1, nitrate; 2, IDS; 3, sulfate; T, formate; R, detector signal (increasing resistance). LE, 2 m$_M$ His-HCl + 2 m$_M$-His + 3 m$_M$ $CaCl_2$ + 0.1% HPMC; TE, 10 m$_M$ ammonium formate. The analysis was performed as in Fig. 1. One analysis took 15–20 min, depending on the sample composition [30].

by an addition of Ca^{++} as complexing (with citric acid) cocounterion to the leading electrolyte [54]. CITP was used for the determination of organic acids in various foodstuff such as wheat flour [55], potatoes [34], tomatoes [56], eggs [57], milk [58], fruits [59], and meat [60]. Glutamic acid is one of the most determined amino acids in foodstuff because of the limiting content as food additive. CITP is well suited [61] for this purpose and can easily replace the more expensive HPLC with fluorimetric detection. Kvasnička *et al.* [62] applied CITP (cationic mode) for the determination of free (added) and bound lysine in animal feed mixtures. This method can be easily used for the simultaneous determination of lysine, arginine, and ornithine in foodstuff.

There are some typical components of meat proteins, *e.g.*, 3-methylhistidine (3-MeHis), N_ε-methyllysine and creatinine. Especially 3-MeHis is a normal constituent of the myofibrillar protein actine and myosine. Levels of protein-bound 3-MeHis in different species of meat are constant. Hitherto, no 3-MeHis residues have been determined in other foodstuff, rich in proteins, as milk, eggs, soy, *etc.*. For this reason the determination of 3-MeHis in meat products can be used as an index of lean meat content. Simultaneous determination of mentioned meat protein markers was published by Kvasnička [63]. Holloway [64] developed a method for simultaneous determination of acidic and basic amino acids. The trick of the developed method consists in derivatization of amino acids with the anhydride of citraconic acid followed by the separation of the negatively charged products at pH 6. Kaniansky and co-workers [65] developed a method for the determination of 2,4-dinitrophenyl derivatives of amino acids in wine. Using visible photometric detection at 405 nm, an LOD of 10^{-8} mol/L was reached.

Water-soluble vitamins were determined by CITP in various foodstuff. Röben and Rubach [66] determined thiamine, pyridoxamine, pyridoxal and nicotinamide as cations. In anionic analysis at high pH they separated riboflavine-5-phosphate, nicotinic acid, folic acid, biotin, and pantothenic acid. The pair ascorbic/dehydroascorbic acid are the most popular vitamins determined by CITP [67–69]. The content of dehydroascorbic acid is usually calculated from differences of the ascorbic acid content before and after reduction with a suitable agent (cysteine, mercaptoethanol, H_2S). The determination of peptides or proteins by CITP is widely used in biochemistry. Stover [70] described a CITP method for the determination of lysozyme from white and ovalbumin from yolk.

There are several naturally occurring antinutrients and toxins that were determined by CITP. Phytic acid (*myo*-inositol-hexaphosphate) was determined in grains and legumes [71], glucosinolates sinalbin and sinigrin were determined in mustard seeds [72], and glycoalkaloids α-chaconine and α-solanine in potato and potato crisps [73]. Fish poisoning by consuming members of the order *Tetraodontiformes* is one of the most violent intoxications from marine species. The gonads, liver, intestines, and skin of pufferfish can contain levels of tetrodotoxin (perhydroquinazoline) sufficient to produce rapid and violent death. The flesh of many pufferfish is not usually dangerously toxic. Kazuko with co-workers [74] developed a CITP method (cationic mode) for the determination of tetrodotoxin in pufferfish in crude water/dioxane extracts of fish meat.

3.2 Food additives

Many chemical substances serve as food additives such as preservatives, artificial sweeteners, flavor enhancers, synthetic dyes, stabilizers, *etc.* The limits (qualitative and/or quantitative) regarding the addition of such compounds vary from country to country. To some foodstuff no additives may be applied (*e.g.*, addition of preservatives to beer). To have control over the addition of compounds in question, CITP offers the determination of such compounds.

Organic acids such as benzoic, sorbic, formic, lactic, or propionic acids [75–78] are commonly used as food preservatives for beverages, jam, mayonnaise and other foodstuff. These compounds could be easily determined by anionic ITP after extraction of solid sample (jam, bread) with water or after dilution of a liquid sample. Potassium pyrosulfite is often added to mustard as both preservative and antioxidant agent. Due to sulfur-containing substances in mustard all classical titrimetric methods other than CITP failed [79]. Free and bound sulfite in mustard was determined at a low pH of the leading electrolyte on the column-coupling instrument. The LOD was as low as 5 mg of SO_2/kg of mustard. EDTA is commonly used as a stabilizer (antioxidant) for mayonnaise and salad dressings. This compound was determined as a complex with Fe(III) by anionic ITP [80] after sample treatment or by on-line coupling CITP-CZE without any sample pretreatment [81].

Karovičová *et al.* [82] developed a CITP method for the determination of some water-soluble synthetic food dyes. These compounds were determined in a methanolic extract of a sample by anionic ITP. Commercial caramel color is mainly manufactured by a sugar-ammonia or sugar-ammonia-sulfite reaction procedure, during which imidazole and pyrazine derivatives are formed. The content of 4-methylimidazole (4-MeI) is limited due to its possible toxicity. The CITP method [83] enables direct determination of 4-MeI at ppm levels (LOD is 5 mg of 4-MeI/kg). Figure 4 shows an isotachopherogram of a sample of caramel color.

Glutamate, guanosine-5'monophosphate and inosine-5'-monophosphate are added to soups, sauces, and salad dressings as flavor enhancers. After extraction of solid samples with hot or cold water or after dilution of liquid samples, these compounds could be very easily determined by anionic ITP in one run at 10 ppm levels [84]. Quinine [85] in tonic water was determined after degassing (removal of carbon dioxide) by cationic ITP analysis. The artificial sweeteners acesulfam K [86], saccharine and cyclamates [87] or aspartame [88] were determined by anionic ITP in low-calory products or in foods for diabetics.

3.3 Food contaminants

CITP was applied for the determination of some food contaminants, *viz.*, residues of biocides, heavy metals, nitrates, and phosphate. The pesticides in question can be divided into three groups according to their behavior in electric fields, *i.e.*, anionic, cationic and nonionic. Anionic CITP was used for the determination of glyfosat and its metabolite aminomethanephosphoric acid in apples with an LOD of 1 mg/kg (without any pretreatment of apple juice) [89]. Triazines as representative of cationic pesticides were determined by cationic ITP in milk [90] and water [91]. Pyrethroides are nonionic pesticides. The determination of such compounds could be carried out

indirectly after alkaline hydrolysis. Phenoxybenzoic, dichlorochrysanthemic or chrysanthemic acids as products of such hydrolysis were determined by anionic ITP [92].

Heavy metals (Cd, Cu, Pb, Zn) and aluminum were determined by cationic ITP [93]. The metals are released from solid samples by nitric acid (mineralization) and after pH adjustment by ammonium acetate they are trapped and concentrated on selective sorbent. In the case of water the mineralization step is omitted and metals are concentrated on selective sorbents (solid-phase extraction). The trapped metals are eluted from sorbents by 1 M HCl, evaporated to dryness, dissolved in water, and analyzed by cationic ITP. Since I experienced that the published leading electrolyte with α-hydroxyisobutyric acid (α-HIBA) as counterion is not sufficiently robust (very small changes in pH caused a dramatic shift in step height of copper and aluminum), I developed another more robust leading electrolyte with glycolic acid as counterion [94]. The isotachopherogram of a model mixture in this system is given in Fig. 5.

4 Conclusions

CITP is a powerful electrophoretic method, especially for the determination of non-UV-absorbing small ions, such as alkali and alkaline earth metals, inorganic and organic acids, *etc.* Due to the concentrating effect of CITP, excellent LODs as low as 10^{-8} mol/L can be reached. Very promising is its on-line combination with CZE where the

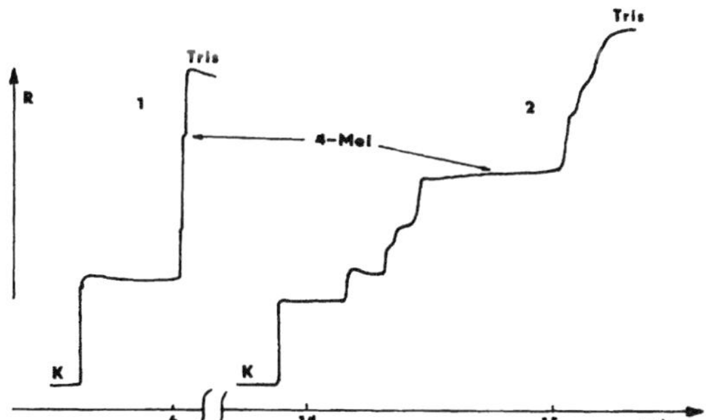

Figure 4. Isotachopherograms of caramel color. 1, Preseparation capillary 160 × 0.8 mm ID; 2, analytical capillary 160 × 0.3 mm ID; LE1 (preseparation capillary), 5 mM KOH + 20 mM HEPES + 0.1% PVP; LE2 (analytical capillary), 1 mM KOH + 4 mM HEPES + 0.1% PVP; TE, 10 mM Tris + 5 mM HAc; R, detector signal (increasing resistance). The analysis was performed on a column coupling isotachopherograph ZKI 01. The applied driving current was 150 µA (preseparation capillary) and 15 µA that was decreased to 5 µA during detection.

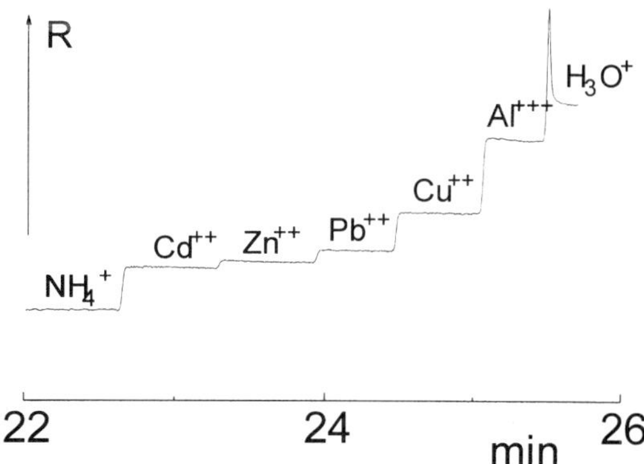

Figure 5. Isotachopherogram of standard of Cd, Zn, Pb, Cu, and Al (0.5 mM each). LE, 30 mM NH₄Ac + 10 mM glycolic acid; TE, 5 mM HAc. Analyses were performed on a single-capillary isotachopherograph IONOSEP 900.1. R, detector signal (increasing resistance). The applied driving current was 150 µA, which was decreased to 50 µA during detection.

CITP step serves as preconcentrating and/or sample clean-up step for the analysis of sample with complex matrices such as foodstuff.

The author is grateful for financial support provided by the grant agency of the Czech Republic (Project No. 525/99/1279).

Received February 3, 2000

5 References

[1] Kohlrausch, F. W. G., *Ann. Phys. Chem. (Leipzig)* 1897, *62*, 209–239.

[2] Martin, A., Everaerts, F. M., *Proc. Roy. Soc. London* 1970, *A 316*, 493–514.

[3] Everaerts, F. M., Beckers, J. L., Verheggen, T. P. E. M., *Isotachophoresis: Theory Instrumentation and Applications*, Elsevier, Amsterdam 1976.

[4] Boček, P., Deml, M., Gebauer, P., Dolník, V., *Analytická kapilární isotachoforéza, Pokroky chemie, Academia Praha, 1987.*

[5] Boček, P., Deml., M., Gebauer, P., Dolník, V., *Analytical Isotachophoresis*, VCH Publishers, Weinheim 1987.

[6] Righetti, P. G., *J. Chromatogr.* 1999, *838*, 1–320.

[7] Kaiser, K. P., Hupf, H., *Dtsch. Lebensm. Rdsch.* 1979, *75*, 300–304.

[8] Kaiser, K. P., Hupf, H., *Dtsch. Lebensm. Rdsch.* 1979, *75*, 346–349.

[9] Klein, H., Stoya, W., *Mitt. Gebiete Lebensm. Hyg.* 1988, *79*, 413–432.

[10] Marák, J., Laštinec, J., Kaniansky, D., Madajová, V., *J. Chromatogr.* 1990, *509*, 287–299.

[11] Kaniansky, D., Marák, J., Madajová, V., Šimuničová, E., *J. Chromatogr.* 1993, *638*, 137–146.

[12] Křivánková, L., Gebauer, P., Thormann, W., Mosher, R. A., Boček, P., *J. Chromatogr.* 1993, *638*, 119–135.

[13] Thormann, W., Arn, D., Schumacher, E., *Electrophoresis* 1985, *6*, 10–18.

[14] Hirokawa, T., Watanabe, K., Yokota, Y., Kiso, Y., *J. Chromatogr.* 1993, *633*, 251–259.

[15] Hirokawa, T., *J. Chromatogr. A* 1994, *686*, 158–163.

[16] Hirokawa, T., Ohta, T., Nakamura, K., Nishimoto, K., Nishiyama, F., *J. Chromatogr.* 1995, *709*, 171–180.

[17] Vacík, J., Muselasová, I., *J. Chromatogr.* 1985, *320*, 199–203.

[18] Kvasnička, F., Parkin, G., Harvey, C., *Int. Sugar Jnl.* 1993, *95*, 451–458.

[19] Kvasnička, F., *Acta Alimentaria Polonica* 1991, *XVII*, 317–325.

[20] Valášková, I., Zelenský, I., Zelenská V., Komarnicky, A., Kaniansky, D., *Collect. Czech. Chem. Commun.* 1990, *502*, 143–153.

[21] Kvasnička, F., Čopíková, J., Sterziková, E., *Listy cukrov.* 1988, *104*, 255–259.

[22] Stover, F. S., *J. Chromatogr.* 1984, *298*, 203–210.

[23] Kaniansky, D., Zelenský I., Valášková, I., Marák, J., Zelenská, V., *J. Chromatogr.* 1990, *502*, 143–153.

[24] Blatný, P., Kvasnička, F., Loučka, R., Šafářová, H., *J. Agric. Food Chem.* 1997, *45*, 3554–3558.

[25] Kvasnička, F., Ph. D. Habilitation, Institute of Chemical Technology, Prague 1999, p. 108.

[26] Boček, P., Medziak, I., Deml, M., Janák, J., *J. Chromatogr.* 1977, *137*, 83–91.

[27] Zelenský, I., Zelenská, V., Kaniansky, D., Lednárová, V., *J. Chromatogr.* 1984, *294*, 317–327.

[28] Meissner, T., Eisenbeiss, F., Jastorff, B., *J. Chromatogr. A* 1999, *838*, 81–88.

[29] Meissner, T., Eisenbeiss, F., Jastorff, B., *J. Chromatogr. A* 1998, *810*, 201–208.

[30] Blatný, P., Kvasnička, F., Application Note No. 50 for isotachophoretic analyzer IONOSEP 900.1, *Recman-laboratorní technika, Czech Republic, 1994.*

[31] Blatný, P., Kvasnička, F., *J. Chromatogr. A* 1994, *670*, 223–228.

[32] Madajová, V., Turcajová, E., Kaniansky, D., *J. Chromatogr.* 1992, *589*, 329–332.

[33] Kováč, J., Farkaš, J., Svoboda, M., *Kvas.prům.* 1988, *35*, 296–299.

[34] Blatný, P., Kvasnička, F., Application Note No. 31 for isotachophoretic analyzer IONOSEP 900.1, *Recman-laboratorní technika, Czech Republic, 1994.*

[35] Matějovič, I., Bielikova, M., *Collect.-Czech. Chem. Commun.* 1988, *53*, 3067–3071.

[36] Zelenský, I., Zelenská, V, Kaniansky, D., *J. Chromatogr.* 1987, *390*, 111–120.

[37] Blatný, P., Kvasnička, F., Kenndler, E., *J. Agric. Food Chem.* 1995, *43*, 129–133.

[38] Janoš, P., *J. Chromatogr.* 1990, *516*, 473–477.

[39] Blatný, P., Kvasnička, F., *J. Chromatogr. A* 1999, *834*, 419–431.

[40] Rubach, K., Offizorz, P., Bremez, Z., *Z. Lebensm. Unters. Forsch.* 1981, *172*, 351–354.

[41] Ginzinger, W., Hohenauer, W., *Milchwirtsch. Ber.* 1984, *80*, 251–255.

[42] Voldřich, M., Hrdlička, J., Opatová, H., *Potr. Vědy* 1988, *6*, 99–103.

[43] Kvasnička, F., Voldřich, M., *Electrophoresis* 2000, *21*, in press.

[44] Farkaš, J., Koval', M., *Kvas. prům.* 1982, *28*, 256–260.

[45] Farkaš, J., Koval', M., *Vinohrad* 1982, *20*, 186–187.

[46] Reijenga, J. C., Verheggen, T. P. E. M., Everaerts, F. M., *J. Chromatogr.* 1982, *245*, 120–125.

[47] Štechová, A., Kvasnička, F., Čopíková, J., Kadlec, P., *Listy cukrov.* 1985, *101*, 223–228.

[48] Štechová, A., Kvasnička, F., Čopíková, J., Kadlec, P., *Listy cukrov.* 1985, *101*, 278–284.

[49] Štechová, A., Čopíková, J., Kvasnička, F., Kadlec, P., *Listy cukrov.* 1986, *102*, 36–42.

[50] Kvasnička, F., Čopíková, J., Sterziková, E., *Listy cukrov.* 1988, *104*, 255–259.

[51] Blatný, P., Kvasnička, F., Application Note No. 13 for isotachophoretic analyzer IONOSEP 900.1, *Recman-laboratorní technika, Czech Republic, 1994.*

[52] Procházka, L., Kvasnička,, F., Štechová, A., *Kvasný prů-mysl* 1986, *32*, 33–36.

[53] Procházka, L., Štechová, A., *Listy cukrov.* 1987, *103*, 91–94.

 CITP in food analysis 2787

[54] Kvasnička, F., Application Note No. 17 for electrophoretic analyzer EA 100, *Villa-Labeco*, Slovakia, 1998.

[55] Blatný, P., Kvasnička, F., Application Note No. 38 for isotachophoretic analyzer IONOSEP 900.1, *Recman-laboratorní technika,* Czech Republic, 1994.

[56] Blatný, P., Kvasnička, F., Application Note No. 41 for isotachophoretic analyzer IONOSEP 900.1, *Recman-laboratorní technika,* Czech Republic, 1994.

[57] Blatný, P., Kvasnička, F., Application Note No. 45 for isotachophoretic analyzer IONOSEP 900.1, *Recman-laboratorní technika,* Czech Republic, 1994.

[58] Blatný, P., Kvasnička, F., Application Note No. 16 for isotachophoretic analyzer IONOSEP 900.1, *Recman-laboratorní technika,* Czech Republic, 1994.

[59] Blatný, P., Kvasnička, F., Application Note No. 43 for isotachophoretic analyzer IONOSEP 900.1, *Recman-laboratorní technika,* Czech Republic, 1994.

[60] Blatný, P., Kvasnička, F., Application Note No. 46 for isotachophoretic analyzer IONOSEP 900.1, *Recman-laboratorní technika,* Czech Republic, 1994.

[61] Kenndler, E., Huber, J. F. K., *Z. Lebensm. Unters. Forsch.* 1980, *171*, 292–296.

[62] Kvasnička, F., Prášil, T., Zbudilová, D., Svobodová, M., *Biol. Chem. Vet. (Praha)* 1986, *XXII*, 471–478.

[63] Kvasnička, F., *J. Chromatogr. A* 1999, *838*, 191–195.

[64] Holloway, C. J., *J. Chromatogr.* 1987, *390*, 97–100.

[65] Kaniansky, D., Madajová, V., Marák, J., Šimuničová, E., Zelenský, I., Zelenská, V., *J. Chromatogr.* 1987, *390*, 51–60.

[66] Röben, R., Rubach, K., *Analytical and Preparative Isotachophoresis*, Walter de Gruyter, Berlin 1984, pp. 109–116.

[67] Rubach, K., Breyer, C., *Dtsch. Lebensm. Rdsch.* 1980, *76*, 228–231.

[68] Kvasnička, F., Humpolíková, P., Volkmerová, D., *Potrav. Vědy* 1988, *6*, 259–269.

[69] Pažitná, G., Sochorová, V., *Průmysl potravin* 1990, *41*, 246–247.

[70] Stover, F. S., *J. Chromatogr.* 1988, *445*, 417–423.

[71] Blatný P., Kvasnička, F.. Kenndler, E., *J. Agric. Food Chem.* 1995, *43*, 129–133.

[72] Klein, H., *Z. Acker Pflanzenbau* 1981, *150*, 349–355.

[73] Kvasnička, F., Price, K. R., Ng, K., Fenwick, G. R., *J. Liq. Chromatogr.* 1994, *17*, 1941–1951.

[74] Kazuko, S., Masaru, O., Tsutomu, Y., Kimitoshi, N., *J. Food. Sci.* 1983, *48*, 665–667.

[75] Rubach, K., Breyer, C., *Getreide Mehl Brot* 1981, *35*, 91–93.

[76] Karovičová, J., Polonský, J., Šimko, P., *Nahrung* 1991, *35*, 543–544.

[77] Madajová, V., Marák, J., Kaniansky, D., Šimuničová, E., *Chem. Listy* 1992, *86*, 381–385.

[78] Pipek, P., Beneš P., Kvasnička, F., Wilming, M., *Fleischerei Technik* 1999, *15*, 41–42.

[79] Kvasnička, F., Ph. D. Habilitation, *Institute of Chemical Technology, Prague* 1999, p. 124.

[80] Ito, Y., Toyoda, M., Suzuki, H., Iwaida, M., *JAOAC* 1980, *63*, 1219–1223.

[81] Kvasnička, F., Míková, K., *J. Food Comp. Anal.* 1996, *9*, 231–242.

[82] Karovičová, J., Polonský, J., Príbela, A., Šimko, P., *J. Chromatogr.* 1991, *545*, 413–419.

[83] Kvasnička, F., *Electrophoresis* 1989, *10*, 801–802.

[84] Kenndler, E., Huber, J. F. K., *Z. Lebensm. Unters. Forsch.* 1980, *171*, 292–296.

[85] Reijenga, J. C., Aben, G. V. A., Lemmens, A. A. G., Verheggen, T. P. E. M., de Bruijn, C. H. M. M., Everaerts, F. M., *J.Chromatogr.* 1985, *320*, 245–252.

[86] Klein, H., Stoya, W., *Ernährung* 1987, *11*, 322–325.

[87] Rubach, K., Offizorz, P., *Dtsch. Lebensm. Rundsch.* 1983, *79*, 88–90.

[88] Kvasnička, F., *J. Chromatogr.* 1987, *390*, 237–240.

[89] Poláková, J., Application Note No. 7 for electrophoretic analyzer EA 100, *Villa-Labeco*, Slovakia, 1998.

[90] Křivánková, L., Boček, P., Tekel, J., Kovačičová, J., *Electrophoresis* 1989, *10*, 731–734.

[91] Stránský, Z., *J. Chromatogr.* 1985, *320*, 219–231.

[92] Dombek, V., *Anal. Chim. Acta* 1992, *256*, 69–73.

[93] Everaerts, F. M., Verheggen, T. P. E. M., Reijenga, J. C., Aben, G. V. A., Gebauer, P., Boček, P., *J. Chromatogr.* 1985, *320*, 263–268.

[94] Kvasnička, F., Ph. D. Habilitation, *Institute of Chemical Technology, Prague* 1999, p. 134.

Review

Luis A. Colón
Glamarie Burgos
Todd D. Maloney
José M. Cintrón
Ramón L. Rodríguez

Department of Chemistry,
State University of New York
at Buffalo,
Buffalo, NY, USA

Recent progress in capillary electrochromatography

Capillary electrochromatography (CEC) continues to captivate many separation scientists. A remarkable activity is apparent from the numerous publications in the literature using CEC. A review of the most recent progress in CEC is presented herein, covering an extensive fraction of the literature on CEC published from the year 1997 until the beginning of 2000. Most of the recent developments have concentrated on column technology.

Keywords: Capillary electrochromatography / Packed capillaries / Electroseparations / Electro-osmotically driven flows / Open tubular columns / Review   EL 4170

Contents

Correspondence: Dr. Luis A. Colón, Department of Chemistry, State University of New York at Buffalo, 578 NS Complex, Buffalo, New York 14260-3000, USA
E-mail: lacolon@buffalo.edu
Fax: +716-645-6963

Abbreviations: ODS, ocatdecyl silane; **OT**, open tubular; **PAH**, polyaromatic hydrocarbon; **PLOT**, porous-layer open-tubular; **SAX**, strong anionic exchanger; **SCX**, strong cationic exchanger

1 Introduction

Capillary electrochromatography (CEC) continues to captivate the attention of many separation scientists. CEC stems from the combination of high performance liquid chromatography (HPLC) and capillary electrophoresis (CE), as a hybrid of the two techniques. This combination allows the use of an electrically driven flow to transport solvent and solutes through a chromatographic column. In principle, CEC makes possible the exploitation of a plug-like flow profile to achieve high efficiency, while retaining the variety of retention mechanisms and selectivity afforded by HPLC. Separation of uncharged solutes is effected by differential interactions of the solute between the mobile and the stationary phases. Charged species, on the other hand, can be influenced by the applied electric field and by interactions with the stationary phase; hence, separation of charged species is a combination of both differential interaction with the stationary phase and differential electromigration. One very important attraction of CEC is the possibility of increasing the limited peak capacity of traditional liquid chromatography. All these features continue to stimulate the exploration of CEC.

0173-0835/00/1818-3965 $17.50+.50/0

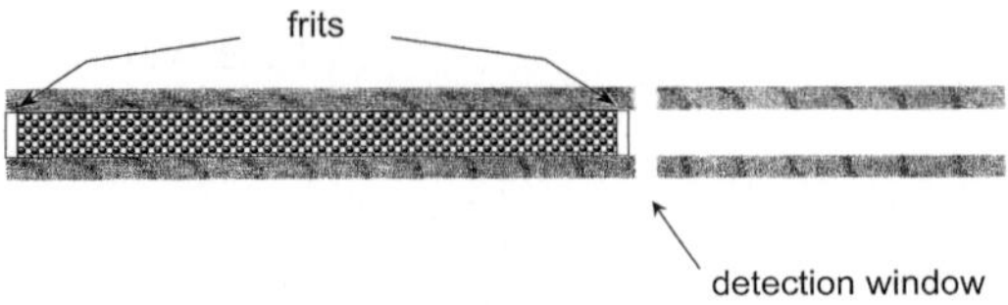

Figure 1. Schematic of a CEC column having a packed and an open segment.

There has been a remarkable increase in activity in the field of CEC during the past few years, which is reflected in the numerous research articles published in the literature. A significant amount of publications has appeared since our last review article on the subject [1]. Other reviews on the basic concepts of CEC have also appeared [2–7]. An introductory book on CEC was published recently [8] and another one is in production [9]. Educational material is also available through CE instrument manufacturers. All this is indeed indicative of the continued interest and growth of CEC.

CEC has been compared with electrokinetic chromatography (EKC) [10, 11]. These two techniques seem to be more competitive than complimentary. Both techniques mainly use chromatographic interactions to effect separation. In CEC, selectivity is achieved by changing the stationary phase (or packing material) while in EKC the pseudostationary phase in the running electrolyte is varied to obtain the same result. Changing the pseudostationary phase can perhaps result in a more convenient operation than changing to a completely new capillary. On the other hand, mass spectrometry (MS) detection is more complicated with EKC because of the additives (pseudophase) included in the running buffer. Further, gradient elution can be incorporated in CEC whereas it is more difficult to implement in EKC. Perhaps a major advantage of CEC over EKC is the infinite elution range and the compatibility with injection solvents and MS.

When a pressure-induced flow is superimposed on the EOF in CEC, the technique can be considered as pseudo-CEC, which can also be referred to as pressure-assisted CEC. Many CEC experiments make use of pressure at both ends of the capillary column, but without pressure differential, to avoid bubble formation; some refer to this as pressurized CEC. The term pressurized CEC must be avoided, however, since it could generate confusion; we therefore suggest using the term pressure-assisted flow CEC, when appropriate. The addition of pressure can aid in different ways. For example, it can provide a certain degree of selectivity not found in common CEC [12]. The use of pressure to assist the EOF

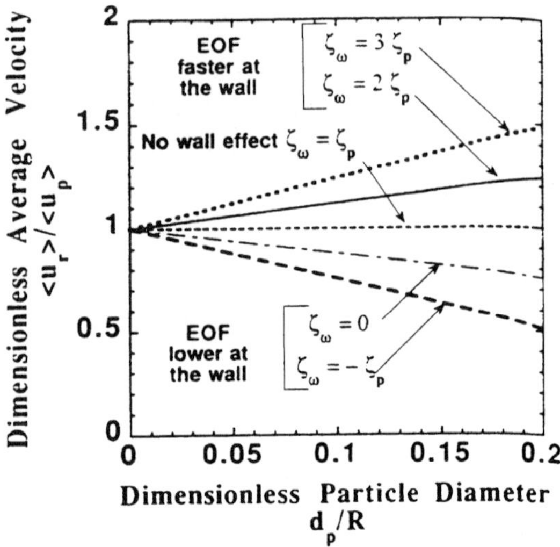

Figure 2. Effect of the packing particle diameter on the average flow velocity for a column with charged walls and packing ($R = 50$ and $\varepsilon = 0.4$). Reprinted from [18], with permission.

also enhances the separation speed [6], minimizes bubble formation, and can facilitate coupling with gradient elution.

This review article expands upon our previous one, appearing in this journal in 1997 [1], and covers a significant fraction of the literature from the second half of 1997 to the early part of 2000. Because of the many review articles appearing since our last review [1], however, we present a literature overview emphasizing the most relevant and recent aspects of CEC. We do not intend to be exhaustive or inclusive, nor do we want to duplicate what has already been covered by previous reviews. To this effect and when appropriate, we refer the reader to specific recent reviews with at-length discussion of subject areas beyond the scope of this article. Following our previous format [1], we cover CEC progress in fundamentals, column technology, and instrumentation. The applications of CEC were recently reviewed in this journal in several ways [13–17], for which reason we do not include such a section here.

2 Fundamental aspects

2.1 EOF

CEC is performed in packed and open tubular columns. The majority of CEC columns used to date, however, are considered packed in nature. Commonly, a packed col-

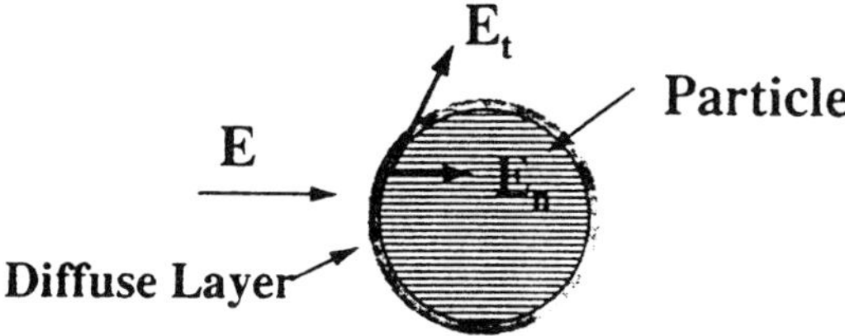

Figure 3. Illustration of the electric field components (normal and tangential) acting on the diffuse layer at a charged curved surface. Reprinted from [18], with permission.

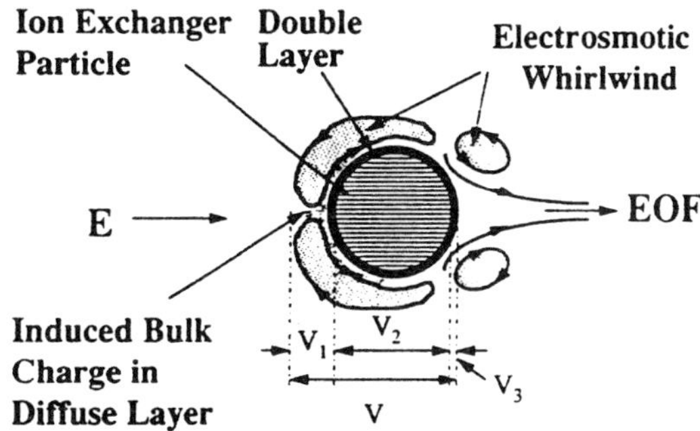

Figure 4. Illustration of the localized induced bulk charge layer and electroosmotic whirlwind around a conductive particle in a relatively low concentration electrolyte solution under the influence of a high electric field. Reprinted from [18], with permission.

umn in CEC is comprised of a packed and open segment (see Fig. 1). The electroosmotic phenomenon in porous media, like the CEC column, is a complex one and at the same time of extreme importance. The EOF is responsible for the bulk transport of the mobile phase and analytes through the capillary column. In the absence of EOF, only species with the appropriate charge will reach the detector. The principles and mechanism of this important parameter in CEC are not completely understood. Theoretical and practical considerations affecting EOF have been treated in the literature [18–29].

The expression commonly used to describe the linear velocity of the EOF (u) is the well-known von Smoluchowski equation given by Eq. (1):

$$u = \frac{\varepsilon_0 \varepsilon_r \zeta E}{\eta} = \mu_{eo} E \tag{1}$$

where ε_0, ε_r, ζ, E, η, and μ_{eo} are, respectively, the permittivity in vacuum, relative permittivity of the medium, zeta potential, applied electric field strength, viscosity of the medium, and the electroosmotic mobility. This expression was developed for the movement of liquid adjacent to a flat, uniformly charged surface as an electric field is applied parallel to the surface. Rice and Whitehead [30] derived the expression for the EOF flow profile in cylindrical capillary columns, assuming that the surface potential is less than 25 mV (see Eq. 2).

$$u(r) = \frac{\varepsilon_0 \varepsilon_r \zeta E}{\eta} \left[1 - \frac{I_0(\kappa r)}{I_0(\kappa a)} \right] \tag{2}$$

In this expression κ, r, and α are, respectively, the reciprocal of the electrical double layer thickness, the distance from the center of the channel, and the radius of the channel; other variables are as in Eq. (1). $I_0(\kappa r)$ is the zero-order modified Bessel function of the first kind. The reciprocal of the double layer thickness, κ, is also known as the Debye-Hückel parameter and given by Eq. (3):

$$\kappa = \frac{1}{\delta} = F\sqrt{\frac{2I}{RT\varepsilon_0\varepsilon_r}} \tag{3}$$

where δ is the Debye length (double layer thickness), F is the Faraday constant, R is the gas constant, T is the temperature (K), and I is the ionic strength.

A theoretical evaluation of Eq. (2) was performed recently by Andrade and Luo [20], confirming the indications of Rice and Whitehead [30] on exercising caution when treating the dependence of the electrokinetic phenomena on the electrokinetic radius ($\kappa\alpha$) at very low values due to the Debye-Hückel approximation. This is the case at low ionic strengths and/or for relatively narrow flow channels. It is apparent that the EOF plug-like profile is lost when the dimensions of the channel diameter and the thickness of the double layer are of similar dimensions.

An in-depth analysis of theories on EOF in porous media was conducted by Rathore and Horváth [18]. They examined the theories of Overbeek [31–33] and Dukhin, Mischuk *et al.* [34–38] as they apply to CEC. From the theory by Overbeek [31–33] for porous or nonporous packing materials of any shape, assuming nonconducting particles with uniform zeta potential and a thin double layer, the average EOF linear velocity in the packed structure in which the column walls and the particles are charged is given as:

$$u = u_p \left[1 + \left(\frac{d_p}{R}\right)\left(\frac{2}{\beta}\right)\left(\frac{\zeta_w}{\zeta_p} - 1\right) \right] \tag{4}$$

where u is the average velocity, u_p is the electroosmotic velocity generated at the packing surface (see Eq. 6 below), d_p is the particle diameter, R is the radius, ζ_w is

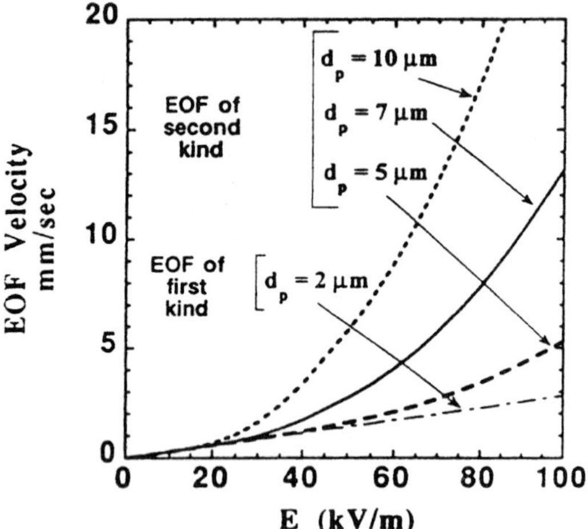

Figure 5. Average EOF velocity *versus* the field strength for different particle diameters with conditions $\sigma_p/\sigma_b = 5 \times 10^2$, $\delta = 500$ Å, $\zeta_w = \zeta_p = 100$ mV, $\varepsilon = 80$, $\varepsilon_o = 8.85 \times 10^{-12}$ C V^{-1} m^{-1}, and $\eta = 10^{-3}$ kg s^{-1} m^{-1}. Reprinted from [18], with permission.

the zeta potential at the capillary wall, ζ_p is the zeta potential of the particle, and β is a dimensionless parameter that depends on the total column porosity (ε_c) and on the structure of the packing and shape of the particles (α). β and u_p are given by

$$\beta = 3\sqrt{\frac{\alpha(1 - \varepsilon_c)}{2}} \tag{5}$$

$$u_p = -\frac{\varepsilon_o\varepsilon_r\zeta_p E}{\eta}\left(\frac{\sigma^*}{\sigma_b}\right) \tag{6}$$

In Eq. (6), σ^*/σ_b is the conductivity ratio, where σ^* and σ_b are the conductivities of the packed column and the open tube filled with the electrolyte solution. The conductivity ratio accounts for the packed bed porosity [18, 24]. Equation (4) takes into consideration the contribution of the wall to the net velocity under CEC conditions. It can be seen that if the zeta potential of the wall is the same as that of the particles, then $u=u_p$. As illustrated in Fig. 2, the "wall effect" is noticeable when both zeta potentials are not equal; it also increases as the particle diameter increases.

The EOF model proposed by Dukhin, Mischuk and co-workers [34–38], termed electroosmosis of the second kind, applied to unexpected high EOF velocities obtained with particles that have a conductivity higher than the

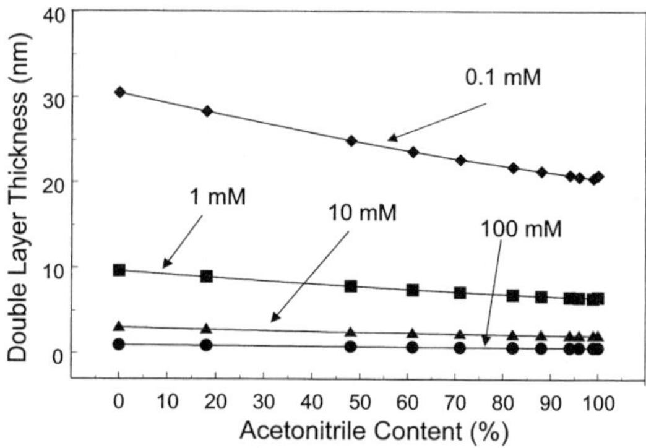

Figure 6. Double layer thickness as a function of acetonitrile concentration, as per Eq. (3) using constants for mixed solvents (as found in [214, 215]).

electrophoretic medium [18]. This unexpected high EOF velocity is assumed to be the result of tangential and normal components of the electric field at the charged surface of the curved particle, inducing bulk charges and hence enhancing EOF (see Fig. 3). Part of the mechanism for enhancing the EOF is attributed to the "electroosmotic whirlwind" around a conductive particle; in this case, the double layer is polarized in an electrolyte solution under the influence of a strong electric field (see Fig. 4). The average linear velocity for a packed bed under Dukhin's model is given by Eq. (6); however, the zeta potential of the particle is replaced by an apparent zeta potential, accounting for the potential drop across the bulk charge layer induced (assumed to be thin in comparison to the particle diameter), and the "wall effect" is negligible [18]. The EOF enhancement is predicted to increase as both the particle size and conductivity are increased, as depicted in Fig. 5. The authors pointed out that the EOF enhancement would take place when the conductivity of the column packing material is higher than the presently used packing in CEC. This enhancement will become more important as ionic exchanger-type of packing materials are used in CEC, which must be considered as column technology advances. Specific experiments designed to obtain evidence to support or refute these theories in CEC are lacking in the literature.

2.1.1 EOF in small flow channels

CEC does not have the pressure limitations frequently found in HPLC. In principle, then, relatively long capillary columns can be packed with small particle diameters to generate high separation efficiencies. The limitation imposed on particle diameter in CEC is determined by the electrical double layer overlapping in the flow channel. As

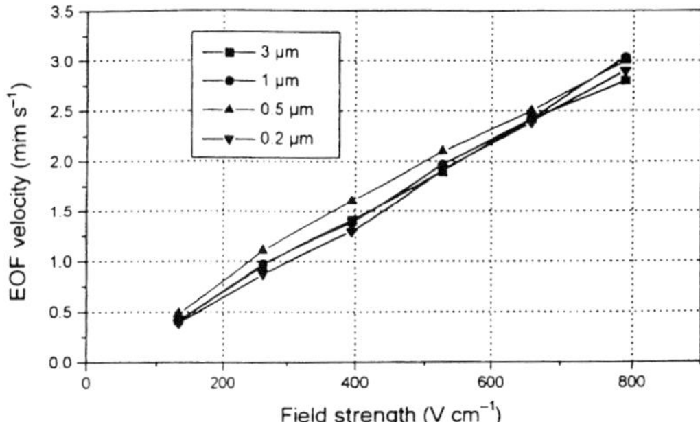

Figure 7. EOF velocity as a function of particle diameter. Reprinted from [44], with permission.

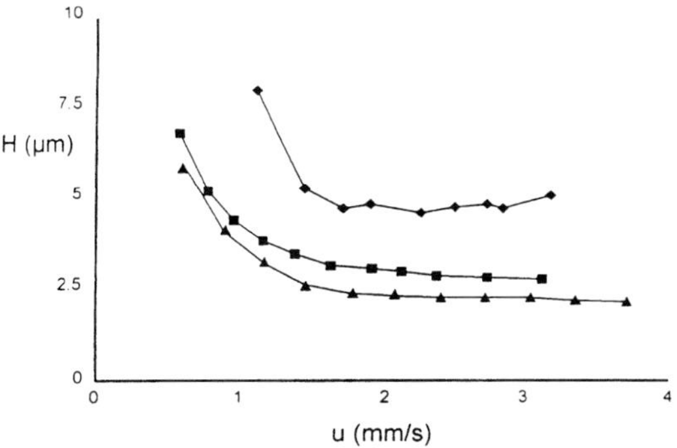

Figure 8. Effect of electrolyte concentration (borate, pH 8.3) on the separation efficiency on Nucleosil 4000-7 C-18 in an acetonitrilel-water (80:20 v/v) mobile phase. Top curve for 0.1 mM, middle curve for 1 mM, and bottom curve for 10 mM. Reprinted from [47], with permission.

Eq. (3) shows, the thickness of the electrical double layer is dependent on the dielectric constant of the mobile phase and the concentration of the electrolyte present (*i.e.*, the ionic strength). Based on the classical double layer theory, as applied to thin capillaries, the flow channel must have a diameter that is at least 20 times the thickness of the electrical double layer ($d_c \geq 20\delta$) in order to minimize the parabolic flow profile [30]; in such a case 80% of the volume transport due to electroosmosis is retained. Below this value, overlapping of the electrical double layer becomes significant and the plug-like profile is lost, decreasing EOF considerably. To maintain 90% of the volume transport due to electroosmosis $d_c \geq 40\delta$ ($\kappa\alpha > 20$), while the total volume transport, as described by Eq. (1), divided by the cross-sectional area of the channel, should be realized at $d_c \geq 100\delta$ ($\kappa\alpha > 50$) [30]. Typical values for δ are less than 10 nm, depending on the electrolyte concentration. Figure 6, for example, shows δ calculations as a function of acetonitrile concentration at different ionic strengths using univalent ions. For δ values of 1–10 nm, the minimum flow channel dimensions required to maintain at least 80% of the transport flow would be 20–200 nm under the conditions of Fig. 6.

The EOF depends on the column packing structure and pore size of the packing material [19–24, 39]. In a packed structure, there are many interconnected channels between particles. These channels depend on the column porosity, which dictates the permeability of the column. Assuming that a CEC column is a collection of capillary tubes and the average channel size between particles corresponds to the diameter of each tube, one can estimate the minimum channel diameter (d_c) in terms of particle size to satisfy the condition of $d_c \geq 20\delta$. This value was taken arbitrarily as the limit to preserve most of the volume transport with plug-like profile [40]. In the past, the flow channel dimensions have been estimated from geometrical considerations of a closed packed structure [41] or from the flow resistance parameter of a packed column

[42]. A relationship between the mean channel diameter and particle size (d_p) was derived by Wan [22], accounting for the particle structure through the interparticle porosity, and is given by:

$$d_c = \frac{0.42 d_p \varepsilon}{1 - \varepsilon} \tag{7}$$

where ε is the interparticle porosity. For a random packing structure, which can be considered a fairly well packed column, the porosity would be 0.4 [43]; hence, $d_c = 0.28\, d_p$. This value for d_c is similar to the one suggested by Knox and Grant [42] (*i.e.*, 0.25 d_p), obtained by the flow resistance parameter of a packed structure. Using Eq. (7), the minimum particle diameter would then be 0.071–0.71 μm for δ of 1–10 nm, for a column with a random packed structure. Unger and co-workers [44] reported no decrease in EOF using particle diameters as low as 0.2 μm, indicating that EOF is independent of the particle size (see Fig. 7). From Eq. (7), values for d_c of about 0.06 μm for 0.2 μm particles are calculated. Under the conditions used by Unger and co-workers, $\delta \sim 3$ nm would correspond to a channel diameter of about 0.06 μm to satisfy the condition of $d_c \geq 20\delta$. According to the above analysis, a reduced EOF would be seen, with diameters below those used by Unger and co-workers. Linear velocities above 2 mm/s have been realized in CEC using sub-μm particles with organic/acqueous mobile phases [44–46]. Nevertheless, one must pay attention at the experimental parameters that control this effect: solvent dielectric constant and viscosity, and the concentration of electrolyte.

If using porous particles in CEC, the possibility of having intraparticle flow exist, depending on the particle pore size. Most commercially available silica-based materials

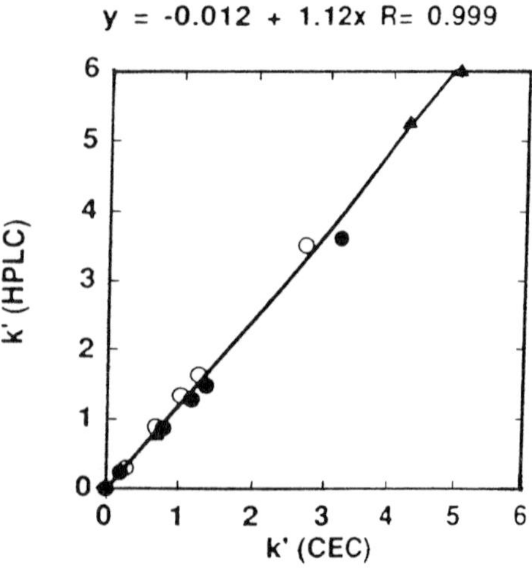

Figure 9. Correlation between the retention factors measured by HPLC and CEC. Reprinted from [51], with permission.

have a pore size close to 10 nm. Given the above discussion on the minimum channel required to support EOF, flow through these channels is unlikely. However, particles with larger pores can support EOF within the pores, exhibiting perfusion through the particles. In fact, EOF transport through the pores of particles has been reported in the literature [24, 47–50]. Remcho and Li [48] first reported perfusive transport through 5–10 µm particles with nominal pore sizes > 2000 Å. In later work, Stol *et al.* [47] also reported EOF transport on wide-pore silica particles. EOF transport through wide-pore silica particles entrapped in a capillary column by the sol-gel process has also been reported [49]. Flow through the pores in the particles is achieved by adjusting the concentration of electrolyte in the mobile phase, which in turn adjusts the thickness of the double layer inside the pore. At high electrolyte concentrations, perfusive flow is observed while at low electrolyte concentrations, double layer interaction occurs and transport through the pores is stopped. Such behavior has been reported by others [24, 47]. Because of an enhanced flow through the pores, mass transfer is improved and higher separation efficiencies are achieved. For example, efficiencies of 430 000 theoretical plates/m were reported for columns packed with 7 µm (C$_{18}$) particles containing pores with a diameter of 4000 Å (nominal value given by manufacturer) [47]. With a decrease in the concentration of the electrolyte in the mobile phase, the efficiency decreased, attributed to double layer interactions that shut down the pores affecting solute mass transfer within the pore, as also observed by others [47, 48]. Such an effect can be seen in Fig. 8. The efficiencies

achieved are comparable to those obtained with small particle sizes [47]. Therefore, the use of wide-pore large particles has been proposed as an alternative to sub-µm materials for the increase in separation efficiency in CEC [47, 48, 51]. When compared to sub-µm material, the wide-pore large particles also offer the advantage of easier packing since agglomeration is minimal. However, particles with large pores have the disadvantage of being fragile and care must be exercised during packing. The increased particle porosity will also decrease sample capacity, and the increase in ionic strength to maintain a thin double layer can lead to heating of the column.

2.1.2 Experimental parameters affecting EOF

The EOF velocity in CEC is affected as several experimental parameters are changed. From Eq. (1) it can be seen that variations in viscosity, dielectric constant, and ionic strength of the mobile phase will have a direct influence on the EOF. A charged surface is required so that the zeta potential can develop properly; hence, the chromatographic material in CEC must posses a charge. The pH of the solution affects the EOF velocity, as the surface is protonated or deprotonated with pH changes. Packed materials with fixed charges should minimize pH effects (see Section 3.2.1). Temperature changes the EOF velocity because of its effect on the viscosity of the system and on the dielectric constant. An increase in the temperature of the mobile phase should increase its velocity [52, 53]. Moreover, because of the zeta potential dependence on the dielectric constant, which is dependent on the temperature, a change in temperature will also affect the zeta potential. The organic modifier in the mobile phase also plays a role, which is difficult to predict since many variables are affected by changes of the organic modifiers: viscosity, dielectric constant, and zeta potential. It thus becomes clear that deconvolution of these parameters is not a trivial task in CEC and that careful manipulation of the chromatographic conditions are needed for its operation. Many reports have appeared in the literature which consider the experimental parameters affecting EOF [18, 19, 26, 44, 52–62].

2.1.2.1 Ionic strength

The ionic strength of the mobile phase can influence the zeta potential as well as the viscosity. It is expected that increasing the ionic strength will result in a decrease of EOF velocity. Recent reports continue to confirm such theoretical predictions [44, 52, 59, 60]. Wan [56] observed that for various phosphate concentrations (0.04, 0.2 and 1 mM) the velocity in CE increased linearly with the field strength as the ionic strength was decreased. However,

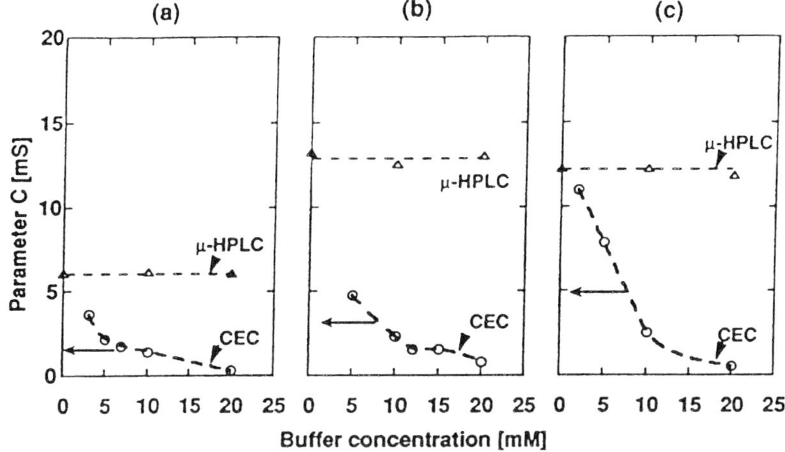

Figure 10. *C*-term *versus* electrolyte concentration as evaluated for HPLC and CEC. Reprinted from [51], with permission.

in CEC the velocities showed positive deviations in plots of the linear velocity as a function of applied field. The heat generation is described by Eq. (8).

$$Q = E^2 \lambda c \varepsilon \gamma \qquad (8)$$

where E, λ, c, ε, and γ are the field strength, equivalent conductivity, electrolyte concentration, column porosity, and tortuosity factor, respectively. The positive deviations in EOF velocity were attributed to local heating because of differences in packing densities. Other laboratories have observed these positive deviations. Knox and Grant [42] also attributed such behavior to heating effects. Horváth and co-workers [18, 19] also found these positive deviations, although Ohm's plot showed a linear relationship, indicating effective heat dissipation. They postulated that possible double layer polarization as described by Dukhin [34] was playing a role. This polarization is predicted to occur when the particles used are more conductive than the electrophoretic medium.

Wan [56] constructed plots of μ_{eo} *versus* the logarithm of the electrolyte molar concentration for CEC and CE systems, showing different behavior for each system. In CE, a linear decrease in the μ_{eo} is observed as the logarithm of the electrolyte concentration increases. However, under CEC conditions the μ_{eo} shows a maximum at intermediate concentration of salt, as was also observed by Knox and Grant [42]. The observed trend under CEC conditions was interpreted by Wan [56] as a double layer overlap at low electrolyte concentrations; this would be more prone to occur where the packing density is high. Similarly, Banholczer and Pyell [61] obtained plots of μ_{eo} *versus* logarithm of the phosphate buffer concentration. A maximum in these plots is observed at phosphate concentrations between 0.4 to 4 mM. The largest variations in

μ_{eo} were at lower phosphate concentrations. Bartle and co-workers [62], on the other hand, found that the linear velocities are comparable at high ionic strengths in both CE and CEC, for ionic strength between 1–20 mM in octadecyl silica (ODS-1) materials. They also observed that at ionic strength lower than 5 mM the linear velocity in CEC starts to decrease moderately. However, for CE the linear velocities continue to increase as the ionic strength decreases. One possible explanation for such results, according to Bartle, is the presence of a double layer overlap inside the pores of the particles.

Dittmann and Rozing [54] studied the effect of ionic strength on the μ_{eo} at 20, 30, and 40°C. At each individual buffer concentration tested, the μ_{eo} was higher for the higher temperatures. This is expected since at higher temperatures the ε/η increases, although the zeta potential can also change with temperature. As predicted by theory, a decrease in the μ_{eo} was observed as the buffer concentration was increased as a result of a more compact double layer; therefore work at low electrolyte concentration and at temperatures above ambient was recommended. Various authors, however, disagree with the recommendation of working at low ionic strengths due to encountered irreproducibility in migration time and in μ_{eo} [60, 61], suggesting the opposite. Furthermore, relatively high ionic strengths have been reported to increase separation efficiency [47, 56, 60].

2.1.2.2 Organic modifier

The vast majority of papers published on CEC have reported on the use of C18-modified silica particles (over 95%) and to a much lesser extent C8 stationary phases, using organic/aqueous mobile phases [63]. For a given particle/stationary phase material, the properties of the

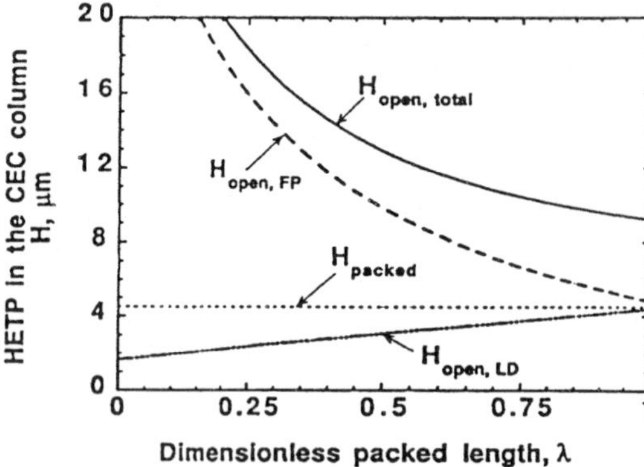

Figure 11. Plate height contributions due to longitudinal diffusion, $H_{open,LD}$, flow profile effects, $H_{open,FP}$, and their sum, $H_{open,total}$, in the open segment of a CEC column as a function of packed bed ($\lambda = L_{packed}/L$). Reprinted from [23], with permission.

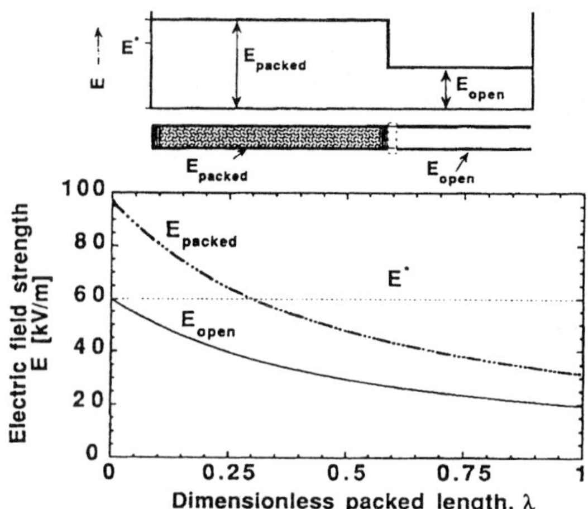

Figure 12. Plots of the actual electric field strength in the packed and open segments as a function of packed bed ($\lambda = L_{packed}/L$). E^* is the apparent or fictitious electric field strength. The conductivity ratio used was 3.1. Reprinted from [23], with permission.

eluent influence the magnitude of the EOF. The EOF is influenced by the type of organic modifier in use [57]. Acetonitrile is the most commonly used organic modifier, although methanol has also been employed. In general, it has been observed that as the percentage of acetonitrile is increased, the EOF velocity increases [44, 52, 54, 62, 64]. The effect of the percent acetonitrile in the mobile phase was studied by Choudhary and Horváth [19] in two sets of experiments: (i) the aqueous buffer was serially diluted with acetonitrile and therefore the ionic strength was not kept constant; and (ii) the electrolyte concentration was kept constant as the organic percentage was increased. It was found that, for CE under constant or varying electrolyte conditions, the EOF velocity decreased by increasing the organic percentage. In CEC, however, keeping the electrolyte concentration constant, the EOF velocity increased with the percentage of acetonitrile. Unexpectedly high EOF velocities were observed at high acetonitrile concentrations [52, 57, 61, 62]. Plots of EOF velocity *versus* % methanol, in contrast, have shown a minimum around 60% methanol [54]. Considering that the ε/η ratio is proportional to the EOF velocity and if all other variables in Eq. (1) remain constant, plots of EOF velocity *versus* organic percentage should follow those of ε/η *versus* organic percentage. The observed increase in EOF velocity with acetonitrile concentration is not the sole result of differences in the ε/η ratio; variations on the charge surface may play an important role [54, 57, 60, 61].

2.2 Retention and selectivity

The selectivity in CEC, as in HPLC, can be altered by changes in the mobile and/or stationary phase. In gen-

eral, the same behavior observed in HPLC is also observed in CEC. In reversed-phase, solutes are retained stronger in mobile phases containing methanol than those containing acetonitrile [54, 65]. The predominant separation mechanism for neutral compounds in ODS columns is that of reversed-phase chromatography, with similar retention behavior in CEC and HPLC [51, 65–67]. This is illustrated in Fig. 9, where the retention factor for a given component in HPLC is plotted against that in CEC collected from different packing materials, showing a linear relationship between CEC and HPLC. Euerby and co-workers [53] reported minimal changes in selectivity in CEC for C8, C18, and phenyl-based stationary phases. Interestingly, they also reported the longest retention for the C8 in the separation of barbiturates. In HPLC, however others have also found small changes in the retention of barbiturates between the C8 and C18 phases [68].

By increasing the ionic strength of the mobile phase, a decrease in the retention factor was observed [44, 59, 60]; an increase in retention time was also reported [44, 59, 60]. A comparison of different electrolytes (phosphate, MES, and Tris) at a constant pH on the separation of polycyclic aromatic hydrocarbons (PAHs) did not reveal any changes in retention and elution order [60]. Banholczer and Pyell [61] did not find any dependence of the retention factor on the concentration of phosphate for the neutral compounds pyrene, acenaphthene, acenaphthylene, ethyl benzoate, and methyl benzoate. According to these authors, the buffer concentration did not affect the distribution coefficient of the analytes between the stationary and the mobile phase. They also indicated the absence of temperature effects in their study since an increase in electrolyte concentration did not decrease the retention factor, which one may expect as temperature increases.

When using strong cationic exchangers (SCX) as stationary phase, one can change the selectivity by manipulating mobile phase components [69]; for example, altering the concentration of a competing ion in the mobile phase, the contributions between ion-exchange chromatography and CE can be controlled. Selectivity in CEC can also be changed by selecting/changing the stationary phase. Columns packed with mixed-mode stationary phases, for example, can provide separation of species based on mechanisms involving hydrophobic and ion-exchange interactions, as well as electrophoretic migration. The selectivity of charged compounds can be fine-tuned by changing the length of the packed segment in a typical CEC column [27]. This is the case because, in the open section, separation is achieved solely by electrophoresis while in the packed section is a combination of chromatographic and electrophoretic mechanisms.

In the case of charged solutes, migration is determined by electrophoretic and chromatographic transfer mechanisms. Stahlberg [70] reported that, in general, the convolution of chromatographic and electrophoretic mechanisms brings about nonlinear effects. Such nonlinearity may cause band broadening. The combination of the electric field and the adsorption, however, can stabilize the eluting band during its migration through the column and these effects can explain some of the focusing observed in CEC.

Careful optimization of experimental parameters leads to a desired selectivity for a specific application. Using a central composite design approach. Miyawa *et al.* [71] optimized several operating parameters (the applied electric field, acetonitrile concentration, and buffer composition) to separate a variety of degradation products from the antibacterial 3-[4-(methylsulfinyl)phenyl]-5S-acetamidomethyl-2-oxazolidinone. Djordjevic *et al.* [72] used a blending of bare silica and reversed-phase material in a column to affect retention. By adjusting the column composition through such blending, they were able to reduce the retention factors and analysis time by a factor of two, when compared to a column packed with reversed-phase material.

2.3 Band broadening

The high efficiencies obtained with CEC are attributed to the characteristic flat flow profile of the EOF [2, 51, 73]. In a recent study, Horváth and co-workers [51] performed a systematic examination of band broadening between electrically (*i.e.*, CEC) and pressure-driven (*i.e.*, micro-HPLC) flows in several packed capillaries under otherwise identical experimental conditions. The contribution of detection and injection to band spreading were found to be negligible. The columns for these experiments were packed with 5 µm Spherisorb ODS 300 Å, 5 µm Spherisorb SCX 300 Å, and 8 µm polymeric 1000 Å. The experimental data were fitted to the simplified van Deemter expression (Eq. 9) to evaluate the eddy diffusion (A) and mass transfer (C) parameters for each mode of flow.

$$H = A + \frac{B}{u} + Cu \tag{9}$$

For the columns tested, the A-term was shown to be smaller in CEC than in HPLC; this is attributed to the EOF plug-like flow profile reducing multipath band dispersion by a factor of 2–4. The C-term in the van Deemter expression was shown to be higher in HPLC than in CEC for the packing materials possessing average pore sizes of $\geq$ 300 Å. The enhancement was not observed for columns containing packing material with an average pore size of 80 Å, ascribed to not having EOF transport through the pores of the particle because of double layer interactions, as discussed above in Section 2.1.1. The intraparticle resistance to mass transfer contributing to the plate height (H_i) was examined with Eq. (10).

$$H_i = \left(\frac{\theta(k_0 + k + k_0 k)^2 d_p}{30 k_0 (1 + k_0)^2 (1 + k)} \right) \left(\frac{D_{eff}}{D_{app}} \right) \left(\frac{u d_p}{D_m} \right) \tag{10}$$

In this equation, θ is the tortuosity of the support, d_p is the particle diameter, D_m is the molecular diffusivity of the solute in the mobile phase, u is the interstitial mobile phase velocity, k is the retention factor, D_{eff} is the effective molecular diffusivity in the pores, and D_{app} is the apparent diffusivity that accounts for transport in the porous particles by diffusion and by intraparticle-convective transport. The term k_0 is given by

$$k = \frac{\varepsilon_i (1 - \varepsilon_i)}{\varepsilon_e} \tag{11}$$

where ε_i and ε_e are the intraparticle and interstitial porosities, respectively. Figure 10 is an indication that intraparticle EOF through large-pore particles is very effective in reducing the C-term. It can be noticed that the C-term decreases as the concentration of the electrolyte in solution is increased. This clearly indicates that mass transfer resistance is significantly lower for electrically driven mobile phase flows than for pressure-driven ones, due to generation of intraparticle EOF inside the porous parti-

Table 1. Equations to characterize EOF and CEC columns

Formula		Description
$\mu_{eo,open} = \dfrac{L_d L}{t_{o,open} V}$	(1)	EOF mobility in an open tube
$\mu_{eo,packed}^* = \dfrac{L_{packed} L}{t_{o,packed} V}$	(2)	Apparent EOF mobility that is evaluated for the packed segment by using the voltage applied to the entire column
$\mu_{eo,packed} = \dfrac{L_e^2}{t_{o,packed} V_{packed}} = \dfrac{\tau^2 L_{packed}^2}{t_{o,packed} V_{packed}}$	(3)	Actual interstitial EOF mobility
$\tau = \dfrac{L_e}{L_{packed}}$	(4)	Tortuosity
$V_{packed} = V\left(1 - \dfrac{i_{packed} L_{open}}{i_{open} L}\right)$	(5)	Potential drop across the packed segment
$L_e = L\left(\sqrt{\dfrac{i_{open}}{i_{packed}}}\right) - L_{open}$	(6)	Length of the actual flow path which a tracer follows through the packed section of the column
$\sigma_{open} = \dfrac{i L_{open}}{V_{open} A_{open}}$	(7)	Conductivity of an electrolyte-filled cylindrical capillary (ionic conduction is the dominant mechanism for ionic migration)
$\sigma_{open} = \dfrac{i_{open} L}{V A_{open}}$	(8)	Conductivity of open segment in the absence of packing
$\sigma_{packed} = \dfrac{i_{open} L i_{pakced} L_{packed}}{V(i_{open} L - i_{packed} L_{open}) A_{open}}$	(9)	Conductivity of packed segment for column having packed and open segments
$\phi = \dfrac{\sigma_{packed}}{\sigma_{open}} = \varepsilon_T^m$	(10)	Conductivity ratio related to electrokinetic porosity (ε_T); m is an empirical constant dependent on the morphology of packing particles; for porosity of the media >0.2 $m = 1.5$
$\sigma_{packed} = \dfrac{i L}{V A_{open}}$	(11)	Conductivity of monolithic structure without open segments
$\mu_{eo,packed}^* = \dfrac{L_d L}{t_2 V}$	(12)	Apparent EOF mobility for monolithic structure without open segment
$\mu_{eo,packed} = \dfrac{L_e^2 L_d}{t_o V L} = \dfrac{\tau^2 L L_d}{t_o V}$	(13)	Actual EOF mobility for monolithic structure without open segment
$L_e = L\left(\sqrt{\dfrac{i_{open}}{i_{packed}}}\right)$	(14)	The equivalent length for monolithic structure without open segment
$E_{packed} = \dfrac{V_{packed}}{L_e}$	(15)	Electric field in the packed segment of a column having packed and open segments
$E_{open} = \dfrac{V_{open}}{L_{open}}$	(16)	Electric field in the open segment of a column having packed and open segments

Other variables in the equations are defined as follows:
L_d distance between the inlet and the point of detection L total length of the capillary t_o migration time of a neutral tracer $t_{o,open}$ migration time of the tracer in an open tube $t_{o,packed}$ migration time of the inner tracer in the packed segment L_{packed} the respective length of the packed section in a CEC column L_{open} the respective length of the open section in a CEC column V total potential difference across the column V_{packed} the respective potential difference across the packed segment of a CEC column V_{open} the respective potential difference across the open segment of a CEC column A_{open} cross-sectional area of the capillary column i current flowing through when potential drop, V, is applied across i_{open} current measured within the capillary in the absence of packing i_{packed} current measured within the capillary in the presence of packing

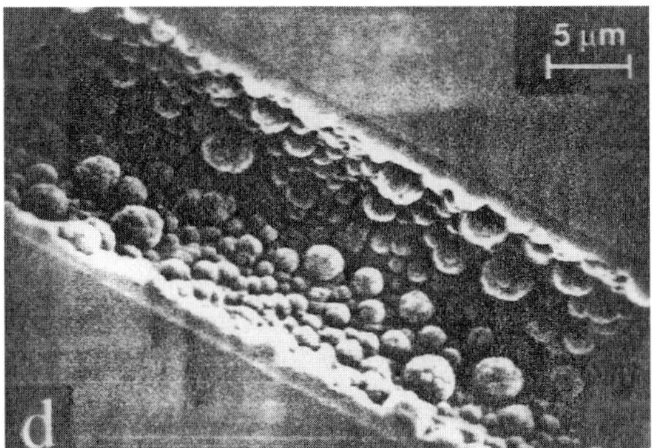

Figure 13. Scanning electron micrograph lof a polymeric porous layer in a capillary column. Reprinted from [85], with permission.

cles. The *A*-term of the van Deemter expression did not change with changes in the electrolyte concentration. The mass transfer enhancement in electrically driven flows is most pronounced for retention factors below one. The authors suggested that the use of relatively large particles (5–10 μm), with appropriate wide pores, can be an alternative to using small particle sizes ($\leq$1μm), as others have also suggested [47, 48].

For a typical CEC packed column, the nonuniformities in EOF velocities in the open segment region immediately after the packed bed causes band broadening [23, 74]. In this region, the characteristic flat flow profile no longer holds, changing to parabolic flow because of an induced pressure required to equalize the differences in EOF velocities, preserving the conservation of volumetric flow. The plate height contribution (H_{open}) for the open segment is then given by [23]:

$$H_{open} = H_{open,LD} + H_{open,FP} =$$

$$= \frac{2D_m}{U^a_{eo,open}} + \frac{a^2}{24D_m u^a_{eo,open}} \left[\left(\frac{P_0 - P_i}{L_{open}} \right) \frac{a^2}{8\eta} \right]^2 \quad (12)$$

where $H_{open,\,LD}$ is the plate height contribution due to longitudinal diffusion, $H_{open,\,FP}$ is due to the induced parabolic flow profile, D_m is the molecular diffusivity of a neutral tracer, $u^\alpha_{eo,\,open}$ is the actual flow velocity in the open segment, a is the radius of the capillary tube, L_{open} is the length of the open segment, η is the viscosity, and $P_o - P_i$ is the intersegmental pressure difference. A graphical representation of Eq. (12) is shown in Fig. 11, indicating that band broadening is appreciable, particularly with long open segments relative to the packed one.

Pyell *et al.* [75] studied band broadening due to sample injection and derived the following expressions to calculate optimum injection parameters:

$$V_{max} = \frac{0.7LL_r(1 + k_s)}{\mu_{eo} t_i \sqrt{N}} \quad (13)$$

$$t_{max} = \frac{0.7LL_r(1 + k_s)}{\mu_{eo} V_i \sqrt{N}} \quad (14)$$

where V_{max}, t_{max}, L, L_T, k_s, μ_{eo}, t_i, V_i, and N are maximum injection voltage, maximum injection time, the end of the column from injection to detection, total length of the column, retention factor for the analyte (sample solvent equal to mobile phase), electroosmotic mobility, time of injection, injection voltage, and number of theoretical plates, respectively. They calculated that for a standard injection in CEC (5 s at 5 kV), using C18 stationary phases, the injection volume would have an impact on band broadening that is larger than the typically tolerated 5%.

2.4 Electrical conductivity in CEC

The actual electroosmotic mobility in a CEC column depends on the column architecture [19] (*i.e.*, packed structure having open and packed segments, open tubular, and monolithic columns). The permeability of a packed column in CEC is a function of the column porosity and tortuosity [22, 24]. The electrical conductivity is indicative of these parameters, given that the EOF does not contribute significantly to the observed conductivity by transporting excess charges at the double layer; under CEC conditions, this contribution accounts for only 5% and can be neglected [22, 24]. Therefore, the use of conductivities has been suggested as a means to examine the architecture of CEC columns [19, 21–26, 28]. The packing structure can be evaluated and/or characterized by measuring the current through the column, which leads to the calculation of the column's conductivity. The electrical conductivity is a structural constant independent of column dimensions, particle size, and field strength [21]. Because of the dual nature of most packed CEC columns (*i.e.*, a packed and unpacked segment), each segment should have its own chracteristic conductivity. This results in different electric field strengths across each segment, hence, different EOF mobilities. To properly obtain the magnitude of the electroosmotic mobility, the actual field strength across the packed section must be known. Often, the potential drop across the entire column (packed and unpacked sections) is used to calculate EOF mobilities, resulting in an apparent EOF mobility and not the actual one (see Fig. 12). Conductivity measurements

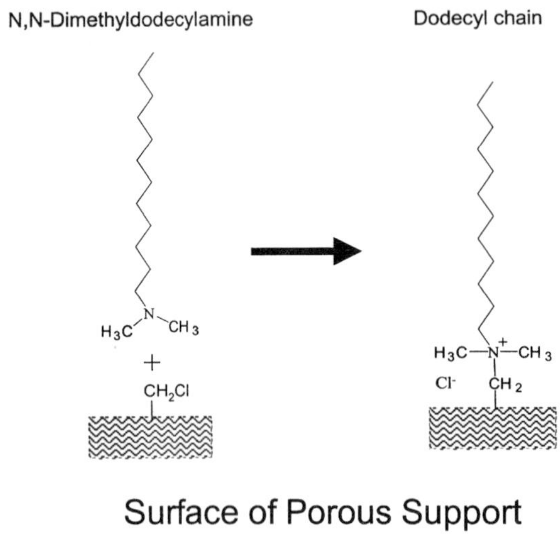

N,N-Dimethyldodecylamine   Dodecyl chain

Surface of Porous Support

Figure 14. Reaction scheme to attach a hydrophobic chain and a positive charge at the surface of the PLOT columns. Adapted from [85], with permission.

allow for the estimation of the actual voltage and EOF mobility across the packed and unpacked segments; it also provides means to obtain tortuosity and electrokinetic porosity of the packed column [19, 24]. Several reports have addressed this issue [23–26, 28]; the reports by Rathore and Horváth *et al.*, appear to be the most exhaustive [23, 24]. They have developed a series of equations that allows the calculation of conductivity, tortuosity, porosity, and electroosmotic mobility for the different column architectures in CEC, using migration data of a neutral tracer and current measurements. The equations are shown in Table 1. In their work, they evaluate the conductivity ratio of open to packed segments, tortuosity, porosity, and the actual EOF mobility for columns containing different stationary phases (*i.e.*, SCX and ODS), thus defining structural properties for the packing used. They reported that the SCX columns showed higher conductivity. In all columns considered, as the concentration of the organic modifier in the mobile phase was increased, the conductivity ratio increased, even though the total conductivity of the column decreased. The conductivity ratio was also high for large-pore packing having high porosity.

In separate studies, Cikalo *et al.* [25, 26] evaluated packed columns by performing current measurements. They reported that under their experimental conditions and at intermediate pH (pH 7.5), the field strength was similar for both packed and unpacked segments of a CEC column. At extreme pH (pH 2.9 and 10.5), the field strength appeared to be larger in the packed section.

Table 2. Most common reversed-phase packing materials in CEC

Packing material	Particle size
Hypersil (ODS 1)	3 µm, 5 µm
Nucleosil (ODS 1, ODS 2)	3 µm, 5 µm
Spherisorb (ODS 1)	3 µm
Zorbax (ODS 2)	6 µm, 7 µm

EOF mobilities were higher at extreme pH values and similar at the intermediate pH studied. They reported no difference in conductivity ratio between C18 and SCX silica-based columns. Henry and co-workers [28] also performed electrical measurements to characterize particle loaded monolithic sol-gel columns. In general, measurements of electrical conductivity can lead to a general protocol by which CEC columns can be evaluated. More experimentation in this area should correlate such measurements with column efficiency.

3 Column technology

Column technology is an active area of exploration in CEC. Column designs have included open tubular columns [1, 2, 77–88], monolithic columns [28, 37, 46, 49, 86, 89–107], packed columns [6, 16, 26, 29, 44, 46, 54, 58, 62, 63, 69, 81, 100, 108–131], and microfabricated structures [132–137] (open or continuous beds). These can be categorized into two major types: (i) open channels and (ii) packed structures. The packed structures can be subdivided into three different types: (i) columns packed with particles, (ii) columns containing a continuous bed that has been fabricated by *in situ* polymerization, creating a "rod-like" monolithic structure; and (iii) columns with entrapped or immobilized particulate material. Discussion of this last type of CEC columns will be included within the monolithic columns since entrapped columns are essentially a combination of packed and monolithic structures, resembling more of a monolith.

3.1 Open tubes

Open tubular (OT)-CEC is the simplest means of performing EC in capillary columns. OT-CEC is similar to CE, with the exception of having a stationary phase attached to the walls of fused-silica capillaries with small dimensions, typically ≤25 µm ID; the walls of the column must also be charged to support EOF. Detailed discussions on theoretical aspects of OT-CEC and on the various types of stationary phases incorporated have been reviewed elsewhere [1, 2, 77, 79]. The reader is referred to such reviews for more details on the subject, including applications. The discussion herein will involve the most recent developments in OT-CEC stationary phases.

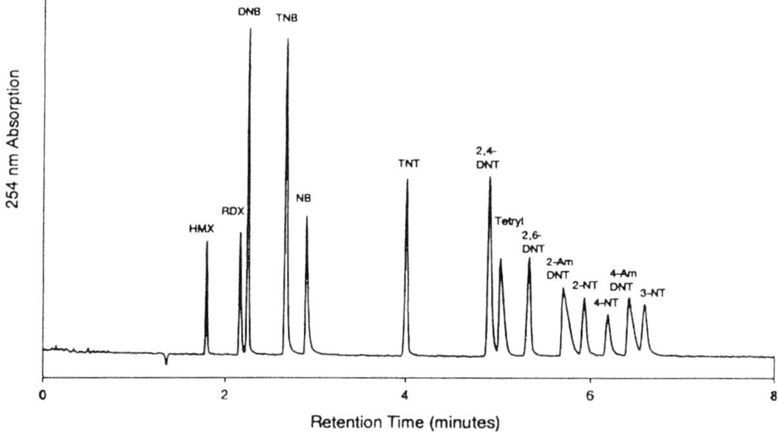

Figure 15. CEC separation of 14 explosives using 105 µm nonporous ODS particles. Reprinted from [118], with permission.

An approach to OT-CEC was developed by Liu *et al.* [78] by adsorption of the stationary phase to the wall of the capillary column. This was performed by rinsing the column with a cationic buffer such as cetyltrimethylammonium bromide (CTAB) or with a basic protein such as lysozyme. Upon adsorption of a surfactant bilayer or protein layer, the capillary was rinsed with running buffer containing no CTAB or protein, depending on the material adsorbed to the column wall. The adsorbed CTAB layer was suitable for separating neutral components (PAHs) in 6 min. Reproducibility of the CTAB phases was < 3%. The adsorbed lysozyme layer was utilized as a chiral stationary phase for the separation of a series of amino acid enantiomer pairs and the drug mephenytoin in less than 5 min. It was indicated that lysozyme is good chiral phase for OT-CEC because of its high isoelectric point (*i.e.*, 11.1). In order to adjust the solute-protein interaction, 2-propanol was used as a modifier to the mobile phase.

The sample capacity of OT columns is well known to be low, due to the column's low surface area. Pesek's group [79–83] continues to work vigorously in the area of OT-CEC after developing an approach to increase the surface area of the OT columns. They use ammonium hydrogen difluoride to etch the inner surface of the capillary column, thereby increasing the surface area. The etching process alters the chemical behavior of the wall. The presence of anodic EOF at a pH < 4.5 is observed, which suggests that a nitrogen containing species from the etching agent has incorporated itself in the surface structure. This seems to be an inherent characteristic of the etching process. After etching, silanization/hydrosilation of the surface is performed to anchor the stationary phase at the surface. Etched columns (20, 50, and 75 µm ID) have been characterized for the separation of cytochrome *c*,

lysozymes, amino acids, and some basic compounds [80–82]. Atomic force microscopy (AFM) was utilized to examine the surface of the column wall after etching. It was determined that etching times of about 4 h at 400°C were necessary to fabricate a somewhat more uniform and regular surface. They have shown that etched and chemically modified OT-CEC columns gave better resolution and greater retention than unetched columns for OT-CEC of a tryptic digest of transferrin [83]. When Polybrene was employed as a stationary phase for a mixture of four proteins, a significant increase in retention was also observed for the etched OT-CEC column when compared to the unetched column.

Using the etched columns, Pesek and co-workers [79] tested the potential of various chiral stationary phases for OT-CEC. The selectors evaluated for their potential use in OT-CEC were lactone, β-cyclodextrin, and naphthylethylamine. Diffuse reflectance infrared Fourier transform (DRIFT) was utilized to characterize the surface of the inner capillary wall by removing the polyimide coating and smashing the capillary column. DRIFT experiments confirmed the presence of the above-mentioned stationary phases attached to the surface of the capillary wall. The chiral phases were tested by separations of tricyclick antidepressants, benzodiazipenes, dansyl amino acids, and several drugs. Each of the bonded phases provided separations of some of the racemic solutes, but the phases were not able to completely resolve all of the enantiomeric mixtures tested. The stability of the columns was shown to be good over 200 injections with no change in resolution or retention. Columns used for these studies had 50 µm ID; mobile phase conditions and composition varied considerably depending on the species being separated. The potential applicability of the liquid crystals

Electrophoresis 2000, *21*, 3965–3993

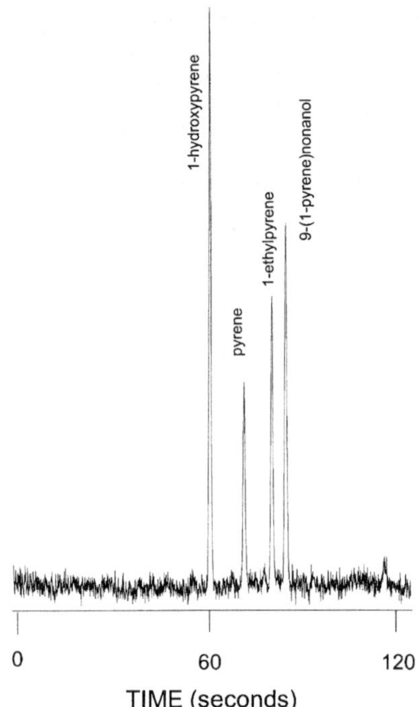

0   60   120

TIME (seconds)

Figure 16. CEC separation of pyrene derivatives in a 50 μm ID capillary packed with 450 nm organosilica particles using 80% acetonitrile – 20% of 50 mM Tris (pH 8) at 27 kV. Total column length, 34 cm; 12.7 cm packed, LIF detection with 325 nm excitation.

cholesterol-10-undeceneoate land 4-cyano-4'-pentoxybiphenyl as stationary phases for OT-CEC using the etched capillaries was also demonstrated [84].

Horváth and co-workers [85] have used porous-layer open-tubular (PLOT) columns to separate basic proteins and peptides. The use of these types of columns was prompted by the high permeability and their loading capacity due to an increased surface area (see Fig. 13). The polymeric porous layer was prepared at the inner wall of a fused-silica capillary previously silanized with 3-(trimethoxysily)propyl methacrylate. The silanization facilitated anchoring of the polymeric layer by covalent bonds *via* the vinyl functions. *In situ* polymerization was performed using vinylbenzyl chloride and divinylbenzene in the presence of 2-octanol as a porogen. The polymeric layer shielded any residual dissociated silanols that could potentially interact with the analyte. The chloromethyl functions at the surface of the porous polymeric support layer were reacted with *N,N*-dimethyldodecylamine to obtain a positively charged chromatographic surface with

fixed C_{12} alkyl chains (see Fig. 14). The fixed charges of the quaternary ammonium groups provided the required surface charge to generate EOF upon application of the electric field. The dodecyl chains provided selective retaining sites for reversible binding of the sample components to be separated in the chromatographic process. The PLOT columns provided high electroosmotic mobility, indicating a high surface concentration of the quaternary ammonium groups. In addition, a decrease in mobility per unit salt concentration was significantly greater than that obtained with packed capillary columns of comparable chromatographic surface. Interestingly, it was also observed that the retention of the solutes in this type of column increased with a concomitant increase in the concentration of the organic modifier in the mobile phase.

Another approach to increase the phase ratio in OT-CEC columns was followed by Sawada and Jinno [86]. They used a polymeric stationary phase anchored to the capillary walls *via* a bifunctional reagent. The stationary phase was fabricated by polymerizing *N-ter*-butylacrylamide and 2-acrylamido-2-methyl-1-propanesulfonic acid, providing for hydrophobicity and charged groups for the generation of EOF. Small neutral compound were separated. The small diffusion of the solutes in such polymeric phases leads to low efficiency. Tan and Remcho [87] have also developed a procedure for attaching thick polymethacrylate films to capillary walls for OT-CEC. The procedure enabled the formation of both linear and cross-linked polymer films within 25 μm ID capillaries. Cross-linker concentrations were optimized for maximum capacity factor and flow velocity for the separation of parabens and high efficiencies were obtained (~200 000–400 000 plates/m). The selectivity of the polymeric stationary phase could also be adjusted by changing monomer/cross-linker concentrations or by incorporating other functional groups within the matrix.

Using a cation-exchange OT-CEC column, Regnier and Xu [88] separated several proteins. Columns were coated with poly(aspartic acid) (PAA) *via* the *in situ* derivatization reaction of polysuccinimide and β-alanine. Separation was based on adsorption of the proteins in combination with EOF and electrophoretic mobility, depending on the pH of the mobile phase in relation to the p*I* of the protein and the ionic strength of the solution. Column efficiencies for the separation of various proteins were 10–100 times greater than in HPLC. They reported that the advantage of using these columns is that isocratic separation of protein mixtures can be accomplished while in HPLC such a separation would require gradient elution.

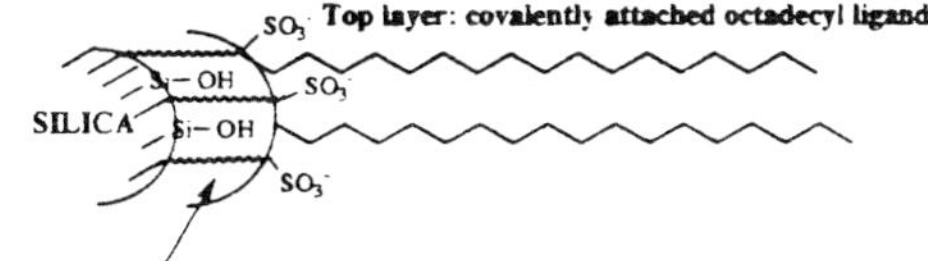

Figure 17. Scheme illustrating the octadecyl sulfonated silica material. Reprinted from [216], with permission.

3.2 Packed columns

The most commonly used columns in CEC are packed capillaries. Generally, they consist of two sections – one packed and one unpacked (or open; see Fig. 1); the packing material is kept in place by means of retaining frits. Traditionally, capillary columns of < 100 µm ID are packed with reversed-phase HPLC spherical packing materials (1.5–10 µm diameter). Reducing the column ID, while keeping the particle diameter constant, does not have an influence on column efficiency [74, 138, 139]. Varying the fraction of the packed section (25–100% of the open) has shown a dependence on the overall observable EOF velocity [25]. The EOF can also be accelerated by means of segmented capillaries [50]. With such capillaries one segment is packed with chromatographic material to effect separation while the other segment is packed with bare silica. The segment packed with bare silica serves as an EOF accelerator. The average velocity of the segmented column increases linearly with the fraction length of the accelerator and is affected by the porosity of the bare silica material.

The packed columns provide higher retention and column capacity than OT columns. For the most part, columns in CEC have been packed with reversed-phase materials. However, stationary phases involving mixed modes, such as SCX or strong anionic exchanger (SAX) with reversed phase, have also been used to pack capillaries for CEC. The subject of packing columns for CEC has been reviewed in detail recently by our laboratory [108], to which the reader is referred for more details on column fabrication and packing methodology. Therefore, our discussion here will focus on the most salient aspects.

3.2.1 Packing materials

3.2.1.1 Silica-based materials

One important characteristic of a packing material for CEC is that it must be charged to support EOF for the bulk transport of the mobile phase and at the same time must possess retentive properties. If EOF is not present, the species reaching the detector will only be those that

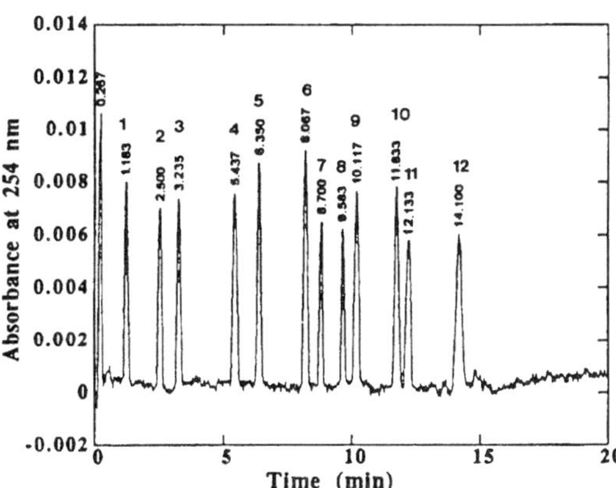

Figure 18. CEC separation of mono-, di-, and triphosphate nucleosides in a column packed with octadecyl sulfonated silica material. Reprinted from [124], with permission.

are charged. For a list on the many HPLC materials employed in CEC the reader is referred to the reviews by Robson *et al.* [109] and by Dermaux and Sandra [110], where the application of each particular material is also stated. Zimina *et al.* [111] also studied the electrochromatographic behavior of several HPLC packing materials. The five most used materials are for reversed-phase chromatography and are listed in Table 2. Materials modified for chiral recognition also continue to be tested for CEC [5, 16, 109, 110, 112–115, 140] and have been reviewed recently [16]. Other materials have also been employed, including fluoropolymers [141], and cellulose-based beads [142, 143].

CEC Hypersil and ODS-1 type have been popular among the reversed-phase materials since they seem to offer the fastest EOF. These materials are HPLC supports that have not been end-cappeda and, therefore, a relatively large amount of silanol groups is left on the surface, which can generate EOF. As the amount of alkyl substitution at the packing surface is increased, the EOF decreases, which results from a concomitant decrease of silanols groups responsible for the generation of the EOF [44, 54]. EOF is mostly generated at the particle surface, since most of the area is provided by the packing material, with negligible contribution from the capillary walls [6, 54, 111, 116]. It has also been shown that EOF is independent of the particle size [44, 62] down to 0.2 µm particles, as predicted by Eq. (1). Porous and nonporous silica materials have been used in CEC. Very high separation efficiencies (>500 000 plates/m) have been generated with columns packed with 1.5 µm nonporous ODS packing materials [58, 117–120]. The use of SDS (below its critical concentration) in the mobile phase has been recommended to prevent bubble formation and stabilize the EOF through dynamic modification of the alkylated surface. Figure 15 illustrates the separation of 14 explosive compounds using such materials [118]. So far, the reports on the use of sub-µm particles [44, 46] do not seem to offer a tremendous improvement over the efficiencies just mentioned, indicating that diffusion may be becoming the predominant source of dispersion. Nevertheless, the small particles do provide for rapid separations. For example, Fig. 16 shows the separation of four pyrene derivatives in less than 90 s using a capillary column packed (12.7 cm) with 450 nm C-8 reversed-phase material.

3.2.1.2 Mixed-mode materials

One major disadvantage of the traditional silica-based materials is the EOF dependence on the pH of the mobile phase. To avoid such dependence, ionic exchangers (*i.e.*, SCX, SAX) are used in CEC as a way to increase the surface charge; they have been used to separate charged organic and inorganic solutes [26, 69, 121–123]. Materials with cationic [26, 29, 54, 63, 69, 124–129] and anionic [100, 130, 131] exchange properties have been under recent consideration in CEC because of their so-called mixed-mode phase character. In principle, they should also provide for a strong, constant, and stable EOF through a wide pH range. The SCX-type of materials used contains sulfonic groups, providing a negative surface charge. A recent study [26] on the behavior of SCX materials in CEC did not find any conclusive evidence to support the popular belief of having a greater EOF when using SCX materials. However, detailed studies are still required to completely elucidate the behavior of such materials in CEC. To allow separation in the reversed-phase mode, packing materials incorporating hydrophobic moieties have been used [29, 44, 54, 63, 124–126, 128]. El Rassi's group [29, 124, 125], for example, synthesized a silica-based stationary phase containing two layers. A sublayer attached to the silica support contained the sulfonic acid groups, while a toplayer contained octadecyl groups chemically attached to the sublayer, named octadecyl-sulfonated silica by the authors (see Fig. 17). A separation of mono-, di-, and triphosphate nucleosides using this type of phase is shown in Fig. 18; the selectivity was significantly different than that of just ODS.

A different approach of using SCX columns for CEC has been taken, in which a column packed with an SCX packing material was dynamically modified with a long-chain quaternary ammonium salt; CTAB was added to the mobile phase [144]. The CTAB ions were adsorbed onto the surface of the SCX packing material, creating a hydrophobic layer useful as a stationary phase for CEC. Reten-

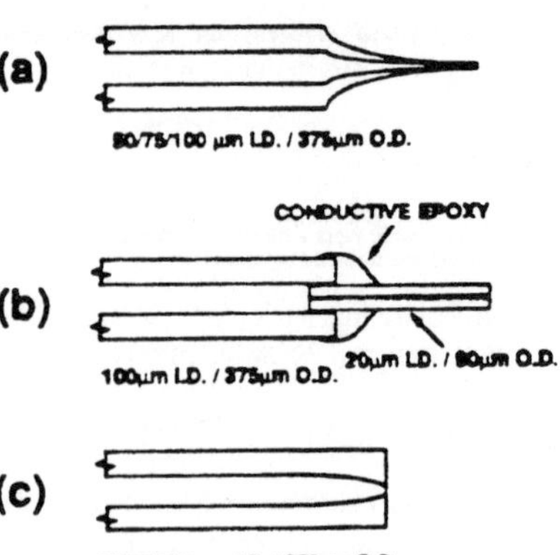

Figure 19. Schematic representation of (a) external taper for column inlet/outlet, (b) restrictor capillary for column inlet/outlet, and (c) internal taper for column outlet. Reprinted from [155], with permission.

tion factors showed reproducibility below 2%. The adsorbed CTAB on SCX material showed excellent stability at high and low pHs as well as the ability to separate a sample mixture containing acidic, neutral, and basic solutes at low pH; tailing was observed for basic compounds [144].

The SAX materials have not been as popular as the SCX ones. These types of materials provide for a positively charged surface, most typically through quaternary amine groups. Columns packed with SAX materials exhibit an anodic EOF. It is worth mentioning the use of SAX packing material in the separation of peptides. Lubman and co-workers [130], for example, used a commercially available mixed-mode stationary phase combined with voltage tuning for the separation of tryptic digests of horse heart myoglobin by pressure-assited CEC with ion trap storage/ reflectron TOF-MS detection. They observed a separation efficiency similar to that of a reversed-phase column, with different elution order of peptides. The mixed-mode phase incorporated C18 and secondary amine groups. High efficiency and reproducible separations of peptides were also reported by Ye *et al.* [131]. They attribute the high column efficiency to capillary electrophoretic stacking and chromatographic focusing phenomena during the injection and separation of the positively charged peptides. These positively charged mixed mode phases are attractive for the separation of proteins since they can operate at very low pH. At low pH, the surface of the material has a high charge density, allowing for the generation of a larger EOF; this positive charge also will aid in the repulsion of positively charged proteins. It has been observed that the sample buffer concentration has a marked effect on the retention factor in the separation of small peptides; as the concentration of the electrolyte was increased, the retention factor decreased [131].

3.2.2 Frits and bubble formation

Despite the numerous advantages associated with packed columns, fabrication of retaining frits continues to be a problem and a critical parameter in column performance [145, 146]. The most frequent method to fabricate retaining frits consists of sintering a portion of the packing material in place by heating. Several disadvantages of frit fabrication *via* this method have been cited [147]. The reproducibility of the frit is difficult. The column becomes fragile at the location of the frit by removal of the polyimide coating. There is band broadening and bubble formation associated with frits, and little or no control of frit porosity. There are indications that bubble formation is not primarily created by self-heating effects [23, 148]. Discontinuities of flow velocities at the interface of the packed and unpacked segment of a typical CEC column lead to bubble formation; a flow-equalizing intersegmental pressure develops at the open and packed segments, as shown by Rathore and Horváth [23]. Furthermore, sintering of the frit can also alter characteristics of the packing material at the frit position, creating nonhomogeneous packing at the frit and resulting in different electrical resistivities when compared to the open and packed segments of the columns combined with different zeta potentials. This can contribute to nonuniformities in EOF and lead to bubble formation at the boundary between the frit and the unpacked segment of the capillary [23, 148]. Variations in EOF can also depend on the functionally of the material being sitered as the frit and the time of heating [149].

The formation of bubbles in CEC has also been correlated to frit fabrication [150]; as the length of the frit is increased, bubble formation at a given potential also increases. Resilanizing of the packing material surface can reduce the likelihood of bubble formation [150]. It has been shown through fluorescence imaging experiments that deactivation of the frits, by means of silanization, also reduces undesirable adsorption of analytes at the frit [151]. Bubble formation continues to be minimized by using columns with high permeable frits, well degassed solvents, working at reduce temperatures (*e.g.*, 15°C) and applying pressure at both ends of the capillary [1]. Pressurization typically used in CEC (100 psi) will cause little, if any, contribution to the entire chromatographic experiment if the particle size of the packing material is very small and/or has long capillary columns.

Despite the difficulties associated with sintered frits, sintering of packing material continues to be the most popular approach to frit fabrication. Nevertheless, alternate protocols have been pursued. One alternative is to use

UV photopolymerization of a glycidyl methacrylate and trimethylpropane trimethacrylate solution (UV radiation, 365 nm for 1 h) [147]. This process has been shown to be reproducible; in addition, there is no weakening of the column at the frit since the polyimide coating is not removed, nor is the stationary phase altered since no heating is involved in the process. Additional advantages of photopolymierzed frits are that the frits can be placed at any point within a given column, and that the porosity of the frit can be controlled through a porogen used in the polymerization process. Cassidy and co-workers [152] recently optimized a silicate polymerization method similar to that of Cortes *et al.* [153], in which a silicate solution is used to make a silica-based porous frit. The use of optimum concentrations for the silicate solutions and short heating times (< 5 s) resulted in short-length (< 75 μm) frits, showing stability over a wide range of acetonitrile/water mixtures, with no bubble formation. Using a 20% solution of polydimethoxysilane (PDMOS) in dry methanol (v/v), Schmid *et al.* [154], produce retaining frits. Completely packed columns having PDMOS frits were connected to an unpacked open capillary column by a piece of PTFE tubing, allowing for on-column spectroscopic detection. The columns showed similar performance to those of sintered frit fabrication protocols under various acetonitrile/ Tris concentrations.

External and internal tapers and restrictors have been fabricated on the fused-silica columns as an alternative to frit formation, illustrated in Fig. 19 [155, 156]. These approaches are shown to be useful in interfacing CEC with MS. The external tapers are fabricated using a micropipette pulling device and are inherently fragile since they are weak points on the column, prone to breakage. The internal tapers, on the other hand, are fabricated by sealing the end of a silica column with a high temperature flame, followed by careful grinding of the end of the capillary to generate a small hole suitable for flow [156]. These are more robust than the external tapers; in addition, they facilitate the packing procedure by eliminating the need to have a temporary retaining frit. A restrictor is a capillary with a small OD that is inserted inside the capillary column, which will contain the packing material and glue in place, as illustrated in Fig. 19. The implementation of these restrictor capillaries can be time-consuming, particularly when trying to insert the small capillary inside the one that will be packed, thus calling for a very skillful operator.

3.2.3 Column packing

Column packing has not changed much since our previous review [1]. The most popular method to pack capillaries continues to be pressure packing. Electrokinetic packing is mostly practiced by commercial sources and packing by centripetal forces has only been performed in our laboratory [138, 157]. Other methods include packing capillaries using CO_2 [158, 159] and gravity [160]. All the methods to pack capillary columns for CEC have been reviewed very recently [108]; sufficient details are provided to guide those who are not experienced in the field. A preliminary comparison of all methods suggests that slurry pressure packing renders column with the lowest efficiency and electrokinetic packing is the simplest and easiest method to implement [108]; nevertheless, a more complete study is required. In the next few paragraphs, we briefly mention those methods not covered in our previous review as well as advances since then [1].

3.2.3.1 Supercritical CO_2 packing

Supercritical CO_2 packing has been utilized to pack columns for CEC [158, 159]. In this approach, a capillary column is attached to a pressure reservoir (typically a 2 mm ID HPLC column) that contains the dry packing material and the other end of the column is connected to a union containing a metallic frit or the column has a temporary frit sintered in it. At the other end of the connecting union, a restrictor is attached to maintain supercritical conditions during the packing process. The packing is delivered by pressurizing with CO_2. The column is immersed in an ultrasonicated bath and the temperature of the bath is maintained above the critical temperature of CO_2 (approximately 50–60°C). The column is typically packed at pressures above the critical pressure of CO_2 (3000–4500 psi). However, it is beneficial to increase the pressure of the CO_2 after the packing is completed while sonicating to aid in compressing the packed bed. To avoid disturbing the packed bed and potentially creating voids, the column is usually allowed to depressurize for at least 6 h, and occasionally overnight. The column is then solvated with water and the retaining frits are constructed on column.

3.2.3.2 Pseudoelectropacking

Recently, Tjaden *et al.* [161] developed a pseudoelectrokinetic packing method to pack capillary columns for CEC. In their packing method, a high electric field is used in conjunction with a hydrodynamic flow to ensure robust, well-packed columns. The column, equipped with a temporary frit, is mounted in a commercially available CE instrument and flushed with a background electrolyte consisting of 75/25k methanol-water/10 mM Tris buffer. The entrance of the capillary column is placed in a slurry that is made of the background electrolyte. A portion of the slurry is pumped into the column. Once the column is filled, an electric field is applied to migrate the packing

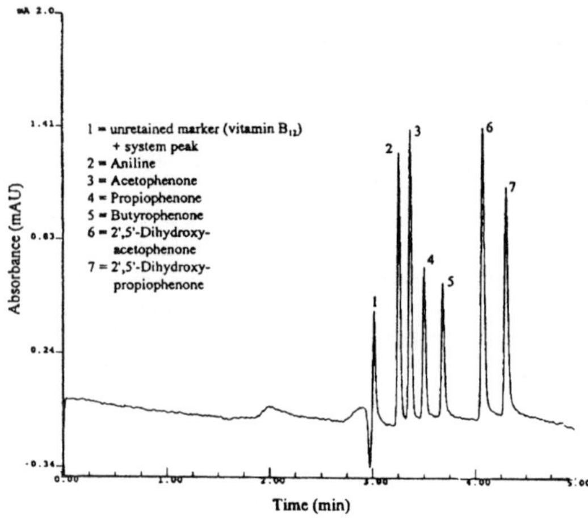

Figure 20. CEC separation of phenones in a macroporous polyacrylamide/poly(ethylene glycol) monolithic column under "isocratic" conditions. (Reprinted from [106], with permission.

material towards the temporary frit, creating a packed bed in approximately 15 min. The columns are then connected to a pump and flushed with water prior to frit fabrication. Columns packed with this method have demonstrated stability over hundreds of runs and no bubble formation was observed while running at high voltages without application of stabilizing pressure to the inlet/outlet vials. Columns were run with 80/20 acetonitrile/water (2 mM Tris) and efficiencies of ~280 000 plates/m were reported.

3.2.3.3 Packing by centripetal forces

Another alternative to pressure packing is to use centripetal forces to pack columns for CEC, achieved by centripetal acceleration of the particles through the capillary column [138, 157]. Particles are slurried in an appropriate slurry solvent (approximately 10–50 mg/mL) and then placed in the slurry reservoir of the packing apparatus. Upon rotation of the packing apparatus, the particles travel into the columns where a bed is formed as they arrive at the temporary frit within the column. Columns are typically packed in 5–15 min, depending on the viscosity of the solvent containing the packing material. For example, ~25 cm of a 50 μm ID capillary column is packed in 15 min at 2000 rpm using a 10 mg/mL slurry of 3 μm ODS packing material in isopropanol. Introducing a drying step during the column fabrication process increases the efficiency and retention characteristics of the columns (approximately 15–20% and ~13%, respectively) [157]. This drying step can also be implemented with any other packing method. The extra drying step will add time

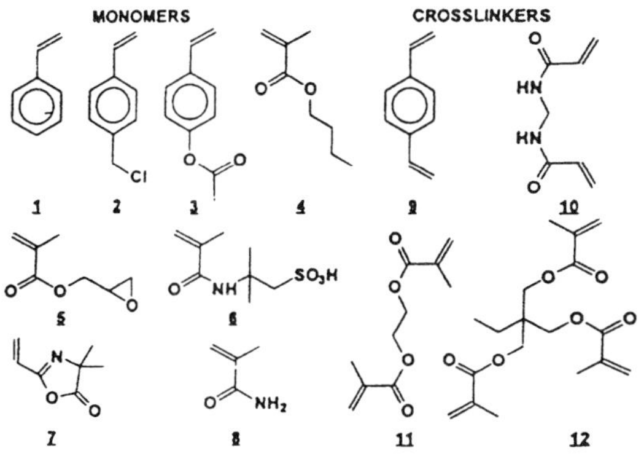

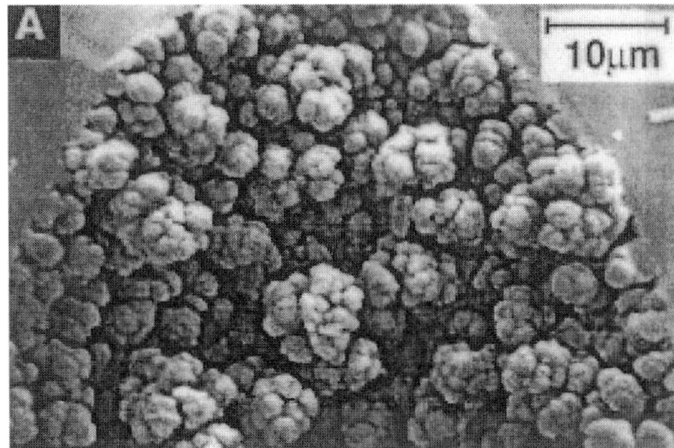

Figure 21. Example of monomers and cross-linking agents to form porous rigid organic monolithic structures. Reprinted from [175], with permission.

Figure 22. Scanning electron micrograph of a styrene-based monolithic structure in a capillary column. Reprinted from [162], with permission.

to the column fabrication process and the minimal increase in column performance may not justify the increase in column fabrication time. However, if a faster, more convenient dry-packing protocol could be developed, it could certainly be advantagenous in packing columns for CEC.

3.2.3.4 Packing by gravity

Packing material can also be delivered into capillary columns by means of gravity [45, 160]. The packing device used to demonstrate the concept was a 1 mL syringe that contains ~100 µL of a 10 mg/mL slurry of 3 µm ODS material in acetone. The syringe was connected to the open end of the capillary column by a piece of PTFE tubing attached to the syringe needle. The syringe plunger has been modified by connecting a screw cap to the end in order to prevent evaporation of the slurry. Capillary columns were filled with acetone and connected to the apparatus. Sedimentation of the packing material is permitted over a 12–24 h period, replacing the slurry every 4–6 h to ensure enough material for packing and to avoid clogging of the column entrance. Once packed, the columns are connected to a pump and flushed with water before fabrication of the retaining frits.

3.3 Monolithic structures

Capillary columns with continuous beds, or monoliths, have attracted considerable attention in CEC because they are alternatives to packed columns [28, 37, 46, 49, 86, 89–107]. The monolithic columns eliminate difficulties encountered with packing columns, particularly the fabrication of retaining frits. In one approach to monolithic columns, monomeric precursors are introduced into the cap-

illary and allowed to polymerize *in situ*, forming a rod-like structure. The monoliths are porous in nature and can be either rigid structures [28, 49, 90–100, 104, 107, 162–165] or soft gels [89, 95, 101, 105, 106, 166–172]. Organic and inorganic polymeric monoliths can be fabricated in this fashion. Luo and Andrade [20] have suggested that a continuous polymeric bed with sub-µm channels is perhaps the ideal packing structure for CEC because the geometric tortuosity is greatly reduced and EOF is more favorable. From the practical point of view, this approach should also simplify column technology and other potential problems in CEC (*e.g.*, bubble formation). An excellent review on the *in situ* polymerization of synthetic organic polymers as monolithic stationary phases for CEC was published recently [173]. Another approach is to immobilize conventional particulate material inside the column [28, 96–99, 159].

3.3.1 Soft gels

The use of acrylamide-based gels inside capillary columns as stationary phase for CEC was pioneered by Fujimoto and co-workers [95, 166, 168–170] and Hjertén *et al.* [89, 167, 171, 172]. The nature of EOF to drive the solution through the column without pressure allows the use of these soft materials in CEC while they are discouraged in HPLC. Typically, acrylamide-type gels are prepared by incorporating a functional monomer that would provide a characteristic surface charge, such as 2-acrylamido-2-propanesulfonic acid and vinylsulfonic acid. This permits the generation of EOF within the pores of the polymeric matrix. Hydrophobicity of these aqueous gels can be increased by incorporating butyl methacrylate or stearyl methacrylate into the polymerization, *via* emulsification [89, 172], or by derivatization with an organic ligand (*e.g.*, C18) after polymerization [167]. Separation efficiencies of

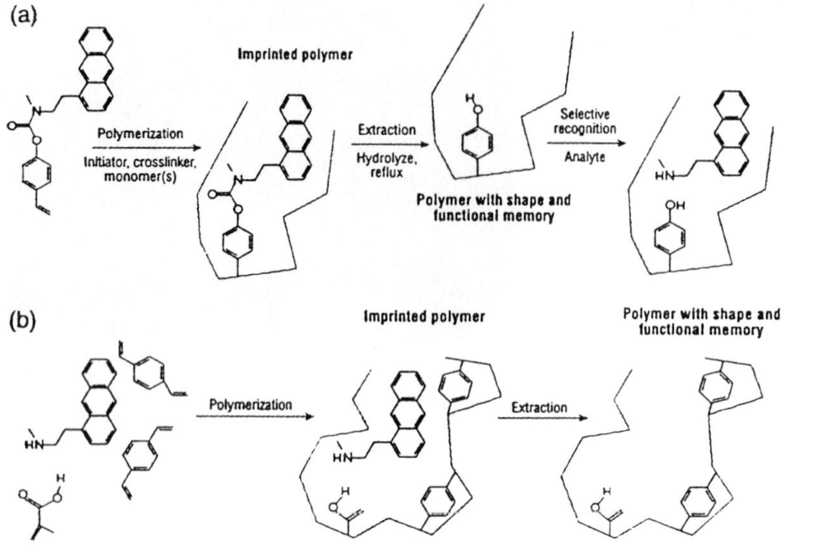

Figure 23. Schematic illustration of (a) covalent and (b) noncovalent imprinting procedures. Reprinted from [165], with permission.

160 000 plates/m for steroids [169] and 180 000 plates/m for naphthalene [89] have been reported. Perhaps the most impressive separations on acrylamide-based monolithic collumns were those reported by Novotny and Palm (300 000–398 000 plates/m) [106]. The monoliths were prepared *in situ* by the copolymerization of acrylamide and bisacrylamide as cross-linker. The polymerizing matrix also included an alkyl acrylate (butyl, hexyl, or lauryl) ligand, acrylic or vinylsulfonic acid for EOF generation, and poly(ethylene glycol), leading to a macroporous polymeric structure. The polymerization process was performed in an aqueous-organic mixture, which also facilitated the incorporation of hydrophobic ligands. An example of a separation of alkyl phenones is shown in Fig. 20. This type of column was also used for the separation of peptides and carbohydrates. Monolithic gel columns prepared with β-cyclodextrin-bonded positively charged polyacrylamide gels have also been shown to be effective for enantiomeric separations; their potential was demonstrated by separating dansyl-DL-amino acids [174].

3.3.2 Rigid organic polymers

Another type of polymeric monoliths is based on polystyrene or polymethacrylate polymers [90–92, 107, 162, 163], which are more rigid in nature. Many different chemistries can be incorporated within the polymeric structure to give a desired selectivity. Frechet and co-workers [90, 92, 107] have shown the feasibility of using polymethacrylate-based monolithic polymers inside capillaries for CEC. The monolith is fabricated by filling a desired portion of the capillary with a polymerizing mixture, which, through radi-

cal polymerization initiated by temperature or UV radiation, give rise to the porous polymer. The porous characteristics of the monolith are controlled by adjusting a ternary porogenic solvent system consisting of water, 1-propanol, and 1,4-butanediol [90]. Numerous monomers can be used in the preparation of the polymeric monolith [175]. Figure 21 shows a few examples of monomers and cross-linking agents.

EOF is generated through sulfonic acid functionalities. The monolithic columns are stable through a wide pH range (pH 2–12). Monolithic columns have been prepared with reversed-phase [91, 92, 164] and chiral [107] characteristics. Efficiencies as high as 210 000 plates/m have been found for monolithic columns with optimized porous properties [164]. Horváth and co-workers [162] prepared monolithic columns for CEC by copolimerizing chloromethylstyrene and divinylbenzene using a suitable porogen in silanized fused-silica capillaries (75 µm ID). The resulting monolith is cross-linked and highly porous. Figure 22 shows a scanning electron micrograph of a styrene-based monolithic structure in a capillary column. It can be seen that the polymeric material exists as tiny clumps that appear to be fused at some locations, creating a rigid, rod-like monolithic support. The porous monolithic support was used either directly or after functinalization of the surface. A positive charge was imparted to the surface by reacting the chloromethyl moieties with *N,N*-dimethyloctylamine; hence the EOF direction was from cathode to anode. The octyl group provided the stationary phase for the separation of peptides by CEC. The columns were used to separate angiotensin-type peptides with routine plate heights of 8 μm. Xiong *et al.* [163] used poly(styrene-co-divinylbenzene-co-methacrylic acid) to fabricate a rigid monolithic structure with similar retention mechanism as reversed-phase packing materials [163].

3.3.3 Molecular imprint polymers (MIPs)

Polymers with molecular imprints have been used as stationary phases for molecular recognition [165]. In the synthesis of polymers with molecular recognition, assemble of monomeric units (*e.g.*, styrene and methacrylic acid) and the desired template molecule take place through molecular interactions. The three-dimensional arrangement is fixed during *in situ* polymerization, producing an imprint of the template molecule. The imprint can be obtained through covalent or noncovalent interactions with the template molecule that is removed after polymerization. The final polymeric structure is rigid and porous in nature. The process is shown schematically in Fig. 23. The performance of molecular-imprinted monolithic polymers, as stationary phase, seem to be more attractive in CEC than in HPLC, where excessive tailing has predominated [104]. Most applications in CEC have focused on chiral separations, because of the unique feature afforded by molecular recognition that allows the preparation of stationary phases with a predetermined selectivity [101, 102, 104, 176]. Figure 24 shows an example for the separation of propranolol enantiomers in a monolithic column where the imprints were fabricated using (*R*)-propranolol as the template molecule. The enantiomer with the strongest interaction with the stationary phase is that recognized by the imprint. In general, the peaks for the components separated in MIP columns are broad, with relatively low efficiencies. MIPs have also been used in OT-CEC, affording higher efficiency than the monoliths [176]. Nevertheless, this type of phases is relatively new and still under development; future work should improve peak shape and efficiency. The subject of molecular imprints as a stationary phase for CEC has been reviewed recently [104, 165].

3.3.4 Silica-based monoliths

Porous silica monolithic structures have been prepared inside capillary columns, mostly by the sol-gel process. In this approach, monomeric precursors are mixed under ambient conditions and introduced into the capillary column. After hydrolysis and polycondensation reactions, a three-dimensional network is created that is highly porous. Tanaka's group [93, 94] has fabricated and evaluated monolithic silica columns fabricated by the sol-gel method in pressure-driven and electrically driven separations. The macroporous silica gel network was prepared by *in situ* hydrolysis and polymerization of tetramethoxysi-

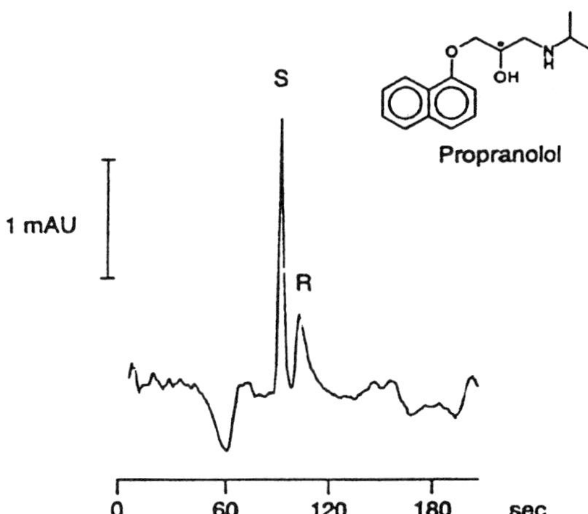

Figure 24. CEC separation of propranolol enantiomer in monolithic column containing imprints of (*R*)-propranolol imprints. Reprinted from [101], with permission.

lane and poly(ethylene oxide) inside a 100 μm ID capillary column, allowed to react overnight. The monolith was then washed with water before treating with aqueous ammonium hydroxide followed by washing with ethanol. Columns were dried in an oven for 24 h at 330°C and then reacted with octadecyldimethyl-*N,N*-diethylaminosilane in toluene for 2 h. The morphology of such a monolith is demonstrated in Fig. 25. A series of small silica rods, with aggregate-like clumps of silica attached, can be seen. The pores within the monolith varied but were as large as 8 μm. Columns were evaluated in pressure-driven and electrically driven modes and exhibited separation performances of 48 000 plates/m and 128 000 plates/m, respectively. Separations of probe compounds are shown in Fig. 26. The rather low plate count for the pressure-driven separations was attributed to the *A*-term contributions of the van Deemter equation; thereby, the monolithic system seems more appropriate for CEC than for HPLC. Honda *et al.* [100] and Fujimoto [95, 170] have also followed a similar procedure to fabricate monolithic columns for CEC. The sol-gel-based monolithic columns showed higher permeability than columns packed with particles of about 3 μm.

3.3.5 Particle entrapment

Another alternative to fabricate monolithic columns is to immobilize conventional packing materials inside the capillary column. In this approach, chromatographic particulate is delivered into the column. The particles, however, remain in place by entrapment, instead of using retaining frits. Since the entire bed is fixed, the problem of emptying

Electrophoresis 2000, *21*, 3965–3993

Figure 25. Scanning electron micrograph of a continuous monolithic silica bed inside a 100 μm ID capillary. Reprinted from [93], with permission.

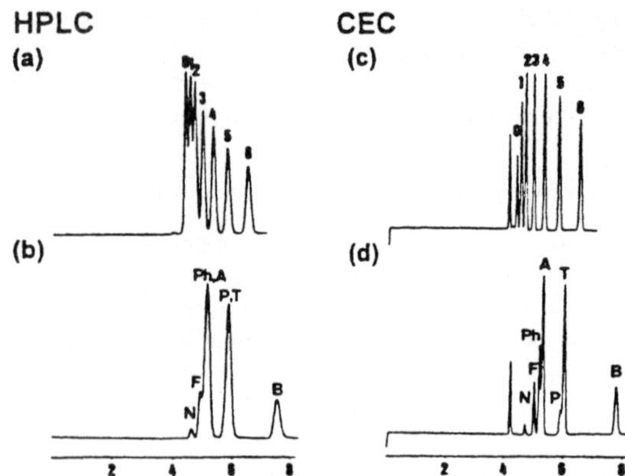

Figure 26. Separation of (a, c) alkylbenzenes and (b, d) PAHs by HPLC (a, b: 90% acetonitrile / 10% water) and CEC (c, d: 90% acetonitrile / 10% 50 mM Tris-HCl, pH 8) in a monolithic silica column. Reprinted from [93], with permission.

the capillary due to a loss of frit, at either the inlet or the outlet end, is avoided. In addition, the potential of gap formation within the packed bed is prevented. All of the above contribute to improved column longevity. Immobilization of the packing material is accomplished by forming a silica-based matrix (*e.g.*, sol-gel processing) [28, 49, 97–99] or sintering the particles by heat [96]. In the sintering method [96], a capillary tube is packed with silica particles containing the stationary phase (*e.g.*, C_{18}). The column is submitted to washings (with 0.1 M $NaHCO_3$ and acetone), drying (with nitrogen gas), and a heating treatment (two steps, 120°C and 360°C) that allows sintering of the particles. The stationary phase may be damaged during the process; therefore, resilanization is required. Monolithic columns fabricated by this method have been reported to be very stable, producing reproducible separations, with theoretical plates of about 125 000 plates/m [96]. The multistep procedure used to fabricate the column is a drawback of this approach because the process can be time-consuming, including the reattachment of the stationary phase.

The packing material can also be entrapped using silicate solutions or sol-gel technology [28, 49, 97–99, 159]. Sol-gel processing was first used to fabricate stationary phases for OT-CEC [177]. Two different procedures have been utilized to entrap the chromatographic particles *via* sol-gel processing. In one procedure, a sol-gel solution (typically a mixture containing alkoxysilanes, ethanol, and hydrochloric acid) is prepared and the particles containing the stationary phase are added, forming a suspension containing the particles [28, 97]. A capillary tube is then filled with the suspension. The packing material is embedded in the sol-gel matrix after drying, also called particle-

loaded monolith. The reported separation efficiencies of columns prepared by this approach are not as high as those of traditionally packed columns. Furthermore, the reproducibility of column preparation and homogeneity of the packing material inside the column seem to be a problem.

The second approach to entrap the chromatographic particles is by introducing the entrapping solution after the column has been packed [49, 98, 99, 159]; the column is then dried, after which the temporary retaining frit (originally used to pack the column) is eliminated and the column cut to the desired length. The method by Remcho and Chirica [98] makes use of silicate sol solutions to fill the pressure-packed capillaries with subsequent heating. The columns are then cured with 0.1 M ammonium hydroxide and dried at 160°C to produce monolithic structures with reported efficiencies similar to those of packed columns, although with reduced retentive characteristics. Lee's group [49, 99, 159], on the other hand, first packed the capillary using a CO_2 slurry packing method and then filled the capillary with a sol solution prepared with tetramethoxysilane and ethyltrimethylsilane. The column filled with the sol solution was dried using supercritical CO_2 for 5 h at 40°C followed by further drying at 120°C and 250°C for the same amount of time. The particles seem to be glued together in the column, as shown in Fig. 27. Efficiencies of 220 000 plates/m and 175 000 plates/m were reported for columns containing sol-gel-bonded 7 μm, 4000 Å ODS and 3 μm, 80 Å ODS, respectively. The immobilizing matrix used in the methods by Remcho and Chirica [98] and Lee *et al.* [49, 99, 159] is less than that of the particle loading approach and since the capillaries are

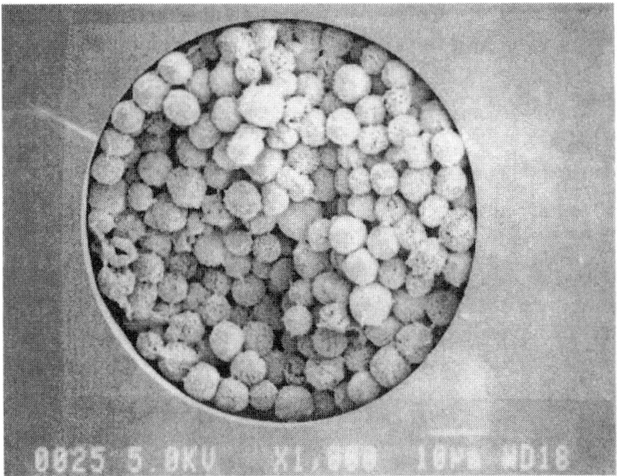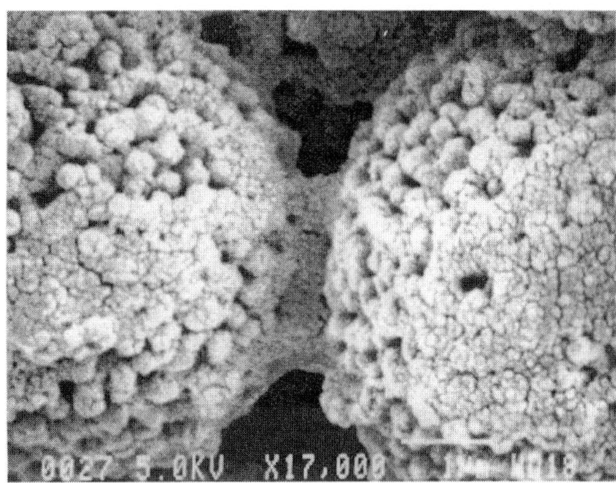

Figure 27. Scanning electron micrographs of sol-gel bonded particles (7 μm, 4000 Å pores). Reprinted from [99], with permission.

packed before entrapment, reproducibility and homogeneity of the packing material seem to be higher. The efficiencies of columns prepared by this approach have also been shown to be higher than those prepared by the particle loading approach.

4 Instrumentation

For the most part, CEC is carried out in instrumentation designed for CE, either commercially available or built in the laboratory. This instrumentation may not be completely adequate for CEC. For example, in most cases, commercially available CE units do not provide for equilibration of a CEC column under pressure-induced flow, although some manufacturers have included a feature with low pressure, which has been used to minimize bubble formation. The generation of gradients in the mobile phase has been accomplished by modification of standard HPLC gradient systems. Reports have appeared, however, that explore the use of integrated instrumental platforms that permit the use of three modes of operation – capillary HPLC, CE, and CEC [178, 179]. Such instrumentation should facilitate method development, making it possible to test different techniques for the separation of complex samples [178]. A few manufacturers are attempting to commercialize a unit with the three modes of operation.

4.1 Gradient elution

The high efficiency afforded by CEC offers the opportunity of increasing peak capacity in LC. The use of gradient elution in CEC can improve peak capacity even further, increasing the potential for analyzing very complex samples. Gradients in the composition of the mobile phase are the most common in CEC, although temperature [51] and voltage [180] gradients are also possible in CEC. The change in composition of the mobile phase in CEC continues to be mostly realized by a gradient delivery HPLC pump. Most applications make use of such an approach [181–184]. The use of gradient steps is also employed and convenient when using a commercial CE unit: here the mobile phase composition is varied by first stopping the applied voltage in order to change the inlet vial mobile phase and then the voltage is applied again [185, 186]. One problem with solvent gradient in CEC is that changes in flow velocity can be expected as the concentration of organic modifier is varied [6, 54, 116]. Most advances in gradient elution CEC in the past few years have focused on constructing a more convenient system and demonstrating their feasibility [182, 187–190]. From these, the system used by Dorsey's group [190] is quite simple and relatively easy to implement, as shown in Fig. 28; although the interface was made of Kel-F polymer, a commercially available plastic "tee" connector may be adapted for the same purposes. One particular development in solvent gradient for CEC was the implementation of an automated gradient system in microchip EC. Ramsey and co-workers [132] demonstrated solvent programming for open-channel EC, obtaining a separation of model compounds in less than 20 s. They used electroosmosis to drive and mix the solvents, generating the gradient similarly to the approach of Zare and co-workers [191] using capillary columns.

In the approach to gradient elution in CEC for capillary columns developed by Novotny and co-workers [192], the gradient of mobile phase is not generated by a pump, but rather by a series of coupled pieces of silica tubing. The

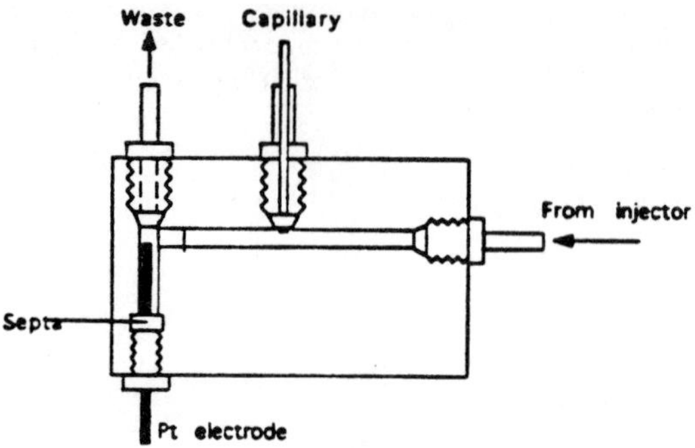

Figure 28. Diagram of an interface for connecting CEC column to a solvent gradient. Reprinted from [90], with permission.

gradient generator was constructed by inserting a piece of 100 μm fused-silica capillary column into the modified injection end of a 400 μm glass capillary (see Fig. 29). The separation column was connected to the other end of the glass capillary. The total volume of the gradient generator is 6.28 μL. The ratio of the ID of the glass capillary to that of the fused-silica capillary is 4:1, creating a ratio of 16:1 with respect to their cross-sectional areas. Therefore, the linear velocity of the mobile phase is 16 times faster in the fused-silica capillary than in that of the glass capillary, while the volumetric flow rates are kept constant. It is the difference in linear velocity that is responsible for the mixing of the two mobile phases and hence, the generation of the gradient. The authors demonstrated the ability to change the shape, curvature, and slope of the gradient by manually adding various amounts of strong mobile phase (high organic content) or by modifying the amount of the organic component present in both the weak and strong mobile phases. Optimized conditions for the gradient system were obtained while the sample was injected into the column that had been equilibrated with the weak mobile phase, followed by injecting 3.5 μL of the strong mobile phase into the gradient generator. These conditions were used to separate a group of alkylphenones, with resolution of all compounds in < 20 min. Reproducibility of the gradient was RSD < 3.5%. A key advantage of this gradient system to currently used CEC gradient interfaces is that no mechanical pumps are used to generate the gradient.

4.2 Detection

The detection schemes readily available in CE have been implemented with CEC, with UV detection being the most common. The detection cell must remain smaller than the peak standard deviation to avoid peak distortion and band broadening. It has been reported that using a cell with increased path length could increase detectability without adversely affecting peak efficiency [6]. A capillary column with 100 μm ID, for example, could use a cell with a volume of about 12 nL, which is in the same order of magnitude of the peak standard deviation [6].

Perhaps the most important advances in detection schemes for CE, hence CEC, are MS and nuclear magnetic resonance (NMR) spectroscopy. These detection schemes offer unique possibilities because of the additional structural information that can be obtained. On-line coupling of NMR with capillary separation techniques offers many possibilities [193] and was recently coupled to CEC [184, 194–197] as well as being used, for example, in metabolite analysis [195, 197] and analgesics [194]. The use of CEC-MS is becoming more popular mainly because of solvent compatibility, in addition to the volumetric flow rates that closely match the requirements of electrospray interfaces available. Banks [198] has reviewed advances in CEC-MS, while Lüdtke and Unger [199] reviewed the potential of CEC as a coupling technique, with MS indicating the main advantages of the technique. In a more specific article, Lubman and co-workers [200] reviewed CEC interfaced to ion-trap storage/reflectron TOF-MS. Many reports on CEC-MS have focused on the characterization of the coupling between the two techniques [155, 201–206]. CEC-MS has been used recently for several applications including analysis of samples containing, for example, peptides and amino acids [130, 156, 183, 204], preservatives is pharmaceutical creams [127], metal ions [207], triglycerides and fatty acids [204, 208], DNA [209–211], tags from combinatorial organic synthesis [212], drug candidates [213], and fatty acids and vitamins [204], among others.

5 Conclusions

Column technology has been the subject most studied in CEC in the past few years, leading to several column architectures for CEC, and there is no indication that this will stop soon. Not much instrumental development has been reported recently. An important issue is the fabrication of separation media specifically for CEC, containing a charge for the proper generation of EOF and groups for the desired chromatographic interactions. Mixed-mode type of stationary phases seem like a viable alternative. These phases also offer the possibility of multiple separation mechanisms; however, more work is needed to fully understand their behavior in CEC. The frit fabrication in packed columns continues to be challenging, requiring skilled personnel. The fritless, monolithic columns are an alternative to the problems associated with frit formation. A standard protocol to fully evaluate and compare col-

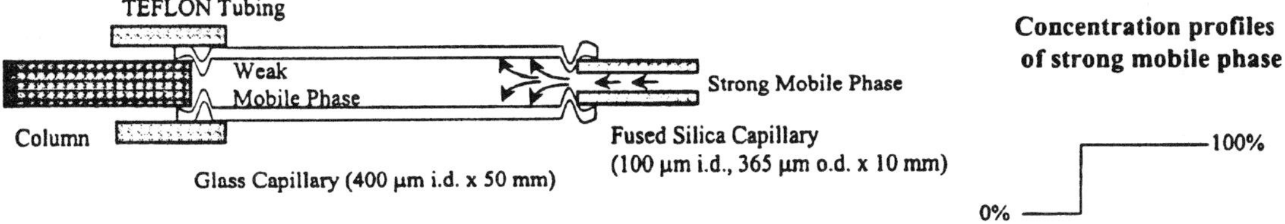

(a) Introduction of strong mobile phase into gradient generator

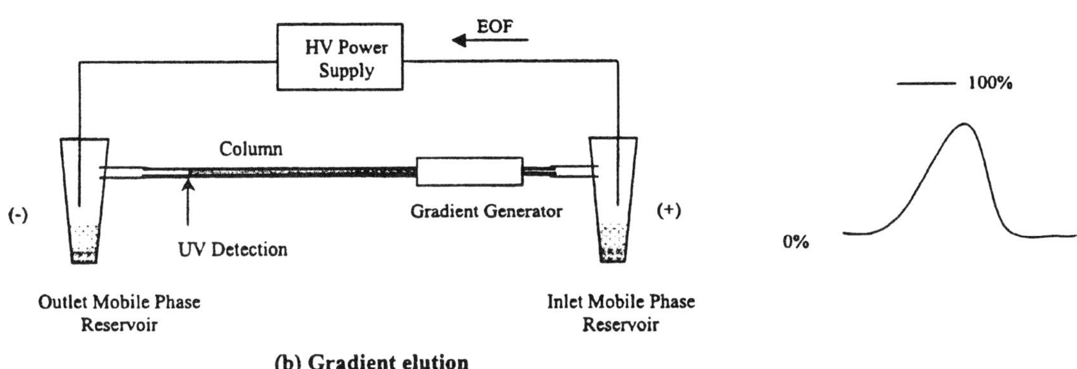

(b) Gradient elution

Figure 29. Diagram of the gradient CEC system. (a) Introduction of a strong mobile phase into the gradient generator; (b) gradient elution. Reprinted from [192], with permission. Copyright John Wiley & Sons 2000.

umns may be of benefit to compare columns fabricated by different laboratories. An area that is expected to grow is that of CEC in microfabricated devices [132–137], offering the possibility of having very well controlled microfabricated channels to perform separations, recently reviewed by Regnier [136]. Fundamental considerations on the generation of EOF and on the separation mechanisms are leading to a better understanding of CEC; however, more research is still required.

The field of CEC has undoubtedly grown considerably in the past few years, with enthusiasm generated in the pharmaceutical industry and the academic world, and research will continue in the foreseeable future. At this time, the technique is not used routinely for chemical analysis and it is difficult to predict if it will ever do so, for the incremental improvement in efficiency and speed. HPLC is a well-established technique that keeps improving; CEC needs more time to mature, with unique applications that will stimulate its use to become routine. At present, CEC is still in the research laboratory and continues to fascinate academicians. Will it stay that way? Time will answer that question.

Financial support by the National Science Foundation under grant CHE-9614947 is acknowledged.

Received June 19, 2000

6 References

[1] Colón, L. A., Reynolds, K. J., Alicea-Maldonado, R., Fermier, A. M., *Electrophoresis* 1997, *18*, 2162–2174.

[2] Colón, L. A., Guo, Y., Fermier, A., *Anal. Chem.* 1997, *69*, 461A–467A.

[3] Altria, K. D., Smith, N. W., Turnbull, C. H., *Chromatographia* 1997, *46*, 664–674.

[4] Dadoo, R., Yan, C., Zare, R. N., Anex, D. S., Rakestraw, D. J., Hux, G. A., *LC·GC* 1997, *15*, 630–632.

[5] Cikalo, M. G., Bartle, K. D., Robson, M. M., Myers, P., Euerby, M. R., *Analyst* 1998, *123*, 87R–102R.

[6] Dittmann, M. M., Rozing, G. P., Ross, G., Adam, T., Unger, K. K., *J. Capil. Electrophor.* 1997, *4*, 201–212.

[7] Vallano, P. T., Remcho, V. T., *J. AOAC Int.* 1999, *82*, 1604–1612.

[8] Krull, I. S., Stevenson, R. L., Mistry, K., Swartz, M. E., *Capillary Electrochromatography and Pressurized Flow Capillary Electrochromatography*, HNB Publishing, New York 2000.

[9] Svec, F., Deyl, Z., *Capillary Electrochromatography*, Elsevier, Amsterdam, 2000.

[10] Quirino, J. P., Terabe, S., *J. Chromatogr. A* 1999, *856*, 465–482.

[11] Palmer, C. P., *J. Chromatogr. A* 1997, *780*, 75–92.

[12] Eimer, T., Unger, K. K., Van Der Greef, J., *TrAC* 1996, *15*, 463–468.

[13] Karcher, A., El Rassi, Z., *Electrophoresis* 1999, *20*, 3280–3296.

[14] Sovocool, G. W., Brumley, W. C., Donnelly, J. R., *Electrophoresis* 1999, *20*, 3297–3310.

[15] El Rassi, Z., *Electrophoresis* 1999, *20*, 3134–3144.

[16] Schurig, V., Wistuba, D., *Electrophoresis* 1999, *20*, 2313–2328.

[17] Smyth, W. F., McClean, S., *Electrophoresis* 1998, *19*, 2870–2882.

[18] Rathore, A. S., Horváth, C., *J. Chromatogr. A* 1997, *781*, 185–195.

[19] Choudhary, G., Horváth, C., *J. Chromatogr. A* 1997, *781*, 161–183.

[20] Luo, Q.-L., Andrade, J. D., *J. Microcol. Sep.* 1999, *11*, 682–687.

[21] Wan, Q.-H., *J. Phys. Chem. B* 1997, *101*, 8449–8453.

[22] Wan, Q.-H., *J. Phys. Chem. B* 1997, *101*, 4860–4862.

[23] Rathore, A. S., Horváth, C., *Anal. Chem.* 1998, *70*, 3069–3077.

[24] Rathore, A. S., Wen, E., Horváth, C., *Anal. Chem.* 1999, *71*, 2633–2641.

[25] Cikalo, M. G., Bartle, K. D., Myers, P., *J. Chromatogr. A* 1999, *836*, 25–34.

[26] Cikalo, M. G., Bartle, K. D., Myers, P., *Anal. Chem.* 1999, *71*, 1820–1825.

[27] Rathore, A. S., Horváth, C., *Anal. Chem.* 1998, *70*, 3271–3274.

[28] Ratnayake, C. K., Oh, C. S., Henry, M. P., *J. High Resolut. Chromatogr.* 2000, *23*, 81–88.

[29] Zhang, M., El Rassi, Z., *Electrophoresis* 1999, *20*, 31–36.

[30] Rice, C. L., Whitehead, R., *J. Phys. Chem.* 1965, *69*, 4017–4024.

[31] Overbeek, J. T. G., Wijga, P. W. O., *Rec. Trav. Chim.* 1946, *65*, 556.

[32] Overbeek, J. T. G., in: Kruyt, H. R. (Ed.), *Colloid Science*, Elsevier, Amsterdam 1952, Chapter V.

[33] Overbeek, J. T. G., in: Kruyt, H. R. (Ed.), *Colloid Science*, Elsevier, Amsterdam 1952; Chapter IV.

[34] Dukhin, S. S., *Adv. Colloid Interface Sci.* 1991, *35*, 173.

[35] Baran, A. A., Babich, Y. A., Tarovsky, A. A., Mischuk, N. A., *Colloids Surf.* 1992, *68*, 141.

[36] Dukhin, S. S., Mischuk, N. A., *J. Membr. Sci.* 1993, *79*, 199.

[37] Mischuk, N. A., Takhistov, P. V., *Colloids Surf. A: Physicochem. Eng. Asp.* 1995, *95*, 119–155.

[38] Mischuk, N. A., Takhistov, P. V., *Colloid J. Russ. Acad. Sci.* 1993, *55*, 244.

[39] Wan, Q.-H., *Anal. Chem.* 1997, *69*, 361–363.

[40] Knox, J. H., Grant, I. H., *Chromatographia* 1991, *32*, 317–328.

[41] Stevens, T. S., Cortes, H. J., *Anal. Chem.* 1983, *55*, 1365–1370.

[42] Knox, J. H., Grant, I. H., *Chromatographia* 1987, *24*, 135–143.

[43] Giddings, J. C., *Dynamics of Chromatography*, Marcel Dekker, New York 1965, Chapter 5.

[44] Adam, T., Lüdtke, S., Unger, K. K., *Chromatographia* 1999, *49*, S49–S55.

[45] Reynolds, K. J., Colón, L. A., *J. Liq. Chromatogr. Relat. Technol.* 2000, *23*, 161–173.

[46] Lüdtke, S., Adam, T., Unger, K. K., *J. Chromatogr. A* 1997, *786*, 229–235.

[47] Stol, R., Kok, W. T., Poppe, H., *J. Chromatogr. A* 1999, *853*, 45–54.

[48] Li, D., Remcho, V. T., *J. Microcol. Sep.* 1997, *9*, 389–397.

[49] Tang, Q., Lee, M. L., *J. High Resolut. Chromatogr.* 2000, *23*, 73–80.

[50] Yang, C., El Rassi, Z. E., *Electrophoresis* 1999, *20*, 18–23.

[51] Wen, E., Asiaie, R., Horváth, C., *J. Chromatogr. A* 1999, *855*, 349–366.

[52] Cahours, X., Morin, P., Dreux, M., *J. Chromatogr. A* 1999, *845*, 203–216.

[53] Euerby, M. R., Johnson, C. M., Smyth, S. F., Gillott, N., Barrett, D. A., Shaw, P. N., *J. Microcol. Sep.* 1999, *11*, 305–311.

[54] Dittmann, M. M., Rozing, G. P., *J. Microcol. Sep.* 1997, *9*, 399–408.

[55] Euerby, M. R., Gilligan, D., Johnson, C. M., Roulin, S. C. P., Myers, P., Bartle, K. D., *J. Microcol. Sep.* 1997, *9*, 373–387.

[56] Wan, Q.-H., *J. Chromatogr. A* 1997, *782*, 181–189.

[57] Wright, P. B., Lister, A. S., Dorsey, J. G., *Anal. Chem.* 1997, *69*, 3251–3259.

[58] Seifar, R. M., Kok, W. T., Kraak, J. C., Poppe, H., *Chromatographia* 1997, *46*, 131–136.

[59] Moffatt, F., Cooper, P. A., Jessop, K. M., *Anal. Chem.* 1999, *71*, 1119–1124.

[60] Crego, A. L., Martinez, J., Marina, M. L., *J. Chromatogr. A* 2000, *869*, 329–337.

[61] Banholczer, A., Pyell, U., *J. Chromatogr. A* 2000, *869*, 363–374.

[62] Cikalo, M. G., Bartle, K. D., Myers, P., *J. Chromatogr. A* 1999, *836*, 35–51.

[63] Smith, N., Evans, M. B., *J. Chromatogr. A* 1999, *832*, 41–54.

[64] Liu, Z., Zou, H., Ye, M., Ni, J., Zhang, Y., *Sepu* 1999, *17*, 245–248.

[65] Moffatt, F., Cooper, P. A., Jessop, K. M., *J. Chromatogr. A* 1999, *855*, 215–226.

[66] Zhang, Y., Shi, W., Zhang, L., Zou, H., *J. Chromatogr. A* 1998, *802*, 59–71.

[67] Wei, W., Wang, Y. M., Luo, G. A., Wang, R. J., Guan, Y. H., Yan, C., *J. Liq. Chromatogr. Relat. Technol.* 1998, *21*, 1433–1443.

[68] Barret, D. A., Brown, V. A., Shaw, P. N., Davies, M. C., Ritchie, H. J., Ross, P., *J. Chromatogr. Sci.* 1996, *34*, 146–156.

[69] Hilder, E. F., Macka, M., Haddad, P. R., *Anal. Commun.* 1999, *36*, 299–303.

[70] Stahlberg, J., *Anal. Chem.* 1997, *69*, 3812–3821.

[71] Miyawa, J. H., Alasandro, M. S., Riley, C. M., *J. Chromatogr. A* 1997, *769*, 145–153.

[72] Djordjevic, N. M., Fowler, P. W. J., Houdiere, F., Lerch, G., *J. Liq. Chromatogr. Relat. Technol.* 1998, *21*, 2219–2232.

[73] Dittman, M. M., Wienand, K., Bek, F., Rozing, G. P., *LC·GC* 1995, *13*, 800–814.

[74] Rebscher, H., Pyell, U., *Chromatographia* 1996, *42*, 171–176.

[75] Pyell, U., Rebscher, H., Banholczer, A., *J. Chromatogr. A* 1997, *779*, 155–163.

[76] Pesek, J. J., Matyska, M. T., Mauskar, L., *J. Chromatogr. A* 1997, *763*, 307–314.

[77] Pesek, J. J., Matyska, M. T., *Electrophoresis* 1997, *18*, 2228–2238.

[78] Liu, Z., Zou, H., Ni, J. Y., Zhang, Y., *Anal. Chim. Acta* 1999, *378*, 73–76.

[79] Pesek, J. J., Matyska, M. T., Menezes, S., *J. Chromatogr. A* 1999, *853*, 151–158.

[80] Pesek, J. J., Matyska, M. T., Cho, S., *J. Chromatogr. A* 1999, *845*, 237–246.

[81] Matyska, M. T., Pesek, J. J., Sandoval, J. E., Parkar, U., Liu, X., *J. Liq. Chromatogr. Relat. Technol.* 2000, *23*, 97–111.

[82] Pesek, J. J., Matyska, M. T., *J. Capil. Electrophor.* 1997, *4*, 213–217.

[83] Pesek, J. J., Matyska, M. T., Swedberg, S., Udivar, S., *Electrophoresis* 1999, *20*, 2343–2348.

[84] Matyska, M. T., Pesek, J. J., Katrekar, A., *Anal. Chem.* 1999, *71*, 5508–5514.

[85] Huang, X., Zhang, J., Horváth, C., *J. Chromatogr. A* 1999, *858*, 91–101.

[86] Sawada, H., Jinno, K., *Electrophoresis* 1999, *20*, 24–30.

[87] Tan, Z. J., Remcho, V. T., *J. Microcol. Sep.* 1998, *10*, 99–105.

[88] Xu, W., Regnier, F. E., *J. Chromatogr. A* 1999, *853*, 243–256.

[89] Liao, J.-L., Chen, N., Ericson, C., Hjertén, S., *Anal. Chem.* 1996, *68*, 3468–3472.

[90] Peters, E. C., Petro, M., Svec, F., Frechet, J. M. J., *Anal. Chem.* 1997, *69*, 3646–3649.

[91] Peters, E. C., Petro, M., Svec, F., Frechet, J. M. J., *Anal. Chem.* 1998, *70*, 2288–2295.

[92] Peters, E. C., Petro, M., Svec, F., Frechet, J. M. J., *Anal. Chem.* 1998, *70*, 2296–2302.

[93] Ishizuka, N., Minakuchi, H., Nakanishi, K., Soga, N., Nagayama, H., Hosoya, K., Tanaka, N., *Anal. Chem.* 2000, *72*, 1275–1280.

[94] Tanaka, N., Nagayama, H., Kibayashi, H., Ikegami, T., Hosoya, K., Ishizuka, N., Minakuchi, H., Nakanishi, K., Cabrera, K., Lubda, D., *J. High Resol. Chromatogr.* 2000, *23*, 111–116.

[95] Fujimoto, C., *J. High Resolut. Chromatogr.* 2000, *23*, 89–92.

[96] Asiaie, R., Huang, X., Farnan, D., Horváth, C., *J. Chromatogr. A* 1998, *806*, 251–263.

[97] Dulay, M. T., Kulkarni, R. P., Zare, R. N., *Anal. Chem.* 1998, *70*, 5103–5107.

[98] Chirica, G., Remcho, V. T., *Electrophoresis* 1999, *20*, 50–56.

[99] Tang, Q., Wu, N., Lee, M. L., *J. Microcol. Sep.* 1999, *11*, 550–561.

[100] Suzuki, S., Kuwahara, Y., Makiura, K., Honda, S., *J. Chromatogr. A* 2000, *873*, 247–256.

[101] Schweitz, L., Andersson, L. I., Nilsson, S., *Anal. Chem.* 1997, *69*, 1179–1183.

[102] Schweitz, L., Andersson, L. I., Nilsson, S., *Chromatographia* 1999, *49*, S93–S94.

[103] Schweitz, L., Andersson, L. I., Nilsson, S., *J. Chromatogr. A* 1997, *792*, 401–409.

[104] Schweitz, L., Andersson, L. I., Nilsson, S., *J. Chromatogr. A* 1998, *817*, 5–13.

[105] Schure, M. R., Murphy, R. E., Klotz, W. L., Lau, W., *Anal. Chem.* 1998, *70*, 4985–4995.

[106] Palm, A., Novotny, M. V., *Anal. Chem.* 1997, *69*, 4499–4507.

[107] Peters, E. C., Lewandowski, K., Petro, M., Svec, F., Frechet, J. M. J., *Anal. Commun.* 1998, *35*, 83–86.

[108] Colón, L. A., Maloney, T. D., Fermier, A., *J. Chromatogr. A* 2000, in press.

[109] Robson, M. M., Cikalo, M. G., Myers, P., Euerby, M. R., Bartle, K. D., *J. Microcol. Sep.* 1997, *9*, 357–372.

[110] Dermaux, A., Sandra, P., *Electrophoresis* 1999, *20*, 3027–3065.

[111] Zimina, T. M., Smith, R. M., Myers, P., *J. Chromatogr. A* 1997, *758*, 191–197.

[112] Lammerhofer, M., Lindner, W., *J. Chromatogr. A* 1998, *829*, 115–125.

[113] Krause, K., Girod, M., Chankvetadze, B., Blaschke, G., *J. Chromatogr. A* 1999, *837*, 51–63.

[114] Wolf, C., Spence, P. L., Pirkle, W. H., Derrico, E. M., Cavender, D. M., Rozing, G. P., *J. Chromatogr. A* 1997, *782*, 175–179.

[115] Wolf, C., Spence, P. L., Pirkle, W. H., Cavender, D. M., Derrico, E. M., *Electrophoresis* 2000, *21*, 917–924.

[116] Dittmann, M. M., Rozing, G. P., *J. Chromatogr. A* 1996, *744*, 63–74.

[117] Seifar, R. M., Heemstra, S., Kok, W. T., Kraak, J. C., Poppe, H., *Biomed. Chromatogr.* 1998, *12*, 140.

[118] Bailey, C. G., Yan, C., *Anal. Chem.* 1998, *70*, 3275–3279.

[119] Zhang, L., Shi, W., Zou, H., Ni, J., Zhang, Y., *J. Liq. Chromatogr. Relat. Technol.* 1999, *22*, 2715–2728.

[120] Zhang, L., Zhang, Y., Zhu, J., Zou, H., *Anal. Lett.* 1999, *32*, 2679–2690.

[121] Kitagawa, S., Tsuji, A., Watanabe, H., Nakashima, M., Tsuda, T., *J. Microcol. Sep.* 1997, *9*, 347–356.

[122] Li, D., Knobel, H. H., Remcho, V. T., *J. Chromatogr. B* 1997, *695*, 169–174.

[123] Breadmore, M. C., Macka, M., Haddad, P. R., *Electrophoresis* 1999, *20*, 1987–1992.

[124] Zhang, M., Yang, C., El Rassi, Z., *Anal. Chem.* 1999, *71*, 3277–3282.

[125] Zhang, M., El Rassi, Z., *Electrophoresis* 1998, *19*, 2068–2072.

[126] Walhagen, K., Unger, K. K., Olsson, A. M., Hearn, M. T. W., *J. Chromatogr. A* 1999, *853*, 263–275.

[127] Adam, T., Kramer, M., *Chromatographia* 1999, *49*, 35–40.

[128] Spikmans, V., Lane, S. J., Smith, N. W., *Chromatographia* 2000, *51*, 18–24.

[129] Wei, W., Luo, G., Yan, C., *Am. Lab.* 1998, *30*, 20C–20E.

[130] Huang, P., Jin, X., Chen, Y., Srinivasan, J. R., Lubman, D. M., *Anal. Chem.* 1999, *71*, 1786–1791.

[131] Ye, M., Zou, H., Liu, Z., Ni, J., *J. Chromatogr. A* 2000, *869*, 385–394.

[132] Kutter, J. P., Jacobson, S. C., Matsubara, N., Ramsey, J. M., *Anal. Chem.* 1998, *70*, 3291–3297.

[133] He, B., Tait, N., Regnier, F., *Anal. Chem.* 1998, *70*, 3790–3797.

[134] Oleschuk, R. D., Shultz-Lockyear, L. L., Ning, Y., Harrison, D. J., *Anal. Chem.* 2000, *72*, 585–590.

[135] Ericson, C., Holm, J., Ericson, T., Hjertén, S., *Anal. Chem.* 2000, *72*, 81–87.

[136] Regnier, F. E., *J. High Resolut. Chromatogr.* 2000, *23*, 19–26.

[137] He, B., Ji, J., Regnier, F. E., *J. Chromatogr. A* 1999, *853*, 257–262.

[138] Fermier, A. M., Colon, L. A., *J. Microcol. Sep.* 1998, *10*, 439–447.

[139] Witowski, S. R., Kennedy, R. T., *J. Microcol. Sep.* 1999, *11*, 723–728.

[140] Wistuba, D., Schurig, V., *Electrophoresis* 1999, *20*, 2779–2785.

[141] Alicea-Maldonado, R., Colón, L. A., *Electrophoresis* 1999, *20*, 37–42.

[142] Maruska, A., Pyell, U., *J. Chromatogr. A* 1997, *782*, 167–174.

[143] Maruska, A., Pyell, U., *Chromatographia* 1997, *45*, 229–234.

[144] Ye, M., Zou, H., Liu, Z., Ni, J., Zhang, Y., *Anal. Chem.* 2000, *72*, 616–621.

[145] van den Bosch, S. E., Heemstra, S., Kraak, J. C., Poppe, H., *J. Chromatogr. A* 1996, *755*, 165–177.

[146] Seifar, R. M., Kraak, J. C., Kok, W. T., Poppe, H., *J. Chromatogr. A* 1998, *808*, 71–77.

[147] Chen, J.-R., Dulay, M. T., Zare, R. N., Svec, F., Peters, E., *Anal. Chem.* 2000, *72*, 1224–1227.

[148] Rebscher, H., Pyell, U., *J. Chromatogr. A* 1996, *737*, 171–180.

[149] Hilder, E. F., Klampfl, C. W., Macka, M., Haddad, P. R., Myers, P., *Analyst* 2000, *125*, 1–4.

[150] Carney, R. A., Robson, M. M., Bartle, K. D., Myers, P., *J. High Resolut. Chromatogr.* 1999, *22*, 29–32.

[151] Behnke, B., Johansson, J., Zhang, S., Bayer, E., Nilsson, S., *J. Chromatogr. A* 1998, *818*, 257–259.

[152] Chen, Y., Gerhardt, G., Cassidy, R., *Anal. Chem.* 2000, *72*, 610–615.

[153] Cortes, H. J., Pfeiffer, C. D., Stevens, T. S., Richter, B. E., *J. High Resolut. Chromatogr.* 1987, *10*, 446–448.

[154] Schmid, M., Bauml, F., Kohne, A. P., Welsch, T., *J. High Resolut. Chromatogr.* 1999, *22*, 438–442.

[155] Lord, G. A., Gordon, D. B., Myers, P., King, B. W., *J. Chromatogr. A* 1997, *768*, 9–16.

[156] Choudhary, G., Horváth, C., Banks, J. F., *J. Chromatogr. A* 1998, *828*, 469–480.

[157] Maloney, T. D., Colon, L. A., *Electrophoresis* 1999, *20*, 2360–2365.

[158] Robson, M. M., Roulin, S., Shariff, S. M., Raynor, M. W., Bartle, K. D., Clifford, A. A., Myers, P., Euerby, M. R., Johnson, C. M., *Chromatographia* 1996, *43*, 313–321.

[159] Tang, Q., Xin, B., Lee, M. L., *J. Chromatogr. A* 1999, *837*, 35–50.

[160] Reynolds, K. J., Colon, L. A., *Analyst* 1998, *123*, 1493–1495.

[161] Stol, R., Mazereeuw, M., Tjaden, U. R., van der Greef, J., *J. Chromatogr. A* 2000, *873*, 293–298.

[162] Gusev, I., Huang, X., Horváth, C., *J. Chromatogr. A* 1999, *855*, 273–290.

[163] Xiong, B., Zhang, L., Zhang, Y., Zou, H., Wang, J., *J. High Resol. Chromatogr.* 2000, *23*, 67–72.

[164] Yu, C., Svec, F., Frechet, J. M. J., *Electrophoresis* 2000, *21*, 120–127.

[165] Remcho, V. T., Tan, Z. J., *Anal. Chem.* 1999, *71*, 248A–255A.

[166] Fujimoto, C., *Anal. Chem.* 1995, *67*, 2050–2053.

[167] Ericson, C., Liao, J.-L., Nakazato, K. I., Hjertén, S., *J. Chromatogr. A* 1997, *767*, 33–41.

[168] Fujimoto, C., Sakurai, M., Muranaka, Y., *J. Microcol. Sep.* 1999, *11*, 693–700.

[169] Fujimoto, C., Fujise, Y., Matsuzawa, E., *Anal. Chem.* 1996, *68*, 2753–2757.

[170] Fujimoto, C., *Analusis* 1998, *26*, M49–M52.

[171] Hjertén, S., Vegvari, A., Srichaiyo, T., Zhang, H.-X., Ericson, C., Eaker, D., *J. Capil. Electrophor.* 1998, *5*, 13–26.

[172] Ericson, C., Hjertén, S., *Anal. Chem.* 1999, *71*, 1621–1627.

[173] Svec, F., Peters, E. C., Sykora, D., Yu, G., Frechet, J. M., *J. High Resolut. Chromatogr.* 2000, *23*, 3–18.

[174] Koide, T., Ueno, K., *J. High Resolut. Chromatogr.* 2000, *23*, 59–66.

[175] Svec, F., Frechet, J. M. J., *Ind. Engineer. Chem. Res.* 1999, *38*, 34–48.

[176] Tan, Z. J., Remcho, V. T., *Electrophoresis* 1998, *19*, 2055–2060.

[177] Guo, Y., Colón, L. A., *Anal. Chem.* 1995, *67*, 2511–2516.

[178] Choudhary, G., Hancock, W., Witt, K., Rozing, G., Torres-Duarte, A., Wainer, I., *J. Chromatogr. A* 1999, *857*, 183–192.

[179] Dasgupta, P. K., Surowiec, K., *LC·GC* 1998, *16*, 44–46.

[180] Xin, B., Lee, M. L., *J. Microcol. Sep.* 1999, *11*, 271–275.

[181] Behnke, B., Metzger, J. W., *Electrophoresis* 1999, *20*, 80–83.

[182] Huber, C. G., Choudhary, G., Horváth, C., *Anal. Chem.* 1997, *69*, 4429–4436.

[183] Huang, P., Wu, J.-T., Lubman, D. M., *Anal. Chem.* 1998, *70*, 3003–3008.

[184] Gfroerer, P., Schewitz, J., Pusecker, K., Tseng, L.-H., Albert, K., Bayer, E., *Electrophoresis* 1999, *20*, 3–8.

[185] Euerby, M. R., Gilligan, D., Johnson, C. M., Bartle, K. D., *Analyst* 1997, *122*, 1087–1088.

[186] Ding, J., Szeliga, J., Dipple, A., Vouros, P., *J. Chromatogr. A* 1997, *781*, 327–334.

[187] Alexander, J. N. I. V., Poli, J. B., Markides, K. E., *Anal. Chem.* 1999, *71*, 2398–2409.

[188] Robson, M. M., Bartle, K. D., Myers, P., *Chromatographia* 1999, *50*, 711–715.

[189] Apffel, A., Yin, H., Hancock, W. S., McManigill, D., Frenz, J., Wu, S.-L., *J. Chromatogr. A* 1999, *832*, 149–163.

[190] Lister, A. S., Rimmer, C. A., Dorsey, J. G., *J. Chromatogr. A* 1998, *828*, 105–112.

[191] Yan, C., Dadoo, R., Zare, R. N., Rakestraw, D. J., Anex, D. S., *Anal. Chem.* 1996, *68*, 2726–2730.

[192] Que, A. H., Kahle, V., Novotny, M. V., *J. Microcol. Sep.* 2000, *12*, 1–5.

[193] Olson, D. L., Lacey, M. E., Sweedler, J. V., *Anal. Chem.* 1998, *70*, 257A–264A.

[194] Gfrorer, P., Schewitz, J., Pusecker, K., Bayer, E., *Anal. Chem.* 1999, *71*, 315A–321A.

[195] Pusecker, K., Schewitz, J., Gfrorer, P., Tseng, L.-H., Albert, K., Bayer, E., Wilson, I. D., Bailey, N. J., Scarfe, G. B., Nicholson, J. K., Lindon, J. C., *Anal. Commun.* 1998, *35*, 213–215.

[196] Pusecker, K., Schewitz, J., Gfroerer, P., Tseng, L.-H., Albert, K., Bayer, E., *Anal. Chem.* 1998, *70*, 3280–3285.

[197] Schewitz, J., Gfrorer, P., Pusecker, K., Tseng, L.-H., Albert, K., Bayer, E., Wilson, I. D., Bailey, N. J., Scarfe, G. B., Nicholson, J. K., Lindon, J. C., *Analyst* 1998, *123*, 2835–2837.

[198] Banks, J. F., *Electrophoresis* 1997, *18*, 2255–2266.

[199] Lüdtke, S., Unger, K. K., *Chimia* 1999, *53*, 498–500.

[200] Wu, J.-T., Qian, M. G., Li, M. X., Zheng, K., Huang, P., Lubman, D. M., *J. Chromatogr. A* 1998, *794*, 377–389.

[201] Bayer, E., Gfrorer, P., Rentel, C., *Angew. Chem.* 1999, *38*, 992–995.

[202] Lane, S. J., Tucker, M. G., *Rapid Commun. Mas Spectrom.* 1998, *12*, 947–954.

[203] Palmer, M. E., Clench, M. R., Tetler, L. W., Little, D. R., *Rapid Commun. Mass Spectrom.* 1999, *13*, 256–263.

[204] Rentel, C., Gfrorer, P., Bayer, E., *Electrophoresis* 1999, *20*, 2329–2336.

[205] Spikmans, V., Lane, S. J., Tjaden, U. R., Van Der Greef, J., *Rapid Commun. Mass Spectrom.* 1999, *13*, 141–149.

[206] Warriner, R. N., Craze, A. S., Games, D. E., Lane, S. J., *Rapid Commun. Mass Spetrom.* 1998, *12*, 1143–1149.

[207] Chen, W. H., Lin, S. Y., Liu, C. Y., *Anal. Chim. Acta* 2000, *410*, 25–35.

[208] Dermaux, A., Medvedovici, A., Ksir, M., Van Hove, E., Talbi, M., Sandra, P., *J. Microcol. Sep.* 1999, *11*, 451–459.

[209] Ding, J., Vouros, P., *Am. Lab.* 1998, *30*, 15–16.

[210] Ding, J., Vouros, P., *Anal. Chem.* 1997, *69*, 379–384.

[211] Ding, J., Barlow, T., Dipple, A., Vouros, P., *J. Am. Soc. Mass Spectrom.* 1998, *9*, 823–829.

[212] Lane, S. J., Pipe, A., *Rapid Commun. Mass Spectrom.* 1998, *12*, 667–674.

[213] Paterson, C. J., Boughtflower, R. J., Higton, D., Palmer, E., *Chromatographia* 1997, *46*, 599–604.

[214] Akerlöf, G., *J. Am. Chem. Soc.* 1932, *54*, 4125.

[215] D'Aprano, A., Fuoss, R. M., *J. Phys. Chem.* 1969, *73*, 100.

[216] Yang, C., El Rassi, Z., *Electrophoresis* 1998, *19*, 2061–2067.

Electrophoresis 2000, *21*, 3905–3918

Review

Rosanne M. Guijt-van Duijn
Johannes Frank
Gijs W. K. van Dedem
Erik Baltussen

Delft University of Technology,
Kluyver Laboratory for
Biotechnology,
Department of Analytical
Biotechnology,
Delft, The Netherlands

Recent advances in affinity capillary electrophoresis

Use of the specificity of (bio)interactions can effectively overcome the selectivity limitation faced in capillary electrophoresis (CE), and the resulting technique usually is referred to as affinity capillary electrophoresis (ACE). Despite the high selectivity of ACE, several important problems still need to be addressed. A major issue in all CE separations, including ACE, is the concentration detection limit. Using UV detection, this is usually in the order of 10^{-6} M whereas laser-induced fluorescence (LIF) detection can provide detection limits down to the sub-10^{-10} M range. However, a marked disadvantage of LIF is that labeling of the analytes is usually required, which might change the interaction behavior of the solutes under investigation. Additionally, labeling reactions at sub-10^{-10} M concentration levels are certainly not trivial and often difficult to perform quantitatively. Alternative and universal detection approaches, particularly mass spectrometric (MS) detection, look very promising but (A) CE-MS techniques are still far from routine application. Important future progress in sensitive detection strategies is likely to increase the use of ACE in the future.

Keywords: Affinity capillary electrophoresis / Binding constants / Review   EL 4216

Contents

Correspondence: Dr. Erik Baltussen, Delft University of Technology, Kluyver Laboratory for Biotechnology, Department of Analytical Biotechnology, Julianalaan 67, NL-2628 BC Delft, The Netherlands
E-mail: h.a.baltussen@tnw.tudelft.nl
Fax: +31-15-278-2355

Abbreviations: ACE, affinity capillary electrophoresis; **AChe**, acetylcholinesterase; **AThCh**, acetylthiocholine; **CAB**, carbonic anhydrase B; **CPM**, coumarinylphenylmaleimide; **FA**, frontal analysis; **FACCE**, FA continuous CE; **FTPF-ACE**, flow through partial filling-ACE; **μTAS**, minaturized total analysis system; **VACE**, vacancy affinity capillary electrophoresis

1 Introduction

Weak noncovalent interactions are widespread in nature and are at the basis of chemical reactions and processes in living cells. An important starting point for a better understanding of living systems is the detailed study of the specific biointeractions that are found throughout nature. In the end this can enable us to better interpret the enormous amount of proteomic and genomic data. Additionally, binding interactions can also be effectively used to set up assays for specific compounds or compound classes. Use of the high selectivity of biological recognition systems allows quantitation of analytes in complex matrices, such as biological samples, with virtually no sample pretreatment.

Affinity capillary electrophoresis (ACE) is an analytical approach in which the migration patterns of interacting molecules in an electrical field are recorded and used to quantitate and identify specific binding and to estimate binding constants [1]. In ACE, a distinction between gel and free-flow electrophoresis has to be made. Gel electrophoresis is often used to determine the size of large biomolecules such as DNA and proteins. Additionally, it has been shown to be a viable option for microchip-based instruments, as demonstrated by the introduction of the Agilent 2100 bioanalyzer. Unfortunately, gel electrophoresis is often operated in the denaturing mode, which is detrimental for the survival of weak biological interactions. ACE in free solutions looks much more promising as denaturing conditions are usually not required and separations can be performed in more or less physiological solutions, thereby preserving biological intertactions. Additionally, ACE does not suffer from the complications resulting from two-phase chromatographic systems and is therefore particularly suited to study complex formation reactions.

Today, a variety of interactants including detergents [2, 3], cyclodextrins [4–7], crown ethers [8], antibiotics [7], peptides and proteins [4, 9] is used and formally all micellar electrokinetic chromatography (MEKC) and chiral separations can be considered to be ACE methods. However, the term ACE generally seems to be more or less reserved for stronger and more specific interactions with well-defined stoichiometry. This review focuses on methodologies and quantitative aspects of ACE according to this restricted definition. However, in case of CE-on-chip, a wider field of applications is considered due to the fewer number of papers on CE in microdevices in general and ACE in particular. Especially microdevices can benefit strongly from the high selectivity of ACE due to their restricted efficiency caused by the lower electric field that can be applied over the short separation channels. Compared to our previous review [10], only later publications will be included and the reader is referred to the earlier review for a complete overview.

2 Principles

2.1 Basic principles

Parameters describing binding interactions and interaction stoichiometries are essential for translation of the data coming out of genomics and proteomics into an understanding of how biological systems work. These fundamental parameters can be estimated by the use of affinity CE [11]. Several ACE techniques have been developed to study molecular interaction and to be able to discriminate between weak and strong affinity complexes. In frontal analysis, the vacancy peak method and the Hummel-Dreyer method, the affinity interaction is directly measured from the peak areas in the electropherogram. This is in contrast with ACE and vacancy affinity capillary electrophoresis (VACE), where the binding parameters are estimated from the change in migration time [12]. Recently, the possibilities and limitations of currently applied ACE methods have been reviewed by Heegaard and Kennedy [1].

Three different interaction phases can be distinguished in ACE: the stationary, pseudostationary, and the mobile phase. First, the interaction can take place at the surface of a coated capillary wall or with a stationary phase present in the capillary. The affinity molecule is either immobilized at the capillary wall, or bound to or imprinted in a column material or gel. When the capillary wall is coated with an affinity molecule, the analyte with affinity for this compound will be delayed compared with migration times in an untreated capillary. This approach is sometimes referred to as open-tubular electrochromatography. Coating of the capillary wall with heparin was used to determine the affinity of heparin for antithrombin III and for the secretory leukocyte proteinase inhibitor [13]. Zhang *et al.* [14] demonstrated the determination of methyltestosterone doping by immobilization of polyclonal antibodies in a hydrogel.

Second, the interaction can take place in semistationary phases like cyclodextrins and micelles analogous to MEKC. The possibilities of different ACE techniques that can be applied to study the interaction of drugs with cyclodextrins or amphiphilic compounds as well as several pseudostationary phases, such as micelles, microemulsions and liposomes, have been recently reviewed [15]. Third, the interactions can take place when both the analyte and the affinity molecule are in free solution. For studying these interactions, two analysis methods have been developed representing the extremes of slowly and rapidly dissociating complexes. For slowly dissociating (strong binding) complexes, the complex will not dissociate during the separation. Here the sample is preincubated, and ACE is only used as a means of separating the bound and unbound analyte. Weak binding complexes (rapid on-off kinetics) will be analyzed under constant equilibrium. A change in migration time corresponding to the time the analyte was bound to the affinity molecule is recorded. This can be determined by mobility change analysis.

2.1.1 Scatchard analysis

In binding studies, the binding and dissociation constants are important parameters. Scatchard analysis is a traditional way to linearize binding data, originating from the time where nonlinear regression programs were unavailable. From a Scatchard plot, the equilibrium binding constant K_B can be determined.

$$K_B = \frac{[LR]}{[L][R]} \tag{1}$$

where [LR] is the concentration of the formed ligand-receptor complex while [L] and [R] are the concentrations of free ligand and receptor, respectively; K_B is expressed in L/mol and all concentrations in mol/L. A small K_B means that the receptor has a low affinity for the ligand, while a large K_B indicates high affinity.

For Scatchard analysis Eq. (1) is written as

$$\frac{[RL]}{[L]} = -K_B[RL] + K_B[R]_{tot} \tag{2}$$

In a Scatchard plot, the *X*-axis is the bound fraction and the *Y*-axis is the ratio of the bound fraction to the concentration of the free ligand (see Fig.1). As indicated in Eq. (2), the K_B value can be determined by calculating the slope.

In ACE, K_B can be determined from the changes in electrophoretic mobility (μ_p in m^2/Vs) of a receptor protein P on complexation with a ligand L. The electrophoretic mobility of a compound is given by

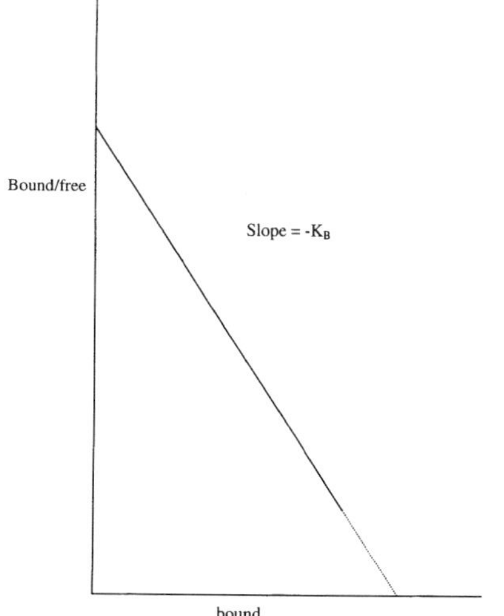

Figure 1. A Scatchard plot.

$$\mu = \frac{l_C l_D}{Vt} \tag{3}$$

where l_C and l_D are the capillary length and the detector in m, respectively, t is the migration time in s, and V is the voltage applied across the capillary in V. The change in μ_p is related to the migration time of a reference ion without affinity for L by

$$\Delta\mu_{P,L} = \frac{l_C l_D}{V\left[\left(\frac{1}{t_{ref}} - \frac{1}{t_p}\right) - \left(\frac{1}{t_{ref,L}} - \frac{1}{t_{P,L}}\right)\right]} \tag{4}$$

where $t_{P,L}$ and $t_{ref,L}$ are the measured migration times of the protein peak and the reference ion at specific ligand concentration, and t_P and t_{ref} are the measured migration times of the protein peak and the reference peak in absence of ligand, in s. In order to determinate K_B using Scatchard analysis, the ratio of the change in mobility $\Delta\mu_{P,L}$ and [L] is plotted *versus* $\Delta\mu_{P,L}$, according to Eq. (5).

$$\frac{\Delta\mu_{P,L}}{[L]} = K_B\Delta\mu_{P,L}^{max} - K_B\Delta\mu_{P,L} \tag{5}$$

In practice, however, this method is limited to migration times that are constant from run to run. A more reliable approach compared to the used mobilities is the use of the mobility ratio, *M*. This method has the advantage of being independent of the capillary length, the applied voltage, and the EOF velocity [16]. The mobility ratio is defined as:

$$M = \frac{\mu_{net}}{\mu_{EOF}} = \frac{(\mu + \mu_{EOF})}{\mu_{EOF}} \tag{6}$$

Combining with Eq. (3) gives

$$M = \frac{t_{EOF}}{t_P + 1} \tag{7}$$

For Scatchard analysis Eq. (8) can be used:

$$\frac{\Delta M_{P,L}}{[L]} = K_b\Delta M_{P,L}^{max} - K_b\Delta M_{P,L} \tag{8}$$

Here, $\Delta M_{P,L}$ is the change in migration ratio as a function of the concentration of the ligand. Using the mobility ratio, binding constants estimated for carbonic anhydrase B (CAB) and vancomycin agree well with those obtained using conventional assay techniques [16].

2.1.2 Mathematics and statistics

Most papers discussing affinity interactions in CE describe analyte migration based on a dynamic equilibrium process between the analyte and a single additive in 1:1 binding stoichiometry. Studying binding interactions of biomolecules in CE, however, often involves interactions with more than one analyte. Even when experimental conditions and concentrations are kept constant, unawareness of the existence of more complicated equilibria when studying secondary equilibria may lead to erroneous results. In general, equilibrium constants cannot be measured independently of other interactions.

In order to determine the multiple equilibrium phenomena more systematically, the migration behavior of an analyte can be described using contour plots and profile plots of multivariate isotherms. Profile plots of the corrected apparent mobility *versus* concentration of additive A give a strong indication of the binding stoichiometry with additive B; an example of a profile plot is demonstrated in Fig. 2. A common intersection point in a profile plot identifies the concentration of additive A where the net electrophoretic mobility of the analyte is independent on the concentration of B. This concentration of A is known as dengsu concentration and indicates 1:1 stoichiometry of B with the analytes and the absence of an interaction. Dengsu means same speed in Chinese, referring to the unaffected net electrophoretic mobility.

The shape of a contour plot (concentration of additive A *versus* concentration of additive B for constant apparent mobility) provides information about the interaction between the additives and can be used to determine whether the analyte can bind both additives at the same time or not. Bowser *et al.* [17] presented a theoretical discussion about the features of multivariate binding isotherms using applied mathematics. Insights in these isotherms provides a better insight in our understanding of how the separation buffer affects the chemical equilibria in CE separations, resulting in the basis for describing multiadditive CE systems. In a second paper, Bowser and Chen [18] describe the effect of analyte-additive interactions on analyte migration behavior when both 1:1 and 1:2 binding stoichiometry are present simultaneously. Equations based on the capacity factors for each individual interaction are derived to account for the effect of both first and second order equilibria. Microscopic equilibrium constants and mobilities are fully taken into account to calculate the macroscopic migration values. Linear transformation of the binding isotherms are compared with respect to the detection of higher order equilibria. Equilibria with more than one analyte can be described using the extended capacity factor. In correlation with chromatography, where the capacity factor k' is defined as

$$k' = \frac{\text{amount of analyte in the stationary phase}}{\text{amount of analyte in the mobile phase}} \qquad (9)$$

The expanded capacity factor describes the ratio of bound and free analyte. The expanded capacity factor for the interaction of analyte A with additive C is given by

$$k'_{AC} = \frac{n_{AC}}{n_A} = \frac{[AC]}{[A]} = K_1[C] \qquad (10)$$

where K_1 is the equilibrium constant of the formation of complex AC. More complex equilibria can be described analogously by defining an expanded k' for each interaction.

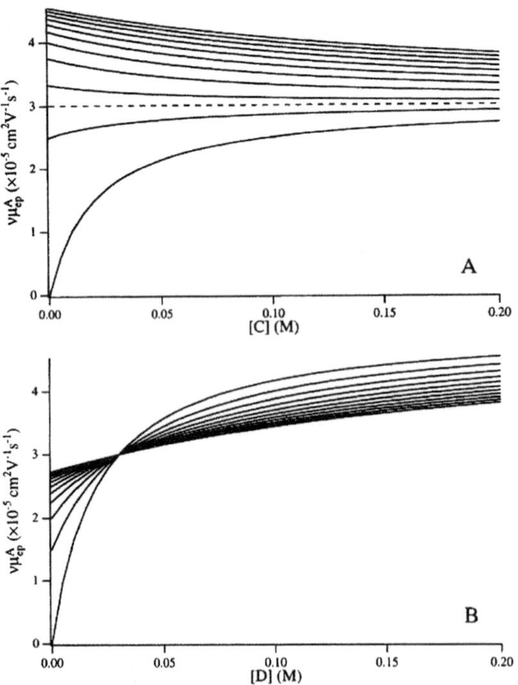

Figure 2. Profile plots of a binding isotherm surface where both additives interact with the analyte with 1:1 binding stoichiometries. The constants are $K_{AC} = 50$ M^{-1}; $K_{AD} = 50$ M^{-1}; $\mu_{ep,A} = 0$ cm^2 V^{-1} s^{-1}; $\mu_{ep,AC} = 3 \times 10^{-5}$ cm^2 V^{-1} s^{-1}; $\mu_{ep,AD} = 5 \times 10^{-5}$ cm^2 V^{-1} s^{-1}. The second additive concentration ranges from 0 to 0.2 M in increments of 0.02 M. The dashed line in (A) is the binding isotherm for C when [D] is at the dengsu concentration. Reprinted from [17], with permission.

2.1.3 Physicochemical parameters

Next to the interest in binding reactions and stoichiometry under physiological conditions, attention is also paid to interactions under nonphysiological conditions. These data provide more detailed information about the physicochemical parameters influencing the binding interaction, and provide a better insight in the binding mechanism. The effect of organic modifiers on protein-drug interactions was studied by Liu *et al.* [19]. The decrease in binding affinity with an increasing amount of acetonitrile in the buffer is most probably caused by protein conformational changes or partial denaturation.

The intrinsic binding constant is expected to be predominantly electrostatic, and therefore dependent on variables as the total protein charge, and thus pH. In 1980, Record *et al.* [20] obtained a relationship between K_{obs} and ligand charge Z at constant ionic strength for DNA (D) and an oppositely charged ligand species L:

$$\log K_{obs} = \log K^0 - Z\psi \log [\text{M}^+] \tag{11}$$

where K^0 is the equilibrium constant for

$$L + D \xleftarrow{\ \ } \xrightarrow{K^0} L–D + Z\psi M^+ \tag{12}$$

where ψ is the fraction of counterion M^+ thermodynamically associated with the DNA (per phosphate unit) and [M^+] is the added monovalent salt. Hallberg and Dubin [21] studied this dependence and found a semilogarithmic relationship; however, the used polyelectrolyte and ligand were charged equally. Here, another model where the free binding energy is linear with Z, without solely invoking the stoichiometric displacement of counter ions, is proposed.

2.2 Frontal analysis

In frontal analysis (FA), the affinity interaction takes place in the sample vial where the analytes are incubated with the affinity ligand. FA can only be used for equilibria where the dissociation rate is relatively slow compared with the separation time. After injection of the sample into the capillary, the separation field is applied and the free and bound fractions are separated. Due to the relatively slow dissociation rate, the formed complexes do not dissociate during separation. The advantages of FA are the low sample consumption and that the chemical equilibrium is not affected; the disadvantage is that it is restricted to equilibria with relatively slow binding kinetics.

2.2.1 Flow through partial filling ACE

In flow through partial filling ACE (FTPF-ACE), the capillary is filled with buffer, and a plug of solution containing ligand is injected at the injection side prior to the injection of a sample containing the receptor and noninteracting standards [22]. When both the ligand and the receptor have been injected in two sequential steps, the separation voltage is applied. During electrophoresis the receptor and standards flow through the domain of the ligand plug, where the receptor interacts with the ligand. A schematic

overview of a FTP-ACE separation is demonstrated in Fig. 3. Analysis of the change in migration time of the receptor, relative to the noninteracting standards, as a function of the concentration of the ligand yields a value for the binding constant. The relative migration time ratio R of the receptor is referenced to two noninteracting standards:

$$R = \frac{(t_r - t'_s)}{(t'_s - t_s)} \tag{13}$$

where, t_r, t_s and t'_s are the measured migration times of the receptor peak and two noninteracting standards, respectively. A Scatchard plot is obtained via equation 14:

$$\frac{\Delta R_{R,L}}{[L]} = K_b \Delta R_{R,L}^{max} - K_b \Delta R_{R,L} \tag{14}$$

Here, $\Delta R_{R,L}$ is the change in relative migration time ratio as a function of the ligand concentration. The use of a relative timescale compensates for fluctuations in the applied voltage in the capillary, EOF velocity, and capillary length resulting in run-to-run comparable results. This

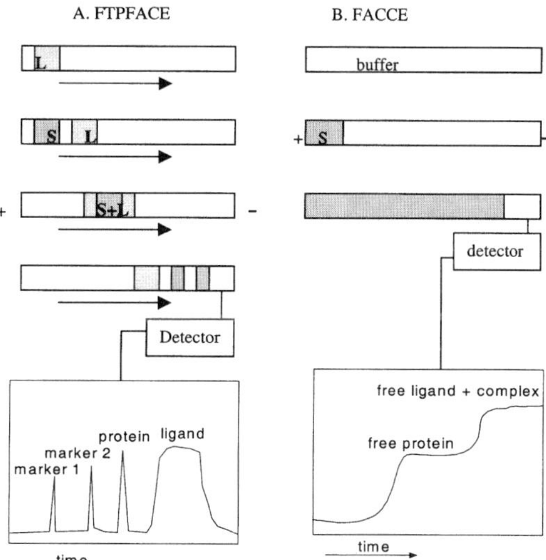

Figure 3. Schematic drawing of separation principles. (A) FTPF-ACE: After hydrostatic injection of a ligand plug (L) and a sample plug (S), the separation voltage is applied. The retardation caused by the protein-ligand interaction is related to the migration time of two noninteracting markers. (B) FACCE: The sample containing protein and ligand is injected continuously into the capillary. In the electropherogram continuous injection plateaus will appear instead of peaks. The first plateau is related to the fastest compound, the protein, while the second plateau is caused by the free ligand and complex passing the detector.

approach is introduced by Mito and Gomez [22] for studying the interaction of CAB with arylsulfonides and of vancomycin with D-Ala-Ala peptides. The obtained K_B values are in agreement with other techniques, while only 7–13 nL of ligand solution per run (typically 1 µL in conventional ACE) is consumed.

2.2.2 FA continuous CE

In contrast to other ACE techniques, FA continuous CE (FACCE) involves continuous sample introduction [23]. The capillary is filled and equilibrated with run buffer prior to sample introduction. The inlet of the capillary is then immersed in the sample vial, and a voltage is applied across the capillary to start the injection and separation process. Sample introduction and separation are integrated into one process. Separated analytes appear as progressive plateaus in the electropherogram. A schematic drawing of a FACCE separation is given in Fig. 3. Advantages of FACCE are (i) lower detection limits caused by increased injection times; (ii) the static equilibrium in the sample vial is measured instead of a dynamic equilibrium in the capillary; (iii) reduction of tedious calibration; one single calibration curve is sufficient to quantify the dependence of plateau height on the free ligand concentration, this in contrast with the Hummel-Dreyer method where multiple calibrations for each run buffer are required; and (iv) the analysis speed; a Scatchard plot can be obtained in a single day instead of weeks. Despite the increased sample consumption compared with conventional CE, the overall consumption of ligand is reduced by the ligand-conserving calibration method.

2.3 VACE

In VACE, the capillary is filled with a solution containing buffer, protein, and drug. The presence of these compounds causes a high detector signal. After injection of a small plug of buffer, the voltage is switched on and two negative peaks will appear in the electropherogram corresponding to drug and protein. These peaks are caused by a local deficiency of free protein and drug, respectively. The area of these peaks is related to the drug and protein concentration, while the migration time is correlated with the degree of complexation. The binding parameters can be estimated by defining r, the ratio of the concentration of bound drug ($[D_B]$), and the total protein concentration ($[P_{Tot}]$), as

$$r = \frac{[D_B]}{[P_{Tot}]} = \sum_{i=1}^{m} \frac{n_i K_i [D_f]}{1 + K_i [D_f]} \qquad (15)$$

where m is the number of identical independent binding sites of the protein and n_i is the number of sites class i.

The free and bound drug concentrations can be calculated from the actual mobility of the drug which ranges between the mobility of the free drug and the mobility of the complex. When a second analyte, the so-called displacer, is added to the system, there will be a competitive interaction when both compounds show affinity for the same active site. The displacement of the bound drug with the displacer can be recognized by a shift in the actual mobility. Due to the displacement, the free drug concentration will increase, resulting in a shift in the actual mobility. An important requirement for the use of VACE as a technique to recognize displacement is that either the mobilities of the target drug and displacer drug differ or the detector does not respond to the displacer drug. Erim and Kraak [12] tested VACE for its suitability to study drug displacement by taking the displacement of warfarin from bovine serum albumin (BSA) by furosemide and phenylbutazine as a model system.

2.4 Fluorescence polarization studies of affinity interactions in CE

In current practice of ACE, the formation and stability of the complex is established by titration experiments, in which a series of solutions containing the substrate and its binding protein are analyzed in various ratios. The emergence of a new peak upon addition of the binding protein to the substrate is evidence of the formation of a complex while the peak areas of the complex and the substrate, respectively, correspond to concentration. These titration experiments have proven to be very successful in studies of binding interactions; however, they are time-consuming and unable to provide unequivocal quantitation of the complex when the complex is not well separated from the unbound molecules.

CE light-induced fluorescence polarization (LIFP) uses fluorescence polarization to distinguish the formed complex from the free analytes [24]. Fluorescence polarization is a measure of the time-averaged rotational motion of molecules. Small molecules rotate faster than larger molecules and therefore exhibit a lower polarization value. The smallest molecule of the binding partners is labeled with a fluorophore. After mixing of the binding partners, the complex is electrophoretically separated from the unbound substrate and detected using LIFP. The value of polarization can be calculated using

$$P = \frac{(I_v - G I_h)}{(I_v + G I_h)} \qquad (16)$$

where I_v and I_h are the fluorescence intensities of the vertically and horizontally polarized components, respectively, and G is an empirical constant correcting for the polarization bias introduced by the optics and the detection system. This method avoids tedious titration curves since the higher polarization displayed by larger molecules is evidence for the binding of the substrate to its antibody. Once the complex has been identified, one of the two polarized components is taken for quantification, where the sensitivity is equal to normal fluorescence detection. This makes CE-LIFP particularly suitable for applications such as screening for specific monoclonal antibodies where speed and convenience are the major concerns when choosing a screening method. Potential unique applications of CE-LIFP include analyses (i) where not baseline separation of the complex and substrate can be achieved; (ii) for screening of antibody and affinity binding reagents; and (iii) for monitoring of fluorescent labeling reactions. A possible disadvantage of CE-LIFP is the influence of the label on the binding constant of the affinity complex.

3 Applications

In pharmaceutical and medical research, affinity interactions of drugs and interfering substances with serum proteins and receptors are important parameters of interest for drug discovery, action, and metabolism. ACE is a fast and low-consuming separation technique for studying molecular interactions, properties that are suitable with the requirements in medical and pharmaceutical research. This explains that most applications of ACE in literature can be found in these areas. A better insight into the working principle of the human body can be obtained by studying interactions at the cellular and molecular level. In general, an event in the body results in the release of a protein that interacts with other proteins and causes a cascade of reactions resulting in a response of the body. For drug development and metabolic research, protein-protein interactions are of enormous importance. ACE techniques are excellently suited to investigate protein-protein interactions, in particular because they can be performed in free solutions, generally under nondenaturing conditions and they require only minimal amounts of sample. Samples do not even need to be pure, which facilitates their availability. Protein-protein interaction has been the subject of two recent reviews [25, 26]. Recently, ACE in protein-protein interaction has been focused on the interaction of the human immunodeficiency virus HIV-1 gp41 or HIV-2 gp36 protein with the putative cellular receptor proteins P45 and P62 from human T- and B-lymphocytes [27, 28]. The interaction between these proteins could be established and characterized as serving as a basis for further exploration of the protein domains involved in the affinity, which is important for our understanding of how the virus enters the human immune system.

Drugs not only interact with proteins, but also with other compounds present in the body. Interactions between different starch degradation products and the β-blocker propanolol were studied using permeation experiments and ACE. In the presence of malto-oligosaccharides, the transport of propanolol across artificial lipid membranes was retarded. ACE of propanolol demonstrated discrete interaction with starch degradation products such as malto-oligosaccharides [29]. This interaction was too small for reliable determination of the binding constants; however, it could be responsible for the retarded transport of propanolol over the membranes. ACE can be successfully applied to improve detection sensitivity or selectivity of analytes in complex samples [30].

3.1 Frontal analysis

3.1.1 Regular frontal analysis

FA is used for relatively slow interactions where the interaction is effected in the sample vial. Advantages of FA are the unaffected chemical equilibrium and the low sample consumption; a disadvantage is the restriction to slow binding kinetics. The development of an analytical method for determination of the free bilirubin content in an undisturbed bilirubin/albumin system is urgently needed in medicine. Bilirubin, a product of heme degradation, is a poorly water-soluble compound with a strong affinity for albumin in the circulation. Free bilirubin is deposited in the brain of jaundiced babies and causes severe damage or even death. Administration of bilirubin-displacing drugs has to be avoided for babies infected with jaundice and for lactating women. An analytical method for determination of the free bilirubin content in an undisturbed bilirubin/albumin system in a newborn's blood sample should be robust and fast, consuming only small sample volumes. Fung *et al.* [31] used CE-FA to determine free bilirubin and to study the interaction of albumin with bilirubin and cobinding with aspirin under clinical conditions without disturbance of the chemical equilibrium between bilirubin and albumin. The advantage of CE-FA in this study is that the chemical equilibrium in the blood is not disturbed, while the low volume consumption is compatible with the low blood sample volume from newborn babies.

CE-FA was also applied to study the enantioselective binding of nilvadipine (NV), a calcium channel blocker, to plasma lipoproteins [32]. It was found that the binding of NV to high-density lipoprotein (HDL), to low-density lipoprotein (LDL) and to oxidized LDL was nonspecific and nonenantioselective. Partition-like binding to the lipid part of these lipoproteins seemed to occur predominantly. The binding affinity of NV to LDL was about seven times stronger than that to HDL, and the oxidation of LDL enhanced the binding affinity significantly. The binding capacity and association constants of β-blockers to individual serum proteins, serum protein mixtures, and human serum were also determined using CE-FA [33]. The values for the association constants were in good agreement with those obtained from traditional binding assays found in the literature. In a similar way, Ding *et al.* [34] determined the affinity and binding stoichiometries of the β-blockers propanolol and verapamil with human serum albumin.

3.1.2 FACCE

FACCE was used to study the binding of proteins to polyelectrolytes [35]. FACCE offers enhanced detection limits and is free from effects due to slow binding kinetics, thus making it suitable for studying equilibrium systems. In addition, FACCE provides efficient quantitative analysis with a single calibration. For β-lactoglobulin-sodium poly-(styrenesulfonate) as model system, FACCE yields reproducible calibration curves and binding isotherms corresponding with values found in the literature [21]. Similarly, the interaction between BSA and hydrophobically modified polyacrylates (HMPA) was investigated, leading to binding isotherms from which it could be concluded that the association is controlled by both the concentration of free BSA and the hydrophobicity of the HMPA [36]. Long-range electrostatic contributions to the stability of the affinity complex have been analyzed using a series of substituted benzenesulfonamide inhibitors of carbonic anhydrase [37]. The use of protein charge ladders together with ACE allowed for the measurement of these forces. However, Monte-Carlo simulations suggest that the experimental resolution should be enhanced to measure the predicted differences between some of the lysine residues involved in the interaction.

3.2 Electrophoretically mediated microanalysis

In electrophoretically mediated microanalysis (EMMA), the capillary is used as a microreactor. The different reagents are injected in such a way that they pass each other during the electrophoretic run, thereby creating migrating reaction zones. EMMA is applied to study the binding of cyclosporin A (CsA), an immunosuppressive drug to prevent rejection after organ transplantation. It binds competitively with and inhibits the action of the enzyme cyclophilin (Cyp). The formed complex interacts with other proteins, hereby surpressing the T-cell proliferation in an unknown way. The therapeutic use of CsA is hampered by disadvantages like toxicity and poor solubility in water. Therefore, the pharmaceutical industry is looking for CsA derivatives with similar efficacy, lower tox-

icity and better solubility. ACE in the equilibrium mixture mode and EMMA provide a simple and effective system for the estimation of binding constants for recombinant human Cyp and CsA and derivatives [38]. The order of magnitude and the comparison between different derivatives can be achieved in a short time and with a minimal comsumption of substances. Furthermore, EMMA provides the advantageous possibility to study the interaction of the formed Cyp-CsA complex with other proteins in order to elucidate the working principle of the immunosuppressant action of CsA.

3.3 Molecular recognition

3.3.1 General approach

ACE appears to be a valuable tool to explore the factors involved in molecular recognition. It has been successfully used to study the forces underlying affinity interactions and to explore systematically the structural elements involved in the interaction. Structural information about molecular recognition can be obtained with ACE when a large number of variants of the interacting molecules is available. Such an application of ACE, for instance, is the investigation of epitopes. Qian and Tomer [39] were able to locate a peptide in the core protein of the human immunodeficiency virus (HIV) that had a strong interaction with the monoclonal antibody against this core protein. Inversely, it is also possible to study the extent to which a fragment of an antibody is still able to interact with its antigen [40]. Similarly, it is possible to analyze which part of a heparin binding glycopeptide recognizes heparin [41].

Ginseng polypeptide is a pharmaceutically active peptide isolated from the *Ginseng* plant. In order to overcome the tedious insulation procedure and to be able to make analogs, chemical synthesis routes have been developed. The binding properties of a chemically synthesized *Ginseng* polypeptide were characterized. ACE was used to explore the effect of substitutions on the binding of D-ribose and adenosine [42]. The interaction appears to be highly specific since most of the substitutions led to an inactive peptide. The binding of hyaluronan to cell surfaces is mediated by CD44 through its oligosaccharide structure. Analysis of the interaction of a series of oligosaccharides with hyaluronan revealed several structural elements that stimulated or inhibited binding [43]. The structural aspects of the interaction between dextrin oligomers and iodine were extensively investigated using ACE [44]. Together with spectroscopic techniques, insight was obtained into the mutual influence that these molecules exert on each other. ACE appeared to be particularly useful to rapidly assess the effects of external parameters on the interaction.

Kedarcin is a highly potent and active chromoprotein in antitumor antibiotics. The antitumor activity of this protein durg is due primarily to the chromophore. The chromophore, however, is labile when in free solution. An apoprotein is involved in the working mechanism to stabilize the chromophore and appears to direct the delivery of the chromophore to the DNA of intact cells. Detailed studies on the molecular interactions between kedarcin chromophore and the apoprotein provide useful information in understanding the mechanism of action. Liu *et al.* [19] studied this interaction using ACE. They varied the organic solvent as buffer component in order to establish a suitable binding assay. The organic solvent was required in order to stabilize the chromophore. From the binding constants, it appeared that the binding affinity between the kedarcin-chromophore and the apoprotein is reduced when the concentration of organic solvent is increased. ACE provides a way of studying interactions under normal and extreme conditions giving some insight into the mechanism of action of the complexation process between the apoprotein and its natural ligand.

3.3.2 Affinophoresis

Affinophoresis is a type of ACE using an affinophore, a soluble ionic carrier bearing affinity ligand(s). Affinophoresis is used to study the binding stoichiometry under different conditions since more affinity ligands can be immobilized at one carrier. Interactions between mannoside ligands coupled to polylysine with pea lectin were studied by affinophoresis to evaluate the contributions of monovalent and divalent interaction [45]. The affinophoresis of divalent lectin with the polyliganded affinophore was investigated using CE. The affinity was larger for affinophores having higher ligand density. The contribution of monovalent and divalent interactions to the binding in the lectin-affinophore complex could be estimated using the experimental data where the mobility of lectin is inhibited by a neutral ligand with a known affinity constant for lectin. There were more divalent complexes for affinophores having higher ligand density.

3.4 High-throughput screening and combinatorial chemistry

In high-throughput screening and combinatorial chemistry, a large number of samples has to be screened for biological activity in a very short time. The low-consuming, high-speed assays that can be performed using ACE seem to be very suitable for this purpose. A model system for the application of ACE for rapid screening of protein-ligand interactions has been developed for porphyrin-protein interaction [46]. Rapid screening of peptide libraries was developed by Caldwell *et al.* [47]. ACE was performed to assess the effect of the peptide dissolved in the electrophoretic buffer on the mobility of erythoprotein receptor. Screening of heparin-binding peptides appears to be easily done with capillaries coated with heparin and heparin sulfate [48]. The method is highly selective and allows the distinction of peptides that differ only in the stereochemistry of a proline residue.

The combination of the high selectivity of ACE and the structural identification of MS make ACE-MS a powerful technique in screening combinatorial libraries [49, 50]. An interesting approach to screen single-component libraries has been reported by Sun *et al.* [51]. In screening interactions with an acceptor molecule such as avidin, the number of injected samples can be reduced using a systematic manner of pooling and the high resolving power of ACE. Screening of peptides with unnatural amino acids and other ligands binding to DNA have also been performed with ACE [52, 53].

3.5 Affinity interaction in selective detection

Zhang *et al.* [14] used affinity interactions twice for the analysis of doping methyltestosterone. To improve separation selectivity, polyclonal antibodies were cross-linked in a reversible hydrogel while in the same assay the detection selectivity was improved using a fluorescently labeled antigen. Separation of the free and bound antigen allows determination of the bound and/or free concentration, which in turn is related to the amount of unlabeled antigen in the sample. Specific detection of methyl testosterone in serum down to the 50 ng/mL levels were achieved using this method. Using the labeled/unlabeled antigen technique, the glaucoma drug dorzolamide could be detected in biological fluids at the picomole level using its affinity for human carbonic anhydrase II [54]. This method to detect trace levels of drugs is also called affinity probe-CE (APCE).

3.6 DNA separations

Small DNA mutations of certain genes are the origin of heritable disorders and cancers. In order to discover and hopefully prevent these disorders, gene mutation assays are of increasing importance in diagnostic and medical fields. Gel electrophoresis is the traditional separation technique for DNA fragments, since the sieving nature of the gel will provide a size-based separation. Capillary affinity gel electrophoresis was applied to separate oligodeoxynucleotide mixtures using sequence-specific and base composition-specific recognition. The affinity interaction is based on the formation of a complex, known as heteroduplex, between the solved oligodeoxynucleotide sample and a nucleic acid analog immobilized on the capillary gel. A selective separation of hexathymidylic acid from a mixture of four homopolymers and five heteropolymers of hexadeoxynucleotides is demonstrated. In the same setup, the selective and sensitive sequence-specific recognition of DNA isomers was demonstrated [55]. A combination of size and sequence-dependent oligonucleotide separation methods using capillary affinity gel electrophoresis has been reviewed by Baba [56].

Contrary to most known affinity systems in CE, which operate in a continuous mode, Muscate *et al.* [57] separate the affinity from the separation step. The first step is the preseparation, removing all nonspecific solutes from the sample. At low temperature, oligonucleotides with complementary sequences in the sample will bind to the immobilized recognition sequence while unrelated oligonucleotides will migrate and leave the gel. In the second step, the bound oligonucleotides are released from the gel. This is achieved at stepwise-elevated temperature; each oligonucleotide will loose binding at its specific temperature. Based on the techniques described above, Katayama *et al.* [58] developed an ACE assay for gene mutation analysis using oligonucleotide polyacrylamide conjugate as pseudoimmobilized ligand. The magnesium ion concentration proved to be a key factor to achieve efficient separation of sequential isomers of oligonucleotides with the same length. The use of oligonucleotides as affinity ligand in contrast with the adenine bases increased the performance and applicability of the technique, as shown by Baba [55]. Different magnesium ion concentrations to control the interaction of the analytes with the affinity ligand enabled many components to be separated at the same time, this in contrast with Muscate *et al.* [57], where a stepwise raising of the temperature was required to separate each oligonucleotide.

Instead of immobilizing the oligonucleotides in a gel, Ozaki *et al.* [59] immobilized the oligonucleotides at the capillary wall. Again, the presence of the magnesium ion, which stabilizes the DNA duplex formation, is essential in this technique. A 100% match disappears from the electropherogram when the Mg^{2+} is present, while even small mismatches result in a consistent peak over the Mg^{2+} concentration range. The matching sequence disappears since it binds tightly to the antisequence immobilized at the wall, while the mismatch stays in solution because it does not bind to the immobilized nonmatching antisequence. In a capillary coated with $(dT)_{12}$, single T-mutants in a $d(A)_{12}$ sequence could be detected using variable concentrations of Mg^{2+}. In the same way, a one base mutant of the c-K-ras codon could be detected when its antisequence was immobilized. This system is potentially useful for practical gene mutations.

4 Affinity interactions in microfabricated devices

This section discusses the application of microfabricated devices in studying affinity interactions. In contrast with the rest of this review, non-bio affinity interactions in CE like MEKC and CEC will be taken into account in this section. Since the introduction of the miniaturized total analysis system (μTAS) principle by Manz *et al.* [60], microfabricated devices for increasingly complex analyses are revolutionizing the area of chemical and biological analysis [61–63]. This revolution is mainly caused by the advantages of miniaturization like reduced reagent consumption, the rapid analysis time, opportunities for parallel screening and the ease of automation. Separation techniques that are based on electrophoresis are attractive for application in micromachined devices. The electrokinetic effects provide both efficient transport and a simple way to control fluid flow in a network of channels. Application of computer-controlled electrokinetic fluid manipulation enables complex enzymatic assays on a chip using a fraction of the conventional amount of reagents. Well-defined sample plugs can be injected reproducibly using computer-controlled fluid manipulation in crossed channels.

In 1997, Hadd *et al.* [64] described a β-galactosidase (β-Gal) assay using resorufin β-D-galactopyranoside (RBG) as substrate. β-Gal catalyzed hydrolysis of the nonfluorescent RBG to resorufin (λ_{em} = 585 nm) and D-galactose. In this microassay, the consumption of enzyme and substrate is reduced four orders in magnitude compared to conventional assays. A mixing ratio of 40% flow from the mixing channel and 30% flow from each of the side channels was used for all experiments. Detection of resorufin was obtained 20 mm downstream from the reaction cross, which corresponded to a reaction volume of 6.5 nL. The reagent flow rate and mixing time were controlled by modifying the electric field strength in the reaction channel, which was 220 V/cm for most experiments. The on-chip determined Michaelis-Menten constant compared well between chip and conventional assays.

A more complex immunoreactor for determination of theophyllin in plasma was demonstrated by Chiem and Harrison [65]. The design of the chip allows 1:1 mixing at junctions 1 and 3. (see the layout in Fig. 4). In this competitive immunoassay of diluted serum, samples containing theophyllin (reservoir 5) and a labeled tracer compound Th* (reservoir 6) are mixed at junction J1. In the mixing coil Th and Th* are mixed and a selective antibody Ab from reservoir 7 is added at junction J3. Injection of the sample into the separation channel takes place after incubation of the Th/Th* mixture with the Ab solution in the second mix-

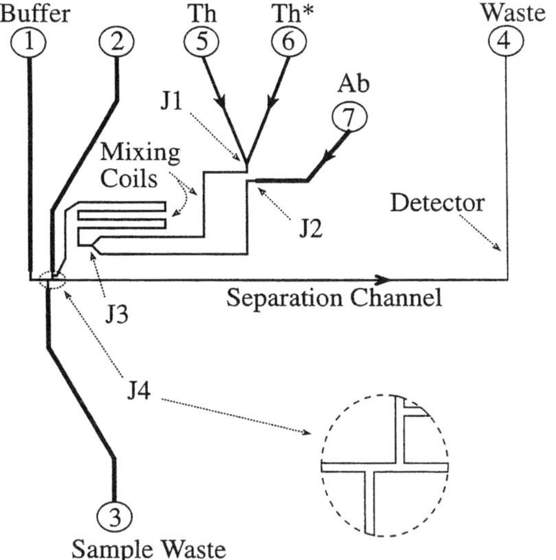

Figure 4. Device layout for competitive immunoassay showing numbering scheme for solution reservoirs and junctions. Reprinted from [65], with permission.

ing coil. The incubation time could be increased by stopping the flow in the reaction coil. The performance of this device is comparable with conventional instrumentation. Therefore, no sacrifice is made by integrating and miniaturizing. This fast and low-sample and reagent-volume-consuming device could form the basis for a new method of automating immunoassays at the point-of-care in clinical and hospital environments.

An assay for multiplex screening of four cationic inhibitors of acetylcholinesterase (AChE) was presented in 1999 [66]. In this chip, four inhibitors of AChE were separated after incubation with AChE. The extent of inhibition of AChE was monitored on-chip by measuring a fluorescent reaction product. A schematic overview of this device with the chemical reactions of the assay is given in Fig. 5. The inhibitor sample was pumped from the reservoir through the valve to the sample waste reservoir. The transportation of the AChE solution was controlled in such a way that a fraction goes to the sample waste and the main stream goes continuously down the separation channel. For injection, the potentials were temporally changed to allow a small plug of inhibitor to enter the separation channel. The enzyme AChE is mixed with the inhibitor mixture while the inhibitors are separated electrophoretically. Depending on the affinity of the inhibitors for AChE, a fraction of each inhibitor will bind to the enzyme, these equilibria taking place in the electrophoretically separated zones of the inhibitor mixture. The unbound AChE fraction

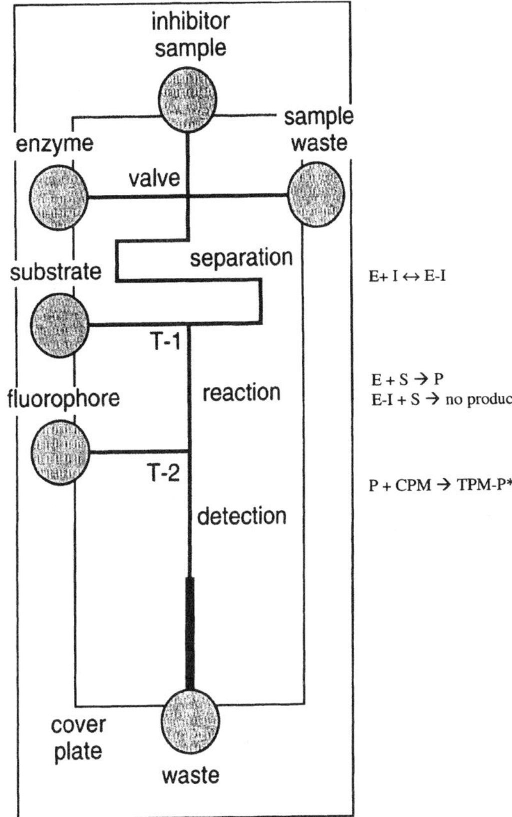

Figure 5. Schematic of the microchip used for analysis of acetylcholinesterase inhibitors. The reactions occurring in the channels are indicated at the right side. The fluid reservoirs were filled with AChE for enzyme, AThCh chloride for substrate and CPM for fluorophore. Reprinted from [66], with permission.

in each zone converts its substrate acetylthiocholine (AThCh) to thiocholine. This product reacts with coumarinylphenylmaleimide (CPM) to form the fluorescent CPM-thiocholine, which is measured in the detection channel.

Figure 6 gives the electropherogram of the assay described above. Assays to determine the K_I of tacrine on-chip compared well with a cuvette assay. Clinically relevant is the microfabricated electrophoresis chip for simultaneous bioassays of glucose, uric acid, ascorbic acid, and acetaminophen [67]. In this assay, glucose, ascorbic acid, acetaminophen, and uric acid are separated and detected using amperometric detection. In the separation channel, glucose oxidase (GOx) converts glucose into gluconic acid and hydrogen peroxide. The neutral peroxide is separated from the negatively charged uric and

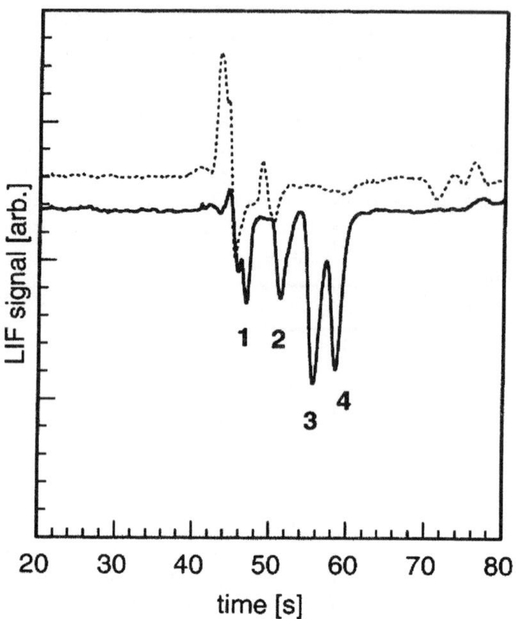

Figure 6. Electrophoretic separation of (1) 8 mM tetrametylammonium chloride, (2) 2 mM tetraethylammonium chloride, (3) 20 nM tacrine, and (4) 4 μM edophonium, detected by inhibition of AChE. The dotted line indicates the fluorescence signal *versus* time of thiocholine (produced off-chip) for injection of the same inhibitor mixture. The thiocholine peak is offset by 5% for comparison. Conditions: 0.2 s injection; 2.5 nM AChE; 500 μM AThCh; 250 μM CPM. Reprinted from [66], with permission.

ascorbic acid, and at the end of the channel these three compounds can be detected amperometrically. The presence of acetaminophen, a common natural interference, in the sample makes this separation system more complex.

In principle, electrophoretic separation of hydrogen peroxide and acetaminophen is impossible since both are neutral. Discrimination between glucose and acetaminophen is obtained using the specific affinity of GOx to convert glucose. In presence of GOx, glucose is converted to gluconic acid and hydrogen peroxide, a neutral and electrochemically detectable analyte. In absence of GOx, the glucose is not converted, and the only detectable neutral species is acetaminophen. Subtraction of the peak area in absence of GOx and the area in presence of GOx gives the peak area that corresponds with the concentration of glucose in the sample. Problems with the run-to-run variations caused by electrokinetic injection are reduced using an internal standard.

Next to bioaffinity, nonbioaffinity interactions such as micelles and packed columns also proved to be useful on a chip. In conventional CE, selectivity can be increased by addition of a surfactant to the separation buffer resulting in an MEKC-type separation. Unfortunately, this often leads to excessive Joule heating caused by increase in separation current. The use of microfabricated devices in micellar separation systems is an attractive way of avoiding this problem. Microfabricated channels have a more efficient heat dissipation than conventional capillaries, and are therefore less sensitive towards Joule heating effects when higher voltages are applied.

Compared with conventional MEKC, MEKC on a chip has proven to be 1–2 magnitudes faster, with higher efficiency and without expense of accuracy and precision [68]. For chip-based MEKC, plate numbers of 60 000 have been obtained for the separation of explosives in a 65 mm channel. The separation was completed within 1 min [69]. Cyclodextrin-modified MEKC (CD-MEKC) is employed to achieve high-speed chiral separations on microchip electrophoresis devices [70]. Chiral separations are an important issue in pharmaceutical and biotechnology fields because the bioactivity and pharmacological effects of stereoisomers are different. Enantiomers do not differ in their physicochemical properties; hence they cannot be separated without the presence of a chiral selector in the separation system. By reason of short separation channels and a relatively high electric field strength in combination with small-volume-defined injection plugs and operating in the counter-EOF mode, fast and efficient separations of FITC-labeled amino acid isomers were achieved. With analysis times ranging from 75 s to 160 s and efficiencies from 7000 up to 28 000 plates (100 000–395 000 plates/m), microchip MEKC provided overall better performance. Compared with conventional MEKC, the threefold better efficiency could be achieved in a tenfold shorter analysis time. The absolute chiral resolution, however, was better in conventional CE since the dynamic process of chiral discrimination is favored in longer separation columns.

Instead of affinity interactions in solution, the affinity phase can also consist of packed particles in a capillary. CEC is a hybrid technique between CE and LC. The high separation efficiency of CE is combined with the affinity differences of the analyte with the column material. Packing particles in a capillary, and especially in a channel, is very difficult. Therefore, light-initiated polymerization processes are very well suited to achieve monolith formation within a specified area [71]. The applicability of photopolymerization has been proven by Yu *et al.* [71] where the calculated efficiency of a CEC separation in a photopolymerized column in a conventional capillary is over 43 000 plates/m for small peptides.

All the ligands described before are of natural origin. In molecular imprinting (MIP), in contrast, the ligand is synthesized by imprinting it in a polymer. To do so, complexes are formed in a mixture of functional monomers, the imprint molecule, and a cross-linking monomer. These complexes, based on noncovalent interactions, such as hydrogen bonding, electrostatic interactions, reversible covalent interactions or metal ion-mediated interactions, are "frozen" during polymerization. After extraction of the imprint molecule, a polymer with specific cavity for this imprint molecule is left [72].

MIPs could also be used as stationary phase in on-chip CEC [72]. In conventional CE, enantiomeric separation of, for example, aromatic amino acids and local anesthetics could be achieved by coating the capillary wall with molecularly imprinted polymers. Separations of unlabeled amino acids achieved in a capillary column filled with grained imprinted polymers is demonstrated [1, 72, 73]. As described above, filling the channel is a bottleneck in CEC. Usually, MIPs are prepared by a bulk polymerization method, resulting in a block of MIP that should be crushed, ground, and sieved to produce packing materials. The applicability of CEC on-chip could be increased when the MIP was made on-chip. The photopolymerization reaction described by Yu [71] could be useful for molecular imprinting on-chip.

5 Conclusions and future directions

ACE has proven to be an adequate technique to study molecular interactions. Important applications can be found in pharmaceutical research and medical diagnostics. The advantages of ACE over other assay techniques are the high separation efficiency and selectivity, the low sample volumes and the use of (almost) unpurified samples. In no other binding assay is it possible to obtain binding constant determinations in solution on nanoliter volumes of unlabeled nonpurified samples. A marked disadvantage of the small sample volumes is the poor detection limit obtained using optical detection techniques. Standard (A)CE systems are equipped with UV detectors, resulting in path-length limited detection in capillaries of < 100 μm. Presently, on-line MS detection appears to be a promising detection method since it is highly sensitive and allows structural elucidation to some extent.

Traditional binding assays like ELISA are tedious and significant amounts of expensive reagents are consumed. The introduction of the μTAS concept provides new opportunities in studying binding interactions in a faster,

cheaper and parallel approach, as well as the use of affinity interactions in order to improve detection and/or separation efficiency. The low volume scale of microseparation systems reduces the sample and reagent consumption several orders in magnitude even compared with normal (A)CE, while the computer-controlled electrokinetic fluid handling enables automation of the whole assay. In the future, applications of ACE-based chips for high-throughput screening and combinatorial chemistry are expected, particularly due to the easy parallelization of such techniques. Affinity interactions in columns packed with (molecularly imprinted) stationary phases are also a growing field of interest. On-chip polymerization reactions to form monolithic beds is a promising approach to obtain a highly selective system in a completely "chemical" way. Revolutions in medicine and diagnostics can be expected when adequate, fast screening systems for studying molecular interactions become available on a routine basis. In medicine, DNA as well as blood protein assays will become essential in diagnosis and prevention of diseases. In high-throughput screening and combinatorial chemistry, parallel ACE chips, coupled with MS, can rapidly provide structural information of the biologically active compounds from a whole library within days/hours instead of weeks to months using conventional assay techniques.

RG gratefully acknowledges the support of STW, the Dutch foundation for Science and Technology, in the framework of the BIOMAS project (DST66.4351).

Received July 7, 2000

6 References

[1] Heegard, N. H., Kennedy, R. T., *Electrophoresis* 1999, *20*, 3122–3133.

[2] Yarabe, H. H., Billiot, E., Warner, I. M., *J. Chromatogr. A* 2000, *875*, 179–206.

[3] El Rassi, Z., *J. Chromatogr. A* 2000, *875*, 207–233.

[4] Fanali, S., *J. Chromatogr. A* 1997, *792*, 1–2, 227–267.

[5] Koppenhoefer, B., Zhu, X. F., Jakob, A., Wuerthner, S., Lin, B. C., *J. Chromatogr. A* 2000, *875*, 135–161.

[6] Fanali, S., *J. Chromatogr. A* 2000, *875*, 89–122.

[7] Fanali, S., Aturki, Z., Desiderio, C., *Enantiomer* 1999, *4*, 229–241.

[8] Kuhn, R., *Electrophoresis* 1999, *20*, 2605–2613.

[9] Vespalec, R., Boček, P., *Electrophoresis* 1999, *20*, 2579–2591.

[10] Rippel, G., Corstjens, H., Billiet, H. H., Frank, J., *Electrophoresis* 1997, *18*, 2175–2183.

[11] Kasička, V., *Electrophoresis* 1999, *20*, 3084–3105.

[12] Erim, F. B., Kraak, J. C., *J. Chromatogr. B* 1998, *710*, 205–210.

[13] Wu, X. J., Linhardt, R. J., *Electrophoresis* 1998, *19*, 2650–2653.

[14] Zhang, X. X., Li, J., Gao, J., Sun, L., Chang, W. B., *Electrophoresis* 1999, *20*, 1998–2002.

[15] Neubert, R. H. H., Schwarz, M. A., Mrestani, Y., Platzer, M., Raith, K., *Pharmaceut. Res.* 1999, *16*, 1663–1673.

[16] Kawaoka, J., Gomez, F. A., *J. Chromatogr. B* 1998, *715*, 203–210.

[17] Bowser, M. T., Kranack, A. R., Chen, D. Y., *Anal. Chem.* 1998, *70*, 1076–1084.

[18] Bowser, M. T., Chen, D. Y., *Anal. Chem.* 1998, *70*, 3261–3270.

[19] Liu, J. P., Abid, S., Hail, M. E., Lee, M. S., Hangeland, J., Zein, N., *Analyst* 1998, *123*, 1455–1459.

[20] Record Jr., M. T., Anderson, C. F., Lohman, T. M. Q., *Rev. Biophys.* 1980, *11*, 103–109.

[21] Hallberg, R. K., Dubin, P. L., *J. Phys. Chem. B* 1998, *102*, 8629–8633.

[22] Mito, E., Gomez, F. A., *Chromatographia* 1999, *50*, 689–694.

[23] Gao, J. Y., Dubin, P. L., Muhoberac, B. B., *Anal. Chem.* 1997, *69*, 2945–2951.

[24] Wan, Q. H., Le, X. C., *Anal. Chem.* 1999, *71*, 4183–4189.

[25] Vergnon, A. L., Chu, Y. H., *Methods Comp. Methods Enzymol.* 1999, *19*, 270–277.

[26] Kajiwara, H., *Anal. Chim. Acta* 1999, *383*, 61–66.

[27] Chen, Y. H., Xiao, Y., Wu, W. C., Wang, Q. G., Luo, G. A., Dierich, M. P., *Immunobiology* 2000, *201*, 317–322.

[28] Xiao, Y., Wu, W. C., Dierich, M. P., Chen, X. H., *Int. Arch. Allergy Immunol.* 2000, *121*, 253–257.

[29] Dongowski, G., Neubert, R. H. H., Platzer, M., Schwarz, M. A., Schnorrenberger, B., Anger, H., *Pharmazie* 1998, *53*, 871–875.

[30] Taga, A., Yabusako, Y., Kitano, A., Honda, S., *Electrophoresis* 1998, *19*, 2645–2649.

[31] Fung, Y. S., Sun, D. X., Yeung, C. Y., *Electrophoresis* 2000, *21*, 403–410.

[32] Mohamed, N. A. L., Kuroda, Y., Shibukawa, T., Nakagawa, T., El Gizawy, S., Askal, H. F., El Kommos, M. E., *J. Pharm. Biomed. Anal.* 1999, *21*, 1037–1043.

[33] McDonnell, P. A., Caldwell, G. W., Masucci, J. A., *Electrophoresis* 1998, *19*, 448–454.

[34] Ding, Y. S., Zhu, X. F., Lin, B. C., *Chromatographia* 1999, *49*, 343–346.

[35] Gao, J. Y., Dubin, P. L., *Biopolymers* 1999, *49*, 185–193.

[36] Porcar, I., Cottet, H., Gareil, P., Tribet, C., *Macromolecules* 1999, *32*, 3922–3929.

[37] Caravella, J. A., Carbeck, J. D., Duffy, D. C., Whitesides, G. M., Tidor, B., *J. Am. Chem. Soc.* 1999, *121*, 4340–4347.

[38] Kiessig, S., Bang, H., Thunecke, F., *J. Chromatogr. A* 1999, *853*, 469–477.

[39] Qian, X. H., Tomer, K. B., *Electrophoresis* 1998, *19*, 415–419.

[40] Lin, S. M., Tang, P., Hsu, S. M., *Electrophoresis* 1999, *20*, 3388–3395.

[41] Heegaard, N. H. H., *J. Chromatogr. A* 1999, *853*, 189–195.

[42] Kajiwara, H., *J. Chromatogr. A* 1998, *817*, 173–179.

[43] Skelton, T. P., Zeng, C. X., Nocks, A., Stamenkovic, I., *J. Cell Biol.* 1998, *140*, 431–446.

[44] Hong, M., Soini, H., Novotny, M. V., *Electrophoresis* 2000, *21*, 1513–1520.

[45] Shimura, K., Kasai, K., *Electrophoresis* 1998, *19*, 397–402.

[46] Lin, M., Wu, N. A., *J. Liq. Chromatogr. Rel. Technol.* 1999, *22*, 2167–2175.

[47] Caldwell, G. W., McDonnel, P. A., Masucci, J. A., Johnson, D. L., Jollife, L. K., *J. Biochem. Biophys. Methods* 1999, *40*, 17–25.

[48] VanderNoot, V. A., Hileman, R. E., Dordick, J. S., Linhardt, R. J., *Electrophoresis* 1998, *19*, 437–441.

[49] Lynen, F., Zhao, Y., Becu, C., Borremans, F., Sandra, P., *Electrophoresis* 1999, *20*, 2462–2474.

[50] Cheng, C. C., Chu, Y. H., *Am. Lab.* 1998, *30*, 79–80.

[51] Sun, S. X., Headrick, J., Staller, T., Sepaniak, M., *J. Microcol. Sep.* 1998, *10*, 653–660.

[52] Li, C. Z., Martin, L. M., *Anal. Biochem.* 1998, *263*, 72–78.

[53] Ihara, T., Ozaki, Y., Maeda, M., Takagi, M., *Anal. Sci.* 1997, *13*, 501–504.

[54] Tim, R. C., Kautz, R. A., Karger, B. L., *Electrophoresis* 2000, *21*, 220–226.

[55] Baba, Y., Sawa, T., Kishida, A., Akashi, M., *Electrophoresis* 1998, *19*, 433–436.

[56] Baba, Y., *J. Biochem. Biophys. Methods* 1999, *41*, 91–101.

[57] Muscate, A., Natt, F., Paulus, A., Ehrat, M., *Anal. Chem.* 1998, *70*, 1419–1424.

[58] Katayama, Y., Arisawa, T., Ozaki, Y., Maeda, M., *Chem. Lett.* 2000, *106–107*.

[59] Ozaki, Y., Katayama, Y., Ihara, T., Maeda, M., *Anal. Sci.* 1999, *15*, 389–392.

[60] Manz, A., Graber, N., Widner, H. M., *Sens. Actuators* 1990, *B1*, 244–248.

[61] Jakeway, S. C., De, M. A., Russell, E. L., *Fresenius J. Anal. Chem.* 2000, *366*, 525–539.

[62] Weller, M. G., *Fresenius J. Anal. Chem.* 2000, *366*, 635–645.

[63] Effenhauser, C. S., Bruin, G. J. M., Paulus, A., *Electrophoresis* 1997, *18*, 2203–2213.

[64] Hadd, A. G., Raymond, D. E., Halliwell, J. W., Jacobson, S. J., Ramsey, J. M., *Anal. Chem.* 1997, *69*, 3407–3412.

[65] Chiem, N. H., Harrison, D. J., *Clin. Chem.* 1998, *44*, 591–598.

[66] Hadd, A. G., Jacobson, S. J., Ramsey, J. M., *Anal. Chem.* 1999, *71*, 5206–5212.

[67] Tian, B., Chatrathi, M. P., Polsky, R., *Anal. Chem.* 2000, *72*, 2514–2518.

[68] von Heeren, F., Verpoorte, E., Manz, A., Thormann, W., *Anal. Chem.* 1996, *68*, 2044–2053.

[69] Wallenburg, S. R., Bailey, C. G., *Anal. Chem.* 2000, in press.

[70] Rodriguez, I., Jin, L. J., Ly, S. F. Y., *Electrophoresis* 1999, *21*, 211–219.

[71] Yu, C., Svec, F., Frechet, J. J., *Electrophoresis* 2000, *21*, 120–127.

[72] Schweitz, L., Andersson, L. I., Nilsson, S., *J. Chromatogr. A* 1998, *817*, 5–13.

[73] Takeuchi, T., Haginaka, J., *J. Chromatogr. B* 1999, *728*, 1–20.

Electrophoresis 2000, *21*, 3919–3930

Review

Dieter Schmalzing[1]
Scott Buonocore[1]
Christine Piggee[2]

[1]Whitehead Institute for
 Biomedical Research,
 Cambridge, MA, USA
[2]Laboratory of Neurotoxicology,
 National Institute of
 Mental Health,
 National Institutes of Health,
 Bethesda, MD, USA

Capillary electrophoresis-based immunoassays

This review covers the progress and developments in the field of capillary electrophoresis immunoassay (CEIA) over the past three years. Because many excellent
descriptions of the principles of these methods are available (*e.g.*, in the reviews listed
in this article), no elementary introduction is given to the field of immunoassays (IAs) or
CEIAs. This report focuses exclusively on experimental results, dividing the CEIA
papers into the categories of direct, indirect, and microchip electrophoretic immunoassays. In the last section, a brief summary of the current status of the CEIA field is presented.

Keywords: Capillary electrophoresis / Immunoassays / Microchips / Review   EL 4171

Contents

1 Introduction

The first data on competitive and direct capillary electro-
phoretis immunoassays (CEIAs) appeared in 1993 and
1994, respectively [1, 2]. After these encouraging initial
reports, development work on many CEIAs was initiated,
covering the entire spectrum of traditional immunoassay
(IA) analytes [3]. CE appears to be well-suited for this
important field of bioanalysis due to its excellent separation power, high analysis speed, small sample volume
requirements, complete automation, sensitive laser-
induced fluorescence detection, and the possibility of multianalyte analysis. On the other hand, several fundamental methodological problems have been uncovered over
the past years, such as relatively poor sensitivity and poor
resolution (at least for the direct format), which still await
resolution.

The goal of this report is to summarize the on-going CEIA
research from 1997 to present. For a description of the
principles of this method, one may consult recent reviews
[3, 4]. Competitive CEIAs will be discussed first since they
are the furthest developed, followed by direct CEIAs and,
finally, the performance of CEIA in electrophoretic microdevices, which is still at an early stage. The article will
close with a short overview on the current status of the
CEIA field.

2 Competitive CEIAs

One area in which competitive CEIAs have found utility is
in biomedical monitoring. Cyclosporin A (CsA) is administered as an immunosuppressive reagent. Because of its
very narrow therapeutic range of 150–450 ng/mL of blood
(150–450 nM), which needs to be tightly controlled, Ye
et al. [5] developed a competitive CEIA for CsA in patient
blood. The immunoreagents consisted of CsA labeled
with fluorescein (Ag*) and a monoclonal antibody (Ab).
The detection system was composed of an argon-ion
laser, postcolumn sheath-flow, and two photomultiplier
tubes (PMTs), which enabled the simultaneous recording
of both horizontally and vertically polarized light for the
spectroscopic differentiation between complexed and free
Ag*. Figure 1 shows electropherograms recorded concurrently with the two PMTs. The signal strength of free Ag*
was identical in both detectors but differed drastically for
Ab-Ag*, allowing for direct peak identification. This mole-
cule-dependent strength in fluorescent signal is due to the

Correspondence: Dr. Dieter Schmalzing, Whitehead Institute
for Biomedical Research, 9 Cambridge Center, Cambridge,
MA 02142, USA
E-mail: schmalzing@wi.mit.edu
Fax: +617-258-7663

Abbreviations: Ab, antibody; **Ab***, fluorescently labeled antibody; **Ab-Ag**, antibody/antigen complex; **Ag**, antigen; **Ag***, fluorescently labeled antigen; **ALP**, alkaline phosphatase; **CEIA**,
capillary electrophoresis immunoassay; **CsA**, cyclosporin A;

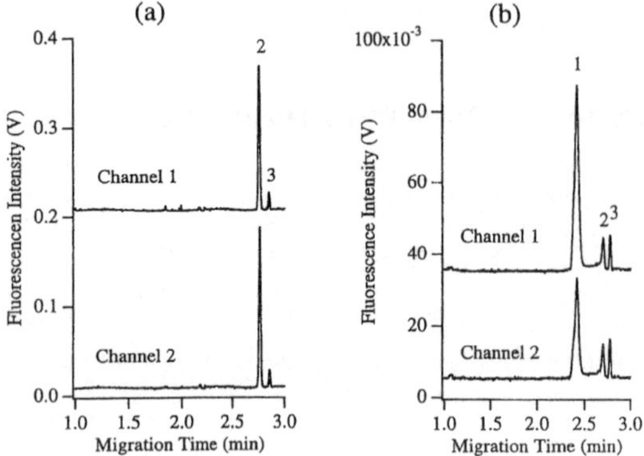

Figure 1. Comparison of the electropherograms obtained with two PMTs for LIF polarization detection. (a) Free fluorescently labeled cyclosporin A (CsA*) reagent and (b) a mixture of anti-CsA and CsA*. Peaks: 1, CsA*-Ab complex; 2, unbound CsA*; 3, free fluorescein dye. Reprinted from [5], with permission.

fact that small molecules such as free Ag* rotate rapidly in solution, have short relaxation times, and consequently do not exhibit significant fluorescence polarization (and *vice versa* for large conglomerates such as Ab-Ag*). The assay was linear between 2.5 and 25 nM CsA. The limit of detection (LOD) was 0.9 nM (S/N = 3), which was more sensitive than traditional immunoassay (IA) methods.

Consequently, this ability to differentiate between free and bound Ag* *via* fluorescence polarization detection was adopted to bypass the need for any electrophoretic separation prior to detection [6]. CE was employed solely to sweep unresolved Ag* and Ab-Ag* past the detector

window (comigration time was 2.6 min). Quantification was therefore based exclusively on the single unresolved peak consisting of the two coeluting species, whose total polarization signal varied with the amount of unlabeled Ag present in the sample.

The continuous on-line monitoring of insulin secretion from single islets of Langerhans has been elegantly described [7]. It is well-known that insulin levels in the body change in pulses. In the past it has been difficult to find techniques which can time-resolve high frequency oscillations of insulin secretion which are, for instance, typical for rats. In this proposed method, perfusate from single rat islets was mixed and reacted on-line with FITC-labeled antigen and unlabeled monoclonal antibody. On-line insulin secretion was stimulated by increasing the administered glucose concentration in a stepwise fashion from 3 to 17 mM. Typically, single islets could be monitored for 1–4 h. Stimulation products were injected every 3 s into the separation column *via* a flow-gated interface (Fig. 2). By using an ultrashort separation column (8 mm effective length) and high field strength (3000 V/cm), bound and free Ag* could be baseline-separated in only 1 s. A representative electropherogram is depicted in Fig. 3. With a sampling rate of approximately 0.1 Hz, the LOD was 0.3 nM of insulin. The authors conclude that this novel method could be easily extended to the *in vitro* monitoring of biologically active chemicals in neurons, brain sections, and the like.

The competitive CEIA was also effectively utilized for disease diagnosis and toxin detection. One group [8] extended their previous CEIA work on screening for scrapie protein, PrP^{SC}, which is associated with transmissible spongiform encephalopathies (TSE) in sheep and goat.

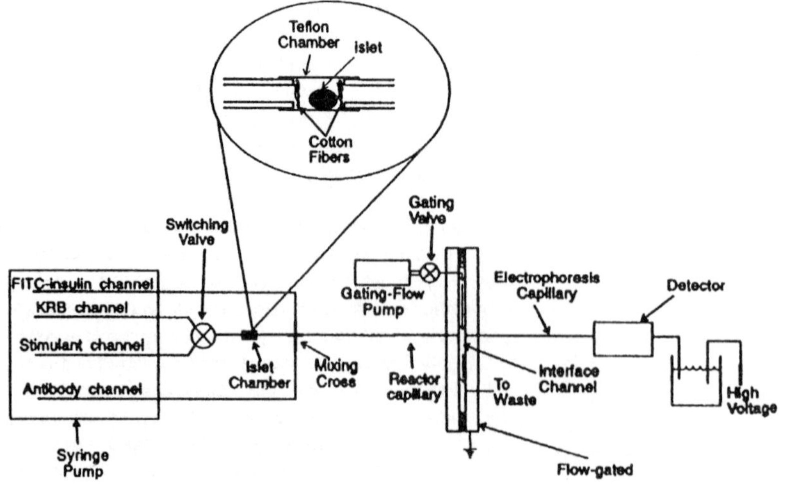

Figure 2. Block diagram of a system for on-line competitive IA. A single islet was perfused in the islet chamber. Perfusate was mixed on-line with FITC-insulin and Ab solution in the mixing cross. The binding reaction took place in the reactor capillary connecting the cross and flow-gated interface. Once at the interface, reaction products were injected into the separation capillary where bound and free FITC-insulin were separated and detected. Reprinted from [7], with permission.

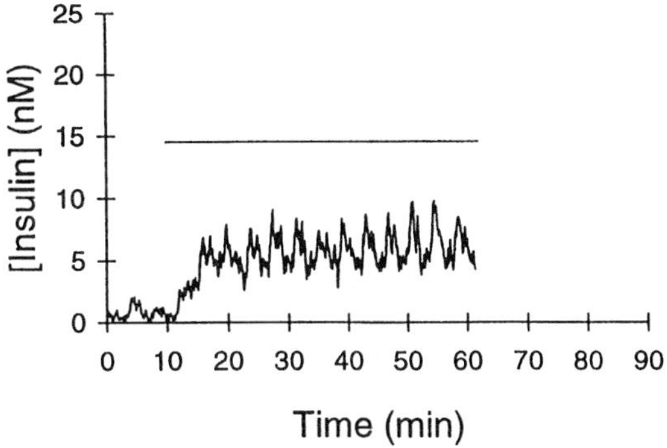

Figure 3. Typical oscillatory insulin secretion of single rat islets during stimulation with 16 mM glucose. The bar indicates the time the islet was stimulated with 16 mM glucose; all other times it was in 3 mM glucose. Reprinted from [7], with permission.

Four distinct peptide sequences from the PrP^{SC} protein were synthesized and labeled with fluorescein to yield the four antigens. Rabbits were immunized to produce a specific antibody for each peptide. After a 10 min off-line incubation, bound and free Ag* were separated by CE in approximately 1 min. Once linear calibration plots were established, extracts of sheep brains were tested for TSE. The assay successfully differentiated between scrapie-positive and normal sheep, and as little as 135 pg of PrP^{SC} could be detected. Purification of PrP^{SC} by HPLC prior to CEIA was found to significantly improve the assay reproducibility. In a subsequent paper, the same group reported on the continuation of their TSE work [9].

Lam *et al.* [10] examined the analysis of the staphylococcal enterotoxin A (SEA). Staphylococcal enterotoxins are proteins with molecular masses of approximately 30 kDa and basic p*I*s. A very low dose of these proteins (< 1 µg/kg body weight) is already sufficient to cause food poisoning in humans. FITC-labeled staphylococcal enterotoxin A (SEA*) was employed as the Ag* and rabbit anti-SEA as the selector. After a 30 min off-line incubation, the separation of free and bound Ag* required approximately 3 min. Multiple peaks were observed for free SEA* (most likely due to multiple FITC labeling); all of them were found to be immunologically active. Utilizing sheath flow to enhance detection, linearity was observed over the range from 0.3 to 6.5 nM of SEA (SD = 5%). The LOD was 0.3 nM SEA in buffer.

A major area of competitive CEIA development has been in the analysis of drugs of abuse. In one example, a competitive CEIA for the urinary analysis of methamphetamine (MA) was demonstrated [11]. Three polyclonal anti-

MA antibodies were evaluated. A methamphetamine derivative, *N*-(4-aminobutyl)methamphetamine, was labeled with fluorescein and used as the Ag*. As shown in Fig. 4, free Ag* and Ab-Ag* were well-separated in 5 min for all three Ab preparations. Responding successfully to a wide variety of antigen levels, the system provided a linear calibration from 50 to 1000 ng/mL and the LOD was 19 ng/mL methamphetamine. Not only was the LOD of the CEIA better than that of enzyme-linked immunosorbent assays (ELISAs) but also the cross-reactivities of the two systems differed despite the fact that both assay systems utilized the same antibodies. These discrepancies were explained by the fact that the ELISA employed a differ-

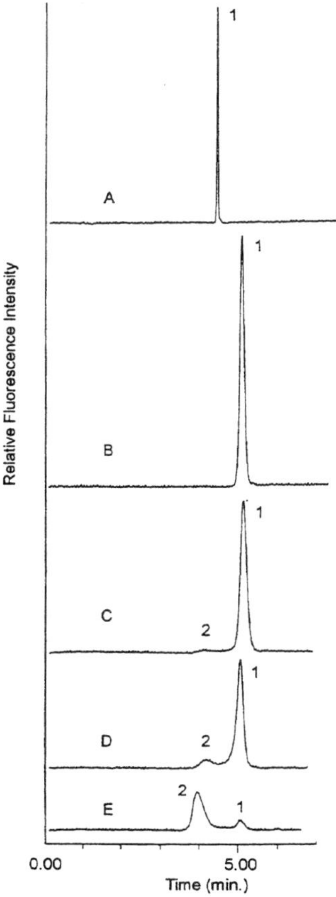

Figure 4. CEIA electropherograms of purified MA-FITC tracer (Ag*) and of screening profiles for anti-BSA activity of various antiserum preparations. (A) Plain Ag* without antiserum; (B) Ag* plus normal goat serum without Ab; (C) Ag* plus anti-MA-BSA(L) antiserum; (D) Ag* plus anti-MA-BSA(H) antiserum; (E) Ag* plus anti-MA-keyhole limpet hemocyanine (KLH) antiserum. Peaks: 1, free MA-FITC; 2, complexed MA-FITC. Reprinted from [11], with permission.

ently labeled tracer. Intra- and interassay precision and recoveries were found to be good. In a subsequent paper, this group [12] extended the application of their assay to various antisera, confirming the broader applicability of this CEIA system for MA analysis.

The feasibility of urinary methadone analysis by competitive CEIA and micellar electrokinetic capillary chromatography immunoassay (MECCIA) was evaluated by Thormann's group [13]. Immunoreagents were taken from a commercially available fluorescence polarization immunoassay (FPIA) methadone kit. Urine samples were incubated off-line for 10 min. Free Ag* and Ab-Ag* were not well-separated by MECCIA; thus quantification under these electrophoretic conditions was difficult. CEIA, on the other hand, allowed for the complete separation of free labeled antigen and complex in approximately 10 min and yielded a 10 ng/mL LOD for methadone in urine. A very similar system was used for the CEIA/MECCIA screening of urinary amphetamine and its analogs [14].

One of the potential advantages of CEIA over most other IA methods is the prospect of simultaneous analysis of several antigens due to the high separation power of CE. In an interesting paper, the same group [15] extended the work discussed in the previous paragraph one important step further. They described a competitive CEIA for the simultaneous urinary analysis of four drugs of abuse: methadone (M), opiates (O), the cocaine metabolite benzoylecgonine (C), and amphetamine/methamphetamine (A). The reagents of four commercially available FPIA kits, each analyzing one of the four substances, were added together in the appropriate ratios with urine samples and incubated for 10 min prior to injection in CE. The same conditions as before were applied for laser-induced fluorescence (LIF) detection and free-zone electrophoresis [13]. Figure 5 shows the multianalyte CEIA separation profiles. Baseline separation of all four analytes was achieved in approximately 12 min. Moreover, the changing peak ratios of M, O, C, and A prove that the CEIA system is responsive to the different amounts of drugs of abuse present in the screened urine samples. The CEIA data were in good agreement with an enzyme-modulated immunotest (EMIT) and FPIA data. The LOD in this multiplexed system was 30 ng/mL. The authors anticipate an even larger multianalyte capacity of their system through

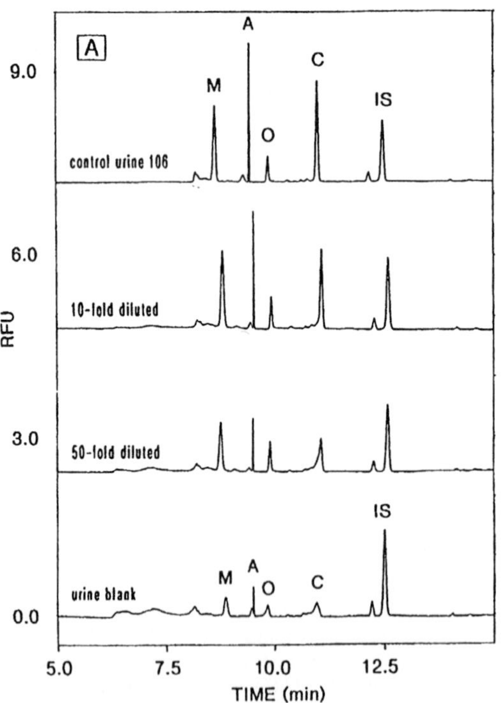
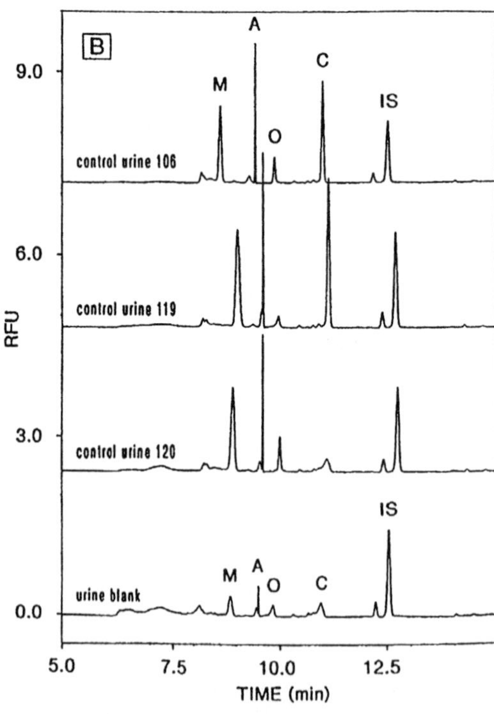

Figure 5. Electropherograms from multianalyte CEIA of four drugs of abuse. System response (A) at different sample dilutions, and (B) in different control urine samples. Peaks: M, methadone; O, opiates; C, cocaine metabolite benzoylecgonine; A, amphetamine/methamphetamine; IS, internal standard. Reprinted from [15], with permission.

the additional implementation of spectroscopic multiplexing, where different antigens will be tagged with different color fluorophores. Another multiplexed CEIA for screening of urinary drugs has been reported recently [16].

A thermally reversible hydrogel containing cross-linked polyclonal antibodies was investigated for rapid CEIA analysis of metabolites of methyltestosterone (MTS) in serum [17]. To simplify CEIA assay methodology and to avoid time-consuming off-line incubation, the system combined on-line complex formation and electrophoretic separation of free from bound Ag* in a single step. The immunogel was prepared from a mixture of *N*-isopropylacrylamide and *N*-acrylsuccinoamide monomers and anti-MTS polyclonal antibodies. Prior to MTS analysis, the separation capillary was filled with the Ab-containing gel; the FITC-labeled antigen was mixed off-line with the sample. No antibody had to be added to the sample before injection. The immunogel analysis resulted in well-shaped peaks and no noticeable complex dissociation during electrophoresis. It was observed that an increase in buffer and gel concentration and run voltage led to reduced migration times and improved resolution. Figure 6 illustrates the effect that increasing hydrogel concentrations

had on run time and separation. A linear calibration curve ($r^2 = 0.9913$) was generated in the range of 20–200 ng/mL of MTS.

Environmental monitoring assays were also transferred to competitive CEIA format. CEIA was applied to the quantification of the pesticide 2,4-dichlorophenoxyacetic acid (2,4-D) in river water [18]. An intact monoclonal antibody was used, and 2,4-D was labeled with FITC. After a 1 h off-line incubation with the antibody, the sample cocktail was filtered through a nylon filter to remove particles and then separated by CE. The separation time was approximately 14 min. Besides the well-separated free Ag* and complex peaks, additional peaks attributed to FITC labeling artifacts were seen. Although they generated a relatively complicated electropherogram, they did not interfere with the quantification. A linear response ($r^2 = 0.99$) was found from 10–600 ppb of 2,4-D. With an LOD of 1.2 nM, the CEIA assay was found to be slightly less sensitive than ELISA.

The first results of a CEIA for the natural toxins potato glycoalkaloids (GAs) were reported [19]. Diluted extracts of sprouted Yukon Gold potato tubers and nonsprouted Yukon Gold tubers were analyzed. The assay components consisted of rabbit polyclonal antiserum and the fluorescein-labeled alkaloid solanidine as the tracer. Solanidine was recognized by the Ab preparation used in this study as the most common alkaloid moiety of potato GAs. While the off-line incubation time was 30 min, CE analysis required approximately 4 min. A slightly basic buffer pH was chosen to ensure strong Ab-Ag binding, high quantum yield, and good solubility of GAs. The detergent SDS was needed as an additive to achieve baseline separation between free tracer and complex since their separation was found to be nontrivial. Future work is planned to validate the method.

3 Direct CEIAs

Most of the research in this field is still directed towards methodology rather than specific applications. This is probably due to the serious experimental difficulties encountered when the Ab*-Ag complex has to be separated from free labeled Ab* by CE for quantification. Surprisingly, the problem is not only observed for small antigens, which by their very nature contribute only a small amount of mass and charge to the Ab*-Ag complex, but also for much larger antigens such as proteins. A good illustration of this fundamental technical problem in direct CEIA was given by Ou and co-workers [20]. When attempting to analyze bovine serum albumin (BSA) by direct CEIA, no complete separation between free and bound antibody could be achieved, rendering quantifica-

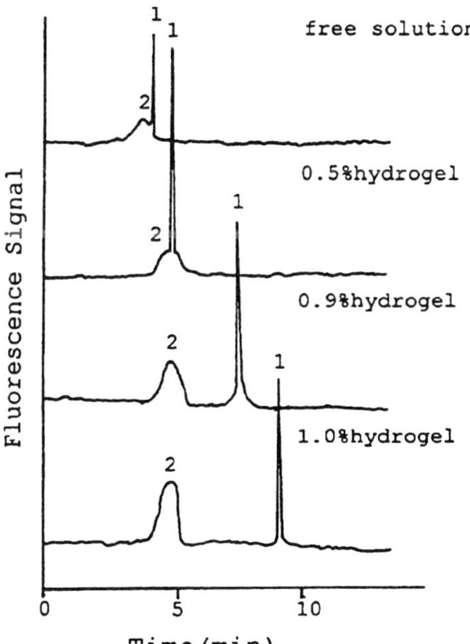

Figure 6. CE separation of bound and free Ag* by different concentrations of hydrogel containing immobilized Abs at 363 V/cm in 100 mM Tris-borate buffer. Peaks: 1, free Ag* 2, bound Ag*; Reprinted from [17], with permission.

tion impossible (Fig. 7). When the same analyte was quantified by competitive CEIA, however, near baseline separation was achieved. A linear calibration curve could then be established and an LOD of 47 nM of BSA was reached in uncoated capillaries.

One very interesting and strikingly simple solution to this direct CEIA separation problem was proposed by the same group [21]. In this novel approach, nondenaturing SDS capillary gel electrophoresis was used to analyze anti-BSA in water by CEIA with UV detection. The use of SDS renders the charges of all proteins the same, allowing them to be sorted exclusively according to their molecular weight. In addition, nondenaturing conditions allow the proteins to keep their native state during analysis. After a 10 min incubation of BSA and anti-BSA (both unlabeled), SDS was added to the sample, and the mixture was pressure-injected. Figure 8 shows that two complex peaks (with estimated molecular masses of 209 and 258 kDa) eluted at approximately 27 and 30 min, respec-

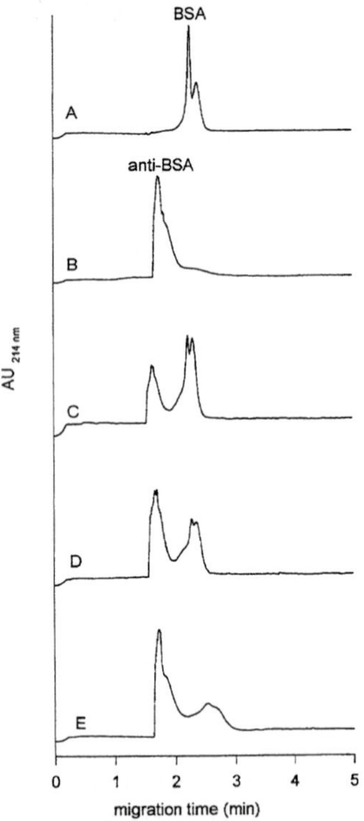

Figure 7. Example of an incomplete separation between free and bound Ab* in direct CEIA. CE-UV electropherograms of (A) 1 mg/mL BSA, (B) 1 mg/mL monoclonal anti-BSA, and mixtures of 1 mg/mL BSA with increasing concentrations of anti-BSA: (C) 0.5 mg/mL, (D) 1 mg/mL, and (E) 1 mg/mL. Reprinted from [20], with permission.

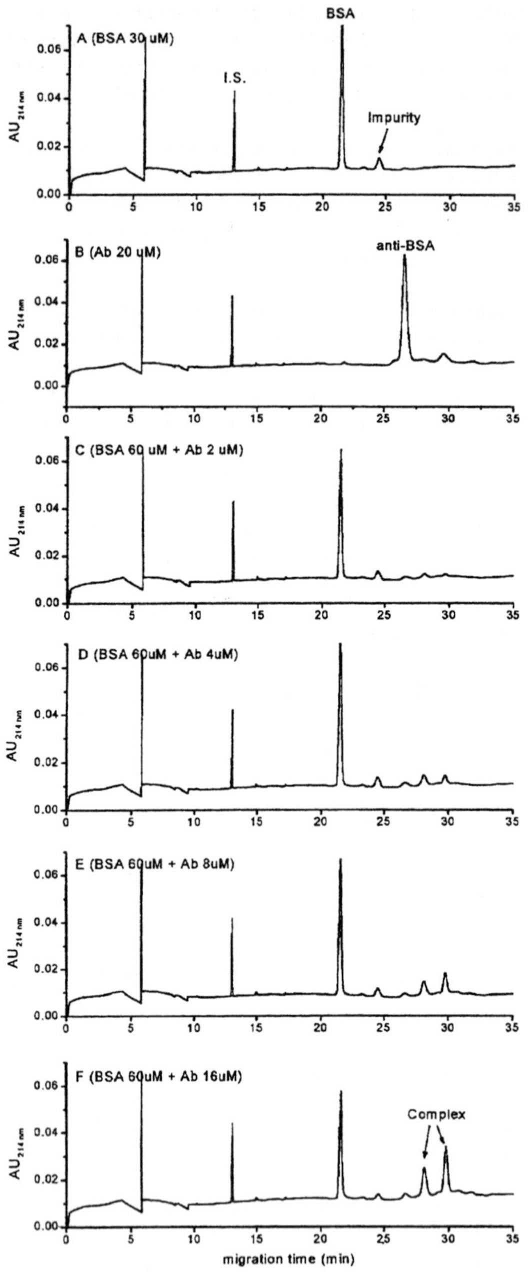

Figure 8. Nondenaturing gel electrophoresis employed for the separation of free and bound Ab in direct CEIA format. UV was used for detection. (A) 30 μM BSA; (B) 20 μM anti-BSA Ab, and mixtures of 60 μM BSA with different concentrations of anti-BSA antibody: (C) 2 μM, (D) 4 μM, (E) 8 μM, and (F) 16 μM. Reprinted from [21], with permission.

tively, well-separated from both free anti-BSA (147 kDa) and unbound BSA (64 kDa). The two complex peaks are likely to represent the binding of one and two BSA molecules to one anti-BSA molecule; their signal strength increased as the amount of BSA was increased. A linear calibration curve was obtained with an LOD of 0.1 µM of anti-BSA. The practicality of the proposed method will ultimately depend on its compatibility with biological fluids, its quantitative behavior, the possibility to use LIF detection, and the reduction of the separation time.

Another possibility to circumvent the difficulties encountered with the CE separation of free Ab* and Ab*-Ag in the direct CEIA format is the use of alternative selectors. The requirements for any type of antibody substitute in CEIA are strong binding affinity and high selectivity. Aptamers, single-stranded DNA or RNA molecules which bind with high specificity and affinity to target molecules, were investigated as potential antibody substitutes [22]. Known to exist for a wide variety of molecules, aptamers are developed through a combinatorial selection process. From the resulting small set of sequences with high binding affinities, the best candidates are then identified and synthesized. In this paper, IgE- and thrombin-binding aptamers were labeled with FITC at the 5′ position and used as selectors in IgE or thrombin CEIA, respectively. Off-line incubation time was 3 min for both systems. For the IgE assay no complex peak was seen. This was attributed to complex dissociation and adsorption to the capillary surface during electrophoresis. It was observed that the binding affinity of the IgG aptamer had decreased after FITC labeling, probably due to a change in the tertiary structure of the aptamer. An intact complex could only be detected after the shortening of the separation column, the increase in buffer pH, and the simultaneous application of vacuum to the capillary outlet to accelerate the migration of the complex. Using this method, a calibration was established that was linear up to approximately 300 nM IgE. The resulting LOD was 46 nM (S/N = 2). Similar results were achieved when serum was used. The anti-thrombin aptamer had an even weaker binding constant; therefore complete baseline separation between the free and bound thrombin aptamer was not possible. A connecting plateau was observed between the two peaks due to complex dissociation during separation (Fig. 9). For the thrombin assay, the LOD was 40 nM based on the complex peak. The authors believe that further development in aptamer technology, particularly in the enhancement of binding strength, could ultimately result in aptamers that are genuine alternatives to antibodies in CEIAs.

Green fluorescent protein (GFP) was explored as an alternative fluorescent labeling reagent in CEIA [23]. An acidic, globular protein with a molecular mass of 30 kDa,

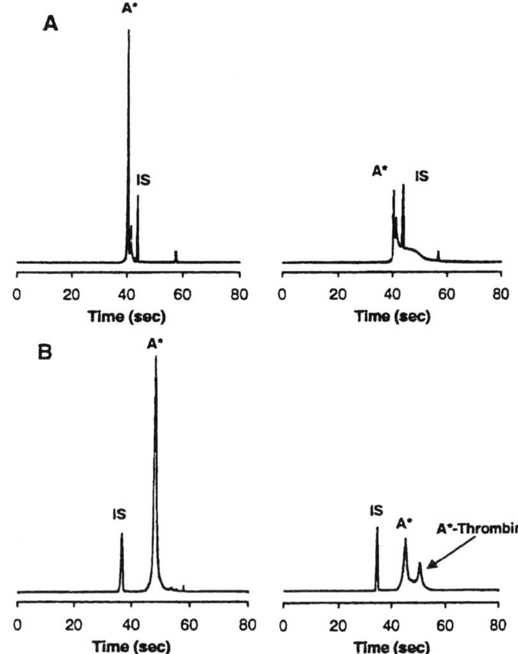

Figure 9. Determination of thrombin using FITC-labeled aptamer. (A) Electropherograms obtained for 2000 nM aptamer A* with 0 (left) and 3000 nM (right) thrombin at pH 7.4 and no flow. A large plateau next to the free FITC-labeled aptamer indicates the presence of the quickly dissociating complex. (B) Electropherograms obtained for the same concentrations at pH 8.2 with the application of vacuum. IS, internal standard. Reprinted from [22], with permission.

GFP exhibits natural fluorescence (excitation/emission = 470/509 nm). The tagging of antibodies, proteins, and peptides with GFP can be achieved by recombinant DNA technology through the fusion of genes encoding GFP and the protein of choice, followed by expression of the fluorescent fusion protein. For feasibility testing, nonfused GFP was used for rapid screening of anti-GFP. An argon-ion laser was used for the excitation of GFP at 488 nm. Complexation between GFP and anti-GFP occurred during a 4 min inubation, and separation by CE took approximately 5 min. Only incomplete separation of complexed and free GFP was achieved. In addition, it was also found that the fluorescence of the free GFP was sensitive to the pH of the incubation solution, while the quantum yield of the bound GFP was unaffected.

To achieve more sensitive detection in CEIA, exploratory work was performed on the combination of LIF with enzymatic signal amplification [24]. A schematic of the assay is presented in Fig. 10. Alkaline phosphatase (ALP) was used for enzymatic tagging of the antibody, and nonfluorescent fluorescein diphosphate (FDP) was added to the separation buffer as the enzyme substrate. ALP catalyzes the conversion of FDP to fluorescein through the cleavage of the two phosphate ester bonds present in FDP, which is the most sensitive fluorogenic ALP substrate currently available. After on-line amplification, fluorescein was detected by LIF. As a test system, ALP-conjugated antibody (goat anti-rat IgG) was used for the analysis of rat IgG. Following 1 h off-line incubation, free and conjugated ALP-Ab were separated in approximately 3 min. An amplification time of a few minutes generated sufficient fluorescein to guarantee good signal strength for the complex.

Two different capillary assay systems, flow-through IA and CEIA, were utilized for the direct analysis of monoclonal anti-BSA [25]. In the flow-through format (which does not require electrophoresis) an Ab capture column filled with protein-G-coated beads was connected to a plain detection capillary. Protein G is known to bind selectively to a wide range of IgG species. First, FITC-BSA was incubated off-line for 20 min with anti-BSA for complex formation. The mixture was then pressure-loaded into the affinity column for the selective capture of the complex. To ensure complete binding of the Ab-Ag* to the short protein G capture segment, a low flow rate was used, and the total amount of loaded complex was never allowed to exceed one-twentieth of the binding capacity of the capture column. After pressure washing for sample matrix removal followed by Ab-Ag* desorption with glycine buffer, the fluorescent complex was moved by pressure to the LIF detection window for quantification. The response was linear in the range from 80 to 800 pM, and the LOD

was 60 pM (S/N = 3). A small fluorescent background peak, which eluted close to the sample peak, dictated the LOD. Under nonoptimized conditions, the total assay time was 30 min. In the CEIA format, the same off-line incubation conditions were used. However, the separation between free Ag* and complex was achieved by electrophoresis instead of an Ab capture column. Separation by CE required approximately 10 min. A linear relationship ($r^2 = 0.999$) was found between sample amount and signal strength for up to 160 nM of anti-BSA. For the CEIA, the LOD was 8 nM as dictated by the detector response. The lower LOD of the first method (100 × more sensitive) was due to the fact that the protein capture column acted not only as an affinity reactor but also as a complex preconcentrator.

In a subsequent paper [26], the same group used a similar CEIA for the analysis of monoclonal anti-BSA in mouse serum. This time, however, BSA was labeled with the near-infrared dye Cy5 instead of the more common

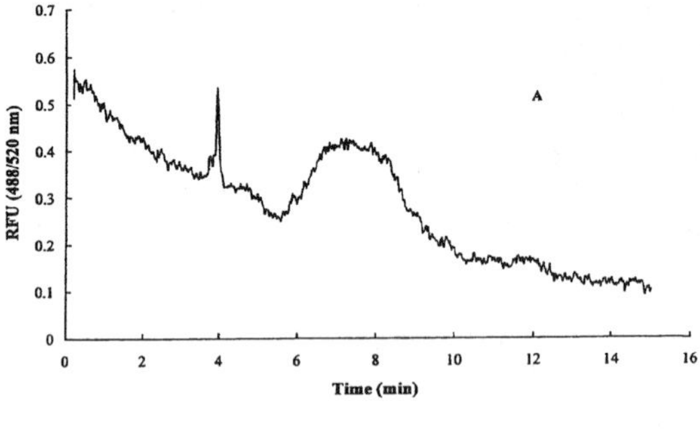

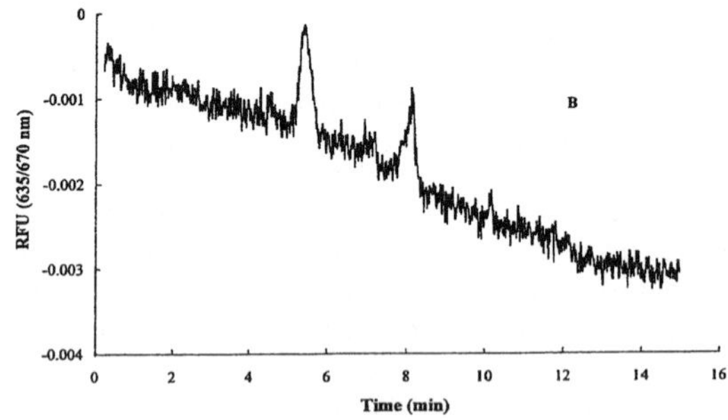

Figure 11. CE-LIF electropherograms illustrating the background fluorescence of blank mouse serum with two different lasers. (A) The 488 nm line of an argon-ion laser was employed for excitation; emitted fluorescence was detected at 520 nm. (B) The 635 nm line of a diode laser was used for excitation; emitted fluorescence was detected at 670 nm. Note the dramatic difference in the absolute signal strength of the background fluorescence. Reprinted from [26], with permission.

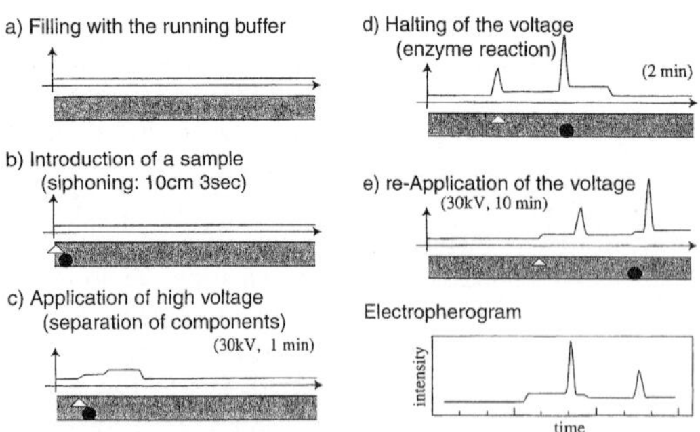

Figure 10. Schematic of the direct enzymatic CEIA procedure with alkaline phosphatase as label and nonfluorescent FDP in the running buffer. The circle and triangle represent the free and complexed ALP-conjugated Ab. Reprinted from [24], with permission.

FITC, permitting the use of a diode laser. Diode lasers, besides being less expensive and more compact than argon-ion lasers, offer the practical advantage of preventing the excitation of background components present in biological fluids, which do not fluoresce at 630 nm. A comparison of the two laser systems in this respect is shown in Fig. 11. As done previously, the off-line incubation time in serum was 20 min. The separation required approximately 9 min on uncoated capillaries. A linear relationship ($r^2 = 0.9997$) was found for 4–40 nM of anti-BSA, and the LOD was approximately 1.2 nM (S/N = 3). When compared to the argon-ion laser used in the previous setup, the improved LOD of the diode laser system was attributed to the significantly reduced background fluorescence. Recovery and precision for this assay were 97.5–104.6% and < 3.5%, respectively.

In addition to the developments in methodology, a few practical applications were also published. Direct CEIA was used for the determination of the minor protein bovine lactoferrin (bLF) in whey [27]. The analysis of bLF in whey is usually hampered by the high concentration of background components such as the major whey proteins and lipids. Two different selectors were investigated: FITC-labeled polyanionic lipopolysaccharide (LPS*), serving as a potential Ab substitute, and FITC-labeled anti-bLF. Although baseline resolution between free and complexed LPS* could be achieved in approximately 12 min, the determination of bLF in whey was not possible using this selector due to its polyanionic character which resulted in nonspecific interactions with other whey components. No separation was achieved for the anti-bLF system. Nevertheless, a decrease of peak area was observed when bLF was added (possibly due to the quenching of Ab fluorescence during complex formation) permitting quantification despite the complete lack of resolution. However, the CEIA quantification agreed in only a few cases with the results achieved by traditional ELISA.

An interesting assay for the rapid and simultaneous monitoring of three recombinant cytokines in biological fluids from patients suffering from malignant melanoma and undergoing cytokine therapy was described by Phillips and Dickens [28]. Instead of performing the typical homogeneous CEIA where both Ab and Ag are in solution, a mixture containing an equal amount of the three selectors was immobilized at the inlet of the capillary for the selective capture of the three cytokines prior to their separation. Antigen binding fragments (Fabs) rather than intact antibodies were chosen as selectors. An excess of Fabs was used to ensure total antigen capture. The selector immobilization contributed the added benefits of on-line analyte extraction and preconcentration. Patient samples were fluorescently labeled with Cy5 prior to their analysis

by LIF detection. A parabolic mirror was placed around the detection window of the capillary to increase the signal strength. After ~ 30 nL aliquots of sample were injected by vacuum into the capillary followed by a 2-min on-line incubation, the capillary was purged with buffer to remove all unbound materials. Recovery and electrophoretic separation of the three recombinant cytokines was accomplished in 5 min with complete baseline resolution. Not only did the assay results compare well to conventional IA, but also the CEIA analysis time of 20 min constituted only a fraction of the time typically required for other IA formats. The authors believe that their method should be applicable to a wide variety of analytes.

Postcapillary affinity detection was applied to the fast screening of antibody heterogeneity in cell cultures and other complicated sample matrices [29]. Fluorescein-labeled fragment B (B*) of protein A was chosen as a universal but specific Ab selector. In this technique, Ab isoforms were vacuum-transferred after their electrophoretic separation across a 20 µm gap from the separation capillary into the reaction capillary. The postcolumn affinity reactor was filled with a buffer solution containing B*. Labeled fragment B and separated Ab variants were then allowed to react before being moved past the LIF detector for quantification. The reaction time was restricted to only 10 s to keep additional peak broadening to a minimum. Depending on the sample type, total analysis time was between 20 and 30 min. No interferences from the sample matrices were observed, and the technique was found to be linear for various Ab preparations. This novel method might allow rapid evaluation of Ab heterogeneity and product consistency for quality control during Ab production with minimal sample preparation.

4 Microchip-based CEIAs

In conjunction with the exponential growth in microfabricated device research, several developmental advances have been realized in mirochip-based CEIAs. Jiang *et al.* [30] investigated the utility of red solid-state diode lasers (635 nm excitation) for direct CEIA analysis on microchips. Solid-state lasers are an interesting alternative to the commonly used gas-phase lasers. The bulky size of the gas-phase laser is incongruous with the compactness of microdevices, whereas the solid-state laser is smaller and, as such, might be more compatible with microelectrophoretic systems, particularly when portable CEIA analyzers are considered. The test system consisted of Cy5-labeled anti-ovalbumin with ovalbumin as the antigen. Following off-line incubation, the complex could not be fully separated from free Ab* under the conditions used. Next, the influence of optical filter sets, laser spot size, confocal pinhole size and microfabricated channel dimen-

    Electrophoresis 2000, *21*, 3919–3930

sions on the detector performance were evaluated with Cy5 dye as the test compound. Under optimized detection conditions, the LOD (S/N = 3) for 13 and 20 µm deep channels was 20 and 9 pM, respectively. These results indicated that solid-state diode lasers could be a useful light source for CEIA.

Another way of rendering microanalyzers simpler and more compact might be by eliminating the light source altogether. Post-separation chemiluminescence detection was employed for CEIA in microchips [31]. It is well-known that chemiluminescence detection, which is widely used in IA, does not require any physical light source since the emitted light is generated by a photochemical process based on the horseradish peroxidase (HRP)-catalyzed reaction of luminol with peroxidase. HRP-derivatized anti-mouse IgG conjugate was used in a test system for the direct CEIA of mouse IgG. The separation buffer contained luminol for the chemiluminescence reaction as well as a high salt and Tween-20 content, both for the reduction of protein adsorption to the surface of the microchannel. A side channel, the "postcolumn reaction channel", intersected at the end of the separation channel to supply peroxide to the analytes after their separation, thereby initiating the photochemical process. The immunoreagents were incubated off-line and injected after 15 min. After a very rapid 15 s separation, the chemiluminescent signal of the free Ab* peak could be detected. However, no complex peak was seen due to immunoprecipitation of the complex in the sample channel. Usually the complex signal is preferred for quantification in the direct format since it is far more responsive than the free Ab* signal to changes in Ag concentration (the Ab* is typically present in large excess). Nevertheless, a linear calibration curve could be established using the free Ab*. The detection limits for the 13 and 40 µm deep channels were 35 and 9 nM, respectively, which were found to be a factor of 100–1000 less sensitive than when conventional LIF detection was employed.

A competitive CEIA for theophylline (Th) in human serum was performed on an integrated electrophoretic microdevice where both incubation and subsequent electrophoretic analysis were performed on-chip [32]. To allow for uniform dispersion of the sample components across the channel, mixing coils were designed to be long and narrow. LIF was used for detection, and a commercially available FPIA kit containing labeled Ag*, Ab, and Th serum calibrators was employed. Figure 12 illustrates how on-chip mixing, reaction, and separation were performed. The on-line immunoreaction, which relied on electroosmotic pumping, required only 1.5 min to nearly reach completion, whereas the separation itself required 1 min. A typical electropherogram of the CEIA achieved

A) On-Chip Mixing & Reaction

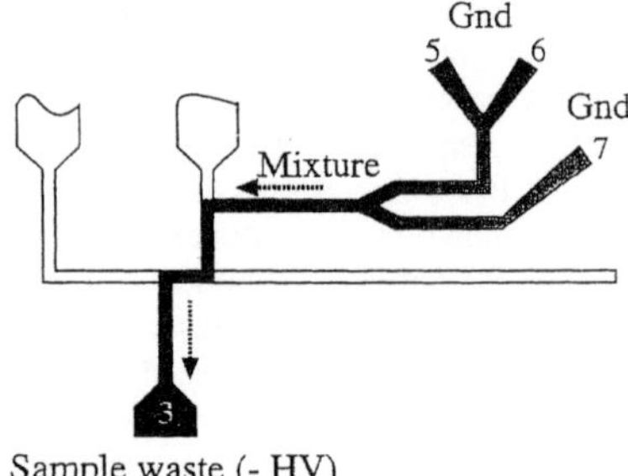

B) Separation

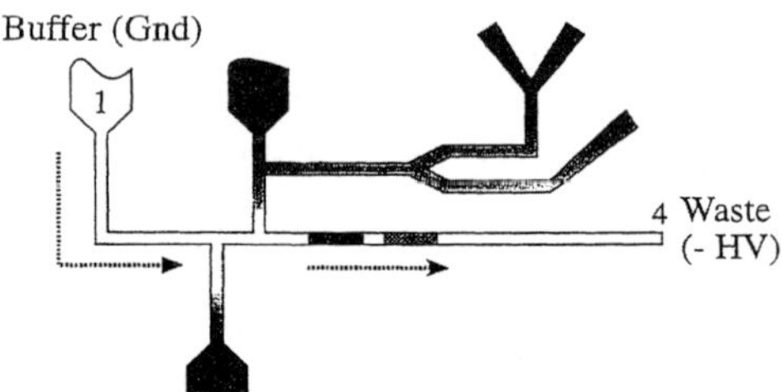

Figure 12. Schematic of on-line microchip competitive CEIA for Th. (A) solutions of Th, Th* and Ab in reservoirs 5, 6, and 7, respectively, were electroosmotically pumped along the shaded area by applying electrical ground (Gnd) to these reservoirs and negative high voltage (–HV) at the sample waste (reservoir 3). This forms a plug of the on-line incubated mixture at the double-T intersection. (B) The reagents and products were separated by switching Gnd to reservoir 1 and –HV to the separation waste (reservoir 4). Reprinted from [32], with permission.

directly after on-chip mixing is shown in Fig. 13. A calibration curve was established in the clinically relevant range of 4–40 mg/L. The measured LOD was 0.26 mg/L of Th. In addition, some recovery values were determined. The same authors also applied an electrophoretic microdevice to the determination of the affinity constant of a monoclonal antibody for fluorescently labeled BSA [33].

5 Review articles

Since 1997, two reviews have been published specifically on CEIAs [3, 4]. The reader is referred to these reviews for a discussion of the principles and theory of CEIA. Several reviews were published that addressed CEIA but covered broader topics such as clinical chemistry [34–37], microchip-CE [38, 39], and affinity CE [40–42]. Finally, one review specifically reported on chemiluminescent IAs [43].

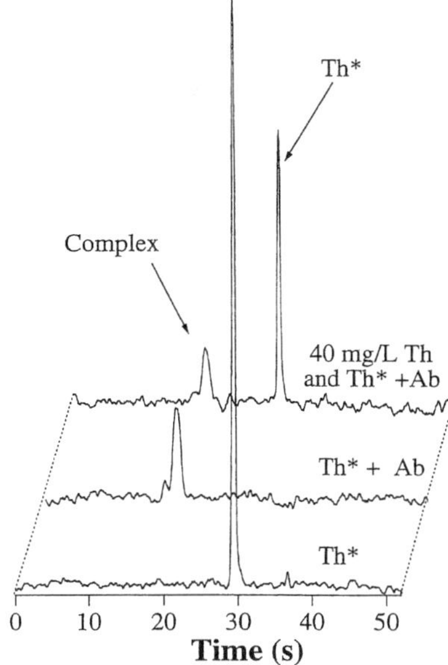

Figure 13. Electropherograms of the competitive Th CEIA with on-chip mixing of immunoreagents. The lower trace results from having Th* in reservoir 6 and buffer in reservoirs 5 and 7. The middle trace is due to the subsequent substitution of buffer with Ab in reservoir 7. The upper trace is caused by the subsequent replacement of buffer with Th sample in reservoir 5. Reprinted from [32], with permission.

6 Conclusions

CEIAs have now been investigated for quite some time. Essentially, any type of candidate molecule for traditional IA is now being tested in the CE format. Applications developed so far include: biomedical monitoring, toxicological assays, disease diagnosis, drug testing, and environmental monitoring. Thus far, the methodology of competitive CEIAs appears to be far more developed than that for the direct format. For the former, a simple protocol has evolved which is followed by most authors. Most commonly, fluorescently tagged antigen and unlabeled, intact antibody or antibody fragments are incubated off-line with the sample containing the antigen to be measured. The typical incubation volume ranges from 10 to 30 μL, and the typical incubation time is on the order of minutes. After separating free Ag* from complex Ab-Ag* on uncoated capillaries by free-zone electrophoresis, detection is usually performed with LIF (mostly in conjunction with FITC labeling and low power argon-ion lasers). The separation between free and bound Ag* is usually trivial, and the best LODs are in the nM range.

Direct CEIAs, on the other hand, have attracted less attention. They lag seriously behind in development despite the fact that they are as important as competitive assays. Moreover, purely from the point of view of assay development, they should be even easier to establish than competitive IAs since they do not require the tedious adjustments of the properties of the immunoreagents (*e.g.*, linker chemistries, type of Ab, type of tracer, *etc.*) for optimal quantification. Their only requirement, notwithstanding high specificity and strong binding, is that the Ab* be present in excess to ensure total Ag capture. Furthermore, direct IAs exhibit larger dynamic ranges than competitive IAs. It was realized early on, however, that the electrophoretic separation between free and bound Ab* is not trivial. At present, no simple and general CEIA technique has been developed as in the case of the competitive format. General assay concepts are highly desirable since they greatly reduce assay development costs. In fact, several of the direct CEIAs discussed in this review achieved only incomplete or no separation, thereby seriously impairing quantification. When good resolution was achieved, it was usually not because of the separation power of free-zone CE but rather due to the use of affinity precolumns [25, 28], unintended immunoprecipitation of the complex inside the capillary [31], or the inversion of the direct format (where the Ag is used as selector for the Ab) [25, 26]. The only novel approach wherein baseline resolution was achieved by introducing electrophoretic mobility differences employed nondenaturing gel CE [21]. This proposed method promises to be elegant, straightforward, and general for larger antigens, once certain present limitations such as poor LOD and comparatively low speed have been addressed. Also, the use of alternative selectors such as aptamers may ultimately become attractive in this respect.

Besides the current restriction of practical CEIA to competitive assays, there is to date no proof of concept that CEIAs can be performed in parallel and fully automated. All analyses thus far were performed manually in a serial fashion. Both automation and higher throughput are necessary features if CEIA is to successfully compete with well-established IA methods. Also, no solution has arisen as yet to improve the current LOD of CEIAs beyond the typical nM range, limiting the applicability of this method to more concentrated samples. Other detection schemes such as red diode lasers, enzyme amplification, and chemiluminescence do not seem to overcome this limitation thus far. Moreover, the experimental data available on such essential parameters as recovery, accuracy, and precision are currently sparse and restricted to small sets of real samples. It remains to be seen how CEIAs will perform under real-world conditions where thousands of samples have to be routinely analyzed. On the other hand, CEIAs compare well to established IA formats in respect to sample volume requirements, speed, ease of multiplexing (electrophoretically and spectroscopically), and the possibility of miniaturization and total process integration through the use of microfabricated devices. It will be interesting to follow over the coming years how CEIA will advance and where it will find its place in this tremendously important field of analytical chemistry.

Received June 23, 2000

7 References

[1] Schultz, N. M., Kennedy, R. T., *Anal. Chem.* 1993, *65*, 3161–3165.

[2] Shimura, K., Karger, B. L., *Anal. Chem.* 1994, *66*, 9–15.

[3] Schmalzing, D., Nashabeh, W., *Electrophoresis* 1997, *18*, 2184–2193.

[4] Bao, J. J., *J. Chromatogr. B* 1997, *699*, 463–480.

[5] Ye, L. W., Le, X. C., Xing, J. Z., Ma, M. S., Yatscoff, R., *J. Chromatogr. B* 1998, *714*, 59–67.

[6] Wan, Q. H., Le, X. C., *J. Chromatogr. A* 1999, *853*, 555–562.

[7] Tao, L., Aspinwall, C. A., Kennedy, R. T., *Electrophoresis* 1998, *19*, 403–408.

[8] Schmerr, M. J., Jenny, A., *Electrophoresis* 1998, *19*, 409–414.

[9] Schmerr, M. J., Jenny, A. L., Bulgin, M. S., Miller, J. M., Hamir, A. N., Cutlip, R. C., Goodwin, K. R., *J. Chromatogr. A* 1999, *853*, 207–214.

[10] Lam, M. T., Wan, Q. H., Boulet, C. A., Le, X. C., *J. Chromatogr. A* 1999, 853, 545–553.

[11] Choi, J., Kim, C., Choi, M. J., *J. Chromatogr. B* 1998, *705*, 277–282.

[12] Choi, J., Kim, C., Choi, M. J., *Electrophoresis* 1998, *19*, 2950–2955.

[13] Thormann, W., Lanz, M., Caslavska, J., Siegenthaler, P., Portmann, R., *Electrophoresis* 1998, *19*, 57–65.

[14] Ramseier, A., Caslavska, J., Thormann, W., *Electrophoresis* 1998, *19*, 2956–2966.

[15] Caslavska, J., Allemann, D., Thormann, W., *J. Chromatogr. A* 1999, *838*, 197–211.

[16] Thormann, W., Caslavska, J., Ramseier, A., Siethoff, C., *J. Microcol. Sep. 2000, 12, 13–24.*

[17] Zhang, X. X., Li, J., Gao, J., Sun, L., Chang, W. B., *Electrophoresis* 1999, *20*, 1998–2002.

[18] Rogers, K. R., Apostol, A. B., Brumley, W. C., *Anal. Lett.* 2000, *33*, 443–453.

[19] Driedger, D. R., LeBlanc, R. J., LeBlanc, E. L., Sporns, P., *J. Agric. Food Chem.* 2000, *48*, 1135–1139.

[20] Ou, J. P., Wang, Q. G., Cheung, T. M., Chan, S. T. H., Yeung, W. S. B., *J. Chromatogr. B* 1999, *727*, 63–71.

[21] Ou, J. P., Chang, S. T. H., Yeung, W. S. B., *J. Chromatogr. B* 1999, *731*, 389–394.

[22] German, I., Buchanan, D. D., Kennedy, R. T., *Anal. Chem.* 1998, *70*, 4540–4545.

[23] Korf, G. M., Landers, J. P., Okane, D. J., *Anal. Biochem.* 1997, *251*, 210–218.

[24] Koizumi, A., Morita, T., Murakami, Y., Morita, Y., Sakaguchi, T., Yokoyama, K., Tamiya, E., *Anal. Chim. Acta* 1999, *399*, 63–68.

[25] Wang, Q. G., Luo, G., Ou, J. P., Yeung, W. S. B., *J. Chromatogr. A* 1999, *848*, 139–148.

[26] Wang, Q. G., Luo, G., Wang, Y. M., Yeung, W. S. B., *Anal. Lett.* 2000, *33*, 589–602.

[27] Riechel, P., Weiss, T., Weiss, M., Ulber, R., Buchholz, H., Scheper, T., *J. Chromatogr. A* 1998, 817, 187–193.

[28] Phillips, T. M., Dickens, B. F., *Electrophoresis* 1998, *19*, 2991–2996.

[29] Kelly, J. A., Lee, C. S., *J. Chromatogr. A* 1997, *790*, 207–214.

[30] Jiang, G. F., Attiya, S., Ocvirk, G., Lee, W. E., Harrison, D. J., *Biosens. Bioelectron.* 2000, *14*, 861–869.

[31] Mangru, S. D., Harrison, D. J., *Electrophoresis* 1998, *19*, 2301–2307.

[32] Chiem, N. H., Harrison, D. J., *Clin. Chem.* 1998, *44*, 591–598.

[33] Chiem, N. H., Harrison, D. J., *Electrophoresis* 1998, *19*, 3040–3044.

[34] Anderson, D. J., Guo, B. C., Xu, Y., Ng, L. M., Kricka, L. J., Skogerboe, K. J., Hage, D. S., Schoeff, L., Wang, J., Sokoll, L. J., Chan, D. W., Ward, K. M., Davis, K. A., *Anal. Chem.* 1997, *69*, R165–R229.

[35] Thormann, W., Aebi, Y., Lanz, M., Caslavska, J., *Forens. Sci. Int.* 1998, *92*, 157–183.

[36] Couderc, F., Causse, E., Bayle, C., *Electrophoresis* 1998, *19*, 2777–2790.

[37] Thormann, W., Wey, A. B., Lurie, I. S., Gerber, H., Byland, C., Malik, N., Hochmeister, M., Gehrig, C., *Electrophoresis* 1999, *20*, 3203–3236.

[38] Colyer, C. L., Tang, T., Chiem, N., Harrison, D. J., *Electrophoresis* 1997, *18*, 1733–1741.

[39] Effenhauser, C. S., Bruin, G. J. M., Paulus, A., *Electrophoresis* 1997, *18*, 2203–2213.

[40] Shimura, K., Kasai, K., *Anal. Biochem.* 1997, *251*, 1–16.

[41] Heegaard, N. H. H., Nilsson, S., Guzman, N. A., *J. Chromatogr. B* 1998, *715*, 29–54.

[42] Heegaard, N. H. H., Kennedy, R. T., *Electrophoresis* 1999, *20*, 3122–3133.

[43] Baeyens, W. R. G., Schulman, S. G., Calokerinos, A. C., Zhao, Y., Campana, A. M. G., Nakashima, K., De Keukeleire, D., *J. Pharm. Biomed. Anal.* 1998, *17*, 941–953.

Electrophoresis 2000, *21*, 4179–4191

Review

Andrei R. Timerbaev[1]
Oleg A. Shpigun[2]

[1]Vernadsky Institute of
Geochemistry and Analytical
Chemistry,
Russian Academy of Sciences,
Moscow, Russia
[2]Chemistry Department,
Lomonosov Moscow State
University,
Moscow, Russia

Recent progress in capillary electrophoresis of metal ions

Advances in the fundamental studies and methodology of capillary electrophoresis (CE) as applied to metal ion analysis over the last two years are reviewed, with the objective of providing the interested reader with a state-of-the-art picture of technique's potentialities in the area. In particular, novel strategies for separation selectivity control and CE system innovations designed to enhance the detection sensitivity are described. In addition, a brief overview of the primary metal analytes and samples for which the technique appears to be best suited is given. The current limitations of the technique regarding most of all the implementation for routine use are considered along with the approaches on how they could be addressed. Finally, some pointers as to the likely trends in the future research are discussed.

Keywords: Capillary electrophoresis / Metal ions / Metal complexes / Organometals / Review

EL 4174

Contents

Correspondence: Prof. Andrei R. Timerbaev, Vernadsky Institute of Geochemistry and Analytical Chemistry, Kosygin St. 19, 117975 Moscow, Russia,
E-mail: rtimer@online.ru
Fax: +7-095-938-2054

Abbreviations: CDTA , 1,2-cyclohexanediaminetetracetic acid; **CL**, chemiluminescence; **ED**, electrochemical detection; **HEDTC**, bis(2-hydroxyethyl)dithiocarbamate; **ICP-MS**, inducitvely coupled plasma – mass spectrometry; **PAR**, 4-(2-pyridylazo)resorcinol; **PDC**, 2,6-pyridinedicarboxylic acid; **RE**, rare-earth elements

1 Introduction

Since the introduction of CE into the domain of inorganic ion analysis in the early 1990s, the separation and determination of metal ions is without doubt the area in which CE is being used to an increasing extent. This fact has been reflected in numerous publications revealing that there is not a single reason why CE might be a method of choice more suitable than, *e.g.*, HPLC, the conventional technique for multimetal species analysis. First, in carrying out such analyses, CE's specific strength is an ability to provide superior resolution, which is achievable in a shorter time, using simpler instrumentation, with lower consumption of materials (and hence fewer disposal problems), and at lower cost than in HPLC. Next, and an often more important benefit of CE is a higher degree of independence to complex matrices that simplifies sample preparation methodologies. Furthermore, CE exerts minor disturbances on the original distribution of metal species in a sample that makes the method highly promising for speciation analysis purposes.

These and other advantages of CE along with a broad range of applications to metal determinations in various matrices have been evaluated in depth in a number of review papers that appeared over the past five years [1–

44] (this listing could be continued; see also Table 1 for the subjects examined). Among these, a comprehensive coverage of contributions dealing with metal ion separations by CE, published up to the beginning of 1997, should be mentioned specifically [19], as one of our major aims was to provide a continuation of that review. More recently, a special issue of Journal of Chromatography A [45] was exclusively devoted to CE of inorganic species, including a diversity of metal analytes. The issue contains a large proportion of the overviews [28–40], the subject of which is closely related or even coincided with that of the present work. In view of these circumstances and in an effort to help the reader assess the latest trends avoiding duplication of material covered in previous reviews, only the literature of the last two years is examined here. A survey of the selected key separation approaches will be given as well as of general methodology used for improving the versatility of the CE technique. New and improved detection designs have been developed by several workers and will be discussed, as will the trend toward the coupling of CE with element-specific detectors. In particular, strong emphasis is placed on major developments and current limitations in real sample analysis. Finally, the possible future trends and prospects in metal ion analysis using CE will be briefly discussed.

2 Reviews

During the two-year review period, a number of important reviews discussing basic and application principles relevant to the separation and quantification of metal ions were published [20–44]. Table 1 gives a summary of major topics covered. The introductory-level book chapter by Fritz [22], one of the pioneers in the area, provides brief background information on fundamentals and basic methodological principles of CE in metal ion analysis. This work would be particularly useful for better orienta-

Table 1. Summary of review papers on CE of metal ions

Main subject	Reference
General principles	[22, 24, 28]
Instrumentation	[22]
Separation strategies	[22, 23, 28, 30, 31, 35, 36]
Detection	[27, 34, 35, 42]
– ICP-MS	[25, 26]
– Electrochemical methods	[32, 33]
– Chemiluminescence	[21]
Sample introduction modes	[29, 34, 36, 42]
Speciation analysis	[20, 25, 30, 36, 43, 44]
Applications	[28, 29, 35]
– Environmental samples	[37, 38, 41]
– Biological samples	[24, 38–40]
– Food and beverages analysis	[39, 40]

tion of those just beginning to work with CE methods development. Janos [28] prepared a general review on the role of complexing and other side equilibria in a CE system in controlling the effective mobilities of the metal ion analytes. The methods for optimization of the separation selectivity by exploiting complex formation were also discussed by Chiari [23]. Okada [31] provided a detailed discussion on the use of polyethers both as complexing ligands and separation medium modifiers incorporated in running solutions to enhance separations of metal ions. Metal ion complexation in its various modes has been extensively treated by Pacakova and co-workers [35] who summarized numerous complexation systems in use in table format. Derivatization into stable complexes with the metal ion analytes continues to be a frequently used approach for the purposes of enhancement of both the separation selectivity and detection sensitivity. A lengthy review by Liu *et al.* [36] is devoted to practical and theoretical considerations, such as the various modes of CE, approaches to derivatization, choice of complexing reagents, and manipulation of sensitivity and selectivity.

According to the amount of published work, going forward to an advanced methodology to lower concentration limits of detection presents one of the main current trends in metal ion CE. Timerbaev and Buchberger [34] prepared an extensive review which compares various nonabsorbance-based detection modes in terms of applicability, effects on separation efficiency, costs and manufacture concerns. Also, thoroughly discussed sensitivity enhancement means were on-capillary and pre-electrophoresis concentration techniques and chemical derivatization. The update of developments involving newly emerging detection methods has recently appeared [42] (note that this review was completed and published in the year of 2000). Over the past years, there has been certain interest in the use of electromigration mode of sample introduction for on-line enrichment of metal ions, and its advances have been reviewed by Krivacsy *et al.* [29].

Turning to specific detection techniques, the reports by Barnes [25] and Sutton and Caruso [26] overviewed the latest achievements in coupling CE with inductively coupled plasma MS (ICP-MS). Both articles are primarily focused on the current state-of-the-art for interface designs between the two types of instruments that appear to be the bottleneck of CE-ICP-MS shift to practical use. Recent progress in the design and application of electrochemical detection (ED) systems was highlighted by Kappes and Hauser [32] and Polesello and Valsecchi [33]. These reviews cover amperometric, potentiometric and conductometric detectors which from both sensitivity and specificity points of view hold a considerable (but different) promise in the field of metal ion analysis based on

CE separation. According to a survey by Zhang *et al.* [21], advances in combining CE with on-line chemiluminescence (CL) detection, an evolving and extremely sensitive technique, should be based on more reliable and possibly simpler detector configurations.

Application to determining different species of a metal was a popular topic for review (see Table 1), with a first review article on element speciation analysis by CE authored by Dabek-Zlotorzynska and her colleagues [20]. This is not surprising given the number of research groups pushing CE into this field where the technique foremost benefit of negligible (or only minimal) changes in speciation during separation and detection plays a key role. The authors of that review centered on a description of various CE methods for speciation studies of individual metals as well as for characterization of metal interactions with biological and naturally occurring macromolecules. The use of chemical modification in CE to obtain quantitative elemental speciation information was the subject of concern in a compilation by Liu and Lee [30]. Most recently, Timerbaev [43] conducted an extensive literature survey on simultaneous separation and determination of different chemical forms of metals. The present issue categorizes the original papers according to the type of specific speciation task solved, that is, differentiation of metal ions in different oxidation states, metal complexes with inorganic and organic ligands, metal oxoanions, organometallic compounds, *etc.*, and contains the most complete summary of reported application studies to date.

A great deal of work over the past few years has been done on CE application to the analysis of real-world samples. Numerous innovative applications, as shown in nearly each aforementioned review paper, witness not only a considerable potential of CE but also the method's premature status in important areas of metal ion analysis. Furthermore, the growing practical utility of CE involving a range of specific samples, such as various environmental matrices [37, 38, 41], biological systems [24, 38–40], and food samples [39, 40], was the subject of respective recent reviews.

3 Basic principles

At present, the fundamental studies in the area can be characterized as having reached a quieter period after the initial steps of progress which had played an important part in the method's emergence. This confirms one of the major trends of metal ion CE, that is, a shift of research interests toward application domains. Havel and co-workers [46, 47] reported on the use of neural networks for computer-aided modeling and optimization of CE systems

involving multiple parameters influencing the migration process. It was demonstrated that the optimization of resolution can be greatly facilitated by combining artificial neural networks with experimental design. Such a combination allows for predicting the optimal separation conditions from a low number of experiments and without knowing an explicit model of the separation process or the mathematical description of operational parameters. The feasibility of the optimization scheme developed has been illustrated for the separation of multicomponent mixtures of metal ions [46] and metal complexes [47]. The development of a valid migration model for micellar CE of metal ions complexed with bis(2-hydroxyethyl)dithiocarbamate (HEDTC) and 1,2-cyclohexanediaminetetraacetic acid (CDTA) was recently reported by Breadmore *et al.* [48]. This model was evolved to relate the effective mobility to the electrophoretic mobility of the anionic analyte, the partition coefficient of the analyte into the micelle and system variables such as concentrations of surfactant and organic modifier. Using nonlinear regression, parameters for the model were derived from experimental data and these parameters were then used to predict effective mobilities of the analytes. As can be seen in Fig. 1, predicted values agreed fairly well with experimental values.

Both the surface of the internal capillary wall and ligand purity need to be characterized in order to identify and minimize a number of problems that alter the separation in metal-complexing CE systems. In detailing the analyte-wall interactions for the anionic metal complexes of large metallochromic ligands, Macka *et al.* [49] compared separations obtained for several capillaries differing in nature

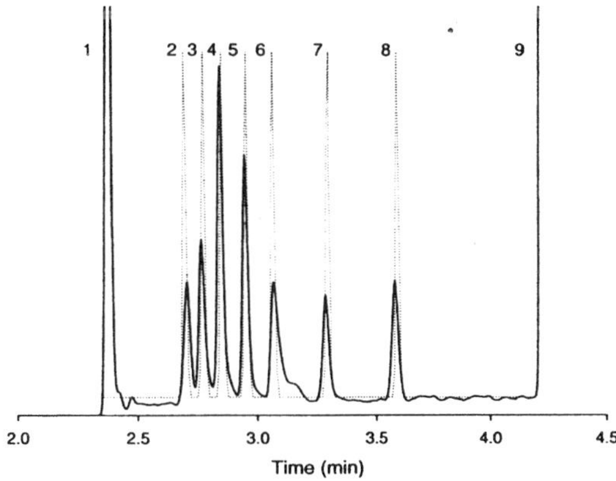

Figure 1. Experimental (solid line) and predicted (broken line) MEKC separations of metal-HEDTC complexes. Peaks: , EOF; 2, Cd(II); 3, Pb(II); 4, Ni(II); 5, Co(II); 6, Bi(III); 7, Cu(II); 8, Hg(II); 9, HEDTC. Reprinted from [48], with permission.

of the capillary wall. Strong adsorption of complexes of uranium (VI) and lanthanides with arsenazo III was observed both on uncoated fused-silica (with a negatively charged wall) and coated capillaries with neutral hydrophilic, positively charged and neutral hydrophobic surfaces. Dynamic coating of the bare fused-silica capillary with Carbowax 20M provided better separation performance owing to suppression of EOF and blockage of wall adsorption sites [49, 50]. Another study by Haddad's group [51], examined the role of ligand purity on the separation of metal ions as kinetically labile complexes with a metallochromic ligand contained within the carrier electrolyte. Using alkaline-earth metal ions-arsenazo I as a model complexing system, it was brought to focus that the presence of complexing (micro)impurities, in the form of various metal ions and competing ligands, causes substantial deterioration of peak shapes. The reagent and other chemicals used should therefore be as free as possible of these kinds of impurities.

Important contributions have been made in the past few years to increase the precision of CE methods. Concerning metal ion separations, Faller and Engelhardt [52] assessed the accuracy and precision of peak integration. The errors introduced by integration software were studied by transferring the same data sets (for eight different cations) to different commercialized integration software. Whereas no differences in RSD values between the various softwares could be observed at S/N > 35, at lower S/N the newly released software produced much better reproducibility of integration.

4 Separation systems

This section reviews the identification and application of capillary electrolyte additives relevant to the control and enhancement of the separation selectivity. Also examined is the utility of novel complexing reagents that assist in the electrophoretic separation of metal ions.

4.1 Surfactants

Haddad and co-workers [47, 48, 53] used a micellar concentration of SDS to realize high-efficiency MEKC separation of a number of transition metal complexes with HEDTC. In view of the dithiocarbamate ligand's inclination to decomposition and adsorption on the capillary wall in acidic media, the choice of background electrolyte was pointed out as an important step in method development. For the CDTA complexes, in contrast, the introduction of SDS to the electrolyte solution has resulted in no significant improvement in the separation from that achieved under regular CE conditions [47, 48]. This observation was in accord with the fact that interaction of the multiply

charged anionic analytes with the micelle is minimal [13]. MEKC with SDS was found superior to CE for the separation of diacetylacetonato-beryllium from acetylacetone [54].

4.2 Polymers

Separations utilizing an ion-exchange mechanism for mobility differentiation are being systematically investigated by a group headed by Yotsuyanagi and Shpigun [55, 56]. The addition of a cationic water-soluble polymer, namely poly(diallyldimethylammonium chloride), provides the condition for selective separation of anionic chelates (as well as some inorganic anions [55]) in the ion-exchange electrokinetic chromatography mode [36]. In particular, the EDTA or CDTA complexes, which are singly or doubly negatively charged, were highly efficiently discriminated through the interaction with a polycationic polymeric "pseudostationary" phase [55]. Extending this work, a spectacular separation of a series of trivalent metal ions as *N,N*-bis(hydroxybenzyl)ethylenediamine-*N,N*-diacetic acid complexes was recently attained [56], as shown in Fig. 2.

4.3 Nonaqueous media

Nonaqueous solvents are attractive for use in CE since they, in particular, allow higher mobilities and potentially

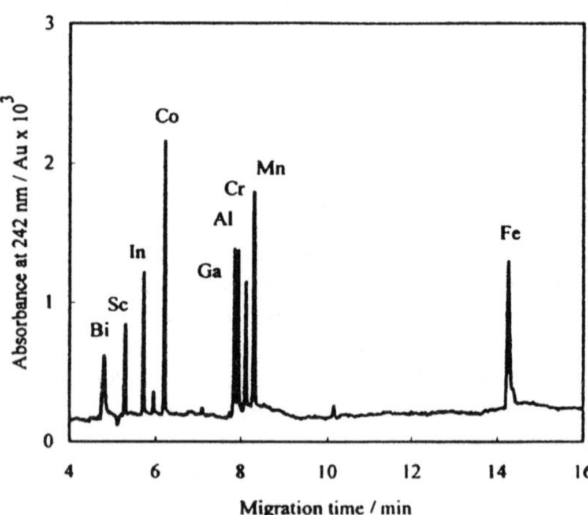

Figure 2. Separation of nine trivalent metal-*N,N*-bis(hydroxybenzyl)ethylenediamine-*N,N*-diacetic acid (HBED) chelates, Capillary, 50 cm × 50 µm ID fused-silica; carrier electrolyte, 10 mM $Na_2B_4O_7$, 50 mM poly(diallyldimethylammonium acetate), 0.1 mM HBED, pH 9.2. CE conditions: voltage, −20 kV; sample introduction, pressure, 15 psi·s; detection, 242 nm. Sample: 2.5×10^{-5} M Bi, Sc, 5×10^{-5} M other metals, 1×10^{-3} M HBED. Reprinted from [56], with permission.

faster separations, permit application of greater field strengths (that also accelerates the separation), feature a different selectivity compared to that normally obtained with water-based electrolytes (due to a change in the solvation of the analytes) [57]. Altria *et al.* [58] proved some of these advantages of nonaqueous (methanol) CE electrolytes for a range of metal cations. In addition, a superior use of ACN-based buffer systems for CE-ED of organometallic species was reported by Matysik and co-authors [59, 60].

4.4 Macrocyclic ligands

Due to their specific recognition of the metal analytes, macrocyclic compounds find an appropriate application as the CE additives to improve the separation resolution and selectivity [61]. The work of Hu *et al.* [62] manifested that a novel class of such compounds, macrocyclic polyamines, could be a valuable complement to the commonly used crown ethers. Furthermore, a specifically studied 1,4,7,10-tetraazacyclotridecane-11,13-dione was sufficient in the separation of alkali metal, alkaline earth metal and ammonium ions (in the same use of crown ethers, another complexing additive is typically required; see, *e.g.*, [63–68]). Reijenaga *et al.* [69] evaluated the potential of cyclofructan (and its permethylated form), which has a circular hexameric structure resembling that of 18-crown-6, as a complexing agent for the CE separation of metal cations. Another type of macrocyclic oligomer, the sulfonated calixarenes, (tested by Arce and co-workers [70]) possess differential selectivity with the different alkaline-earth metal cations when included in the electrolyte solution.

4.5 Mixed complexing systems

Inducing a secondary complexing equilibrium may serve as a supportable source of separation selectivity. An electrolyte system containing α-hydroxyisobutyric acid as the main complexing agent and malonic acid as the auxiliary one was developed by Hirokawa and co-authors [71] for simultaneously separating 15 rare-earth element (RE) ions. The specific role of the latter carboxylic acid appears to enlarge the migration window between the two adjacent peaks of lanthanide ions, Tb and Dy, in order to accommodate the peak of yttrium. Macka *et al.* [50] applied a binary mixture of a strong reagent, arsenazo III, and a weaker citrate to separate lanthanides, uranium(VI) and thorium(IV). Addition of the second reagent competing with the chelating ligand for the metal ions was found to be essential to achieve better selectivity and peak shapes. The authors came to the conclusion that rapid complexation kinetics is a necessary prerequisite for the competing ligand to bring an improvement in the separation. The same separation principle was later utilized to resolve V(V), Nb(V), and Ta(V) as ternary complexes with 4-(2-pyridylazo)resorcinol (PAR) and tartaric acid [72]. Acetate was found to act as the auxiliary ligand favoring more complete and more stable complexation of Th(IV), U(VI) and RE with a chelating reagent; hence their more efficient and reproducible CE separation [73]. Pozdniakova and Padarauskas [74] employed a mixture of *o*-phenanthroline and CDTA to selectively complex and then separate Fe(II) and Fe(III) as a cationic and an anionic complex, respectively.

4.6 Precapillary complexing reagents

Several attractive reagents were newly introduced for the separation (and direct photometric detection) of specific metal groups over the review period (see also the following subsection). Thiourea, a well-known complex-forming reagent for precious metals, was shown to be suitable for separating selected platinum metals [75]. In a later related paper, Hamacek and Havel [76] separated and determined Pd(II) and Pt(II) [or Pt(IV)] in the form of thiocyanate complexes. Cheng and co-workers [77] examined the separation of thorium, uranium and RE, often co-existing in nature, after complexation with 2-(2-arsenophenylazo)-1,8-dihydroxy-7-(4-chloro-2,6-dibromophenylazo)-naphthalene-3,6-disulfonic acid [73] and 2-(2-arsenophenylazo)-1,8-dihydroxy-7-(2,4,6-tribromophenylazo)-naphthalene-3,6-disulfonic acid. They utilized these analogues complexing agents in counterelectroosmotic mode and in a CE mode with reversed EOF (by adding CTAB), respectively, but even at optimized conditions, all of the RE studied such as La(III), Ce(III), Pr(III), Nd(III), Sm(III) or La(III), Ce(III), Dy(III), Ho(III), migrated as one peak. Ammonium bis(carboxymethyl)dithiocarbamate enabled a baseline separation of a group of heavy metals with increased sensitivities owing to an SPE preconcentration step [78] (see also Table 2). The pH-dependent solubility of the metal complexes formed makes them advantageous for SPE prior to CE. Of eight derivatives of *o*-phenanthroline investigated by Yokoyama *et al.* [79], 4,7-dimethyl-1,10-phenanthroline was found to be best for separating (and determining) transition metal ions; separately investigated bathophenanthroline disulfonic acid [80] turned out to be less efficient for this purpose. 2,4,6-Tri(2'-pyridyl)-1,3,5-triazine was proved to be suitable for the selective CE determination of Fe(II), the on-capillary complexation mode being preferable [81], while 2,6-pyridinedicarboxylic acid (PDC) as the carrier electrolyte additive worked well to confer a negative charge and a different mobility on transition metal cations [82]. Li and Wang [83] developed and patented an electrolyte buffer (and the corresponding buffer kit) for analyzing inorganic cations by CE, comprising nitrilotriacetic acid and other complexing agents.

Table 2. Advanced detection and sensitivity enhancement techniques for metal ion analysis by CE[a]

Method	Typical sensitivity (M)	Enhancement[b]
Laser-based fluorescence		
– Direct	10^{-7}–10^{-6}	10
– Indirect	10^{-5}–10^{-6}	–
Chemiluminescence	10^{-6}–10^{-12}	10–10^{-6}
Atomic emission spectroscopy	10^{-6}–10^{-8}	10–100
Conductivity	10^{-6}–10^{-7}	5–10
Amperometry	10^{-7}–10^{-8}	10–100
Potentiometry	10^{-5}–10^{-6}	–
Mass spectrometry		
– ICP	10^{-6}–10^{-10}	10–10^{-5}
– Electrospray	10^{-5}–10^{-6}	–
Sample stacking		
– Field amplification	–	50–1000
– Transient ITP	–	10–500
Solid-phase extraction	–	100–1000
Supported liquid membranes	–	50–500
Chemical derivatization for direct absorbance detection	10^{-6}–10^{-7}	10

a) Adopted from [42], with permission
b) Relative to standard CE methodology based on UV absorbance detection

4.7 Ion-pairing effects

Cationic ion-pairing agents are convenient electrolyte additives to adjust selectivity for negatively charged metal complexes *via* ion-pairing mechanism [14]. For example, the tetrabutylammonium cation added to the electrophoretic buffer induced the drastic changes in mobilities of metal anionic chelates of 1-nitroso-2,7-dihydroxynaphthalene-3,6-disulfonic acid [84] and 2-nitroso-1-naphthol-5-sulfonic acid [85]. Further, a recent report by Liu *et al.* [86] demonstrated the superiority of using mixed tetraalkylammonium additives for the separation of some transition metal complexes of PAR [86].

5 Detection

Improvements in performance of the existing detection methods and the development of new detector designs received more literature coverage than any other single topic during the review period. The paramount motive seems to be the growing understanding that without substantial gains in sensitivity, CE will face serious problems in being accepted as a routine tool in trace metal analysis. Table 2 gives a comparison of sensitivity capacities of various CE detection techniques and sample enrichment means.

5.1 MS

The interfacing of CE to ICP-MS, undoubtedly the most perspective detection methodology for quantification of trace metals and their forms, was explored, in depth, by several groups [87–97]. These studies revealed that the major challenges in CE-ICP-MS coupling, such as a post-capillary detector design (with accompanying problems in applying a high voltage across the capillary and keeping the electrode grounded), the low flow rate of CE separations limiting the choice of a nebulizer, and separation buffer incompatibilities with the ionization process, have been overcome to a different degree. As a result, this hyphenated technique offers detection limits typically better than 10^{-7} M. However, applications to real samples are still missing. As for electrospray ionization-MS, no published examples on its recent use as a detector for CE relative to metal species are known to the authors (not counting an off-line combination investigated by Schramel *et al.* [98]).

5.2 Electrochemical detection

The growing popularity of ED has become evident within the last few years. Principally, there are three ED systems suitable for sensitive detection of metal ions, *i.e.*, amperometric, conductometric and potentiometric detection. For all of these, investigations into simpler and more robust cell arrangements continued to be a main trend. Amperometric detection appears to be by far the most suited for CE of the metal analyte ions because a high-sensitivity response can be expected for a wider range of metal groups. The progress in detector design – from a constant voltage [59, 60, 99] to a pulsed [100] mode and then to a fast-scan cyclic voltammetry system [101] – has resulted in impressive sensitivity figures of merit. For example, in the latter mode providing analyte preconcentration, Cassidy and co-workers [100] attained detection limits as low as 5×10^{-9} M for the metal ions shown in Fig. 3. Most recent advances in amperometric detection involved a simplified detector cell with a single (not electrophoretic) electrode [102], a field-portable CE instrument which incorporates, in a single unit, both amperometric and potentiometric detection [103], and a sensor for nonelectroactive cations employing a graphite electrode modified with a mixed-valent ruthenium-iron cyanide [104]. Whereas the positioning of conductivity sensor at the outlet of the capillary became the accepted (and first commercialized) configuration of conductivity detector in inorganic analysis, more efforts to develop an easy-to-assemble contactless conductometric CE detector were reported for metal cations [105–107]. Conductivity detector of this design exhibits greater ease of use and robustness, and according to a most recent essay [107], its sensitivity

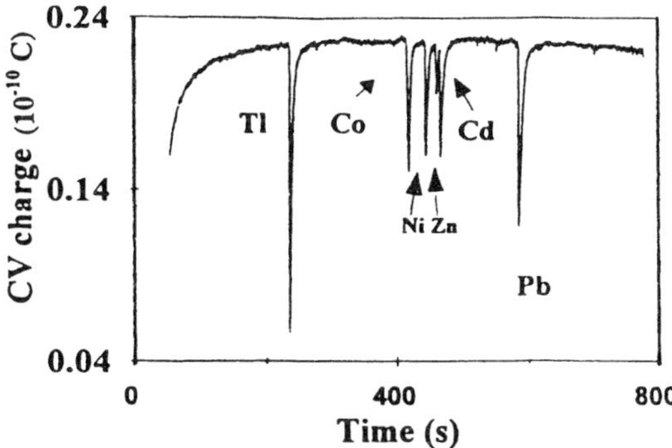

Figure 3. Electropherogram of metal ions with fast-scan cyclic voltammetry detection. Capillary, 60 cm × 25 µm ID fused-silica; carrier electrolyte, 30 mM creatinine, 8 mM α-hydroxyisobutyric acid, pH 4.8. CE conditions: voltage, 20 kV; sample introduction, electromigration, 5 kV for 10 s. Detection conditions: preconcentration, 220 ms; scan rate, 275 V/s; potential, −900 to 300 mV. Metal concentration, 1×10^{-7} M each. Reprinted from [101], with permission.

could be lowered down to single-digit µg/L levels. On the other hand, improvements in the end-capillary mode for minimizing the interference from the separation voltage and simplifying the measurement electrode design [108] should be mentioned. Potentiometric detection of alkali and alkaline earth metals using coated-wire ion-selective electrodes was also recently addressed [66, 103, 109], the former of these contributions being demonstrated a compact, portable device for on-site analysis.

5.3 Fluorescence detection

According to the recent literature, this method, in different modes of derivatization, arrangement of the separation system, and inducing and measuring the fluorescence signal, holds promise for a wider implementation in highly sensitive detection of metal ions. For example, building on a postcapillary reaction with an established fluorescence reagent, 8-hydroxyquinoline-5-sulfonic acid, Zhu and Kok [110] and Kutter *et al.* [111] developed attractive detection schemes in which, after separation, the metal ions were complexed, respectively, in a laboratory-built reactor and on-chip, and subsequently detected. Indirect detection mode offers an alternative means for detection of usually nonfluorescent metal cations. Quinine sulfate as the fluorescent carrier coion demonstrated low micromolar concentration detection limits for a range of RE cations [112, 113].

5.4 Chemiluminescence detection

In terms of sensitivity, CL perhaps represents the ultimate detection scheme. A variety of metal ions or their complexes catalyze the reaction of hydrogen peroxide and luminol in alkaline solution and can thus form the basis of direct CL assay following the electrophoretic separation [21, 114]. Metal ions showing no direct catalytic effect (for instance, RE) can be subject to CL detection owing to their ability to enhance CL emission produced by the same reaction catalyzed by cobalt(II) [115]. However, one must be aware that this secondary enhancement behavior is much less sensitive.

5.5 Other detection techniques

Another element-specific technique, atomic emission spectroscopy, based on an ICP source, can be hyphenated with CE [116] but less favorably from the viewpoint of sensitivity. The first operation of a thermal lens spectroscopic detector enabling a 30-fold better detection limit than conventional absorption detection was demonstrated by Seidel and Faubel [117]. Application of radioactivity detection [112] has been shown to be practical for the separation of metal ions characteristically associated with nuclear fusion. Other detection principles reported over this review period, based on the measurement of refractive index [118] and X-ray fluorescence [119], should undergo further refinement to provide lower detection limits (which are presently in the 10^{-4}–10^{-5} M range). Also, the construction, operation, and performance of a postcapillary reactor that facilitates UV-visible detection of trace metals were detailed by Hardy *et al.* [120].

6 Sample introduction modes

This section addresses briefly the various sample injection approaches that can be adopted to concentrate the target metal ions on-line, minimize the unfavorable injection biases and expand the analytical possibilities of using CE. Stacking of metal analyte ions in a hydrodynamically injected large sample plug remains the most frequently applied technique useful for lowering the detection limits (see [49, 87, 111, 116, 121] among others and also Table 2). Several groups investigated the utility of an effective modification of conventional sample stacking procedure (sometimes called large-volume sample stacking) using voltage reversal between stacking and separation to remove most of the water matrix [122–124]. Specifically, Liu and Lee [124] reported sub-µg/L sensitivities for a range of metal and organometal species resulting from up to 1500-fold on-capillary enrichment of their charged complexes. Alternatively, the high concentration factors can be achieved by electrostacking supported by an isotachophoretic effect [33, 41]. In this way, over 700-fold improvements in the detection limits for a series of lanthanides were reached, H^+ and tris(hydroxymethyl)aminomethane being used as leading and terminating ions, respectively [112, 113].

Kuban and Karlberg [125] developed an interesting sample introduction scheme for the aim of simultaneous determination of cations and anions. Two portions of the same sample solution were subsequently introduced into the first (anodic) and second (cathodic) ends of a separation capillary so that after applying the voltage, the cations and the anions started to migrate against each other toward the detection window placed approximately at the middle of the capillary. Thus, with a suitable background electrolyte (see the legend to Fig. 4), as many as 22 cations and anions could be separated in a single 5-min run using one capillary and just one detector. A similar approach for the separation of both cations and anions, with the only notable exception that the capillary ends were inserted in the sample vials at the same time and electromigration injection was performed, was independently proposed by Padarauskas *et al.* [126]. (Note that the same analytical task can be accomplished by precapillary or on-capillary transformation of metal cations into negatively charged complexes, *e.g.*, with EDTA [127, 128] and PDC [82], respectively, and their separation together with

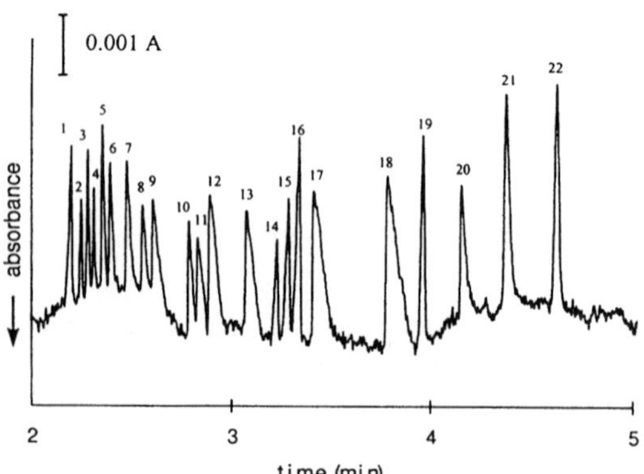

Figure 4. Separation of 22 cations and anions. Capillary: 50 cm × 50 µm ID fused-silica; carrier electrolyte, 6 mM 4-aminopyridine, 2.7 mM H_2CrO_4, 3×10^{-5} M CTAB, pH 8.0. CE conditions: voltage, 20 kV; sample introduction, gravity, 40 s at 10 cm (cathode end); 20 s at 5 cm (anodic end); time between the injections, 80 s; detection, 262 nm. Peaks: 1, $S_2O_3^{2-}$; 2, Br^-; 3, Cl^-; 4, SO_4^{2-}; 5, NO_2^-; 6, NO_3^-; 7, WO_4^{2-}; 8, MoO_4^{2-}; 9, citrate; 10, maleate; 11, fumarate; 12, F^-; 13, HPO_4^{2-}; 14, Cs^+; 15, K^+; 16, NH_4^+; 17, HCO_3^-; 18, acetate; 19, Na^+; 20, Ca^{2+}; 21, Mg^{2+}; 22, Li^+. Reprinted from [125], with permission.

the anions using an anionic separation mode). The development of a flow injection analysis (FIA)-CE sampling system based on hydrodynamic injection was the aim of a recent study by Kuban *et al.* [129]. Such a system offers a possibility of introducing constant sample volumes, regardless of variations in sample conductivity, without the need of addition of an internal standard or conductivity corrections. This merit favors the options of FIA-CE for implementation of automated sample pretreatment procedures.

7 Applications

Since this volume implies chiefly CE methodology and theory items rather than practical applications, only a concise description of the utilization of CE for metal ion analysis will be presented in this section, focusing on procedures which are (or could be) of real use and some ways to avoid and overcome problems in practice. Among application areas where CE has attracted much interest in the past few years, the environmental field is a dominant one. Most of the reports on environmental matrices involve various aquatic samples. New procedures for the determination of alkali and alkaline earth metal ions from single rain, cloud, or fog drops and ice crystals [130, 131], aerosols and depositions [132], drinking water [133], mineral and tap water [127, 128, 134], river water [66], groundwater [135], underground water [136], and seawater [137], Ca, Mg, Zn, and Al from tap and surface water [110], for iron(II) quantification in rain and tap water [117] and natural waters [81], and for measuring the content of Pb(II) in seawater [124] were reported. Usually these samples do not require any pretreatment other than filtration and occasionally dilution. However, the need to preserve Fe(II) in surface waters prior to CE analysis had to be responded to by the addition of concentrated sulfuric acid [81]. Similarly, natural water samples were acidified to pH 2 to avoid the interference of the fluoride ion with the determination of aluminum [110]. Also, in the event of a great disparity in concentration between the analyte ions, special care should be paid on optimizing the nature and concentration of both the complexing agent and the chromophore [136, 138]. The analysis of solid environmental samples with CE has received relatively less attention compared to water analysis. Apparently, less enthusiasm of researches is at the bottom of more tedious and insofar less furnished for the coupling with CE sample preparation procedures necessary to bring the metal species into solution and alleviate matrix effects. In particular, Fung *et al.* developed several appropriate electrolyte systems for the determination of water leachable and total alkali, alkaline earths, and zinc [63, 64] and water and acid leachable transition metals [64, 139] from various aiborne particulate matters. Another heavy metal ion, lead(II), was quantified in the leachate of road dust by Kappes and Hauser [102]. In spite of interferences from the complex matrix of vehicle exhaust particulates, satisfactory accuracy was obtained for Pd(II) determination in a reference material [76].

In view of the current sensitivity limitations of CE method, samples with relatively high concentration levels of analytes are more amenable to direct analysis. This makes CE especially useful for the determination of metal ions in industrial matrices and technical materials. For example, the contents of Ni, Co, and Cu in nickel ores were determined precisely using zinc as an internal standard [140]. Similar samples analyzed successfully by CE encompass RE ores [71, 73] and geological standards [141], duralmin alloys [79], and platinum metal concentrates [75]. In an interesting application, Kelly and his group [142–144] systematically applied CE to study corrosion processes taking place in aircraft aluminum alloys. The method's flexibility and robustness in conjunction with a specifically developed extraction procedure allowed sample solutions of small volume to be fully analyzed for their composition regarding the cationic type that causes or is produced by corrosion. Figure 5 demonstrates the possibility of such an analysis from a very small blister (100 nL). Soga and Ross [82] reported on the determination of nickel, along

with several kinds of supporting additives, in a plating bath solution. CE also provides an effective means for the analysis of various drugs [76, 145], salts [85, 146, 147], explosives [148], *etc.*

The prospects for CE in bioinorganic analysis also seem to be bright, although the number of published accounts remains comparatively small to date [65, 68, 104, 110, 122, 149–151]. It should be emphasized that sample pretreatment steps appear to be most critical in analyses required. Bächmann and co-workers [150] demonstrated for the first time the applicability of CE for the analysis of a single plant cell vacuole. Due to judiciously designed and optimized sampling, sample handling, and injection procedures, it was possible to measure with a high reliability inorganic cation concentrations in 10 pL subsamples. Muse *et al.* [151] used a series of certified reference materials of plant origin to prove the method's potential in routine determinations of common alkali and alkaline earth metals as well as manganese performed after microwave-assisted digestion of the samples. A closed-vessel architecture of the sample decomposition step reduced substantially the risk of losses and contamination. Only zinc was determined in hair samples as a 2-(5'-bromo-2'-pyridylazo)-5-diethylaminophenol complex [122]; the quantification of other heavy metals existing at relatively low levels in hair was not feasible, and large-volume sample stacking was not successful. Other bioapplication papers mentioned deal mostly with body fluids (blood serum, vitreous humor, *etc.*) analyzed for alkali and alkaline earth metal contents.

8 Speciation studies

The potential and practical usefulness of CE for metal speciation purposes have been evaluated in many publications in the last two years [43]. That review revealed that CE techniques are extremely powerful in discriminating the speciation pattern of various environmentally and biologically important metals such as mercury [99, 123, 124, 152], lead [94, 123, 124, 152], cadmium [91, 92], uranium [153], aluminum [110, 152, 154], chromium [74, 96, 97, 116, 155], tin [89], iron [59, 60, 74, 89, 152, 156], copper [91, 152], zinc [91, 157], nickel [89], vanadium [74, 158], platinum [87], iridium [159], *etc.* To date, however, relatively little use has been made of CE in speciation measurements of real samples. Despite the method's two clear benefits, which are the minimal impact of the analytical system on the original distribution of sample species and good tolerance to complex matrices, still challenging remains the development of a CE methodology that works well for direct analysis of trace metal species. Other current problems in CE applying to real-world speciation analysis, such as the shortage of sample prepara-

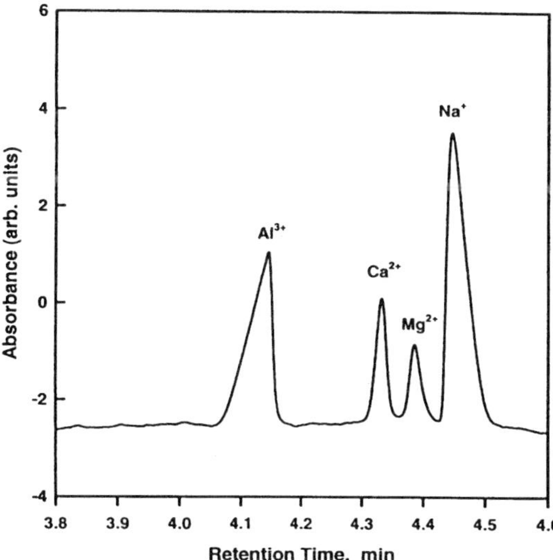

Figure 5. Electropherogram from 2 mm diameter red blister on a vinyl-coated aluminum alloy after 13 days of exposure to 0.6 M NaCl. Capillary, 60 cm × 75 µm ID fused-silica; carrier electrolyte, 1 mM guanidine hydrochloride, pH 3.1 (100 mM HNO₃). CE conditions: voltage, 17 kV; sample introduction, gravity, 30 s at 10 cm; detection, 185 nm. Reprinted from [142], with permission.

tion procedures ensuring species stability and avoiding matrix effects, possible changes in speciation during electrophoresis, the lack of performance testing and application of quality assurance/quality control approaches, among others, were brought into focus and critically evaluated by Timerbaev [44]. The potential sources in analysis errors specific to the CE-ICP-MS combination were discussed by Olesik *et al.* [89] and Majidi and Miller-Ihli [92]. Notwithstanding, several advanced procedures applied to the analysis of atmospheric aerosol [74], surface and tap water [110], soil solutions [87, 154], plant material [110], heavy fuel ash [74, 158], industrial waste water [74], *etc.*, prove the ability of CE to render reliable speciation information. As an example, Fig. 6 shows an electropherogram obtained for iron speciation in an aqueous extract of atmospheric aerosol.

9 Miscellaneous

Systematic studies on complex formation between alkali metal ions and crown ethers and ion association reactions of resulting complexes with the hydrophobic anions, carried out by Motomizu and co-workers [160–164], are mentioned as an interesting supplemental area of CE research. Similar applications of CE on the association of various Co(III) coordination compounds and inorganic anions [165], the coordination of carboxylate ions to Cu(II) complexed with *o*-phenanthroline and bipyridine derivatives [166], and the complex formation between Cu(II)-Alizarin complexone and various amines [167] have been reported. Some previously cited reports also deal with the determination of the complexation and ion-association constants based on electrophoretic mobility measurements [69, 84].

10 Conclusions

A successful decade of development in the field of metal ion determinations has brought CE the status of method of choice for many analytes and areas of application. The continuing efforts by many research groups have led to vast improvements in basic principles and methodology, selectivity and detection sensitivity, performance stability, and sample analysis. To this end, the persistent refinement in commercial instruments during the formative era of CE has resulted in a situation where many analysts, who looked at the technique in its early days and concluded that it was not for them, could today well find it worthwhile to re-examine CE due to its great changes in the past years. The method's growing acceptance by practicing analytical chemists has made this publication particularly timely. This overview concludes with more than 150 references that also illustrate that the rate of development and acceptances of CE for separations of metal ions in terms of the number of publications appearing annually (*cf.* Fig. 1 in [19]) is becoming even greater.

In the authors' opinion, the following main directions along which advancements in CE will take place are likely in the years ahead. No matter what viewpoint on CE is correct, whether the method has to be considered as complementary to or competitive with HPLC, the latter being a generally accepted analytical tool in the field, the prospects of CE would largely be related to application areas in which it has decided advantages over HPLC. Future work should therefore be increasingly centered on the analysis of complex samples and speciation analysis. However, a real breakthrough in both areas would only be possible following substantial improvements in detection sensitivity and sample preparation methodology. Thus, one can envision that considerable research emphasis will also be placed on more powerful detection systems, sample enrichment steps prior to CE, and a combination thereof, as well as on more efficient means to pretreat liquid and especially solid samples.

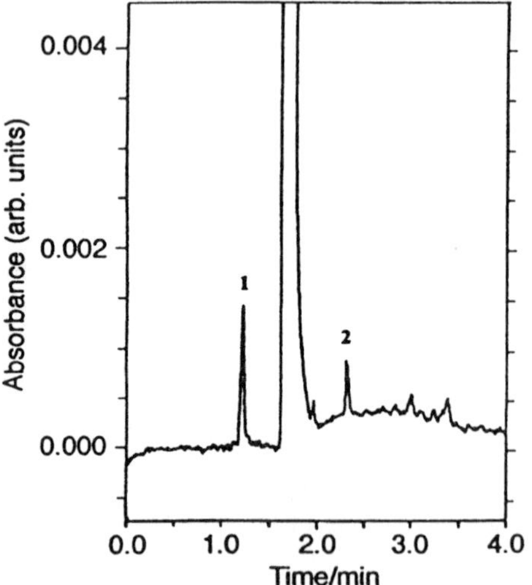

Figure 6. Speciation of iron in aqueous atmospheric aerosol extract. Capillary, 47 cm × 75 µm ID fused-silica; carrier electrolyte, 100 mM borate buffer, pH 9.0. CE conditions: voltage, 30 kV; sample introduction, pressure, 12 s at 3.43 kPa; detection, 254 nm. Peaks: 1, Fe(II); 2, Fe(III). Reprinted from [74], with permission.

The authors express their sincere gratitude to Dr. Olga Semenova of Vienna University for her assistance in conducting the comprehensive literature search.

Received July 13, 2000

11 References

[1] Timerbaev, A. R., *J. Capil. Electrophor.* 1995, *2*, 14–23.

[2] Guan, F., *Fenxi Huaxue* 1995, *23*, 111–116.

[3] Timerbaev, A. R., *J. Capil. Electrophor.* 1995, *2*, 165–174.

[4] Manabe, T., Otsuka, K., Baba, Y., Suzuki, S., Kakehi, K., *Kyapirari Denki Eido* 1995, 142–243.

[5] Jimidar, M., Yang, Q., Smeyers-Verbeke, J., Massart, D. L., *Trends Anal. Chem.* 1996, *15*, 91–102.

[6] Song, L., Qu, Q., Yu, W., *Sepu* 1996, *14*, 98–101.

[7] Chernoglazov, V. M., Nesterenko, P. N., *Russian Chem. J.* 1996, *40*, 100–110.

[8] Guan, F., *Fenxi Huaxue* 1996, *24*, 109–114.

[9] Belen'kii, B. G., Belov, Y. V., Kasalainen, G. E., *J. Anal. Chem.* 1996, *51*, 752–769.

[10] Timerbaev, A. R., *Electrophoresis* 1997, *18*, 185–195.

[11] Haddad, P. R., *J. Chromatogr. A* 1997, *770*, 281–290.

[12] Roland-Gosselin, P., Caudy, P., Parvez, H., Parvez, S., *Prog. HPLC-HPCE* 1997, *5*, 445–471.

[13] Haddad, P. R., Macka, M., Hilder, E. F., Bogan, D. P., *J. Chromatogr. A* 1997, *780*, 329–341.

[14] Timerbaev, A. R., *J. Chromatogr. A* 1997, *792*, 495–518.

[15] Buchberger, W., in: Shintani, H., Polansky, J. (Eds.), *Handbook of Capillary Electrophoresis Applications*, Blackie, London, UK 1997, pp. 509–516.

[16] Erim, F. B., in: Shintani, H., Polansky, J. (Eds.), *Handbook of Capillary Electrophoresis Applications*, Blackie, London, UK 1997, pp. 517–530.

[17] Buchberger, W., in: Shintani, H., Polansky, J. (Eds.), *Handbook of Capillary Electrophoresis Applications*, Blackie, London, UK 1997, pp. 531–549.

[18] Haber, C., in: Landers, J. P. (Ed.), *Handbook of Capillary Electrophoresis,* CRC, Boca Raton, FL 1997, pp. 425–447.

[19] Macka, M., Haddad, P. R., *Electrophoresis* 1997, *18*, 2482–2501.

[20] Dabek-Zlotorzynska, E., Lai, E. P. C., Timerbaev, A. R., *Anal. Chim. Acta* 1998, *359*, 1–26.

[21] Zhang, Y., Huang, B., Cheng, J., *Fenxi Ceshi Xuebao* 1998, *17*, 81–84.

[22] Fritz, J. S., in: Camillery, P. (Ed.), *Capillary Electrophoresis. Theory and Practice*, CRC Press, Boca Ranton, FL 1998, pp. 249–272.

[23] Chiari, M., *J. Chromatogr. A* 1998, *805*, 1–15.

[24] Timerbaev, A. R., Buchberger, W., in: Deyl, Z., Miksik, I., Tagliaro, F., Tesarova, E. (Eds.), *Advanced Chromatographic and Electromigration Methods in BioSciences*, Elsevier, Amsterdam 1998, pp. 963–1012.

[25] Barnes, R. M., *Fresenius J. Anal. Chem.* 1998, *361*, 246–251.

[26] Sutton, K. L., Caruso, J. A., *LC.GC* 1999, *17*, 36–45.

[27] Morin, P. H., *Analusis* 1999, *27*, 107–119.

[28] Janos, P., *J. Chromatogr. A* 1999, *834*, 3–20.

[29] Krivacsy, Z., Gelencser, A., Hlavay, J., Kiss, G., Sarvari, Z., *J. Chromatogr. A* 1999, *834*, 21–44.

[30] Liu, W., Lee, H. K., *J. Chromatogr. A* 1999, *834*, 45–63.

[31] Okada, T., *J. Chromatogr. A* 1999, *834*, 73–87.

[32] Kappes, T., Hauser, P. C., *J. Chromatogr. A* 1999, *834*, 89–101.

[33] Polesello, S., Valsecchi, S. M., *J. Chromatogr. A* 1999, *834*, 103–116.

[34] Timerbaev, A. R., Buchberger, W., *J. Chromatogr. A* 1999, *834*, 117–132.

[35] Pacakova, V., Coufal, P., Stulik, K., *J. Chromatogr. A* 1999, *834*, 257–275.

[36] Liu, B.-F., Liu, L.-B., Cheng, J.-K., *J. Chromatogr. A* 1999, *834*, 277–308.

[37] Fukushi, K., Takeda, S., Chayama, K., Wakida, S.-I., *J. Chromatogr. A* 1999, *834*, 349–362.

[38] Valsecchi, S. M., Polesello, S., *J. Chromatogr. A* 1999, *834*, 363–385.

[39] Sadecka, J., Polonsky, J., *J. Chromatogr. A* 1999, *834*, 401–417.

[40] Blatny, P., Kvasnicka, F., *J. Chromatogr. A* 1999, *834*, 419–431.

[41] Timerbaev, A. R., Dabek-Zlotorzynska, E., van den Hoop, M. A. G. T., *Analyst* 1999, *124*, 811–826.

[42] Timerbaev, A. R., *J. Capil. Electrophor.* 1998, *5*, 185–192.

[43] Timerbaev, A. R., *Talanta* 2000, *52*, 573–606.

[44] Timerbaev, A. R. (Ed.), *Anal. Chim. Acta* 2001, in press.

[45] Jones, P., *J. Chromatogr. A* 1999, *834*, 1–459.

[46] Havel, J., Pena, E. M., Rojas-Hernandez, A., Doucet, J.-P., Panaye, A., *J. Chromatogr. A* 1998, *793*, 317–329.

[47] Havel, J., Breadmore, M., Macka, M., Haddad, P. R., *J. Chromatogr. A* 1999, *850*, 345–353.

[48] Breadmore, M. C., Macka, M., Haddad, P. R., *Anal. Chem.* 1999, *71*, 1826–1833.

[49] Macka, M., Nesterenko, P., Haddad, P. R., *J. Microcol. Sep.* 1999, *11*, 1–9.

[50] Macka, M., Nesterenko, P., Andersson, P., Haddad, P. R., *J. Chromatogr. A* 1998, *803*, 279–290.

[51] Macka, M., Paull, B., Bogan, D. P., Haddad, P. R., *J. Chromatogr. A* 1998, *793*, 177–185.

[52] Faller, T., Engelhardt, H., *J. Chromatogr. A* 1999, *853*, 83–94.

[53] Hilder, E. F., Macka, M., Haddad, P. R., *Analyst* 1998, *123*, 2865–2870.

[54] Takaya, M., *J. Chromatogr. A* 1999, *850*, 363–368.

[55] Krokhin, O. V., Adamov, A. V., Hoshino, H., Shpigun, O. A., Yotsuyanagi, T., *J. Chromatogr. A* 1999, *850*, 269–276.

[56] Krokhin, O. V., Hoshino, H., Shpigun, O. A., Yotsuyanagi, T., *Chem. Lett.* 1999, 903–904.

[57] Valko, I. E., Siren, H., Riekkola, M., *LC.GC Int.* 1997, *10*, 190–196.

[58] Altria, K. D., Wallenberg, M., Westerlund, D., *J. Chromatogr. B* 1998, *714*, 99–104.

[59] Matysik, F.-M., *J. Chromatogr. A* 1998, *802*, 349–354.

[60] Matysik, F.-M., Nyholm, L., Markides, K. E., *Fresenius J. Anal. Chem.* 1999, *363*, 231–235.

[61] Lamb, J. D., Smith, R. G., *Compr. Supramol. Chem.* 1996, *10*, 79–112.

[62] Hu, S., Fu, E., Li, P. C. H., *J. Chromatogr. A* 1999, *844*, 439–446.

[63] Fung Y. S., Lau, K. M., Tung, H. S., *Talanta* 1998, *45*, 619–629.

[64] Fung, Y. S., Tung, H. S., *Proc. Annu. Meet. – Air Waste Manage. Assoc.* 1998, *91*, TP31B06/1- TP31B06/15.

[65] Ferslew, K. E., Hagardorn, A. N., Travis Harrison, M., McCormick, W. F., *Electrophoresis* 1998, *19*, 6–10.

[66] Kappes, T., Hauser, P. C., *Anal. Commun.* 1998, *35*, 325–329.

[67] Colombara, R., Massaro, S., Tavares, M. F. M., *Anal. Chim. Acta* 1999, *388*, 171–180.

[68] Maruszak, W., Trojanowicz, M., *Chem. Anal. (Warsaw)* 1999, *44*, 437–453.

[69] Reijenaga, J. C., Verheggen, T. P. E. M., Chiari, M., *J. Chromatogr. A* 1999, *838*, 111–119.

[70] Arce, L., Segura Carretero, A., Rios, A., Cruces, C., Fernandez, A., Valcarcel, M., *J. Chromatogr. A* 1998, *816*, 243–249.

[71] Mao, Q., Hashimoto, Y., Manabe, Y., Ikuta, N., Nishiyama, F., Hirokawa, T., *J. Chromatogr. A* 1998, *802*, 203–210.

[72] Liu, B.-F., Liu, L.-B., Chen, H., Cheng, J.-K., *J. Chromatogr. A* 2000, in press.

[73] Liu, B.-F., Liu, L.-B., Cheng, J.-K., *Talanta* 1998, *47*, 291–299.

[74] Pozdniakova, S., Padarauskas, A., *Analyst* 1998, *123*, 1497–1500.

[75] Khramov, A. N., Havel, J., *Scripta Fac. Sci. Nat. Univ. Brun.* 1998, *28*, 63–73.

[76] Hamacek, J., Havel, J., *J. Chromatogr. A* 1999, *834*, 321–327.

[77] Liu, B.-F., Liu, L.-B., Cheng, J.-K., *Anal. Chim. Acta* 1998, *358*, 157–162.

[78] Rudnev, A., Spivakov, B., Timerbaev, A., *Chromatographia* 2000, *52*, 99–102.

[79] Yokoyama, T., Akamatsu, T., Ohri, K., Zenki, M., *Anal. Chim. Acta* 1998, *364*, 75–81.

[80] Yokoyama, T., Nakano, K., Zenki, M., *Anal. Chim. Acta* 1999, *396*, 117–123.

[81] Dahlen, J., Karlsson, S., *J. Chromatogr. A* 1999, *848*, 491–502.

[82] Soga, T., Ross, G. A., *J. Chromatogr. A* 1999, *834*, 65–71.

[83] Li, S. F. Y., Wang, T., *PCT Int. Appl.* 1998.

[84] Wan, T., Wang, Y., Yan, Y., Oshima, M., *J. Wuhan Univ. Technol. Mater. Sci. Ed.* 1999, *14*, 40–45.

[85] Wan, T., Wang, Y., Yan, Y., *Huaxue Shijie* 1999, *40*, 322–325.

[86] Liu, B.-F., Liu, L.-B., Cheng J.-K., *J. Chromatogr. A* 1999, *848*, 473–484.

[87] Lustig, S., Michalke, B., Beck, W., Schramel, P., *Fresenius J. Anal. Chem.* 1998, *360*, 18–25.

[88] Kirlew, P. W., Castilliano, M. T. M., Caruso, J. A., *Spectrochim. Acta B* 1998, *53*, 221–237.

[89] Olesik, J. W., Kinzer, J. A., Grunwald, E. J., Thaxton, K. K., Olesik, S. V., *Spectrochim. Acta B* 1998, *53*, 239–251.

[90] Kirlew, P. W., Caruso, J. A., *Appl. Spectrosc.* 1998, *52*, 770–772.

[91] Majidi, V., Miller-Ihli, N. J., *Analyst* 1998, *123*, 803–808.

[92] Majidi, V., Miller-Ihli, N. J., *Analyst* 1998, *123*, 809–813.

[93] Sutton, K. L., B'Hymer, C., Caruso, J. A., *J. Anal. Atom. Spectrom.* 1998, *13*, 885–891.

[94] Schaumlöffel, D., Prange, A., *Fresenius J. Anal. Chem.* 1999, *364*, 452–456.

[95] Day, J. A., Sutton, K. L., Soman, R. S., Caruso, J. A., *Analyst* 2000, *125*, 819–823.

[96] Chen, W.-H., Lin, S.-Y., Liu, C.-Y., *Anal. Chim. Acta* 2000, *410*, 25–35.

[97] Stewart, I. I., Olesik, J. W., *J. Chromatogr. A* 2000, *872*, 227–246.

[98] Schramel, O., Michalke, B., Kettrup, A., *J. Chromatogr. A* 1998, *819*, 231–242.

[99] Lai, E. P. C., Zhang, W., Trier, X., Georgi, A., Kowalski, S., Kennedy, S., MdMuslim, T., Dabek-Zlotorzynska, E., *Anal. Chim. Acta* 1998, *364*, 63–74.

[100] Wen, J., Cassidy, R. M., Baranski, A. S., *J. Chromatogr. A* 1998, *811*, 181–192.

[101] Wen, J., Baranski, A. S., Cassidy, R. M., *Anal. Chem.* 1998, *70*, 2504–2509.

[102] Kappes, T., Hauser, P. C., *Analyst* 1999, *124*, 1035–1039.

[103] Kappes, T., Schnierle, P., Hauser, P. C., *Anal. Chim. Acta* 1999, *393*, 77–82.

[104] Fu, C., Wang, L., Fang, Y., *Anal. Chim. Acta* 1999, *391*, 29–34.

[105] Zemann, A. J., Schnell, E., Volgger, D., Bonn, G. K., *Anal. Chem.* 1998, *70*, 563–567.

[106] Da Silva, J. A. F., Do Lago, C. L., *Anal. Chem.* 1998, *70*, 4339–4343.

[107] Mayrhofer, K., Zemann, A. J., Schnell, E., Bonn, G. K., *Anal. Chem.* 1999, *71*, 3828–3833.

[108] Zhao, H., Dadoo, R., Reay, R. J., Kovacs, G. T. A., Zare, R. N., *J. Chromatogr. A* 1998, *813*, 205–208.

[109] Kappes, T., Hauser, P. C., *Anal. Chem.* 1998, *70*, 2487–2492.

[110] Zhu, R., Kok, W. T., *Anal. Chim. Acta* 1998, *371*, 269–277.

[111] Kutter, J. P., Ramsey, R. S., Jacobson, S. C., Ramsey, J. M., *J. Microcol. Sep.* 1998, *10*, 313–319.

[112] Klunder, G. L., Andrews Jr., J. E., Church, M. N., Spear, J. D., Russo, R. E., Grant, P. M., Andresen, B. D., *J. Radioanal. Nucl. Chem.* 1998, *236*, 149–153.

[113] Church, M. N., Spear, J. D., Russo, R. E., Klunder, G. L., Grant, P. M., Andresen, B. D., *Anal. Chem.* 1998, *70*, 2475–2480.

[114] Zhang, Y., Gong, Z., Zhang, H., Cheng, J., *Anal. Commun.* 1998, *35*, 293–296.

[115] Zhang, Y., Cheng, J., *J. Chromatogr. A* 1998, *813*, 361–368.

[116] Chan, Y. Y., Chan, W. T., *J. Chromatogr. A* 1999, *853*, 141–149.

[117] Seidel, B. S., Faubel, W., *Fresenius J. Anal. Chem.* 1998, *360*, 795–797.

[118] Swinney, K., Pennington, J., Bornhop, D. J., *Microchem. J.* 1999, *62*, 154–163.

[119] Ringo, M. C., Huhta, M. S., Shea-McCarthy, G., Penner-Hahn, J. E., Evans, C. E., *Nucl. Instr. Methods Phys. Res. B* 1999, *149*, 177–181.

[120] Hardy, S., Jones, P., Riviello, J. M., Avdalovic, N., *J. Chromatogr. A* 1999, *834*, 309–320.

[121] Timerbaev, A. R., Semenova, O. P., Fritz, J. S., *J. Chromatogr. A* 1998, *811*, 233–239.

[122] McClean, S., O'Kane, E., Coulter, D. J. M., McClean, S., Smyth, W. F., *Electrophoresis* 1998, *19*, 11–18.

[123] Liu, W., Lee, H. K., *J. Chromatogr. A* 1998, *796*, 385–395.

[124] Liu, W., Lee, H. K., *Anal. Chem.* 1998, *70*, 2666–2675.

[125] Kuban, P., Karlberg, B., *Anal. Chem.* 1998, *70*, 360–365.

[126] Padarauskas, A., Olsauskaite, V., Schwedt, G., *J. Chromatogr. A* 1998, *800*, 369–375.

[127] Kobayashi, J., Shirao, M., Nakazawa, H., *J. Liq. Chromatogr. Rel. Technol.* 1998, *21*, 1445–1456.

[128] Kuban, P., Kuban, P., Kuban, V., *J. Chromatogr. A* 1999, *836*, 75–80.

[129] Kuban, P., Pirmohammadi, R., Karlberg, B., *Anal. Chim. Acta* 1999, *378*, 55–62.

[130] Tenberken, B., Bächmann, K., *Atmos. Environ.* 1998, *32*, 1757–1760.

[131] Tenberken-Pötzsch, B., Schwikowski, M., Gäggeler, H. W., *J. Chromatogr. A* 2000, *871*, 391–398.

[132] Tam, W. F. C., Tanner, P. A., Law, P. T. R., Bächmann, K., Pötzsch, S., *Anal. Chim. Acta* 2000, in press.

[133] Qiu, J., Fu, X., *Lihua Jianyan, Huaxue Fence* 1998, *34*, 342–343.

[134] Padarauskas, A., Olsauskaite, V., Paliulionyte, V., *Anal. Chim. Acta* 1998, *374*, 159–165.

[135] Hiissa, T., Siren, H., Kotiaho, T., Snellman, M., Hautojärvi, A., *J. Chromatogr. A* 1999, *853*, 403–411.

[136] Motellier, S., Petit, S., Decambox, P., *Anal. Chim. Acta* 2000, *410*, 11–23.

[137] Wang, T., Lee, H. K., Li, S. F. Y., *J. Liq. Chromatogr. Rel. Technol.* 1998, *21*, 2485–2496.

[138] Cahours, X., Morin, P., Dreux, M., *J. Chromatogr. A* 1998, *810*, 209–220.

[139] Fung, Y. S., Tung, H. S., *Electrophoresis* 1999, *20*, 1832–1841.

[140] Yongtan, Y., Liang, B., Kang, J., Qu, Q., *Yankuang Ceshi* 1998, *17*, 275–278, 289.

[141] Verma, S. P., Garcia, R., Santoyo, E., Aparicio, A., *J. Chromatogr. A* 2000, *884*, 317–328.

[142] Kelly, R. G., Yuan, J., Weyant, C. M., Lewis, K. S., *J. Chromatogr. A* 1999, *834*, 433–444.

[143] Lewis, K. S., Yuan, J., Kelly, R. G., *J. Chromatogr. A* 1999, *850*, 375–380.

[144] Cooper, K. R., Kelly, R. G., *J. Chromatogr. A* 1999, *850*, 381–389.

[145] Erim, F. B., Akm-Senel, K., *Fresenius J. Anal. Chem.* 1998, *362*, 418–421.

[146] Yang, Y., Kang, J., Li, J., Ou, Q., *Sepu* 1998, *16*, 433–435.

[147] Yang, Y., Kang, J., You, J., Ou, Q., *Anal. Lett.* 1998, *31*, 1955–1964.

[148] Kishi, T., Nakamura, J., Arai, H., *Electrophoresis* 1998, *19*, 3–5.

[149] Lopez, C. E., Castro, J. M., Gonzalez, V., Gonzalez, E., Perez, J., Seco, H. M., Fernandez, J. M., *J. Chromatogr. Sci.* 1998, *36*, 352–356.

[150] Bazzanella, A., Lochmann, H., Tomos, A. D., Bächmann, K., *J. Chromatogr. A* 1998, *809*, 231–239.

[151] Muse, J. O., Dabas, P. C., Carducci, C. N., *J. Liq. Chromatogr. Rel. Technol.* 1999, *22*, 2741–2753.

[152] Norden, M., Dabek-Zlotorzynska, E., *Proceedings of the Workshop on Long-lived Radionuclide Chemistry in Nuclear Waste Treatment*, Paris, France, OECD Publ. 1998, pp. 49–61.

[153] Scapolan, S., Ansoborlo, E., Moulin, C., Madic, C., *J. Alloys Comp.* 1998, 271–273, 106–111.

[154] Göttlein, A., *Eur. J. Soil. Sci.* 1998, *49*, 107–112.

[155] Himeno, S., Nakashima, Y., Sano, K., *Anal. Sci.* 1998, *14*, 369–373.

[156] Liu, C. Y., Chen, W. H., *J. Chromatogr. A* 1998, *815*, 251–263.

[157] Malik, A. K., Faubel, W., *Talanta* 2000, *52*, 341–346.

[158] Pozdniakova, S., Padarauskas, A., *Chemija* 1998, 240–244.

[159] Sanchez, J. M., Salvado, V., Havel, J., *J. Chromatogr. A* 1999, *834*, 329–340.

[160] Takayanagi, T., Iwachido, T., Motomizu, S., *Bull. Chem. Soc. Jpn.* 1998, *71*, 1373–1379.

[161] Takayanagi, T., Motomizu, S., *Chem. Lett.* 1999, *523–524*.

[162] Manege, L. C., Takayanagi, T., Oshima, M., Iwachido, T., Motomizu, S., *Bull. Chem. Soc. Jpn.* 1999, *72*, 1301–1306.

[163] Takayanagi, T., Manege, L. C., Motomizu, S., *J. Microcol. Sep.* 2000, *12*, 113–119.

[164] Manege, L. C., Takayanagi, T., Oshima, M., Motomizu, S., *Analyst* 2000, *125*, 699–703.

[165] Rasmussen, B. W., Bjerrum, M. J., *J. Chromatogr. A* 1999, *836*, 93–105.

[166] Yokoyama, T., Tsuji, H., Zenki, M., *Anal. Chim. Acta* 2000, *409*, 55–64.

[167] Yokoyama, T., Tashiro, K., Murao, T., Yanase, A., Nishimoto, J., Zenki, M., *Anal. Chim. Acta* 1999, *398*, 75–82.

Review

Holger Becker[1]
Claudia Gärtner[2]

[1]Jenoptik Mikrotechnik,
 Jena, Germany
[2]amt – Application Center
 for Microtechnology Jena,
 Jena, Germany

Polymer microfabrication methods for microfluidic analytical applications

A growing number of microsystem technology (MST) applications, particularly in the field of microfluidics with its applications in the life sciences, have a need for novel fabrication methods which account for substrates other than silicon or glass. We present in this paper an overview of existing polymer microfabrication technologies for microfluidic applications, namely replication methods such as hot embossing, injection molding and casting, and the technologies necessary to fabricate the molding masters. In addition, techniques such as laser ablation and layering techniques are examined. Methods for bonding and dicing of polymer materials, which are necessary for complete systems, are evaluated.

Keywords: Polymer microfabrication / Microfluidics / Miniaturized total analysis system / Review

EL 3734

Contents

Correspondence: Dr. H. Becker, Jenoptik Mikrotechnik GmbH, Prüssingstr. 41, D-07745 Jena, Germany
E-mail: holger.becker@jenoptik.com
Fax: +49-3641-653-562

Abbreviations: COC, cycloolefin copolymer; **LIGA**, German acronym for Lithographie (lithography), Galvanoformung (electroplating), Abformung (molding); **MST**, microsystem technology; **PA**, polyamide; **PBT**, polybutyleneterephthalate; **PC**, polycarbonate; **PDMS**, poly(dimethylsiloxane); **PE**, polyethylene; **PEEK**, polyetheretherketone; **PMMA**, polymethylmethacrylate; **POM**, polyoxymethylene; **PP**, polypropylene; **PPE**, polyphenylene ether; **PS**, polystyrene; **RMS**, root mean square; **μ-TAS**, miniaturized total analysis system

1 Introduction

The life sciences currently experience a revolution which is only comparable with the foundation of the information society due to the introduction of microelectronic circuits. The application of microsystem technologies (MST or, in the American community, MEMS, for "microelectromechanical systems" or "micromachine" in Japan), for instance in the research of the human genome, the drug discovery process in the pharmaceutical industry, clinical diagnostics, or in analytical chemistry is not only increasing the performance of conventional methods in these disciplines and changing our understanding of the analytical process but, much more important, it allows totally new access to information on the molecular level [1]. The underlying concept of this development is called μ-TAS, the "miniaturized total chemical analysis system" and arose historically from analytical chemistry. Already in the 70s, in a remarkable effort, Terry *et al.* [2] miniaturized a gas chromatography system and integrated the complete system on a silicon wafer. Tragically, this work went

0173-0835/00/0101-0012 $17.50+.50/0

unnoticed for more than a decade, until the conceptual μ-Tas paper was published in 1990 [3]. This article triggered an avalanche of developments and discoveries, which led to a truly exponential growth of this field, first only in academic research, but since the mid 90s also on a commercial basis [4]. In the beginning, as microsystem technologies often use the techniques developed in microelectronics, the devices were fabricated on a silicon wafer, using standard photolithography and subsequent wet etching to produce microchannels on a planar substrate [5]. Electrokinetic pumping was established as the method of choice for transporting liquid samples in these microchannels; the development focused on various types of glass [6–10] or quartz [11–13] as a substrate material, as the conductivity of silicon proved problematic for the application of the high voltage needed for electroosmotic flow (for recent reviews see [14, 15]). The fabrication methods, however, remained mainly the same; isotropic wet etching, using hydrofluoric acid (HF) or KOH as an etchant. However, for the ongoing commercialization of this technology, these fabrication processes represent certain disadvantages.

(i) Cost of substrate material: As many microfluidic devices have a comparatively large footprint (typically several cm^2) to achieve either a long separation channel length or a high parallelization of their functions, the cost of the substrate material is an important factor in high volume production. Polymers like polymethylmethacrylate (PMMA) are of the order of 0.2–2 cents per cm^2, while boro-float glass (*e.g.*, Corning Pyrex) are of the order of 10–20 cents/cm^2, boro-silicate glass (*e.g.*, schott B270) around 5–15 cents/cm^2, and photostructurable glass (*e.g.*, Schott Foturan) around 20–40 cents/cm^2. But while the size of a microchip could become smaller and smaller in microelectronics, due to the progress in circuit engineering and lithographical techniques, the area of a microfluidic chip often cannot be decreased without a loss in performance (*i.e.*, by making a curved channel instead of a straight one) or compatibility with existing systems (*i.e.*, footprint of a microwave plate or pipetting robot). (ii) Many steps (cleaning, resist coating, photolithography, development, wet etching) and harmful wet chemistry (*e.g.*, HF) are involved. Despite the fact that these steps are well known from the microelectronics industry and can be fully automated, each device goes through this fabrication process serially, which increases the fabrication time, and therefore the cost, as well as the risk of fabrication errors. In addition, the costs are significant due to the reagents involved as well as their waste disposal. (iii) Limitations in geometrical design due to the isotropicity of the etching process, which allows only shallow, mainly semicircular channel cross sections in glass substrates, are another disadvantage. For many applications,

however, channel cross sections (such as high aspect ratio square channels, channels with a defined but arbitrary wall angle, or channels with different heights) are desirable. These cannot achieved with standard microfabrication methods in glass or quartz substrates. In silicon, due to the advanced silicon dry etch processes [16], a larger variety of geometries, particularly vertical trenches with a high aspect ratio, can be produced; however, the costs per unit substrate area are significant. (iv) Surface chemistry: Particularly for continuous flow systems, the surface chemistry of silicon substrates poses a problem, as biomolecules (oligonucleotides, DNA, or proteins, *etc.*) tend to create a bond to silicon surface groups and therefore stick to the silicon surfaces. This can be prevented with a surface coating (*e.g.*, silanization), but carries with it the risks of an additional process step.

Polymers as substrate material offer a possible solution to these fabrication challenges and lend themselves to the mass fabrication of microfluidic devices. Their wide range of material properties, their normally low costs and the development of suitable polymer microfabrication methods in the past years have attracted enormous interest, particularly as this opens up the road to a high-volume production of disposable microfluidic devices, which allows the successful commercialization of the μ-TAS concept.

In this review, we have limited ourselves to analytical microfluidic devices. Other aspects of microfluidics, such as micropumps, valves, and mixers, as well as chemically modified surfaces (DNA-array technologies) have consciously been omitted. The same holds true for a large number of other polymer fabrication technologies which so far have not been applied to microfluidics (such as extrusion, reaction injection molding, vacuum casting, injection compression molding, *etc.*) and have therefore been left out. This paper is organized as follows: After a brief introduction into polymer characteristics, the various microfabrication methods are explained. They are classified into two groups, (i) the replication methods, which dominate the commercial sector of microfluidics, and (ii) the serial/individual device techniques, which allow the rapid fabrication of prototype volumes. A chapter on additional steps for device completion (like dicing and bonding) is added to complete the overview on manufacturing technologies.

2 Polymer materials

Polymers are the most promising materials for microsystem technology since they are applicable for mass replication technologies such as injection molding and hot embossing, as well as for methods for rapid prototyping, like casting or laser micromachining. The polymers differ in their mechanical properties, optical characteristics, temperature stability, resistance against chemicals such as acids, alkalines or organic solutions, and there are also biodegradable polymers; therefore, there is a suitable polymer material for nearly every application.

2.1 General properties

Polymers are macromolecular substances with a relative molecular mass between 10 000 and 100 000 Da and more than 1000 monomeric units. The polymerization process is started by an initiator substance or a change in the physical parameters (light, pressure, temperature). Due to the great length of the polymer chain, polymers are bulk materials. In most cases polymers are amorphous or in some cases microcrystalline, where the length of the polymer chains is larger than the size of the crystallites. The chain length of the polymer molecules varies in bulk material; therefore, a polymeric material does not have an exactly defined melting temperature. Instead, a melt interval exists in which the viscosity changes strongly and the material turns into a highly viscous mass. The decomposition temperature is another characteristic point above which the thermal cracking of the material starts and the material ceases to function.

Many of these highly molecular substances solidify after cooling under the so-called glass transition temperature, T_g, and the solid phase that results is as hard as glass and brittle. For fabrication processes, this is one of the important parameters. If the temperature is increased above T_g, the material becomes plastic-viscous and can be molded. It is important for the molding process to cool the material below T_g before demolding; otherwise the geometric stability of the molded component can suffer due to the relaxation during the demolding and the resulting entropy elasticity. Softeners can be used to reach lower glass transition temperatures for the respective material. The elasticity, impact strength, and expansion of the polymer increase and the hardness decreases due to the softeners.

Polymers can be classified into the following three categories according to their molding behavior, which is determined by the interconnection of the monomer units in the polymer chain. (i) *Thermoplastic polymers.* These consist of unlinked or weakly linked chain molecules. At a temperature above the glass transition temperature, these materials become plastic and can be molded into specific shapes, which they will retain after cooling below T_g. (ii) *Elastomeric polymers.* These are also very weakly cross-linked polymer chains. If an external force is applied, the molecular chains can be stretched, but relax

and return to their original state (higher entropy) once the external force is removed. Elastomers also do not melt before reaching their decomposition temperature. (iii) *Duroplastic polymers.* In these materials, the polymer chains are cross-linked stronger, and therefore a molecular movement for a change in shape is not easily possible. These materials therefore have to be cast into their final shape. They are harder and more brittle than thermoplastic materials and soften only very little before the temperature reaches their decomposition temperature.

Meanwhile a wide variety of polymer materials have been used for microfabrication processes: Standard polymer materials such as polyamide (PA), polybutyleneterephthalate (PBT), polycarbonate (PC), polyethylene (PE), polymethylmethacrylate (PMMA), polyoxymethylene (POM), polypropylene (PP), polyphenylene ether (PPE), polystyrene (PS) and polysulfone (PSU); engineering plastics like liquid crystal polymer (LCP), polyetheretherketone (PEEK) and polyetherimide (PEI), as well as biodegradable materials such as polylactide. To date, PMMA and PC are the most popular polymer materials for microfabrication *via* hot embossing and injection molding. A cycloolefin copolymer (COC), is currently being tested in the hot embossing process. This new material is extremely promising for applications in chemical engineering and molecular biotechnology since it has high chemical stability and is optically transparent [17].

Tables 1 and 2 summarize the physical and chemical properties of the most commonly used thermoplastic polymers for micromolding. In comparison to glass, the resistance to chemicals, the aging, and the mechanical as well as UV stability can restrict the use of polymers for certain applications. Also, an increased fluorescence at shorter wavelengths (around 400 nm, *e.g.*, for dyes like fluorescein) in comparison to glass can be ascertained, which can lead to reduced sensitivity in case of laser-induced fluorescence (LIF) detection. Table 3 gives an overview of the typical materials used for micromolding and a judgement about their behavior in the injection molding process. Nevertheless, the characteristics depend on the product, polymer type, and the respective conditions in the injection molding process. For the hot embossing process, filling and separation of the mold are less critical than for injection molding.

To date there has been no special development of plastics for the micromolding process since even a mass production of one million pieces only leads to perhaps one ton of polymer material – an amount much too small for any special effort. Therefore, already commercially available standard polymers are used for micromolding processes. However, the existing applications prove their prin-

Table 1. Basic physical properties of molding polymer materials

Thermoplastic materials for micromolding	Density $(\times 10^3$ kg/m^3)[a]	Glass temperature T_g (°C)	Permanent temperature of use (°C)	Thermal conductivity λ (W m^{-1} K^{-1})	Linear expansion coefficient α (10^{-6} K^{-1})	Heat distortion temperature measurement method: Vicat Method B (°C)
Polyamide 6 (PA 6)	1.13	60	80–100	0.29	80	180
Polyamide 66 (PA 66)	1.14	70	80–120	0.23	80	200
Polycarbonate (PC)	1.2	150	115–130	0.21	65	148–150
Polyoxymethylene (POM)	1.41–1.42	–60	90–110	0.23–0.31	90–110	154–160
Cycloolefin copolymer (COC)	1.01[b]	138[b]	Not available	Not available	60[b]	123[b]
Data for the cycopentadien-norbornen copolymer Zeonex						Measurement method: ASTM D648
Polymethylmethacrylate (PMMA)	1.18–1.19	106	82– 98	0.186	70– 90	80–110
Polyethylene low density (PE-LD)	≤ 0.92	–10	70[c]	0.349	140	40
Polyethylene high density (PE-HD)	≤ 0.954	–	90[d]	0.465	200	60– 65
Polypropylene (PP)	0.896–0.915	0–10	100	0.22	100–200	90–100
Polystyrene (PS)	1.05	80–100	70[e]	0.18	70	78– 99

Data taken from [18].
a) [19]
b) Product information leaflet for Zeonex
c) For Lupolen 1800a of BASF
d) For Lupolen 6031 of BASF
e) For polystyrol 159 K of BASF

cipal suitability for microfabrication. To our knowledge only thermoplastic and elastomeric materials have been used to date for the fabrication of microfluidic devices.

2.2 Photoresists

Photoresists are another class of polymers used in microsystem technology and for microfluidic systems. Irradiation, with electrons, ions, X-rays, UV or visible light, leads to a photochemical reaction of the resist material, which is coated onto a carrier substrate (typically either silicon or another polymer). In the case of the so-called positive tone resists (or singly positive resists) the solubility of the irradiated areas increases; in negative tone resists, it decreases. The irradiated or nonirradiated areas, respectively, will be removed by a developer. For deep X-ray lithography the standard resist material is PMMA which acts as a positive tone resist. Polyactide or copolymers of lactide and glycolide can be used as resist material for X-ray lithography [17]. In the longer wavelength range, SU-8 is a resist material developed by IBM for UV-lithography which allows the fabrication of structures with heights of more than 1000 µm (see Section 4.2).

3 Replication technologies

The secret of the commercial success of polymer microfabrication in microfluidics lies in the establishment of a low-cost manufacturing process. Replication technologies are good for this application because the principles behind these processes are already well known in the macroworld, and, in the case of injection molding, represent a standard technology for macroscopic polymer component manufacturing. The underlying principle is the replication of a microfabricated mold tool, which represents the negative (inverse) structure of the desired polymer structure. The (expensive) microfabrication step is therefore only necessary once for the fabrication of this master structure, which then can be replicated many times into the polymer substrate. In addition to the cost advantages, this replication also offers the benefit of the freedom of design: the master can be fabricated with a large number of different microfabrication technologies (see below), which allow various geometries to be realized.

Nevertheless, some restrictions apply to these replication techniques. (i) As the master has to be removed from the

Table 2. Basic chemical properties of molding polymer materials

Thermoplastic materials for micromolding	Solvent resistance	Acid and alkaline resistance	Trade name
Polyamide 6 (PA 6)	Resistant against: ethanol, benzene, aromatic and aliphatic hydrocarbons, mineral oils, fats, ether, ester, ketones	Not resistant against: diluted and concentreated mineral acids, formic acid	Perlon Durethan (Bayer) Ultramid (BASF)
Polyamide 6.6 (PA 6.6)	Same as above	Same as above	Nylon Nylind (Du Pont) Celanese (Ticona) Ultramid (BASF)
Polycarbonate (PC)	Resistant against: water, benzene, mineral oils Conditionally resistant against: alcohols, ether, ester	Resistant against: diluted mineral acids	Makrolon (Bayer)
Polyoxymethylene (POM)	Resistant against: fuels, mineral oils, usual solvents	Not resistant against: anorganic acids, acetic acid, oxidating solvents	Hostaform (Hoechst)
Cycloolefin copolymer (COC)[a]	Resistant against: acetone, methylethylketone, methanol, methanol, isopropanol	Resistant against: diluted and concentrated mineral acids and alkalines, 30% H_2O, 40% formaldehyde, detergents in water	Topas (Ticona) Zeonex (Nippon Zeon)
Cycopentadien-norbornen copolymer Zeonex	Not resistant against: ether, aromatic and aliphatic hydrocarbons, methylmethacrylate	Not resistant against: concentrated HNO_3	
Polymethylmethacrylate (PMMA)	Resistant against: water, mineral oils, fuel, fatty oils	Resistant against: up to 20%, diluted acids, diluted alkalines, NH_3	Plexiglas (Röhm) Lucryl (BASF) Perspex (ICI)
Polyethylene (PE)	Resistant against: alcohols, benzene, toluene, xylene	Resistant against: NH_3, diluted HNO_3, H_2SO_4, HCL, KOH, NaOH	Lupolen (BASF)
Polypropylene (PP)	Resistant against: diluted solutions of salts, lubricating oils, chlorinated hydrocarbons and alcohols	Resistant against: most diluted acids and alkalines	Hostalen (Hoechst)
Polystyrene (PS)	Resistant against: alcohols, polar solvents Not or hardly resistant against: ether, benzene, toluene, chlorinated hydrocarbons, acetone, ethereal oils	Resistant against: diluted and concentrated acids (except HNO_3) and alkalines	Polystyrol (BASF)

Data for solvent, acid and alkaline resistance taken from [18].
a) Product information leaflet for Zeonex

molded structure, undercuts (*i.e.*, structures in the polymer with overhanging edges) can not be fabricated by these means. (ii) The primary success, the lifetime of the mold tool, and the achievable aspect ratios depend very strongly on the surface quality of the mold tool. In general, the smoother the tool surface, the lower the frictional forces on the tool as well as the polymer microstructure in the demolding step. Typically, surface roughness values of better than 100 nm root mean square (RMS) are necessary for a good and reliable replication. This puts certain limitations on the fabrication methods for the mold tool. (iii) The interface chemistry between tool material and substrate polymer is also a critical factor. If the two materials form any kind of chemical or physical bond during the replication step, this adds to the forces during the demolding step. Release agents, which are often used in

Table 3. Molding behavior of polymer materials

Thermoplastic materials for micromolding	Reproduction	Filling	Separation
Polyamide (PA)	++	+	+
Polycarbonate (PC)	+	+	−
Polyetheretherketone (PEEK)	++	−	−
Polyoxymethylene (POM)	++	++	+
Polymethylmethacrylate (PMMA)	+	o	−
Polypropylene (PP)	+	o	+
Polystyrene (PS)	+	+	−

++ very good, + good, o average, − bad
Data taken from [20]

the macroworld to help the mold release of complex structures, are often not suited for microfluidic devices, as on the one hand they might diffuse into the polymer matrix and contaminate the sample if they are released into the liquid being pumped through the channel, and on the other hand they tend to increase the autofluorescence of the polymer. With these criteria in mind, the following section describes the technologies available for the mold tool fabrication.

3.1 Master fabrication

Various techniques have been used in the past for the fabrication of molding tools. We will concentrate on the methods relevant for microfluidic structures.

3.1.1 Micromachining methods

As conventional machining has also undergone miniaturization efforts, modern micromachining technologies (sawing, cutting, milling, turning) are capable of producing mold tools with structures in the range of a few 10 μm. Their particular advantage is the wide range of materials which can be machined, particularly stainless steel, which is not accessible with other microfabrication technologies and which offers good mold insert lifetimes. In addition, the development times for micromachined tools can be shorter as no mask fabrication and lithography step is involved. Especially comparatively simple channel structures with straight walls are well-suited geometries for these techniques. Channel crossings, high aspect ratio structures, very deep holes or very small structures, however, cannot be fabricated, or only with major drawbacks, with these methods. Special mention should be made of microelectrodischarge machining (μ-EDM), which allows the fabrication of quasi three-dimensional structures in conducting materials. Here the material is removed due

to the high energy electric discharge between an electrode and the workpiece. The technique offers a high degree of flexibility in terms of materials and geometries, but produces a comparatively rough surface. This method nevertheless has potential for further development. Very simple structures have reportedly been fabricated by replicating thin chromel wires [22, 23].

3.1.2 Electroplating methods

The most commonly used methods for master fabrication employ an electroplating step, resulting in a replication master made out of nickel or a nickel alloy like NiCo or NiFe. The process starts with photolithography, in which a photoresist-coated substrate with a conducting electroplating starting layer is exposed to light and subsequently developed, so that the areas to be electroplated are free of resist. This structure is then placed into a galvanic bath, where, due to the migration of metal ions between the bath and the conducting starting layer, the metal starts to grow in the resist structure. After the resist structure is overgrown with the metal, the resist and starting layer can be dissolved and the resulting metal structure (so-called shim) can be processed further. Conventional photoresist technologies allow structural heights of the order of 10–40 μm; for higher structures special thick resists (EPON SU-8) have to be used, which can result in heights of up to 1 mm. Other techniques to produce electroplating forms are the LIGA technique, which is a German acronym for Lithographie (lithography), Galvanoformung (electroplating), Abformung (molding), where thick PMMA layers are exposed with synchrotron radiation [25] and the laser-LIGA process [26], where the synchrotron radiation is replaced by pulsed UV-light, which ablates the polymer material. These techniques have in common that the surface roughness is quite small (LIGA down to about 10 nm RMS) and that the resulting nickel tool has a good surface chemistry for most polymers. Drawbacks are the comparatively slow growth rate of nickel in the electroplating process (a typical rate is typically between 10 and 100 μm/h), the high stress levels in thick nickel layers, which tend to bend the master, and the radial dependency of the growth rate, which can result in a different height of the nickel structure in the middle and at the rim of a nickel wafer.

3.1.3 Silicon micromachining

As silicon itself has suitable material properties for a mold tool (high level of stiffness, high heat conductivity) and a large variety of silicon surface micromachining techniques exist as well as commercially available services, attempts have been made to use silicon as a tool directly. Several

Table 4. Overview of the different molding technologies

Process	Materials	Tool costs	Cycle time	Forces and temperatures	Automation	Geometry	Minimum dimensions Aspect ratios
Hot embossing	Thermoplastics	Low-medium	Medium-long (3–10 min)	High (kN)	Little	Planar, *e.g.* wafers, plates	nm (nanoimprint)
	Duraplastic thin films			Around T_g (100–200°C)			50 small areas, 5 wafer scale
Injection molding	Thermoplastics	High	Short-medium (0.3–3 min)	High	Yes	Bulk, spherical	Some 10 µm
	Duroplastics			Above melting (150–400°C)			50 small areas, 5 larger areas
Casting	Elastomers Epoxies	Low	Long (min-h)	No forces Room temperature –80°C	Little	Planar	nm About 1

micromachining technologies have been under investigation, the simplest one being wet etching of silicon. A wet etching step of 100-silicon results in a structure with a wall angle of 54.7 degrees, which therefore forms a trapezoidal channel. The slant in the wall allows a good mold release and the surface roughness of the wet etching process of well-oriented monocrystalline silicon wafers is excellent [22, 27, 28]. Obviously, the channel cross section in this case is limited to this shape, although some isotropic etching techniques for silicon also exist. This limits the achievable aspect ratio. Dry etching methods, *i.e.*, reactive ion etching (RIE), advanced silicon etch (ASE) or the Bosch process (for a review see [16]), however, allow deep structures with vertical walls to be fabricated, although with a surface roughness which strongly depends on the etch rate. Fast etches produce a very rough surface, which for practical reasons limits the achievable depth of the structures. Typical depths range between 10 and 40 µm for a conventional dry etch and up to more than 200 µm in case of an ASE process, with etch rates of the order of 1–5 µm/min. All silicon tools, however, have the common problem of a potential stiction problem with many polymers due to their surface chemistry. A combination of silicon etching and electroplating exists, which is called DEEMO (deep etching, electroplating, molding) [29], where the micromachined silicon is used as a base for an electroplating step equivalent to the electroplating of the photoresist described above.

3.2 Hot embossing

Currently the most widely used replication process to fabricate channel structures for microfluidic applications is hot embossing [22, 23, 28, 30–33]. The microfabrication process of hot embossing itself is rather straightforward [34]. After fabrication of the master, it is mounted in the embossing system together with the planar polymer substrate. Both are heated separately in a vacuum chamber to a temperature just above the glass transition temperature, T_g, of the polymer materials, which is typically of the order of 50–150°C (see Table 1). The vacuum is necessary to prevent the formation of air bubbles due to entrapment of air in small cavities. It also allows water vapor (which, during the process, is driven out of the polymer substrate) to be removed. Additionally it increases the lifetime of nickel tools, as it prevents corrosion of the nickel at these elevated temperatures. The tool is brought into contact with the substrate and then embossed with a controlled force. Typical embossing forces are of the order of 0.5–2 kN per cm^2. Still applying the embossing force, the tool-substrate sandwich is then cooled to just below T_g.

To minimize thermally induced stresses in the material as well as replication errors due to the different thermal expansion coefficients of tool and substrate, this thermal cycle should be as small as possible. After reaching the lower cycle temperature, the embossing tool is mechanically driven apart from the substrate which now contains the desired features. This is usually the most critical step, as now the highest forces act on the polymer microstructure, particularly if a structure with vertical walls and a high aspect ratio is desired. Therefore, an automated mold release is crucial for a higher production yield. Overall cycle time of the embossing process for materials such as PMMA is of the order of 5–7 min. The microstructured polymer wafer can now be processed further. Figure 1 shows a diagram of a hot embossing machine; Fig. 2, a channel structure fabricated with hot embossing on a PMMA substrate; and Fig. 3 shows a separation of *Hae*III ΦX174 DNA fragments. The high structural resolution achievable with hot embossing can be seen in Fig. 4, with Fig. 4a showing the embossing tool, fabricated with an

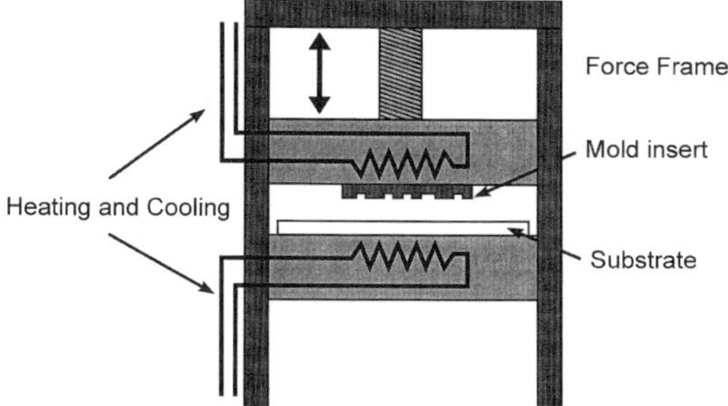

Figure 1. Diagram of a hot embossing machine.

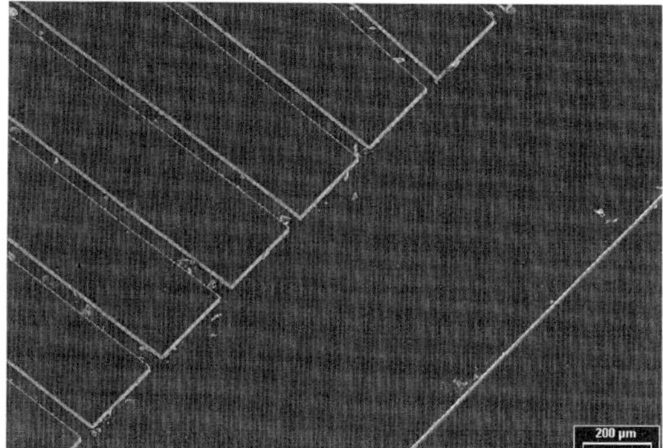

Figure 2. Microchannel structure for capillary electrophoresis fabricated with hot embossing in a PMMA substrate.

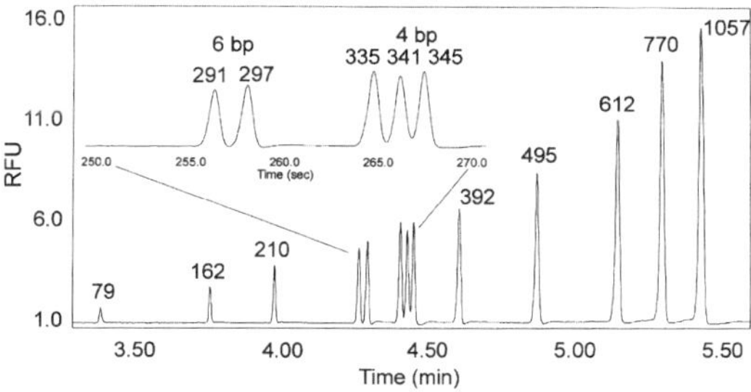

Figure 3. Separation of *Hae*III ΦX174 DNA fragments in a hot embossed PMMA chip. Data kindly supplied by A. Paulus (Acalara BioSciences).

advanced silicon etch process, and Fig. 4b showing the resulting channel structure in PMMA. This structure represents a two-dimensional channel array with microchannels of less than 1 μm in width.

3.3 Injection molding

In the macroscopic world, injection molding is one of the widespread standard processes to fabricate polymer

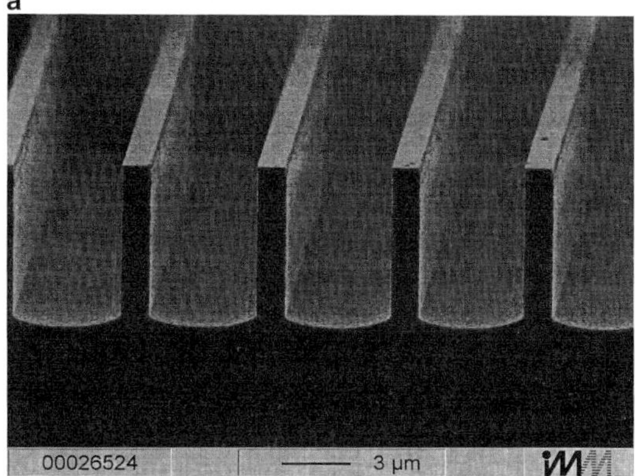

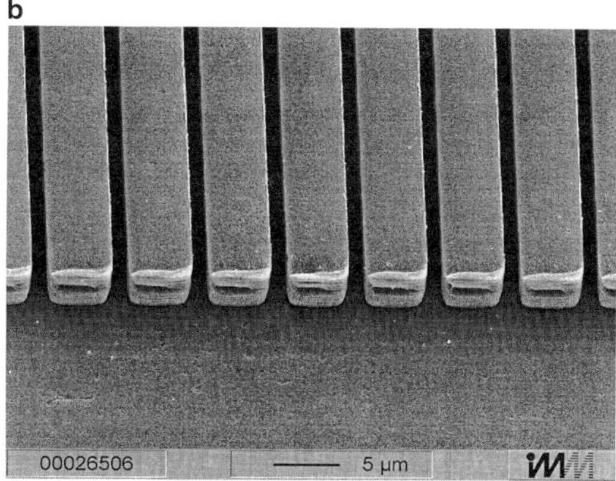

Figure 4. (a) Embossing tool fabricated with an advanced silicon etch; (b) resulting channel array structure.

parts. It is used to form almost any geometry from a large variety of thermoplastic materials, and almost any plastic part with dimensions in the millimeter to centimeter range will be manufactured with this technology. It is therefore not surprising that attempts have been made to apply this process to microsystem technologies [35, 36] for the fabrication of microfluidic devices [27, 37]. Figure 5 shows a cross section of an injection-molding machine with the different process steps. The process starts with the raw polymer material, which is used in granular form. These granules are fed into the cylinder, a heated screw, where the pellets start to melt. This melt is then transported forward towards the mold cavity. Typical temperatures in this ragion range from 200°C (for polymers such as PMMA and PS) over ~ 280°C (for PC) up to ~ 350°C vor materials like PEEK. The molten material is then injected under a high pressure (typically 60–100 MPa = 600–1000 bar) into the evacuated cavity, which contains the mold insert

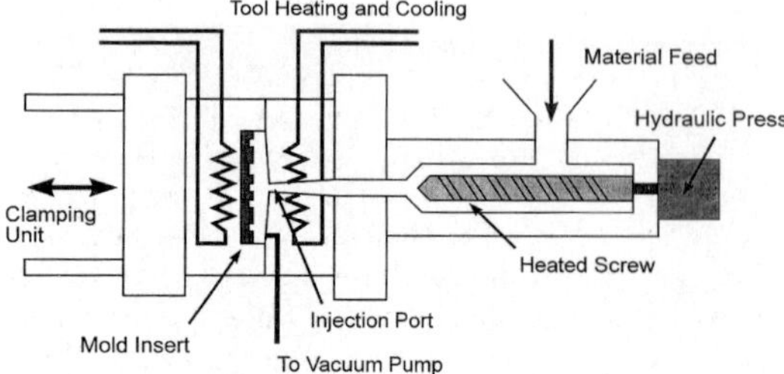

Figure 5. Schematic diagram of an injection molding machine.

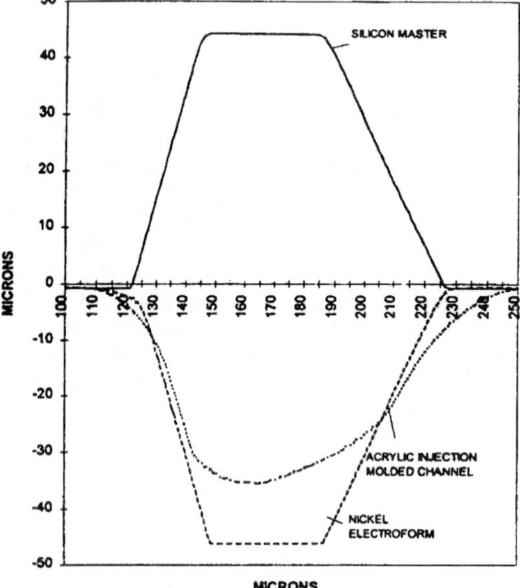

Figure 6. Profilometer scan of an injection molded channel. Reprinted from [27] with permission.

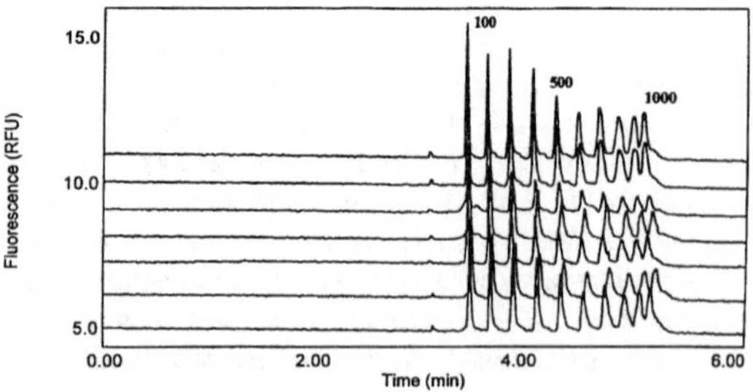

Figure 7. Separations of a DNA sizing ladder in parallel capillaries fabricated with injection molding. Reprinted from [37], with permission.

as the master structure. In macroscopic systems, the cavity can be held at a temperature below the solidification temperature of the polymer (usually between approximately 60–120°C; so-called cold-cavity process). This allows rapid fabrication: cycle times of only several seconds are standard for most applications. If the structures become smaller, and therefore less material has to be injected into the cavity and the surface-to-volume ratio increases, the cavity has to be heated closer to the melting point of the polymer material to allow the polymer to flow into all small structures of the mold insert. The cavity will then be cooled to allow the ejection of the microstructured part. This process, called variotherm, allows the fabrication of smaller structures than the cold-cavity process, but increases the cycle time due to the heating and cooling. Typical cycle times for microinjection molding are of the order of 1–3 min. As a large thermal gradient between the injection temperature and the ejection temperature of the polymer exists, as well as the phase transition between the liquid and the solid phase, volume changes and thermal shrinkage have to be taken into account in the master fabrication. Figure 6 shows a profilometer scan across an injection-molded PMMA microchannel structure [27]. The deviation of the replicated polymer structure from the master is due to a nonoptimized injection molding process. Figure 7 shows a separation of a DNA sizing ladder in parallel capillaries [37].

3.4 Casting

A process that has found widespread use mainly in the academic world is the casting of silicone-base elastomers. The earliest mention of a miniaturized separation device based on polymer methods was reported in a patent by Ekström *et al.* [38] in 1990. They use a cast silicone rubber to form microchannels. This elastomeric layer was then placed between two glass plates for mechanical support and channel sealing. Casting generally offers flexible and low-cost access to planar microchannel structures [39], and the material involved, mostly poly(dimethylsiloxane) (PDMS) of type Sylgard 184, offers good optical properties with high transparency above 230 nm and little autofluorescence [40]. In this microfabrication method, a mixture of the elastomer precursor and its curing agent are poured over the molding templates. These templates are made, *e.g.*, by silicon surface micromachining [40, 41], by photostructuring a printed circuit board [42, 43] or by lithographically patterning a photoresist layer [44], and might be surface-modified for a better mold release. After curing, which typically takes several hours, the soft elastomer copy can simply be peeled off the mold. The so-formed microstructures can simply be placed against a planar surface, *e.g.*, a glass slide [40, 45], a plastic sheet [44], or a printed circuit

Figure 8. Injection region of a cast PDMS CE chip. Reprinted from [40], with permission.

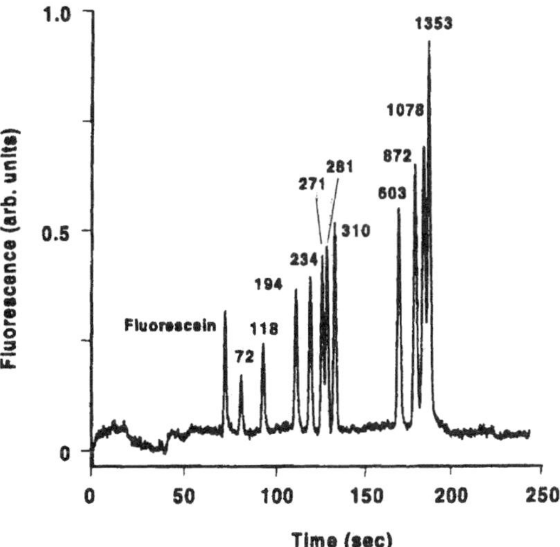

Figure 9. Separation of *Hae*III ΦX174 DNA fragments in a PDMS chip. Reprinted from [40], with permission.

board containing electrodes [43], to form closed channels. Figure 8 shows the injection region of a CE device, replicated from a silicon mold [40]. A good structural replication can be observed. As an example of the performance, Fig. 9 shows the separation of *Hae*III ΦX174 DNA fragments achieved with this structure. With this technique the group of Whitesides [46] has demonstrated the fabrication of structures with nm features.

3.5 Summary of replication technologies

The above-mentioned replication methods represent the commercial pathways for the microfabrication of fluidic devices. They have the potential for high volume production and allow μ-TAS to become disposable devices. Table 4 summarizes the properties of these technologies.

4 Direct techniques

In contrast to the above-mentioned replication techniques, which allow the repetitive production of a polymer device from a single mold, several techniques exist, which micromachine each single device individually. On one hand this allows a rapid fabrication of single devices, as no previous master fabrication step is involved; on the other hand the fabrication throughput is limited by the fabrication time of each individual device.

4.1 Laser-based technologies

A widely used technology for the fabrication of microfluidic devices is laser ablation [47, 48]. In this process, the energy of a laser pulse is used to break bonds in a polymer molecule and to remove the decomposed polymer fragments from the ablation region. A typical laser ablation setup consists of an eximer laser, which delivers light pulses at 193 nm (ArF) or 248 nm (KrF) with typical pulse frequencies of 10–100 Hz (ArF) to several kHz (KrF), a mask or aperture, and an *xy*-table on which the substrate is mounted. The mask defines the ablated region, while the complete pattern is made by moving the substrate on the *x-y* stage underneath the mask. Depending on the energy available per laser pulse and on the substrate material, typical ablation rates per laser pulse range between some hundred nanometers [49, 50] and 5 μm [49]. With this technology, a wide range of polymeric materials, including PMMA, PS, PC polyethyleneterephthalate (PET), cellulose acetate, polyimide and photoresists have been structured [47–50]. The accuracy of the process in *x-y* direction is mainly given by the energy distribution in the beam (which is normally constant to about 5% across the mask or aperture) and the quality of the *x-y* stage.

Figure 10. Laser-ablated microchannel. Picture kindly supplied by N. Rivzi, Exitech Ltd.

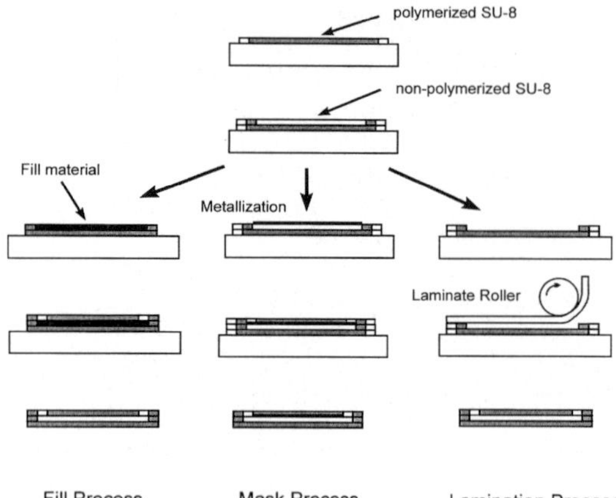

Figure 11. Fabrication methods for microchannels using optical lithography with SU-8.

Typically they are of the order of a few microns. Depth control, which ultimately also determines the wall roughness, is of the order of 0.1 µm. Due to the interaction of the laser light and the polymer material, however, certain modifications are induced in the surface chemistry in comparison to the untreated material [48]. Nevertheless, electroosmotic flow could be achieved in laser-ablated channels [48]. Figure 10 shows the cross section of such a laser-ablated network of curved microchannels. Another possibility to create microchannels with laser ablation, which albeit results in larger channels (typically several 100 µm in width), is the fabrication of "gaskets", *i.e.*, the outline of the microchannel is cut out with the laser from a thin polymer foil, and the remaining material is simply removed with tweezers [51–53]. To complete a microfluidic device, this gasket is then placed between two flat polymer or glass plates.

4.2 Optical lithography in deep resists

Many microfabrication technologies involve the use of a lithography step; the idea of directly patterning a photosensitive polymer material to form microchannels is thus obvious. After the development of SU-8, an epoxy polymer for high aspect ratio UV-LIGA [24, 54], its application for the fabrication of microfluidic devices was quickly proven [55]. Such thick resists allow (in contrast to the normal resists in photolithography, which usually have a thickness in the range of 0.5–3 µm) large structural heights of up to several hundred µm with a single spin coating step. Therefore, microchannels with typical heights of some 10 µm can be fabricated. Several fabrication methods based on the exposure of SU-8 and similar resists have been proposed [56] (Fig. 11). All three proc-

esses have as common first steps the deposition of a base layer of SU-8, which is processed by exposure with UV light and post-bake. Then a second SU-8 layer is spun on top and processed identically.

(i) The so-called fill process, where a sacrificial layer of another polymer (*e.g.*, Araldite) is used to fill a channel which has been lithographically defined in the second step and has been formed after development of the SU-8. After filling the channel, another layer of SU-8 is spun on top, exposed, and baked. The sacrificial polymer is dissolved, thus creating a closed microchannel system (compare also to the process described in Section 4.4).

(ii) In the mask process the second SU-8 layer is not developed. Instead a metal layer, which acts as a shadow mask, is deposited on top of the second SU-8 layer. A third layer is then spun on this stack and illuminated. Afterwards, the resist stack is developed, which leads to a dissolution of the material in the shadow region, thus forming the channel. This dissolution, however, is a rather slow process and it takes many hours for a channel of 10 mm length to form at channel cross sections of 50 by 50 µm [56].

(iii) In the lamination process, the first steps are identical with the fill process. To close the channel, however, a layer of a dry film of SU-8 is laminated on top of the stack (see also Section 6.1.1). This method has the advantage of comparatively short processing times, as no dissolution steps have to take place. Also, completely sealed cavities can be fabricated this way.

Figure 12 shows a cross section of a microchannel made of SU-8. A big advantage is the fact that the closing of the channel structure is included in the fabrication method. SU-8 is a material which is not easy to process, has a rather large internal stress and, once developed, is hard to remove from structures.

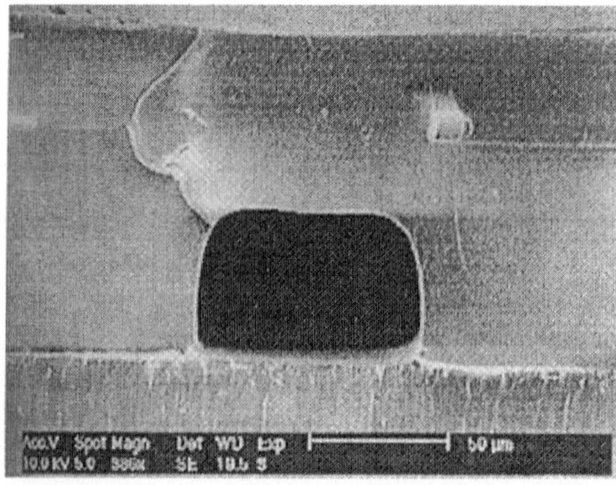

Figure 12. Cross section of a microchannel made of SU-8. Reprinted from [56], with permission.

4.3 Stereolithography

A method which allows real three-dimensional microfabrication is stereolithography. In this method, a photocuring liquid polymer is exposed to focused laser light. In the focal point, the polymer cures and forms a solid. By moving the container with the polymer relative to the focal point, a structure is built up, one volume element after another. Typically the laser focus is scanned in *x-y* direction, while the container is moved in *z* direction, thus forming a layer by layer structure. An application of this microfabrication method for microfluidics has been reported by Ikuta *et al.* [57], who fabricated a fluidic system with several channels and chambers. The main disadvantage of this method lies in its slow buildup of the structure by volume elements, which therefore allows only limited device volumes. A single device can take several hours. On the other hand, no masks or other steps are involved, and the structure is directly formed according to computer-aided design (CAD) data, which control the movement of laser and container. It is therefore especially suited for rapid prototyping applications.

4.4 Layering techniques

A method that, by and large, is compatible with integrated circuit (IC) technologies is the growth of thin layers of

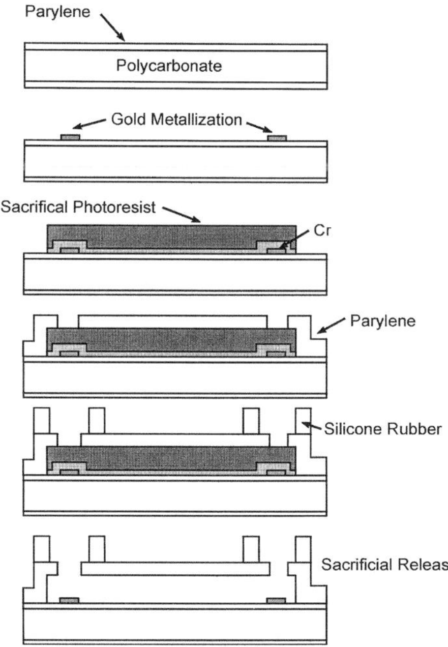

Figure 13. Diagram of the fabrication process for layered structures.

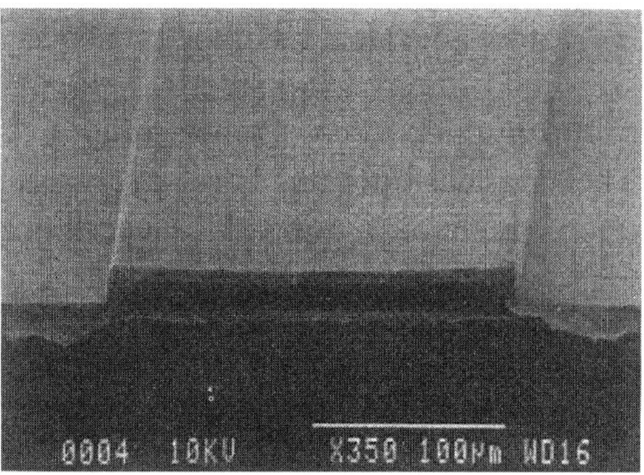

Figure 14. Scanning electron microscope (SEM) picture of a channel fabricated with the process shown in Fig. 15. Reprinted from [59], with permission.

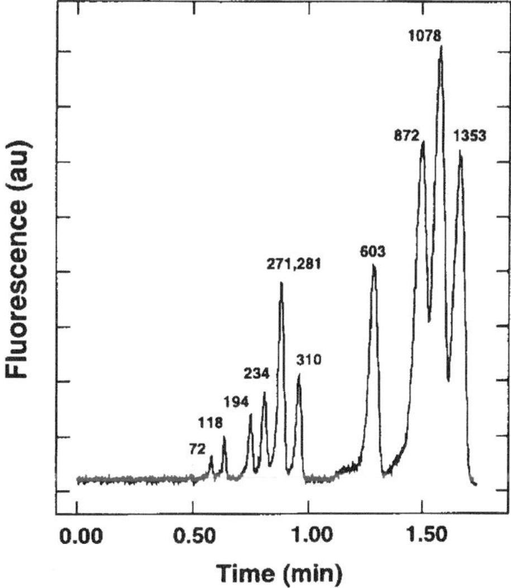

Figure 15. Separation of DNA fragments in a parylene microchannel. Reprinted from [59], with permission.

polymers on planar substrates and the use of sacrificial layers to create open volumes between these layers. This technique has been described by Webster *et al.* [58, 59] and Mastrangelo *et al.* [60] and is schematically shown in Fig. 13. A thin layer of parylene is deposited onto a substrate, either silicon or polycarbonate. Onto this layer, metal can be deposited using evaporation or sputtering methods to form electrode structures. This structure is then covered with a thin chromium layer and a layer of a sacrifical polymer. The thickness of this sacrifical polymer defines the channel height. The next step consists of depositing an additional layer of parylene on top of this structure to form the channel walls and cover. This represents a big advantage because the closing of the channel structure is naturally included in this fabrication method. *Via* holes are defined in the parylene which allow fluid connections in the completed system as well as access of the etchant for the now following sacrificial etch, mostly with acetone. Due to the comparatively small channel cross section, this etch can take several hours as the dissolved material can only be transported by diffusion. Cast silicone structures on top of the parylene layer then define fluid reservoirs. Figure 14 shows a cross section of such a microchannel and Fig. 15 a separation of DNA fragments achieved in such a structure.

5 Additional manufacturing technology for complete devices

In addition to the pure fabrication processes of microchannels or channel networks, several other issues have to be addressed to account for a completed microfluidic device. This includes, most importantly, the closing of the microchannels to form capillaries, but also refers to the dicing of devices, *via* hole fabrication and possible inclusion of metal structures. All these steps are necessary to create a fully functional microsystem.

5.1 Bonding

In order to form capillaries, the microchannels which are normally open after the fabrication step have to be closed, obviously without clogging the channels, changing their physical parameters or altering their dimensions. This

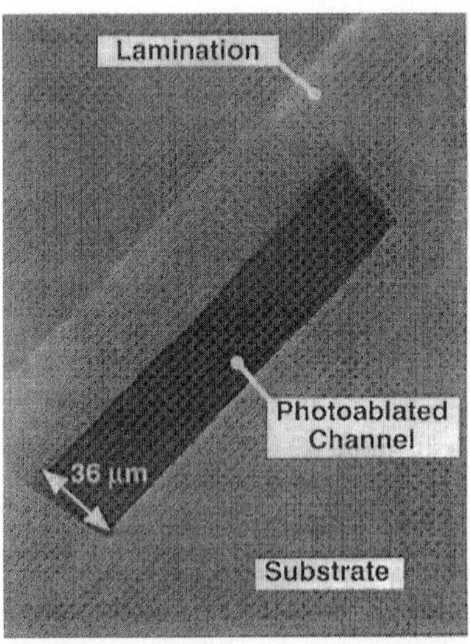

Figure 16. Laser-ablated microchannel laminated with a PET/PE foil. Reprinted from [48], with permission.

often represents a big challenge for higher volume fabrication methods. Several methods have been reported in the literature.

5.1.1 Lamination

In the lamination process, a thin PET foil (typical thickness about 30 µm) coated with a melting adhesive layer (typical thickness 5–10 µm) is rolled onto the structure with a heated roller [48, 62]. The adhesive layer melts in this process and combines the lid foil with the channel plate. This method is widespread in the macroworld for encapsulation of paper and polymers in a polymer film and works well for larger channels. With very small channels, however, the adhesive tends to block the channel. Due to the difference in the materials used, an inhomogeneous interface between lid and channel plate is created, which leads to a sudden change of parameters such as refractive index at the interface. Figure 16 shows the cross section of such a microchannel, closed with a PET/PE laminate [48].

5.1.2 Gluing

Similar to lamination, conventional gluing can be used to join channel plate and lid [38, 62], whereby the same problems arise, mainly the risk of blocking the channels.

5.1.3 Application of heat and pressure

Several teams have reported a sealing of structures by heating up the polymer and applying a force to close the channels [37]. Care has to be taken not to damage the structures; therefore this method is advisable mainly for designs with comparatively small structured areas in comparison to the whole chip surface.

5.1.4 Laser welding

Polymers can be joined by local melting due to heat generated by a laser. This has been successfully demonstrated in the fabrication of micropumps [63], but so far no reports have been published on microchannel applications. The main reason for this is the fact that all welding lines have to be drawn out with the laser, which in the case of microchannels amounts to comparatively large distances and therefore long welding times.

5.1.5 Ultrasonic welding

A method well known in the macroworld is ultrasonic fusion of two polymer layers, where a local melting of the polymer is achieved by the energy density of an ultrasonic sound wave. To date, no application of this technique is known in microfluidics. Common for all these bonding technologies is the need for very clean processing conditions, as especially particle contamination reduces the bond quality and yield dramatically. Clean-room processing is therefore highly advised.

5.2 Dicing

Dicing the devices can be realized either with a conventional rotating saw (no wafer saw, the polymer tends to smear the cutting wire) or with lasers, mainly CO_2 lasers with little energy (typically 50 W). The method of choice depends on the material used and the accuracy required.

5.3 Electrode fabrication

If electrode structures for amperometric detection ([64]; Rossier *et al.*, submitted) or the application of the separation voltage are included, the normal deposition methods such as sputtering and thermal or electron beam evaporation can be used. A limitation exists in the achievable dimensions of the electrode structures, as the deposition can only be done easily with a shadow mask, which restricts the electrode width to about 40 μm and above. Photolithographic processes are difficult to carry out on already structured polymer chips; therefore only the layering technique (Section 4.4) lends itself easily to the fabrication of smaller metal structures. Alternatively, laser-ablated microchannels can be filled with a conducting ink and act as electrodes (Rossier *et al.*, submitted).

6 Outlook and conclusions

The fabrication of polymer microfluidic devices is a rather young field which nevertheless shows enormous growth. Todays, hardly any conference in the field of μ-TAS lacks a session on polymeric devices, which is in stark contrast to the situation only three years ago, when at μ-TAS '96 only a single paper [41] addressed this issue. The driving force behind this development, on the one hand, is certainly the commercialization of microfluidics with its applications in genomics, drug discovery, and diagnostics. These areas all demand a high number of devices at low cost. Ultimately the devices will be used as disposables. On the other hand, μ-TAS is becoming emancipated from its historical fabrication roots, which came from microelectronics, and is developing fabrication technologies which are better suited for the above-mentioned applications in the life sciences.

What are the main trends in this exciting field? First, several groups are working on the integration of sample preparation issues and/or detection systems onto fluidic chips.

This is a development towards a "real" miniaturized system. Second, the first applications with a demand for a very high number of devices (several million a year) are emerging, which will certainly stimulate development in the fabrication technologies. Third, more attention will be directed to the materials themselves, as the range of polymeric material is much wider than the scope currently under investigation; the tuning of material parameters for specific applications will become an important topic, as well as potential surface modifications. We have only seen the beginning of these developments, but commercialization will prove a tremendous stimulus. In addition, more and more academic groups are realizing the great potential for simple and fast in-house production of design prototypes with polymer fabrication methods. Polymer-based systems will certainly become "household" items in the years to come.

Received June 7, 1999

7 References

[1] Manz, A., Becker, H. (Eds.), *Microsystem Technology in Chemistry and Life Science*, Springer, Heidelberg 1998.

[2] Terry, S. C., Jermann, J. H., Angell, J. B., *IEEE Trans. Electron. Devices* 1979, *ED-26*, 1880–1886.

[3] Manz, A., Graber, N., Widmer, H. M., *Sensors Actuators* 1990, *B 1*, 244–248.

[4] http://www.the-scientist.library.upenn.edu/yr1997/sept/latta_p1_970915.html

[5] Manz, A., Fettinger, J. C., Verpoorte, E., Lüdi, H., Widmer, H. M., Harrison, D. J., *Trends Anal. Chem.* 1991, *10*, 144–149.

[6] Harrison, D. J., Fluri, K., Seiler, K., Fan, Z., Effenhauser, C., Manz, A., *Science* 1993, *261*, 895–897.

[7] Fan, Z., Harrison, D. J., *Anal. Chem.* 1994, *66*, 177–184.

[8] Effenhauser, C. S., Manz, A., Widmer, H. M., *Anal. Chem.* 1993, *65*, 2637–2642.

[9] Jacobson, S. C., Hergenröder, R., Koutny, L. B., Ramsey, J. M., *Anal. Chem.* 1994, *66*, 2369–2373.

[10] Jacobson, S. C., Hergenröder, R., Moore, J. A. W., Ramsey, J. M., *Anal. Chem.* 1994, *66*, 4127–4132.

[11] Jacobson, S. C., Moore, A. W., Ramsey, J. M., *Anal. Chem.* 1995, *67*, 2059–2063.

[12] Jacobson, S. C., Ramsey, J. M., *Electrophoresis* 1995, *16*, 481–486.

[13] Becker, H., Lowack, K., Manz, A., *J. Micromech. Microeng.* 1998, *8*, 24–28.

[14] Becker, H., Manz, A., in: Baltes, H., Göpel, W., Hesse, J., (Eds.), *Sensors Update*, Vol. 3, VCH Weinheim 1998, pp. 208–238.

[15] Kopp, M. U., Crabtree, H. J., Manz, A., *Curr. Opin. Chem. Biol.* 1997, *1*, 410–419.

[16] Jansen, H. V., Gardeniers, J. G. E., de Boer, M. J., Elwenspoek, M. C., Fluitman, J. H. J., *J. Micromech. Microeng.* 1996, *6*, 14–28.

[17] Ehrfeld, W., Hessel, V., Löwe, H., Schulz, C., Weber, L., *Microsystem Technol.* 1999, *5*, 105–112.

[18] Merkel, T., *Taschenbuch der Werkstoffe*, Fachbuchverlag, Köln 1994.

[19] *CRC Handbook of Chemistry and Physics*, 73rd Ed., CRC Press, Boca Raton, FL 1993.

[20] Weber, L., Ehrfeld, W., *Kunststoffe – plast europe*, 1998, *88*, 60–63.

[21] Weck, M., Fischer, S., Vos, M., *Nanotechnology* 1997, *8*, 145–148.

[22] Martynova, L., Locasico, L. E., Gaitan, M., Kramer, G. W., Christensen, R. G., MacCrehan, W. A., *Anal. Chem.* 1997, *69*, 4783–4789.

[23] Locascio, L. E., Gaitan, M., Hong, J., Eldefrawi, M., *Proc. Micro-TAS '98*, Banff, Canada 1998, pp. 367–370.

[24] Despont, M., Lorenz, H., Fahrni, N., Brugger, J., Renaud, P., Vettiger, P., *Proc. MEMS '97*, Nagoya, Japan 1997, pp. 518–523.

[25] Ehrfeld, W., Münchmeyer, D., *Nucl. Instrum. Methods* 1991, *A303*, 523–532.

[26] Arnold, J., Dasbach, U., Ehrfeld, W., Hesch, K., Löse, H., *Appl. Surf. Sci.* 1995, *86*, 251–258.

[27] McCormick, R. M., Nelson, R. J., Alonso-Amigo, M. G., Benvegnu, D. J., Hooper, H. H., *Anal. Chem.* 1997, *69*, 2626–2630.

[28] Becker, H., Heim, U., *Proc. MEMS '99*, Orlando, FL 1999, pp. 228–231.

[29] Elders, J., Jansen, H. V., Elwenspoek, M., Ehrfeld, W., *Proc. MEMS '95*, Amsterdam 1995, pp. 238–244.

[30] Niggemann, M., Ehrfeld, W., Weber, L., *Proc. SPIE Micromachining and Microfabrication Process Technology IV*, Vol. 3511, Santa Clara, CA 1998, pp. 204–213.

[31] Becker, H., Dietz, W., Dannberg, P., *Proceedings Micro-TAS '98*, Banff, Canada 1998, pp. 253–256.

[32] Becker, H., Dietz, W., *Proceedings SPIE Microfluidic Devices and Systems*, Santa Clara, CA 1998, pp. 177–182.

[33] Konrad, R., Ehrfeld, W., Hartmann, H. J., Jacob, P., Pommersheim, R., Sommer, I., *Proc. 3rd Int. Conference on Microreaction Technologies*, Frankfurt April 18–21, 1999, in press.

[34] Heckele, M., Bacher, W., Müller, K. D., *Microsystem Technol.* 1998, *4*, 122–124.

[35] Piotter, V., Hanemann, T., Ruprecht, R., Haußelt, J., *Microsystem Technol.* 1997, *3*, 129–133.

[36] Weber, L., Ehrfeld, W., Freimuth, H., Lacher, M., Lehr, H., Pech, B., *Proc. SPIE Micromachining and Microfabrication Process Technology II*, vol. 2879, Austin, TX 1996, pp. 156–167.

[37] Paulus, A., Williams, S. J., Sassi, A. P., Kao, P. K., Tan, H., Hooper, H. H., *Proc. SPIE Microfluidic Devices and Systems*, Vol. 3515, Santa, Clara, CA 1998, pp. 94–103.

[38] Ekström, B., Jacobsen, G., Öhman, O., Sjödin, H., International Patent WO 91/16966, 1990.

[39] Qin, D., Xia, Y., Rogers, J. A., Jackman, R. J., Zhao, X. M., Whitesides, G. M., in: Manz, A., Becker, H. (Eds.), *Microsystem Technology in Chemistry and Life Science*, Springer, Heidelberg 1998, pp. 2–20.

[40] Effenhauser, C. S., Bruin, G. I., Paulus, A., Ehrat, M., *Anal. Chem.* 1997, *69*, 3451–3457.

[41] Effenhauser, C. S., Bruin, G. I., Paulus, A., Ehrat, M., *Anal. Methods Instrum.* 1996, Special issue μ-TAS '96, pp., 124–125.

[42] Fielden, P. R., Baldock, S. J., Goddard, N. J., Pickering, L. W., Prest, J. E., Snook, R. D., Treves Brown, B. J., Vaireanu, D. I., *Proc. Micro-TAS '98*, Banff, Canada 1998, pp. 323–326.

[43] Baldock, S. J., Bektas, N., Fielden, P. R., Goddard, N. J., Pickering, L. W., Prest, J. E., Snook, R. D., Treves Brown, B. J., Vaireanu, D. I., *Proc. Micro-TAS '98*, Banff, Canada 1998, pp. 359–362.

[44] Hosokawa, K., Fujii, T., Endo, I., *Proc. Micro-TAS '98*, Banff, Canada 1998, pp. 307–310.

[45] Bruno, A. E., Baer, E., Völkel, R., Effenhauser, C. S., *Proc. Micro-TAS '98*, Banff, Canada 1998, pp. 281–285.

[46] Kim, E., Xia, Y., Whitesides, G. M., *Nature* 1995, *376*, 581–584.

[47] Pethig, R., Burt, J. P. H., Parton, A., Rizvi, N., Talary, M. S., Tame, J. A., *J. Micromech. Microeng.* 1998, *8*, 57–63.

[48] Roberts, M. A., Rossier, J. S., Bercier, P., Girault, H., *Anal. Chem.* 1997, *69*, 2035–2042.

[49] Schwarz, A., Rossier, J. S., Bianchi, F., Reymond, F., Ferrigno, R., Girault, H. H., *Proc. Micro-TAS '98*, Banff, Canada 1998, pp. 241–244.

[50] Becker, H., Klotzbücher, T., *Proc. 3rd Int. Conference on Microreaction Technologies*, Frankfurt, April 18–24, 1999, in press.

[51] Weigl, B. H., Yager, P., *Science* 1999, *283*, 346–347.

[52] Martin, P. M., Matson, D. W., Bennett, W. D., Hammerstrom, D. J., *Proc. SPIE Microfluidic Devices and Systems*, Vol. 3515, Santa Clara, CA 1998, pp. 172–176.

[53] Matson, D. W., Martin, P. M., Bennett, W. D., Kurath, D. E., Lin, Y., Hammerstrom, D. J., *Proc. Micro-TAS '98*, Banff, Canada 1998, pp. 371–374.

[54] Lorenz, H., Despont, M., Fahrni, N., Brugger, J., Renaud, P., Vettiger, P., *Sensors Actuators* 1998, *A 64*, 33.

[55] Guerin, L. J., Bossel, M., Demierre, M., Calmes, S., Renaud, P., *Proc. 1997 IEEE Int. Conf. Solid-State Sens. Actuators*, Chicago, IL 1997, pp. 1419–1422.

[56] Renaud, P., van Lintel, H., Heuschkel, M., Guerin, L., *Proc. Micro-TAS '98*, Banff, Canada 1998, pp. 17–22.

[57] Ikuta, K., Maruo, S., Fukaya, Y., Fujisawa, T., *Proc. MEMS '98*, Heidelberg 1998, pp. 131–136.

[58] Webster, J. R., Burke, D. T., Mastrangelo, C. H., *Proc. 1997 IEEE Int. Conf. Solid-State Sens. Actuators*, Chicago, IL 1997, pp. 503–506.

[59] Webster, J. R., Burns, M. A., Burke, D. T., Mastrangelo, C. H., *Proc. Micro-TAS '98*, Banff, Canada 1998, pp. 249–252.

[60] Man, P. F., Jones, D. K., Mastrangelo, C. H., *Proc. IEEE MEMS '97*, Nagoya, Japan 1997, pp. 311–316.

[61] Mastrangelo, C. H., Burns, M. A., Burke, D. T., *Proc. IEEE*, 1998, *86*, pp. 1769–1787.

[62] Soane, D. S., Soane, Z. M., Hooper, H. H., Alonso-Amigo, M. G., International Patent Application WO 98/45693, 1998.

[63] Kämper, K. P., Döpper, J., Ehrfeld, W., Oberbeck, S., *Proc. MEMS '98*, Heidelberg 1998, pp. 432–437.

[64] Woolley, A. T., Lao, K., Glazer, A. N., Mathies, R. A., *Anal. Chem.* 1998, *70*, 684–688.

Review

J. Cooper McDonald[1]
David C. Duffy[2]
Janelle R. Anderson[1]
Daniel T. Chiu[1]
Hongkai Wu[1]
Olivier J. A. Schueller[1]
George M. Whitesides[1]

[1]Department of Chemistry
and Chemical Biology,
Harvard University,
Cambridge, MA, USA
[2]Gamera Bioscience,
Medford, MA, USA

Fabrication of microfluidic systems in poly(dimethylsiloxane)

Microfluidic devices are finding increasing application as analytical systems, biomedical devices, tools for chemistry and biochemistry, and systems for fundamental research. Conventional methods of fabricating microfluidic devices have centered on etching in glass and silicon. Fabrication of microfluidic devices in poly(dimethylsiloxane) (PDMS) by soft lithography provides faster, less expensive routes than these conventional methods to devices that handle aqueous solutions. These soft-lithographic methods are based on rapid prototyping and replica molding and are more accessible to chemists and biologists working under benchtop conditions than are the microelectronics-derived methods because, in soft lithography, devices do not need to be fabricated in a cleanroom. This paper describes devices fabricated in PDMS for separations, patterning of biological and nonbiological material, and components for integrated systems.

Keywords: Microfluidics / Rapid prototyping / Poly(dimethylsiloxane) / Review   EL 3710

Contents

Correspondence: Dr. George M. Whitesides, Department of Chemistry and Chemical Biology, Harvard University, 12 Oxford Street, Cambridge, MA, 02138 USA
E-mail: gwhitesides@gmwgroup.harvard.edu
Fax: +617-495-9857

Abbreviations: CAD, computer-aided design; **PDMS**, poly(dimethylsiloxane)

1 Introduction

Microfluidics, the manipulation of liquids and gases in channels having cross-sectional dimensions on the order of 10–100 µm, will be a central technology in a number of miniaturized systems that are being developed for chemical, biological, and medical applications. These applications can be categorized into four broad areas: miniaturized analytical systems, biomedical devices, tools for chemistry and biochemistry, and systems for fundamental research. In order for these systems to be successful, they must have the attributes that are required for the particular application – *e.g.*, optical properties and surface chemistry – and they must also be fabricated in materials that are inexpensive and rugged and use processes that are amenable to manufacturing. Here, we review the design, fabrication, and applications of microfluidic devices in one material – poly(dimethylsiloxane) (PDMS) – that shows particular promise in the fabrication of systems for biological and water-based applications. We begin by discussing the motivation underlying the development of microfluidic systems. We then describe the fabrication of microfluidic systems in PDMS by casting the polymer against models that are usually created by using photolithography. We emphasize systems fabricated by using a technique we call rapid prototyping [1, 2], which combines high-resolution commercial printing, photolithography, and soft lithography, and allows microfluidic systems to be designed and fabricated rapidly and inexpensively. We then discuss applications of microfluidic systems in PDMS that have been developed in separations, patterning of biological and nonbiological materials, and components for integrated systems.

1.1 Motivation behind the development of microfluidic devices

Microfluidic systems have the potential for wide application (Table 1). Miniaturization of devices for use in these areas leads to many benefits, including decreased cost in manufacture, use, and disposal; decreased time of analysis; reduced consumption of reagents and analytes; reduced production of potentially harmful by-products; increased separation efficiency; and increased portability. In addition, some studies are difficult or impossible in larger-scale devices. For example, microfluidic channels can approximate the size and flow conditions ($\sim$ 10 μm, 0.1 cm/s) found *in vivo* in capillaries [3]; use of research and diagnostic devices of the same sizes and with similar

elasticity as found in biology could lead to more accurate information and greater understanding of physiology. In addition, smaller channels increase resolution while decreasing the overall size of the device, but small channels also make detection more demanding, are susceptible to blockages from particles, and are more sensitive to adsorption of species on the surface [4].

1.2 Historical background of the development of microfluidic systems

The first microfluidic device was a miniaturized GC developed at Standord University in the 1970s [5]. Although this miniaturized GC system was not developed further, the growth of molecular biology, especially genomics, has stimulated the development of technology for the analysis of complex mixtures of macromolecules, especially DNA and proteins, in aqueous solutions by CE and LC. Microfluidic systems that analyzed aqueous solutions developed originally in four laboratories: those of Manz [4, 6–9], Harrison [10–15], Ramsey [16–21], and Mathies [22–25]. Most of these early systems were fabricated by technology derived from microelectronics – photolithography and etching in silicon and glass – because these technologies were available and highly developed. Silicon is, however, a relatively expensive material; it has the further

Table 1. Potential applications of microfluidic devices

Area	Application
Miniaturized analytical systems	
Genomics and proteomics	Rapid, high density sequencing [22, 23, 25], DNA fingerprinting, combinatorial analysis, forensics, gene expression assays, integration of fluidics with DNA arrays
Chemical/biological warfare defense	Early detection and identification of pathogens and toxins; early diagnosis; triage
Clinical analysis	Rapid analysis of blood and bodily fluids [72, 73], point of care diagnostics based on immunological [14, 28, 74] or enzymatic assays [21], electrochemical detection, and cell counting [57]
High throughput screening	Combinatorial synthesis and assaying for drugs. Toxicological assays [8, 75]
Environmental testing	*In situ* analysis of environmental contamination [76]
Biomedical devices	
Implantable devices	Devices for *in vivo* drug delivery [36], *in vivo* monitoring for disease and conditions
Tools for chemistry and biochemistry	
Small-scale organic synthesis	Combinatorial synthesis [77]
Sample preparation	Purification of biological samples for further analysis [15]
Amplification of nucleic acids/sequences	PCR [9, 17, 31, 78, 79], RT-PCR
Systems for fundamental research	
Systems with which to study the flow of fluids	Studies on EOF and laminar flow in small channels [80], studies of diffusion
Studies of chemical reactions	Enzyme-substrate
Biomimetic systems	Development of machines that mimic biological functions
Systems to study small amounts of sample	Detection of single molecules [53, 58, 81]

disadvantage that it is opaque in the visible/UV region of the spectrum, thus making it unsuitable for systems that use optical detection. Glass is transparent, but because it is amorphous, vertical side walls are more difficult to etch than in Si. Although batch processing of both silicon and glass is possible, the commonly used processes for sealing these materials require that each device be made in a cleanroom environment. These sealing processes also typically require high voltages or temperatures. Glass and oxidized silicon, however, have desirable surface characteristics: they possess a negative charge and support electroosmotic flow (EOF), and the channels are fabricated by etching, which cleans the surfaces as it produces the channels. Glass systems have proved especially successful when applied to separating and sequencing DNA [22–26], but when used with proteins, adsorption can be a problem.

1.3 New materials for the fabrication of devices

Since the early work in the field, there has been a rapid expansion [27–37] into new types of materials, especially polymers [38–45]. Polymers, in contrast to silicon and glass, are inexpensive; channels can be formed by molding or embossing rather than etching; and devices can be sealed thermally or by using adhesives. The disadvantages of polymers are that more care must be taken to control their surface chemistry than with glass or silicon; they are often incompatible with organic solvents and low molecular weight organic solutes; and they are generally incompatible with high temperatures.

Our own work in microfluidics has focused on polymer systems made of PDMS. PDMS is an excellent material for the fabrication of microchannel systems for use with biological samples in aqueous solutions for a number of reasons: (i) features on the micron scale can be reproduced with high fidelity in PDMS by replica molding; (ii) it is optically transparent down to 280 nm so it can be used for a number of detection schemes (*e.g.*, UV/Vis absorbance and fluorescence); (iii) it cures at low temperatures; (iv) it is nontoxic; mammalian cells can be cultured directly on it; and devices made from it can be implanted *in vivo*; (v) it can be deformed reversibly; (vi) it can seal reversibly to itself and a range of other materials by making molecular (van der Waals) contact with the surface, or it can seal irreversibly after exposure to an air plasma by formation of covalent bonds (see Section 2.4); (vii) its surface chemistry can be controlled by reasonably well-developed techniques; and (viii) because it is elastomeric, it will conform to smooth, nonplanar surfaces, and it releases from delicate features of a mold without damaging them or itself.

2 Fabrication of microfluidic systems in PDMS

New methods and materials for fabricating microfluidic systems are needed because etching in Si and glass is too expensive and time-consuming. Some important issues to consider in a method of fabrication are the speed at which designs can be reduced to working devices and evaluated, the design parameters such as channel size and geometry, and the availability of required components, *e.g.*, for injection, separation, or detection. In this section we discuss methods for fabrication of devices in PDMS, how to address certain aspects of fabrication – sealing and surface chemistry – when using PDMS, and the design of devices in PDMS.

2.1 Soft lithography

Our approach to the fabrication of microfluidic devices is based primarily on the techniques of soft lithography [2, 46, 47], specifically rapid prototyping and replica molding. Soft lithography is a suite of nonphotolithographic methods for replicating a pattern. An elastomeric structure with the patterns embedded as a bas-relief on the surface acts as the pattern transfer agent. These methods have the characteristic that routine access to a cleanroom is not necessary when producing most structures relevant to microfluidics (20–100 μm). They also enable pattern transfer to curved materials. We use these techniques in the fabrication of channels in bulk polymer.

2.2 Rapid prototyping

Rapid prototyping begins with creating a design for a device in a computer-aided design (CAD) program. A high-resolution commercial image setter then prints this design on a transparency. This transparency serves as the photomask in contact photolithography to produce a positive relief of photoresist on silicon wafer (Fig. 1). We refer to this positive relief as a "master"; it is used for the casting of PDMS devices. A master in SU-8 photoresist, a photocurable epoxy, [48] on a silicon wafer is durable and can be used indefinitely; failure usually occurs from the user breaking the fragile silicon wafer or from the photoresist releasing from the wafer. Replication of the master as one piece in a hard polymer, *e.g.*, structural polyurethane or epoxy, can further extend its lifetime.

The hallmark of rapid prototyping is the reduction in time and cost for a cycle of design, fabrication, and testing of new ideas compared to methods that use a chrome mask in the photolithographic step. The chrome mask that the transparency replaces can be 20–100 times more expensive and can take weeks compared to hours to obtain.

    Electrophoresis 2000, *21*, 27–40

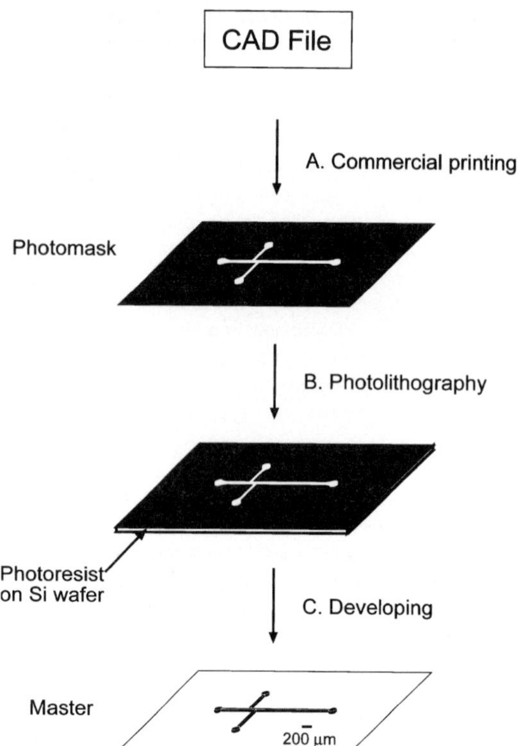

Figure 1. Scheme for rapid prototyping and replica molding of microfluidic devices in PDMS. A design for channels is created in a CAD program. (A) This file is printed on a high-resolution transparency. (B) The transparency then serves as the photomask in contact photolithography. (C) Dissolving away the unpolymerized photoresist leaves a positive relief that serves as a master.

The drawback is that the resolution of the transparency is lower (> 20 μm) than that of a chrome mask (~ 500 nm). Access to image setters with higher resolution (> 3386 dpi) would decrease the size of features obtainable with rapid prototyping. The channel or capillary diameter for most applications, however, ranges from 50–100 μm, which is within the capacity of rapid prototyping. For devices that require features smaller than 20 μm, a chrome mask needs to be used. The resolution of the transparency also leads to two walls with rough edges (Fig. 2). Experiments show that this edge roughness does not compromise resolution for 50×50 μm channels compared to separations in fused silica capillaries with circular cross sections [1].

2.3 Replica molding

Once a master is fabricated, we form channels in PDMS by replica molding. Replica molding is simply the casting

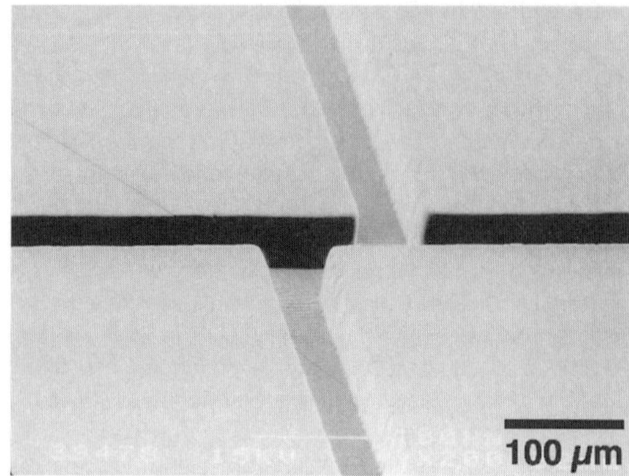

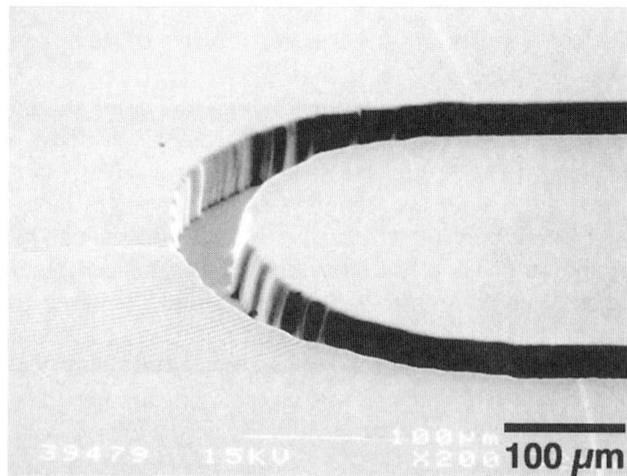

Figure 2. Scanning electron micrograph of 50×50 μm^2 channels of a miniaturized CE device that was created by molding PDMS against a photolithographic master. The top image shows straight sections of the channel. The bottom image shows a curve in the channel. The roughness of the vertical side walls arises from the resolution (3386 dpi) of the transparency used to create the channels. This edge roughness is more pronounced in the curved section. Reprinted from [1], with permission.

of prepolymer against a master and generating a negative replica of the master in PDMS, *i.e.*, ridges on the master appear as valleys in the replica (Fig. 3). The PDMS is cured in an oven at 60°C for 1 h, and the replica is then peeled from the master. Access holes for channels and reservoirs for buffer can be added in the replication step by appropriate placement of posts on the master or punched out of the cured layer by using a borer. Although we use masters produced in photoresist from rapid prototyping, masters can be fabricated by many techniques including etching in silicon, electroforming metal, or con-

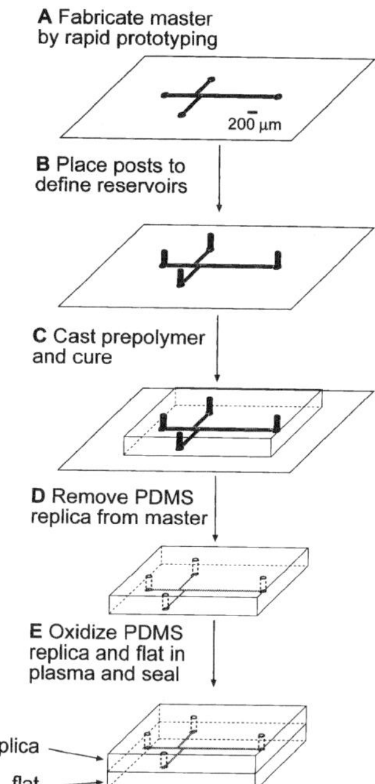

Figure 3. Scheme describing replica molding of microfluidic devices. (A) A master is fabricated by rapid prototyping. (B) Posts are placed on the master to define reservoirs. (C) The prepolymer is cast on the master and cured. (D) The PDMS replica is removed from the master. (E) Exposing the replica and an appropriate material to an air plasma and placing the two surfaces in conformal contact makes a tight, irreversible seal. Reprinted from [1], with permission.

ventional machining of other hard materials. The composition of the master used in production may depend on the production run. Masters made from metal or other hard materials may be used in manufacturing when the production run is large because of their durability; the expense of the master becomes negligible after many uses. For prototyping new devices and for limited run production, however, metal or silicon molds are time-consuming and expensive to make, especially in a program of research and development where several iterations are necessary for the development of a final design.

2.4 Sealing

Molding provides a PDMS replica that contains three of the four walls necessary for enclosed channels. Sealing the replica to a flat surface provides the fourth wall. This flat material can be PDMS, to give channels in which all four walls are made from the same material, or another material. Sealing occurs in two ways: (i) reversible, con-

formal sealing with a flat surface, and (ii) irreversible sealing to certain substrates upon exposure of both surfaces to an air plasma [49]. Reversible sealing occurs because PDMS is flexible and can conform to minor imperfections in a "flat" surface making van der Waals contact with this surface. This method of sealing is watertight and fast and occurs at room temperature. Simply peeling the PDMS off the flat surface breaks this reversible seal. It does not withstand high pressures (> 5 psi) in the capillaries. Removal of the PDMS leaves little or no residue on the other material, and resealing can occur numerous times without degradation in the PDMS.

PDMS comprises repeating units of $-O-Si(CH_3)_2-$. Exposing a PDMS replica to an air plasma introduces polar groups on the surface. We believe the plasma introduces silanol groups (Si-OH) at the expense of methyl groups (Si-CH$_3$) [49–51]. We believe these silanol groups then condense with appropriate groups (OH, COOH, ketone) on another surface when the two layers are brought into conformal contact. For PDMS and glass, this reaction yields Si-O-Si bonds after loss of a water. These covalent bonds form the basis of a tight, irreversible seal: attempting to break the seal results in failure in the bulk PDMS [1, 49]. The seal withstands pressures of 30–50 psi. It is possible to seal PDMS irreversibly to the surfaces of a number of materials: PDMS, glass, Si, SiO$_2$, quartz, silicon nitride, polyethylene, polystyrene, and glassy carbon [1]. This method, however, does not work with all polymers, *e.g.*, Saran, polyimide, poly(methylmethacrylate), and polycarbonate [1].

2.5 Surface chemistry

One of the most important issues in the selection of an appropriate material for a device is surface chemistry. Unmodified PDMS presents a hydrophobic surface. Channels in hydrophobic PDMS are difficult to wet with aqueous solutions, are prone to the adsorption of other hydrophobic species, and easily nucleate bubbles. Exposure to plasma oxidation, however, renders the surface hydrophilic because of the presence of silanol groups. Aqueous solutions then easily wet these oxidized channels. In addition, the presence of silanol groups on the walls of the channels provides ionizable groups (SiOH $\leftrightarrow$ SiO$^-$ + H$^+$) that, when in contact with neutral or basic solutions, support a strong EOF towards the cathode that can be used for capillary zone electrophoresis (CZE). These negatively charged channels have greater resistance to adsorption of hydrophobic and negatively charged analytes than unmodified PDMS, but some proteins still adsorb on the surface [1]. Adsorption of charged and neutral polymers [1] and covalent attachment of trichlorosilanes [49, 50] on oxidized PDMS can also modify the sur-

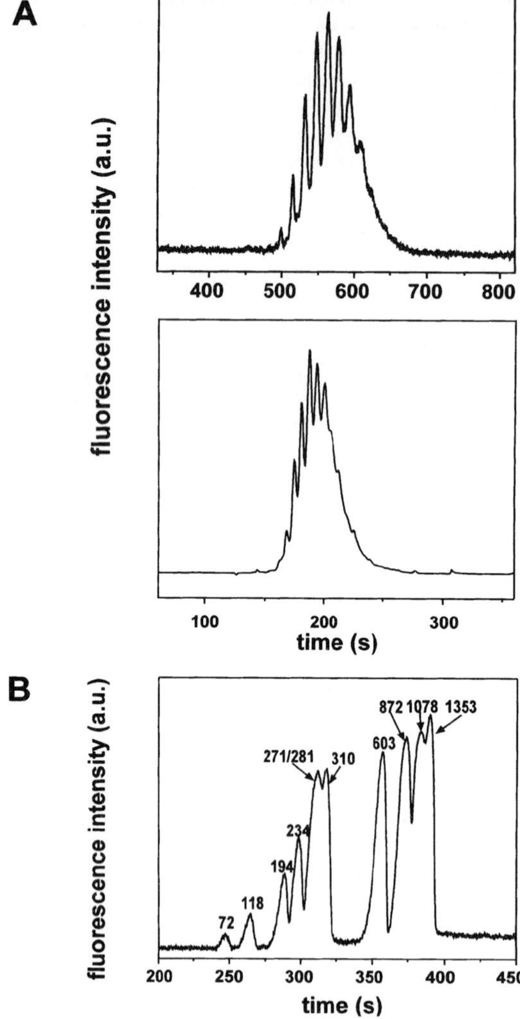

A

B

Figure 4. (A) Electropherograms of a fluorescently labeled charge ladder of bovine carbonic anhydrase. The charge ladder was formed by modifying the ε-amino groups of lysine residues of the protein first with 0.5 eq. of 5-carboxyfluorescein, succinimidyl ester and then with 20 eq. of acetic anhydride. The sample injected was 3 mg/mL of modified protein. The running buffer was 25 mM Tris-192 mM glycine, pH 8.4. In the miniaturized CE device (top), the oxidized PDMS channel was 42 cm long with a cross-sectional area of 50 × 50 μm, the sample was detected 35 cm from the point of injection, and the separation voltage was 5 kV. In the commercial CE (Beckman P/ACE 5000; bottom), the 50 μm ID fused silica capillary was 37 cm long, the detector was 30 cm from the point of injection, and the separation voltage was 15 kV. The resolution obtained with the PDMS device compared favorably to the resolution obtained with the commercial instrument. (B) Electropherogram of the φX-174/*Hae*III DNA restriction fragments intercalated with YOYO-1 fluorescent dye separated in an oxidized PDMS channel that had a cross-sectional area of 50 × 50 μm and was 28 cm long. The sample was detected 21 cm from 6 the injection point, and the separation voltage was 5 kV. The separation buffer contained 40 mM Tris, 40 mM acetic acid, 1 mM Na_2EDTA, 10 μM 9-aminoacridine, and 0.75% w/v hydroxypropyl cellulose (pH 8). The DNA concentration was 100 ng/μL; the base pair to YOYO-1 ratio was 1:10. This high concentration of DNA caused the relatively poor resolution in the separation [82], but when lower concentrations were used, the smaller fragments were no longer detectable with our instrument. The number of base pairs in each fragment is indicated. Adapted from [1].

face properties of PDMS. The oxidized surface of PDMS is unstable in air and reverts to being hydrophobic in ~30 min. Keeping the oxidized PDMS in contact with a polar liquid, however, protects the surface although the long-term stability of the oxidized layer is unknown [1, 49, 51].

2.6 Design of microfluidic systems

Each application requires different components in a device, *e.g.*, those for injection, separation, detection, heating, mixing, and post-treatment. The design of a device must therefore take into account how to fabricate these components and how to move samples from one component to the next. In general, components with few moving parts are desirable since these parts complicate fabrication and can break or become clogged. Usually EOF or pressure is used to move fluids, although some workers have used ultrasonic, bubble, and rotary pumps [52]. PDMS is an excellent material for EOF and pressure pumping since its surface can be charged and, when sealed irreversibly, can withstand high pressure. In addition, since PDMS is elastomeric, membrane pumps and check valves are easy to incorporate into a system.

3 Microfluidic systems in PDMS

Several microfluidic systems in PDMS have been developed. Most applications have been in separations, patterning of biological and nonbiological materials on substrates by using channels in PDMS, and components for integrated analytical systems. We consider each of these areas in turn.

3.1 Separations

3.1.1 Miniaturized CE systems

The first miniaturized CE system in PDMS was developed by Effenhauser *et al.* [53]. This system used a commercially obtained positive relief of silicon as the master in replica molding. The device sealed reversibly against a flat piece of PDMS. The walls of the channel were not modified and therefore were hydrophobic and uncharged; the channels thus supported at most a very weak EOF, and negatively charged molecules migrated towards the anode. These workers used a polymer sieving matrix to separate restriction fragments of DNA and fluorescently labeled peptides. They also explored single DNA molecule (many fluorophores) detection limits with λ DNA achieving 50% detection. The greatest advantage of this device was that it could be disassembled, easily cleaned, and reused.

We developed a PDMS device for CE whose properties differ significantly from that of Effenhauser *et al.* [1]. The device was fabricated by using rapid prototyping and was irreversibly sealed by using plasma oxidation. We have successfully separated fluorescently labeled amino acids, and protein charge ladders using CZE, and DNA restriction fragments using a sieving matrix. Figure 4 shows an electropherogram of a charge ladder of carbonic anhydrase obtained in these devices. A charge ladder is a set of modified proteins obtained by successive acylation of lysine amino groups [54, 55]. These separations utilized several surface modifications of PDMS. For the separation of amino acids and negatively charged proteins, no surface modifications other than plasma oxidation were necessary, although for a separation of an insulin charge ladder, a zwitterion had to be added to the running buffer to reduce adsorption of the protein and its derivatives onto the walls [56]. For the separation of a charge ladder of lysozyme – a positively charged protein – the channels were coated with Polybrene®, a polymer containing quaternary amines, to make them positively charged. This modification reduced the adsorption of the positively charged protein to the walls of the channels and reversed the direction of EOF to be toward the anode. The separation, however, also required the addition of a zwitterionic component to the buffer [56]. The sieving matrix used in

the separation of restriction fragments of DNA eliminated EOF. The device could be used to separate different samples by flushing the channels between uses. Blockages of the channels, however, were difficult to remove since the device was irreversibly sealed.

3.1.2 Sorting of cells

Bakajin *et al.* [57] used a PDMS device to sort white blood cells. They fabricated a lattice of small channels of varying lengths that mimic the size restrictions imposed on cells by capillaries *in vivo*. They used PDMS because devices fabricated in glass and silicon dioxide were too adhesive for the cells studied. Using this device, they separated two classes of white blood cells, T-lymphocytes and granulocytes, and showed that the cells' passage through the lattice depended strongly on size and nuclear morphology.

3.1.3 Sizing and sorting of DNA

Chou *et al.* [58] developed a microfluidic device for the sizing and sorting of restriction fragments of DNA based on single DNA molecule detection. The device was molded against a silicon master and had features ranging from 5 to 100 μm. The channels were only 3 μm deep, and posts were necessary in the 100 μm sections to prevent the PDMS from sagging and blocking the channels. They rendered the channels hydrophilic by soaking the channels in dilute HCl, which hydrolyzes some of the Si-O-Si bonds, and sealed the device reversibly to a glass cover slip. The reversible seal allowed all data to be taken with the same device. The device sized DNA on the basis of fluorescence intensity from an intercalated dye; individual DNA molecules that were detected had between 500 and 5000 dye molecules intercalated. DNA was pumped through the channels by a mixture of capillary action and EOF. Individual DNA molecules were detected as they passed through the narrow portion (5 μm) of the channel. The size of each molecule was determined by the intensity of the fluorescence. Counting the number of times that each fluorescence intensity (corresponding to a specific number of base pairs) was detected gave the concentration of each restriction fragment. Detection took place at a T-junction; each fragment could therefore go one of two directions after being detected. The fragments could be sorted by steering them down one of the possible paths using electric fields.

3.2 Patterning of biological and nonbiological materials on substrates

Controlled deposition or removal of material from substrates finds application in many areas. Patterned deposition of materials in small (< 100 μm) features is important

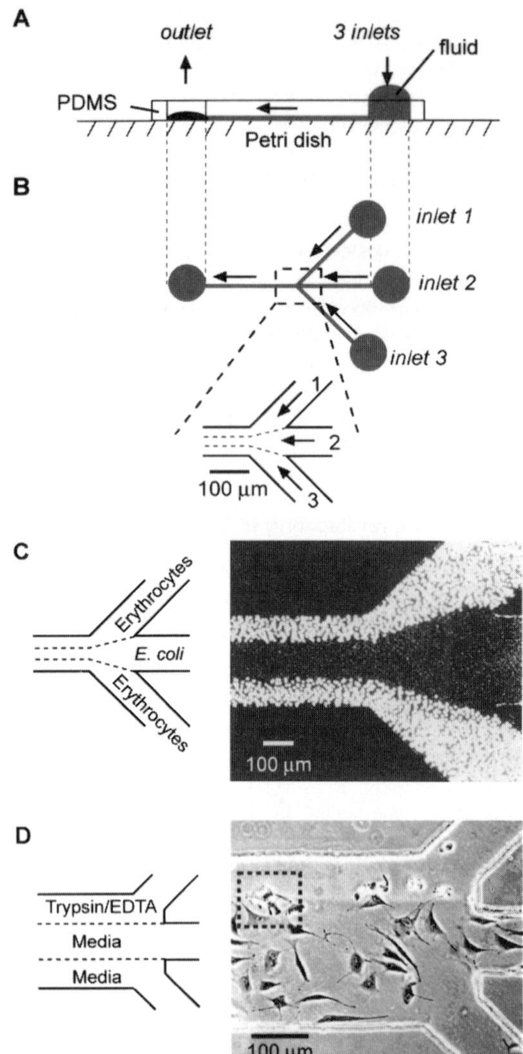

Figure 5. Laminar flow patterning of biological materials. (A) Side view of a PDMS membrane containing microscopic channels and reversibly sealed to a polystyrene Petri dish. Fluids were pumped by gentle aspiration at the outlet or by pressure from the inlets. (B) Top view of the channels. The closeup shows the paths that the flows from the three inlets take in the main channel. (C) Two different types of cells, chicken erythrocytes and *E. coli*, can be patterned next to each other. A suspension of chicken erythrocytes was placed in inlets 1 and 3, and PBS in inlet 2, and allowed to flow by gravitational force for 5 min followed by a 3 min wash with PBS; these flows patterned the outer lanes. Next, a suspension of *E. coli* in inlet 2 and PBS in inlets 1 and 3 were allowed to flow for 10 min followed by a 3 min wash with PBS. This flow patterned the middle lane. Cells adhered by nonspecific adsorption and were visualized by fluorescence microscopy. (D) Patterned detachment of bovine capillary endothelial cells with trypsin/EDTA. Cells adhered and spread in a fibronectin-coated region. Trypsin/EDTA and media were then allowed to flow from the designated inlets. Trypsin/EDTA caused the detachment of cells where it flowed. The patterning is well defined, and it was possible to detach only portions of a cell (dashed box). The cells were visualized by phase contrast microscopy. Adapted from [63].

for applications of miniaturized systems in biochemistry and cell biology. Many assays are based on attaching ligands to a surface and determining the locations of binding of species of interest. Patterned attachment of cells is also important in cell-based sensors where the reaction of cells to stimuli in specific areas of a device is necessary for detection of species of interest. There are several options for patterning materials from solutions, including ink-jet printing, stamping, patterning from fluids through stencils, and patterning in capillaries. We focus on patterning in capillaries by using microfluidic systems. This method relies on the reversible sealing of a network of channels in PDMS directly on the surface to be patterned.

Fluids containing the patterning species flow through these channels and pattern the surface through covalent attachment, adsorption, or dissolution of material already present.

3.2.1 Patterning of cells and proteins

Delamarche *et al.* [59, 60] first reported the use of channels in a PDMS replica that was reversibly sealed to an underlying substrate to pattern proteins on the substrate. These workers used channels with cross sections of 1.5 µm $\times$ 3 µm to pattern mm^2 areas; the channels allowed for simultaneous patterning of different proteins

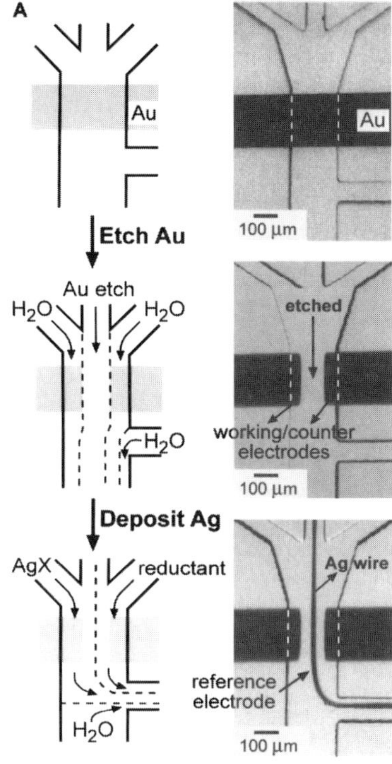

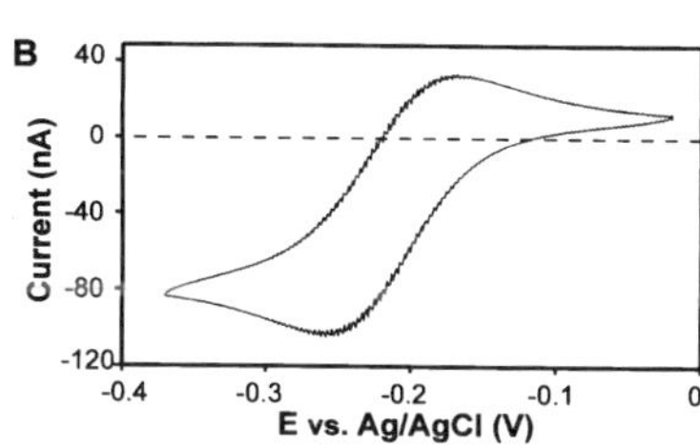

Figure 6. (A) Scheme for the fabrication of a three-electrode detector by using laminar flows. Two gold electrodes are fabricated by selectively etching an evaporated stripe of gold. The silver reference electrode was fabricated at the interface of flows of silver ion and reductant. (B) Cyclic voltammogram of 5 nL of 2 mM $Ru(NH_3)_6Cl_3$ in 0.1 M NaCl as recorded by the three-electrode system (scan rate = 100 mV/s). Adapted from [64].

in well-defined areas adjacent to each other. They also demonstrated the use of these patterns in bioassays. One important aspect of this work was the wettability of the PDMS surface and the propensity for the surface to adsorb proteins. In nonflowing systems, the substrate was patterned only about 100 µm from the inlet. In that distance, all of the protein had adsorbed onto the walls of the channel (three walls of PDMS and one wall of sub-

strate) leaving no reagent to couple to the surface. The workers overcame the problem by creating flowing systems that replenished the supply of protein available to couple to the surface. They also used bovine serum albumin as a blocking agent to prevent protein adsorption on the PDMS walls; the albumin also has the benefit of rendering hydrophobic PDMS slightly hydrophilic [60].

Folch and Toner [61] and Folch *et al.* [62] used PDMS microchannels to pattern proteins and cells. They used deeper channels (20–30 µm) than Delamarche *et al.* [59, 60] in order to slow transport of material. In one experiment, they nonspecifically patterned collagen or fibronectin on a variety of substrates and then seeded cells on these patterns. A second culture with a different cell type was possible simply by seeding the areas not already patterned with cells that need no specific protein to promote adhesion [61]. They also used the microchannels to pattern cells directly on a substrate [62]. Using this method, they could pattern two different types of cells simultaneously [62].

Takayama *et al.* [63] have also used microfluidic channels to pattern proteins and cells. Their method differed from previous approaches, however, in that they used a single channel to pattern multiple species simultaneously. This approach relies on the fact that flow in these channels is laminar. The nature of a flow of fluid is characterized by the Reynolds number, $R_e = \rho v l/\eta$, where ρ is the density of the fluid, v is the velocity of the flow, l characterizes the shape and dimensions of the channel, and η is the viscosity of the fluid. Flows that have $R_e < 2000$ are laminar. When two or more streams with low R_e are combined into a single stream, the combined streams flow side by side without turbulent mixing; mixing occurs only by diffusion. By combining two or more streams containing different patterning materials, stripes of these materials can be placed adjacent to one another on a substrate (Fig. 5). These workers exploited laminar flow to pattern proteins and cells side by side in well-defined areas (Fig. 5). This method allowed for the selective patterning of a substrate and/or for the selective patterning of cells and even portions of cells. They used these methods to pattern *Escherichia coli* and chick erythrocytes side by side. In addition, they performed an enzymatic assay that used trypsin/EDTA to cleave fibronectin and thus detach cells or portions of cells in the affected area.

3.2.2 Patterning of nonbiological materials

Kenis *et al.* [64] have used multiple parallel flows to pattern nonbiological materials within channels. They used this technique to fabricate a three-electrode system that could potentially be used as a detector or a sensor

(Fig. 6). First they evaporated a thin stripe of gold on a glass slide. They then placed a PDMS replica on the slide with the channels oriented perpendicular to the stripe of gold. A portion of the gold was then etched by flowing three streams, two water flows flanking a central flow of gold etchant, over the gold. The removal of the gold from the center of the channel made two gold electrodes, a working and a counter electrode, separated by a distance controllable by the relative rates of flow of the three streams. A third, reference electrode was made by flowing two streams – one of silver ion and one of reductant – in the channel. Diffusion mixed the two flows at the interface, and silver metal plated onto the channel; the electrode was then treated with HCl to make an Ag/AgCl reference electrode. A cyclic voltammogram of $Ru(NH_3)_6Cl_3$ in water was obtained with these electrodes. This laminar flow-based method of fabrication both fabricates devices inside preformed channels and eliminates the need for an additional alignment step.

3.3 Components

While several systems have been developed for use in patterning and separations, the addition of certain components – especially in detection and fluid handling – can help to extend the range of applications for these sys-

tems. Most systems currently use fluorescence detection, but to reduce the complexity of the fabrication of a system, other methods of detection that are simple and compact, and reduce the need for labeling, may be needed. In addition, fine control of fluid motion by using pumps and valves, for example, will be necessary in order to effectively inject, pretreat, separate, and post-treat samples under analysis. Here we outline a sensor, two detectors, and a switch that have been developed in PDMS; these components illustrate the range of functional structures that can be fabricated easily in PDMS by using rapid prototyping.

3.3.1 Cell-based biosensor

DeBusschere *et al.* [65] developed a prototype cell-based biosensor using both PDMS and silicon. The PDMS portion comprises the fluid channels, interconnects, injection ports, and sensing chambers. The sensor relies on the reaction of cells exposed to irritants compared to a reference cell population. The cells are electrically active, and when the detecting population of cells is exposed to a toxin or irritant, they will react differently than the reference populaltion that has not been exposed to any contaminants. A difference in the electrical signals between the two populations of cells is used as the basis of detec-

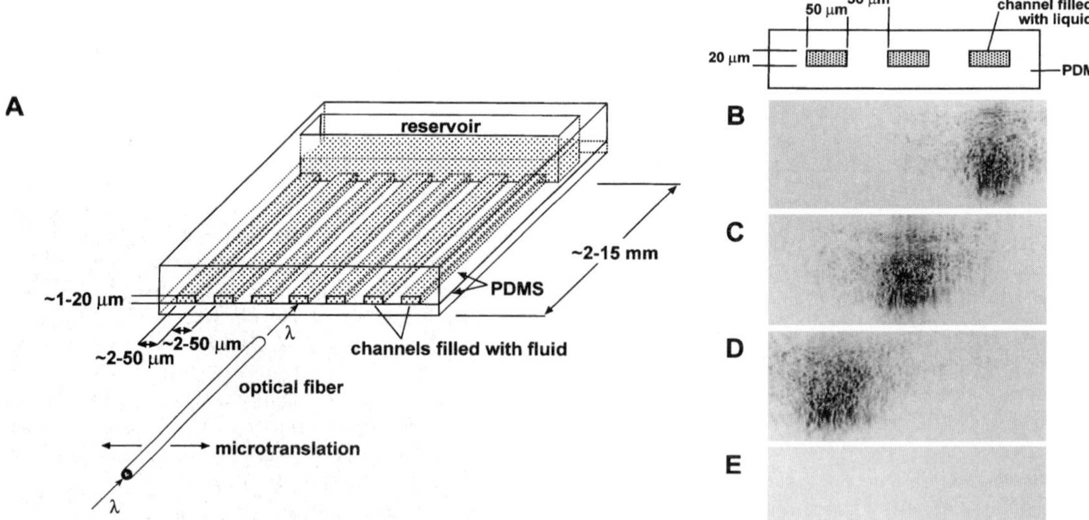

Figure 7. Schematic diagram of the optical coupling to the waveguides. (A) An optical fiber is used to couple light from a laser into the fluid-containing channels. High index fluids in the channels guide light by total internal reflection. A CCD camera images the output of the waveguides. (B)–(D) Photographs of the output of $20 \times 50~\mu m^2$ channels as a function of the location of the coupling fiber. Selection of a particular waveguide is accomplished by changing the lateral position of the fiber. Light is coupled into and guided through the channels when the end of the fiber is directly facing the end of a channel. (E) No output is seen when the fiber is not aligned with one of the channels. Adapted from [68].

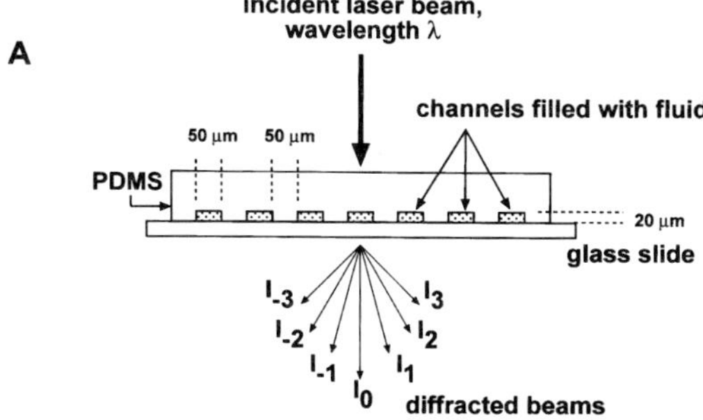

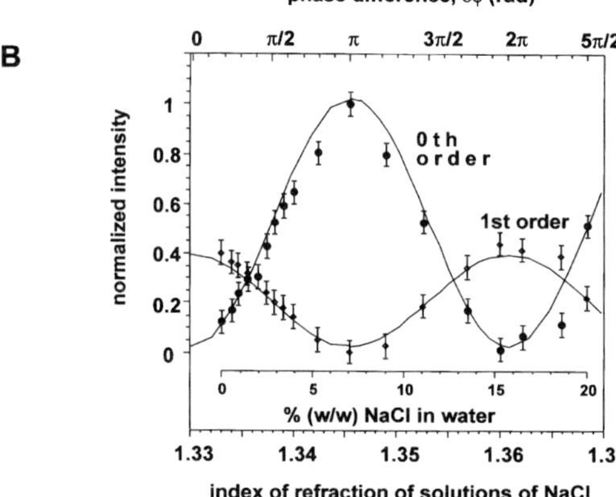

Figure 8. (A) An array of parallel channels embedded in PDMS is sealed against a glass slide. When filled with fluids that differ in index of refraction from PDMS, the array acts as a diffraction grating. If the fluid does not absorb the incident light, only the phase of the light is affected. If the fluid absorbs the incident light, both the phase and amplitude of the light change. Diffraction results from the interference of light passing through the PDMS and light passing through the channels. The changes in phase and amplitude modulate the relative intensities of the diffracted beams. (B) Variation in the intensities of the 0th (filled circles) and 1st (open diamonds) order beams of the diffraction pattern from a microfluidic grating as a function of the concentration of NaCl in solutions filling the microchannels. Adapted from [67].

tion. PDMS is especially suitable in this application because it is highly permeable to oxygen and carbon dioxide, so that the cells are able to respire normally even when not directly exposed to the atmosphere.

3.3.2 Detectors

One alternative to detecting analytes by fluorescence is to detect changes in index of refraction of the liquid in the channels caused by the presence of analytes. Detection of index of refraction has the following advantages: it is universal, *i.e.*, it does not require labeling of analytes; it is concentration dependent; and it is optically simple and less expensive than fluorescence [66]. Some disadvantages of the method are the dependence of refractive index on the temperature and poor detection limits achieved thus far [67]. Two components that use index of refraction to manipulate beams of light are liquid-core waveguides and diffraction gratings.

A component that exploits differences in index of refraction is a liquid-core waveguide (Fig. 7) [68]. These waveguides were fabricated by sealing channels molded in PDMS, either reversibly or irreversibly, against an appropriate material. When a liquid with a higher index of refraction than PDMS ($n = 1.41$) fills the channel, incident light is guided along the channel by total internal reflection. Changes in the index of refraction of the liquid core will change the output of the waveguide; this change provides the basis for a simple detector.

Schueller *et al.* [67] report the fabrication of a microfluidic diffraction grating. The grating is based on a network of parallel channels in PDMS (Fig. 8). These channels were 50 μm wide and 20 μm deep and spaced by 50 μm. Filling the channels with liquids that differ in refractive index from PDMS produces a phase grating: light incident on the grating will be diffracted because of the phase differences caused by the different indexes of refraction. Use of a dye in the channels that absorbs the light of interest leads to an additional modulation of the amplitude. Filling the channels with different liquids easily reconfigures the grating. Since the diffraction and amplitude patterns are sensitive to the composition of the liquid filling the channels, this device could find use as a detector or as an actuator for wavefront engineering or beam steering.

3.3.3 Microfluidic switch

Duffy *et al.* [69] demonstrated a microfluidic deflection switch. This switch could be used to control the direction of flow in complex microfluidic devices that require splitting of flows into or out of a channel. They fabricated this switch using rapid prototyping and sealed the devices irreversibly using plasma oxidation. The fluidic deflection switch is a forked channel with control channels at the junction that steer the fluid down one of the paths of the fork (Fig. 9). The direction of EOF or hydrodynamic flow emanating from the control channels determines which path the flow from the main channel will take. This switch could be used in microfluidic systems where fractionation of species is desired. The switch contains no moving

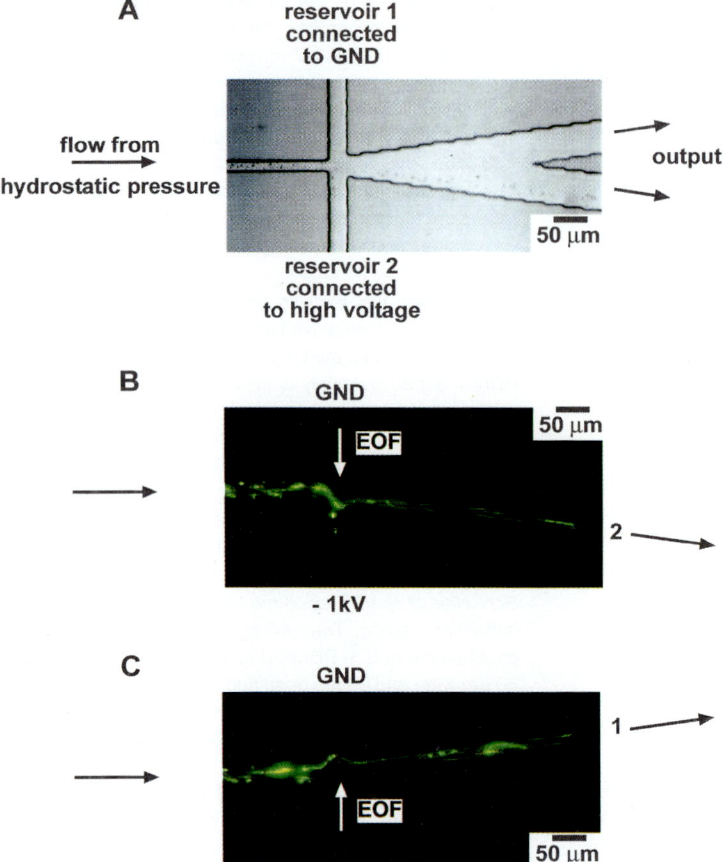

Figure 9. Microfluidic deflection switch actuated by EOF. (A) Optical micrograph of the switch. (B) Fluorescent micrograph of 1 µm diameter beads labeled with fluorescein. A negative potential (–1 kV) applied to control reservoir 2 deflected the beads into output channel 2. (C) A positive voltage (+ 1 kV) applied to control reservoir 2 deflected the beads into output channel 1. The beads flowed from the inlet to the outlet under hydrostatic pressure. Reprinted from [69], with permission.

parts that can fail or become clogged by small particles. Careful control of the magnitude of the flow in the control channel must be taken to prevent the flow from the main channel from being diluted by fluid from the control channels or being directed into the control channel.

4 Conclusions

While PDMS has many desired attributes as a material for fabrication of microfluidic systems, it also has some limitations. While it is compatible with aqueous media and some alcohols, most organic solvents are soluble in bulk PDMS and swell the polymer. In addition, small organic analytes with appreciable solubility in water may also dissolve in the bulk PDMS. While PDMS reproduces masters with high fidelity, some geometries tend to collapse because of the elasticity of the material [60, 62, 70]. Other properties of PDMS that could be advantageous or detrimental depending on the circumstances are its poor thermal conductivity ($\sim$ 0.2 Wm/K) [71] and high permeability to gases. A low thermal conductivity can lead to a rise in temperature due to resistive heating in electrophoresis; this rise in temperature causes broadening of peaks and lowers the resolution of the system. Conversely, a low thermal conductivity could find use in systems such as an incubator where an elevated temperature is desired. Permeability to gases like water vapor, oxygen, and ammonia could be disadvantageous, for example, when evaporation of water in channels can occur in devices in storage. The gas permeability could be useful in systems requiring gas, but not liquid, exchange.

Although it is still in the early stages of development in this field, we believe that PDMS is an extremely promising material for the fabrication of microfluidic systems that use aqueous media. Since PDMS can be molded at low temperatures and by using rapid prototyping, chemists and biologists working on a benchtop can make devices quickly and easily. PDMS is robust, and devices made in it, unlike hard materials, can be dropped and bent without loss of performance. PDMS is ideal for the fabrication of devices that require more than one material in fabrication, since it can seal to a variety of materials. In addition, fabrication of 3-D devices will be straightforward since layers of PDMS with channels, reservoirs, valves, and through holes can easily be aligned and sealed.

This research was supported by DARPA, NSF ECS-9729405, and NIH GM51559. J. C. M. thanks the NSF for a predoctoral fellowship. We wish to thank our colleagues who participated in this research.

Received July 12, 1999

5 References

[1] Duffy, D. C., McDonald, J. C., Schueller, O. J. A., Whitesides, G. M., *Anal. Chem.* 1998, *70*, 4974–4984.

[2] Qin, D., Xia, Y., Whitesides, G. M., *Adv. Mater.* 1996, *8*, 917–919.

[3] Tortora, G. J., Grabowski, S. R., *Principles of Anatomy and Physiology*, Harper Collins, New York 1996.

[4] Manz, A., Harrison, D. J., Verpoorte, E., Widmer, H. M., *Adv. Chromatogr.* 1993, *33*, 1–65.

[5] Terry, S. C., Jerman, J. H., Angell, J. B., *IEEE Trans. Electron. Devices* 1979, *ED-26*, 1880–1886.

[6] Seiler, K., Harrison, D. J., Manz, A., *Anal. Chem.* 1993, *65*, 1481–1488.

[7] Manz, A., Graber, N., Widmer, H. M., *Sens. Actuators B* 1990, *1*, 244–248.

[8] Manz, A., Becker, H., *1997 International Conference on Solid-State Sensors and Actuators*, Chicago, IL, June 1997, pp. 915–918.

[9] Kopp, M. U., Mello, A. J. d., Manz, A., *Science* 1998, *280*, 1046–1048.

[10] Fluri, K., Fitzpatrick, G., Chiem, N., Harrison, D. J., *Anal. Chem.* 1996, *68*, 4285–4290.

[11] Salimi-Moosavi, H., Tang, T., Harrison, D. J., *J. Am. Chem. Soc.* 1997, *119*, 8716–8717.

[12] Harrison, D. J., Fluri, K., Seiler, K., Fan, Z., Effenhauser, C. S., Manz, A., *Science* 1993, *261*, 895–897.

[13] Liang, Z., Chiem, N., Ocvirk, G., Tang, T., Fluri, K., Harrison, D. J., *Anal. Chem.* 1996, *68*, 1040–1046.

[14] Chiem, N., Harrison, D. J., *Anal. Chem.* 1997, *69*, 373–378.

[15] Li, P. C. H., Harrison, D. J., *Anal. Chem.* 1997, *69*, 1564–1568.

[16] Khandurina, J., Jacobson, S. C., Waters, L. C., Foote, R. S., Ramsey, J. M., *Anal. Chem.* 1999, *71*, 1815–1819.

[17] Waters, L. C., Jacobson, S. C., Kroutchinina, N., Khandurina, J., Foote, R. S., Ramsey, J. M., *Anal. Chem.* 1998, *70*, 5172–5176.

[18] Kutter, J. P., Jacobson, S. C., Matsubara, N., Ramsey, J. M., *Anal. Chem.* 1998, *70*, 3291–3297.

[19] Jacobson, S. C., Culbertson, C. T., Daler, J. E., Ramsey, J. M., *Anal. Chem.* 1998, *70*, 3476–3480.

[20] Ermakov, S. V., Jacobson, S. C., Ramsey, J. M., *Anal. Chem.* 1998, *70*, 4494–4504.

[21] Hadd, A. G., Raymond, D. E., Halliwell, J. W., Jacobson, S. C., Ramsey, J. M., *Anal. Chem.* 1997, *69*, 3407–3412.

[22] Simpson, P. C., Roach, D., Woolley, A. T., Thorsen, T., Johnston, R., Sensabaugh, G. F., Mathies, R. A., *Proc. Natl. Acad. Sci. USA* 1998, *95*, 2256–2261.

[23] Woolley, A. T., Mathies, R. A., *Anal. Chem.* 1995, *67*, 3676–3680.

[24] Woolley, A. T., Lao, K., Glazer, A. N., Mathies, R. A., *Anal. Chem.* 1998, *70*, 684–688.

[25] Liu, S., Shi, Y., Ja, W. W., Mathies, R. A., *Anal. Chem.* 1999, *71*, 566–573.

[26] Burns, M. A., Mastrangelo, C. H., Sammarco, T. S., Man, F. P., Webster, J. R., Johnson, B. N., Foerster, B., Jones, D., Fields, Y., Kaiser, A. R., Burke, D. T., *Proc. Natl. Acad. Sci. USA* 1996, *93*, 5556–5561.

[27] van den Berg, A., Bergveld, P. (Ed.), *Micro Total Analysis Systems*, Kluwer, Boston 1995.

[28] Colyer, C. L., Tang, T., Chiem, N., Harrison, D. J., *Electrophoresis* 1997, *18*, 1733–1741.

[29] Effenhauser, C. S., Bruin, G. J. M., Paulus, A., *Electrophoresis* 1997, *18*, 2203–2213.

[30] Freemantle, M., *Chem. Eng. News* 1999, *77*, 27–36.

[31] Burns, M. A., Johnson, B. N., Brahmasandra, S. N., Handique, K., Webster, J. R., Krishnan, M., Sammarco, T. S., Man, P. M., Jones, D., Heldsinger, D., Mastrangelo, C. H., Burke, D. T., *Science* 1998, *282*, 484–487.

[32] Yao, S., Anex, D. S., Caldwell, W. B., Arnold, D. W., Smith, K. B., Schultz, P. G., *Proc. Natl. Acad. Sci. USA* 1999, *96*, 5372–5377.

[33] Weigl, B. H., Yager, P., *Science* 1999, *283*, 346–347.

[34] Effenhauser, C. S., Bruin, G. J. M., Paulus, A., *Electrophoresis* 1997, *18*, 2203–2213.

[35] Kricka, L. J., *Clin. Chem.* 1998, *44*, 2008–2014.

[36] Santini, J. T., Cima, M. J., Langer, R., *Nature* 1999, *397*, 335–338.

[37] Ramsey, J. M., Jacobsen, S. C., Knapp, M. R., *Nature Med.* 1995, *1*, 1093–1096.

[38] Ogura, M., Agata, Y., Watanabe, K., McCormick, R. M., Hamaguchi, Y., Aso, Y., Mitsuhashi, M., *Clin. Chem.* 1998, *44*, 2249–2255.

[39] Rohlícek, V., Deyl, Z., Miksik, I., *J. Chromatogr. A* 1994, *662*, 369–373.

[40] Bayer, H., Engelhardt, H., *J. Microcol. Sep.* 1996, *8*, 479–484.

[41] Schützner, W., Kenndler, E., *Anal. Chem.* 1992, *64*, 1991–1995.

[42] Roberts, M. A., Rossier, J. S., Bercier, P., Girault, H., *Anal. Chem.* 1997, *69*, 2035–2042.

[43] McCormick, R. M., Nelson, R. J., Alonso-Amigo, M. G., Benvegnu, D. J., Hooper, H. H., *Anal. Chem.* 1997, *69*, 2626–2630.

[44] Martynova, L., Lacascio, L. E., Gaitan, M., Kramer, G. W., Christensen, R. G., MacCrehan, W. A., *Anal. Chem.* 1997, *69*, 4783–4789.

[45] Ford, S. M., Davies, J., Kar, B., Qi, S. D., McWhorter, S., Soper, S. A., Malek, C. K., *J. Biomech. Eng.* 1999, *121*, 13–21.

[46] Xia, Y., Whitesides, G. M., *Angew. Chem., Int. Ed.* 1998, *37*, 550–575.

[47] Qin, D., Xia, Y., Rogers, J. A., Jackman, R. J., Zhao, X., Whitesides, G. M., *Top. Curr. Chem.* 1998, *194*, 1–20.

[48] Shaw, J. M., Gelorme, J. D., LaBianca, N. C., Conley, W. H., Holmes, S. J., *IBM J. Res. Develop.* 1997, *41*, 81–94.

[49] Chaudhury, M. K., Whitesides, G. M., *Langmuir* 1991, *7*, 1013–1025.

[50] Chaudhury, M. K., Whitesides, G. M., *Science* 1991, *255*, 1230–1232.

[51] Morra, M., Occhiello, E., Marola, R., Garbassi, F., Humphrey, P., Johnson, D., *J. Colloid Interface Sci.* 1990, *137*, 11–24.

[52] Kovacs, G. T. A., *Micromachined Transducers Sourcebook*, McGraw-Hill, New York 1998.

[53] Effenhauser, C. S., Bruin, G. J. M., Paulus, A., Ehrat, M., *Anal. Chem.* 1997, *69*, 3451–3457.

[54] Carbeck, J. D., Colton, I. J., Gao, J., Whitesides, G. M., *Acc. Chem. Res.* 1998, *31*, 343.

[55] Colton, I. J., Anderson, J. R., Gao, J., Chapman, R. G., Isaacs, L., Whitesides, G. M., *J. Am. Chem. Soc.* 1997, *119*, 12701–12709.

[56] Mammen, M., Gomez, F., Whitesides, G. M., *Anal. Chem.* 1995, *67*, 3526–3535.

[57] Bakajin, O., Carlson, R., Chou, C. F., Chan, S. S., Gabel, C., Knight, J., Cox, T., Austin, R. H., *Solid-State Sensor and Actuator Workshop*, Hilton Head Island, SC, June 1998, pp. 116–119.

[58] Chou, H.-P., Spence, C., Schere, A., Quake, S., *Proc. Natl. Acad. Sci. USA* 1999, *96*, 11–13.

[59] Delamarche, E., Bernard, A., Schmid, H., Michel, B., Biebuyck, H., *Science* 1997, *276*, 779–781.

[60] Delamarche, E., Bernard, A., Schmid, H., Bietsch, A., Michel, B., Biebuyck, H., *J. Am. Chem. Soc.* 1998, *120*, 500–508.

[61] Folch, A., Toner, M., *Biotechnol. Prog.* 1998, *14*, 388–392.

[62] Folch, A., Ayon, A., Hurtado, O., Schmidt, M. A., Toner, M., *J. Biomech. Eng.* 1999, *121*, 28–34.

[63] Takayama, S., McDonald, J. C., Ostuni, E., Liang, M. N., Kenis, P. J. A., Ismagilov, R. F., Whitesides, G. M., *Proc. Natl. Acad. Sci. USA* 1999, *96*, 5545–5548.

[64] Kenis, P. J. A., Ismagilov, R. F., Whitesides, G. M., *Science* 1999, *285*, 83–85.

[65] DeBusschere, B. D., Borkholder, D. A., Kovacs, G. T. A., *Solid-State Sensor and Actuator Workshop*, Hilton Head Island, SC, June 1998, pp. 358–362.

[66] Burggraf, N., Krattiger, B., de Mello, A. J., de Rooij, N. F., Manz, A., *Analyst* 1998, *123*, 1443–1447.

[67] Schueller, O. J. A., Duffy, D. C., Rogers, J. A., Brittain, S. T., Whitesides, G. M., *Sens. Actuators A* 1999, in press.

[68] Schueller, O. J. A., Zhao, X.-M., Whitesides, G. M., Smith, S. P., Prentiss, M., *Adv. Mater.* 1999, *11*, 37–41.

[69] Duffy, D. C., Schueller, O. J. A., Brittain, S. T., Whitesides, G. M., *J. Micromech. Microeng.* 1999, *9*, 211–217.

[70] Delamarche, E., Biebuyck, H. A., Schmid, H., Michel, B., *Adv. Mater.* 1997, *9*, 741–746.

[71] Dow Corning Corporation, Midland, MI.

[72] Lauks, I. R., *Acc. Chem. Res.* 1998, *31*, 317–324.

[73] Peled, N., *Pure Appl. Chem.* 1996, *68*, 1837–1841.

[74] Kricka, L. J., Wilding, P., *Pure Appl. Chem.* 1996, *68*, 1831–1836.

[75] Braxton, S., Bedilion, T., *Curr. Opin. Biotech.* 1998, *9*, 643–649.

[76] van den Berg, A., Grisel, A., Verney-Norberg, E., van der Schoot, B. H., Koudelka-Hep, M., de Rooij, N. F., *Sens. Actuators B* 1993, *13*, 396–399.

[77] Konrad, R., Ehrfeld, W., Freimuth, H., Pommersheim, R., Schenk, R., Weber, L., *Microreaction Technology: Proceedings of the First International Conference on Microreaction Technology* 1998, pp. 348–356.

[78] Northrup, M. A., Benett, B., Hadley, D., Landre, P. L., Lehew, S., Richards, J., Stratton, P., *Anal. Chem.* 1998, *70*, 918–922.

[79] Cheng, J., Waters, L. C., Fortina, P., Hvichia, G., Jacobson, S. C., Ramsey, J. M., Kricka, L. J., Wilding, P., *Anal. Biochem.* 1998, *257*, 101–106.

[80] Brody, J. P., Yager, P., Goldstein, R. E., Austin, R. H., *Biophys. J.* 1996, *71*, 3430–3441.

[81] Dörre, K., Brakmann, S., Brinkmeier, M., Han, K., Riebesee, K., Schwille, P., Stephan, J., Wetzer, T., Lapczyna, M., Stuke, M., Bader, R., Hinz, M., Seliger, H., Holm, J., Eigen, M., Rigler, R., *Bioimaging* 1997, *5*, 139–152.

[82] Zhu, H., Clark, S. M., Benson, S. C., Rye, H. S., Glazer, A. N., Mathies, R. A., *Anal. Chem.* 1994, *66*, 1941–1948.

Electrophoresis 2000, *21*, 3931–3951

3931

Review

Gerard J. M. Bruin

Novartis Pharma AG,
Drug Metabolism &
Pharmacokinetics,
Basel, Switzerland

Recent developments in electrokinetically driven analysis on microfabricated devices

This review is devoted to the rapid developments in the field of microfluidic separation devices in which the flow is electrokinetically driven, and where the separation element forms the heart of the system, in order to give an overview of the trends of the last three years. Examples of microchip layouts that were designed for various application areas are given. Optimization of mixing and injection strategies, designs for the handling of multiple samples, and capillary array systems show the enormous progress made since the first proof-of-concept papers about lab-on-a-chip devices. Examples of functional elements for on-chip preconcentration, filtering, DNA amplification and on-chip detection indicate that the real integration of various analytical tasks on a single microchip is coming into reach. The use of materials other than glass, such as poly(dimethylsiloxane) and polymethylmethacrylate, for chip fabrication and detection methods other than laser-induced fluorescence (LIF) detection, such as mass spectrometry and electrochemical detection, are described. Furthermore, it can be observed that the separation modes known from capillary electrophoresis (CE) in fused-silica capillaries can be easily transferred to the microchip platform. The review concludes with an overview of applications of microchip CE and with a brief outlook.

Keywords: Microfluidics / Chips capillary electrophoresis / Micro-total analysis systems / Separation systems / Proteomics / Review EL 4163

Contents

Correspondence: Dr.Gerard J. M. Bruin, Novartis Pharma AG, Drug Metabolism & Pharmacokinetics, WKL-135.2.22, Basel, Switzerland
E-mail: gerardus_j_m.bruin@pharma.novartis.com
Fax: +41-61-6964317

Abbreviations: COMOSS, collocated monolith support structures; **PDMS**, poly(dimethylsiloxane)

1 Introduction

Sinces the introduction of the concept of micro-total analysis systems (μ-TAS) in 1989 [1], interest in analytical chip-based microdevices has increased strongly. The concept of μ-TAS proposes the integration of the different steps of an analytical process into a miniaturized flow system, thereby creating the right conditions for faster, automated analysis. One of the earliest examples of μ-TAS systems was the introduction of CE separations in etched microchannels in glass chips [2]. The numerous papers about capillary electrophoretic separations in microfabricated channels that have been published since the first proof-of-concept papers demonstrate the enormous analytical potential for miniaturized separation devices.

There are several arguments in favor of a size reduction of analytical separation devices. They include better performance in terms of increased separation speed and thus higher sample throughput, reduced costs (particularly in the case of disposable devices), and possibilities for parallel processing in small working areas. The use of microfabricated analytical devices by the latter is made extremely attractive when high sample throughput is needed. In recent years, it has become generally accepted that large-scale, high-throughput analytical methods are needed in various emerging application fields of the life sciences, especially in the fields of genomics and proteomics. This statement also strongly applies to the pharmaceutical field, where these analytical issues play a major role as well with regard to combinatorial chemistry and drug discovery.

This review is restricted to the description of microfluidic separation devices in which the flow is electrokinetically driven, and where the separation element forms the heart of the system. Other examples of analytical chip-based microsystems, such as gene microarray systems (so-called DNA chips), are outside the scope of this review and have been covered in other reviews ([1, 3, 4] and references therein). Note that the main part of the review is completely devoted to developments over the last three years. References to earlier work in chip CE can be found in several reviews and book chapters on analytical microdevices [1, 3–16].

2 Microchip layouts – mixing, focusing, multiplexing, faster separations, modeling and more

An attractive feature in utilizing microfabricated structures is the integration of different kinds of fluid manipulation on a single device. The first chip designs consisted of microchannels laid out in cross geometries, with a straight separation channel intersected by a second channel for sample injection. Four fluid reservoirs were positioned at the ends of the channels, two for introduction of sample and background electrolyte, the other two serving as waste reservoirs. One of the clear trens over the last three years has been towards more complex microfluidic networks, incorporating a large variety of functional elements in chip format.

2.1 Mixing and pumping devices

One of the key issues that arises with increasing complexity of the chip design is efficient reagent mixing [17–20]. It was shown by Jacobsen *et al.* [17] that in a compact chip architecture and with a single voltage source, parallel mixing with different mixing ratios of sample to

buffer can be accomplished by using a series of T-intersections (see Fig. 1A). The channels had similar cross-sectional areas, which made channel resistances proportional to length. Applying the same potential to both the sample and buffer reservoirs allowed accurate and fast mixing in analysis channels A1–A7 with a sample fraction of 0 (only buffer) in A1 and a fraction of 1 (only sample) in A7, and liquid streams with different mixing ratios of sample and buffer in A2–A6. The mixing ratio depended on the relative lengths of the buffer channels, B1–B6, and sample channels S2–S7. A scheme for serial electrokinetic mixing was also demonstrated.

In another paper [18], the operation of a so-called electroosmotically induced hydraulic pump was reported. Due to a coating on the channel wall of the side arm of a T-intersection, the electroosmotic flow could be strongly reduced in this channel compared to the uncoated main channel. When potential was applied between the reservoirs at the top of the main channel and the end of the side arm, pressure was produced at the T-intersection. This resulted in hydraulically induced pumping of fluid in the electric field-free channel of the T. Mass transport of cations and anions through the T-intersection was investigated from a theoretical and experimental point of view. Depending on the experimental conditions chosen, a continuous separation of cations from anions could be generated.

Precise mixing of fluids was demonstrated in a paper on the use of gradient elution in MEKC by Kutter *et al.* [19]. The device for electrokinetically controlled mixing of pure buffer and acetonitrile or methanol is shown in Fig. 1B. Gradients were realized by changing the voltages at reservoirs "Solvent 1" and "Solvent 2" over time in order to change the mixing ratio at the T according to the programmed gradient shape. Linear, concave and convex gradients with both modifiers could be performed within 10 s. The width of the migration window, analysis time, selectivity and resolution showed the influence of isocratic and gradient solvent changes in the MEKC separation of five neutral coumarin dyes. Electroosmotic pumping of organic solvents and reagents has been demonstrated by Salimi-Moosavi *et al.* [20]. The precise mixing of two reactants, dissolved either in acetonitrile or methanol, showed the potential for on-chip combinatorial synthesis in parallel channels, automated on a microscale.

2.2 Integrated devices

Several examples of increasingly integrated devices combining various steps of a biochemical analysis have been reported. Completely integrated thermal cell lysis, PCR amplification, electrophoretic analysis and labeling for fluorescence detection on a single microchip has been dem-

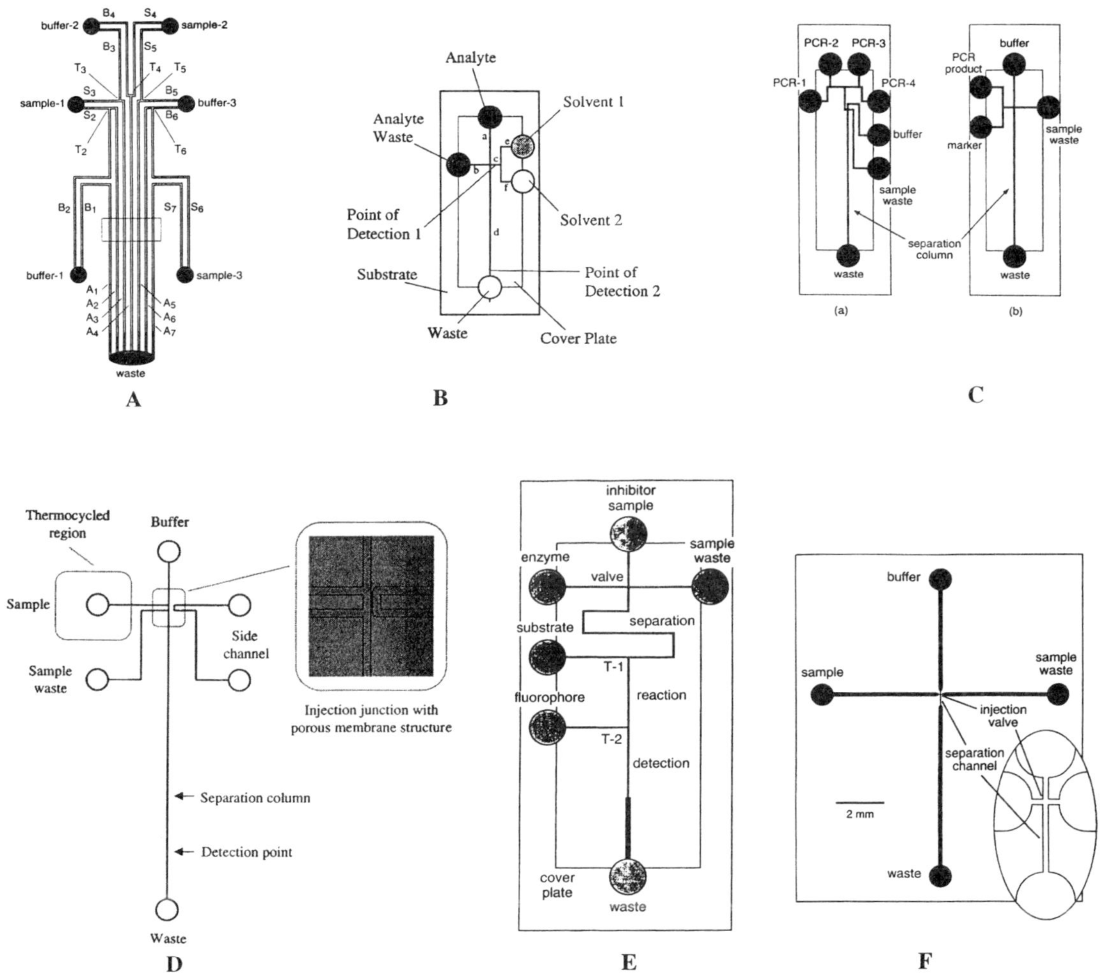

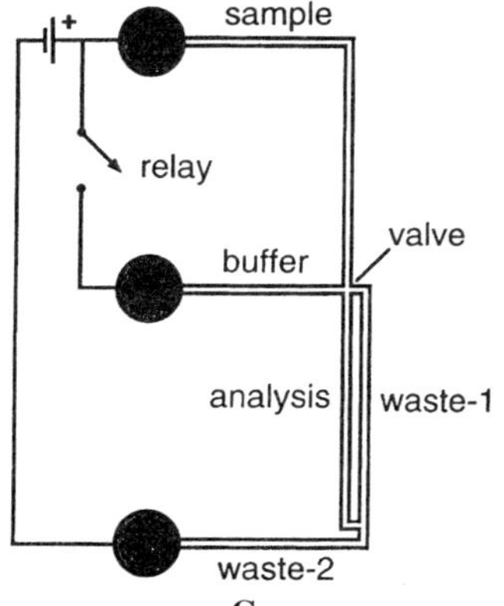

Figure 1. Examples of chip layouts for different applications. (A) Microchip for parallel electrokinetic mixing. Sample, buffer and analysis channels are labeled "S", "B" and "A", respectively; from [17], with permission. (B) Microchip for isocratic and gradient elution in MEKC; from [19], with permission. (C) a: Microchip for PCR amplification of multiple DNA samples; b: microchip for sizing PCR products with a DNA sizing marker; from [22], with permission. (D) Microchip for on-chip DNA preconcentration and integrated rapid PCR; from [23], with permission. (E) Microchip for analysis of acetylcholinesterase inhibitors; from [25], with permission. (F) Microchip for high-speed electrophoretic separations. Inset: enlargement of the injection valve and separation channel; from [28], with permission. (G) Microchip for gated valving with only one voltage source; from [32], with permission.

onstrated by Ramsey's group [21, 22]. Figure 1C depicts the schematic of the microchip used for PCR amplification and electrophoretic analysis of four DNA samples. It was clearly demonstrated how DNAs were amplified from three separate PCR reservoirs. The products could be either individually sampled by sequentially applying voltage to each of the four resevoirs or by simultaneous application of voltage to all four reservoirs. Resulting electropherograms can be found in Fig. 2. To allow an accurate size determination of PCR products, coelectrophoresis of a known DNA marker and PCR products can be performed in the five-port device as shown in Fig. 1C. In a follow-up paper [23], an element for preconcentration was combined with an on-chip assembly for DNA amplification (the preconcentration aspect will be discussed below). The thermal cycling assembly consisted of dual Peltier thermoelectric elements. The use of long injection times enabled the detection of PCR products after running only ten cycles. The chip layout can be found in Fig. 1D.

Another integrated microfluidic system for DNA amplification and CE was described by Lagally *et al.* [24]. The system was comprised of microfluidic valves and vents for loading of sample in closed PCR chambers of only 280 nL. The use of thin-film heaters allowed very fast cycle times of only 30 s. After amplification, the amplified DNA could be injected directly and separated in the CE channel. The authors foresee that applications such as

performing multiple PCR reactions and multiplex PCR reactions in parallel on a single device and thermal cycling-based DNA sequencing will come into reach. A device with even more integrated steps of an analytical process was designed for a microfluidic assay of acetylcholine esterase (AchE) inhibitors, in which flow injection analysis and electrophoretic separation of AchE were combined (see Fig. 1E) [25]. Substrate, inhibitor, enzyme, and derivatizing agent were mixed using computer-controlled electrokinetic transport. Another microchip-based enzyme assay was reported for protein kinase A [26].

The integration of complex, multiple fluidic manipulations on a compact microchip makes the requirements for the separation step with regard to resolving power often more demanding. In such cases where higher separation efficiencies are needed, the separation channel length can be enhanced while keeping the electric field constant. This can be achieved by using a serpentine geometry. The influence of turns in such geometries has been studied by Culbertson *et al.* [27]. They describe the peak broadening generated by a turn. Due to differences in distance traveled and electric field strengths at the outside *versus* the inside of a turn, additional dispersion in the total band broadening is generated. A predictive model was developed and validated for various chip designs, differing in channel width and curve radius. The model was based on the ratio of the transverse analyte diffusion time, t_D, to the analyte transit time, t_t. The influence of analyte velocity, diffusion coefficient, channel width, and curve radius on t_D/t_t and hence on separation efficiency, was studied.

When the goal is not separation efficiency improvement, but rather high-speed electrophoresis for ultrahigh throughput or millisecond time-scale kinetics for chemical reactions then a design as depicted in Fig. 1F can be used [28]. Due to the resistivity of the wide channels (440 μm, 7 μm depth) being nearly 17 times lower than that of the narrow channels (26 μm, 7 μm depth), separation fields as high as 53 kV/cm could be achieved with an applied potential of 8.6 kV. A mixture of rhodamine B and dichlorofluorescein was electrophoretically separated in 0.8 ms, a world record in separation sciences. The main source for band broadening was probably Joule heating, especially at higher field strengths.

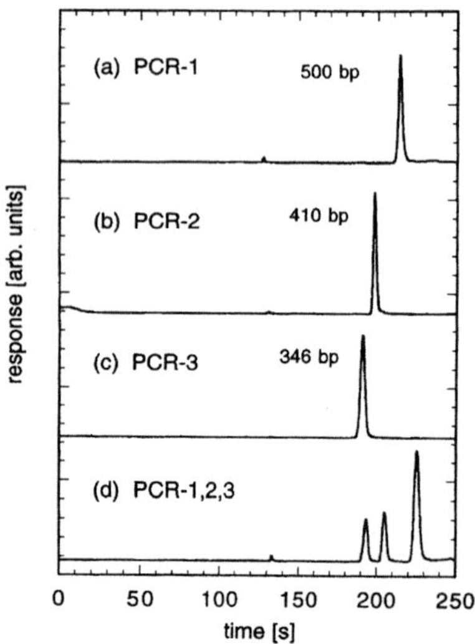

Figure 2. Electropherograms of PCR products made from three DNA samples and analyzed on the microchip from Fig. 1C. (a) λ DNA 500 bp, (b) plasmid DNA 410 bp, (c) genomic DNA 346 bp, (d) combined analysis of (a)–(c). From [22], with permission.

2.3 Injection

The complexity of fluid manipulation by voltage control increases with the number of channels and intersections. Especially the injection of a well-defined, reproducible injection plug into the separation channel is of paramount importance to find the right tradeoff between separation efficiency and detection sensitivity. As in all separation methods, sample plugs which are too broad lead to lower separation efficiencies, but can also result in a gain in sensitivity. The effect of the electric field distribution in a simple cross microchip during the sample loading step and the sample dispensing steps was investigated by Alarie *et al.* [29]. Electric fields in the sample, waste, buffer and analysis channel were defined in such a way that samples pinched to different degrees were created. Strongly pinched samples give smaller spatial variances than medium or weakly pinched sample. The electric field distribution during the dispensing step also plays a role, although to a lesser extent. With higher fields in the sample and sample waste channels during dispensing, which results in smaller amounts of sample dispensed in the analysis channel, smaller spatial variances were found. This study can be regarded as an extension of earlier work on electrokinetic focusing in microfabricated channel structures [30]. Applications for electrokinetic focusing include flow cytometry and ultrasensitive fluorescence detection (see Section 5.1).

The effects of electric field distribution during sample loading, injector geometry, and sample matrix were qualitatively described for a double-T injector [31]. Various factors that could influence flow within channels and intersections, such as the electrical impedance of channels involved, diffusion, leakage flow arising from the Venturi effect, pH and ionic strength mismatches of the buffer and sample matrix, were mentioned. All of these factors contribute to the amount of sample injected into the analysis channel. Long enough injection times allow the slowest moving sample constituents to arrive at and fill the complete injection volume, thereby leading to a geometrically defined injection plug which is truly representative of the sample. A clear conclusion was that matching the ionic strength and pH of sample and buffer will produce the most predictable results.

Increased integration of multiple functions onto one microchip should be done in such a way that the output of one high voltage power supply can be distributed to multiple points. Otherwise, the hardware needed to regulate the applied voltages in the different parts of the architecture could become too complex. This issue has been addressed by Jacobsen *et al.* [32]. By using a microchip as depicted in Fig. 1G and only a single fixed voltage source and a high-voltage switch, precise loading and dispensing of a sample was investigated. The injection time was defined by the time that the voltage switch was open to remove the potential at the buffer reservoir. The analysis and waste channels merge in a common waste-2 channel and waste reservoir, thereby leading to a compact architecture. Replicate sample profiles for injection

times of 0.2, 0.4 and 0.8 s showed a peak area reproducibility of better than 0.5% RSD ($n = 20$).

2.4 Capillary array systems

The application of sequential fast separations on microchips has the potential to increase sample throughput considerably. However, a real breakthrough in throughput can be expected by performing separations in parallel using microfabricated capillary array systems as demonstrated in three papers from the Mathies group [33–35]. After a first device [33], which permitted analysis of 12 samples in parallel, a second design with 96 sample wells and 48 separation channels was published (Fig. 3A) [34]. The whole assembly consisted of a sealed micromachined glass chip, a poly(dimethylsiloxane) (PDMS) gasket containing holes for reservoirs placed on top of it, and an electrode array then placed on top of the PDMS elastomer for voltage application (Fig. 3B). Detection was by using a laser-induced confocal fluorescence scanner. The layout of the sample injector was designed in such a way that, after facile sample loading with an 8-tip pipetter, sequential analysis of two samples per microchannel could be done. As an example to demonstrate the power of this design, the analysis of 96 PCR products could be carried out in less than 8 min, or, in other words, less than 5 s per sample. An even more sophisticated design (Figs. 3C, D) with 96 individual separation channels connected to 96 sample reservoirs together with a rotary confocal fluorescence detection system solved most limitations of the design in Fig. 3A [35]. Especially the potential problem of cross contamination due to the analysis of two different samples in the same channel was overcome. A 96-capillary array loader that allowed direct transfer of samples from a microtiter plate to the sample reservoirs on the microseparation device was also described (see Fig. 3D). The latter system has been one of the few examples in literature so far that addresses the problem of "connecting microchips to the outside world" for efficient sample handling. High-throughput genotyping separations with a separation throughput of 0.5–0.6 sample/s were demonstrated. Another multichannel device with 48 separation channels combined with a transmission imaging spectrograph, consisting of a dual laser excitation source and a sensitive charge-coupled device, was described by Simpson *et al.* [36]. Examples of the separation of single-stranded DNA fragments up to 500 bases in length were shown.

2.5 On-chip preconcentration

As already mentioned, microfabrication offers the advantage of integration of small volume compartments for different steps in a chemical analysis. Besides well-known sample stacking, several new examples of on-chip pre-

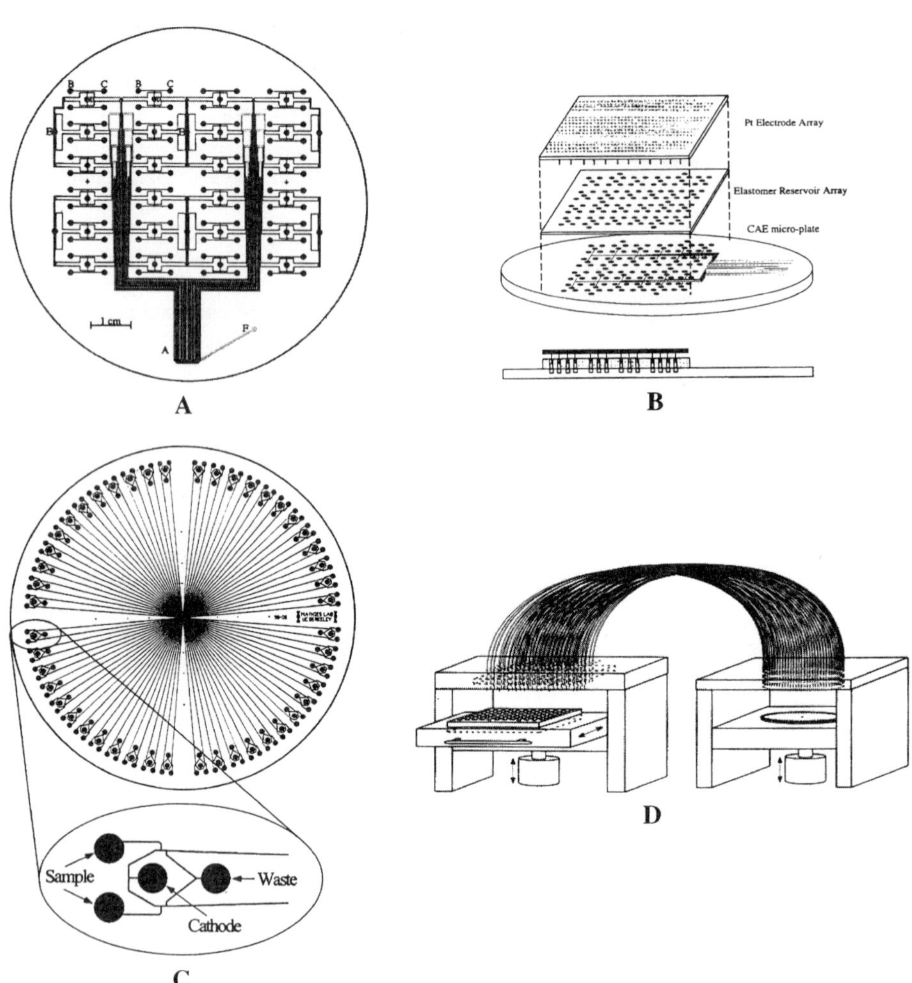

Figure 3. (A) Chip device for the 96-sample / 48-separation channel array. A, detection region; B, C, injection reservoirs. (B) Schematic of the PDMS reservoir array and electrode array loading assembly. From [34], with permission. (C) Chip device for the 96-channel radial capillary array. The substrate is 10 cm in diameter. (D) Schematic of the 96-sample capillary array loader. Pressurization is used to transfer samples from the microtiter wells to the sample reservoirs of the radial microplate. From [35], with permission.

concentration have been published during the last three years. Solid-phase extraction of a neutral coumarin dye was investigated in a chip layout that is similar to the one in Fig. 1B for solvent programming [37]. The only difference was that the channel walls were modified with octadecyl silica (ODS), which served the enrichment process. Sample enrichment in the C18-coated channel was achieved with buffers with low acetonitrile content. Elution of the sample was simply carried out by switching to a buffer with a higher acetonitrile content. The gain in con-

centration was 80-fold. A different approach for solid-phase extraction was described by Oleschuk *et al.* [38]. They packed a 330 pL cavity, which was formed by two weirs within a sample channel, with ODS-coated silica beads. The beads could be moved into and out of the cavity by electroosmotic flow. A concentration enhancement of about 500 of a 1 nM solution of a nonpolar analyte was obtained after loading the sample, followed by a buffer flush and subsequent elution of the sample in acetonitrile. The design with the weirs circumvented the problem of frit preparation for retaining packing material. An electrochromatographic separation of the analyte and fluorescein was shown as well.

An alternative design for sample preconcentration and subsequent electrophoretic analysis can be found in Fig. 1D [23, 39]. Here, two adjacent channels were connected electrically by a narrow layer of a porous membrane, consisting of polysilicate. Preconcentration of DNA samples by up to two orders of magnitude prior to separation could be achieved by applying voltage between the analyte and side reservoirs for up to 250 s. A nonlinear dependence of concentration enhancement with time was found.

A point of debate certainly will be how much integration is possible or needed for a chemical or biochemical analysis. A more modular approach, in which micromachined injection and reactor devices are combined with classical separation steps in capillaries or HPLC columns, still makes sense, as demonstrated in several papers [40–43]. Viewed the other way around, a classical off-chip sample pretreatment with further on-chip separation and detection could in some cases be more economical than integrating all steps onto a single device. A modular concept also plays an important role in the coupling of microfabricated separation devices with mass spectrometry (see Section 5.3).

2.6 Filtering devices

Microdevices typically have channel depths of 10–40 µm and widths of 60–200 µm, which makes blockage by particles a realistic problem. The filtering of liquids in micromachined capillary systems can be accomplished by lateral percolation [44]. The purpose was to filter cells, dust, and other particles, found in solvents or arising from sample crystallization or microbial growth, from liquid entering the channels on the microdevice. A filtering design for the isolation of white blood cells from whole blood was described by Wilding *et al.* [45]. Various designs were tested. Efficient trapping of white blood cells could be done in microfilters, consisting of a 3.5 µm gap between the top of a silicon dam and the glass used to cover the chip. This filter type was integrated into a microdevice for both cell filtration and PCR amplification. After the PCR amplification of a 202 bp sequence of exon 6 of the dystrophin gene directly from genomic DNA in trapped white blood cells, the amplification product was removed from the chip and analyzed with conventional CE.

2.7 Electroosmotic flow control

Controlling the electroosmotic flow (EOF) is of primary importance for a reproducible and precise fluid control in microfabricated fluid networks. A possible way to achieve this is by applying an additional perpendicular external voltage across the wall of the microchannel [46, 47]. It was shown that electroosmosis could be altered with external voltages smaller than 120 V. External voltages were applied by two electrodes, positioned 50 µm away from the inner surface of a 30 µm wide channel [46]. A similar approach was demonstrated in microchannels surrounded by a thin, free-standing and fragile nitride wall (390 nm thick). A so-called three-capacitor model, used by others to describe the effect of radial fields in conventional CE, showed that modulation of the zeta-potential with low voltages required extremely thin walls. Manipulation of electroosmotically driven fluids by changing the zeta potentials was demonstrated [47].

2.8 Modeling of electrokinetic transport

Modeling of identified parameters for an adequate description of flow in microchannel networks is increasingly recognized as important for chip design. Initial computer simulations of electrokinetic transport in microchannels have been reported by three groups [48–50]. Patankar and Hu [48] developed a scheme to simulate electroosmotic flow in a cross-channel, thereby evaluating the behavior of a sample plug during loading. It was shown that (i) the injected plug is less distorted when injection at higher field strength is applied; (ii) application of a higher electric potential in the side channels (in most cases the analysis and buffer channel) than at the center of intersection gives well-defined, focused plugs without leaking; and (iii) the latter effect can also be obtained by applying pressure at the side reservoirs, although with greater sample plug distortion. Similar simulations of fluid distribution for flow distribution at a T-junction showed the generations of pressure drops at the intersection by the inhomogeneous distribution of the EOF in the different channels [49].

With a 2-D model by Ermakov *et al.* [50], which considered EOF, electrophoretic motion and diffusion, typical features of sample transport, (*e.g.*, electric focusing during injection and mixing of fluid streams in a T-channel structure), could be simulated and compared with experimental results from earlier studies. Simulations were performed under various operating conditions. Experimental and simulated images of electrokinetic focusing agreed nicely, although small differences were found with respect to the concentration profile across the channels involved. In case of sample mixing in the T-intersection, where diffusion is the only source of mixing, the influence of mixing time (channel length, electric field strength) and the width of the mixing channel on concentration homogeneity across the cross section were investigated. More homogenous mixing can be obtained by taking narrower mixing channels and a lower field strength.

3 Polymer-based microfluidic devices

To date, most of the work on chips in the field of CE has been done successfully with various glass substrates. Glass has many advantages, such as its optical transparency, generation of high EOF rates, and favorable clean surface characteristics after fabrication by etching. However, in the case of biopolymer separations, which constitute a sizable number of potential applications, adsorption of the analyte at the negatively charged channel walls can be a serious problem. This is particularly true in case of proteins. Besides this, there are also some disadvantages in the fabrication process: (i) Due to the amorphous character, vertical side walls are difficult to etch. Mostly semicircular or trapezoidal cross sections are created. (ii) Sealing of glass plates normally requires a cleanroom environment and high voltages or temperatures. This hinders fabrication of chips with integrated temperature-sensitive material. Low temperature bonding for fabrication of microchip devices could possibly solve this problem [51, 52]. (iii) The fabrication time is still slow and the material is fragile. In addition, though batch fabrication of devices is possible, that is, several can be made at one time, glass chips still tend to be expensive, due to their relatively large size. Batch fabrication for photolithograpically produced devices becomes economical when many devices can be made on a single wafer, as in the case of microelectronic chips.

These disadvantages have led to much research activity into new materials, especially polymers, and fabrication procedures for the production of microchips [53]. The fabrication methods include laser ablation, replication, injection molding, and hot embossing. A few examples are briefly discussed below.

A UV laser ablation method was used in order to produce channels in polystyrene, polycarbonate, cellulose acetate and poly(ethylene terephthalate) [54]. Sealing was accomplished using a film lamination technique. Differences in hydrophilicity and surface roughness were observed between the polymers studied. Cathodic EOF was generated and characterized as a function of electric field, pH, and ionic strength of the buffers used. No electrophoretic separations have been shown in these channels so far.

The use of polymethylmethacrylate (PMMA) for fabrication of microcapillary electrophoresis devices was described by several groups. Imprinting by using a small-diameter wire to create an impression of simple linear channel designs in the PMMA at elevated temperature showed promising results [55–59]. Repetitive, highly efficient separations of DNA fragments and an immunoassay with human anti-goat antibody and goat IgG were possi-

ble in cross channel configurations [55, 57]. More complex configurations were fabricated by imprinting methods using an inverse image of devices micromachined on a silicon wafer. An imprinting method for PMMA or the copolymer Vivak at room temperature improved the device yield from about 10 to above 100 devices per master [56]. For the latter copolymer, it was shown that the EOF can be increased after alkaline hydrolysis and reversed after flushing with CTAB solutions [58]. Micromachining in PMMA using X-ray lithography has also been demonstrated [59].

The master or molder can also be replicated by a technique called injection molding [60, 61]. In this process, the polymer is melted and injected against the molder in the molding chamber. Injection molding allows high production rates. This approach has been used for single channels and channel arrays in an acrylic copolymer resin and applied for the analysis of various DNA sequencing mixtures and dsDNA fragments [60, 61].

After the first paper on a miniaturized CE system in PDMS by Effenhauser *et al.* [62], several groups have focused on the procedures used to fabricate chips in this elastomer, and on the separation characteristics under typical electrophoretic conditions and applications [63–65]. Also, the handling of picoliter liquid samples by air pressure using microfabricated vent valves instead of using EOF in PDMS has been described [66].

Among the tested fabrication methods for microstructures, the molding process for polymer systems made of PDMS is the easiest to date [62]. Besides that, the material itself has several physicochemical characteristics that make it well suited for use in miniaturized electrophoretic separation systems. Whitesides' group [64] used masters produced in photoresist for rapid prototyping, but masters can also be fabricated by other techniques, such as in silicon etching or conventional machining of other hard materials. Channels in PDMS are easily formed by replica molding, which is simply the generation of a negative replica of the master in PDMS by pouring the prepolymer over the master and subsequent curing at atmospheric pressure at somewhat elevated temperatures for a few hours.

A schematic representation of replica molding can be found in Fig. 4. In the last step, the PDMS replica containing the channel network was sealed irreversibly with a second flat slab of PDMS. In the example shown, this is due to the oxidation of both surfaces in an oxygen plasma discharge. Oxidation results in a stronger seal than with untreated PDMS and it seems to be easier to fill oxidized PDMS channels with liquids [63, 64]. Furthermore higher cathodic EOFs were generated, which is consistent with a

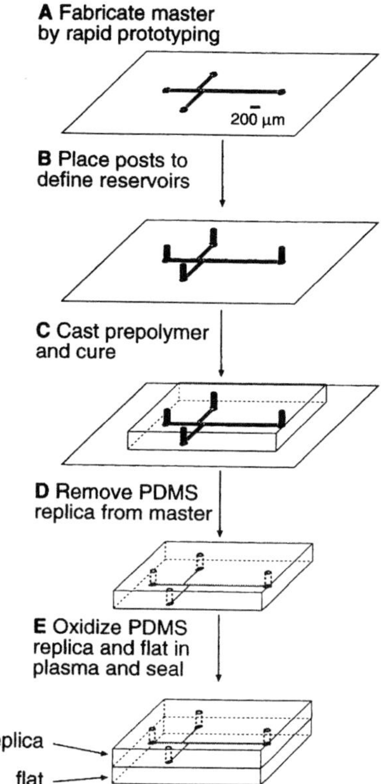

Figure 4. Schematic diagram illustrating how microfabricated structures are made in PDMS. From [63], with permission.

more negatively charged surface of PDMS after oxidation. Highly efficient charge ladders of bovine carbonic anhydrase II, insulin, and lysozyme were separated in oxidized PDMS channels and the results compared favorably in most cases with separations in fused-silica capillaries. (A charge ladder is a set of derivatives that differ in the number of charged groups by partial acetylation of the ε-amino groups of the lysine residues). It was shown that with added surfactants, such as sulfonates and Polybrene, adsorption of positively charged lysozyme can be greatly reduced and the EOF reversed, respectively (as in fused-silica capillaries) [63].

The electrokinetic fluid behavior in untreated PDMS was investigated in detail by Ocvirk *et al.* [65]. The effect of pH on EOF in untreated and thus hydrophobic PDMS was very similar to the pH effect in fused silica capillaries. It was argued that the sigmoidal shape of the curve of electroosmotic velocity as a function of pH over the range pH 3–8, and observed electroosmotic mobilities in the same

order of magnitude as in fused-silica capillaries, could be attributed to silica filler into the prepolymer formulation. Higher reproducibility and stability of EOF than in oxidized PDMS were observed. With careful and comprehensive voltage control in all reservoirs both during injection and separation, excellent efficiency (N = about 64 000 for fluorescein), repeatability of injection volumes (RSD = 1.5%), and migration times (RSD = 0.9%) were realized. Besides that, fluid control in devices with narrower channels seemed to be less sensitive to leakage effects. An extensive overview on the pros and cons of PDMS for the fabrication of microfluidic devices was recently published by McDonald *et al.* [64]. In this review, examples of devices for separation, patterning of biological material (cells, proteins) and components for integrated systems (sensors, detectors, microswitches) were given.

4 Capillary electrochromatography on chips

To date, an interesting but hardly explored research area in microfabricated structures is the chromatographic separation mode, the most applied method in separation sciences. As in a capillary, several possibilities exist to perform capillary electrochromatography (CEC) on a microchip. Since 1997 several papers have appeared that show CEC (i) in open channels [67], (ii) with stationary phases constructed by micromachining [68, 69], and (iii) using continuous beds [70].

4.1 CEC in open channels

Kutter *et al.* [67] investigated open-channel electrochromatography in combination with gradient elution in the same type of chip layout as the one depicted in Fig. 1B, used earlier for solvent mixing in MEKC [67]. All channels with depths between 2.9 and 10.2 µm were coated with octadecylsilanes to attach a reversed-phase stationary phase at the wall. Separations of four neutral coumarin dyes under different isocratic and gradient conditions were demonstrated. Under optimized conditions, the four peaks eluted within a 6 s window, with the chromatogram being completed in about 20 s. A steep gradient, taking only 5 s to go from 29 to 50% acetonitrile, was generated for this particular separation. A strong influence of channel depth on efficiency was observed, especially for channels deeper than about 5 µm. Decreased channel depth results in lower plate heights, while the minima of the H-u curves shift toward higher linear velocities. This influence of the channel depth on separation performance was theoretically confirmed in a paper by Zhang and Regnier [71]. They analyzed the impact of channel geometry on separation efficiency in open channel electrochromatography by using a three-dimensional random walk model.

The resulting, simulated plate height curves agreed well with the experimental data from [67]. Further, contributions of injection plug length, cross-sectional channel area, and aspect ratio of rectangular channels to the total band broadening were studied. Although simulations indicate that plate heights less than half of the channel depth might be obtained, it must be admitted that channels with 2 µm or less in depth are difficult to fabricate. Furthermore, handling could be problematic, due to their increased susceptibility to clogging.

4.2 CEC with microfabricated support structures

A totally new concept for performing CEC on microchips was introduced by He *et al.* [68, 69]. They produced the so-called collocated monolith support structures

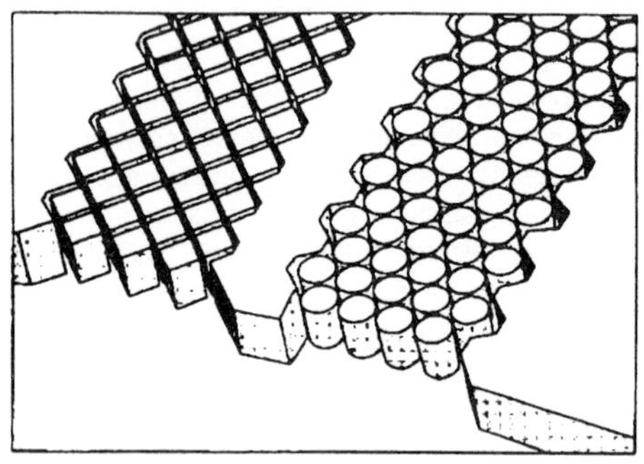

A

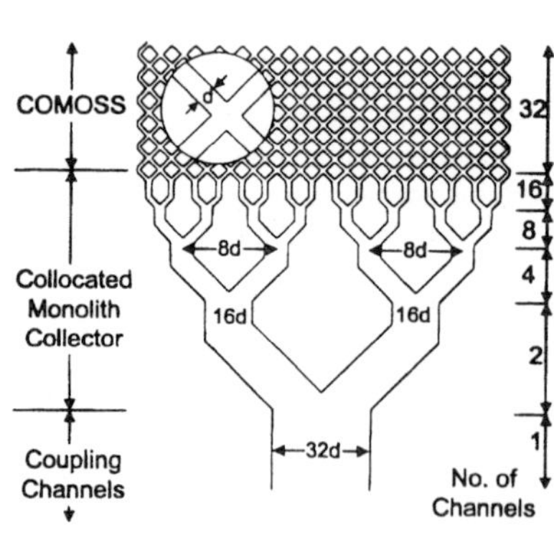

B

Figure 5. (A) Illustration of micromachined cylindrical and cubic COMOSS structures that are suitable for liquid chromatography. (B) Configuration of the inlet splitter: a design that reflects the 2^n architecture of the splitter with constant cross-sectional area. From [68], with permission.

(COMOSS) separated by rectangular channels (1.5 µm wide and 10 µm deep) by deep reactive ion etching in quartz. Reversed-phase stationary phases were bonded to the structures by using standard coating procedures. An illustration of the fabricated structures plus an inlet distribution network of channels is depicted in Fig. 5A and B. In the structure shown here, potential wall effects due to stagnant mobile-phase pools can be strongly reduced by a proper column-wall interface architecture. Inlet and outlet channel dimensions were defined such that the combined cross-sectional area of all channels at any point along the length of the inlet, separation, column, and outlet were equal. Otherwise, unwanted pressure drops with accompanying bubble formation could be generated, resulting in decreased separation performance. Compared with the problems encountered when packing channels with particles on a microchip, the advantages of channels with COMOSS are numerous. To mention only a few: (i) channel dimensions are independent of the packing process and dimensions of the monolith structure; (ii) the support is attached to the wall of the channel (no frit preparation necessary); and (iii) in principle several columns can be constructed on a single wafer. A first application has been demonstrated with this approach [69]. A tryptic digest of ovalbumin was performed with a C18 COMOSS column operated in the CEC mode and compared with CZE of the same digest with an uncoated COMOSS column. Inferior performance (under nonoptimized CZE conditions) was found in the latter instance.

4.3 CEC with continuous polymer beds

It is generally accepted that efficient packing of particles into the narrow channels of a chip is very complicated, although first attempts to pack channels for planar CEC look promising [72]. Fabrication of frits in planar devices is as difficult as in conventional capillaries if not more so. The design of appropriate fluidic connections between the macroscopic world and the chip for introduction of particle slurries can also pose a challenge, particularly if high pressures are required for packing. Therefore, as in electrochromatography in capillaries, the use of separation media consisting of continuous polymer beds is also being explored. A photo-initiated free radical polymerization for the preparation of porous polymer monoliths in UV-transparent fused-silica capillaries has been used [73]. The experiments served as a model for preparation within channels of microchips in a later phase. Photopolymerization appears to be an interesting option because the bed formation can be achieved in a well-defined space on the microchip. The developed monolithic polymers were characterized with regard to pore size distribution, flow velocity, and electrochromatographic separation performance. Although the actual method transfer from capillaries to chip systems remains, this early work confirms the potential of the method.

Ericson *et al.* [70] already fabricated continuous beds for various chromatographic modes in an interesting experimental setup with UV detection. Classical polymerization methods with TEMED and ammonium persulfate as initiators were used to prepare the polymer beds in the microchip. The system allowed pressure- and electrically driven chromatography with gradient elution capability. Pressure-driven anion-exchange chromatography of proteins and electrochromatography of neutral compounds in straight channels and channels with a serpentine geometry (about 30 cm long!) were demonstrated. A comparison of H-u curves for a chip column and a fused-silica capillary filled with a continuous bed derivatized with C3 and SO3- ligands showed no significant differences. This indicated formation of uniform polymer beds in both cases. Besides that, the plate height was nearly independent of the flow velocity above 0.5 mm/s, allowing fast separations. "Wall effects" are minimized, due to the covalent attachment of the bed to the channel wall and the good homogeneity in the region close to the wall, as shown in scanning electron microscopy (SEM) pictures.

5 Detection

5.1 Optical detection

Laser-induced fluorescence (LIF) is still the most preferred detection method on-chip because of its high sensitivity. Careful optimization of parameters such as pinhole size, channel depth, excitation efficiency and laser spot size in a typical detection set up for confocal epifluorescence microscopy allows detection of concentrations as low as 300 fM fluorescein [74]. With the same setup in combination with a red diode laser (see Fig. 6) a concentration detection limit of 9 pM Cy-5 was obtained, which corresponds to detection of 900 molecules in a probe volume of 1.6 pL [75].

Ultrasensitive LIF-detection of separated rhodamine 6G (R6G) and rhodamine B with concentration detection limits of 1.7 pM and 8.5 pM, respectively, and even single molecule detection of R6G after injection of solutions in the low pM range has been achieved [76]. Single molecule detection of DNA has been reported by Haab and Mathies as well [77]. A method was developed for physical and electrokinetic focusing of the sample stream to improve both mass and detection limits. Physical focusing can be realized by tapering the separation channel shortly before the detection spot, which physically constricts the sample to a smaller width. This, in combination with electrokinetic focusing, enabled the reduction of the sample stream

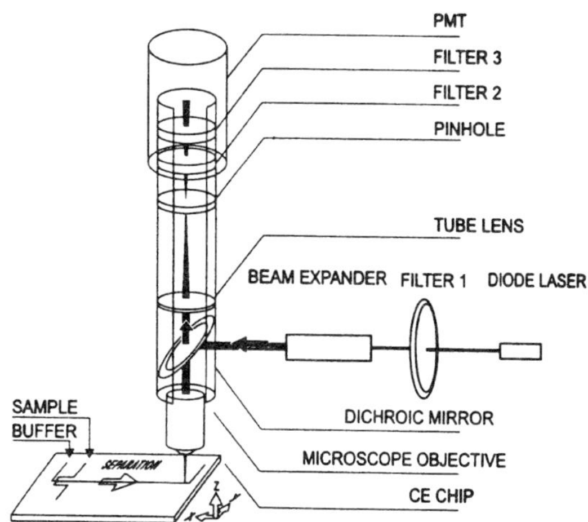

Figure 6. Basic confocal epifluorescence setup for high-sensitivity detection. From [75], with permission.

width from 62 to a minimum value of 13 µm. A threefold increase of detection efficiency was achieved by using focused single molecule counting in the separation of a 100–1000 bp DNA ladder. Electrokinetic focusing was also applied to demonstrate flow cytometry of fluorescently labeled and unlabeled latex particles on a microfabricated device [78].

Another method to obtain S/N enhancement is the application of cross correlation techniques [79]. Here, multiple injections are made in a random sequence. This technique has good prospects in microfabricated electrophoresis devices because no moving parts during injection are involved, thereby enabling reproducible and rapid injections. Peaks with intensities below the detection limit in the "normal" fluorescence detection mode could be visualized in the so-called correlogram using cross correlation methods. A proof-of-principle of a new detection concept was represented in a paper by Crabtree *et al.* [80]. The method, named Shah convolution Fourier transform detection (SCOFT), converts a time-domain electropherogram into a frequency-domain plot by means of Fourier transformation. A sample was injected into a separation channel, which was masked from an expanded and focused laser beam except at a series of 55 micromachined 300 µm wide transparent detection slits. For one compound injected, the detector signal shows regularly spaced peaks along the time axis which, when the data are transformed, results in a single frequency peak. The principle was also demonstrated for a two-component mixture, producing two frequency peaks. The influence of baseline drift and other parameters on the signals obtained in the plots of the frequency-domain data were addressed. The authors stated that SCOFT could possibly yield better and faster resolution and could also be applied to detection methods other than LIF.

UV-absorbance is the most widely used detection method in CE and HPLC. This is in sharp contrast to the situation for microchip-based electrophoresis devices, where UV-absorbance is in most cases too insensitive due to the small channel depths. To solve the problem of the very short light path length, a multireflection cell for enhanced UV-absorbance detection was microfabricated and tested [81]. Aluminum mirrors were patterned above and below the separation channel, with the entrance and exit apertures for ingoing and outgoing light positioned 200 µm apart. The influence of incident beam angle, channel depth, and the thickness of the glass plates used on the number of reflections, the effective path length, the noise and thus ultimately on the sensitivity enhancement was studied both theoretically and experimentally. A 5- to 10-fold sensitivity enhancement was found compared with single pass devices with insignificant contribution from detection to the total band broadening. Although the fabrication of this kind of cells was complicated, the work clearly demonstrated the potential of UV-absorbance for CE chips.

5.2 Electrochemical and other detection methods

The large optical detection devices needed for high sensitivity partly reduce one of the advantages of CE chip sytems, namely the small dimensions of the rest of the experimental setup. The large dimensions of a nonintegrated detector might complicate the development of portable analysis devices. Woolley *et al.* [82] showed that a real integration of the detector within the microfabricated CE chip is feasible in the case of electrochemical detection. The detector consisted of plasma-sputtered Pt working and counter electrodes and a reference electrode that was positioned in an access hole as close as possible (300 or 600 µm) to the working electrode. Sloping baselines and decreased signals were observed at higher separation voltages with a 600 µm distance between reference and working electrode, whereas these effects were much less pronounced with a 300 µm distance. Separations of dopamine, epinephrine and catechol were demonstrated with detection limits between 3.7 and 12 µM. Indirect electrochemical detection for separation and detection of DNA was also demonstrated. A different integrated electrochemical detector, in which a gold working electrode was sputtered directly onto the separation channel outlet, also showed good characteristics [83]. The detection limit for dopamine was 1 µM and separation voltages up to 3 kV were applied without a noticeable

influence on baseline drift or noise. The same group also constructed a thick film electrochemical detector [84]. The detector consisted of planar screen-printed carbon line electrode mounted perpendicular to the flow direction. This detector was also used in an electrophoresis chip for simultaneous quantitation of glucose, uric acid and ascorbic acid [85].

In a setup for continuous channel electrophoretic separations, a microfabricated electrochemical array detection scheme was described by Gavin and Ewing [86, 87]. The detector consists of 100 platinum electrodes, each 95 µm wide with 5 µm spacing between them. The separations take place in a wide, rectangular channel, making the setup design different from most of the work reviewed here. Samples are continuously injected by a fused-silica capillary moving across the width of the channel. For fundamentals, details and applications of the separation technique used in their work, termed channel electrophoresis, see [86–88].

The need for a universal and sensitive detection protocol is still obvious. The potential of refractive index detection was shown in a paper that presented the application of a holographic-based refractive index detector to the separation of carbohydrates [89]. Although the detection limits were very high (10 mM carbohydrate), the paper can be regarded as a first step towards a universal, small volume detector for microfabricated separation systems. Also, the construction of a refractive index detector based on backscatter interferometry has been reported [90]. Raman spectroscopy combined with on-chip ITP of two herbicides has also been described, thereby further illustrating the search for alternative detection schemes and the application of alternative CE separation modes [91].

5.3 Mass spectrometry

Over the past three years, the coupling of microfabricated devices to electrospray ionization-mass spectrometry (ESI-MS) has attracted increased attention. The combination of handling small sample volumes and fast separations on the one side and sensitive, nearly universal detection with structure elucidation of separated compounds on the other is too attractive to leave unexplored. The small size of a microfluidic device for sample pretreatment, *e.g.*, preconcentration, tryptic digestion and/or separation, and the bulky MS instrument seem to exclude a coupling of both. However, due to the close match between flow rates required for ESI-MS and flow rates genrated by CE chips, MS is a promising alternative for LIF detection.

In several papers on this topic, the microfabricated devices, mostly with an array of parallel channels and several sample reservoirs for sequential analysis, were used much more as a sample delivery system rather than as a platform for separation [92–99]. The microfabricated devices were made of glass or quartz, but also the use of a PDMS device was reported [98]. The work focused on creating stable electrospray conditions, reducing background noise, and finding the right coating chemistry for efficient transport of compounds to the MS inlet. The systems described were used for sequential automated analysis of proteins and tryptic digests with continuous sample infusion. For instance, selected protein spots, separated by 2-D gel electrophoresis, were subjected to tryptic digestion and further analyzed using the microfabricated device combined with MS. Generation of MS/MS spectra together with the use of appropriate sequence databases resulted in distinguishable sequences of peptides and fast protein identification [94–96, 98].

In a number of applications, detection of the minute amounts of samples by the mass spectrometer required the use of sample preconcentration methods. This can be done by solid-phase extraction as shown by Figeys and Aebersold [95]. They used a small C18 cartridge, incorporated in the μESI interface at the end of a transfer capillary, for the separation by frontal analysis of low concentrations of peptides and tryptic digests. The chip was used for the supply of a solvent gradient by differential electroosmotic pumping of an aqueous phase and organic solvent. A detection limit of 0.1 nM could be obtained.

The first on-chip separations prior to mass spectral analysis were reported in 1999 by the groups of Harrison [100], Karger [101], and Ramsey [102]. Although previous papers had shown that direct ionization from the outlet separation channel at the chip surface with Taylor cone formation is possible, this is most likely not optimal. Droplet formation at the surface causes too large a dead volume, thereby preventing such a design for on-chip CE-ESI-MS. With short transfer capillaries that were attached to the microdevice, an external electrospray could be incorporated. Typical schematic representations of chip CE with ESI-MS are depicted in Fig. 7A and B for a conventional sheath flow configuration and for a disposable nanoelectrospray emitter, respectively. Figure 7C illustrates the different interfaces for chip CE-ESI-MS as described above, into more detail. It was shown by Harrison's group that it is possible and absolutely necessary to generate very low dead volume couplings between the chip separation channel and the transfer capillary [100, 103]. This was done by using flat-tipped drills to create holes at the end of the glass device, in the plane of the separation channel. An example of the analysis of four synthetic peptides using a nanoelectrospray emitter was given. The separation performances ranged from 500–

3500 theoretical plates. The use of coated chip devices and a longer transfer capillary, also coated, together with the sheath flow interface, gave better separation performance (N = 2600–58 000) for different peptide standards, due to reduced adsorption at the wall. Sensitivity was improved by stacking procedures [100]. The detection sensitivity and separation performance were substantially improved for the nanoelectro-spray emitter by using a gold-coated nanoelectrospray tip and wall-coated channels and capillary [104]. Low nanomolar detection limits and efficiencies of 2400–10 500 for a separation length of 7 cm were thus made possible.

An expanded application area of chip CE combined with nano-QqTOF-MS systems will be the emerging field of proteomics. This was shown convincingly for the identification of membrane proteins that form a subset of the *Haemophilus influenza* proteome [105]. Here, chip CE-MS was used as the last step after a 1-D gel electrophoresis, followed by excision and digestion of selected protein spots without sample cleanup of extracted tryptic peptides. Li *et al.* [105] addressed the issue of enhancing the sample loading capacity of gel-isolated proteins, which will often be necessary to meet the sensitivity requirements for analysis of excised spots at femtomole levels. On-chip sample stacking with polarity switching for the removal of sample buffer resulted in detection limits in the subnanomolar range. A second approach for preconcentration was the use of a disposable adsorption concentrator, with a single layer of a C18 membrane in it, and connected to the chip by a transfer capillary. Low nM detection limits were achieved. The chip CE-nanoelectrospray combination provided a mass accuracy of less than 5 ppm when using an internal standard. Zhang *et al.* [106] showed efficient CE separations of angiotensin peptides in a device with a liquid junction as shown in Fig. 8. The fabrication and use of a microdevice with an integrated nebulizer and the outlet of the separation channel as the direct electrospray exit was also reported. In a design by Lazar *et al.* [102], the nanoelectrospray emitter was placed in an opening that was drilled perpendicular to the planar surface of the chip, thereby deviating somewhat from the devices described above. A novel device fabricated from polycarbonate using laser micromachining techniques for IEF was reported by Wen *et al.* [107]. The device, with reservoirs for sample, analyte and catholyte, sheath gas and sheath liquid, incorporated a sharply pointed nanoelectrospray emitter tip. The performance and stability was demonstrated with a mixture of model proteins.

Sample cleanup of complex biological samples, such as crude cell lysates, will be important for the generation of high-quality spectra as shown by Xiang *et al.* [108]. They

Electrophoresis 2000, *21*, 3931–3951

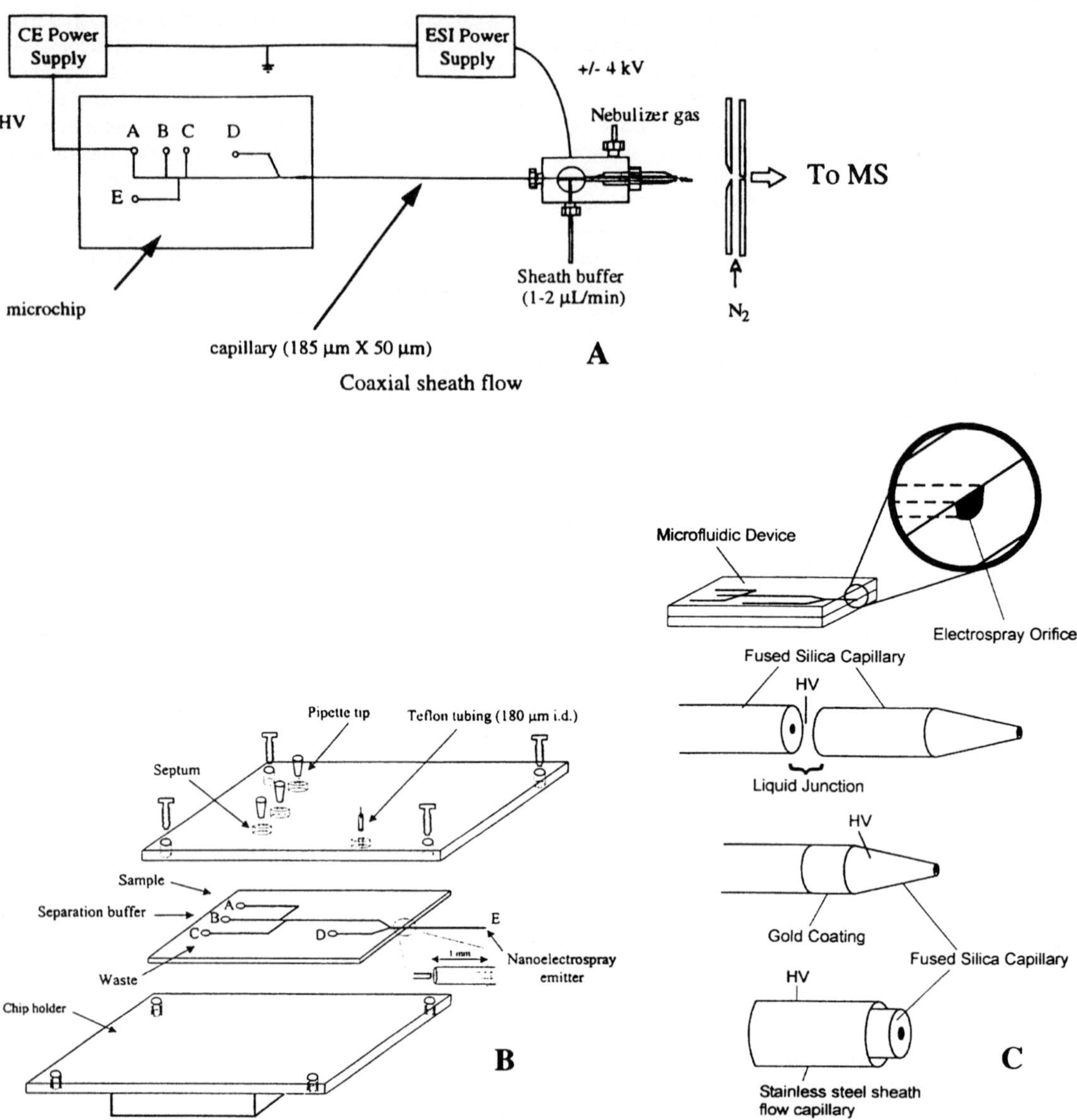

Figure 7. Schematic representation of a chip CE configuration using (A) a sheath flow ESI-MS interface; from [100], with permission; (B) a disposable nanoelectrospray emitter; from [104], with permission. (C) Schemes of different electrospray interfaces for chip ESI-MS. From top to bottom: spraying directly from an exposed channel at the edge of a chip; liquid junction interface; gold-coated capillary interface; coaxial sheath flow configuration; from [109], with permission.

developed a microfabricated dual-microdialysis device for removing both high-molecular-weight and low-molecular-weight compounds. The device consists of two microdialysis membranes (one high M_r cutoff, one low M_r cutoff), sandwiched between three polymer layers with serpentine flow channels for sample introduction, dialysis and a counterflow of buffer. After clean-up, the sample flows directly into a nanoelectrospray tip for online ESI-MS. Selective removal of high M_r compounds, such as BSA, with accompanying improvement of signal-to-noise ratio depending on the M_r cutoff value of the membrane was demonstrated. An extensive review about the coupling of

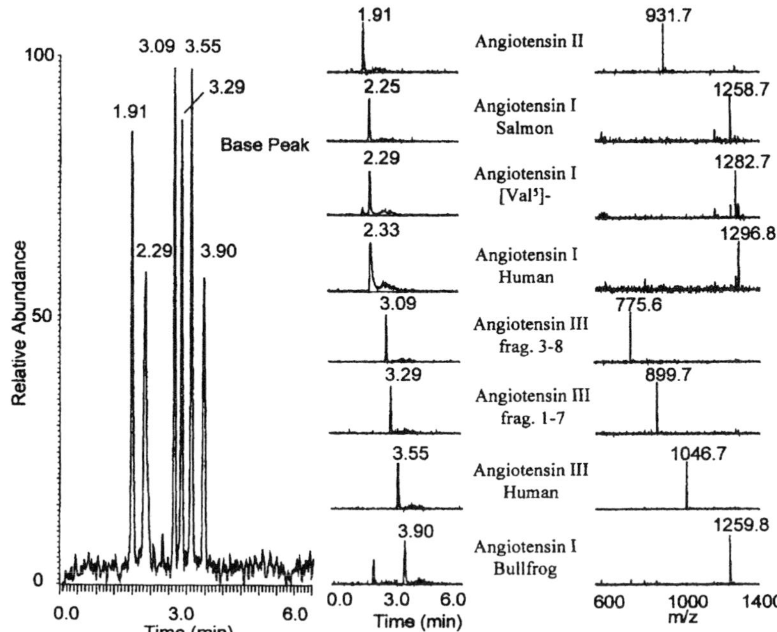

Figure 8. Chip CE-ESI-MS analysis of a mixture of angiotensin peptides. The signals from base peak monitoring (left), selected ion monitoring (middle), and single-scan mass spectra corresponding to peak maximums. From [106], with permission.

microfluidic devices and mass spectrometers, and its potential role in proteomics and drug discovery has been published previously [109].

6 Applications

Table 1 gives an overview of applications published in the period from 1997 until June 2000. The majority of applications is devoted to the analysis of DNA, proteins and peptides. The analysis of DNA, mostly in combination with LIF detection, is in a more mature phase than applications in the field of proteins and peptides. This rapid development in DNA analysis is partly due to the Human Genome Project and the numerous research projects resulting from it. It is generally expected that the microfabricated separation devices will be used in situations with a need for ultrahigh throughput, for two reasons. First, the time per analysis can be strongly reduced without the loss of separation performance. Second, the possibilities for parallel sample processing are nearly unlimited. Examples of DNA analysis on microchips have confirmed this optimism so far. For instance, it was demonstrated that genotyping for a single locus can be done in 30 s and analysis of PCR samples containing four loci can be done in less than 2 min in a highly optimized system without replacement of the separation matrix [119]. This is a 10- to 100-fold increase in separation speed compared to capillary or slab-gel electrophoresis.

The demand for DNA sequencing has substantially increased, and will continue to do so in the near future. A steady performance improvement of microfabricated electrophoresis devices for DNA sequencing can be observed [118–123]. DNA sequencing of 400–500 bases can now be done in less than 20 min using microfabricated electrophoresis devices under optimal conditions [120, 121]. Recently, read lengths of 640 bases in about 30 min with a base-calling accuracy of 98.5% on a 11.5 cm long separation channel could be achieved [122]. Work is in progress to run multiple channels in parallel, which would result in an even more dramatic increase of DNA sequencing rates per microplate. The numerous papers on genotyping by microchip CE, in single separation channels, or in arrays of separation channels, and combined with either on-chip or off-chip DNA amplification, also show the potential of using microchip CE for pharmacogenetics and for the analysis of genes responsible for hereditary diseases [33–35, 113–115].

LIF detection of proteins and peptides was and still is somewhat limited since most of these biopolymers are not natural fluorophors and their derivatization is not always straightforward. A driving force for the analysis of proteins and peptides using microfabricated systems is the demonstrated ability to couple these devices with mass spectrometers. The emergence of the relatively new application field of proteomics coincides with a rapid growth in the use of microfluidic devices for life science

Table 1. Applications of CE on microchips

Application	Detection mode	CE mode	References
DNA			
Sizing	LIF	Sieving in polymer solutions	[21, 22, 33, 35, 39, 57, 60-63, 77, 110–115]
	Electrochemical		[82]
Oligonucleotides	LIF	Sieving	[116, 117]
Genotyping	LIF	Sieving	[33–35, 115]
Sequencing	LIF	Sieving	[36, 118–122]
Integrated PCR/CE	LIF	Sieving	[21–24, 124]
Proteins, peptides			
Model proteins	LIF	Sieving	[125]
Model proteins	UV	CEC	[70]
Model proteins	MS	CZE	[92, 99, 106, 108]
Model proteins	UV/LIF	IEF	[126, 127]
Model proteins	MS	IEF	[107]
Tryptic digests	LIF	CEC	[69]
Tryptic digests	MS	CZE	[93–98, 100, 101, 104–106]
Model peptides	LIF	CZE	[62]
Model peptides	MS	CZE	[100–102, 104–106]
Antihuman immunoglobulin G	LIF	CZE	[128]
Human serum proteins	LIF	CZE	[129]
Protein charge ladders	LIF	CZE	[63]
Enzyme assays			
Protein kinase A	LIF	CZE	[26]
Acetylcholinesterase	LIF	CZE	[25]
β-Galactosidase	LIF	CZE	[130]
Glucose oxidase	LIF	CZE	[85]
Immuno assays			
Thyroxine	LIF	CZE	[131]
Goat IgG	LIF	CZE	[55]
Chiral separations			
Amino acids	LIF	CZE	[132, 133]
Miscellaneous			
Catecholamines	Electrochemical	CZE	[82–86]
Sugars	Refractive index	CZE	[89]
Explosives	Indirect LIF	MEKC	[134]
Amino acids	Indirect LIF	CZE	[135]
Photographic developer solutions	Indirect LIF	CZE	[136]
Herbicides	Raman spectroscopy	ITP	[91]
Alkylphenones	UV	CEC	[70]
Biological cells	Light scattering		[137]
Porphyrins	LIF	CZE	[138]

Table 1. continued

Application	Detection mode	CE mode	References
Latex particles	LIF	Flow cytometry	[78]
Metal cations	LIF	CZE	[139]
Amphetamine and analogs	LIF	MEKC	[140, 141]
Single molecule detection	LIF	CZE, sieving	[62, 76, 77]
Amino acids	LIF	CZE	[63]
Coumarin dyes	LIF	MEKC, CEC	[19, 37, 67]

applications. It appears that chip-based technologies could be well-positioned "to be there at the right time at the right place" with respect to protein and peptide analysis.

Another trend that can be observed is the conversion from the fused-silica capillary platform to the microchip format for several CE applications. This allows good comparison of the separation performance between the two analytical configurations [63, 135]. Microchip CE has the luxury of benefiting from the CE method development over the last 20 years. Comparisons of capillary and microchip electropherograms can be found in Fig. 9. The first example shows the indirect detection of nine amino acids in a fused-silica capillary with an effective separation length of 30 cm and an electric field of 224 V/cm (Fig. 9A), and the same separation on a microchip with a separation length of 5.5 cm and a field of 183 V/cm (Fig. 9B). Comparable resolution was observed with much shorter migration times. Reoptimization of separation conditions from those determined with the fused-silica capillary was unnecessary, thereby illustrating the relative ease of transferring the method from one format to the other [135]. The second example shows the separation of a charge ladder of insulin obtained in an oxidized PDMS channel with an effective separation length of 21 cm and an electric field of 36 V/cm (Fig. 9C), compared with that obtained in a fused-silica capillary of 20 cm separation length and with an electric field of 67 V/cm (Fig. 9D). The first seven peaks represent the seven regioisomers upon modification with acetic anhydride of the two α-amino groups in Phe and Gly and the Lys ε-amino group. Slightly better resolution was obtained in the PDMS device than in the fused-silica capillary[63].

Table 1 also clearly shows that a wide variety of samples other than DNA, peptides and proteins have been analyzed by chip-based CE. Furthermore, most of the CE modes, known from the fused-silica capillary format, such as ITP, CEC, IEF and capillary electrophoresis (CGE), have already been applied for specific applications, as described in more detail in Section 2. For instance, Hofmann *et al.* [126] described first attempts to perform capillary isoelectric focusing (CIEF) on glass chips. With EOF-driven mobilization, which occurs simultaneously with focusing, six Cy5-labeled peptides were nearly baseline-resolved in a simple 7 cm long, 200 µm wide, straight channel in less than 7 min. Mao and Pawliszyn [127] also studied some basic parameters for IEF on glass chips. Low molecular mass p*I* markers and myoglobin were mixed with carrier ampholyte solution and injected *via* a connection capillary by pressure. The focusing process was recorded by real-time absorption imaging detection. Good resolution, sensitivity and reproducibility were only obtained in coated channels without EOF. Also, IEF combined with ESI-MS for the separation of some standard proteins has been reported [107].

7 Outlook

In this review, a broad overview of the possibilities of electrokinetically driven analysis on microfabricated devices is presented. Especially with real integrated chips for sample preparation, separation, reaction, and detection and with chips designed to handle multiple samples (multiplexing), most of the papers have convincingly shown the proof of concept for very different analytical applications, such as enzyme and immunoassays, on-chip PCR-CE, and ultrahigh throughput analysis, during the last three years. The use of novel materials and fabrication methods, in addition to microfabrication based on photolitho-graphy, will probably further expand the prospects to create tailor-made chip devices for specific applications.

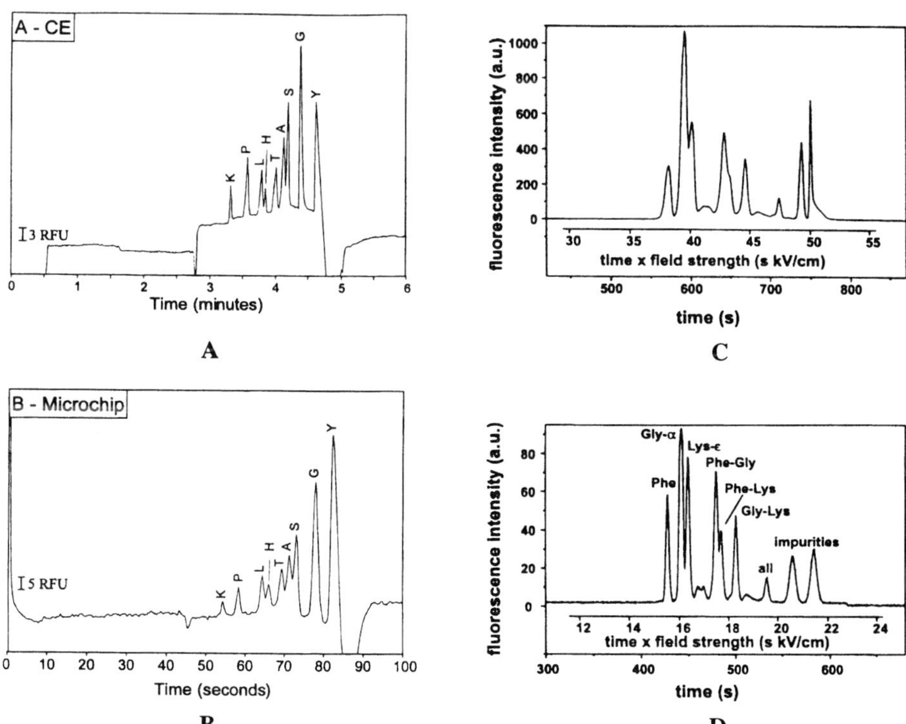

Figure 9. Comparison of electropherograms from microchips and fused-silica capillaries. Indirect detection of nine amino acids in (A) a fused-silica capillary (l_{eff} = 30 cm, E_{sep} = 224 V/cm) and (B) in a glass microchip (l_{eff} = 5.5 cm, E_{sep} = 183 V/cm). BGE: 1.0 mM sodium carbonate and 0.5 mM fluorescein at pH 11.0. From [135], with permission. Electropherograms of a charge ladder of insulin in (C) an oxidized PDMS channel and (D) a fused-silica capillary. The charge ladder was created by modifying insulin with 8 equiv of fluorescent 5-carboxyfluorescein succinimidyl ester and then with acetic anhydride. BGE: 25 mM Tris-192 mM Gly containing the surfactant 3-quinuclidinepropanesulfonate to reduce adsorption of insulin to the PDMS wall. From [63], with permission.

Electrophoresis 2000, *21*, 3931–3951

The relatively fast introduction of the first commercial instrument for DNA and RNA analysis with a glass chip for electrophoretic separations, developed by Agilent together with Caliper Technologies, and the activities of other microchip companies show the optmism among instrument suppliers [112]. Probably, the availability of commercial, more integrated devices (for instance with on-chip sample preparation) is only a matter of time. The problem of handling the numerous samples to be analyzed on one microchip still seems to be somewhat underestimated in the literature and could be one of the bottlenecks for the realization of a completely automated lab-on-the-chip. The introduction of micropumps on chip devices to drive fluids through the microchannels to overcome problems associated with EOF characteristics could be worthwile. This would allow a greater degree of freedom with regard to the chemistry on the microchip.

Although most of the applications shown so far have been in the field of genomics-based analysis, the application field could very well shift in other directions. Particularly the coupling of microfluidic devices for sample preparation and/or separation with new mass spectrometric techniques could be of great significance in the field of proteomics. This field deals with a complete description of proteins encoded by a certain genome plus the understanding of the external influence of diseases and drugs on protein expression. The resolving power of 2-D gel electrophoresis, the technique used routinely in proteomics for protein analysis, is impressive and probably hard to surpass by microfabricated devices. However, for further automated, high-throughput analysis of digests of excised proteins from the gels, analytical microdevices for sample preparation and separation prior to MS combined with ESI-MS or MALDI could become the method of choice in the near future. The application of chip-based CE and other integrated approaches could also play a role in the pharmaceutical industry, in conjunction with the development of combinatorial chemistry. After the first step of lead identification, huge numbers of samples have to be identified and further screened. In this application field the use of multiplexed chips could also have several advantages.

Although the use of microfluidic devices is now in a phase in which the robustness of the technique has to be tested with "real life" samples, there is no doubt that the impact of the results obtained so far will be significant for further developments in analytical (bio)chemistry. A higher degree of integration of chemical manipulation functions onto chips will, in the medium term, lead to increasingly automated devices that will certainly simplify the analyst's job. More fundamentally, the viewpoint of scientists on laboratory methodology and infrastructure in the past is on the verge of a major change. The use of microfabricated devices for chemical and biological applications represents a paradigm shift in how activities in these domains are pursued. Beyond this, these technologies will help open up new and previously unimaginable possibilities for scientists to probe, study, and ultimately understand our chemical and biological world.

Sabeth Verpoorte (University of Neuchatel, Switzerland) is acknowledged for critically reading the manuscript and the helpful comments.

Received June 12, 2000

8 References

[1] Manz, A., Becker, H. (Eds.) *Microsystem Technology in Chemistry and Life Science*, Springer Verlag, Berlin 1998.

[2] Harrison, D. J., Fluri, K., Seiler, K., Fan, Z., Effenhauser, C. S., Manz, A., *Science* 1993, *261*, 895–897.

[3] Sanders, G. H. W., Manz, A., *Trends Anal. Chem.* 2000, *19*, 364–378.

[4] Kricka, L. J., *Clin. Chem.* 1998, *44*, 2008–2014.

[5] Effenhauser, C. S., Bruin, G. J. M., Paulus, A., *Electrophoresis* 1997, *18*, 2203–2213.

[6] Jacobsen, S. C., Ramsey, J. M., in: Landers, J. P. (Ed.), *Handbook of Capillary Electrophoresis*, CRC Press, Boca Raton, FL 1997, pp. 827–839.

[7] Colyer, C. L., Tang, T., Chiem, N., Harrison, D. J., *Electrophoresis* 1997, *18*, 1733–1741.

[8] Campaña, A. M. G., Baeyens, W. R. G., Aboul-Enein, H.Y., Zhang, X., *J. Microcol. Sep.* 1998, *10*, 339–355.

[9] Effenhauser, C. S., *Topics Curr. Chem.* 1998, *194*, 51–82.

[10] Fintschenko, Y., van den Berg, A., *J. Chromatogr. A* 1998, *819*, 3–12.

[11] Regnier, F. E., He, B., Lin, S., Busse, J., *Trends Biotechnol.* 1999, *17*, 101–106.

[12] Wilding, P. Kricka, L. J., *Trends Biotechnol.* 1999, *17*, 465–468.

[13] Dolník, V., Liu, S., Jovanovich, S., *Electrophoresis* 2000, *21*, 41–54.

[14] Carrilho, E., *Electrophoresis* 2000, *21*, 55–65.

[15] Figeys, D., Pinto, D., *Anal. Chem.* 2000, *72*, 330A–335A.

[16] Kutter, J. P., *Trends Anal. Chem.* 2000, *19*, 352–363.

[17] Jacobsen, S. C., McKnight, T. E., Ramsey, J. M., *Anal. Chem.* 1999, *71*, 4455–4459.

[18] Culbertson, C. T., Ramsey, R. S., Ramsey, J. M., *Anal. Chem.* 2000, *72*, 2285–2291.

[19] Kutter, J. P., Jacobsen, S. C., Ramsey, J. M., *Anal. Chem.* 1997, *69*, 5165–5171.

[20] Salimi-Moosavi, H., Tang, T., Harrison, D. J., *J. Am. Chem. Soc.* 1997, *119*, 8716–8717.

[21] Waters, L. C., Jacobsen, S. C., Kroutchinina, N., Khandurina, J., Foote, R. S., Ramsey, J. M., *Anal. Chem.* 1998, *70*, 158–162.

[22] Waters, L. C., Jacobsen, S. C., Kroutchinina, N., Khandurina, J., Foote, R. S., Ramsey, J. M., *Anal. Chem.* 1998, *70*, 5172–5176.

[23] Khandurina, J., McKnight, T. E., Jacobsen, S. C., Waters, L. C., Foote, R. S., Ramsey, J. M., *Anal. Chem.* 2000, *72*, 2995–3000.

[24] Lagally, E. T., Simpson, P. C., Mathies, R. A., *Sens. Actuators B* 2000, *63*, 138–146.

[25] Hadd, A. G., Jacobsen, S. C., Ramsey, J. M., *Anal. Chem.* 1999, *71*, 5206–5212.

[26] Cohen, C. B., Chin-Dixon, E., Jeong, S., Nikiforov, T. T., *Anal. Biochem.* 1999, *273*, 89–97.

[27] Culbertson, C. T., Jacobsen, S. C., Ramsey, J. M., *Anal. Chem.* 1998, *70*, 3781–3789.

[28] Jacobsen, S. C., Culbertson, C. T., Daler, J. E., Ramsey, J. M., *Anal. Chem.* 1998, *70*, 3476–3480.

[29] Alarie, J. P., Jacobsen, S. C., Culbertson, C. T., Ramsey, J. M., *Electrophoresis* 2000, *21*, 100–106.

[30] Jacobsen, S. C., Ramsey, J. M., *Anal. Chem.* 1997, *69*, 3212–3217.

[31] Shultz-Lockyear, L. L., Colyer, C. L., Fan, Z. H., Roy, K. I., Harrison, D. J., *Electrophoresis* 1999, *20*, 529–538.

[32] Jacobsen, S. C., Ermakov, S. V., Ramsey, J. M., *Anal. Chem.* 1999, *71*, 3273–3276.

[33] Woolley, A. T., Sensabaugh, G. F., Mathies, R. A., *Anal. Chem.* 1997, *69*, 2181–2186.

[34] Simpson, P. C., Roach, D., Woolley, A. T., Thorsen, T., Johnston, R., Sensabaugh, G. F., Mathies, R. A., *Proc. Natl. Acad. Sci. USA* 1998, *95*, 2256–2261.

[35] Shi, Y., Simpson, P. C., Scherer, J. R., Wexler, D., Skibola, C., Smith, M. T., Mathies, R. A., *Anal. Chem.* 1999, *71*, 5354–5361.

[36] Simpson, J. W., Ruiz-Martinez, M. C., Mulhern, G. T., Berka, J., Latimer, D. R., Ball, J. A., Rothberg, J. M., Went, G. T., *Electrophoresis* 2000, *21*, 135–149.

[37] Kutter, J. P., Jacobsen, S. C., Ramsey, J. M., *J. Microcol. Sep.* 2000, *12*, 93–97.

[38] Oleschuk, R. D., Shultz-Lockyear, L., Ning, Y., Harrison, D. J., *Anal. Chem.* 2000, *72*, 585–590.

[39] Khandurina, J., Jacobsen, S. C., Waters, L. C., Foote, R. S., Ramsey, J. M., *Anal. Chem.* 1999, *71*, 1815–1819.

[40] van der Moolen, J. N., Poppe, H., Smit, H. C., *Anal. Chem.* 1997, *69*, 4220–4225.

[41] Eijkel, J. C. T., Prak, A., Cowen, S., Craston, D. H., Manz, A., *J. Chromatogr. A* 1998, *815*, 265–271.

[42] Litborn, E., Emmer, Å., Roerade, J., *Electrophoresis* 2000, *21*, 91–99.

[43] Kopp, M. U., de Mello, A. J., Manz, A., *Science* 1998, *280*, 1046–1048.

[44] He, B., Tan, L., Regnier, F., *Anal. Chem.* 1999, *71*, 1464–1468.

[45] Wilding, P., Kricka, L. J., Cheng, J., Hvichia, G., Shoffner, M. A., Fortina, P., *Anal. Biochem.* 1998, *257*, 95–100.

[46] Polson, N. A., Hayes, M. A., *Anal. Chem.* 2000, *72*, 1088–1092.

[47] Schasfoort, R. B. M., Schlautmann, S., Hendrikse, J., van den Berg, A., *Science* 1999, *286*, 942–945.

[48] Patankar, N. A., Hu, H. H., *Anal. Chem.* 1998, *70*, 1870–1881.

[49] Bianchi, F., Ferrigno, R., Girault, H. H., *Anal. Chem.* 2000, *72*, 1987–1993.

[50] Ermakov, S. V., Jacobsen, S. C., Ramsey, J. M., *Anal. Chem.* 1998, *70*, 4494–4504.

[51] Wang, H. Y., Foote, R. S., Jacobsen, S. C., Schneibel, J. H., Ramsey, J. M., *Sens. Actuators B* 1997, *45*, 199–207.

[52] Chiem, N., Lockyear-Shultz, L., Andersson, P., Skinner, C., Harrison, D. J., *Sens. Actuators B* 2000, *63*, 147–152.

[53] Becker, H., Gärtner, C., *Electrophoresis* 2000, *21*, 12–26.

[54] Roberts, M. A., Rossier, J. S., Bercier, P., Girault, H., *Anal. Chem.* 1997, *69*, 2053–2042.

[55] Martynova, L., Locascio, L. E., Gaitan, M., Kramer, G. W., Christensen, R. G., MacCrehan, W. A., *Anal. Chem.* 1997, *69*, 4783–4789.

[56] Xu, J., Locascio, L., Gaitan, M., Lee, C. S., *Anal. Chem.* 2000, *72*, 1930–1933.

[57] Chen, Y.-H., Chen, S.-H., *Electrophoresis* 2000, *21*, 165–170.

[58] Wang, S.-C., Perso, C. E., Morris, M. D., *Anal. Chem.* 2000, *72*, 1704–1706.

[59] Ford, S. M., Kar, B., McWhorter, S., Davies, J., Soper, S. A., Klopf, M., Calderon, G., Saile, V., *J. Microcol. Sep.* 1998, *10*, 413–422.

[60] McCormick, R. M., Nelson, R. J., Alonso-Amigo, M. G., Benvegnu, D. J., Hooper, H. H., *Anal. Chem.* 1997, *69*, 2626–2630.

[61] Paulus, A., Williams, S. J., Sassi, A. P., Kao, P., Tan, H., Hooper, H. H., *SPIE* 1998, *3515*, 94–103.

[62] Effenhauser, C. S., Bruin, G. J. M., Paulus, A., Ehrat, M., *Anal. Chem.* 1997, *69*, 3451–3457.

[63] Duffy, D. C., McDonald, J. C., Schueller, O. J. A., Whitesides, G. M., *Anal. Chem.* 1998, *70*, 4974–4984.

[64] McDonald, J. C., Duffy, D. C., Anderson, J. R., Chiu, D. T., Wu, H., Schueller, O. J. A., Whitesides, G. M., *Electrophoresis* 2000, *21*, 27–40.

[65] Ocvirk, G., Munroe, M., Tang, T., Oleschuk, R., Westra, K., Harrison, D. J., *Electrophoresis* 2000, *21*, 107–115.

[66] Hosokawa, K., Fujii, T., Endo, I., *Anal. Chem.* 1999, *71*, 4781–4785.

[67] Kutter, J. P., Jacobsen, S. C., Matsubara, N., Ramsey, J. M., *Anal. Chem.* 1998, *70*, 3291–3297.

[68] He, B., Tait, N., Regnier, F., *Anal. Chem.* 1998, *70*, 3790–3797.

[69] He, B., Ji, J., Regnier, F., *J. Chromatogr. A* 1999, *853*, 257–262.

[70] Ericson, C., Holm, J., Ericson, T., Hjertén, S., *Anal. Chem.* 2000, *72*, 81–87.

[71] Zhang, X., Regnier, F. E., *J. Chromatogr. A* 2000, *869*, 319–328.

[72] Ceriotti, L., Verpoorte, E., de Rooij, N. F., in: van den Berg, A., Olthuis, W., Bergvelt, P. (Eds.), *Micro Total Analysis Systems 2000*, Kluwer Academic Publishers, Dordrecht, Netherlands 2000, pp. 225–228.

[73] Yu, C., Svec, F., Fréchet, J. M. J., *Electrophoresis* 2000, *21*, 120–127.

[74] Ocvirk, G., Tang, T., Harrison, D. J., *Analyst* 1998, *123*, 1429–1434.

[75] Jiang, G., Attiya, S., Ocvirk, G., Lee, W. E., Harrison, D. J., *Biosens. Bioelectron.* 2000, *14*, 861–869.

[76] Fister III, J. C., Jacobsen, S. C., Davis, L. M., Ramsey, J. M., *Anal. Chem.* 1998, *70*, 431–437.

[77] Haab, B. B., Mathies, R. A., *Anal. Chem.* 1999, *17*, 5137–5145.

[78] Schrum, D. P., Culbertson, C. T., Jacobsen, S. C., Ramsey, J. M., *Anal. Chem.* 1999, *71*, 4173–4177.

[79] Fister III, J. C., Jacobsen, S. C., Ramsey, J. M., *Anal. Chem.* 1999, *71*, 4460–4464.

[80] Crabtree, H. J., Kopp, M. U., Manz, A., *Anal. Chem.* 1999, *71*, 2130–2138.

[81] Salimi-Moosavi, H., Jiang, Y., Lester, L., McKinnon, G., Harrison, D. J., *Electrophoresis* 2000, *21*, 1291–1299.

[82] Woolley, A. T., Lao, K., Glazer, A. N., Mathies, R. A., *Anal. Chem.* 1998, *70*, 684–688.

[83] Wang, J., Tian, B., Sahlin, E., *Anal. Chem.* 1999, *71*, 3901–3904.

[84] Wang, J., Tian, B., Sahlin, E., *Anal. Chem.* 1999, *71*, 5436–5440.

[85] Wang, J., Chatrathi, M. P., Tian, B., Polsky, R., *Anal. Chem.* 2000, *72*, 2514–2518.

[86] Gavin, P. F., Ewing, A. G., *Anal. Chem.* 1997, *69*, 3838–3845.

[87] Gavin, P. F., Ewing, A. G., *J. Microcol. Sep.* 1998, *10*, 357–364.

[88] Ewing, A. G., Gavin, P. F., Hietpas, P. B., Bullard, K. M., *Nature Med.* 1997, *3*, 97–99.

[89] Burggraf, N., Krattiger, B., de Mello, A. J., de Rooij, N. F., Manz, A., *Analyst* 1998, *123*, 1443–1447.

[90] Swinney, K., Markov, D., Bornhop, D. J., *Anal. Chem.* 2000, *72*, 2690–2695.

[91] Walker III, P. A., Morris, M. D., Burns, M. A., Johnson, B. N., *Anal. Chem.* 1998, *70*, 3766–3769.

[92] Xue, Q., Foret, F., Dunayevsky, Y. M., Zavracky, P. M., McGruer, N. E., Karger, B. L., *Anal. Chem.* 1997, *69*, 426–430.

[93] Xue, Q., Dunayevsky, Y. M., Foret, F., Karger, B. L., *Rapid Commun. Mass Spectrom.* 1997, *11*, 1253–1256.

[94] Figeys, D. Ning, Y., Aebersold, R., *Anal. Chem.* 1997, *69*, 3153–3160.

[95] Figeys, D., Aebersold, R., *Anal. Chem.* 1998, *70*, 3721–3727.

[96] Figeys, D., Gygi, S. P., McKinnon, G., Aebersold, R., *Anal. Chem.* 1998, *70*, 3728–3734.

[97] Pinto, D. M., Ning, Y., Figeys, D., *Electrophoresis* 2000, *21*, 181–190.

[98] Chan, J. H., Timperman, A. T., Qin, D., Aebersold, R., *Anal. Chem.* 1999, *71*, 4437–4444.

[99] Ramsey, R. S., Ramsey, J. M., *Anal. Chem.* 1997, *69*, 1174–1178.

[100] Li, J., Thibault, P., Bings, N. H., Skinner, C. D., Wang, C., Colyer, C. L., Harrison, J., *Anal. Chem.* 1999, *71*, 3036–3045.

[101] Zhang, B., Liu, H., Karger, B. L., Foret, F., *Anal. Chem.* 1999, *71*, 3258–3264.

[102] Lazar, I., Ramsey, R. S., Sundberg, S., Ramsey, J. M., *Anal. Chem.* 1999, *71*, 3627–3631.

[103] Bings, N. H., Wang, C., Skinner, C. D., Colyer, C. L., Thibault, P., Harrison, D. J., *Anal. Chem.* 1999, *71*, 3292–3296.

[104] Li, J., Kelly, J. F., Chernushevich, I., Harrison, D. J., Thibault, P., *Anal. Chem.* 2000, *72*, 599–609.

[105] Li, J., Wang, C., Kelly, J. F., Harrison, D. J., Thibault, P., *Electrophoresis* 2000, *21*, 198–210.

[106] Zhang, B., Foret, F., Karger, B. L., *Anal. Chem.* 2000, *72*, 1015–1022.

[107] Wen, J., Lin, Y., Xiang, F., Matson, D. W., Udseth, H. R., Smith, R. D., *Electrophoresis* 2000, *21*, 191–197.

[108] Xiang, F., Lin, Y., Wen, J., Matson, D. W., Smith, R. D., *Anal. Chem.* 1999, *71*, 1485–1490.

[109] Oleschuk, R. D., Harrison, J. D, *Trends Anal. Chem.* 2000, *19*, 379–388.

[110] Mueller, O., Hahnenberger, K., Dittmann, M., Yee, H., Dubrow, R., Nagle, R., Ilsley, D., *Electrophoresis* 2000, *21*, 128–134.

[111] Chou, H.-P., Spence, C., Scherer, A., Quake, S., *Proc. Natl. Acad. Sci. USA* 1999, *96*, 11–13.

[112] Various home pages of instrument manufacturers: www.agilent.com; www.calipertech.com; www.aclara.com

[113] Munro, N. J., Snow, K., Kant, J. A., Landers, J. P., *Clin. Chem.* 1999, *45*, 1906–1917.

[114] Hofgärtner, W. T., Hühmer, A. F. R., Landers, J. P., Kant, J. A., *Clin. Chem.* 1999, *45*, 2120–2128.

[115] Cheng, J., Waters, L. C., Fortina, P., Hvichia, G., Jacobsen, S. C., Ramsey, J. M., Kricka, L. J., Wilding, P., *Anal. Biochem.* 1998, *257*, 101–106.

[116] Ueda, M., Kiba, Y., Abe, H., Arai, A., Nakanishi, H., Baba, Y., *Electrophoresis* 2000, *21*, 176–180.

[117] Soper, S. A., Ford, S. M., Xu, Y., Qi, S., McWhorter, S., Lassiter, S., Patterson, D., Bruch, R. C., *J. Chromatogr. A* 1999, *853*, 107–120.

[118] Backhouse, C., Caamano, M., Oaks, F., Nordman, E., Carillo, A., Johnson, B., Bay, S., *Electrophoresis* 2000, *21*, 150–156.

[119] Schmalzing, D., Koutny, L., Adourian, A., Belgrader, P., Matsudaira, P., Ehrlich, D., *Proc. Natl. Acad. Sci. USA* 1997, *94*, 10273–10278.

[120] Schmalzing, D., Adourian, A., Koutny, L., Ziaugra, L., Matsudaira, P., Ehrlich, D., *Anal. Chem.* 1998, *70*, 2303–2310.

[121] Schmalzing, D., Tsao, N., Koutny, L., Chisholm, D., Srivastava, A., Adourian, A., Linton, L., McEwan, P., Matsudaira, P., Ehrlich D., *Genome Res.* 1999, *9*, 853–858.

[122] Salas-Solano, O., Schmalzing, D., Koutny, L., Buonocore, S., Adourian, A., Matsudaira, P., Ehrlich, D., *Anal. Chem.* 2000, *72*, 3129–3137.

[123] Liu, S., Shi, Y., Ja, W. W., Mathies, R. A., *Anal. Chem.* 1999, *71*, 566–573.

[124] Burns, M. A., Johnson, B. N., Brahmasandra, S. N., Handique, K., Webster, J. R., Krishnan, M., Sammarco, T. S., Man, P. M., Jones, D., Heldsinger, D., Mastrangelo, C. H., Burke, D. T., *Science* 1998, *282*, 484–487.

[125] Yao, S., Anex, D. S., Caldwell, W. B., Arnold, D. W., Smith, K. B., Schultz, P. G., *Proc. Natl. Acad. Sci. USA* 1999, *96*, 5372–5377.

[126] Hofmann, O., Che, D., Cruickshank, K. A., Müller, U. R., *Anal. Chem.* 1999, *71*, 678–686.

[127] Mao, Q., Pawliszyn, J., *Analyst* 1999, *124*, 637–641.

[128] Rodriguez, I, Zhang, Y., Lee, H. K., Li, S. F. Y., *J. Chromatogr. A* 1997, *781*, 287–293.

[129] Colyer, C. L., Mangru, S. D., Harrison, D. J., *J. Chromatogr. A* 1997, *781*, 271–276.

[130] Hadd, A. G., Raymond, D. E., Halliwell, J. W., Jacobsen, S. C., Ramsey, J. M., *Anal. Chem.* 1997, *69*, 3407–3412.

[131] Schmalzing, D., Koutny, L. B., Taylor, T. A., Nashabeh, W., Fuchs, M., *J. Chromatogr. B* 1997, *697*, 175–180.

[132] Hutt, L. D., Glavin, D. P., Bada, J. L., Mathies, R. A., *Anal. Chem.* 1999, *71*, 4000–4006.

[133] Rodriguez, I., Jin, L. J., Li, S. F. Y., *Electrophoresis* 2000, *21*, 211–219.

[134] Wallenborg, S. R., Bailey, C. G., *Anal. Chem.* 2000, *72*, 1872–1878.

[135] Munro, N. J., Huang, Z., Finegold, D. N., Landers, J. P., *Anal. Chem.* 2000, *72*, 2765–2773.

[136] Sirichai, S., de Mello, A. J., *Analyst* 2000, *125*, 133–137.

[137] Li, P. C. H., Harrison, D. J., *Anal. Chem.* 1997, *69*, 1564–1568.

[138] Zhang, Y., Lee, H. K., Li, S. F. Y., *Talanta* 1998, *45*, 613–618.

[139] Kutter, J. P., Ramsey, R. S., Jacobsen, S. C., Ramsey, J. M., *J. Microcol. Sep.* 1998, *10*, 313–319.

[140] Ramseier, A., von Heeren, F., Thormann, W., *Electrophoresis* 1998, *19*, 2967–2975.

[141] Wallenborg, S. R., Lurie, I. S., Arnold, D. W., Bailey, C. G., *Electrophoresis* 2000, *21*, 3257–3263.

Review

Vladislav Dolník
Shaorong Liu
Stevan Jovanovich

Molecular Dynamics,
Sunnyvale, CA, USA

Capillary electrophoresis on microchip

Capillary electrophoresis and related techniques on microchips have made great strides in recent years. This review concentrates on progress in capillary zone electrophoresis, but also covers other capillary techniques such as isoelectric focusing, isotachophoresis, free flow electrophoresis, and micellar electrokinetic chromatography. The material and technologies used to prepare microchips, microchip designs, channel geometries, sample manipulation and derivatization, detection, and applications of capillary electrophoresis to microchips are discussed. The progress in separation of nucleic acids and proteins is particularly emphasized.

Keywords: Capillary electrophoresis / Microchip / Microdevices / Microfabrication / Micromachining / Review

EL 3712

Contents

1 Introduction

Capillary electrophoresis on microchips is an emerging new technology that promises to lead the next revolution in chemical analysis. It has the potential to simultaneously assay hundreds of samples in a matter of minutes or less. The rapid analysis combined with massively parallel analysis arrays should yield ultrahigh throughput. Microchips typically consume only picoliters of samples. These samples may potentially be prepared on-board for a complete integration of sample preparation and analysis functions. These features make microchips an attractive technology for the next generation of capillary electrophoresis instrumentation.

The first instrument on a microchip was an integrated gas chromatograph [1]. While this device never commercially succeeded, it nonetheless initiated the application of micromachining technology to build chemical analysis devices. By the early 90s, chemical analysis on microchips had been demonstrated for many capillary electrophoresis applications [2–10]. Microfabricated devices have so far separated fluorescently labeled amino acids [11], DNA restriction fragments [4, 6, 12], PCR products, short oligonucleotides [13], and sequencing ladders (reviewed in [15–26]).

Capillary electrophoresis on microchips is based upon microfabrication techniques developed in the semiconductor industry. Microchannels are fabricated in microchips using photolithography or micromolding to form channels for sample injection and capillary electrophore-

Correspondence: Dr. Vladislav Dolník, Molecular Dynamics, 928 East Arques Avenue, Sunnyvale, CA, 94086, USA
E-mail: vdolnik@mdyn.com
Fax: +408-737-4808

Abbreviation: CAE, capillary array electrophoresis

0173-0835/00/0101-0041 $17.50+.50/0

    Electrophoresis 2000, *21*, 41–54

sis separation. After all solutions, including samples, are loaded, samples are typically transferred electrokinetically into an injector region. The samples are then separated by applying high voltage while a potential is applied to the sample and waste reservoirs to prevent the sample from bleeding [13]. The analytes are detected by laser-induced fluorescence (LIF) or other methods. The small injection plugs, high fields, and short separation lengths produce separation times measured in seconds or minutes. Because fairly standard microfabrication technology is used, mass production of microchip devices for capillary electrophoresis should be economically feasible. The following sections give an overview of fabrication methods, microchip designs, and applications of capillary electrophoresis to microchips.

2 Materials and fabrication technologies

CE chips are mainly fabricated using various glass substrates [6, 27–30], from inexpensive soda lime glass to high quality quartz. Glass substrates are the most common substrates because of their good optical properties, well-understood surface characteristics, and well-developed microfabrication methods, adapted from the microelectronics industry. Recently, various polymer materials have been used to fabricate microchips for CE separations [31–34]. Polymer microchips are of increasing interest because their potentially low manufacturing costs may allow them to be disposable. Fabrication procedures for these materials are quite different from those for glass. The following sections describe typcial fabrication procedures for both types of materials.

2.1 Fabrication procedures for glass materials

Structures on glass substrates are usually generated using standard photolithographic technologies [6, 27–29]. Figure 1 presents a diagram of such a procedure. First a sacrificial etching mask layer is attached to the surface to be structured. The most often used sacrificial mask is Cr/Au [5, 27, 29, 35]. The thin (usually 100–500 Å) film of Cr is used to enhance the adhesion between the substrate and the gold layer, which is the real sacrificial mask. The Cr/Au film is effective for almost all types of glass etchants, particularly hydrogen fluoride (HF)/HNO$_3$ which attacks other etch masks, such as amorphous silicon. An alternative mask is amorphous silicon [36, 37]. Amorphous silicon normally adheres to glass better than Cr/Au and works well for concentrated HF as an etchant. High quality and very deep channels (> 70 µm) can be etched with few defects [38]. A layer of photoresist can also be used as a mask to etch shallow channels [6, 12, 39, 40]. In this case, the photoresist serves two functions,

a sacrificing mask and a regular photoresist, to transfer photomask patterns.

The pattern to be etched is then transferred to the wafer. First, a layer of photoresist is spin-coated on top of the mask layer. Photoresist is a polymer that becomes soluble (for positive photoresist) or insoluble (for negative photoresist) in developer solutions after exposure to light. In the next step, the photoresist spun on the microchip is exposed in the region defined by a photomask, typically using an aligner. The photomask is a plate with a user-designed pattern that is transparent while the background is opaque (or *vice versa*) to the exposition light. After the microchip is baked to harden unexposed resist, the exposed photoresist is dissolved with a developer solution. The sacrificial mask layer of the exposed region is removed using the appropriate etchants. During this time, the sacrificial layer underneath the unexposed photoresist remains intact.

After pattern transfer and development, the portions of the microchip that are to be etched have been unmasked and are now ready for chemical etching. HF is used as the primary etchant and can be prepared in various solutions including HF/NH$_4$F, HF/HNO$_3$, and concentrated HF. The etching rate of HF for glass is readily controllable if the temperature is controlled. The etching progress can be monitored with a profilometer. Following etching of the microchannels, the photoresist and sacrificing mask layer are stripped, and the access holes drilled (not shown in Fig. 1). The access holes can be drilled on the etched substrate or on another blank glass wafer. When holes are drilled on the etched substrate, aligning two substrates for bonding is much easier. Finally, the substrate is bonded to another piece of substrate to form a finished microchip. Thermal diffusion is the most often used method for glass bonding [3, 41]. Other methods such as chemical activated bonding [42] and adhesive annealing [43, 44] are also used.

2.2 Fabrication procedures for polymer materials

Methods for the fabrication of plastic microchips include laser ablation [31], injection molding [32, 45], silicone rubber casting [33], and hot embossing [34].

2.2.1 Laser ablation

The photoablation process involves absorption of a short-wavelength laser pulse to break covalent bonds in long-chain polymer molecules with production of a shock wave that ejects decomposed polymer fragments [46]. Many commercially available polymers can be photoablated, including polycarbonate, poly(methyl methacrylate) (PMMA), polystyrene, nitrocellulose, poly(ethylene terephthalate) (PET or Melinex), and poly(tetrafluoroethylene) (Teflon) [31, 47]. The laser energy can be specially patterned using a mask with the subsequent generation of microcavities and channels in various geometries or by controlling position of the laser with *x-y* stages. The resulting structures are generally characterized as having little thermal damage, straight vertical walls, and well-defined depth [46, 48]. However, laser ablation does not lend itself well to mass production.

2.2.2 Injection molding, embossing and casting

The formation of microchannels and other structures using molding methods generally involves two primary steps: (i) fabrication of a molder (also known as a master), and (ii) channel pattern transfer from the molder to polymer substrates. Various techniques can be used to produce a molder depending upon the channel dimensions and precision requirements. For large structures (> 100 μm), traditional computer numerical control (CNC) machining of materials like stainless steel can be sufficiently accurate. For smaller features (< 100 μm), a silicon wafer or a thick photoresist structure is etched and then electroplated with a metallic material such as nickel or nickel-cobalt. For very small channels with high aspect ratios, lithography, electroplating and molding (LIGA), followed by electroplating [49] is the method of choice to produce the molder; LIGA is a process that uses a synchrotron to produce X-rays for photolithography of an X-ray resist, followed by electroplating to form a molder.

Replication of the molder to produce microchips can be accomplished by injection molding, embossing, or casting. In the injection molding process, polymer is melted and injected against the molder in the molding chamber. Molded devices are released every 5–10 s. Injection molding allows very high-throughput production with low production costs [32]. In the embossing process [50], the embossing tool and the polymer substrate are heated separately under vacuum (this extends the lifetime of the molder) to a temperature just above the glass transition temperature of the polymer material. The tool is then brought into contact with the substrate and embossed. Embossing can take several minutes per device and can be a useful tool in rapid prototyping devices. In the casting process, polymer material such as poly(dimethylsiloxane) (PDMA) is usually poured on the top of the molder and then cured/hardened [33] at atmospheric pressure and temperature. Elevating the temperature can accelerate the curing. Casting is the simplest among these three molding processes, but requires contact with the molder for minutes or hours.

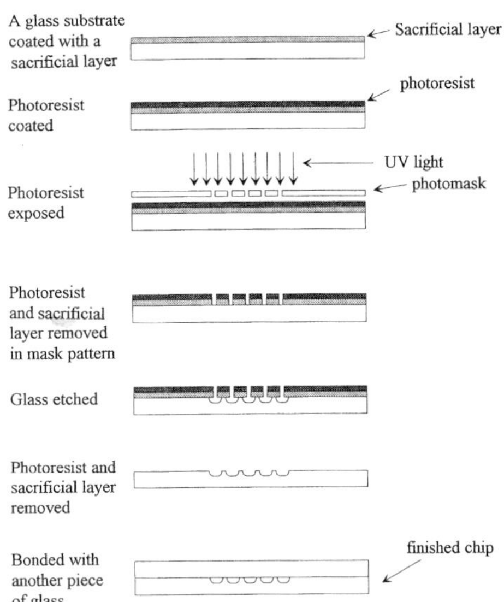

Figure 1. Photolithographic process for making chips.

2.2.3 Groove enclosure

Thermal lamination is normally used to seal grooves to form channels [31, 32]. Channels formed using this method can have two different surface types, three walls of the polymer substrate and one wall of the laminated film. If the materials are not well matched or appropriately post-modified, plug flow is disrupted due to the different characteristics of the two surfaces and the separation is degraded. Another bonding technique is to anneal the molded plate directly to another plate at room temperature [33, 51]. Very strong bonding (possibly some covalent bonding) was reported after two PDMA surfaces were treated with oxygen plasma [51]. Clean PDMA surfaces also strongly bonded to other surfaces such as glass and plastics [33]. When a PDMA-molded plate is bonded with a thin slab of PDMA, four equivalent walls are formed [33]. Channels with four equivalent walls can also be made with other polymeric materials [50]; however, details of the bonding procedure have not been described.

3 Microchip format

The design of microchips for capillary electrophoresis (CE) has undergone significant development from simple single-channel structures to increasingly complex ones. Design rules for many channel geometries and separation lengths are being developed and computer-aided design tools are becoming commercially available. Current designs of CE chips allow reactions on-chip and separations in multiple channels.

3.1 Microchip layout

During the past several years, significant progress has been made in the design of CE chips (Fig. 2). The first CE chips used single microchannels fabricated on relatively large-scale substrates (14.8 × 3.9 cm) [3] (Fig. 2a). Several years later, miniaturized chips appeared containing two crossed channels and four reservoirs for sample, waste, and cathode and anode electrolytes [4]. Separation channel lengths can be increased by introducing serpentine turns into the separation channels (Fig. 2b) [5]. To further extend the separation path, a synchronized cyclic CE chip has been designed: the channel forms a

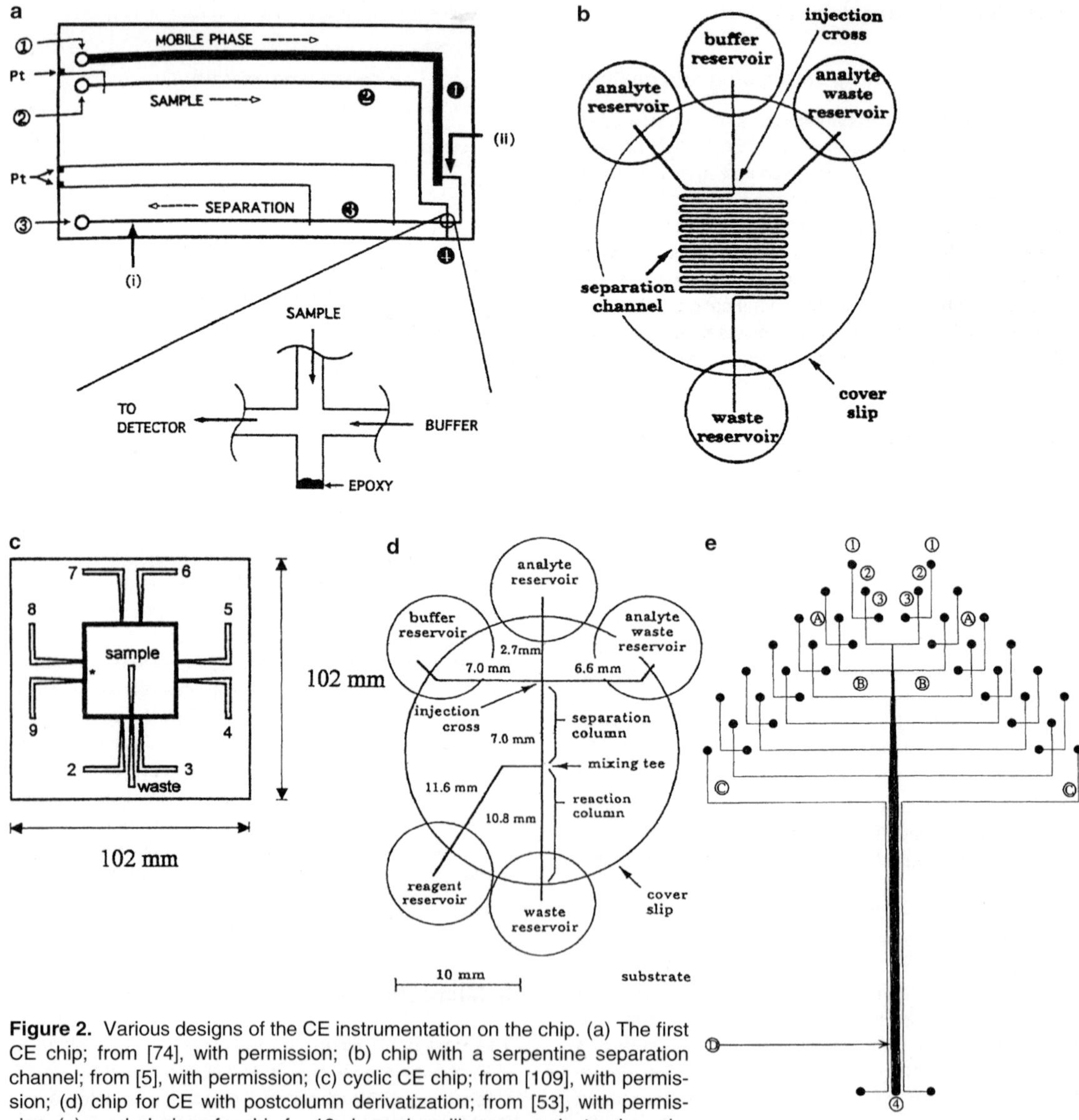

Figure 2. Various designs of the CE instrumentation on the chip. (a) The first CE chip; from [74], with permission; (b) chip with a serpentine separation channel; from [5], with permission; (c) cyclic CE chip; from [109], with permission; (d) chip for CE with postcolumn derivatization; from [53], with permission; (e) mask design of a chip for 12-channel capillary array electrophoresis; from [12], with permission.

loop with two pairs of electrodes connected. Switching voltage between pairs of electrodes enables separands to migrate in cycles around the loop and thus to increase the migration path (Fig. 2c) [52]. Microreactors can also be added to perform on-line reactions such as sample preparation (see Section 6.1) or postcolumn derivatization (see Section 4.2) (Fig. 2d) [53]. Some other chip layouts constructed include ones with a rectangular channel where the sample is loaded by a sampling capillary [54] and free-flow electrophoresis microchips [55]. These are just few examples; naturally more single-channel variants are in use.

To simultaneously analyze large number of samples, arrays of separation channels have been designed on microchips [12, 26, 36]. Multichannel microchips present unique design constraints. The geometry of multiple separation channels can be constrained by substrate size, requirements for straight channels, and by the detection method. Sample and waste wells can consume significant real estate and usually need to be laid out with geometries that match the loading device. Finally, all channels may be constrained to have similar geometries and equal distances from the cathode to the anode and from the injection region to the detector. One solution to many of these constraints is a multichannel microchip demonstrated for genotyping (Fig. 2e) [36]. This microchip uses 48 parallel separation channels to separate 96 samples in two sets of 48 samples in less than 8 min. A microchip with 96 radial separation channels has also been designed, as described in [15].

3.2 Separation channel length

The change of the CE column format from long capillaries to shorter microfabricated channels on microchips brings new opportunities, limitations, and challenges. Miniaturization, smaller injection plugs, and shorter separation paths enable rapid separations without significant peak broadening due to diffusion. If constant selectivity can be maintained, shorter separation paths can be used since resolution is proportional to the square root of the migration path when other contributors to peak broadening other than diffusion are ignored [56–59]. However, the shorter separation channels on microchips makes the length of the sample plug more critical. The contribution of the sample plug length to peak broadening can be calculated. When the conductivity of the sample zone equals that of the background electrolyte, the channel length L_{req} required for the full separation of separands A and B equals

$$L_{req} = L_{sample} \frac{\mu_B}{\mu_A - \mu_B}$$

where L_{sample} is the length of the sample plug and μ_A and μ_B are mobilities of the separands A and B [60]. To achieve separation in a separation channel of length L_{req}, the sample plug must not be longer than L_{sample}. The length of the sampling plug can be manipulated by electric field amplification [61–66], preconcentration by isotachophoresis [67–69], and, in micellar electrokinetic chromatography, by sweeping [70, 71]. In CE in sieving media, the interface between free electrolyte and polymer solution can also reduce the length requirements for the separation channel [72].

If the approaches described above are insufficient, a longer channel can simply be used alternatively. Microchips have been prepared with long straight channels of 20 cm and more [73]. Channel length can also be increased by using a serpentine geometry. An initial study showed that two rectangular corners in a 8.5 mm long channel did not increase zone dispersion significantly and it was predicted that a 50 cm long serpentine capillary channel could be fabricated in less than a 1 cm^2 area [74]. A real chip with a channel length of 165 mm has been produced in an area of less than 1 cm^2 [5]. Research examining zone dispersion, however, has shown that the impact of turns is more complex than previously thought [75, 76]. A zone of analyte migrating around the turn in a serpentine capillary is dispersed due to differences in the migration path at the inner and the outer perimeter of the turn. The amount of dispersion due to channel turns depends on the magnitude of the diffusion coefficient of the analyte. A one-dimensional model based on the ratio of the transverse analyte diffusion time to the analyte transit time around a turn allows prediction of the amount of the dispersion caused by turns [75].

3.3 Channel geometry

Microdevices for CE typically have channel depths of 15–40 μm and widths of 60–200 μm [4, 77, 78], although channels with depths of less than 10 μm have been reported [79, 80]. The small cross section of the separation channels and large thermal mass of the microchip allows Joule heat to be dissipated efficiently. Thus, high electric fields (over 2 kV/cm) can be applied on microchips [74]. An electric field as high as 53 kV/cm has been used in a specially designed channel to achieve sub-millisecond separations [80]. In free-air convection, small radial temperature gradients of 2–4°C from the center of the channel to the wall are formed. With forced cooling, the center-to-wall temperature difference is less than 1°C [37].

4 Sample manipulation and derivatization

Sampling, sample concentration by stacking, precolumn or postcolumn derivatization can be performed on-line in CE microchips. Samples are typically loaded from microtiter plates into sample wells using standard pipettors.

4.1 Sample injection and zone manipulation

Integrated sample injection is typically used to produce the small sample size (measured in picoliters) required for CE on microchips. The integrated injectors are usually either cross-channel injectors, formed by orthogonally intersecting the separation channel with a channel connecting the sample to waste, or twin-T injectors, where the two arms of the sample to waste channel are offset to form a larger injector region. Integrated sample injectors permit volume-defined electrokinetic sample injection [3, 5, 81, 82] of short injection plugs with reproducibility of peak heights better than 4.1% [83]. Band broadening of the plugs can occur from leakage at the injection channel intersection [27, 84]. Electrical biasing of the different reservoirs, such as in a pinched injection, minimizes the bleeding. When a modular injector on a microchip is connected to a fused separation capillary, the injection repeatability is also improved [85]. To enhance the detection limit, a stack injection can be applied to concentrate samples, as demonstrated for dansylated amino acids [86]. A major sample component can make CE analysis of trace sample components difficult or impossible. This type of interference can be significantly reduced or eliminated through zone manipulation [87]. Manipulation of zones during separation can be used to redirect them for fraction collection or other assays. Collection of zones by stopped-flow methods results in significant zone dilution. On microchips, selected zones can be redirected from the separation channel to a side channel electrically. The dilution of the isolated sample components hardly exceeded a factor of four [88].

4.2 On-line sample derivatization

To increase detection sensitivity of analytes without a strong inherent analytical signal, they are typically labeled with a fluorophore and detected by LIF. The labeling can be precolumn, *i.e.*, made prior to the CE separation, or postcolumn, when already separated analytes are labeled and detected. For precolumn sample derivatization, a 1 nL microreactor has been constructed [28]. Electrical control of background electrolyte, sample, and reagent stream allows a precise manipulation of the fluids within the channel manifold. Halftimes of the reaction between amino acids, arginine and lysine, and *o*-phthaldialdehyde are about 5 s with detection limits of 0.5–0.83 fmol [28].

For postcolumn sample derivatization, a microreactor was integrated after CE separation channels on a microchip [35, 53, 89]. The reactor geometry causes about 10% contribution to peak broadening. An on-chip postcolumn reaction of *o*-phthaldialdehyde and amino acids gives a separation efficiency of about 83 000 theoretical plates [35]. For chemiluminescence detection, a method of postcolumn labeling has been developed using horseradish peroxidase-catalyzed reaction of luminol with peroxide [82].

5 Detection

5.1 Optical detection

Sensitive detection schemes are essential in microfabricated devices for CE due to the extremely small size of the detection cell. LIF is so far the most popular detection scheme for CE chips because of its sensitivity [11, 28, 52, 53, 90–95]. A common LIF detection system that can be applied to microchips uses a confocal detection system, based on that described by Mathies and Huang [96]. A laser provides a coherent, collimated beam that is reflected by a dichroic beamsplitter into a high numerical aperture objective. The objective focuses the laser to a small beam waist inside the microchannel to excite fluorescently labeled analyte. The fluorescence emitted from fluorescently labeled analyte is collected and collimated by the objective, and passes back through the dichroic beamsplitter. In this direction, the beamsplitter reflects laser light and passes through the longer wavelength fluorescence light. An achromatic lens focuses the light onto the entrance of a spatial filter (confocal aperture) which is "confocal" with the microscope objective, *i.e.*, only light emitted from the focal region in the microchip passes through the pinhole; light scattered from the surface of the microchip and fluorescent light originating outside the channels are rejected. This increases the S/N ratio and can produce a sensitive detection system. The light exiting the spatial filter is directed to one or more detectors, such as photomultiplier tubes (PMTs) or a charge-coupled device (CCD) array. Filter sets and additional beamsplitters can be inserted to spectrally separate different emission channels. The output from the PMTs or CCD can be preamplified, digitized, and acquired by a computer for processing.

LIF detection of an array of microchannels in microchips presents additional requirements compared to capillaries or a single channel. The laser can be scanned over the microchannels, which reduces the duty cycle, or continuous illumination can be used, which reduces laser power density at each microchannel. Either method can reduce the S/N ratio. The quick transit of the samples past the detector in microchips also requires higher sampling rates than capillary array electrophoresis (CAE). Development of a robust LIF detector for multichannel microchips remains a challenge, although it has been demonstrated for genotyping [36].

Absorption detectors can be used with electrophoresis chip applications, but detection sensitivity is a concern because of the limited optical path length of the microfabricated channel. With the development of stable low-current detection electronics, detection limits in the low micromolar range are readily achieved in conventional CE, where the optical path length ranges from 50 to 100 μm [97]. To further improve the detection sensitivity on microchips, a 140 μm U-shaped detection cell was fabricated that could detect 6 μM of fluorescein [92]. To incorporate this cell into a UV/Vis detector, two additional channels were fabricated on both side of the U-cell to host optical fibers to guide the source light to the cell and the transmitted light to a photodetector [92].

5.2 Mass spectrometry

Mass spectroscopy (MS) is a powerful tool in analytical chemistry. It performs the separation, detection, and identification of a broad range of compounds, including polypeptides and nucleic acids. MS has been successfully coupled to CE with various electrospray microdevices. CE on microchips and electrospray ionization (ESI)/MS have recently been connected [98–104]. These same devices are used to load samples into MS using electroosmotic flow without any CE separation. Multiple samples can be placed on the microdevice and analyzed in an automatic mode. Recently a pneumatic nebulizer has been introduced to generate a stable sample flow for electrospray, eliminating the need for an electrospray tip and allowing microfabrication of the device in a single etching step [105]. An integrated microfluidic system for MS analysis of proteins has been constructed where the sample flow and data analysis are completely under computer control [98–100]. However, voltage is used to pump the liquid sample by electroosmotic flow to the electrospray and MS without any electrophoretic separation.

5.3 Electrochemical detection

While LIF provides sensitivity even to the single molecule limit, the detection system is much larger than the microfabricated analysis devices. This reduces the benefits of miniaturization [106]. Electrochemical methods provide an alternative detection approach to address this issue. When electrodes are microfabricated, they generally result in higher sensitivity and quicker response times [8, 76, 107]. Gavin and Ewing [8, 107] made a microfabri-

cated electrochemical array detector for continuous electrophoretic separations in narrow channels. An array of 100 platinum microelectrodes at 5 μm spacing (95 μm wide, 1.2–2 mm long, 0.2 μm high) was placed on the microchip at the exit of the narrow channels. This setup allows electrochemical detection of neurotransmitter separations [107]. An integrated electrochemical detector for DNA analysis on a microchip has been constructed [106]. An integrated electrochemical detector for DNA analysis on a microchip has been constructed [106] with a 10 μm wide working electrode fabricated 30 μm from the end of the separation channel. To minimize the interference of the high separation voltage, an off-column detection scheme is employed after the separation channel. A ΦX174 *Hae*III restriction digest is detected in an indirect detection mode with a detection limit of 28 zmol [106].

5.4 Other detection methods

Some other detection methods have been applied to CE chips including Raman spectroscopy and holographic refractive index detection. Raman spectroscopy has been demonstrated as a suitable detector for microchip [108] for the isotachophoretic separation of pesticides (paraquat and diquat). Raman spectra were generated with a 2 W, 532 nm NdYVO$_4$ laser and collected at 8 cm^{-1} resolution with the data acquisition rate of 2–5 spectra/s. The 980 cm^{-1} Raman band for the counter ion, sulfate, was used as an internal standard to correct for instrument variations. Working sample concentrations ranged from 10^{-5} to 10^{-7} M [108]. A holographic refractive index detector has been constructed for CE on microchips [109]. It was tested by analysis of model saccharides at a concentration of 33 mM. The detection limit is rather poor, but the method has potential as a universal detector.

6 Applications

A development of applications for a new technology is always a confirmation that the technology is maturing. Electrophoresis on microchips has now been applied to the analysis of nucleic acids, proteins, and other types of samples in research applications. Commercial applications are expected to appear soon.

6.1 Nucleic acids

One of the leading applications of CE on microchips is the analysis of nucleic acids. Microchips have analyzed oligonucleotides and RNA, and genotyped and sequenced DNA. The analyses are extremely rapid, from less than a minute for oligonucleotides [13] to less than 20 min for DNA sequencing [14]. DNA is typically detected with LIF, but electrochemical detection has been applied as well

[106]. Microchips are being developed for commercial applications such as genotyping medically important loci and sequencing genomic human DNA. In addition, efforts are being made to integrate sample preparation and CE analysis on microchips.

6.1.1 Nucleic acid sizing

Microchips have been applied to size short oligonucleotides (10–25 bases) [13] and restriction fragments such as ΦX174 *Hae*III DNA [6, 32, 33, 51, 110], and ribosomal RNA [111]. Separation of ΦX174 *Hae*III DNA fragments from 70 to 1000 bp is completed in 120 s using an array of microchannels [6].

6.1.2 Genotyping

Genotyping by microchip CE is a rapidly developing application allowing a quick identification of genes responsible for hereditary diseases, such as hemochromatosis and Duchenne/Becker muscular dystrophy, and eventually for pharmacogenetics. In 1997, CAE on microchips analyzed restriction fragment markers from the HLA-H gene, a candidate gene for the diagnosis of hereditary hemochromatosis. Twelve samples were analyzed in parallel with a LIF scanner and an intercalating dye, thiazole orange [12]. Recently, 96 hemochromatosis samples were analyzed in less than 8 min on a microchip with 96 sample reservoirs and 48 analysis capillaries using a LIF scanner [36]. Locus-specific, multiplex PCR products specific for deletions causing Duchenne/Becker muscular dystrophy have been separated on a silicon-glass microchip [112]. Single-channel CE microchips with a 2.6 cm separation length and a replaceable polyacrylamide matrix have been used to analyze fluorescently-labeled CTTv PCR samples (containing the four loci CSF1PO, TPOX, THO1, and vWA) and short tandem repeats [113]. This body of work firmly establishes the feasibility of using CE microchips for genotyping. Commerical high-speed, high-throughput genotyping applications are now under development.

6.1.3 DNA sequencing

DNA sequencing with CAE on microchips is an area of intense interest, particularly with the exciting developments in the Human Genome Program. DNA sequencing with microchips was first demonstrated in 1995 in the Mathies lab [39]. Glass microchips with a denaturing polyacrylamide sieving matrix were used to separate DNA sequencing fragment ladders fluorescently labeled with energy transfer dye primers. Single base resolution reached 150–200 bases in 10–15 min with an effective

separation channel distance of 3.5 cm [39]. To achieve longer read-lengths, a microchip with long separation channels has been constructed [73]. Using a single-color detector, a DNA sequencing sample can be separated in less than 14 min with a 11.5 cm single-channel CE microchip using a linear polyacrylamide matrix. The resolution should be sufficient to achieve read-lengths of 400 bases [57]. An alternative approach to extend read-lengths on microchips is to use shorter channels and optimize the separation conditions. Using a four-color detector and a linear polyacrylamide matrix, read-lengths of over 500 bases were achieved in about 20 min in a single channel run at 99.4% accuracy (Fig. 3) [14]. This demonstrated the feasibility of high-speed, high-throughput four-color DNA sequencing using CE on microchips [14].

6.1.4 Integrated nucleic sample preparation and analysis

Integrating sample preparation and analysis is one of the prime goals of the micro-total-analysis system (μ-TAS) approach. Applications of μ-TAS to nucleic acid sample preparation and integrated analysis were first published in 1996 [40, 114–116]. One integrated device performs an automated restriction digest, and injects and separates the restriction fragments by CE [114]. Another reported developing components for thermocapillary pumping, thermal-cycling chambers, gel electrophoresis, and on-board DNA detection [115]. PCR reactors microfabricated in silicon have been coupled to CE microchips to create an integrated DNA analysis system [40, 116]. The microfabricated PCR device has rapid thermal cycling characteristics and *Salmonella* genomic DNA can be amplified and analyzed in under 45 min [40]. These reports demonstrated that high-speed DNA analyses could be integrated with sample preparation on microchips.

An integrated monolithic microchip was used recently for cell lysis, multiplex PCR amplification, and electrophoretic analysis [117]. PCR targets such as bacteriophage λ, *Escherichia coli* genomic DNA, and plasmid DNA can be amplified from *E. coli* cells on a microchip and then electrophoretically separated in less than 3 min [117]. A different approach performed individual sample preparation reactions on individual microchips and then analyzed them on a CE microchip [112]. In this study, random nonspecific amplifications of the human genome were performed on one microchip and then aliquots were loaded into another microchip for a locus-specific, multiplex PCR for Duchenne/Becker muscular dystrophy. The amplicons were then analyzed both by traditional CE and by CE on microchips, with similar results. This segmented approach demonstrated that complex sample preparation and analysis can be performed on microchips [112].

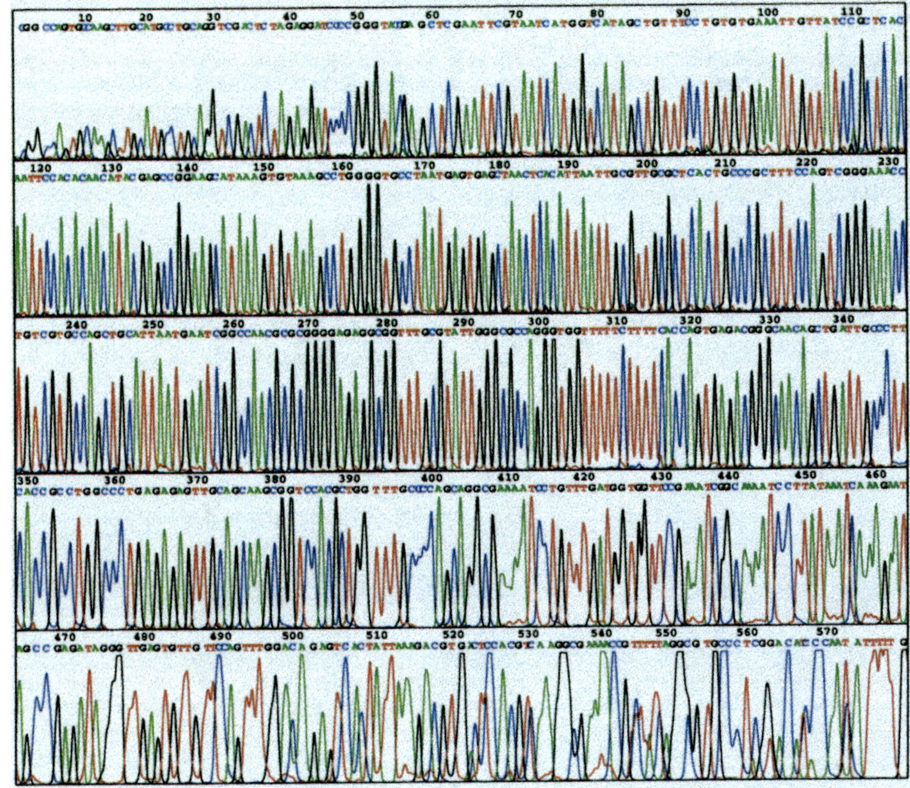

Figure 3. DNA sequencing by CE on microchip. The M13 sequencing trace has been processed and base-called with BaseFinder software [135]. The accuracy was 99.4% out to 500 bases. From [14], with permission.

Finally, an integrated silicon device has been microfabricated that prepares DNA samples, separates fragments, and detects samples with an on-board integrated photodiode [118]. This *tour de force* microchip is capable of measuring and moving samples in microchannels, with heaters, temperature sensors, and a fluorescence photodiode detector. Strand displacement amplification reaction was applied to amplify target DNA, mixed with intercalating dye, separated on cross-linked polyacrylamide gel, and detected with the photodiode, all on board the microchip. The appropriate fragment was detected in 4 min [118]. These results demonstrate the enormous potential for integration of sample preparation, analysis, and detection on CE microchips.

6.2 Peptides and proteins

Separation of proteins has always been an important part of the analysis of biological systems and more efforts have been devoted lately to separating proteins on microchips by CE. The rapid separation of proteins using zone electrophoresis, SDS electrophoresis, and isoelectric focusing have now been demonstrated. While the reports show the basic feasibility of protein separation, they are only a first glimmer of the potential of this powerful technology. Capillary SDS electrophoresis on a microchip has been reported recently. Fluorescein-labeled proteins (calmodulin, α-lactalbumin, pepsinogen, ovalbumin, BSA, and β-galactosidase) were separated in a 5 cm channel filled with SDS 24–200 (a sieving matrix from Beckman-Coulter) and detected by LIF detection. With an applied voltage from 1 to 5 kV, the separation time ranged from 35 to 200 s. Under optimized conditions, plate heights of *ca.* 1 μm can be obtained, whereas plate heights with the same proteins in capillary exceed 10 μm [119].

Two microchip devices for isoelectric focusing of proteins have recently been published [120, 121]. Capillary IEF with detection on-line requires zone mobilization, bringing protein zones into the detector. The three most common methods of mobilization, chemical mobilization, pressure mobilization, and electroosmotic mobilization, of separated proteins were compared. It was found that electroosmotically driven mobilization, which occurs simultaneously with the focusing, is the most suitable technique for miniaturization because of its high speed, compatibility of electroosmotic flow, and minimal instrumentation requirements [120]. Further, Cy5-labeled peptides can be focused in less than 30 s in a 7 cm $\times$ 200 μm $\times$ 10 μm channel with a total analysis time of less than 5 min. The maximum peak capacity is about 30–40 peaks [120]. The other approach used UV imaging detection that eliminates the requirement of zone mobilization. The separation channel is 4 cm $\times$ 100 μm $\times$ 10 μm. The detection limit is 2.4 ng for myoglobin at the optical path length of 10 μm [121].

Capillary zone electrophoresis (CZE) has been applied to analyze several proteins. A mixture of model human proteins (albumin, α_1-antitrypsin, transferrin, and IgG), simulating the electrophoretic pattern of human serum proteins, has been separated by CZE on a microchip after labeling with 2-toluidinonaphthalene-6-sulfonate. Real-world samples of human serum proteins, however, did not provide the expected set of five traditional peaks due to the poor labeling of several serum proteins [122, 123]. Further, CE of antibodies was performed on microchips. Anti-human IgG labeled with fluorescein isothiocyanate (FITC) has been analyzed on a microchip with an effective length of 2.8 cm in less than 16 s [124]. Monoclonal mouse anti-BSA IgG was measured in mouse ascites fluid by a direct microchip-based CE immunoassay [125, 126]. The calibration curve is linear up to at least 135 mg/L. The method can measure the antigen-antibody interactions between BSA and anti-BSA and calculate stoichiometry and equilibration constants. A free-flow electrophoretic system has been constructed for continuous separation and micropreparation of proteins on microchips [55, 94]. The instrument was tested using BSA, bradykinin, and ribonuclease A as a model mixture. It can separate proteins and collect individual fractions for further use as shown on rat plasma [55, 94].

6.3 Other applications

An effort has been devoted to using CE microchips for the analysis of other types of compounds in addition to nucleic acids and polypeptides. Immunoassays, enzyme assays, micellar electrokinetic chromatography (MEKC) and isotachophoresis have been adapted to microchips and applied to assay herbicides, amino acids, biogenic amines, ions, and drugs.

Immunoassays followed by CE separations have been developed to analyze low molecular weight compounds in capillaries and in microchips (see review [127]). A competition immunoassay has been developed to determine the concentration of theophylline, a drug for the treatment of asthma, in serum samples [125, 128]. A sample containing unlabeled theophylline is mixed with known amounts of fluorescently labeled theophylline and theophylline antibody. The unlabeled theophylline molecules in the sample compete with the labeled molecules for the limited amount of antibody. As the content of theophylline in the sample increases, less labeled theophylline is bound to the antibody, resulting in a decreased signal in the theophylline-antibody complex and an increased signal for the free labeled theophylline. CE separation on microchips with LIF detection can separate and quantify the free theophylline and theophylline-antibody complex peaks. The limit of detection for theophylline in diluted serum is 1.25 μg/L with a separation time of less than 50 s [125, 128]. Immunoassays followed by CE on microchips have measured serum thyroxin [129] and serum cortisol [29]. For thyroxin, using fluorescein-labeled thyroxin and a polyclonal antibody preparation as assay reagents, serum thyroxin could be electrophoretically analyzed in about 15 s [129]. Serum cortisol can be measured with a microchip-based electrophoretic immunoassay over a range of clinical interest (10–600 μg/L) without sample preconcentration [29].

MEKC on microchips has been applied to analyze biogenic amines (putrescine, cadaverine, spermidine, spermine, histamine, tyramine, tryptamine, and phenylethylamine) after derivatization with FITC. The analysis time is about 80 s, the run-to-run reproducibility of migration times varies between 0.15 and 0.54%, day-to-day reproducibility of migration times is between 0.40 and 0.80%, and the detection limit varies between 2.94 μM (putrescine) and 6.57 μM (spermine). The method was applied to real samples by analyzing biogenic amines in soy sauce [130].

Amphetamine and its analogs, methamphetamine, 3,4-methylenedioxymethamphetamine, and β-phenylethylamine, were analyzed in human urine after FITC derivatization by CZE and MEKC in synchronized cyclic mode (see Section 3.2) [131]. Using solid-phase extraction to concentrate the urine, the limit of identification is about 10 mg/L, a value that is currently too high for practical applications. The synchronized cyclic mode of MEKC has also been applied to separate a mixture of FITC-labeled amino acids [132].

Isotachophoresis on a microchip has been combined with Raman spectroscopy detection to assay the herbicides paraquat and diquat. Working sample concentrations ranged from 10^{-5} to 10^{-7} M [108]. Magnesium and calcium can be determined by electrophoresis on chip after sample stacking and on-chip complexation. After a gated injection, the sample is concentrated in the stacking channel and derivatized on-chip with 8-hydroxyquinolin-5-sulfonic acid. The limit of detection is 18 ppb (0.45 µM) for calcium and 0.5 ppb (21 nM) for magnesium [133].

7 Conclusions and outlook

CE on microchips is a rapidly emerging, new analytical technology. The fundamental feasibility of applying microfabrication to CE is now well established and numerous applications have been demonstrated. The adaptation of CE to microchips has many advantages. Integrated injectors produce well-defined sample plugs that can be resolved in short path lengths. The short plugs, good dissipation of Joule heating, and high field strengths result in extremely rapid separations that consume only picoliter sample volumes [13]. Microchips with both sample preparation and separation functionalities [117] raise the possibility of integrating sample preparation and analysis on a single device. This integration could allow matching the sample preparation and analysis requirements with only picoliter volumes of samples being prepared. Photolithographic micromachining methods make the task of producing arrays of structures on microchips as simple as altering computer-aided design (CAD) drawings and should make manufacturing less expensive than traditional CAE. The applications described above illustrate that these advantages of microchips can potentially be translated into ultrafast computer-controlled CE separations for a wide range of analytes.

To achieve this potential, microchip technology must overcome many challenges. The interface between the microworld of the chip and the outside world needs to be improved with novel hardware solutions and interconnections amenable to microfluidic sample handling and eventual automation. Microchips need to be integrated in a systems level solution to sample preparation and analysis, and smaller sample volumes prepared [134] if the potential savings of volumetric reduction are to be realized. This exciting technology will need to be hardened until it is ready for commercial applications with more robust methods and instrumentation developed that require less human intervention. Microchips must be developed that can either be reused, for glass microchips, or be disposable, for plastic microchips. If these and other challenges can be overcome, CE on microchips will become a major analytical technology.

The rapid progress that CE on microchips has made in the last decade suggests the impact it may have in the next decade. CE on microchips has the potential to be the first widespread analytical application of fundamentally new microfluidic and microchip technologies. Microchip-based CE instruments are expected to be developed that will automatically perform rapid electrophoresis applications with extremely high throughputs and minimal sample preparation costs. In the coming years, microchip-based technology may revolutionize many industrial CE applications including drug screening, single nucleotide polymorphism (SNP) analysis, resequencing for pharmacogenetics, proteomics, and diagnostics.

The authors gratefully acknowledge funding from the National Institute of Standards and Technology (Grant number 70NANB5H0131) and the National Institutes of Health (grant numbers 5 R01HG01775-03, 1 R43HG02980-01, and 2 R44HG01563-02).

Received July 29, 1999

8 References

[1] Terry, S. C., Jerman, J. H., Angell, J. B., *IEEE Trans Electron Devices* 1979, *26*, 880–886.

[2] Manz, A., Graber, N., Widmer, H. M., *Sens. Actuators B* 1990, *1*, 244–248.

[3] Harrison, D. J., Manz, A., Fan, Z. H.. Ludi, H., Widmer, H. M., *Anal. Chem.* 1992, *64*, 1926–1932.

[4] Jacobson, S. C., Hergenroder, R., Koutny, L. B., Ramsey, J. M., *Anal. Chem.* 1994, *66*, 1114–1118.

[5] Jacobson, S. C., Hergenroder, R., Koutny, L. B., Warmack, R. J., Ramsey, J. M., *Anal. Chem.* 1994, *66*, 1107–1113.

[6] Woolley, A. T., Mathies, R. A., *Proc. Natl. Acad. Sci. USA* 1994, *91*, 11348–11352.

[7] He, B., Regnier, F., *J. Pharm. Biomed. Anal.* 1998, *17*, 925–932.

[8] Gavin, P. F., Ewing, A. G., *J. Am. Chem. Soc.* 1996, *118*, 8932–8936.

[9] Cheng, J., Shoffner, M. A., Hvichia, G. E., Kricka, L. J., Wilding, P., *Nucleic Acids Res.* 1996, *24*, 380–385.

[10] Cheng, J., Shoffner, M. A., Mitchelson, K. R., Kricka, L. J., Wilding, P., *J. Chromatogr. A* 1996, *732*, 151–158.

[11] Harrison, D. J., Fan, Z. H., Seiler, K., Manz, A., Widmer, H. M., *Anal. Chim. Acta* 1993, *283*, 361–366.

[12] Woolley, A. T., Sensabaugh, G. F., Mathies, R. A., *Anal. Chem.* 1997, *69*, 2181–2186.

[13] Effenhauser, C. S., Paulus, A., Manz, A., Widmer, H. M., *Anal. Chem.* 1994, *66*, 2949–2953.

[14] Liu, S. R., Shi, Y. N., Ja, W. W., Mathies, R. A., *Anal. Chem.* 1999, *71*, 566–573.

[15] Freemantle, M., *C&EN* 1999, *77*, 27–36.

[16] Regnier, F. E., He, B., Lin, S., Busse, J., *Trends Biotech.* 1999, *17*, 101–106.

[17] Campana, A. M. G., Baeyens, W. R. G., Aboulenein, H. Y., Zhang, X. R., *J. Microcol. Sep.* 1998, *10*, 339–355.

[18] Mastrangelo, C. H., Burns, M. A., Burke, D. T., *Proc. IEEE* 1998, *86*, 1769–1787.

[19] Mitchelson, K. R., Cheng, J., Kricka, L. J., *Trends Biotech.* 1997, *15*, 448–458.

[20] Effenhauser, C. S., Bruin, G. J. M., Paulus, A., *Electrophoresis* 1997, *18*, 2203–2213.

[21] Kopp, M. U., Crabtree, H. J., Manz, A., *Curr. Opinion Chem. Biol.* 1997, *1*, 410–419.

[22] Jacobson, S. C., Ramsey, J. M., in: Landers, J. P. (Ed.), *Handbook of Capillary Electrophoresis*, 2nd Ed., CRC Press, Boca Raton, FL 1997, pp. 827–839.

[23] Effenhauser, C. S., *Topics Curr. Chem.* 1998, *194*, 51–82.

[24] Kricka, L. J., *Clin. Chem.* 1998, *44*, 2008–2014.

[25] MacTaylor, C. E., Ewing, A. G., *Electrophoresis* 1997, *18*, 2279–2290.

[26] Paulus, A., *Amer. Lab.* 1998, *30*, No. 8, 59–62.

[27] Fan, Z. H., Harrison, D. J., *Anal. Chem.* 1994, *66*, 177–184.

[28] Jacobson, S. C., Hergenroder, R., Moore, A. W., Ramsey, J. M., *Anal. Chem.* 1994, *66*, 4127–4132.

[29] Koutny, L. B., Schmalzing, D., Taylor, T. A., Fuchs, M., *Anal. Chem.* 1996, *68*, 18–22.

[30] Jacobson, S. C., Moore, A. W., Ramsey, J. M., *Anal. Chem.* 1995, *67*, 2059–2063.

[31] Roberts, M. A., Rossier, J. S., Bercier, P., Girault, H., *Anal. Chem.* 1997, *69*, 2035–2042.

[32] McCormick, R. M., Nelson, R. J., Alonso-Amigo, M. G., Benvegnu, J., Hoopwer, H. H., *Anal. Chem.* 1997, *69*, 2626–2630.

[33] Effenhauser, C. S., Bruin, G. J. M., Paulus, A., Ehrat, M., *Anal. Chem.* 1997, *69*, 3451–3457.

[34] Martynova, L., Locascio, L. E., Gaitan, M., Kramer, G. W., Christensen, R. G., MacCrehan, W. A., *Anal. Chem.* 1997, *69*, 4783–4789.

[35] Fluri, K., Fitzpatrick, G., Chiem, N., Harrison, D. J., *Anal. Chem.* 1996, *68*, 4285–4290.

[36] Simpson, P. C., Roach, D., Woolley, A. T., Thorsen, T., Johnston, R., Sensabaugh, G. F., Mathies, R. A., *Proc. Natl. Acad. Sci. USA* 1998, *95*, 2256–2261.

[37] Liu, K. L. K., Davis, K. L., Morris, M. D., *Anal. Chem.* 1994, *66*, 3744–3750.

[38] Simpson, P. C., Woolley, A. T., Mathies, R. A., *Biomed. Microdevices* 1998, *1*, 7–26.

[39] Woolley, A. T., Mathies, R. A., *Anal. Chem.* 1995, *67*, 3676–3680.

[40] Woolley, A. T., Hadley, D., Landre, P., Demello, A. J., Mathies, R. A., Northrup, M. A., *Anal. Chem.* 1996, *68*, 4081–4086.

[41] Manz, A., Harrison, D. J., Verpoorte, E., Fettinger, J. C., Ludi, H., Widmer, H. M., *J. Chromatogr.* 1992, *593*, 253–258.

[42] Nakanishi, H., Abe, H., Nishimoto, T., Arai, A., *Bunseki Kagaku* 1998, *47*, 361–368.

[43] Wang, H. Y., Foote, R. S., Jacobson, S. C., Schneibel, J. H., Ramsey, J. M., *Sens. Actuators B* 1997, *45*, 199–207.

[44] Ewing, A. G., Gavin, P. F., Hietpas, P. B., Bullard, K. M., *Nature Med.* 1997, *3*, 97–99.

[45] Ekstrom, B., Jacobson, G., Ohman, O., Sjodin, H., *US Patent 5,376,252*, 1994.

[46] Reyna, L. G., Sobehart, J. R., *J. Appl. Phys.* 1994, *76*, 4367–4371.

[47] Srinivasan, R., Braren, B., *Chem. Rev.* 1989, *89*, 1303–1316.

[48] Niino, H., Yabe, A., *Appl. Surf. Sci.* 1993, *69*, 1–6.

[49] Ehrfeld, W., Munchmeyer, D., *Nucleic Instrum. Methods A* 1991, *303*, 523–532.

[50] Becker, H., Dietz, W., Dannberg, P., in: Harrison, D. J., Van den Berg, A. (Eds.), *Micro Total Analysis Systems '98*, Kluwer Academic Publishers, Boston, MA 1998, pp. 253–356.

[51] Duffy, D. C., McDonald, J. C., Schueller, O. J. A., Whitesides, G. M., *Anal. Chem.* 1998, *70*, 4974–4984.

[52] Burggraf, N., Manz, A., Effenhauser, C. S., Verpoorte, E., Derooij, N. F., Widmer, H. M., *J. High Res. Chromatogr.* 1993, *16*, 594–596.

[53] Jacobson, S. C., Koutny, L. B., Hergenroder, R., Moore, A. W., Ramsey, J. M., *Anal. Chem.* 1994, *66*, 3472–3476.

[54] Hietpas, P. B., Bullard, K. M., Gutman, D. A., Ewing, A. G., *Anal. Chem.* 1997, *69*, 2292–2298.

[55] Raymond, D. E., Manz, A., Widmer, H. M., *Anal. Chem.* 1994, *66*, 2858–2865.

[56] Luckey, J. A., Norris, T. B., Smith, L. M., *J. Phys. Chem.* 1993, *97*, 3067–3075.

[57] Schmalzing, D., Adourian, A., Koutny, L., Ziaugra, L., Matsudaira, P., Ehrlich, D., *Anal. Chem.* 1998, *70*, 2303–2310.

[58] Dolník, V., *Electrophoresis* 1997, *18*, 2353–2361.

[59] Kim, Y., Yeung, E. S., *J. Chromatogr. A* 1997, *781*, 315–325.

[60] Martin, A. J. P., Everaerts, F., *Proc. Roy. Soc. London* 1970, *316*, 493–514.

[61] Burgi, D. S., *Anal. Chem.* 1993, *65*, 3726–3729.

[62] Burgi, D. S., Chien, R. L., *Anal. Biochem.* 1992, *202*, 306–309.

[63] Burgi, D. S., Chien, R. L., *Anal. Chem.* 1991, *63*, 2042–2047.

[64] Burgi, D. S., Salomon, K., Chien, R. L., *J. Liq. Chromatogr.* 1991, *14*, 847–867.

[65] Chien, R. L., Burgi, D. S., *J. Chromatogr.* 1991, *559*, 141–152.

[66] Chien, R. L., Burgi, D. S., *J. Chromatogr.* 1991, *559*, 153–161.

[67] Dolník, V., Cobb, K. A., Novotny, M., *J. Microcol. Sep.* 1990, *2*, 127–131.

[68] Kanansky, D., Marák, J., *J. Chromatogr.* 1990, *498*, 191–204.

[69] Foret, F., Šustáček, V., Boček, P., *J. Microcol. Sep.* 1990, *2*, 229–233.

[70] Quirino, J. P., Terabe, S., *Anal. Chem.* 1999, *71*, 1638–1644.

[71] Quirino, J. P., Terabe, S., *Science* 1998, *282*, 465–468.

[72] Klepárník, K., Malá, Z., Boček, P., *J. Chromatogr. A* 1997, *772*, 243–253.

[73] Davidson, C., Balch, J., Brewer, L., Kimbrough, J., Swierkowski, S., Nelson, D., Madabhushi, R., Pastrone, R., Lee, A., McCready, P., Adamson, A., Bruce, R., Mariella, R.,

Carrano, A., *DOE Human Genome Program Contractor-Grantee Workshop VI*, Santa Fe, NM 1997.

[74] Seiler, K., Harrison, D. J., Manz, A., *Anal. Chem.* 1993, *65*, 1481–1488.

[75] Culbertson, C. T., Jacobson, S. C., Ramsey, J. M., *Anal. Chem.* 1998, *70*, 3781–3789.

[76] Gavin, P. F., Ewing, A. G., *J. Microcol. Sep.* 1998, *10*, 357–364.

[77] Jacobson, S. C., Hergenroder, R., Koutny, L. B., Ramsey, J. M., *Anal. Chem.* 1994, *66*, 2369–2373.

[78] Liu, Y. M., Sweedler, J. V., *Anal. Chem.* 1996, *68*, 2471–2476.

[79] Harrison, D. J., Fluri, K., Seiler, K., Fan, Z. H., Effenhauser, C. S., Manz, A., *Science* 1993, *261*, 895–897.

[80] Jacobson, S. C., Culbertson, C. T., Daler, J. E., Ramsey, J. M., *Anal. chem.* 1998, *70*, 3476–3480.

[81] Effenhauser, C. S., Manz, A., Widmer, H. M., *Anal. Chem.* 1993, *65*, 2637–2642.

[82] Mangru, S. D., Harrison, D. J., *Electrophoresis* 1998, *19*, 2301–2307.

[83] Evans, C. E., *Anal. Chem.* 1997, *69*, 2952–2954.

[84] Seiler, K., Fan, Z. H. H., Fluri, K., Harrison, D. J., *Anal. Chem.* 1994, *66*, 3485–3491.

[85] vanderMoolen, J. N., Poppe, H., Smit, H. C., *Anal. Chem.* 1997, *69*, 4220–4225.

[86] Jacobson, S. C., Ramsey, J. M., *Electrophoresis* 1995, *16*, 481–486.

[87] Dolník, V., Deml, M., Boček, P., in: Holloway, C. J. (Ed.), *Analytical And Preparative Isotachophoresis*, Walter de Gruyter, Berlin 1984, pp. 55–62.

[88] Effenhauser, C. S., Manz, A., Widmer, H. M., *Anal. Chem.* 1995, *67*, 2284–2287.

[89] Harrison, D. J., Fluri, K., Chiem, N., Tang, T., Fan, Z. H., *Sens. Actuators B* 1996, *33*, 105–109.

[90] Flster, J. C., Jacobson, S. C., Davis, L. M., Ramsey, J. M., *Anal. Chem.* 1998, *70*, 431–437.

[91] Flanagan, J. H., Owens, C. V., Romero, S. E., Waddell, L., Kahn, S. H., Hammer, R. P., Soper, S. A., *Anal. Chem.* 1998, *70*, 2676–2684.

[92] Liang, Z. H., Chiem, N., Ocvirk, G., Tang, T., Fluri, K., Harrison, D. J., *Anal. Chem.* 1996, *68*, 1040–1046.

[93] Ocvirk, G., Tang, T., Harrison, D. J., *Analyst* 1998, *123*, 1429–1434.

[94] Raymond, D. E., Manz, A., Widmer, H. M., *Anal. Chem.* 1996, *68*, 2515–2522.

[95] Zhang, Y., Lee, H. K., Li, S. F. Y., *Talanta* 1998, *45*, 613–618.

[96] Mathies, R. A., Huang, X. C., *Nature* 1992, *359*, 167–169.

[97] Bruin, G. J. M., Stegman, G., van Asten, A. C., Xu, X., Kraak, J. C., Poppe, H., *J. Chromatogr.* 1991, *559*, 163–181.

[98] Figeys, D., Aebersold, R., *J. Biochech. Eng. Trans. ASME* 1999, *121*, 7–12.

[99] Figeys, D., Aebersold, R., *Anal. Chem.* 1998, *70*, 3721–3727.

[100] Figeys, D., Ning, Y. B., Aebersold, R., *Anal. Chem.* 1997, *69*, 3153–3160.

[101] Ramsey, R. S., Ramsey, J. M., *Anal. Chem.* 1997, *69*, 1174–1178.

[102] Xu, N. X., Lin, Y. H., Hofstadler, S. A., Matson, D., Call, C. J., Smith, R. D., *Anal. Chem.* 1998, *70*, 3553–3556.

[103] Xue, Q. F., Foret, F., Dunayevskiy, Y. M., Zavracky, P. M., Mcgruer, N. E., Karger, B. L., *Anal. Chem.* 1997, *69*, 426–430.

[104] Xue, Q. F., Dunayevskiy, Y. M., Foret, F., Karger, B. L., *Rapid Commun. Mass Spectrom.* 1997, *11*, 1253–1256.

[105] Foret, F., Liu, H., Zhang, B., Felten, C., Karger, B. L., in: Harrison, D. J., Van Den Berg, A. (Eds.), *Micro Total Analysis Systems '98*, Kluwer Academic Publishers, Boston, MA 1998, pp. 35–38.

[106] Woolley, A. T., Lao, K. Q., Glazer, A. N., Mathies, R. A., *Anal. Chem.* 1998, *70*, 684–688.

[107] Gavin, . F., Ewing, A. G., *Anal. Chem.* 1997, *69*, 3838–3845.

[108] Walker, P. A., Morris, M. D., Burns, M. A., Johnson, B. N., *Anal. Chem.* 1998, *70*, 3766–3769.

[109] Burggraf, N., Krattiger, B., deMello, A. J., deRooij, N. F., Manz, A., *Analyst* 1998, *123*, 1443–1447.

[110] Effenhauser, C. S., Bruin, G. J. M., Paulus, A., *Electrophoresis* 1997, *18*, 2203–2213.

[111] Ogura, M., Agata, Y., Watanabe, K., McCormick, R. M., Hamaguchi, Y., Aso, Y., Mitsuhashi, M., *Clin. Chem.* 1998, *44*, 2249–2255.

[112] Cheng, J., Waters, L. C., Fortina, P., Hvichia, G., Jacobson, S. C., Ramsey, J. M., Kricka, L. J., Wilding, P., *Anal. Biochem.* 1998, *257*, 101–106.

[113] Schmalzing, D., Koutny, L., Adourian, A., Belgrader, P., Matsudaira, P., Ehrlich, D., *Proc. Natl. Acad. Sci. USA* 1997, *94*, 10273–10278.

[114] Jacobson, S. C., Ramsey, J. M., *Anal. cCem.* 1996, *68*, 720–723.

[115] Burns, M. A., Mastrangelo, C. H., Sammarco, T. S., Man, F. P., Webster, J. R., Johnson, B. N., Foerster, B., Jones, D., Fields, Y., Kaiser, A. R., Burke, D. T., *Proc. Natl. Acad. Sci. USA* 1996, *93*, 5556–5561.

[116] Woolley, A. T., Northrup, M. A., Mathies, R. A., *Abst. Paper Amer. Chem. Soc.* 1996, *212*, 155.

[117] Waters, L. C., Jacobson, S. C., Kroutchinina, N., Khandurina, J., Foote, R. S., Ramsey, J. M., *Anal. Chem.* 1998, *70*, 158–162.

[118] Burns, M. A., Johnson, B. N., Brahmasandra, S. N., Handique, K., Webster, J. R., Krishnan, M., Sammarco, T. S., Man, P. M., Jones, D., Heldsinger, D., Mastrangelo, C. H., Burke, D. T., *Science* 1998, *282*, 484–487.

[119] Yao, S., Anex, D. S., Caldwell, W. B., Arnold, D. W., Smith, K. B., Schultz, P. G., *Proc. Natl. Acad. Sci. USA* 1999, *96*, 5372–5377.

[120] Hofmann, O., Che, D. P., Cruickshank, K. A., Muller, U. R., *Anal. Chem.* 1999, *71*, 678–686.

[121] Mao, Q., Pawliszyn, J., *Analyst* 1999, *124*, 637–641.

[122] Colyer, C. L., Mangru, S. D., Harrison, D. J., *J. Chromatogr. A* 1997, *781*, 271–276.

[123] Colyer, C. L., Tang, T., Chiem, N., Harrison, D. J., *Electrophoresis* 1997, *18*, 1733–1741.

[124] Rodriguez, I., Zhang, Y., Lee, H. K., Li, S. F. Y., *J. Chromatogr. A* 1997, *781*, 287–293.

[125] Chiem, N., Harrison, D. J., *Anal. Chem.* 1997, *69*, 373–378.

[126] Chiem, N. H., Harrison, D. J., *Electrophoresis* 1998, *19*, 3040–3044.

[127] Schmalzing, D., Nashabeh, W., *Electrophoresis* 1997, *18*, 2184–2193.

[128] Chiem, N. H., Harrison, D. J., *Clin. Chem.* 1998, *44*, 591–598.

[129] Schmalzing, D., Koutny, L. B., Taylor, T. A., Nashabeh, W., Fuchs, M., *J. Chromatogr. B* 1997, *697*, 175–180.

[130] Rodriguez, I., Lee, H. K., Li, S. F. Y., *Electrophoresis* 1999, *20*, 118–126.

[131] Ramseier, A., vonHeeren, F., Thormann, W., *Electrophoresis* 1998, *19*, 2967–2975.

[132] vonHeeren, F., Verpoorte, E., Manz, A., Thormann, W., *Anal. Chem.* 1996, *68*, 2044–2053.

[133] Kutter, J. P., Ramsey, R. S., Jacobson, S. C., Ramsey, J. M., *J. Microcol. Sep.* 1998, *10*, 313–319.

[134] Kricka, L. J., *Clin. Chem.* 1998, *44*, 2008–2014.

[135] Giddings, M. C., Severin, J., Westphall, M., Wu, J. Z., Smith, L. M., *Genome Res.* 1998, *8*, 644–665.

Review

Emanuel Carrilho

Instituto de Química de São Carlos, Universidade de São Paulo
São Carlos - SP, Brazil

DNA sequencing by capillary array electrophoresis and microfabricated array systems

To comply with the current needs for high-speed DNA sequencing analysis, several instruments and innovative technologies have been introduced by several groups in recent years. This review article discusses and compares the issues regarding high-throughput DNA sequencing by electrophoretic methods in miniaturized systems, such as capillaries, capillary arrays, and microchannels. Initially, general features of several capillary array designs (including commercial ones) will be considered, followed by similar analyses with microfabricated array electrophoretic devices and how they can contribute to the success of large sequencing projects.

Keywords: High-throughput DNA sequencing / Microchannel array electrophoresis / Instrumentation / Capillary electrophoresis / Review   EL 3760

Contents

Correspondence: Prof. Emanuel Carrilho, Universidade de São Paulo, Instituto de Química de São Carlos, Av. Dr. Carlos Botelho, 1465, Cx. Postal 780, 13560-970 São Carlos - SP, Brazil
E-mail: emanuel@iqsc.sc.usp.br
Fax: +55-16-273-9983

Abbreviations: AFM, atomic force microscopy; **CAE**, capillary array electrophoresis; **µCAE**, microfabricated CAE; **CGE**, capillary gel electrophoresis; **ET dyes**, energy transfer dyes; **HEC**, hydroxyethyl cellulose; **HGP**, Human Genome Project; **LPA**, linear poly(acrylamide), **PDMA**, poly(dimethylacrylamide); **PEO**, poly(ethyleneoxide)

1 Introduction

The Human Genome Project (HGP) was initiated in 1990 with the goal to sequence the human genetic code with three billion base pairs in 15 years [1, 2]. Since then, the publicly funded Human Genome Research Institute (HGRI) has set five-year plans (1993/98 and 1998/2003) to guide the project to achieve its specific goals [3]. In October 1998, at the beginning of the second half of the project, there were a little over 180 million base pairs (Mbp) of human DNA in public databases (*i.e.*, less than 6% of the whole human genome) and a sequencing rate capacity of 90 Mbp/year [4]. Obviously these numbers are not satisfactory to comply with the newly revised goals for the second 5-year plan (1998–2003) which is aiming to produce sequencing data at a rate of 500 Mbp of finished sequences per year. Such optimistic goals were motivated by the continuous improvements on the commercial, slab-gel-based automated sequencers and their sequencing throughput. However, the project will be even more accelerated with the validation and use of capillary electrophoresis for DNA sequencing and the spreading of multiple channel instrumentation.

Assuming that all HGP goals can be met with the new capillary array technologies, all human genetic codes could be unveiled by the end of 2001. However, the human genome is only one of almost one hundred genomes being sequenced around the world. Also, *de novo* sequencing, sequencing of other organisms, comparative genomics, single nucleotide polymorphism (SNP), and other genome-related issues are examples of applications that will continue to seek new advances in high-throughput sequencing technologies.

2 Competing technologies for high-throughput DNA sequencing

Although the absolute majority of all DNA sequencing is currently generated by slab gel electrophoresis, there are a number of technologies that could produce high throughput data. Important advances in ionization of biopolymers in the gas phase during the last decade allowed mass spectrometry (MS) to be considered a promising technique for fast, accurate, high-throughput analysis. One mode of sequencing by MS is to use matrix-assisted laser desorption/ionization time-of-flight MS (MALDI-TOF-MS) [5], and another is electrospray ionization-MS (ESI-MS) [6]. Because very large DNA molecules take only a few microseconds to reach the detector, MALDI-TOF has become an attractive tool for high-throughput DNA sequencing [7, 8]. MS is indeed a powerful technique for DNA analysis; however, limitations in sensitivity and efficient ionization of large molecular sizes must be overcome before MS becomes a high-throughput DNA sequencing tool [9].

Sequencing by hybridization (SBH) is another technique that claims to have the potential to fulfill the high throughput demands of the HGP [10]. The concept, introduced at the end of the last decade by three separate groups [11–13], is based on hybridization of the DNA to an oligonucleotide probe of known sequence. Only the perfect match of the probe sequence to the target DNA will strongly bind the strands together. Obviously, the stability of the complex depends not only on the sequence itself, but also on the temperature, ionic strength, and concentration [14]. The target DNA can be probed against a large array of short oligonucleotides with all sequence combinations possible. For example, a complete array of 10-mer oligonucleotides would be made of 1 048 576 (or 4^{10}) probes. It is postulated that such a number would suffice to walk through an unknown genome [11–13]. The sequence is determined by detecting all the hybridized DNA and by overlapping all the known sequences from the probes. Although it is difficult to control all parameters to yield ideal hybridization conditons for all sequences, the application of this technology for sequence verification and mutation analysis is strongly predicted [10].

Revolutionary throughputs of sequencing data could be achieved, in the theory, using an enzymatic approach to the question. Two interesting projects could be mentioned. One proposed monitoring the fluorescence of single nucleotides released by enzymatic digestion of a strand of DNA attached to a magnetic bead. As the exonuclease digestion proceeds, a flowing stream drags the labeled nucleotides to a detection cell where the fluorescence is detected [15]. The other enzymatic method utilized an RNA polymerase [16]. One fragment of DNA was attached to a particle that was held in place by atomic force microscopy (AFM). As the polymerase pulled down the DNA strand in the synthesis of RNA, the force applied by the enzyme was monitored by AFM and the force variation was proportional to the sequence. Although both studies are still underway, their feasibility is clear. More important than the fast sequencing information provided by such enzymatic methods (several hundreds of sequenced bases per second) will be the ability to sequence very long stretches of DNA, thus breaking the 1000–1200 nucleotide barrier obtainable with electrophoretic methods [17–19].

3 Capillary electrophoresis for automated DNA sequencing

Fast and completely automatable capillary electrophoresis appears to be a realistic option to increase the throughput of the HGP. The high surface-to-volume ratio of a small capillary tube can efficiently dissipate the heat produced during electrophoresis; therefore, the electric field can be higher than that practiced in slab gel electrophoresis, only limited by the onset of the biased reptation model (BRM) [20–22]. Capillary electrophoresis has been extensively investigated as a practical tool for DNA sequencing. Initially introduced as capillary gel electrophoresis (CGE) [23–28], it has proven to be a successful analytical technique for DNA sequencing, capable of fast and sensitive separations. However, CGE still had some features that were not compatible with a high-throughput regime. Among them were the short column lifetime as well as injection-related problems [29, 30].

DNA sequencing using noncross-linked polymer solutions was introduced by Karger *et al.* in 1993 [31] and it became a major breakthrough, solving earlier problems faced by CGE. Through the use of replaceable linear polymer solutions, it was possible to reuse the same capillary many times with a fresh load of polymer solution (the separation medium) for each sample. Linear poly (acrylamide) (LPA) has been utilized by several groups [32–36]. The promise that DNA sequencing by replaceable polymer solutions could bring new means of speed and automation prompted other groups to search for alternative matrices [37–41], but none have simultaneously yielded fast analysis time and a length-of-read as good as LPA [42].

DNA sequencing using replaceable polymer solutions offers a variety of analytical conditions to be tailored to optimize analysis time and/or length of read. For example, one might use short capillaries and high electric fields to separate 300 bases in less than 5 min [22, 43], or use elevated temperatures to minimize compressions and the onset of the BRM [44] and lower the electric field to extend the read-length to more than a thousand bases [18, 45].

Compared to slab gel electrophoresis, CE with polymer solutions is about 8–10 times faster per lane. Unfortunately, this is not sufficient to compete in throughput due to the parallel nature of slab gel instruments. To turn it into a parallel instrument, capillary array electrophoresis (CAE) was introduced in 1992 [46]. The search for a comprehensive instrument was attempted by several groups seeking an instrument capable of fast, automated, sensitive, and, most important of all, rugged operation [47–54].

3.1 CAE

The concept of a high-throughput DNA analyzer based on CE can only be materialized using an array of capillaries. The parallel feature of slab gels is unmatched for high-number sample processing. Given the nature of this challenge, an ideal instrument should be based on capillaries with replaceable polymer solutions capable of separating over a thousand bases in a single run per capillary per hour of operation. Also, the capillary column should be stable and maintain the same separation efficiency for several hundred runs without any change in the instrument.

Designing and constructing a CAE instrument is not a simple task. In many aspects there are many solutions to a problem; however, the choice of each one will depend on the direct relationship of all elements, for example, the illumination of the columns. It is preferable for the light to be evenly distributed among all capillaries in a continuous fashion because scanning systems with illumination of one capillary at a time could be less efficient and sensitive if a very large number of capillaries is involved. However, the laser power of a continuous system has to be many times more powerful than a scanning system one. The same is valid for the detection of the electrophoretic bands; an optical design should be able to continuously sample all capillaries with high frequencies (2–5 Hz per column), required for fast separations. To accomplish this in scanning detection is not straightforward; however, detection systems which use imaging devices to view the entire array at once often deal with issues such as cross-talk between channels.

4 Important sequencing issues

4.1 Sample preparation

Advances in all steps of the sequencing process contributed substantially to the current state of the art and to permit actual sequencing rates by the large genomic centers. Issues related to sample preparation were certainly the most relevant of all. Improvements in enzyme processivity [55], fluorescent-dye chemistry [56], sequencing strategy [57], and sample clean-up and injection (for CE analysis) [45, 58] permitted rugged production of sequencing samples.

4.1.1 Enzymes

The improvement in DNA polymerase enzymes has greatly contributed to the quality of the sequencing reactions and sequencing data. Initially, isothermal DNA polymerases were used in manual and automated DNA sequencing [59]. The reactions were performed at physiological temperatures (~37°C) for a few minutes (–20 min). These enzymes (T4 or T7 DNA polymerases) evenly incorporated all four terminators, even the dye-labeled ones. However, these polymerases were very sensitive to temperature and could degrade easily. With the discovery of the polymerase chain reaction (PCR) [60] through the use of a heat-stable DNA polymerase *Thermus aquaticus* (*Taq* polymerase), the ability to perform sequencing reactions (cycle-sequencing) with reduced amounts of DNA template compared to isothermal enzymes became possible [61]. The major drawback of cycle-sequencing using *Taq* polymerase compared to PCR was a severe discrimination of the enzyme to ddNTPs over dNTPs. A single substitution of one amino acid in the primary sequence of the enzyme completely changed this effect [55] and the rate of ddNTP incorporation was substantially equalized to that of dNTPs. Further development on the activity of the enzymes and on the fluorescent labels attached to ddNTPs [62] allowed for significant improvement of the sequencing data. This new dye-terminator kit produces more even peak heights, which facilitates automated base-calling.

4.1.2 *Fluorescent-dye chemistry*

A significant advance in dye-primer chemistry was the introduction of energy transfer (ET) dyes [56, 63]. They consisted of two dyes per primer, one being a common donor and other the acceptor dye. The common donor could be a fluorescein derivative (FAM) or a cyanine (Cy5) [64] at the 5'-end. The second dye, the discriminating one, was located about ten bases away, with the separation between the dyes optimized for ET efficiency and minimum electrophoretic mobility shifts. The four acceptors are those commonly used in dye-primer chemistry: FAM, JOE, TAMRA and ROX [65].

The major advantages of ET dyes are that they can be almost evenly excited by a single wavelength (488 nm) and the electrophoretic mobility shifts are minimal*. Another chemical class of fluorescent dyes, Bodipy, can be used to produce similar primers. Bodipy dyes seem to offer narrower spectral bandwith and better quantum efficiency [66]. The combination of cycle sequencing with modified *Taq* polymerase with ET dyes yielded reactions with stronger signals even using lower amounts of templates. Another advantage in using ET dyes was instrument-related: since only one wavelength was needed to excite all four dyes, the instrument could be simpler in its optical design.

Fluorescence decay, or time-resolved fluorescence, has been investigated as one alternative to conventional laser-induced fluorescence DNA sequencing [67, 68]. It offers sensitive instrumentation, no need for excitation-scattering rejection, and easier deconvolution of overlapped peaks; the major difficulty, however, is to design four appropriate dyes with different decay spectra [69]. One-lane, four-color DNA sequencing using fluorescence lifetime detection was successfully demonstrated by Lieberwirth *et al.* [68] after introduction of three newly synthesized fluorescent labels. The new detection scheme yielded separations up to 660 bp with base-calling accuracy over 90% without need for mobility correction. Also, all fluorescently labeled primers were excited with a semiconductor laser centered in 630 nm.

4.2 Sequencing strategies

Shotgun and primer-walking sequencing approaches are the most common methods for genomic sequencing. While the first generates random sequencing, the latter can complete gaps in the sequence left by the shotgun approaches. A new variation of the shotgun method was introduced by Venter *et al.* [70] upon shotgunning a whole genome at once. This strategy depended enormously on computational resources to align all generated sequences, but the results were rewarded with the sequencing of the *Haemophilus influenzae* genome in only 18 months [57].

4.3 Injection

The transfer of all knowledge of the sequencing protocols from slab gel format to CE was not straightforward. In gel electrophoresis a substantial aliquot (few microliters) of the sample is directly loaded on the wells. In CGE the hydrodynamic loading was not feasible due to the presence of a physical gel inside the column. The alternative was the electrokinetic loading of the sample into the capillaries. Even after the substitution of CGE by replaceable matrices, electrokinetic injection is the most suitable method of sample introduction.

Electrokinetic injection is limited by the salt content of the sample, thus requiring some sort of desalting. Ethanol precipitation, which is typically used, provided a low enough salt content, but the procedure is not highly reproducible. Karger *et al.* [45, 58] presented a complete study about sample purification and injection. They proposed a clean-up protocol to reduce the salt content to a low millimolar level and to successfully remove the DNA template, also proved to have a deleterious effect on the separation performance. After the clean-up, a tenfold improvement in the signal-to-noise ratio (S/N) was obtained from a fraction of the full reaction. Signal improvement was also shown by Swerdlow *et al.* [71] with successful stacking of the sequencing sample on replaceable polymer solutions. Improvement in sample introduction and signal strength is important because it allows reducing the cost of the sequencing reaction. However, the lowest sample volume was limited to a few microliters, since a minimum volume was necessary to produce efficient electrical contact between the electrodes and the capillary. In this regard, microchips can show an important advantage. They are able to handle and manipulate smaller volumes than required by the conventional protocols.

Sample injection in microdevices is normally executed electrokinetically using a cross channel, with the sample channel perpendicular to the separation channel. An interesting feature in sequencing reactions is the exponential decay in the peak height profile, *i.e.*, shorter DNA fragments are present in larger amounts than the longer ones. In microdevices it is possible to design a cross channel of variable length by placing the opposite sample channels far apart, in a Z or double-T format. The distance between the opposite channels defines the sample volume or the amount injected. Upon continuous injection of the sequencing sample through this section, shorter and more mobile fragments are partially eliminated from the injection volume, whereas the less mobile, longer sequencing fragments are continuously introduced in the separation channel. This procedure produces a more even signal distribution throughout the separation, significantly improving the S/N at the end of the separation [72].

4.4 Channel coating and polymer solution

Control the EOF has been an important topic of study in CE. Several approaches have been reported to control or suppress the EOF for open-tube CE in an variety of ways

---

* Due to differences in charge and size, the dyes impart a differential migration pattern to the DNA. The effect is most pronounced for small fragments (< 200 bases).

[73, 74]. Covalently bound hydrophilic coatings [75–78] have been the most successful approach to date, yielding stable columns with near-zero EOF. However, a few papers focused on EOF suppression for use with polymer solutions in DNA sequencing [37, 41, 79].

The passiveness of the capillary wall is achieved by efficient silanol masking of the capillary surface. The better the silanol coverage, the better the EOF suppression because protection of the silanol groups reduces the electrical double layer [80]. Additionally, the solution viscosity at the wall is greatly enhanced by the presence of polymers. Even if a small number of ionizable silanol groups are left out after coating, the high viscosity medium should offer enough resistance to the flow.

It is well known that proteins tend to adsorb onto the capillary walls; indeed, this has recently been proven by direct probing of the surface using AFM [81]. DNA molecules tend not to adsorb on the silica surface due to electrostatic repulsion between the negatively charged wall and the phosphodiester backbone of DNA. Nevertheless, DNA analysis cannot be carried out in free zone electrophoresis due to the lack of charge-to-mass selectivity, thus requiring a size-separation matrix [82]. Although dsDNA has been successfully separated in both coated and uncoated capillaries with ultradilute cellulose solutions [83], ssDNA presents a more hydrophobic character than dsDNA and could possibly interact with the silica wall. Because of this, ssDNA analysis is normally performed in polymer solutions with coated capillaries.

Polymers such as linear poly(acrylamide) (LPA) [31, 84], hydroxyethyl cellulose (HEC) [38], poly(ethyleneoxide) (PEO) [37] and poly(dimethylacrylamide) (PDMA) [41] have been used with some success as separation matrices for DNA sequencing. Some matrices (LPA and HEC) required a hydrophilic layer deposited on the internal wall of the capillary, while others (PEO and PDMA) were self-coating, not requiring immobilized coatings. Hydrophilic polymers are better solvated by the aqueous buffer, and their adsorption energy is not sufficiently high to maintain the polymer molecules adsorbed to the capillary walls. Although the presence of the polymer suppresses the EOF by decreasing the silica zeta potential, if the polymer solution is not viscous enough (low concentration), the EOF can mobilize the bulk matrix out of the column. On the other hand, water is not a good solvent for PEO or PDMA, thus enhancing the interactions of the polymer molecules with the hydrophobic silica tube. The adsorption energy of the polymer molecules to the wall is such that it efficiently restrains the establishment of the EOF in an electric field. However, the more hydrophobic the polymer is, the stronger the interaction between the separa-

tion matrix with the larger DNA fragments, and this interaction can increase the analysis time and decrease the number of resolvable bases.

4.5 Base-calling software

Important issues in DNA sequencing are the computational resources and the bioinformatic tools. Computers and software technically run all operations, from sample preparation to sequence alignment. Massive use of algorithms is necessary for data acquisition and storage, spectral deconvolution, dye-trace recognition (color sorting), peak finding, mobility shift correction, base calling, and sequence alignment. Of particular interest is the base-calling software. Many logical and mathematica approaches have been used to transform the analog/digital signal from the fluorescence detector onto finished sequences; graph-theoretic [85], adaptive equalization [86], and expert systems [87] are a few. The base-calling algorithm can extend the sequencing results beyond the separation power of electrophoresis. Visually, the length-of-read (LOR) [42] is limited to resolution values around 0.5. Expert software can read beyond this resolution limit with high accuracy, significantly extending the LOR, and therefore, the cost and number of runs [87].

5 Instrumentation

5.1 Commercial capillary arrays

After the first study exploiting the potential of a multiple capillary instrument was reported by Mathies *et al.* [46], several investigators endeavored to produce a capillary array that could significantly improve the sequencing rate compared to the state-of-the-art slab-gel-based sequencers. Today, there are three commercial versions of CAE instruments and several other groups are still trying to bring new alternatives in system design. At the XII International Symposium on High Performance Capillary Electrophoresis & Related Techniques (HPCE'99), a section was dedicated to high-throughput DNA sequencing. In this section, the latest information on all three commercial instruments was presented. All CAE sequencers feature 96 capillaries: MegaBACE 1000 from Molecular Dynamics (Sunnyvale, CA, USA), ABI 3700 from PE Biosystems (Foster City, CA, USA) and SCE9600 from SpectruMedix (State College, PA, USA). Their actual design originated from prototypes of Mathies *et al.* [88], Kambara and Tkokahashi [47], and Yeung and Ueno [49], respectively. The major features of these instruments, as presented at the conference, are listed in Table 1.

Molecular Dynamics' instrument, based on confocal detection, consisted of a microscope objective to focus the laser light inside the capillaries and, at the same time,

Electrophoresis 2000, *21*, 55–65

Table 1 Principal features of commercial CAE instruments

Feature	Molecular Dynamics	Perkin-Elmer	SpectruMedix
Capillaries	LPA coated capillaries; stable over 200 runs	Bare fused silica with dynamic coating; stable for over 300 runs	Bare fused silica with dynamic coating with hydrophobic polymer
Matrix	2–4% LPA	5% PDMA	2–6% mix of PEO of varied molecular mass (up to 8 MDa)
Illumination	Confocal MO, scanning system; 488 and 532 nm	Liquid sheath with postcolumn illumination; 488 nm	Side illumination with 488 nm; capillaries immersed in refractive index matching liquid
Detection	MO Confocal light collection, 4 photomultiplier tubes (PMTs) and filter set for color separation	Cooled CCD operating at 4 Hz, grating for light dispersion: 522–700 nm	CID with F#1.4 camera lens; multicolor detection
Sequencing data	Up to 800 bp, over 500 bp on average	600 bp in 120 min;	500 bp with 98% accuracy
Turnaround time	<2 h	~2.6 h	~2 h
Other	42°C	50°C	–

MO, microscope objective

collect the emitted light from the center of the column. To collect the data from each and every one of the capillaries, a scanning system is used. This procedure can be a limiting step when fast sampling rates are required, and imply mechanical stress of moving parts. The confocal system utilizes a set of four photomultiplier tubes with proper filters and dichroic beam splitters. Coated columns with LPA were used [79], and although initially the separation was carried out with HEC as separation media [38], it was changed to low viscosity LPA solution for a significant improvement in read length.

Perkin-Elmer's system employed postcolumn detection with liquid sheath flow [48, 89]. Briefly, the capillary bundle is aligned inside a quartz cuvette. Along the dead space between the columns and the walls of the cuvette, a buffer solution is pumped through the cell. The liquid sheath flowing outside the capillary drags down the electrophoretic bands eluting from the column, tapering them to a smaller diameter. A laser beam crosses all flow streams and excites the fluorescent molecules. Light collection was made at 90° from the laser plane, which reduced the scattered light. The fluorescent light is dispersed through a diffraction grating and imaged onto a cooled CCD camera.

On-column detection is the approach of the SpectruMedix instrument [49, 90]. The laser beam crosses all 96 capillaries that are immersed in a refractive index-matching liquid between glass plates. Due to the glass plates, the laser light is confined, and losses are minimized. The fluorescent light is collected at a right angle from the laser axis by an F# 1.4 lens and detected by a CCD camera able to perform multicolor detection. An expansion to 384 capillaries is foreseen, followed by an exchange in the camera lens.

5.2 Homemade capillary arrays

Despite the availability of commercial instruments, several other groups are still working on the development and design of CAE sequencers, mainly schemes concerning illumination and detection. Kambara's group [89, 91] has shown several versions of its CAE instrument. The first two designs used the liquid sheath approach. The color discrimination was accomplished using narrow bandpass filters stacked together. The filtered light was then split into four and refocused onto a CCD camera. The image seen by the camera consisted of four colored lines of 24 spots each. In both approaches, replacing the polymer solution was not easily possible, at least from the detection side. For this reason, Kambara *et al.* [51] began testing side illumination with detection on column. Just recently, they improved the lens effect approach by intercalating each capillary with a glass rod of the same external diameter as the capillaries. With this design, the authors improved the light transmission with side illumination, although only 48 capillaries were used [92].

Another elegant approach for a multiple capillary instrument was the use of optical fibers for illumination and collection of the signal in a 90° arrangement, as shown by Quesada and Zhang [50]. The collection fibers directed the fluorescent light to a spectrograph for full spectrum detection, but the base-calling program used only small regions of the spectrum according to the dye-chemistry used. For example, with dye-primer chemistry, only four slices sufficed for the base coding; with old dye-terminator chemistry, nine slices were required for proper dye discrimination. With recent advances in ET dyes, this task would be much simpler. This instrument allowed matrix replacement, and the low light collection efficiency of the optical fibers was compensated by the use of an intensified CCD. However, the optical fiber illumination required a high-powered laser for high-sensitivity analysis. For this reason, the authors switched to the use of a single beam to illuminate all capillaries, using the capillary walls as lenses [93], a similar approach as that of Anazawa *et al.* [51].

The instrument first developed in Karger's group was an instrument capable of simultaneous replacement of the matrix solution in 12 capillaries, as well as simultaneous and continuous illumination of all capillaries. The fluorescent light was collected at 90° relative to the capillaries, dispersed by a ruled grating, and then refocused back inside a thermoelectrically cooled CCD camera for fast, low-noise, and high-sensitivity detection. This prototype was developed initially to help understand the important issues of a rugged CAE instrument. Although the instrument lacked lenses with good imaging properties, it was suitable to produce new sequencing data from a small region of the mouse genome [94]. Also, it was helpful in addressing issues on the illumination of the capillaries and detection optics, as well as several issues related to software development. The main challenge was to produce similar separation performances, *i.e.*, separation of 1000 bases in both platforms, in the single capillary instrument as well as in the capillary array. After several improvements in the illumination (use of a laser line generator instead of a beam splitter) and detection arrangements (camera lenses instead of light condensers and transmission grating instead of a ruled grating) they have shown over 1000 bases of sequencing information with ~ 99% accuracy on 48 capillaries simultaneously in less than 60 min, even for nonstandard DNA such as human DNA from chromosome 17 [87].

Since the introduction of the concept of CAE, Mathies *et al.* [95] continued to push its development. To break the actual limit of 96 capillaries, they developed a confocal system with the capillaries aligned in a circular array. The microscope objective spins inside a drum, interrogating each one of the columns at a time. Currently they have shown sequencing data from 128 columns, but a larger number of capillaries could be easily accommodated in this geometry. Despite the criticism of the possible lack of sensitivity due to small sampling time, they have shown

that the actual exposure and measurement time is close to optimal for the fluorescence detection of the electrophoretic bands [96].

The latest CAE instrument proposal comes from Heller *et al.* [97]. In their paper, besides the extensive review of all configurations published to date, the authors described the rationale that guided their decisions. For illumination of all 96 capillaries, a laser line generator was used to produce a uniform intensity profile over the array, yet using a low power laser (~ 30 mW). The detection set-up was complex, with several optical elements, such as lenses, filters, and holographic transmission grating. The image of the array is formed onto a back-illuminated CCD chip. As mentioned earlier, imaging systems often suffer from cross-talk between capillaries. Indeed, this system presented 1–2% cross-talk of the main signal, and even though this level is not significant, it could easily be corrected by software processing [54]. Although four color DNA sequencing was not shown, given the instrumental set-up presented, it is possible to foresee this application.

5.3 Microfabricated CAE (μCAE)

A totally different platform to perform CAE is by using microchips, or micromachined or microfabricated devices. Relatively well known in electronics, the microelectromechanical systems (MEMS) are being brought to chemistry as a compact platform to perform chemical analysis in the last decade [98]. When applied to DNA analysis, particularly to DNA sequencing, such devices consist of reservoirs (for buffer and sample) and channels in which the electrophoresis and separation of the sequencing fragments occur [99]. Some advantages of these systems include high lane density and compactness. Because of its size, a large number of separation channels can be placed close together, generating more electrophoretic lanes per unit of length. This factor facilitates detection, whether using optical imaging or scanning lenses.

Two of the major protocols to prepare the microchips should be mentioned; one uses conventional photolithography on glass or silicon substrates [100], the other uses molded plastic over a micromachined template [101]. Some of the polymers used for fabrication of microchips are poly(dimethylsiloxane) (PDMS) [101, 102] and poly(methylmethacrylate) (PMMA) [103]. Details and procedures on the fabrication of micromachined devices can be found elsewhere [104].

Although several papers have been published about separations in microfabricated devices, especially PCR-amplified fragments, few have dealt with DNA sequencing samples [72, 105, 106]. In one section at HPCE'99, the results of two groups were presented. Ehrlich *et al.* [106] evaluated the relationship of sequencing read length with separation length and applied voltage in microdevices. In a single-lane microchip, they separated up to 400 bases in 14 min at 200 V/cm, using 4% replaceable LPA in poly(-acrylamide) coated microchannels. The system was not fully compatible with 4-color DNA sequencing since only one wavelength could be detected.

In their second approach to DNA sequencing on micro-chips, Mathies *et al.* [72, 105] significantly improved the separation performance after changing the LPA polymeri-zation protocol according to Carrilho *et al.* [18]. Initially, a 9% LPA solution was used and ~430 bases could be sep-arated in 10 min using single-color detection and a sepa-ration distance of 3.5 cm. Four-color DNA sequencing was obtained in a single-lane microchip with confocal arrangement. Single-base resolution was detected up to 150 bases in only 540 s with 97% accuracy [105]. When a 4% LPA solution, 160 V/cm, and 7 cm long channel was used by Liu *et al.* [72], up to 600 bases after the primer were called by the sequencing software in 20 min. The best sequencing results using four-color detection were 500 bases, with an accuracy of 99.4%. Another important study by these authors was to investigate the influence of the size of the sample injection channel on the separation performance and injection time as means to enhance the S/N of longer DNA fragments.

Mathies [107] has also shown several microchip designs to fit an increasing number of channels. A conventional-size glass chip can normally fit 12 channels, capable of analyzing up to 48 samples in a special arrangement. As an innovation, a 6 inch circular glass plate carried 96 radial channels, converging at the center of the chip. For detection in this microfabricated device, a spinning confo-cal system was used, which detected the fluorescence from each channel, one at a time.

6 High throughput sequencing

High-throughput DNA sequencing by CE or microfabri-cated channels faces a scientific dilemma. Separation sci-entists always try to improve the separation of their ana-lytes to the maximum extent possible, or at least to baseline resolution. Separation of DNA sequencing frag-ments should follow such premises; however, baseline separation of more than a 1000 compounds is a major challenge. Due to the nature of the analysis (qualitative, not quantitative analysis), resolution over 0.7 is rarely important. Besides, the actual development of different base-calling software often allows separations with much poorer electrophoretic resolution. Indeed, what really mat-ters is the right migration order, not how far apart the

peaks are; therefore, for sequencing purposes, the reso-lution of the separation is less important than the confi-dence levels of the bases called.

6.1 Results per lane

For proper evaluation of high throughput CAE DNA sequencing, the number of bases called per lane (a lane being either a capillary tube or a microfabricated chan-nel), should also be taken into consideration. After investi-gating several combinations of capillary coatings and replaceable matrices for DNA sequencing by CE, it was possible to identify some of the most relevant results.

As discussed earlier, more hydrophobic polymers, such as PEO and PDMA, can use bare fused-silica capillaries with successful EOF suppression. However, the lifetime of the capillaries was not unlimited, and PEO columns lasted only around 30 runs before reconditioning [37]. PDMA columns, when mixed with a small amount of poly(vinylpyrrolidone), can last up to a few hundred runs with proper rinsing and activation of the surface every time that the resolution falls below a certain threshold value. In both cases, the sequencing average 500–600 bases in roughly 2 h. LPA matrices were always used with coated channels. Two of the most successful ones are LPA and poly(vinylacohol) (PVA) coatings, with both being able to last over a few hundred runs. In the first case, sequencing reads averaged 600 bases in less than 2 h, and, at lower electric fields, it reached 800 bases [108]. The combination of PVA-coated capillaries with a mixture of high and low molecular mass LPA as replacea-ble matrix routinely yielded read-lengths of over 1000 bases in less than 1 h with 98% base-calling accuracy. At lower electric fields it could read even further, reaching 1300 bases in 2 h. This was the longest read-length known to date, obtained in capillary format [87].

In the microchip format, all papers described the modified use of Hjertén's procedure [75] for coating the channels with LPA. In all cases, the read-lengths were shorter than those obtained in capillaries with LPA separation matrix. However, because the separation distance in microchips was shorter than in capillaries (about 25% of the separa-tion length) the sequencing rate was actually higher (if calculated in terms of bases/min). Table 2 summarizes some of the main sequencing numbers of a few represen-tative systems of channel/polymer chemistry in both the capillary and microchip platforms.

6.2 Results per instrument

The general throughput for all CAE instruments described here are nearly the same, since they all have 96 columns and similar read lengths of 500–600 bases in an average run. Assuming 24 h operation, the capillary-based CAE can produce up to 0.5 Mbp/day. Elevating this mark 4-fold may be possible by extending the read-lengths further than the 1000-base level in a 1 h turnaround time or fitting more capillaries in the array. The first option is feasible and was already demonstrated by Karger, while increasing the number of columns represented a more demanding task.

Microfabricated devices could, in principle, offer better throughput with 96 channels. It is not clear yet whether it is better to take advantage of reusing the microchip or if is better to use a new one every time; however, in both cases, the set-up time has to be taken in account. Based on the best results shown on microchips so far and a 10 min set-up, roughly 1200 bases per channel/h could be possible. In a 24 h cycle, such devices with 96 channels could yield up to 2.7 Mbp/day.

6.3 System integration; from matrix replacement to sample preparation

In the pursuit of very high throughput by CAE, one of the most important issues is the integration of sample preparation, separation, and data analysis in a closed loop system for continuous 24 h nonattended operation. All commercial CAE instruments have this claim in their specifications. Automation of μCAE, however, is still under development and it is not clear whether the microchips will be disposable or reusable. Regardless of this fact, microchips are the most suitable platform to perform multiple tasks and sample handling before electrophoresis.

Reuse of capillaries and microchips requires the separation matrix to be replaced by means of pumping a new load of the polymer solution inside the channels. This is normally achieved by moving the capillary tips from the buffer reservoir to a pressurization vessel. This movement is easily achieved given the capillary flexibility and dimension. Replacing polymer solutions in microchips can be performed by using pressurized syringes (or a

Table 2. DNA sequencing performance of several combinations of channel type and separation chemistry

Platform	Channel	Medium	Read length	Time (min)	Accuracy (%)	Ref.
Capillary	Coated	LPA	1000	53	98	[109]
Capillary	Noncoated	PDMA	600	120	94	[41]
Capillary	Noncoated	PEO	460	120	98	[110]
Chip	Coated	LPA	400	14	-	[106]
Chip	Coated	LPA	600	20	95	[72]

manifold thereof) loaded with the polymer solution. A certain degree of complexity is involved in systems with disposable microchips.

To fully profit from the high throughput of such miniaturized systems, the integration of the sample preparation to the sequencing instrument has recently been demanded. Some papers have shown a variety of couplings for fluid handling with capillaries and valves in order to produce, purify, and inject the sequencing sample [110–112]. Some others used total integration on microchips, performing PCR reactions, sample purification and separation, demonstrating a potential application to DNA sequencing samples [113].

7 Future trends

Although the conception of an ideal high-throughput DNA sequencing instrument is possible by adding together the best features of all systems, the attainment of the whole package is not trivial and many factors are involved, such as cost, size, performance, compatibility, *etc.*. However, there is still space for improvement in many areas such as channel and polymer chemistry, microfabrication tools, optical arrangements, system integration, and software for managing, base calling, sequence assembly, *etc.*

Received August 15, 1999

8 References

[1] Watson, J. D., *Science* 1990, *248*, 44–49.

[2] Cantor, C. R., *Science* 1990, *248*, 49–51.

[3] Collins, F., Galas, D., *Science* 1993, *262*, 43–46.

[4] Collins, F. S., Patrinos, A., Jordan, E., Chakravarti, A., Gesteland, R., Walters, L., *Science* 1998, *282*, 682–689.

[5] Karas, M., Hillenkamp, F., *Anal. Chem.* 1988, *60*, 2301–2303.

[6] Fenn, J. B., Mann, M., Meng, C. K., Wong, S. F., Whitehouse, C. M., *Mass Spectrom. Rev.* 1990, *9*, 37–70.

[7] Karas, M., Bahr, U., *Trends Anal. Chem.* 1990, *9*, 321–325.

[8] Spengler, B., Pan, Y., Cotter, R. J., Kan, L. S., *Rapid Commun. Mass Spectrom.* 1990, *4*, 99–102.

[9] Henry, C., *Anal. Chem.* 1997, *69*, 243A–246A.

[10] Lipshutz, R., Morris, D., Chee, M., Hubbel, E., Kozal, M., Shah, N., Shen, N., Yang, R., Fodor, S., *BioTechniques* 1995, *19*, 442–448.

[11] Bains, W., Smith, G. J., *J. Theoret. Biol.* 1988, *135*, 303–307.

[12] Lysov, Y., Yu, P., Khorlin, A. A., Khrapko, K. R., Shick, V. V., Florentiev,, V. L., Mirzabekov, A. D., *Proc. USSR Acad. Sci.* 1988, *303*, 1508–1511.

[13] Drmanac, R., Labat, I., Brukner, I., Crkvenjakov, R., *Genomics* 1989, *4*, 114–128.

[14] Lysov, Y. P., Chernyi, A. A., Balaeff, A. A., Beattie, K. L., Mirzabekov, A. D., *J. Biomol. Struct. Dyn.* 1994, *11*, 797–812.

[15] Ambrose, W. P., Goodwin, P. M., Jett, J. H., Johnson, M. E., Martin, J. C., Marrone, B. L., Schecker, J. A., Wilkerson, C. W., Keller, R. A., Haces, A., Shih, P. J., Harding, J. D., *Berichte Bunsen Ges. Physikal. Chemie* 1993, *1535–1542.*

[16] Andoh, T., Nishizawa, M., Hida, T., Ariyoshi, Y., Takahashi, T., Ueda, R., *Oncology Res.* 1996, *8*, 229–238.

[17] Grothues, D., Voss, H., Stegemann, J., Wiemann, S., Sensen, C., Zimmermann, J., Schwager, C., Erfle, H., Rupp, T., Ansorge, W., *Nucleic Acids Res.* 1993, *21*, 6042–6044.

[18] Carrilho, E., Ruiz-Martinez, M. C., Berka, J., Smirnov, I., Goetzinger, W., Miller, A. W., Brady, D., Karger, B. L., *Anal. Chem.* 1996, *68*, 3305–3313.

[19] Kim, Y., Yeung, E. S., *J. Chromatogr. A* 1997, *781*, 315–325.

[20] Noolandi, J., Slater, G. W., Lim, H. A., Viovy, J. L., *Science* 1989, *243*, 1456–1458.

[21] Slater, G. W., Noolandi, J., *Biopolymers* 1989, *28*, 1781–1791.

[22] Yan, J. Y., Best, N., Zhang, J. Z., Ren, H. J., Jiang, R., Hou, J., Dovichi, N. J., *Electrophoresis* 1996, *17*, 1037–1045.

[23] Cohen, A. S., Najarian, D. R., Paulus, A., Guttman, A., Smith, J. A., Karger, B. L., *Proc. Natl. Acad. Sci. USA* 1988, *85*, 9660–9663.

[24] Cohen, A. S., Najarian, D. R., Karger, B. L., *J. Chromatogr.* 1990, *516*, 49–60.

[25] Swerdlow, H., Gesteland, R., *Nucleic Acids Res.* 1990, *18*, 1415–1419.

[26] Drossman, H., Luckey, J. A., Kostichka, A. J., D'Cunha, J., Smith, L. M., *Anal. Chem.* 1990, *62*, 900–903.

[27] Rocheleau, M. J., Dovichi, N. J., *J. Microcol. Sep.* 1992, *4*, 449–453.

[28] Luckey, J. A., Norris, T. B., Smith, L. M., *J. Phys. Chem.* 1993, *97*, 3067–3075.

[29] Swerdlow, H., Dew-Jager, K. E., Brady, K., Grey, R., Dovichi, N. J., Gesteland, R., *Electrophoresis* 1992, *13*, 475–483.

[30] Figeys, D., Renborg, A., Dovichi, N. J., *Electrophoresis* 1994, *15*, 1512–1517.

[31] Ruiz-Martinez, M. C., Berka, J., Belenkii, A., Foret, F., Miller, A. W., Karger, B. L., *Anal. Chem.* 1993, *65*, 2851–2858.

[32] Best, N., Arriaga, E., Chen, D. Y., Dovichi, N. J., *Anal. Chem.* 1994, *66*, 4063–4067.

[33] Grossman, P. D., *J. Chromatogr. A* 1994, *663*, 219–227.

[34] Manabe, T., Chen, N., Terabe, S., Yohda, M., Endo, I., *Anal. Chem.* 1994, *66*, 4243–4252.

[35] Wu, C. H., Quesada, M. A., Schneider, D. K., Farinato, R., Studier, F. W., Chu, B., *Electrophoresis* 1996, *17*, 1103–1109.

[36] Fang, Y., Zhang, J. Z., Hou, J. Y., Lu, H., Dovichi, N. J., *Electrophoresis* 1996, *17*, 1436–1442.

[37] Fung, E. N., Yeung, E. S., *Anal. Chem.* 1995, *67*, 1913–1919.

[38] Bashkin, J., Marsh, M., Barker, D., Johnston, R., *Appl. Theor. Electrophor.* 1996, *6*, 23–28.

[39] Gelfi, C., Perego, M., Libbra, F., Righetti, P. G., *Electrophoresis* 1996, *17*, 1342–1347.

[40] Menchen, S., Johnson, B., Winnik, M. A., Xu, B., *Electrophoresis* 1996, *17*, 1451–1459.

[41] Madabhushi, R. S., *Electrophoresis* 1998, *19*, 224–230.

[42] Quesada, M. A., *Curr. Opin. Biotech.* 1997, *8*, 82–93.

[43] Muller, O. M., Minarik, M., Foret, F., *Electrophoresis* 1997, *19*, 1436–1444.

[44] Kleparnik, K., Foret, F., Berka, J., Goetzinger, W., Miller, A. W., Karger, B. L., *Electrophoresis* 1996, *17*, 1860–1866.

[45] Salas-Solano, O., Ruiz-Martinez, M. C., Carrilho, E., Kotler, L., Karger, B. L., *Anal. Chem.* 1998, *70*, 1528–1535.

[46] Huang, X. C., Quesada, M. A., Mathies, R. A., *Anal. Chem.* 1992, *64*, 967–972.

[47] Kambara, H., Takahashi, S., *Nature* 1993, *361*, 565–566.

[48] Bay, S., Strake, H., Zhang, J. Z., Elliot, L. D., Dovichi, N. J., *J. Capil. Electrophor.* 1994, *1*, 121–126.

[49] Ueno, K., Yeung, E. S., *Anal. Chem.* 1994, *66*, 1424–1431.

[50] Quesada, M. A., Zhang, S. P., *Electrophoresis* 1996, *17*, 1841–1851.

[51] Anazawa, T., Takahashi, S., Kambara, H., *Anal. Chem.* 1996, *68*, 2699–2704.

[52] Kheterpal, I., Scherer, J. R., Clark, S. M., Radhakrishnan, A., Ju, J., Ginther, C. L., Sensabaugh, G. F., Mathies, R. A., *Electrophoresis* 1996, *17*, 1852–1859.

[53] Madabhushi, R. S., Vainer, M., Dolnik, V., Enad, S., Barker, D. L., Harris, D. W., Mansfield, E. S., *Electrophoresis* 1997, *18*, 104–111.

[54] Carrilho, E., Miller, A. W., Ruiz-Martinez, M. C., Kotler, L., Kesilman, J., Karger, B. L., in: Cohn, G. E., Soper S. A. (Eds.) *Ultrasensitive Biochemical Diagnostics II,* Proceedings of SPIE 1997, Vol. 2985, pp. 1–16.

[55] Tabor, S., Richardson, C. C., *Proc. Natl. Acad. Sci. USA* 1995, *92*, 6339–6343.

[56] Ju, J., Ruan, C., Fuller, C. W., Glazer, A. N., Mathies, R. A., *Proc. Natl. Acad. Sci. USA* 1995, *92*, 4347–4351.

[57] Fleischmann, R. D., Adams, M. D., Venter, J. C., *Science* 1995, *269*, 469–512.

[58] Ruiz-Martinez, M. C., Salas-Solano, O., Carrilho, E., Kotler, L., Karger, B. L., *Anal. Chem.* 1998, *70*, 1516–1527.

[59] Tabor, S., Richardson, C. C., *Proc. Natl. Acad. Sci. USA* 1987, *84*, 4767–4771.

[60] Mullis, K. B., Faloona, F. A., *Methods Enzymol.* 1987, *155*, 335–350.

[61] Murray, V., *Nucleic Acids Res.* 1989, *17*, 8889.

[62] Protocol P/N 403041 Rev. A. ABI Prism dRhodamine Terminator Cycle Sequencing Kit With AmpliTaq DNA Polymerase, FS. PE Applied Biosystems, Foster City, CA 1997.

[63] Ju, J., Kheterpal, I., Scherer, J. R., Ruan, C., Fuller, C. W., Glazer, A. N., Mathies, R. A., *Anal. Biochem.* 1995, *231*, 131–140.

[64] Hung, S. C., Ju, J., Mathies, R. A., Glazer, A. N., *Anal. Biochem.* 1996, *243*, 15–27.

[65] Ju, J., Glazer, A. N., Mathies, R. A., *Nature Med.* 1996, *2*, 246–249.

[66] Metzker, M. L., Lu, J., Gibbs, R. A., *Science* 1996, *271*, 1420–1422.

[67] Soper, S. A., Legendre, B. L., Williams, D. C., *Anal. Chem.* 1995, *67*, 4358–4365.

[68] Lieberwirth, U., Arden-Jacob, J., Drexhage, K. H., Herten, D. P., Müller, R., Neumann, M., Schulz, A., Siebert, S., Sagner, G., Klingel, S., Sauer, M., Wolfrum, J., *Anal. Chem.* 1998, *70*, 4771–4779.

[69] Soper, S. A., *Pharmaceutical and Biomedical Applications of Capillary Electrophoresis*, Pergamon Press, Oxford 1996, pp. 181–228.

[70] Venter, J. C., Smith, H. O., Hood, L., *Nature* 1996, *381*, 364–366.

[71] Xiong, Y., Park, S. R., Swerdlow, H., *Anal. Chem.* 1998, *70*, 3605–3611.

[72] Liu, S. R., Shi, Y. N., Ja, W. W., Mathies, R. A., *Anal. Chem.* 1999, *71*, 566–573.

[73] Landers, J. P., Oda, R. P., Madden, B. J., Spelsberg, T. C., *Anal. Biochem.* 1992, *205*, 115–124.

[74] Hayes, M. A., Kheterpal, I., Ewing, A. G., *Anal. Chem.* 1993, *65*, 27–31.

[75] Hjertén, S., *J. Chromatogr.* 1985, *347*, 191–198.

[76] Nashabeh, W., El Rassi, Z., *J. Chromatogr.* 1991, *559*, 367–383.

[77] Schmalzing, D., Piggee, C. A., Foret, F., Carrilho, E., Karger, B. L., *J. Chromatogr. A* 1993, *652*, 149–159.

[78] Goetzinger, W., Karger, B. L., PCT Int. Appl. WO9623220, August, 1996.

[79] Dolnik, V., Xu, D., Yadav, A., Bashkin, J., Marsh, M., Tu, O., Mansfield, E., Veiner, M., Madabhushi, R., Barker, D., Harris, D., *J. Microcol. Sep.* 1998, *10*, 175–184.

[80] Rieger, P. H., *Electrochemistry*, 2nd Ed., Chapman & Hall, New York 1994.

[81] Bonvent, J. J., Barberi, R., Bartolino, R., Capelli, L., Righetti, P. G., *J. Chromatogr. A* 1996, *756*, 233–243.

[82] Grossman, P. D., Soane, D. S., *Biopolymers* 1991, *31*, 1221–1228.

[83] Barron, A. E., Sunada, W. M., Blanch, H. W., *Electrophoresis* 1995, *16*, 64–74.

[84] Zhang, J., Fang, Y., Hou, J. Y., Ren, H. J., Jiang, R., Roos, P., Dovichi, N. J., *Anal. Chem.* 1995, *67*, 4589–4593.

[85] Berno, A. J., *Genome Res.* 1996, *6*, 80–91.

[86] Proakis, J. G., *Digital Communications* 3rd Ed., McGraw-Hill, New York 1995.

[87] Karger, B. L., *Proceedings of the 12th International Symposium on High Performance Capillary Electrophoresis & Related Microscale Techniques*, Palm Springs, CA 1999, p. 39.

[88] Huang, X. C., Quesada, M. A., Mathies, R. A., *Anal. Chem.* 1992, *64*, 2149–2154.

[89] Takahashi, S., Murakami, K., Anazawa, T., Kambara, H., *Anal. Chem.* 1994, *66*, 1021–1026.

[90] Li, Q., Yeung, E. S., *Appl. Spectrosc.* 1995, *49*, 825–833.

[91] Chen, D. Y., Swerdlow, H. P., Harke, H. R., Zhang, J. Z., Dovichi, N. J., *J. Chromatogr.* 1991, *559*, 237–246.

[92] Anazawa, T., Takahashi, S., Kambara, H., *Electrophoresis* 1999, *20*, 539–546.

[93] Quesada, M. A., Dhadwal, H. S., Fisk, D., Studier, F. W., *Electrophoresis* 1998, *19*, 1415–1427.

[94] Cai, W., Cao, W., Wu, L., Exley, W. E., Waneck, G. L., Karger, B. L., Warner, C. M., *Immunogenetics* 1996, *45*, 97–107.

[95] Scherer, J. R., Kheterpal, I., Radhakrishnan, A., Ja, W. W., Mathies, R. A., *Electrophoresis* 1999, *20*, 1508–1517.

[96] Kheterpal, I., Mathies, R. A., *Anal. Chem.* 1999, *71*, 31A–37A.

[97] Behr, S., Matzig, M., Levin, A., Eickhoff, H., Heller, C., *Electrophoresis* 1999, *20*, 1492–1507.

[98] Manz, A., Harrison, D. J., Verpoorte, E. M. J., Fettinger, J. C., Ludi, H., Widmer, H. M., *Chimia* 1991, *45*, 103–105.

[99] Woolley, A. T., Mathies, R. A., *Proc. Natl. Acad. Sci. USA* 1994, *91*, 11348–11352.

[100] Jacobson, S. C., Hergenroder, R., Koutny, L. B., Warmack, R. J., Ramsey, J. M., *Anal. Chem.* 1994, *66*, 1107–1113.

[101] Duffy, D. C., McDonald, J. C., Schueller, O. J. A., Whitesides, G. M., *Anal. Chem.* 1998, *70*, 4974–4984.

[102] Effenhauser, C. S., Bruin, G. J. M., Paulus, A., Ehrat, M., *Anal. Chem.* 1997, *69*, 3451–3457.

[103] Ford, S. M., Kar, B., McWhorter, S., Davies, J., Soper, S. A., Klopf, M., Calderon, G., Saile, V., *J. Microcol. Sep.* 1998, *10*, 413–422.

[104] Kovacs, G. T. A., Petersen, K., Albin, M., *Anal. Chem.* 1996, *68*, 407A–412A.

[105] Woolley, A. T., Mathies, R. A., *Anal. Chem.* 1995, *67*, 3676–3680.

[106] Schmalzing, D., Adourian, A., Koutny, L., Ziaugra, L., Matsudaira, P., Ehrlich, D., *Anal. Chem.* 1998, *70*, 2303–2310.

[107] Simpson, P. C., Roach, D., Woolley, A. T., Thorsen, T., Johnston, R., Sensabaugh, G. F., Mathies, R. A., *Proc. Natl. Acad. Sci. USA* 1998, *95*, 2256–2261.

[108] Bashkin, J., Rank, D., Tu, O., Amjadi, M., Aplaon, D., Hoang, C., Solomon, N., Ernst, N., Li, D., Ellis, T., He, C., Yogev, P., Kuo, J.–D., Karbelashvilii, M., Franklin, H., McArdle, B., Mamone, T., Fuller, C., Mardis, E., Snyan, J., Hillier, L., Wilson, R., Lewis, M., *Proceedings of the 12th International Symposium on High Performance Capillary Electrophoresis & Related Microscale Techniques*, Palm Springs 1999, p. 39.

[109] Salas-Solano, O., Carrilho, E., Kotler, L., Miller, A. W., Goetzinger, W., Sosic, Z., Karger, B. L., *Anal. Chem.* 1998, *70*, 3996–4003.

[110] Tan, H., Yeung, E. S., *Anal. Chem.* 1998, *70*, 4044–4053.

[111] Tan, H., Yeung, E. S., *Anal. Chem.* 1997, *69*, 664–674.

[112] Swerdlow, H., Jones, B. J., Wittwer, C. T., *Anal. Chem.* 1997, *69*, 848–855.

[113] Burns, M. A., Johnson, B. N., Brahmasandra, S. N., Handique, K., Webster, J. R., Krishnan, M., Sammarco, T. S., Man, P. M., Jones, D., Heldsinger, D., Mastrangelo, C. H., Burke, D. T., *Science* 1998, *282*, 484–487.

Review

András Guttman[1]
Zsolt Rónai[1,2]

[1]Novartis Agricultural
 Discovery Institute,
 La Jolla, CA, USA
[2]Institute of Medical Chemistry,
 Molecular Biology and
 Pathobiochemistry,
 Semmelweis University,
 Medical School,
 Budapest, Hungary

Ultrathin-layer gel electrophoresis of biopolymers

Emerging need for large-scale, high-resolution analysis of biopolymers, such as DNA sequencing polymerase chain reaction, (PCR) product sizing, single nucleotide polymorphism (SNP) hunting and analysis of protein molecules necessitated the development of automated and high-throughput gel electrophoresis based methods enabling rapid, high-performance separations in a wide molecular weight range. Scaling down electric field mediated separation processes supports higher throughput due to the applicability of higher voltages, thus speeding up analysis time. Indeed, efforts in miniaturization resulted in faster, easier, less costly and more convenient analyses, fulfilling the needs of the emerging biotechnology industry for microscale and massively parallel assays. The two primary approaches in miniaturizing electrophoresis dimensions are the capillary and microslab formats. This latter one evolved towards ultrathin-layer gel electrophoresis which is, except from the thickness of the separation platform, slightly in the upper side of the scale, resulting in considerably easier handling. Ultrathin-layer gel electrophoresis combines the advantages of conventional slab-gel electrophoresis (multilane format) and capillary gel electrophoresis (rapid, high-efficiency separations). It is readily automated, automatic versions of it have been extensively used for large-scale DNA sequencing in the Human Genome Project and more recently became popular in high throughput DNA fragment analysis. Ultrathin-layer techniques are the first step towards the wider use of electrophoresis microchips in perfecting a user-friendly interface between the user and the microdevice.

Keywords: Ultrathin-layer gel electrophoresis / Microelectrophoresis / Microgel / DNA / Protein / High-throughput separation / Review   EL 4196

Contents

Correspondence: Dr. András Guttman, Novartis Agricultural Discovery Institute, 3115 Merryfield Row, La Jolla, CA 92121, USA
E-mail: andras.guttman@nadii.novartis.com
Fax: +858-812-1097

Abbreviation: TBE, Tris-borate-EDTA buffer

1 Introduction: Early miniaturization attempts

Miniaturization of electric field mediated separation systems was first attempted as early as a half a century ago in order to find faster and higher resolution electrophoresis techniques and media [1]. In the 1950's, Edstrom first described picogram-scale electrophoresis separation of nucleic acids along a silk thread [2]. A decade later, Matioli and Niewisch [3] developed a polyacrylamide fiber based microelectrophoresis method for the separation of hemoglobin variants from individual erythrocytes. They attained very rapid separations by applying electric fields as high as 1000 V/cm. Narrow bore capillary tubes filled with appropriate gels also provided good separation performance offering a more generally applicable technique [4]. In the late 60's, Catsimpoolas [5] reported on a microversion of isoelectric focusing (IEF) in cylindrical poly-

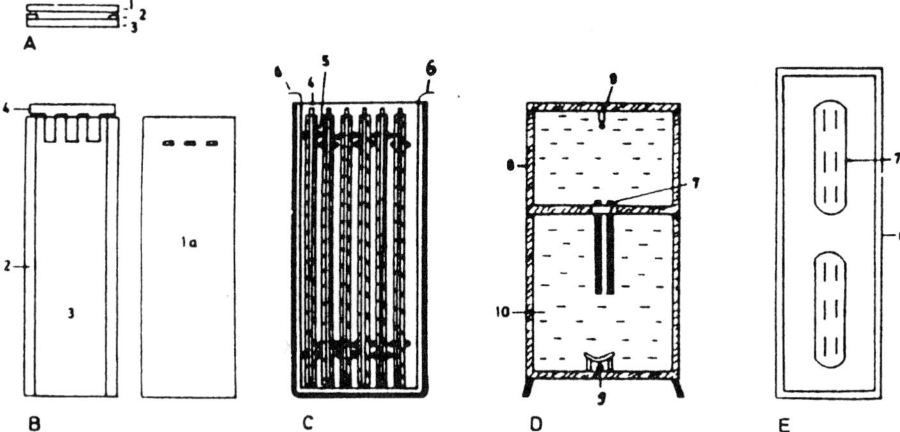

Figure 1. Early microslab-gel electrophoresis equipment (size: 25 mm × 75 mm). Gel cell: (A) cross view, (B) top view. 1, Cover slide, (1a: cover slide with rectangular holes); 2, glass strips; 3, back slide; 4, Teflon comb. (C) Beaker with gel cells. 5, O-ring; 6, plastic net strip. (D), (E) Electrophoresis apparatus. From [12], with permission.

acrylamide gels, which method was later refined by other groups [6]. In the early 70's, Dames and Maurer [7] described the preparation of gel concentration gradients in 1–50 µL capacity capillary tubes, thus increasing the relevance of microgel-based separations. In the same time, Neuhoff [8] separated DNA and RNA molecules on 2 mm long polyacrylamide gels in narrow bore glass tubes and called it microelectrophoresis. His progress was ultimately hampered by the inability to control extra Joule heat during the application of the extremely high separation voltages. The relationship between the dimensions of the electrophoresis separation platform and the accumulations of Joule heat are now fully understood. First Karger and co-workers [9] later Grossmann *et al.* [10] concluded that one of the key factors influencing the extent of Joule heat is the thickness of the separation platform. Indeed, electrophoresis in capillary dimensions proved to be much less susceptible to effects of Joule heat, because of the ability to dissipate heat *via* the large surface areas typical of capillary columns and ultrathin-layer platforms. Obtaining high surface to volume ratios in planar (ultrathin-layer) format electrophoretic systems require very thin layers, preferably reaching thickness of the capillary dimensions. Separation platforms no thicker than 0.25 mm allow the application of high electric field strengths (up to hundreds of V/cm), therefore, enabling rapid and efficient separations of biopolymers [11].

Maurer and Dati [12] fabricated an early microslab-gel apparatus employing regular microscope slides of 75 mm × 25 mm × 0.75 mm (Fig. 1), and used it for elec-

trophoretic separation of human serum protein samples. The same group also used microslab flat gels for immunoelectrophoresis. A thin-layer gel system was also successfully used for efficient IEF [13] followed by miniaturization of 2-D protein electrophoresis almost a decade later by Ruchel [14] and Poehling and Neuhoff [15]. One of the reasons that these microslab electrophoresis methods were not exploited to a greater extent after the initial demonstrations, was probably due to the inadequacy of imaging techniques at the time, *i.e.*, to capture high-resolution separation images in such minute scales. Other problems with possible leaking, formation of air bubbles in the separation medium and the hardship of sample application into the microgels represented significant difficulties, as well. Even in many current applications, the necessary decrease in sample volume required by the ultrathin-layer format has limited its use, due to the lack of appropriate injection methodology. However, recently introduced techniques such as membrane-mediated loading [16] and capillary sample injection [17] successfully addressed this issue.

2 Recent advances in ultrathin-layer gel electrophoresis of DNA

Electric field mediated separation of DNA molecules, such as sequencing ladders, restriction digestion fragments, PCR products, *etc.*, is usually accomplished in agarose, polyacrylamide or composite agarose-polyacrylamide gels [18]. At slightly basic separation conditions (7<pH<9), DNA molecules are negatively charged, so they are loaded at the cathode end of the separation platform and migrate towards the anode under the influence of the electric field. Using denaturing separation conditions the electrophoretic mobility of DNA fragments is primarily determined by their size, while under nondenaturing separation conditions it is also influenced by the sequence-dependent secondary structure [19]. Crosslinked polyacrylamide gels are regularly used to attain high-resolution separation of single-stranded DNA molecules from several bases up to a thousand bases. Agarose gels are usually employed to analyze double-stranded (ds) DNA molecules ranging from hundreds of base pairs (bp) to tens of thousands of base pairs. Albeit, some of the large, automated DNA sequencing systems have recently been reported to accommodate applications other than sequencing (*e.g.*, genotyping and STR profiling [20] using cross-linked polyacrylamide gels and fluorescently labeled primers), the configuration of these devices does not accommodate the use of large pore size agarose gels as separation medium. In addition, all the reported methods on those commercially available automated sequencing systems require preseparation covalent fluorophore labeling.

Conventional gel electrophoresis based separation of DNA fragments relies on the complex relationship of several parameters, including gel pore size, electric field strength, ionic moiety, buffer composition and conductivity, as well as on the type of DNA fragment migration through the porous gel matrix. Smaller DNA fragments tumble through the pores of much larger average radii that of the gyration radius of the fragments. Thus, they become size-separated on the basis of time required to find their path through the porous gel matrix [21]. Larger fragments, *i.e.*, whose radii of gyration are larger than that of the average pore size of the gel, become elongated towards the electric field in order to pass through the smaller pores. This phenomenon is referred to as reptation and mainly induced through increases in either the gel concentration (*i.e.*, decreasing the pore size) or the applied electric field strength and in extreme instances, it may even result in size independent migration of the analyte DNA molecules [22].

The most recent advances in the electrophoretic separation of nucleic acids came from the exploration of novel separation matrices. Linear polymers, such as noncross-linked polyacrylamide [23], derivatized celluloses [24] and polyethylene oxides [25], have all been demonstrated to be effective in electric field mediated size separation of single- and double-stranded DNA molecules. The advantages of these noncross-linked polymers have been demonstrated almost entirely in high-performance capillary electrophoresis applications [26], albeit, it had been

shown that very high concentration linear polymers can also be used in planar separation format [27, 28]. Agarose gel-filled capillaries were extensively studied by Boček and Chrambach [29] for the separation of dsDNA molecules. Chemically modified agarose gels or composite agarose – noncross-linked polymer gels, capable of resolving several base pair differences in DNA fragments of several hundreds of base pairs in length, have also been developed [30]. The employment of noncross-linked, linear polymers for DNA fragment analysis applications may be advantageous in several respects. First, it has been shown that noncross-linked polymers can be supplied in a desiccated, dry form, providing long shelf life [31]. Second, planar form linear polymers can be rehydrated to any of the range of final gel concentrations, buffer compositions and/or ionic strengths [32]. Note that lower viscosity noncross-linked polymers are easily replaced even within capillary formats [33], therefore, ultrathin-layer platforms supporting repetitive work can be readily used with these matrices.

Visualization of nucleic acids separated by gel electrophoresis was usually accomplished by covalent labeling with fluorescent tags, such as fluorescein, tetramethylrhodamine, Texas Red *etc.* [34]. This approach is extensively used for high-sensitivity DNA analysis, provided the analyte is covalently labeled by the appropriate fluorophore prior to the separation process (*e.g.*, DNA sequencing applications). Another approach employs noncovalent affinity binding (*e.g.*, intercalation) dyes for *"in migratio"* labeling of dsDNA fragments during their electrophoresis separation. Noncovalent, instantaneous, fluorophore labeling of DNA fragments by intercalation expands the detection sensitivity and separation potential of gel electrophoresis [35], since unlabeled DNA fragments can be readily visualized as they become labeled during the separation process. In addition, the complexation phenomenon usually increases the separation selectivity, resulting in higher resolution [36]. In theory, the complexing dye (ligand:L^+), intercalates between the strands of double-stranded DNA (P^{n-}) molecules and due to its positive charge (at the separation pH of 8.4), it results in decreased electrophoretic mobility of the DNA-ligand complex ($PL_m^{(n-m)-}$) by reducing its charge-to-mass ratio:

$$P^{n-} + mL^+ \Leftrightarrow PL_m^{(n-m)-} \tag{1}$$

$$K = \frac{[PL_m^{(n-m)-}]}{[P^{n-}][L^+]^m} \tag{2}$$

where K is the formation constant of the complex, m is the number of the positively charged fluorescent dye molecules in the complex, and n is the total number of negative charges on the DNA molecules (in this instance, equal to the number of phosphate groups). Double-stranded DNA molecules have approximately one ethidium bromide binding site for five base pairs, slightly influenced by the dye/polymer ratio and salt concentration but most importantly, there is no reported base composition selectivity of this binding [37]. While, for example, a commonly used intercalator, ethidium bromide, affects the mobility and resolution of double-stranded DNA molecules, it does not bind to single-stranded oligonucleotides and therefore has no significant effect on their migration and separation. The appropriate concentration and type of the fluorophore complexing dye should be optimized for the excitation laser used in the illumination/detection system.

2.1 Polyacrylamide gel-based systems

2.1.1 DNA sequencing

The first powerful DNA sequencing techniques were reported in the late 70's by Maxam and Gilbert (chemical method) [38] and Sanger *et al.* (chain termination method) [39]. Both of these methods depended on subsequent high-resolution polyacrylamide gel electrophoresis separation to resolve oligonucleotides, each differing by one nucleotide in size. The chain termination sequencing method proved to be simpler, quicker and ready for easy automation. Originally, the technique employed radioactive labeling of DNA fragments, generated in four sets of sequencing reactions terminated by one of the corresponding dideoxynucleotides, and consequently size separation by polyacrylamide gel electrophoresis. Using cross-linked polyacrylamide sieving medium and denaturing conditions (*e.g.*, 7 M urea and 50°C), the migration velocity of the apparently identical mass-to-charge ratio of DNA fragments is almost strictly dependent on their size (chain length); therefore, the labeled fragments propagate as discrete bands with different velocities [40]. The detection of the separated radioactively labeled fragments was accomplished by postseparation autoradiography [41], and the sequence information was obtained by manually evaluating the patterns of the different lanes corresponding to the G-, A-, T-, and C-terminated reactions. Although this was the first viable method for large-scale DNA sequencing, it was tedious and labor-intensive.

In their original publication in 1986, Hood and co-workers [42] reported on fluorescent detection in DNA sequence analysis using polyacrylamide gels, initiating the since large-scale development in automated DNA sequencing technology. They used fluorophore labels attached to the sequencing primers in a way, that four different color fluorophores were used in each of the reactions (G, A, T and

C). The combined reaction mixtures were then separated and evaluated by high-resolution ultrathin-layer polyacrylamide gel electrophoresis. The fluorescently labeled DNA fragments were illuminated by a narrow-band light source (laser), focused into a small spot onto the separation gel at the wavelength that was optimal to excite the fluorophore. This technique usually employed one or two laser line sources being focused onto the separation gel and a photomultiplier-based detection mechanism on the same or the opposite side of the separation platform [43]. A translation stage moved the detector head across the detection zone providing continuous collection of the emitted fluorescent light from the migrating bands [44].

At the end of the 80's, other groups also reported on so-called automated DNA sequencing, *i.e.*, automated reading of the sequence information (base calling) of the separated DNA fragments [45, 46]. All these techniques used either one, or a set of four fluorophores (labeled primers and or labeled dideoxynucleotide terminators) to label the corresponding G, A, T and C reactions, and the resulting fragments were separated by ultrathin-layer polyacrylamide slab-gel electrophoresis. For example, Ansorge and de Mayer [47] employed tetramethyl-rhodamine labeled primers in the sequencing reaction and detected by laser-induced fluorescence based method during the separation process obtaining 0.1 fmol per band sensitivity using thinlayer (0.2 mm) polyacrylamide gels. The same group used porous-comb/membrane-mediated loading technology for simultaneous loading of 200 DNA sequencing samples on vertical and horizontal ultrathin gels [48].

In high-throughput automated DNA sequencing, the thickness of ultrathin-layer gels ranges from 0.050 to 0.25 mm. Novel discontinuous buffer systems proved to be useful during ultrathin-layer gel electrophoresis, especially when high voltages were applied [49]. Using a 0.5 mm thick unbuffered stacking gel connected to a 0.05 mm thick buffered separation gel resulted in easy loadability and excellent separation of DNA sequencing fragments. This novel discontinuous gel-buffer system also allowed sample preconcentration and seemed to be beneficial by avoiding band compression of GC rich regions. Stein *et al.* [50] introduced a simple miniaturized gel system suitable for DNA sequencing. The small ultrathin polyacrylamide gels were run horizontally in a standard minigarose apparatus (Fig. 2). Their sample loading system permitted volumes of standard sequencing reactions as small as 0.1 μL to be analyzed. In order to increase separation efficiency, Brumley and Smith [51] circulated coolant directly under the glass plates of the ultrathin-layer horizontal polyacrylamide gel electrophoresis apparatus which resulted in heat exchange that was approximately

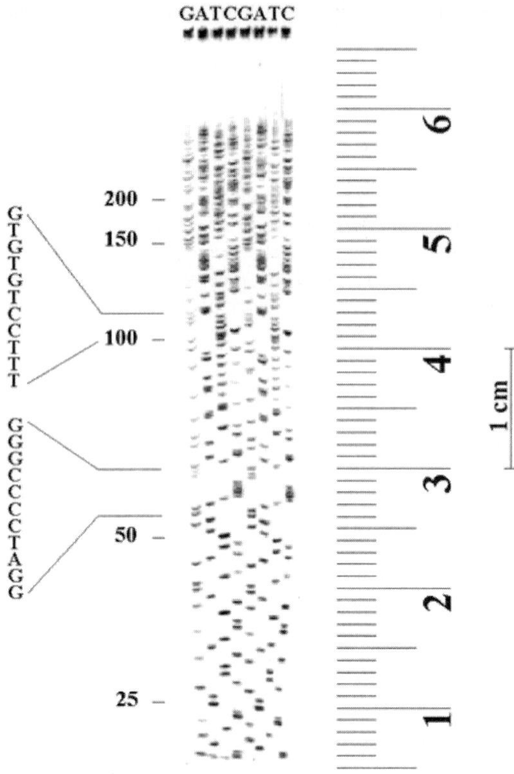

Figure 2. Autoradiogram of ^{35}S-labeled sequencing reaction separated on a miniaturized polyacrylamide gel. Samples were loaded in duplicate (GATC GATC). Separation time was 7 min. The autoradiogram was printed alongside a transparent ruler with a centimeter scale. From [50], with permission.

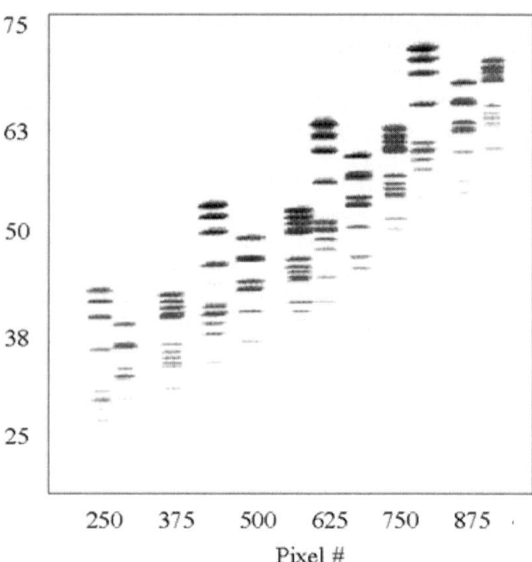

Figure 3. Ultrathin-layer gel electrophoresis separations of digested φX174, pUC18 and pBR322 DNA using capillary-assisted injection method. Conditions: 6% polyacrylamide gel (0.057 mm thickness), 16 cm total separation length (11.5 cm to the detector); applied field strength, E = 125 V/cm; injection capillary, 42 cm (0.1 mm ID); injections: 3 s at 15 kV every 3 min. From [56], with permission.

nine times more efficient than commonly used passive thermal transfer methods. When the system was used in conjunction with ^{32}P-based autoradiography, the DNA bands appeared substantially sharper than those obtained in conventional electrophoresis.

2.1.2 Double-stranded DNA fragment analysis

In the early 90's, Heller and Tullis [52] developed an instrumental micromethod that significantly reduced the linear dimensions necessary for electrophoretic separation of dsDNA fragments. They used high-concentration cross-linked polyacrylamide gels of 12–18%T / 6%C and obtained fast and effective separations of DNA fragments in very short running distances (<2 mm). The complete separation patterns were instantly detected and imaged

by means of an epifluorescent microscope. Later MacDonell and Roszak [53] introduced high-concentration linear (noncross-linked) polyacrylamide gels (up to 14%) for successful separation of nucleic acids in planar ultrathin-layer format having 0.1 mm gel thickness. However, their lack of ability to apply electric field strengths higher than several volts per cm resulted in relatively long separation times of 30–40 min. Horizontal ultrathin-layer polyacrylamide gel electrophoresis was implemented by van den Berg [54] in a multizonal format for large-scale analysis of PCR products in order to reveal short sequence repeat (SSR) polymorphisms. Detection of dsDNA fragments in the range of 200–3000 bp was accomplished by silver staining and up to 400 PCR samples were analyzed in 2 h.

Ewing's group [55] introduced a novel ultrathin-layer method for DNA fragment analysis, which combined the parallel processing capabilities of slab gels with the advantages of sample introduction obtained with a single capillary column. Ultrathin-layer slab gels were fabricated by using 57 μm spacers between quartz plates, and a single capillary was used to introduce dsDNA fragments into the gel-filled ultrathin-layer separation platform (Fig. 3). They obtained spatially resolved detection using an argon ion laser based fluorescence excitation and CCD detection system [56]. The capillary sample introduction approach allowed multiple samples to be rapidly deposited on the edge of the ultrathin slab gels for consequent separation. They applied the method for large-scale analysis of PCR amplified STR samples. Ultrathin-layer nondenaturing polyacrylamide gel and ethidium bromide labeling with laser-induced fluorescence was used to separate and detect the amplified fragments. This technique exhibited a great potential to increase the throughput of STR analysis techniques [57].

2.2 Agarose gel-based ultrathin-layer systems

Agarose gels are characterized as large pore size, high mechanical strength, and biologically inert separation matrices [58]. Despite of numerous refinements in electric field mediated agarose gel-based separation techniques over the past decades, this process is still not efficient enough and hardly automated. The time to prepare, load and separate DNA fragments by conventional agarose gel electrophoresis, plus staining and visualization times are all added up. Although attempts were made to automate this technology [59], DNA analysis by agarose gel electrophoresis in most laboratories is still done in a conventional way: using submarine gels and postseparation visualization techniques. The recently introduced, high-sensitivity fluorescent dyes provided the possibility of employing regular or electronic CCD cameras to take pictures of transilluminated gels for data evaluation and archiving [60]. By all means, the existing technology of separating DNA fragment molecules using conventional gel electrophoresis is a task that requires multiple labor-intensive steps, such as gel casting, sample loading, imaging, documentation and data evaluation. As agarose is the medium of choice for the separation of relatively large dsDNA molecules with conventional slab-gel electrophoresis, it can also be used to separate DNA in capillary dimensions as was reported earlier by Compton and Brownlee [61]. Since then, other research groups [62] have reported promising results with agarose-filled capillaries using purified grades of low electroendosmosis (EEO) agaroses to avoid electroosmotic flow mediated disruption of the separation in capillary dimensions.

Recent publications of the Guttman group [63, 64] reported the development and implementation of a novel, automated ultrathin-layer agarose gel electrophoresis system, equipped with integrated scanning laser-induced fluorescence/avalanche photodiode detection, for large-scale DNA fragment analysis. Figure 4 shows the picture of their automated ultrathin-layer agarose gel electro-

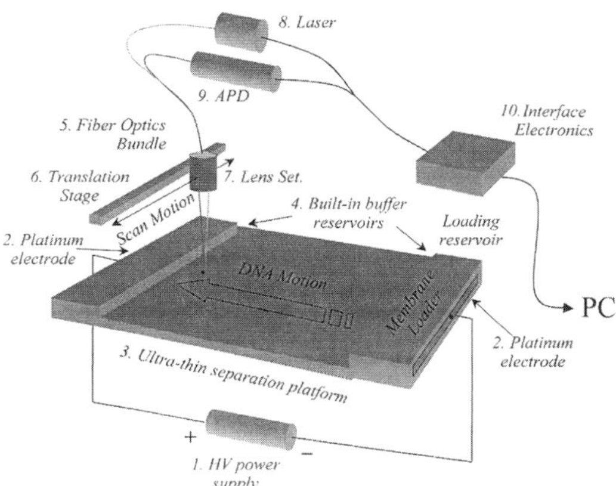

Figure 4. Block diagram of the automated ultrathin-layer agarose gel electrophoresis system. From [63], with permission.

phoresis setup, consisting of a high-voltage power supply, ultrathin-layer separation cassette with built-in buffer reservoirs and a fiberoptic bundle based scanning detection system. A lens set, connected to the illumination/detection block *via* fiberoptic bundle, scans across the ultrathin-layer separation gel by means of a high-speed translation stage. A laser excitation source (532 nm frequency doubled Nd-yttrium-aluminium-garnet (YAG) laser) and an avalanche photodiode are connected to the central excitation fiber and the surrounding collecting fibers of the fiberoptic bundle, respectively [65]. Interface electronics is used to digitize the analog output of the detector and to connect the system to a personal computer. The horizontal assembly also included a positional heat sink to hold the gel-filled cartridge and to eliminate local heat-spot generated separation irregularities by means of homogeneous dissipation of any extra heat over the gel surface during the separation.

The flat bottom and top plates of the ultrathin-layer separation cassette were joined and secured in a parallel manner spaced 190 μm apart. Buffer reservoirs were permanently fixed to both ends. For filling the ultrathin-layer separation platform, a syringe is used with a sealing applicator nozzle that matches the top of the buffer reservoirs [66]. To introduce the melted agarose into the separation cassette, the nozzle is placed at the top of one of the reservoirs, and the gel is pumped into platform. In order to prevent premature solidification of the separation matrix, the separation cassette was preheated in a microwave oven for 45 s (40–50°C) prior to the introduction of the melted agarose gel. For visualization, 10–50 nM ethidium bromide was added to the melted agarose solution just

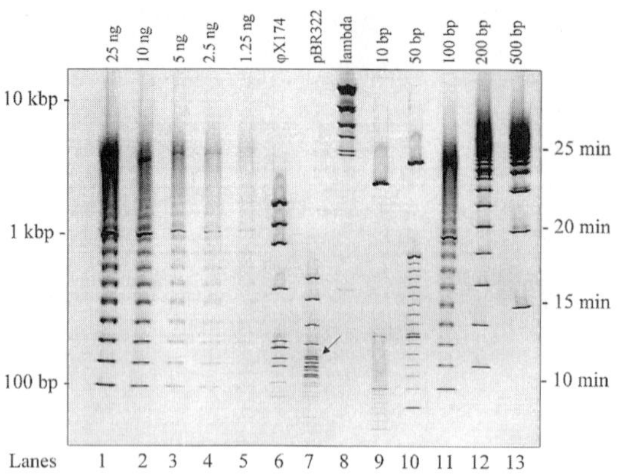

Figure 5. Automated ultrathin-layer agarose gel electrophoresis separation of DNA sizing ladders and restriction digest fragment mixtures. Lanes (1)–(5) serial dilution of a 100 bp ladder (25, 10, 5, 2.5, and 1.25 ng total DNA injected); (6) φX174 DNA *Hae*III restriction fragment mixture; (7) pBR322 DNA *Msp*I restriction fragment mixture (arrow labels separation of 238/242 bp fragments); (8) lambda DNA *Hin*dIII restriction fragment mixture; (9)–(13) 10, 50, 100, 200, and 500 bp DNA sizing ladders. Conditions: effective separation length, 6 cm; separation matrix, 2% agarose gel in 0.5 × Tris-borate-EDTAl (TBE) buffer containing 25 nM ethidium bromide; running buffer, 0.5 × TBE; applied voltage, 750 V; temperature, 25°C; injection, membrane-mediated (0.5 µL/tab). From [63], with permission.

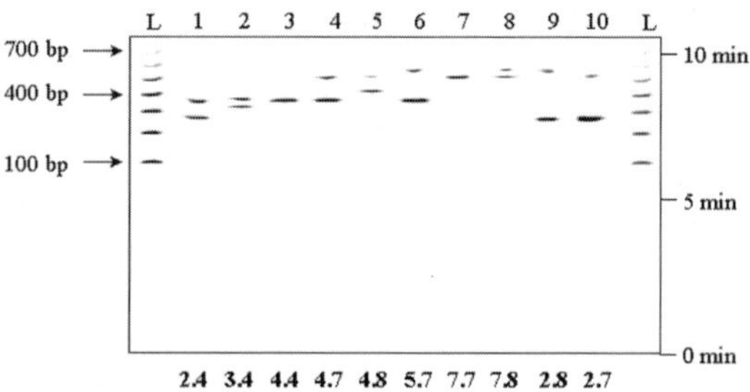

Figure 6. High-throughput genotyping of dopamine D4 receptor short tandem repeats by ultrathin-layer agarose gel electrophoresis. The number of the repeat sections are given under the separation picture (*e.g.*, 2.4 means 2 × 48 bp and 4 × 48 bp). Conditions: 1% agarose – 2% linear polyacrylamide composite gel matrix in 0.5 × TBE buffer containing 50 nM ethidium bromide; running buffer, 0.5 × TBE; applied voltage, 750 V; effective separation length, 4 cm; current, 8 mA; ambient temperature. From [72], with permission.

before filling into the cassette. The inner surface of the ultrathin-layer agarose gel electrophoresis cassette was coated with linear polyacrylamide in order to avoid the formation of an electric double layer and concomitant electroosmotic flow generation [67].

Membrane-mediated sample loading provided a reliable and easy loading mechanism for ultrathin-layer gels, and it was applicable for both vertical and horizontal formats [68]. The samples were spotted manually or automatically (robots) onto the surface of the loading membrane tabs, outside of the separation/detection platform. The sample-spotted membrane was then placed into the injection (cathode) side of ultrathin-layer separation platform, in intimate contact with the straight gel edge. Under the influence of the electric field, the sample components migrated into the gel. There was no need to form individual injection wells in the separation gel and loading was accomplished easily on the bench-top, outside of the separation platform. This novel sample injection method is readily automated [69] and can be applied to most high-

throughput thinlayer slab-gel electrophoresis based DNA analysis applications (*e.g.*, automated DNA sequencing [70]).

Figure 5 exhibits simultaneous high-performance separation of various dsDNA fragment mixtures over a broad size range of 20 bp–23 130 bp, using a single agarose gel composition. The migrating bands were visualized by *"in migratio"* ethidium bromide staining (50 nM). The first five lanes depict the dilution series of the 100 bp DNA ladder representing 25, 10, 5, 2.5, and 1.25 ng total amount of DNA injected (lanes 1–5), corresponding to 0.863, 0.30, 0.17, 0.08, and 0.04 ng per band, respectively. Based on these results, the limit of detection (LOD) of the automated high-performance ultrathin agarose gel electrophoresis system using laser-induced fluorescence/avalanche photodiode detection was found to be 0.04 ng DNA per band. Lane 6 shows a separation of the φX174 DNA *Hae*III restriction digest mixture ranging from 72 to 1353 bp. Lane 7 depicts a rapid (<17 min) and extremely high-resolution separation of the pBR322 DNA *Msp*I restriction digest mixture. The high separation efficiency of the system enabled to obtain baseline resolution of the four base pair difference of the 242 and 238 bp fragments (see arrow). Lane 8 exhibits the separation of the lambda DNA *Hin*dIII restriction digest fragments on the same gel, ranging up to 23 130 bp in size. Lanes 9–13 depict the separations of 10, 50, 100, 200, and 500 bp DNA sizing ladders, respectively. The high resolving power of the system is demonstrated here by the nice separation of the ten base pair ladder (lane 9) ranging from 20 to 320 bp. Similar, high-resolution separations can be observed for the other ladders.

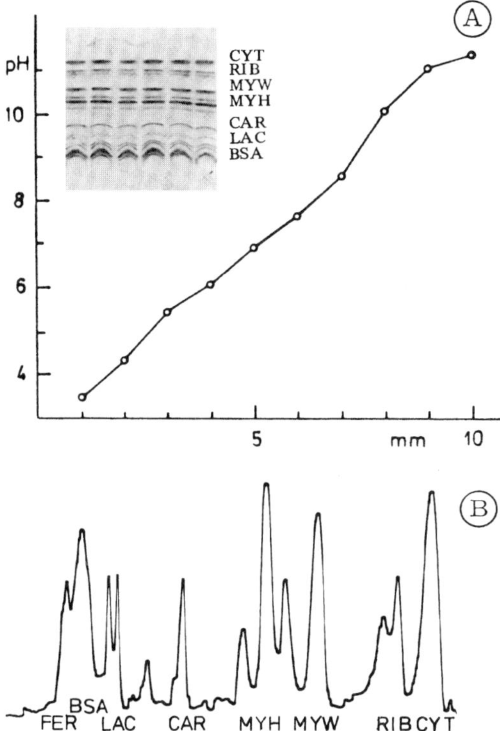

Figure 7. Ultrathin-layer IEF of pH marker proteins (FER, ferritin, BSA, bovine serum albumin; LAC, β-lacto-globulin; CAR, carbonic anhydrase; MYH, horse myoglo-bin; MYW, sperm whale myoglobin; RIB, ribonuclease; CYT, cytochrome *c*). Conditions: Gel thickness, 20 µm, pH 4–9 carrier ampholytes; separation distance, 1 cm; focusing voltage, 800–1000 v/cm; focusing time, 1 min (A) pH gradient (inset: picture of the stained separation gel). (B) Electropherogram. From [79], with permission.

Figure 6 depicts an example of rapid and simultaneous analysis of PCR-generated genotyping samples using ultrathin-layer agarose gel electrophoresis. It has been reported that such human behavioral traits as novelty seeking, hyperactivity disorders and substance abuse may possibly be associated with the presence of a certain 48 bp variable number tandem repeat polymorphism of the human dopamine D4 receptor gene (DRD4) [71]. A highly optimized genotyping method was elaborated, allowing simultaneous PCR amplification with large differ-ences in amplicon sizes (379 bp *vs.* 667 bp) using nano-grams of DNA template. High-throughput separation and instant visualization of the PCR products were accom-plished within 10 min by automated ultrathin-layer aga-rose gel electrophoresis analysis. Rapid detection of a

wide range of possible genotypes, including such rear heterozygotes as the 2 × 48 bp and 8 × 48 bp repeats (referred to as 2.8 in Fig. 6) proved the reliability of the method and its ability to detect the longer alleles in the presence of shorter alleles [72]. The high sensitivity of the laser-induced fluorescence detection for the ethidium bro-mide stained DNA fragments was especially important where deoxynucleotide analogues were used during PCR amplification. Automated ultrathin-layer agarose gel elec-trophoresis provided dependable and high-throughput genotyping that led to better understanding of genetic fac-tors in normal and pathological human behavior.

3 Analysis of proteins by ultrathin-layer gel electrophoresis

Applications of ultrathin-layer gel based electrophoresis methods for the analysis of protein molecules were almost exclusively limited to IEF and sodium dodecyl sul-fate-polyacrylamide gel electrophoresis (SDS-PAGE), as they also represent the most frequently used two tech-niques in regular electrophoresis based protein analysis. The most common separation matrix of choice was cross-linked polyacrylamide, however, the use of noncross-linked polymers, such as linear polyacrylamide, high mo-lecular weight dextran and polyethylene oxide have also been reported [73], and recently, special agarose gels were also introduced [74]. The quite adequate mechani-cal strength even at low gel concentrations, biological inertness and stability in the pH range of 4–9 made aga-rose a popular electrophoresis separation matrix. The rel-atively large temperature difference between the gelling and the melting temperatures of agarose allowed its use even at electric fields as high as > 40 V/cm, especially under ultrathin-layer separation conditions, where dissi-pation of extra Joule heat is efficient. In order to increase both the mechanical stability and the resolving power of agarose gels, composite separation matrices were intro-duced [75], usually comprising agarose mixed with other polymers, such as cross-linked or linear polyacrylamide or polyethylene oxide.

Besides the extensive miniaturization efforts of the two main applications (IEF and SDS-PAGE), limited attempts were made on separating native proteins in ultrathin-layer format. Maly and Toranelli [76] separated lactate dehydrogenase isoenzymes on native, ultrathin-layer (0.15 mm) polyacrylamide gels. They incorporated a mini-ature size wells into the gel and accomplished all analyti-cal steps under paraffin oil to prevent exposure to air. With this method, the heterogeneous distribution pattern of the lactate dehydrogenase isoenzymes of various mammals was demonstrated from several nanograms of microdissected liver tissue samples. In another approach,

Electrophoresis 2000, *21*, 3952–3964

ultrathin gels were covalently bound onto tiny glass plates and native proteins were separated under basic and acidic conditions on the bound polyacrylamide gels. The whole procedure required 100–120 min from gel preparation to the end of the electrophoresis [77].

3.1 IEF

After the early attempts of Radola's group [78, 79] (Fig. 7), ultrathin-layer IEF of proteins was revitalized during the late 80's. Budowlee and Eberhardt [80] used thin-layer polyacrylamide gel IEF for hemoglobin typing and clearly resolved the A, F, S, C forms and a number of other, rare variant allelic products. Their method proved to be excellent for rapid screening of a large number of blood samples. Others reported that IEF in ultrathin-layer agarose gels resulted in equally reliable performance for phenotyping erythrocyte acid phosphatase (EAP) as the well established regular polyacrylamide slab-gel based IEF methodl [81]. They focused the sample proteins within a thin (0.168 mm) agarose gel and obtained adequate results in 90 min, providing a cost-effective phenotyping method. Inczedy-Marcsek and co-workers [82] performed protein mapping by ultrathin-layer horizontal electrophoresis and IEF to study protein extraction. The 0.12–0.36 mm thick polyacrylamide gel layer was deposited onto tiny glass plates (*e.g.*, microscope slides). The method allowed them to analyze 1 ng tissue culture specimens. Immobilized pH gradients were extensively studied in ultrathin gel formats featuring very short focusing times [83]. Ultrathin-layer polyacrylamide IEF was also used in two-dimensional analysis of plant and fungal proteins.

Marlow *et al.* [84] reported on the use of 0.2 mm semirigid backing (polyester) supported IEF gels as the first dimension. This ultrathin-layer IEF gel quickly dried on the backing after the focusing step. Then narrow strips were cut from the dried gel and easily applied to the second dimension. Another interesting agricultural application was published by van den Berg [85], who used ultrathin-layer IEF for inbred testing of seed proteins from tomato variants. With this new methodology, he found several genetic variants among open-pollinated and hybrid varieties. The throughput of his micromethod was as high as almost 800 seed sample analysis per day. Yakhyayev *et al.* [86] used cellophane foil to support ultrathin-layer IEF gels. As the polyacrylamide gel was firmly attached to the cellophane foil, it provided good protection from mechanical damage and enabled easy handling. Since cellophane is permeable to ions, the use of this combination alleviated difficulties of the removal of the thin gels from the support. Using a similar approach, proteins were also separated under nondenaturing conditions and transferred onto a nitrocellulose membrane followed by enzyme assay based detection.

3.2 Ultrathin-layer SDS-gel electrophoresis

Electrophoresis in polyacrylamide or other gels containing ionic detergents, such as SDS has proven to be a powerful tool for size separation of protein molecules, estimation of their molecular mass and assessment of their purity [87]. In the presence of a thiol-reducing agent, the disulfide bridges are cleaved and the detergent (SDS) dissociates proteins into their constituent subunits and binds to the polypeptide chains in a way that similar charge-to-mass ratios of the resulting complexes are obtained [88]. In gel electrophoresis, the sieving medium (polymer network structure) separates the solute molecules according to their size [89]. In SDS-PAGE of proteins a linear relationship is observed when the logarithm of the molecular mass of standard polypeptide chains are plotted against their electrophoretic mobilities [90]. The method was found to be reliable and reproducible in the molecular mass range of ten to several hundred kDa generally within 5–10% of those obtained by mass spectrometry (MS) [91]. Conventionally a variety of cross-linked polyacrylamide gels were used as sieving matrices in rod or slab formats to separate the SDS - polypeptide complexes [92]. In the past several years, gel electrophoresis based analysis of proteins has been successfully utilized in capillary dimensions. Early attempts to utilize capillary gel electrophoresis for protein separations by SDS-PAGE involved highly concentrated cross-linked polyacrylamide as sieving medium [93] using UV detection. Later, both a lower concentration of cross-linked polyacrylamide [94] and linear (noncross-linked) polyacrylamide were employed in fused-slica capillaries as sieving material [95]. Nonpolyacrylamide-type low-viscosity polymers such as dextrans or polyethylene oxides were also utilized to obtain size separations [96]. These latter ones feature excellent UV transparency at 214 nm.

Visualization of the separated SDS-protein complexes in slab-gel electrophoresis is usually performed after the separation process. Most commonly used staining methods are Coomassie Brillant Blue and silver staining. Both require up to several hours of extra time after the separation process. Furthermore, Coomassie Brillant Blue staining is not of high sensitivity and silver staining requires several labor-intensive steps and expensive chemicals, which are relatively unstable and toxic. Recently various high-sensitivity fluorescent labeling methods were introduced, such as staining with ethidium bromide, Nile red or the novel Sypro™ dyes which provide detection limits similar to silver staining (1–2 ng protein/band). Covalent fluorophore labeling with various high-sensitivity fluorescent dyes were also introduced to enhance detectability of proteins separated by SDS-capillary gel electrophoresis. Such preseparation covalent labeling was reported earlier

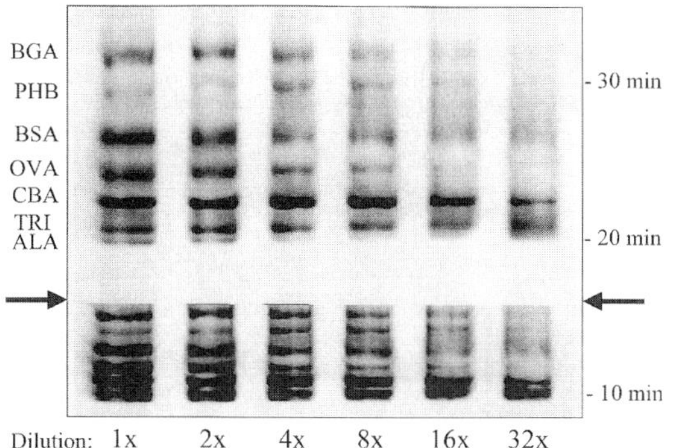

Figure 8. Ultrathin-layer SDS-gel electrophoresis of a seven-protein mixture of α-lactalbumin (ALA), trypsin inhibitor (TRI), carbonic anhydrase (CBA), ovalbumin (OVA), bovine serum albumin (BSA), phosphorylase B (PHB) and β-galactosidase (BGA) (0.2 mg/mL each) injected in a set of serial double dilutions (1, 2, 4, 8, 16 and 32-fold) and detected at 2.5 cm (lower panel) and 5.5 cm (upper panel) from the loading site. Arrows at 16 min show the time when the scanning distance was changed without interrupting the application of the electric field strength. Separation conditions: gel, 1% agarose, 2% linear polyacrylamide (LPA, M_r 700 000–1 000 000) in 50 mM Tris, 50 mM N-tris(hydroxymethyl)methyl-3-amino-propanesulfonic acid (TAPS), 0.05% SDS (pH 8.4); separation buffer, 50 mM Tris, 50 mM TAPS, 0.05% SDS (pH 8.4); separation voltage, 420 V; current, 5 mA; gel thickness, 190 µm; temperature, 25°C; sample loading, 0.2 µL into 2.5 × 4 × 0.19 mm injection wells. Sample buffer contained 0.05% SDS and 1 × Sypro Red. From [99], with permission.

by Dovichi and co-workers [97] and more recently by Hunt and Nasabeh [98], who used 5-TAMRA.SE as a very sensitive approach for monitoring consistency of biotechnology products. Guttman and co-workers [99] have introduced noncovalent, instant labeling in automated ultra-thin-layer SDS-gel electrophoresis. SDS-protein complexes were stained immediately prior to electrophoresis with a novel high sensitivity fluorophore Sypro Red™. During labeling optimization process, they found that the surfactant (SDS) concentration had a significant effect on noncovalent protein staining. When the regular amount of 1–2% SDS was used in the sample buffer, apparently a large amount of staining dye complexed with the SDS micelles. This resulted in a very intensive SDS-dye front, overlapping the lower molecular mass protein bands (< 30 000 Da). Best results were obtained using 0.05–0.1% final SDS concentration in the sample buffer.

Ultrathin-layer SDS-gel electrophoresis, a combination of the multilane format slab-gel electrophoresis and the high separation efficiency (performance) capillary SDS-gel

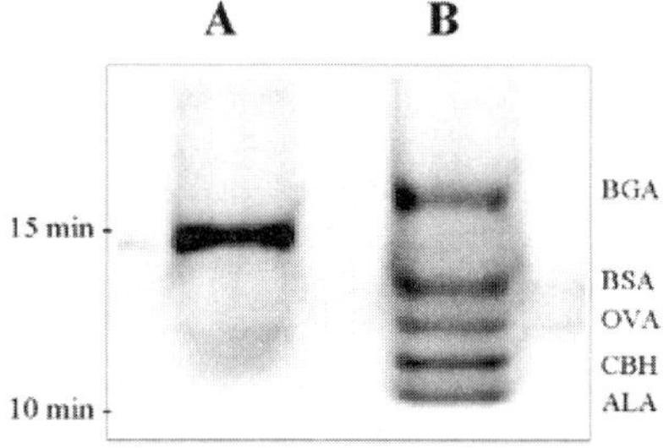

Figure 9. Lane (A) The analysis of phosphorylase *b* (PHB); (B) separation of a five-protein test mixture of α-lactalbumin (ALA), carbonic anhydrase (CBA), ovalbumin (OVA), bovine serum albumin (BSA) and β-galactosidase (BGA). Conditions are the same as in Fig. 8; effective separation length, 3.5 cm; protein concentration, 0.2 mg/mL each; sample buffer contained 0.05% SDS and 1 × Sypro Red. From [99], with permission.

electrophoresis featured rapid, high-throughput and high-resolution analysis of proteins in the moleclar mass range of 14–116 kDa [100]. Figure 8 depicts a typical automated ultrathin-layer SDS-gel electrophoresis separation of a seven-protein test mixture (α-lactalbumin (ALA), trypsin inhibitor (TRI), carbonic anhydrase (CBA), ovalbumin (OVA), bovine serum albumine (BSA), phosphorylase B (PHB), and β-galactosidase (BGA), 0.2 mg/mL each). The sample was loaded onto the ultrathin-layer SDS gel electrophoresis platform in a set of serial double dilutions up to 32-fold (1-, 2-, 4-, 8-, 16-, 32-fold). The migrating bands were first detected at 2.5 cm from the injection site (lower panel). The figure shows that almost complete separation of all proteins were obtained in less than 15 min even with this very short effective separation distance. When all the bands passed through the detection zone at 2.5 cm, the scanning detector head (similar to that described under Section 2.2) was moved to 5.5 cm from the loading site without the interruption of the run. In this way, the same proteins were detected again (upper panel) at 5.5 cm with higher resolution. One of the most important features of this consecutive double distance detection shown in Fig. 8 is the apparently high migration time fidelity of the detected bands at the 5.5 cm distance (average measured and calculated migration times erred only 0.562%). This actually means that one can accurately predict the migration time at any distance from the loading well based on the mobility values calculated from the migration time results of the first detection point, allowing precise collection of any detected proteins for further characterization by microsequencing or MS. Please note, that even in 32-fold dilution, the trypsin inhibitor and the carbonic anhydrase bands were still clearly visible. In this instance the detected amount of these proteins corresponded to LOD = 1.25 ng/band (60 and 42 fmol for the trypsin inhybitor and carbonic anhydrase, respectively).

Automated ultrathin-layer SDS-gel electrophoresis can be readily applied for the analysis of protein mixtures and for molecular mass estimation of the proteins in the sample. Lane A in Fig. 9 depicts the automated ultrathin-layer SDS gel electrophoresis analysis of phosphorylase B using noncovalent fluorescent labeling. A standard curve was constructed for the molecular mass estimation of the unknown by plotting the logarithmic molecular masses of the five test proteins (lane B) against their electrophoretic mobilities. The best fitting of the standard curve was attained by applying a second order polynomial function resulting in an extremely high confidence level (r^2 = 0.9999) for this relationship. This is in contrast to previous reports on linear relationship between the logarithmic molecular mass and electrophoretic mobility. The slight curvature of this calibration plot was probably caused by the noncovalent attachment of the negatively charged staining dye that increased the charge and concomitantly, the overall electrophoretic velocity of the complex. Based on the calibration curve obtained, the molecular mass of the phosphorylase B band in lane A was estimated to be 97.250, representing only a 0.25% error compared to the literature value of 97 400 [101].

4 Conclusions and outlook

Automated ultrathin-layer gel electrophoresis is a novel combination of the well-established methodologies of slab-gel electrophoresis and capillary gel electrophoresis. This review summarized the history and recent applications of this technology for high-performance analysis of biopolymers. Ultrathin-layer systems readily accommodate fluorophore labeling during the electrophoresis separation process (*"in migratio"*), such as by complexation with novel, high-sensitivity staining dyes, in addition to the use of conventional preseparation labeling. Sample loading onto the ultrathin separation platform is easily accomplished by membrane-mediated loading technology, which also enables robotic spotting of multiple samples. The ultrathin-layer format provides a multilane separation platform (a plurality of virtual channels) with excellent heat dissipation characteristics allowing the application of high voltages necessary to obtain rapid and extremely efficient analysis of biopolymers. Detection of the separated bands is accomplished in real time by continuous scanning at just a few centimeters from the injection site of the multilane separation platform. It is important to note here, that the detection of the migrating biopolymers is accomplished in a timely basis, similar to capillary electrophoresis (CE) and high-performance liquid chromatography

Table 1. Comparison of separation performance and sensitivity of automated ultrathin-layer agarose gel electrophoresis and manual submarine slab-gel electrophoresis systems

Feature	dsDNA fragment analysis	
	Automated	Manual
Gel preparation and loading	10 min	35 min
Separation	25 min	60 min
Staining/destaining	None	30 min
Imaging/evaluation	None	15 min
Total time	40 min	155 min
Number of lanes	32	14
Time per sample	1.09 min	10 min
Sensitivity (ethidium bromide)	0.04 ng/band	0.2 ng/band[a]
Separation range (single gel)	20–25 000 bp	50–500, 500-5K, 5K–50K
Resolution	4 bp at 240 bp in 12 min	4 bp at 240 bp in 2 h
Required buffer volume	25 mL	250 mL
Required gel volume	2 mL	20 mL

a) With UV transillumination and CCD camera detection imaging
From [63], with permission.

(HPLC), which is in contrast to the traditional spatial based detection in slab-gel electrophoresis. This guarantees that all solute molecules in the sample mixture travel the same distance from the injection point to the detection site.

Ultrathin-layer gel electrophoretic methods provide a number of advantages. While gel and buffer compositions are the same as those used in the conventional procedures, the efficiency is increased due to the reduced cross-sectional area. The ability to apply higher voltage gradients is due to the more efficient heat dissipation, which in turn results in shorter analysis times. Microanalytical methods also represent savings in reagent and sample consumption as well as in operating costs. Table 1 compares the separation performance and detection sensitivity of conventional submarine type agarose gel electrophoresis to the ultrathin-layer agarose gel electrophoresis platform, exhibiting significantly faster and higher sensitivity analysis for the latter in DNA fragment analysis.

Recently emerging microfluidics based analytical techniques [102] brought the promise to further speed and throughput of electric field mediated separations. Microfluidic devices equipped with cross-channel injectors enable precisely controlled loading of picoliter amounts essential for short separation distances [103], also comprising an inherent capability of multiplexing. Integrated microfabricated device technology opens up new horizons in bioseparations for the biotechnology industry. Entering the age of genomics and proteomics, we expect to see a paradigm shift towards miniaturized, high-resolution separation techniques used in integrated and automated fashion to solve formidable separation problems and provide means for ultrahigh-throughput analysis. Most current separation protocols for DNA and protein analysis, already in use in molecular biology and biotechnology labs, can be readily transferred to microfabricated devices. Another real strength of miniaturization is the possibility to integrate existing methods/functionalities in a way that allows for sample preparation, reactions, analysis and even fraction collections carried out on a single microchip (lab-on-a-chip). Microfabricated devices are intrinsically acquiescent to full automation enabling large-scale analyses with considerably less human intervention than conventional techniques, resulting in significant savings in time, labor and costs [104].

The authors greatly appreciate the help of Csaba Barta during the preparation of the manuscript.

Received August 7, 2000

5 References

[1] Neuhoff, V., *Electrophoresis* 2000, *21*, 3–11.

[2] Edstrom, J. E., *Biochim. Biophys. Acta* 1956, *22*, 378–383.

[3] Matioli, G. T., Niewisch, H. B., *Science* 1965, *150*, 1824–1826.

[4] Grossbach, U., *Biochim. Biophys. Acta.* 1965, *107*, 180–182.

[5] Catsimpoolas, N., *Anal. Biochem.* 1968, *26*, 480–482.

[6] Quentin, C. D., Neuhoff, V., *Int. J. Neurosci.* 1972, *4*, 17–24.

[7] Dames, W., Maurer, H. R., in: Allen, R. C., Maurer, H. R. (Eds.), *Electrophoresis and Isoelectric Focusing in Polyacrylamide Gel*, Walter de Gruyter, Berlin, Germany 1974.

[8] Neuhoff, V., *Micromethods in Molecular Biology*, Springer-Verlag, New York, NY 1973.

[9] Nelson, R. J., Paulus, A., Cohen, A. S., Guttman, A., Karger, B. L., *J. Chromatogr.* 1989, *480*, 111–127.

[10] Grossman, P. D., Menchen, S., Hershery, D., *Gene Anal. Tech. Appl.* 1992, *9*, 9–16.

[11] Brumley, R. L., Smith, L. M., *Nucleic Acids Res.* 1991, *19*, 4121–4126.

[12] Maurer, H. R., Dati, F. A., *Anal. Biochem.* 1972, *46*, 19–32.

[13] Radola, B. J., *Biochim. Biophys. Acta* 1969, *194*, 335–338.

[14] Ruchel, R., *J. Chromatogr.* 1977, *132*, 451–468.

[15] Poehling, H. M., Neuhoff, V., *Electrophoresis* 1980, *1*, 90–98.

[16] Cassel, S., Guttman, A., *Electrophoresis* 1998, *19*, 1341–1346.

[17] Suljak, S. W., Thompson, L. A., Ewing, A. G., in: Harrison, D. J., van den Berg, A. (Eds.), *Micro Total Analysis Systems '98*, Kluwer, Dordrecht, The Netherlands 1998.

[18] Rickwood, D., Hames, B. D., *Gel Electrophoresis of Nucleic Acids*, Oxford University Press, UK 1990.

[19] Chrambach, A., *Practice of Quantitative Gel Electrophoresis*, VCH, Deerfield Beach, FL 1985.

[20] Ballard, L. W., http://www.medstv.unimelb.edu.au/ABRFNews/1997/September1997/

[21] Ogston, A. G., *Trans. Faraday. Soc.* 1958, *54*, 1754–1757.

[22] Lumpkin, O. J., DeJardin, P., Zimm, B. H., *Biopolymers* 1985, *24*, 1573–1593.

[23] Heiger, D. N., Cohen, A. S., Karger, B. L., *J. Chromatogr.* 1990, *516*, 33–48.

[24] Schwartz, H. E., Ulfelder, K., Sunzeri, F. J., Busch, M. P., Brownlee, R. G., *J. Chromatogr.* 1991, *559*, 267–283.

[25] Chang, H. T., Yeung, E. S., *J. Chromatogr. B* 1995, *669*, 113–123.

[26] Guttman, A., in: Landers, J. P. (Ed.), *Handbook of Capillary Electrophoresis*, CRC Press, Boca Raton, FL 1994.

[27] Bode, H. J., *Electrophoresis '79*, Walter de Gruyter & Co., New York, NY 1980, p. 39.

[28] Tietz, D., Gottlieb, M. H., Fawcett, J. S., Chrambach, A., *Electrophoresis* 1986, *7*, 217–222.

[29] Boček, P., Chrambach, A., *Electrophoresis* 1991, *12*, 1059–1061.

[30] Soto, D., Sukumar, S., *PCR Methods Appl.* 1992, *2*, 96–98.

[31] Schwatz, H., Guttman, A., *Separation of DNA by Capillary Electrophoresis*, Primer #5, Beckman Instruments, Fullerton, CA 1995.

[32] Allen, C. R., Graves, G., Budowle, B., *BioTechniques* 1989, *7*, 736–744.

[33] Guttman, A., *US Patent 5,370,777*, 1994.

[34] Haugland, R. P. H., in: Spence, M. T. Z. (Ed.), *Handbook of Fluorescent Probes and Research Chemicals*, Molecular Probes, Eugene, OR 1996.

[35] Guttman, A., in: Swadesh, J. (Ed.), *HPLC, Practical and Industrial Applications*, CRC Press, Boca Raton, FL 1997.

[36] Guttman, A., Cooke, N., *Anal. Chem.* 1991, *63*, 2038–2042.

[37] LePecq, J. B., Paleotti, C., *J. Mol. Biol.* 1967, *27*, 87–106.

[38] Maxam, A. M., Gilbert, W., *Proc. Natl. Acad. Sci. USA* 1977, *74*, 560–564.

[39] Sanger, F., Nicklen, S., Coulson, A. R., *Proc. Natl. Acad. Sci. USA* 1977, *74*, 5463–5467.

[40] Freifelder, D., *Physical Biochemistry*, W. H. J. Freeman & Co, New York, NY 1982.

[41] Sambrook, J., Fritch, E. F., Maniatis, T., *Molecular Cloning*, Cold Spring Harbor Lab. Press, Plainview, NY 1987.

[42] Smith, L. M., Sanders, J. Z., Kaiser, R. J., Hughes, P., Dodd, C., Connell, C. R., Heiner, C., Kent, S. B., Hood, L. E., *Nature* 1986, *321*, 674–679.

[43] Adams, M. D., *Nature* 1994, *368*, 474–475.

[44] Ishino, Y., Mineno, J., Inoue, t., Fujimiya, H., Yamamoto, K., Tamura, T., Homma, M., Tanaka, K., Kato, I., *BioTechniques* 1992, *13*, 936–943.

[45] Prober, J. M., Trainor, G. L., Dam, R. J., Hobbs, F. W., Robertson, C. W., Zagursky, R. J., Cocuzza, A. J., Jensen, M. A., Baumeister, K., *Science* 1987, *238*, 336–341.

[46] Ansorge, W., Sproat, B. S., Stegemann, J., Schwager, C., *J. Biochem. Biophys. Methods* 1986, *13*, 315–323.

[47] Ansorge, W., de Maeyer, L., *J. Chromatogr.* 1980, *202*, 45–53.

[48] Erfle, H., Ventzki, R., Voss, S., Rechmann, S., Benes, V., Stegemann, J., Ansorge, W., *Nucleic Acids Res.* 1997, *25*, 2229–2230.

[49] Carninci, P., Volpatti, F., Schneider, C., *Electrophoresis* 1995, *16*, 1836–1845.

[50] Stein, A., Hill, S. A., Cheng, Z., Bina, M., *Nucleic Acids Res.* 1998, *15*, 452–455.

[51] Brumley, R. L., Smith, L. M., *Nucleic Acids Res.* 1991, *19*, 4121–4126.

[52] Heller, M. J., Tullis, R. H., *Electrophoresis* 1992, *13*, 512–520.

[53] MacDonell, M. T., Roszak, D. B., *GATA* 1993, *10*, 10–15.

[54] van den Berg, B. M., *Electrophoresis* 1997, *18*, 2861–2864.

[55] Ewing, A. G., Gavin, P. F., Hietpas, P. B., Bullard, K. M., *Nature Med.* 1997, *3*, 97–99.

[56] Hietpas, P. B., Bullard, K. M., Gutman, D. A., Ewing, A. G., *Anal. Chem.* 1997, *69*, 2292–2298.

[57] Bullard, K. M., Hietpas, P. B., Ewing, A. G., *Electrophoresis* 1998, *9*, 71–75.

[58] FMC, MetaPhor™ Agarose. FMC Bioproducts Application Note, 1992.

[59] Gombocz, E., Cortez, E., *Appl. Theor. Electrophor.* 1995, *4*, 197–209.

[60] Southerland, J. C., in: Chrambach, A., Dunn, M. J., Radola, B. J. (Eds.), *Advances in Electrophoresis* VCH Publishers, Weinheim, Germany 1993.

[61] Compton, S. W., Brownlee, R. G., *BioTechniques* 1988, *6*, 432–440.

[62] Boček, P., Chrambach, A., *Electrophoresis* 1992, *13*, 31–34.

[63] Guttman, A., *LC.GC* 1999, *17*, 1020–1026.

[64] Guttman, A., Barta, C., Szoke, M., Sasvari-Szekely, M., Kalasz, H., *J. Chromatogr. A* 1998, *828*, 481–487.

[65] Trost, P., Guttman, A., *Anal. Chem.* 1998, *70*, 3930–3935.

[66] Guttman, A., *Trends Anal. Chem.* 1999, *18*, 694–702.

[67] Hjertén, S., *J. Chromatogr.* 1985, *347*, 191–198.

[68] Guttman, A., *Anal. Chem.* 1999, *71*, 3598–3602.

[69] Stanchfield, J. E., Batey, D. W., Poster Presentation at *Genome Mapping and Sequencing Symposium*, Cold Spring Harbor, May 13–17 1998, p. 214.

[70] Gerstner, A., Sasvari-Szekely, M., Kalasz, H., Guttman, A., *BioTechniques* 2000, *28*, 628–630.

[71] van Tol, H. H. M., Wu, C. M., Guan, H. C., Ohara, K., Bunzow, J. r., Civelli, O., Kennedy, J., Seeman, P., Niznik, H. B., Jovanovic, V., *Nature* 1992, *358*, 149–152.

[72] Ronai, Z., Guttman, A., Nemoda, Z., Staub, M., Kalasz, H., Sasvari-Szekely, M., *Electrophoresis* 2000, *21*, 2058–2061.

[73] Bode, H. J., *FEBS Lett.* 1976, *65*, 56–58.

[74] Gerstner, A., Csapo, Z., Sasvari-Szekely, M., Guttman, A., *Electrophoresis* 2000, *21*, 834–840.

[75] Lengyel, T., Guttman, A., *J. Chromatogr. A* 1999, *853*, 511–518.

[76] Maly, I. P., Toranelli, M., *Anal. Biochem.* 1993, *214*, 379–388.

[77] Heukeshoven, J., Dernick, R., *Electrophoresis* 1992, *13*, 654–659.

[78] Radola, B. J., *Biochim. Biophys. Acta* 1973, *295*, 412–428.

[79] Kinzkofer, A., Radola, B. J., *Electrophoresis* 1981, *2*, 174–183.

[80] Budowlee, B., Eberhardt, P., *Hemoglobin* 1986, *10*, 161–172.

[81] Frank, W. E., Stolorow, M. D., *J. Forens. Sci.* 1986, *31*, 1089–1094.

[82] Inczedy-Marcsek, M., Lindner, E., Hassler, R., Zwack-Megele, G., Roisen, F., Yorke, G., *Acta. Hystochem. Suppl.* 1988, *36*, 377–394.

[83] Pascali, V. L., Dobosz, M., d'Aloja, E., *Electrophoresis* 1988, *9*, 514–519.

[84] Marlow, G. C., Wurst, D. E., Loschke, D. C., *Electrophoresis* 1988, *9*, 693–704.

[85] van den Berg, B. M., *Electrophoresis* 1990, *11*, 824–829.

[86] Yakhyayev, A. V., Voronkova, I. M., Sukhanov, V. A., *Electrophoresis* 1991, *12*, 680–682.

[87] Andrews, A. T., *Electrophoresis* Claredon Press, Oxford 1986.

[88] Weber, K., Osborn, M., *J. Biol. Chem.* 1969, *244*, 4406–4412.

[89] Hames, B. D., Rickwood, D. (Eds.), *Gel Electrophoresis of Proteins*, IRL, Washington, DC 1983.

[90] Reynolds, J. A., Tanford, C., *Proc. Natl. Acad. Sci. USA* 1970, *66*, 1002–1007.

[91] Guttman, A., *Electrophoresis* 1996, *17*, 1333–1341.

[92] Chrambach, A., *The Practice of Quantitative Gel Electrophoresis*, VCH, Deerfield Beach, FL 1985.

[93] Cohen, A. S., Karger, B. L., *J. Chromatogr.* 1987, *397*, 409–417.

[94] Tsuji, K., *J. Chromatogr.* 1991, *550*, 823–830.

[95] Widhalm, A., Schwer, C., Blass, D., Kenndler, E., *J. Chromatogr.* 1991, *546*, 446–451.

[96] Ganzler, K., Greve, K. S., Cohen, A. S., Karger, B. L., Guttman, A., Cooke, N. C., *Anal. Chem.* 1992, *64*, 2665–2671.

[97] Craig, D. B., Polakowski, R. M., Wong, J. C. Y., Ahmadzeh, H., Stathakis, C., Dovichi, N. J., *Electrophoresis* 1998, *19*, 2175–2178.

[98] Hunt, G., Nasabeh, W., *Anal. Chem.* 1999, *71*, 2390–2397.

[99] Csapo, Z., Gerstner, A., Sasvari-Szekely, M., Guttman, A., *Anal. Chem.* 2000, *72*, 2519–2525.

[100] Guttman, A., Ronai, Z., Csapo, Z., Gerstner, A., Sasvari-Szekely, M., *J. Chromatogr. A* 2000, in Press.

[101] Sigma Catalog, Sigma Chemical Co., St Louis, MO 1997, p. 1895.

[102] Jacobson, S. C., Ramsey, J. M., in: Khaledi, M. G. (Ed.), *High Performance Capillary Electrophoresis*, John Wiley & Sons, New York, NY 1998.

[103] Dolnik, V., Liu, S., Jovanovich, S., *Electrophoresis* 2000, *21*, 41–54.

[104] Manz, A., Becker, H. (Eds.), *Microsystem Technology in Chemistry and Life Sciences*, Springer-Verlag, Berlin, Germany 1999.

Review

Andreas Chrambach
Sergey P. Radko

Section on Macromolecular
Analysis, Laboratory of
Cellular and Molecular
Biophysics, National
Institute of Child Health and
Human Development,
National Institutes of Health,
Bethesda, MD, USA

Size-dependent retardation and resolution by electrophoresis of rigid, submicron-sized particles, using buffered solutions in presence of polymers: A review of recent work from the authors' laboratory

Capillary zone electrophoresis (CZE) was conducted in buffered solutions of polyacrylamide (PA) and polyethylene glycol (PEG) to find the degree and the manner in which separation and resolution of submicron-sized rigid spherical polystyrene sulfate and carboxylate particles were affected by the presence of those polymers. In resolving pairs of representative particles, maximal resolution was observed at or near the entanglement threshold concentration, c^*, of the polymer. The value of that maximum represents a several-fold increase in resolution. Since c^* can be calculated from intrinsic viscosity, and the latter from the molecular weight of the polymer (and some constants available in the literature), optimally resolving polymer conditions become predictable. The maximum can also be experimentally determined by measuring intrinsic viscosity and calculating c^*, or by either systematically varying the concentration of a polymer of constant molecular weight or by varying the molecular weight of a polymer at constant concentration. An optimally resolving field strength is superimposed on the maximally resolving condition of polymer concentration and weight.

Keywords: Capillary zone electrophoresis / Submicron-sized particles / Review   EL 3682

Contents

Correspondence: Dr. A. Chrambach, National Institutes of Health, Bldg. 10, Rm. 9D50, Bethesda, MD 20892-1580, USA
E-mail: acc@cu.nih.gov
Fax: +301-402-0263

Abbreviations: EDL, electric double layer; **PA,** polyacrylamide

1 Introduction

Within the past year, considerable advances have been made in our laboratory concerning the mechanisms by which submicron-sized particles separate in capillary zone electrophoresis (CZE) [1–4]. However, the emphasis on mechanism has necessarily led to relatively complex formulations which obscure the practical relevance of those insights. The present review of that work aims to remedy and to some degree invert that disbalance between theory and practice in order to communicate the findings more effectively to the majority of readers working in practical separation science. We restrict ourselves to CZE conducted in buffers and buffered polymer solutions since, in general, submicron-sized charged particles will be unable to migrate into gels, while they can freely penetrate into the solutions contained in a capillary. To obviate electroosmosis, capillaries internally coated with polyacrylamide were used exclusively.

1.1 The nature of submicron-sized particles under discussion

The experimental work under review deals with polystyrene latex particles that can be considered rigid and spherical. This restriction serves to avoid the increased complexity of systems comprising both flexible polymer networks and flexible particles such as those that change their conformation as a function of pH, ionic strength, and polymer concentration in the fashion of DNA fragments greater than 300 bp in length. Polystyrene size standards are the only particles in that category that are available commercially in a large assortment of sizes. They are charged by sulfate or carboxylate groups and have at a representative ionic strength of electrophoresis (about 0.01 M) radii, R, 10–100 times larger than the thickness of the electric double layer (EDL) surrounding them. They have variable surface charge densities and can either belong to the type of particles with a smooth surface, *i.e.*, which are incompatible with flow below the surface, or that of particles with surfaces covered by a layer of flexible polymers and thus allowing for flow beneath the surface of that layer ("hairy" or soft particles). It is assumed that sulfated or carboxylated polystyrene approaches the former particle type while "carboxyl modified" polystyrene and most biological particles belong to a relatively "hairy" particle type.

1.2 The nature of the polymers used

In principle, all hydrophilic uncharged polymers lend themselves to form restrictive networks above some entanglement threshold concentration, c^*, *i.e.*, in the semidilute concentration regime. However, theoretical calculations of the screening length, ξ, which may be viewed as the average distance between the overlap points of polymer chains in a network or as its mesh size, depend on numerical coefficients (in Eqs. 7–9 of [3]) which vary with the method by which they were found and to the degree to which the idealized condition of a "good solvent" for a particular polymer is met by aqueous solutions. Among water-soluble polymers tested to date – *i.e.*, polyacrylamide (PA), agarose, dextran, hydroxymethylpropyl cellulose, hydroxyethyl cellulose, polyvinyl alcohol, polyvinyl pyrrolidone, polyethylene glycol – that condition

seems to hold at least for the first and last of the above series at a temperature of 25°C. The uncertainty in the calculation of ξ due to the solvent dependence of the numerical coefficients is, however, a moot issue when we are concerned with the order of magnitude of the ratio of particle radius over screening length of the polymer network, R/ξ, rather than its precise value. A further restriction of our investigations and conclusions derives from the limited and arbitrary molecular weight range and degree of homogeneity in which commercial polymers of the above-listed types are available. Finally, it must be remembered that limiting the studies reviewed here to the semidilute concentration regime, *i.e.*, a concentration range giving rise to a polymer network in solution, is arbitrarily due to some analogy of those networks to the gels previously studied [5].

Retardation and resolution of particles in the submicron to micron size range in both the dilute and the concentrated polymer regimes have been empirically observed but have not been made the subject of systematic and theoretical studies to date. In relation to the particle size range reviewed here, in excess of 15 nm in radius, the mesh sizes of the networks formed in semidilute solutions of commercially available polymers are smaller, maybe even much smaller, than the particle size. For instance, expressed in relation to the particle radius, the average mesh size of a polymer network in a semidilute concentration range (0.3–1%) of polyacrylamide of 1 million molecular weight is characterized by the ratio of R/ξ with representative values of 3–8, 8–20, and 13–31 for particles of 55, 140, and 215 nm radius [3].

Polymer size is governed by the desirability to conduct electrophoresis near the entanglement threshold concentration, c^* (see below). With relatively large polymers, c^* is a low concentration, and, accordingly, operation around that concentration is free of the problems of irreproducibility of mobility and peak width that we observed at concentrations beyond 7–8%. Correspondingly, the use of relatively small polymers with high values of c^* invites the use of high polymer concentrations, such that those problems are encountered. Moreover, at those high concentrations, the concentration range between dilute and concentrated regime – the semidilute regime – may be inconveniently narrow. Nonetheless, the data on resolution (Section 3) indicate that there may be advantages to resolution in electrophoresis using low polymer molecular weights.

1.3 Particle-size-dependent retardation in polymer-free solutions

In application to macromolecular-sized particles, with radii less than 10 nm and a high degree of diversity in surface net charge, mobility in free solution is a very shallow function of particle size and does not provide size separations to a practical degree. This is not necessarily the case as the particle size is increased into the submicron and micron range [1]. In that size range, electrophoretic separations based on size differences can be practical even in the absence of polymers, provided that the particles share a similar surface net charge density and that the separation is conducted at a medium ionic strength. Possibly, a high field strength may also contribute to size separations in the absence of polymers [1]. Although those conditions may hold in a few applications of CZE, a far more general approach to electrophoretic size separations in the submicron and micron size range is to employ polymer solutions as retarding media.

1.4 Separation in semidilute polymer solutions, based on size differences or rigid spherical particles with radii less than 15 nm

Consideration of the particle size range below 15 nm, which includes all rigid spherical macromolecules, serves only as the background for this review of electrophoresis of the submicron and micron size range. These "small" particles display a retardation in polymer solution which resembles that in gels and fundamentally differs in mechanism and properties from the retardation of the larger particles. These properties are: (i) a concave plot of the logarithm of relative mobility, μ/μ_o, vs. polymer concentration (Ferguson plot), at least in PA and PEG solutions, which can be linearized by plotting the 0.75th power of the polymer concentration (Figs. 2 of [6] and [7]) with the resulting slope designated as retardation coefficient, K_R; (ii) a linear plot with positive slope of K_R vs. particle radius ("*R*-plot"; Fig. 1 of [8]; Fig. 4 of [7]; Fig. 3 of [6]); (iii) independence of field strength (Fig. 1 of [8]) and near-independence of the molecular weight of the polymer (Fig. 1 of [6]); (iv) availability of at least four different theoretical treatments which account for the retardation in the terms of a fit of the particle into the available spacings of the polymer network [9–11] or in terms of hydrodynamic interactions [12]. The available spacings can be related to the mesh size (screening length), ξ, a function of polymer concentration, which is independent of polymer molecular weight.

2 Separation of semidilute polymer solutions, based on size differences of rigid spherical particles with radii larger than 15 nm

In contrast to smaller particles, the size separation of particles with a radius larger than 15 nm, conducted in polymer solutions, remains a *terra incognita*, lacking the guiding light of theory and, therefore, predictability. Moreover,

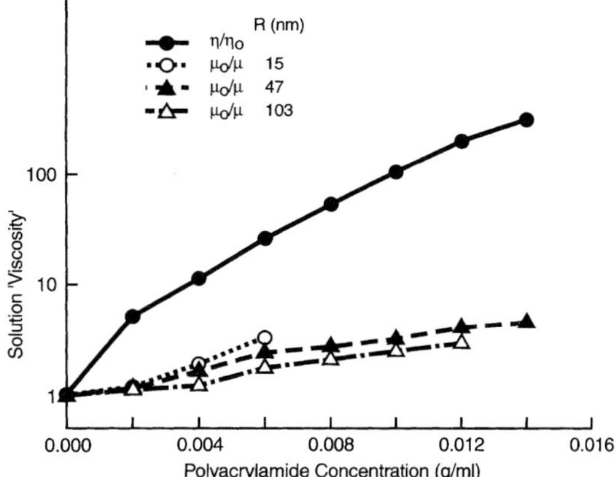

Figure 1. Microviscosity (μ_o/μ) surrounding submicron-sized polystyrene sulfate particles during CZE, compared to zero shear macroscopic relative viscosity (η/η_o). Mobilities are measured in CZE (150 μm ID capillary, 40 cm effective length, 270 V/cm, 25°C, initial zone length $\simeq$ 6 mm). Adapted from Fig. 2B of [2].

it appears that, at least in polyacrylamide solutions of high molecular weight, retardation decreases with particle size to about 300 nm but increases upon increase of the particle radius to 800 nm, and ondulates again with a further increase in *R* (Fig. 2 of [8]), suggesting the existence of at least 2–3 different dominant retardation mechanisms. Qualitatively similar data for PEG exist (Fig. 5 of [13]). The data reviewed here mostly refer to the first of these retardation mechanisms, related to particles less than 300 nm in radius.

Retardation of electrophoretic migration by polymeric media can be viewed in different ways: (i) In the traditional manner of gel electrophoresis, interpreted by mechanistic models which are unverifiable by experimental measurements, or (ii) in terms of microviscosity, *i.e.*, a parameter originally introduced to characterize the viscous properties of the medium over distances commensurate with the size of the particles. The advantage of operating with such a parameter is, first, that it can be compared with the macroscopic (bulk) viscosity which is accessible to experimental measurement, and second, that the phenomenological description and mechanistic concepts developed for macroscopic viscous flow may be applied to the study of particle transport in polymer solutions.

According to classical electrokinetic theory, the mobility of a rigid nonconducting sphere in a medium of viscosity η is given by

$$\mu = (2\varepsilon\zeta/3\eta)f/\kappa R) \tag{1}$$

Electrophoresis 2000, *21*, 259–265

where ε is the dielectric permittivity, ζ the zeta potential, f(κR) Henry's function and $1/\kappa$ the thickness of the EDL. Since in the calculation of relative mobility, the values of μ and μ_o at moderate concentrations of neutral polymers derive from the identical medium except for the viscosity, all of the above parameters except η cancel in the ratio of μ/μ_o. Thus, a microviscosity can be measured from

$$\mu_o/\mu = \eta/\eta_o \tag{2}$$

$$\eta = (\mu_o/\mu)/\eta_o$$

We will discuss retardation below in both manners (i) and (ii).

2.1 Ferguson and *R*-plots

The Ferguson plot, $\log(\mu/\mu_o)$ *vs.* polymer concentration, in the particle size range under study, slightly deviates from linearity but may be linearized as a first approximation, at least in the concentration range up to 1%. However, the fundamental difference of the plot from that of macromolecules, both when derived from gel electrophoresis and the polymer solution electrophoresis of smaller particles, is the fact that the slope of the linearized plot, *i.e.*, retardation, decreases with increasing particle size. This fact gives rise to an *R*-plot, K_R *vs. R*, with a negative slope (Fig. 3 of [2]). In terms of microviscosity, the viscous friction experienced by particles in that size range decreases with particle size. Figure 1 shows a plot of the microviscosity, calculated by measurement of mobility in CZE, of three particles differing in size *vs.* polymer concentration, superimposed on the corresponding plot of zero shear (relative) macroviscosity of the medium measured by viscometry.

2.2 Field strength dependence

Resistance to migration of particles in the size range under consideration decreases with electric field strength.

That decrease can be viewed either as a decrease of microviscosity (defined as viscosity in the vicinity of the particle surface) with particle translational velocity (Fig. 2A) or as a decrease of the retardation coefficient, K_R, with field strength (Fig. 2B). The separation, *i.e.*, K_R differences, between pairs of the three particles tested can either clearly improve or deteriorate as the field strength is lowered.

2.3 Dependence on polymer molecular weight

As the M_r of the polymer is increased, the retardation increases in inverse relation to particle size (Fig. 3). The rate at which retardation increases with M_r (polymer) appears to be directly related to particle size.

2.4 Dependence of retardation on sample concentration

Retardation of the representative particles of 55–215 nm radius also decreases with sample concentration. Translating that concentration into values of distance between particles, it appears that the rate of decrease is indistinguishable for particles of the three sizes. The decrease of retardation with sample concentration is therefore not affected by an increase in the particle size and can for that reason not be due to the presence of an aggregated form of the particle. Aggregation leading to a system of interacting size isomers is also excluded by the fact that band width was found not to increase to any appreciable extent with sample concentration. Thus, the mechanism underlying the effect of sample concentration on retardation remains unknown.

2.5 Retardation of submicron-sized particles viewed in terms of microscopic viscosity

The mechanisms by which particles in the size range of 15–300 nm radius are retarded may rest on a decreased viscosity at the interface between the migrating particle

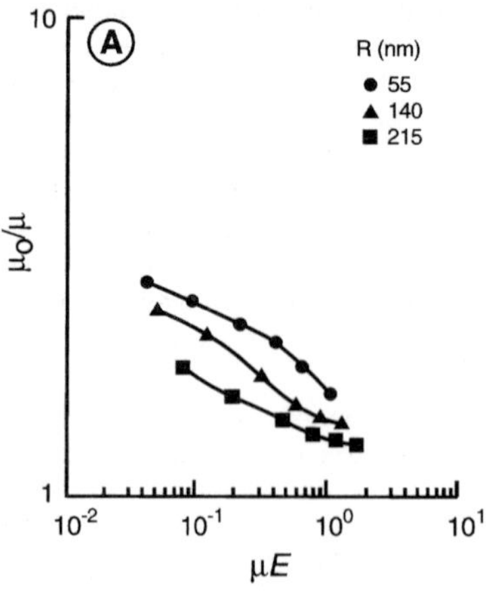

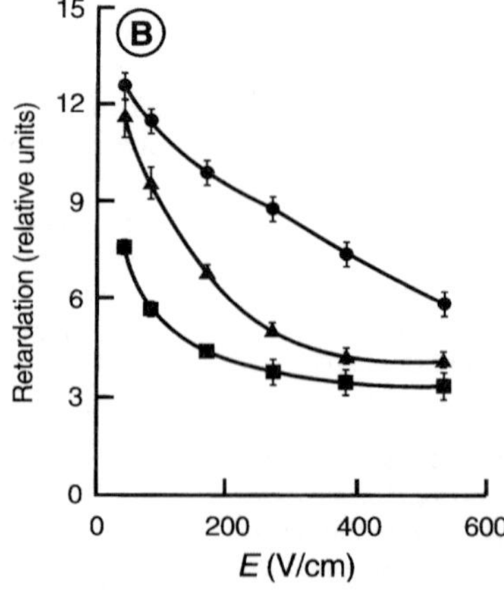

Figure 2. Microviscosity as a function of (A) particle velocity and (B) retardation as a function of electric field strength. PA, M_r (0.7–1.0) $\times$ 10^6. (A) PA concentration, 0.8%. Figure 9 of [3]. (B) Fig. 7 of [3].

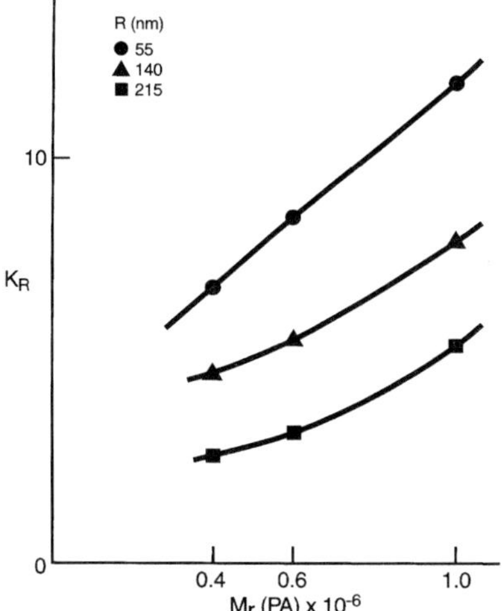

Figure 3. Retardation as a function of the molecular weight of the polymer. K_R designates retardation in relative units. Adapted from Fig. 10A of [3]. Particle translational velocity, 0.5 mm/s.

and the polymer network in the surrounding solution. In more general terms: once the particle size sufficiently exceeds the screening length (average mesh size) of the polymer network, the concept of a particle traveling through a network by finding spaces large enough for entrance into them fails. At the scale of increasing R/ξ, the retardation may be expected to progressively become a function of the interaction between the particle and the polymer network as a viscous continuum. When a particle undergoes a translational motion due to diffusion or sedimentation, viscous friction experienced by the particle derives from the commensurate spatial region surrounding it. By contrast, the viscous friction in the electrophoresis of large particles at the scale of κR derives entirely from the EDL surrounding the particle, the thickness of which is an explicit function of the ionic strength. In fact, the viscosity η in Eq. (1) is the viscosity of the solution within the EDL rather than that of the bulk solution [14]. Thus, the events occurring within the region of the EDL with length of $1/\kappa$ (which is of the order of several nm at $I = 0.01$) are thought to be crucial to particle retardation. It was found that (i) microviscosity decreases with increasing particle size in the range of 15–300 nm radius at a given electric field strength (particle translational velocity); (ii) microscopic velocity decreases with electric field strength (particle translational velocity) for a particle of a given size; (iii) macroscopic viscosity exceeds microscopic viscosity by at least several times even at the lowest electric field strength (Fig. 1). The causes of that dramatic reduction in resistance to migration near the particle surface remain speculative. Two possible mechanisms have been advanced: (i) a shear-like deformation of the network in the vicinity of the particle surface, and (ii) a decrease of polymer segment density in the direction toward the particle surface due to the loss of configurational entropy of the polymer coils (formation of a "depletion layer"). The retardation behavior of submicron-sized particles in polymer solutions is believed to be due to a complex interplay between these mechanisms.

3 Resolution based on size differences of rigid spherical particles with radii larger than 15 nm

The results summarized in Section 2, defining separation of particles in the submicron size range in polymer-containing buffers, are insufficient to define the usefulness of polymers in practice since they neglect the role of peak width and peak spreading. This neglect is overcome by measuring peak width and calculating resolution, defined as

$$Res = (\Delta\mu/\mu_{ave})\,/N/16)^{0.5} \tag{3}$$

where $\Delta\mu$ designates the mobility difference between two peaks, μ_{ave} their average mobility, and N the number of theoretical plate equivalents, which relates to peak width at half-height, w, measured in spatial units, as

$$N = 5.55\,x^2/w^2 \tag{4}$$

when x is the migration distance (effective capillary length in CZE).

3.1 Resolution as a function of polymer concentration

Resolution in three representative cases exhibits a maximum close to the entanglement threshold concentration, c^* (Fig. 4). The finding suggests a practical way to select an optimally resolving polymer concentration in the neighborhood of c^* which can be determined experimentally as

$$c^* \simeq 1/[\eta] \tag{5}$$

or can be calculated for a polymer of known molecular weight since

$$[\eta] = K\,M_r^a \tag{6}$$

(Mark-Houwink-Sakurada equation), where K and a are listed in the literature for most relevant polymers.

3.2 Resolution as a function of the M_r of the polymer

Resolution increases upon the approach toward c^* by a decrease of the molecular weight of the polymer at a con-

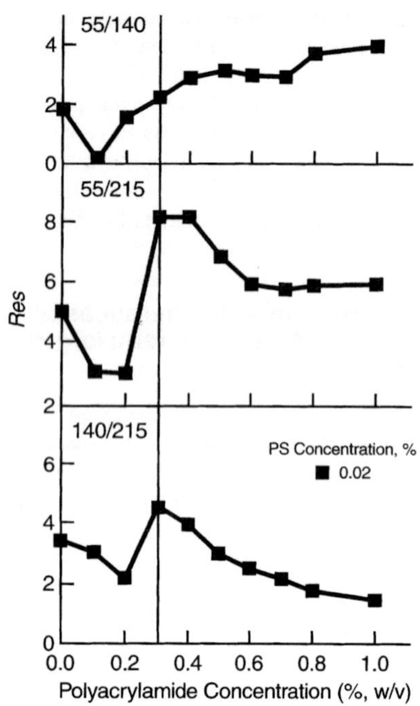

Figure 4. Resolution as a function of polymer concentration. The vertical solid line denotes the position of c^*. PA, M_r (0.7–1.0) $\times$ 10^6. From Fig. 6 of [4]. Resolution of particles the radii (nm) of which are shown numerically.

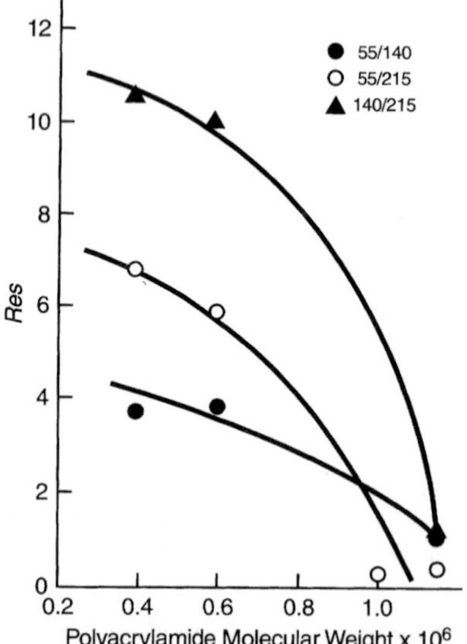

Figure 5. Resolution as a function of the molecular weight of the polymer. PA concentration, 1%; field strength, 270 V/cm (Fig. 7B of [4]).

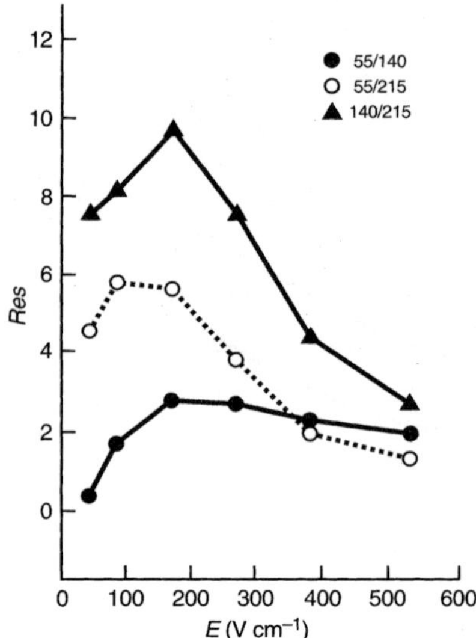

Figure 6. Resolution as a function of field strength. PA concentration, 0.4%; M_r of PA (0.7–1.0) $\times$ 10^6 (Fig. 9 of [4]).

stant polymer concentration (Fig. 5). Although the decrease of resolution with the approach of polymer weight toward "zero" is not experimentally shown, such a decrease and a corresponding maximum of resolution must exist in analogy to the maximum approached at constant polymer weight by variation of the polymer concentration (Fig. 4). A practical limit to the decrease in the molecular weight of the polymer toward its c^* is set due to the high value of such c^* and the perturbations encountered in CZE at polymer concentrations above 7–8% (see Section 1.2).

3.3 Resolution as a function of field strength

Resolution in the representative cases exhibits a maximum around 200 V/cm (Fig. 6). This is not due to peak spreading when the Joule heat exceeds the heat dissipation capacity of the apparatus since it has previously been shown [3] that, in the CZE apparatus used, no appreciable Joule heating can be observed, even at 500 V/cm, in a buffer of 0.01 M ionic strength, at 25°C. This is also not due to the parabolic temperature profile which was shown not to contribute to peak spreading to any appreciable degree at field strengths up to 400 V/cm. Since Joule heating and temperature profile effects on peak width were ruled out for CZE in a capillary of 150 μm diameter, these effects are *de majore* of no consequence in application of CZE to smaller capillary diameters.

4 Conclusions

The conclusions are drawn from the results of CZE in only two polymers, PA and PEG, and only a few of the commercially available molecular sizes of each. They are based on plots with a minimum of data points, and for both reasons, must be considered preliminary. Nonetheless, they are illuminating an area of separation science that Section 1 has accurately described as a *terra incognita*. The most important conclusion of the work reviewed here is that the addition of polymers to the electrophoretic buffer improves resolution of submicron-sized rigid spherical particles. The degree of improvement is less than one order of magnitude, but nonetheless several-fold. The optimally resolving condition was found at or near the entanglement threshold concentration, c^*, which can be predicted by calculation on the basis of the molecular weight of the polymer (related to intrinsic viscosity) or can be approached experimentally either by varying systematically the concentration of a polymer of constant molecular weight, or by varying the molecular weight of the polymer at a constant concentration. Applying relatively small polymers with high c^* to resolution, a practical limit to reducing the molecular weight of the polymer is set due to perturbations of CZE when the polymer concentration exceeds 7–8%. An optimally resolving electric field strength at about 200 V/cm is suggested by the available data. It can be expected to exist in general but must be found in each application by electrophoresis at the optimal polymer concentration near c^*.

Received June 11, 1999

5 References

[1] Radko, S. P., Chrambach, A., *J. Chromatogr. B* 1999, *722*, 1–10.

[2] Radko, S. P., Chrambach, A., *Electrophoresis* 1998, *19*, 2423–2431.

[3] Radko, S. P., Chrambach, A., *Macromolecules* 1999, *32*, 2617–2628.

[4] Radko, S. P., Chrambach, A., *J. Chromatogr. A* 1999, *848*, 443–455.

[5] Chrambach, A., *The Practice of Quantitative Gel Electrophoresis*, VCH, Weinheim 1985.

[6] Radko, S. P., Chrambach, A., *J. Phys. Chem.* 1996, *100*, 19461–19465.

[7] Radko, S. P., Chrambach, A., *Biopolymers* 1997, *42*, 183–189.

[8] Radko, S. P., Chrambach, A., *Electrophoresis* 1996, *17*, 1094–1102.

[9] Ogston, A. G., Preston, P. N., Wells, J. D., *Proc. Royal Soc. A London* 1973, *333*, 297–316.

[10] Langevin, D., Rondelez, F., *Polymer* 1978, *19*, 875–882.

[11] Slater, G. W., Treurniet, J. R., *J. Chromatogr. A* 1997, *772*, 39–48.

[12] Cukier, R. I., *Macromolecules* 1984, *17*, 252–255.

[13] Radko, S. P., Chrambach, A., *Appl. Theor. Electrophor.* 1995, *5*, 79–88.

[14] Donath, E., Krabi, A., Allan, G., Vincent, B., *Langmuir* 1996, *12*, 3425–3430.

Review

Methal N. Albarghouthi
Annelise E. Barron

Polymeric matrices for DNA sequencing by capillary electrophoresis

We review the wide range of polymeric materials that have been employed for DNA sequencing separations by capillary electrophoresis. Intensive research in the area has converged in showing that highly entangled solutions of hydrophilic, high molar mass polymers are required to achieve high DNA separation efficiency and long read length, system attributes that are particularly important for genomic sequencing. The extent of DNA-polymer interactions, as well as the robustness of the entangled polymer network, greatly influence the performance of a given polymer matrix for DNA separation. Further fundamental research in the field of polymer physics and chemistry is needed to elucidate the specific mechanisms by which DNA is separated in dynamic, uncross-linked polymer networks.

Keywords: DNA sequencing / Capillary electrophoresis / Polymer solutions / Matrices / Review

EL 4173

Contents

Correspondence: Dr. Annelise E. Barron, 2145 Sheridan Road, Room E136, Department of Chemical Engineering, Northwestern University, Evanston, IL 60208, USA
E-mail: a-barron@northwestern.edu
Fax: +847-491-3728

Abbreviations: AAP, *N*-acryloylaminopropanol; **HEC,** hydroxyethylcellulose; **LOR,** length of read; **LPA,** linear polyacrylamide; **PDMA,** poly-*N,N*-dimethylacrylamide; **PEO,** polyethylene oxide; **PNIPA,** poly-*N*-isopropylacrylamide; **TAPS,** *N*-tris(hydroxymethyl)methyl-3-aminopropanesulfonic acid; **TBE,** Tris-borate-EDTA buffer; **TTE,** Tris-TAPS-EDTA buffer

1 Introduction

Genome sequencing projects reveal the genetic makeup of an organism by reading off the sequence of the DNA bases that encode the fundamental information necessary for the life of the organism. Already revolutionizing biology, genome research provides a vital thrust to the increasing productivity and pervasiveness of the life sciences. Availability of the genome sequences of humans, animals, plants and microorganisms will have unprecedented impact on many disciplines, including health care, biological and biomedical research, forensics, biotechnology, agriculture, and the environment. To date, there are over 50 complete genomes available in public databases, and over 350 ongoing genome projects [1]. All genome projects and their applications will continue to seek new advances in high-throughput and cost-effective sequencing technologies. Therefore, the development of automated, high-throughput, cost-effective DNA sequencing technologies has gained huge momentum in both academic and industrial communities.

The focus of this review is upon polymeric materials that have been tested for DNA sequencing by capillary electrophoresis (CE). Hypothesized mechanisms of DNA separation in polymer solutions are briefly reviewed, along with a discussion of some experimental results that provide better insight into these mechanisms. The empirical approaches that have been taken towards the selection and performance optimization of polymers for DNA sequencing are discussed so as to provide guidelines for the selection of appropriate materials and conditions for resolving DNA sequencing fragments by CE.

2 DNA sequencing

The present state-of-the-art method for determining the base sequence of DNA relies upon the Sanger dideoxy-terminated reaction [2]. The method is based on controlled interruption of the enzymatic replication of ssDNA by DNA polymerase enzyme. The modern Sanger cycle sequencing reaction produces a fluorescently labeled, nested, single-base ladder of ssDNA fragments ranging in chain length from just a few bases to a few thousand bases. This set of DNA fragments is separated according to chain length using gel electrophoresis, allowing the order of bases to be read. Originally, the sequence was determined using autoradiography after electrophoresis was complete, and required up to 24 h. This type of "snapshot" detection severely limited the capacity and throughput of sequencing systems. High-throughput DNA sequencing became a reality with the commercial introduction of an apparatus for automated "finish line" detection of DNA fragments labeled with laser-induced fluorescence (LIF) dyes.

The majority of DNA sequencing has been carried out by slab-gel electrophoresis. Commercial sequencing instruments based on ultrathin cross-linked polyacrylamide slab-gel electrophoresis, such as the ABI PRISM 377™ (PE Biosystems, Foster City, CA, USA), produce about 600 bases per sample at a top rate of 200 bases/h/lane, and can only provide over 750 bases at the expense of extending run times to 10–12 h [3]. Modern sequencing gels have the advantage of running up to 96 samples in parallel, which increases the throughput of the instrument. However, slab-gel electrophoresis suffers from many limitations. Gel pouring is a manual process, and there is always a possibility of trapping air bubbles in the gel that can harm the separation efficiency. Furthermore, the Joule heating effect, which arises from the passage of electric current through the gel, places upper limits on the voltage that can be applied, resulting in long run times. Finally, with slab-gel systems, sample loading is done manually, obstructing complete automation of the sequencing process.

CE is an attractive alternative to slab-gel electrophoresis and is rapidly becoming the dominant technique in DNA sequencing centers [4, 5] due to its capacity for full automation. In CE, DNA separation is achieved in a fused-silica capillary, 25–100 µm in diameter. The high ratio of surface area to volume of the small capillary tube serves to efficiently dissipate the heat produced during electrophoresis, allowing the use of higher electric fields, which decreases the run time and improves DNA resolution. Although CE provides much faster separation than slab gels, it is necessary to operate multiple capillaries in par-

allel to compete with the throughput of slab-gel electrophoresis systems, which can run up to 96 samples in parallel. After the first study exploiting the potential of a multiple-capillary instrument was reported by Mathies *et al.* [6, 7], several investigators endeavored to produce a capillary array system that could significantly improve the sequencing rate relative to slab-gel-based instruments (see [8] for review). Today, there are two commercial versions of capillary array electrophoresis instruments, each featuring 96 capillaries and relying on LIF detection: the MegaBACE 1000™ from Molecular Dynamics (Sunnyvale, CA, USA) and the ABI PRISM 3700™ from PE Biosystems.

The MegaBACE 1000™ instrument echoes the original Mathies design [6, 7] that was based on confocal LIF detection consisting of a microscope objective to focus the laser light inside the capillaries and, at the same time, to collect emitted light from the center of the columns. To collect data from each and every capillary at the required frequency, a scanning system is used. The instrument uses polyacrylamide-coated capillaries with a covalent coating attachment and a high molar mass, linear polyacrylamide-based LongRead® matrix. Over 500 bases per capillary can be sequenced in 2 h, yielding a minimum throughput of 250 bases/h/capillary [9].

The ABI PRISM 3700™ employs postcolumn LIF detection with liquid sheath flow [10, 11]. In this detection system, the anodic end of the capillary bundle is aligned inside a quartz cuvette. Along the dead space between the columns and the walls of the cuvette, a polymer solution is pumped through the cell. The liquid sheath flowing outside the capillary entrains the DNA bands as they elute from the capillary, tapering them to a smaller diameter. A laser beam crosses all of the flow streams and excites the fluorescence of the dyes with which the DNA strands are labeled. Fluorescent light is dispersed through a diffraction grating and imaged onto a charge-coupled device (CCD) camera. The 3700™ instrument uses a separation matrix that is based on relatively low molar mass, linear polydimethylacrylamide, commercially known as POP®, for "performance-optimized polymer". The zero-shear viscosity of this matrix is relatively low compared to the aforementioned LongRead® matrix. The system can achieve a read length of 550 bases per capillary with a run time of approximately 4 h [12]. The same polymer is used for "dynamic coating" of the internal capillary wall by physical adsorption. Hence, arrays of bare fused-silica capillaries can be used in this instrument, reducing array cost. Typically, more than 100 consecutive runs can be carried out in such a "dynamically coated" capillary array before performance begins to degrade, after which time the capillaries can be "regenerated" with an acid rinse.

Electrophoresis 2000, *21*, 4096–4111

The next generation of automated, high-throughput DNA sequencing instruments most likely will be based on microfluidic devices or microchips (here, we are referring not to DNA hybridization chips, but to electrophoresis microchips). DNA sequencing on a microchip was first demonstrated in 1995 by Mathies and Woolley [13]. In the first unoptimized system of DNA sequencing on a microfluidic device, single-base resolution reached 150–200 bases in 10–15 min in an *in situ*-polymerized linear polyacrylamide (LPA) matrix, with an effective channel length of 3.5 cm. Later, optimized separation conditions that utilized a 3% LPA matrix, polymerized outside the chip, yielded read lengths of over 500 bases in about 20 min in a single channel run at 99.4% sequencing accuracy [14]. The shorter run time (20 min as compared to 2–4 h for capillaries) is a consequence of the narrow injection zones achievable on chips that have cross-flow or "T" injection geometries. Hence, even a 3.5 cm channel can deliver a 500-base read length, a surprising result for many people, providing a potential sequencing rate of 1500 bases/h/lane if the system could be automated. This work demonstrated the feasibility of high-speed DNA sequencing by CE on microchips and showed the tremendous potential for increased throughput that these systems offer. However, this breadboard system did not enable automated replacement of the separation matrix; automating that particular function may be more difficult to accomplish on microfluidic devices than it is in capillary arrays.

3 Mechanism of DNA electrophoresis in polymer solutions

An understanding of the mechanism of size-based separation of DNA chains in polymeric media is critical for improving and optimizing the performance of sieving matrices and electrophoresis conditions. By some, the mechanism of DNA separation in uncross-linked polymer solutions has been described as being essentially the same as that in traditional slab-gel electrophoresis, as modeled by the Ogston model and the reptation models, recently reviewed by Heller [15]. In the Ogston model, a DNA molecule is assumed to migrate as a rigid spherical particle through a network of rigid, linear, infinitely long rods, and to diffuse laterally until it encounters a "pore" large enough to allow its passage. Thus smaller molecules migrate faster through the matrix as they have access to a larger fraction of the pores. The Ogston model predicts a linear dependence of log (μ), where μ is the electrophoretic mobility, on matrix concentration for small DNA fragments in the limit of low electric fields [16, 17]. According to this model, however, molecules with radii of gyration larger than the average pore size should not penetrate the matrix at all, in contrast with the actual

behavior of DNA molecules in gel electrophoresis. Hence, the assumptions underlying the Ogston model are invalid for larger DNA [18]. Furthermore, fundamental assumptions of this model about the nature of the separation matrix are extremely far from the reality of a dynamic entangled polymer network.

In the reptation model, the DNA molecule travels through a matrix as a flexible chain that reptates "snake-like" through a fictitious "tube" in the polymer network. The model predicts, in the limit of zero electric field, that the electrophoretic mobility is inversely proportional to DNA size. This dependence is indeed observed under some conditions for slab gels, over a certain "medial" range of DNA size. However, experimental observations that DNA larger than 20 kbp migrate with a constant electrophoretic mobility led to a modification to the reptation model, the biased reptation model (BRM) [19, 20]. This model postulates that when the electric field strength is large, DNA chains will orient in the direction of the electric field, strongly biasing the random walk in the forward direction so that DNA assumes a stretched, rod-like conformation. DNA molecules in this oriented conformation no longer take a tortuous path through the gel, thus the size-based separation ability of the gel is lost for large DNA and at high field strengths. The biased reptation with fluctuations (BRF) model was proposed to improve the original BRM by incorporating fluctuations in the theoretical tube length [21].

Polymer solutions differ from cross-linked gels in that the physical entanglements between the chains have a finite lifetime. The group of Viovy and Duke [17] adopted the physical concept of "constraint release" to describe this phenomenon through a modification of the BRF model. Such constraint release occurs when the polymers, which form the tube, are allowed to reptate away even while the DNA remains within the tube. Cottet *et al.* [22] described the dynamics of the entangled polymer network by the reptation or relaxation time which, for a given polymer, increases with increasing polymer concentration or molar mass but decreases with increasing temperature. In order to achieve uniform mobility and good resolution of electrophoresing DNA, the relaxation time should be greater than the residence time of the DNA molecule segment in the blob [22, 23].

The Ogston model and reptation models are applicable at low electric field strengths such as those typically used in slab-gel electrophoresis, but would seem to be inapplicable for CE in uncross-linked polymer solutions, where much higher field strengths are commonly applied. This points to a fundamental inconsistency of the application of the reptation model to DNA separation by CE. Polymer motion by "reptation" is postulated to prevail in a fixed polymer network, in which chains are sufficiently constrained, each in a reptation "tube", that end-on motion is their only possible mode of movement besides small, lateral segmental motions (the formation of "chain defects") [24]. This type of motion has been seen by microscopy to occur in entangled DNA solutions under equilibrium conditions of zero applied field [25].

Mathematically, the initial distribution of chain conformations for computer simulation of reptative polymer motion is assumed to be either Gaussian or self-avoiding, depending upon solvent conditions, and hence is derived by equilibrium-based, statistical mechanical arguments [26, 27]. These assumptions are only appropriate in the case of zero external applied force, and hence are violated in the presence of a strong electric field [28, 29]. Under an external field, the initial distribution of chain conformations (for the DNA) must be heavily influenced by kinetics, and hence cannot be determined accurately from statistical mechanical arguments. Hence, the reptation model *per se* cannot be applied meaningfully to high-field electrophoresis. Although it has had significant success for interpreting slab-gel electrophoresis, it may not be a useful framework with which to consider the mechanism of separation for CE in uncross-linked polymer solutions. While these mechanisms may be applicable at extremes of low electric field and high polymer concentrations, these are generally not conditions of interest for genomic sequencing.

The difference in stiffness of ssDNA (4 nm persistence length) and dsDNA (45 nm persistence length) molecules, as well as the availability of bases in ssDNA molecules to hydrogen bond with the surrounding water molecules, urea (when used as denaturant) and polymer chains have been shown to affect the electrophoretic migration of DNA and have led to the identification of new separation regimes [15, 30]. Furthermore, the separation mechanisms postulated by both the Ogston and reptation models require polymer concentration to be in the entangled regime for DNA separation to be achieved. However, Barron *et al.* [31–33] have shown that separation of DNA fragments is possible in dilute polymer solutions at concentrations well below the entanglement threshold. To explain this experimental observation, the authors proposed a transient entanglement coupling mechanism for DNA separation. It was hypothesized that DNA fragments migrating through the polymer solution entangle or collide with individual polymer molecules and are forced to drag them through the solution, resulting in a decrease in DNA electrophoretic mobility. Larger DNA molecules were postulated to have a higher probability of encountering

and entangling polymer molecules, hence to experience a greater reduction in mobility. Variations of this mechanism are likely to apply to both dilute and entangled polymer solutions, as the CE separations of DNA fragments obtained at concentrations above and below the polymer entanglement threshold were similar [32, 34].

Direct observations of dsDNA molecules electrophoresing through polymer solutions [35–38] shed new light on the mechanism of separation and the interactions between DNA and polymer molecules during electrophoresis. DNA molecules electrophoresing through high molecular mass polymer solutions, both above and below the entanglement threshold, undergo conformational changes from a compact, globular conformation to U-shaped conformations that appear to drag the entangling polymer molecules. However, in low molecular mass polymer solutions DNA migrates primarily in a globular conformation, not forming entanglements. Three types of DNA-polymer interactions have been proposed to account for this behavior [37, 38]: (i) U-shape collisions, in which the DNA entangles with the polymer obstacle and releases itself in a pulley-like motion; (ii) brief collisions, in which the DNA entangles with the polymer for a brief period of time before the polymer releases itself from the DNA; and (iii) transient nonentangling collisions, where the DNA in globular form pushes polymer molecules out of its path. Nonentangling collisions, modeled by Sunada and Blanch [38], occur between small DNA and small polymer molecules and explain the separation of small DNA molecules, whilst the entangling collisions, modeled by Hubert *et al.* [39], are necessary to separate large DNA molecules by entangling with large polymer molecules. Based on these mechanisms of separation, one can explain the experimentally observed improvements of separation of a wider size range of DNA molecules in solutions containing high and low molecular mass polymer, both above and below the entanglement threshold [40–43]. DNA was not observed to reptate, but to cycle in conformation between a spherical glob and a deformed U-shape.

It is clear that the mechanism of DNA electrophoretic separation in uncross-linked polymer solutions is more complicated than that in cross-linked gels, due to the dynamic nature of the network and due to the fact that separation is possible even in the absence of the polymer network in dilute solutions. At present, theoretical work on the mechanism of DNA separation is most useful for providing a physical framework with which to qualitatively interpret experimental results. None of these models are quantitatively predictive at this time. However, considering the entangling and nonentangling collisions which have been seen to take place over a wide range of polymer concentrations, it should be possible for theorists to formulate a single model of DNA electrophoresis in polymer solutions, taking into account the properties of DNA, both in single- and double-stranded form, and polymer molecules under the various electrophoretic conditions. This is an exquisitely complicated problem in polymer physics and will, no doubt, keep theorists occupied for a long while.

4 Quantitative analysis of DNA separation

A quantitative measure of electrophoretic separation performance is the resolution, R_s, which can be expressed as:

$$R_s = 0.59 \frac{x_2 - x_1}{FWHM} = \frac{1}{4} \frac{\Delta\mu}{\mu_{av}} N_P^{1/2} \qquad (1)$$

where x_i is the center of peak i, *FWHM* is the peak full width at half maximum (this is derived based on the assumption that both peaks have the same width), $\Delta\mu$ is the difference in electrophoretic mobility for two fragments, μ_{av} is the average mobility between two fragments, and N_P is the number of theoretical plates. The term $\Delta\mu/\mu_{av}$ is a measure of the selectivity or peak spacing in a separation and reflects the sieving power of the matrix. The term N_P is a measure of separation efficiency or peak width due to the dispersion characteristics of the electrophoresis system.

An alternative quantitative measure of the performance of a sequencing separation matrix and instrument is the length-of-read, LOR. The LOR is defined as the point at which the peak spacing is equal to the peak width giving a resolution value of 0.59 [16]. This point can be determined graphically from the point of intersection of the plots of peak width and peak spacing *versus* DNA size. Another definition of the LOR is associated with the use of base-calling software used with DNA sequencing instruments. It is defined as the number of bases that can be called accurately, and requires single-base resolution of DNA fragments. Expert computer programs that are able to call bases accurately at resolution as low as 0.25 have been reported [44].

5 ssDNA *versus* dsDNA

DNA molecules in aqueous solution can exist in native double-stranded form or in denatured, single-stranded form, depending upon conditions such as solvent and temperature. DNA sequencing is necessarily performed under denaturing conditions, because the sequencing reactions deliver ssDNA molecules with a partly elongated complementary DNA strand attached. Also, ssDNA

strands have the propensity to form intrastrand base pairs and secondary structure. In particular, genomic DNA often has "trouble regions" of repetitive sequence that are difficult to fully denature and hence difficult to sequence. Full denaturation is needed to enable correct and reliable size-based separation depending strictly upon chain length. For DNA sequencing applications, DNA denaturation is typically achieved by using high concentrations of urea (> 4 M, more typically 7–8 M), formamide (~ 10%) and high temperatures (up to 70°C), or any combination of the three.

For the separation of PCR products and restriction fragments, DNA molecules are separated under nondenaturing conditions in the double-stranded form. For such separations, single-base resolution is not a critical requirement, allowing the use of less concentrated, and hence less viscous, polymer solutions, and permitting easier filling and replacement of polymer matrix in the capillaries. Van der Schans *et al.* [45] reported that when DNA is separated in single-stranded form, the selectivity of separation is increased, which results in significantly higher resolution of DNA fragments up to 500 bases. Using 4% linear polyacrylamide solution, baseline resolution of fragments differing in size by four base pairs in the 200 base pair range was achieved under denaturing conditions, while the separation failed under native conditions.

In a series of articles, Heller [15, 46, 47] compared the separation of ssDNA and dsDNA (< 700 bases) using similar conditions except that electrophoretic runs under denaturing conditions were carried out at 50°C, in the presence of 4 M urea, whilst for nondenaturing conditions the temperature was 25°C. In all cases, peak spacing decreases with increasing DNA size, but increases with increasing polymer concentration. However, separations of ssDNA fragments give larger peak spacings than separations of dsDNA fragments (< 700 bases or base pairs). Peak widths are smaller under denaturing compared to nondenaturing conditions. As a result, the resolution of ssDNA fragments is much higher than that for dsDNA of the same length. The use of strongly alkaline conditions (pH 11) has been reported to denature DNA fragments in hydroxyethylcellulose (HEC) solution and to improve the resolution of DNA fragments compared to that of dsDNA at neutral pH [48]. Although separation of DNA has been shown to be possible in dilute solutions, concentrated, entangled polymer solutions gave superior separations of DNA as compared to dilute solutions [48]. Therefore, for high-resolution applications, such as DNA sequencing, relatively concentrated, entangled polymer solutions are needed.

6 Polymer matrices used in DNA sequencing

Several different types of water-soluble polymers have been used in DNA sequencing by CE. The choice of polymer has often been arbitrary and empirical, primarily because the mechanism of DNA separation in uncross-linked polymer solutions is not fully understood. The chemical structures of the polymers that have been tested for this application are shown in Table 1. In addition to chemical properties, the physical properties of polymers used for CE are critically important, as they control the attributes of the entangled polymer network and hence influence the predominant mechanism and time-scale of DNA-polymer and polymer-polymer interactions [23]. To compare the performance of different polymers as DNA sequencing matrices, a number of important issues need to be evaluated simultaneously: resolving power, speed of separation, viscosity of polymer solution and suppres-

Table 1. Structures of polymers used for DNA sequencing by CE

Polymer	Chemical structure
LPA	H_2N–CO–$(CH_2$–$CH)_n$
PDMA	$(CH_3)_2N$–CO–$(CH_2$–$CH)_n$
PEO	$(O$–CH_2–$CH_2)_n$
PVP	pyrrolidone–$(CH_2$–$CH)_n$
PEG (end-capped)	$F_{2m+1}C_m$–$(O$–CH_2–$CH_2)_n$–C_mF_{2m+1} $m = 6$ or 8
poly(AAP)	HO–CH_2–CH_2–CH_2–NH–CO–$(CH_2$–$CH)_n$
HEC	cellulose with hydroxyethyl groups

sion of electroosmotic flow. Resolving power reflects the ability of the polymer matrix to achieve long LORs, which is a highly desired feature for genome sequencing projects, though less important for many other biological and medical applications of DNA sequencing. Long LORs achieved within reasonable time limits, minimize sample preparation and computational effort required assembling the sequencing data into finished sequence. Thus, long LORs increase the throughput and reduce the cost of sequencing.

The viscosity of a polymer matrix is an important factor considered in designing and engineering automated CE sequencing instruments. Low-viscosity polymer solutions may be pumped into the capillary and replaced using a practically achievable and robust pressurizing system, and the requisite matrix loading step may significantly contribute to the turnaround time and cost of the instruments. More subtly, the dependence of polymer solution viscosity on the rate of applied shear is an important attribute of DNA sequencing matrix, as it controls the rate of flow of the solution through the microchannel under a given applied force. Another factor that contributes to cost is the ability of the polymer to adsorb on the internal capillary wall and suppress electroosmotic flow, thus eliminating the need for a covalently bound capillary coating. The lifetime of the coating, as well as the ease of its regeneration, impact the overall performance and cost of sequencing.

6.1 Linear polyacrylamide

Replaceable LPA, introduced for DNA sequencing in 1993 [49], has become one of the most widely employed separation media in DNA sequencing by CE due to its excellent performance in terms of LOR and separation time. Carrilho *et al.* [50] investigated the effect of molecular mass and concentration of LPA on DNA sequencing. For a given high LPA molecular mass, it was found that higher solution concentrations (4% LPA) improved the resolution of DNA fragments smaller than 450 bases, while a lower concentration (2% LPA) gave better resolution of DNA fragments larger than 450 bases. Others reported similar results on the effect of LPA concentration on separation performance [51, 52]. Selectivity for low base numbers was higher for more concentrated polymer solutions, and decreased with base number for all polymer solutions. In addition, the rate of change of selectivity with base number was significantly greater for high-concentration LPA solutions. For any polymer concentration, peak efficiency decreased with increasing DNA fragment size; however, there was no significant difference in efficiency for a specific fragment with changing LPA concentration, suggesting that simply increasing the polymer concentration will not serve to improve the resolution of DNA peaks.

Using a 4% LPA solution, the selectivity of bases up to 500 bases was higher with matrices formulated with high-molecular mass LPA as opposed to low-molecular mass LPA. For fragments larger than 500 bases, the selectivity merged to a common low value for all matrices. Migration time was found to be more dependent on polymer concentration than on polymer molecular mass. For each polymer solution, a maximum in efficiency was observed, and this maximum was linearly shifted to longer DNA fragments with increased LPA molecular mass. Moreover, the absolute value of the maximum efficiency increased with LPA molecular mass. Also, lowering the electric field strength from 200 V/cm to 150 V/cm improved the selectivity and shifted the onset of biased reptation to longer DNA fragments, extending the sequencing LOR at the expense of run time. Thus, the use of a low concentration of high-molecular mass LPA extended the LOR. Using 2% high-molecular mass LPA (> 5.5 MDa) solution, a column temperature of 50°C, and 150 V/cm field strength, sequencing analysis of more than 1000 bases was achieved with 96.8% accuracy in 80 min. A similarly spectacular performance was obtained by high-molecular mass (9 MDa) LPA prepared by inverse emulsion polymerization [53].

The performance of LPA sequencing performance was further improved by fine-tuning of polymer molecular mass distribution, LPA solution composition, electric field strength, run temperature, internal capillary diameter, dye chemistry, sample clean-up protocols [41], and base-calling software. The final conditions comprised a 2.5% LPA mixture consisting of 2% high-molecular mass (9 MDa) and 0.5% low-molecular mass (50 kDa) LPA operated at 60°C and 200 V/cm. The new conditions produced an LOR of 1000 bases in 55 min at 98–99% accuracy. An increase of LPA concentration by 0.5% with low-molecular mass LPA improved the separation of DNA fragments below 100 bases, without causing a significant reduction in the resolution of the large fragments. This was explained by the fact that short-chain polymers interact preferentially with small DNA molecules, whereas long-chain polymers interact mainly with large DNA fragments. More likely, small-DNA separation is insensitive to polymer molecular mass, but instead is dependent on overall polymer concentration, whereas large DNA can only be separated by large polymers [34].

An increase in the run temperature shifts the onset of biased reptation to longer DNA fragments [20], and this counterbalances the effect of increased electric field strength on the onset of biased reptation (in terms of DNA chain length). Moreover, the use of elevated temperature helps to minimize compressions that result from DNA secondary structure, improving resolution and reducing separation time [51, 54, 55]. Table 2 summarizes the effect of LPA solution composition, temperature and electric field strength on electrophoretic separation of DNA fragments as determined by Karger's group [44]. Optimal results were obtained using an LPA matrix comprising 0.5% 270 kDa and 2% 17 MDa LPA. This formulation delivered an LOR of 1300 bases (average 1249 bases) at 98.5% accuracy in 2 h at 70°C and 125 V/cm. The final resolution was 0.25. This work also involved some improvements in the base-calling software and the DNA Sanger cycle sequencing reaction.

Wu *et al.* [56] studied the separation of ssDNA sequencing reaction products up to 393 bases in size in polyacrylamide solutions. It was found that the electrophoretic mobility of DNA molecules up to 393 bases in size increased with decreasing polyacrylamide concentration, leading to a faster separation of DNA fragments. The molecular mass and polydispersity of the polymer did not seem to have a significant effect on the mobility. However, increasing the average polyacrylamide chain length reduced the dispersion of electrophoresing DNA and increased the sharpness of the bands, implying an increase in efficiency of separation.

Using low-viscosity (150 cP), 6.2% low-molecular mass LPA (339 kDa) solution, Grossman [57] resolved DNA sequencing fragments up to 580 bases long in 110 min at 218 V/cm. Under the run conditions, the selectivity decreased with fragment size, whilst the efficiency of separation increased, and did not exhibit a maximum value

Table 2. Comparison of migration times and LORs for different compositions of LPA matrices at different temperatures and electric field strengths [44]

LPA	Temp. (°C)	Electric field (V/cm)	Migration time for base 1019 (min)	LOR at at 98.5% accuracy (bases)
2% 17 MDa + 0.5% 270 kDa	70	125	105.4	1249
2% 10 MDa + 0.5% 270 kDa	70	125	101.0	1190
2% 17 MDa + 0.5% 50 kDa	70	125	100.0	1083
2% 10 MDa + 0.5% 50 kDa	70	125	99.5	965
2% 10 MDa + 0.5% 50 kDa	60	200	55.6	1013
2% 10 MDa	50	150	81.0	951
2% 10 MDa + 0.5% 270 kDa	70	250	44	927
2% 10 MDa + 0.5% 270 kDa	70	200	55.6	1042
2% 10 MDa + 0.5% 270 kDa	70	150	80.5	1127
2% 10 MDa + 0.5% 270 kDa	70	100	131.0	1172

as was observed by Karger's group using high-molecular mass LPA [50, 58]. Empirical models, derived by Manabe *et al.* [52], predicted that a high electric field and a high LPA concentration are needed to achieve high separation efficiency, whilst a low electric field and a low polymer concentration are required to achieve long LORs. Using 9%T *in situ* polymerized LPA and an electric field strength of 100 V/cm, single-base resolution of ssDNA sequencing fragments up to 520 bases was obtained. Zhang *et al.* [51] separated DNA sequencing fragments up to 640 bases in 2 h using a 5%T polyacrylamide solution at 60°C and 150 V/cm.

Although LPA matrices have a high sieving capacity for DNA fragments, they have a few drawbacks. One of the main problems of high-molecular mass LPA is the high viscosity of its solutions, a result of the high hydrophilicity of the polymer. A saving grace is that LPA solutions behave as non-Newtonian (shear-thinning) fluids, such that the viscosity of high-molecular mass LPA comes close to that of concentrated low-molecular mass LPA when the shear force is increased [53]. Nonetheless, the viscosity of the 2%, 9 MDa LPA was *ca.* 260 000 cP at zero shear and about 27 000 cP at a shear rate of $1.32\,\mathrm{s}^{-1}$ [53]. High-molecular mass LPA also requires careful preparation and handling, because the method of dissolution can affect the mass distribution through chain scission, which can adversely affect the subsequent separation of DNA sequencing fragments. Chemical instability towards alkaline hydrolysis and the neurotoxicity of acrylamide monomer are other drawbacks of LPA [59, 60]. Another problem with using LPA for CE is that it cannot be used with bare fused-silica capillaries. The capillary has to be precoated to suppress electroosmotic flow. Capillary coatings have the propensity to become fouled after a number of DNA sequencing runs, and this leads to an eventual deterioration of the separation performance.

6.2 Polyethylene oxide

Polyethylene oxide (PEO) is another polymer that has been extensively studied for DNA sequencing [42, 43, 61, 62]. It was first introduced in 1995 by Fung and Yeung [61]. In 1997, Kim and Yeung [42] reported the separation of a single-color DNA sequencing ladder up to 1000 bases (resolution of raw data = 0.5 at 966 bases) in a mixture of 1.5% high-molecular mass PEO (8 MDa) and 1.4% low-molecular mass PEO (0.6 MDa) at 75 V/cm. A separation time of 7 h was needed to achieve this separation, which is very long compared to the 1 h run time needed for LPA to achieve the same LOR [58]. Increasing the concentration of the high-molecular mass PEO to 2.0% decreased the size of the largest DNA fragments that could be resolved. The influences of separation volt-

age, column length, and PEO composition were also studied. In agreement with findings for other polymers [50, 63], PEO performance increased with lower field strengths and longer separation distances. The selectivity up to 320 bases was independent of the electric field strength and decreased with increasing field strength for DNA larger than 320 bases. The efficiency of the separation increased to a maximum at around 250 bases and then decreased with increasing base number. Regarding the PEO composition, it was found that when low-molecular mass PEO was blended with high-molecular mass PEO, the separation of smaller DNA fragments improved significantly, a result similar to that obtained for other polymers [40, 44, 58]. For best sequencing performance, the total PEO concentration was found to be between 2.5 and 3.0%.

PEO has the interesting and favorable property of being "self-coating" through adsorption to the capillary wall [61]. This potentially eliminates the need for covalent wall passivation, as is required for LPA matrices, thereby reducing the material cost of DNA sequencing. PEO-coated capillaries had to be flushed with HCl and PEO solutions before being reusable and this required about 2 h. It was further observed that PEO capillaries could not be used for more than 30 runs without reconditioning. This was tentatively attributed to the formation of PEO aggregates on the wall surface, and was increased by increasing the PEO concentration. The adsorbed PEO coating was shown to reduce the electroosmotic flow velocity by more than one order of magnitude compared to the bare capillary at pH 7.0. However, at pH 8.2, the reduction in electroosmotic flow was only 30–50% [64].

6.3 Poly-*N,N*-dimethylacrylamide

Madabhushi [65] introduced the acrylamide derivative poly-*N,N*-dimethylacrylamide (PDMA) for DNA sequencing. The author considered low viscosity of the sieving matrix to be a higher priority than the attainment of a long LOR. Low viscosity translates to the ability to use lower pressure for filling and flushing the capillary, thus providing faster and safer operation. An LOR of 600 bases was achieved in 125 min using a 6.5% low-molecular mass (98 kDa) PDMA, the viscosity of which was claimed to be just 75 cP, many orders of magnitude lower than that of LPA solutions. The increase in PDMA molecular mass beyond 98 kDa slightly improved the resolution but adversely increased the viscosity of the solution. PDMA also exhibits two advantages over LPA: (i) it is resistant to hydrolysis [59]; (ii) it possesses self-coating properties, *i.e.*, it physically adsorbs to the capillary wall and virtually eliminates electroosmotic flow. The dynamic coating process required no special conditioning, *e.g.*, no HCl wash, and capillaries could be used directly after dynamic coating. The capillary was typically ready for use after 1 min of sieving matrix replacement by low pressure. It was reported that at least 100 sequencing runs could be performed in a single capillary using this simple procedure without any adverse effect. In four-color sequencing, 600 bases could be analyzed with a final resolution of 0.59. The accuracy achieved was more than 98% for the first 450 bases and greater than 96.5% for the 600 bases. The run was performed at 42°C and 160 V/cm.

Heller [15, 30, 47, 63] investigated the use of PDMA for ssDNA separation. Peak spacing was found to decrease with increasing DNA size, but increased with increasing polymer concentration. Moreover, for a given PDMA concentration, the peak spacing increased with increasing PDMA chain length. The peak width decreased, reflecting an increase in the efficiency of separation, with increasing polymer concentration and polymer chain length, but it did not change as a function of DNA size. 5% solutions of low-molecular mass (216 kDa) or high-molecular mass (1.15 MDa) PDMA achieved single-base resolution up to 500 bases, but beyond that point the resolution failed dramatically. The use of lower PDMA concentration (1 and 2%) resulted in lower resolution of DNA but the rate of decrease of resolution with DNA base number was smaller than that using 5% solution. As with other polymers, a reduction in electric field strength improved the resolution of the DNA fragments.

The main disadvantage of PDMA is that it is relatively hydrophobic compared to LPA [59]. This increases the extent of hydrophobic interaction between the polymer and the hydrophobic DNA-labeling dyes, resulting in changes in DNA mobility. This can lead to peak shifting and band broadening, thus affecting the resolution of DNA fragments and increasing the chances of errors in base calling. This increased hydrophobicity also causes the PDMA chains to adopt more compact coils in aqueous solution at a given chain length, reducing the average number of chain entanglements and producing a less robust entangled network than, for example, LPA, which is more hydrophilic.

6.4 Polyvinyl pyrrolidone

Using a mixture of PDMA and polyvinyl pyrrolidone (PVP), Madabhushi *et al.* [66] demonstrated the feasibility of four-color sequencing in a bare silica capillary. Later, Gao and Yeung [67] explored the use of PVP as a DNA sieving matrix. The polymer is soluble in water and aqueous buffers, forming a homogenous, low-viscosity (27 cP at 4.5%) solution within less than 1 h. The polymer exhibited a self-coating property and reduced electroosmotic

flow to negligible levels. The low viscosity of the polymer solution reduced the time needed for regeneration of the column surface to 3 min, compared to 2 h for PEO. Single-color sequencing with 7% PVP of 1 MDa molecular mass showed separation of up to 350 bases. The separation was extended to more than 500 bases when a 5% solution of high-molecular mass PVP extracted from the 1 MDa material was used. The sequencing time was 83 min in a 50 cm effective column length at 150 V/cm and room temperature.

6.5 Polyethylene glycol with fluorocarbon tails

A novel class of material was reported by Menchen *et al.* [68] who modified polyethylene glycol (PEG) by end-capping it with a fluorocarbon tail. This material self-assembles in water into equilibrium network structures with a well-defined mesh size. Above a critical concentration, the hydrophobic end groups associate into micelles in water to form extended networks. These micellar systems form flowable gel-like networks that permit electrophoretic DNA sequencing in a capillary column. In addition, the network structure is disrupted under high shear, allowing relatively easy replacement of the sieving matrix. DNA electrophoretic mobility appeared not to be adversely affected by the hydrophobicity of the polymer tails. The use of longer PEG chains decreased the band width and improved the resolution of DNA sequencing fragments. Polymer concentration influenced the peak-to-peak spacing. Longer LORs were obtained by using the polymer at concentrations above the critical micelle concentration. Increasing the temperature from 22°C to 55°C destabilized the polymer network and resulted in severe band broadening and loss of resolution. Optimum sequencing results were obtained from a 6% solution of a 1:1 mixture of C_6F_{13} end-capped and C_8F_{17} end-capped PEG of 35 kDa molecular mass. An LOR of 450 bases was obtained at 200 V/cm. The run time was not reported.

6.6 Hydroxyethylcellulose

The use of HEC as a sieving matrix of ssDNA was explored by Bashkin *et al.* [69]. The best performance was obtained at 190 V/cm using 2% HEC of 90–105 kDa molecular mass. Sharp, well-resolved peaks of one-color sequencing reaction products were observed to around 570 bases in 70 min. Higher voltages shifted the onset of the biased reptation to smaller DNA fragments. Purification of the HEC was needed to remove oxidized forms of cellulose and other charged impurities that affected the performance of the crude HEC sieving matrix. The use of entangled HEC solution at pH 11 significantly improved the efficiency and hence the resolution of ssDNA fragments in denaturing buffer [48]. The drawback of these HEC matrices that prevented their use in a commercial high-throughput sequencing instrument such as the MegaBACE 1000^{TM}, was that in a practical setting with realistic DNA sequencing samples, LORs of *ca.* 450 bases were typical. Inconsistent performance of HEC batches stems from the fact that this is a naturally derived, not synthetic, polymer.

Using one-color DNA sequencing data, Dolník and Gurske [70] derived empirical equations for the relationship between selectivity per base and sieving matrix concentration and between the mobility slope and matrix concentration using HEC (300 kDa). The same authors proposed the use of the inflection slope, *i.e.*, the slope of the log-log mobility curve at its inflection point, as a quantitative parameter of sieving performance. With increasing polymer concentration, the inflection slope decreased and moved to shorter DNA length and eventually reached a plateau. Kheterpal and Mathies [71] reported a comparison of the performance of replaceable LPA, HEC, and a mixture of PEO and HEC under identical conditions for DNA sequencing using identical separation conditions. Although experimental details were not given, LPA was said to provide the longest LOR (~ 1000 bases) compared with ~ 600 bases in HEC and < 300 bases in the HEC-PEO mixture. Mixtures of PEO and HEC were reported by Kim and Yeung [42] to give unreadable resolution of DNA sequencing fragments.

6.7 Poly-*N*-acryloylaminopropanol

Righetti's group [60, 72, 73] explored the utility of a novel acrylamido monomer, namely *N*-acryloylaminopropanol (AAP). The resistance to alkaline hydrolysis of poly(AAP) was shown to be 500 times higher than that of polyacrylamide. Moreover, AAP is more hydrophilic than acrylamide. Single-color DNA sequencing was performed at 50°C and 200 V/cm using a 10.9%T poly(AAP). An LOR of 385 bases was achieved in about 100 min at a final resolution of 0.5. An increase in run temperature from 25°C to 50°C was efficient in resolving compression of DNA zones. The use of formamide was detrimental to gel stability and sieving performance. The performance of an *in situ* cross-linked poly(AAP) gel, with linear poly(AAP) as an additive, was also evaluated. The gel was formed by introducing into the capillary a mixture of the 5.45% AAP monomer, 1.1% dihydroxyethylenebisacrylamide cross-linker and 5.45% prepolymerized poly(AAP). A rise in temperature from 25 to 50°C reduced the LOR from 400 bases to 303 bases and adversely affected the efficiency of the separation and the resolution of the DNA fragments.

6.8 Poly(*N,N*-dimethylacrylamide-co-*N,N*-diethylacrylamide)

Sassi *et al.* [74] observed that for uncross-linked, thermoresponsive polymer solutions, the reversible volume-phase transition at the lower critical solution temperature (LCST) results in a precipitous viscosity drop, by at least one order of magnitude. Taking advantage of this property exhibited by hydrophobically modified polymers, the authors introduced a new approach towards solving the problem of obtaining a single-base resolution of DNA fragments in entangled polymer solutions and ease of handling and loading of polymer solutions into the capillary. Thermoresponsive polymers can be loaded into the capillary at high temperatures ($\geq$ LCST) at which the viscosity is low, and the viscous, entangled network can be regained by cooling the polymer solution below its LCST at which DNA separation can be achieved.

Using copolymers of *N,N*-dimethylacrylamide (DMA) and *N,N*-diethylacrylamide (DEA), it was shown that these copolymers could show single-base resolution of DNA sequencing products up to 150 bases in a solution of 6% copolymer. The copolymer had a viscosity of 10 cP at 70°C, making it easy to load and replace in the capillary. We have developed different formulations of linear DEA/DMA copolymers for use in DNA sequencing (Buchholz *et al.*, in preparation). Our results show that 420 bases can be sequenced at 98.5% accuracy in 75 min, using a copolymer of 53% DEA, 47% DMA copolymer (4.0 MDa molecular mass). The electrophoretic run was performed at 44°C and 140 V/cm. Furthermore, it was found that increasing the DEA composition decreased the LOR due to increasing polymer hydrophobicity, which caused band broadening and peak shifting. A comparison of the sequencing performance, rheological properties and ability to suppress electroosmotic flow of the various polymers used for DNA sequencing by CE is given in Tables 3, 4 and 5, respectively.

7 Polymers for separation of dsDNA

The separation of dsDNA fragments has a number of important biological applications, such as the separation of PCR products, restriction fragments, and plasmids. For such applications, single-base resolution is not a necessity. Thus, many of the polymers that have been used for DNA sequencing have provided excellent separation of dsDNA [32, 34, 40, 43, 62, 67, 73–77]. A variety of polymers has also been successfully explored for the separation of dsDNA, and may potentially be applicable for DNA sequencing under the appropriate conditions. For several of these polymers, the challenge has been to achieve sufficiently high-molecular mass polymers to form robust, entangled networks as needed for long-read DNA sequencing.

Table 3. Comparison of performance of polymer solutions for DNA sequencing by CE

Polymer	Mol. mass (kDa)	Conc. (wt%)	Electric field (V/cm)	Effective length (cm)	Temp. ($^{\circ}$C)	Buffer[a]	LOR (bases)	Time (min)	Resolution at LOR	Reference
LPA	> 5500	2	150	30	50	TTE, 7 M urea	1000	80	0.4	[50]
LPA	9000	2	150	30	50	TTE, 7 M urea	1000	80	0.4	[53]
LPA	9000 + 50	2 0.5	200	30	60	TTE, 7 M urea	1000	55	0.25	[58]
LPA	17 000 + 270	2 0.5	125	30	70	TTE, 7 M urea	1300	120	0.25	[44]
LPA	1000	8	100	12	Room	TBE, 3.5 M urea, 30% formamide	393	90	nr[a]	[56]
LPA	339	6.2	218	40	Room	180 mM Tris-H_3PO_4, 8 M urea	580	109	0.75	[57]
LPA	nr[b]	9 T	100	40	Room	100 mM Tris-borate, 7 M urea	520	430	0.55	[52]
LPA	nr[b]	5 T	150	39	60	TBE, 3.5 M urea	640	118	nr[b]	[51]
PEO	8000 + 600	1.5 1.4	75	70	Room	TBE, 3.5 M urea	1000	420	0.5	[42]
PDMA	98	6.5	160	40	42	100 mM TAPS, 8 M urea	600	125	0.59	[65]
PDMA	1145	5	210	36	50	0.5 × TBE, 4 M urea	500	34[c]	0.5	[47]
PVP	1000	5	150	50	Room	TBE, 3.5 M urea	500	83	nr[b]	[67]
PEG[d]	35	6	200	36	22	100 mM TAPS, 6.6 M urea	450	nr[b]	0.59	[68]
HEC	97	2	190	41	Room	TBE, 6 M urea, 10% formamide	570	70	0.5	[69]
Poly(AAP)	nr[b]	10.9 T	200	40	50	TBE, 7 M urea	385	100	0.5	[55]
Poly(DMA/DEA)	nr[b]	6 T	170	45	Room	TBE, 7 M urea	150	60	nr[b]	[74]

a) TBE buffer: 89 mM Tris, 89 mM borate, 2 mM EDTA; TTE buffer: 50 mM Tris, 50 mM TAPS (*N*-tris(hydroxymethyl)methyl-3-aminopropanesulfonic acid) and 2 mM EDTA
b) No value reported
c) Estimated from electrophoretic mobility data from [15]
d) 1:1 mixture of C_6F_{13} end-capped and C_8F_{17} end-capped PEG 35 kDa

Table 4. Comparison of viscosities of polymer solutions used for DNA sequencing

Polymer	Composition (wt%)	Mol. mass (kDa)	Buffer[a]	Temp. ($^{\circ}$C)	Viscosity (cP)	Reference
LPA	6.2	339	180 mM Tris-H_3PO_4, 8 M urea	25	150	[57]
LPA	2	9000	TTE, 7 M urea	25	27 000[b]	[53]
LPA	2	9000	TTE, 7 M urea	25	3700[c]	[53]
LPA	6	nr[d]	TBE, 3.5 M urea	25	4900	[61]
PEO	1.5 + 1.4	600 8000	TBE, 3.5 M urea	Room	1200	[61]
PDMA	6.5	98	100 mM TAPS, 8 M urea	30	75	[65]
PDMA	6.5	200	100 mM TAPS, 8 M urea	30	1200	[65]
PEG[e]	6	35	100 mM TAPS, 6.6 M urea	nr[d]	10 000	[68]
HEC	2	97	TBE, 6 M urea, 10% formamide	25	5000	[69]
PVP	4.5	1000	TBE	20	27	[67]
Poly(DMA/DEA)	6	nr[d]	TBE, 7 M urea	70	10	[74]

a) Buffer compositions as in Table 3
b) Viscosity measured at 1.32 s^{-1} shear rate
c) Viscosity measured at 22 s^{-1} shear rate
d) No value reported
e) 1:1 mixture of C_6F_{13} end-capped and C_8F_{17} end capped PEG 35 kDa

Table 5. Suppression of electroosmotic flow by different DNA sequencing polymers

Polymer	Mol. mass (kDa)	Conc. (wt%)	Buffer[a]	Temp. (°C)	Field strength (V/cm)	Electrophoretic mobility (10^{-4} cm^2/V s)	Reference
None	–	–	100 mM glycine	30	200	6.29	[65]
LPA	150	0.01	100 mM glycine	30	200	2.43	[65]
PDMA	98	0.01	100 mM glycine	30	200	0.20	[65]
PVP	40	0.01	100 mM glycine	30	200	0.87	[65]
PVP	360	0.01	100 mM glycine	30	200	0.47	[65]
PEG	35	0.01	100 mM glycine	30	200	1.44	[65]
PNIPA	50	0.01	100 mM glycine	30	200	0.34	[65]
PDMA	nr[b]	0.01	TTE	nr[b]	nr[b]	0.17	[80]
PDMA	nr[b]	0.05	TTE	nr[b]	nr[b]	0.107	[80]
PVP	1000	0.1	TBE	Room	250	0.27	[67]
PVP	1000	1.0	TBE	Room	250	0.06	[67]
PEO	8000	0.2	Phosphate, pH 7.0	Room	nr[b]	0.2	[64]
PEO	8000	0.2	Phosphate, pH 8.2	Room	nr[b]	2.5	[64]

a) Buffer composition as in Table 3
b) No value reported

Several novel *N*-substituted acrylamides, which combine two important properties needed for DNA fragment separation, hydrophilicity and resistance to alkaline hydrolysis, have been proposed by Righetti and co-workers [59, 60, 72]. Chiari *et al.* [59] introduced *N*-acryloylethoxyethanol (AAEE) as a monomer that could be polymerized *in situ* yielding a sieving matrix for dsDNA that gave excellent performance in separation of DNA fragments ranging from few hundreds to 23 000 bp [78]. Furthermore, poly-(AAEE) offered excellent performance as a covalent coating of the silica capillary wall [79]. However, while the polymer offered good performance in CE, the monomer suffered from two main limitations. First, the yield of the monomer synthesis was low, at less than 18% [59]. Second, it was found that the monomer exhibited a unique instability. Upon storage, even in the presence of polymerization inhibitors, it could suddenly auto-polymerize [60], which greatly limited its shelf life.

N-(acrylaminoethoxy)ethyl-β-ᴅ-glucopyranose (AEG) is another monomer bearing hydrophilic glucose units on *N*-substituted acrylamido derivatives, synthesized by Chiari *et al.* [75] *via* a chemo-enzymatic pathway. Radical polymers of AEG were found to have low viscosity in solution relative to polyacrylamide. Moreover, the viscosity of poly(AEG) solution increases weakly with polymer concentration, allowing the replacement in the capillary of more concentrated solution. Poly(AEG) was reported to have a dsDNA sieving capacity fully comparable with that of polyacrylamide, but migration was faster in the former, possibly due to the formation of a more flexible network with less robust entanglements.

Chiari *et al.* [80] proposed another novel low-viscosity glucose-containing copolymer of acrylamide and allyl-β-ᴅ-glucopyranose (AG). The viscosity of the copolymer solution decreased as the sugar content of the copolymer increased. In spite of its low molecular mass (209 kDa), the copolymer obtained by polymerizing a 1:1 molar ratio of AG:acrylamide showed the ability to sieve oligonucleotides and dsDNA ranging in size from a few tens up to 600 bases. The matrix showed good resolving power of DNA restriction fragments. However, single-base resolution needed for DNA sequencing application was not demonstrated.

Another class of material that has been extensively studied for the separation of dsDNA fragments is the group of polysaccharides, such as glucomannan [81], dextran [46, 82], agarose solutions [83], cellulose derivatives, including methylcellulose (MC) [84], HEC [16, 31, 32], hydroxypropylcellulose (HPC) [34, 84, 85] and hydropropylmethylcellulose (HPMC) [84, 86, 87], and composite agarose/HEC [88]. HEC polymers in water assume a stiff and extended conformation [31]. Cellulose derivatives are typically used to separate DNA fragments at concentrations of 0.2–1%. Due to the fact that stiff and extended molecules become entangled more easily than flexible, random-coil polymers and, furthermore, form more robust entanglements, the operative concentration of cellulosic polymers is considerably lower than that generally used for LPA. Baba *et al.* [84] studied the performance of cellulose derivatives, MC, HPC and HPMC, as sieving networks for DNA separation by CE. Solutions of MC, HPC, and HPMC at a concentration of 0.5% provided separation of large fragments (from 1000 to 23000 bp) of a 1 kbp DNA ladder. However, a higher concentration (0.7%) was needed to improve the resolution of smaller fragments from 75 to 500 bp. Using HPC (1 MDa) solution, Mitnik *et al.* [85] showed that increasing the electric field strength increased the separation speed and improved the resolution of small DNA fragments. Resolution of large fragments was lost at electric fields greater than 500 V/cm. In agreement with other findings, the resolution of DNA fragments larger than a certain size (600 bp) degraded with increasing polymer concentration whilst the resolution of smaller fragments improved.

Barron *et al.* [34] evaluated the separation of dsDNA in HEC, HPC and LPA in the dilute and entangled solution regimes. HEC, which is a stiffer polymer than HPC and LPA, provided better resolution of large DNA fragments (> 603 bp) than the other two polymers. This result pointed out the importance of polymer chain stiffness (or persistence length) for the resolution of large DNA fragments. Moreover, it was shown that the three polymers, HEC, HPC and LPA, show similar trends with respect to the effect of molecular mass and polymer concentration on the sieving performance. The authors also investigated the effect of polymer polydispersity on DNA separation performance using 1 MDa LPA of different polydispersities. It was concluded that polymer polydispersity had only a small effect on the separation of small DNA fragments whereas the separation of large DNA fragments is significantly improved by the use of low-polydispersity, high-molecular mass polymer solutions. All three of these polymers separate large dsDNA fragments (> 600 bp) well in the unentangled regime, if countermigration CE in uncoated capillaries in employed. In coated capillaries, lower resolution was obtained [89] because the effective separation distance is shortened in this mode of separation.

Thermoresponsive copolymers of poly-*N*-isopropylacrylamide (PNIPA), densely grafted with short PEO chains, were applied for dsDNA separation by CE [90, 91]. The copolymer had a random-coil conformation at low concentrations and low temperatures (< 31°C). At temperatures above 31°C, the copolymer underwent a broad coil-to-globule transition accompanied by a small decrease in viscosity (~ 10%), with the PNIPA chains inside the core and the hydrophilic PEO on the surface. With a further increase in temperature (> 35°C), the PNIPA backbone chains became more hydrophobic so that the interchain aggregation became very strong at high concentrations, resulting in a dramatic increase in viscosity. As a separation medium for dsDNA restriction fragments at temperatures less than 31°C, the copolymer provided a resolution comparable to that of LPA with the advantage of a shorter

migration time. Moreover, the copolymer PNIPA-g-PEO showed a self-coating ability for the capillary inner wall. At temperatures above 31°C, the copolymer chains collapsed to the globular state, destroying the chain network and thus losing their DNA-separating ability. These matrices are likely to be too hydrophobic in nature (due to the PNIPA portion of the chains) to be useful for high-efficiency DNA sequencing.

Triblock copolymers of PEO_{99}-PPO_{69}-PEO_{99} (PPO being polypropylene oxide), also known as Pluronic F127®, showed a thermoresponsive behavior and capillary coating ability similar to that of PNIPA-g-PEO [92–96]. Increasing the temperature from 4 to 25°C, 18–30% F127 aqueous solution was transformed from a low-viscosity solution (40 cP at 4°C) to a transparent, self-supporting, gel-like liquid crystalline phase. The attainment of this gel-like state is thermally reversible. The crystalline phase consists of micelles, each composed of a dense core of PPO and a hydrated PEO shell. DNA fragments in the size range of 50–3000 bp were separated in a 20% Pluronic solution. Increasing the polymer concentration increased the number of micelles formed, resulting in slower DNA migration. This improved the resolution of the fragments but reduced the separable DNA size range to an upper limit of 1000 bp. Increasing the temperature led to the formation of more compact micelles, and reduced the interaction between the DNA fragments and the copolymer resulting in loss of sieving power. DNA separation in Pluronic liquid crystals is envisioned to occur by mechanisms other than the presently hypothesized ones for CE.

8 Conclusions

The intensive search for and optimization of polymer matrices applied for DNA sequencing seems to indicate that the optimal matrix has yet to be found. The systematic approach that has been undertaken to optimize the performance of these matrices have revealed that important properties for an "ideal" matrix for DNA sequencing by CE are: (i) high resolving power; (ii) high hydrophilicity; (iii) high average molecular mass to separate large DNA fragments; (iv) sufficient concentration to separate small DNA fragments; (v) hydrolytic stability at pH 7–8.5; (vi) relatively low viscosity for easy capillary loading and matrix replacement; and (vii) dynamic adsorption onto a bare capillary wall for suppression of electroosmotic flow.

For DNA sequencing applications, the use of concentrated, entangled polymer solutions is needed to provide single-base resolution. The delivery of long sequencing LORs is greatly dependent upon the stability and robustness of the entangled polymer network in solution. Good resolution of electrophoresing DNA molecules can be obtained as long as the DNA-polymer interactions responsible for the size-based separation of DNA molecules do not disrupt the polymer-polymer entanglements and hence the polymer network. Hence, the extent and strength of entanglements between polymer chains in solution control the robustness of the network. Hydrophilic, high-molecular mass polymer chains such as LPA adopt an extended conformation in aqueous solution, allowing more frequent entanglements along the chains. An increase in polymer hydrophobicity, such as in PDMA, PVP and PEG, results in the adoption of more compact coil sizes in aqueous solution and the formation of less robust polymer networks, that do less well in resolving large DNA fragments than LPA. Side chains pendant on the polymer backbone also boost the entanglements between the polymer chains, as in the case of HEC. For PEO, the necessity for low field strengths (75 V/cm) and long run times most likely reflects the inherent flexibility of the unsubstituted PEO chains and hence the "delicate" nature of its entangled polymer networks. DNA molecules, traveling under powerful electric fields such as 150 V/cm, are simply not sufficiently hindered by PEO-PEO entanglements to be separated by chain entanglement mechanisms. A good analogy is to American football: large DNA molecules are like huge linebackers, shoving polymer chains out of their way as they barrel towards the anode. Strong points of entanglement that force DNA chains to pause in their headlong rush provide the necessary change in dependence of DNA molecular friction coefficient for separation. Although the presence of side chains on the matrix polymers has a positive effect on the robustness of the polymer network, bulky side chains, such as aminopropanol in AAP, can increase the difficulty of obtaining high-molecular mass polymers due to steric hindrance in the polymerization.

High polymer concentrations improve the separation of small DNA sequencing fragments (< 500 bases), whilst low concentrations of high-molecular mass polymers are better in resolving large DNA fragments (> 500 bases). Small DNA fragments are not long enough to interact sufficiently by entangling with the polymer chains of the matrix. Increasing the polymer concentration increases the frequency of small DNA-polymer collisions that lead to separation. For large DNA fragments, large polymer chains are needed to maintain the robustness of the entangled polymer network and prevent facile dragging of the polymer chains by migrating DNA chains.

Thermoresponsive polymers are potentially suitable candidates as DNA sequencing matrices for CE and microchips, allowing easy filling and replacement in the capillary at one temperature and providing high resolution of DNA sequencing reaction products at another, though typically at the cost of somewhat shorter LORs due to the hydrophobicity of these polymers. Further experimental and theoretical investigations are needed to elucidate the specific mechanisms by which polymer chemical and physical properties influence matrix performance for DNA sequencing separation. In light of the new insights researchers have gained into the impact of polymer properties on DNA sequencing performance, the door is still open to exploit and invent new polymers that will improve overall DNA sequencing throughput and reduce the cost-per-base rate.

Financial support was provided by the National Institutes of Health (NIH), Grant R01HG01970-01. We are grateful to Professor Monica Olvera de la Cruz (Northwestern University) for helpful discussions of DNA separation mechanisms.

Received July 5, 2000

9 References

[1] http://geta.life.uiuc.edu/~nikos/genomes.html

[2] Sanger, F., Nicklen, S., R., Coulson, A. R., *Proc. Natl. Acad. Sci. USA* 1977, *74*, 5463–5467.

[3] http://www.pebio.com/ab/about/dna/377/377a1a.html

[4] Marshal, E., Pennisi, E., *Science* 1998, *280*, 994–995.

[5] Brennan, M., Zurer, P., *Chem. Eng. News* 2000, *78*, 11.

[6] Huang, X. H. C., Quesada, M. A., Mathies, R. A., *Anal. Chem.* 1992, *64*, 2149–2154.

[7] Huang, X. H. C., Quesada, M. A., Mathies, R. A., *Anal. Chem.* 1992, *64*, 967–972.

[8] Carrilho, E., *Electrophoresis* 2000, *21*, 55–65.

[9] http://www.mdyn.com/products/MegaBACE/default.htm

[10] Swerdlow, H., Zhang, J. Z., Chen, D. Y., Harke, H. R., Grey, R., Wu, S., Dovichi, N. J., *Anal. Chem.* 1991, *63*, 2835–2841.

[11] Kambara, H., Takahashi, S., *Nature* 1993, *361*, 565–566.

[12] Mullikin, J. C., McMurray, A. A., *Science* 1999, *283*, 1867–1868.

[13] Woolley, A. T., Mathies, R. A., *Anal. Chem.* 1995, *67*, 3676–3680.

[14] Liu, S. R., Shi, Y. N., Ja, W. W., Mathies, R. A., *Anal. Chem.* 1999, *71*, 566–573.

[15] Heller, C., *Electrophoresis* 1999, *20*, 1962–1977.

[16] Grossman, P. D., Soane, D. S., *Biopolymers* 1991, *31*, 1221–1228.

[17] Viovy, J. L., Duke, T., Caron, F., *Contemp. Phys.* 1992, *33*, 25–40.

[18] Slater, G. L., Mayer, P., Drouin, G., *Methods Enzymol.* 1996, *270*, 272–295.

[19] Slater, G. W., Noolandi, J., *Biopolymers* 1986, *25*, 431–454.

[20] Lumpkin, O. J., Dejardin, P., Zimm, B. H., *Biopolymers* 1985, *24*, 1573–1593.

[21] Duke, T., Viovy, J. L., Semenov, A. N., *Biopolymers* 1994, *34*, 239–247.

[22] Cottet, H., Gareil, P., Viovy, J. L., *Electrophoresis* 1998, *19*, 2151–2162.

[23] Bae, Y. C., Soane, D., *J. Chromatogr.* 1993, *652*, 17–22.

[24] de Gennes, P. G., *Phys. Today* 1983, *36*, 33–39.

[25] Smith, D. E., Perkins, T. T., Chu, S., *Phys. Rev. Lett.* 1995, *22*, 4146–4149.

[26] deGennes, P. G., *J. Chem. Phys.* 1971, *55*, 572–579.

[27] Doi, M., Edwards, S. F., *The Theory of Polymer Dynamics*, Clarendon Press, Oxford 1986.

[28] Olivera de la Cruz, M., Deutsch, J. M., Edwards, S. F., *Phys. Rev. A* 1986, *33*, 2047–2055.

[29] Shaffer, E. O., Olivera de la Cruz, M., *Macromolecules* 1989, *22*, 1351–1355.

[30] Heller, C., *Electrophoresis* 1998, *19*, 3114–3127.

[31] Barron, A. E., Soane, D. S., Blanch, H. W., *J. Chromatogr.* 1993, *652*, 3–16.

[32] Barron, A. E., Blanch, H. W., Soane, D. S., *Electrophoresis* 1994, *15*, 597–615.

[33] Barron, A. E., Sunada, W. M., Blanch, H. W., *Biotechn. Bioengineer.* 1996, *52*, 259–270.

[34] Barron, A. E., Sunada, W. M., Blanch, H. W., *Electrophoresis* 1996, *17*, 744–757.

[35] Shi, X., Hammond, R., Morris, M. D., *Anal. Chem.* 1995, *67*, 1132–1138.

[36] Carlsson, C., Larsson, A., Jonsson, M., Norden, B., *J. Am. Chem. Soc.* 1995, *117*, 3871–3872.

[37] Sunada, W. M., Blanch, H. W., *Biotechnol. Progr.* 1998, *14*, 766–772.

[38] Sunada, W. M., Blanch, H. V., *Electrophoresis* 1998, *19*, 3128–3136.

[39] Hubert, S. J., Slater, G. W., Viovy, J. L., *Macromolecules* 1996, *29*, 1006–1009.

[40] Bunz, A. P., Barron, A. E., Prausnitz, J. M., Blanch, H. W., *Ind. Engineer. Chem. Res.* 1996, *35*, 2900–2908.

[41] Ruiz-Martinez, M. C., Salas-Solano, O., Carrilho, E., Kotler, L., Karger, B. L., *Anal. Chem.* 1998, *70*, 1516–1527.

[42] Kim, Y., Yeung, E. S., *J. Chromatogr. A* 1997, *781*, 315–325.

[43] Madabhushi, R. S., Vainer, M., Dolnik, V., Enad, S., Barker, D. L., Harris, D. H., Mansfield, E. S., *Electrophoresis* 1997, *18*, 104–111.

[44] Zhou, H., Miller, A. W., Sosic, Z., Buchholz, B., Barron, A. E., Kotler, L., Karger, B. L., *Anal. Chem.* 2000, *72*, 1045–1052.

[45] van der Schans, M. J., Kuypers, A. W., Kloosterman, A. D., Janssen, H. J., Everaerts, F. M., *J. Chromatogr. A* 1997, *772*, 255–264.

[46] Heller, C., *Electrophoresis* 1998, *19*, 1691–1698.

[47] Heller, C., *Electrophoresis* 1999, *20*, 1978–1986.

[48] Liu, Y., Kuhr, W. G., *Anal. Chem.* 1999, *71*, 1668–1673.

[49] Ruiz-Martinez, M. C., Berka, J., Belenkii, A., Foret, F., Miller, A. W., Karger, B. L., *Anal. Chem.* 1993, *65*, 2851–2858.

[50] Carrilho, E., Ruiz-Martinez, M. C., Berka, J., Smirnov, I., Goetzinger, W., Miller, A. W., Brady, D., Karger, B. L., *Anal. Chem.* 1996, *68*, 3305–3313.

[51] Zhang, J. Z., Fang, Y., Hou, J. Y., Jiang, R., Roos, P., Dovichi, N. J., *Anal. Chem.* 1995, *67*, 4589–4593.

[52] Manabe, T., Chen, N., Terabe, S., Yohda, M., Endo, I., *Anal. Chem.* 1994, *66*, 4243–4252.

[53] Goetzinger, W., Kotler, L., Carrilho, E., Ruiz-Martinez, M. C., Salas-Solano, O., Karger, B. L., *Electrophoresis* 1998, *19*, 242–248.

[54] Kleparnik, K., Foret, F., Berka, J., Goetzinger, W., Miller, A. W., Karger, B. L., *Electrophoresis* 1996, *17*, 1860–1866.

[55] Lindberg, P., Righetti, P. G., Gelfi, C., Roeraade, J., *Electrophoresis* 1997, *18*, 2909–2914.

[56] Wu, C. H., Quesada, M. A., Schneider, D. K., Farinato, R., Studier, F. W., Chu, B., *Electrophoresis* 1996, *17*, 1103–1109.

[57] Grossman, P. D., *J. Chromatogr. A* 1994, *663*, 119–227.

[58] Salas-Solano, O., Carrilho, E., Kotler, L., Miller, A. W., Goetzinger, W., Sosic, Z., Karger, B. L., *Anal. Chem.* 1998, *70*, 3996–4003.

[59] Chiari, M., Micheletti, C., Nesi, M., Fazio, M., Righetti, P. G., *Electrophoresis* 1994, *15*, 177–186.

[60] SimoAlfonso, E., Gelfi, C., Sebastiano, R., Citterio, A., Righetti, P. G., *Electrophoresis* 1996, *17*, 723–731.

[61] Fung, E. N., Yeung, E. S., *Anal. Chem.* 1995, *67*, 1913–1919.

[62] Chang, H. T., Yeung, E. S., *J. Chromatogr. B* 1995, *669*, 113–123.

[63] Heller, C., *Electrophoresis* 2000, *21*, 593–602.

[64] Preisler, J., Yeung, E. S., *Anal. Chem.* 1996, *68*, 2885–2889.

[65] Madabhushi, R. S., *Electrophoresis* 1998, *19*, 224–230.

[66] Madabhushi, R. S., Menchen, S. M., Efcavitch, J. W., Grossman, P. D., *US Patent No. 5,567,292*, 1996.

[67] Gao, Q., Yeung, E. S., *Anal. Chem.* 1998, *70*, 1382–1388.

[68] Menchen, S., Johnson, B., Winnik, M. A., Xu, B., *Electrophoresis* 1996, *17*, 1451–1459.

[69] Bashkin, J., Marsh, M., Barker, D., Johnston, R., *Appl. Theor. Electrophor.* 1996, *6*, 23–28.

[70] Dolník, V., Gurske, W. A., *Electrophoresis* 1999, *20*, 3373–3380.

[71] Kheterpal, I., Mathies, R. A., *Anal. Chem.* 1999, *71*, 31A–37A.

[72] SimoAlfonso, E., Gelfi, C., Sebastiano, R., Citterio, A., Righetti, P. G., *Electrophoresis* 1996, *17*, 732–737.

[73] Gelfi, C., SimoAlfonso, E., Sebastiano, R., Citterio, A., Righetti, P. G., *Electrophoresis* 1996, *17*, 738–743.

[74] Sassi, A. P., Barron, A., AlonsoAmigo, M. G., Hion, D. Y., Yu, J. S., Soane, D. S., Hooper, H. H., *Electrophoresis* 1996, *17*, 1460–1469.

[75] Chiari, M., Riva, S., Gelain, A., Vitale, A., Turati, E., *J. Chromatogr. A* 1997, *781*, 347–355.

[76] Braun, B., Blanch, H. W., Prausnitz, J. M., *Electrophoresis* 1994, *15*, 1994–1997.

[77] Gelfi, C., Perego, M., Libbra, F., Righetti, P. G., *Electrophoresis* 1996, *17*, 1342–1347.

[78] Chiari, M., Nesi, M., Righetti, P. G., *Electrophoresis* 1994, *15*, 616–622.

[79] Chiari, M., Nesi, M., Sandoval, J. E., Pesek, J. J., *J. Chromatogr. A* 1995, *717*, 1–13.

[80] Chiari, M., Damin, F., Melis, A., Consonni, R., *Electrophoresis* 1998, *19*, 3154–3159.

[81] Izumi, T., Yamaguchi, M., Yoneda, K., Isobe, T., Okuyama, T., Shinoda, T., *J. Chromatogr.* 1993, *652*, 41–46.

[82] Ganzler, K., Greve, K. S., Cohen, A. S., Karger, B. L., Guttman, A., Cooke, N. C., *Anal. Chem.* 1992, *64*, 2665–2671.

[83] Boček, P., Chrambach, A., *Electrophoresis* 1992, *13*, 31–34.

[84] Baba, Y., Ishimura, N., Samata, K., Tsuhako, M., *J. Chromatogr.* 1993, *653*, 329–335.

[85] Mitnik, L., Salome, L., Viovy, J. L., Heller, C., *J. Chromatogr. A* 1995, *710*, 309–321.

[86] Shihabi, Z. K., *J. Chromatogr. A* 1999, *853*, 349–354.

[87] Han, F., Huynh, B. H., Ma, Y., Lin, B., *Anal. Chem.* 1999, *71*, 2385–2389.

[88] Siles, B. A., Anderson, D. E., Buchanan, N. S., Warder, M. F., *Electrophoresis* 1997, *18*, 1980–1989.

[89] Barron, A. E., Sunada, W. M., Blanch, H. W., *Electrophoresis* 1994, *15*, 64–74.

[90] Liang, D. H., Song, L. G., Zhou, S. Q., Zaitsev, V. S., Chu, B., *Electrophoresis* 1999, *20*, 2856–2863.

[91] Liang, D. H., Zhou, S. Q., Song, L. G., Zaitsev, V. S., Chu, B., *Macromolecules* 1999, *32*, 6326–6332.

[92] Rill, R. L., Locke, B. R., Liu, Y., Van Winkle, D. H., *Proc. Natl. Acad. Sci. USA* 1998, *95*, 1534–1539.

[93] Wu, C. H., Liu, T. B., Chu, B., *Electrophoresis* 1998, *19*, 231–241.

[94] Wu, C., Liu, T., Chu, B., Schneider, D., Graziano, V., *Macromolecules* 1997, *30*, 4574–4583.

[95] Liang, D. H., Chu, B., *Electrophoresis* 1998, *19*, 2447–2453.

[96] Rill, R. L., Liu, Y., Van Winkle, D. H., Locke, B. R., *J. Chromatogr A* 1998, *817*, 287–295.

Review

Beate Maichel
Ernst Kenndler

Institute of Analytical
Chemistry, University of
Vienna, Austria

Recent innovation in capillary electrokinetic chromatography with replaceable charged pseudostationary phases or additives

Recent developments of separation of neutral analytes in capillary systems with the mobile phase driven by the electroosmotic flow (EOF) and charged additives acting as a pseudostationary phase are reviewed. As pseudostationary phases a number of additives are used. Soluble polymers, either anionic or cationic, were applied as alternatives to micelles. Monomeric charged additives are also intended to form associates with the analytes, leading to selective retention and separation in a similar way as the polymeric pseudostationary phases. Dendrimers, spherical macromolecules with highly branched chains and charged terminal groups, are successfully applied for the separation of lipophilic analytes. Polymers with covalently stabilized structures are introduced in the form of permanent micelles and are therefore insensitive to the mobile phase composition, enlarging the applicability of micellar electrokinetic capillary chromatography (MEKC).

Keywords: Capillary electrophoresis / Electrokinetic chromatography / Pseudostationary phases / Polymeric phases / Tetraalkylammonium ions / Dendrimers / Review
EL 4075

Correspondence: Prof. Ernst Kenndler, Institute of Analytical Chemistry, University of Vienna, Währingerstr. 38, A-1090 Vienna, Austria
E-mail: ernst.kenndler@univie.ac.at
Fax: +43-1-4277-9523

Abbreviations: BBMA, butyl acrylate-butyl methacrylate-methacrylic acid copolymer; **BMAC**, butyl methacrylate-methacryloyloxyethyltrimethyl ammonium chloride copolymer; **BNA**, (±)-1,1′-binaphthyl-2,2′-diamine; **BNP**, (±)-1,1′-binaphthyl-2,2′-diyl hydrogen phosphate; **BOH**, (±)-1,1′-2,2′-binaphthol; **Brij-S**, lauryl poly(oxyethylene) sulfate; **DOSS**, sodium dioctyl sulfosuccinate; **Elvacite 2669**, poly-(methyl methacrylate-ethacrylate-methacrylic acid); **LSER**, linear solvation energy relationship; **PAA**, polyallylamine-supported phases; **PAH**, polynuclear aromatic hydrocarbon; **PDADMA**, poly(diallyldimethylammonium); **PEI**, polyethyleneimine; **poly SUA**, poly sodium 10-undecylenate; **poly LL-SUILV**, poly sodium *N*-undecanoyl LL-isoleucyl-valinate; **poly LL-SULL**, poly sodium *N*-undecanoyl LL-leucyl-leucinate; **Poly R-478**, poly(vinylamine)sulfonate anthrapyridone; **poly SUP**, poly sodium *N*-undec-10-ene-1-oyl-aminoethyl-2-phosphonate; **poly SUS**, poly sodium 10-undecenylsulfate; **poly SUT**, poly sodium *N*-undec-10-ene-1-oyl-taurate; **poly L-SUT**, poly sodium *N*-undecanoyl L-threoninate; **poly L-SUV**, poly sodium *N*-undecanoyl L-valinate; **poly D-SUV**, poly sodium *N*-undecanoyl D-valinate; **poly LL-SUVV**, poly sodium *N*-undecaoyl LL-valyl-valinate; **TFAE**, (±)-trifluoro-1-(9-anthryl)ethanol

Contents

    254   0173-0835/00/1515-3160 $17.50+.50/0

1 Introduction

A number of separation methods are being carried out in the capillary format based on migration of the analytes induced by an electric field. There are various criteria for a distinction of these methods. In a first instance, we might differentiate them concerning the electrical property of the analytes, either ionic or neutral. In the case of electrically charged analytes the separation technique is capillary electrophoresis, with zone electrophoresis being the most popular type. Here the different migration in the electric field is determined by the effective ionic mobility of the solutes. The present review will not treat electrophoretic techniques, although a clear distinction is not possible in all cases between these methods and those belonging to the second class, because in some cases a mixed separation mechanism takes place.

With this second class of methods noncharged analytes are separated; it uses electroosmosis as the nonselective flow-generating phenomenon. The electroosmotic flow (EOF) here plays the role of the pressure drop in classical chromatography. Electrochromatography and electrokinetic chromatography (EKC) belong to this class. The straightforward analog to high performance liquid chromatography, electrochromatography, applies a porous bed as stationary phase, and the background electrolyte (BGE) solution is the mobile phase. Another analogous method, micellar electrokinetic capillary chromatography (MEKC), introduced by Terabe in the early 1980s, uses moving charged micelles instead of fixed porous particles as a pseudostationary phase. The separation principle is then based on the partition of the commonly neutral solutes that were transported by the EOF into the micelles. It is decisive that the EOF and the micelles have different migration velocities, but not necessarily different migration directions. MEKC became popular due to its easy applicability and the high separation efficiency achievable. As a further expansion of MEKC, microemulsions were used instead of micelles, but the separation principle is the same.

Some disadvantages of MEKC, and of capillary electrochromatography, led to the search for alternatives. One is based on the use of additives other than the micelle forming ones, which can be regarded as pseudophases in the widest sense as well (*cf.* other reviews [1–7]). It is obvious that soluble charged polymeric additives could be considered as such. An advantage of these materials over micelles is that their application does not require a certain lower, critical concentration (which depends on the BGE), below which micelles are not formed. It is obvious that monomers, on the low molecular size end, can hardly be considered as a pseudostationary phase; however, there

is no clear limit in molecular weight or chain length of the additive that restricts, at least formally, the type of technique. Therefore, charged monomeric additives are treated in this review together with charged polymeric ones.

Soluble neutral polymers, used as additives for the separation of charged analytes, are not the topic of this paper, as they are applied in electrophoretic techniques rather than in EKC. Charged cyclodextrins, applied for the separation of neutral solutes as well, are also not subject of this article. Probably somewhat arbitrarily, we do not consider them as a pseudophase, because normally they form well-defined complexes with the analytes (in contrast to phases). For the same reason we do not include calixarenes and resorcarenes.

It is therefore the topic of the present paper to review the application of charged polymeric additives and dendrimers, respectively, used as pseudostationary phases in a kind of EKC. Also included in this survey are charged monomeric additives, which might also interact with the solutes in a way similar to polymers. The review is finalized by the description of charged polymers that form permanent micelles, although these additives do not strictly belong to the nonmicellar type of pseudostationary phases.

2 Polymeric pseudostationary phases

An overview about the papers dealing with polymeric pseudostationary phases is given in Table 1. We first classify these polymers according to their charge, then according to their chemical composition.

2.1 Anionic polymers

2.1.1 Acrylate copolymers

To date, several acrylate copolymers (Fig. 1A–C) have been employed as pseudostationary phases in EKC for the separation of both neutral and charged analytes [8–16]. Ozaki *et al.* [8] and Terabe *et al.* [9] first reported the use of butyl acrylate-butyl methacrylate-methacrylic acid copolymer (BBMA; Fig. 1A), an acrylate block copolymer with negatively charged carboxylate groups, for the separation of substituted benzenes and naphthalenes as well as ingredients of cold medicines. Concerning naphthalene compounds, BBMA exhibited a significantly different selectivity than that of SDS micelles in MEKC [8]. Ozaki *et al.* [10] also demonstrated that the combination of a high-molecular-mass surfactant with a chiral selector, *i.e.*, β-cyclodextrin, is superior to cyclodextrin-modified MEKC. In contrast to SDS monomers, BBMA is too large to be

Table 1. Polymeric pseudostationary phases

Pseudostationary phase	Analytes	References
Anionic polymers		
(i) Acrylate copolymers		
BBMA	Substituted benzenes and naphthalenes, cold medicine ingredients	[8, 9]
	Naphthalene derivatives	[10]
	Phenyltrimethylammonium chloride, 1-naphthylamine, quinine sulfate, tetraphenyl phosphonium chloride, octaoxyethylenedodecanol	[11]
	Standard mixtures, pharmaceuticals, industrial surfactants	[12]
BBMA + β-cyclodextrin	Dansylated DL-amino acids	[10]
Elvacite 2669	PAHs, alkyl phenones, fullerenes	[13]
	Substituted benzenes, PAHs	[14]
	Organophosphorous pesticides	[15]
	Corticosteroids	[16]
(ii) PAA		
PAA-C$_8$ PAA-C$_{10}$ PAA-C$_{12}$ PAA-C$_{16}$	PAHs, alkyl phenyl ketones	[18]
PAA-C$_{12}$ PAA-C$_{16}$	PAHs, alkyl phenyl ketones	[20]
PAA-C$_{10}$/C$_{16}$ (Mixed phase)	PAHs, alkyl phenyl ketones	[19]
(iii) Other polymeric anionic pseudostationary phases		
Linear polyacrylamide (hydrolyzed)	Phenols	[21]
Poly R-478	Aminoaromatic and nitroaromatic compounds	[25]
Polymers based on silicone backbone	Alkyl phenyl ketones, substituted benzene and naphthalene compounds	[26]
(iv) Cationic polymers		
BMAC	Naphthalene derivatives	[10]
PEI	Proteins, peptides	[27]
	Nitrophenols, chlorophenols	[28]
	Phenols	[29, 30]
PDADMA	Substituted benzenes, 1-naphthol	[31]
	Phenols, benzyl alcohol, benzene	[32]
	Substituted benzenes and naphthalenes	(Maichel *et al.*, submitted)

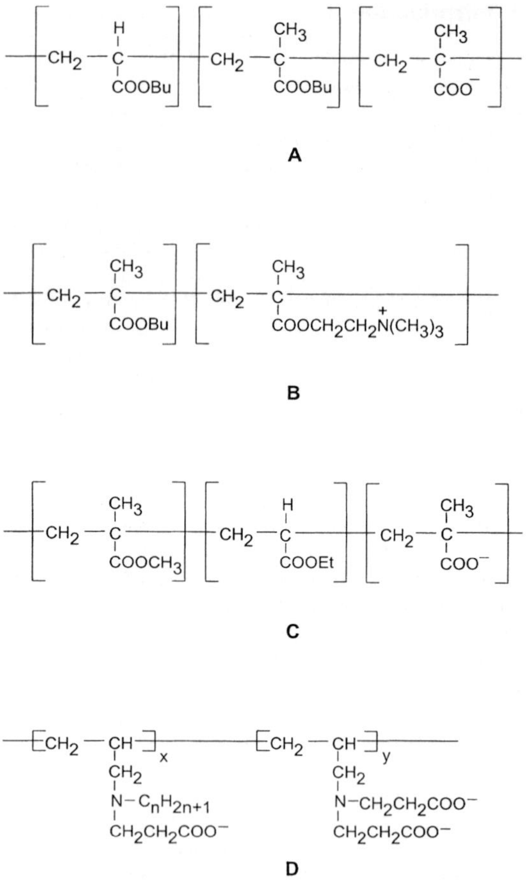

Figure 1. Structure of acrylate copolymers and PAAs. (A) BBMA; (B) BMAC; (C) Elvacite 2669; (D) PAA.

co-included into the cavity of the cyclodextrin with the analyte enantiomers, and thus does not compete with complex formation. As polymeric pseudostationary phases will not dissociate in an electrospray ionization interface or interfere in the low-molecular-mass region of a mass spectrum, they are predestined bo be applied with mass spectrometric detection. Several papers [11, 12, 17] showed that EKC with BBMA as pseudostationary phase was successfully coupled with electrospray ionization mass spectrometry (ESI-MS).

Separation of a wide range of hydrophobic compounds (polynuclear aromatic hydrocarbons (PAHs), alkyl phenyl ketones, fullerenes) was achieved with an anionic high-molecular-mass surfactant of similar structure [13] as BBMA. The structural formula of poly-(methyl methacrylate-ethylacrylate-methacrylic acid; Elvacite 2669) is given in Fig. 1C. Linear solvation energy relationships

(LSER) were used to describe the retention characteristics of aromatic test analytes for this polymer [14]. In comparison with several conventional micelle-forming surfactants, no fundamental difference in selectivity was found for the polymeric pseudostationary phase. Further applications of Elvacite 2669 comprised the separation of organophosphorus pesticides [15] and neutral corticosteroids of closely similar structure [16]. Results obtained by dynamic light scattering and capillary electrophoresis showed that, above a certain concentration, single polymer chains can adsorb on the fused-silica wall and enhance the EOF [16].

2.1.2 Polyallylamine-supported phases

Tanaka *et al.* [18–20] prepared a new type of polymeric pseudostationary phase by attaching alkyl and carboxylate groups on commercially available polyallylamine. These polyallylamine-supported phases (PAAs, Fig. 1D) could be synthesized with various chain lengths (C_8–C_{16}) and different degrees of alkylation in order to separate PAHs and alkyl phenyl ketones. Although bearing amino groups in the chain, the polymer exhibits net negative charges due to the excess of carboxylate functions. At higher concentrations of organic modifier (40–60% methanol) an expansion of the migration time window was observed [18]. With longer alkyl groups this expansion occurred at even higher methanol levels. It was stated

that a structural change of the polymeric phase causes a greater electrophoretic mobility of the polymer relative to the EOF, thus leading to a wider migration time range.

Consequently, mixtures of PAAs with different alkyl chain lengths were used to control the migration time window and modify the selectivity [19]; see the example in Fig. 2. While neither PAA with decyl groups (PAA-C_{10}) nor PAA with hexadecyl groups (PAA-C_{16}) in 35% acetonitrile could separate a mixture of 16 PAHs completely, a mixed pseudostationary phase consisting of both PAA-C_{10} and PAA-C_{16} was successful in separating compounds 1–13. It can be seen that PAA-C_{16} provided better separation for compounds with relatively small *k* values, whereas PAA-C_{10} contributed more to the separation of stronger retarded analytes. Recently, the influence of the degree of alkylation of PAA phases, the content of organic solvent and the polymer concentration on the selectivity and electrophoretic mobility of the pseudostationary phase were examined in detail [20]. It was reported that intermolecular aggregation of the pseudostationary phase could occur under poorly solvated conditions, especially at higher polymer concentrations.

2.1.3 Other anionic polymers

Another approach to obtain a charged pseudostationary phase was performed with linear polyacrylamide, which was partially hydrolyzed in order to introduce negative

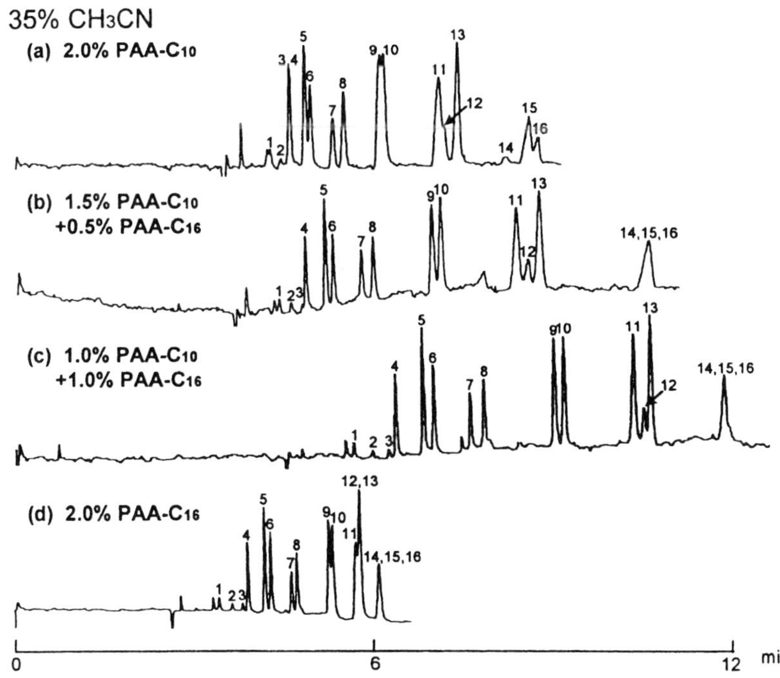

Figure 2. Separation of 16 PAHs with (A) PAA-C_{10}, 2.0% w/v; (B) PAA-C_{10}, 1.5% w/v + PAA-C_{16}, 0.5% w/v; (C) PAA-C_{10}, 1.0% w/v + PAA-C_{16}, 1.0% w/v; and (D) PAA-C_{16} 2.0% w/v, in 20 mM borate buffer-acetonitrile (65:35 v/v) mixtures (pH 9.3). Other conditions: field strength, 400 Vcm^{-1}, capillary (50 µm ID, 375 µm OD), 48 cm total and 33 cm effective length; ambient temperature (air circulation); detection at 254 nm. Peaks: 1, naphthalene; 2, acenaphthylene; 3, acenaphthene; 4, fluorene; 5, phenanthrene; 6, anthracene; 7, fluoranthene; 8, pyrene; 9, chrysene; 10, benz[a]anthracene; 11, benzo[b]fluoranthene; 12, benzo[k]fluoranthene; 13, benzo[a]pyrene; 14, dibenz[a,h]anthracene; 15, indenol[1,2,3-cd]pyrene; 16, benzo[ghi]perylene. Reproduced from [19], with permission.

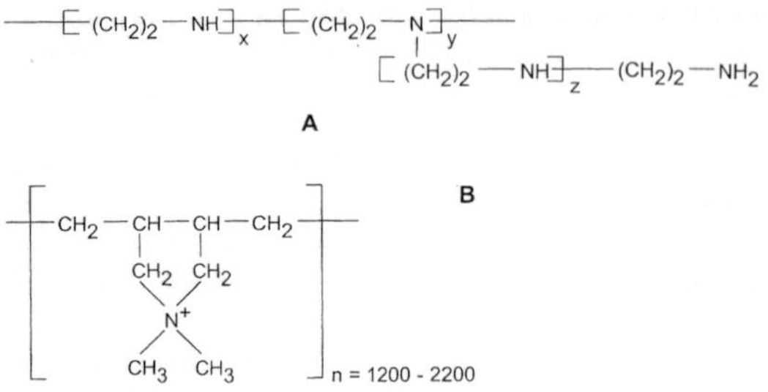

    Electrophoresis 2000, *21*, 3160–3173

A

B

n = 1200 - 2200

Figure 3. Structure of cationic polymeric additives. (A) PEI; (B) PDADMA.

charge [21]. Separation of neutral compounds was possible, but low stability of the phase and high salt concentration of the BGE hindered further applications. For the separation of aromatic analytes based on charge transfer interaction, π-electron-rich anions were applied as carriers. Besides some monomeric additives [22–24], poly-(vinylamine)sulfonate anthrapyridone (poly R-478), a polymeric dye, offered alternative selectivities to SDS micelles when separating aminoaromatic and nitroaromatic compounds [25].

Chen and Palmer [26] recently investigated the possibility of employing polymers based on a silicone backbone in EKC. The application of these polymeric phases promised to adapt previously developed stationary-phase chemistries for EKC. Among several kinds of silicone phases that were synthesized, only partially hydrolyzed poly(bis(3-cyanopropyl)) siloxane provided selectivity for substituted benzene and naphthalene compounds. This pseudostationary phase exhibited greater electrophoretic mobility and different selectivity than conventional SDS micelles. However, the limited aqueous solubility of silicone-based polymers represented a major problem.

2.2 Cationic polymers

2.2.1 Polyethyleneimine

Polyethyleneimine (PEI; see Fig. 3A) is a branched polymer containing primary, secondary, and tertiary amine

groups. Strong adsorption of PEI on the capillary wall creates a positively charged coating and therefore reverses the EOF towards the anode. High-molecular-mass PEI was applied as buffer additive to control the separation selectivity of proteins and peptides [27]. Erim [28] separated partially charged nitro- and chlorophenols using PEI as pseudostationary phase. In this work, seemingly a mixed mechanism consisting of both capillary zone electrophoresis and chromatographic interaction with the polymeric phase takes place.

EKC of neutral mono- and oligophenols with replaceable PEI as separation medium was reported by Maichel *et al.* [29]. The solutes were retarded mainly according to the number and position of hydroxy groups without a significant influence of methylation of the analytes. Furthermore, the effect of organic solvents on the separation selectivity in this system was investigated [30]. Addition of methanol or acetonitrile reduced the resolution of phenolic compounds, indicating a decrease in retardation.

2.2.2 Poly(diallyldimethylammonium)

Poly(diallyldimethylammonium) (PDADMA) represents a new separation medium that was initially applied as a dynamic coating material. It consists of a diallyl backbone with strongly basic quaternary ammonium groups (Fig. 3B). Without forming micelles, easily replaceable PDADMA provided selectivity for neutral analytes, mainly substituted benzenes [31]. Based on an isotachophoretic regime, Potocek *et al.* [31] determined the electrophoretic mobility of the pseudostationary phase that is essential for the calculation of capacity factors (retention factors). The linear free-energy relationship (LFER) model allowed a characterization of solute-polymer interactions on the basis of these data. The resulting solvation parameters for the polymeric pseudostationary phase (4% PDADMA) compared with a cationic (CTAB) and an anionic micellar phase (SDS) are given in Table 2. It can be seen that mainly π and n electron interactions as well as the hydrogen-bond basicity of the polymer determine the retention in the PDADMA separation system. Cavity formation is a

Table 2. Solvation parameters of the PDADMA polymer solution obtained by linear multiple regression

Parameter	PDADMA	CTAB	SDS	Type of interaction
c	−0.81 (0.07)	−1.67 (0.11)	−1.82 (0.07)	None; regression constant
m	–	3.40 (0.10)	2.99 (0.07)	Cavity formation
r	0.75 (0.09)	0.61 (0.06)	0.46 (0.05)	Interaction with n- or π-electrons
s	−0.35 (0.07)	−0.55 (0.07)	−0.44 (0.05)	Dipole-dipole and dipole-induced dipole
a	0.19 (0.05)	0.58 (0.06)	−0.30 (0.05)	Hydrogen bond basicity
b	–	−3.08 (0.10)	−1.88 (0.08)	Hydrogen bond acidity

Data taken from [31]. Values in parenthesis are standard deviations.

relevant contribution to interaction in micellar phases, whereas with a polymeric pseudostationary phase it is not significant.

The retention characteristics of ten neutral aromatic solutes in PDADMA were compared with those in systems consisting of ionic monomeric additives of a similar chemical structure (Maichel *et al.*, submitted). These monomeric cations were tetraalkylammonium ions with short alkyl chains. Such monomeric additives also introduced retention of the analytes, and the separation selectivity was roughly the same as in the PDADMA system. However, changes in the migration sequence could be observed. Recently, apparent diffusion coefficients of aromatic solutes were determined with different concentrations of PDADMA [32]. The change of the apparent diffusion coefficients due to association of the analytes with the polymer made it possible to derive capacity factors and distribution constants. Comparison with distribution constants obtained by EKC [31] showed good agreement. Note in this context that diffusion coefficients (and mobilities) of nonassociated solutes are often nearly independent of the macroscopic viscosity of polymer solutions, even in highly viscous media [33].

2.2.3 Cationic polymer with acrylate structure

In an uncoated capillary, a negatively charged pseudostationary phase migrates opposite to neutral solutes that are transported by the cathodic EOF. This counterflow is a prerequisite for separation. With a cationic polymer the EOF is reversed (due to coating with the polymer), and uncharged analytes move against the positively charged pseudostationary phase towards the anode. This was in fact observed by Ozaki *et al.* [10] when comparing anionic BBMA with a cationic acrylate copolymer, butyl methacrylate-methacryloyloxyethyltrimethyl ammonium chloride copolymer (BMAC; Fig. 1B).

3 Monomeric ionic additives

We refer here to those monomeric ions that are constituents of the BGE with the intention to interact with the analytes and to influence their retention. It is obvious that a monomeric electrolyte is normally present in the BGE solution, but it has other goals, namely to adjust the pH, or the ionic strength. However, this electrolyte may also often interact with the solutes. A well-known example is borate, which forms anionic complexes with *cis*-diols. Such ionic additives, which form specific complexes, are not considered here; for this reason charged cyclodextrins are also not discussed. A survey of monomeric ionic additives that are discussed in this review is given in Table 3.

Table 3. Monomeric ionic additives

Additive	Analytes	References
Tetraalkylammonium ions		
Tetrahexylammonium	PAHs	[24, 34, 35]
Tetraheptylammonium	PAHs, substituted benzenes	[36]
Tetraoctylammonium Tetradecylammonium	PAHs, alkylbenzenes	[37]
Tetradecylammonium	Vitamin K_1, parabens	[39]
	Fat-soluble vitamins (vitamin A palmitate, vitamin E acetate, vitamin D_3)	[40]
	Fat-soluble vitamins and water-soluble vitamins (B_1, B_2, B_6, B_{12}, C, nicotine amide, pantothenic acid)	[41]
Tetramethylammonium, tetraethylammonium	Substituted benzenes and naphthalenes	(Maichel *et al.*, submitted)
Other monomeric ionic additives		
DOSS	PAHs, substituted benzenes and naphthalenes	[42]
Laurylpoly(oxyethylene) sulfate (sulfonated Brij-30)	PAHs, substituted benzenes, testosterones	[43]
	PAHs, phenols, basic drugs	[44]
Tetraphenyl porphyrintetrasulfonate	Nitroaromatic compounds	[22]
	Nitroaromatic and phenolic compounds	[23]
Cardiogreen	Nitroaromatic and phenolic compounds	[23]
Tropylium ion	PAHs	[24]
2,4,6-Triphenylpyrylium ion	PAHs	[24]

3.1 Tetraalkylammonium ions

In 1986 already, Walbroehl and Jorgenson [34], reported the separation of neutral analytes with a monomeric cationic additive. A tetrahexylammonium salt was added to a BGE with a high content of acetonitrile (50% v/v or more) and separation of several PAHs was achieved. Selectivity was attributed to so-called solvophobic interaction with the tetrahexylammonium ions that formed positively charged species with the analytes. As the EOF and the solutes migrated in the same direction, the larger and more hydrophobic analytes with the strongest interaction were eluted first.

The same separation principle was employed in combination with UV-laser-induced fluorescence detection (UV-LIF) [35]. With tetrahexylammonium ions in a mixed acetonitrile/water medium, subattomole detection limit in analysis of PAHs was reached. Another quaternary ammonium ion, tetraheptylammonium, provided selectivity for a wide range of neutral organic compounds [36]. In this work, the effect of type and concentration of the ammonium salt, solvent composition, and pH on the electroosmotic and electrophoretic migration was investigated.

Muijselaar *et al.* [37] used both tetraoctylammonium and tetradecylammonium ions for the separation of highly hydrophobic compounds. Resolution of some geometric PAH isomers demonstrated the potential of these additives. It was noted that both tetraoctylammonium as well as tetradecylammonium can form micelles in organic media [38] and that cationic hemimicelles may reverse the EOF when dynamically coating the capillary wall. The authors stated that in the downstream mode, where the analytes migrate towards the cathode and are detected before the EOF, the associate-forming species is mainly the single ammonium ion. On the other hand, a change in solvent composition of only 10% v/v can drastically influence the migration behavior. In the anionic or upstream mode, the migration order is reversed and the pseudostationary phase contains both single ions and micellar aggregates [37].

Small tetraalkylammonium ions were used as additive to compare the separation selectivity with a system consisting of PDADMA as pseudostationary phase (see Section 2.2.2; Maichel *et al.*, submitted). It was pointed out that the main advantage of polymeric additives over monomeric ones with similar chemical structure is the larger retention window. This fact was related to the reduction of the mobility of the analyte-additive associate in case of monomeric additives, in contrast to polymeric pseudophases. Association constants with the additives were determined and found to be in the range between 5 and 170.

Different types of quaternary ammonium ions and counterions were investigated in nonaqueous capillary electrophoresis for the separation of highly hydrophobic solutes [39]. Not only the properties of the organic solvent but also the type of counterion was seen to have a pronounced effect on the separation window. As an application, quantitative determination of vitamin K_1 and alkyl parabens in a pharmaceutical product was performed with tetradecylammonium bromide in propylene carbonate. Pedersen-Bjeergaard *et al.* [40] focused on the separation of nonionic fat-soluble vitamins by means of so-called hydrophobic interaction EKC. Three vitamins were separated in decreasing order of hydrophobicity prior to the EOF with a BGE containing tetradecylammonium ions

in 80% v/v acetonitrile. In a subsequent work, the analysis of vitamin formulations using the same pseudostationary phase was further developed [41]. In addition to the separation of lipophilic vitamins, resolution of some water-soluble vitamins (B_1, B_2, B_{12}, nicotinamide) was achieved.

3.2 Other monomeric ionic additives

An anionic surfactant additive, sodium dioctyl sulfosuccinate (DOSS), was employed for the separation of PAHs and fairly hydrophilic solutes by Shi and Fritz [42]. Different overall mobilities of the neutral organic compounds resulted from formation of association "complexes" with the additive that was not supposed to form micelles. Using a countermigration separation mode, the elution order was reversed compared to that with tetraheptylammonium ions [36]. The same group obtained a novel pseudostationary phase by sulfonating Brij-30 [43]. The negatively charged lauryl poly(oxyethylene) sulfate (Brij-S) with the formula $C_{12}H_{25}(OCH_2CH_2)_4OSO_3^-$ exhibited excellent selectivity for PAHS, substituted benzenes and testosterones. Even structural isomers of methylated anthracenes as well as acetophenone and deuterated acetophenone could be separated in the countermigration mode.

Furthermore, Brij-S was studied for the analysis of PAHs, phenols, and basic drugs under acidic conditions [44]. At an apparent pH of 2.4 the EOF was very low compared to the electrophoretic mobility of the analyte-Brij-S associates. With reversed polarity, solutes migrated in order of decreasing hydrophobic character. Whereas separations with counterflowing EOF and pseudostationary phase required quite long analysis times at basic pH [43], fast and efficient separations were demonstrated at low pH. Figure 4 shows the difference between these two separation modes. Separation of PAH compounds under acidic conditions (Fig. 4B) was accomplished in 20 min. On the contrary, analysis of similar solutes required nearly double that time with reversed elution order (Fig. 4A).

Molecules with highly delocalized π-electron systems are well suited as additives for charge transfer interaction with aromatic analytes [22–24]. For instance, tetraphenyl porphyrintetrasulfonate and Cardiogreen were used for the separation of nitroaromatic and phenolic compounds [22, 23]. However, since these dyes absorb in the UV range, a partial filling technique had to be employed. Application of a hydrodynamic counterflow prevented the pseudostationary phase from reaching the detector, but as a consequence the plug-like flow profile was distorted and efficiency was reduced [22].

Two planar organic cations were also reported to form charge-transfer complexes with PAHs in a purely nonaqu-

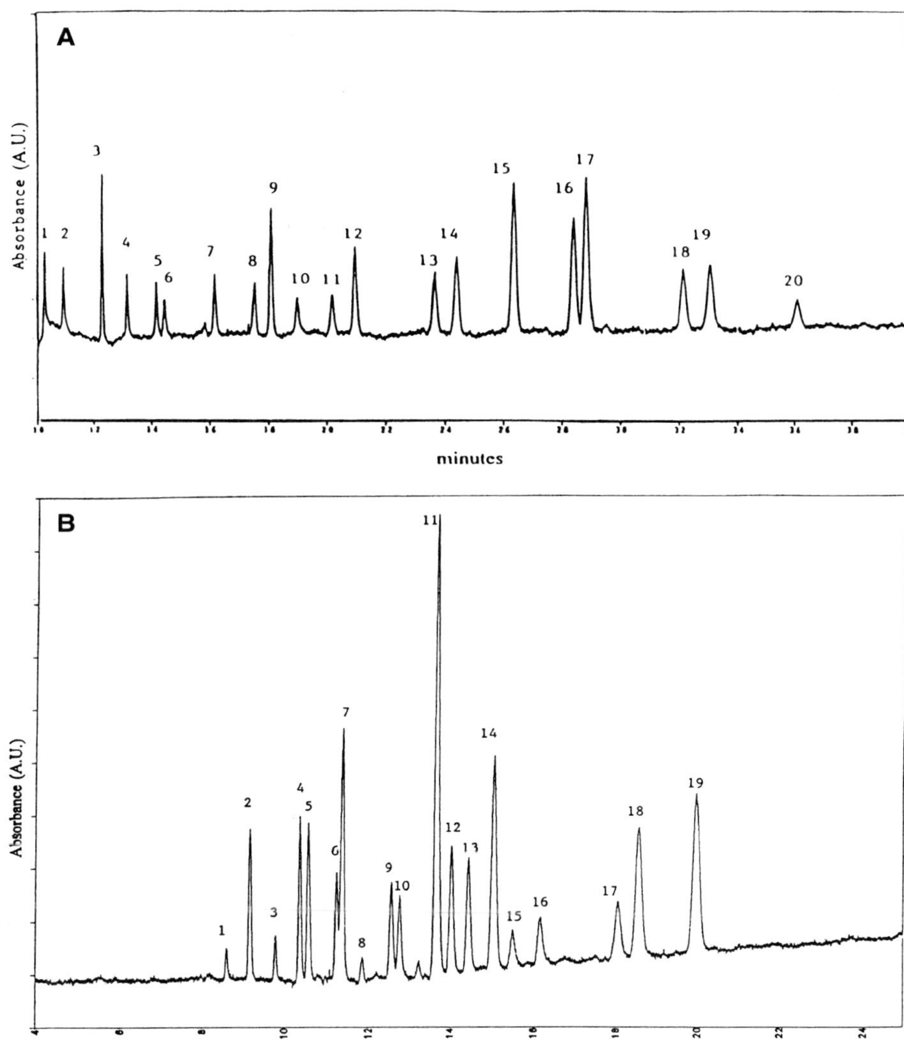

Figure 4. Separation of PAH compounds with a BGE of 40 mM sulfonated Brij-S, 8 mM sodium borate, 40% acetonitrile, pH 9.0. Other conditions: uncoated fused-silica capillary (50 μm ID), 75 cm length; voltage, 20 kV; current, 68 μA; hydrostatic injection, 30 s; detection at 254 nm; temperature, 25°C. Peaks: 1, acetophenone; 2, nitrobenzene; 3, 5,6-benzoquinoline; 4, benzophenone; 5, azulene; 6, naphthalene; 7, acenaphthylene; 8, acenaphthene; 9, fluorene; 10, 3-aminofluoranthene; 11, benz[a]anthracene; 12, anthracene; 13, fluoranthene; 14, pyrene; 15, 2,3-benzofluorene; 16, chrysene; 17, 2,3-benzphenathrene; 18, perylene; 19, benzo[a]pyrene; 20, benzo[ghi]perylene. Reproduced from [43], with permission. (B) Separation of PAH compounds in acidic condition with 30 mM sulfonated Brij-S, 10 mM HCl, 20% acetonitrile, 20% v/v 2-propanol, pH 2.4. Other conditions: uncoated fused-silica capillary (50 μm ID), 45 cm total and 37.5 cm effective length; voltage, −22 kV; current, 50 μA; hydrostatic injection, 10 s; detection at 254 nm; temperature, 25°C. Peaks: 1, rubrene; 2, 9,10-diphenylanthracene; 3, benzo[ghi]perylene; 4, benzo[a]pyrene; 5, benzo[e]pyrene; 6, benz[a]anthracene; 7, chrysene; 8, 2,3-benzofluorene; 9, pyrene; 10, fluoranthene; 11, anthracene; 12, phenanthrene; 13, 1-hydroxypyrene; 14, fluorene; 15, acenaphthene; 16, 1-methylnaphthalene; 17, naphthalene; 18, azulene; 19, benzophenone. Reproduced from [44], with permission.

eous medium [24]. Tropylium and 2,4,6-triphenylpyrylium ions were dissolved in acetonitrile to serve as pseudostationary phases. Quantitative structure-migration relationships (QSMRs) led to the assumption that, besides charge transfer interactions, van der Waals forces are responsible for interaction between the species.

4 Dendrimers

Dendrimers are spherical macromolecules with highly branched chains and charged terminal groups (Figs. 5A, B). During synthesis, their size and chemical structure can be determined and different functional groups can be introduced. Besides the variable selectivity of starburst dendrimers, the good stability in organic media favors their application as carriers in EKC. A summary of dendritic phases reported in literature is given in Table 4. The first report of dendrimers as pseudostationary phases in EKC was by Tanaka *et al.* [45]. Starburst dendrimers (poly(amidoamines)) with ethylenediamine core and carboxylate end groups provided a different selectivity for uncharged aromatic compounds than MEKC systems with cetyltrimethylammonium chloride or SDS. Later, sim-

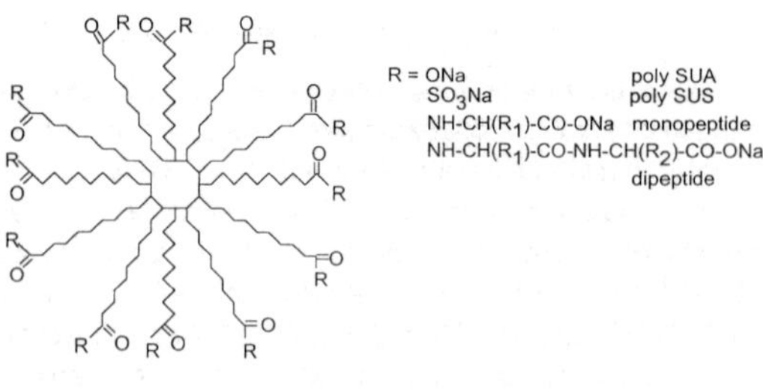

Figure 5. Structure of dendrimers and micelle polymers. (A) Amide-based dendrimer (second generation); (B) diaminobutane dendrimer; (C) micelle polymers (general structure).

Table 4. Dendrimers

Pseudostationary phase	Analytes	References
Poly(amidoamine) (PAMAM) dendrimers with carboxylic acid terminal		
Ethylenediamine core	Substituted benzene and naphthalene compounds, PAHs.	[45]
	Aromatic amino acids	[46]
Amide-based core	Alkyl parabens, cough medicine ingredients	[47]
	Alkyl parabens, xanthines	[48]
p-Xylylenediamine core	Substituted benzene and naphthalene compounds, PAHs	[50]
Diaminobutane-based poly(propyleneimine) dendrimer	Substituted benzene compounds	[49]
Alkylated PAMAM dendrimers with carboxylic acid terminal		
Ethylenediamine core C_8, C_{12}	PAHs, benzene and naphthalene derivatives, alkyl phenyl ketones	[52]
p-Xylylenediamine core C_8	Substituted benzene and naphthalene compounds, PAHs	[50]
	PAHs, benzene and naphthalene derivatives, alkyl phenyl ketones	[51]
C_{12}	PAHs, benzene and naphthalene derivatives, alkyl phenyl ketones	[51]
C_4, C_8, C_{12}	PAHs, benzene and naphthalene derivatives, alkyl phenyl ketones	[52]
Sulfonic acid terminal	Dimethylphenols	[53]
Methoxy group terminal	PAHs, naphthalene derivatives	[54]

ilar dendrimers were explored for the separation of several aromatic amino acids [46].

Kuzdzal *et al.* [47] applied polyacidic dendrimers with an amido-based core (Fig. 5A) for the analysis of a homologous series of alkyl parabens and cough medicine ingredients under aqueous conditions. Furthermore, they studied interactions between these dendrimers and neutral solutes, mainly alkyl parabens, by means of EKC [48]. Determination of capacity factors and distribution coefficients allowed the calculation of thermodynamic parameters. Diaminobutane-based poly(propyleneimine) dendrimers have a more hydrophilic interior than SDS micelles because of tertiary amines that are present in the macromolecule [49]. It was reported that with these dendrimers more polar phenolic compounds were more strongly retarded than less polar solutes.

Tanaka *et al.* [50] focused on the structural selectivity provided by starburst dendrimers that were synthesized with ammonia and *p*-xylylenediamine cores. Similar to polymer gel packing materials in reversed-phase liquid chromatography, these dendrimers showed selectivity for the carbon skeletons of aromatic analytes, but not for their functional groups. In order to increase the hydrophobic selectivity of dendrimer-supported pseudostationary phases, Tanaka *et al.* [50] modified starburst dendrimers with alkyl chains. Octyl-modified dendrimers provided increased retention and better efficiency than the parent dendrimer without modification. In a subsequent work, different selectivity of octyl- and dodecyl-modified starburst dendrimers was observed [51]. The dodecyl derivative was found to be superior when separating hydrophobic compounds in a full range of water-methanol (up to 90%) mixtures by EKC.

For further investigations of alkylated dendrimers, dendritic pseudostationary phases with different cores (*p*-xylylenediamine and ethylenediamine) and varying alkyl chain length were prepared [52]. Figure 6 illustrates the separation of benzene and naphthalene derivatives with (A) an unmodified dendrimer and (B) SDS micelles as well as (C) butyl-modified, (D) octyl-modified, and (E) dodecyl-modified dendrimer. The unmodified parent dendrimer (Fig. 6A) only recognizes the carbon backbone of the analytes, whereas the selectivities of the alkylated phases (Fig. 6D and E) were closer to that provided by SDS micelles (Fig. 6B). The selectivity of alkylated dendrimers was primarily determined by the chain length of the alkyl groups, and no pronounced influence of the generation of the dendrimer or the structure of its core was found. Considering the composition of the BGE, a nearly linear relation between methanol content and logarithm of the capacity factor was observed.

Gray and Hsu [53] reported the use of dendrimers with terminal groups other than carboxylate. They intended to

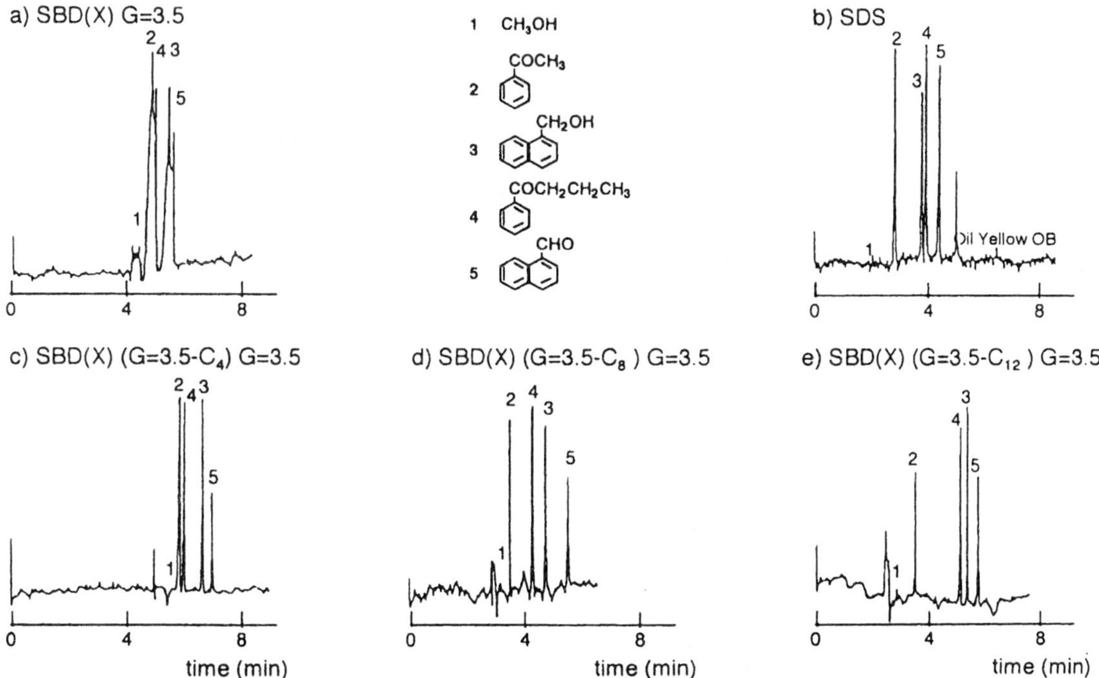

Figure 6. Comparison of the effect of carriers for the separation of benzene and naphthalene derivatives. (A) Starburst dendrimer with *p*-xylylenediamine core (SBD(X)) generation (G) 3.5, 5 mM, pH 10.3; (B) SDS, 30 mM, pH 9.4; (C) SBD(X)(G = 3.5-C₄) G = 3.5, 5 mM, pH 10.4; (D) SBD(X)(G = 3.5-C₈) G = 3.5, 5 mM, pH 10.1; (E) SBD(X)(G = 3.5-C₁₂) G = 3.5, 5 mM, pH 10.4. Other conditions: fused-silica capillary (50 μm ID and 375 μm OD), 48 cm total and 33 cm effective length; field strength: (A), (C), (E) 300 Vcm⁻¹; (D) 250 Vcm⁻¹; 20 mM borate buffer, detection at (B) 210 nm or (A), (C), (E) 254 nm; ambient temperature. Peaks: 1, methanol; 2, acetophenone; 3, 1-naphthalenemethanol; 4, phenyl propyl ketone; 5, 1-naphthaldehyde; 6, Oil Yellow OB. Reproduced from [52], with permission.

imitate the structure of an SDS micelle by introducing sulfonic acid terminals to a poly(amidoamine) starburst dendrimer. Separation of positional isomers of neutral dimethylphenols was mainly attributed to hydrogen bonding of the phenols with amine groups in the dendrimer. A novel dendrimer for EKC was presented by Haynes *et al.* [54]. The diaminobutane dendrimer was synthesized with methoxy end groups and contained tertiary amines and carbonyl groups (Fig. 5B). In contrast to formerly described dendrimers, this pseudostationary phase is positively charged at pH values below 9. Separation of naphthalene derivatives and PAHs was achieved, but the diaminobutane dendrimer offered no selectivity for smaller benzyl derivatives. Addition of organic modifier (methanol) reduced the selectivity.

5 Micelle polymers

5.1 Achiral separations

The great variety of so-called micelle polymers can be seen in Table 5. Palmer and co-workers [55, 56] first reported the use of a micelle polymer in EKC (Fig. 5C). Poly sodium 10-undecylenate (poly SUA) was synthesized by free radical initiation from sodium 10-undecenylate. Due to its covalently stabilized structure, poly SUA could be applied in the presence of relatively high amounts of organic modifier. Addition of up to 50% methanol or 45% acetonitrile made it possible to adjust the retention of hydrophobic analytes such as alkyl phthalates and PAHs, which are difficult to analyze by conventional MEKC. Moy *et al.* [57] achieved separation of 16 PAH environmental contaminants in presence of 20% tetrahydrofuran. Recently the effect of free-radical initiators of different hydrophobicity on the synthetic yield and chemical selectivity of poly SUA was studied [58].

However, the applicability of the undecylenate polymer was limited by its carboxylate head group whose ionization led to a low electrophoretic mobility and solubility of the phase at pH values below 8. In order to overcome these problems and to mimic the chemistry of an SDS micelle, Palmer and Terabe [59, 60] synthesized a polymerized surfactant with a sulfate head group, namely, poly sodium 10-undecenylsulfate (poly SUS). Compared to conventional SDS micelles, both poly SUA and poly SUS exhibited better selectivity towards more polar analytes and greater electrophoretic mobility, thus providing wider migration time ranges. Major problems connected with the free radical initiation were reported to be low synthetic yields and contamination of the product with sodium sulfate [60]. Warner and co-workers [61, 62] avoided these disadvantages by using ^{60}Co γ-irradiation to initiate polymerization instead. They reported the separation of

priority PAH pollutants [61] and monomethylbenz[a]anthracene isomers [62]. With addition of poly SUS to the BGE the analysis of tricyclic antidepressant drugs and β-adrenergic blocker drugs by ESI-MS could be improved [63].

In the course of a further investigation of the chemical selectivity of micelle polymers the synthesis and characterization of two new pseudostationary phases with taurate or phosphonate head groups was reported [64]. Poly sodium *N*-undec-10-ene-1-oyl-taurate (poly SUT) and poly sodium *N*-undec-10-ene-1-oyl-aminoethyl-2-phosphonate (poly SUP) provided a significantly improved stability and solubility over previous structures. However, LSER results indicated that the selectivity of undecenylate micelle polymers is not strongly affected by changes in the head group chemistry.

5.2 Chiral separations

Modification of these polymerized surfactant aggregates with peptide head groups introduced chiral selectivity and thus represented a decisive step forward in the development of new enantioselective phases in EKC [4, 6, 65–75]. The first report of a monopeptide micelle polymer was published by Wang and Warner [65]. With poly sodium *N*-undecanoyl, L-valinate (poly L-SUV), separations of (±)-binaphthol and DL-laudanosine were demonstrated, being superior to those obtained with the corresponding monomeric sufactant. Besides the elimination of the dynamic equilibrium of the additive between the monomeric and the micellar form, this improvement was explained by the rigid and compact structure of the polymerized surfactant. As the solutes could not penetrate as deeply into the core of the micelle polymer, an enhanced rate of mass transfer of the solutes led to reduced peak broadening compared to normal micelles [75]. In contrast, Dobashi *et al.* [66] did not observe better chiral selectivity for 3,5-dinitrobenzoylated amino acid isopropylesters with poly L-SUV than that obtained with conventional micelles of sodium dodecanoyl-L-valine. The combination of poly D-SUV and γ-cyclodextrin [73, 75] provided significantly improved enantioselectivity because of the elimination of inclusion of surfactant monomers into the cyclodextrin. Besides a great number of enantioseparations with poly SUV [65–71, 73], the application of monopeptide surfactants with threonine [69], leucine [70–72] and alanine [71] as head groups was likewise reported.

Another approach to further develop micelle polymers was the introduction of multiple chiral centers into the head group. Shamsi *et al.* [69] prepared the first polymeric dipeptide surfactant, namely, poly sodium *N*-undecanoyl LL-valyl-valinate (poly LL-SUVV). They compared

Table 5. Micelle polymers

Pseudostationary phase	Analytes	References
Achiral separations		
Poly SUA	Alkyl phthalates, PAHs, alkyl benzenes	[55, 56]
	PAHs	[57]
	Substituted benzene and naphthalene compounds, PAHs	[60]
	Alkyl phenyl ketones, substituted benzene and naphthalene compounds	[58]
Poly SUS	Substituted benzene and naphthalene compounds, PAHs, cold medicine ingredients	[59, 60]
	PAHs	[61]
	Tricyclic antidepressant drugs, β-adrenergic blocker drugs	[63]
	Monomethylbenz[a]anthracene isomers	[62]
Poly SUT	Alykl phenyl ketones, PAHs, cold medicine ingredients	[64]
Poly SUP	Alkyl phenyl ketones, PAHs, cold medicine ingredients	[64]
Chiral separations		
(i) Monopeptide surfactants		
Poly L-SUV	(±)-1,1′-bi-2-naphthol (BOH), DL-laudanosine	[65]
	3,5-Dinitrobenzoylated amino acid isopropylesters	[66]
	Coumarinic anticoagulant drugs, BOH, (±)-1,1′-binaphthyl-2,2′-diamine (BNA), Tröger's base, paveroline drugs	[67]
	(±)-Propranolol, (±)-alprenolol, (±)-1,1′-binaphthyl-2,2′-diyl hydrogen phosphate (BNP), (±)-trifluoro-1-(9-anthryl)ethanol (TFAE)	[68]
	Phenylthiohydantoin-DL-amino acids	[69]
	BOH, BNP	[70, 71]
Poly D-SUV + γ-cyclodextrin	BOH, BNP, DL-laudanosine, (±)-verapramil	[73]
Poly L-SUT	Phenylthiohydantoin-DL-amino acids	[69]
Poly sodium N-undecanoyl -leucinate (Poly L-SUL)	BOH, BNP	[70, 71]
	BOH, BNP, (±)-propranolol, (±)-alprenolol	[72]
Poly D-SUL	BOH, BNP, BNA, (±)-propranolol, (±)-alprenolol	[72]
Poly sodium N-undecanoyl L-alaninate (Poly L-SUAla)	BOH, BNP	[71]

Table 5. continued

Pseudostationary phase	Analytes	References
(ii) Dipeptide surfactants		
Poly LL-SUVV	(±)-propranolol, (±)-alprenolol, BNP, TFAE	[68]
	BOH, BNP	[70, 71]
Poly sodium N-undecanoyl LL-leucyl-leucinate (Poly LL-SULL)	BOH, BNP	[70, 71]
	BOH, BNP, BNA, (±)-propranolol, (±)-alprenolol	[72]
Poly DD-SULL Poly LD-SULL Poly DL-SULL	BOH, BNP, BNA, (±)-propranolol, (±)-alprenolol	[72]
Poly sodium N-undecanoyl L-leucyl-glycinate (Poly L-SULG)	BOH, BNP, BNA, (±)-propranolol, (±)-alprenolol	[72]
Poly sodium N-undecanoyl L-glycyl-leucinate (Poly L-SUGL)		
Poly sodium N-undecanoyl LL-valyl-leucinate (Poly LL-SUVL)	BOH, BNP	[70, 71]
Poly sodium N-undecanoyl LL-leucyl-valinate (Poly LL-SULV)	BOH, BNP	[70, 71]
	Binaphthyl derivatives, β-blockers, glutethimide, aminoglutethimide, benzodiazepines, TFAE	[74]
Poly LL-SUILV	Binaphthyl derivatives β-blockers, glutethimide, aminoglutethimide, benzodiazepines, TFAE	[74]
All possible combinations of the L-form of alanine, valine, and leucine and the achiral glycine (except glycine-glycine)	BOH, BNP	[71]

the monopeptide surfactant poly L-SUV with the dipeptide surfactant poly LL-SUVV as pseudostationary phase for the separation of basic, acidic, and neutral enantiomers. For both cationic and anionic analytes the chiral selectivity was improved when applying the phase containing two chiral centers (*i.e.*, poly LL-SUVV). The single amino acid polymer, poly L-SUV, exhibited slightly better resolution for a neutral racemate, but only with longer analysis time and reduced efficiency. The influence of amino acid order on chiral selectivity in dipeptide surfactants was studied in two papers of Billiot *et al.* [70, 71].

Polymeric diastereomeric dipeptide surfactants were used to determine the site of chiral recognition [72]. It was concluded that the depth of penetration of the analyte into the hydrophobic core of the pseudostationary phase is decisive for chiral resolution and that this depth is governed by the hydrophobicity of the analyte and electrostatic interactions. With isoleucine and valine as head group, a polymeric dipeptide surfactant containing three chiral centers, namely, poly sodium *N*-undecanoyl LL-isoleucyl-valinate (poly LL-SUILV), was recently synthesized and evaluated for enantioseparation [74]. Compared to a two-chiral-center dipeptide surfactant, poly LL-SULV, poly LL-SUILV did not show clear superiority concerning chiral recognition. It was assumed that the sterical hindrance in poly LL-SUILV could both assist and limit stereoselectivity. Note that a current review [6] covers chiral EKC using dipeptide polymeric surfactants in great detail.

6 Conclusions

Many additives have been employed as an alternative to capillary electrochromatography (which uses porous particles as in HPLC as stationary phase in the column) or to classical MEKC (with charged monomeric detergents applied above the critical concentration, thus forming micelles). They are used because of their advantages over capillary electrochromatography: they are replaceable and therefore permit identical experimental conditions to be reproduced after each run without the problem of undesired column modification due to contaminants. It is possible to establish reproducible EOF by rinsing the system with chemically aggressive solutions (that would destroy the stationary phase in capillary electrochromatography). They offer large flexibility concerning the variation of the separation selectivity, in principle after each run. They are mechanically uncomplicated, *e.g.*, they do not need any frits or similar attachments to keep the stationary phase particles inside the capillary. Compared to classical MEKC they are not limited to certain lower concentrations, and have no restrictions concerning their stability in presence of organic solvent modifiers. However, micelle polymers avoid such limitations, and therefore significantly enlarge the applicability of MEKC in this respect.

Received February 29, 2000

7 References

[1] Palmer, C. P., *J. Chromatogr. A* 1997, *780*, 75–92.

[2] Palmer, C. P., Tanaka, N., *J. Chromatogr. A* 1997, *792*, 105–124.

[3] Riekkola, M.-L., Wiedmer, S. K., Valko, I. E., Siren, H., *J. Chromatogr. A* 1997, *792*, 13–35.

[4] Shamsi, S. A., Warner, I. M., *Electrophoresis* 1997, *18*, 853–872.

[5] de Boer, T., de Zeeuw, R. A., de Jong, G. J., Ensing, K., *Electrophoresis* 1999, *20*, 2989–3010.

[6] Haddadian, F., Shamsi, S. A., Warner, I. M., *Electrophoresis* 1999, *20*, 3011–3026.

[7] Quirino, J. P., Terabe, S., *J. Chromatogr. A* 1999, *856*, 475–482.

[8] Ozaki, H., Terabe, S., Ichihara, A., *J. Chromatogr.* 1994, *680*, 117–123.

[9] Terabe, S., Ozaki, H., Tanaka, Y., *J. Chin. Chem. Soc.* 1994, *41*, 251–257.

[10] Ozaki, H., Ichihara, A., Terabe, S., *J. Chromatogr. A* 1995, *709*, 3–10.

[11] Ozaki, H., Itou, N., Terabe, S., Takada, Y., Sakairi, M., Koizumi, H., *J. Chromatogr. A* 1995, *716*, 69–79.

[12] Ozaki, H., Terabe, S., *J. Chromatogr. A* 1998, *794*, 317–325.

[13] Yang, S., Bumgarner, J. G., Khaledi, M. G., *J. High Resol. Chromatogr.* 1995, *18*, 443–445.

[14] Yang, S., Bumgarner, J. G., Khaledi, M. G., *J. Chromatogr. A* 1996, *738*, 265–274.

[15] Aguilar, M., Farran, A., Serra, C., Sepaniak, M. J., Whitaker, K. W., *J. Chromatogr. A* 1997, *778*, 201–205.

[16] Wiedmer, S. K., Tenhu, H., Vastamäki, P., Riekkola, M.-L., *J. Microcol. Sep.* 1998, *10*, 557–565.

[17] Rundlett, K. L., Armstrong, D. W., *Anal. Chem.* 1996, *68*, 3493–3497.

[18] Tanaka, N., Nakagawa, K., Iwasaki, H., Hosoya, K., Kimata, K., Araki, T., Patterson Jr., D. G., *J. Chromatogr. A* 1997, *781*, 139–150.

[19] Tanaka, N., Nakagawa, K., Hosoya, K., Palmer, C. P., Kunugi, S., *J. Chromatogr. A* 1998, 802, 23–33.

[20] Tanaka, N., Nakagawa, K., Nagayama, H., Hosoya, K., Ikegami, T., Itaya, A., Shibayama, M., *J. Chromatogr. A* 1999, *836*, 295–303.

[21] Potocek, B., Maichel, B., Gaš, B., Chiari, M., Kenndler, E., *J. Chromatogr. A* 1998, *798*, 269–273.

[22] Kutter, J., Welsch, T., *J. High Resol. Chromatogr.* 1995, *18*, 741–744.

[23] Welsch, T., Kolb, S., Kutter, J. P., *J. Microcol. Sep.* 1997, *9*, 15–20.

[24] Miller, J. L., Khaledi, M. G., Shea, D., *Anal. Chem.* 1997, *69*, 1223–1229.

[25] Kolb, S., Kutter, J. P., Welsch, T., *J. Chromatogr. A* 1997, *792*, 151–156.

[26] Chen, T., Palmer, C. P., *Electrophoresis* 1999, *20*, 2412–2419.

[27] Cifuentes, A., Poppe, H., Kraak, J. C., Erim, F. B., *J. Chromatogr. B* 1996, *681*, 21–27.

[28] Erim, F. B., *J. Chromatogr. A* 1997, *768*, 161–167.

[29] Maichel, B., Potocek, B., Gaš, B., Chiari, M., Kenndler, E., *Electrophoresis* 1998, *19*, 2124–2128.

[30] Maichel, B., Potocek, B., Gaš, B., Kenndler, E., *J. Chromatogr. A* 1999, *853*, 121–129.

[31] Potocek, B., Chmela, E., Maichel, B., Tesarova, E., Kenndler, E., Gaš, B., *Anal. Chem.* 2000, *72*, 74–80.

[32] Maichel, B., Gaš, B., Kenndler, E., *Electrophoresis* 2000, *21*, 1505–1512.

[33] Shimizu, T., Kenndler, E., *Electrophoresis* 1999, *20*, 3364–3372.

[34] Walbroehl, Y., Jorgenson, J. W., *Anal. Chem.* 1986, *58*, 479–481.

[35] Nie, S., Dadoo, R., Zare, R. N., *Anal. Chem.* 1993, *65*, 3571–3575.

[36] Shi, Y., Fritz, J. S., *J. High Resol. Chromatogr.* 1994, *17*, 713–718.

[37] Muijselaar, P. G., Verhelst, H. B., Claessens, H. A., Cramers, C. A., *J. Chromatogr. A* 1997, *764*, 323–329.

[38] Hinze, W. L., in: Hinze, W. L., Armstrong, D. W. (Eds.), *Ordered Media in Chemical Separations, American Chemical Society, Washington, DC 1987, Vol. 342, pp. 2–82.*

[39] Tjornelund, J., Hansen, S. H., *J. Chromatogr. A* 1997, *792*, 475–482.

[40] Pedersen-Bjergaard, S., Rasmussen, K. E., Tilander, T., *J. Chromatogr. A* 1998, *807*, 285–295.

[41] Naess, O., Tilander, T., Pedersen-Bjergaard, S., Rasmussen, K. E., *Electrophoresis* 1998, *19*, 2912–2917.

[42] Shi, Y., Fritz, J. S., *Anal. Chem* 1995, *67*, 3023–3027.

[43] Ding, W., Fritz, J. S., *Anal. Chem.* 1997, *69*, 1593–1597.

[44] Ding, W., Fritz, J. S., *Anal. Chem.* 1998, *70*, 1859–1865.

[45] Tanaka, N., Tanigawa, T., Hosoya, K., Kimata, K., Araki, T., Terabe, S., *Chem. Lett.* 1992, *XX*, 959–962.

[46] Gao, H., Carlson, J., Stalcup, A. M., Heineman, W. R., *J. Chromatogr. Sci.* 1998, *36*, 146–154.

[47] Kuzdzal, S. A., Monnig, C. A., Newkome, G. R., Moorefield, C. N., *J. Chem. Soc. Chem. Commun.* 1994, *XX*, 2139–2140.

[48] Kuzdzal, S. A., Monnig, C. A., Newkome, G. R., Moorefield, C. N., *J. Am. Chem. Soc.* 1997, *119*, 2255–2261.

[49] Muijselaar, P. G. H. M., Claessens, H. A., Cremers, C. A., Jansen, J. F. G. A., Meijer, E. W., de Brabander-van den Berg, E. M. M., van der Wal, S., *J. High Resol. Chromatogr.* 1995, *18*, 121–123.

[50] Tanaka, N., Fukutome, T., Tanigawa, T., Hosoya, K., Kimata, K., Araki, T., Unger, K. K., *J. Chromatogr. A* 1995, *699*, 331–341.

[51] Tanaka, N., Fukutome, T., Hosoya, K., Kimata, K., Araki, T., *J. Chromatogr. A* 1995, *716*, 57–67.

[52] Tanaka, N., Iwasaki, H., Fukutome, T., Hosoya, K., Araki, T., *J. High Resol. Chromatogr.* 1997, *20*, 529–538.

[53] Gray, A. L., Hsu, J. T., *J. Chromatogr. A* 1998, *824*, 119–124.

[54] Haynes III, J. L., Shamsi, S. A., Dey, J., Warner, I. M., *J. Liq. Chromatogr.* 1998, *21*, 611–624.

[55] Palmer, C. P., McNair, H. M., *J. Microcol. Sep.* 1992, *4*, 509–514.

[56] Palmer, C. P., Khaled, M. Y., McNair, H. M., *J. High Resol. Chromatogr.* 1992, *15*, 756–762.

[57] Moy, T. W., Ferguson, P. L., Grange, A. H., Matchett, W. H., Kelliher, V. A., Brumley, W. C., Glassman, J., Farley, J. W., *Electrophoresis* 1998, *19*, 2090–2094.

[58] Palmer, C. P., Tellman, K. T., *J. Microcol. Sep.* 1999, *11*, 185–191.

[59] Palmer, C. P., Terabe, S., *J. Microcol. Sep.* 1996, *8*, 115–121.

[60] Palmer, C. P., Terabe, S., *Anal. Chem.* 1997, *69*, 1852–1860.

[61] Shamsi, S. A., Akbay, C., Warner, I. M., *Anal. Chem.* 1998, *70*, 3078–3083.

[62] Akbay, C., Warner, I. M., Shamsi, S. A., *Electrophoresis* 1999, *20*, 145–151.

[63] Lu, W., Shamsi, S. A., McCarley, T. D., Warner, I. M., *Electrophoresis* 1998, *19*, 2193–2199.

[64] Tellman, K. T., Palmer, C. P., *Electrophoresis* 1999, *20*, 152–161.

[65] Wang, J., Warner, I. M., *Anal. Chem.* 1994, *66*, 3773–3776.

[66] Dobashi, A., Hamada, M., Dobashi, Y., Yamaguchi, J., *Anal. Chem.* 1995, *67*, 3011–3017.

[67] Agnew-Heard, K. A., Sánchez Peña, M., Shamsi, S. A., Warner, I. M., *Anal. Chem.* 1997, *69*, 958–964.

[68] Shamsi, S. A., Macossay, J., Warner, I. M., *Anal. Chem.* 1997, *69*, 2980–2987.

[69] Yarabe, H. H., Shamsi, S. A., Warner, I. M., *Anal. Chem.* 1999, *71*, 3992–3999.

[70] Billiot, E., Macossay, J., Thibodeaux, S., Shamsi, S. A., Warner, I. M., *Anal. Chem.* 1998, *70*, 1375–1381.

[71] Billiot, E., Agbaria, R. A., Thibodeaux, S., Shamsi, S., Warner, I. M., *Anal. Chem.* 1999, *71*, 1252–1256.

[72] Billiot, E., Thibodeaux, S., Shamsi, S., Warner, I. W., *Anal. Chem.* 1999, *71*, 4044–4049.

[73] Wang, J., Warner, I. M., *J. Chromatogr. A* 1995, *711*, 297–304.

[74] Haddadian, F., Billiot, E. J., Shamsi, S. A., Warner, I. M., *J. Chromatogr. A* 1999, *858*, 219–227.

[75] Williams, C. C., Shamsi, S. A., Warner, I. M., *Adv. Chromatogr.* 1997, *37*, 363–423.

Review

Christopher P. Palmer[1]

[1]Department of Chemistry,
New Mexico Institute of Mining
and Technology,
Socorro, NM, USA

Polymeric and polymer-supported pseudostationary phases in micellar electrokinetic chromatography: Performance and selectivity

Several types of synthetic ionic polymers have been employed as pseudostationary phases in electrokinetic chromatography. The polymers have been shown to have some significant advantages and different chemical selectivity relative to conventional surfactant micelles. Polymeric phases are effective for the separation and analysis of hydrophobic and chiral compounds, and may be useful for the application of mass spectrometric detection. Additionally, the polymeric phases often demonstrate unique selectivity relative to micellar phases, and can be designed and synthesized to provide desired selectivity. This review covers efforts to develop and characterize the performance, characteristics, and selectivity of synthetic polymeric pseudostationary phases since their introduction in 1992. Some ideas for the future development of polymeric pseudostationary phases and the role they may play in electrokinetic separations are presented.

Keywords: Micellar electrokinetic chromatography / Polymer surfactants / Polymer micelles / Pseudostationary phase / Selectivity / Review

EL 4183

Contents

Correspondence: Dr. Christopher P. Palmer, Department of Chemistry, New Mexico Institute of Mining and Technology, Socorro, NM 87801, USA
E-mail: palmer@nmt.edu
Fax: +505-835-5364

Abbreviations: AMPS, 2-acrylamido-2-methylpropane sulfonic acid; **BBMA**, butyl acrylate-butyl methacrylate-methacrylic acid; **BMAC**, butylmethacrylate methacryloyloxyethyltrimethyl ammonium chloride; **LMAm**, lauryl methacrylamide; **LSER**, linear solvation energy relationship; **PAA**, polyallylamine; **PAH**, polynuclear aromatic hydrocarbon; **PDADMA**, polydiallyldimethylammonium **PEI**, polyethyleneimine; **SBD**, starburst dendrimer; **SUA**, sodium undecylenate; **SUP**, sodium-*N*-undec-10-en-1-oyl aminoethyl-2-phosphonate; **SUS**, sodium undecenyl sulfate; **SUT**, sodium *N*-undec-10-en-1-oyl taurate

1 Introduction

Introduced in 1984 by Terabe *et al.* [1], micellar electrokinetic chromatography (MEKC) utilizes a micellar pseudostationary phase in a capillary electrophoresis format to effect the separation of nonionic compounds and/or to alter the selectivity of the electrophoretic separation of ionic compounds. Typically, negatively charged micelles formed from anionic surfactants such as sodium dodecyl sulfate (SDS) make up the pseudostationary phase, which migrates at a rate slower than that of the electroosmotic flow. The rate of migration of an analyte depends on its partition coefficient between the micelles and the aqueous phase. This has proven to be a powerful tool for the separation and analysis of a variety of analytes [2–5]. The micellar phase is not truly stationary, but migrates through the capillary, typically in the same direction as the analytes, with a migration time t_{mc}. This results in a limited migration range for separation of the analytes, and modified equations for the retention factor, k:

$$k = \frac{t_r - t_0}{t_0 \left(1 - \frac{t_r}{t_{mc}}\right)} \tag{1}$$

and for the resolution between two analytes, R_s:

$$R_S = \frac{\sqrt{N}}{4} \frac{\alpha - 1}{\alpha} \frac{k_2}{k_2 + 1} \left[\frac{1 - \left(\frac{t_0}{t_{mc}}\right)}{1 + \left(\frac{kt_0}{t_{mc}}\right)}\right] \tag{2}$$

In both equations t_0 is the migration time of a nonionic species that does not interact with the pseudostationary phase, and t_r is the migration time of the analyte. In Eq. (2), α is the chemical selectivity between two analytes (defined as the ratio of the retention factors), and N is the number of theoretical plates. Both equations reduce to the conventional equations when the t_{mc} becomes infinite, corresponding to a conventional stationary phase. The effect of the mobility of the pseudostationary phase on resolution (Fig. 1) is the product of the two retention factor terms in Eq. (2), plotted as a function of the retention factor for various ratios of t_{mc}/t_0. For typical pseudostationary phases with t_{mc}/t_0 of 4–5, it can be seen that the resolution suffers dramatically relative to conventional chromatography at all retention factors, and especially at high retention factors. Relative to conventional HPLC, this is more than compensated by the improvement in efficiency, except for very highly retained hydrophobic compounds.

Given the plot in Fig. 1, it is clear that pseudostationary phases should have several properties in order to allow optimum resolution. To permit adjustment of the retention factor to within a fairly narrow optimum range, they should be stable and soluble under a range of analytical conditions. To provide a wide migration range, they should have high electrophoretic mobility. Phases should be available with a wide range of chemical structures to provide desired chromatographic selectivity. To permit efficient application of secondary media such as cyclodextrins and to minimize Joule heating, they should have very low or zero CMC. To maintain high plate numbers, they should be monodisperse, at least with respect to chemical interactions with solutes and electrophoretic mobility, and provide fast mass transfer for analytes between the pseudostationary phase and buffer medium. Ideally, a pseudostationary phase should permit mass spectrometric detection.

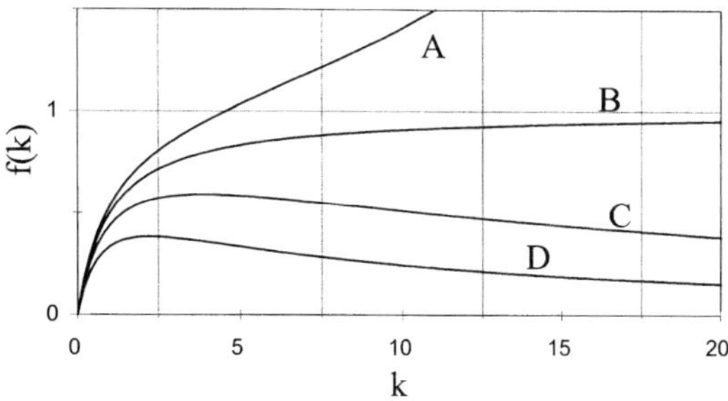

Figure 1. Plot of the product of the two retention factor terms in Eq. (2) as a function of the retention factor (A) $t_{mc}/t_0 = -30$ (negative value indicates that the pseudostationary phase has a net velocity opposite to electroosmotic flow); (B) $t_{mc}/t_0 = \infty$ (infinite value is equivalent to conventional chromatography); (C) $t_{mc}/t_0 = 15$; (D) $t_{mc}/t_0 = 5$.

Conventional micellar pseudostationary phases, while they do provide highly efficient separations, cannot meet all of the criteria noted above because they are self-assemblies in a state of equilibrium with the free surfactant in the surrounding buffer medium. The micellar equilibrium is characterized by the CMC of free surfactant and the aggregation number, or the number of surfactant monomers assembled in a single micelle. The CMC and aggregation number vary with the analytical conditions: they are affected by changes in the temperature, salt concentration, pH, surfactant concentration, and the concentration and nature of buffer additives. The equilibrium status of the micellar phase limits the flexibility of the technique in terms of the choice of analytical conditions. The surfactants must have a relatively low CMC, limiting the choice of surfactants considerably. The surfactant concentration must be high relative to the CMC, or irreproducibility will result. The effect of organic additives on the CMC and structure of micelles adds complications for the adjustment of retention factors for the separation of hydrophobic compounds.

The use of conventional surfactants also limits the applicability of MEKC for mass spectrometric detection. Unless the surfactant is somehow removed, the presence of a high concentration of low molecular mass surfactant leads to large signals in the low molecular mass region of the mass spectra, interfering with most MEKC analyses. High concentrations of surfactants also limit the sensitivity of the electrospray ionization process [6].

Several types of alternative pseudostationary phases have been developed and employed in efforts to address the limitations of micellar phases. Neutral pseudophases such as cyclodextrin polymers [7–9] and polyvinyl pyrrolidone [10–15] have been employed to provide selectivity

Electrophoresis 2000, *21*, 4054–4072

for ionic compounds. Proteins [16–22] and charged cyclo-dextrins [23–27] have also been employed for chiral separations. Calixarenes [28–35] are stable structures that permit the separation of hydrophobic compounds and enantiomers, but which are limited by background UV absorbance. Dendrimers and modified dendrimers have also been utilized as monomolecular pseudostationary phases.

This review concentrates on the use of synthetic ionic polymers and dendrimers as pseudostationary phases. Many of these polymers are amphiphilic. Because these polymers have similar properties to conventional surfactant micelles (solubilization, surface active properties, structure determined by hydrophobic effect), they are often referred to as polysoaps or micelle polymers. The fundamental difference between micelle polymers and conventional micelles is that in the polymer the size and primary structure of the phase is fixed by covalent bonds, rather than by hydrophobic association and self-assembly. Two reviews have covered the origins, structure, and properties of micelle polymers and polymer surfactants in detail [36, 37]. This review concentrates on studies that have characterized the structure, selectivity, and performance of these phases. The review is organized by the structures of the polymers. Two reviews of polymeric phases and the chemical selectivity of polymeric phases were published three years ago [38, 39].

Polymers provide stable pseudostationary phases with zero CMC. Because the CMC is zero, the phases can be used at virtually any concentration. The primary chemical structure and concentration of the phase do not change with changes in the analytical conditions. The structures can be used in the presence of relatively high amounts of organic modifier and with mass spectrometric detection [40–43]. The effects of organic modifiers in the run buffer can be studied without altering the primary covalent pseudostationary phase structure. Chiral separations can be achieved by using cyclodextrins [44, 45] or by using chiral-substituted polymers [45, 56]. Dendrimers may have advantages over some types of polymer surfactants in that they are relatively monodisperse.

In the development and characterization of polymeric pseudostationary phases, it has often been noted that these phases afford unique selectivity relative to micelles of SDS [38, 57–61] or of the monomer [62]. The fundamental difference between polymers and conventional micelles is the presence of covalent bonds between the hydrophobic unit structures in the polymer, whereas in micelles the monomers self-associate through hydrophobic interactions. Covalent stabilization results in a more structured phase with greater steric constraints than

micellar phases. This greater structural rigidity may lead to unique structural selectivity. However, this rigidity may also diminish the ability of the polymer to create suitable hydrophobic domains for the solvation of some hydrophobic compounds, and may limit certain interactions through steric hindrance.

Relative to conventional micelles, there are more variables which may be used to alter the selectivity of polymeric pseudostationary phases. The requirement of self-association is eliminated, so polymers without significant hydrophobic character can be employed. The pendent group and ionic head group chemistry can be varied while keeping the backbone structure of the polymer constant. Modification of a polymer backbone to achieve unique and varied selectivity is comparable to liquid chromatography, where a given support material can be modified to provide stationary phases with dramatically different selectivity. In electrokinetic chromatography, mixtures of polymeric phases with different selectivity can be used to achieve intermediate and predictable selectivity, or to adjust the effective migration range [63, 64].

Several approaches have been taken to characterize the selectivity of pseudostationary phases. The simplest approach is to compare the migration order, or relative migration times, of the solutes on the polymeric phase relative to another pseudostationary phase, typically SDS micelles. Alternatively, if the migration time of the polymer can be measured, the retention factors and the selectivity between analytes can be determined. The methylene selectivity (α_{CH2}, the selectivity between two compounds in a homologous series which differ only by the presence of a methylene group) is often reported. This is generally accepted as a measure of the hydrophobicity of the pseudostationary phase, with greater methylene selectivity indicating greater hydrophobicity. In many studies the logarithm of the retention factor for a series of compounds on one pseudostationary phase is plotted *vs.* the logarithm of the retention factor on a second pseudostationary phase. When the selectivity of the phases is the same, the points are expected to fall on a straight line. If the selectivity is different, a scatter plot is expected. Differences is log(k) are directly related to selectivity:

$$\log(k_2) - \log(k_1) = \log\left(\frac{k_2}{k_1}\right) = \log(\alpha) \tag{3}$$

and thus these plots are good indicators of overall selectivity. Further information can be deduced from the plots by observing the relative position of various classes of analytes. Linear solvation energy relationship studies are perhaps the most comprehensive approach to character-

Table 1. Selected selectivity descriptors of polymeric pseudostationary phases

Phase	$\alpha_{CH2}(H_2O)$	$\alpha_{CH2}(\%$ organic)	Cohesivity
SDS	2.29	1.27 (60% MeOH) 1.25 (50% ACN)[b]	3.95 (0.39)[a]
LiPFOS	NA	NA	2.44 (0.39)[a]
SUA[b]	1.99	NA	NA
SUS[b]	2.20	1.36 (40% ACN)	NA
SUP[c]	1.77	NA	3.86 (0.91)
SUT[c]	1.78	NA	2.85 (0.73)
Acrylate C_9	1.49	NA	NA
Acrylate C_{13}	2.18	NA	NA
Acrylate C_{18}	2.57	NA	NA
Elvacite 2669	NA	NA	3.00 (0.31)[a]
PAA-C_{12}	2.45	1.32 (60% MeOH)	NA
PAA-C_{16}	2.70 (20% MeOH)	1.67 (60% MeOH)	NA
AMPS/LMAm[d]	2.54	NA	NA
Siloxane[e]	1.99	NA	3.01 (0.48)
SBD G=1–3[f]	1.09	NA	NA
SBD G=4[f]	1.32	NA	NA
SBD G=3.5	1.0[g]	NA	NA
SBD-C_8 G=3.5	1.99	NA	NA
SBD-C_{12} G=3.5	2.32	1.57 (60% MeOH)	NA
PDMDA	NA	NA	(0)[h]

Higher values of methylene selectivity indicate more hydrophobic phases, and lower numbers for cohesivity represent more cohesive phases.
Unless otherwise noted, data are taken from the collection in [38].
a) [77], b) [69]; c) [72]; d) Shi, Watson and Palmer, in press; e) [90]
f) G denotes generation, calculated from data in [95]
g) Assumed from lack of separation for homologs in [97]
h) [94], reported as not significantly different from water
Abbreviations: LiPFOS, lithium perfluorooctane sulfonate; NA, not available; SBD, starburst dendrimer

izing the interactions between solutes and pseudostationary phases, but this approach has been employed in relatively few instances to date. Some of the descriptive parameters for many of the pseudostationary phases studied to date are presented in Table 1.

2 Micelle polymers

The term micelle polymer refers to polymerized surfactants of the tail-end geometry which are polymerized in solutions in excess of the CMC. These polymers may be considered to have structures similar to micelles, but lack the dynamic nature of micelles. The chemistry employed to date in MEKC involves the polymerization of ω-undecylenic acid and homologs or derivatives thereof. The various structures of the monomers are shown in Fig. 2.

2.1 Poly(sodium 10-undecylenate)

The first successful reports of the use of a micelle polymer were those of Palmer *et al.* [62, 65]. These authors used a micelle polymer of sodium-10-undecylenate (SUA,

Fig. 2A) as a pseudostationary phase to achieve the separation of alkyl phthalates and polynuclear aromatic hydrocarbons (PAHs) in buffers modified with up to 50% methanol or 45% acetonitrile. The polymer was observed to have unique selectivity relative to micelles of SDS or of the monomer, as determined by the relative migration times. The polymer was more retentive of polar compounds, and less retentive of nonpolar compounds. The electrophoretic mobility of polySUA increases substantially between 30 and 40% acetonitrile, and plots of log k *vs.* percent acetonitrile are nonlinear [65]. The retention factors of all of the analytes studied were affected to nearly the same extent, resulting in no dramatic change in the selectivity of the polymer. The increase in mobility and apparent change in interaction with analytes were interpreted to mean that the structure and solvation of the polymer are dynamic and that a change in the conformation of the polymer at high acetonitrile concentrations induces greater mobility and alters the chemical interactions with the solutes. It is possible that in the absence of organic modifier the polymer maintains a collapsed or entangled structure which minimizes interactions between

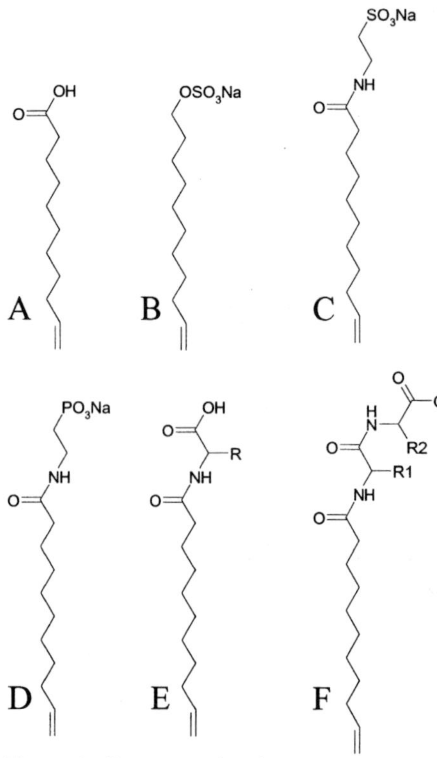

Figure 2. Structures of various monomers employed to synthesize micelle polymers. (A) SUA; (B) SUS; (C) SUT; (D) SUP; (E) chiral amino acid modified monomer, R=H, CH_3, $CH(CH_3)_2$, $CH_2CH(CH_3)_2$, or $CH(OH)(CH_3)$; (F) dipeptide modified monomer, R1=H, CH_3, $CH(CH_3)_2$, or $CH_2CH(CH_3)_2$ and R2=H, CH_3, $CH(CH_3)_2$, or $CH_2CH(CH_3)_2$.

water and the hydrophobic alkyl chains, while in the presence of acetonitrile the polymer can assume a more open structure with the alkyl chains solvated by the acetonitrile.

In an effort to determine whether the chemical selectivity of the polymer is affected by the presence of the chemical initiator in the core of the structure, Palmer and Tellman [66] polymerized the material with three different initiators of different hydrophobicity. The polymer initiated with very hydrophobic 2,2'-azobis(2,4-dimethylvaleronitrile) had a significantly lower molecular weight, and was not soluble in aqueous buffers. The overall selectivity, as determined by a plot of the logarithms of the capacity factors on two phases initiated with different initiators, is not significantly different (r^2 = 0.998, m = 0.962). The end groups on the interior of the polymer are not a significant factor in determining the selectivity of these phases. Interaction with the end-group functionality in the rigid core of the polymer

micelles is probably sterically restricted. Moy *et al.* [67] compared the performance of this polymer to other electrokinetic approaches for the separation of PAH. They found polySUA material provided the best separations. Using THF as an organic modifier, they were able to separate all sixteen priority pollutant PAHs. Additionally, they predicted that by using electrokinetic chromatography with polySUA as the second dimension in a gel permeation chromatography (GPC)-EKC apparatus, they would be able to separate up to 1000 compounds. The system was applied to a soil extract with good results. The polymer could not be used, however, with LIF detection at 257 nm.

2.2 Poly(sodium 10-undecenyl sulfate)

In part to eliminate the problems with the carboxylate head group, and in part to prepare a micelle polymer with the same head group chemistry as SDS, Palmer and Terabe [68, 69] synthesized and employed poly(sodium 10-undecenyl sulfate) (SUS, Fig. 2B), the sulfate analog of poly(sodium undecylenate). Shamsi *et al.* [70, 71] have also utilized this polymer for the separation of PAH and monomethylbenz[a]anthracenes. Like its carboxylate counterpart, this polymer provides efficient and selective separations of a variety of compounds in aqueous and modified aqueous buffers [68–71]. The chemical selectivity of the polymer was studied using a series of substituted benzene and naphthalene compounds, and the structural selctivity was studied using PAH [69].

A comparison of the separation of substituted benzene and naphthalene compounds achieved with the sulfate polymer and SDS micelles is shown in Fig. 3. Changes in migration order indicate the different selectivity of the polymer and the SDS micelles. Plots of log k on polySUA and polySUS *vs.* log k using SDS micelles indicate that both polymers interact more strongly with compounds having amine or hydroxyl groups, implying that they are more polar or are better hydrogen bond acceptors or donors. A plot of log k on polySUA *vs.* log k on the SUS phase shows that the selectivity of the two polymers is virtually identical. Together these results illustrate that the differences in selectivity between the polymers and SDS micelles are not due to the structure of the ionic head group.

The stronger interaction of polySUA and polySUS with polar compounds, and the lower methylene selectivity of these polymers relative to SDS micelles (Table 1), are indications of lower overall hydrophobicity. This may be the result of the smaller size and shorter alkyl chain length relative to SDS micelles. The fact that polySUA initiated with hydrophobic initiators had both lower molecular

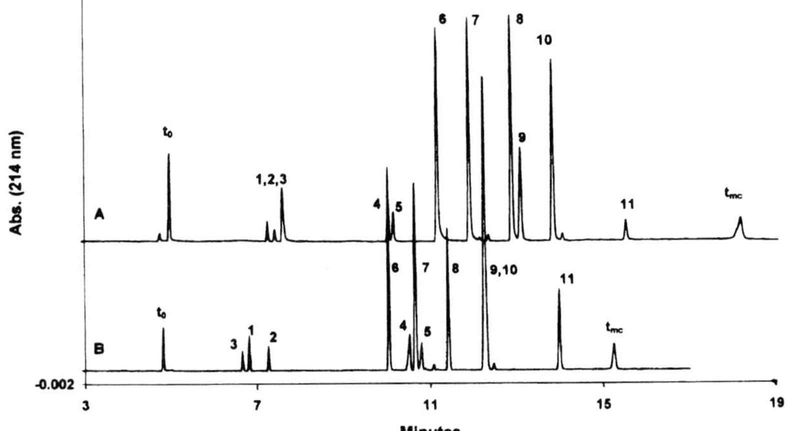

Figure 3. Separation of substituted aromatic compounds using (A) polySUS and (B) SDS micelles. 1, Nitrobenzene; 2, anisole; 3, *p*-nitroaniline; 4, *o*-xylene; 5, *m*-xylene; 6, naphthylamine; 7, naphthalene methanol; 8, acenaphthenol; 9, naphthalene; 10, naphthalene ethanol; 11, diphenyl ether. Reprinted from [69], with permission.

weight and lower methylene selectivity is evidence that the size of the polymer may play a role in determining the strength of the hydrophobic interactions. Steric constraints may prevent formation of a hydrophobic domain capable of solvating the hydrophobic compounds studied.

The SUS polymer is also useful in buffers modified with organic solvents. Shamsi *et al.* [70] separated all sixteen PAHs using polySUS in 57% acetonitrile. The selectivity of the polymer for hydrophobic compounds is also interesting [69, 71]. Relative to micelles of SDS, the SUS polymer retains greater methylene selectivity in buffers modified with acetonitrile and methanol [69]. In 60% methanol the structurally rigid PAHs (rings connected at more than one point) have less affinity for the polymer phase relative to SDS micelles, while more flexible compounds (rings connected by a bridging bond) are more highly attracted to the polymer phase. However, in 40% acetonitrile the opposite result is observed [69]. The methanol results are understandable given the smaller size and more rigid structure of the polymer, which may render it less able to accommodate large inflexible molecules. The acetonitrile results are more difficult to understand, and imply that the stronger solvent may lead to a larger hydrophobic region due to better solvation or to a more structured conformation in the alkyl chains or polymer backbone. Akbay *et al.* [71] attempted to correlate retention of monomethylbenz-[a]anthracene isomers with the length and length-to-breadth ratio (L/B). Unlike liquid chromatography, they found a better correlation with length than with L/B. These authors also noted no significant change in the partial specific volume of the phase in 35% acetonitrile relative to water, indicating that any change in the conformation of the polymer in acetonitrile-modified buffers must be subtle.

2.3 Sodium-*N*-undec-10-en-1-oyl taurate and sodium-*N*-undec-10-en-1-oyl aminoethyl-2-phosphonate

In a further effort to characterize the effect of the ionic head group on the selectivity of the micelle polymers, Tellman and Palmer [72] synthesized and characterized analogs with amide bonds and sulfonate and phosphonate ionic groups (SUT, SUP, Fig. 2C and D). PolySUT did provide significantly higher electrophoretic mobility, and both of the polymers could be employed in low pH (2.5) buffers. As shown in Fig. 4, the selectivity of the phosphonate phase is significantly different from that of the sulfonate phase for certain classes of compounds. However, very little difference in selectivity was observed between polySUA, polySUS, polySUT, and polySUP for the separation of the majority of substituted aromatic compounds studied. The selectivity of the phosphonate phase was also similar to the sulfonate phase for cold medicine ingredients. Linear solvation energy relationship studies showed some significant differences in the cohesiveness of polySUP and polySUT, but otherwise no significant differences were observed. Both phases were found to be more cohesive than SDS micelles (Table 1). While the ionic head group does have some effect on performance, the effect on the selectivity appears to be minor.

2.4 Chiral micelle polymers

In contrast to the conclusion drawn for the general selectivity of micelle polymers, many reports have now demonstrated that micelle polymers with chiral head groups do provide chiral selectivity, that the selectivity can be reversed, and that the interactions between solutes and the polymers can be probed by changing the nature of the

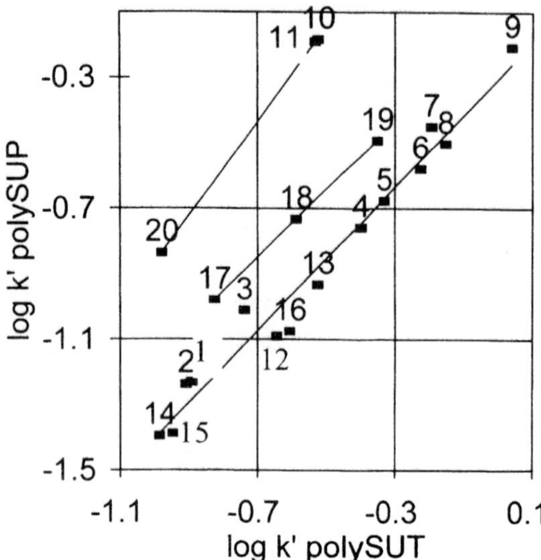

Figure 4. Comparison of the selectivity of polySUP to polySUT using 20 compounds. 1, Anisole; 2, nitrobenzene; 3, *p*-nitroaniline; 4, naphthylamine; 5, naphthylene methanol; 6, acenaphthenol; 7, naphthalene; 8, naphthalene methanol; 9, diphenyl ether; 10, *p*-xylene; 11, *m*-xylene; 12, 4-chloroaniline; 13, chlorobenzene; 14, benzene; 15, acetanilide; 16, methylbenzoate; 17, acetophenone; 18, propiophenone; 19, butyrophenone; 20, ethylbenzene. Three best fit lines: 10, 11 and 20; $r^2 = 0.99995$; 17, 18, and 19, $r^2 = 1.000$; others, $r^2 = 0.972$.

head group. Polymers with the structures shown in Fig. 2E and F have been synthesized with a variety of amino acids in both the L- and D-forms. These polymers have now been applied to the separation of a variety of chiral analytes. The motivation for the use of polymeric phases has predominately been that the elimination of the dynamic equilibrium between the pseudostationary phase and surfactant monomers would improve the chiral resolution. Poly(sodium undecenoly-L-valine) and a cationic amide of this compound have been synthesized, polymerized, and employed for chiral separations. The two structures were patented for used in chiral separations by electrokinetic chromatography in 1992 [73, 74].

Wang and Warner [46] investigated and published their results on the use of this chiral polymer in 1994. Substantially improved separations of (±)-1,1'-bi-2-naphthol were observed when poly(sodium undecenoyl-L-valine) was employed relative to the monomer surfactant. The migration order of the analytes was reversed when poly(sodium-undecenoyl-D-valine) was employed as a pseudostationary phase. In combination with γ-cyclodextrin, the additive effects of the chiral polymer and the chiral recog-

nition of the cyclodextrin provided greatly improved separations [45]. The micelle polymer cannot be included in the cyclodextrin cavity and does not interfere with the chiral recognition of the cyclodextrin. An advantage of the chiral micelle polymer is that it can be employed in buffer media modified with organic solvents. Addition of up to 40% methanol did improve some chiral separations, but addition of acetonitrile had a detrimental effect. The authors report better separations at pH 10 than at pH 9, which they attribute to a more open structure of the polymer at high pH [75] that leads to better interactions. A poly(vinyl alcohol)-coated capillary was employed to eliminate any adsorption of the micelle polymer on the surface of the capillary. This improved the separation for two paveroline derivatives, but eliminated all selectivity for a third paveroline derivative.

Dobashi *et al.* [55] concentrated on the separation of dinitrobenzoyl amino acid isopropyl esters and compared the separations achieved with micelles formed from chiral surfactants to those obtained with the chiral micelle polymer [55]. Using the polymer, however, the selectivities were not as good as those obtained when conventional micelles of sodium dodecanoyl-L-valine were employed. Additionally, peak tailing could only be eliminated by the addition of SDS to the separation buffer. They concluded that the increased order of the polymer relative to the micelles does not prevent binding of the substrate molecules and that an ordered interfacial region where enantiomer binding and recognition can occur exists in either case. The lower selectivity observed with the micelle polymer was attributed to spaces between the surfactant monomers and penetration of water into the interior of the micelle polymer.

Warner's group [51] has since embarked on an ambitious program to characterize the structure of these polymers, and the nature of the chemical interactions that contribute, or do not contribute, to chiral resolution. The single amino acid valinate polymer has been compared to the corresponding dipeptide surfactant, poly(sodium undecenyl-(LL)-valine-valine). The dipeptide polymer demonstrated a significant improvement in chiral selectivity relative to the single amino acid polymer for three out of four analytes studied [51].

To better characterize the mechanism of interaction, a large variety of dipeptide polymers with different amino acids and different order of attachment have been synthesized and studied [52–54]. It was shown that the amino acid order has a significant impact on the chiral selectivity of the surfactant [54]. A structural explanation of the results was proposed in which it was argued that the conformation of the dipeptide surfactants is governed by hydrophobic interactions and steric constraints [53]. The

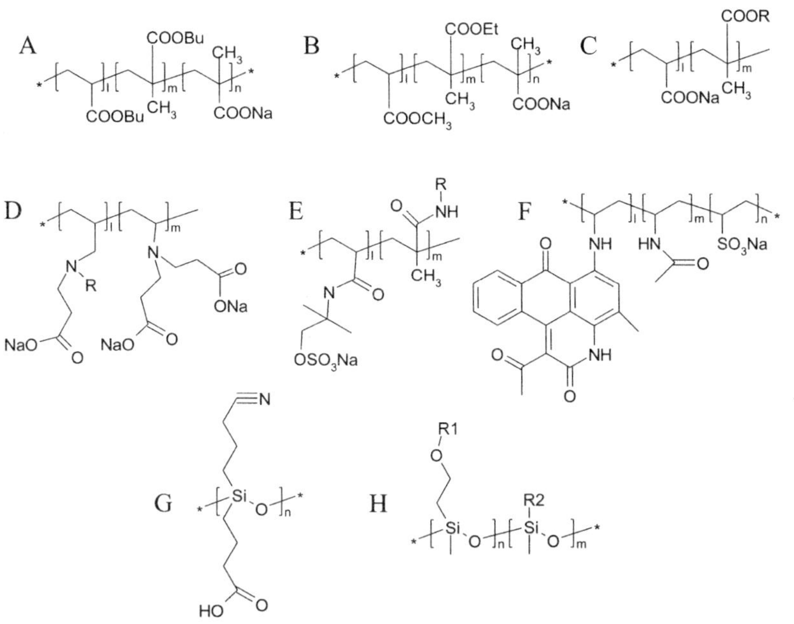

Figure 5. Structures of anionic copolymers. (A) BBMA; (B) Elvacite 2669; (C) acrylate copolymers, R = C_9, C_{13}, or C_{18}; (D) PAA, R = C_{10}, C_{12}, C_{16}; (E) AMPS/LMAm, R = C_{12}; (F) polymeric dye, $l/m/n$ = 2 /4/4; (G) and (H) siloxane polymers, R1 = s ulfate or sulfonate terminated ethers or polyethers, R2 = methyl or pentyl groups.

preferential binding site of various solutes has also been investigated using diastereomeric dipeptide surfactants [52]. The depth of penetration of the analyte into the polymer, governed by hydrophobic and electrostatic interactions, was thought to be the deciding factor in determining the selectivity observed. The role of the second chiral center was investigated by comparing the resolution of a variety of solutes using dipeptide surfactants with one chiral center with that of dipeptide surfactants with two chiral centers [50]. In some cases, trends could be observed indicating that the interactions are primarily governed by electrostatic and hydrophobic interactions. In many cases, however, no preferential site or mechanism of interaction could be determined.

Warner's group [49] has also studied two of the polymeric dipeptide surfactants by analytical ultracentrifugation. Comparing poly(sodium undecenyl-L-valinate) with poly(-sodium undecenyl-L-threoninate) showed that while the apparent average molecular mass of the valinate remained constant with temperature, that of the threoninate varied from 12 000 to 20 000 g/mol between 20 and 40°C [49]. This variation strongly suggests intermolecular aggregation, leaving open the possibility that even the 12 000 g/mol measurements represent upper limits of associated lower oligomers.

3 Anionic polymers

Polymer surfactants with the structures shown in Fig. 5 have been employed for MEKC separations of cold medi-cine ingredients [17, 76], substituted benzenes [17, 76, 77], substituted naphthalenes [17, 76], and hydrophobic compounds (PAHs, *n*-alkyl phenyl ketones, fullerenes) [78]. The polymers have also been employed with cyclodextrins for the separation of dansyl amino acids [44] and for MEKC with mass spectrometric detection [42, 43]. The chemical selectivity of these polymers has also been studied in some detail [77].

3.1 Acrylate copolymers

Ozaki *et al.* [44, 76] were the first to report the use of an acrylate copolymer as a pseudostationary phase. They employed butyl acrylate-butyl methacrylate-methacrylic acid (BBMA, Fig. 5A) copolymers for the separation of benzene derivatives, cold medicine ingredients, and naphthalene derivatives. In comparison with SDS micelles, BBMA had similar selectivity for the cold medicine ingredients and benzene derivatives, but significantly different selectivity for the substituted naphthalene compounds. Polymer structure and pH were also shown to affect the selectivity and performance of the polymer. From pH 4 to 7 electrophoretic mobility of the polymer increased considerably due to increased ionization of the carboxylate groups. At the same time, the retention factors for naphthalene compounds decreased, also due to increases in surface charge. Increases in the fraction of methacrylic acid in the copolymer had similar effects: at higher fractions where the surface charge is greater, the electrophoretic mobility was greater and the retention factors were

Electrophoresis 2000, *21*, 4054–4072

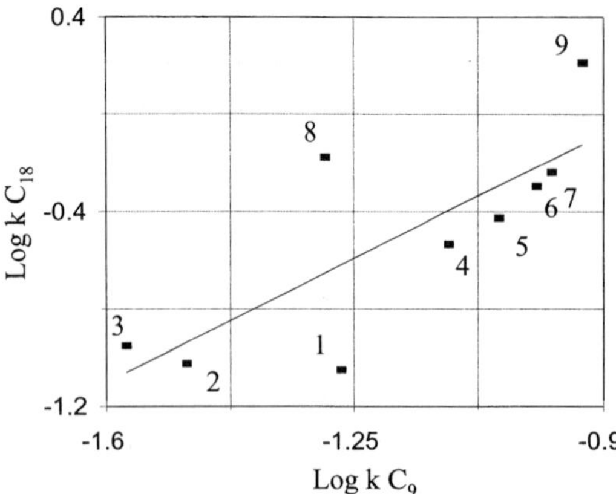

Figure 6. Selectivity comparison of C_{18} acrylate copolymer with C_9 acrylate copolymer. 1, *p*-Nitroaniline; 2, nitrobenzene; 3, anisole; 4, naphthalene methanol; 5, acenapthenol; 6, naphthylamine; 7, naphthalene ethanol; 8, *o*-xylene; 9, naphthalene. $r^2 = 0.61$.

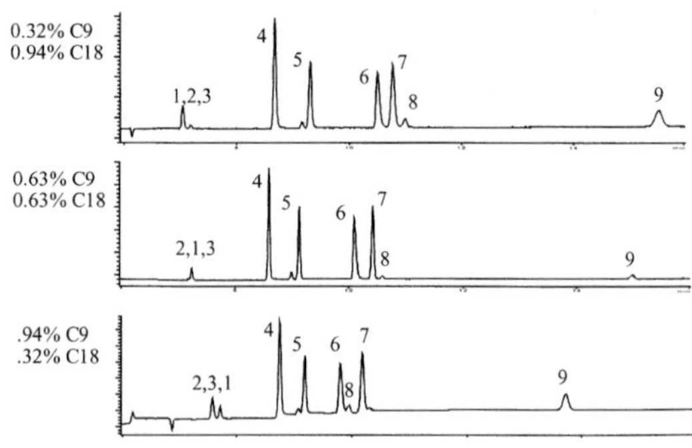

Figure 7. Change in selectivity as C_{18} acrylate copolymer is mixed with C_9 acrylate copolymer in different ratios. Selectivity differences are most apparent for solutes 1, 2, 3, 6, 7, 8, and 9. Analytes numbered as in Fig. 6.

lower. It was concluded by the authors that it is more suitable to change the polymer chemistry than to change the pH, since at pHs below 4 the polymer precipitates and since changes in pH are accompanied by changes in the electrophoretic mobility. Significantly, three different molecular masses of the BBMA polymer gave essentially the same separations of naphthalene derivatives. As long as the polymer chemistry was constant, the molecular mass did not affect the electrophoretic mobility of the polymer or the retention factors of the solutes.

The effects of the addition of methanol and a nonionic surfactant, octaoxyethylenedodecanol, $(EO)_8R_{12}$, have also been investigated [44]. Addition of methanol to the run buffer was found to reduce the retention factors of substituted naphthalene compounds, as would be expected from reductions in hydrophobic interactions. Minor selectivity changes were also noted. Addition of $(EO)_8R_{12}$ was found to increase the retention factors of the same compounds, while the migration range was diminished. The nonionic surfactant evidently forms comicelles with the polysoap, rather than forming independent nonionic micelles.

Ozaki *et al.* [44] also demonstrated the utility of BBMA in combination with cyclodextrins for the chiral separation of dansylated amino acids. In combination with 10 mM β-CD, nine of ten pairs of dansylated amino acids were successfully separated, and eight had separation factors greater than those observed with SDS and 60 mM β-CD. The superior results relative to SDS can be attributed to the absence of monomeric surfactant, which can

be coincluded in the cyclodextrins [79–82], reducing chiral selectivity. The BBMA cannot be coincluded in the cyclodextrin cavity, owing to its large size. It was also demonstrated that it is important to purify the polymer of low molecular mass impurities, as these impurities can also interfere with the separation.

Yang *et al.* [78] have used a similar polysoap, poly(methyl methacrylate-ethyl acrylate-methacrylic acid) (Elvacite 2669, Fig. 5B) as a pseudostationary phase for the separation of hydrophobic compounds [78], and have used linear solvation energy relationships (LSERs) to characterize the chemical selectivity of this polymer relative to several conventional micelles [77]. LSER studies using 60 aromatic test solutes were used to measure the relative cohesiveness, hydrogen bond acceptor strength, and hydrogen bond donor strength of Elvacite 2669. Cohesiveness, or resistance to cavity formation, and hydrogen bond donor strength were found to be the most important contributors to the selectivity of Elvacite 2669, while the hydrogen bond acceptor strength plays a minor but significant role. This is similar to the retention behavior of SDS and sodium cholate micelles. The polymer was found to resist cavity formation nearly as much as fluorocarbon micelles (Table 1). This might be expected from the more rigid covalent structure of the polysoap, which would require greater energy to rearrange in order to solubilize larger analytes. The polymer was found to have intermediate hydrogen bond donor strength and the hydrogen bond acceptor strength was relatively high. The migration behavior of the 60 test compounds was studied relative to SDS micelles by plotting log *k* with the polymer *vs.* log *k* with SDS micelles. Hydrogen bond donor compounds or strong dipolar compounds were found to interact more

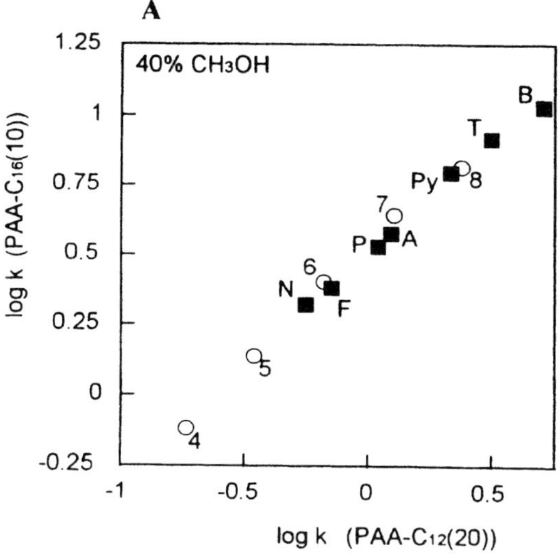
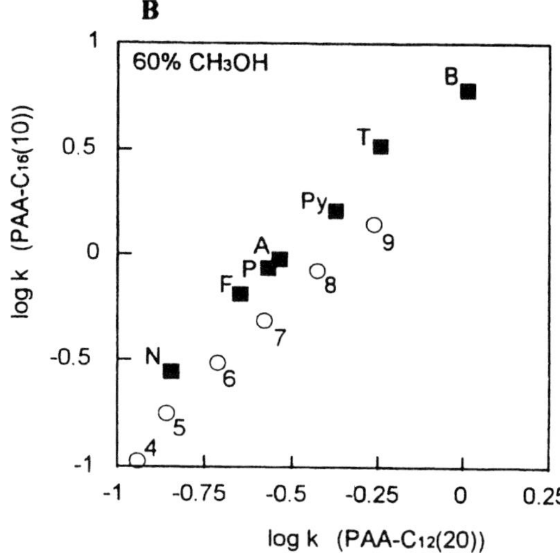

Figure 8. Selectivity comparison of PAA-C$_{16}$ with PAA-C$_{12}$ for alkyl phenyl ketones and PAH. 4–9 refers to the number of carbons in the alkyl phenyl ketone chain; N, naphthalene; F, fluorene; P, phenanthrene; A, anthracene; Py, pyrene; T, triphenylene; B, benzo[a]pyrene. Reprinted from [38], with permission.

strongly with the strong hydrogen bond-accepting Elvacite 2669, while hydrogen bond acceptor compounds were found to interact more strongly with the hydrogen bond donating SDS micelles. In many cases, the compounds that interact more strongly with Elvacite 2669 are the same compounds that interact strongly with SUA and SUS polymer micelles. Elvacite 2669 was also studied for the separation of organophosphorus pesticides in methanol and acetonitrile modified buffers [83]. Plate numbers and resolution were lower with the polymer than with cholate micelles. Retention factors were lower, and resolution was the same or less than with SDS micelles. Overall, the polymer was not competitive with the cholate phase.

Wiedmer *et al.* [84] have studied the behavior of Elvacite 2669 for the separation of hydrophobic compounds in buffers modified with methanol. The viscosity, the migration times of some neutral hydrophobic compounds, and the light-scattering properties of polymer solutions were studied as a function of the concentration of methanol and the polymer. The results showed the structure and behavior of Elvacite 2669 to be highly dependent on the methanol-water ratio in the buffer. Significant intermolecular aggregation was observed. Significant interaction of the Elvacite 2669 with the capillary wall was also observed at polymer concentrations in excess of 0.5%.

Palmer [85] has studied a series of acrylate copolymers with differing alkyl chain lengths and molecular weights; for structures, see Fig. 5C. All polymers had the same acrylate/alkyl acrylate mole ratio and approximately the

same molecular mass, and thus changes in selectivity were solely due to the differences in alkyl chain length. As shown in Table 1, the methylene selectivity increased with increased alkyl chain length. Relative to SDS micelles, the polymers progressed from having greater overall interaction with polar compounds (C9) to having greater overall interaction with nonpolar or hydrophobic compounds (C18). The exception to this rule was amine compounds, which invariably interact more strongly with the polymer phases. The amines are hydrogen bond acceptors, and the strong interaction indicates that the polymers are in general stronger hydrogen bond donors than SDS micelles. The C13 polymer had a selectivity most similar to SDS micelles. Figure 6 shows a plot of log *k* on the C9 phase *vs.* that on the C18 phase, demonstrating that significant differences in selectivity are realized by varying the alkyl chain length. In several instances (*p*-nitroaniline, nitrobenzene, anisole and *o*-xylene), the migration order of analytes was reversed as the alkyl chain length progressed from nine to eighteen carbons.

Palmer [63] has shown that mixtures of the C9 and C18 phases provide intermediate and predictable selectivity. Figure 7 shows the change in selectivity on going from 75% C18 to the 75% C9 phases. Using a simple model based on no intermolecular aggregation, it was possible to predict the mobility of analytes to within 10% absolute error. Plots of log *k* *vs.* carbon number for homologous series of alkyl phenyl ketones and alkyl benzoates in aqueous buffers, which are generally linear [86], were not

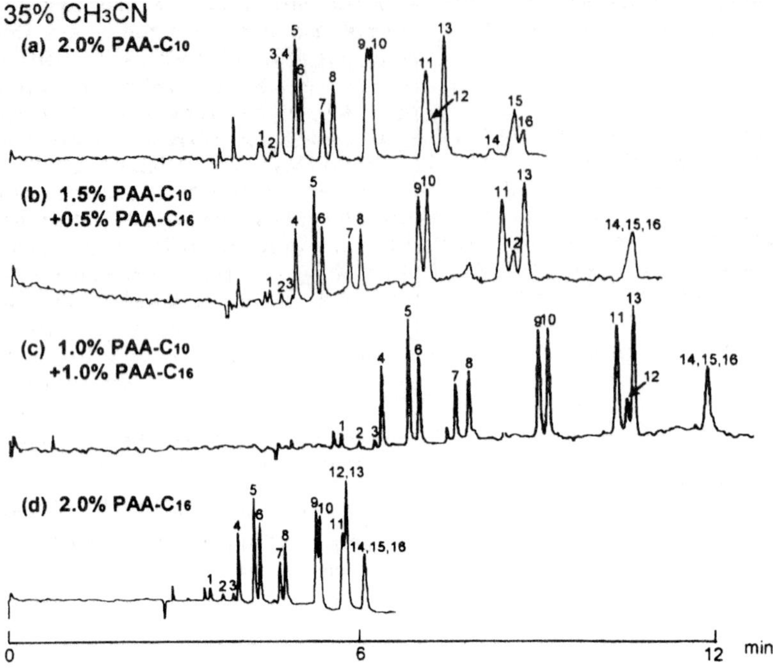

Figure 9. Separation of 16 PAHs with (a) PAA-C_{10}, 2% w/v; (b) PAA-C_{10}, 1.5% w/v + PAA-C_{16}, 0.5% w/v; (c) PAA-C_{10}, 1.0% w/v + PAA-C_{16}, 1.0% w/v; (d) PAA-C_{16}, 2.0% w/v. 1, Naphthalene; 2, acenaphthylene; 3, acenaphthene; 4, fluorene; 5, phenanthrene; 6, anthracene; 7, fluoranthene; 8, pyrene; 9, chrysene; 10, benz[α]anthracene; 11, benzo[*b*]fluoranthene; 12, benzo[*k*]fluoranthene; 13, benzo[α]pyrene, 14, dibenz[α *h*]anthracene; 15, indeno[1,2,3-*cd*]pyrene; 16, benzo[*ghi*]perylene. Re-printed from [64], with permission.

linear for the C13 and C9 polymers. Negative deviations were observed for homologs with longer alkyl chains (four to six carbons), and the deivations were more severe with the C9 phase than with the C13 phase. The C18 phase did yield linear plots. Evidently, homologs with longer alkyl chains are not well solvated by polymers with shorter alkyl chains, implying limited ability of these polymers to create a large hydrophobic domain.

3.2 Polyallylamine-supported phases

Tanaka *et al.* [57, 64, 87] have studied polyallylamine (PAA) supported pseudostationary phases (Fig. 5D). These polymers were synthesized with varying alkyl chain lengths, and different degrees of alkylation. The polymers were studied in methanol-modified buffers for the separation of alkyl phenyl ketones and PAH. The methylene selectivity of PAA modified with dodecyl chains is similar to that of SDS micelles in both aqueous media and 60% methanol. However, the methylene selectivity of PAA modified with hexadecyl chains is higher than that observed with SDS in both 20 and 60% methanol. The methylene selectivity of the hexadecyl-modified polymer is similar to that observed by Palmer with acrylate copolymers with octadecyl chains (Table 1) [85]. Plots of log k *vs.* carbon number for the alkyl-phenyl ketones are not always linear, especially at intermediate methanol con-

centrations. The nonlinearity for higher carbon numbers must be due to the inability of the polymer to create a large hydrophobic domain capable of solvating long hydrocarbon chains. This is especially true at intermediate concentrations of methanol, where only part of the alkylated polymer is solvated by methanol. As presented in Fig. 8, the selectivity of hexadecyl-modified PAA is very similar to that for dodecyl-modified PAA for the separation of PAH and alkyl-phenyl ketones in 40% methanol, but rather different in 60% methanol. This appears to be due to a change in the selectivity of the hexadecyl phase, which shows strong preference for the PAH in 60% methanol, but not in 40% methanol.

As was observed with polySUA in acetonitrile-modified buffers, the electrophoretic mobility of these polymers was observed to increase dramatically at a particular concentration of methanol as a modifier. This increase in electrophoretic mobility provided a wide migration range, and meant that separations of hydrophobic compounds could be optimized in a similar manner to reversed-phase liquid chromatography. Additionally, as was observed with the SUA polymer, plots of log k *vs.* percent methanol were nonlinear, indicating a change in the retention mechanism. These results can be correlated with the results of dynamic light scattering studies, which have shown that there is a bimodal distribution of relaxation times in aque-

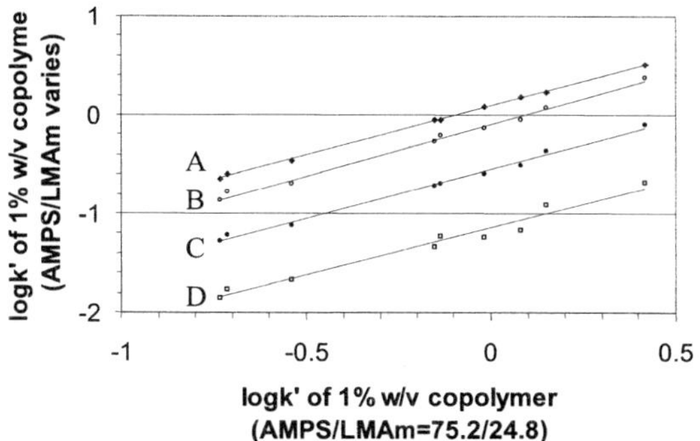

Figure 10. Selectivity plots for AMPS/LMAm copolymers with different monomer ratios relative to AMPS/LMAm with the ratio 75.2/24.8 (A) 75.1/24.9, r^2=0.999; (B) 85/15, r^2=0.994; (C) 91/9, r^2=0.992; (D) 97/3, r^2=0.970. Analytes are not numbered, but are the same analytes as in Fig. 6.

ous solutions, indicating intermolecular association [87]. In 40% methanol, a narrow and presumably unimolecular distribution in observed.

The degree of substitution of the backbone with dodecyl (10–20%) and hexadecyl (5–25%) chains was also studied [87]. Greater substitution leads to a narrower migration range, greater sample capacity, and greater methylene selectivity. Mixtures of PAA polymers with decyl and hexadecyl chains were used to modify the migration window and the selectivity of separation of PAH in methanol- and acetonitrile-modified buffers [64]. An example of the results is presented in Fig. 9. The mixture of two phases provided adequate peak capacities for both early and late eluting compounds. A model was presented which assumed independent contribution of the two phases to solute partition. This model adequately predicted the observed selectivity, but gave up to 40% error in the t_r/t_0 values for later eluting compounds.

3.3 2-Acrylamido-2-methylpropanesulfonic acid (AMPS) copolymers

Shi and Palmer (submitted) have recently reported studies on the use of copolymers of AMPS and lauryl methacrylamide (LMAm, Fig. 5E). This polymer structure takes advantage of the strong acidity and stability of the sulfonic acid head group, and should permit the introduction of a variety of functional groups to alter the chemical selectivity [88, 89]. The initial study has investigated the effect of the ratio of the two monomers, and found that (as with PAA phases) increasing the degree of substitution with lauryl groups reduces the migration range, increases sample capacity, and increases methylene selectivity. A

reasonable compromise between selectivity, migration range, and sample capacity was found at approximately 20–25% lauryl methacrylamide. The electrophoretic mobility of polymers of this structure is not significantly different from SDS, and the methylene selectivity is greater (Table 1). No differences in chemical selectivity between polymers with different ratios of monomers were observed, as can be seen in Fig. 10.

3.4 Polymeric dye

An aromatic polymeric dye (Fig. 5F) has been employed for the separation of nitro and amino aromatic compounds in buffers modified with up to 70% methanol [59]. The dye showed significantly different chemical selectivity from SDS micelles, but due to strong UV absorbance a partial filling technique with counterpressure had to be used. This led to poor efficiency for early eluting compounds, but improved the resolution for late eluting compounds. The results indicate that polymers with aromatic functionality could be used to provide unique selectivity with alternative modes of detection.

3.5 Siloxane polymers

Chen and Palmer [90] and Peterson and Palmer ([91] and in press) Peterson and Palmer, submitted have begun to investigate the applicability of siloxane polymers as pseudostationary phases. Use of siloxane polymers should permit application to electrokinetic chromatography of the wealth of stationary phase chemistry based on silica and siloxane chemistry developed in the past 30–40 years. The first of these studies investigated several siloxane polymers, but found only one with sufficient aqueous solubility for application in electrokinetic chromatography (Fig. 5G) [90]. This polymer had severely limited sample capacity, but very different selectivity from SDS micelles. LSER studies showed that this polymer has significantly different selectivity relative to SDS micelles in every regard. As with other polymeric phases, the cohesiveness of the polymer was found to be greater than SDS micelles.

More recent studies have overcome the problem of limited aqueous solubility by introducing ionic groups to commercially available siloxane backbones (Fig. 5H) ([91]; Peterson and Palmer, in press). These polymers all have strongly acidic sulfate or sulfonate head groups, and can be synthesized with varied side chain chemistry to alter the selectivity. Five such polymers have been introduced. Each has a different selectivity relative to SDS, and there are some differences in selectivity relative to one another. Polymers substituted with ionic functionality at all silicone centers show significant peak asymmetry for more hydro-

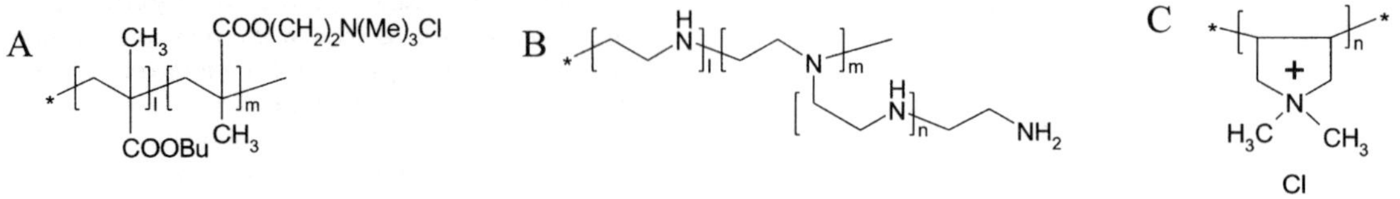

Figure 11. Structures of cationic polymers. (A) BMAC; (B) PEI; (C) PDADMA; *n*=1200–2200.

phobic analogs due to a lack of sufficient hydrophobic domains to provide sufficient sample capacity (Peterson and Palmer, in press). These polymers also show significant nonequilibrium band broadening, possibly due to slow mass transfer. However, substitution of at least 15% of the silicon centers with hydrophobic groups appears to alleviate or eliminate both problems.

4 Cationic polymers

Ozaki and Ichihara [44] were the first to report the use of a cationic polymer as a pseudostationary phase. They used butyl methacrylate methacryloyloxyethyltrimethyl ammonium chloride (BMAC, Fig. 11A). The cationic polymer adsorbed to the capillary wall, reversing electroosmotic flow. Due to limited solubility, the polymer could only be used in buffers modified with 20% isopropyl alcohol. The selectivity of the polymer is not very different from that of anionic BBMA.

Polyethyleneimine (PEI, Fig. 11B) was first employed as a pseudostationary phase for the separation of phenols in 1997 [92]. Strong interactions, which were at least partly ionic in nature, were observed between the phenols and PEI. Strong interactions were also observed, however, at pHs where the phenols are not ionized. Maichel *et al.* [60] and Potocek *et al.* [61] have also reported the use of PEI and polydiallyldimethyl-ammonium (PDADMA; Fig. 11C) [93].

Adsorption of PEI to the capillary wall reverses and enhances electroosmotic flow and thus provides faster analysis. The electroosmotic flow is reproducible. The selectivity of PEI differs from that of other pseudostationary phases studied [61]. The polymer provided separations of phenols with migration time based on the number of hydroxyl groups. However, no selectivity was observed between mono-, di- and trimethyl-substituted phenols, demonstrating a lack of selectivity based on lipophilicity. Addition of methanol or acetonitrile to the separation buffer reduced the affinity of the phenols for the phase, which appears contrary to interactions based on polarity alone [60]. Both organic solvents were found to be detrimental to separations of phenols with a PEI pseudostationary phase.

PDADMA also reproducibly reverses the electroosmotic flow. Like PEI, the phase appears to interact more strongly with solutes having greater polarity or a greater number of polar groups. LSER studies showed the phase to have very different selectivity characteristics from other phases studied. The polymer has a cohesiveness term not different from water (Table 1), which makes it a very cohesive phase relative to micelles and other polymeric phases, but which means that no selectivity should be observed based on solute size alone. Additionally, the hydrogen bond donor strength of the phase is not significantly different from water, which makes this a very acidic phase relative to others. The hydrogen bond acceptor term is positive, like cationic micelles. The selectivity of the PDADMA phase is dominated by a term describing interactions with *n*- and π-electrons [93].

5 Dendrimers

Dendrimers are highly branched polymers that are synthesized in generations from a core using multistep repetitive syntheses. The resulting macromolecules have well-defined branches and are said to have very specific molecular masses and uniform sizes. The polymers differ from linear polymers in that they do not have entangled chains but do have numerous chain-ends that can be functionalized. Unlike micelles, dendrimers become more sterically hindered toward the exterior of the molecule, and the interior may be hydrophobic or hydrophilic. Dendritic molecules have been shown to provide unique selectivity. The selectivity is often affected by the presence of internal functionality, such as amine groups. It has also been shown that the selectivity can be altered through modification of the exterior of the dendrimer with alkyl chains of varying length. Figure 12 shows the general structure of some of the dendrimers used to date as pseudostationary phases, as well as the structure of modified dendrimers.

Kuzdzal *et al.* [94, 95] employed amide-based dendrimers with carboxylic acid terminus for the separation of alkyl parabens and Robitussen cold medicine ingredients with good selectivity and efficiency. No organic modifier was required to separate alkyl parabens up to butyl paraben. Good separations were obtained at pH 10, but pH 8 and 6

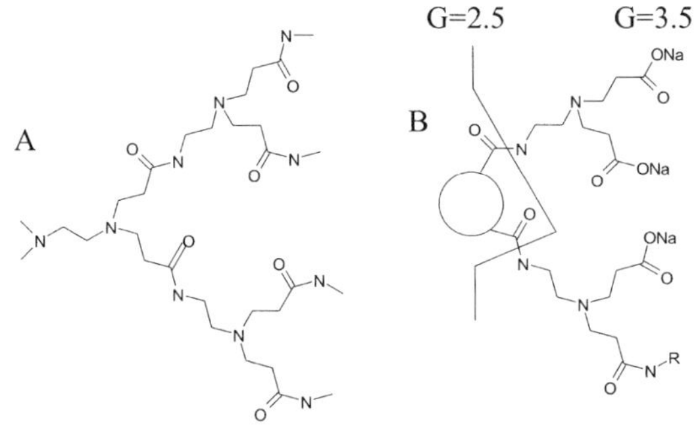

Figure 12. Structures and notations for dendrimers and alkyl-modified dendrimers. (A) General structure of the dendrimers with ethylenediamine core (dendrimers with xylenediamine core have also been employed); (B) SBD-C_n, G=3.5, R=C_4, C_8 or C_{12}.

did not provide useful separations due to increased cationic behavior and reduced electrophoretic mobility of the dendrimer. The migration times were observed to increase significantly with increased size of the dendrimer from first to third generations. The third generation dendrimer provided less efficiency than lower generations. Van't Hoff studies were used to characterize the enthalpy and entropy of the interactions. The enthalpy was found to be less favorable for higher generation dendrimers, but the entropy was more favorable. Dendrimers of different generations give similar retention in concentration ratios in approximate agreement with the calculated phase ratio.

Tanaka *et al.* [96–99] have studied both standard and modified poly(amidoamine) dendrimers. The selectivity of unmodified dendrimers was found to differ significantly from that of conventional micelles [96, 97]. The dendrimers were found to separate benzene derivatives from naphthalene derivatives, but did not separate well the compounds with a difference in functional groups. The contrast in selectivity relative to SDS-MEKC is demonstrated in Fig. 13. Very little selectivity was observed for alkyl benzenes using the dendrimers (methylene selectivity $\approx$ 1). PAHs were separated in 40% methanol and the selectivity was compared to SDS micelles in 60% methanol [97]. The dendrimer preferentially retained the more rigid, compact hydrocarbons relative to SDS micelles. This is likely due to the hydrophilic and rigid structure of the highly branched polymers. Modifying these dendrimers with alkyl chains changes their selectivity, and enhances their utility [96, 98]. Butyl-, octyl- and dodecyl- and tetradecyl-modified dendrimers have been synthesized at different generations (Fig. 12). Tetradecyl-modified dendrimers have limited solubility in aqueous and methanol-

modified buffers at similar degrees of alkylation as the others. Another generation can be added to the dendrimer after alkylation, yielding a polymer with the alkyl groups partially embedded in the core.

All of the modified dendrimers retain greater recognition of the backbone structure of analytes, and thus provide unique selectivity relative to SDS micelles. This is demonstrated in Fig. 13, where the alkylated dendrimers show selectivity between that of SDS micelles and the unmodified dendrimers. There is a progression from dendrimer-like selectivity to SDS micelle-like selectivity as the chain length of the alkyl modifiers becomes greater. Alkylated dendrimers provided differentiation between the benzene and naphthalene derivatives than SDS micelles, and greater differentiation between functional groups than the parent dendrimer. Dendrimers modified with dodecyl chains show greater utility than those modified with octyl chains, due to greater recognition of analyte functionality and hydrophobicity. Likewise, as the alkyl chain length is increased, greater methylene selectivity is obtained (Table 1). The dodecyl-modified dendrimer shows similar methylene selectivity to SDS micelles in aqueous buffers, and retains greater methylene selectivity in methanol-modified buffers. The methylene selectivity in aqueous media correlates well with the hydrophobicity as measured by the fluorescence of pyrene. The separation of alkyl phenyl ketones using SDS micelles shows an abnormal elution profile in 40% methanol, and very narrow migration windows above 40% methanol, while the dodecyl-modified dendrimer shows consistent separation in up to 80% methanol [98]. No dramatic selectivity differences were observed between the ethylenediamine and *p*-xylenediamine cores, or as a function of the degree of alkylation [96].

PAHs were also separated using the alkyl-derivatized dendrimers [98]. The octyl dendrimer was used in up to 80% methanol and, like the parent dendrimer, planar rigid PAH were preferentially retained relative to SDS micelles. The dodecyl derivative was employed in up to 90% methanol, and displayed selectivity more like that of SDS micelles. Sixteen priority pollutant PAHs were separated using the dodecyl-modified dendrimer in 65% methanol. Plots of log *k* *vs.* percent methanol on the dodecyl-modified dendrimer were nearly linear, implying a stable structure without conformational or structural change in the presence of high concentrations of methanol [98]. However, an increase in the electrophoretic mobility of the modified dendrimers is observed around 40–60% methanol [96]. This is similar to what was reported with SUA-modified PAA phases, and indicates a conformational or structural change in the polymer backbone or in the attached alkyl groups at higher organic modifier concentrations.

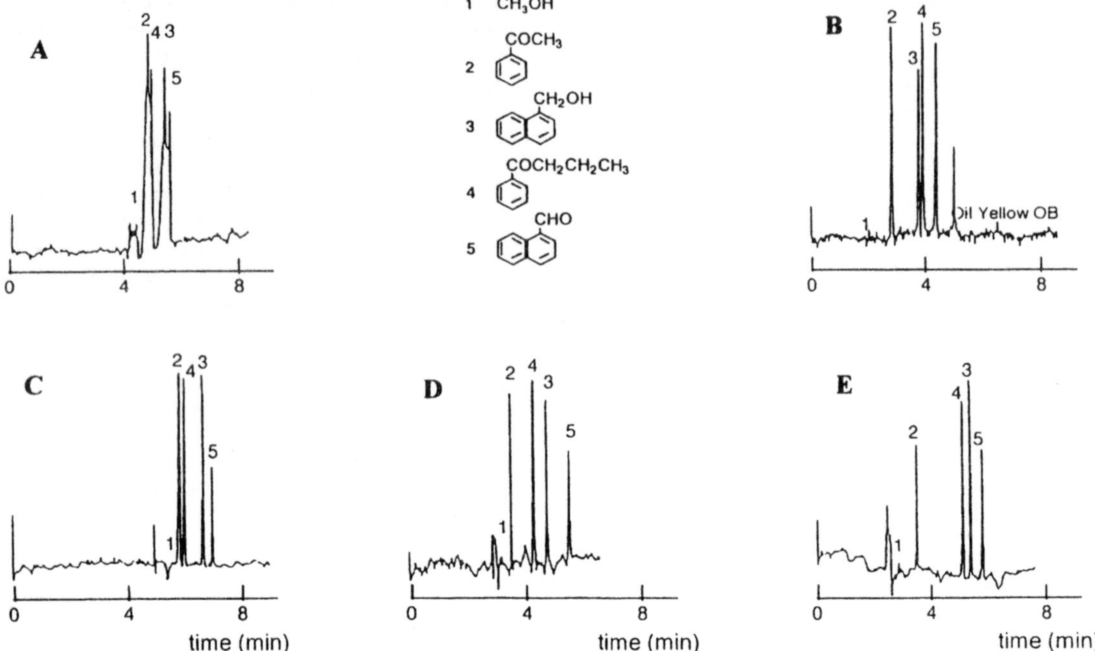

Figure 13. Comparison of SBD pseudostationary phases for the separation of benzene and naphthalene compounds using 5 mM solutions of G=3.5 dendrimers of different pendent alkyl chain length. (A) Unmodified dendrimer; (B) 30 mM SDS; (C) SBD-C$_4$; (D) SBD-C$_8$; (E) SBD-C$_{12}$. 1–5, as noted; 6, Oil Yellow OB. Reprinted from [96], with permission.

Muijselaar *et al.* [100] have studied the selectivity of a diaminobutane-based poly(propylenimine) dendrimer for the separation of substituted benzene compounds. The dendrimer was found to have substantially different selectivity relative to micelles of SDS. The differences in selectivity are explained by the greater hydrogen bond-accepting capabilities of the internal tertiary amines, which led to greater interaction with hydrogen bond-donating compounds such as hydroquinone and resorcinol. Stathakis *et al.* [101] have used anionic and cationic poly(amidoamine) dendrimers as buffer additives to alter the selectivity of separations of chicken sarcoplasmic proteins. Very low concentrations (10^{-4}% w/v) of anionic dendrimer were observed to improve resolution substantially. A cationic dendrimer also improved resolution, but at the cost of very long analysis times. The improvement in separation was attributed to either ionic interactions or hydrophobic partitioning. In either case, separation efficiency was rather poor.

Commercial dendrimers modified with sulfonic acid terminal groups have been used to separate two-dimethyl-phenol isomers [102]. The optimal separation using dendrim-

ers was better than with SDS micelles. Better separation efficiency was observed with higher concentrations of dendrimer. Gao *et al.* [103] have studied the properties and utility of commercially available poly(amidoamine) dendrimers with ethylene diamine cores for the separation of phenylglycine, tyrosine, phenylalanine, ditrydroxyphenylalanine (DOPA), homophenyalanine and methylDOPA. Characterization of the materials by capillary electrophoresis and MALDI-TOF showed that the dendrimer materials were actually a complex mixture rather than uniform monomolecular materials. Relatively poor selectivity and high baseline noise was observed at pH 7. At pH 2.5 the dendrimers were still anionic and the cationic analytes were carried toward the anode with strong ionic association and with better selectivity. Reduced background absorbance and better baselines were also observed at the lower pH.

6 Additional applications

Three studies have reported the utility of polymeric phases for application with mass spectrometric detection with mixed results [41–43]. In the first study, up to 2% purified BBMA was used with ESI mass spectrometry for the separation and detection of several analytes, including sulfamides [42]. A stable electrospray was obtained with 2% BBMA, but the signal was diminished. Loss of signal was not as great as with SDS, and the polymer did not interfere with detection at *m/q* less than 1000. In a second report, Ozaki and Terabe [43] utilized BBMA and ESI-mass spectrometry for the analysis of pharmaceutical compounds. In this case, they found that BBMA did not adversely affect the signal at concentrations below 0.5%, but signal was diminished by nearly 50% at a concentration of 1%. This was better than SDS micelles, but was considered severe enough that a partial filling approach was used. The polymer of SUS has also been used for mass spectrometric detection of tricyclic antidepressants and β-adrenic blocker drugs [41]. The signal was severely diminished at a polySUS concentration of 0.5%, but better separations and a sufficient signal were obtained at a concentration of 0.1%. The authors point out that to utilize this approach, a compromise between signal and resolution may have to be made. The suppression of signals by SDS in ESI is related to the large surface excess of the surfactant [6]. Pseudostationary phases that are not surface-active should perform better for this application.

Increases in sensitivity have been reported using normal stacking with polymeric pseudostationary phases [104]. The use of BBMA and polySUA led to better stacking efficiency and sharper analyte peaks than SDS micelles. The authors propose that this better performance may be related to the more stable molecular structure of the polymeric phases.

7 Conclusions and future directions

The results to date have demonstrated the utility of polymeric phases as pseudostationary phases for MEKC. Some advantages over conventional micelles have been demonstrated, particularly for the analysis of hydrophobic compounds and chiral compounds. In spite of early concerns about mass transfer and polydispersity and their effects on efficiency, separations using the polymers do not necessarily suffer from reduced plate counts. An exception may be hydrophilic polymers lacking sufficient hydrophobic domains to solvate hydrophobic molecules. These polymers show some tendency to suffer from sample overload and, in the case of siloxane polymers, possibly slow mass transfer. Overall, however, the chromatographic performance and chemical selectivity of the polymeric pseudostationary phases is good, with the polymers often providing unique selectivity and broad migration range.

The chemical selectivity of the polymeric pseudostationary phases is invariably different from that of SDS micelles. It seems likely that the differences in selectivity

can be traced to the more rigid structure of the polymer surfactants, which affects structural selectivity and limits the ability of the polymer to create a large and unstructured hydrophobic domain favorable for the solvation of flexible hydrophobic compounds such as alkyl phenyl ketones. As shown in Table 1, the methylene selectivity of the polymeric phases is often lower than SDS micelles in aqueous systems. Where the methylene selectivity is the same or higher, the polymer usually has longer alkyl chains. The methylene selectivity is, however, maintained to a greater extent in organic-modified buffers due to the greater stability of the polymeric phases. Another common theme evident in Table 1 is that the polymers have greater resistance to cavity formation than SDS or even fluorocarbon micelles. This is also most likely a result of the more rigid structure of the polymeric materials.

Polymers of sodium undecylenate, Elvacite 2669, modified polyallylamine polymers and alkyl-modified dendrimers have all been observed to undergo a conformational or structural change in organic-modified media. Although the primary structure of the polymers is covalently stabilized, a balance between hydrophobic and polar interactions determines the conformation of the polymers in solution. The polymers can also associate *via* intermolecular interactions. Adding organic solvents alters the solvation of the polymer, permitting the polymer to assume a different conformation and disrupting intermolecular interaction. This change in structure affects the migration times and migration range, and yields nonlinear plots of log *k* *vs.* percent organic modifier, but does not dramatically alter the chemical selectivity. However, it is possible that a structural change alters the structural selectivity of the SUS polymer, causing different selectivity in acetonitrile *vs.* methanol-modified buffers.

The critical parameters which affect the performance and selectivity of the polymers studied to date include the alkyl chain length and density of alkyl chains on the polymer backbone, the chemistry and density of the ionic groups, the nature of the polymer backbone, and the composition of the buffer medium. The chemistry of the ionic head group made little or no difference in the selectivity of polymers of SUA, SUS, SUT and SUP, or between BBMA and BMAC, but substitution with chiral head groups does permit chiral separations. The chemistry and structure of the polymer backbone can have a dramatic effect on selectivity, as was demonstrated with dendritic polymers *vs.* linear polymers and siloxane phases *vs.* SDS micelles.

The degree of substitution of the polymer backbone affects the capacity, mobility, and methylene selectivity, but does not have an effect on the overall selectivity of

Electrophoresis 2000, *21*, 4054–4072

the polymers. The molecular weight of the phases appears to have minimal effects on the selectivity or affinity of the phases, and thus polydispersity is not considered to be a critical factor in determining the efficiency of the separations. It has been demonstrated with polymer micelles, acrylate copolymers, dendrimers, siloxane polymers and polyallylamine phases that polymeric pseudostationary phases with unique selectivities can be synthesized using a single polymer backbone. Furthermore, mixtures of polymers with different selectivities can be used to achieve separations with intermediate and more-or-less predictable selectivity.

One of the advantages of the MEKC and polymer EKC approaches is that pseudostationary phases are easily introduced and replaced in the capillary or a micromachined channel. Phases with different selectivity can be easily introduced and employed to achieve desired separations. Additionally, the use of polymeric pseudostationary phases with appropriate selectivity and mobility, combined with optimization of the buffer composition, allows one to fully utilize the advantage of electrokinetic chromatography of having wide peak spacing at the beginning and a relatively narrow peak spacing at the end of a separation. This characteristic of electrokinetic chromatography is due to the mobility of the pseudostationary phase, and permits the separation of a mixture having a wide range of hydrophobic properties in a short time without the need for gradient elution.

The studies to date have employed a relatively small number of polymer structures for a limited number of separations. The number and types of polymers employed, however, is steadily increasing. Further work in this area should concentrate on the introduction and characterization of new polymeric pseudostationary phases, fundamental characterization of the interactions between polymers and analytes as well as the effect of polymer structure and solvation on these interactions, refinement of the use of mass spectrometric detection for MEKC, and application of polymers in areas where packed capillaries are difficult to prepare or maintain.

It has been demonstrated that it is possible to synthesize polymers with altered and unique selectivities. This avenue should be further explored with the synthesis and introduction of polymers with unique selectivity and high electrophoretic mobility. An obvious avenue of work is the synthesis and application of additional polymers with chiral selectivity. As a long-term effort, it may be possible to employ molecularly imprinted polymers for extremely selective separations. Polymeric phases can currently be seen as specialty phases, which will see application only in instances where unique chemical selectivity or high pseudostationary phase stability is a requirement. This is

due at least in part to the lack of commercial sources for most of the more selective and efficient polymeric phases. The commercial availability of dendrimers may make their use more commonplace, but as is evidenced by the work of Gao *et al.* [103], better quality control of these materials will be necessary before widespread application can be expected.

Received May 30, 2000

8 References

[1] Terabe, S., Otsuka, K., Ichikawa, A., Tsuchiya, A., Ando, T., *Anal. Chem.* 1984, *56*, 111–113.

[2] St. Claire III, R. L., *Anal. Chem.* 1996, *68*, R569–R586.

[3] Vindevogel, J., Sandra, P., *Introduction to Micellar Electrokinetic Chromatography*, Huthig, Heidelberg 1992.

[4] Beale, S. C., *Anal. Chem.* 1998, *70*, R279–R300.

[5] Quirino, J. P., Terabe, S., *J. Chromatogr. A* 1999, *856*, 465–482.

[6] Rundlett, K. L., Armstrong, D. W., *Anal. Chem.* 1996, *68*, 3493–3497.

[7] Nishi, H., Nakamura, K., Nakai, H., Sato, T., *J. Chromatogr. A* 1994, *678*, 333–342.

[8] Ingelse, B. A., Everaerts, F. M., Desiderio, C., Fanali, S., *J. Chromatogr. A* 1995, *709*, 89–98.

[9] Fanali, S., Aturki, Z., *Electrophoresis* 1995, *16*, 1505–1509.

[10] Schützner, W., Fanali, S., Rizzi, A., Kenndler, E., *J. Chromatogr.* 1993, *639*, 375–378.

[11] Schützner, W., Caponecchi, G., Fanali, S., Rizzi, A., Kenndler, E., *Electrophoresis* 1994, *15*, 769–773.

[12] Blatny, P., Fischer, C.-H., Kenndler, E., *Fresenius J. Anal. Chem.* 1995, *352*, 712–714.

[13] Schützner, W., Fanali, S., Rizzi, A., Kenndler, E., *Anal. Chem.* 1995, *67*, 3866–3870.

[14] Blatny, P., Fisher, C.-H., Rizzi, A., Kenndler, E., *J. Chromatogr. A* 1995, *717*, 157–166.

[15] Schützner, W., Fanali, S., Rizzi, A., Kenndler, E., *J. Chromatogr. A* 1996, *719*, 411–420.

[16] Wu, N., Barker, G. E., Huie, C. W., *J. Chromatogr. A* 1994, *659*, 435–442.

[17] Terabe, S., Ozaki, H., Tanaka, Y., *J. Chin. Chem. Soc.* 1994, *41*, 251–257.

[18] Tanaka, Y., Terabe, S., *J. Chromatogr. A* 1995, *694*, 277–284.

[19] Kilár, F., *Electrophoresis* 1996, *17*, 1950–1953.

[20] Schmid, M. G., Gubitz, G., Kilár, F., *Electrophoresis* 1998, *19*, 282–287.

[21] Tanaka, Y., Terabe, S., *Chromatographia* 1999, *49*, 489–495.

[22] Fanali, S., Desiderio, C., *J. High Resolut. Chromatogr.* 1996, *19*, 322–326.

[23] Izumoto, S., Nishi, H., *Bunseki Kagaku* 1998, *47*, 739–746.

[24] Bachmann, K., Bazzanella, A., Haag, I., Han, K. Y., *Fresenius J. Anal. Chem.* 1997, *357*, 32–36.

[25] Spencer, B. J., Zhang, W. B., Purdy, W. C., *Electrophoresis* 1997, *18*, 736–744.

[26] Zhu, W. H., Vigh, G., *Anal. Chem.* 2000, *72*, 310–317.

[27] Tanaka, Y., Yanagawa, M., Terabe, S., *J. High Resolut. Chromatogr.* 1996, *19*, 421–433.

[28] Hamilton, K. Y., Warner, I. M., *Abstracts of Papers of the American Chemical Society* 1999, p. 92-Anyl.

[29] Hamilton, K. Y., Ruguti, J. K., Pena, M. S., Zhang, Y. L., Warner, I. M., *Abstracts of Papers of the American Chemical Society* 1998, p. 4-Pres.

[30] Sepaniak, M. J., Whitaker, K. W., Sun, S., Fox, S. B., *Abstracts of Papers of the American Chemical Society* 1997, p. 151-Coll.

[31] Zhao, T., Hu, X. B., Cheng, J. K., Lu, X. R., *J. Liq. Chromatogr. Relat. Technol.* 1998, *21*, 3111–3124.

[32] Bazzanella, A., Bachmann, K., Milbradt, R., Bohmer, V., Vogt, W., *Electrophoresis* 1999, *20*, 92–99.

[33] Bazzanella, A., Morbel, H., Bachmann, K., Milbradt, R., Bohmer, V., Vogt, W., *J. Chromatogr. A* 1997, *792*, 143–149.

[34] Zhao, T., Hu, X., Cheng, J., Lu, X., *Anal. Chim. Acta* 1998, *358*, 263–268.

[35] Warner, I. M., *Abstracts of Papers of the American Chemical Society* 1997, p. 161-Coll.

[36] Anton, P., Koberle, P., Laschewsky, A., *Makromol. Chem.* 1993, *194*, 1–27.

[37] Laschewsky, A., *Adv. Polymer Sci.* 1995, *124*, 3–85.

[38] Palmer, C. P., Tanaka, N., *J. Chromatogr. A* 1997, *792*, 105–124.

[39] Palmer, C. P., *J. Chromatogr. A* 1997, *780*, 75–92.

[40] Ozaki, H., *Bunseki Kagaku* 1999, *48*, 1023–1024.

[41] Lu, W. Z., Shamsi, S. A., Mccarley, T. D., Warner, I. M., *Electrophoresis* 1998, *19*, 2193–2199.

[42] Ozaki, H., Itou, N., Terabe, S., Takada, Y., Sakairi, M., Koizumi, H., *J. Chromatogr. A* 1995, *716*, 69–79.

[43] Ozaki, H., Terabe, S., *J. Chromatogr. A* 1998, *794*, 317–325.

[44] Ozaki, H., Ichihara, A., Terabe, S., *J. Chromatogr. A* 1995, *709*, 3–10.

[45] Wang, J., Warner, I. M., *J. Chromatogr. A* 1995, *711*, 297–304.

[46] Wang, J., Warner, I. M., *Anal. Chem.* 1994, *66*, 117–123.

[47] Williams, C. C., Shamsi, S. A., Warner, I. M., *Adv. Chromatogr.* 1997, *37*, 363–423.

[48] Haddadian, F., Billiot, E. J., Shamsi, S. A., Warner, I. M., *J. Chromatogr. A* 1999, *858*, 219–227.

[49] Yarabe, H. H., Shamsi, S. A., Warner, I. M., *Anal. Chem.* 1999, *71*, 3992–3999.

[50] Billiot, E., Warner, I. M., *Anal. Chem.* 2000, *72*, 1740–1748.

[51] Shamsi, S. A., Macossay, J., Warner, I. M., *Anal. Chem.* 1997, *69*, 2980–2987.

[52] Billiot, E., Thibodeaux, S., Shamsi, S., Warner, I. M., *Anal. Chem.* 1999, *71*, 4044–4049.

[53] Billiot, E., Agbaria, R. A., Thibodeaux, S., Shamsi, S., Warner, I. M., *Anal. Chem.* 1999, *71*, 1252–1256.

[54] Billiot, E., Macossay, J., Thibodeaux, S., Shamsi, S. A., Warner, I. M., *Anal. Chem.* 1998, *70* 1375–1381.

[55] Dobashi, A., Hamada, M., Dobashi, Y., *Anal. Chem.* 1995, *67*, 3011–3017.

[56] Haddadian, F., Shamsi, S. A., Warner, I. M., *Electrophoresis* 1999, *20*, 3011–3026.

[57] Tanaka, N., Nakagawa, K., Iwasaki, H., Hosoya, K., Kimata, K., Araki, T., Patterson, D. G., *J. Chromatogr. A* 1997, *781*, 139–150.

[58] Potocek, B., Maichel, B., Gaš, B., Chiari, M., Kenndler, E., *J. Chromatogr. A* 1998, *798*, 269–273.

[59] Kolb, S., Kutter, J. P., Welsch, T., *J. Chromatogr. A* 1997, *792*, 151–156.

[60] Maichel, B., Potocek, B., Gaš, B., Kenndler, E., *J. Chromatogr. A* 1999, *853*, 121–129.

[61] Maichel, B., Potocek, B., Gaš, B., Chiari, M., Kenndler, E., *Electrophoresis* 1998, *19*, 2124–2128.

[62] Palmer, C. P., Mcnair, H. M., *J. Microcol. Sep.* 1992, *4*, 509–514.

[63] Palmer, C. P., *International Symposium on Capillary Chromatography and Electrophoresis, Park City, UT* 1999.

[64] Tanaka, N., Nakagawa, K., Hosoya, K., Palmer, C. P., Kunugi, S., *J. Chromatogr. A* 1998, *802*, 23–33.

[65] Palmer, C. P., Khaled, M. Y., Mcnair, H. M., *J. High Resolut. Chromatogr.* 1992, *15*, 756–762.

[66] Palmer, C. P., Tellman, K. T., *J. Microcol. Sep.* 1999, *11*, 185–191.

[67] Moy, T. W., Ferguson, P. L., Grange, A. H., Matchett, W. H., Kelliher, V. A., Brumley, W. C., Glassman, J., Farley, J. W., *Electrophoresis* 1998, *19*, 2090–2094.

[68] Palmer, C. P., Terabe, S., *J. Microcol. Sep.* 1996, *8*, 115–121.

[69] Palmer, C. P., Terabe, S., *Anal. Chem.* 1997, *69*, 1852–1860.

[70] Shamsi, S. A., Akbay, C., Warner, I. M., *Anal. Chem.* 1998, *70*, 3078–3083.

[71] Akbay, C., Warner, I. M., Shamsi, S. A., *Electrophoresis* 1999, *20*, 145–151.

[72] Tellman, K. T., Palmer, C. P., *Electrophoresis* 1999, *20*, 152–161.

[73] Hara, S., Dobashi, A., Japan, Patent # 92, 149, 205, 1992.

[74] Hara, S., Dobashi, A., Japan, Patent # 92, 149, 206, 1992.

[75] Chu, D. Y., Thomas, T. K., *Macromolecules* 1991, *24*, 2212–2216.

[76] Ozaki, H., Terabe, S., Ichihara, A., *J. Chromatogr. A* 1994, *680*, 117–123.

[77] Yang, S. Y., Bumgarner, J. G., Khaledi, M. G., *J. Chromatogr. A* 1996, *738*, 265–274.

[78] Yang, S. Y., Bumgarner, J. G., Khaledi, M. G., *J. High Resolut. Chromatogr.* 1995, *18*, 443–445.

[79] Dharmawardana, U. R., Christian, S. D., Tucker, E. E., Taylor, R. W., Scamehorn, J. F., *Langmuir* 1993, *9*, 2258–2263.

[80] Palepu, R., Reinsborough, V. C., *Can. J. Chem.* 1988, *66*, 325–328.

[81] Satake, L., Yoshida, S., Hayakawa, K., Maeda, T., Kusumoto, Y., *Bull. Chem. Soc. Jpn.* 1986, *59*, 3991–3993.

[82] Satake, L., Ikenou, T., Takeshita, T., Hayakawa, K., Maeda, T., *Bull. Chem. Soc. Jpn.* 1985, *58*, 2746–2750.

[83] Aguilar, M., Farran, A., Serra, C., Sepaniak, M. J., Whitaker, K. W., *J. Chromatogr. A* 1997, *778*, 201–205.

Electrophoresis 2000, *21*, 4054–4072

[84] Wiedmer, S. K., Tenhu, H., Vastamakì, P., Riekkola, M.-L., *J. Microcol. Sep.* 1998, *10*, 557–565.

[85] Palmer, C. P., *HPCE '97*, Anaheim, CA 1997.

[86] Tanaka, N., Thornton, E. R., *J. Am. Chem. Soc.* 1977, *99*, 7300–7307.

[87] Tanaka, N., Nakagawa, K., Nagayama, H., Hosoya, K., Ikegami, T., Itaya, A., Shibayama, M., *J. Chromatogr. A* 1999, *836*, 295–303.

[88] Morishima, Y., Kobayashi, T., Furui, T., Nozakura, K., *Macromolecules* 1987, *20*, 1707–1712.

[89] Morishima, Y., Kobayashi, T., Nozakura, K., *Polymer J.* 1989, *21*, 267–274.

[90] Chen, T., Palmer, C. P., *Electrophoresis* 1999, *20*, 2412–2419.

[91] Peterson, D., Palmer, C. P., *Electrophoresis* 2000, *21*, 3174–3180.

[92] Erim, F. B., *J. Chromatogr. A* 1997, *768*, 161–167.

[93] Potocek, B., Chmela, E., Maichel, B., Tesarova, E., Kenndler, E., Gaš, B., *Anal. Chem.* 2000, *72*, 74–80.

[94] Kuzdzal, S. A., Monnig, C. A., Newkome, G. R., Moorefield, C. N., *J. Am. Chem. Soc.* 1997, *119*, 2255–2261.

[95] Kuzdzal, S. A., Monnig, C. A., Newkome, G. R., Moorefield, C. N., *J. Chem. Soc. Chem. Commun.* 1994, 2139–2140.

[96] Tanaka, N., Iwasaki, H., Fukutome, T., Hosoya, K., Araki, T., *J. High Resolut. Chromatogr.* 1997, *20*, 529–538.

[97] Tanaka, N., Fukutome, T., Tanigawa, T., Hosoya, K., Kimata, K., Araki, T., Unger, K. K., *J. Chromatogr. A* 1995, *699*, 331–341.

[98] Tanaka, N., Fukutome, T., Hosoya, K., Kimata, K., Araki, T., *J. Chromatogr. A* 1995, *716*, 57–67.

[99] Tanaka, N., Tanigawa, T., Hosoya, K., Kimata, T., Araki, T., Terabe, S., *Chem. Lett.* 1992, *959–962*.

[100] Muijselaar, P. G. H. M., Claessens, H. A., Cramers, C. A. J. J. F. G. A., Meijer, E. W., de Brababander-van den Berg, E. M. M., van der Wal, S., *J. High Resolut. Chromatogr.* 1995, 121–123.

[101] Stathakis, C., Arriaga, E. A., Dovichi, N. J., *J. Chromatogr. A* 1998, *817*, 233–238.

[102] Gray, A. L., Hsu, J. T., *J. Chromatogr. A* 1998, *824*, 119–124.

[103] Gao, H., Carlson, J., Stalcup, A. M., Heineman, W. R., *J. Chromatogr. Sci.* 1998, *36*, 146–154.

[104] Quirino, J. P., Terabe, S., *J. Chromatogr. A* 1997, *781*, 119–128.

Review

Hanfa Zou
Mingliang Ye

National Chromatographic
R & A Center, Dalian Institute
of Chemical Physics,
The Chinese Academy of
Sciences,
Dalian, China

Capillary electrochromatography with physically and dynamically absorbed stationary phases

Adsorption is always considered a troublesome effect in capillary electrophoresis (CE) and capillary electrochromatography (CEC). However, the adsorption effect can also be exploited to prepare or optimize the stationary phase in CEC. Compared with the chemical synthesis of new stationary phase materials for CEC, this method is simpler and more convenient. This review is focused on CEC with physically and dynamically adsorbed stationary phases. Separation of some acidic, basic and neutral solutes as well as enantiomers in CEC with dynamically adsorbed stationary phases are presented. The theory for the migration of charged solutes and the stationary phases currently used in CEC are also briefly reviewed.

Keywords: Capillary electrochromatography / Stationary phase / Physical adsorption / Dynamic coating / Additive agent / Review

EL 4195

Contents

Correspondence: Dr. Hanfa Zou, National Chromatographic R & A Center, Dalian Institute of Chemical Physics, The Chinese Academy of Sciences, Dalian 116011, China
E-mail: zouhfa@mail.dlptt.ln.cn
Fax: +86-411-3693407

Abbreviations: CSP, chiral stationary phase; **DMS-CEC**, CEC on dynamically modified stationary phase; **HAC-TEA**, acetic acid-triethylamine; **ODS**, octadecyl silica; **OT-CEC**, open-tubular CEC; **PC-CEC**, packed-column CEC; **SAX**, strong anion-exchange phase; **S-CD**, sulfated β-CD; **SCX**, strong cation-exchange phase

1 Introduction

Capillary electrochromatography (CEC), which was first proposed by Pretorius *et al.* [1] in 1974, has experienced explosive growth in recent years. CEC uses electroosmotic flow (EOF) to transport the mobile phase through a capillary column packed with stationary phase material or coated with stationary phase on the capillary wall. According to the existing state of the stationary phase, CEC can be classified into two branches – open-tubular CEC (OT-CEC) and packed-column CEC (PC-CEC) [2, 3]. In OT-CEC, the stationary phase is attached to the capillary wall, while in PC-CEC the stationary phase material is packed into the capillary column. Neutral solutes will be separated mainly based on their different interaction with the stationary and mobile phases, while charged solutes will be separated based on both chromatographic and electrophoretic mechanisms. In fact, CEC is the hybrid technique of high-performance liquid chromatography (HPLC) and capillary electrophoresis (CE).

Adoption of EOF in CEC results in two important advantages for CEC over conventional HPLC. First, the nearly flat profile of EOF reduces the band broadening caused by trans-channel diffusion and eddy diffusion, and thus higher column efficiency can be obtained in CEC than that in HPLC. Second, the EOF is independent of particle size and does not generate back pressure; therefore smaller

Electrophoresis 2000, *21*, 4073–4095

particles and longer columns can be used in CEC than in HPLC, and the plate number can be further increased. Compared with CE, many kinds of stationary phases can be used in CEC, and thus higher selectivity can be obtained with this mode. As a miniaturized separation technique, the rewards for using CEC include high separation efficiency, low eluent and sample consumption, high mass sensitivity, and low operational cost. All of these features have made CEC an attractive separation technique.

Adsorption at the capillary wall was usually considered a troublesome effect in capillary zone electrophoresis (CZE), especially for the separations of proteins and peptides. The adsorption of solute on the capillary wall may cause many serious problems, for example, tailing of the peak, instability of the baseline, variance of migration time, loss of efficiency and recovery, and even reduction of the capillary life. Therefore, it is necessary to minimize the adsorption effect in CE. The adsorption in CEC is also considered a problem. Gillott *et al.* [4] found that the repeatability for separation of basic compounds in CEC with bare silica at neutral pH was not good. The elution time of the bases was found to increase with repetitive injections due to their adsorption onto the surface of the silica. The same phenomenon was also found in separation of anionic compounds in CEC with strong anion-exchange (SAX) packing [5]. The EOF was observed to decrease with repetitive injections of anionic samples due to the adsorption of these solutes onto the positively charged groups. Therefore, the adsorption effect in CEC should also be avoided. However, it was recently reported that the adsorption effect could be exploited to prepare a stationary phase both for OT-CEC and PC-CEC [6–9]. Compared with the preparation of a stationary phase by chemical synthesis, the attractive advantage of this method is its simplicity of the preparation procedure.

Chromatography with dynamically adsorbed stationary phases on silica in HPLC was first reported by Ghaemi and Wall [10]. In this system, the HPLC column was packed with bare silica and a certein amount of long-chain quaternary ammonium salts such as cetyltrimethylammonium bromide (CTAB) was added into the mobile phase [11]. CTAB ions have considerable affinity to ionized silanol groups because of the strong electrostatic attraction between the positive quaternary ammonium group and the negative silanol group. Therefore, CTAB ions will be adsorbed onto the silica surface in such a way that the C_{16} carbon chain is pointing away from the surface. The CTAB ions adsorbed on the silica surface form an apolar layer similar to that of a chemically bonded octadecyl silica (ODS) material. The obvious difference is that the apolar layer is formed through a dynamic adsorption proc-

ess, since CTAB ions adsorbed will constantly interchange with CTAB ions present in the mobile phase. The separation mechanism for neutral compounds in this system is based on the reversed-phase partition between the adsorbed stationary phase and mobile phase, while that for ionic compounds is relatively complex [11]. Anionic compounds are retained as ion pairs by a reversed-phase mechanism, while cationic compounds are retained partly by a reversed-phase mechanism and partly by cation exchange. The bare silica is modified by the dynamic adsorption of CTAB. This method is also termed chromatography on dynamically modified silica [11].

Although the performing of OT-CEC with a dynamically adsorbed stationary phase was reported in 1993 [12], the idea to perform PC-CEC with a dynamically adsorbed stationary phase was recently realized [13–15]. This novel mode of CEC was established similar to that in HPLC, the packing materials can be bare silica, strong cation-exchange (SCX) phase, SAX phase, the dynamically modifying agents can be surfactants like CTAB, and ionic chiral selectors like sulfated β-CD (S-CD) [13–15]. The dynamically adsorbed agents, such as CTAB, S-CD, can be used for separation of neutral, acidic and basic compounds as well as enantiomers. The original stationary phase is modified by the dynamic adsorption of these agents. This novel mode of CEC is also called CEC on dynamically modified stationary phase (DMS-CEC).

Several reviews [2, 3, 16–19] have been published in recent years covering the general technique of CEC. And some special aspects of CEC, such as the analysis of acidic compounds [20], chiral separation [21–23] and the packing materials in CEC [24, 25] were also reviewed by several authors. Here we will focus on CEC with physically and dynamically adsorbed stationary phases. The theory for the migration of charged solutes in CEC and the stationary phase used in CEC are also briefly reviewed. Separation of the neutral, basic, and acidic compounds as well as enantiomers are demonstrated in this review.

2 Retention of solutes in CEC

The chromatographic retention factor k' is always used in describing the migration process in HPLC, and can also be used for neutral solutes in CEC because their retention is based on a purely chromatographic mechanism. However, it is not suitable to be used for charged solutes in CEC because of the coupled electrophoretic migration. Therefore, it is necessary to define an equivalent parameter to describe the migration process in CEC [16]. This parameter should include both an electrophoretic and a chromatographic process because they are the characteristic components in CEC. Rathore and Horváth [26] have defined the electrochromatographic retention factor (k^*) as the ratio of the separative and nonseparative virtual migration length for describing the migration process. The relationship between this parameter with the retention factor caused by chromatography alone, the electrophoretic mobility of the solute and the electroosmotic mobility was obtained from a unified theory using virtual migration distances [26]. The retention factor in HPLC can be easily calculated from the migration time of solute and the void time, which can be directly obtained from a chromatogram. Similarly, the k^* in CEC was also defined by the following equation by some authors [27–29]:

$$k^* = (t_r - t_0)/t_0 \tag{1}$$

where t_r is the migration time of a solute, and t_0 is the migration time of a neutral and chromatographically unretained solute. The electrochromatographic retention factor k^* has been used to describe the migration process of charged solutes in both reversed-phase CEC [27, 29] and ion-exchange CEC [30, 31].

The separation mechanism of charged solutes in HPLC [32, 33] and micellar electrokinetic capillary electrochromatography (MEKC) [34, 35] has been studied by phenomenological approach and the results have been experimentally verified. This approach assumes that the migration parameter of an ionic compound such as capacity factor (k'), migration velocity (μ_m) and electrophoretic mobility (μ_{ep}) is the weighted average of the parameter of the compound in different forms. This approach can also be used to study the migration behavior of charged solutes in CEC [28]. Figure 1 shows the migration process of charged solutes in CEC, which indicates that both chromatographic and electrophoretic mechanisms make contribution to the migration behavior of the charged solutes. According to the definition of the capacity factor (k') in HPLC, the fractions of solute A⁻ in the stationary phase (F_s) and in the mobile phase (F_m) can be expressed as:

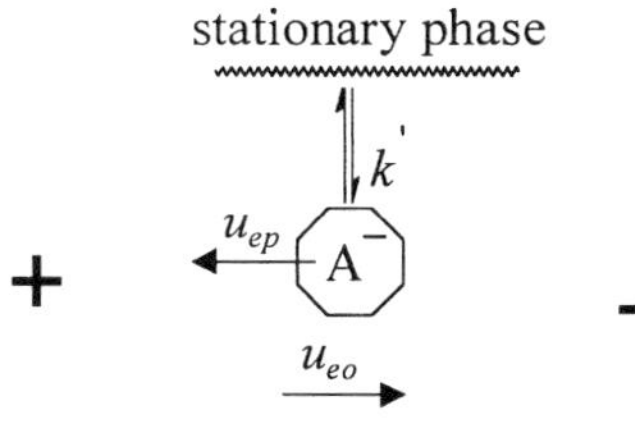

Figure 1. The migration process of charged solutes in CEC. Reprinted from [28], with permission.

$$F_s = k'/(1 + k') \tag{2}$$

$$F_m = 1/(1 + k') \tag{3}$$

The migration velocity of the solute in the mobile phase is the sum of the migration velocities arising from the EOF and the electrophoretic migration and thus is given by:

$$u_m = u_{eo} + u_{ep} = \mu_{eo}E + \mu_{ep}E \tag{4}$$

where u_{eo}, u_{ep} are the linear velocities of EOF and electrophoresis, μ_{eo} and μ_{ep} are the EOF mobility and electrophoretic mobility, and E is the applied electrical field strength. The charged solutes in the stationary phase may also migrate because of the applied electrical field. If the migration of the solute in the stationary phase is considered, then the migration velocity of the solute in CEC (u_{mig}) is the weighted average of the migration velocities in the mobile phase (u_m) and in the stationary phase (u_s) and can be given by

$$u_{mig} = F_s u_s + F_m u_m \tag{5}$$

Substituting Eq. (2)–(4) into Eq. (5), we have

$$u_{mig} = \frac{k' u_s}{1 + k'} + \frac{u_{ep} + u_{eo}}{1 + k'} \tag{6}$$

t_r and t_0 in Eq. (1) can be expressed by

$$t_r = L_{id}/u_{mig} \tag{7}$$

$$t_0 = L_{id}/u_{eo} \tag{8}$$

where L_{id} is the length of capillary from injection end to detection window. Substituting Eq. (7) and (8) into Eq. (1), k^* is given by

$$k^* = u_{eo}/u_{mig} - 1 \tag{9}$$

Substituting Eq. (6) into Eq. (9), we have

$$k^* = \frac{1}{\dfrac{1 + \mu_{ep}/\mu_{eo}}{1 + k'} + \dfrac{k'}{1 + k'}\mu_s/\mu_{eo}} - 1 \tag{10}$$

where μ_s is the electrophoretic mobility of the charged solute in a stationary phase. It is reasonable to assume that the solute migration in a stationary phase is much slower than that in a mobile phase, thereby μ_s in Eq. (10) can be neglected and the following equation can be obtained:

$$k^* = \frac{k' - \mu_{ep}/\mu_{eo}}{1 + \mu_{ep}/\mu_{eo}} \tag{11}$$

This equation is similar to that reported by Wu *et al.* [27], but does not agree with that reported by Rathore and Horváth [26]. The reason may be the different definition of the CEC retention factor.

It can be seen from Eq. (11) that the k^* value reflects the concurrence of the chromatographic and electrophoretic processes. For neutral solutes, μ_{ep} is zero and k^* becomes k', reflecting a purely chromatographic process. However, for charged unretained solutes, k' is zero, reflecting a purely electrophoretic process. If the solutes are charged and retained in the stationary phase, the two processes will both be involved. The value of k' in HPLC with the exception of size-exclusion chromatography is always positive because the retention time of solutes is always greater than the void time. Because of the sign associated with the electrophoretic and electroosmotic mobilities, the value of k^* in CEC can be either positive or negative. For example, if the signs of μ_{ep} and μ_{eo} are the same and the value of k' is not too great, the sign of the k^* can be negative. In other words, if the direction of the electrophoretic migration of a charged solute is the same as that of the EOF and the retention of the solute in the stationary phase is not too strong, the migration time of the solute can be smaller than the void time. Some basic compounds in reversed-phase CEC [4, 36, 37] and some acidic compounds in SAX-CEC [31] were reported to migrate out before the t_0 marker. But if the retention of the charged solute on the stationary phase is relatively strong, the influence of the electrophoretic mechanism on the separation is insignificant. It was reported that the log k^* of peptide and the logarithm of buffer concentration (log [c]) in ion-exchange CEC obey excellent linear relationships just like the relationship of log k' and log [c] in ion-exchange HPLC [30], which indicated that the separation of peptides in ion-exchange CEC was mainly based on the ion-exchange mechanism.

3 CEC stationary phases

3.1 OT-CEC

The stationary phase most often used in OT-CEC is a hydrophobic stationary phase or a reversed-phase stationary phase, where the alkyl chain or hydrophobic polymer is bonded to or coated onto the capillary wall. The first OT-CEC column was fabricated by drawing large bore soda-lime glass tubes, and treating the inner surface with NaOH solution, and then chemically bonding the capillary inner wall with octadecylsilane [38]. A series of hydrocarbons were separated in this system by reversed-phase partitioning mechanism. In order to increase the surface area and phase ratio, Bruin *et al.* [39] reported

the preparation of an OT-CEC column by laying down a porous silica layer on the capillary wall, and then attaching functional groups onto the prepared layer by chemical bonding. Polycyclic aromatic hydrocarbons were separated in this system, and a plate height as small as 1.2 µm was observed for unretained components. Pesek *et al.* [40, 41] demonstrated the separation of tetracyclines and proteins by OT-CEC with etched fused-silica capillaries modified with a C_{18} phase. The etching procedure was performed to increase the surface area of the capillary and to provide more capacity. In brief, the preparation procedure for the above columns involved two steps: (i) increasing the surface area by etching or laying down a porous layer and (ii) attaching functional groups onto the prepared layer through chemical bonding. This two-step procedure is lengthy. To alleviate this drawback, an organic-inorganic hybrid glass material fabricated by the sol-gel process to prepare the stationary phase for OT-CEC [42–44] was introduced. The unique preparation procedure combines the synthesis of a bonded phase and a supporting porous glass matrix in a single step. The OT-CEC column was prepared by the acid-catalyzed hydrolysis and condensation of a precursor mixture containing tetraethoxysilane and *n*-octyltriethoxysilane or tridecafluoro-1,1,2,2-tetrahydrooctyl-1-triethoxysilane. This approach increases the column surface area leading to a higher phase ratio and, at the same time, reduces the overall column preparation time. Separation efficiencies over 280 000 plates/m were typical in this system.

Another way to increase the phase ratio is by adoption of the polymeric stationary phase. Pfeffer and Yeung [45] presented the preparation of an OT-CEC column by coating with a polymer solution of PS-264. However, the velocity for the EOF in this column is very low because of the full coverage of the silanol groups by the stationary phase. In order to increase the EOF velocity, CTAB was added to the mobile phase to increase the zeta potential on the surface of the capillary wall. Tan and Remcho [46] have reported the preparation of bonded linear polymethacrylate stationary phases for OT-CEC. Due to the use of a linear polymer film, the silanol group on the capillary wall was still exposed to the mobile phase, and the EOF in this column was still strong. The separation of four benzoate derivatives was achieved in 8 min, and an efficiency of 270 000 plates/m was observed for the unretained compound. A porous-layer OT-CEC column was introduced by Huang *et al.* [47] for the separation of basic proteins and peptides. The porous layer was highly linked and prepared by *in situ* polymerization of vinylbenzyl chloride and divinylbenzene. The chloromethyl groups at the surface of the porous polymeric support layer reacted with *N,N*-dimethyldodecylamine to form a positively charged chromatographic surface with fixed C_{12} alkyl chains, which resulted in anodic EOF during CEC separation. Although the polymeric stationary phases usually provide good column stability, high phase ratio, and sufficient retention, the drawback of these stationary phases is their poor column efficiency due to the slow diffusivity of solutes in the retentive layers.

OT-CEC has been gradually extended to the separation of chiral compounds. The majority of stationary phases used were CD-based chiral selectors [48–52]. For example, the column coated with Chirasil-DEX (mono-6-*O*-octamethylenepermethyl-β-CD chemically linked to dimethylpolysiloxane) can be used to separate enantiomers of the underivatized nonsteroidal antiinflammatory drugs (NSAIDs) via OT-CEC mode [50]. More recently, it was demonstrated that the same single column could also be used to run enantiomeric separation in gas chromatography, supercritical fluid chromatography, and liquid chromatography besides OT-CEC [51]. It was reported that molecularly imprinted polymer (MIP) was also covalently linked to the capillary wall for enantiomeric separation in OT-CEC [53, 54]. The advantage of this system is that the selectivity can be predetermined. Chiral separation can also be achieved with a cellulose-coated column in OT-CEC mode [55]. Hofstetter *et al.* [56] presented the preparation of a chiral OT-CEC column by chemically bonding the protein, bovine serum albumin (BSA), to the inner capillary surface. Enantiomer separations of a number of dinitrophenyl (DNP)-amino acids and 3-hydroxy-1,4-benzodiazepines were successfully achieved in OT-CEC mode. Recently, Liu *et al.* [6–8] reported the preparation of a chiral OT-CEC column by simple physical adsorption of a basic chiral selector such as protein, peptide, and amino acid to the capillary wall. A series of enantiomers were successfully resolved in this system.

3.2 PC-CEC

The stationary phase material used in HPLC serves only to affect the separation, while it has dual roles in PC-CEC. Specifically, in addition to affecting separation, the stationary phase material is also the vehicle that enables the EOF. The majority of packing materials is silica-based, and the EOF with these packings results from the negative charge on the packing surface due to the ionization of silanol groups. The mobile phase in CEC is driven by EOF, and the analysis time is influenced by the velocity of EOF. Some packing materials used in HPLC can not generate relatively strong EOF, and consequently they are not suitable to be used in CEC due to the long analysis time. For example, the base-deactivated stationary phase (BDS) used in HPLC for separation of basic compounds is undesirable to be used in CEC bacause the silanol groups on the packing surface is endcapped, which results in serious reduction of EOF. Older type HPLC stationary phases have generally been employed in PC-CEC as they are unendcapped and possess a large number of acidic silanol groups. Most of the papers published to date on PC-CEC have reported the use of ODS and, to a much lesser extent, C_8 stationary phases [57]. The magnitude of EOF in these systems is determined by the charge density of the packing surface. If the packing surface possesses more silanol groups, more charge density is expected to be generated due to the ionization of more silanol groups, and thus a strong EOF will be formed.

In order to generate high EOF velocity, Yang and El Rassi [58] introduced a specially designed ODS phase with 75% of the surface silanols unreacted. However, the ionization of silanol groups depends of the pH value of the mobile phase. At high pH, more silanol groups ionized and a higher EOF velocity can be generated, while at low pH value, the ionization is suppressed and the EOF is very poor. It is well known that the pH value is one of the important parameters in HPLC to adjust the separation selectivity. However, the use of this method in CEC is limited because of the dependence of EOF on the pH value. In order to overcome this problem, several groups have reported the use of some new stationary phases with both sulfonic acid groups and alkyl chains bonded on silica [20, 59–66]. The alkyl chains act as stationary phases to retain solutes while the sulfonic acid groups result in high EOF. Because both reversed-phase groups and lon-exchange groups are immobilized on the packing, these specially developed CEC materials are also called mixed-mode stationary phases. High linear velocities can be obtained with the mixed-mode stationary phase throughout a wide pH range because the sulfonic acid groups are ionized over a wide pH range. It was reported that the EOF with this stationary phase was largely independent of the mobile-phase pH (2.3–7.8) and was substantially higher than the conventional C_8, C_{18}, and phenyl phases [60].

However, the synthesis of the mixed-mode stationary phase is relatively complex and time-consuming. As reported by Zhang and El Rassi [61], the synthesis process involved three steps to bond three coating layers on the packing surface. First the hydrophilic sublayer was bonded with γ-glycidoxypropyltrimethoxysilane. Next, a sulfonated layer was sandwiched through covalent bonding between the sublayer and an octadecyl top layer. An alternative way to use the mixed-mode stationary phase is to pack the CEC column with a physical mixture of SCX and ODS [67]. Figure 2 shows the effect of pH on EOF

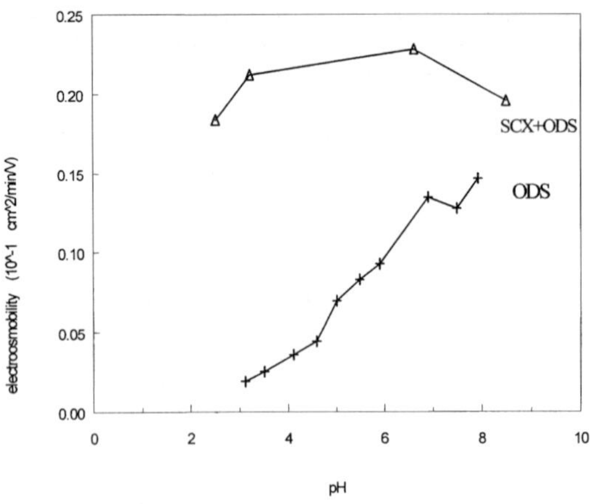

Figure 2. Effect of pH on EOF mobility in CEC with ODS-packed column and mixed stationary phases packed column. Conditions: mobile phase, V_{ACN}/V_{H2O} = 70/30, 3 mM phosphate buffer; stationary phase, (ODS), 5 μm Spherisorb ODS; (SCX + ODS), 5 μm Spherisorb SCX/ODS 1/2.5 w/w. Reprinted from [67], with permission.

mobility with an ODS-packed column and mixed packing column (MP column). The EOF mobility in the ODS-packed column decreases rapidly with decreasing pH value because the ionization of silanol groups is suppressed, while the influence of the pH value on the EOF mobility in an MP column is small, and the mobility is also substantially greater than that in an ODS-packed column.

Ion-exchange phases such as SCX [30, 57, 68–70] and SAX [31, 71, 72] phases have also been used in CEC. A series of ionic compounds like tricyclic antidepressants [69], peptides [30, 70], benzoic acids [31], and some inorganic anions [71, 72] have been separated in ion-exchange CEC. Bare silica was also used as cation-exchange stationary phase for the separation of basic compounds in CEC [4, 73]. Abnormal high efficiency was observed in ion-exchange CEC [30, 57, 69, 71], and the reason for the focusing of the peaks is still not known. Although ionic compounds can be separated in ion-exchange CEC with high efficiency, the ability of this system to separate neutral solutes is poor because the hydrophobicity of the packing is low. Recently, it was reported that the hydrophobicity of cation-exchange packing, like bare silica, and SCX can be increased by dynamic adsorption of CTAB [13, 14]. Neutral solutes can be separated in this system *via* a reversed-phase partitioning mechanism.

Chiral separation is one of the important application fields for CEC. Many chiral stationary phases (CSPs) including CDs [74–78], proteins [9, 79, 80], cellulose derivatives [81], macrocyclic antibiotics [82, 83], quinine-based chiral stationary phase [84], molecular imprinting phases [85, 86] and so on, were used in PC-CEC. Several reviews have summarized the application of CEC in chiral separations [21–23]. The majority of CSPs are silica-based, and the EOF towards the cathode also results from the ionization of silanol groups. Since anions migrate against EOF due to the involvement of the electrophoresis mechanism, the chiral separation of acidic compounds in this system is very difficult. It was reported that acidic enantiomers could not be resolved in CEC with α_1-acid glycoprotein as stationary phase because of their counterdirectional electrophoretic migrations [79]. The packing material with net positive charge on the surface is desirable for the analysis of acidic enantiomers because the anodic EOF can then be generated. Lammerhofer and Lindner [84] have presented CEC enantioseparations of anionic *N*-derivatized α-amino acid employing a capillary column packed with a weak anion-exchange type CSP based on a quinine-derived chiral selector. Koide and Ueno [78] also demonstrated enantiomeric separation of acidic compounds by CEC with β-CD-bonded positively charged polyacrylamide gels. Recently, it was found that some anionic enantiomers like tryptophan, fenoprofen, ketoprofen, and warfarin could be separated in PC-CEC with protein adsorbed on the SAX packing surface [9]. The surface of the packing was still positively charged after the adsorption of protein and anodic flow could still be generated; therefore the anionic solutes could be eluted in reasonable time.

Typical PC-CEC columns are prepared by packing 1.5–5 μm packing materials into 50–100 μm capillary column. In order to prevent the packing materials from exiting the capillary, two on-column end-frits are prepared by sintering packing materials *in situ*. There are several drawbacks to using frits in CEC. These include the extreme difficulty in their preparation, their lack of reproducibility, their tendency to act as a catalyst for bubbles, their unpredictable influence upon EOF and band-spreading [87]. An attractive alternative for the packed columns is continuous-bed columns, whereby the problems related to the frits are eliminated because no end-frits are required in these column. Various continuous-bed columns have been reported for CEC. Generally speaking, they can be classified into three types: organic porous polymer [22, 88, 89], inorganic porous polymer [24, 90, 91], and fusion of individual particles [87, 92–94]. The preparation and performance of continuous-bed columns in CEC have been summarized in several review articles in more detail [22, 24, 25].

4 CEC with physically adsorbed stationary phase

4.1 OT-CEC

Liu and co-workers [6–8] have reported that the adsorption effect could be used to prepare a stationary phase for OT-CEC. Cationic surfactants and basic proteins, peptides, and amino acids were successfully absorbed to the capillary wall. Neutral solutes could be separated on the adsorbed surfactant stationary phase and enantiomers could be separated on the adsorbed CSP. The inner wall surface of the fused-silica capillary is negatively charged due to the dissociation of silanol groups under neutral pH value. Thus, positively charged compounds are readily adsorbed onto the capillary wall. The cationic surfactant with a long alkyl chain, such as CTAB, can be adsorbed to the capillary wall by the electrostatic attraction. Thus, a hydrophobic bilayer will be formed on the capillary wall. Because the CTAB molecule has a long alkyl chain containing 16 carbons, the adsorbed bilayer may act as a reversed-phase stationary phase, just as the role of C_{18} or C_8 in HPLC. The procedure for the preparation of the physically adsorbed CTAB stationary phase was as follows [6–8]: the capillary was first rinsed with phosphate buffer containing some amount of CTAB, so that the CTAB will be adsorbed to the capillary wall. Then, the residue of CTAB was flushed out with a mobile phase without CTAB, and the separation was performed by using the mobile phase containing no CTAB. As shown in Fig. 3, separation of five alkyl-substituted benzenes was achieved successfully in this system. Since all of the five test solutes are neutral and the mobile phase contains no micelles, they can not be separated *via* electrophoresis;

thus, their separation must be based on the reversed-phase partitioning between the mobile phase and the adsorbed CTAB bilayer. Because the physically adsorbed CTAB can easily be desorbed from the capillary wall under the electric field, the adsorbed stationary phase could last only one injection. Therefore, a repreparing procedure was necessary before each run in order to make up the loss of CTAB. The repeatability for this system was relatively good, and the relative standard deviations (RSDs) of the migration time for the five test solutes were less than 3%.

Yang and Hage [95] observed that the adsorbed layer of human serum albumin (HSA) on a capillary wall acted as a stationary phase when they investigated the separation mechanism in chiral CZE using HSA as a buffer additive. The HSA is an acidic protein (the isoelectric point, p*l*, is about 4.5), which is negatively charged at a neutral pH buffer condition, and the capillary wall is also negatively charged under this condition. Therefore, the amount of HSA adsorbed on the capillary wall was relatively low due to the charge repulsion and only the hydrophobic interaction was responsible for the adsorption of HSA. In order to increase the amount of the adsorbed chiral selector, basic proteins, cytochrome *c* [7] and lysozyme [6–8] were selected. These proteins are positively charged at neutral pH, and they can be adsorbed onto the negatively charged capillary wall by electrostatic attraction besides the hydrophobic attraction. Therefore, the adsorbed amount for a basic protein should be more than that for an acidic protein.

In CZE, in order to minimize the adsorption effect, an extreme pH value, high buffer concentration and rather large diameter of the capillary were employed. However, in order to obtain a large phase ratio by adsorption of more protein onto the capillary wall, a buffer with neutral pH value and low salt concentration and a capillary with a small inner diameter were adopted for the preparation of the adsorbed stationary phase for OT-CEC [6]. Five pairs of enantiomers, tryptophan, PTH-aspartic acid, PTH-threonine, dansyl-leucine and mephenytoin, were successfully resolved in OT-CEC with adsorbed lysozyme as the stationary phase [6]. It was also found that some basic peptides (such as Lys-Tyr and Lys-Ser-Tyr) and basic amino acids *i.e.*, L-Lys) could also be adsorbed to the capillary wall as the stationary phase for chiral separation [7]. Figure 4 shows the chiral separation of amino acid racemates by OT-CEC with physically adsorbed stationary phases. The highest efficiency for chiral separation was up to 590 000 plates/m and the average plate number was more than 250 000 plates/m.

Unlike CTAB, the adsorption of basic proteins and peptides onto the capillary wall is very strong. The reprepara-

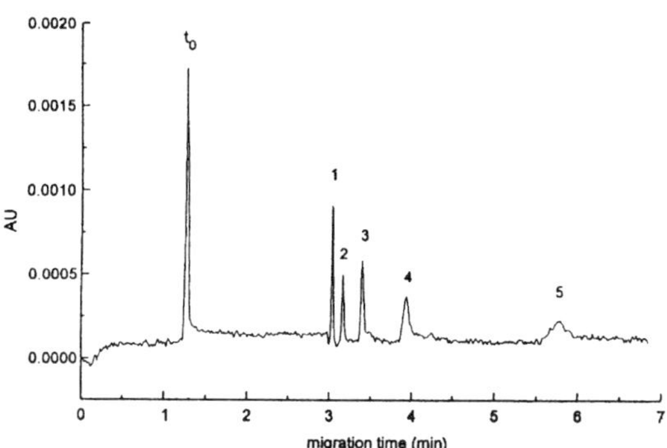

Figure 3. Separation of alkyl benzenes by OT-CEC with adsorbed CTAB bilayer as stationary phase. Solutes: 1, benzene; 2, toluene; 3, ethylbenzene; 4, propylbenzene; 5, butylbenzene. Reprinted from [6], with permission.

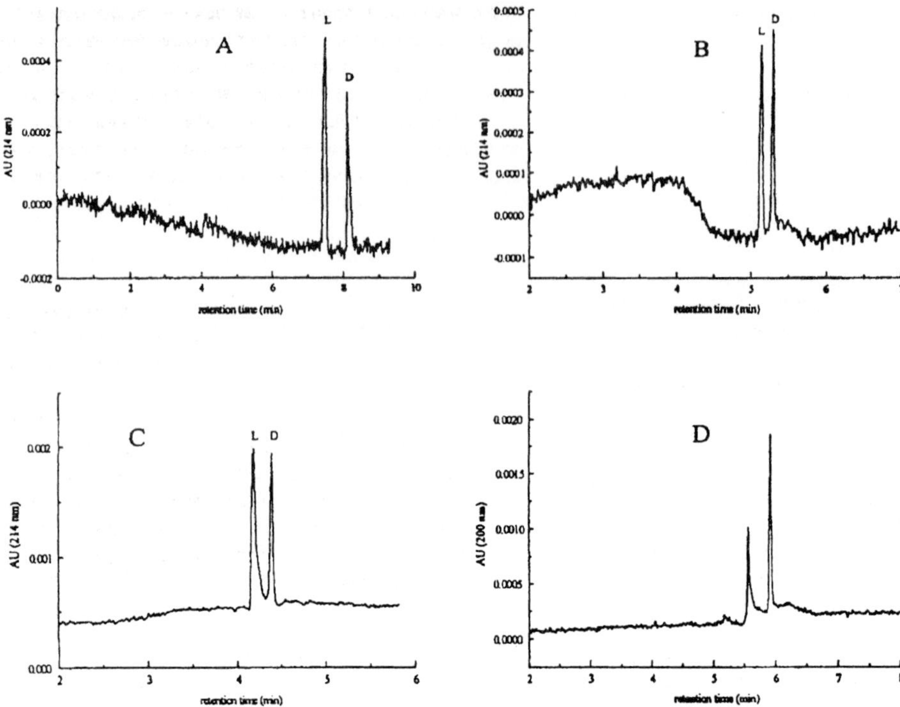

Figure 4. Chiral separation of amino acid racemates by OT-CEC with physically adsorbed stationary phases. Solutes: (A)–(C) DL-tyrosine (0.5 mg/mL); (D) DL-phenylalanine (0.1 mg/mL). Adsorbed stationary phase: (A) cytochrome *c*; (B) Lys-Tyr; (C) Lys-Ser-Tyr; (D) L-lysine. Reprinted from [7], with permission.

tion procedure before each run is unnecessary because the desorption of the chiral stationary phase is insignificant. The column can be used for a long time without loss of chiral selectivity. For example, the chiral selectivity of cytochrome *c* stationary phase could last at least 30 days if the column was preserved well [7]. One of the advantages of this method is that the chiral stationary phase could be prepared again if the chiral selectivity is completely lost [7]. First, the stationary phase was broken down and washed out with methanol (and/or surfactant solution), followed by dilute nitric acid solution. With reactivation of the capillary by a normal activation procedure, a new stationary phase could be obtained again with the same preparation method. Compared with attaching protein onto the capillary wall by chemical bonding [56], adsorption of proteins on the stationary phase surface is simple and easier to be accomplished.

4.2 PC-CEC

Protein stationary phases such as serum albumin, α_1-acid glycoprotein, avidin, *etc.*, could be used for the resolution of acidic enantiomers in HPLC [96–98]. However, almost all protein phases used in HPLC are immobilized on a silica surface, and the charge on these packings is negative. Thus, cathodic EOF will be generated in a CEC column packed with these materials. The generation of cathodic EOF in protein-based PC-CEC was verified by the chiral separations with α_1-acid glycoprotein [79] and HSA phases [80]. The resolution of the acidic enantiomer in this system is problematic in that the ionized acid with negative charge will electrophoretically migrate towards the anode, which is against the direction of EOF; thereby they will migrate out over a long period of time or perhaps can not be loaded into the column by electrokinetic injection if the electrophoretic mobility is greater than that of the EOF. It was observed that some acidic enantiomers, like fenoprofen, ibuprofen, ketoprofen and naproxen, could not be resolved in CEC with α_1-acid glycoprotein as the stationary phase [80]. Therefore, it is necessary to find a way to resolve acidic enantiomers in PC-CEC with protein stationary phases. It is obvious that the separation of an acidic enantiomer with a protein stationary phase can not be performed in ion-suppressed mode, as often used in an achiral system [20], because the structure of the protein will be destroyed at extremely low pH and the enantioselectivity may also be lost.

It is also not suitable to use a buffer additive in this system to reverse the direction of EOF as reported in the β-CD packed CEC column [74]. The reason is the ability of the biopolymers to strongly bind so many different ligands, and the significant effects that additives could have on the binding of analytes. Therefore, it is difficult to solve this problem by simply changing the composition of the mobile phase. A possible way to solve this problem is to develop special protein packing for CEC. In order to generate anodic EOF, the net charge of this special packing should be positive. Although special packings were developed for reversed-phase CEC, these specially developed protein packings have not been reported until now. To synthesize this kind of packing may be complex and time-consuming. An alternative way is to prepare such a packing by adsorption of a protein on SAX material. Ye and co-workers [9] found that BSA adsorbed to SAX could still display enantioselectivity in CEC. The merit of this system is that the EOF direction is from cathode to anode due to the positively charged surface, which is preferable for the separation of acidic enantiomers. The anionic enantiomers, tryptophan, ketoprofen, fenoprofen and warfarin, were successfully resolved in this system.

The procedure for the preparation of the protein-adsorbed stationary phase on the SAX packing in CEC is similar to that in HPLC [99]. The capillary column packed with SAX material is first flushed with phosphate buffer (pH 6.5) containing some amount of BSA for a long enough time. The acidic BSA protein, which is negatively charged at this pH value, will be adsorbed to the packing surface by electrostatic attraction. After that, the residual BSA is flushed out by phosphate buffer. Then the column is conditioned with the mobile phase for a long enough time before running. Figure 5 shows the chromatograms of DL-tryptophan before and after the adsorption of BSA. As can be seen, the column does not display enantioselectivity before the adsorption of BSA, while DL-tryptophan can be baseline-separated after the adsorption. The column separation efficiencies for D-tryptophan and L-tryptophan were 54 500 and 40 200 plates/m, respectively. Because some positively charged groups on the packing surface are neutralized by the adsorbed protein, the EOF was found to be reduced by 26.3%. BSA is a biopolymer, which can not access all the positively charged groups; thereby only part of the positive charges were neutralized and the net charge on the packing surface was still positive. Therefore, the EOF direction after protein adsorption was still from cathode to anode.

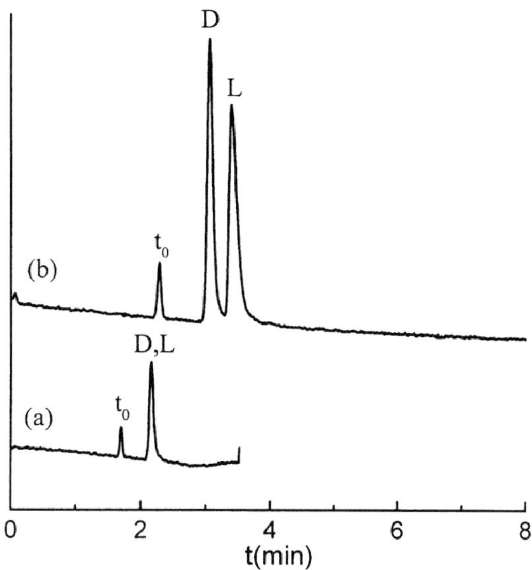

Figure 5. Separation of DL-tryptochan in CEC with (a) SAX stationary phase and (b) adsorption of BSA on SAX. Conditions: column packed with 5 μm SAX; applied voltage, 10 kV; mobile phase, 10% acetonitrile, 5 mM phosphate buffer (pH 6.5). Reprinted from [9], with permission.

One important aspect of this method is its stability because the adsorbed protein may be lost after a long running time. It was observed that the repeatability was relatively good at a low organic modifier content [9]. The RSDs for void time (t_0), selectivity (α) and capillary electrochromatographic retention factor (k^*) of D-tryptophan and L-tryptophan were, respectively, 0.9, 0.5, 0.8 and 0.7% for 21 consecutive runs when the mobile phase containing 7% acetonitrile was used. The decrease of k^* was not found for the consecutive runs, which means that the desorption of protein was negligible under this condition. However, when the acetonitrile content increased to 30%, the desorption of protein was obvious. This was verified by the fact that the k^* for the fenoprofen enantiomers decreased with repetitive runs [9]. Therefore, this system is only stable at low organic modifier content.

It was found that the retention of acidic drugs like ketoprofen, fenoprofen and warfarin in this system was very strong [9]. None could be eluted in 1 h with a mobile phase containing 15% acetonitrile. In order to reduce the strong binding of drugs on the protein, *n*-hexoic acid was added in the mobile phase as the displacer. Figure 6 shows the chromatograms for the separation of warfarin enantiomers with different *n*-hexoic acid concentrations present in the mobile phase. Because *n*-hexoic acid competes with warfarin for the binding site in the protein, the

Electrophoresis 2000, *21*, 4073–4095

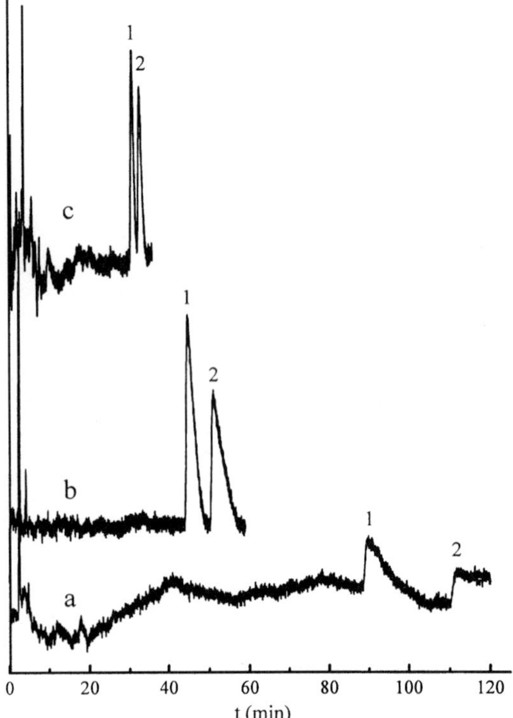

Figure 6. Chiral separation of warfarin at different *n*-hexoic acid concentrations in CEC with adsorption of BSA on SAX packing. Conditions: mobile phase contained (a) 0 mM; (b) 2 mM; (c) 10 mM *n*-hexoic acid. Reprinted from [9], with permission.

retention of warfarin decreased rapidly. Without addition of *n*-hexoic acid, the more retarded warfarin enantiomer was eluted at 112 min. But when 2 mM of *n*-hexoic acid was added in the mobile phase, the migration time was reduced to 52 min, while the selectivity was only decreased by 8%. When the *n*-hexoic acid concentration increased to 10 mM, the warfarin enantiomers were separated in 33 min. Ketoprofen and fenoprofen did not migrate out within 1 h when the mobile phase did not contain *n*-hexoid caid, but their enantiomers were separated in 15 and 60 min, respectively, by addition of 10 mM *n*-hexoic acid into the mobile phase. The SAX packing with a pore size of 120 Å was used to adsorb protein. The surface area of this packing was too large, which resulted in a great amount of adsorbed protein, and consequently strong retention of the enantiomers was observed. The situation is expected to change when packing material with a larger pore size is adopted. The S-CD could also be physically adsorbed to the SAX packing as the stationary phase for chiral separation [15]. However, compared

with that of the protein, the adsorption of S-CD was not strong. The desorption of S-CD occurred after consecutive runs, which resulted in poor repeatability of the system. In order to improve the repeatability, S-CD was added in the mobile phase to dynamically modify the packing. This aspect will be discussed later in more detail.

5 CEC with dynamically adsorbed stationary phase

5.1 OT-CEC

The first example of using dynamically adsorbed stationary phases in CEC was performed in OT-CEC mode [12]. The system was established by coating the capillary wall with a hydrophobic stationary phase, OV-17, and adding CTAB in the mobile phase. The hydrophobic alkyl chain of CTAB is expected to adsorb onto the hydrophobic surface of the coated capillary, allowing the charged end of the molecule to be exposed to the aqueous mobile phase. The charged groups of the adsorbed CTAB act as the ion-exchange site and anions with similar mobility were successfully separated in this system due to the ion-exchange mechanism. The fact that the EOF was reversed in this system, indicates that the surface of the coated capillary was positively charged, which strongly supports the adsorption of CTAB. Compounds were shown to have a decrease in retention when the concentration of the buffer ions increased. This phenomenon is very similar to that in ion-exchange chromatography, which also means the retention of the anions is based on the ion-exchange mechanism.

It was also reported that anions having similar electrophoretic mobilities could be successfully separated in OT-CEC mode with the addition of the ion-pairing agent, tetrabutylammonium hydroxide, to the mobile phase [100]. The addition of the ion-pairing agent to the mobile phase to enhance retenion and resolution has been widely used in HPLC for separation of charged species on reversed-phase columns [101–104]. The ion-pair model and dynamic ion-exchange model typically explain the retention behavior in ion-pair HPLC [76]. The ion-pair model presumes that an ion pair between the ion-pairing agent and the sample ion is formed in the mobile phase. The ion-pair then partitions into the stationary phase. The dynamic ion-exchange model presumes that the ion-pairing agent is dynamically adsorbed onto the hydrophobic stationary phase, where it behaves as an ion-exchanger. The retention and separation of solutes in this mode are very similar to those in ion-exchange chromatography. The dynamic adsorption of ion-pairing agent onto the hydrophobic stationary phase can be well verified in OT-

CEC by the fact that the direction of EOF was reversed with addition of ion-pairing agent in the mobile phase [12, 45, 100]. According to the direction of the EOF, the packing surface is positively charged. Thus, the ion-exchange mechanism inevitably contributes to the migration behavior of anions in OT-CEC. The results of these experiments support the dynamic ion-exchange model for retention and separation [12, 100].

Since adsorption of a surface active agent, such as CTAB and tetrabutylammonium hydroxide, onto the coated capillary is based on hydrophobic interaction, the effect of organic modifier on the amount of adsorbed agent is significant. The EOF velocity is directly related to the surface charge. Peffer and Yeung [45] found that the EOF rapidly decreased linearly with increasing acetonitrile concentration. One of the main reasons is that at low organic modifier content, more CTAB molecules partition into the stationary phase, thus leading to more positive charges on the capillary wall and stronger EOF. Because of the desorption of CTAB at high acetonitrile concentration, the retention of anions also decreased quickly due to the decrease of ion-exchange sites [12].

5.2 PC-CEC

The idea of performing chromatography on the dynamically modified stationary phase in PC-CEC (DMS-CEC) was recently realized [13, 14]. Bare silica was selected as the packing material in the first applications of DMS-CEC [13], and the system was established similar to that in HPLC [11]. The capillary column with a diameter of 100 µm was packed with 5 µm bare silica, and some amount of CTAB was added to the mobile phase to dynamically modify the packing surface. The packing surface is negatively charged under neutral pH conditions because of the ionization of the silanol groups. The CTAB molecule with its permanent positive group is expected to be adsorbed by electrostatic attraction. Thus, a hydrophobic layer would be formed on the packing surface. The separation of neutral solutes in this system is based on the different partitioning between the dynamically adsorbed hydrophobic layer and the mobile phase. The typical chromatogram for the separation of formamide and five alkyl benzene homologous solutes is shown in Fig. 7. The number of theoretical plates obtained was about 71 500/m, which was not very good for a CEC column packed with 5 µm particles. The reason may be the small capacity of the dynamically modified silica and its mass transfer limitation. It is well known that there is a linear relationship between the logarithm of the capacity factor (log k') of homologous solutes and their carbon numbers (Nc) in reversed-phase HPLC. Figure 8 shows the log k' *vs. Nc* plots for five alkyl benzene homologs in DMS-CEC

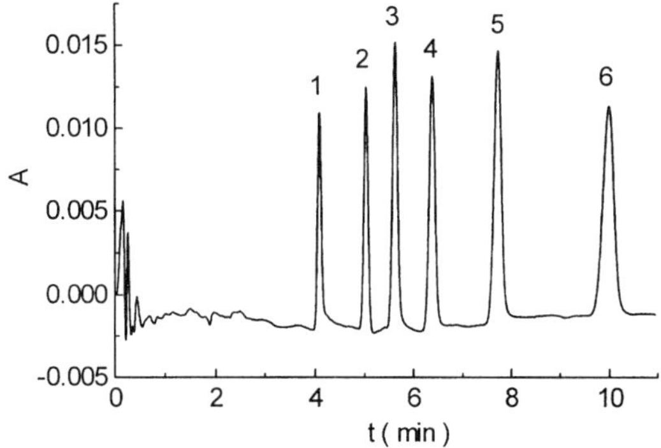

Figure 7. Separation of aromatic compounds by DMS-CEC with bare silica packing. Conditions: packing material, 5 µm Spherisorb-silica gel; mobile phase, 5 mM phosphate buffer (pH 7.5) containing 60% methanol and 2 mM CTAB; applied voltage, 20 kV. Solutes: 1, formamide; 2, benzene; 3, toluene; 4, ethyl benzene; 5, *n*-propyl benzene; 6, *n*-butyl benzene. Reprinted from [13], with permission.

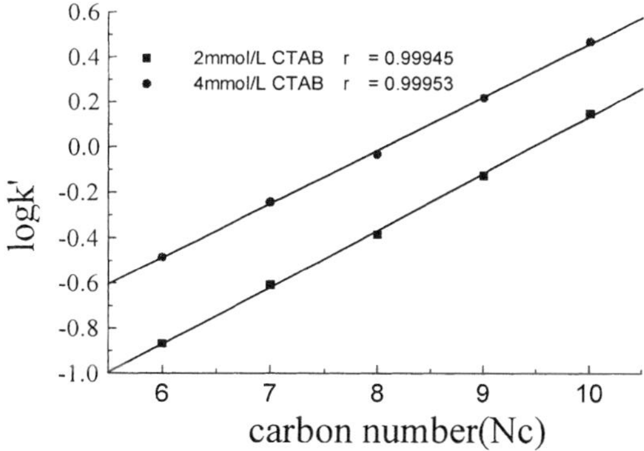

Figure 8. Linear relationship of log k' and carbon number of solutes (Nc) in DMS-CEC with bare silica packing. Conditions and solutes as in Fig. 7. Reprinted from [13], with permission.

with the mobile phase containing 2 mM and 4 mM CTAB. As can be seen from Fig. 8, log k' and Nc obey excellent linear relationships ($r > 0.999$), and the positive value of the slope means that the retention value in DMS-CEC increased with the hydrophobicity of the solute. This result strongly supported the fact that the separation mechanism of neutral solutes in this system is based on reversed-phase partitioning.

In order to confirm that the separation of the six test solutes was based on the hydrophobic interaction with the asdorbed CTAB layer, the same solutes were also analyzed by two other modes [13]. One was the CZE mode where the running buffer used was the same as in the mobile phase in DMS-CEC. Another one was also CEC with silica packing, but the mobile phase did not contain CTAB. The chromatograms obtained by the three modes are shown in Fig. 9. It can be seen from Fig. 9 that only *n*-propylbenzene and *n*-butylbenzene were partially sep-

arated in CZE mode. Since the neutral solutes had no charge and no micelle was formed in the running electrolyte in CZE because the CTAB concentration was lower than the critical micellar concentration, the separation of *n*-propylbenzene and *n*-butylbenzene was obtained from their partitioning between the mobile phase and the adsorbed CTAB double layer as in OT-CEC with CTAB-adsorbed stationary phase. The siloxane bridge of bare silica has some hydrophobicity [105]. The partial separation of the five alkylbenzenes in CEC with silica-packed column as shown in Fig. 9b indicates that the hydrophobicity of the bare silica is still not strong enough to be used as reversed-phase stationary phase. The baseline separation of the six solutes in DMS-CEC demonstrates that the hydrophobicity of the silica surface can be enhanced rapidly by the dynamic adsorption of CTAB.

It is well known that the ionization of silanol groups on the bare silica surface is suppressed at low pH, and consequently the charge density on the packing surface decreases seriously. Thus, the amount of CTAB adsorbed also decreases rapidly, which results in decreasing hydrophobicity of the packing surface. Therefore, the ability of this system to separate neutral solutes also decreases. The separation of five neutral solutes in DMS-CEC packed with bare silica with a mobile phase at pH 2.21 is shown in Fig. 10b. It can be seen from Fig. 10b that the five solutes, which can be baseline-separated at

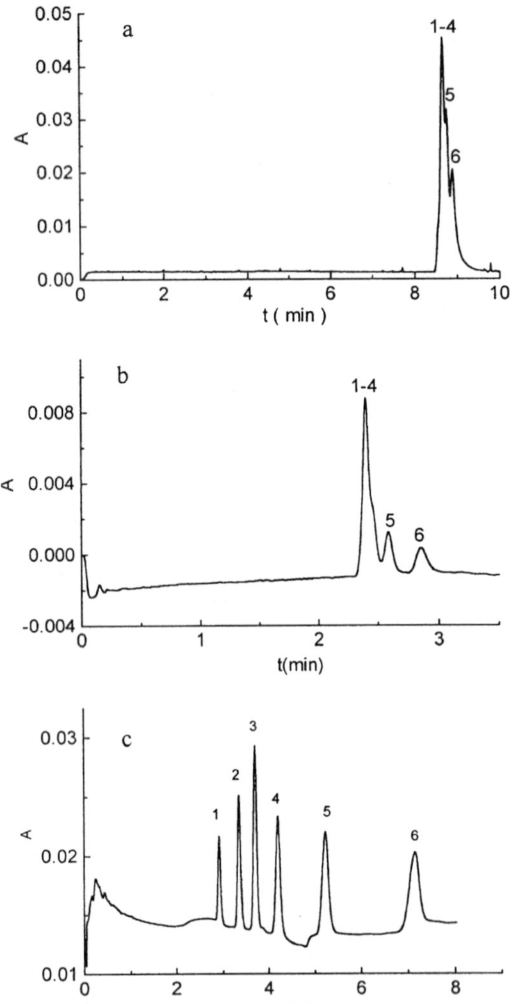

Figure 9. Separation of test solutes in CZE, CEC and DMS-CEC. Conditions: (a) CZE: running buffer, 5 mM phosphate buffer (pH 7.5), 50% methanol and 2 mM CTAB. (b) CEC: packing material, 5 µm Spherisorb-silica gel; mobile phase, 5 mM phosphate buffer (pH 7.5), 50% methanol; (c) DMS-CEC: 5 µm Spherisorb-silica gel; mobile phase, 5 mM phosphate buffer (pH 7.5), 50% methanol, 2 mM CTAB. Solutes as in Fig. 7. Reprinted from [13], with permission.

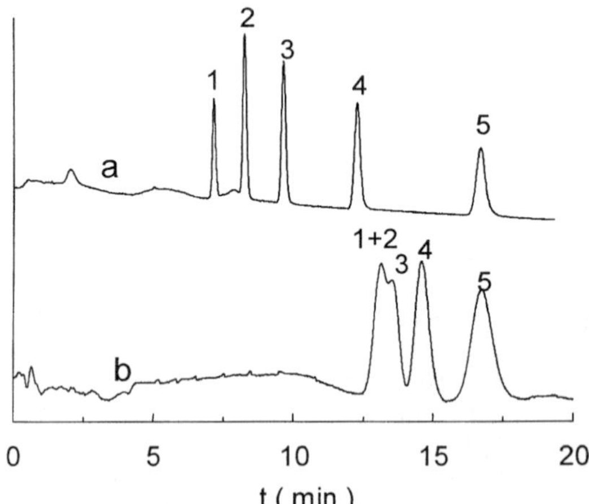

Figure 10. Separation of five neutral solutes in DMS-CEC with different packing materials. Conditions: (a) 5 µm Spherisorb-SCX; (b) 5 µm Spherisorb-silica gel; applied voltage, 15 kV; mobile phase, methanol/25 mM CTAB/100 mM NaH_2PO_4 buffer (pH 2.21)/water = 60/8/10/26. Solutes: 1, benzene; 2, toluene; 3, ethyl benzene; 4, *n*-propyl benzene; 5, *n*-butyl benzene. Reprinted from [14], with permission.

high pH, are only partially separated at low pH. The EOF velocity also decreases significantly due to the suppressing of ionization of silanol. In order to solve this problem, SCX was selected instead of bare silica as the packing [14]. The separation of the same five solutes with the same mobile phase in DMS-CEC with SCX packing is shown in Fig. 10a. The five solutes are completely separated, and the results are much better than in bare silica packed column. Furthermore, the EOF in the SCX column was much stronger than in the bare silica column. The reason for the better performance in the latter case is that the strong sulfonic acid groups on the SCX packing are still ionized at low pH value.

In these two cases, DMS-CEC is based on the dynamic adsorption of CTAB. A dynamic equilibration of CTAB partitioning into the mobile phase and the stationary phase is formed, and the amount of CTAB partitioning into the stationary phase is strongly dependent on the concentration of CTAB in the mobile phase. The influence of the concentration of CTAB on the retention of the solute was investigated in DMS-CEC with both bare silica and SCX phases [13, 14]. The effect of the CTAB concentration on k' values of five neutral solutes in DMS-CEC with SCX packing is shown in Fig. 11. It can be seen from Fig. 11 that the effect of CTAB on capacity factors is very similar to the form of Langmuir adsorption. At a low CTAB concentration, k' values increase quickly, with increasing of CTAB concentration, while at a high CTAB concentration, k' values increase relatively moderately. For example, an increase of the CTAB concentration from 0 to 1 mM resulted in an 8.7-fold increase of k' (*n*-butyl benzene). But an increase of the CTAB concentration from 1 mM to 4 mM only resulted in an 0.24-fold increase of k'. The increase of the retention of neutral solutes while increasing the CTAB concentration was also found in DMS-CEC with bare silica packing [13]. These results mean that an increase of CTAB concentration will increase hydrophobicity of the stationary phase, and consequently the retention of solutes. The increase of hydrophobicity is due to more CTAB adsorbed onto the packing surface at a high CTAB concentration. The increase of the amount of CTAB adsorbed was also demonstrated by the decrease of EOF with increasing CTAB concentration due to the neutralization of the negative charge on the packing surface by CTAB [13, 14].

Another important parameter to adjust the selectivity and retention is the organic modifier content of the mobile phase [13, 14]. The effect of the methanol fraction on k' of five neutral solutes in DMS-CEC with SCX packing is shown in Fig. 12. There are two reasons for the decrease in k' value with increasing methanol fraction. First, since the separation of neutral solutes in this system was based on the reversed-phase partitioning mechanism, an increasing the methanol fraction would increase the elution strength of the mobile phase, resulting in the decrease of k' values of the neutral solutes. Second, the amount of CTAB adsorbed on the SCX packing surface would decrease with increasing methanol fraction, and thus the hydrophobicity of the packings decreased, which also resulted in the decrease of the k' values for neutral solutes. Similar results were also found in DMS-CEC with bare silica packing [13]. The EOF velocity in DMS-CEC

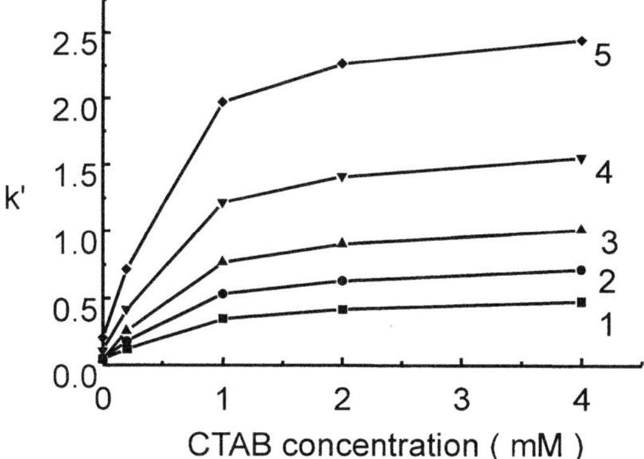

Figure 11. Effect of CTAB concentration on capacity factors (k') of neutral solutes in DMS-CEC with SCX packing. Conditions: packing material, 5 μm Spherisorb-SCX; applied voltage, 15 kV; mobile phases, 60% methanol, 10 mM NaH$_2$PO$_4$ buffer (pH 2.21), 0–4 mM CTAB. Solutes as in Fig. 10. Reprinted from [14], with permission.

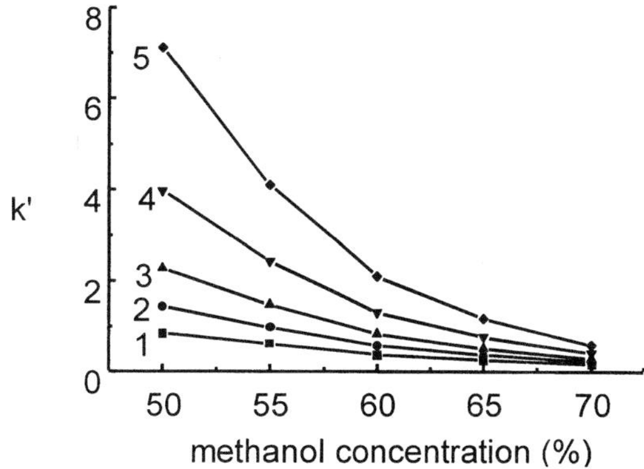

Figure 12. Effect of methanol fraction on capacity factor (k') of neutral solutes. Conditions: packing material, 5 μm Spherisorb-SCX; applied voltage, 15 kV; mobile phases, 50–70% methanol, 2 mM CTAB, 10 mM NaH$_2$PO$_4$ buffer (pH 2.21). Solutes as in Fig. 10. Reprinted from [14], with permission.

was also found to increase with increasing methanol concentration. One of the reasons is that the negative charge increases due to the desorption of CTAB at high organic modifier. In conclusion, in order to shorten the analysis time, it is preferable to increase the organic modifier content, while in order to increase the resolution, the separation should be performed at a low organic modifier content.

As in OT-CEC [12], a surfactant can also be adsorbed onto the hydrophobic stationary phase as dynamic ion-exchange sites. The dynamic adsorption of an anionic SDS surfactant on an ODS phase in PC-CEC was examined by several groups [106–110]. The adsorption of SDS resulted in the increase of a negative charge on the ODS packing, which in turn resulted in increasing the EOF. The retention of basic compounds was found to increase with addition of SDS to the mobile phase [106, 111]. The reason may be that the retention of the solute due to the dynamic ion-exchange mechanism becomes stronger when the negative charge density on the packing surface increases. Huber *et al.* [112] found that the retention of a basic PTH-amino acid like PTH-Arg increased with increasing pH in reversed-phase CEC with a mobile phase in the absence of SDS. In contrast to their report, Seifar *et al.* [109] found that the relative velocities of basic PTH-amino acids increased with increasing pH in reversed-phase CEC with a mobile phase containing SDS. There may exist three different mechanisms to influence the elution behavior of basic PTH-amino acids in the above-mentioned CEC system: (i) their positive charge at lower pH gives them an electrophoretic mobility in the same direction as the EOF; (ii) retention is possibly based on the hydrophobic interaction with the ODS stationary phase; and (iii) retention may be based on electrostatic interaction with the negative groups on the stationary phase. For the first two processes, it is expected that an increase of the mobile-phase pH decreases the average charge on the basic compounds, and will result in longer elution times: the effective cationic mobility will decrease and the hydrophobic interaction will increase with lower effective charge. For the third mechanism it is expected that the decreased average charge at high pH leads to a smaller interaction with the stationary phase, and in turn to shorter elution times. Apparently, the electrostatic interaction with the negatively charged groups in the stationary phase was the most important mechanism in this system.

Similar results for the effect of pH on basic solutes were also observed for the separation of related opiate compounds by reversed-phase CEC with the addition of SDS to the mobile phase [106]. At low pH range, the six opiates appeared before the EOF marker with a mobile

phase without SDS; however, all solutes appeared after the EOF marker with mobile phase in the presence of 5 mM SDS. The stronger retention after the addition of SDS may be contributed by an ion-pair mechanism. The adsorption of SDS on the hydrophobic stationary phase was verified by the fact that the EOF increased with the addition of SDS. These results also support the dynamic ion-exchange model for retention and separation in ion-pair chromatography.

6 Relationship between CEC with physically and dynamically adsorbed stationary phase

The difference between CEC on physically adsorbed stationary phase (PAS-CEC) and CEC on dynamically adsorbed stationary phase, or DMS-CEC, is the composition of the mobile phase. In the former mode, the additive is adsorbed on the capillary wall or the packing surface, and the separation is performed by using a mobile phase without addition of the adsorbed agent. However, in the latter mode, the column is conditioned with the mobile phase containing a surface-active agent for a sufficient amount of time, and the separation is also performed with this mobile phase based on the dynamically adsorbed stationary phase on the surface of the capillary inner wall or packing. In PAS-CEC, if the adsorption of the agent is not strong enough, the desorption of the adsorbed stationary phase is quite obvious. In order to obtain good repeatability, a re-preparing procedure is necessary before each run in order to make up the loss of the desorbed stationary phase. However, this approach is inconvenient and time-consuming for PC-CEC. In other words, if the affinity of the adsorbed agent to the packing surface is not strong enough, the agent is preferably not to be used in PAS-CEC with packed column. An agent with strong adsorption is more desirable for use in PAS-CEC, like the adsorption of BSA on SAX packing. In the case of poor adsorption, the dynamic adsorption to prepare the stationary phase is more desirable. In the latter case, the adsorbed agent is added to the mobile phase to yield the equilibrium of the agent between the mobile phase and the capillary wall surface or the packing surface; thus the amount of the adsorbed stationary phase does not change throughout the experiment, and consequently good repeatability can be obtained.

The repeatability for PAS-CEC and DMS-CEC was compared with the adsorption of S-CD on SAX packing for chiral separation [15]. The void time and the migration times for DL-tryptophan for 17 consecutive runs in PAS-CEC and DMS-CEC are shown in Fig. 13. We see the void time in PAS-CEC increases with repetitive runs. The reason is that the adsorbed S-CD on the packing surface was desorbed gradually, which resulted in the decrease of the net negative charge on the packing surface and in turn the decrease of EOF. The migration time of the enantiomers of tryptophan in PAS-CEC also increased with increasing number of runs. The repeatability for the separation can be improved significantly in DMS-CEC mode. The void time and migration time of the enantiomers of tryptophan almost do not change with the 17 consecutive runs in this system, as shown in Fig. 13. Good repeatability for the void time and the migration times of DL-tryptophan was obtained with RSDs of 0.53, 0.62, and 0.69%, respectively. The reason for the good repeatability is that the equilibration of the added agent between the mobile phase and the stationary phases will be formed, and the amount of the adsorbed agent on the packing surface will not decrease during the experiments.

7 Application of dynamic adsorption in CEC

7.1 Analysis of ionic compounds

In theory, additional selectivity can be obtained in the separation of ionic compounds in CEC because of the involvement of electrophoretic mechanisms. But actually the separation of ionic compounds in CEC is problematic. The major difficulty for separation of basic compounds by

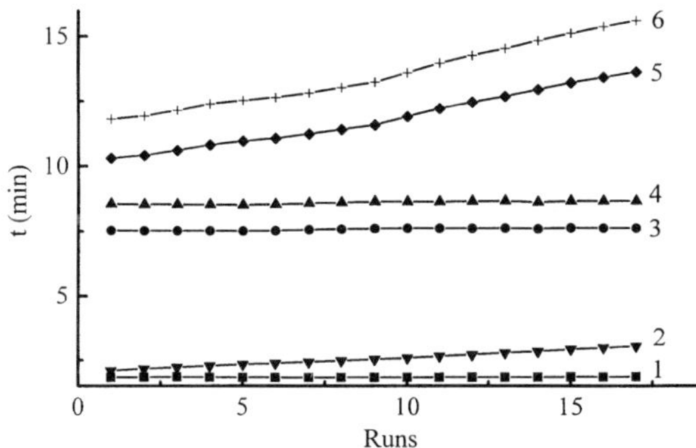

Figure 13. Repeatability of the void time and migration time of tryptophan enantiomers in CEC with physical and dynamic adsorption of S-CD. Conditions: packing material, 5 µm SAX phase; applied voltage, 10 kV; mobile phase for CEC with dynamically modified stationary phase, 10% methanol and 2 mg/mL S-CD in 20 mM acetic acid-triethylamine (HAC-TEA) buffer (pH 4.0); mobile phase for CEC with physically adsorbed stationary phase, 10% methanol in 20 mM HAC-TEA buffer (pH 4.0). Lines (1), (3) and (4), void time and migration time of L-tryptophan and D-tryptophan in CEC with dynamically modified stationary phase; (2), (5), (6) times in CEC with physically adsorbed stationary phase. Reprinted from [15], with permission.

reversed-phase CEC is peak tailing [113], which results from the strong secondary interactions of the base with the ionized silanol groups on the packing surface. Smith and Evens [69] reported a discouragingly poor peak shape of strong basic compounds on the silica-based ODS stationary phase. A typical way to solve this problem in HPLC is by end-capping the free silanol groups on the packing material. However, end-capped C_8- or C_{18}-bonded silica does not represent an acceptable alternative in CEC due to the significant decrease in EOF. Another way to solve this problem in HPLC is by adding competing base in the mobile phase. The competing base has stronger affinity to the ionized silanol groups, and consequently the interaction between the solute base and the silanol groups will be suppressed. This method was also used in CEC for the separation of basic compounds [29, 36, 37, 59, 106, 111, 114]. For example, Gillot *et al.* [37] demonstrated the effect of adding triethylamine or triethanolamine in CEC to improve the peak symmetry of pharmaceutical compounds. The separation of neutral solutes by DMS-CEC packed with bare silica was also based on reversed-phase partitioning. It was found that the peak tailing of basic compounds could be effectively eliminated in this system because the surfactant used had a stronger affinity to the silanol group than the common basic compound had [13].

Figure 14a and b show the chromatograms for the separation of formamide (neutral solute, peak 1) and *o*-toluidine (basic solute, peak 2) on the reversed-phase CEC and DMS-CEC. As can be seen from Fig. 14, the neutral solute has a good peak symmetry in both systems, while the basic solute *o*-toluidine has a poor peak shape in the reversed-phase CEC but a good peak shape in DMS-CEC. Figure 15 shows the chromatogram for the separation of small basic peptides. It can be seen that all basic peptides have good peak symmetry. Basic compounds are retained partly by a reversed-phase mechanism and partly by cation-exchange in DMS-CEC. The cation-exchange effect is, however, of only minor importance for most cationic solutes due to the high affinity of CTAB to silica and to the relatively high concentration of CTAB presented in the eluent.

The difficulty for separation of acidic compounds in reversed-phase CEC is that these solutes can not be loaded into the CEC column or migrate out slowly because the acidic solutes in the ionic form will migrate against the direction of the EOF. Huber *et al.* [112] found that some acidic PTH-amino acids were subject to counterdirectional electrophoretic migration and did not enter the ODS column at pH 7.55. In order to resolve anionic enantiomers, Li and Lloyd [74] have reported that the direction of EOF in CEC with β-CD CSP can be reversed

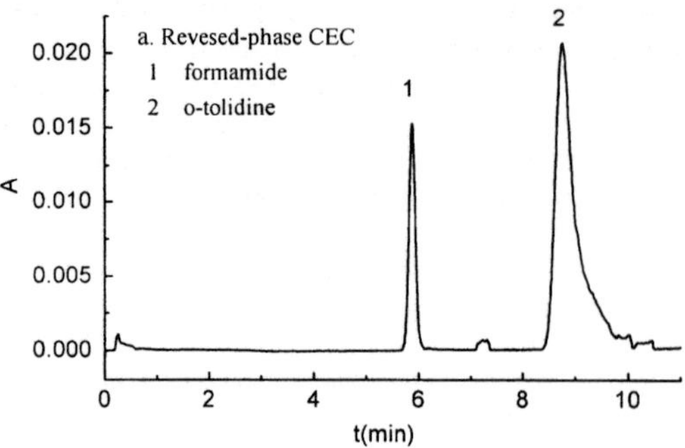

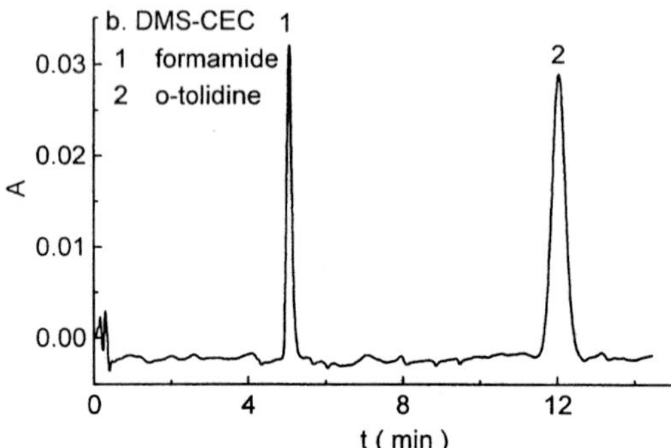

Figure 14. Chromatogram of basic solute in reversed-phase CEC and DMS-CEC. Conditions: (a) reversed-phase CEC, 5 µm Spherisorb-ODS I; mobile phase, 1 mM phosphate buffer (pH 7.5) containing 70% methanol; applied voltage, 15 kV; (b) DMS-CEC, 5 µm Spherisorb-silica gel; mobile phase, 5 mM phosphate buffer (pH 7.5) containing 55% methanol and 2 mM CTAB; applied voltage, 20 kV. Reprinted from [13], with permission.

by the use of triethylammonium acetate as background electrolyte. The reversal of EOF in reversed-phase CEC with the addition of a surface-active agent has not been reported until now. The typical way to analyze acidic compounds in reversed-phase CEC is in ion-suppressed mode where low pH electrolytes are used as the eluent [20]. The acidic compounds are in neutral form at low pH, and thereby they would elute with the EOF.

The original studies were performed on the conventional HPLC packing [115, 116]. For example, a number of acidic pharmaceuticals have been resolved on a standard C_{18} column at pH 2.3 [116]. The disadvantage of this method is that the EOF is low because ionization of the surface silanol groups is suppressed at low pH, which results in long analysis times. Using the special stationary phase materials, which were prepared by coating silica particles with a mixture of sulfonic acid groups and alkyl chain moieties, can solve this problem. It is preferable to

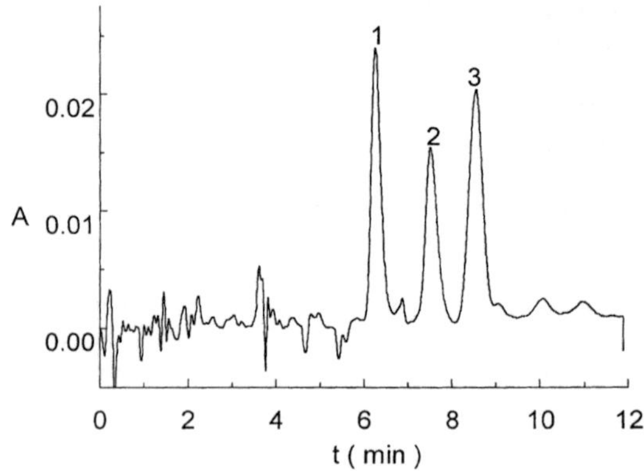

Figure 15. Separation of three basic peptides by DMS-CEC. Conditions: packing material, 5 µm Spherisorb-silica gel; mobile phase, 10 mM phosphate buffer (pH 7.65) containing 60% methanol and 2 mM CTAB; separation voltage, 15 kV. Solutes: 1, Arg-Gly; 2, Arg-Trp; 3, Lys-Tyr-Ser. Reprinted from [13], with permission.

separate acidic compounds in ion-suppressed mode with these packings [20, 59, 60], because the EOF in this system is strong at low pH due to the dissociation of the sulfonic acid groups over a wide pH range. Thus, acidic compounds can be separated in a relatively short time. Altria *et al.* [20] have separated eight acidic drugs with a mobile phase of pH 1.5 on a 7 cm packed column in less than 8 min with such a specific stationary phase compared to an analysis time of 15 min on a standard ODS stationary phase. The separation of acidic compounds in ion-suppressed mode was also performed in DMS-CEC with SCX packing [14]. The EOF in this system was also relatively strong because the sulfonic acid group still ionized at low pH value. Five organic acids can be separated in reasonable time as shown in Fig. 16. The separation of the weak acids in this system is mainly based on reversed-phase partitioning, while that for strong acids, besides the reversed-phase partitioning, the ion-pair and electrophoretic mechanisms, also contribute to the migration behavior.

Simultaneous separation of acidic, basic and neutral compounds is a challenging topic in CEC. Lurie *et al.* [59] achieved the simultaneous separation of acidic, basic and neutral compounds by reversed-phase CEC with Hypersil-C_8 as the stationary phase and low-pH buffer containing hexylamine as the mobile phase. Although basic compounds can be separated with high efficiency by SCX-CEC [30, 57, 69], the ability for separation of neutral solutes is poor. However, the hydrophobicity of the ion-exchange packing like SCX phase can be enhanced by dynamic adsorption of CTAB as in DMS-CEC, and the

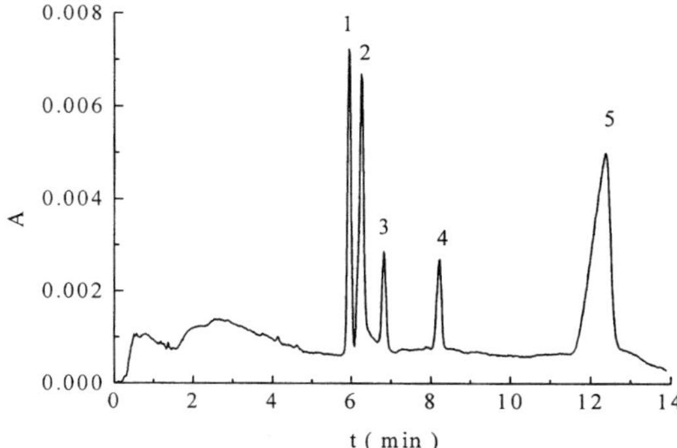

Figure 16. Separation of acidic compounds in DMS-CEC with SCX packing at low pH of the eluent. Conditions: packing material, 5 μm Spherisorb-SCX; mobile phase, methanol/25 mM CTAB/100 mM NaH_2PO_4 buffer (pH 2.21)/ water = 60/8/10/26; applied voltage, 15 kV. Solutes: 1, phenylacetic acid (pK_a 4.28); 2, benzoic acid (pK_a 4.20); 3, *o*-toluic acid (pK_a 3.91); 4, *p*-nitrobenzoic acid (pK_a 3.43); 5, phthalic acid (pK_a 2.95). Reprinted from [14], with permission.

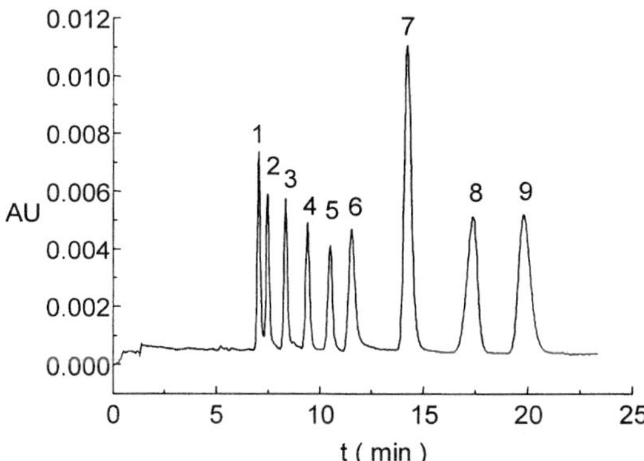

Figure 17. Simultaneous separation of acidic and neutral compounds on the DMS-CEC with SCX packing at low pH value. Conditions: packing material, 5 μm Spherisorb-SCX; applied voltage, 15 kV; mobile phase, methanol/ 25 mM CTAB/100 mM NaH_2PO_4 buffer (pH 2.21)/water = 55/8/10/27. Solutes: 1, phenylacetic acid (pK_a 4.28); 2, benzoic acid (pK_a 4.20); 3, *o*-toluic acid (pK_a 3.91); 4, toleuene; 5, *p*-nitrobenzoic acid (pK_a 3.43); 6, ethyl benzene; 7, naphthalene; 8, phthalic acid (pK_a 2.95); 9, biphenyl. Reprinted from [14], with permission.

neutral compounds can be separated by reversed-phase partitioning mechanism [14]. Acidic compounds exist in neutral form at low pH, and they can also be separated by partitioning mechanism under this condition. Figure 17 shows the simultaneous separation of acidic and neutral compounds by DMS-CEC with SCX packing and a mobile

phase of low pH value. Excellent separation of all the nine compounds was accomplished. Under these conditions, chromatography played a major role in the separation process of the weakly acidic and neutral compounds, but the chromatography and electrophoresis processes may both contribute to the separation of strong acids.

Figure 18 shows the simultaneous separation of neutral, acidic and basic compounds at low pH value. Although all of the six compounds were well separated, peak tailing for basic compounds was observed, which may probably be caused by the strong electrostatic interaction between the positive charge of the basic compounds and the negative charge on the SCX packing. But it was found that the peak tailing of the basic solutes could be eliminated by performing CEC at high pH value. Figure 19 shows simultaneous separation of basic and neutral compounds by the DMS-CEC. It can be seen that all of those neutral and basic compounds are well separated mainly based on the reversed-phase partition mechanism, and very good peak symmetry for the basic compounds has been observed.

7.2 Chiral separation

Enantiomer separation based on a dynamically adsorbed CSP was reported recently [15]. The capillary column was packed with SAX material, and S-CD was used as an additive in the mobile phase, which was dynamically adsorbed onto the packing surface to form a layer of β-CD

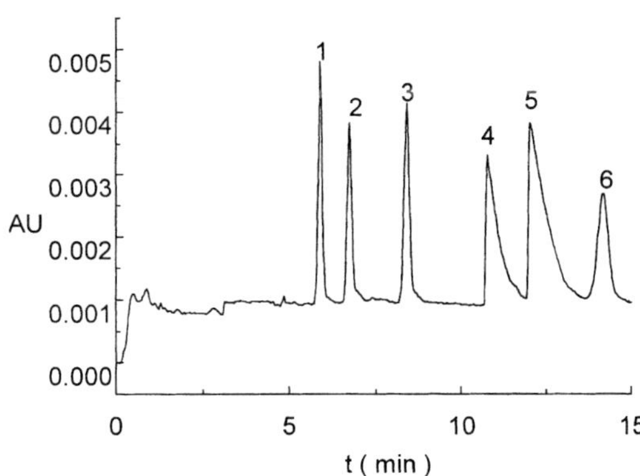

Figure 18. Simultaneous separation of acidic, basic and neutral solutes on the DMS-CEC with SCX packing at low pH value. Conditions: packing material, 5 μm Spherisorb-SCX; applied voltage, 15 kV; mobile phase, methanol/ 25 mM CTAB/100 mM NaH_2PO_4 buffer (pH 2.21)/water = 60/8/10/22. Solutes: 1, phenylacetic acid (pK_a 4.28); 2, *o*-toluic acid (pK_a 3.91); 3, ethyl benzene; 4, quinoline (pK_b 9.15); 5, 2-methyl-quinoline (base); 6, *n*-butyl benzene. Reprinted from [14], with permission.

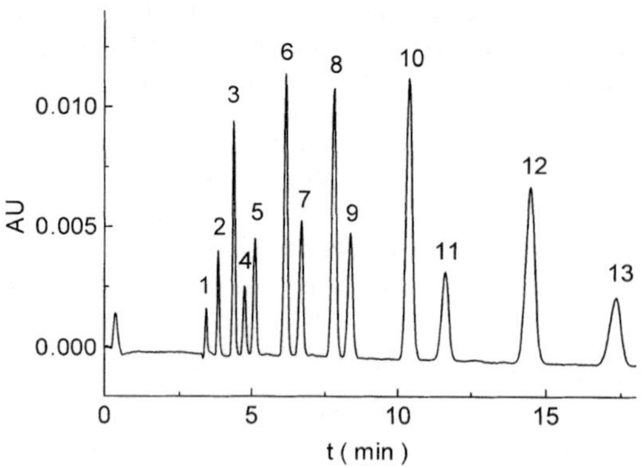

Figure 19. Simultaneous separation of neutral and basic solutes at high pH. Conditions: packing material, 5 μm Spherisorb-SCX; mobile phase, methanol/25 mM CTAB/ 100 mM NaH$_2$PO$_4$ buffer (pH 7.65)/water = 60/8/10/22. Solutes: 1, formamide; 2, pyridine (pK_b 8.83); 3, aniline; 4, uinoline (pK_b 9.15); 5, 2-methyl-quinoline (base); 6, phenetole; 7, toluene; 8, bromobenzene; 9, ethyl benzene; 10, naphthalene; 11, *n*-propyl benzene; 12, biphenyl; 13, *n*-butyl benzene. Reprinted from [14], with permission.

stationary phase. Separation of enantiomers was achieved by their different interaction with the adsorbed stationary phase. The enantiomers of tryptrophan, praziquantel, atropine, metoprolol, and verapamil were successfully separated in this system with a column efficiency varying from 36 000 to 412 000 plates/m. It was found that higher column efficiency could be obtained for the strongly retained solutes. For example, the efficiencies for atropine, metoprolol and verapamil were relatively high, and about 300 000 plates/m were obtained. Figure 20 shows the separation of enantiomers of tryptophan, atropine and verapamil in a single run. Each pair of enantiomers was well resolved with high efficiency.

Most chiral separations in this system were achieved at an S-CD concentration of 2.0 mg/mL. The typical substitution of S-CD is 7–11 moles/mole β-CD. If the substitution is assumed to be 9, the mole concentration of S-CD is calculated to be 0.974 mM. The CD derivatives are often used as running buffer additive for chiral separation in CE. In almost all cases, the CD concentrations are higher than 1 mM, and for some cases the concentration even higher than 50 mM [117]. Compared with CE, the CD concentration required in DMS-CEC is much lower. In fact, it was reported that the adsorption of S-CD on the SAX packing surface was already saturated at a concentration of 2.0 mg/mL [15]. That means that the concentration can be further lowered. In DMS-CEC, the adsorbed chiral selectors are responsible for the chiral separation.

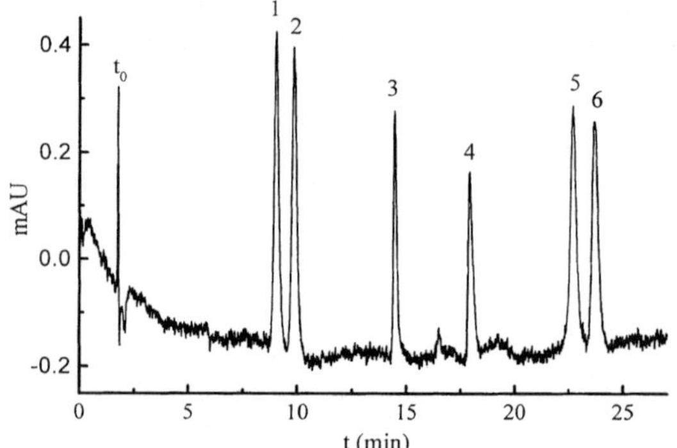

Figure 20. Chiral separation of trytophan, atropine and verapamil in a single run by SAX-CEC with dynamically modified S-CD. Conditions: packing material, SAX packing; mobile phase, 30% methanol and 2.0 mg/mL S-CD in 20 mM HAC-TEA buffer (pH 4.0); applied voltage, 14 kV. Solutes: 1, L-trytophan, 2, D-trytophan; 3 and 4, atropine; 5 and 6, verapamil. Reprinted from [15], with permission.

If the adsorption is very strong, the concentration of the surface active agent in the mobile phase can be very low. A special case is that the adsorption is extremely strong so that the desorption can be ignored, and the separation could be performed with the absence of the chiral selector in the mobile phase. An example is the chiral separation in PC-CEC with SAX phase having physically adsorbed BSA [9]. The lower concentration of the chiral selector required in the mobile phase in DMS-CEC than that in CE permits the use of expensive chiral selector in CEC. In CE mode, because both the chiral selector and the target enantiomers are added to the running buffer, their mobilities should be different in order to achieve chiral separation. However, in DMS-CEC, this restriction is avoided because the chiral selector is adsorbed on the packing surface. Thus, some enantiomers that can not be resolved in CE may be resolved in DMS-CEC mode. Therefore, DMS-CEC can be a complementary technique to chiral separation by CE. More ionic chiral selectors, such as chiral surfactants used in CE, could also be used in CEC as dynamic modification agents. Therefore, the applications of DMS-CEC for chiral separation will be broadened in the future.

7.3 Modification of EOF

One of the important application areas for the dynamic adsorption of additives in CEC is to modify the EOF. The first application of dynamic adsorption for modifying EOF was reported in OT-CEC. Pfeffer and Yeung [45] have reported that the EOF velocity in OT-CEC with a hydrophobic stationary phase can be enhanced with the dynamic adsorption of surfactant. The capillary column was coated with the cross-linked polymer PS-264; the EOF velocity in this system was very poor due to the fairly complete coverage of the silanol groups by the polymer stationary phase. When CTAB was added in the mobile phase, the surface of the capillary would be expected to carry a positive charge due to the adsorption of CTAB. Thus relatively strong EOF can be generated in this system.

Generally, the surface-active agent is an ionic compound, so that when it is adsorbed to the capillary inner surface or packing surface, the zeta potential of the surface changes, which results in the change of EOF. It was reported that EOF characteristics in CEC packed with ODS, bare silica, β-CD packing, SCX packing and SAX packing can be influenced by dynamic modification with SDS [106–110], spermine [118] or CTAB [13], triethylammonium [74], CTAB [14], and sulfated β-CD [9], respectively. The force for these agents dynamically adsorbed to the packing surface can be either electrostatic attraction or hydrophobic interaction based on the property of the packing surface and the EOF modifier agent.

The generation of EOF in PC-CEC occurs at the packing surface rather than at the capillary walls. The magnitude of the EOF is expected to decrease due to nonalignment of the flow channels in the packed bed with the capillary axis, and by lack of electrodrive within the particle pores [119]. However, if the property of the packing surface in PC-CEC is similar to that of the capillary wall, the law of EOF in PC-CEC should also be similar to that of CE. But it was found that the modifications of EOF in CEC with bare silica by dynamically adsorbed spermine [118] or CTAB [13] were not as effective as that in CE. Figure 21 shows the influence of CTAB concentration on the EOF in CE and CEC with bare silica. It can be seen from Fig. 21 that the EOF mobility from anode to cathode in CE obviously decreased with increasing CTAB concentration, and the EOF was reversed when the CTAB concentration exceeded about 0.9 mM, and then the EOF mobility from cathode to anode increased with further increasing of CTAB concentration. This is because a monolayer of CTAB is adsorbed onto the capillary wall at a low concentration of CTAB, with first increase of CTAB concentration, the negative charge on the capillary wall will be neutralized by CTAB, resulting in decreasing EOF mobility. But, when the CTAB concentration was increased to 0.9 mM, almost all the negatively charged groups were covered by CTAB, and the EOF was almost suppressed. Further increasing in the CTAB concentration yielded a bilayer of CTAB on the capillary wall, and the net charge changed from negative to positive, which resulted in the reversal of EOF. The net positive charge density in-

creased with increasing CTAB concentration, resulting in the increase of EOF. As can be seen from Fig. 21, the effect of the CTAB concentration on EOF in CEC with bare silica packing was quite different from that in CE. First, the EOF direction does not change throughout the range of the CTAB concentration investigated, which is always from anode to cathode. Second, the CTAB concentration has less influence on EOF mobility in CEC than that in CE.

The effect of CTAB on EOF was found to have similar tendencies in DMS-CEC packed with SCX material [14]. The EOF mobility was observed to decrease with increasing CTAB concentration, and the reversal of EOF was not found at a CTAB concentration below 4 mM. This result can be explained by the fact that a CEC column packed with silica or SCX particle had a higher surface area than the capillary wall, and therefore more negative groups in the packing surface could be neutralized in CEC by CTAB than in CE. In other words, a CTAB concentration of 4 mM was not enough to neutralize all negative groups and only a monolayer of CTAB was formed on the surface. The reversal of EOF may occur at a higher concentration of CTAB, but it was found difficult to perform CEC at a CTAB concentration higher than 4 mM due to the unstable baseline and current which may be caused by the bubble formation [13]. Although the adsorption of CTAB in CEC with bare silica and SCX is based on electrostatic attraction, the influence of methanol concentration on the amount of adsorbed CTAB is obvious. The EOF in these system was found to decrease with increasing methanol content, which means that less CTAB is adsorbed by the packing surface at high methanol content [13, 14].

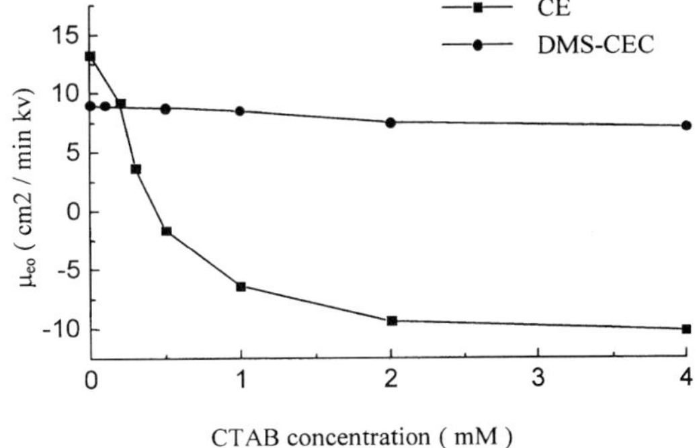

Figure 21. Influence of the CTAB concentration on electroosmotic mobility (μ_{eo}) in the CE and DMS-CEC with bare silica. Conditions: mobile phase, 5 mM phosphate buffer (pH 7.5) containing 50% methanol and various concentrations of CTAB. (a) DMS-CEC: column packed with bare silica; applied voltage, 25 kV. (b) CE: applied voltage, 20 kV. Reprinted from [13], with permission.

It was reported that the EOF in CE could be altered by addition of amine bases [120–122]. The reason is that the negatively charged silanol groups on the capillary wall are dynamically covered by the amine base, resulting in the alteration of the zeta potential. The EOF in reversed-phase CEC also results from the ionization of the silanol group on the packing surface; the negative charges can also be neutralized by the adsorption of amine base, which in turn result in decreasing the EOF. The minimization of EOF occurred after addition of triethylammonium [37] or hexylamine [59] in the mobile phase, respectively. But Hilhorst *et al.* [36] found that the addition of the amine base to the mobile phase had little effect on the EOF. In conclusion, the influence of the amine base on EOF in reversed-phase CEC is not significant compared with that in CE. The reversal of EOF by amine base in CE is common [120, 121], while the reversal of EOF in reversed-phase CEC is not reported. One of the possible reason is that a high organic modifier content is always used in reversed-phase CEC, preventing the adsorption of double layers of amine base.

The EOF modifiers in CE are all amine bases because the adsorption is based on electrostatic attraction. But the EOF modifier in reversed-phase CEC can be an anionic surfactant because this can be adsorbed to the packing surface by hydrophobic interaction. Seifar and co-workers [107] first reported that the EOF in CEC packed with 1.5 µm ODS-modified nonporous silica spheres could be enhanced by the addition of SDS in the mobile phase. Other researchers also reported similar results [106, 108–110]. This is because the C_{12} alkyl chain of the SDS molecule is adsorbed onto the reversed-phase packing by hydrophobic interaction, and consequently the charge density on the packing surface is increased, resulting in the EOF increase. Since the adsorption of SDS is based on hydrophobic interaction, the influence of acetonitrile on the amount of SDS adsorbed is inevitable. The EOF mobility increased with increasing acetonitrile content in reversed-phase CEC with mobile phase in the absence of SDS [123], but the mobility did not change with the increase of the acetonitrile content in the mobile phase in the presence of 5 mM SDS [107]. This is because the adsorption of SDS decreases at a high acetonitrile content, which results in the decrease of the charge density on the packing surface, and therefore the tendency for EOF to increase was counteracted by the decrease of the zeta potential. This indicates that the effect of SDS on EOF is more significant at a low organic modifier content. The attractive advantage of this method to enhance EOF mobility is that the EOF hardly depends on the pH value at low organic modifier concentration.

Lim *et al.* [111] have reported that the EOF mobility increased with increasing pH value when the mobile phase did not contain SDS, but the retention time of the EOF marker was almost the same at different pH values when SDS was present in the mobile phase. The EOF was also found to be much stronger in the latter case, which mainly results from the negative group on the adsorbed SDS on the packing surface. The sulfonic acid groups in the SDS molecule are ionized over a wide range of pH, which leads to an EOF independent of pH values. The experiment was conducted with an acetonitrile concentration less than 30%, but when the acetonitrile concentration was increased to 60%, the situation changed. The influence of the pH value on the EOF mobility was more obvious at high concentrations of an organic modifier as was reported by Seifar *et al.* [107]. The reason is that the amount of SDS partitioning into the stationary phase is low at a high organic content, and consequently the negative charge resulting from the ionization of silanol groups, which depends strongly on pH value, can not be ignored.

In addition to enhancing the EOF in reversed-phase CEC, the addition of SDS to the mobile phase is also helpful to stabilize the current and EOF [107]. Bailey and Yan [110] reported that due to the particularly nonwetting characteristics of the ODS packing and the highly aqueous mobile phase necessary for a baseline resolution, the CEC separation of explosives was occasionally hampered by the drying out of the capillary. This situation can be changed by adding of SDS to the mobile phase. This is because adsorption of surfactant decreases the surface tension at the solid-liquid interface, leading to increased wetting on the surface of the packing. Lim *et al.* [111] also found that the addition of 5 mM SDS is effective for improving the separation efficiency as well as for preventing bubble formation in reversed-phase CEC.

The reversal of EOF in PC-CEC was only found in two chiral separation applications. The EOF in CEC packed with β-CD was reversed by the addition of triethylammonium in the mobile phase [74], whereas in CEC packed with the SAX phase it was reversed by the addition of sulfated β-CD [15]. The reversal of EOF in the β-CD packed column is because the bilayer of triethylammonium was adsorbed onto the silica by electrostatic attraction, which resulted in the reversal of the zeta potential at the solid-liquid interface. It was reported that the adsorption of triethylammonium was saturated on the packing surface at the triethylammonium concentration of 5 mM. A similar or higher concentration of amine base was also added in the mobile phase of reversed phase CEC [29, 36, 37, 59]. But the reversal of EOF was not found in reversed-phase CEC. The reason for the reversal of EOF in β-CD-packed CEC is because double layers of triethylammonium are adsorbed, but in the case of SAX-packed CEC only a monolayer of S-CD is adsorbed [15]. The substitution of S-CD is 7–11; this means there are 7–11 sulfonic acid groups on the CD ring. When S-CD is added in the mobile phase, the positive charge on the packing surface is neutralized by one side of sulfonic acid groups on the CD ring, while the other side of sulfonic acid groups results in the reversal of the EOF.

A new generation of columns that holds great potential for CEC consists of a continuous bed inside the column. This eliminates the problems in column packing and frit construction. Compared with HPLC, the continuous bed for CEC applications must assume the role of a vehicle that enables strong EOF, in addition to affecting the separation. In order to generate strong EOF, a functional monomer containing charged group, such as a sulfonic acid group [88, 124, 125] or ammonium group [126] must be incorporated into the polymerization mixture. Therefore, the experience acquired earlier with the preparation of continuous bed for HPLC is not directly transferable to the preparation of capillary columns for CEC [125]. The EOF in continuous bed CEC can be enhanced or generated by the addition of surface-active agent in the mobile phase as that in OT-CEC. It was reported that the EOF in CEC using fluoropolymer as the chromatographic support material increased with addition of trifluoroacetic acid [127]. The enhancement of EOF was attributed to a dynamic coating of the particles by trifluoroacetic acid.

Surfactants such as SDS and CTAB can modify EOF in CEC with hydrophobic continuous beds. SDS or CTAB will be dynamically adsorbed on the packing surface by hydrophobic interaction, which results in changing the zeta potential of the surface, and in turn the change of EOF. Therefore, the necessity to incorporate a charged group into the polymer can be avoided, and the method used in HPLC for the preparation of a continuous bed can be directly transferred to the preparation of a CEC column. In addition, adjusting the EOF is more convenient by using a surfactant. The EOF in this system can be enhanced by the addition of SDS or can be reversed by the addition of CTAB. The cathodic flow is desirable to simultaneously separate basic and neutral solutes, and therefore SDS could be added to increase the EOF velocity. However, the anodic flow is preferable to simultaneously separate acidic and neutral solutes, so that CTAB could be added for reversing and enhancing the EOF. CEC in this field is still being studied in our group.

8 Conclusions

The majority of the stationary phases used in CEC are based on HPLC materials, which are not optimum for CEC. Some stationary phases such as base-deactivated silica are not suitable for use in CEC because of the poor EOF generated. Most standard HPLC stationary phases are not suitable for the analysis of ionic compounds because of the presence of silanol groups on the packing surface. Although some new stationary phase materials were synthesized for CEC, this method is tedious and time-consuming. One alternative method to prepare the stationary phase is by physical adsorption. The attractive advantage of this method is the simplicity of the preparation procedure. As for long-term stability, the stationary phase prepared by adsorption is not good.

The property of the stationary phase materials can be conveniently modified by the dynamic adsorption of some agents. The hydrophobicity of hydrophilic stationary phases can be increased by the adsorption of surfactants, and it can be used as reversed-phase stationary phases for the separation of neutral solutes. The reversed-phase stationary phases can also be transferred to the ion-exchange phase by the adsorption of a charged surfactant, which can be used to separate anions with similar mobility in OT-CEC. Achiral stationary phases can be transferred to chiral phases by adsorption of chiral agents. In addition, the zeta potential on the stationary phase is also changed after the adsorption, which results in the alteration of the EOF. In conclusion, the most important advantages of this method are (i) easy adjustment of the chromatographic property of the stationary phase, and (ii) easy control of the magnitude and direction of the EOF. Compared with the synthesis of new stationary phase materials for CEC, the use of dynamically adsorbed stationary phases is simpler and more convenient. The application of dynamic adsorption in CEC is still not popular now, and further studies of the choices of the adsorbed agents and stationary phases are necessary. Since some of the problems encountered in CEC can be solved by dynamically modifying the stationary phases, this mode should be regarded as a complementary technique for CEC.

The financial support from the Natural Science Foundation of Liaoning Province, China, to Dr. Hanfa Zou is gratefully thanked. Dr. Hanfa Zou is the recipient of the Excellent Young Scientist Award from the National Natural Science Foundation of China (No. 29725512).

Received July 7, 2000

9 References

[1] Pretorious, V., Hopkins, B. J., Schieke, J. D., *J. Chromatogr.* 1974, *99*, 23–30.

[2] Colón, L. A., Reynolds, K. J., Alicea-Maldonado, R., Fermier, A. M., *Electrophoresis* 1997, *18*, 2162–2174.

[3] Crego, A. L., González, A., Marina, M. L., *Crit. Rev. Anal. Chem.* 1996, *26*, 261–304.

[4] Gillott, N. C., Euerby, M. R., Johnson, C. M., Barrett, D. A., Shaw, P. N., *Chromatographia* 2000, *51*, 167–173.

[5] Lei, Z., Ye, M., Zou, H., Wu, R., Ni, J., *J. Chin. Anal. Chem.* 2000, in press.

[6] Liu, Z., Zou, H., Ni, J., Zhang, Y., *Anal. Chim. Acta* 1999, *378*, 73–76.

[7] Liu, Z., Zou, H., Ye, M., Ni, J., Zhang, Y., *Electrophoresis* 1999, *20*, 2891–2897.

[8] Liu, Z., Zou, H., Ye, M., Ni, J., Zhang, Y., *Chin. J. Chromatogr. A* 1999, *17*, 245–248.

[9] Ye, M., Zou, H., Lei, Z., Wu, R., Liu, Z., Ni, J., *J. Chromatogr. A* 2000, in press.

[10] Ghaemi, Y., Wall, R. A., *J. Chromatogr.* 1979, *174*, 51–59.

[11] Helboe, P., Hansen, S. H., Thomsen, M., *Adv. Chromatogr.* 1989, *28*, 195.

[12] Garner, T. W., Yeung, E. S., *J. Chromatogr.* 1993, *640*, 397–402.

[13] Ye, M., Zou, H., Liu, Z., Ni, J., Zhang, Y., *J. Chromatogr. A* 1999, *855*, 137–145.

[14] Ye, M., Zou, H., Liu, Z., Ni, J., Zhang, Y., *Anal. Chem.* 2000, *72*, 616–621.

[15] Ye, M., Zou, H., Lei, Z., Wu, R., Liu, Z., Ni, J., *Electrophoresis* 2000, *21*, in press.

[16] Colón, L. A., Guo, Y., Fermier, A., *Anal. Chem.* 1997, *69*, 461A–467A.

[17] Altria, K. D., Smith, N. W., Turnbull, C. H., *Chromatographia* 1997, *46*, 664–674.

[18] Cikalo, M. G., Bartle, K. D., Robson, M. M., Myers, P., Euerby, M. R., *Analyst* 1998, *123*, 87R–102R.

[19] Altria, K. D., *J. Chromatogr. A* 1999, 856, 443–463.

[20] Altria, K. D., Smith, N. W., Turnbull, C. H., *J. Chromatogr. B* 1998, *717*, 341–353.

[21] Hatajik, T. D., Brown, P. R., *J. Capil. Electrophor.* 1998, *5*, 143–151.

[22] Schweitz, L., Andersson, L. I., Nilsson, S., *J. Chromatogr. A* 1998, *817*, 5–13.

[23] Wistuba, D., Schurig, V., *J. Chromatogr. A* 2000, *875*, 255–276.

[24] Fujimoto, C., *Trends Anal. Chem.* 1999, *18*, 291–301.

[25] Svec, F., Peters, E. C., Sykora, D., Yu, C., Fréchet, J. M. J., *J. High Resolut. Chromatogr.* 2000, *23*, 3–18.

[26] Rathore, A. S., Horváth, C., *J. Chromatogr. A* 1996, *743*, 231–246.

[27] Wu, J. T., Huang, P., Li, M. X., Lubman, D. M., *Anal. Chem.* 1997, *69*, 2908–2913.

[28] Ye, M., Zou, H., Liu, Z., Zhu, Z., Ni, J., Zhang, Y., *Science in China* 1999, *42*, 639–648.

[29] Walhagen, K., Unger, K. K., Olsson, A. M., Hearn, M. T. W., *J. Chromatogr. A* 1999, *853*, 263–275.

[30] Ye, M., Zou, H., Liu, Z., Ni, J., *J. Chromatogr. A* 2000, *869*, 385–394.

[31] Ye, M., Zou, H., Liu, Z., Ni, J., *J. Chromatogr. A* 2000, *887*, 223–231.

[32] Foley, J. P., May, W. E., *Anal. Chem.* 1987, *59*, 102–109.

[33] Melander, C. W., Molnár, I., *Anal. Chem.* 1977, *49*, 142–154.

[34] Khaledi, M. G., Smith, S. C., Strasters, J. K., *Anal. Chem.* 1991, *63*, 1820–1830.

[35] Strasters, J. K., Khaledi, M. G., *Anal. Chem.* 1991, *63*, 2503–2508.

[36] Hilhorst, M. J., Somsen, G. W., de Jong, G. J., *J. Chromatogr. A* 2000, 872, 315–321.

[37] Gillott, N. C., Euerby, M. R., Johnson, C. M., Barrett, D. A., Shaw, P. N., *Anal. Commun.* 1998, *35*, 217–220.

[38] Tsuda, T., Nomura, K., Nakagawa, G., *J. Chromatogr.* 1982, *248*, 241–247.

[39] Bruin, G. J. M., Tock, P. P. H., Kraak, J. C., Poppe, H., *J. Chromatogr.* 1990, *517*, 557–572.

[40] Pesek, J. J., Matyska, M. T., *J. Chromatogr. A* 1996, *736*, 255–264.

[41] Pesek, J. J., Matyska, M. T., Cho, S., *J. Chromatogr. A* 1999, 845, 237–246.

[42] Guo, Y., Colón, L. A., *J. Microcol. Sep.* 1995, *7*, 485–491.

[43] Guo, Y., Colón, L. A., *Anal. Chem.* 1995, *67*, 2511–2516.

[44] Narang, P., Colón, L. A., *J. Chromatogr. A* 1997, *773*, 65–72.

[45] Pfeffer, W. D., Yeung, E. S., *Anal. Chem.* 1990, *62*, 2178–2182.

[46] Tan, Z. J., Remcho, V. T., *Anal. Chem.* 1997, *69*, 581–586.

[47] Huang, X., Zhang, J., Horváth, C., *J. Chromatogr. A* 1999, *858*, 91–101.

[48] Armstrong, D. W., Tang, Y., Ward, T., Nichols, M., *Anal. Chem.* 1993, *65*, 1114–1117.

[49] Mayer, S., Schurig, V., *J. High Resolut. Chromatogr.* 1992, *15*, 129–131.

[50] Mayer, S., Schurig, V., *J. Liq. Chromatogr.* 1993, *16*, 915–931.

[51] Schurig, S., Jung, M., Mayer, S., Fluck, M., Negura, S., Jakubetz, H., *J. Chromatogr. A* 1995, *694*, 119–128.

[52] Szemán, J., Ganzler, K., *J. Chromatogr. A* 1994, *668*, 509–517.

[53] Schweitz, L., Andersson, L. I., Nilsson, S., *Anal. Chem.* 1997, *69*, 1179–1183.

[54] Tan, Z. J., Remcho, V. T., *Electrophoresis* 1998, *19*, 2055–2060.

[55] Francotte, E., Jung, M., *Chromatographia* 1996, *42*, 521–527.

[56] Hofstetter, H., Hofstetter, O., Schurig, V., *J. Microcol. Sep.* 1998, *10*, 287–291.

[57] Smith, N., Evans, M. B., *J. Chromatogr. A* 1999, *832*, 41–54.

[58] Yang, C., El Rassi, Z., *Electrophoresis* 1998, *19*, 2061–2067.

[59] Lurie, I. S., Conver, T. S., Ford, V. L., *Anal. Chem.* 1998, *70*, 4563–4569.

[60] Euerby, M. R., Johnson, C. M., Bartle, K. D., *LC.GC* 1998, *16*, 386–394.

[61] Zhang, M., El Rassi, Z., *Electrophoresis* 1998, *19*, 2068–2072.

[62] Zhang, M., El Rassi, Z., *Electrophoresis* 1999, *20*, 31–36.

[63] Zhang, M., Yang, C., El Rassi, Z., *Anal. Chem.* 1999, *71*, 3277–3282.

[64] Huang, P., Jin, X., Chen, Y., Srinivasan, J. R., Lubman, D. M., *Anal. Chem.* 1999, *71*, 1786–1791.

[65] Dittmann, M. M., Rozing, G. P., *J. Microcol. Sep.* 1997, *9*, 399–408.

[66] Euerby, M. R., Johnson, C. M., Smyth, S. F., Gillott, N., Barrett, D. A., Shaw, P. N., *J. Microcol. Sep.* 1999, *11*, 305–311.

[67] Zhang, L., Zhang, Y., Shi, W., Zou, H., *J. High Resolut. Chromatogr.* 1999, *22*, 666–670.

[68] Cikalo, M. G., Bartle, K. D., Myers, P., *Anal. Chem.* 1999, *71*, 1820–1825.

[69] Smith, N. W., Evans, M. B., *Chromatographia* 1995, *41*, 197–203.

[70] Choudhary, G., Horváth, C., *J. Chromatogr. A* 1997, *781*, 161–183.

[71] Li, D., Knobel, H. H., Remcho, V. T., *J. Chromatogr. B* 1997, *695*, 169–174.

[72] Kitagawa, S., Tsuji, A., Watanabe, H., Nakashima, T., Tsuda, T., *J. Microcol. Sep.* 1997, *9*, 347–356.

[73] Wei, W., Luo, G. A., Hua, G. Y., Yan, C., *J. Chromatogr. A* 1998, *817*, 65–74.

[74] Li, S., Lloyd, D. K., *J. Chromatogr. A* 1994, *666*, 321–335.

[75] Lelièvre, F., Yan, C., Zare, R. N., Gareil, P., *J. Chromatogr. A* 1996, *723*, 145–156.

[76] Zou, H., Zhang, Y., Lu, P., *Ion-Pair High Performance Liquid Chromatography*, Henan Science and Technology Press, Zhengzhou (China), 1994, pp. 60–114.

[77] Wistuba, D., Czesla, H., Roeder, M., Schurig, V., *J. Chromatogr. A* 1998, 815, 183–188.

[78] Koide, T., Ueno, K., *J. High Resolut. Chromatogr.* 2000, *23*, 59–66.

[79] Li, S., Lloyd, D. K., *Anal. chem.* 1993, *65*, 3684–3690.

[80] Lloyd, D. K., Li, S., Ryan, P., *J. Chromatogr. A* 1995, *694*, 285–296.

[81] Masruska, A., Pyell, U., *J. Chromatogr. A* 1997, *782*, 167–174.

[82] Carter-Finch, A. S., Smith, N. W., *J. Chromatogr. A* 1999, *848*, 375–385.

[83] Wikstrom, H., Svensson, L. A., Torstensson, A., Owens, P. K., *J. Chromatogr. A* 2000, *869*, 395–409.

[84] Lammerhofer, M., Lindner, W., *J. Chromatogr. A* 1998, *829*, 115–125.

[85] Schweitz, L., Andersson, L. I., Nilsson, S., *J. Chromatogr. A* 1997, *792*, 401–409.

[86] Lin, J. M., Uchiyama, K., Hobo, T., *Chromatographia* 1998, *47*, 625–629.

[87] Ratnayake, C. K., Oh, C. S., Henry, M. P., *J. High Resolut. Chromatogr.* 2000, *23*, 81–88.

[88] Fujimoto, C., Fujise, Y., Matsuzawa, E., *Anal. Chem.* 1996, *68*, 2753–2757.

[89] Peters, E. C., Petro, M., Svec, F., Fréchet, J. M. J., *Anal. Chem.* 1998, *70*, 2288–2295.

[90] Asiaie, R., Huang, X., Farnan, D., Horváth, C., *J. Chromatogr. A* 1998, *806*, 251–263.

[91] Fujimoto, C., *J. High Resolut. Chromatogr.* 2000, *23*, 89–92.

[92] Dulay, M. T., Kulkarni, R. P., Zare, R. N., *Anal. Chem.* 1998, *70*, 5103–5107.

[93] Tang, Q., Xin, B., Lee, M. L., *J. Chromatogr. A* 1999, *837*, 35–50.

[94] Tang, Q., Lee, M. L., *J. High Resolut. Chromatogr.* 2000, *23*, 73–80.

[95] Yang, J., Hage, D. S., *Anal. Chem.* 1994, *66*, 2719–2725.

[96] Noctor, T. A. G., Felix, G., Wainer, I. W., *Chromatographia* 1991, *31*, 55–59.

[97] Menzel, S. S., Geisslinger, G., Brune, K., *J. Chromatogr.* 1990, *532*, 295–303.

[98] Mano, N., Oda, Y., Asakawa, N., Yoshida, Y., Sato, T., Miwa, T., *J. Chromatogr. A* 1994, *687*, 223–232.

[99] Jacobson, S. C., Guiochon, G., *J. Chromatogr.* 1992, *590*, 119–126.

[100] Pfeffer, W. D., Yeung, E. S., *J. Chromatogr.* 1991, *557*, 125–136.

[101] Bieganowska, M. L., Petruczynik, A., Gadzikowska, M., *J. Chromatogr.* 1990, *520*, 403–410.

[102] Molnar, I., Knauer, H., Wilk, D., *J. Chromatogr.* 1980, *201*, 225–240.

[103] Iskandarani, Z., Pietezyk, D. J., *Anal. Chem.* 1982, *54*, 1065–1071.

[104] Inczedy, J., Szokoli, F., *J. Chromatogr.* 1990, *508*, 309–317.

[105] Bij, K. E., Horváth, C., Melander, W. R., Nahum, A., *J. Chromatogr.* 1981, *203*, 65–84.

[106] Seifar, R. M., Kraak, J. C., Poppe, H., Kok, W. T., *J. Chromatogr. A* 1999, *832*, 133–140.

[107] Seifar, R. M., Kok, W. T., Kraak, J. C., Poppe, H., *Chromatographia* 1997, *46*, 131–136.

[108] Seifar, R. M., Heemstra, S., Kok, W. T., Kraak, J. C., Poppe, H., *J. Microcol. Sep.* 1998, *10*, 41–49.

[109] Seifar, R. M., Kraak, J. C., Kok, W. T., Poppe, H., *J. Chromatogr. A* 1998, *808*, 71–77.

[110] Bailey, C. G., Yan, C., *Anal. Chem.* 1998, *70*, 3275–3279.

[111] Lim, J., Zare, R. N., Bailey, C. G., Rakestraw, D. J., Yan, C., *Electrophoresis* 2000, *21*, 737–742.

[112] Huber, C. G., Choudhary, G., Horváth, C., *Anal. Chem.* 1997, *69*, 4429–4436.

[113] Majors, R. E., *LC.GC* 1998, *16*, 96–103

[114] Helboe, T., Hansen, S. H., *J. Chromatogr. A* 1999, *836*, 315–324.

[115] Euerby, M. R., Gilligan, D., Johnson, C. M., Bartle, K. D., *Analyst* 1997, *122*, 1087–1088.

[116] Euerby, M. R., Gilligan, D., Johnson, C. M., Roulin, S. C. P., Myers, P., Bartle, K. D., *J. Microcol. Sep.* 1997, *9*, 373–387.

[117] Gübitz, G., Schmid, M. G., *J. Chromatogr. A* 1997, *792*, 179–225.

[118] Govindaraju, K., Ahmed, A., Lloyd, D. K., *J. Chromatogr. A* 1997, *768*, 3–8.

[119] Knox, J. H., Grant, I. H., *Chromatographia* 1991, *32*, 317–328.

[120] Kaneta, T., Tanaka, S., Taga, M., *J. Chromatogr. A* 1993, *653*, 313–319.

[121] Lucy, C. A., Underhill, R. S., *Anal. Chem.* 1996, *68*, 300–305.

[122] Cohen, N., Grushka, E. J., *J. Chromatogr. A* 1994, *678*, 167–175.

[123] Bosch, S. E., Heemstra, H., Kraak, J. C., Poppe, H., *J. Chromatogr. A* 1996, *755*, 165–177.

[124] Yu, C., Svec, F., Fréchet, M. J., *Electrophoresis* 2000, *21*, 120–127.

[125] Peters, E. C., Petro, M., Svec, F., Fréchet, M. J., *Anal. Chem.* 1997, *69*, 3646–3649.

[126] Ericson, C., Hjertén, S., *Anal. Chem.* 1999, *71*, 1621–1627.

[127] Alicea-Maldonado, R., Colón, L. A., *Electrophoresis* 1999, *20*, 37–42.

Review

Pier Giorgio Righetti[1]
Cecilia Gelfi[2]
Alessandra Bossi[1]
Erna Olivieri[1]
Laura Castelletti[1]
Barbara Verzola[1]
Alexander V. Stoyanov[3]

[1]University of Verona,
Department of Agricultural &
Industrial Biotechnologies,
Verona, Italy
[2]ITBA, CNR, L.I.T.A.,
Segrate (Milano), Italy
[3]Institute of Physical Chemistry,
Russian Academy of Sciences,
Moscow, Russia

Capillary electrophoresis of peptides and proteins in isoelectric buffers: An update

Capillary electrophoresis in acidic, isoelectric buffers is a novel methodology allowing fast protein and peptide analysis in uncoated capillaries. Due to the low pH adopted and to the use of dynamic coating with cellulose derivatives, silanol ionization is essentially suppressed and little interaction of macromolecules with the untreated wall occurs. In addition, due to the low conductivity of quasi-stationary, isoelectric buffers, high-voltage gradients can be applied (up to 800 V/cm) permitting fast peptide analysis with a high resolving power due to minimal diffusional peak spreading. Four such buffers are here described: cysteic acid (Cys-A, p*I* 1.85), iminodiacetic acid (IDA, p*I* 2.23), aspartic acid (Asp, p*I* 2.77) and glutamic acid (Glu, p*I* 3.22). A number of applications are reported, ranging from food analysis to the study of folding/unfolding transitions of proteins.

Keywords: Isoelectric buffers / Capillary electrophoresis / Peptide analysis / Review   EL 4176

Contents

1 Introduction

Zone electrophoresis in isoelectric buffers has had a long gestation period, since the seminal papers by Mandecki and Hayden (dealing with slab-gel electrophoresis of

---

Correspondence: Prof. Pier Giorgio Righetti, University of Verona, Department of Agricultural & Industrial Biotechnologies, Strada Le Grazie No. 15, 37134 Verona, Italy
E-mail: righetti@mailserver.unimi.it
Fax: +39-045-8027901

Abbreviations: Acp, acylphosphatase; **Cys-A**, cysteic acid; **HEC**, hydroxyethylcellulose; **IDA**, iminodiacetic acid; β-**ME**, β-mercaptoethanol

DNA in histidine buffers) [1], by Bier *et al.* [2, 3] (proposing preparative free-flow electrophoresis in a cycloserine buffer) and by Hjertén *et al.* [4] (the latter dealing with CZE of proteins in isoelectric Lys and other amphoteric buffers). Starting in 1997, we developed extensively the technique in the CZE format and we give here a list of our major contributions:

(i) Methodological papers. Our first report dealt with the use of isoelectric Asp (p*I* 2.77 at 50 mM concentration) for peptide mapping [5]. Other suitable buffers we described comprise iminodiacetic acid (IDA, p*I* 2.23) [6], cysteic acid (p*I* 1.85 at 100 mM concentration) [7] and Glu (p*I* 3.22) [8]. A report on their properties and their correct use appeared in Current Protocols [9].

(ii) Theoretical papers. These two reports [10, 11] deal extensively with the fundamental properties of such isoelectric buffer and offer guidelines for their selection.

(iii) Reviews. A few reviews on this topic have already appeared [12–15]. The present one is an update of our paper published in 1997 from the pages of this journal [12].

(iv) Applications. A number of methodological developments, covering some areas of food analysis and biomedical aspects, have already appeared, dealing with: wheat cultivar discrimination *via* analysis of gliadins [16]; identification of maize lines *via* CZE of zeins [17, 18]; tryptic maps of α- and β-globin chains [19], and detection of neutral mutations in γ- and β-globin chains [20, 21]; determination of cow's milk and ripening time in nonbovine and mixed cheeses [22, 23].

(v) Studies on protein folding/unfolding. Although we have only a couple of papers on this subject [24, 25], it appears that this technique has much to offer in these kinds of studies, since it is fast, reliable and very sensitive. Because it can physically separate misfolded polypeptide chains, it would seem ideally suited for those studies regarding protein aggregation due to misfolding, such as amyloidosis and prion proteins.

(vi) Studies on oligonucleotide and DNA fragment separations. Although not the subject of the present review, these studies are of high interest, in that they demonstrate unique separations of DNA fragments ranging in size from a few nucleotides up to 6–700 bp (26–32). Most of the work here reported was conducted in isoelectric His buffer, except for [32], in which 60-mers demonstrating the separation of the β-39 missense mutation in thalassemia were separated in free solution (*i.e.*, in the absence of sieving liquid polymers) in isoelectric IDA buffer in 7 M urea (apparent pH 3.3). In His buffer, in the analysis of native, ds-DNA, it was reported that His would massively coated the DNA double helix, leading to more rigid and expanded coils, to the point at which even small fragments would be sieved in dilute polymer solutions in which size discrimination would normally be lost in conventional buffers [31].

The present review will thus be limited to the topics illustrated above and will not cover general applications of CZE to peptide/protein analysis. For that, the readers are referred to extensive reviews covering peptide applications [33], biotechnological proteins [34], proteins in general [35], and proteins in food analysis [36].

2 Some general, theoretical and experimental aspects

If peptide/protein separation is sought in uncoated capillaries, a possible solution is to work at a pH close to the neutralization point of the silica wall, determined as pH 2.3. Given an average pK value of the silanols of 6.3, at pH 2.3 essentially all silanols should be protonated, thus should be unable to adsorb proteinaceous samples by an ion-exchange mechanism [37]. We have tested and evaluated four such acidic, amphoteric buffer: cysteic acid (pI 1.85), IDA (pI 2.23); Asp (pI 2.77) and Glu (pI 3.22). Their properties are listed in Table 1. Please note the following:

(i) The pI is not an absolute value, but a limiting pH value: at vanishing concentrations of the amphotere in solution, the pH will tend to the pI of pure water (pH 7.0); at high enough concentration, the pH will tend to the true pI of the amphotere. For cysteic acid (Cys-A) and IDA, the given pI value corresponds to a 100 mM solution; for Asp and Glu, it corresponds to a 50 mM solution.

Table 1. Properties of amphoteric, acidic, isoelectric buffers

Electrolyte	Chemicals formulas	pK_1	pK_2	pK_3	pI	$pI - pK_{prox}$
Cys-A	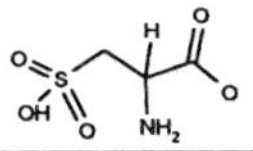	1.55	2.15	8.7	1.85	0.3
IDA	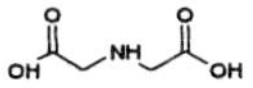	1.73	2.73	–	2.23	0.5
Asp	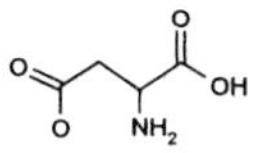	1.88	3.66	9.9	2.77	0.89
Glu	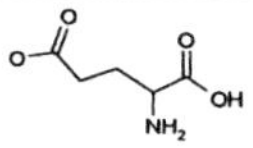	2.19	4.25	9.5	3.22	1.03

(ii) The ($pI - pK_{prox}$) difference represents the hall mark of the amphotere. Very small values (up to 1.0) indicate a very good species, possessing a high buffering capacity at pH = pI. Larger values (2 and greater) indicate a poor compound, unable to properly buffer at its pI value. The pK_{prox} symbol indicates a protolytic group active in the proximity of the pI value. In the case of Table 1, it refers to either pK_1 or pK_2. Note that it is quite irrelevant what pK value will be displayed by the amino group in the molecule (here referred to as pK_3); any pK value above 6.0 will be compatible with the given pIs.

(iii) Solubility. Cys-A and IDA display very high solubilities in water, whereas Asp is soluble only up to 50 mM (at 25°C).

(iv) Compatibility with organic solvents. Cys-A and IDA are compatible with a number of organic solvents, such as trifluoroethanol (TFE) and hexafluoropropanol (HFP), which impart unique electrophoretic properties to peptides. On the contrary, Asp is compatible with only 10% TFE.

(v) Conductivity. The lower the pI, the higher the conductivity of the amphotere, due to dissociation of bulk water. Thus Cys-A and IDA should preferably be used in presence of organic solvents (such as TFE and HFP) or in 7 M urea, which, in addition to modulating the mobility of peptides/proteins, also act as conductivity quenchers.

(vi) Addition of liquid polymers. Although the acidic pH values adopted should ensure absence of interaction of peptide/proteins to the silica wall, in reality this risk always exists. Thus, in all the buffer formulations reported, small amounts of liquid polymers (typically hydroxyethylcellulose (HEC), 27 000 Da molecular mass, at a level of 0.5%) are always added. HEC is dissolved at concentrations well below the entanglement threshold, thus it does not exert any sieving of the macromolecular analytes. On the contrary, it is very effective in forming a "dynamic

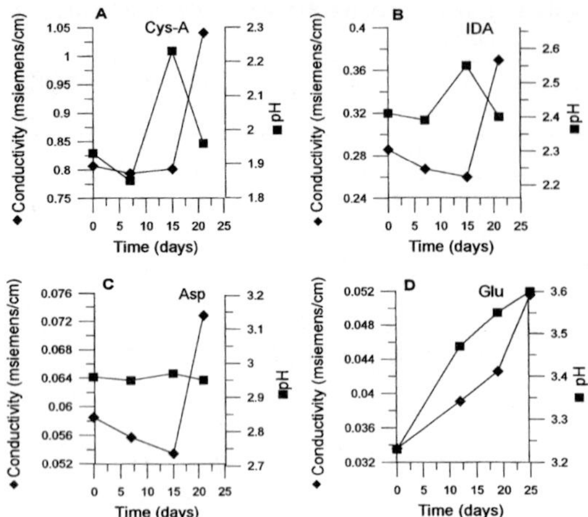

Figure 1. Conductivity and pH profiles of (A) Cys-A, (B) IDA, (C) Asp and (D) Glu upon ageing for three weeks at room temperature. The four amphoteres were dissolved at 50 mM concentration in plain water as a solvent. Reprinted from [7], with permission.

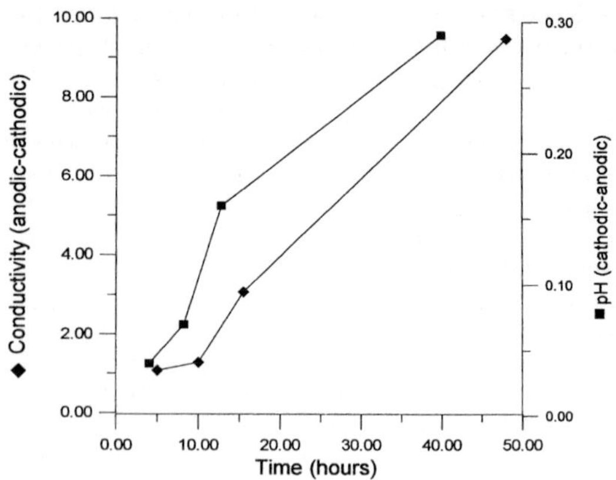

Figure 2. Conductivity and pH profiles of a 40 mM Asp, 6 M urea and 0.5% HEC solution used for up to 50 h as anolyte and catholyte. Electrophoretic conditions: 20 kV at room temperature without any change of electrolytes. The data are expressed as the difference in conductivity and pH measured in the two electrodic vessels. Reprinted from [7], with permission.

coating" onto the silica wall, thus minimizing binding of polypeptides.

(vii) Use of neutral and alkaline amphoteres. Although His and Lys could be excellent amphoteric ions for such separations, we have chosen not to deal with them, as mentioned above. His has been excellent for oligonucleotide separations [26–31] any Lys has been reported for analysis of native proteins [4]. However, in both cases, one should use coated capillaries, thus considerably adding to the burden and expenses of the technique.

Another important aspect of the use of isoelectric buffers regards their shelf life when stored at room temperature, dissolved as pure species in plain water as a solvent. In this case, the final pH of such solutions will approach the p*I* value of each compound. This study was deemed necessary due to the fact that quite a number of them have strongly acidic p*I* values, thus they could potentially degrade at such local pH values. As indicators of potential degradation, pH and conductivity changes as a function of storage time were evaluated. Figure 1 gives these variations for the four compounds of Table 1, in order of increasing p*I* values: Cys-A (A), IDA (B), Asp (C), and Glu (D). In all cases, these changes have been monitored for a period of up to three weeks. There seems to be a general trend emerging in all four cases: the pH and conductivity of all solutions remain stable for a period of *ca.* ten days, then there appear to be fluctuations in both parameters. The pH changes, in some cases, appear to be very minute (*e.g.*, Fig. 1C, Asp) or to be nonmonotonic, so it

would seem that they could be neglected. On the contrary, the conductivity increases steadily for all solutions analyzed, and this increment (in some cases quite marked) appears to be substantial after a period of two weeks. This process seems to be markedly accelerated when running CZE in presence of 6 M urea.

Another practical point of interest to users of CZE is how often one should replenish the anolyte and catholyte solutions, so as to avoid electrode polarization and marked changes of buffer pH and conductivity due to buffering ion and counterion depletion. This point, in principle, should not be critical when using isoelectric buffers, since, due to the fact that they should not bear a net charge, their depletion at either electrode should be negligible. Figure 2 explores potential pH and conductivity variations as a function of time of operation of the CZE instrument under a constant electric field of 20 000 V, when using Asp as the sole buffering species in all compartments (anode, cathode, capillary lumen). The data are expressed as difference in pH and conductivity between the two electrode vessels. It is noted that, upon continuous operation, the pH keeps increasing at the cathode, whereas the conductivity keeps augmenting at the anode. The meaning of that: there must be a net transport of Asp from the cathodic to the anodic vessels. If that is the case, the pH should keep increasing at the cathode and decreasing a the anode. Decrements of pH in the anodic vessel should bring about increments of conductivity. This is to be expected. In fact, at progressively more acidic pH values, a fraction of the amphotere should be nonisoelectric, in order to neutralize the excess positive charges brought about by bulk water ionization. For example, in a solution of Asp at pH 3.0, there must be 1 mM proton concentration which must be neutralized by a 1 mM fraction of Asp being nonisoelectric, but negatively charged. This excess, nonisoelectric molar fraction will constantly migrate towards the anode, creating an unbalance in concentration of Asp in the two electrodic vessels. This phenomenon can be neglected during the first 10 h of operation, but after that it becomes quite appreciable.

As a conclusion of the above findings, the following guidelines should apply to CZE in isoelectric, acidic buffers: (i) The buffers, especially if made in presence of urea, should be freshly prepared every week and discarded after this period. (ii) The anolyte and catholyte solutions should be refreshed after each run when using the most acidic buffer (Cys-A); as the p*I* increases, the electrodic solution can be progressively used for a few sequential runs before replenishing; *e.g.*, in case of Asp, this buffer can be used for a full day of operation before being discarded. (iii) As a corollary of the above, monitoring the electrodic vessels for pH and conductivity changes will give a sure indication of when to refresh these solutions.

3 Examples of some separations

3.1 Identification of maize lines *via* CZE of zeins

Variety identification in maize lines has been achieved by CZE of zeins (the prolamins or seed storage proteins in maize) [17, 18]. The running buffer consisted of 40 mM isoelectric Asp, in presence of 6 M urea and 0.5% HEC (apparent pH 3.8; p*I* 2.77 in the absence of urea). A total of 31 different zein peaks were mapped out of a total of 21 different maize lines. Each line exhibited typically seven to twelve peaks, with some lines showing up to 20 zein bands. Due to slightly changing elution times, caused by a lack of reproducibility of the electroendoosmotic flow in uncoated silica surfaces, correct peak assignment and alignment among different runs was obtained by multivariate statistical analysis. This CZE method compared well, both in resolution and total number of peaks, with current protocols adopted for screening of maize inbreds, which consist of isoelectric focusing in agarose gels. Figure 3 gives typical CZE profiles of three related lines, the W64A+ and its fl2 and o2 variants.

3.2 Determination of cow's milk and ripening time in nonbovine cheese

CZE in isoelectric, acidic buffers (50 mM IDA, pH = p*I* 2.3) was successfully reported also for analyzing adulteration of goat and ewe cheeses with cow's milk. The cheese

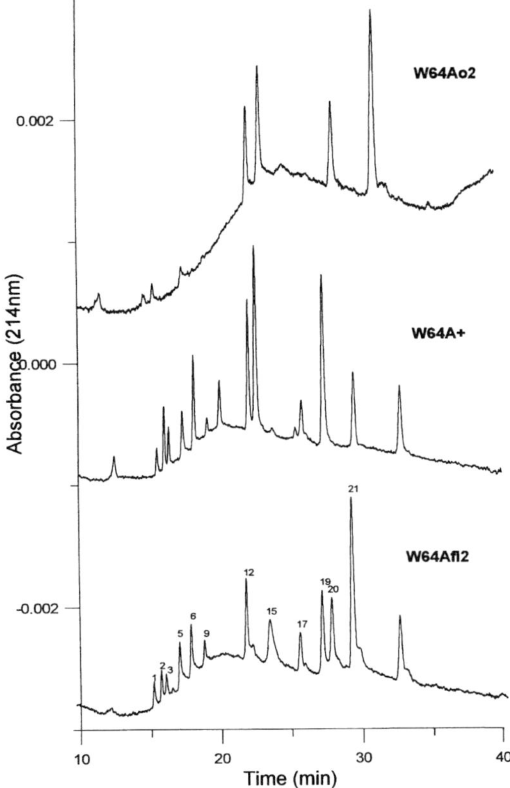

Figure 3. Representative CZE runs of zeins, obtained by 70% ethanol, 2% β-ME extraction of ground endosperm, from three related lines of maize. Run performed in 40 mM Asp buffer, 6 M urea and 0.5% HEC (apparent pH 3.7). Conditions: 50 μm ID, 30 cm long capillary, run at 800 V/cm and 30°C. Detection at 214 nm. From top to bottom: W64Ao2; W64A+; W64Afl2. Peak numbering as obtained by multivariate statistical analysis. Reprinted from [18], with permission.

samples were extracted with a 20:80 v/v ethanol-water mixture in presence of 3 M urea and 1% β-mercaptoethanol (β-ME) for 1 h. After centrifugation and lipid extraction, the samples were dissolved in 50 mM IDA, 6 M urea and 0.5% HEC and analyzed by CZE at 700 V/cm. A total of 18 characteristic peaks were resolved among the three types of cheeses and 18 variables were defined as their respective areas. There was an excellent similarity among the electrophoretic patterns obtained with cheeses of a given type of milk, while cheeses made with different types of milk were easily distinguishable. Most peaks were common to all cheeses, but the profile was different depending of the type milk used. Principal component analysis, linear discriminant analysis and partial least

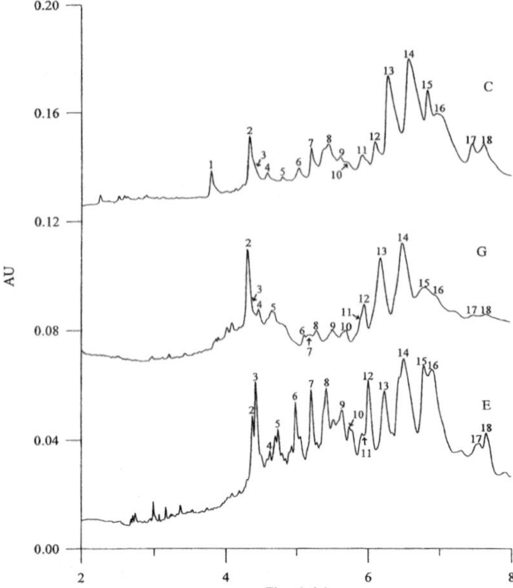

Figure 4. Electropherogram of cow (C), goat (G) and ewe (E) cheeses, showing the 18 peaks selected as variables (60 mAUFS for each electropherogram). CZE conditions: 50 mM IDA, 6 M urea and 0.5% HEC; 700 V/cm. Reprinted from [22], with permission.

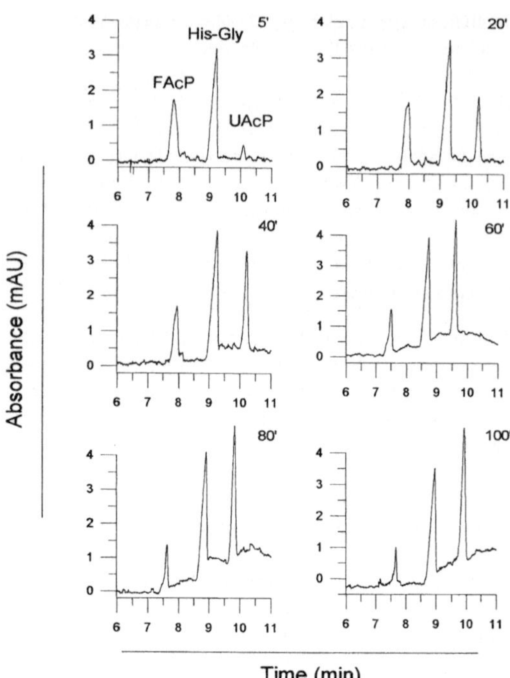

Figure 5. Unfolding kinetics of AcP in 7 M urea solutions. At the indicated times (5 to 100 min incubation outside the capillary) the samples were injected and run under the following conditions. Waters Quanta 4000E instrument, equipped with a 75 μm ID uncoated capillary 36 cm total length; 28 cm to the observation window. Protein samples were routinely injected into the capillary for 6 s. Electrophoresis was performed in a capillary containing 50 mM Glu and 1 mM TEPA (pH 3.3). Running conditions: 10 kV at 18 μA, 20°C. Detection at 214 nm. Sample concentration: 0.5 mg/mL of AcP and 0.2 mg/mL of His-Gly (used as an internal standard). FAcP: folded AcP; UAcP: unfolded AcP. Reprinted from [25], with permission.

squares regression (PLS) were used for statistical analysis of the data obtained by CZE. In particular, by using PLS multivariate regression, the contents of cow's milk in presumably pure goat and ewe cheeses, as well as in binary and ternary mixtures, could be predicted with relative standard deviations of *ca.* 6–7%. In addition, the ripening time in goat and ewe cheeses could also be predicted [22, 23]. Figure 4 gives the typical profiles of protein/peptides extracted from cow's (C), goat (G) and ewe's (E) cheeses.

3.3 Monitoring equilibria and kinetics of protein folding/unfolding

CZE in acidic, isoelectric buffers appears to be also an excellent method for studying conformational transitions of protein due to denaturing agents. A nice example comes from the study of acylphosphatase (AcP), a small, globular enzyme of 98 residues, exhibiting slow unfolding kinetics [25]. The sample was run in 50 mM isoelectric glutamic acid (pH = p*I* 3.2) added with 1 mM oligoamine (tetraethylenepentamine; TEPA) for quenching protein interaction to the capillary wall (final pH 3.3). Muscle AcP, in this buffer, exhibited a free solution mobility of 2.63×10^{-4} cm^2 V^{-1} s^{-1}. By studying the unfolding kinetics, as a func-

tion of time of incubation in 7 M urea, it was possible to measure the rate constant of the unfolding reaction, estimated to be 0.00030 ± 0.00006 s^{-1} (see Fig. 5). The same measurements, when repeated *via* spectroscopic monitoring of intrinsic fluorescence, gave a value of 0.00034 ± 0.00002 s^{-1}, thus in excellent agreement with CZE data. By equilibrium unfolding CZE studies, it was possible to construct the typical sigmoidal transition of unfolding *vs.* urea molarity: the midpoint of this transition, at which the folded and unfolded states should be equally populated, was estimated to be at 4.56 M urea (see Fig. 6). Similar experiments by fluorimetric analysis gave a value of 4.60 M area as midpoint of the unfolding curve.

4 Silica and proteins: les liaisons dangereuses?

This is a most intriguing question, to which unfortunately there is no easy answer. The silica wall seems to act like a Leviathan (or Moloch) towards proteins: if it can adsorb them, it surely will, no matter how careful is the experimenter in trying to avoid it. There are innumerable recipes and protocols described for minimizing protein interaction with fused silica (for a review, see [38]). In addition, we have recently described newer protocols, which help assessing the phenomenon and give figures of merit for a number of dynamic coating procedures, such as with amines and oligoamines [39], with adsorption of neutral polymers [40] and *via* shielding with surfactants (both neutral and zwitterionic) [41]. There seems to be no universal remedy and a perfect cure for the phenomenon.

In case of proteins, it is believed that the biggest offender is ion-ion interaction between the negatively charged silanols (at suitable operative pH values, in general above pH 3) and positively charged protein surfaces, here too at suitable pH values (in general when the pH of the background electrolyte is below the p*I* of the protein, which thus acquires a net positive charge; but beware, even when a protein is 1 or 2 pH units above the p*I* plenty of interaction can occur). These interactions are, in general, distance-dependent, with the energies being inversely proportional to the distance r (or a given power of it) separating the two groups. As the power of the inverse distance dependency increases, the interaction approaches zero more rapidly as r increases, and thus becomes a short-range interaction. The interaction energy between two charges varies as $1/r$, which qualifies it as a long-range interaction. At the other extreme are the induced dipole-induced dipole (of dispersion) interactions, with a dependence on $1/r^6$, which classifies them in the domain of very short-range interactions. This last type of interaction describes the natural tendency of atoms to attract, regardless of charge and polarity, because of the polarizability of their electron clouds. Given the above, it is highly probable that ion-ion interactions predominate on the general mechanism of protein deposition onto silica surfaces, although in should be considered that such interaction becomes shielded in a polar medium, and is therefore weakened.

In addition, it is known that, close to the silica surface, there is a monomolecular layer of water tightly bound to the wall, which should further shield such ion-ion interactions. It is quite probable though that, under conditions of high charges on both the silica and protein surfaces, cooperative phenomena prevail over shielding effects and adsorption takes place. But why should adsorption occur

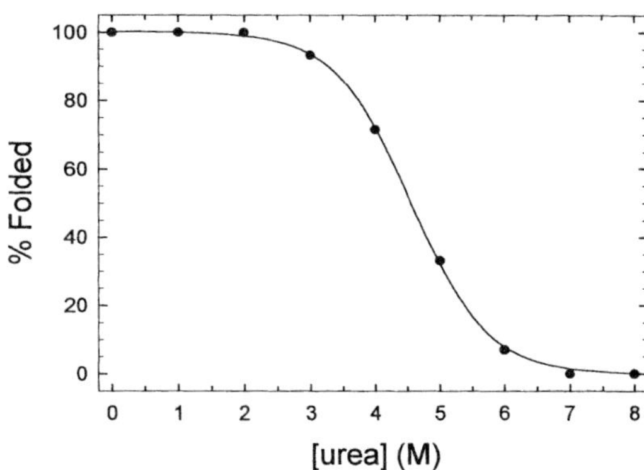

Figure 6. Equilibrium unfolding in different urea molarities. AcP was first incubated for 3 h at room temperature at the different urea molarities and then injected in CZE. The decrement of peak area at the various urea concentration was plotted *vs.* urea molarities. The midpoint of the denaturation curve, at which the folded and unfolded states are about equally populated, was found at 4.56 M urea. Reprinted from [25], with permission.

even under our conditions of operating in a strongly acidic milieu? (We did not want to disclose this, in the hope that the readers would come to the end of the article, but, yes, there is non-negligible adsorption of protein/large peptides even in our system). When very few (and sparse) silanols are ionized, cooperativity of binding cannot be invoked. The mechanism of adsorption could then be due to residual van der Waals interactions. Although these interactions, as we have seen, are extremely weak (they decay with a power law of $1/r^6$), the situation is more complex than that. This power law refers to interactions between two point-like objects. When dealing with two surfaces (silica on one side, protein on the other) the power law becomes $1/r^2$, indicating a much stronger interaction. Thus, it is quite possible that the adsorption mechanism in our case is driven by such residual van der Waals forces, and then it is amplified by additional interaction mechanisms once the protein is deposited onto the silica surface.

Paradoxically, adsorption is truly magnified if the protein under study is not native, but present in solution as a random coil, thus fully denatured. This has been the case in most of our separations in very acidic isoelectric buffers, where proteins are in purpose denatured in 6–8 M urea, in some cases for solubilizing them, in others to prevent native/denatured equilibria. When proteins are fully denatured, plenty of hydrophobic residues are exposed to the solvent, but we cannot hypothesize hydrophobic interaction between such residues and the silica wall as the mechanism responsible for their adsorption, since the silica wall is, by its own nature, highly hydrophilic. Thus, it is quite possible that here too the first mechanism of attraction to the wall is *via* van der Waals forces. Once the denatured protein is adsorbed onto the silica surface, it will offer now a hydrophobic moiety which will be available to additional protein molecules to be adsorbed onto. Thus, layers and layers of proteins could built up, justifying the massive adsorption mechanism which we often notice when dealing with denatured proteins (and which we cannot simply find when running native proteins).

Although the additives we have suggested help considerably in lessening the phenomenon (HEC, but also tetraethylene pentamine and spermine are, individually, strongly effective additives), they do not abolish it. So, the only other remedy is to watch the decay of separation performance of a given capillary upon repeated injections of a given sample. When the separation parameters are not any longer acceptable, the capillary should be washed free of adsorbed proteinaceous (but also other components of the sample matrix) material. What is the best washing procedure? If one aims at keeping the residual electroendoosmotic flow constant and reproducible, one should avoid rough washing cycles with NaOH or with HCl. The best procedure we have found is a frontal elution protocol, consisting in loading SDS micelles at the cathodic side and driving them electrophoretically into the capillary lumen. SDS is the most efficient sweeper ever invented [39–41].

5 Conclusions

We hope we have given to the readers a panoramic view of what could be achieved when separating proteins/peptides in acidic, isoelectric buffers. The technique is simple, reliable and truly user-friendly. It has the non-negligible advantage, among others, of not requiring any covalent coating procedure, except for transient, dynamic coatings. It can be helpful in a number of applications, with the proviso that not (or minimal) charge modulation can be implemented, since we work at pH values largely suppressing carboxyl ionization, where proteins/peptides will exhibit maximum positive charge. In case of coincident pH/mobility curves of our analytes, the best pH values allowing fanning out of the titration curves will be typically in the pH 5–8 region. If the readers have profitably enjoyed this review, then we hope we do not have to close it with the verses of Martial (in his famous Epigrams):

Versus scribere me parum severos,
Nec quos praelegat in schola magister,
Corneli, quereris...,

Quare deposita severitate,
Parcas lusibus et iocis rogamus,
Nec castrare velis meos libellos
(Cornelius, you always complain of my nonserious verses, which could not possibly be declaimed in front of scholars.... mitigate your temper, absolve these verses and do not try to castrate our sentences).

This work was supported in part by a grant from MURST (Coordinated Project 40%, Folding/Unfolding of proteins, 1999) and by ASI (Agenzia Spaziale Italiana, grant No. I/R/28/00) to PGR.

Received May 31, 2000

6 References

[1] Mandecki, W., Hayden, M., *DNA* 1988, *7*, 57–62.

[2] Bier, M., Long, T., *J. Chromatogr.* 1992, *604*, 73–83.

[3] Bier, M., Ostrem, J., Marquez, R. B., *Electrophoresis* 1993, *14*, 1011–1018.

[4] Hjertén, S., Valtcheva, L., Elenbring, K., Liao, J. L., *Electrophoresis* 1995, *16*, 584–594.

[5] Nembri, F., Righetti, P. G., *J. Chromatogr. A* 1997, *772*, 203–211.

[6] Bossi, A., Righetti, P. G., *Electrophoresis* 1997, *18*, 2012–2018.

[7] Bossi, A., Righetti, P. G., *J. Chromatogr.* 1999, *840*, 117–129.

[8] Bossi, A., Olivieri, E., Castelletti, L., Gelfi, C., Hamdan, M., Righetti, P. G., *J. Chromatogr. A* 1999, *853*, 71–82.

[9] Righetti, P. G., Bossi, A., Gelfi, C., *Curr. Prot. Protein Sci.* 1999, 10.3.1–10.3.16.

[10] Stoyanov, A. V., Righetti, P. G., *J. Chromatogr. A* 1997, *790*, 169–176.

[11] Stoyanov, A. V., Righetti, P. G., *Electrophoresis* 1998, *19*, 1674–1676.

[12] Righetti, P. G., Gelfi, C., Perego, M., Stoyanov, A. V., Bossi, A., *Electrophoresis* 1997, *18*, 2145–2153.

[13] Righetti, P. G., Bossi, A., Gelfi, C., *J. Capil. Electrophor.* 1997, *4*, 47–59.

[14] Righetti, P. G., Bossi, A., *Anal. Chim. Acta* 1998, *372*, 1–19.

[15] Righetti, P. G., Bossi, A., Olivieri, E., Gelfi, C., *J. Biochem. Biophys. Methods* 1999, *40*, 1–16.

[16] Capelli, L., Forlani, F., Perini, F., Guerrieri, N., Cerletti, P., Righetti, P. G., *Electrophoresis* 1998, *19*, 311–318.

[17] Righetti, P.G., Olivieri, E., Viotti, A., *Electrophoresis* 1998, *19*, 1738–1741.

[18] Olivieri, E., Viotti, A., Lauria, M., Simò-Alfonso, E., Righetti, P. G., *Electrophoresis* 1999, *20*, 1595–1604.

[19] Capelli, L., Stoyanov, A. V., Wajcman, H., Righetti, P. G., *J. Chromatogr. A* 1997, *791*, 313–322.

[20] Righetti, P. G., Saccomani, A., Gelfi, C., Stoyanov, A., *Electrophoresis* 1998, *19*, 1733–1737.

[21] Saccomani, A., Gelfi, C., Wajcman, H., Righetti, P. G., *J. Chromatogr. A* 1999, *832*, 225–238.

[22] Herrero-Martinez, J. M., Simò-Alfonso, E. F., Ramis-Ramos, G., Gelfi, C., Righetti, P. G., *Electrophoresis* 2000, *21*, 633–640.

[23] Herrero-Martinez, J. M., Simò-Alfonso, E. F., Ramis-Ramos, G., Gelfi, C., Righetti, P. G., *J. Chromatogr. A* 2000, *878*, 261–271.

[24] Stellwagen, E., Gelfi, C., Righetti, P. G., *J. Chromatogr. A* 1999, *838*, 131–138.

[25] Verzola, B., Chiti, F., Manao, G., Righetti, P. G., *Anal. Biochem.* 2000, *282*, 239–244.

[26] Stoyanov, A. V., Gelfi, C., Righetti, P. G., *Electrophoresis* 1997, *18*, 717–723.

[27] Perego, M., Gelfi, C., Stoyanov, A. V., Righetti, P. G., *Electrophoresis* 1997, *18*, 2915–2920.

[28] Gelfi, C., Perego, M., Righetti, P. G., Cainarca, S., Firpo, S., Ferrari, M., Cremonesi, L., *Clin. Chem.* 1998, *44*, 906–913.

[29] Gelfi, C., Mauri, D., Perduca, M., Stellwagen, N. C., Righetti, P. G., *Electrophoresis* 1998, *19*, 1704–1710.

[30] Stellwagen, N. C., Gelfi, C., Righetti, P. G., *J. Chromatogr. A* 1999, *838*, 179–189.

[31] Magnusdottir, S., Gelfi, C., Hamdan, M., Righetti, P. G., *J. Chromatogr. A* 1999, *859*, 87–98.

[32] Gelfi, C., Viganò, A., Carta, P., Manchi, P., Cossu, G., Righetti, P. G., *Electrophoresis* 2000, *21*, 780–784.

[33] Kašička, V., *Electrophoresis* 1999, *20*, 3084–3105.

[34] Lagu, A., L., *Electrophoresis* 1999, *20*, 3145–3155.

[35] Dolnik, V., *Electrophoresis* 1999, *20*, 3106–3115.

[36] Frazier, R. A., Ames, J. M., Nursten, H. E., *Electrophoresis* 1999, *20*, 3156–3180.

[37] Bello, M. S., Capelli, L., Righetti, P. G., *J. Chromatogr. A* 1994, *684*, 311–320.

[38] Chiari, M., Nesi, M., Righetti, P. G., in: Righetti, P. G. (Ed.), *Capillary Electrophoresis in Analytical Biotechnology*, CRC Press, Boca Raton, FL 1995, pp. 1–336.

[39] Verzola, B., Gelfi, C., Righetti, P. G., *J. Chromatogr. A* 2000, *868*, 85–89.

[40] Verzola, B., Gelfi, C., Righetti, P.G., *J. Chromatogr. A* 2000, *874*, 293–303.

[41] Castelletti, L., Verzola, B., Gelfi, C., Stoyanov, A., Righetti, P. G., *J. Chromatogr. A* 2000, *894*, 281–289.

Electrophoresis 2000, *21*, 3994–4016

Review

Frank Steiner
Marc Hassel

University of the Saarland,
Instrumental/Environmental
Analysis, Saarbrücken,
Germany

Nonaqueous capillary electrophoresis: A versatile completion of electrophoretic separation techniques

Nonaqueous capillary electrophoresis (NACE) is the application of a conductive electrolyte dissolved in either one organic solvent or a mixture of several organic solvents to carry out zone electrophoresis or related techniques in fused-silica capillaries. A complete review on the fundamentals, the optimization of analytical methods, practical considerations, and applications is given. To explain the differences to CE in aqueous media, a brief summary on solvent properties and molecular interactions in solutions introduces the reader into these fields. The use of additives to tune separation selectivity by means beyond a pure zone-electrophoretic mechanism is discussed in detail for organic media. Special detection techniques providing high potential for NACE are presented. Data on the precision of NACE methods and a list of relevant applications are included. More specialized applications like the determination of physicochemical constants in NACE or the setup of a semipreparative mode are described.

Keywords: Nonaqueous capillary electrophoresis / Organic solvents / Electroosmotic flow / Additives / Detection / Review   EL 4198

Contents

1 Introduction

For the first time in the history of capillary electrophoresis (CE), Walbroehl and Jorgenson [1] in 1984 published the results of an analysis in a pure organic solvent. They separated five quinoline-type compounds in acetonitrile (ACN) applying a tetraethylammonium perchlorate/hydrochloric acid electrolyte. Although they already considered nonaqueous capillary electrophoresis (NACE) to deserve further investigation, nobody followed their recommendation in the 1980s. Several publications on the addition of organic modifiers to the aqueous buffer in capillary zone electrophoresis (CZE) were published in those years [2–7], followed by recent contributions to systematic studies on water-methanol [8], water-ethanol [9], water-1-propa-

Correspondence: Dr. Frank Steiner, University of the Saarland, Instrumental/Environmental Analysis, P.O. Box 151150, 66041 Saarbrücken, Germany
E-mail: f.steiner@mx.uni-saarland.de
Fax: +49-681-3022963

Abbreviations: AN, acceptor number; **CTC**, chlortetracycline; **DMA**, *N,N*-dimethylacetamide; **DN**, donor number; **EDL**, electric double layer; **FA**, formamide; **NACE**, nonaqueous capillary electrophoresis; **NMF**, *N*-methylformamide; **OTC**, oxytetracycline; **TC**, tetracycline

nol [10] and water-ACN [11] mixtures by Sarmini and Kenndler. In the early papers on partially organic electrolytes in CZE, a modifier content of 40% was rarely exceeded. Some authors reported on problems with system robustness in terms of varying migration times and current breakdown at organic contents higher than 40% [6, 7]. Working at high alcohol contents in the running buffer was considered to slow down separations due to a tremendous decrease in analyte and electroosmotic mobility [7]. The method further struggled with the fact that common knowledge on electrochemistry, molecular interactions, solvation phenomena and protolysis in organic solvents is not comprehensive. However, the reputation of NACE obviously changed after 1993, when Benson *et al.* [12] published the results of a separation of pyrazoloacridine from metabolic degradation products and synthesis impurities in neat methanol. They reported better results for this system than for any investigated water-methanol buffer. With that, rapid development of NACE began and the acceptance of the method was enhanced from year to year. In the meanwhile, the outstanding potential of NACE for certain applications became unambiguous. It was summarized by Riekkola *et al.* [13] in the following words: "...possibly the most attractive feature of organic solvents is that their physical and chemical properties are very different from each other and from water, allowing the important characteristics of separations to be controlled on a wider scale than with water alone. Solvent properties affect the acid-base behavior of analytes in a way that pK_a values can differ up to many orders of magnitude. Hence, enormous variations in electrophoretic mobilities can be achieved and separations not possible in aqueous CE can be performed with excellent selectivities".

A further benefit of NACE over aqueous CE is the increased solubility of several pharmaceutical compounds and long-chain surfactants in organic solvents. This fact extends the applicability of CE to more hydrophobic ionic compounds. Throughout the experiences with mixed aqueous organic solvents, NACE was usually considered a borderline case when working at high organic modifier content. From a different point of view, it is the water that needs to be considered as the borderline case, with the highest solvent strength (interaction with polar solutes) in a large pool of solvents providing a vast variety of physicochemical properties.

2 Early contributions to NACE and fundamental theoretical considerations

In 1994, Sahota and Khaledi [14] reported on the application of pure formamide in CE. They pointed out its solubilization power for many supporting electrolytes due to its

high dielectric constant and the low specific conductance resulting from its high viscosity. Excellent linearity of current *vs.* applied voltage even at 0.1 M K_2HPO_4 content was reported up to 30 kV using a capillary of 57 cm length and 75 µm ID. The authors found out that these facts enabled them to work with wider bore capillaries and/or higher electrolyte concentrations at a tolerable level of Joule heating, thus improving loadability and detectability. Even in such capillaries, the analysis speed and efficiency could be increased, since higher field strengths could be applied. Sahota and Khaledi mentioned the hydrolysis instability and reduced UV transparency as intrinsic drawbacks for formamide. A relatively slow separation of peptides indicated low analyte mobilities in this high-viscosity medium.

An early paper focusing on the versatility of electrophoretic selectivities in NACE was published in 1994 by Bjørnsdottir and Hansen [15]. They reported on the separation of amines having a very similar charge-to-mass ratio in various organic solvents (formamide, *N*-methylformamide, *N,N*-dimethylformamide, dimethyl sulfoxide, methanol, and ACN) using volatile buffers (ammonium acetate). Their observed changes in selectivity were ascribed to a change in the basicity of the tricyclic antidepressants when varying the solvent properties for both donating and accepting hydrogen bonds. This can be achieved very efficiently in methanol-ACN mixtures at different ratios. The authors further reported tremendous changes in the electroosmotic flow as a function of the solvent. An increase by a factor of ten was mentioned when changing from methanol to ACN, showing a maximum in a mixture of 75:25 v/v ACN-methanol. These observations were related to changes in the dielectric constant, viscosity, and zeta potential at the inner capillary wall, but no detailed discussion was given. A later publication from the Hansen group [16] provided more details on the relationship of physicochemical parameters of solvents and their influence on the separation of drug substances.

Jansson and Roeraade [17] gave for the first time a comprehensive introduction into the fundamental theory of CE with respect to the application of organic media. Three equations were derived describing the dependence of ion mobility, electroosmotic mobility and generated theoretical plates per unit time on properties of analyte and solvent.

$$\mu_{ion} = \frac{2}{3}\left(\frac{\varepsilon_0\varepsilon_r\zeta_{ion}}{\eta}\right) \tag{1}$$

$$\mu_{eo} = -\left(\frac{\varepsilon_0\varepsilon_{r,wall}\zeta_{wall}}{\eta_{wall}}\right) \tag{2}$$

$$\frac{N}{t} \propto \frac{r}{kT\eta}(2\zeta_{ion} - \zeta_{wall})^2(\varepsilon_0\varepsilon_r E)^2 \qquad (3)$$

In Eq. (1), μ_{ion} describes the mobility of the analyte ion, ε_r the relative dielectric constant of the medium, ε_0 the permittivity of the vacuum, ζ_{ion} the zeta potential of the analyte ion and η the viscosity of the solvent. In Eq. (2), $\varepsilon_{r,wall}$ and η_{wall} are the relative dielectric constant and viscosity in the electric double layer (EDL) close to the capillary wall and ζ_{wall} is the zeta potential of the capillary wall. The authors mentioned the fact that the dielectric constant and viscosity in the EDL are different from their bulk values and that the ε_r value is considered to be substantially lower in the EDL. Since these data are usually not available, inserting the bulk values (from literature) can be considered sufficient when suitability of different solvents for CE is compared. A linear increase of the ion mobility and the electroosmotic mobility with increasing ε_r/η ratio can be deduced from Eqs. (1) and (2), thus speeding up analysis when working in a coelectroosmotic mode. In Eq. (3), N is the number of theoretical plates, t the migration time of the analyte to the detector, r the Stokes radius of the analyte, k the Boltzmann's constant, T the absolute temperature and E the applied electric field strength. This demonstrates that for a given analyte and field strength the separation efficiency increases with the ε_r^2/η ratio of the solvent.

Jansson and Roeraade [17] achieved data on CE measurements in water-*N*-methylformamide (NMF) mixtures from 0 to 100%. NMF exhibits a high ε_r/η ratio and a very high ε_r^2/η ratio due to its extremely elevated relative dielectric constant ($\varepsilon_r = 182$) which is more than twice that of water. Values on EOF, apparent pH (pH*), electric conductivity, and analyte mobility were plotted *vs.* the content of NMF. In pure NMF the field strength was varied up to 1000 V/cm and no influence on the EOF mobility observed. Very weak Joule heating effects occurred due to the low electric conductivity, which was about 15% of the value measured in water when a 50 mM H$_3$BO$_3$ buffer was applied. Electropherograms of carboxylic acid separations and the separation of strongly basic compounds like propranolol were shown. The authors mentioned that no peak distortion occurred, even when the concentration of the analyte was significantly higher than 1/100 of the background electrolyte concentration. Several publications on the application of NACE for the determination of pharmaceutical compounds practically insoluble in aqueous buffers appeared in 1994. Tomlinson *et al.* [18–20] reported on the analysis of haloperidol and Ng *et al.* [21] on the analysis of tamoxifen and metabolites.

3 Characteristics of solvents applied in NACE

To discuss the vast field of interactions that occur in different organic solvents with very different physicochemical properties, a classification of solvents has been undertaken. In the context of nonaqueous CE, a concise but very convenient classification of solvents was given by Riekkola *et al.* [13]. This classification mainly considers the ability of a solvent to accept or to transfer protons and to undergo an eventual autoproteolysis, which is a characteristic ability of so-called amphiprotic solvents, whereby they distinguished between: (i) amphiprotic solvents with about equally good proton donor and acceptor capabilities (*e.g.*, water and alcohols); (ii) amphiprotic solvents with predominantly proton donor (acidic) properties (*e.g.*, acetic acid); (iii) amphiprotic solvents with predominantly proton acceptor (basic) properties (*e.g.*, formamide and *N*-substituted formamides); (iv) aprotic solvents which are not capable of autoproteolysis but able to accept protons (*e.g.*, ACN or dimethyl sulfoxide); and (v) inert solvents that are neither capable of autoproteolysis nor of accepting or donating protons (*e.g.*, hexane). Pure inert solvents are not relevant for the preparation of running buffers in electrophoresis since they are practically unable to dissolve ions. More detailed types of solvent classification further considered the dielectric constant, thus dividing the amphiprotic classes into dipolar and apolar solvents [22].

An easy-to-read overview on the influence of solvent properties on separations in NACE, especially selectivity, efficiency, and detectability was given by Valkó *et al.* [23]. They refer to the equations given in Section 2 and summarize the related parameters of common solvents (Table 1). It can be deduced from the ε/η and the ε^2/η ratios that NMF and acetonitrile are favorable in terms of electrophoretic and electroosmotic velocity when compared to water. In NMF, extremely high efficiencies should be generated according to theoretical prediction. Conversely, formamide should be considered unfavorable since the ε^2/η ratio predicts a smaller amount of plates/min that for aqueous systems. However, Ward and Khaledi [24] reported higher efficiencies for the separation of oligosaccharides in formamide than in aqueous solutions in a recent paper. The observed effect was related to differences in Joule heating and wall adsorption effects, although the given discussion was not always very straightforward. As can be seen in Table 1, the amount of autoprotolysis differs greatly for various solvents. This leads to scales of pH completely different from water, individually defined as pH* for a distinct solvent. In terms of its influence on the separation in CE, the pH* in an organic solvent cannot be directly compared to the equiv-

Table 1. Physicochemical parameters of selected organic solvents and water

Solvent	η (cP)	ε	ε/η (cP^{-1})	ε^2/η (cP^{-1})	pK_{auto}	Polarity	T_{boil} (°C)
Water	0.89	80	89.9	7191	14.0	10.2	100
FA	3.30	111	33.6	3733	16.8	9.6	210
NMF	1.65	182	110.3	20075	10.7	6.0	182
DMF	0.80	36.7	45.9	1683	29.4	6.4	153
DMA[a]	0.78	37.8	48.5	1831	24.0		166
DMSO	2.00	46.7	23.4	1090	33.3	7.2	189
ACN	0.34	37.5	110.3	4136	*	5.8	82
Methanol	0.54	32.7	60.6	1980	17.2	5.1	65
Ethanol	1.07	24.6	23.0	566	18.9	4.3	78

a) DMA, *N,N*-dimethylacetamide
* No detectable autoprotolysis
Reprinted from [23], with permission.

alent pH in water or to the pH* in another organic solvent. Dissociation constants can vary significantly at nominally equal pH* in different media which, on the other hand, is one of the reasons for the power of NACE to tune selectivity of compounds with similar structures.

A deeper insight into molecular interactions and their relation to solvent parameters has been given by Bowser *et al.* [25] in their review on analyte-additive interctions in NACE. They comprehensively explained the process of dissolving species in liquid phases and the formation of aggregates in solvents as well as the driving forces of these phenomena. The authors basically made divisions into solvophobic interactions, related to the solvent strength, electrostatic interactions and donor-acceptor interactions. The phenomenon of solvophobic interactions is important when dissolving of nonpolar analytes occurs. Unlike polar analytes, they interact very weakly with solvent molecules, making solvation energetically unfavorable. Considering changes in enthalpy and entropy it can be deduced that strong solvents for nonpolar analytes must exhibit weak solvent-solvent interactions to minimize the energy required to disrupt the solvent when the dissolving occurs. Since strong solvent-solute interactions only take place in polar media, solvent strength as a general scale should be related to the ability of a solvent to interact with polar analytes. When less polar analytes are dissolved in polar media, analyte-analyte interaction can minimize the required amount of solvation and thus the unfavorable interruption of the strong solvent-solvent interactions. These nonpolar interactions are named solvophobic and will be further discussed in Section 4.2 to explain attractive forces between analytes and additives.

A general question is: how to quantify polarity or the forces between solvent molecules? The authors explained that common parameters like dielectric con-

stant and dipole moment can not completely describe the interactions occurring in solutions since they do not account for, *e.g.*, donor-acceptor interactions, steric factors or solvent density. They considered the cohesion energy density P_S a more appropriate measure for solvent-solvent interactions and thus solvophobic interactions. The cohesion energy is defined as

$$P_s = \frac{E_{vap}}{V_M} = \delta^2 \tag{4}$$

where E_{vap} is the heat of vaporization and V_M is the molar volume. It is also common to apply the so-called solubility parameter δ, which is equal to the square root of P_S. The values of P_S for a series of solvents are listed in Table 2. It can be seen that water has the highest cohesion energy density in this list, indicating that solvophobic interactions are generally weaker in organic solvents than in aqueous solutions.

The term "electrostatic interactions" was used by Bowser *et al.* [25] to summarize ion-ion, ion-dipole, and dipole-dipole interactions. Since the dielectric constant determines the strength of electrostatic interactions, it is a measure for the ability of a solvent to stabilize an unpaired ion or dipole through solvation. It can be concluded that electrostatic interaction is generally stronger in solvents with lower dielectric constants. This is the main driving force for, *e.g.*, water and the formamides to dissolve salts, thereby interrupting their strong Coulomb interactions occurring in the solid state. The authors further mentioned the significant influence of dissolved ions on the rotational freedom of the oriented solvent molecules in their vicinity. Thus an area with a very low dielectric constant ($\varepsilon \approx 2$ for most of the solvents) was reported to be created, directly surrounding the dissolved ion. For the bulk properties of diluted electrolyte solutions, these effects can be neglected. However, they have a significant influence on interactions between analytes and buffer additives added to optimize separations. Another effect mentioned by the authors is the decrease of the dielectric constant in the strong electric field in an electrophoresis device due to reorientation of solvent molecules under this influence.

A more specialized third class of interactions reported by Bowser *et al.* [25] is related to donor-acceptor interactions. Their consideration extends the electrostatic interactions to the cases where an electron transfer occurs. If this happens, an additional so-called coordination comes into play. To describe the ability of a solvent to accept or to transfer electrons in a quantitative manner, the acceptor number (*AN*) and the donor number (*DN*) were introduced. Proton transfer in so-called hydrogen-bonding is included in this type of interaction but simply regarded from the opposite point of view. Besides the hydrogen bond donors (*e.g.*, hydroxyls), in which the donated hydrogen is an electron acceptor, lewis acids with unfilled orbitals (*e.g.*, metal cations or metal halogenides) exist as a second type of electron acceptor. Electron donors are molecules with free electron pairs, usually present on oxygen, sulfur, or nitrogen atoms. A quantitative measure for the donor ability of the solvent was introduced by Gutmann [26] through the donor number (*DN*). The higher the *DN*, the stronger the ability of a solvent as electron donor. *DN* can be determined from the formation enthalpy of $SbCl_5$-solvent complexes. From comparing the ^{31}P-NMR shifts of triethylphosphine oxide when bound to the solvent molecules, a scale of acceptor numbers (*AN*) has been defined [26–28]. *AN* is a dimensionless number since normalized to the measured value for $SbCl_5$, which was arbitrarily set at 100. Solvents being able to act as hydrogen bond donors (*e.g.*, water, methanol, formamide) generally have higher *AN* values than those without that ability (*e.g.*, DMSO, ACN, DMF). The values of *DN* and *AN* are listed in Table 2 for some CE-relevant solvents. Donor-acceptor interaction can play a marked role in interaction between analytes and buffer additives. This will be further discussed in the related section (Section 4.2).

Besides these detailed considerations given above, two basic facts should be mentioned at the end of this section. The ability of a medium to dissolve ionic species is gener-

ally crucial for its applicability to CE. Several organic solvents are able to dissolve ions to a certain extent. Nevertheless, the application of organic media usually implies some restrictions on the choice of electrolytes. Common buffers in aqueous CE like, *e.g.*, phosphate and borate, are not applicable in some nonaqueous media (ACN, alcohols). A further difficulty is the fact that pH* values in organic solvents cannot be directly measured with a pH meter. Hence the simple and fast control of this very important influencing parameter on the separation is difficult compared to aqueous CE.

4 Method development and method robustness in NACE

4.1 Selectivity optimization in NACE *via* solvent variation

It has been discussed in Section 3 that the alteration of solvent properties has a significant influence on solute-solvation and ion-ion interactions. The variation of the solvent can change solute migration and separation selectivity tremendously. Consequently, the mixing of solvents to varying ratios turned out to be a powerful tool in CE separation optimization. Even neglecting distinct influences of solvent properties on interaction phenomena, changes in solute mobility result from differing dielectric constants and viscosities of the medium due to mixing of solvents, which can be described by the von Smolukowski equation [29] as follows:

$$\mu_{(normalized)} = \mu_{(SolventA)} \, [\varepsilon/\eta]_{(mixture)}/[\varepsilon/\eta]_{(SolventA)} \qquad (5)$$

where $\mu_{(normalized)}$ describes the mobility of the solute ion in the solvent mixture normalized to the mobility in solvent A ($\mu_{(SolventA)}$), ε the dielectric constant, and η the viscosity of the mixture or solvent A, respectively. Salimi-Moosavi and Cassidy [30] applied Eq. (5) to predict changes in migration velocity of alkanesulfonates and alkyl sulfates in mixtures of methanol and ACN. They related the $[\varepsilon/\eta]_{(mixture)}$ to the percentage of each solvent. This approach considers the ζ potential of the analyte ions to be independent of the solvent composition. In Fig. 1 the theoretical values for C_2SO_3 (12) and C_8SO_4 (13) are plotted together with the experimental results of the authors for sulfonates and sulfates of different chain lengths. For the alkyl sulfates the results comply with the theory in a satisfying way. However, in the case of the alkanesulfonates, the mobility decreased tremendously when an acetonitrile content higher than 50% was applied, whilst theory predicts an augmentation. This phenomenon was interpreted by significant differences in the tendency to form ion pairs between the different functional groups of the analytes. Generally, the addition of ACN, an aprotic solvent with

Table 2. Physical properties affecting analyte – additive interactions in solvents commonly used in CE

Solvent	P_s (J mL^{-1})	ε	*DN* (kJ mol^{-1})	*AN*
Water	2291	78	138[a]	54.8
FA	1638	111	151[a]	39.8
n-Methylformamide	992	182	205[a]	32.1
Methanol	928	33	126[a]	41.3
Ethanol	693	25	134[a]	37.1
ACN	655	38	59	18.9
DMF	613	37	111	16.0
DMSO	602	47	121	19.3
DMA	488	38	116	13.6
Acetone[b]	435	21	71	12.5
Diethyl ether[b]	280	4	80	3.9

a) These *DN* were measured indirectly and represent the donating ability of the bulk solvent.
b) These solvents should not be used in CE due to their low flash points and high volatility. They were included in the table so a wider range of solvents could be compared.

Reprinted from [25], with permission. Values for ε may differ slightly from Table 1 according to data given in the literature source.

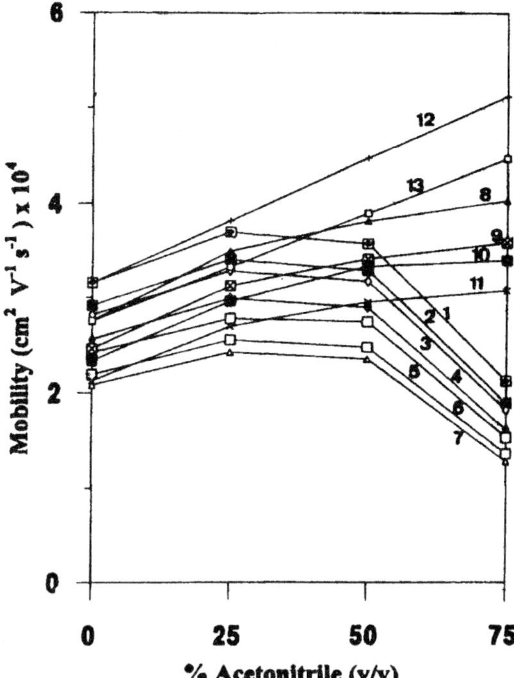

Figure 1. Effect of the addition of ACN on the mobilities of alkanesulfonates and alkyl sulfates in methanol. Peak identification: 1, C_2SO_3; 2, C_4SO_3; 3, C_5SO_3; 4, C_8SO_3; 5, $C_{10}SO_3$; 6, $C_{14}SO_3$; 7, $C_{16}SO_3$; 8, C_8SO_4; 9, $C_{12}SO_4$; 10, $C_{14}SO_4$; 11, $C_{18}SO_4$; 12, C_2SO_3 (theoretical); 13, C_8SO_4 (theoretical). Experimental conditions: electrolyte, 0.01 mol/L p-toluenesulfonic acid sodium salt and 0.005 mol/L p-toluenesulfonic acid. Reprinted from [30], with permission.

weaker solvent strength, promotes the formation of ion pairs, thus decreasing analyte mobilities. Figure 1 clearly demonstrates the power of NACE to generate a distinct selectivity in electrophoresis. In a solvent mixture of ACN-MeOH, 75:25 v/v, a marked selectivity between the two groups of analytes could be achieved, thus enabling separations never obtainable in aqueous CE. However, at ACN concentrations above 75% the analytes were no longer soluble.

Solvent mixing has become an efficient tool to optimize selectivity in NACE. Systematic studies on this have been published by the Cassidy group for other analytes as well. Salimi-Moosavi and Cassidy [31] reported on alkali-alkaline earth cation selectivity in methanol-ACN mixtures depending on the nature of the acid to adjust pH*. The observed effects were related to different solvation phenomena of ions when the solvent was altered. Significant

changes in the electrophoretic selectivity of the alkali and alkaline earth cations were reported when the ACN content in the methanol solution was increased from 0 to 60% and an imidazole/acetic acid electrolyte was applied. The mobility of the alkali metals and ammonium increased at higher ACN content whereas the alkaline earth metal mobility decreased and magnesium and calcium changed their migration order. This indicates opposite changes in solvation phenomena when comparing monovalent and divalent ions.

A paper dealing with the selectivity of amino acids in methanol and ethanol depending on buffer composition and pH* was also published by Salimi-Moosavi and Cassidy [32]. Separations of dansylated and underivatized amino acids are compared. The buffers were composed from tetraethylammonium hydroxide and either acetic acid or mineral acids like HCl or $HClO_4$. The general phenomenon ion-interaction, summarizing any ion association, ion-pairing or ion-cluster formation depending on varying solvation phenomena when the solvent is altered, was considered to explain the observed changes in selectivity. The best results were reported in merely protic solvents (methanol, ethanol) whereas no significant changes upon addition of ACN were found. A systematic study of the migration behavior upon the addition of ethanol into methanol was carried out. Changes in pH* had a strong effect on analyte selectivity, especially on the migration behavior of Dns-Glu, which was the only nonneutral amino acid in the mixture. Especially the mixing of methanol and ACN for selectivity optimization was reported by a series of authors [33–38]. Mixtures of methanol and ACN are of interest since methanol is a protic solvent whereas ACN is an aprotic solvent. Hence, the donor ability of the solvent can be controlled by the ratio of the mixture.

The use of tetrahydrofuran (THF) in nonaqueous CE is rarely reported, most likely due to the fact that a certain content of THF in the running buffer is hazardous to common CE instrumentation. Song et al. [39] applied THF contents up to 20% in methanol to tune a separation of quinolizidine alkaloids. The mixture contained the two N-oxides oxymatrine (OMT) and oxysophocarpine (OSC) as a critical pair to separate. At a content of 10% THF, OMT and OSC could be baseline-separated, which was explained by a dipole-dipole interaction of THF with the N-oxide moiety. An example for the application of ternary mixtures has been published by Tjørnelund and Hansen [40]. They added up to 6% N,N-dimethyl formamide into a methanol-ACN mixture to optimize the separation of tetracycline antibiotics. The addition of up to 6% DMF into MeOH-MeCN (48:52 v/v) had a more pronounced influence on the separation than the variation of the MeOH-

ACN ratio from 42:52 to 52:42 at the constant amount of 6% DMF. Changes in the composition of the ternary solvent to optimize the electrophoretic resolution affected the speed of electroosmosis rather than the analyte selectivity. However, this should not be considered as a general rule. A further application of ternary mixtures has been published by Thibon *et al.* [41]. They separated zinc dialkyl dithiophosphates in a medium composed of MeOH-ACN-THF at the ratio 70:15:15.

4.2 The use of additives in NACE

In aqueous CE the interactions between analyte and additive are mainly of a solvophobic nature (see Section 3), which enables the use of cyclodextrins for chiral separations and surfactants to form micelles for the separation of uncharged or hydrophobic molecules. In solvents of high cohesion energy density, the formation of such aggregates is definitely more favorable since it implies less interruption of their strong solvent-solvent interactions. Conversely, the additives mentioned above are mostly not as effective in organic solvents as they are in water. In a review, Bowser *et al.* [25] discussed the effects of the solvent on several types of analyte-solvent interactions. It was mentioned that at a high solvent strength (see Section 3), strong solvent-solvent interactions are found and therefore both the solvent-analyte and the solvent-additive interaction are strong for nonpolar analytes and additives. At a high solvent strength of the medium, nonpolar solutes (analytes and additives) interact with each other rather than with the solvent molecules. Changing to lower solvent strength (organic media) leads to a reduction of the following interactions: analyte-analyte, additive-additive and analyte-additive, which results in a better solvation of both analyte and additive. Thus micelles do not form in most of organic solvents and the formation constant of the analyte-additive complex is lower.

4.2.1 Chiral separations

The most common type of additives for chiral separation is the class of cyclodextrins. Wang and Khaledi [42] carried out chiral separations in formamide (FA), NMF and DMF using β-cyclodextrin (β-CD). They demonstrated that the formation constants between the chiral selector and the analytes were considerably altered when changing from aqueous to nonaqueous buffer systems. The formation constant between β-CD and enantiomers decreases with decreasing cohesion energy density (see Section 3). The strongest interaction was found in water followed by FA, where the formation constants were 2–4 orders of magnitude lower than in water. In NMF and DMF, the formation constants were too small to be determined appropriately. Thus at first sight, FA seems to be the best choice for nonaqueous chiral separations. In recent studies, both Li *et al.* [43] and Ren *et al.* [44] used this solvent and β-CD to separate chiral drugs and modified amino acids, respectively. Valkó *et al.* [45] demonstrated with their separation of dansyl-amino acids that not only the formation constant is crucial for a chiral separation, but that solubility also plays an important role. Because of the poor solubility of β-CD in water, the enantiomers could not be separated in an aqueous buffer. In the FA/β-CD buffer, high selectivity but poor efficiency was obtained, whereas in NMF the analytes were separated fast and efficiently with an 8-fold β-CD concentration compared to the FA buffer.

Different selectivities from those applying neutral CDs can be generated with charged CDs. Especially when enantiomers are separated with oppositely charged CDs, the selectivity can be enhanced due to a higher difference in the mobility of the free enantiomer from the coordinated one. The use of cationic CDs such as quaternary ammonium β-CD (QA-β-CD) in solvents like water, FA, NMF, MeOH, and DMSO was reported by Wang and Khaledi [46] for the separation of profens and several chiral acids. Chiral separation of profens and some amino acids was solely achieved in FA. But most of the acidic enantiomers were separated in all of these solvents. Furthermore, the addition of QA-β-CD caused a reversal of the EOF in all of the investigated solvents.

Tacker *et al.* [47] demonstrated separations of 40 basic analytes with the oppositely charged heptakis(2,3-dimethyl-6-sulfato)-β-CD (HDMS-β-CD) in acidic methanol. The advantage of methanol over FA or NMF is its UV-transparency at 214 nm. Moreover, in addition to the solvophobic analyte-selector interactions, the charges of the sulfate groups seem to exhibit electrostatic interactions, allowing the use of media with low cohesion energy density when their dielectric constant is sufficiently low. Vincent and Vigh [48] observed very high selectivities for separations of weak bases with negatively charged heptakis(2,3-diacetyl-6-sulfato)-β-CD whereas neutral and acidic analytes were coordinated very weakly by the chiral selector. Wang and Khaledi separated basic chiral pharmaceuticals in FA using CDs having a degree of sulfation of four [49]. Compared to neutral CDs, a lower concentration of $CD(SO_4)_4$ was required and stronger analyte-selector binding, due to an increased electrostatic contribution, was observed. The effect of solvent nature on the analyte-selector formation constant was studied in water, FA, NMF and DMF. It was shown that for the test solutes the binding constants increased with increasing polarity of the solvents. Thus the hydrophobic analyte-selector interaction still remained the predominant binding force.

A promising way to enhance the effectiveness of additives in NACE is to concentrate on substances being able to form electrostatic interactions. Focusing on electrostatic interactions, Mori *et al.* [50] separated the enantiomers of primary amino compounds in FA with a chiral crown ether ((+)-18-crown-6 tetracarboxylic acid) acting as electron donor. The donor-acceptor interactions in FA, showing high *AN* and *DN* (see Section 3), are rather low but this is counterbalanced by the fact that there are six binding sites (oxygens) available at the additive. Furthermore, note that the analyte-polyether binding is still more intense than in water, where no separation was achieved using the same additive.

Stalcup and Gahm [51] took advantage of the electrostatic ion-ion interaction to separate the enantiomers of *N*-modified amino acids with quinine where the protonated amino group acted as electron acceptor. When analytes, like the substances separated in this work, have an aromatic ring, a second (π-π) interaction takes place between the quinoline group of the selector and the aromatic group of the analyte. Piette *et al.* [52] used quinine and modified quinine (tert-Bu carbamoylated quinine) to separate chiral *N*-protected amino acids. Best enantiomeric separation was achieved with the modified quinine in an ethanol-methanol mixture (60:40 v/v). Taking advantage of a similar binding mechanism, Bjørnsdottir *et al.* [53] used a camphorsulfonate enantiomer for the separation of amino drug enantiomers. Like quinine, camphorsulfonate provides two binding sites. The sulfonate and the keto group acted as electron donor, while the amino groups of the analyte acted as electron acceptor.

4.2.2 Separation of neutral analytes

In aqueous media, the most often applied electroseparative technique for the separation of neutral analytes is micellar electrokinetic chromatography (MEKC). With increasing hydrophobicity of the analytes, the micelle-analyte interaction becomes too strong, resulting in a loss of resolution and an increased analysis time. A solution of this problem can be found in increasing the amount of organic solvent in the buffer system. But at a certain level of organic amount, as already mentioned above, micelles no longer form in most of the organic solvents and MEKC becomes impossible. Taking advantage of ion-dipole interactions in NACE, neutral substances can be supplied with an electrical charge using complexing agents like inorganic ions and can thus be separated in electrophoresis.

Walbroehl and Jorgenson [2] first demonstrated the use of tetrahexylammonium to separate hydrophobic polycyclic aromatic hydrocarbons (PAHs) in NACE. They found that the separation was improved by the addition of water to ACN up to 50%, which supports the theory of hydrophobic interactions between the hydrophobic PAHs and the alkyl chains of the additive. Tjørnelund and Hansen [54] used tetraalkylammonium and alkyltrimethylammonium ions in ACN and propylene carbonate, respectively, for the determination of the water-insoluble vitamin K_1 in a pharmaceutical product as well as for the separation of parabenes. However, 0.5% of protic solvents like methanol or acetic acid in the running buffer made separations impossible. Propylene carbonate (a nonstable solvent in the presence of water or alcohols) is an aprotic, dipolar medium like ACN but dissolves much better the long-chain trimethylammonium ions. The analyte-additive interactions were reported to be of a hydrophobic nature as well as in [2]. Bowser *et al.* [25] proposed that the positive charge of the ammonium group induced a dipole in the aromatic ring of the analyte, thus generating electrostatic interactions. In a recent paper, a mixture of the aprotic, dipolar ACN and ethylene carbonate was used to separate several alcohols with lanthanide triflates as additives [55].

Miller *et al.* [56, 57] separated PAH employing planar organic cations like 2,4,6-triphenylpyrylium or tropylium in ACN. The binding mechanism was interpreted as a positive charge at the additive inducing a dipole on the PAHs, resulting in a charge transfer complex. The complex stability could be related to the polarizability of the π-electron cloud of the PAHs. Due to the fact that the polarizability of the PAHs depends on the number of rings, the complex stability and thus the electrophoretic mobility can be related to their molecular mass. A further attempt to separate PAHs was undertaken by Li and Fritz [58]. They used anionic surfactants like sodium tetradecyl sulfate (STS) in methanol. The analyte-additive interactions were assumed to be hydrophobic. It was found that the addition of water to the buffer had two effects. With an increasing amount of water on one hand, hydrophobic interactions were promoted, resulting in higher mobilities of the analytes, whilst the ε/η ratio [23] decreased, leading to lower mobilities. Taking account of these opposite effects, an optimum water-methanol mixture was found at 25:75 v/v. Okada [107] separated several polyethers, taking advantage of their ability to form complexes with small cations in low dielectric constant solvents like methanol. In a solvent with a high dielectric constant such as water, the donor-acceptor interactions are too weak for binding to take place.

4.2.3 The use of additives for the separation of charged analytes

In the separation of charged analytes, additives can be used to obtain new selectivities. Tjørnelund and Hansen [59, 60] increased the separation selectivities of tetracycline antibiotics by adding magnesium ions to the NMF running buffer. The analytes were separated as metal chelates where the complexation intensified the fluorescence of the tetracyclines, so that laser-induced fluorescence (LIF) detection could be applied. Bellini *et al.* [61] used the cationic surfactant trimethyloctadecyl ammonium bromide to refine resolution in the separation of aromatic compounds, providing a carboxylic moiety along with other functional groups in a mixture of methanol-ACN (50:50 v/v). Through an ion pairing mechanism the negative charge of the carboxylic groups of the analytes was partially compensated and the addition of the octadecyl chain from the surfactant resulted in a higher hydrophobicity of the complex. The newly obtained selectivities allowed the separation of the ten compounds.

As already mentioned in Section 4.2.2, ionic additives can be used to separate uncharged polyethers. Conversely, polyethers can be used to separate charged compounds. Bowser *et al.* used Brij 35 to enhance selectivity and efficiency in separations of porphirin acid monomers [62] and oligomers [63] in methanol. An electrostatic ion-dipole interaction was discussed with the carboxylic groups acting as electron donor and the ether function acting as electron acceptor. The low acceptor properties of the ether was supposed to be counterbalanced by the multiplicity of available binding sites and the low dielectric constant of methanol. Wright and Dorsey [64] applied the principles of argentation chromatography to NACE. Several sulfonamides were separated in ACN with silver ions as additive. Compounds such as *N*-containing heterocyclics were found to build selective charge transfer complexes with Ag^+ improving selectivity of the separation. In Section 4.2.1, CDs were presented as chiral selectors. But in electrophoretic separations they can also be used to improve resolution of nonchiral compounds. Morin *et al.* [65] separated hydrophobic phytostyrol derivatives in a methanol-ACN-acetic acid based buffer. Due to hydrophobic interactions, the selectivity was improved by adding neutral heptakis-(2,6-di-*O*-methyl)-β-CD.

4.3 EOF

Electroosmosis can be considered as the pump of the buffer in CE. On the one hand, it is directly related to the speed of analysis since the EOF mainly contributes to the resulting speed of analyte migration through the capillary. On the other hand, the mobility of the EOF is an important parameter to optimize resolution in CE. When the analytes migrate in the same direction as the EOF, a shorter separation path length is simulated, thus decreasing resolution. Analyte migration against the EOF generates a longer virtual separation path length and increases resolution.

4.3.1 Electroosmosis in different media

It has been mentioned in Section 2 (Eq. 2) that solvent parameters have a marked influence on electroosmosis. Wright *et al.* [66] reported the magnitude of electroosmosis in five organic solvents (ACN, methanol, DMF, DMSO, and FA) in the absence of a background electrolyte as well as in deionized water. Although studies like this might be considered of minor practical relevance, since buffering is essential for the application of an electrophoretic method, these data are of great theoretical value. Measuring in buffered eluents additionally implies that it is hard to predict whether the solvent properties or special interaction of the buffer with the capillary surface in a distinct medium becomes predominant. The authors applied acetone or benzene as inert markers to measure μ_{eo} according to their solubility in the solvent to study. The EOF was found to be equal to or greater than the values obtained with common buffer additives (CAPS, borate, phosphate, acetate) in aqueous CE. While the EOF velocity varied in aqueous buffers upon the pH (4–10.9) by a factor of 2.5, one complete order of magnitude was covered in the varying unbuffered organic solvents. The EOF decreased in the following sequence: ACN > water > MeOH > DMF > FA > DMSO. For ACN/water and methanol/water mixtures, similar experiments were carried out. The authors further calculated zeta potentials from these data, as will be discussed in Section 7.

Valkó *et al.* [67] also reported on EOF studies in neat water and organic solvents covering a wider range of solvents than the publication referred to above. Water was the EOF marker in most of the solvents except water itself and solvents not miscible with water. In these cases methanol was used. Detection was carried out by UV at 254 nm, obviously mainly due to changes in refractive index. In their paper, a detailed and comprehensive introduction into the theory of EOF generation and the parameters that electroosmosis depends upon is given. Values for the electroosmotic mobilities of all solvents showing measurable electroosmosis are depicted in Table 3. In the two inert solvents chloroform and hexane no relevant EOF was observed. For acetic acid a very low EOF was reported which was related to excessive protonation of the surface silanols. All standard deviations given in Table 3 were calculated from five repetitive runs with 30 s conditioning steps in-between. The calculated zeta potentials given in this table will be discussed in Section 7. Valkó *et al.* [67] give a detailed explanation for the remarkably fast electroosmosis in aprotic (or aprotogenic) solvents like ACN. Due to the lack of autoprotolysis, in these solvents virtually no anions exist at a certain purity level. Solvent molecules can accept protons from dissociating surface silanols and the negative charge of the capillary wall is

Table 3. Electroosmotic mobility (μ_{eo}) and zeta potentials of the wall (ζ_{wall}) in a fused-silica capillary at 25°C in solvents without the addition of ionic species

Solvent	μ_{eo} (m²/Vs)	ζ_{wall} (mV)	RSD (%)[a]
ACN	1.8149×10^{-7}	−195.2	0.19
Acetone	1.2046×10^{-7}	−198.8	2.61
2-Butanone	7.6546×10^{-8}	−186.0	0.60
Water	7.6147×10^{-8}	− 97.8	1.89
Deuterium oxide	6.5917×10^{-8}	−104.6	0.34
DMF	5.8630×10^{-8}	−144.7	0.11
NMF	5.0878×10^{-8}	− 52.0	0.44
Methanol	4.4608×10^{-8}	− 84.0	0.86
DMSO	3.2168×10^{-8}	−155.3	0.69
Ethanol	1.4644×10^{-8}	− 72.6	1.07
FA	1.3368×10^{-8}	− 45.6	1.84
Ethyl acetate	1.1264×10^{-8}	− 98.8	2.74
Tetrahydrofuran	1.0536×10^{-8}	− 74.1	0.23
1-Propanol	4.1273×10^{-9}	− 46.7	2.71
Morpholine	3.4407×10^{-9}	−104.0	0.17

a) $N = 5$
Reprinted from [67], with permission.

solely balanced by this. Hence, ion strengths are extremely low and the potential distribution close to the capillary wall is completely different from amphiprotic solvents (especially when autoprotolysis is high). In the latter case, the typical relatively dense double layers of both anions and cations in the liquid phase exhibit a more effective shielding of the negative surface potential. Furthermore, Valkó *et al.* described studies on temperature dependence of the EOF, thickness of electrical double layers, and EOF behavior in mixed aqueous-organic media.

In real applications of NACE, the addition of a buffer is mandatory. Systematic studies on the influence of the kind of electrolyte, adjustment of pH* and the addition of additives in nonaqueous media on the resulting EOF, including a revealing explanation of the given results, can rarely be found in the literature. In fact, this turned out to be a complex and problematical field. The theoretical background is related to the considerations on solvent-solute and solute-solute interactions given above. Salimi-Moosavi and Cassidy [31] reported on EOF in methanol-ACN mixtures depending on the nature of the acid to adjust pH*. They discussed the individual solvation effects and their influence on the electric double layer at the capillary wall. HCl, HClO$_4$ and acetic acid at 0.01 mol/L were applied to prepare the background electrolyte. To occasionally increase pH*, triethylamine was added. Significant differences in the speed of electroosmosis depending on the solvent composition, the nature of the

acid anion and the pH* could be observed (Fig. 2 and 3). Their obtained values for μ_{eo} applying triethylamine electrolytes with acetic acid, HCl, or HClO$_4$ are depicted in Fig. 2. A further study without the addition of triethylamine was reported (Fig. 3). Generally, under stronger acidic conditions the EOF is mainly anodic, which indicates an adsorption of solvated protons and/or triethylammonium ions onto the capillary surface. The influence of the ACN/methanol ratio is interpreted through an additional adsorption of the distinct buffer anion to the capillary wall and varying ion-interaction phenomena in the liquid phase. ACN solvates acetate and chloride weaker than perchlorate, thus supporting their adsorption and inducing a more positive zeta potential.

Similar to the application of aqueous buffers, the influence of the ionic strength of the buffer on the thickness of the electrical double layer and thus on the zeta potential and

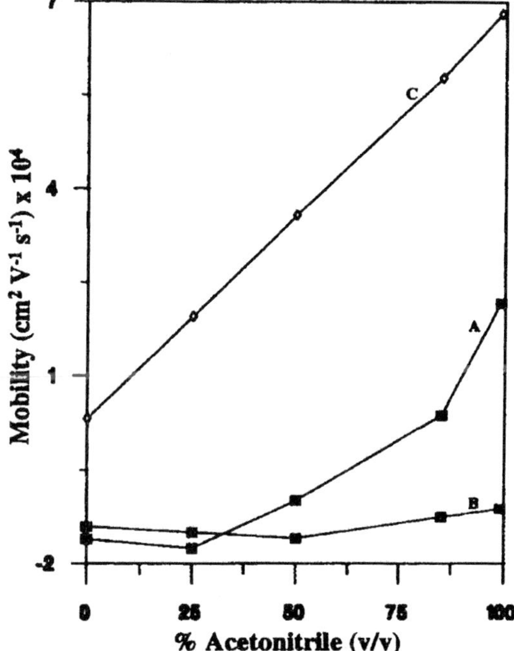

Figure 2. Effect of the addition of ACN to methanol electrolytes on the EOF. Experimental conditions: electrolyte, (A) 0.01 mol/L tetraethylammonium chloride and 0.01 mol/L HCl; (B) 0.01 mol/L tetraethylammonium perchlorate and 0.01 mol/L HClO$_4$; (C) 0.01 mol/L tetraethylammonium acetate and 0.01 mol/L acetic acid in methanol; 50 cm capillary (75 µm ID) with injection-to-detection length of 43 cm; separation voltage, 20 kV, either positive or negative; indirect detection at 214 nm. Reprinted from [31], with permission.

EOF can be suppressed by adding polyvinyl alcohol (PVA) or polyethylene glycol (PEG) to the running buffer [68]. As already mentioned in Section 4.2.1, a reversal of the EOF can be obtained by adding quaternary CDs to the running buffer [46]. A polymeric surfactant that yields an EOF reversal in aqueous as well as in nonaqueous media is hexadimethrine bromide (HDB). This cationic polymer was used to reverse the EOF for a faster analysis of hydroxy- and dihydroxy benzoic acid [69] and for the separation of acetylsalicylic acid and three of its metabolites [70].

A way to avoid the problems coming up with the reduced surfactant-wall interaction could be the use of permanent coatings instead of dynamic ones. Belder *et al.* [71, 72] investigated the performance and the stability of permanent PVA and PEG coatings in methanol-based buffers. In PVA-coated capillaries, EOF was suppressed as expected while in PEG-coated capillaries a reversed EOF could be observed. A combination of dynamic and permanent coating was used to separate positively derivatized fatty acids in methanol by Gallaher and Johnson [73]. To prevent analyte wall adsorption, the capillary was trimethylsilylated. To this permanent coating, a cationic surfactant was added.

4.4 Robustness of methods in NACE

In quantitative routine analysis, the obtainable precision of CE in most cases does not reach the standards set by HPLC. This can be related to the scale of the method, the injection technique, the complexity of parameters influencing separation, the integration of the often triangular peaks, and the elevated baseline noise in CE detection compared to normal bore HPLC. Experimental data on these problems as well as on the influence of capillary equilibration prior to an aqueous CE run have been published by Faller and Engelhardt [74]. Unlike HPLC, where the eluent pumps are nowadays very accurate, electroosmotic pumping applied in CE and CEC is still a matter of concern. Wright *et al.* [66] reported on the precision of the EOF in pure ACN. They achieved relative standard deviations (RSD values) of less than 2% over 25 runs within three nonconsecutive days. Even when the capillary was exposed to other organic solvents between the ACN runs, RSD < 3% for N = 15 was reported. Although these results are promising, they are of limited practical relevance since no buffer was applied for these studies. Wright *et al.* further reported that the change from buffered aqueous systems to pure ACN required a long equilibration time, resulting in worse reproducibility than the values given above. It can be deduced that the use of the same capillary for aqueous and nonaqueous systems or for the use of different electrolytes is not recommendable.

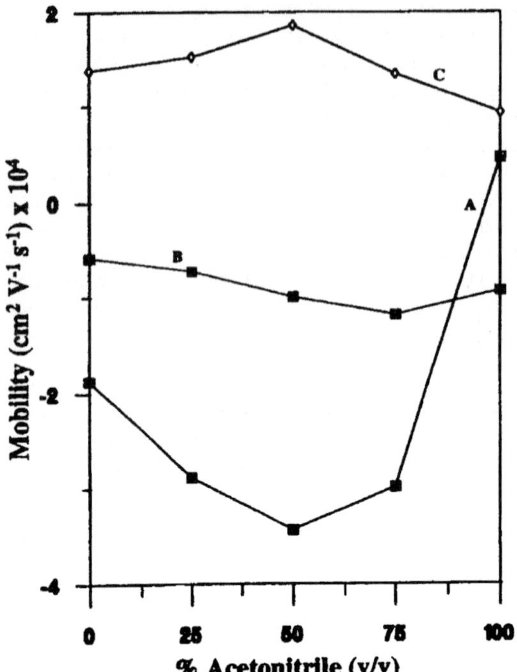

Figure 3. Effect of the addition of ACN to methanol on the EOF. Experimental conditions: electrolyte, 0.02 mol/L imidazole and 0.03 mol/L (A) HCl, (B) HClO$_4$, (C) acetic acid. Other conditions as for Fig. 2. Reprinted from [31], with permission.

EOF is pronounced in NACE. Altria and Bryant [34] varied the concentration of sodium acetate in an ACN/methanol (50:50 v/v) mixture over a range from 0.5 to 10 mM. The augmentation of the buffer concentration resulted in a decrease of EOF by more than a factor of 2. A reversal of the EOF upon the addition of quaternary ammonium β-CD has already been referred to in a section on the application of additives in NACE (Section 4.2.1).

4.3.2 Wall modification for EOF control

For an application in aqueous systems, a large number of surfactants is known to supply the capillary wall with a dynamic coating for EOF suppression, EOF reversal, or the suppression of analyte-wall interaction. Unlike aqueous media, the adsorption of solutes to the capillary walls is rarely observed in nonaqueous media. This means that coatings for the suppression of analyte-wall interaction are not needed in most cases. Due to this reduced interaction, it is more difficult for surfactants to get fixed to the capillary wall. Thus, most of the surfactants used in aqueous buffers are ineffective in nonaqueous buffers. The

Compared to aqueous CE, the lower heat capacities, thermal conductivities, and boiling points of many organic solvents must be considered. Thus, the precision of peak areas (even when normalized to migration time) is strongly instrument-dependent in NACE. Leung *et al.* [33] applied an ACN-methanol-acetic acid (49:50:1) medium for the analysis of basic drugs. Operating without temperature control and with loosely covered sample vials, they obtained RSD values for migration-time-normalized peak areas of more than 15%. Repeatability of quantification could be improved to an RSD below 9% when working with internal standard. Applying a different instrument with septum-closed vessels and temperature control, normalized peak areas for the same substances showed 0.7–1.2% RSD. For the routine analysis of pharmaceutical compounds in an ACN/MeOH medium, Altria and Bryant [34] reported RSD values on the order of 1% for both migration times and peak areas relative to internal standards. The validation of a NACE method (NMF) for the determination of oxytetracycline in ointment has been published by Tjørnelund and Hansen [75]. A correlation coefficient (linearity) higher than 0.999 and a variation coefficient (precision, sensitivity) lower than 3.6% was obtained for concentrations in the range of 0.2–3.0 mg/mL.

In another publication, Tjørnelund and Hansen [76] reported on an important aspect for the ruggedness of NACE methods, namely the effect of small amounts of water in the "nonaqueous" running buffer on the separation characteristics. Tricyclic antidepressants were selected as cationic substances, benzoic acid and tolfenamic acid as anionic ones. They measured the water content of commercially available solvents (ACN, dimethylsulfoxide, methanol, and NMF) prior to adding water from a sub-percent level up to 10% to the buffer prepared with 25 mM ammonium acetate and 1 M acetic acid. The water contents of the original solvents determined by Karl-Fischer titration were found to be between 0.03 and 0.08%. The largest changes in analyte mobility and EOF were found at small percentages of water in NMF. Generally, efficiency, peak tailing, and selectivity were practically not affected when adding water up to the 1% level. Thus, the authors conclude that small variations in the water content are of minor importance to separation reproducibility in NACE.

5 Detection in NACE

Similar to aqueous CE, the measurement of UV absorption is the most common detection technique in NACE. Unfortunately, UV detection is not very sensitive due to the short light path through the capillary. Sensitivity can be considerably enhanced by employing LIF detection, but in most cases this requires a time-consuming derivatization of the analyte to introduce a fluorophore. Another successful attempt to improve sensitivity is amperometric detection. Some solutes like inorganic anions that do not absorb at wavelengths of UV light can be detected by indirect UV measurements. The coupling of CE to MS has not yet reached the routine stage attained in HPLC-MS. Nevertheless, it is about to become a promising tool in NACE. Thermooptical detection, a method not very common to CE, is used by Li *et al.* [77] in the separation of tricyclic antidepressants in a methanol-ACN buffer. The principle of this technique can be described as follows: within a sample a pulsed laser beam induces a temperature rise proportional to the laser power and to the absorbance of the sample zone at the wavelength of the laser. The rise in temperature causes a change in the refractive index and can thus be measured optically. The detection limits were estimated to be in the range of 0.1–0.5 μmol/L.

5.1 Amperometric detection

Salimi-Mossavi and Cassidy [78] were the first to report amperometric detection in NACE for a separation of inorganic anions in methanol. Amperometric detection occurred at a 25 μm microdisc electrode yielding a detection limit of 10^{-8} mol/L for some anions. Matysik [79] constructed an electrochemical cell for amperometric detection in aqueous CE. The cell consisted of a cylindrical glass body containing a working electrode, a ground electrode and a reference electrode. The working electrode was a carbon microdisk electrode (Ø 30 μm) consisting of a carbon fiber sealed in a glass. The microdisk electrode was aligned with the CE capillary. The ground electrode was a platinum wire and the reference electrode a silver/silver chloride electrode separated from the main compartment of the cell by a porous ceramic frit. During the experiment, the cell was filled with the CE running buffer. Later, the same cell was used for detection in NACE [80]. The difference to the system described above was the replacement of the carbon fiber in the working electrode by a platinum wire. The diameter of the microdisk was shrunk to 25 μm. The detection cell, working in the three electrode mode, was characterized using ferrocene and (ferrocenylmethyl)trimethylammonium perchlorate ([FcMTMA]ClO$_4$) as analytes in ACN which had already proved to be an ideal solvent for performing electrochemical measurements [81]. The limit of detection for [FcMTMA]ClO$_4$ turned out be 6×10^{-8} mol/L. In such a system, several water-insoluble dyes could be separated with a detection limit of 7.4 ng/mL compared to 0.9 μg/mL obtained with UV detection for the analyte basic green 1. The design of the described cell was further modified as shown in Fig. 4 (from [82]). The glass cylinder was replaced by a PTFE body and the ground electrode consisted of one of the guiding stainless steel tubes. The sil-

Electrophoresis 2000, *21*, 3994–4016

ver/silver chloride reference electrode was fixed in the cap on the top of the cell. As in the previously described system, the cell was filled with the same CE run buffer reported previously [80]. With this detector, Matysik could determine the nicotine content of tobacco with a detection limit of 13 ng/mL.

Another example for the usefulness of ACN in electrochemical detection is demonstrated by Luong *et al.* [83] with a separation of chlorinated phenolic compounds. Detection occurred with an end-column detector. Again platinum and silver/silver chloride electrodes were used. The design of the cell is shown in Fig. 5. The detection limits ranged from 30 to 500 nmol/L, which was 3–8 times lower than those yielded in aqueous buffers.

5.2 Indirect detection

For indirect detection, a UV absorbing background electrolyte is added to the buffer. The charge of the absorbing electrolyte ion must be of the same sign as that of the analyte ions. At the position where the analyte ions migrate, the background electrolyte is partially displaced because of the required electroneutrality. To obtain symmetric peaks, it is mandatory that the background electrolyte has a similar electrophoretic mobility to the analyte ions. In this mode, the concentration of the analytes is related to the transmittance of the zone passing the detector. The sensitivity of indirect UV detection depends on the molar absorptivity of the background electrolyte and is comparable to direct UV detection.

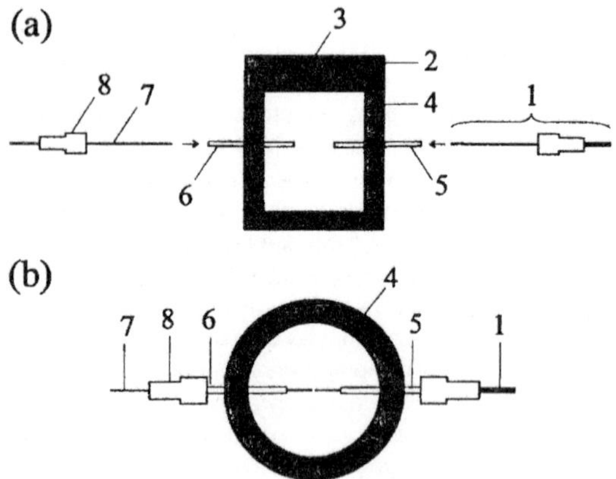

Figure 4. Schematic of the electrochemical detector. (a) Side view, (b) top view. 1, Working electrode; 2, PTFE cap; 3, hole for inserting the reference electrode; 4, PTFE cell body; 5, guiding stainless steel tube for the working electrode; 6, guiding stainless steel tube for the separation capillary; 7, separation capillary; 8, PTFE adapter holding the separation capillary. Reprinted from [82], with permission.

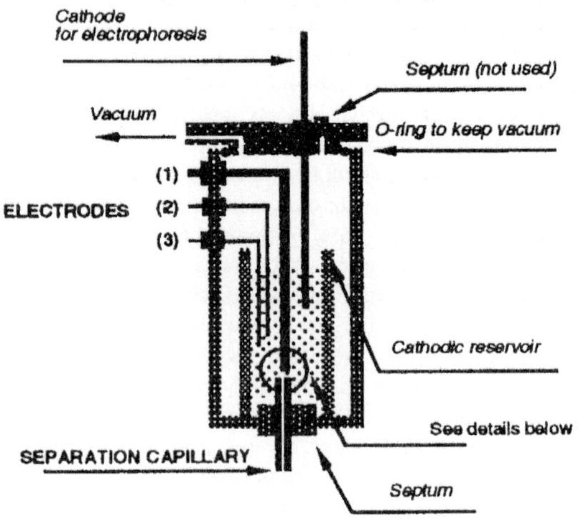

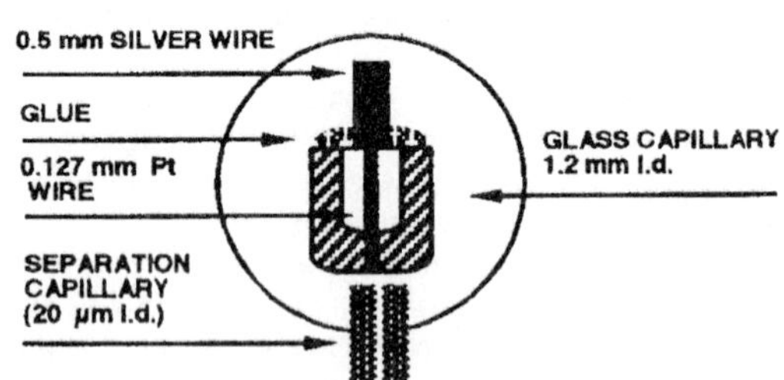

Figure 5. Amperometric detection cell with details of the detecting electrode. Components are not drawn to scale. Electrodes: (1) working: 0.127 mm platinum wire; (2) counter: 0.5 mm platinum wire; (3) reference: 0.5 mm Ag/AgCl. Reprinted from [83], with permission.

Salimi-Moosavi and Cassidy [84] evaluated the application of indirect UV detection of inorganic anions in methanol. Chromate, phthalate and benzoate background electrolytes in methanol were tested. Chromate provided the highest molar absorptivity but was not stable in methanol, so phthalate with a slightly lower sensitivity was the electrolyte of choice. The detection limits were in the range of $2 \times 10^{-5} - 3.4 \times 10^{-5}$ mol/L, which was about four orders of magnitude higher than that obtained with the amperometric detection described above. Suzuki *et al.* [85] also used phthalate in a methanol-DMF mixture for the quantitation of inorganic counterions of water-insoluble drugs. The method was applied to determine the chloride content in azelastine hydrochloride.

Indirect UV detection of underivatized amino acids was compared to direct UV detection of dansylated amino acids by [86]. For indirect detection, dimethylbenzylamine (DMBA) was used as the background electrolyte in acidic methanol to yield detection limits in the range of 9×10^{-6} – 4×10^{-5} mol/L, which was one order of magnitude higher than for direct detection of the dansylated amino acids. However, the response factors were more stable and the peak symmetries were higher for the dansylated amino acids than for the underivatized amino acids when applying the indirect method, which demonstrates in this case the advantage of a prior derivatization.

Anionic surfactants such as alkyl sulfates, ether sulfates, and sulfonates could be separated in a methanol-based buffer by Heinig *et al.* [87]. The influence of different background electrolytes was tested and it was shown that for the analytes with low electrophoretic mobilities, dodecylbenzene sulfonate or octylbenzene sulfonate yielded good peak shapes whereas *p*-hydroxylbenzoate or naphthalene sulfonate was useful to separate surfactants with higher mobilities. The detection limit for sodium dodecyl sulfate was 4.6×10^{-6} mol/L. Drange and Lundanes [88] used anthrachinone-2-carboxylic acid in an NMF-dioxane (3:1 v/v) mixture to separate long-chained fatty acids (C_{14}-C_{26}) at a detection limit of about 25 μmol/L. Salimi-Moosavi and Cassidy [30] detected alkyl sulfates and alkanesulfonates in MeOH/ACN by applying *p*-toluenesulfonic acid for the indirect detection at 214 nm.

5.3 Fluorescence detection

As mentioned earlier in this section, fluorescence detection and especially LIF can considerably improve detection sensitivity. A further potential of fluorescence detection is the increased selectivity. During the prior derivatization of the analytes, most of the impurities or matrix substances are not provided with a fluorophore and do not interfere with analyte detection. Ward and Khaledi [89] investigated the influence of nonaqueous solvents on LIF detection. They found out that NMF yielded the best detection limits, and enhanced the fluorescence signal by a factor of two compared to an aqueous buffer. It was proposed that the enhancement of fluorescence intensity can be correlated to the viscosity and the polarity of the solvent.

Tjørnelund and Hansen [90] investigated the fluorescence behavior of tetracycline antibiotics. It was found that complexation with magnesium ions considerably increased fluorescence intensity of the tetracyclines. The use of organic solvents further intensified fluorescence. In DMF, the factor of enhancement was reported to be 34 when compared to the signal in water and in NMF the intensity

was multiplied by 9. Nevertheless, the solvent of choice in this method was NMF because the solubility of ions like magnesium was insufficient in DMF. As already mentioned in Section 4.3.2, Gallaher and Johnson [73] separated fatty acids in methanol. Prior to separation, the analytes were labeled with a near-infrared fluorescence dye (excitation at 780 nm; emission at 850 ± 20 nm). It was pointed out in Section 5.2 that analytes without any chromophore can be detected by indirect UV detection. A similar principle can be applied in fluorescence detection. Wang *et al.* [91] used indirect fluorescence detection in the separation of fatty acids in a basic methanol-ACN buffer. The analytes were detected at a level in the range of 10–20 μmol/L.

5.4 NACE coupled to mass spectrometry

Among the commonly discussed topics of sensitivity and versatility of a detection method, another important aspect is their ability for peak identification. Electrospray ionization (ESI) as an upcoming atmospheric pressure ionization technique in mass spectrometry (MS) allowed the powerful coupling of a method giving highly efficient separation (CE) to a method providing peak identification and the possibility of structural elucidation of the analytes. Smith *et al.* [92] and Lee *et al.* [93] were the first to report CE-MS coupling. Tomlinson *et al.* [94, 95] described the effect of organic solvents in CE-MS in the separation of therapeutic drugs and their metabolites. Probably due to the high volatility and low surface tension of organic solvents like methanol or ACN, the use of these media in ESI-MS enhances ionization, resulting in an improved detection limit compared to separations in aqueous buffer systems. Lu *et al.* [96] demonstrated that even the addition of the ion-suppressing surfactant SDS to a methanolic buffer improved resolution and sensitivity in a separation and identification of tamoxifen from its metabolites compared to separations in aqueous buffers. Yang *et al.* [97] described the development of CE-MS in a review. The beneficial effects of NACE-MS on the separation of lipophilic peptides and the elucidation of metabolisms of therapeutic drugs are demonstrated. Further publications reported on CE-MS separations of phospholipids [98], tricyclic antidepressants [99], haloacetic acids [100], basic pharmaceuticals and hydrohobic sulfonic acids [101]. All the above applications have in common that the adoption of the EOF in CE (in the range of several nL/min) to the flow required by the spray device to maintain a stable electrospray current (in the range of several μL/min) occurred by the addition of a so-called sheath flow. The advantage of this technique is that an addition of certain additives (dissolved in the sheath liquid) can be done just after the completed CE separation. In several cases, this can tremendously facilitate ionization and improve the stability of the system. The inconvenient aspect is the increasing detection limit, due to a dilution in a range of 1:10. A promising way to improve detection limits could be the application of an online nanospray MS without the addition of a sheath flow.

6 Electrical current characteristics and application of wide-bore capillaries in NACE

Lister *et al.* [102] reported on current measurements in NACE and nonaqueous CEC in various organic solvents and water in the absence of a background electrolyte. Homemade CE equipment with a picoampere meter at the low voltage end was used to carry out the measurements. For the solvents investigated, the resulting currents increased as follows: ACN < H_2O < MeOH $\approx$ DMSO < DMF << FA. The resulting currents varied between 5.1 nA for ACN and 5980 nA for FA at a voltage of 30 kV applied to a capillary of 50 cm × 50 µm ID. The low value for ACN was related to its high purity and extremely low autoprotolysis while the high currents for FA were attributed to impurities contained in this rather unstable solvent. Lister *et al.* [102] concluded that the magnitude of current is a function of solvent purity rather than of the properties of the neat solvent. They mentioned traces of ionic contaminants, water absorbed from the atmosphere, and hydrolyzed solvent molecules. Concerning the possible sources of charge carriers in pure solvents, they refer to a publication on electrospray MS [103]. Since the currents in pure solvents were found to be three orders of magnitude smaller than in common aqueous buffers for CE, Lister *et al.* proposed the use of low electrolyte concentrations combined with increased capillary diameters in order to improve detectability without being troubled by increased Joule heating. However, this strategy might suffer from poor analyte mass loadability of the system and the impossibility to make use of efficient electrostacking in the injection zone.

Valkó *et al.* [104] presented an approach to overcome the problems of CE when it comes to semipreparative tasks. They investigated the application of CE with wide-bore capillaries and nonaqueous media. When the capillary diameter *r* is increased, the maximum sample load is increased by a factor of r^2. Since the resulting current is also proportional to r^2, the generated amount of Joule heat becomes critical for separation efficiency, method precision, and stability when working with wide-bore capillaries. The use of NACE may be a solution to this problem. Valkó *et al.* measured the current in capillaries of 50, 100, 150, and 200 µm ID as a function of the applied voltage up to values of 30 kV. Under these conditions, they compared the current characteristics when applying an

aqueous buffer (water – acetic acid; 99:1 v/v, with 20 mM ammonium acetate) and a nonaqueous buffer (ethanol – acetonitrile – acetic acid, 50:49:1 v/v/v, with 20 mM ammonium acetate). In the case of the aqueous buffer and the 150 µm capillary, the integrated upper power limitation interrupted the system at 25 kV. In the 200 µm capillary the current broke down at 12 kV, most probably due to bubble formation in the solvent close to the boiling point.

As can be deduced from Fig. 6, the linearities of the current curves for the capillaries larger than 50 µm were very poor, thus indicating considerable Joule heat generation and the deviations resulting from Ohm's law. When the nonaqueous buffer was applied, a good linearity for the current even in the 200 µm capillary was achieved and the currents were only one-fourth to one-third of the level measured in the aqueous medium under otherwise identical conditions. The authors further reported on a separation of organic acids in the nonaqueous system. The obtained efficiency was slightly lower for the 200 µm capillary compared to the narrower ones. The separation selectivity was altered upon varying the capillary ID. A possible explanation given by the authors is the strong influence of temperature on the apparent pH. Since the volume flow increases with r^2 (like the mass load and current), differences in the buffer level at the inlet and outlet vial develop much faster than in narrow-bore CE, resulting in an additional hydrodynamic flow which can no longer be neglected because its linear velocity also increases with r^2. When a 200 µm capillary was applied, the separation efficiency was tremendously decreased even when the level at the inlet was only 2 mm lower than at the outlet. The authors figured out that the influence of siphoning on plate numbers is more pronounced than its effect on the migration time. The amount of siphoning flow *F* can be calculated by Eq. (6)

$$F = \frac{\rho g d r^2}{8 L \eta} \tag{6}$$

where ρ is the solvent density, *g* the acceleration due to gravity, *d* the difference between inlet and outlet level, η the solvent viscosity, and *r* and *L* are the radius and length of the capillary. Thus, the ratio of solvent density to solvent viscosity ρ/η is crucial for the contribution of an additional pressure-driven flow. The values for ρ/η are given in Table 4. From this it can be deduced that FA and DMSO are ideal for wide-bore applications, whereas methanol and ACN are not the best choice.

A possible solution for the problems related to the siphoning effect is suppression of the EOF *via* the application of coated capillaries, at least if the initial levels can be balanced. Further applications of wide-bore capillaries in NACE [105] and approaches to suppress the siphoning

[106] have also been published by Riekkola and co-workers. Although the application of NACE appears very advantageous in terms of current and heat generation at first sight, one possible difficulty needs to be mentioned. Altria and Bryant [34] pointed out that organic solvents generally exhibit lower thermal conductivities than water,

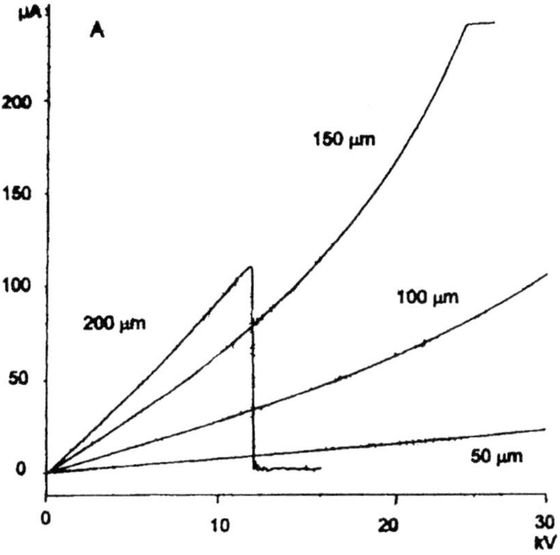

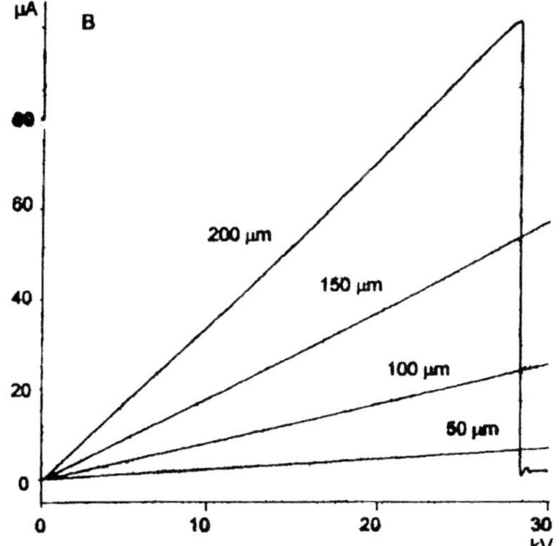

Figure 6. Ohm plots in 58.5 cm long fused-silica capillaries. The capillaries were filled with (A) water-acetic acid (99:1 v/v) containing 20 mM ammonium acetate and (B) ethanol-ACN-acetic acid (50:49:1 v/v/v) containing 20 mM ammonium acetate. The voltage was increased linearly from 0 to 30 kV and the capillary was thermostated at 20°C. Reprinted from [104], with permission.

Table 4. Density (ρ), dynamic viscosity (η), and ρ/η ratio of selected solvents and the linear velocity of the hydrodynamic flow

	ρ (g/cm^3)	η (mPa s)	ρ/η (s/cm^2)	$v_{1.200}$ (μm/min)
FA	1.1292	3.3020	0.3420	751
DMSO	1.0958	1.9960	0.5490	1206
n-Methylformamide	0.9988	1.6500	0.6053	1329
Ethanol	0.7850	1.0780	0.7282	1599
Acetic acid	1.0437	1.1550	0.9036	1985
Water	0.9970	0.8903	1.1199	2460
DMF	0.9440	0.8020	1.1770	2585
Methanol	0.7866	0.5445	1.4447	3173
ACN	0.7766	0.3450	2.2510	4944
Acetone	0.7844	0.3040	2.5803	5667

Hydrodynamic flow generated by 1 mm difference in the inlet and outlet buffer levels in 200 μm ID capillaries ($v_{1,200}$) at 25°C; total capillary length, 33.5 cm. Reprinted from [104], with permission. Values for η may differ slightly from Table 2 according to data given in the literature source.

thus reducing heat dissipation at the capillary wall. In addition, some of the nonaqueous media have significantly lower boiling points than water (see Table 1), *e.g.*, methanol (65°C) and ACN (82°C), and smaller heat capacities. Outgassing of the solvent and current breakdown may therefore result at significantly lower electric performances than would occur in water. Altria and Bryant recommended adjusting the ionic strengths of the buffer and the applied field strength in order to avoid currents higher than 45 μA.

7 Determination of physicochemical constants in NACE

Okada [107–109] published a series of papers on the application of NACE as a powerful method to measure physicochemical parameters in organic solvents and contribute to studies on fundamental solution chemistry. It is commonly accepted that separation techniques are more sensitive to study solution chemistry than spectrometric or electrochemical methods. Since chromatography always involves a stationary phase, retention equilibria are not solely affected by solution equilibria. This problem does not arise in CE. In his publication on NACE measurements of monodisperse polyoxyethylene (POE) and crown ethers in methanol in the presence of alkali ions, ammonium, or protonated amines, weak complex formation equilibria are studied quantitatively [107]. Methanol was chosen as solvent because it appeared to be favorable for polyether complexation of cations. This phenomenon is unlikely to occur in solvents with a strong ability to solvate hard cations (*e.g.*, water) where the ligands cannot replace solvent molecules in the solvation shell. In

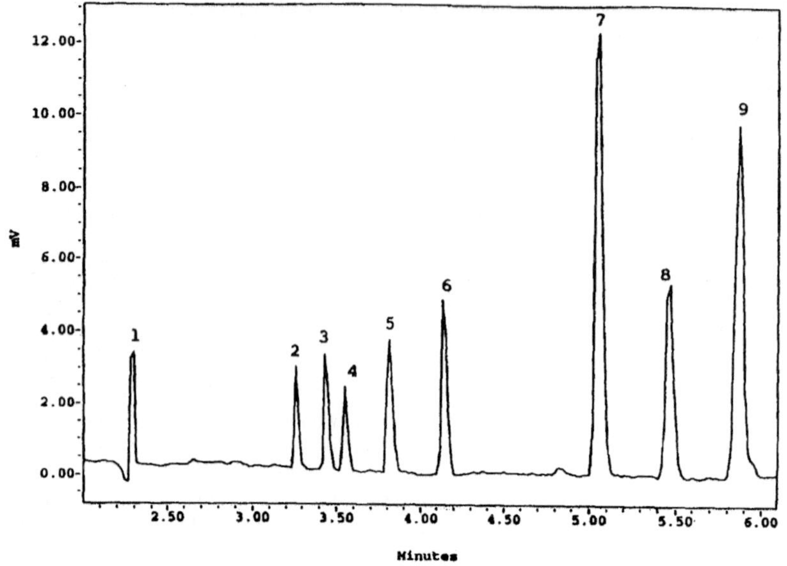

Figure 7. Electropherogram of a mixture of nine morphine analogs (100 ppm each). Peaks: 1, methadone; 2, pentazocine; 3, ethoheptazine; 4, levallorphan; 5, meperidine; 6, fentanyl; 7, nalorphine; 8, diphenoxylate; 9, nikethamide. Running medium, ACN-methanol-acetic acid (49:50:1)-20 mM ammonium acetate; capillary, 59.6 cm × 74 μm ID; applied voltage, 30 kV. Reprinted from [110], with permission.

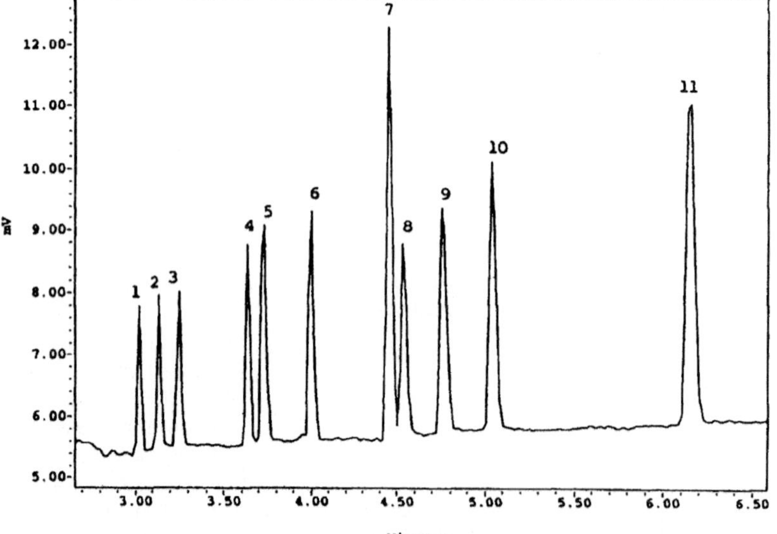

Figure 8. Electropherogram of a mixture of eleven antihistamines (100 ppm each). Peaks: 1, tripelennamine; 2, pyrilamine; 3, promethazine; 4, dimenhydrinate; 5, pheniramine; 6, chlorpheniramine; 7, pyrrobutamine; 8, methaphenilene; 9, cyclizine; 10, chlorcyclizine; 11, buclizine. Running medium, ACN-methanol-acetic acid (49:50:1)-20 mM ammonium acetate; capillary, 62 cm × 75 μm ID; applied voltage, 30 kV. Reprinted from [110], with permission.

solvents whose dielectric constants were too low (*e.g.*, dioxane), complex formation is interfered with by ion-pair formation. The author could demonstrate that NACE is suited to determine complexation constants in the range of 5–20 which is hardly possible by any other physicochemical method.

In a further publication on the separation of Brønsted acids as heteroconjugated anions [108], Okada studied interactions of phenols, carboxylic acids and alcohols with anions like ClO_4^-, BF_4^-, NO_3^-, $CH_3SO_3^-$ and Cl^- in ACN as solvent. He figured out that the electrophoretic mobility in ACN in the presence of such anions are not due to proton transfer, but merely the result of the formation of a so-called heteroconjugated anion. This describes a complex $[X^-...HA]$ of anion X^- and a weak Brønsted acid HA (*e.g.*, phenol, alcohol), not dissociated in ACN, through a hydrogen bond association. Heteroconjugated anion formation constants and molar ionic conductivities of these heteroconjugates were calculated for a series of phenols with

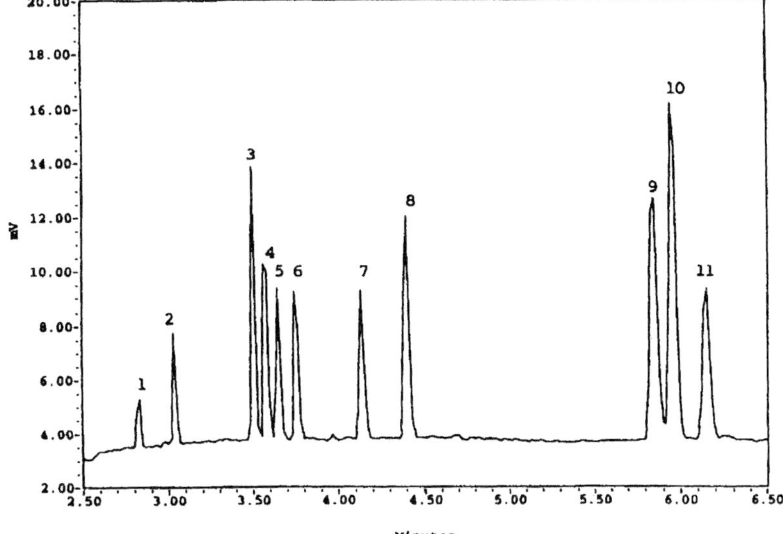

Figure 9. Electropherogram of a mixture of eleven antipsychotic drugs (100 ppm each). Peaks: 1, promethazine; 2, ethopropazine; 3, methotrimeprazine; 4, promazine; 5, thioridazine; 6, mesoridazine; 7, molindone; 8, thiothixene; 9, reserpine; 10, deserpidine; 11, benzquinamide. Running medium, ACN-methanol-acetic acid (49:50:1)-20 mM ammonium acetate; capillary, 59.6 cm × 75 μm ID; applied voltage, 30 kV. Reprinted from [110], with permission.

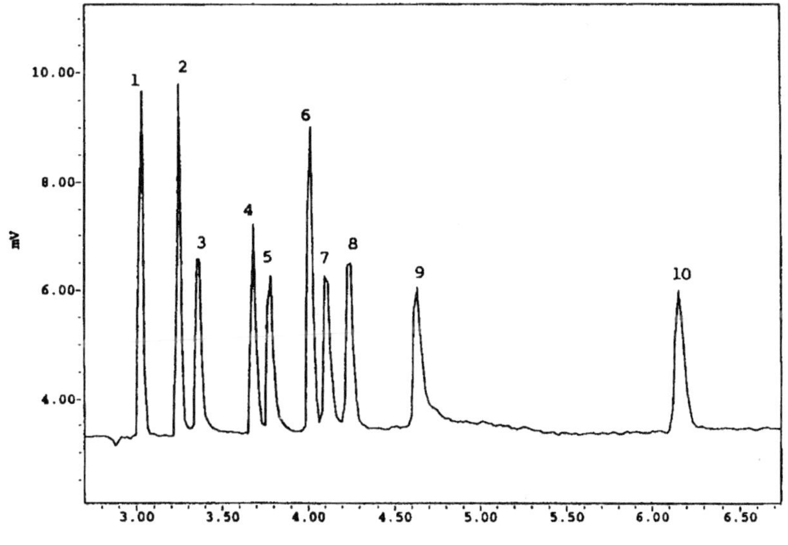

Figure 10. Electropherogram of a mixture of ten stimulants (75 ppm each). Peaks: 1, xylometazoline; 2, oxymetazoline; 3, amphetamine; 4, ephedrine; 5, phenylpropanolamine; 6, benzphetamine; 7, phenylephrine; 8, nylidrin; 9, phenmetrazine; 10, phendimetrazine. Running medium, ACN-methanol-acetic acid (49:50:1)-10 mM ammonium acetate; capillary, 58.2 cm × 75 μm ID; applied voltage, 25 kV. Reprinted from [110], with permission.

chloride and nitrate ions and several aromatic carboxylic acids with chloride ions. In his review on NACE and nonaqueous ion-exchange chromatography, Okada [109] summarized the results on the separation of polyethers and crown ethers *via* complex formation with cations, another determination of complex formation constants in methanol and heteroconjugated anion formation of weak acids in ACN. In the case of polyether complexation, he mentioned that these methods could stimulate the development of new ligands and analytical methods.

CE can be considered a very fast method to determine zeta potentials of a surface exposed to an electrolyte solution, provided that the dielectric constant ε_r and the viscosity η of the EDL in the system are known. The zeta potential is the potential that the diffuse double layer adopts at the so-called plane of shear relative to the bulk solution. Within EDL the flow velocity drops down when approaching the capillary wall and equals zero at a certain distance which defines the plane of shear. Wright *et al.* [66] and Valkó *et al.* [67] calculated zeta potentials of

Table 5. NACE applications in analysis of food, pharmaceuticals, biological fluids, plant extracts, environment, and technical products

Solvents	Electrolytes	Analytes	Ref.
Applications in food			
NMF-dioxan (1:1 v/v)	40 mM Tris, 2.5 mM anthraquinon 2-2-carboxylic acid	Free saturated long-chained fatty acids (n-C$_{14}$-n-C$_{26}$; separation of dimeric and trimeric acids and hydrogenated fish oil.	[88]
NMF	500 mM magnesium acetate tetrahydrate	Tetracycline (TC), oxytetracycline (OTC), chlortetracycline (CTC), demeclocycline, 4-epitetrycycline, anhydrotetracycline, 4-epianhydrotetracycline, and desmethyltetracycline; TC, OTC and CTC in milk and plasma	[90]
Propylene carbonate	Tetraalkylammonium ions, long-chain trimethylammonium ions 20 mM Tetradecylammonium bromide (Vitamin K$_1$ and preservatives)	Phenanthrene, β-naphthol; preservatives: methylparaben, ethylparaben and propylparaben, thiourea (EOF marker) and vitamin K$_1$	[54]
Applications with pharmaceuticals			
10–100% MeOH	Ammonium acetate, acetic acid	Haloperidol and synthetic putative metabolites, pyrazoloacridine and mifentidine	[18]
Mixture of MeOH and H$_2$O	Ammonium acetate, acetic acid	Haloperidol, cimetropium and mifentidine	[19]
MeOH	5 mM ammonium acetate, 100 mM acetic acid	Haloperidol and its synthetic putative metabolites, pyrazoloacridine and its synthetic putative metabolites, mifentidine and its synthetic putative metabolites	[20]
MeOH and mixture of MeOH and ACN	Ammonium acetate, tetrabutylammonium bromide, tetrabutylammonium hydrogensulfate and tetrapentylammonium bromide	Tamoxifen and four phase I metabolites	[21]
MeOH, ACN, mixture of MeOH and ACN, FA, NMF, DMF, DMA, DMSO	25 mM ammonium acetate, 0–1 M acetic acid or 100 mM sodium acetate. Application: 25 mM ammonium acetate, 1 M acetic acid in MeCN	Imipramine, di-desmethylimipramin, desmethalimipramine, methylimipramine and imipramine-*N*-oxide; maprotiline, amitriptyline, litracene, protriptyline and nortriptyline; application: imipramine *N*-oxide and impurities	[15]
MeOH:MeCN:DMF (45:49:6 v/v/v)	25 mM ammonium acetate, 10 mM citric acid and 118 mM methynesulfonic acid	Tetracycline and three degradation products; tetracycline, oxytetracycline, doxycycline, desmethyltetracycline and chlortetracycline	[40]
MeOH:MeCN (1:1 v/v)	20 mM ammonium acetate, 1 M acetic acid	Morphine analogs, antihistamines, antipsychotics and stimulants	[33]
FA, NMF or DMF	Citric acid or acetic acid mixed with Tris chiral selectors: β-CD, γ-CD and derivatized β-CD. Addition of long--chain alkyl ammonium salts investigated.	Racemic mixtures of: chlophedianol, chlorcyclizine, ethopropazine, mianserin, nefopam, primaquine, propiomezine, trihexyphenidyl, trimeprazine, trimipramine and thioridazine	[42]
NMF, FA and mixtures of both	25–200 mM β-CD, 10 mM NaCl	Dansylated amino acids	[45]

Table 5. continued

Solvents	Electrolytes	Analytes	Ref.
NMF	5–100 mM β-CD and 10 mM NaCl	Dansylated amino acids	[111]
MeOH	Ammonium acetate, acetic acid, quinine	N-3,5-dinitrobenzylated amino acids, (±)-1,1'binaphthyl-2,2'-diylhydrogen phosphate and N-[1-(1-napthyl)ethyl]phthalomic acid.	[51]
ACN	(±)-Camphorsulfonic acid potassium or sodium salt, 1 M acetic acid 0.2 M Tween-20	Atenolol, bisoprolol, bunitrolol, metroprolol, pindolol, propranolol, salbutamol, ephedrine, epinephrine, cisapride and synthetic impurities.	[53]
FA	Tetra-n-butylammonium perchlorate. Chiral selector: (+)-18-crown-6-tetracarboxylic acid	1-Naphthylethylamine, 1-phenylethylamine, phenylalanine, DOPA, thryptophan, norephedrine, noradrenaline and 2-amino-1,2-diphenylethanol 1	[50]
FA, NMF, DMF, DMA, DMSO, MeOH; ACN and mixtures of MeOH and MeCN	25 mM ammonium acetate, 1 M acetic acid	Morphine, codeine, normorphine, thebaine, noscapine and papaverine; application:morphine in opium tincture	[112]
MeOH:MeCN (75:25 v/v)	25 mM ammonium acetate, 1 M acetic acid	Morphine	[113]
NMF	500 mM magnesium acetate tetrahydrate	Oxytetracycline in an ointment	[59]
Mixtures of MeOH and ACN	Ammonium acetate, ammonium chloride, acetic acid, trifluoroacetic acid, formic acid, methyne sulfonic acid	Cis-trans (Z-E) isomers of chlorprothixene, thiothixene, clopenthixol, flupenthixol, flupenthixol decanoate, clomiphene and diastereomers: L-Ala-L-Phe, L-Ala-D-Phe; quinine, quinidine, cinchonine and cinchonidine	[114]
Mixtures of MeOH and ACN	Sodium acetate	A range of penicillins, caphalosporins and nonsteroidal anti-inflammatory drugs	[34]
MeOH	20 mM CAPS and 0–40 mM Brij-35	Mesoporphyrin, coporhyrin, pentaporphyrin, hexacarboxylporphyrin, heptacarboxylporphyrin and uroporphyrin	[62]
MeOH/ACN (30:70 v/v)	Ammonium acetate, acetic acid	Cimetidine, associated impurities and degradation products	[35]
DMF	Citric acid, Tris	Promethazine, dioxypromethazine, chlorpromazine, trifluoperazine	[115]
NMF	NaOH	Pyridinyl-methyl-sulfinyl-benzidimidazoles	[116]
MeOH/ACN (40:60 v/v)	Ammonium acetate	Baclophen, indoprofen, ibuprofen, fenoprofen, indomethacin, ketoprofen, suprofen, diclofenac, mefenamic acid	[38]

Applications within biological fluids

Solvents	Electrolytes	Analytes	Ref.
10–10% MeOH in H_2O	20 mM ammonium acetate, 1% acetic acid	Pyrazoloacridine, two metabolites and a synthetic degradation product in urine	[14]
NMF	50 mM magnesium acetate tetrahydrate	TC, OTC, CTC, demeclocycline, 4-epitetracycline, anhydrotetracycline, 4-epianhydrotetrycycline, and desmethyltetracycline; TC, OTC and CTC in cow milk and human plasma	[60]
MeOH	5 mM ammonium acetate, 10 mM acetic acid	Mifentidine and three metabolites in rat liver homogenate	[95]

Table 5. continued

Solvents	Electrolytes	Analytes	Ref.
MeOH/ACN (50:50 v/v)	50 mM ammonium acetate, 159 mM sodium acetate and 0.002% w/v hexydimethrine bromide	Acetylsalicyclic acid and three metabolites: salicyclic acid, salicyluric acid and gentisic acid in plasma and urine	[70]
Applications with environmental background			
MeOH/THF/ACN (70:15:15 v/v/v)	Trimetylammoniumhydroxide, acetic acid	Zinc dialkyl dithiophosphates in lubricant oils	[41]
MeOH/ACN	Ammonium acetate, acetic acid	Priority pollutant phenols (nitro-, chloro-, amino-phenols)	[117]
Application to natural materials like plant extracts			
MeOH/THF (90:10 v/v)	Ammonium acetate, acetic acid	Matrine, sophoridine, sophocarpine, lehmannine, sophoramine, oxymatrine, oxysophocarpine, cytisine, aloperine, dauricine, scopolamine in *Sophora flavescens, Sophora alopecuroides* and *Sophora tonkinensis*	[39]
MeOH/ACN	Ammonium acetate, acetic acid	Atropine, littorine, apoatropine, ipratropium, scopolamine, *N*-butylscopolamine, methylscopolamine, homatropine in genectically transformed root cultures of *Datura candida* × *Datura aurea*	[37]

Modified from a table in [110].

fused-silica capillaries in different media from the measured values of the EOF, which was already reported in Section 4.3. They assumed the EDL values for ε_r and η to be identical to the data for pure solvents taken from literature. The calculated ζ_{wall} values reported in [67] are listed in Table 3. As can be deduced from this table, zeta potentials only differ by a factor of four over all the investigated solvents, whereas the obtained EOF velocities covered almost two orders of magnitude. No correlation between ζ_{wall} and the resulting electroosmosis was found. This indicated the predominant influence of the ε_r/η ratio on the amount of EOF. The reason for the high zeta potentials found in aprotic solvents like ACN and acetone have already been discussed in Section 4.3. Among the ζ_{wall} potentials reported by Wright *et al.* [66], the values for ACN (207 mV), water (99 mV) and methanol (108 mV) were in good agreement with those calculated by Valkó *et al.*, whereas the values for DMF, FA, and DMSO were remarkably different. Variations in the amount of hydrolysis products and impurities of the solvents might be the reason for these variations.

8　Analytical applications of NACE

In 1996, Leung *et al.* [33] reported on the analysis of basic drugs in NACE by applying a very common solvent mixture. They separated morphine analogs, antihistamines, antipsychotics, and stimulants in an ACN – methanol – acetic acid (49:50:1) medium. The structures of all compounds are given in their publication. Although Leung *et al.* did not demonstrate the application of their method to real samples, their work can be considered a early convincing evidence of the power of NACE to achieve very fast and efficient separations of analytes exhibiting similar structures. As can be seen in Fig. 7–10, the drug substances of each class were combined in mixtures yielding 9–11 components. All these mixtures could be separated within 6.5 min at sufficient electrophoretic resolution. A review paper on published applications in NACE focusing on the analysis of food, pharmaceutical and biological samples was published by Bjørnsdottir *et al.* [110] in 1998. They listed applications in the fields mentioned above in a table and specified the solvents and electrolytes that had been applied. For this general review on NACE, their list was extended to further fields of application and more recent publications were included. Many of the related references have already been cited in this review, but Table 5 gives a complete listing all at one glance.

Received August 20, 2000

9　References

[1] Walbroehl, Y., Jorgenson, J. W., *J. Chromatogr.* 1984, *315*, 135–143.

[2] Walbroehl, Y., Jorgenson, J. W., *Anal. Chem.* 1986, *58*, 479–481.

[3] Fujiware, S., Honda, S., *Anal. Chem.* 1987, *59*, 487–490.

[4] VanOrman, B. B., Liversidge, G. G., McIntire, G. L., *J. Microcol. Sep.*, 1990, *2*, 176–180.

[5] Schwer, C., Kenndler, E., *Anal. Chem.* 1991, *63*, 1801–1807.

[6] Schützner, W., Kenndler, E., *Anal. Chem.* 1992, *64*, 1991–1995.

[7] Janini, G. M., Chan, K. C., Barnes, J. A., Muschik, G. M., Issaq, H. J., *Chromatographia* 1993, *35*, 479–502.

[8] Sarmini, K., Kenndler, E., *J. Chromatogr. A* 1998, *806*, 325–335.

[9] Sarmini, K., Kenndler, E., *J. Chromatogr. A* 1998, *811*, 201–209.

[10] Sarmini, K., Kenndler, E., *J. Chromatogr. A* 1998, *818*, 209–215.

[11] Sarmini, K., Kenndler, E., *J. Chromatogr. A* 1999, *833*, 245–259.

[12] Benson, L. M., Tomlinson, A. J., Reid, J. M., Walker, D. L., Ames, M. M., Naylor, S., *J. High Resol. Chromatogr.* 1993, *16*, 324–326.

[13] Riekkola, M.-L., Wiedmer, S. K., Valkó, I. E., Sirén, H., *J. Chromatogr. A* 1997, *792*, 13–35.

[14] Sahota, R. S., Khaledi, M. G., *Anal. Chem.* 1994, *66*, 1141–1146.

[15] Bjørnsdottir, I., Hansen, S. H., *J. Chromatogr. A* 1995, *711*, 313–322.

[16] Bjørnsdottir, I., Tjørnelund, J., Hansen, S. H., *J. Capil. Electrophor.* 1996, *3*, 83–87.

[17] Jansson, M., Roeraade, J., *Chromatographia* 1995, *40*, 163–169.

[18] Tomlinson, A. J., Benson, L. M., Naylor, S., *J. Capil. Electrophor.* 1994, *1*, 127–135.

[19] Naylor, S., Benson, L. M., Tomlinson, A. J., *J. Capil. Electrophor.* 1994, *1*, 181–189.

[20] Tomlinson, A. J., Benson, L. M., Naylor, S., *LC.GC Int.* 1995, *8*, 210–216.

[21] Ng, C. L., Lee, H. K., Li, S. F. Y., *J. Liq. Chromatogr.* 1994, *17*, 3847–3857.

[22] Popovych, O., Tomkins, R. P. T., *Nonaqueous Solution Chemistry*, John Wiley & Sons, New York 1981, pp. 32–43.

[23] Valkó, I. E., Sirén, H., Riekkola, M.-L., *LC.GC Int.* 1997, *10*, 190–196.

[24] Ward, V. L., Khaledi, M. G., *J. Chromatogr. A* 1999, *869*, 203–219.

[25] Bowser, M. T., Kranak, A. R., Chen, D. D. Y., *Trends Anal. Chem.* 1998, *17*, 424–434.

[26] Gutmann, V., *Electrochim. Acta* 1979, *21*, 661–670.

[27] Burger, K., *Solvation, Ionic and Complex Formation Reactions in Non-Aqueous Solvents: Experimental Methods for Their Investigation*, Elsevier, New York 1983.

[28] Mayer, U., *Pure Appl. Chem.* 1979, *51*, 1697–1712.

[29] Shaw, D. J., *Introduction to Colloid and Surface Chemistry*, Butterworths & Co., London 1985, p. 173.

[30] Salimi-Moosavi, H., Cassidy, R. M., *Anal. Chem.* 1995, *68*, 293–299.

[31] Salimi-Moosavi, H., Cassidy, R. M., *J. Chromatogr. A* 1996, *749*, 279–286.

[32] Salimi-Moosavi, H., Cassidy, R. M., *J. Chromatogr. A* 1997, *790*, 185–193.

[33] Leung, G. N. W., Tang, H. P. O., Tso, T. S. C., Wan, T. S. M., *J. Chromatogr. A* 1996, *738*, 141–154.

[34] Altria, K. D., Bryant, S. M., *Chromatographia* 1997, *46*, 122–130.

[35] Ellis, D. R., Palmer, M. E., Tetler, L. W., Eckers, C., *J. Chromatogr. A* 1998, *808*, 269–275.

[36] Tjørnelund, J., Bazzanella, A., Lochmann, H., Bächmann, K., *J. Chromatogr. A* 1998, *811*, 221–217.

[37] Cherkaoui, S., Mateus, L., Christen, P., Veuthey, J.-L., *Chromatographia* 1999, *49*, 54–60.

[38] Cherkaoui, S., Veuthey, J.-L., *J. Chromatogr. A* 2000, *874*, 121–129.

[39] Song, J.-Z., Xu, H.-X., Tian, S.-J., But, P. P.-H., *Chromatogr. A* 1999, *857*, 303–311.

[40] Tjørnelund, J., Hansen, S. H., *J. Chromatogr. A* 1996, *737*, 291–300.

[41] Thibon, V. R. A., Bartle, K. D., Abbott, D. J., McCormack, K. A., *J. Microcol. Sep.* 1999, *11*, 71–80.

[42] Wang, F., Khaledi, M. G., *Anal. Chem.* 1996, *68*, 3460–3467.

[43] Li, Y., Xie, L. J., Liu, H. W., Hua, W. T., *Chin. Chem. Lett.* 1999, *10*, 303–306.

[44] Ren, X., Huang, A., Wang, T., Sun, Y., Sun, Z., *Chromatographia* 1999, *50*, 625–628.

[45] Valkó, I. E., Sirén, H., Riekkola, M.-L., *J. Chromatogr. A* 1996, *737*, 263–272.

[46] Wang, F., Khaledi, M. G., *J. Chromatogr. A* 1998, *817*, 121–128.

[47] Tacker, M., Glukovskiy, P., Cai, H., Vigh, G., *Electrophoresis* 1999, *20*, 2794–2798.

[48] Vincent, J. B., Vigh, G., *J. Chromatogr. A* 1998, *816*, 233–241.

[49] Wang, F., Khaledi, M. G., *J. Chromatogr. B* 1999, *732*, 187–197.

[50] Mori, Y., Ueno, K., Umeda, T., *J. Chromatogr. A* 1997, *757*, 328–332.

[51] Stalcup, A., Gahm, K. H., *J. Microcol. Sep.* 1996, *8*, 145–150.

[52] Piette, V., Lämmerhofer, M., Lindner, W., Crommen, J., *Chirality* 1999, *11*, 622–630.

[53] Bjørnsdottir, I., Hansen, S. H., Terabe, S., *J. Chromatogr. A* 1996, *745*, 37–44.

[54] Tjørnelund, J., Hansen, S. H., *J. Chromatogr. A* 1997, *792*, 475–482.

[55] Li, S., Weber, S. G., *J. Am. Chem. Soc.* 2000, *122*, 3787–3788.

[56] Miller, J. L., Khaledi, M. G., Shea, D., *Anal. Chem.* 1997, *69*, 1223–1229.

[57] Miller, J. L., Khaledi, M. G., Shea, D., *J. Microcol. Sep.* 1998, *10*, 681–685.

[58] Li, J., Fitz, J. S., *Electrophoresis* 1999, *20*, 84–31.

[59] Tjørnelund, J., Hansen, S. T., *J. Pharm. Biomed. Anal.* 1997, *15*, 1077–1082.

[60] Tjørnelund, J., Hansen, S. T., *J. Chromatogr. A* 1997, *779*, 235–243.

[61] Bellini, M. S., Deyl, Z., Mikisik, I., *Forensic Sci. Int.* 1998, *92*, 185–199.

[62] Bowser, M. T., Sternberg, E. D., Chen, D. D. Y., *Electrophoresis* 1997, *18*, 82–91.

[63] Bowser, N. T., Sternberg, E. D., Chen, D. D. Y., *Anal. Biochem.* 1996, *241*, 143–150.

[64] Wright, P. B., Dorsey, J. G., *J. High Resolut. Chromatogr.* 1998, *21*, 498–504.

[65] Morin, P., Daguet, D., Coic, J. P., Dreux, M., *J. Chromatogr. A* 1999, *837*, 281–287.

[66] Wright, P. B., Lister, A. S., Dorsey, J. G., *Anal. Chem.* 1997, *69*, 3251–3259.

[67] Valkó, I. E., Sirén, H., Riekkola, M.-L., *J. Microcol. Sep.* 1999, *11*, 199–208.

[68] Heinig, K., Vogt, G. W., *Fresenius J. Anal. Chem.* 1997, *358*, 500–505.

[69] Tjørnelund, J., Bazzanella, A., Lochmann, H., Bächmann, K., *J. Chromatogr. A* 1998, *811*, 221–217.

[70] Hansen, S. H., Jensen, M. E., Bjørnsdottir, I., *J. Pharm. Biomed. Anal.* 1998, *17*, 1155–1160.

[71] Belder, D., Elke, K., Husmann, H., *J. Microcol. Sep.* 1999, *11*, 209–213.

[72] Belder, D., Elke, K., Husmann, H., *J. Chromatogr. A* 2000, *868*, 63–71.

[73] Gallaher Jr., D. L., Johnson, M. E., *Anal. Chem.* 2000, *72*, 2080–2086.

[74] Faller, T., Engelhardt, H., *J. Chromatogr. A* 1999, *853*, 83–94.

[75] Tjørnelund, J., Hansen, S. H., *J. Pharm. Biomed. Anal.* 1997, *15*, 1077–1082.

[76] Tjørnelund, J., Hansen, S. H., *Chromatographia* 1997, *44*, 163–169.

[77] Li, X.-F., Liu, C.-S., Roos, P., Hansen Jr., E. B., *Electrophoresis* 1998, *19*, 3178–3182.

[78] Salimi-Moosavi, H., Cassidy, R. M., *Anal. Chem.* 1995, *67*, 1067–1073.

[79] Matysik, F.-M., *J. Chromatogr. A* 1996, *742*, 229–234.

[80] Matysik, F.-M., *J. Chromatogr. A* 1998, *802*, 349–354.

[81] Frey, A. J., in: Kissinger, P. T., Heinemann, W. R. (Eds.), *Laboratory Techniques in Electroanalytical Chemistry*, Marcel Dekker, New York 1996, Chapter 15.

[82] Matysik, F.-M., *J. Chromatogr. A* 1999, *853*, 27–34.

[83] Luong, J. H. T., Hilmi, A., Mguyen, A.-L., *J. Chromatogr. A* 1999, *864*, 323–333.

[84] Salimi-Moosavi, H., Cassidy, R. M., *Anal. Chem.* 1995, *67*, 1067–1073.

[85] Suzuki, N., Ishihama, Y., Kajima, T., Asakawa, N., *J. Chromatogr. A* 1998, *829*, 411–415.

[86] Salimi-Moosavi, H., Cassidy, R. M., *J. Chromatogr. A* 1997, *790*, 185–193.

[87] Heinig, K., Vogt, C., Werner, G., *J. Capill. Electrophor.* 1996, *3*, 261–270.

[88] Drange, E., Lundanes, E., *J. Chromatogr. A* 1997, *771*, 301–309.

[89] Ward, V. L., Khaledi, M. G., *J. Chromatogr. B* 1998, *718*, 15–22.

[90] Tjørnelund, J., Hansen, S. H., *J. Chromatogr. A* 1997, *779*, 235–243.

[91] Wang, T. L., Wei, H. P., Li, S. F. Y., *Electrophoresis* 1998, *19*, 2187–2192.

[92] Smith, R. D., Olivares, J. T., Nguyen, N. T., Udseth, H. R., *Anal. Chem.* 1988, *60*, 436–441.

[93] Lee, E. D., Mueck, W., Henion, J. D., Covey, T. R., *J. Chromatogr.* 1988, *458*, 313–321.

[94] Tomlinson, A. J., Benson, L. M., Naylor, S., *LC.GC* 1994, *12*, 122–130.

[95] Tomlinson, A. J., Benson, L. M., Naylor, S., *J. High Resolut. Chromatogr.* 1994, *17*, 175–177.

[96] Lu, W., Poon, G. K., Carmichael, P. L., Cole, R. B., *Anal. Chem.* 1996, *68*, 668–674.

[97] Yang, Q., Benson, L. M., Johnson, K. L., Naylor, S., *J. Biochem. Biophys. Methods* 1999, *38*, 103–121.

[98] Raith, K., Wolf, R., Wagner, J., Neubert, R. H. H., *J. Chromatogr.* 1998, *802*, 185–188.

[99] Liu, C.-H., Li, X.-F., Pinto, D., Hansen Jr., E. B., Cerniglia, C. E., Dovichi, N. J., *Electrophoresis* 1998, *19*, 3183–3189.

[100] Werner, A., Buchberger, W., *Fresenius J. Anal. Chem.* 1999, *365*, 604–609.

[101] Senior, J., Rolland, D., Tolson, D., Chantzis, S., De Biasi, V., *J. Pharm. Biomed. Anal.* 2000, *22*, 413–421.

[102] Lister, A. S., Dorsey, J. G., Burton, D. E., *J. High Resolut. Chromatogr.* 1997, *20*, 523–528.

[103] Blades, A. T., Ikonomov, M. G., Kebarle, P., *Anal. Chem.* 1991, *63*, 2109–2114.

[104] Valkó, I. E., Porras, S. P., Riekkola, M.-L., *J. Chromatogr. A* 1998, *813*, 179–186.

[105] Porras, S. P., Jussila, M., Sinervo, K., Riekkola, M.-L., *Electrophoresis* 1999, *20*, 2510–2518.

[106] Jussila, M., Palonen, S., Porras, S. P., Riekkola, M. L., *Electrophoresis* 2000, *21*, 586–592.

[107] Okada, T., *J. Chromatogr. A* 1995, *695*, 309–317.

[108] Okada, T., *J. Chromatogr. A* 1997, *771*, 275–284.

[109] Okada, T., *J. Chromatogr. A* 1998, *804*, 17–28.

[110] Bjørnsdottir, I., Tjørnelund, J., Hansen, S. H., *Electrophoresis* 1998, *19*, 2179–2186.

[111] Valkó, I. E., Sirén, H., Riekkola, M.-L., *Chromatographia* 1996, *43*, 242–246.

[112] Bjørnsdottir, I., Hansen, S. H., *J. Pharm. Biomed. Anal.* 1995, *13*, 1473–1481.

[113] Bjørnsdottir, I., Hansen, S. H., *J. Pharm. Biomed. Anal.* 1997, *15*, 1083–1089.

[114] Hansen, S. H., Bjørnsdottir, I., Tjørnelund, J., *J. Chromatogr. A* 1997, *792*, 49–55.

[115] Wang, R., Lu, X., Xin, H., Wu, M., *Chromatographia* 2000, *51*, 29–36.

[116] Tivesten, A., Folestad, S., Schönbacher, V., Svensson, K., *Chromatographia* 1999, *49*, 7–11.

[117] Morales, S., Cela, R., *J. Chromatogr. A* 1999, *846*, 401–411.

Electrophoresis 2000, *21*, 4112–4135

Review

Gerald Gübitz
Martin G. Schmid

Institute of Pharmaceutical
Chemistry,
Karl-Franzens University,
Graz, Austria

Recent progress in chiral separation principles in capillary electrophoresis

This review summarizes recent developments in the field of chiral separations by electromigration techniques including capillary zone electrophoresis (CZE), capillary gel electrophoresis (CGE), isotachophoresis (ITP), electrokinetic chromatography (EKC), and capillary electrochromatography (CEC). This overview focuses on the development of new chiral selectors and the introduction of new techniques rather than applications of already established selectors and methods. The mechanisms of the different chiral separation principles are discussed.

Keywords: Enantiomer separation / Capillary electrophoresis / Review   EL 4172

Contents

Correspondence: Dr. Gerald Gübitz, Institute of Pharmaceutical Chemistry, Karl-Franzens University, Universitätsplatz 1, A-8010 Graz, Austria,
E-mail: guebitz@kfunigraz.ac.at
Fax: +43-316-380-9846

Abbreviations: 18C6H4, 18-crown-6-tetracarboxylic acid; **AGP**, α_1-acid glycoprotein; **AIBN**, 2,2'-azobis(isobutyronitrile); **BNHP**, (±)-1, 1'-binaphthyl-2,2'-diyl hydrogen phosphate; **CHARM**, charged resolving agent migration model; **CM-β-CD**, carboxymethyl-β-CD; **CSP**, chiral stationary phase; **DNP**, dinitrophenyl; **Dns**, dansylated; **EMO**, enantiomer migration order; **FA**, formamide; **FMOC**, 9-fluorenylmethoxycarbonyl; **HP-β-CD**, hydroxypropyl-β-CD; **NMF**, *N*-methylformamide; **NSAID**, nonsteoridal antiinflammatory drug; **QA-β-CD**, quaternary ammonium-β-CD; **S-β-CD**, sulfated β-CD

1 Introduction

About 40% of the drugs in use are known to be chiral. It is well established that the pharmacological activity is mostly restricted to one of the enantiomers. In several cases unwanted side effects or even toxic effects can occur with the second enantiomer. Even if the side effects are not that drastic, the inactive enantiomer has to be metabolized and represents an unnecessary burden for the organism. The administration of the pure pharmacologically active enantiomers is therefore of great importance. The development of methods for enantiomer separation for controlling synthesis, for enantiomeric purity check, and for pharmacodynamic studies is attracting increasing interest. In addition to chromatographic methods, CE and more recently CEC are becoming increasingly attractive.

Several comprehensive review articles have appeared in recent years dealing with general aspects of chiral CE and applications of different chiral separation principles to various compound classes [1–21]. This review will therefore focus on new techniques and new chiral selectors covering the literature from 1997 to date. The reader is referred to several specialized reviews on applications. Earlier articles are cited only if they explain mechanisms or contain the first mention of given separation principles. Chiral separation by CE can be performed either indirectly, using a chiral derivatization agent forming diastereomeric pairs, which can be resolved under achiral conditions, or directly, using chiral selectors as additives to the electrolyte. In CEC, a recent technique, similarly to HPLC, chiral stationary phases or chiral mobile phase additives can be applied.

2 Cyclodextrins (CDs)

CDs represent the most frequently used chiral selectors. Native CDs are cyclic oligosaccharides consisting of six (α-CD), seven (β-CD) or eight (γ-CD) glucopyranose units with a truncated cone providing a hydrophobic cavity. Due to the presence of hydroxyl groups the outside of the CD is hydrophilic. Chiral recognition is based on inclusion of the bulky hydrophobic group of the analyte into this hydrophobic cavity of the CD and lateral interactions of the hydroxyl groups at the C-2 and C-3 at the upper rim of the CD, such as hydrogen bonds and dipole-dipole interactions with the analyte. Several neutral and charged CD derivatives have been synthesized. The use of CDs as chiral selectors is the subject of several selective reviews [12, 20, 22].

2.1 Neutral CDs

Various neutral derivatives of CDs such as heptakis-*O*-methyl-β-CD (M-β-CD), heptakis(2,6-di-*O*-methyl)-β-CD (DM-β-CD), heptakis(2,3,6-tri-*O*-methyl)-β-CD (TM-β-CD), hydroxyethyl-β-CD (HE-β-CD), and hydroxypropyl-β-CD (HP-β-CD) have been synthesized and applied to a great variety of compounds [12, 13, 20]. Numerous applications but only a few new developments have been reported in the past three years. Since most of the CD derivatives represent mixtures of different products showing different substitution patterns, separations are often difficult to reproduce. There is a new trend to separate the pure derivatives from mixtures or to synthesize single isomers and to produce selectively substituted derivatives.

Miura *et al.* synthesized selectively methylated [23] and acetylated [24] CDs and evaluated the different products for their ability to resolve Dns-amino acids, naphthalenesulfonylamino acids and naphthoylamino acids. The authors have shown that there are remarkable differences in selectivity between the methylated and acetylated CDs due to the difference in bulkiness and hydrogen-bonding ability. Also, the position of the methyl or acetyl group

was found to play an important role in chiral recognition ability. Furthermore, the chiral recognition of the different selectors varied for the different analytes depending on their bulkiness. These findings are certainly worthwhile contributions to the developments of chiral selectors and to the clarification of the separation mechanism.

Aturki *et al.* [25] recently introduced a new, uncharged CD derivative, cyanoethylated β-CD, and checked its enantioselectivity with a series of basic and acidic drugs. Acidic drugs were found to be more strongly complexed than basic analytes and lower CD concentrations were required. Interestingly, the authors observed an inversion of migration order for naproxen with this selector compared to trimethyl-β-CD used in previous studies [26]. ^{1}H-NMR studies indicated a strong interaction between the methyl group and proton of the aromatic moiety of naproxen and the cyanoethylated-β-CD. Chiari *et al.* [27] synthesized a new vinylpyrrolidine-β-CD copolymer by radical copolymerization of vinylpyrrolidone and methacroyl-β-CD. This polymer consists of an alkyl backbone and pendant β-CD units. The chiral recognition ability of this selector was evaluated with a mixture of sympathomimetic drugs. This selector showed significant advantages over native β-CD.

New ethylcarbonate-β- and γ-CDs were introduced by Zerbinati *et al.* [28] and were used for the resolution of racemic dichlorprop herbicides. These new CD derivatives were found to be superior to native CDs, M-β-CD, HP-β-CD and a newly synthesized C-6 capped β-CD. Recently, the synthesis of an interesting compound, an L-Ala-Crown(3)-L-Ala capped β-CD, was reported [29]. The structure of 6^A, 6^D,-dideoxy-*N*,*N*'-3,6,9-trioxa-undecanoyl-(LL)-bis-alanyl-β-CD was confirmed by COSY and ROESY NMR spectra. This new selector was tested by means of the chiral separation of Dns-amino acids.

2.2 Negatively charged CDs

A broad spectrum of negatively charged CDs was investigated and applied to different drug classes, preferentially to basic and neutral drugs. The improved selectivity compared to neutral CDs is mainly attributed to the countercurrent mobility. Sulfated β-CDs (S-β-CD), sulfobutyl-(SBE-β-CD) and sulfoethyl ether-β-CD (SEE-β-CD) are the most frequently used charged CDs [12, 20, 21]. The preparation of sulfobutyl-γ-CD was first described by Jung and Francotte [30] and Francotte *et al.* [31]. A reversal of the elution order for several enantiomers was observed on changing from γ-CD to SBE-γ-CD. Rickard *et al.* [32] investigated the influence of the degree of substitution (d.s.) of SBE-β-CDs on the separation and peak shapes using duloxetine as a model compound. From a commercially available SBE-β-CD, fractions of derivatives with different d.s. were isolated and investigated. Other strongly negative charged CD derivatives are sulfated CDs. Commercially available sulfated CDs consist of numerous isomers which differ in their degree and site of substitution. Batch-to-batch variations in composition lead to high variations in mobility and selectivity and so to poor separation reproducibilities. To eliminate these drawbacks, single isomer charged CDs were synthesized.

A family of single isomer β- and γ-CD derivatives was introduced by the group of Vigh [33–35]. These authors prepared derivatives completely sulfated in 6-position and completely substituted on their larger rims with hydrophylic groups (heptakis(2,3-dihydroxy-6-sulfato)-β-CD) [33], moderately hydrophobic groups (heptakis(2,3-diacetyl-6-sulfato)-β-CD) [34] or hydrophobic methyl functional groups (heptakis(2,3-dimethyl-6-sulfato)-β-CD) [35]. More recently, an octakis(2,3-diacetyl-6-sulfato)-γ-CD was synthesized and evaluated [36]. The authors have shown that neutral, basic, zwitterionic and even acidic enantiomers can be separated. As predicted by the charged resolving agent migration, (CHARM) model developed by Williams and Vigh [37], the enantiomeric migration order can be reversed in several cases by increasing the selector concentration.

Highly sulfated CDs with a d.s. of 10, developed by the Beckman company (Palo Alto, CA, USA), were applied to the separation of di- and tripeptides and a broad spectrum of pharmaceuticals. Carboxylfunctional CDs such as carboxymethyl-β-CD (CM-β-CD) [38, 39], carboxyethyl-β-CD (CE-β-CD) [39], and succinyl-β-CD (Succ-β-CD) [39, 40] were introduced and applied to a large variety of compounds [12, 13, 20, 21]. Only a few authors report the use of phosphated CDs [41–43]. Juvancz *et al.* [43] compared the enantioselectivity of α-, β- and γ-CD phosphates for tocainide, metoprolol, and diisopyramide.

2.3 Positively charged CDs

Cationic CDs such as 6-[(3-aminoethyl)amino]-6-deoxy-β-CD [44], 6^A-methylamino-β-CD, 6^A, 6^D-dimethylamino-β-CD [45], a heptasubstituted methylamino-β-CD [46], and mono(6-amino-6-deoxy)-β-CD [47] were the first described cationic CDs to be applied to the chiral separation of different acidic and neutral compounds [12, 13, 20, 21]. O'Keefe *et al.* [48] developed a polycationic CD derivative, heptakis (6-hydroxyethylamino-6-deoxy-β-CD). Using a reversed polarity, acidic compounds such as nonsteroidal antiinflammatory drugs (NSAIDs), Dns-amino acids, and phenoxypropionic acid herbicides were resolved adapting the optimal pH for each compound class.

A new hepta-substituted single isomer cationic β-CD (heptakis (6-methoxyethylamine-6-deoxy)-β-CD) was syn-

thesized by Haynes *et al.* [49] and checked for its separation behavior towards NSAIDs and phenoxypropionic acid herbicides. The resolution was found to be strongly dependent on selector concentration and pH. This selector showed improved enantioselectivity for the compounds tested compared to other cationic CDs. The synthesis of several new 6^I-deoxy-6^I-alkyl- and arylamino derivatives of β-CD [50] and their analytical characterization by CE [51] has recently been reported. These new selectors apparently have not yet been applied to chiral separation.

Histamine-modified cationic β-CDs as chiral selectors were recently introduced by Galaverna *et al.* [52] and applied to the enantiomer separation of Dns-amino acids and carboxylic acids as well as hydroxy acids [53]. The authors compared a 6-deoxy-6-*N*-histamino-β-CD, a monosubsituted, positively charged β-CD, bearing a histamine moiety linked to the C-6 of a glucose unit in the upper rim *via* the amino group and a 6-deoxy [4-(2-aminoethyl) imidazolyl]-β-CD bearing the histamine moiety linked to the C-6 *via* the imidazolyl group. Since the latter selector produced only poor resolutions, the authors conclude that the proximity of the positive charge to the cavity plays an important role in chiral recognition. The proposed complexation mechanism, the inclusion of the aromatic ring of the analyte into the cavity and lateral electrostatic interactions between the carboxyl group and the protonated amino group of the CD was confirmed by electrospray ionization-mass spectrometry and two-dimensional nuclear magnetic resonance (2-D NMR) ROESY experiments.

CDs containing quaternary ammonium groups show a pH-independent electrophoretic mobility. Only very low selector concentrations are necessary to resolve acidic enantiomers because of the strong ionic interactions. 2-Hydroxy-3-trimethylammoniopropyl-β-CD was investigated by several groups [54–58] and applied to the chiral separation of basic, neutral, and acidic compounds. Bunke and Jira [55, 58] showed that there is a reversal of the electroosmotic flow (EOF) with this selector under the conditions applied. As demonstrated with tropic acid, the elution order of the enantiomers can be reversed by changing from an uncoated to a coated capillary, changing the pH from 5 to 2.5, and reversing the polarity of the voltage. Another quaternary ammonium-β-CD (QA-β-CD) of undefined structure, which is commercially available (CerestarUSA, Hammond, IN, USA), was applied to the chiral separation of various acidic analytes [59–62]. This selector was also shown to be applicable in nonaqueous solvents [61]. The authors investigated formamide (FA), *N*-methylformamide (NMF), methanol, dimethyl sulfoxide and water as solvents. In all solvents the EOF was reversed above a certain selector concentration. While most of the Dns-amino acids investigated were resolved in all solvents, NSAIDs of the profen type were only resolved in formamide.

2.4 Amphoteric CDs

Lelievre *et al.* [63] synthesized a new zwitterionic selector, mono-(6-glutamylamino-6-deoxy)-β-CD (Glu-β-CD) which was evaluated using neutral, basic, and acidic drugs [63]. The authors investigated this selector at different pH values and observed that enantioselectivity differs depending on the pH. Tanaka and Terabe [64] investigated a commercially available QA-β-CD and an amphoteric β-CD (AM-β-CD). Both CD derivatives were analyzed by CE and MS to determine their composition. QA-β-CD was found to consist of six components having from one to six quaternary ammonium groups. AM-β-CD, which has quaternary ammonium groups and several carboxyl groups, showed many peaks in CE, but the composition could not be completely identified. QA-β-CD was applied to the chiral separation of various carboxylic acids and AM-β-CD to Dns-amino acids. Detailed information and applications of charged CDs can be found in recent reviews [12, 13, 20, 21].

2.5 CDs and nonchiral additives

CDs have often been used in combination with achiral surfactants such as sodium dodecyl sulfate (SDS). This principle was introduced by Terabe *et al.* [65] and called CD-mediated micellar electrokinetic chromatography (CD-MEKC). Negatively charged micelles migrate in the direction opposite to the EOF while uncharged CDs migrate with the same velocity as the EOF. Partition of hydrophobic analytes between the bulk solution, the CD and the micelle phase occurs, causing retention of the analyte. This principle enabled the chiral separation of a broad spectrum of compounds [12, 13, 21]. The enantiomer migration order can be reversed by changing from CD-CZE to CD-MEKC [66–69]. The addition of borate to CDs as a complexing agent was found to be effective [70–73] for the chiral separation of diols. Since resolution was not obtained without borate, the formation of mixed CD-borate diol complexes was assumed. Recently, this principle was applied to the chiral separation of propranolol metabolites with diol structure [74].

3 Carbohydrates

3.1 Neutral polysaccharides

In addition to CDs, linear neutral and charged carbohydrates were also found to be applicable as selectors for chiral separations. Fundamentals and applications up to 1997 are discussed in several reviews [13, 16, 75]. Malto-

dextrins dextrose equivalent (DE) <20) and higher molecular mass dextrins were successfully applied as chiral selectors. D'Hulst and Verbeke [76] found that the chiral recognition ability depends on the DE (percent reducing sugars) and the degree of polymerization (DP, number of saccharide monomers within the oligopolysaccharide chain). They proposed the requirement of α (1→4) linkage in polysaccharide for chiral recognition, since (1→6)-linked polysaccharides (dextrans) did not exhibit enantioselectivity for NSAIDs [76, 77]. These findings are in contradiction to later results obtained with other analytes [78, 79].

Chankvetadze *et al.* [80] reported the investigation of water-soluble, native polysaccharides such as amyloses of different DP values and laminaran and pullulan. The authors described also the application of derivatized polysaccharides, methylcellulose and hydroxypropylcellulose as well as carboxymethyl amylose, using (±)-1,1'-binaphthyl-2,2'-diyl hydrogen phosphate (BNHP) as a model compound.

Later, Chankvetadze *et al.* [81] investigated a series of noncyclic malto- and oligosaccharides as chiral selectors using BNHP as a model analyte. The authors showed that enantioselectivity is dependent on how the monosaccharide units are linked, and the DP. A change of linkage position between two D-glucose units containing α-linkage from (1→4) to (1→6) resulted in a decrease in chiral recognition ability, and a change of (1→4) linkage to (1→1) linkage reversed the enantiomeric migration order of BNHP. In the case of disaccharides, a change of α-linkage to β-linkage resulted in disappearance of chiral recognition ability (in the case of maltose and cellobiose) and reversal of migration order for maltotriose and cellotriose.

It should be mentioned that atropisomeric binaphthyl derivatives are very frequently used as model analytes since they are resolved by nearly all chiral selectors. Even monosaccharides have recently been shown to exhibit enantioselectivity for BNHP [82]. Several monosaccharides, such as D-glucose, D-mannose and some of their derivatives were found to be applicable as selectors for the chiral separation of BNHP, while galactose, sorbose, fructose, ribose, arabinose, and xylose failed to resolve BNHP. The authors postulated the following structural requirements for a chiral recognition: (i) oxygen atom at C-1 is in α-configuration; (ii)-OH group at C-4 is configurated downward; (iii) the presence of a -CH$_2$OH or -CH$_3$ group at C-5.

The chiral recognition mechanism for polysaccharide-based selectors is still not completely clear. In the case of dextrins the formation of a helical structure with hydrophobic character is assumed to be responsible for the binding of hydrophobic molecules. Lateral binding forces such as hydrogen bondings and dipole-dipole interactions with hydroxy groups of the sugar molecules are to be taken into account [16, 83, 84]. The hypothesis that the chiral recognition ability of dextrins mimics the β-CD cavity [83, 84] was disproved by Chankvetadze *et al.* [85] with the observation that the BNHP- enantiomers showed opposite migration order on maltooligosaccharides and β-CD. Hong *et al.* [86] performed CE complexation studies to elucidate the binding mechanisms between analytes and amylodextrins. Fluorescently tagged amylodextrin oligomers were injected together with model analytes of different structures (S-(+)ibuprofen, warfarin, ketoprofen and furosemide). The separation patterns of the oligomers were found to be altered by the presence of these pharmaceuticals as buffer additives.

3.2 Charged polysaccharides

Negatively charged polysaccharides such as heparin, chondroitin sulfate C, chondroitin sulfate A, dextran sulfate, λ-carrageenan have been shown to be suitable selectors for the chiral separation of bases [16, 75]. Recently, several new glycosaminoglycans have been investigated as chiral selectors: dermatan sulfate (DS, chondroitin sulfate B) was shown to exhibit enantioselectivity for a broad spectrum of basic drugs including β-sympathomimetics, β-blockers, antihypertensives, antimalarials, *etc.* [87]. DS is a complex, polydispersed, sulfated polysaccharide consisting of (1→4)-*O*-(α-idopyranosyluronic acid)-(1→3)-*O*-(2-acetamido-2-deoxy-β-D-galacto-pyranosyl-4-sulfate) disaccharide units. Semisynthetic chondroitins on the basis of oversulfated galactosamino-glycans were recently prepared by the same group [88]

and successfully applied to the chiral separation of basic drugs belonging to different classes.

Tsukamoto *et al.* [89] introduced two new glycosaminoglycans, a fucose-containing glycosaminoglycan (FGAG) and depolymerized holothurian glycosaminoglycan (DHG) and applied them to the chiral separation of a variety of basic drugs. These glycosaminoglycans were found to be complementary to other selectors. Tolperison and eperison, for example, which could not be separated with α- and β-CD or heparin, were separated with FGAG and DHG. The authors have shown that the resolution depends on the molecular mass and the selector concentration. Armstrong's group [90] recently investigated the use of pentosan polysulfate (PPS) as a chiral selector. PPS is a semisynthetic sulfated glycosaminoglycan composed of D-xylopyranose units connected through β-(1→4)-linkage. The applicability of this chiral selector was demonstrated by the chiral separation of tryptophan derivatives and several drugs. Figure 1 gives examples for the application of this new chiral selector.

Phinney *et al.* [91] investigated citrus pectins (polygalacturonic acid sodium and potassium salts and esterified pectins) as chiral selectors for the separation of basic drugs including antihistamines and antimalarial drugs as well as broncho- and vasodilators. The authors showed that esterification of the galacturonic acid has a deleterious effect on enantioresolution. Nishi *et al.* [92] introduced several positively charged polysaccharides. Diethylaminoethyl dextran, and the aminoglycoside antibiotics streptomycin sulfate, kanamycin sulfate and fradiomycin sulfate were applied for the resolution of some acidic analytes. Since then, there have been no further reports on the use of positively charged polysaccharides.

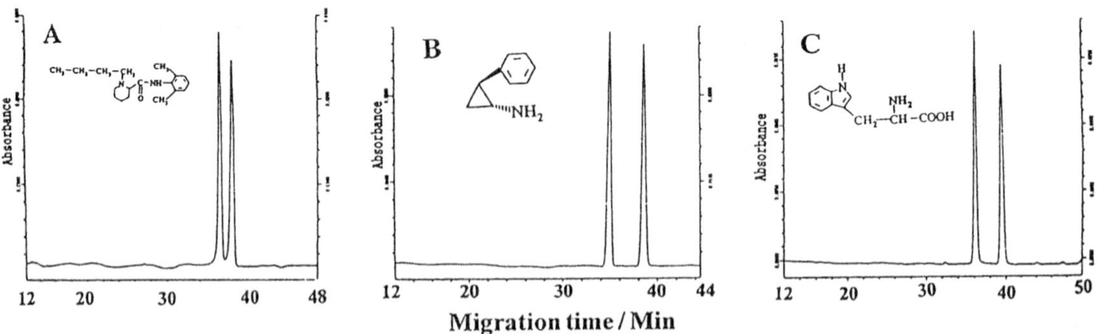

Figure 1. Separation of racemic (A) bupivacaine, (B) *trans*-2-phenylcyclopropylamine and (C) tryptophan. Running solution, 1% w/w pentosan polysulfate in 10 mM phosphate buffer (pH 2) containing (A) 20%, (B) 30%, and (C) 50% v/v methanol. Voltage applied, 28 kV; capillary, 57 cm (50 cm to the detector) × 75 µm ID; temperature, 23°C; detection wavelength, 214 nm. Reprinted from [90], with permission.

4 Chiral crown ethers

Crown ethers are cyclic polyethers which form host-guest complexes with earth metal ions and primary ammonium cations. Chiral crown ethers can stereoselectively include compounds containing primary amino groups. The only chiral crown ether used in CE to date is 18-crown-6-tetracarboxylic acid, (18C6H4) which was synthesized by Behr *et al.* [93]. According to Kuhn *et al.* [94], the primary interactions for complexation are hydrogen bonds between the three amine hydrogens and the oxygens of the macrocyclic ether in a tripod arrangement. In addition, the carboxylic acids groups are arranged perpendiculary to the plane of the macrocyclic ring, forming a chiral barrier, which divides the space available for the substituents at the chiral center into two domains. Thus, two diastereomeric inclusion complexes are formed. Furthermore, ionic-, dipole-dipole interactions or hydrogen bonds between the carboxylic groups and polar groups of the analytes may act as additional supporting interactions.

This selector was introduced by Kuhn *et al.* [95] for the chiral separation of amino acids by CE and found application, among others, in the chiral separation of sympathomimetics [94, 96], dipeptides [97, 98], various amino acid derivatives [99, 100], and different drugs containing primary amino groups [101]. Mori *et al.* [102] described the chiral separation of various drugs using 18C6H4 in non-aqueous medium. Comprehensive surveys of applications of chiral crown ethers to various compounds have recently been provided by Kuhn [103] and Verleysen and Sandra [21].

Verleysen *et al.* [99] tested 18-crown-6-tetracarboxamide, an intermediate in the synthesis of 18C6H4, for its ability to resolve primary amines by CE; however, this compound did not show any stereoselectivity. Tanaka *et al.* [104] recently described a partial filling technique using 18C6H4 in combination with MS detection to prevent entry of the nonvolatile selector into the nozzle of the CE-MS interface and the orifice plate. This technique was performed with racemic 3-aminopyrrolidine and α-amino-ε-caprolactam, which absorb UV only weakly. The combination of 18C6H4 and achiral crown ethers with CDs is discussed in Section 12.

5 Chiral calixarenes

Calixarenes represent a new type of chiral selectors. They are macrocyclic compounds consisting of benzene rings linked by methylene groups forming a hydrophobic cavity which is able to form host-guest complexes. Peña *et al.* [105, 106] synthesized water-soluble (*N*-L-alaninoacyl)calix[4]arene and (*N*-L-valinoacyl)calix[4]arenes.

These calixarenes possess four amino acid residues on the lower rim which permit chiral recognition. The authors resolved BNHP, (±)-1,1'-bi-2-naphthol (BINOL) and (±)-1,1'-binaphthyl-2,2'-diamine (BNA) as model compounds with these chiral calixarenes. Grady *et al.* [107] synthesized a (*S*)-di-naphthylprolinol calix[4]arene and coated the wall of the capillary with it for its hydrophobic properties. The authors demonstrated the applicability of this approach for chiral separations using 2-phenylglycinol as a model compound. Fluorescence quenching experiments with Stern-Volmer plots confirm that the (*R*)-enantiomer of phenylglycinol interacts more strongly with the calixarene.

6 Macrocyclic antibiotics

Macrocyclic antibiotics as chiral selectors were introduced by Armstrong *et al.* [108]. They have several asymmetric centers and many functional groups, allowing multiple interactions with the analytes. In addition to ionic interactions, hydrogen bonding, dipole-dipole, π-π, hydrophobic interactions and steric repulsion are assumed to take effect [109]. Macrocyclic antibiotics possess hydrophobic pockets which can include hydrophobic moieties; due to the presence of pendant polar arms, hydrogen bonds can be formed.

Three classes of antibiotics have been introduced as chiral selectors: Ansamycins such as rifamycin B, rifamycin SV; the glycopeptides vancomycin, ristocetin and teicoplanin and the aminoglucoside antibiotics streptomycin, fradiomycin and kanamycin. A comprehensive description of the properties of these selectors and their applications to the chiral separation of a broad spectrum of compounds is given in several specialized reviews [19, 110, 111]. While rifamycin B showed enantioselectivity for basic compounds, rifamycin SV and the glycopeptide antibiotics were found to be suitable for the chiral separation of acidic compounds.

Since these macrocyclic antibiotics contain aromatic moieties and have a strong UV absorption up to 250 nm, detection is only possible at wavelengths higher than 250 nm. Alternatively, indirect detection can be used. To overcome the detection problems, countercurrent processes have been applied [112, 113]. In this approach, a coated capillary was used to suppress the EOF and a suitable pH provided the selector (vancomycin or ristocetin A) and the analytes with opposite charges. Thus, the positively charged selector moves to the cathode, clearing the detection window, and the analytes can be detected without interferences at the anode. Fanali's group [114–116] used a partial filling method together with the countercurrent mode using vancomycin or teicoplanin as chiral selectors. Countercurrent approaches were also used to allow an interference-free coupling with MS using vancomycin as a chiral selector [117, 118].

Subsequently, some new macrocyclic antibiotics of the glycopeptide type were investigated: Strege *et al.* [119] introduced the glycopeptide antibiotic A 82846B, which was also investigated in countercurrent mode for the chiral separation of profens [120]. LY307599, a derivative of A 82846B, which differs from the latter compound by an additional biphenyl moiety, was evaluated using flurbiprofen [121]. Actaplanin A [122] represents a further member of this family which also found application to the chiral separation of NSAIDs of the profen type.

Avoparcin, a new glycopeptide antibiotic was recently introduced by Armstrong's group [123]. It consists of a mixture of several structurally similar glycopeptide analoges of which α-avoparcin and β-avoparcin are the major components. The authors evaluated the chiral recognition ability of this selector using Dns-amino acids and NSAIDs and compared the results with those obtained with ristocetin A, teicoplanin and vancomycin. Fanali *et al.* [124] evaluated a new antibiotic, Hepta-tyr (MDL 63,246). Hepta-tyr is a semisynthetic small glycopeptide antibiotic, which belongs to the teicoplanin family. Using the partial filling method, this selector was applied to the chiral separation of acidic compounds, among them NSAIDs, anticoagulants and herbicides of the phenoxypropionic acid type using a boric acid-acetic acid-phosphoric acid buffer, pH 5 / acetonitrile mixture as electrolyte.

A further member of the glycopeptide family, A 35512B, was recently investigated by Risley *et al.* [125]. Its good water solubility allows the use of aqueous buffer solutions as mobile phases. The enantioselectivity of this selector was checked using Dns-amino acids. Increasing selector concentration was shown to increase resolution. The influence of pH on resolution was investigated over a range from 6 to 8. In general, the resolution increased with decreasing pH. Baseline resolution for eleven out of thirteen Dns-amino acids tested could be achieved using just an aqueous phosphate buffer, whereas 2-methoxyethanol had to be added to resolve Dns-methionine and Dns-threonine.

To reduce migration time, Kang *et al.* [126] added hexadimethrine bromide, a polycationic polymer, to the run buffer to reverse the EOF. The cationic polymer adsorbs to the capillary wall and adsorption of vancomycin is avoided. The acidic analytes migrate in a co-EOF mode, thereby drastically reducing the migration time. The applicability of this approach is demonstrated with the chiral separation of 9-fluorenylmethoxycarbonyl (FMOC)-

amino acids and ketoprofen. Proof of the hypothesis that the enantiorecognition of macrocyclic antibiotics takes place at the D-Ala-D-Ala binding site, located in the aglycon core, was given by Carotti *et al.* [127]. Teicoplanin was applied in a mixture with D-Ala-D-Ala or L-Ala-L-Ala in the electrolyte using *p*-methoxymandelic acid as a model analyte. When D-Ala-D-Ala was present, no chiral resolution was obtained, whereas with L-Ala-L-Ala there was no significant change in resolution.

7 Proteins

The well-known phenomenon that drugs bind stereoselectively to proteins led to investigations to use proteins as chiral selectors. Several proteins have been successfully applied in the form of chiral stationary phases in HPLC. A great variety of proteins, such as bovine serum albumin, human serum albumin, α_1-acid glycoprotein, avidin, conalbumin, cellulase, ovomucoid, cellobiohydrolase and casein were used as chiral selectors in CE for a broad spectrum of compounds. For literature covering applications up to the year 1997 and detailed information the authors recommend specialized reviews [14, 128, 129]. Proteins can be positively or negatively charged depending on the pH applied. Their charges give them electrophoretic mobility and they can be used for the separation of basic and acidic analytes.

Fanali *et al.* [130] reported on the use of pepsin as a chiral selector using a partial filling technique. A polyacrylamide-coated capillary was used to suppress the EOF and to avoid adsorption of the protein on the capillary wall. The authors evaluated this selector using several β-blockers and some other basic drugs. It was shown that with increasing selector concentration, resolution generally increased, and that the pH investigated in the range between 4 and 7 had to be adapted individually for each compound. The addition of organic modifiers resulted in a decrease in resolution. Tanaka and Terabe [131] studied the enantioselectivity of egg white avidin, succinylated avidin and streptavidin using a partial filling technique. While the basic avidin was found to be useful for the chiral separation of acidic analytes, succinylated avidin showed enantioselectivity for basic analytes. Streptavidin, a neutral nonglycosylated protein, showed enantioselectivity for both acidic and basic analytes. The authors showed that chiral recognition ability was lost when biotin was added as it forms a strong complex with avidin in which the stereoselective binding sites are blocked (Fig. 2).

De Lorenzi *et al.* [132, 133]l evaluated purified quail egg white riboflavin binding protein as a chiral selector for HPLC and CE. Riboflavin-binding proteins are acidic proteins with a molecular weight of about 36 000 containing

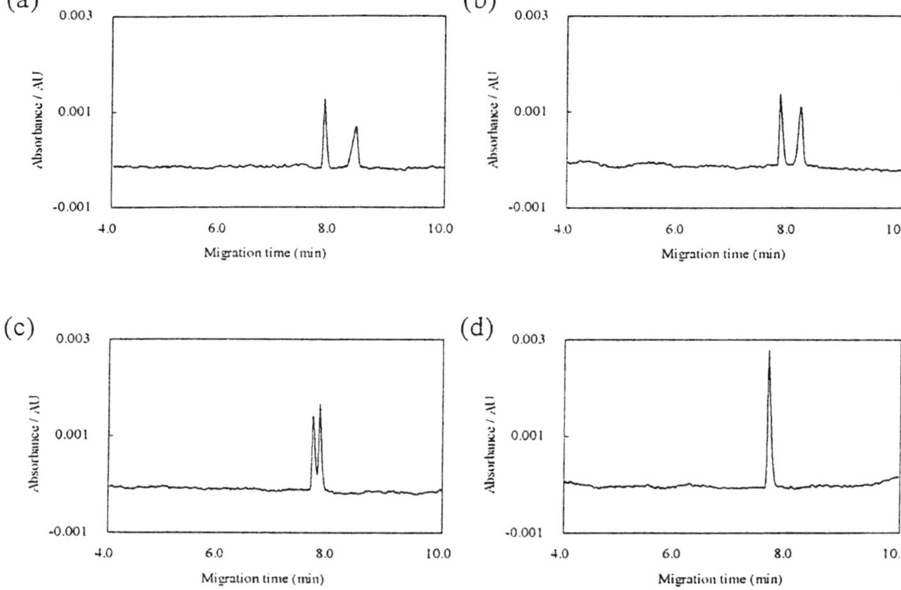

Figure 2. Effect of biotin binding to avidin on enantiomer separation of racemic 4-fluoromandelic acid. Conditions: capillary, 36 cm × 50 µm ID polyacrylamide-coated; running buffer, 50 mM phosphate buffer (pH 4.0); separation solution, 100 µM avidin in running buffer containing (a) 0 µM, (b) 100 µM, (c) 200 µM, (d) 400 µM D-biotin (zone injection, 6.9 kPa × 190 s); applied voltage, −12 kV; sample injection, 6.9 kPa × 2s; capillary temperature, 25°C, detection wavelength, 210 nm. Reprinted from [131], with permission.

about 14% carbohydrate. Complete and partial filling techniques were compared [133] using several basic drugs such as oxazepam, oxprenolol, prilocaine, bupivacaine, *etc.*, as model compounds. Mano *et al.* [134] recently reported the use of native flavoprotein isolated from chicken egg white and chemically modified flavoproteins as chiral selectors. NSAIDs of the profen type, oxprenolol, proglumide and aminoglutethimide, were used to check the chiral recognition ability. The chiral recognition region on the flavoprotein is assumed to consist of an α-helix structure. Studies with the chemically modified flavoproteins led to the conclusion that there are ionic interactions between an amino group and a carboxyl group of the protein and the carboxyl group of the model compound ketoprofen and π-π interaction of the tryptophan moiety of the protein and the aromate of the analyte.

8 Ligand-exchange CE

The first application of the principle of ligand-exchange in CE was reported by Zare's group. They used L-histidine- [135] or aspartam/Cu(II) [136] complexes for the chiral separation of Dns-amino acids. Desiderio *et al.* [137] re-

solved hydroxy acids using L-Pro-, L-Hypro- or aspartam / Cu(II) complexes. Sootornniyomkij *et al.* [138] investigated the role of the metal ion on the resolution of Dns-amino acids using aspartame as chiral selector. The first approach for direct separation of underivatized amino acids using L-Pro or L-Hypro / Cu(II) complexes as chiral selectors was published by Schmid and Gübitz [139]. Using MEKC with SDS added to the electrolyte led to a reversal of the enantiomer migration order (EMO). This is probably caused by hydrophobic and electrostatic interactions with the negatively charged micelles for which the D-enantiomer is more easily accessible.

Chen *et al.* [140] applied a similar principle to the separation of the positional isomers and enantiomers of fluorophenylalanine and tyrosine using L-Hypro with or without SDS. A reversal of the EMO was also observed by these authors when SDS above a certain concentration was added to the electrolyte. The same group [141] reported the separation of sixteen positional and optical isomers of tryptophan derivatives checking SDS, *n*-decyl sulfate (SDeS) and *n*-tetradecyl sulfate (STS) as micelle-forming surfactants. In another paper [142], the mechanism for the flow reversal observed by addition of SDS, SDeS and

STS is discussed and the influence of other surfactants such as Tween-20 and cetyltrimethylammonium bromide (CTAB) as well as organic solvents on the separation is studied. Tween-20 was found to improve resolution at the expense of retention time; the addition of CTAB resulted in a change of the direction of the EOF, but dit not improve resolution. Organic modifiers did not inverse the EMO but caused a decrease in resolution.

Recently, this group [143] published a method for the determination of the critical micelle concentration (CMC) on the basis of ligand-exchange MEKC. This approach is based on the above-mentioned observation that the point of the reversal of the EMO is identical to the CMC. These authors did not observe a reversal of the EMO in the case of hydroxy acids when SDS was added using *trans*-L-Hypro-Cu(II) as selector [144]; this is in contrast to amino acids. A reversal, however, was obtained when the cationic surfactant CTAB was used instead of SDS due to a reversal of the EOF. Interestingly, a reversal of the EMO was also observed when *trans*-L-Hypro was changed for *cis*-L-Hypro. In recent studies, this group [145] showed that the EMO for amino acids also depends on the type of ligand used. While L-Pro and *trans*-L-Hypro showed the same EMO, there was a reversal of the EMO with *cis*-L-Hypro, probably due to the fact that *cis*-L-Hypro represents a tridentate ligand. As was expected, the EMO could also be reversed by changing the chirality of the selector.

Yuan *et al.* [146] reported the use of L-arginine / Cu(II) for the resolution of Dns-amino acids. The resolution was lost when Cu(II) was replaced by Ni(II), Co(II) or Zn(II). When L-Arg was substituted by L-Glu, L-Ala and L-Asp, also no resolution was obtained.

N-alkyl-hydroxyproline such as *N*-(2-hydroxyoctyl)- and *N*-(2-hydroxypropyl)-L-4-hydroxyproline derivatives have recently been synthesized and applied as Cu(II) complexes to the chiral separation of underivatized aliphatic and aromatic amino acids and dipeptides [147, 148]. Compared to L-Hypro, these selectors showed improved resolution with significantly lower selector concentrations. *N*-(2-hydroxyoctyl)-L-4-hydroxyproline was also shown to be applicable for the chiral resolution of sympathomimetic drugs [149] as well as hydroxy acids and β-blockers [150]. Figure 3 shows the chiral resolution of ephedrine, octopamine and orciprenaline. Under these conditions, orciprenaline showed only partial resolution.

More recently, Karbaum and Jira [151] have shown that ligand-exchange CE (LECE) is also possible in nonaqueous solvents. The authors used L-proline / Cu(II) in 25 mM ammonium acetate / 1 M acetic acid in methanol to resolve several aromatic amino acids. The use of L-isoleucine instead of L-proline decreased enantioselectivity

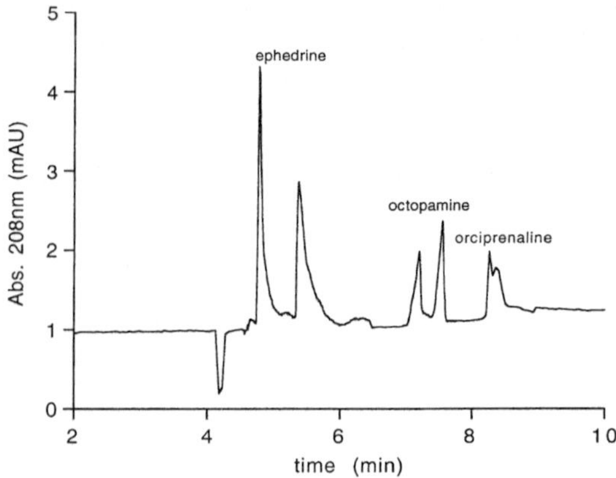

Figure 3. Electropherogram of the chiral separation of (±)-ephedrine, (±)-octopamine and (±)-orciprenaline. Conditions: capillary, 70 cm (26 cm to the detector) × 50 μm ID; electrolyte, 10 mM HO-L-Hypro, 5 mM Cu(II) at pH 12; voltage applied, 15 kV; temperature, 25°C; detection wavelength, 208 nm; injection: 30 mbar × 6 s. Reprinted from [149], with permission.

except for kynurenine and histidine, which could not be separated with L-Pro. The preparation of monolithic phases for LE-CEC [152] is discussed in Section 14.3.

9 Chiral ion-pairing reagents

Although no reports exist to date of successful application of chiral ion-pairing reagents using aqueous BGEs, several ion-pairing reagents have been applied successfully in nonaqueous CE. Obviously, water interferes with the intermolecular interactions responsible for chiral recognition and largely suppresses the formation of intermolecular hydrogen bonds. Exceptions are the use of chiral and achiral counterions as supporting agents for separations with CDs in dual systems [153] (see Section 12). The first ion-pairing reagent reported for CE separations was (+)-*S*-camphor-10-sulfonic acid, which was applied for the chiral separation of basic compounds [154]. To date, this approach has only been successful with compounds with a β-amino alcohol structure. Acetic acid in acetonitrile containing Tween-20 was used as electrolyte. In addition to ionic interactions, dipole-dipole interactions and hydrogen bondings are responsible for chiral recognition. Tween is assumed to act as a hydrophobic pseudostationary phase. Stalcup and Gahm [155] used quinine as an ion-pairing reagent in nonaqueous medium for the chiral resolution of acidic compounds using acetic acid ammonium acetate methanol as BGE.

Lämmerhofer and Lindner [156] and Piette *et al.* [157] investigated quinine and *tert*-butyl carbamoylated quinine as chiral ion-pairing reagents for the enantioseparation of

N-protected amino acids testing ammonium acetate in methanol and ammonia-octanoic acid in an ethanol-methanol mixture as nonaqueous BGEs. Recently, Piette *et al.* [158] screened several cinchona alkaloids and derivatives for their ability to act as chiral counterions for the separation of different *N*-protected amino acids in nonaqueous systems. The authors observed rather poor enantioselectivity when using the natural cinchona alkaloids quinine, quinidine, cinchonine, and cinchonidine. Significantly higher resolution values were obtained with *tert*-butyl carbamoylated quinine, *tert*-butyl carbamoylated quinidine, dinitrophenyl carbamoylated quinine and cyclohexyl carbamoylated quinine as counterions and ethanol-methanol mixtures with ammonia-octanoic acid as BGEs. The best enantioseparations were achieved for Dns-amino acids. Resolution factors up to 78 were observed. The enantioselectivities and the enantiomer migration orders were found to be strongly dependent on the type of selector. In addition to the primary ionic interactions, hydrogen bonding, dipole-dipole, charge transfer (π-π), hydrophobic and steric interactions are to be taken into account.

10 Chiral surfactants

The principle of MEKC using chiral surfactants as selectors was introduced by Terabe *et al.* [159]. Surfactants are amphiphilic molecules composed of a polar head group and a hydrophobic tail. Above a certain concentration (CMC) micelles are formed. These micelles are used as pseudostationary phases in MEKC. The chiral separation of analytes is based on their partition coefficients between the chiral micelle phase and the electrolyte bulk phase.

Different classes of surfactants were applied as chiral selectors: bile salts, saponines, long-chain *N*-alkyl-L-amino acids and *N*-alkanoyl-L-amino acids, *N*-dodecoxycarbonyl amino acids, alkylglycoside surfactants and polymeric amino acid-based surfactants [160, 161]. New carbamate-type surfactants were synthesized by Ding and Fritz [162] from amino acids (L-Leu, L-Val, L-Ile, L-Ser) and alkylchloroformates with chain lengths of C4 to C11. These new selectors were evaluated using propranolol, atenolol, ketamine, laudanosine, nefopam, benzoin and hydrobenzoin as model compounds. The enantioselectivity was found to depend on the amino acids used and the chain length. A significant improvement in resolution was observed when a dual selector system consisting of the carbamate surfactant and a sulfonated-β-CD is used.

Recently, polymeric dipeptide surfactants have been introduced [163–165]. The monomers were synthesized by coupling the *N*-hydroxysuccinimide ester of 10-undecylenic acid to the dipeptide. A survey of the evaluation of different dipeptide polymeric surfactants was recently published by Haddadian *et al.* [166]. The authors have shown that the order of the amino acids in the dipeptide has a marked influence on the enantioselectivity. The applicability of a variety of dipeptide polymeric surfactants for the chiral separation of acidic, basic and neutral analytes was demonstrated. A significant improvement in resolution compared to polymeric amino acid-based surfactants, previously published [161, 167], was observed. The sites of interaction of the analytes with the polar head group of the surfactant is dependent on the depth of penetration of the analyte into the hydrophobic micellar core of the dipeptide chiral pseudostationary phase (CPSP). The depth of penetration is determined by the hydrophobicity of the analytes, as well as by electrostatic interactions of the enantiomers with the surfactant. Since the *N*-terminal amino acids are located close to the hydrophobic core and the *C*-terminal amino acids at the interface of the micelle, hydrophobic enantiomers which penetrate more deeply into the hydrophobic core will interact mainly with the *N*-terminal amino acids. Hydrophilic and cationic enantiomers also interact mostly with the *C*-terminal amino acids. Moderately hydrophobic enantiomers can interact with both amino acids of the CPSP. The authors demonstrated that the enantiomeric migration order strongly depends on the amino acid sequence in the dipeptide CPSP.

Two isomeric D-glucopyranoside-based surfactants were synthesized by Tickle *et al.* [168]. To better understand the possible mechanism of chiral discrimination, the authors calculated the minimum energy conformations of micelles produced by these surfactants. They also investigated the microstructures by transmission electron microscopy. Ju and El Rassi [169, 170] recently presented new chiral glycoside surfactants, cyclohexyl-butyl-β-D-maltoside, cyclohexyl-pentyl-β-D-maltoside and cyclohexyl-pentyl-β-D-maltoside. These cyclohexyl-alkyl-β-D-maltosides have a chiral maltose polar head group and a cyclohexyl-alkyl hydrophobic tail. They form neutral micelles in aqueous media. Since the neutral micelles migrate at the velocity of the EOF, they are only useful for the enantiomer separation of charged enantiomers. These chiral selectors were evaluated using Dns-amino acids, dinitrophenyl (DNP)-amino acids, tryptophan derivatives, and BNHP.

Recently, two new amphiphilic aminosaccharide derivatives, namely 2-carboxy-1-(carboxymethyl)ethyl-6-*O*-[4-carboxy-2*R*-tetradecanoylaminobutanoyl]-2-deoxy-3-*O*-tetradecanoyl-2-tetradecanoyl-amino-α-D-glucopyranoside and 2-carboxy-1-(carboxymethyl)ethyl-6-*O*-[3-carboxy-3*S*-tetradecanoyloxypropanoyl]-2-deoxy-3-*O*-tetradecanoyl-2-tetradeclanoylamino-α-D-glucopyranoside were synthesized by Horimai *et al.* [171]. These negatively charged selectors were applied above their CMC in borate buffer, pH 9.5, for electrokinetic separation of Dns-amino acids and new quinolon antibacterial agents. The authors found that increase of the selector concentration or BGE concentration led to an increase in enantioselectivity. The decrease of electrostatic repulsion and the increase of hydrophobic interaction between the negatively charged pseudostationary phase and the analyte are assumed to be important driving forces for the chiral discrimination.

Bunke *et al.* [172] investigated (–)-*N*-dodecyl-*N*-methylephedrinium bromide (DMEB) as selector for the chiral separation of the atropisomeric methaqualone. Since DMEB is a hydrophobic quaternary ammonium salt, it interacts with the capillary wall of the uncoated capillary and the EOF is reversed. Resolution of methaqualone was only obtained with high concentrations of DMEB. Methaqualone is included in the positively charged micelle and due to the higher mobility of the EOF moves in the direction of the EOF to the anode. When a neutrally coated capillary was used, the analyte moved to the cathode due to the charge of the micelles. An enantioseparation was also observed in this case, but with a lower separation factor.

11 Nonaqueous medium

Nonaqueous CE (NACE) is becoming increasingly popular. Recently, Karbaum and Jira [173] checked several solvents for their suitability for NACE and discussed its advantages. Nonaqueous solvents show several advantages regarding solubility of chiral selectors or samples and reduce unwanted interactions with the capillary wall. An increase in selectivity can often be observed in nonaqueous solvents. Different forms of chemical equilibria in aqueous and nonaqueous systems can lead to different selectivities. Weak interactions which are disrupted by water can become effective in nonaqueous systems. One example for that is the enantiomer separation by ion-pair formation. In nonaqueous solvents, less Joule heating is produced and since higher voltage can be applied, retention times are shorter. Furthermore, nonaqueous solvents are better compatible with CE-MS coupling than aqueous BGEs. First applications of nonaqueous solvents in chiral CE separations were reported by Valkó *et al.* [174] and Wang and Khaledi [175] using CDs in FA, DMF and NMF as solvents.

Wang and Khaledi [176] have recently reviewed the use of NACE for chiral separations. CDs are the most frequently used chiral selectors in nonaqueous solvents. In addition to neutral CDs, charged CDs were used in NACE. Only a few examples will be mentioned here: comparison of the separation of basic compounds in aqueous and nonaqueous systems using S-β-CD in FA showed that there is a significant reduction of band broadening in nonaqueous medium [177]. Vincent and Vigh [178] used the single isomer heptakis(2,3-diacetyl-6-sulfato)-β-CD in pure methanol for the chiral separation of a great variety of basic drugs with remarkable efficiency. The use of a QA-β-CD in pure organic solvents for the chiral separation of amino acids derivatives and some profens was reported by Wang and Khaledi [179].

Mori *et al.* [102] described the application of a chiral crown ether (18C6H4) to the chiral separation of several compounds with primary amino groups in FA. The use of chiral ion-pair reagents such as (+)-*S*-camphorsulfonate [154], quinine [157], and quinine derivatives [157, 158] is discussed in Section 9. In a recent paper, Karbaum and Jira [151] demonstrated the applicability of the principle of chiral ligand-exchange in NACE using L-Pro / Cu(II) in ammonium acetate / acetic acid / methanol. This new approach found application to the separation of underivatized amino acids and drugs with amino alcohol structure. Nonaqueous CEC (NACEC) is a recent trend. Krause *et al.* [180] prepared a helically chiral poly(diphenyl-2-pyridylmethyl) methacrylate coated on aminopropyl-silanized silica. Chiral separation of neutral compounds was carried out in pure methanol. Tobler *et al.* [181] reported the application of a chiral stationary phase (CSP) based on *tert*-butylcarbamoylquinine immobilized on silica gel to NACEC using acetic acid/triethylamine in acetonitrile/methanol as mobile phase (see also Section 14.2).

12 Dual selector systems

The combination of two chiral selectors was found to improve resolution in many cases. Sometimes no separation is obtained using only one of the chiral selectors. The dual selector systems most frequently used are different CDs. The combination of neutral native and CD derivatives significantly enhanced resolution in many cases. Often neutral and charged CDs were combined and improved or even enabled resolution. Since specialized reviews [182, 183] deal with the use of dual CD systems, this will not be detailed here.

The combination of CDs with chiral surfactants such as bile salts [184–187], decanoyl-*N*-methyl glucanoid [188] and poly(sodium-*N*-undecenyl-D-valinate) [189] was found to improve resolution. Synergistic effects were observed when 18C6H4 was used in combination with neutral CDs [190–192]. Recently, Verleysen and Sandra [193] investigated the use of 18C6H4 in combination with negatively charged CDs and the application of a partial filling method [193]. The behavior of primary amines using

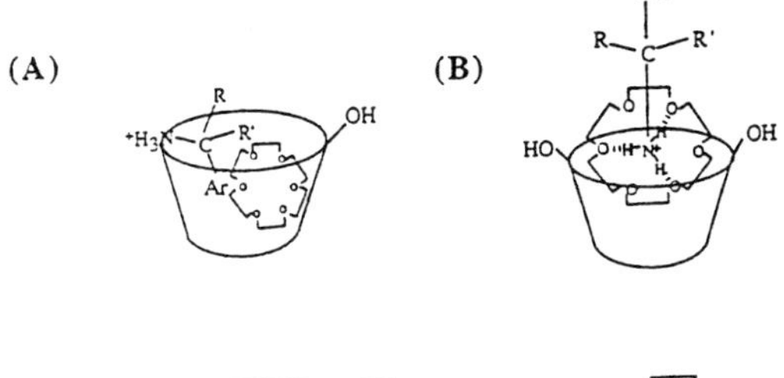

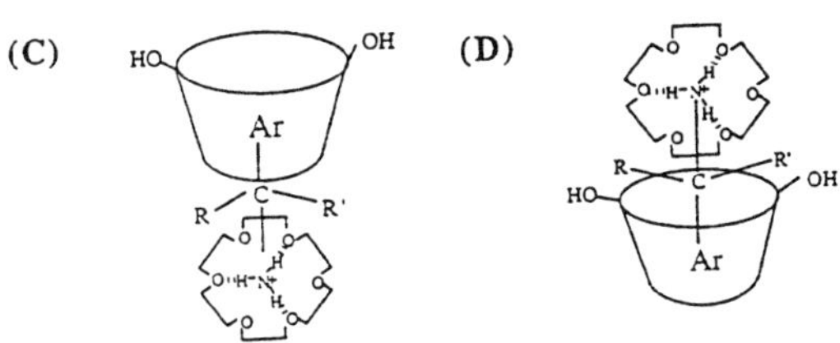

Figure 4. Some possible "three-body" complexes between a hypothetical chiral amine, a CD and 18-crown-6. "D" is thought to be the most important and dominant complex. Reprinted from [198], with permission.

coated and uncoated capillaries at high and low pH was investigated. The synergistic effect of the selector combination is explained and a possible separation mechanism discussed.

The addition of nonchiral crown ethers to CDs was also found to have supporting effects on chiral separations

[194–197]. The authors showed that several primary amines, which could not be separated by CDs alone, were separated after addition of 18-crown-6. Similar investigations were reported by Armstrong *et al.* [198], who discussed several possibilities of "three body" complexes between an amine, a CD, and 18-crown-6 (Fig. 4). The chiral recognition mechanism is explicated by showing the solution equilibria steps (Fig. 5).

Bunke *et al.* [199] observed that the chiral resolution of the drug cyclodrine is only possible with a mixture of β-CD and (+)-*S*-camphor-10-sulfonic acid (CSA). Neither with β-CD nor with CSA alone was resolution obtained. Jira *et al.* [200] investigated several chiral and achiral ion-pairing reagents regarding their supporting effect on the chiral separation using CDs. The authors found out that a significant improvement in resolution is observed regardless of whether the counterion is chiral or not. The influence of (+)- or (–)-camphorsulfonic acid, alkylsulfonic acids and alkanoic acids of different chain length as well as sodium cyclamate on the chiral resolution of basic compounds using different neutral CDs was investigated. On the other hand, basic counterions such as quinine and (*S*)-hyoscyamine were found to improve the resolution of acid compounds using CDs. The combination of the chiral ligand-exchange principle with host-guest complexation was realized by Horimai *et al.* [201] using a dual selector system consisting of γ-CD and Zn(II)-D-phenylalanine. This dual selector system was applied to the chiral separation of new quinolone drugs.

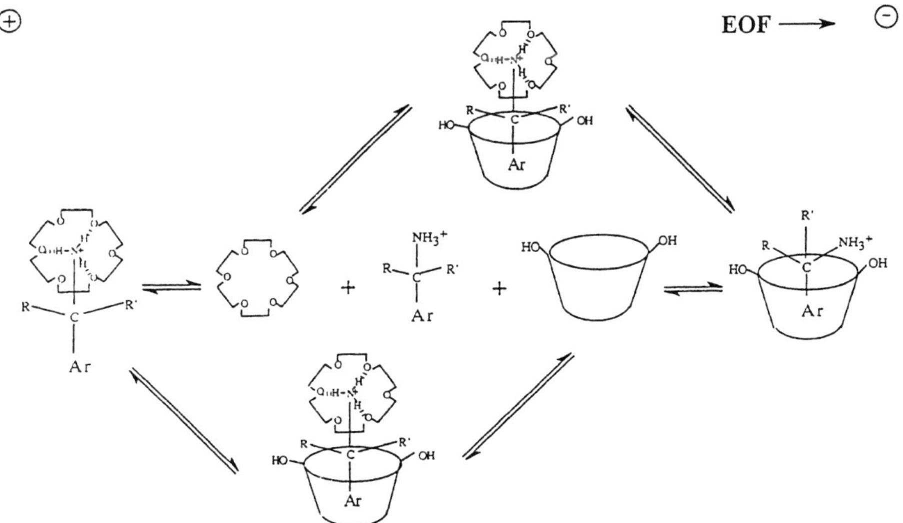

Figure 5. Schematic showing the most important solution equilibria steps in the formation of the "three-body" complex. Reprinted from [198], with permission.

13 Diverse selectors

(+)-(5R,8S,10R)-1-allylterguride, a semisynthetic product, derived from ergot alkaloides, was shown to be a suitable chiral selector for the separation of hydroxy acids [202] and herbicidal enantiomers [203]. A further publication by this group [204] deals with the determination of the formation constant of 1-allyl terguide with mandelic acid by measuring the effective mobility of acidic analytes at varying concentrations. The use of poly-terguide in CEC [205] is discussed in Section 14.1. Nair *et al.* [206] reported the use of D(+)-tubocurarine chloride as a chiral selector. This compound, a macrocyclic bis(benzylisoquinoline) alkaloid, was evaluated by means of the chiral separation of 18 carboxylic acids using methanol / phosphate buffer as electrolyte. Two groups [207, 208] reported the use of cyclohexapeptides prepared by combinatorial synthesis for chiral separations. Several cyclic peptide libraries were prepared and checked for their ability to resolve DNP amino acids.

14 CEC

CEC represents a hybrid method between CE and HPLC making use of the efficiency of CE and the selectivity of stationary phases. Different variants of these recent techniques have been developed for chiral separations. In chiral open tubular capillary electrochromatography (OT-CEC), the chiral selector is covalently attached or coated on the inner surface of a capillary. In contrast, in packed CEC (P-CEC) either an achiral stationary phase in combination with a chiral mobile phase or a CSP can be used. A new alternative to silica-based packed capillaries is the use of monolithic chiral stationary phases prepared by *in situ* polymerization procedures.

14.1 Open tubular capillaries

Several specialized reviews report the use of CEC for chiral separations [209–211]. The principle of chiral CEC was introduced by Schurig's group [212–216]. Permethylated β-CD was covalently linked *via* an octamethylene spacer to a dimethylpolysiloxane and coated on the capillary wall (Chirasil-Dex). These phases were applied to the chiral separation of a broad spectrum of drugs. The possibility of unified chromatography using the same capillary coated with Chirasil-Dex for GC, capillary HPLC, supercritical fluid chromatography (SFC) and CEC was demonstrated [216]. Armstrong *et al.* [217] published a similar approach using a permethylated allyl-substituted β-CD coupled to an organohydrosiloxane polymer. The preparation of a wall-coated capillary using an acrylamide polymer containing a β-CD derivative was described by Szeman and Ganzler [218].

A comprehensive overview of the applications of cyclodextrins in chiral electrochromatography was recently published by Schurig and Wistuba [219]. Francotte and Jung [220] used neat cellulose derivatives, such as 3,5-dimethylphenylcarbamoyl cellulose and *p*-methylbenzoyl cellulose coated on silica gel and applied the same capillaries for CEC and OT-HPLC. A simple approach for preparation of chiral open tubular capillaries was reported by Liu *et al.* [221, 222]. Proteins, peptides and amino acids were adsorbed on the capillary wall by simply rinsing the capillary with a buffer containing the selector. These phases were evaluated using amino acids and some other compounds and showed remarkable results. Surprisingly, the lifetime of the capillaries was between one and several weeks.

Hofstetter *et al.* [223] reported the preparation of an open tubular capillary by immobilizing BSA *via* a silane to the capillary wall. The enantioselectivity was studied using DNP-amino acids and 3-hydroxy-1,4-benzodiazepines as model analytes. Hong *et al.* [224] described a similar approach using α_1-acid glycoprotein (AGP), which was bound to the capillary wall by 3-glycidoxypropyltrimethoxysilane. Promethazine and benzocaine were used as test analytes. Regarding stability, the authors noted that one capillary showed only a small decrease in enantioselectivity after 50 days of use at room temperature. Sinibaldi *et al.* [205] coated capillaries with poly-terguride derived from ergot alkaloids and, on this, positively charged phase acidic analytes such as amino acid derivatives and flurbiprofen were resolved.

Molecularly imprinted polymers are used in CEC, either coated as a thin film to the capillary wall, or in the form of packed capillaries, or as monolithic phases. Several recent reviews [225–229] have been devoted to this basic technique. Generally, a chiral print molecule is loosly bound or entrapped in a polymer. After the print molecule is removed, a chiral imprint of this template remains, which is highly enantioselective for the same or very closely related molecules. Remcho and Tan [226] used L-Dns-phenylalanine as print molecule with methacrylic acid and 2-vinylpyridine as functional monomers and ethylene dimethacrylate as cross-linker in an *in situ* polymerization technique. To obtain a thin film at the capillary wall, the capillary was evacuated to effect shrinking of the polymer. The CSP showed high enantioselectivity for Dns-phenylalanine; efficiency, however, was rather low. Brüggemann *et al.* [230] reported the preparation of very thin coatings. S(+)-2-phenylpropionic acid was used as a print molecule. Due to the extremely strong inclusion of the S(+) enantiomer only the R(−) enantiomer was detected; the S(+) enantiomer eluted as a very broad peak which was not detectable.

14.2 Packed capillaries

14.2.1 Achiral stationary phases with chiral mobile phases

Lelièvre *et al.* [231] used a reversed-phase capillary applying HP-β-CD as an additive to the mobile phase and compared the results with those obtained on a capillary packed with an HP-β-CD phase for the chiral separation of chlorthalidone and mianserin. Wei *et al.* [232] described the use of bare silica packing in combination with HP-β-CD as a mobile phase additive for the chiral separation of phenylephrine and synephrine. Deng *et al.* [233] resolved the enantiomers of salsolinol using a C18-packed capillary and β-CD as an additive to the mobile phase. 1-Heptane sulfonate was added as an ion-pairing reagent to improve the retention behavior of the analyte. Lämmerhofer and Lindner [234] resolved amino acid derivatives on an octadecyl silica (ODS) stationary phase using a quinine derivative as an ion-pairing reagent in the mobile phase.

14.2.2 Chiral stationary phase

Early investigations with packed capillaries deal with the use of β-CD [235] AGP- [236] and HSA- [237] HPLC-grade CSPs. Wistuba *et al.* [238] recently developed a packing material based on permethylated β-CD immobilized to 5 μm (mercaptopropyl) methyl-silica gel (Chirasil-Dex-silica 2) and evaluated this material by means of the chiral separation of barbiturates. The authors compared CEC, capillary HPLC and pressure-supported CEC using the same capillary in a CE system combined with a gradient pump. The authors have shown that with pressure-supported CEC, significantly shorter separation times and sharper peaks are obtained without significant loss of separation selectivity. Recently, Wistuba and Schurig [239] reported the preparation of a polysiloxane-linked permethyl-β-CD thermally immobilized on silica (Chirasil-Dex silica 1). The drawback of polymer-coated silica is that the EOF is reduced because the silanol groups are blocked; this can be overcome by adding unmodified bare silica [239]. In a recent review, Schurig and Wistuba [219] compare the results obtained with open tubular CD-coated capillaries and packed capillaries with CD-CSPs using both CEC and capillary LC.

Krause *et al.* [240] reported the use of poly-*N*-acryloyl-L-phenylalanine ethyl ester covalently bonded to silica (Chiraspher) or cellulose tris(3,5-dimethylphenylcarbamate) coated on silica for CEC, capillary HPLC and pressure-supported CEC and demonstrated the applicability of the different approaches for chiral separations of a broad spectrum of drugs. Recently, Krause *et al.* [180] synthesized a helically chiral poly(diphenyl-2-pyridyl-methyl) methacrylate, which was coated to wide-pore

aminopropyl-silanized silica. These capillaries were used for capillary HPLC and nonaqueous CEC. The anodic EOF is assumed to be generated by not totally covered aminoalkyl groups of the aminopropyl-silanized silica gel.

Different groups have synthesized CEC-CSPs based on macrocyclic antibiotics [241–243]. Dermaux *et al.* [241] packed a capillary with a 5 μm vancomycin phase which is also used in HPLC and tested it with the chiral separation of warfarin and hexobarbital using triethylamine acetate (TEAA)/acetonitrile mixtures as mobile phase. Wikström *et al.* [242] prepared a new vancomycin CSP using a three-step *in situ* immobilization procedure on a packed diol-silica capillary. While vancomycin was used in CE for the chiral separation for acidic compounds, this vancomycin CSP could be applied to a broad spectrum of neutral, acidic, and basic compounds using a reversed-phase and polar organic mode. The resolution values for thalidomide in the reversed-phase mode were 2.5 (80 000 plates per meter) and in the polar organic mode 2.5 (120 000 plates per meter), respectively. Carter-Finch and Smith [243] packed a capillary with a teicoplanin CSP (Chirobiotic T 5 μm) and resolved on this phase tryptophan and dinitrobenzoyl leucine enantiomers using acetonitrile/phosphate buffer, pH 7, as mobile phase.

Wolf *et al.* [244] investigated two Pirkle-type CSPs. An (*S*)-naproxen-derived CSP and a (3*R*,4*S*)-Whelk-O-CSP based on 3 μm silica were packed into 100 μm ID capillaries. Excellent enantioselectivity was obtained on these CSPs for several test analytes with different structures using a MES (pH 6)/acetonitrile mixture as mobile phase with remarkable efficiency (up to 200 000 plates per meter). The chiral recognition mechanism is based on dipole-dipole stacking and π-π interactions. Lämmerhofer and Lindner [245] developed an anion exchange CSP based on *tert*-butyl quinine carbamate immobilized on 5 μm silica. This CSP was applied both for the HPLC and CEC separation of *N*-(9-fluorenylmethoxycarbonyl) amino acids and *N*-(3,5-dinitrobenzoyloxycarbonyl) amino acids. Ionic interactions and π-π interactions are assumed to be the main forces responsible for chiral recognition. Recently, the same group reported nonaqueous CEC using a CSP consisting of the same selector immobilized on 3 μm silica [181]. Acetonitrile/methanol mixtures containing acetic acid and triethylamine were used as mobile phases for the chiral resolution of amino acid derivatives. Compared to the previous published results, faster separations and improved efficiencies were obtained under these conditions. FMOC-Leu was separated in less than 10 min with a resolution factor of 6.9 and about 100 000 plates per meter.

Mayer *et al.* [246] and Francotte and Zhang [247] recently developed a new CSP based on a 3,5-dimethylphenylcarbamoyl cellulose phase immobilized on 5 or 7 µm silica gel of 4000 Å pore size. Acetonitrile/phosphate or citrate buffers were used as mobile phases. This CSP was run both in the capillary LC and CEC mode. CEC showed much better column efficiency and enantioselectivity under similar conditions. Figure 6 shows typical electrochromatograms of some selected model analytes. An improvement in efficiency could certainly be obtained by using a silica gel of smaller particle size. Lin *et al.* [248] prepared molecularly imprinted polymers by copolymerization of methacrylic acid or 2-vinylpyridine as functional monomers and ethylene glycol dimethacrylate as cross-linker in the presence of 2,2'-azobis(isobytyronitrile) (AIBN) using Dns-L-leucine as print molecule. After sieving, the particles were packed into the capillary. The enantioselectivity was good for Dns-leucine but only low for other Dns-amino acids. In another paper these authors described the use of L-phenylalanine anilide as print molecule [249]. This CSP showed chiral recognition ability for phenylalanine and also slightly for other aromatic amino acids.

14.3 Monolithic CSPs

A recent trend in CEC is to move away from packed columns on a silica basis since reproducible packing of capillaries and the preparation of frits by sintering a packing zone requires some experience. Both the silica particles and the frits are sources of air bubbles. Air bubbles can be prevented by applying pressure at both ends of the capillary, but this requires special equipment. The preparation of monolithic phases on silica or polymer basis is a recent technique and certainly one of the most challenging approaches for the development of CSPs for CEC.

The technique of preparation of monolithic phases by *in situ* polymerization in the column was introduced by Hjertén *et al.* [250] in the 1980s already. To fix the polymer on the capillary wall, the capillary is first treated with γ-methacryloxypropyltrimethoxysilane, creating a double bond for copolymerization with the monomer solution. The monomer solution consists of a monomer, a cross-linker and a charge-forming agent. An allylated selector can be added. After degassing, the polymerization is initiated and the solution is drawn into the capillary. Intake must be stopped before the zone reaches the detection window. After 12 h the polymerization is complete.

Koide and Ueno [251] prepared monolithic CSPs by incorporating β-CD polymers such as poly β-CD and CM-β-CD in a polyacrylamide gel and separated terbutaline and benzoin enantiomers in these phases. Later, the same group [252, 253] prepared a positively charged monolithic CSP, to which allyl carbamoylated β-CD (AC-β-CD) was covalently bonded. A mixture of AC-β-CD, acrylamine, *N,N'*-methylenebisacrylamide and *N*-(2-acrylamidoethyl)-triethylammonium iodide in Tris / boric acid buffer containing *N,N,N',N'*-tetramethylethylenediamine and ammonium peroxodisulfate was filled into a capillary pretreated with γ-methacryloxypropyltrimethoxysilane. These monolitic CSPs showed good enantioselectivity for acidic compounds (*e.g.*, Dns-amino acids, warfarin, phenoxypropionic acid) and neutral compounds (benzoin and 1-(1-naphthalene) ethanol) with high efficiency (up to 150 000 plates per meter).

Peters *et al.* [254] prepared a "moulded" monolithic CSP by copolymerization of the chiral monomer 2-hydroxyethyl methacrylate (*N*-L-valine-3,5-dimethylanilide) carbamate with ethylene dimethylacrylate, 2-acrylamido-2-methyl-1-propanesulfonic acid and butyl or glycidyl methacrylate in the presence of a porogenic solvent. This phase has hydrophylic properties which can be enhanced further by hydrolysis of the glycidyl moieties. The chiral recognition ability of this monolithic CSP was demonstrated by means of *N*-(3,5-dinitrobenzoyl)leucine diallylamide. The column efficiency was found to be 61 000 plates per meter.

Recently, Schmid *et al.* [152] synthesized a monolithic ligand-exchange CSP using methacrylamide as a monomer, piperazine diacrylamide as a cross-linker, vinylsulfonic acid as a charge-providing agent and *N*-(2-hydroxy-3-allyloxypropyl)-L-4-hydroxyproline as a chiral selector. The polymer chains so yielded form a homogeneous network of interconnected nodules consisting of microparticles. The voids or channels between the nodules allow

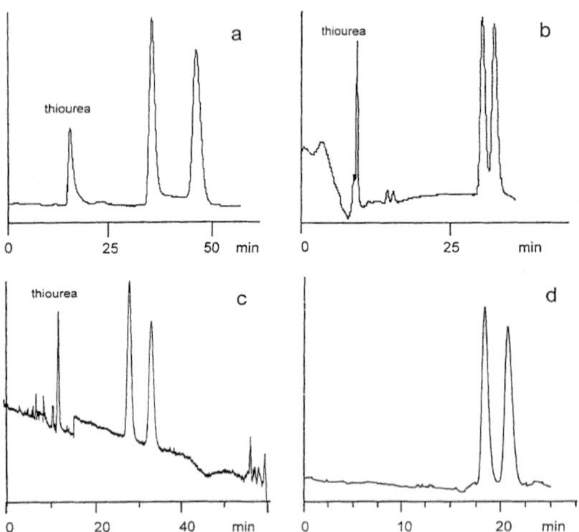

Figure 6. Electropherogram of the enantiomeric separation of (a) lormetazepam, (b) *trans*-stilbene oxide, (c) indapamide, (d) benzoin on a 3,5-dimethylphenylcarbamoyl cellulose phase. Reprinted from [246], with permission.

high permeability and thus low back pressure. These chiral continuous beds are inexpensive and easy to prepare. This CSP was evaluated *via* a selection of underivatized amino acids. The same capillary was used for CEC, capillary LC and pressure-supported CEC (Fig. 7). Faster separations were obtained by "short-end injection". In this case the effective length of the continuous bed was 8.5 cm and the enantiomers of phenylalanine were baseline-resolved within 4 min.

Schweitz *et al.* [255] prepared a monolithic imprinted polymer for chiral CEC separation by filling the capillary pretreated with γ-methacryloxypropyltrimethoxysilane with a prepolymerization mixture of print molecule, functional and cross-linking monomers (methacrylic acid and trimethylolpropane trimethacrylate), radical initiator AIBN and solvent (toluene) and polymerization was performed by placing the capillary under a UV source. The print molecule is removed by flushing the capillary with acetonitrile. An imprint remains, which shows high enantioselectivity for the same or very similar molecules. As print molecules (*R*)-propranolol [255, 256] or (*S*)-metoprolol [255] were used for chiral separation of β-blockers and (*S*)-ropivacaine for the resolution of local anesthetics. Lin *et al.* [257] developed a thermally induced *in situ* polymerization procedure for the preparation of an imprinted polymer using D-phenylalanine as a print molecule. Phenylalanine was baseline-resolved; tyrosine and phenylglycine showed only partial resolution.

15 ITP and IEF

Only a few publications deal with the application of ITP to chiral separations, although ITP was the first technique among electroseparation techniques applied for chiral separations [258–260]. The development of a powerful isotachophoretic sample pretreatment system coupled on-line to CZE was reported by Danková *et al.* [261]. ITP provides a preconcentration- and sample clean-up step enabling the interference-free determination of tryptophan enantiomers in a 90-component model mixture and urine. Figure 9 shows the electropherograms obtained with different working modes explicated in Fig. 8 using α-CD as chiral selector in the CZE system. The authors have clearly demonstrated that all interfering compounds were removed by transferring a narrow sample zone between two spacer constituents to the CZE system. With this approach, it was possible to detect 25 nmol/l L(–)Trp in the presence of the 200-fold amount of D(+)Trp. Hoffman *et al.* [262] developed a continuous flow ITP system for the purification of *R*-(–)-methadone on an mg scale using HP-β-CD as a chiral selector [262]. Recently, Kaniansky *et al.* [263] published a new approach for the preparative separation of enantiomers by ITP. A column-coupling system was used consisting of a preseparation column of

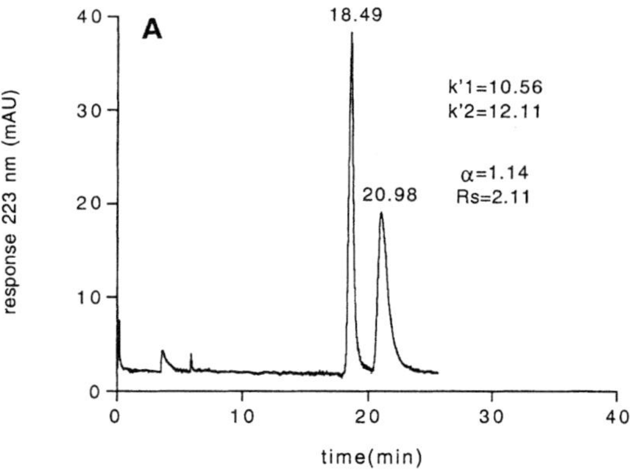

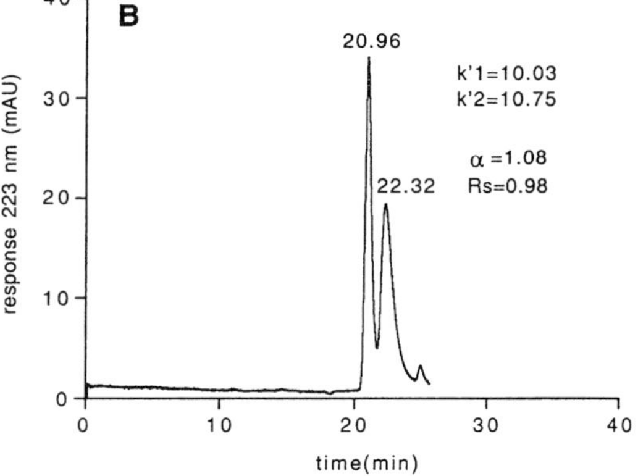

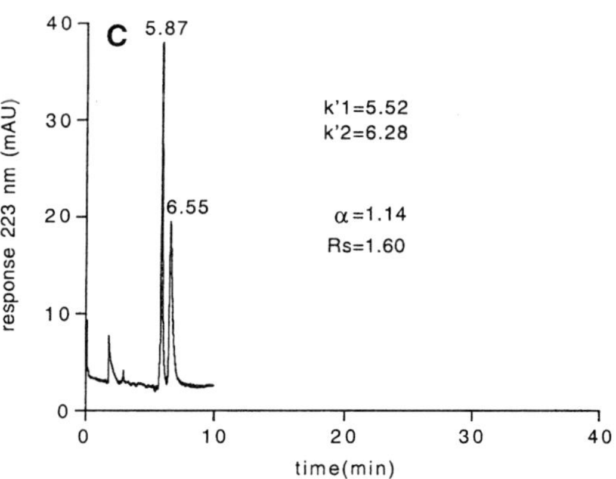

Figure 7. Chiral separation of Phe comparing (A) CEC, (B) pressure-driven micro-HPLC and (C) pressure-supported CEC on a monolithic LE-CSP. Conditions: stationary phase, monolithic ligand-exchange CSP (26 cm × 75 µm); mobile phase, 50 mM sodium dihydrogen phosphate / 0.1 mM Cu(II), pH 4.6; injection, 10 kV × 6 s; (A) 30 kV, (B) 12 bar, (C) 30 kV and 12 bar. Reprinted from [152], with permission.

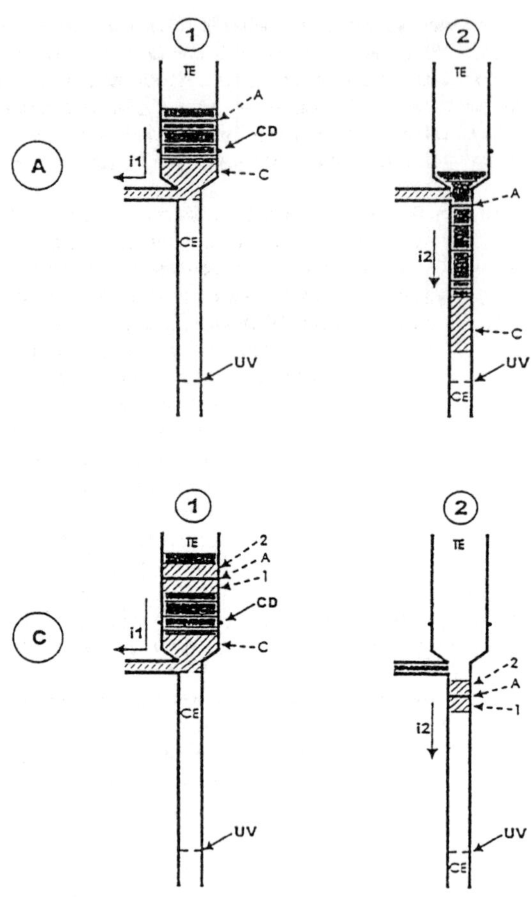

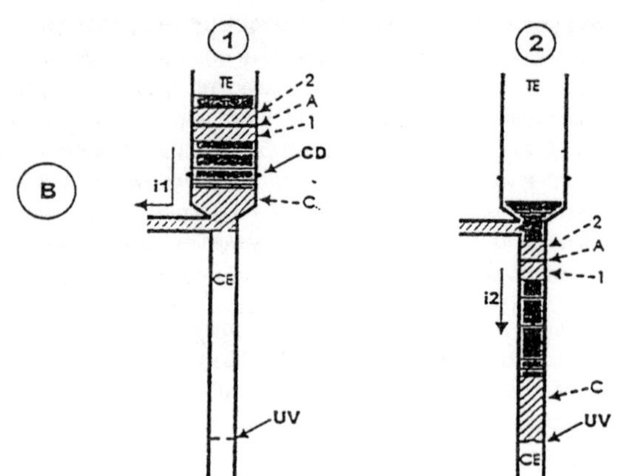

Figure 8. Working modes of the ITP-CZE combination in the column-coupling CE equipment. (A) CZE combined with a transient ITP sample stacking; (B) the same as in (A), only the sample contained a pair of discrete spacers (1, 2) for a spatial separation of the analyte from the matrix constituents; (C) CZE combined with a postcolumn ITP sample clean-up. Characteristic situations in the ITP and CZE stages before the final CZE separations. (A1), (B1) final situations after the separations in the (first) ITP stage of the ITP-CZE runs; (C1) a situation in the ITP stage before a postcolumn ITP clean-up (this included electromigration removals of the sample constituents present in the ITP stack and migrating in front of a front spacer (1, in C1 and C2) and behind the rear spacer (2, in C1 and C2)); (A2)–(C2) starting situations in the (second) CZE stage of the ITP-CZE runs. Symbols: A, analyte (tryptophan enantiomers); C, carbonate zone; 1,2 front and rear spacing constituents, respectively; CD, conductivity detector; UV, photometric absorbance detector. Reprinted from [261], with permission.

1.0 mm ID and a trapping column of 0.8 mm ID. β-CD was used as chiral selector separating DNP-leucine enantiomers as a model analyte. The authors have shown that up to 14 mg of pure DNP-norleucine enantiomers could be obtained in one preparative run. A recent paper deals with the use of transient ITP-CZE for chiral separations [264]. A counterflow is generated by applying counterpressure at the capillary. The system was applied to the separation of clenbuterol enantiomers using dimethyl-β-cyclodextrin as chiral selector. Preparative IEF was used by Gluckhovsky and Vigh [265] for the separation of Dns-phenylalanine enantiomers on mg/h scale.

16 Indirect separation

Little work has been done in this field in recent years. Thorsén *et al.* [266] described the synthesis of a new chiral precolumn derivatization reagent for amino compounds, (+) and (−) 1-(9-anthryl)-2-propyl chloroformate.

This reagent was applied to the chiral separation of 17 amino acids and some small peptides by MEKC using SDS in borax buffer in combination with LIF detection. Kleidernigg and Lindner [267] synthesized (1*R*,2*R*)- and (1*S*-2*S*)-*N*-[(2-isothiocyanato)cyclohexyl]-6-methoxy-4-quinolinylamide) as chiral derivatization reagent for amino acids. The diastereomeric derivatives were separated by RP-HPLC and quasi-CEC using PVP as a pseudostationary phase.

Another chiral derivatization reagent for amino acids and peptides, *R*-(−)- or *S*-(+)-4-(3-isothiocyanatopyrrolidin-1-yl)-7-nitro-2,1,3-benzoxadiazole was described by Liu *et al.* [268]. The resulting derivatives were separated by micelle-mediated CE using a nonionic surfactant Triton X-100 in acetate buffer, pH 4. The fluorescent derivatives were detected by LIF detection using an argon-ion laser. An exciting application of a new derivatization procedure with the goal of determining amino acids and their enantiomeric ratio in micrometeorites was recently reported by Trambouze-Vandenabeele *et al.* [269]. The amino acids

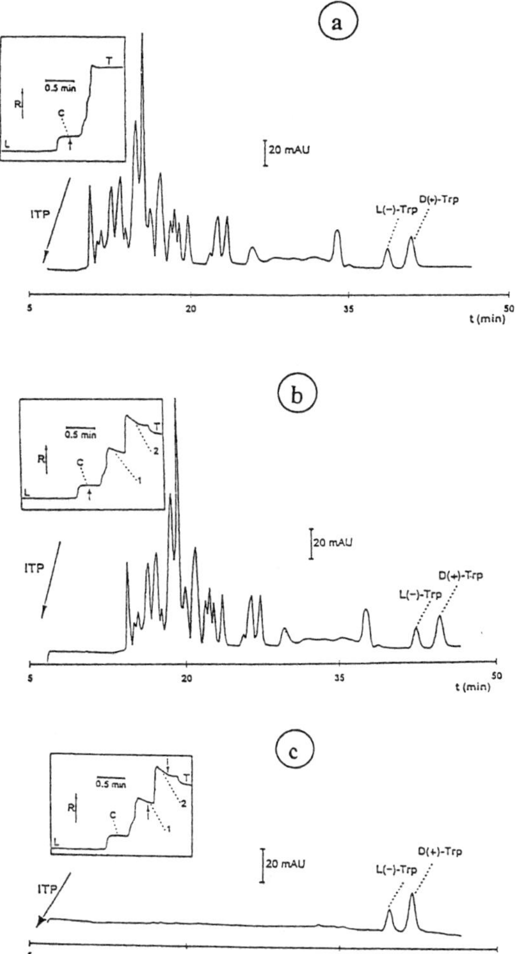

Figure 9. Electropherograms from the separations of tryptophan enantiomers present in a 90 component model mixture of organic acids using various working modes of the ITP-CZE combination. (a)–(c) Electropherograms corresponding to the working modes (A)–(C), respectively, shown in Fig. 8. L(–)-Trp and D(+)-Trp were present in the injected mixture (30 µL) at 0.5 µmol/L and 0.8 µmol/L concentrations, respectively. The concentrations of organic acids in the model mixture were in the range of 1–5 µmol/L. Arrows on the isotachopherograms mark the sample fraction taken for final CZE separations. Glycine (1) and β-alanine (2) added to the sample at 1 mmol/L concentrations served as discrete spacing constituents in the ITP stage of the separation. Reprinted from [261], with permission.

were derivatized using a chloroethylnitrosourea of ε-benzoyloxycarbonyl-lysine-*t*-butyl ester and FITC. The separation of the diastereomeric derivatives was carried out with borate buffer in the presence of the nonionic surfactant Brij-58 using LIF. The low detection limits (about 10 pM for isovaline) make the method suitable to detect trace amounts of amino acid enantiomers in micrometeorites.

17 Miscellaneous

Reversal of the EMO is often of importance, especially with enantiomeric purity check of drugs. The enantiomer present as an impurity in a sample of the active enantiomer should always appear as the first peak to avoid being covered by the tailing of the major component. Chankvetadze *et al.* [270] recently published an excellent survey of various possibilities of reversing the EMO and discussed the mechanisms of the different approaches. The simplest possibility is to use selectors possessing opposite chirality. This is not possible, however, with natural selectors. It was found that different CDs and CD-derivatives can exhibit different chiral recognition ability for the enantiomers. The EMO can be reversed by changing from a neutral to a charged CD. Also, oligosaccharides of different type or chain length were found to respond differently to the enantiomers. Changing the mobility of the analyte by varying the pH is another way of reversing the EMO. On the other hand, the same effect can be obtained by changing the mobility of the chiral selector or by changing the relation of the mobilities of the analyte and the selector and reversing the polarity. Suppressing or reversing the EOF are further means for reversing EMO. Simply the change from an uncoated to a coated capillary may reverse the EMO. EOF can be reversed, for example, by using hydrophobic quaternary ammonium compounds, which form a positively charged layer at the capillary wall. In some cases, EMO was even seen to depend on selector concentration. The combination of different chiral selectors is another tool for reversing the EMO and last but not least, charged micelle forming achiral surfactants are frequently added to reverse EMO.

A new technique, synchronous cyclic CE, was recently introduced by Zhao and Jorgenson [271]. The high resolving power was demonstrated by the application of this technique to isotopic and chiral separations. Over 100 million plates are reported to be reached with this system. Chiral separation of (α-hydroxybenzyl)methyltrimethylammonium and (2-hydroxy-1-phenyl)ethyltrimethylammonium, compounds which could not be separated with any other system, was achieved with this technique using β-CD as chiral selector. The combination of flow injection (FI) with CE for the chiral separation of intermediate enantiomers in chloramphenicol synthesis was described by Liu and Fang [272]. CD derivatives were used as chiral selectors. The samples were injected continuously by FI without current interruption, achieving a five times higher sample throughput with improved precision compared to electrokinetic injection.

Palmarsdottir *et al.* [273] described the development of a supported liquid membrane technique using a microcolumn LC interface for on-line coupling with a CZE system for sample pretreatment and preconcentration. In this system, bambuterol in subnanomole concentrations was detected in human plasma. Using permanently charged CDs, Vigh's group [37] developed the CHARM model, which is based on the consideration of simultaneous protonation and complexation equilibria. This model is helpful for selecting the appropriate operating conditions based on rational predictions. Another approach to give assured predictions for optimal separation is the combination of artificial neural networks of experimental design to optimize the separation of amino acid enantiomers using α-CD as a chiral selector [274].

18 Microchips

Two years ago, the feasibility of using microfabricated lab-on-a-chip systems for the chiral separation of labeled amino acids was demonstrated. Hutt *et al.* [275] resolved FITC-labeled amino acids on a microfabricated CE chip to show how such devices can analyze for extinct or extant signs of life in extraterrestrial environments. The test system consisted of a folded electrophoresis channel (19.0 cm long $\times$ 150 µm wide $\times$ 20 µm deep) that was photolithographically fabricated in a 10 cm diameter glass wafer sandwich; laser-excited confocal fluorescence was used for detection, providing subattomole sensitivity. They used an SDS / γ-CD, pH 10.0, carbonate electrophoresis buffer and a separation voltage of 550 V/cm at 10°C. In only 4 min, baseline resolution was observed for Val, Ala, Glu, and Asp enantiomers. Another approach was published by Rodriguez *et al.* [276] again using CD-modified MEKC for the chiral separation of FITC-amino acids on a microfabricated device. In their experiments, analysis time ranged from 75 s for basic amino acids to 160 s for acidic amino acids. Efficiency up to 395 000 plates/m was reached, promising that the use of such devices could lead to a challenging new field.

Received June 26, 2000

19 References

[1] Snopek, J., Jelinek, I., Smolkova-Keulemansova, E., *J. Chromatogr.* 1988, *452*, 571–590.

[2] Kuhn, R., Hofstetter-Kuhn, S., *Chromatographia* 1992, *34*, 505–512.

[3] Valko, I. E., Billiet, H. A. H., Corstjens, H. A. L., Frank, J., *LC.GC Int.* 1993, *6*, 420–427.

[4] Fanali, S., Cristalli, M., Vespalec, R., Boček, P., in: Chrambach, A., Dunn, M. J., Radola, B. J. (Eds.), *Advances in Electrophoresis*, VCH, Weinheim, New York 1994, pp. 1–86.

[5] Novotny, M., Soini, H., Stefansson, M., *Anal. Chem.* 1994, *66*, 646A–655A.

[6] Terabe, S., Otsuka, K., Nishi, H., *J. Chromatogr. A* 1994, *666*, 295–319.

[7] Vespalec, R., Boček, P., *Electrophoresis* 1994, *15*, 755–762.

[8] Nishi, H., Terabe, S., *J. Chromatogr. A* 1995, *694*, 245–276.

[9] Chankvetadze, B., Endresz, G., Blaschke, G., *Chem. Soc. Ref.* 1996, *25*, 141–153.

[10] Fanali, S., *J. Chromatogr. A* 1996, *735*, 77–121.

[11] Chankvetadze, B., *J. Chromatogr. A* 1997, *792*, 269–295.

[12] Fanali, S., *J. Chromatogr. A* 1997, *792*, 227–267.

[13] Gübitz, G., Schmid, M. G., *J. Chromatogr. A* 1997, *792*, 179–225.

[14] Lloyd, D. K., Aubry, A. F., Delorenzi, E., *J. Chromatogr. A* 1997, *792*, 349–369.

[15] Lurie, I. S., *J. Chromatogr. A* 1997, *792*, 297–307.

[16] Nishi, H., *J. Chromatogr. A* 1997, *792*, 327–347.

[17] Riekkola, M. L., Wiedmer, S. K., Valko, I. E., Siren, H., *J. Chromatogr. A* 1997, *792*, 13–35.

[18] Vespalec, R., Boček, P., *Electrophoresis* 1997, *18*, 843–852.

[19] Ward, T. J., Oswald, T. M., *J. Chromatogr. A* 1997, *792*, 309–325.

[20] Fanali, S., *J. Chromatogr. A* 2000, *875*, 89–122.

[21] Verleysen, K., Sandra, P., *Electrophoresis* 1998, *19*, 2798–2833.

[22] Vigh, G., Sokolowski, A. D., *Electrophoresis* 1997, *18*, 2305–2310.

[23] Miura, M., Funazo, K., Tanaka, M., *Anal. Chim. Acta* 1997, *357*, 177–185.

[24] Miura, M., Kawamoto, K., Funazo, K., Tanaka, M., *Anal. Chim. Acta*, 1998, *373*, 47–56.

[25] Aturki, Z., Desiderio, C., Mannina, L., Fanali, S., *J. Chromatogr. A* 1998, *817*, 91–104.

[26] Fanali, S., Desiderio, C., Aturki, Z., *J. Chromatogr. A* 1997, *772*, 185–194.

[27] Chiari, M., Desperati, V., Cretich, M., Crini, G., Janus, L., Morcellet, M., *Electrophoresis* 1999, *20*, 2614–2618.

[28] Zerbinati, O., Trotta, F., Giovannoli, C., Baggiani, C., Giraudi, G., Vanni, A., *J. Chromatogr. A* 1998, *810*, 193–200.

[29] Corradini, R., Uccella, G., Galaverna, G., Dossena, A., Marchelli, R., *Tetrahedron Lett.* 1999, *40*, 3025–3028.

[30] Jung, M., Francotte, E., *J. Chromatogr. A* 1996, *755*, 81–88.

[31] Francotte, E., Brandel, L., Jung, E., *J. Chromatogr. A* 1997, *792*, 379–384.

[32] Rickard, E., Bopp, R., Skanchy, D., Chetwyn, K., Pahlen, B., Stobaugh, J., *Chirality* 1996, *8*, 108–121.

[33] Vincent, J. B., Kirby, D. M., Nguyen, T. V., Vigh, G., *Anal. Chem.* 1997, *69*, 4419–4428.

[34] Vincent, J. B., Sokolowski, A. D., Nguyen, T. V., Vigh, G., *Anal. Chem.* 1997, *69*, 4226–4233.

[35] Cai, H., Nguyen, T. V., Vigh, G., *Anal. Chem.* 1998, *70*, 580–589.

[36] Zhu, W., Vigh, G., *Anal. Chem.* 2000, *72*, 310–317.

[37] Williams, B. A., Vigh, G., *J. Chromatogr. A* 1997, *777*, 295–309.

[38] Schmitt, T., Engelhardt, H., *J. High Resol. Chromatogr.* 1993, *16*, 525–529.

[39] Schmitt, T., Engelhardt, H., *Chromatographia* 1993, *37*, 475–481.

[40] Schmid, M. G., Wirnsberger, K., Gübitz, G., *Pharmazie* 1996, *51*, 852–861.

[41] Tanaka, Y., Yangawa, M., Terabe, S., *J. High Resol. Chromatogr.* 1996, *19*, 421–433.

[42] Ischibuchi, K., Izumoto, S.-I., Nishi, H., Sato, T., *Electrophoresis* 1997, *18*, 1007–1012.

[43] Juvancz, Z., Jicsinszky, L., Markides, K. E., *J. Microcol. Sep.* 1997, *9*, 581–589.

[44] Terabe, S., *Trends Anal. Chem.* 1989, *8*, 129–134.

[45] Nardi, A., Eliseev, A., Boček, P., Fanali, S., *J. Chromatogr.* 1993, *638*, 247–253.

[46] Fanali, S., Desiderio, C., Aturki, Z., *J. Chromatogr. A* 1997, *772*, 185–194.

[47] Lelièvre, F., Gareil, P., Jardy, A., *Anal. Chem.* 1997, *69*, 385–392.

[48] O'Keefe, F., Shamsi, S. A., Darcy, R., Schwinte, P., Warner, I. M., *Anal. Chem.* 1997, *69*, 4773–4782.

[49] Haynes III, J. L., Shamsi, S. A., O'Keefe, F., Darcey, R., Warner, I. M., *J. Chromatogr. A* 1998, *803*, 261–271.

[50] Sebesta, R., Salisova, M., *Enantiomer* 1999, *4*, 271–277.

[51] Mikus, P., Kaniansky, D., Sebesta, R., Salisova, M., *Enantiomer* 1999, *4*, 279–287.

[52] Galaverna, G., Corradini, R., Dossena, A., Marchelli, R., Vecchio, G., *Electrophoresis* 1997, *18*, 905–911.

[53] Galaverna, G., Corradini, R., Dossena, A., Marchelli, R., *Electrophoresis* 1999, *20*, 2619–2629.

[54] Roussel, C., Favrou, A., *J. Chromatogr. A* 1995, *704*, 67–74.

[55] Bunke, A., Jira, T., *Pharmazie* 1996, *51*, 672–673.

[56] Jakubetz, H., Juza, M., Schurig, V., *Electrophoresis* 1997, *18*, 897–904.

[57] Schulte, G., Chankvetadze, B., Blaschke, G., *J. Chromatogr. A* 1997, *771*, 259–266.

[58] Bunke, A., Jira, T., *J. Chromatogr. A* 1998, *798*, 275–280.

[59] Chankvetadze, B., Endresz, G., Schulte, G., Bergenthal, D., Blaschke, G., *J. Chromatogr. A* 1996, *732*, 143–150.

[60] Tanaka, Y., Terabe, S., *J. Chromatogr. A* 1997, *781*, 151–160.

[61] Wang, F., Khaledi, M. G., *J. Chromatogr. A* 1998, *817*, 121–128.

[62] Wang, F., Khaledi, M. G., *Electrophoresis* 1998, *19*, 2095–2100.

[63] Lelievre, F., Gueit, C., Gareil, P., Bahadi, Y., Galons, H., *Electrophoresis* 1997, *18*, 891–896.

[64] Tanaka, Y., Terabe, S., *J. Chromatogr. A* 1997, *781*, 151–160.

[65] Terabe, S., Miyashita, Y., Shibata, O., Barnhart, E. R., Alexander, L. R., Patterson, D. G., Karger, B. L., Hosoya, K., Tanaka, N., *J. Chromatogr.* 1990, *516*, 23–31.

[66] Chankvetadze, B., Schulte, G., Blaschke, G., *Enantiomer* 1997, *2*, 157–179.

[67] Katayama, H., Ishihama, Y., Asakawa, N., *J. Chromatogr. A* 1997, *764*, 151–156.

[68] Wan, H., Engström, A., Blomberg, E. G., *J. Chromatogr. A* 1996, *731*, 283–292.

[69] Amini, A., Beijersten, I., Pettersson, C., Westerlund, D., *J. Chromatogr. A* 1996, *737*, 301–313.

[70] Stefansson, M., Novotny, M., *J. Am. Chem. Soc.* 1993, *115*, 11573–11580.

[71] Jira, T., Bunke, A., Schmid, M. G., Gübitz, G., *J. Chromatogr. A* 1997, *761*, 269–276.

[72] Schmid, M. G., Wirnsberger, K., Jira, T., Bunke, A., Gübitz, G., *Chirality* 1997, *9*, 153–156.

[73] Noroski, J. E., Mayo, D. J., Moran, M., *J. Pharm. Biomed. Anal.* 1995, *13*, 45–52.

[74] Pak, C., Marriott, P. J., Carpenter, P. D., Aniet, R. G., *J. High Resol. Chromatogr.* 1998, *21*, 640–644.

[75] Sutton, R. M. C., Sutton, K. L., Stalcup, A. M., *Electrophoresis* 1997, *18*, 2297–2304.

[76] D'Hulst, A., Verbeke, N., *J. Chromatogr.* 1992, *608*, 275–287.

[77] Quang, C., Khaledi, M. G., *J. High Reso. Chromatogr.* 199, *17*, 609–612.

[78] Nishi, H., *J. Chromatogr. A* 1996, *735*, 345–351.

[79] Nishi, H., Izumoto, S., Nakamura, K., Nakai, H., Sato, T., *Chromatographia* 1996, *42*, 617–630.

[80] Chankvetadze, B., Saito, M., Yashima, E., Okamoto, Y., *J. Chromatogr. A* 1997, *773*, 331–338.

[81] Chankvetadze, B., Saito, M., Yashima, E., Okamoto, Y., *Chirality* 1998, *10*, 134–139.

[82] Nakamura, H., Sano, A., Sumii, H., *Anal. Sci.* 1998, *14*, 375–378.

[83] Soini, H., Stefansson, M., Riekkola, M. L., Novotny, M., *Anal. Chem.* 1994, *66*, 3477–3484.

[84] Kano, K., Minami, K., Horiguchi, K., Ishimura, T., Kodera, M., *J. Chromatogr. A* 1995, *694*, 307–313.

[85] Chankvetadze, B., Saito, M., Yashima, E., Okamoto, Y., *Chirality* 1998, *10*, 134–139.

[86] Hong, M., Soini, H., Baker, A., Novotny, M. V., *Anal. Chem.* 1998, *70*, 3590–3597.

[87] Gotti, R., Cavrini, V., Andrisano, V., Mascellani, G., *J. Chromatogr. A* 1998, *814*, 205–211.

[88] Gotti, R., Cavrini, V., Andrisano, V., Mascellani, G., *J. Chromatogr. A* 1999, *845*, 247–256.

[89] Tsukamoto, T., Ushio, T., Haginaka, J., *J. Chromatogr. A* 1999, *864*, 163–171.

[90] Wang, X., Lee, J.-T., Armstrong, D. W., *Electrophoresis* 1999, *20*, 162–170.

[91] Phinney, K. W., Jinadu, L. A., Sander, L. C., *J. Chromatogr. A* 1999, *857*, 285–293.

[92] Nishi, H., Nakamura, K., Nakai, H., Sato, T., *Chromatographia* 1996, *43*, 426–430.

[93] Behr, J.-P., Girodeau, J.-M., Heyward, R. C., Lehn, J.-M., Sauvage, J.-P., *Helv. Chim. Acta* 1980, *63*, 2096–2111.

[94] Kuhn, R., Erni, F., Bereuter, T., Häusler, J., *Anal. Chem.* 1992, *64*, 2815–2820.

[95] Kuhn, R., Stöcklin, F., Erni, F., *Chromatographia* 1992, *33*, 32–36.

[96] Höhne, E., Krauss, G.-J., Gübitz, G., *J. High Resol. Chromatogr.* 1992, *15*, 698–700.

[97] Kuhn, R., Riester, D., Fleckenstein, B., Wiesmüller, K.-H., *J. Chromatogr. A* 1995, *716*, 371–379.

[98] Schmid, M. G., Gübitz, G., *J. Chromatogr. A* 1995, *709*, 81–88.

[99] Verleysen, K., Vandijck, J., Schelfaut, M., Sandra, P., *J. High Resol. Chromatogr.* 1998, *21*, 323–331.

[100] Verleysen, K., Van den Bosch, T., Sandra, P., *Electrophoresis* 1999, *20*, 2650–2655.

[101] Nishi, H., Nakamura, K., Nakai, H., Sato, T., *J. Chromatogr. A* 1997, *757*, 225–235.

[102] Mori, Y., Ueno, K., Umeda, T., *J. Chromatogr. A* 1997, *757*, 328–332.

[103] Kuhn, R., *Electrophoresis* 1999, *20*, 2605–2613.

[104] Tanaka, Y., Otsuka, K., Terabe, S., *J. Chromatogr. A* 2000, *875*, 323–330.

[105] Peña, M. S., Zhangh, Y., Thibodeaux, S., Mc Laughlin, M. L., de le Peña, A. M., Warner, I. M., *Tetrahedron Lett.* 1996, *37*, 5841–5844.

[106] Peña, M. S., Zhangh, Y., Warner, I. M., *Anal. Chem.* 1997, *69*, 3239–3242.

[107] Grady, T., Joyce, T., Smyth, M. R., Harris, S. J., Diamond, D., *Anal. Commun.* 1998, *35*, 123–125.

[108] Armstrong, D. W., Rundlett, K. L., Chen, J. R., *Chirality* 1994, *6*, 496–509.

[109] Gasper, M. P., Berthod, A., Nair, U. B., Armstrong, D. W., *Anal. Chem.* 1996, *68*, 2501–2514.

[110] Desiderio, C., Fanali, S., *J. Chromatogr. A* 1998, *807*, 37–56.

[111] Armstrong, D. W., Nair, U. B., *Electrophoresis* 1997, *18*, 2331–2342.

[112] Ward, T. J., Dan III, C., Brown, A. P., *Chirality* 1996, *8*, 77–83.

[113] Oswald, T. M., Ward, T. J., *Chirality* 1999, *11*, 663–668.

[114] Fanali, S., Desiderio, C., *J. High Resol. Chromatogr.* 1996, *19*, 322–326.

[115] Desiderio, C., Polcaro, C. M., Padiglioni, P., Fanali, S., *J. Chromatogr. A* 1997, *781*, 503–513.

[116] Carotti, A., Di Gioia, F., Cellamare, S., Fanali, S., *J. High Resol. Chromatogr.* 1999, *22*, 315–329.

[117] Fanali, S., Desiderio, C., Schulte, G., Heidmeier, S., Strichmann, D., Chankvetadze, B., Blaschke, G., *J. Chromatogr. A* 1998, *800*, 69–76.

[118] Fanali, S., Desiderio, C., Schulte, G., Heidmeier, S., Strichmann, D., Chankvetadze, B., Blaschke, G., *J. Chromatogr. A* 1998, *800*, 77–82.

[119] Strege, A., Huff, B. E., Risley, D. S., *LC.GC* 1996, *14*, 144–150.

[120] Reilly, J., Risley, D. S., *LC.GC* 1998, *16*, 170–171.

[121] Sharp, V. S., Risley, D. S., McCarthy, S., Huff, B. E., Strege, M. A., *J. Liq. Chromatogr. Rel. Technol.* 1997, *20*, 887–898.

[122] Trelli-Seifert, L. A., Risley, D. S., *J. Liq. Chromatogr. Rel. Technol.* 1998, *21*, 299–313.

[123] Ekborg-Ott, K. H., Zientara, G. A., Schneiderheinze, J. M., Gahm, K., Armstrong, D. W., *Electrophoresis* 1999, *20*, 2438–2457.

[124] Fanali, S., Aturki, Z., Desiderio, C., Bossi, A., Righetti, P. G., *Electrophoresis* 1998, *19*, 1742–1751.

[125] Risley, D. S., Trelli-Seifert, L., McKenzie, Q. J., *Electrophoresis* 1999, *20*, 2749–2753.

[126] Kang, J. W., Yang, Y. T., You, J. M., Ou, Q. Y., *J. Chromatogr. A* 1998, *825*, 81–87.

[127] Carotti, A., Di Gioia, F., Cellamare, S., Fanali, S., *J. High Resol. Chromatogr.* 1999, *22*, 315–321.

[128] Hage, D. S., *Electrophoresis* 1997, *18*, 2311–2321.

[129] Haginaka, J., *J. Chromatogr. A* 2000, *875*, 235–254.

[130] Fanali, S., Caponecchi, G., Aturki, Z., *J. Microcol. Sep.* 1997, *9*, 9–14.

[131] Tanaka, Y., Terabe, S., *Chromatographia* 1999, *49*, 489–495.

[132] De Lorenzi, E., Massolini, G., Lloyd, D. K., Monaco, H. L., Galbusera, C., Caccialanza, G., *J. Chromatogr. A* 1997, *790*, 47–64.

[133] De Lorenzi, E., Massolini, G., Quaglia, M., Galbusera, C., Caccialanza, G., *Electrophoresis* 1999, *20*, 2739–2748.

[134] Mano, N., Oda, Y., Ishihama, Y., Katayama, H., Asakawa, N., *J. Liq. Chromatogr. Rel. Technol.* 1998, *21*, 1311–1332.

[135] Gassmann, E., Kuo, J. E., Zare, R. N., *Science* 1985, *230*, 813–814.

[136] Gozel, P., Gassman, E., Michelsen, H., Zare, R. N., *Anal. Chem.* 1987, *59*, 44–49.

[137] Desiderio, C., Aturki, Z., Fanali, S., *Electrophoresis* 1994, *15*, 864–869.

[138] Soontornniyomkij, B., Scandrett, K., Pietrzyk, D. J., *J. Liq. Chromatogr. Rel. Technol.* 1998, *21*, 2245–2263.

[139] Schmid, M. G., Gübitz, G., *Enantiomer* 1996, *1*, 23–27.

[140] Chen, Z., Lin, J. M., Uchiyama, K., Hobo, T., *J. Chromatogr. A* 1998, *813*, 369–378.

[141] Chen, Z., Lin, J. M., Uchiyama, K., Hobo, T., *Chromatographia* 1999, *49*, 436–443.

[142] Chen, Z., Lin, J. M., Uchiyama, K., Hobo, *J. Microcol. Sep.* 1999, *11*, 534–540.

[143] Chen, Z., Lin, J. M., Uchiyama, K., Hobo, *Anal. Chim. Acta* 2000, *403*, 173–178.

[144] Chen, Z., Lin, J. M., Uchiyama, K., Hobo, *Proceedings of 22nd International Symposium on Capillary Chromatography*, Gifu, Japan, Nov. 1999.

[145] Chen, Z., Lin, J. M., Uchiyama, K., Hobo, *Anal. Sci.* 2000, *16*, 131–137.

[146] Yuan, Z., Yang, L., Zhangh, S., *Electrophoresis* 1999, *20*, 1842–1845.

[147] Végvári, Á., Schmid, M. G., Kilár, F., Gübitz, G., *Electrophoresis* 1998, *19*, 2109–2112.

[148] Schmid, M. G., Rinaldi, R., Dreveny, D., Gübitz, G., *J. Chromatogr. A* 1999, *846*, 157–163.

[149] Schmid, M. G., Laffranchini, M., Dreveny, D., Gübitz, G., *Electrophoresis* 1999, *20*, 2458–2461.

[150] Schmid, M. G., Lecnik, O., Sitte, U., Gübitz, G., *J. Chromatogr. A* 2000, *875*, 307–314.

[151] Karbaum, A., Jira, T., *J. Chromatogr. A* 2000, *874*, 285–292.

[152] Schmid, M. G., Grobuschek, N., Tuscher, T., Gübitz, G., Végvári, Á., Machtejevas, E., Hjertén, S., *Electrophoresis* 2000, *21*, 3141–3144.

[153] Karbaum, A., Bunke, A., Jira, T., *J. Chromatogr. A* 1998, *798*, 281–288.

[154] Bjornsdottir, I., Hansen, S. H., Terabe, S., *J. Chromatogr. A* 1996, *745*, 37–44.

[155] Stalcup, A. M., Gahm, K. H., *J. Microcol. Sep.* 1996, *8*, 145–150.

[156] Lämmerhofer, M., Lindner, W., *J. Chromatogr. A* 1999, *839*, 167–182.

[157] Piette, V., Lämmerhofer, M., Lindner, W., Crommen, J., *Chirality* 1999, *11*, 622–630.

[158] Piette, V., Fillet, M., Lindner, W., Crommen, J., *J. Chromatogr. A* 2000, *875*, 353–360.

[159] Terabe, S., Otsuka, K., Ichikawa, K., Tsuchiya, A., Ando, T., *Anal. Chem.* 1984, *56*, 111–113.

[160] Cammileri, P., *Electrophoresis* 1997, *18*, 2322–2330.

[161] Palmer, C. P., Tanaka, N., *J. Chromatogr. A* 1997, *792*, 105–124.

[162] Ding, W., Fritz, J. S., *J. Chromatogr. A* 1999, *831*, 311–320.

[163] Shamsi, S. A., Warner, I. M., *Anal. Chem.* 1997, *69*, 2980–2987.

[164] Billiot, E., Macossay, J., Tibodeaux, S., Shamsi, S. A., Warner, I. M., *Anal. Chem.* 1998, *70*, 1375–1381.

[165] Haddadian, F., Billiot, E. J., Shamsi, S. A., Warner, I. M., *J. Chromatogr. A* 1999, *858*, 219–227.

[166] Haddadian, F., Shamsi, S. A., Warner, I. M., *Electrophoresis* 1999, *20*, 3011–3026.

[167] Shamsi, S. A., Warner, I. M., *Electrophoresis* 1997, *18*, 853–872.

[168] Tickle, D., George, A., Jennings, K., Camillary, P.,. Kirby, A. J., *J. Chem. Soc. Perkin Trans.* 1998, *3*, 467–474.

[169] Ju, M., El Rassi, Z., *Electrophoresis* 1999, *20*, 2766–2771.

[170] Ju, M., El Rassi, Z., *J. Liq. Chromatogr. Rel. Technol.* 2000, *23*, 35–45.

[171] Horimai, T., Arai, T., Sato, Y., *J. Chromatogr. A* 2000, *875*, 295–305.

[172] Bunke, A., Jira, T., Beyrich, T., *Pharmazie* 1997, *52*, 10–11.

[173] Karbaum, A., Jira, T., *Electrophoresis* 1999, *20*, 3396–3401.

[174] Valkó, I. E., Sirén, H., Riekkola, M.-L., *Chromatographia* 1996, *43*, 242–246.

[175] Wang, F., Khaledi, M. G., *Anal. Chem.* 1996, *68*, 3460–3467.

[176] Wang, F., Khaledi, M. G., *J. Chromatogr. A* 2000, *875*, 277–293.

[177] Wang, F., Khaledi, M. G., *J. Chromatogr. B* 1999, *731*, 187–197.

[178] Vincent, J. B., Vigh, G., *J. Chromatogr. A* 1998, *816*, 233–241.

[179] Wang, F., Khaledi, M. G., *J. Chromatogr. A* 1998, *817*, 121–128.

[180] Krause, K., Chankvetadze, B., Okamoto, Y., Blaschke, G., *Electrophoresis* 1999, *20*, 2772–2778.

[181] Tobler, E., Lämmerhofer, M., Lindner, W., *J. Chromatogr. A* 2000, *875*, 341–352.

[182] Lurie, I. S., *J. Chromatogr. A* 1997, *792*, 297–307.

[183] Fillet, M., Hubert, P., Crommen, J., *J. Chromatogr. A* 2000, *875*, 123–134.

[184] Okafo, G. N., Bintz, C., Clarke, S. E., Camilleri, P., *J. Chem. Soc. Chem. Commun.* 1992, 1189–1192.

[185] Lin, M., Wu, N., Barker, G. E., Sun, P., Huie, C. W., Hartwick, R. A., *J. Liq. Chromatogr.* 1993, *16*, 3667–3674.

[186] Okafo, G. N., Camilleri, P., *J. Microcol. Sep.* 1993, *5*, 149–155.

[187] Aumatell, A., Wells, R. J., *J. Chromatogr. A* 1994, *688*, 329–337.

[188] Smith, J. T., Nashabeh, W., El Rassi, Z., *Anal. Chem.* 1994, *66*, 1119–1133.

[189] Wang, J., Warner, I. M., *J. Chromatogr. A* 1995, *711*, 297–304.

[190] Kuhn, R., Steinmetz, C., Bereuter, T., Haas, P., Erni, F., *J. Chromatogr. A* 1994, *666*, 367–373.

[191] Lin, J. M., Nakagama, T., Hobo, T., *Chromatographia* 1996, *42*, 559–565.

[192] Kuhn, R., Wagner, J., Walbroehl, Y., Bereuter, T., *Electrophoresis* 1994, *15*, 828–834.

[193] Verleysen, K., Sandra, P., *J. High Resol. Chromatogr.* 1999, *22*, 33–38.

[194] Huang, W. X., Fazio, S. D., Vivilecchia, R. V., *J. Chromatogr. A* 1997, *781*, 129–137.

[195] Huang, W. X., Xu, H., Fazio, S. D., Vivilecchia, R. V., *J. Chromatogr. B* 1997, *695*, 157–162.

[196] Huang, W. X., Xu, H., Fazio, S. D., Vivilecchia, R. V., *HPCE '98*, Orlando, FL 1998, poster presentation.

[197] Huang, W. X., Xu, H., Fazio, S. D., Vivilecchia, V., *J. Chromatogr. A* 2000, *875*, 361–369.

[198] Armstrong, D. W., Chang, L. W., Chang, S. S. C., *J. Chromatogr. A* 1998, *793*, 115–134.

[199] Bunke, A., Jira, T., Gübitz, G., *Pharmazie* 1995, *50*, 570–571.

[200] Jira, T., Bunke, A., Karbaum, A., *J. Chromatogr. A* 1998, *798*, 281–288.

[201] Horimai, T., Ohara, M., Ichinose, M., *J. Chromatogr. A* 1997, *760*, 235–244.

[202] Ingelse, B. A., Reijenga, J. C., Claessens, H. A., Everaerts, F., *J. High Resol. Chromatogr.* 1996, *19*, 225–228.

[203] Ingelse, B. A., Reijenga, J. C., Flieger, M., Everaerts, F. M., *J. Chromatogr. A* 1997, *791*, 339–342.

[204] Ingelse, B. A., Reijenga, J. C., Everaerts, F. M., *J. Chromatogr. A* 1997, *772*, 179–184.

[205] Sinibaldi, M., Vinci, M., Frederici, F., Flieger, M., *Biomed. Chromatogr.* 1997, *11*, 307–310.

[206] Nair, U. B., Armstrong, D.-W., Hinze, W. L., *Anal. Chem.* 1998, *70*, 1059–1065.

[207] Jung, G., Hofstetter, H., Feiertag, S., Stoll, D., Hofstetter, O., Wiesmüller, K.-H., Schurig, V., *Angew. Chem.* 1996, *35*, 2148–2150.

[208] Chiari, M., Desperati, V., Manera, E., Longhi, R., *Anal. Chem.* 1998, *70*, 4967–4973.

[209] Gübitz, G., Schmid, M. G., *Enantiomer* 2000, *5*, 5–11.

[210] Wistuba, D., Schurig, V., *J. Chromatogr. A* 2000, *875*, 255–276.

[211] Dermaux, A., Sandra, P., *Electrophoresis* 1999, *20*, 3027–3065.

[212] Mayer, S., Schurig, V., *J. High Resol. Chromatogr.* 1992, *15*, 129–131.

[213] Mayer, S., Schurig, V., *J. Liq. Chromatogr.* 1993, *16*, 915–931.

[214] Mayer, S., Schleimer, M., Schurig, V., *J. Microcol. Sep.* 1994, *6*, 43–48.

[215] Mayer, S., Schurig, V., *Electrophoresis* 1994, *15*, 835–841.

[216] Schurig, V., Jung, M., Mayer, S., Fluck, M., Negura, S., Jakubetz, H., *J. Chromatogr. A* 1995, *694*, 119–128.

[217] Armstrong, D. W., Tang, Y. B., Ward, T., Nichols, M., *Anal. Chem.* 1993, *65*, 1114–1117.

[218] Szeman, J., Ganzler, K., *J. Chromatogr. A* 1994, *668*, 509–517.

[219] Schurig, V., Wistuba, D., *Electrophoresis* 1999, *20*, 2313–2328.

[220] Francotte, E., Jung, M., *Chromatographia* 1996, *42*, 521–527.

[221] Liu, Z., Zou, H., Yi Ni, J., Zhang, Y., *Anal. Chim. Acta* 1999, *378*, 73–76.

[222] Liu, Z., Zou, H., Ye, M., Ni, J., Zhang, Y., *Electrophoresis* 1999, *20*, 2891–2897.

[223] Hofstetter, H., Hofstetter, O., Schurig, V., *J. Microcol. Sep.* 1998, *10*, 287–291.

[224] Hong, F., Zhang, X.-X., Chang, W.-B., Ci, Y.-X., *Anal. Chim. Acta* 1998, *373*, 207–212.

[225] Schweitz, L., Andersson, L. I., Nilsson, S., *J. Chromatogr. A* 1998, *817*, 5–3.

[226] Remcho, V. T., Tan, Z. J., *Anal. Chem. News & Features* 1999, 248A–255A.

[227] Sellergren, B., *Trends Anal. Chem.* 1997, *16*, 310–320.

[228] Owens, P. K., Karlsson, L., Lutz, E. S. M., Andersson, L. I., *Trends Anal. Chem.* 1999, *18*, 146–154.

[229] Takeuchi, T., Haginaka, J., *J. Chromatogr. B* 1999, *728*, 1–20.

[230] Brüggemann, O., Freitag, R., Whitcombe, M. J., Vulfson, E. N., *J. Chromatogr. A* 1997, *781*, 43–53.

[231] Lelièvre, F., Yan, C., Zare, R. N., Gareil, P., *J. Chromatogr. A* 1996, *723*, 145–156.

[232] Wei, W., Luo, G. A., Xiang, R., Yan, C., *J. Microcol. Sep.* 1999, *11*, 263–269.

[233] Deng, Y., Zhang, J., Tsuda, T., Yu, P. H., Boulton, A. A., Cassidy, R. M., *Anal. Chem.* 1998, *70*, 4586–4593.

[234] Lämmerhofer, M., Lindner, W., *J. Chromatogr. A* 1999, *839*, 167–182.

[235] Li, S., Lloyd, D. K., *J. Chromatogr. A* 1994, *666*, 321–335.

[236] Li, S., Lloyd, D. K., *Anal. Chem.* 1993, *65*, 3684–3690.

[237] Lloyd, D. K., Li, S., Ryan, P., *J. Chromatogr. A* 1995, *694*, 285–296.

[238] Wistuba, D., Czesla, H., Roeder, M., Schurig, V., *J. Chromatogr. A* 1998, *815*, 183–188.

[239] Wistuba, D., Schurig, V., *Electrophoresis* 1999, *20*, 2779–2785.

[240] Krause, K., Girod, M., Chankvetadze, B., Blaschke, G., *J. Chromatogr. A* 1999, *837*, 51–63.

[241] Dermaux, A., Lynen, P., Sandra, P., *J. High Resol. Chromatogr.* 1998, *21*, 575–576.

[242] Wikström, H., Svensson, L. A., Torstensson, A., Owens, P. K., *J. Chromatogr. A* 2000, *869*, 395–409.

[243] Carter-Finch, A. S., Smith, N. W., *J. Chromatogr. A* 1999, *848*, 375–385.

[244] Wolf, C., Spence, P. L., Pirkle, W. H., Derrico, E. M., Cavender, D. M., Rozing, G. P., *J. Chromatogr. A* 1997, *782*, 175–179.

[245] Lämmerhofer, M., Lindner, W., *J. Chromatogr. A* 1998, *829*, 115–125.

[246] Mayer, S., Briand, X., Francotte, E., *J. Chromatogr. A* 2000, *875*, 331–339.

[247] Francotte, E., Zhang, T., PCT Int. Pat. Appl. WO 9704011.

[248] Lin, J.-M., Uchiyama, K., Hobo, T., *Chromatographia* 1998, *47*, 625–629.

[249] Lin, J.-M., Nakagama, T., Uchiyama, K., Hobo, T., *Chromatographia* 1996, *43*, 585–591.

[250] Hjertén, S., Liao, J.-L., Zhang, R., *J. Chromatogr. A* 1989, *473*, 273–275.

[251] Koide, T., Ueno, K., *Anal. Sci.* 1998, *14*, 1021–1023.

[252] Koide, T., Ueno, K., *Anal. Sci.* 1999, *15*, 791–794.

[253] Koide, T., Ueno, K., *J. High Resol. Chromatogr.* 2000, *23*, 59–66.

[254] Peters, E. C., Lewandowski, K., Petro, M., Svec, F., Frechet, J. M. J., *Anal. Commun.* 1998, *35*, 83–86.

[255] Schweitz, L., Andersson, L. I., Nilsson, S., *Anal. Chem.* 1997, *69*, 1179–1183.

[256] Schweitz, L., Andersson, L. I., Nilsson, S., *Chromatographia* 1999, *49*, 93–94.

[257] Lin, J.-M., Nakagama, T., Wu, X. Z., Uchiyama, K., Hobo, T., *Fresenius J. Anal. Chem.* 1997, *357*, 130–132.

[258] Snopek, J., Jelinek, I., Smolkova-Keulemansova, E., *J. Chromatogr.* 1988, *438*, 211–218.

[259] Jelinek, I., Snopek, J., Smolkova-Keulemansova, E., *J. Chromatogr.* 1988, *439*, 386–393.

[260] Snopek, J., Jelinek, I., Smolkova-Keulemansova, E., *J. Chromatogr.* 1989, *472*, 308–313.

[261] Danková, M., Kaniansky, D., Fanali, S., Iványi, F., *J. Chromatogr. A* 1999, *838*, 31–43.

[262] Hoffmann, P., Wagner, H., Weber, G., Lanz, M., Caslavska, J., Thormann, W., *Anal. Chem.* 1999, *71*, 1840–1850.

[263] Kaniansky, D., Simunicova, E., Ölvecka, E., Ferancova, A., *Electrophoresis* 1999, *20*, 2786–2793.

[264] Toussaint, B., Hubert, P., Tjaden, U. R., van der Greef, J., Crommen, J., *J. Chromatogr. A* 2000, *871*, 173–180.

[265] Glukhovsky, P., Vigh, G., *Anal. Chem.* 1999, *71*, 3814–3820.

[266] Thorsén, G., Engström, A., Josefsson, B., *J. Chromatogr. A* 1997, *786*, 347–354.

[267] Kleidernigg, O., Lindner, W., *J. Chromatogr. A* 1998, *795*, 251–261.

[268] Liu, Y. M., Schneider, M., Sticha, C. M., Toyooka, T., Sweedler, J. V., *J. Chromatogr. A* 1998, *800*, 345–354.

[269] Trambouze-Vandenabeele, O., Grenier-Loustalot, M. F., Albert, M., Despois, D., Dobrijevic, M., Commeyras, A.,

Geffard, G., Couderc, F., Bayle, C., *9th International Symposium on Luminescence Spectrometry in Biomedical and Environmental Analysis*, Montpellier, France, May 2000.

[270] Chankevetadze, B., Schulte, G., Blaschke, G., *Enantiomer* 1997, *2*, 157–179.

[271] Zhao, J., Jorgenson, J. W., *J. Microcol. Sep.* 1999, *11*, 439–449.

[272] Liu, Z. S., Fang, Z. L., *Anal. Chim. Acta* 1997, *353*, 199–205.

[273] Palmarsdottir, S., Mathiasson, L., Jönsson, J. A., Edholm, L.-E., *J. Capil. Electrophor.* 1996, *3*, 255–260.

[274] Dohnal, V., Farkova, M., Havel, J., *Chirality* 1999, *11*, 616–621.

[275] Hutt, L. D., Glavin, D. P., Bada, J. L., Mathies, R. A., *Anal. Chem.* 1999, *71*, 4000–4006.

[276] Rodriguez, I., Jin, L. J., Li, S. F. Y., *Electrophoresis* 2000, *21*, 211–219.

Review

Dorothee Wistuba
Volker Schurig

Institute of Organic Chemistry,
University of Tübingen,
Tübingen, Germany

Recent progress in enantiomer separation by capillary electrochromatography

Enantiomer separation by electrochromatography (CEC) can be performed in three modes: (i) open-tubular capillary electrochromatography (o-CEC), in which the chiral selector is physically adsorbed coated, and thermally immobilized or covalently attached to the internal capillary wall; (ii) packed capillary electrochromatography (p-CEC), in which the capillary is either filled with chiral modified silica particles or with an achiral packing material, and a chiral selector is added to the mobile phase; and (iii) monolithic (rod)-capillary electrochromatography (rod-CEC) in which the chiral stationary phase (CSP) consists of a single piece of porous solid. We present an overview on methods and new trends in the field of electrochromatographic enantiomer separation such as CEC with either nonaqueous mobile phases or stationary phases with incorporated permanent charges, or with packing beds consisting of nonporous silica particles or particles with very small internal diameters .

Keywords: Enantiomer separation / Capillary electrochromatography / Open-tubular capillary electrochromatography / Packed-capillary electrochromatography / Rod-capillary electrochromatography / Review

EL 4243

Contents

Correspondence: Dr. Dorothee Wistuba, Institute of Organic Chemistry, University of Tübingen, Auf der Morgenstelle 18, 72076 Tübingen, Germany
E-mail: dorothee.wistuba@uni-tuebingen.de
Fax: +49-07071-295538

Abbreviations: AGP, α_1-acid-glycoprotein; **AIBN**, azo-isobutyronitrile; **CSP**, chiral stationary phase; **DNP**, dinitrophenyl; **HTAP-β-CD**, β-cyclodextrin-2-hydroxy-3-trimethylammoniumpropyl ether; **MIP**, molecularly imprinted polymer; **NAQ-CEC**, capillary electrochromatography with nonaqueous mobile phase; **o-CEC**, capillary electrochromatography using open-tubular capillary; **ODS**, octadecyl silica; **p-CEC**, capillary electrochromatography using packed capillaries; **rod-CEC**, capillary electrochromatography using monolithic (rod) capillaries; **SPE-β-CD**, 6-*O*-(sulfo-*n*-propyl)-β-cyclodextrin; **TEAA**, triethylammonium acetate

1 Introduction

Enantiomer separation by capillary electrochromatography (CEC) [1–5] has received considerable attention in recent years. The first enantiomer separation by CEC was described by Mayer and Schurig in 1992 [6]. They used wall-coated open tubes with Chirasil-Dex, a permethyl-β-cyclodextrin-modified dimethylpolysiloxane as chiral stationary phase (CSP) for the enantiomer separation of 1,1′-binaphthyl-2,2′-diylhydrogenphosphate and 1-phenylethanol. Later, this method (open-tubular CEC, o-CEC) was used for the enantiomer separation of various racemates on CSPs such as native cyclodextrins, modified cyclodextrins, cellulose, BSA, an alkaloid derivative, molecularly imprinted polymers (MIPs), peptides and proteins [7–21]. Enantiomer separation by CEC with packed capillaries (p-CEC) was first demonstrated in 1993 by Li and Lloyd [22]. The enantiomers of benzoin, hexobarbital, pentobarbital, ifosfamide, cyclophosphamide, diisopyramide, metoprolol, oxprenolol, alprenolol, and propranolol were separated on a capillary filled with an α_1-acid glycoprotein (AGP) stationary phase. In the last three years, a large number of CSPs – all well-known in HPLC – were tested in p-CEC. Saccharides, "Pirkle" phase, macrocyclic antibiotics, anion-exchange-type stationary phases, polyacrylamide derivatives and MIPs were used as chiral selectors [22–39]. The recent trend in CEC is the application of monoliths (or rods) on silica or polymer basis (rod-CEC). This promising technology represents an attractive alternative to p-CEC because the use of retaining frits – the cause of many problems – is unnecessary. To date only a few chiral monoliths for enantiomer separation in CEC have been described [1]. The first generation of chiral monolithic separation systems in CEC was based on continuous rods of imprinted polymers [21, 40]. A chiral cavity was produced by adding a chiral print molecule to a polymerization mixture and removing it after the polymerization. Another approach is the direct copolymerization of a chiral selector such as 2-hydroxyethyl methacrylate (*N*-L-valine-3,5-dimethylanilide) carbamate [41], quinidine carbamate [42] or *N*-(2-hydroxy-3-alloxypropyl)-L-4-hydroproline [43]. Charged cross-linked polyacrylamide gels with incorporated or covalently bonded cyclodextrin derivatives were successfully used for enantiomer separation by CEC [44, 45]. A chiral silica-based monolith was prepared by sintering a packed silica bed at high temperature and derivatizing the formed silica skeleton with a permethyl-β-cyclodextrin containing dimethylpolysiloxane (Chirasil-Dex) [46]. Chiral silica-entrapped columns for CEC were made by trapping particles of a chiral MIP packing, or an *S*-*N*-3,5-dinitrobenzoyl-1-naphthylglycine or *S*-*N*-3,5-dinitrophenylaminocarbonylvaline-modified stationary phase in a network of silica [47].

2 o-CEC

In o-CEC, the chiral selector is immobilized or physically adsorbed to the inner surface of a capillary. So far, capillaries with chiral surfaces have found only limited attention in the field of electrochromatographic enantiomer separation. The main drawback of o-CEC is the low sample loading capacity. This problem can be overcome by

etching the internal wall of the capillary in order to increase the surface area up to 1000-fold. The preparation methods of capillaries for o-CEC may involve (i) chemical bonding of the stationary phase, (ii) physical coating of the selector, and (iii) molecular imprinting techniques.

2.1 Saccharides (cyclodextrins, cellulose)

Cyclodextrins and their derivatives are used most frequently as chiral selectors in chromatography and capillary electrophoresis. Mayer and Schurig [3, 6–9] prepared capillaries coated with Chirasil-Dex, a dimethylpolysiloxane-linked permethylated β-cyclodextrin and separated the enantiomers of 1-phenylethanol, 1,1′-binaphthyl-2,2′-diylhydrogenphosphate, a number of nonsteroidal anti-inflammatory drugs and barbiturates by o-CEC. Figure 1

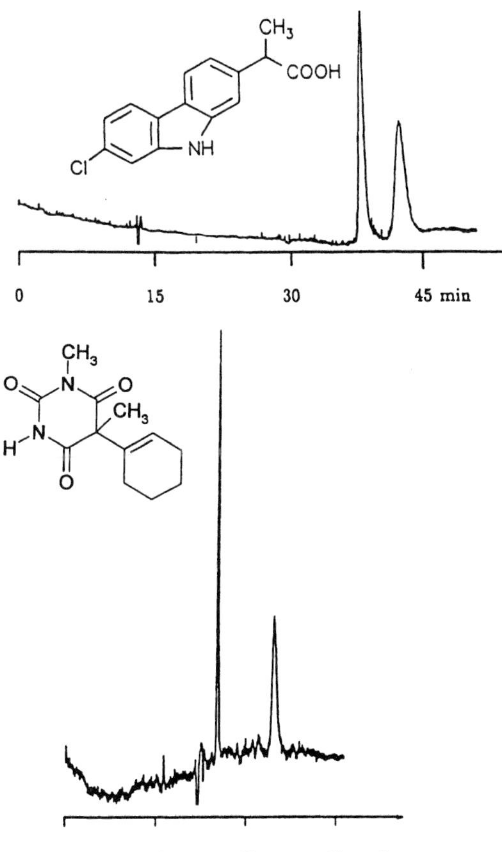

Figure 1. Enantiomer separation of carprofen and ibuprofen by o-CEC on a 80 cm × 50 µm ID capillary coated with immobilized Chirasil-Dex. Buffer, 20 mM Tris, pH 7; applied voltage, 30 kV. Reprinted from [8], with permission.

shows the enantiomer separation of carprofen and hexobarbital. Chirasil-Dex (see Fig. 2) was wall-coated and thermally inmmobilized to yield a nonextractable and buffer-resistant CSP, which was stable under neutral and

acidic (pH 2.5) conditions for several months [8]. An about 30% reduced electroosmotic flow (EOF) was observed, caused by the shielding of the free silanol groups of the inner surface of the capillary [8]. This led to longer elution times compared with untreated capillaries. The film thickness influences both the plate number, N, and the resolution, R_s. While an increase in film thickness leads to a decrease in plate number, a reduced sample capacity was observed with thin films (0.1 µm). The best results were obtained with a film thickness of 0.2 µm [6]. The influence of the organic modifiers methanol, 2-propanol, and acetonitrile were investigated. While the plate number increases with increasing amounts of all investigated modifiers, a reduction of the resolution at a high modifier concentration was observed [49].

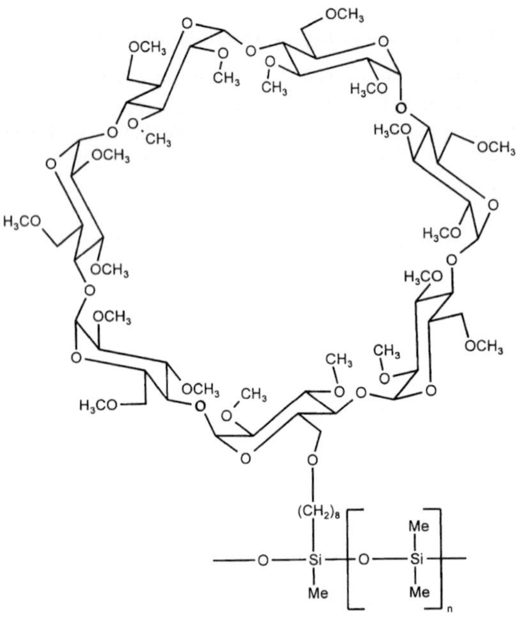

A direct comparison of o-CEC and o-LC with a Chirasil-Dex-coated capillary in a single instrumental setup was described by Jakubetz *et al.* [49]. As expected, they found a higher resolution (R_s) in the CEC mode (*e.g.*, mephobarbital: $R_{s(LC)} = 3.4$, $R_{s(CEC)} = 6$; carprofen: $R_{s(LC)} = 1.8$, $R_{s(CEC)} = 4.1$; 1-(2-naphthyl)ethanol: $R_{s(LC)} = 1.8$, $R_{s(CEC)} = 2$) but often long elution times. The high stability of the CSP allows a unified enantioselective approach by using a single Chirasil-Dex-modified capillary (100 cm × 50 µm ID) in four different chromatographic modes, GC, supercritical fluid chromatography (SFC), o-LC, and o-CEC [49–51]. Figure 3 shows the enantiomer separation of 1-(2-naphthyl)ethanol by GC, SFC, o-LC, and o-CEC. o-CEC has a long breakthrough time because of the low EOF but the retention factors are very low ($\ll 1$). This means that the enantiomers undergo only few encounters

Figure 2. Structure of Chirasil-Dex (dependent upon the reaction conditions, the spacer may be attached to the 2- or 6-position of the cyclodextrin).

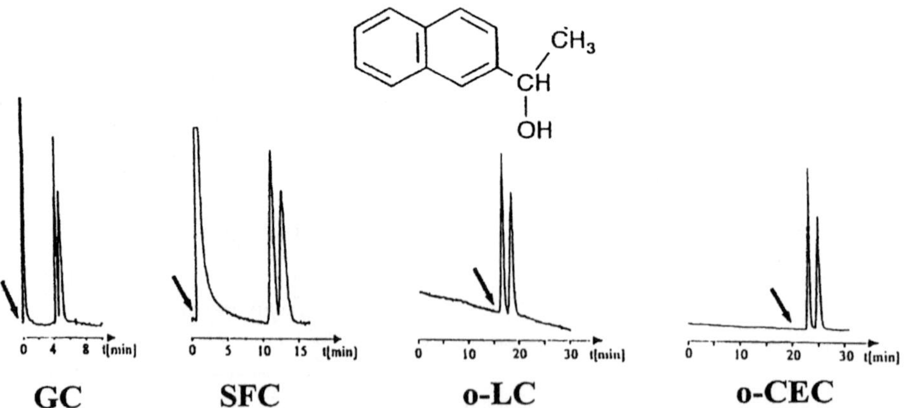

Figure 3. Unified enantioselective chromatography. Enantiomer separation of 1-(2-naphthyl)-ethanol on a 100 cm (80 cm effective) × 50 ID µm capillary coated with Chirasil-Dex by GC, SFC, o-LC, and o-CEC. Conditions for GC: 120°C, 1.1 bar He; SFC: 55°C, 68 bar CO_2; o-LC: 35°C, 0.14 bar, 20 mM phosphate buffer, pH 7; o-CEC: 30 kV, 10 mM borate/phosphate buffer, pH 7.5. The arrows in the chromatograms indicate the breakthrough time. Reprinted from [49], with permission.

with the CSP. On the other hand, in the GC mode the retention factors are large, the breakthrough time of only a few seconds very short, and thus the enantiomers spend most of their time in the CSP. The total analysis time is very long in o-CEC and o-LC because of the long breakthrough time. Switching from the o-CEC to the o-LC mode, peak broadening occurs due to the parabolic flow profile of the pressure-driven method. This leads to a loss of efficiency and hence also of resolution. The use of unified instrumental design to carry out the different chromatographic methods in a single apparatus as described by Ishii *et al.* [52] is desirable in the future. Armstrong *et al.* [10] demonstrated a related approach by using an immo-

bilized stationary phase containing a permethylated β-cyclodextrin covalently linked to dimethylpolysiloxane in a star-type mode for enantiomer separation by GC, SFC, and o-CEC. For o-CEC and SFC an identical column was used while the dimensions of the GC capillary were quite different.

In the usual o-CEC mode, the chiral selector is immobilized on the inner surface of the capillary. The addition of an independent second chiral selector to the mobile phase leads to the concept of a dual chiral recognition two-phase system. Mayer *et al.* [9] and Jakubetz *et al.* [53] investigated the enantiomer separation of hexobarbi-

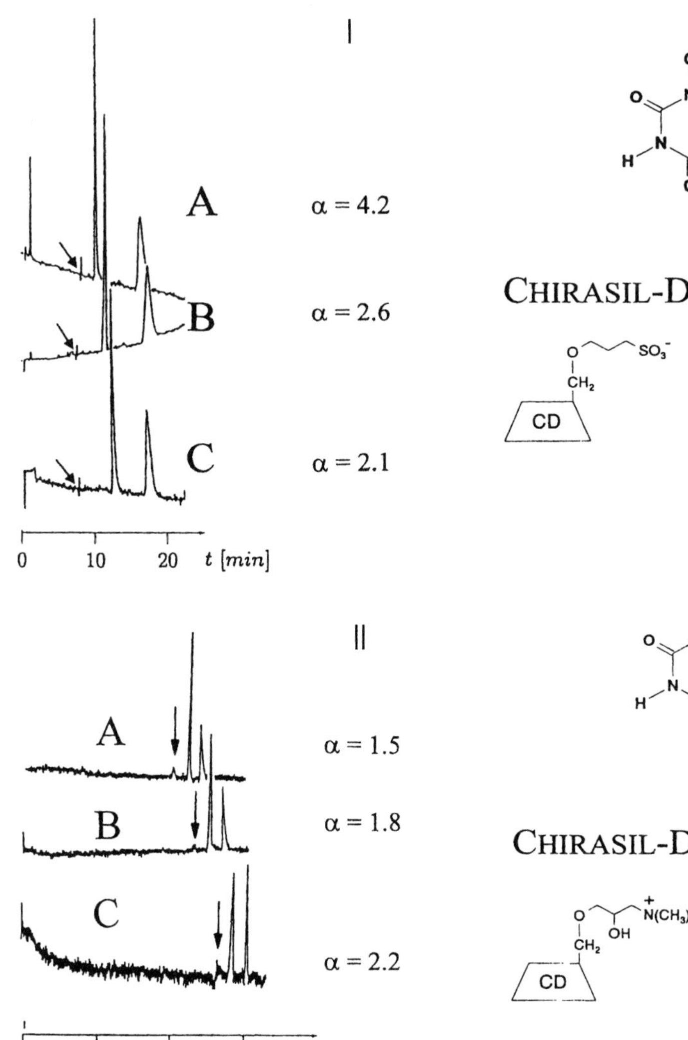

Figure 4. Dual chiral recognition in the enantiomer separation of hexobarbital on an 80 cm × 50 μm ID Chirasil-Dex-coated capillary. (I) Decrease of the separation factor α due to compensation of enantioselectivity; 30 kV, 20 mM borate/phosphate buffer, pH 7. (A) Without additive; (B) 0.5 mM additive; (C) 1.0 mM additive. Additive, SPE-β-cyclodextrin. (II) Increase of the separation factor α due to enhancement of enantioselectivity; 30 kV, 20 mM borate/phosphate buffer, pH 7/acetonitrile 90:10 v/v. (A) Without additive; (B) 1.0 mM additive; (C) 5.0 mM additive. Additive, HTAP-β-cyclodextrin. The arrows in the chromatograms indicate the breakthrough time. Reproduced from [53], with permission.

Electrophoresis 2000, *21*, 4136–4158

tal on a Chirasil-Dex-coated capillary in the presence of either a negatively charged cyclodextrin (6-*O*-(sulfo-*n*-propyl)-β-cyclodextrin, abbreviated SPE-β-cyclodextrin) or a positively charged cyclodextrin (β-cyclodextrin-2-hydroxy-3-trimethylammoniumpropyl ether chloride, abbreviated HTAP-β-cyclodextrin) mobile phase additive. The chiral recognition of hexobarbital by SPE-β-cyclodextrin (*R* before *S*) and by HTAP-β-cyclodextrin (*S* before *R*) is opposite to that verified by using an uncoated capillary and the anionic or cationic selector as buffer additive. The elution order of the hexobarbital enantiomers on a Chirasil-Dex-coated capillary is *R* before *S*. The addition of a charged cyclodextrin derivative to the buffer system of a Chirasil-Dex modified capillary leads to a decrease of the chiral separation factor α in the case of SPE-β-cyclodextrin and in the case of HTAP-β-cyclodextrin to an increase

of α (see Fig. 4). As is evident from Fig. 4, only a partial compensation of the chiral separation factor α is observed since the enantioselective recognition is mainly governed by the stationary phase (Chirasil-Dex). If the enantioselectivity is mainly determined by the chiral mobile-phase additive, an increase of the amount of added selector may result in a reversal of the elution order. As shown in Fig. 5, this phenomenon has been observed during the enantiomer separation of 1,1'-binaphthyl-2,2'-diylhydrogenphosphate by o-CEC on a Chirasil-Dex-modified capillary after the addition of different amounts of SPE-β-cyclodextrin to the mobile phase [9].

Sezemán and Ganzler [11] described a coating procedure for the immobilization of β- and γ-cyclodextrin for o-CEC. First, a polyacylamide layer with free double bonds was

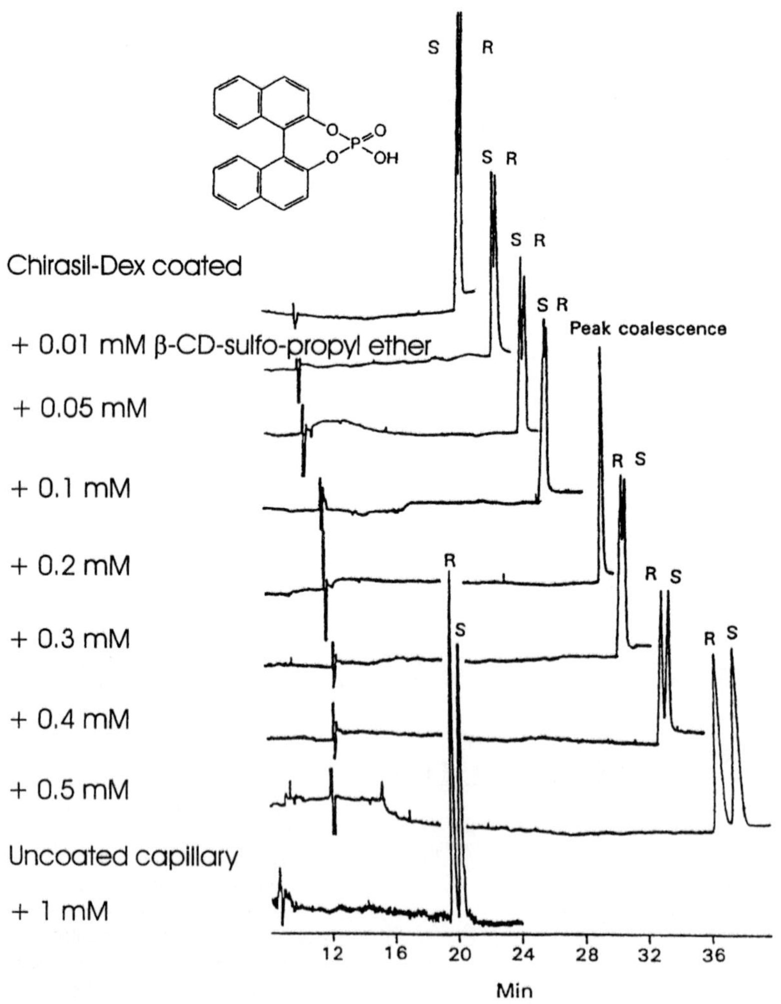

Figure 5. Enantiomer separation of 1,1'-binaphthyl-2,2'-diylhydrogenphosphate on an 80 cm × 50 μm ID Chirasil-Dex-coated capillary; 30 kV, 20 mM borate/phosphate buffer, pH 7; additive, SPE-β-cyclodextrin. Reproduced from [9], with permission.

bound to the capillary and in a second step the cyclodextrin was attached *via* the double bonds of the polyacrylamide in a reaction catalyzed by Ce(IV) salts. With this system, the enantiomer separation of epinephrine is feasible. In order to increase the inner surface of the capillaries, Pesek *et al.* [17] etched the capillaries with ammonium hydrogen difluoride prior to coating and immobilization of 2-hydroxy-3-methacryloyloxypropyl-β-cyclodextrin. Partial resolution of the enantiomers of several benzodiazepines was achieved. Replacing the cyclodextrin selector by a chiral lactone or a naphthylamine, a partial separation of tricyclic depressants, doxepine nortriptyline, clomipramine and derivatized amino acids was achieved. However, the peaks obtained are very broad and accordingly the resolution R_s was very low.

3,5-Dimethylphenylcarbamoyl cellulose (DMPCC) and *para*-methylbenzoyl cellulose (PMBC) are well-known CSPs in HPLC. Francotte and Jung [12] reported on the enantiomer separation by o-CEC and both normal-phase and reversed-phase o-LC in capillaries modified with non-immobilized neat cellulose derivatives as mentioned above. Figure 6 shows the enantiomer separation of 1-(9-anthryl)-2,2,2-trifluoroethanol by o-LC and o-CEC at different temperatures in a capillary coated with DMPCC. The thickness of the coating had a great influence on column performance. The optimal coating thickness of 0.025 μm was considerably lower than for Chirasil-Dex columns (see above). In order to shorten retention, acetonitrile (up to 25%) was used as organic modifier. Unfortunately, the EOF was not constant during continued use of the capillaries and also the lifetime of the capillaries was rather low.

2.2 Peptides and proteins

To obtain capillaries with peptide or protein stationary phases for o-CEC, two different coating procedures proved to be useful: the chiral selector was either covalently bound or physically adsorbed to the internal surface of a capillary. Hofstetter *et al.* [16] described the direct bonding of bovine serum albumin (BSA) on the capillary surface, resulting in a stable stationary phase suitable for enantiomer separation of several dinitrophenyl (DNP)-amino acids. Oxazepam and lorazepam elute in the form of typical interconversion profiles with the appearance of a plateau between the peaks of enantiomers caused by rapid enantiomerization (see Fig. 7). When stored at 4°C, the capillaries are operable up to one year.

Liu *et al.* [13–15] utilized the normally undesirable effect of adsorption of basic peptides and proteins onto the untreated capillary surface during CE runs for a simple preparation of CSPs for o-CEC. The capillaries were rinsed with a buffer system containing proteins or peptides with high isoelectric points such as lysozyme, cyctochrome *c*, Lys-Tyr and Lys-Ser-Tyr, resulting in a physically adsorbed layer stable for about 30 days. This system allowed the enantiomer separation of mephenytoin, fenoprofen, several derivatized and underivatized amino acids with resolutions R_s up to 4.1.

2.3 Anion-exchange-type stationary phases

Poly-terguride, an alkaloid-based chiral selector, was coated and immobilized on the internal wall of capillaries

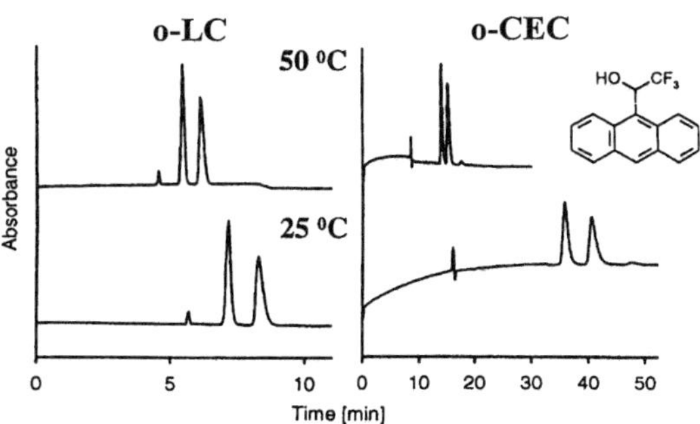

Figure 6. Enantiomer separation of 1-(9-anthryl)-2,2,2-trifluoroethanol by o-LC and o-CEC on a 50 cm × 50 ID μm capillary coated with 0.025 μm 3,5-dimethylphenylcarbamoyl cellulose. o-LC: n-hexane with 4% v/v 2-propanol; 35 mbar. o-CEC: 40 mM phosphate buffer, pH 7; 30 kV. Reprinted from [12], with permission.

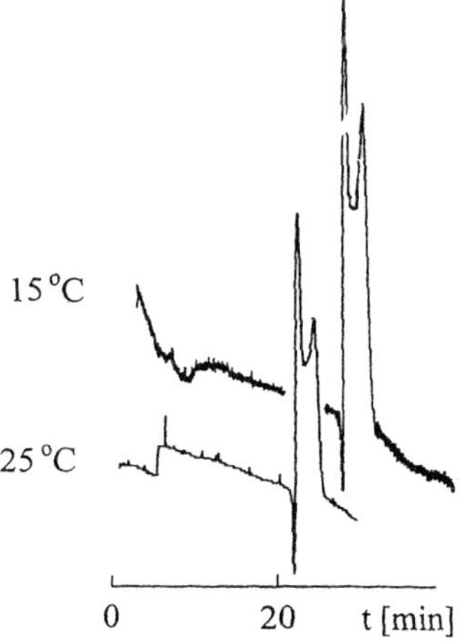

Figure 7. Enantiomer separation of oxazepam by o-CEC displaying enantiomerization with a 60 cm × 50 μm ID capillary derivatized with BSA. Conditions: 50 mM phosphate buffer, pH 8.0; 7.5 kV. Reprinted from [16], with permission.

for the enantiomer separation of flubufen and dansyl-amino acids by o-CEC [18]. In the pH range of 2.5–4.0 the ergolinic skeleton is positively charged (anodic EOF), thus the negatively charged analytes were mainly driven by the EOF and only partially by their anodic electrophoretic mobility. Migration of the enantiomers was affected by the concentration of the organic modifier (acetonitrile), the ionic strength, and the pH of the buffer.

2.4 Molecular imprinted polymers (MIPs)

MIPs were synthesized by mixing functional monomers in the presence of a print molecule. After polymerization, the covalently or noncovalently bonded print molecule was removed to yield a cross-linked polymer possessing a specific recognition cavity complementary to the template [21, 54]. The highly selective recognition properties of this polymer equal those of antibodies or receptors. The use of enantiomerically pure print molecules results in chiral MIPs which were suitable for enantiomer separation in CEC. An advantage of MIP is that the choice of a chiral selector is not based on the "trial and error" concept and that the elution order is predictable. The imprinted enantiomer is more strongly retained and it is therefore eluted last. Drawbacks of this technique are low efficiencies and limited chiral recognition properties. Consequently, only the enantiomers of the print molecule or of molecules with similar structures are separable. For o-CEC, a thin film of MIP was synthesized *in situ* in the capillary and covalently attached to the inner surface.

Tan and Remcho [20] reported on the *in situ* polymerization of methacrylic acid and 2-vinylpyridine (functional monomers), ethylene dimethacrylate, or trimethylol propane trimethacrylate (cross-linker), and L-dansyl-phenylalanine (print molecule) in a functionalized capillary. The polymerization was initiated thermally and vacuum was applied in order to remove the residual solvent and to shrink the polymer into a thin film on the capillary wall. Baseline resolution of the enantiomers of dansyl-phenylalanine was achieved. High efficiency was found for the nonimprinted D-enantiomer (248 600 theoretical plates/m) but only very low efficiency was achieved for the imprinted enantiomer (8000 theoretical plates/m) which eluted as an extremely broad peak (see Fig. 8). Brüggemann *et al.* [19] prepared MIP films by derivatizing the internal surface of the capillary in a first step with 3-methacryloxypropyltrimethoxysilane. In a second step, the capillary was filled with the functional monomer *trans*-3-(3-pyridyl)-acrylic acid, the cross-linker ethyleneglycoldimethacrylate or divinylbenzene, the initiator azo-bis-cyclohexanecarbonitrile and the print molecule *S*-2-phenylpropionic acid in a porogenic solvent. Unfortunately, a direct enantiomer separation of the print molecule is not possible because

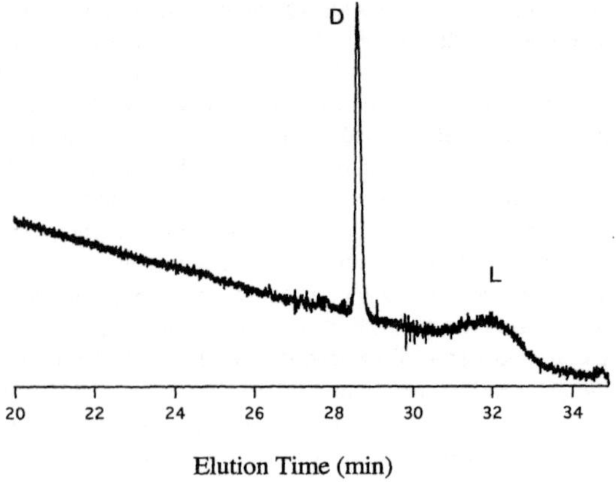

Elution Time (min)

Figure 8. Enantiomer separation of a mixture of D- and L-dansyl-phenylalanine by o-CEC on an MIP-coated capillary (85 cm × 25 µm ID). Conditions: acetonitrile/phosphate buffer (10 mM, pH 7) 10:1 (v/v), 30 kV. Reprinted from [20], with permission.

of the very strong interaction of the print molecule with the MIP resulting in a very broad peak which can not be distinguished from the baseline noise.

3 p-CEC

Capillaries used in p-CEC were filled with typical HPLC materials and the separation bed was fixed by frits (sintered silica). Detection occurs in the empty part of the column. An advantage of this method is that a great number of stationary phases with different selectivities are available. For enantiomer separation, any of the well-known CSPs such as cyclodextrins [23, 58], modified cyclodextrins [25, 27, 31], cellulose derivatives [32, 61–64], "Pirkle"-phases [26, 67], macrocyclic antibiotics [29, 33, 34], quinine-based anion-exchange-type CSP [28, 68–70], polyacrylamide derivatives [32], and MIPs [21, 54, 71–73] have been used. A serious problem in p-CEC is the formation of air bubbles in the capillary. This deleterious effect, mainly caused by differences in the EOF at the interface between the frit and the unpacked part of the capillary or by Joule heating, leads to baseline noise or even a breakdown of the current. To prevent air bubble formation, the separation can be performed under a slight overpressure (up to 12 bar) carried out by pressurizing the inlet and the outlet buffer vials. Alternatively, the flow system can be pressurized by coupling an HPLC pump to the inlet vial. The EOF is complemented by a pressure-driven flow so that this method represents a hybrid form of pure p-CEC and p-LC. This technique is called pressurized CEC, pressure-supported CEC, pressure-assisted

CEC (PEC), pseudoelectrochromatography (pEC), or electro-HPLC [55].

Most research groups used the common chiral HPLC materials as CSPs in p-CEC. To exploit the full potential of CEC, *de novo* designed stationary phases should be developed. In achiral CEC many attempts have been made to design new materials to increase the EOF and enhance selectivity and efficiency. The particle size has been reduced (3 µm and smaller), permanent charges have been introduced by using ion-exchange materials and mixed-mode phases, and nonporous materials or materials with wide pore size have been used [56]. To obtain highly efficient and selective CSPs specially designed for p-CEC, more research must be carried out in this area. A promising new approach – also in the field of chiral analysis – is the use of a nonaqueous mobile phase [57]. This method (abbreviated NAQ-CEC) is attractive if solubility or stability problems in aqueous media exist. Yet at present the understanding of chiral analysis in nonaqueous mobile phases is still incomplete. Enantiomer separation by p-CEC may be performed in two ways: (i) the chiral selector is covalently bound to the silica particle of the packing bed, or (ii) the chiral selector is added to the mobile phase of a capillary packed with silica or modified silica.

3.1 Saccharides

Cyclodextrin and cellulose derivatives are suitable as CSPs in p-CEC. While cyclodextrin derivatives are either covalently linked to the silica packing material or added to the mobile phase of a capillary packed with bare silica, octadecyl silica (ODS) or diol-silica, cellulose derivatives were always coated to silica.

3.1.1 Cyclodextrins and modified cyclodextrins

3.1.1.1 Cyclodextrins covalently linked to silica

Li and Lloyd [23] described the enantiomer separation of some neutral (hexobarbital, benzoin) and anionic (derivatized amino acids) compounds on a native β-cyclodextrin stationary phase. The EOFs in these packed capillaries were drastically influenced by the nature of the applied buffer system. Using triethylammonium acetate (TEAA) as a background electrolyte instead of phosphate buffer, the direction of the EOF was reversed. This effect was explained by adsorption of triethylammonium cations to the free silanol groups of the silica surface. The reversed EOF (anodic) is desirable for analysis of anions because of the additive effect of the electroosmotic and electrophoretic velocity. Acetonitrile or alcohols such as methanol or ethanol were used as organic modifiers, with superior

enantioselectivity being found for methanol. Increasing amounts of methanol lead to reduced retention factors but only slight changes in elution time. Enhancement of the TEAA concentration results in a decrease of the electroosmotic mobility, indicating saturation of the silanol groups with cations. These results were compared with those obtained in an open-tube system using β-cyclodextrin as a chiral buffer additive. The open-tubular capillaries show similar efficiencies and also a reversal of the direction of the EOF when using TEAA instead of phosphate buffer. Lelièvre *et al.* [25] used silica-linked hydroxypropyl-β-cyclodextrin as CSP for the enantiomer separation of chlorthalidon and mianserin. The influence of the organic modifier acetonitrile was studied in the range of 15–30%. With increasing acetonitrile concentration not only the separation factor α and resolution R_s decrease but the elution time also decreases.

Permethyl-β-cyclodextrins as chiral selector were used by Wistuba and Schurig [27, 31] in two approaches: (i) the permethylated β-cyclodextrin was covalently attached to silica *via* a thio ether spacer (Chira-Dex-silica; see Fig. 9) [27], or (ii) the permethylated β-cyclodextrin was bound *via* an octenyl spacer to dimethylpolysiloxane, which was subsequently immobilized to silica (Chirasil-Dex-silica; see also Fig. 2) [31]. To avoid bubble formation and to speed up the analysis, pressure-supported CEC was

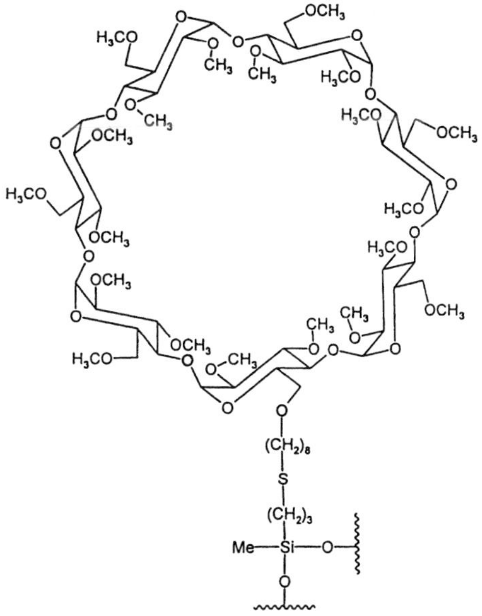

Figure 9. Structure of Chira-Dex-silica (dependent upon the reaction conditions, the spacer may be attached to the 2- or 6-position of cyclodextrin).

utilized for the enantiomer separation of barbiturates, benzoin, α-methyl-α-phenylsuccinimide, γ-phenyl-γ-butyrolactone, methylthiohydantoin (MTH)-proline, methyl mandelate, 1-(2-naphthyl)ethanol, glutethimide, mecoprop methyl, diclofop methyl, and fenoxaprop ethyl. The flow systems were usually pressurized with 10–30 bar by coupling an HPLC pump to the inlet vial. As shown in Fig. 11, the elution time in the enantiomer separation of mephobarbital is dominated by a substantial contribution of the EOF. Comparing p-LC at 10 bar with p-CEC at 10 bar and 20 kV using the same capillary, an approximately ten times longer elution time was found in the LC mode with Chira-Dex-silica and an approximately three times longer elution time in the case of Chirasil-Dex-silica at comparable chiral separation factors α and resolutions R_s. The connection of an HPLC pump to the electrophoretic systems allows switching between the LC and the CEC mode. Thus, a direct comparison of the two methods with one single capillary (unified approach) is feasible. The enantiomer separation of mephobarbital by p-CEC shows a higher efficiency and resolution at comparable chiral separation factor and elution time on both Chira-Dex and Chirasil-Dex columns (see Fig. 10).

The influence of nature and ionic strength of the buffer was investigated. An enhancement of the ionic strength of the phosphate buffer leads to a decrease of the linear flow velocity, the resolution, and the chiral separation factor. The use of the zwitterionic buffer 2-(*N*-morpholino) ethanesulfonic acid (MES) leads to a better stability of the baseline as compared to the phosphate buffer. Acetoni-

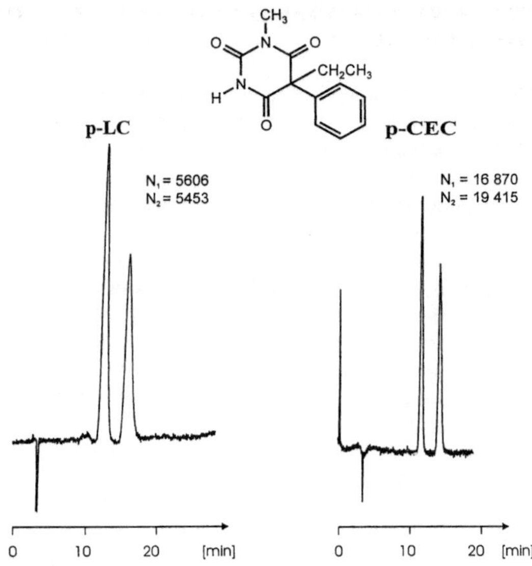

Figure 11. Enantiomer separation of mephobarbital by p-LC and p-CEC. Conditions: column, 23.5 cm × 100 µm ID capillary packed with Chira-Dex-silica; 5 mM phosphate buffer, pH 7.0/methanol 4:1 v/v. CEC: 20 kV, 10 bar; LC: 140 bar. Reprinted from [27], with permission.

trile and methanol were used as organic modifier, with the best enantioselectivity being found for methanol. With increasing amounts of organic modifier, the theoretical plate number *N* increases, while the resolution R_s, the chiral separation factor α, and the elution time decrease. For the elution of various substances on polymer-coated Chirasil-Dex-silica, higher concentrations of the organic modifier are required than with Chira-Dex-silica. The technique of polymer-coating seems to be unsuitable for preparing columns for CEC, because shielding of the free silanol groups by thermal cross-linking and by chemical reaction with residual Si-H groups in the dimethylpolysiloxane matrix remaining after synthesis of Chirasil-Dex reduces the EOF. Addition of bare silica to Chirasil-Dex-silica before packing of the capillary led to an enhancement of the EOF. Differences in selectivity between Chira-Dex-silica and Chirasil-Dex-silica were found in some cases, *e.g.*, the enantiomers of 1-(2-naphthyl)ethanol were resolved only on Chirasil-Dex-silica [31].

Zhu *et al.* [58] used capillaries packed with native β-cyclodextrin CSP for the enantiomer separation of benzoin and mephenytoin. A very short elution time (3.4 min) was found for mephenytoin with a resolution of R_s of 2.6. Recently, Zhang and El Rassi [59] designed a CSP deviced especially for CEC separations. To generate a strong EOF, they prepared a stationary phase containing a hydrophilic sulfonated sublayer to which a chiral top

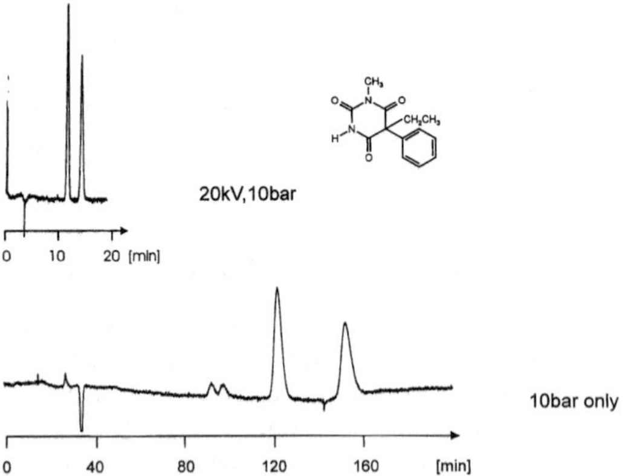

Figure 10. Influence of pressure support on the enantiomer separation of mephobarbital by p-CEC. Conditions: column, 23.5 cm × 100 µm ID capillary packed with Chira-Dex-silica; phosphate buffer (5 mM, pH 7.0)/methanol 4:1 v/v; upper electropherogram: 20 kV, 10 bar; lower chromatogram: 10 bar. Reprinted from [27], with permission.

layer of hydroxypropyl-β-cyclodextrin was immobilized. This allows a rapid enantiomer separation of anionic analytes such as dansyl-amino acids and phenoxy acid herbicides which migrate electrophoretically against the direction of the cathodic EOF. By varying the organic modifier content, the pH, and the ionic strength of the mobile phase, the enantioselectivity and the resolution can be optimized.

3.1.1.2 Cyclodextrins as chiral buffer additive

Capillaries with bare silica, ODS, or diol-silica stationary phase were used in chiral analysis by adding a chiral selector to the mobile phase. A drawback of this method is the consumption of the often very expensive chiral selector. However, with this technique it is possible to change the selector while using a single column. In a relatively short time, a variety of chiral selectors can be screened, thus omitting the time-consuming immobilization procedure. Lelièvre *et al.* [25] demonstrated this method by adding hydroxypropyl-β-cyclodextrin to the mobile phase of an ODS-packed column. Chlorthalidone enantiomers were resolved with resolutions between 0.7 and 1.4, depending on the concentration of the organic modifier acetonitrile. The main factors responsible for the enantiomer separation were described as follows: (i) differences in the stability constants of inclusion complexes formed between the enantiomers and the hydroxypropyl-β-cyclodextrin, (ii) differences in the partition of the inclusion complexes between the mobile and the stationary phase, and (iii) differences in the partition of the free enantiomers between the mobile phase and a CSP arising from the adsorption of hydroxypropyl-β-cyclodextrin. Thesis (iii) was supported by the long equilibration time necessary before an enantiomer separation takes place. Compared with the enantiomer separation of chlorthalidone on an immobilized hydroxypropyl-β-cyclodextrin stationary phase (see Section 3.1.1.1), a longer elution time but a higher resolution was found.

Wei *et al.* [65] also applied hydroxypropyl-β-cyclodextrin as chiral buffer additive but a bare silica instead of an ODS-packed column was used for the enantiomer separation of phenylephrine and synephrine. Probably, the chiral selector is adsorbed onto the silica support and may assist enantiomer separation. In order to optimize the separation, the influence of hydroxypropyl-β-cyclodextrin concentration, the pH of the mobile phase, the buffer concentration, and the separation temperature on the enantiomer separation of phenylephrine and synephrine have been investigated. A comparison between CEC and CZE, both with hydroxypropyl-β-cyclodextrin as mobile phase additive, was carried out. Longer elution times and lower resolutions were found in CZE. A similar approach was

demonstrated by Zhang and El Rassi [60]. They converted a diol-silica stationary phase *in situ* to a dynamically coated CSP by adding hydroxypropyl-β-cyclodextrin to the mobile phase. With this method, the enantiomer separation of neutral and anionic analytes such as organochlorine pesticides and dansyl-amino acids was realized. Neutral hydroxypropyl-β-cyclodextrin acts simultaneously as chiral selector and as "charge-to-mass ratio reducer" for the anionic analytes, thus speeding up their transport with the relatively weak cathodic EOF. The enantioselectivity can be manipulated by varying the concentration of the chiral selector, the nature and content of the organic modifier, the pH, and the ionic strength of the mobile phase.

Deng *et al.* [66] developed a theoretical model with regard to enantioselectivity and resolution for CEC and compared it with the experimental data obtained for the enantiomer separation of salsolinol using an ODS-packed column and β-cyclodextrin as mobile-phase additive. The theoretical model shows that the advantage of the combination of electrophoretic and partitioning mechanisms in CEC is the increase in enantioselectivity. For pressure-supported CEC, whereby the solvent is mainly driven by pressurized flow, the increased enantioselectivity was significantly reduced. Zhu *et al.* [58] used a capillary packed with ODS particles and dimethyl-β-cyclodextrin as buffer additive for the enantiomer separation of propranolol. The later-eluting enantiomer shows a very broad peak shape which was explained by a strong interaction between the positively charged propranolol enantiomer and the SiO⁻ groups on the silica surface.

3.1.2 Cellulose derivatives

Cellulose derivatives such as cellulose tris(3,5-dimethylphenylcarbamate), amylose tris(3,5-dimethylphenylcarbamate), and cellulose tris(4-methylbenzoate) (see Fig. 12) coated on silica were also suitable as CSPs in p-CEC [32, 61–64]. The group of Blaschke [32, 61, 62] coated wide-pore aminopropylsilica with the above-mentioned polysaccharides and separated enantiomers in aqueous as well as nonaqueous media. The enantiomers of indapamide were resolved under aqueous conditions (citrate buffer/acetonitrile) on the cellulose tris(3,5-dimethylphenylcarbamate) stationary phase [32]. Enantiomer separation of thalidomide and its hydroxylated metabolites was performed on a mixed phase consisting of cellulose-tris(3,5-dimethylphenylcarbamate) and amylose-tris(3,5-dimethylcarbamate) with nonaqueous mobile phase [61]. An extensive study on enantiomer separation in NAQ-CEC using tris(3,5-dimethylphenylcarbamate), amylose tris(3,5-dimethylphenylcarbamate), and cellulose tris(4-methylbenzoate) stationary phase was carried out by Girod *et al.* [62]. They resolved the enantiomers of *trans-*

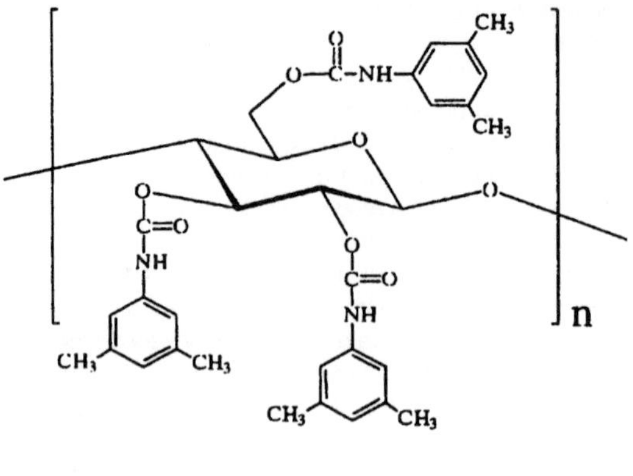

Figure 12. Structures of polysaccharide-type CSPs. (a) Cellulose tris(3,5-dimethylphenylcarbamate) (Chiracel OD); (b) amy-lose tris(3,5-dimethylphenylcarbamate) (Chiralpak AD); (c) cellulose tris(4-methylbenzoate) (Chiralcel OJ). Reprinted from [62], with permission.

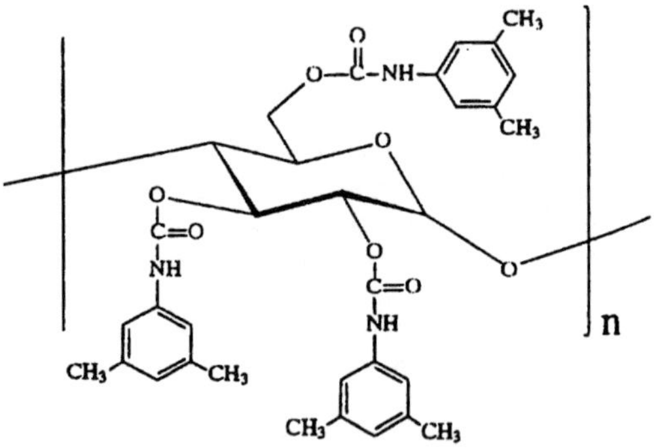

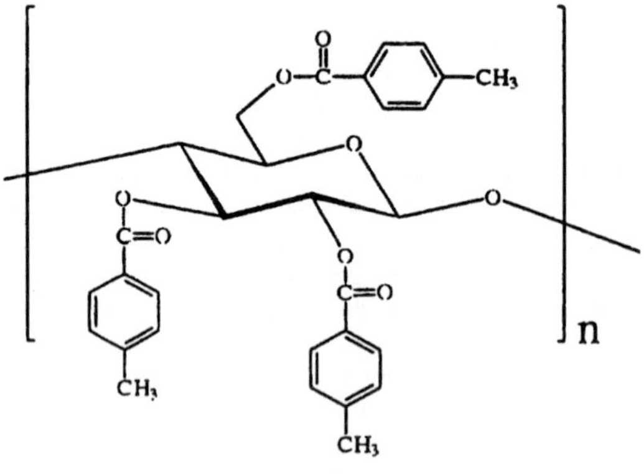

stilbene oxide, econazole, 2,2′-diamino-6,6′-dimethylbiphenyl, glutethimide, aminoglutethimide, piprozolin, etozolin, Troeger's base, indapamide, piprozolin, and metomidate. Methanolic or ethanolic ammonium acetate solutions were used as mobile phase. The anodic EOF decreased drastically when changing from methanol to ethanol and thus the elution time increased (see Fig. 13). But at comparable linear flow velocities, higher enantioselectivities α and thus higher resolutions R_s occur in ethanolic solution. The EOF depended strongly upon the pH*. Increasing EOF was observed in the pH* range of 3.0–5.0. At higher pH* values the EOF decreases. This effect is not fully understood to date. Comparing capillary LC with p-CEC, a moderate improvement in efficiency was observed in the p-CEC mode.

Mayer *et al.* [63] described the enantiomer separation of lorazepam, α-1-hydroxyethyl-naphthalene, benzoin, indapamide, and *trans*-stilbene oxide on cellulose tris(3,5-dimethylphenylcarbamate) immobilized onto macroporous silica gel under aqueous conditions. For lorazepam, a plateau between the peaks of enantiomers, typical for an enantiomerization process, was found. Changes in buffer concentration, amount of the modifier acetonitrile, and the pH of the buffer influence the EOF and the resolution R_s. An enhancement of the resolution can be achieved by decreasing the acetonitrile content. A low acetonitrile concentration, however, leads to an increased elution time. The buffer concentration has a great influence on the EOF. High buffer concentration effected high linear electroosmotic velocities and low resolutions. Comparison of the efficiency and the resolution at similar mobile phase velocity in the nano-LC and the p-CEC mode clearly shows the superiority of the electro-driven method. Recently, Otsuka *et al.* [64] also reported on the enantiomer separation on cellulose tris(3,5-dimethylphenylcarbamate) CSP. Several racemic substrates such as acidic, neutral, and basic drugs were resolved into their enantiomers. As shown in Fig. 14, the enantiomer separation of pindolol is very fast (8 min) and shows a high chiral separation factor ($\alpha = 2.32$).

3.2 Proteins and glycoproteins

Proteins and glycoproteins often possess a strong detector response. This effect can lead to problems when using these compounds as chiral buffer additives in CZE or as a coating layer in o-CEC. Another disadvantage in CZE is the high adsorption affinity of proteins or glycoproteins to the internal capillary surface which results in instability of the baseline, peak tailing, and changes in elution time. To overcome these problems, p-CEC with proteins immobilized on silica can be used. Li and Loyd used both acidic α-glycoprotein (AGP) [22] and human serum albumin

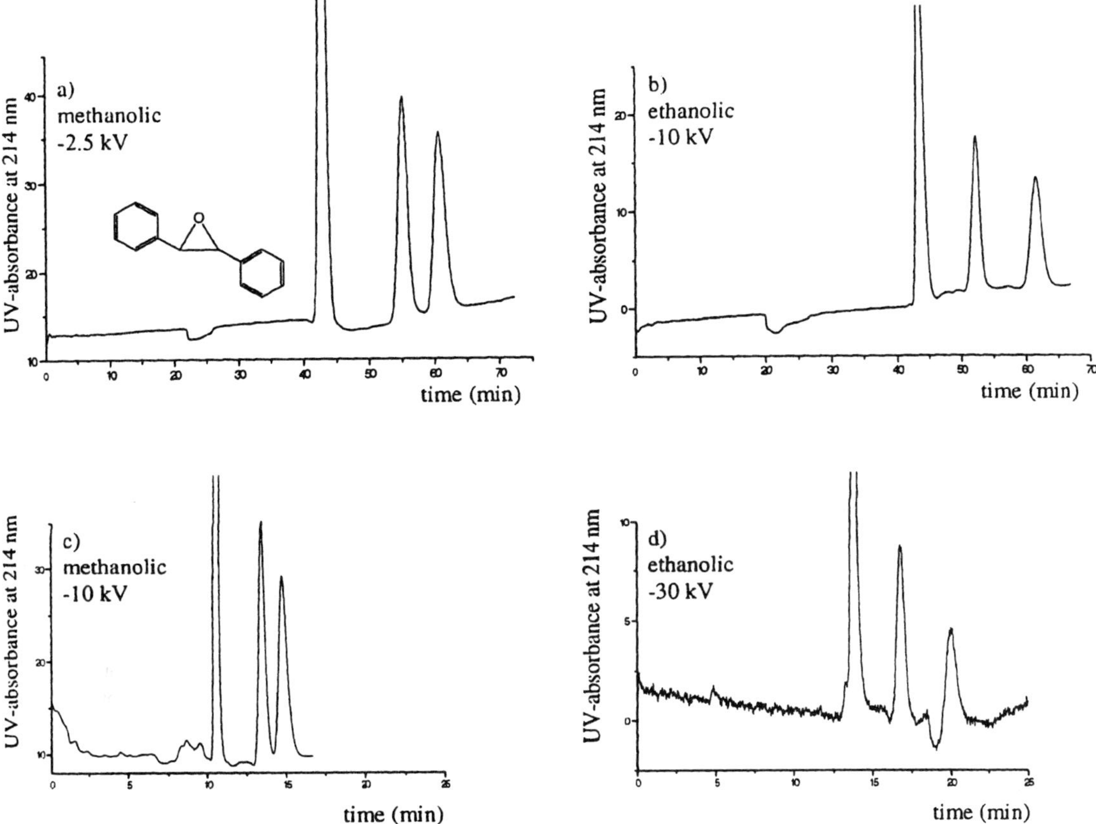

Figure 13. Enantiomer separation of *trans*-stilbene oxide in (a, c) methanolic and (b, d) ethanolic ammonium acetate solution. Conditions: column, 22 cm × 100 μm ID capillary with amylose tris(3,5-dimethylphenylcarbamate) stationary phase; 10 mM methanolic or ethanolic ammonium acetate solution, pH* 7.7. Reprinted from [62], with permission.

(HSA) [24] stationary phases for enantiomer separation by p-CEC. AGP attached to silica was packed in capillaries of 50 μm ID and the enantiomers of benzoin, hexobarbital, pentobarbital, ifosfamide, cyclophosphamide, disopyramide, metoprolol, oxprenolol, and propranolol were separated. Compared with HPLC, the efficiencies observed in the p-CEC mode were generally higher but not as high as those in CZE. In the pH range normally used, AGP was negatively charged (isoelectric point: 2.7), resulting in a cathodic EOF. Methanol, ethanol, 1-propanol, 2-propanol and acetonitrile can be utilized as organic modifiers. The best resolution was obtained with 2-propanol. The enantioselectivity was influenced by changing the amount of organic modifier, but not through variation of the buffer concentration. Columns packed with HSA immobilized on silica support were used to separate the enantiomers of benzoin and temazepam [24]. The best enantioselectivity was observed with 2-propanol or aceto-

nitrile as organic modifier. Efficiencies obtained for benzoin and temazepam are very low (15 000 plates/m and 7000 plates/m, respectively) and comparable with those obtained by HPLC. For both protein stationary phases (AGP, HSA), broad and tailing peaks with relative low column efficiency were observed. Better results may be expected if the relatively large particle size (AGP, 5 μm; HSA, 7 μm) is reduced.

3.3 "Pirkle" phases

The feasibility of enantiomer separation of more than 30 neutral analytes was demonstrated by Wolf *et al.* [26, 67] on *S*-naproxen-derived and 3*R*,4*S*-Whelk-*O* CSPs (see Fig. 15) bonded to silica particles (3 μm). The influence of several buffer systems (MES, acetate, phosphate), different amounts of the organic modifiers methanol and acetonitrile, temperature, applied voltage, and pH on chromato-

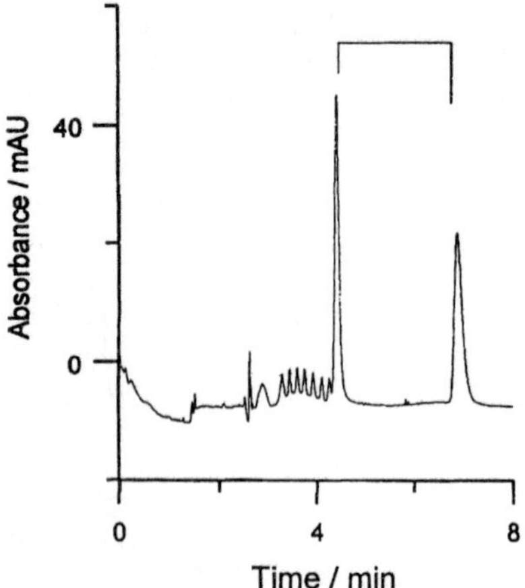

Figure 14. Enantiomer separation of pindolol by p-CEC. Conditions: column, 24 cm × 100 µm ID capillary with cellulose tris(3,5-dimethylphenylcarbamate) stationary phase; 10 mM Na$_2$HPO$_4$/acetonitrile 1:1 v/v, 25 kV, 8 bar (both inlet and outlet vials). Reprinted from [64], with permission.

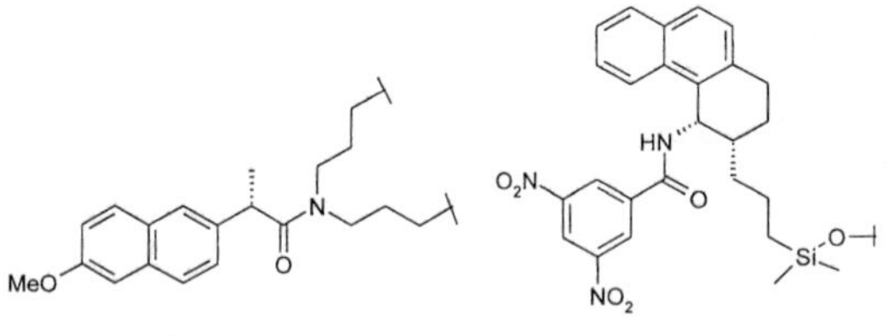

Figure 15. Structure of *S*-naproxen-derived CSP and 3*R*,4*S*-Whelk-O CSP.

graphic performance was studied. Short analysis times (often lower than 10 min) combined with very high enantioselectivities α and high efficiencies were observed, and excellent resolution factors of R_s up to 48 were achieved [67]. The zwitterionic buffer MES was the preferred mobile phase, providing slightly higher enantioselectivities, increased retention factors, and higher resolutions but lower efficiencies than phosphate buffer. The organic modifiers strongly influence the enantioselectivity and efficiency, whereby methanol induced better enantioselectivities and resolutions but longer elution times and lower efficiencies than does acetonitrile. A typical electropherogram is shown in Fig. 16.

3.4 Macrocyclic antibiotics

Macrocyclic antibiotics, *e.g.*, vancomycin and teicoplanin (see Fig. 17), proved to be very successful as stationary phase in HPLC. Using them in p-CEC, Dermaux *et al.*

[29] demonstrated the first enantiomer separation on the macrocyclic antibiotic vancomycin bonded to silica. The enantiomers of warfarin and hexobarbital were separated with about 30% higher efficiency than with HPLC. The organic modifier acetonitrile can be used up to an amount of 30%. With increasing acetonitrile content, the elution time decreased drastically but so did the enantioselectivity. Wikström *et al.* [34] also used a vancomycin CSP for p-CEC. Vancomycin is covalently attached to a diol silica-packed column in an *in situ* immobilization and enan-

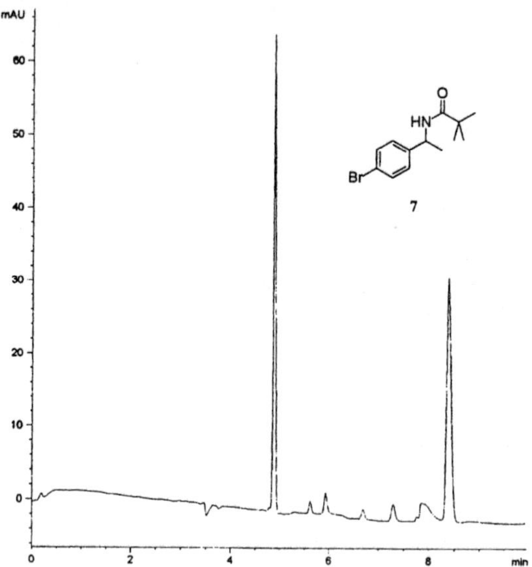

Figure 16. Enantiomer separation of *N*-[1-(4-bromo-phenyl]-2,2-dimethyl-propionamide on 3*R*,4*S*-Whelk-O CSP in p-CEC. Conditions: column, 30.3 cm × 100 µm ID; 25 mM MES buffer, pH 6.0/acetonitrile 1:3.5 v/v, 25 kV. Reprinted from [26], with permission.

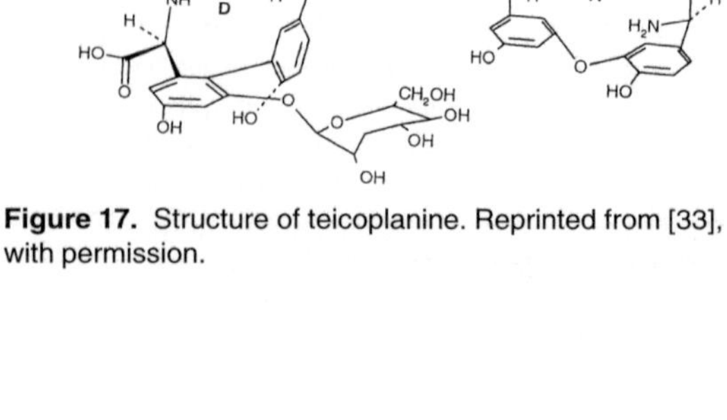

Figure 17. Structure of teicoplanine. Reprinted from [33], with permission.

tiomer separation was performed in both the reversed-phase and polar organic mode. The enantiomers of thalidomide (Fig. 18) were separated in the reversed-phase CEC mode (acetonitrile/0.1% TEAA at pH 6.5, 30:70 v/v)

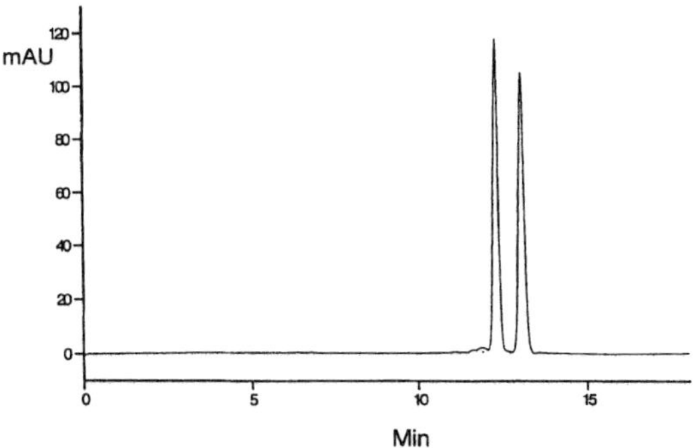

Figure 18. Enantiomer separation of thalidomide on a vancomycin CSP in the polar organic mode. Conditions: column, 35.5 cm × 75 µm ID; methanol/acetonitrile/acetic acid/triethylamine 80:20:0.2:0.2 v/v/v/v, 20 kV, 10 bar. Reprinted from [34], with permission.

with a resolution of 2.5 and 80 000 plates/m and in the polar organic CEC mode (methanol/acetonitrile/acetic acid/triethylamine, 80:20:0.2:0.2 v/v/v/v), also with a resolution of 2.5 but with 120 000 plates/m. Basic analytes including racemic β-adrenergic blocking agents were resolved in their enantiomers under polar organic conditions. Screening of a number of neutral, acidic and basic analytes in the reversed-phase mode led to only minor success. In the enantiomer separation of tryptophan and dinitrobenzoyl-leucine, carried out by Carter-Finch and Smith [33] on a teicoplanin stationary phase by p-CEC, short elution times (less than 6 min) and resolutions up to 2.39 were found. Increased enantioselectivity but a longer elution time were observed for the organic modifiers methanol and ethanol when compared to acetonitrile. The influence of temperature on the resolution was investigated.

3.5 Anion-exchange-type stationary phases

Lämmerhofer and Lindner [28, 68–70] described the use of weak anion-exchanger-type selectors based on chiral quinine carbamate (see Fig. 19) either added as chiral

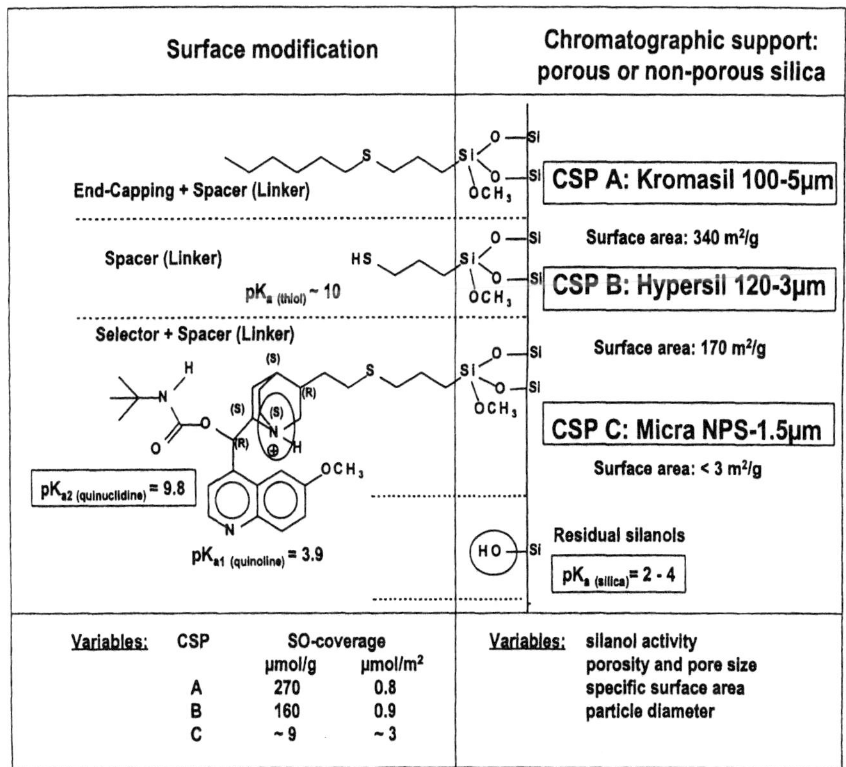

Figure 19. Structure of quinine carbamate-derived CSPs (WAX-type CSPs). Reprinted from [70], with permission.

selector to the buffer system of ODS packed capillaries [68] or linked covalently to silica supports (porous: 3 or 5 µm; nonporous: 1.5 µm) [28, 69, 70]. A number of derivatized amino acids were resolved into their enantiomers under aqueous [28, 68, 70] or nonaqueous conditions [69]. The *tert.*-butyl carbamoyl quinine selector-modified silica surface has a zwitterionic character due to the charge of the positively chargeable chiral selector ($pK_{a1(quinoline)}$ = 3.9; $pK_{a2(quinuclidine)}$ = 9.8) and the silanol groups (see Fig. 19). At pH values below the isoelectric point of the modified surface, the selector is positively charged, whereas the silanol groups are negatively charged, resulting in an anodic EOF; pH values above the isoelectric point lead to a cathodic EOF. A negatively charged analyte can be moved by the EOF and/or the electrophoretic forces which can either be directed with or against the EOF. According to the mobile phase conditions selected (organic modifier, pH) and the different stationary phases (amount of selector loading) used, three different modes have been observed: (i) coelectrophoretic elution of the negatively charged solutes with the anodic EOF in the negative polarity mode, (ii) counterelectrophoretic elution with the cathodic EOF in the positive polarity mode, and (iii) electrophoretically dominated elution in the negative polarity mode with a cathodic EOF (see Fig. 20). The feasibility of enantiomer separation in all three modes was realized [70]. Primarily, an ion-pairing interaction between positively charged chiral selector and negatively charged analyte takes place. Typical organic modifiers were acetonitrile and methanol. The buffer concentration influences the elution time (increasing buffer concentration causes a decrease of elution time) but essentially not the enantioselectivity [28]. In the CEC mode, the same

high enantioselectivity as in pressure-driven methods could be achieved while the theoretical plate numbers (up to 120 000 per/m) were about two to three times higher than in the HPLC mode [28].

Under nonaqueous conditions the use of high buffer concentrations and the application of high electric fields to obtain short elution times is feasible [69]. For example, the enantiomer separation of Fmoc-Leu could be achieved in less than 10 min with a resolution of 6.9 and about 100 000 theoretical plates/m. The mobile phase consists of methanol, acetonitrile, or a mixture of both, and a background electrolyte of acetic acid and triethylamine. The influence of electrolyte concentration, acetonitrile-methanol ratio, and temperature was investigated. The same group reported the use of a quinine-derived chiral selector as a buffer additive for enantiomer separation of derivatized amino acids by enantioselective ion-pair formation with a capillary packed with ODS particle [68]. The separations could be performed under aqueous or nonaqueous conditions. The enantioselectivity, elution time, and elution order can be controlled by two different modes: (i) electrophoretically dominated mode at high background electrolyte concentration leads to high efficiency (two to three times higher than in LC) but only moderate enantioselectivity (about the same as in LC); detection occurs with negative electric polarity; and (ii) electroosmotically dominated mode at low electrolyte concentration, whereby the chiral ion-pair agent itself served as background electrolyte and the electrophoretic mobility of the anionic analyte is overcompensated by the cathodic eluent mobility; detection occurs with positive electric polarity. A higher enantioselectivity, but much lower effi-

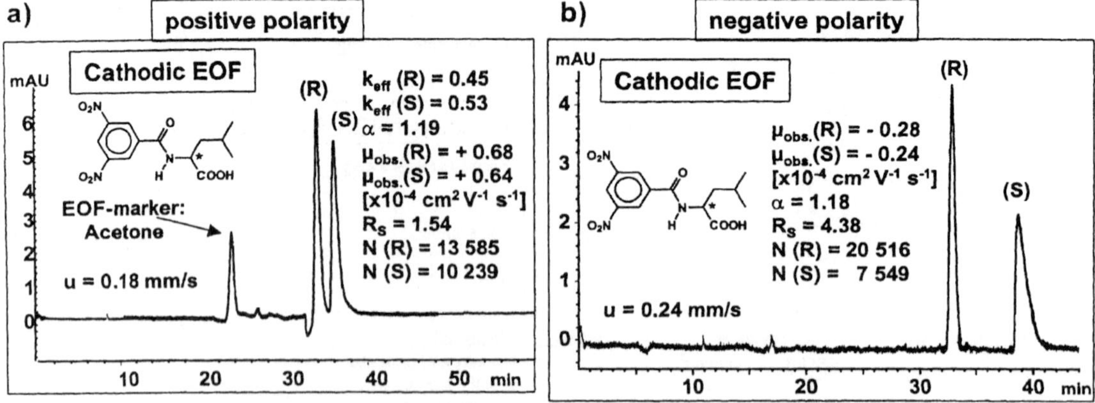

Figure 20. Enantiomer separation of DNB-leucine on WAX-type CSP C. (a) Electroosmotically *vs.* (b) electrophoretically dominated elution mode. Conditions: column, 25 cm × 100 µm ID; organic modifier/0.1 M MES, pH 6.0, 80:20 v/v (adjusted with triethylamine). (a) Organic modifier, acetonitrile; 6 kV; (b) organic modifier, methanol; −15 kV. Reprinted from [70], with permission.

ciency (even inferior to LC) was found. An inversion of the elution order was observed when comparing the electrophoretically with the electroosmotically dominated mode. Under nonaqueous condition, the concentration of the chiral ion-pair agent could be enhanced.

3.6 Chiral polymers

CSPs based on poly-*N*-acryloyl-L-phenylalanine ethyl ester (Chiraspher) covalently attached to silica [32] or helically chiral poly(diphenyl-2-pyridylmethyl methacrylate) coated to aminopropyl-silica [30] (see Fig. 21) were successfully used for enantiomer separation of a number of substances such as bendroflumethiazide [32], Tröger's base, benzoin acetate, methylbenzoin, and *trans*-stilbene oxide [30]. The enantiomer separation of bendroflumethiazide on Chiraspher was accomplished with both organic modifiers, methanol and acetonitrile, whereby a higher enantioselectivity was observed with acetonitrile [32]. Enantiomer separation on helically chiral poly(diphenyl-2-pyridylmethyl methacrylate) coated to aminopropyl-silica was performed under nonaqueous conditions (ammonium acetate in methanol). An anodic EOF caused by the positive charge of the incompletely coated aminopropyl-silica and of the pyridyl group was observed. The feasibility of enantiomer separation was demonstrated with both pressure-supported as well as pure CEC [30].

3.7 MIPs

The functionality of MIPs and their application in o-CEC was described in Section 2.4. Capillaries packed with MIPs were also used for enantiomer separation by p-CEC

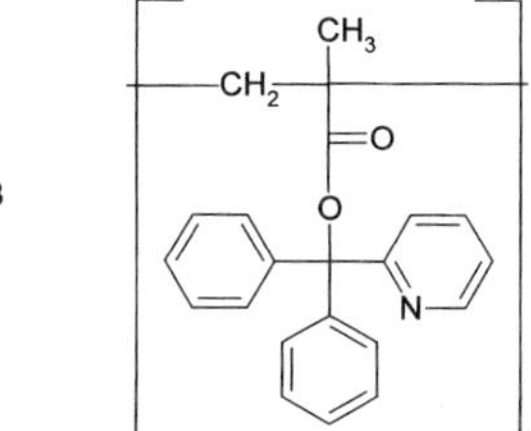

Figure 21. Structure of (A) poly-*N*-acryloyl-*L*-phenylalanine ethyl ester (Chiraspher) and (B) poly(diphenyl-2-pyridylmethyl methacrylate (PDPM).

[21, 54, 71–73]. The capillaries for p-CEC were prepared by synthesizing the MIP in bulk, followed by grinding and sieving the particles ($< 10\ \mu$m) which were then slurry-packed by the conventional HPLC packing procedure. Drawbacks of this method are the necessity of retaining frits and the fact that the particles obtained by grinding and sieving are highly irregular in shape and have a relatively wide size distribution [54]. The feasibility of enantiomer separation of several derivatized amino acids was demonstrated by Lin *et al.* [71–73]. They found that, with dansyl-L-leucine as print molecule, methacrylic acid and/or 2-vinylpyridine as functional monomers, ethylene glycol dimethacrylate as cross-linking monomer, and azo-isobutyronitrile (AIBN) as reaction initiator, only the enantiomer separation of dansyl-DL-leucine ($R_s = 1.20$) was satisfactory [71]. Using L-phenylalanine or L-phenylalanine anilide as print molecule, the enantiomer separation of a number of aromatic amino acids was feasible [72, 73]. Comparing the results with those achieved in HPLC, a higher resolution combined with a better peak shape was observed. The influence of the water content of the mobile phase on the enantiomer separation was investigated, maximum enantioselectivity being obtained with a water content of about 5%. A typical mobile phase consists of 90% acetonitrile, 5% acetic acid, and 5% water. The influence of the column temperature on the separation factor and the resolution was also studied [73].

4 rod-CEC

In the past few years, a multiplicity of different monolithic columns for HPLC and CEC have been described. This requires a definition of the term "monolithic stationary phase". Gusev *et al.* [74] offers the following definition: "A monolithic stationary phase is a continuous unitary porous structure prepared by *in situ* polymerization or consolidation inside the column tubing and, if necessary, the surface is functionalized to convert it into a sorbent with the desired chromatographic binding properties." Monolithic capillaries represent an alternative to packed capillaries for CEC separations, offering several advantages: retaining frits, the cause of many problems, are not necessary; movement of charged separation media in an electrical field is not possible; the columns are highly robust and stable; long capillaries can be prepared without packing problems; and preparation is often easier than packing small particles into a narrow capillary. The main problem in preparing monolithic columns is to find conditions for the polymerization that lead to channels with good flow-through properties. The monolithic stationary phases can be classified into two types: (i) organic polymer-based monoliths prepared by polymerization of organic monomers in presence of a porogen [75, 76], (ii) inorganic-based monoliths [77] prepared by a sol-gel process [78–81] by entrapping particles in inorganic gels [47, 82–84] or sintering [85–88] silica beds. Numerous papers have demonstrated the successful use of monolithic columns in the area of achiral separations. The research activity in the field of rod-CEC for chiral analysis, which requires highly efficient and selective separation media, has expanded in the past few years.

4.1 Organic porous polymer continuous bed

Organic polymer continuous beds for enantiomer separation are prepared either by (i) polymerization of organic monomers in the capillary *in situ* in the presence of a chiral print molecule (molecular imprinting technique) [21, 35–40], or by (ii) direct copolymerization of a chiral selector or a monomer bearing a chiral selector [41–43, 89–91]. Note, that the first generation of chiral monolithic columns was prepared by the first method (i), while research activities in the field of the second method have expanded rapidly in the last year.

4.1.1 MIP monoliths

MIPs can be used as the stationary phase in all three approaches discussed here: (i) o-CEC (see Section 2.4), (ii) p-CEC (see Section 3.7), and (iii) rod-CEC. Imprinted monolithic columns were prepared by polymerization of a mixture consisted of the imprint molecule, the initiator (*e.g.*, AIBN), the functional monomer methacrylic acid and/or 2-vinylpyridine, and the cross-linker monomer (*e.g.*, trimethylolpropane trimethacrylate) in a capillary *in situ*. The polymerization process was initiated by heating or irradiation. To attach the MIPs covalently to the internal surface of the capillary, a pretreatment of the capillary, *e.g.*, with [(methacryloxy)propyl]trimethoxysilane, is necessary. Schweitz *et al.* [39] reported on the first enantiomer separation with monolithic MIP capillaries by CEC in 1997. The polymerization reaction was carried out at –20°C under a UV source. To achieve good flow-through properties through the large pores (1–20 µm diameter) of the aggregates of µm-sized particles, the polymerization process was interrupted before completion by removing the UV source. The resulting pores are large enough to allow direct usage of the material as a monolithic phase in CEC. It need not be crushed and sieved as in the pioneering days of MIP technology. Using *R*-propranolol or *S*-metoprolol as print molecules, the enantiomers of propranolol or metoprolol, respectively, may be separated. Note that the enantiomers of propranolol were completely resolved in less than 2 min. The enantiomer with the same absolute configuration as the imprint molecule is always eluted last.

The same group reported on the use of *S*-ropivacaine as print molecule and separated the enantiomers of ropivacaine and the structural analogs mepivacaine and bupivacaine [37]. To obtain a superporous monolith, 1–25% iso-octane in toluene was used as a porogenic agent. The composition and pH of the electrolyte influence the CEC separations. With increasing pH the selectivity increased but the efficiency decreased. A high amount of acetonitrile (*e.g.*, 80%) was used as organic modifier. Higher temperatures (60°C) were advantageous for enantiomer separation, leading to shorter elution times, higher resolutions and higher plate numbers. In 1999, Schweitz *et al.* [36] reported on the enantiomer separation of several structural analogs of propranolol (pindolol, prenalterol, atenolol) on a MIP-based column prepared with *R*-propranolol as imprinting molecule (see Fig. 22). Lin *et al.* [35] used L-phenylalanine anilide as print molecule in a thermally induced *in situ* polymerization reaction and separated racemic phenylalanine on the resulting stationary phase.

4.1.2 Monoliths with covalently attached chiral selectors

Peters *et al.* [41] described a chiral monolithic stationary phase for CEC produced by direct copolymerization of the chiral monomer 2-hydroxyethyl methacrylate (*N*-L-valine-3,5-dimethylanilide) carbamate with ethylene di-

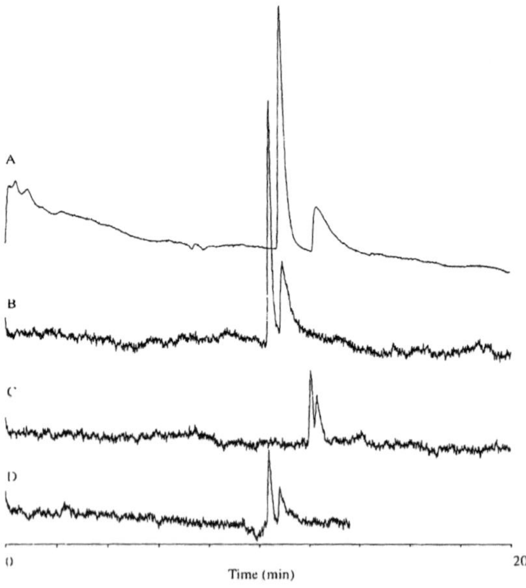

Figure 22. Enantiomer separation of (A) propranolol, (B) pindolol, (C) prenalterol, and (D) atenolol on an MIP-based monolithic column (90 cm × 75 µm ID) by rod-CEC (imprinting molecule, *R*-propranolol). Reprinted from [36], with permission.

methacrylate, 2-acrylamido-2-methyl-1-propanesulfonic acid, and butyl or glycidyl methacrylate in the presence of a porogenic solvent in the capillary. The enantiomers of *N*-(3,5-dinitrobenzoyl)leucine diallylamide were separated with a resolution of 2.0 and a theoretical plate number/m of 61 000. The hydrophilicity of the stationary phase strongly influences the enantioselectivity and efficiency of the separation. An enhanced hydrophilicity could be achieved by hydrolyzing the epoxide group of the glycidyl methacrylate moiety. The copolymerization of a quinidine-functionalized monomer [42] together with a co-monomer (hydroxyethyl methacrylate) and a cross-linker (ethylene dimethacylate) in the presence of a porogenic solvent resulted in a chiral monolith suitable for enantiomer separation of *N*-3,5-dinitrobenzoyloxycarbamyl leucine (DNZ-leucine; see Fig. 23) and *N*-2,4-dinitrophenyl valine (DNP-valine). High resolution factors (up to 7.13) and theoretical plate numbers (up to 242 400) were obtained. The quinine selector bears charges required for generating an EOF (the addition of a charged co-monomer to the poymerization mixture is not necessary) and at the same time undergoes enantioselective interaction with the analyte. Schmid *et al.* [43] separated the enantiomers of underivatized amino acid by ligand-exchange CEC using a continuous bed prepared by *in situ* polymerization of methacrylamide, piperazine diacrylamide, vinylsulfonic acid and *N*-(2-hydroxy-3-alloxypropyl)-L-4-hydroproline. The enantiomer separation of phenylalanine was performed by the following three methods: (i) pure rod-CEC, (ii) pressure-supported rod-CEC, and (iii) nano-HPLC using a single column. While pure rod-CEC shows the highest resolution, the shortest elution time was obtained with pressure-supported CEC.

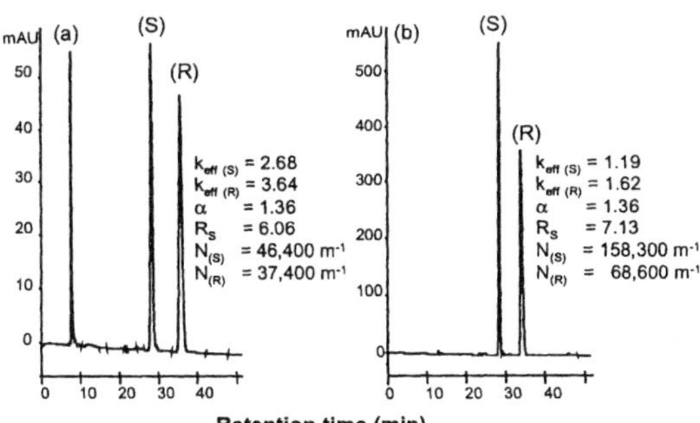

Figure 23. Enantiomer separation of dinitrobenzoyloxycarbamyl (DNZ)-leucine on a quinidine-functionalized chiral monolith. Conditions: column, 25 cm × 100 µm ID; 0.4 M acetic acid and 4 mM triethylamine in 80:20 mixtures of (a) acetonitrile/methanol and (b) acetonitrile/water; − 15 kV, 50°C. Reprinted from [42].

4.1.3 Homogeneous gels

Homogeneous polyacrylamide-based gels prepared *in situ* in the capillaries are quite different from continuous beds, also called continuous polymer rods, silica rods, or monoliths. Homogeneous gels are visually transparent or only slightly opalescent separation media. Continuous beds, on the other hand, consist of small particles which are covalently linked and attached to the internal capillary wall [45]. Both kinds of monoliths need no frits to keep the separation bed in place. Charged polyacrylamide-based gels generate an EOF and can be used as electrochromatographic monolithic separation media. Negatively or positively charged gels bearing covalently attached or incorporated chiral selectors could be used as a stationary phase for enantiomer separation by CEC [44, 45, 89, 90]. Koide and Ueno [44, 90] used charged polyacrylamide gels with either covalently bonded allyl carbamoylated β-cyclodextrin derivatives or incorporated polymeric β-cyclodextrins [89]. Incorporation of monomeric β-cyclodextrin was unsuccessful because the cyclodextrin was eluted from the capillary by EOF [89]. The β-cyclodextrins were covalently attached either to a negatively [90] or a positively [44] charged polyacrylamide gel (see Fig. 24) prepared by adding acrylamido-2-methylpropanesulfonic acid or *N*-(2-acrylamidoethyl)triethylammonium iodide to the polymerization mixture. With the negatively charged cyclodextrin-modified gel, the enantiomer separation of neutral (benzoin) and cationic (terbutaline, propranolol) compounds is feasible.

Capillaries filled with positively charged polyacrylamide gels generating a reversed EOF were used for the enantiomer separation of acidic *i.e.*, dansyl-amino acids (see Fig. 25), phenylmercapturic acid, warfarin, 2-phenoxypropionic acid, *N*-FMOC-valine, and neutral *i.e.*, benzoin, 1-(1-naphthalene)ethanol) compounds. High efficiencies

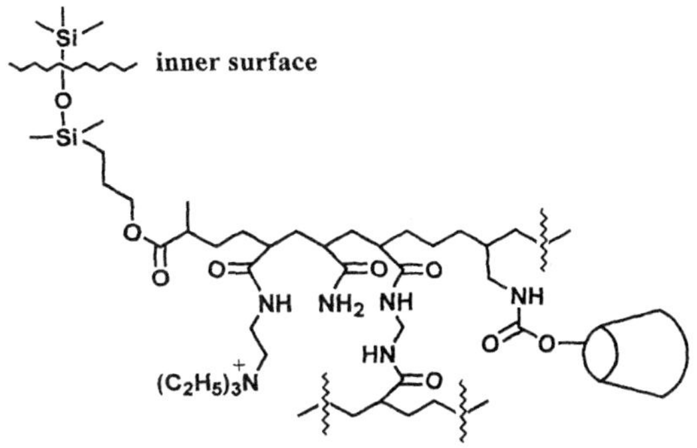

Figure 24. Schematic structure of allyl carbamoylated β-cyclodextrin-bonded positively charged polyacrylamide gels. Reprinted from [44], with permission.

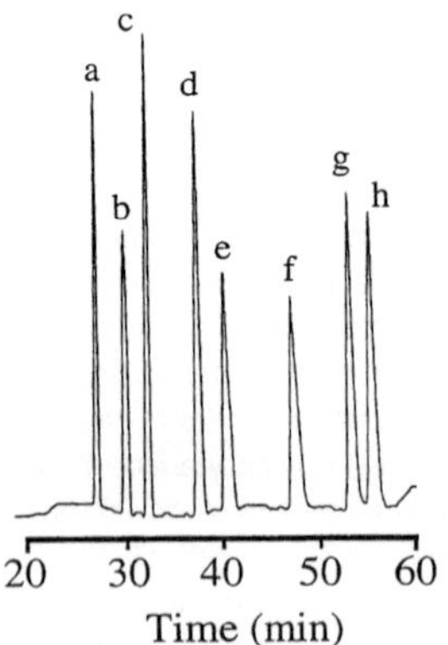

Figure 25. Enantiomer separation of four dansyl-amino acids: (a), (b) dansyl-serine; (c), (d) dansyl-valine; (e), (f) dansyl-leucine; (g), (h) dansyl-tryptophan with an allyl carbamoylated β-cyclodextrin-bonded positively charged polyacylamide gel-filled capillary. Conditions: column, 35 cm × 75 µm ID; 300 mM Tris/boric acid, pH 8.1; 10 kV. Reprinted from [44], with permission.

(up to 150 000 plates/m) and resolution factors of up to 7.1 could be observed. Typically, Tris-boric acid buffers with different pH values and concentrations but without an organic modifier were used as the mobile phase. The gel-filled capillaries were stable and had a satisfactory within-run and between-run reproducibility. Recently, Végvári *et al.* [45] described a similar chirally charged gel for CEC prepared by copolymerization with 2-hydroxy-3-alloxy-propyl-β-cyclodextrin. A number of chiral acidic, neutral and basic drugs (hexobarbital, mephobarbital, warfarin, tropicamid, ibuprofen, propranolol, mephenytoin, hydrobenzoin) could be fully or partially resolved. Enantioseparations on both anionic or cationic gels were performed with high resolution factors (up to 8.18) and theoretical plate numbers (up to 570 000). The resolution is independent of the electroendosmotic velocity, as is the theoretical plate number of the first eluted enantiomer, whereas with increasing velocity the theoretical plate number of the second eluted enantiomer increases. Enantiomer separation with neutral cyclodextrin-containing polyacrylamide gels by capillary gel electrophoresis (CGE) has already been described by Guttman *et al.* [91], Cruzado and Vigh [92] and Lin *et al.* [93].

4.2 Inorganic porous continuous bed

Chiral silica-based monoliths have been prepared either (i) by embedding particles of chiral chromatographic packing materials in a silica network [47, 48], or (ii) by fusion of porous silica packing material in a capillary by way of *in situ* sintering and subsequent modification with

a chiral selector [46]. Chirica and Remcho [47] entrapped the particles of an L-dansyl-phenylalanine imprinted polymer stationary phase in silicate by flushing a packed capillary with potassium silicate in water and subsequent heating from 40 to 160°C over a period of several days. The feasibility of enantiomer separation of dansyl-phenylalanine was demonstrated. Recently, Kato *et al.* [48] modified silica particles (5 µm) with either *S-N*-3,5-dinitrobenzoyl-1-naphthylglycine (phase 1) or *S-N*-3,5-dinitrophenylaminocarbonyl-valine (phase 2) and embedded them in a sol-gel matrix (see Fig. 26). In order to stabilize the EOF, bare silica was added to the sol-gel solution consisting of tetraethoxysilane, ethanol, hydrochloric acid, and the chiral particles. The enantiomers of thirteen amino acids and three nonprotein amino acids, derivatized with the fluorogenic reagent 4-fluoro-7-nitro-2,1,3-benzoxadiazole (NBD) were separated with resolution factors up to 4.45 for phase 1 and 1.17 for phase 2 (see Fig. 27). Phase 1 was found to be superior to phase 2 because of its higher resolution and lower plate height (about two times lower compared to HPLC). The effect of the organic modifier concentration was illustrated for the enantiomer separation of NBD-alanine. With increasing acetonitrile, the elution time increases rapidly, but the resolution decreases. High buffer concentrations shorten the elution time and lower the resolution. At low pH values, NBD-alanine was rapidly eluted as a result of its electrophoretic velocity (the EOF is very small or negligible), while at high pH values the analyte was not eluted after 1 h (the direction of the EOF is opposite to the electrophoretic mobility).

Wistuba and Schurig [46] prepared a chiral monolithic stationary phase by sintering the silica bed of a packed capil-

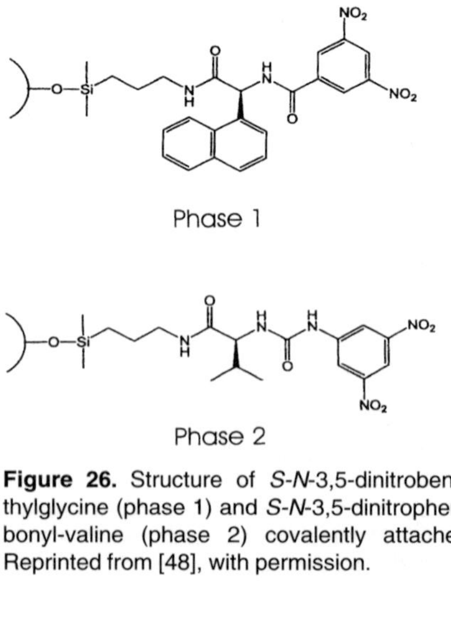

Figure 26. Structure of *S-N*-3,5-dinitrobenzoyl-1-naphthylglycine (phase 1) and *S-N*-3,5-dinitrophenylaminocarbonyl-valine (phase 2) covalently attached to silica. Reprinted from [48], with permission.

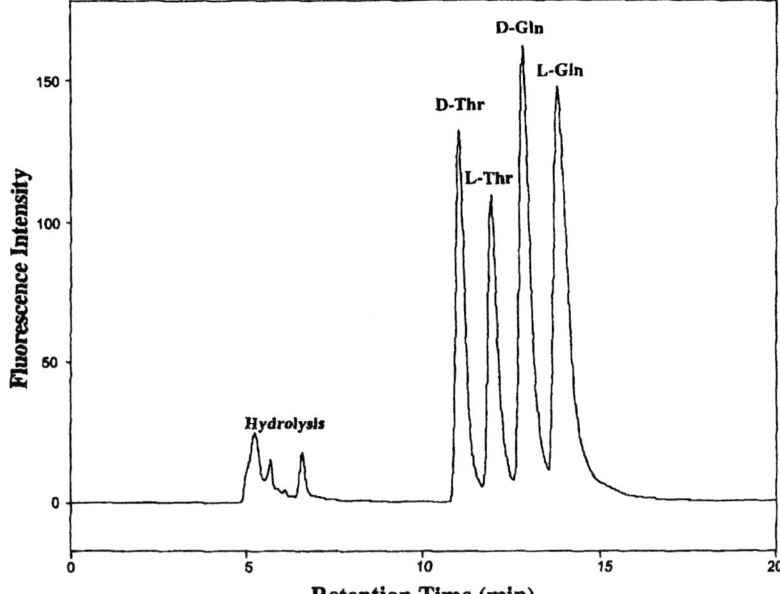

Figure 27. Enantiomer separation of DL-glutamine and DL-threonine on a silica particle (modified with *S-N*-3,5-dinitrobenzoyl-1-naphthylglycine)-loaded monolithic column. Conditions: column, 15 cm × 75 µm ID; 5 mM phosphate buffer, pH 2.5/acetonitrile 30:70 v/v; 0.83 kV/cm. Reprinted from [48], with permission.

lary at high temperatures (380°C) and subsequent modification of the resulting silica monolith with Chirasil-Dex, a permethylated β-cyclodextrin covalently linked *via* an octamethylene spacer to dimethylpolysiloxane (see Fig. 2) by polymer coating. The Chirasil-Dex monolith is a stable and robust separation medium which allows separation runs with an applied voltage of 30 kV and a pressure of more than 400 bar. The enantiomer separation of a number of chiral analytes such as barbiturates, benzoin, α-methyl-α-phenylsuccinimide, MTH-proline, mecoprop methyl, fenoxaprop methyl, carprofen, and ibuprofen was possible (see Fig. 28).

The influence of pressure support on the electrochromatographic enantiomer separation of mephobarbital was demonstrated. The elution time of the first eluted enantiomer was shortened from 26.3 min to 17.4 min by applying a pressure of 12 bar. The theoretical plate number varied from 88 400 (pure CEC) over 65 200 (6 bar pressure support) to 50 600 (12 bar pressure support). While the chiral separation factor remained constant, the resolution factor decreased with increasing pressure support (R_s = 1.97 for pure CEC run, R_s = 1.81 for 6 bar pressure support, and R_s = 1.72 for 12 bar pressure support). Comparing the enantiomer separation of hexobarbital in the LC mode (12 bar) and in the pressure-supported CEC mode (12 bar, 25 kV), an about 4.5 longer elution time was found for the pressure-driven method at comparable chiral separation factor and resolution. Adjusting for equal elution time for hexobarbital in an LC and in a CEC run, a

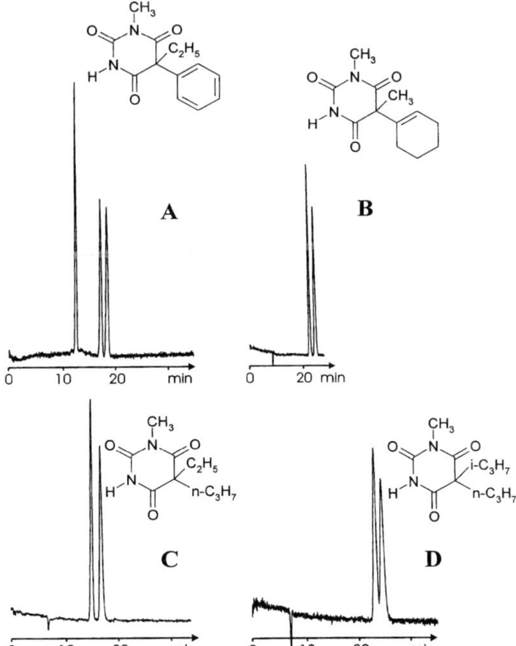

Figure 28. Enantiomer separation of barbiturates on a Chirasil-Dex monolith by CEC. Conditions: column, 20 cm × 100 µm ID; (A) 20 mM MES, pH 6.0/methanol 50:50 v/v; (B)–(D) 20 mM MES, pH 6.0/methanol 70:30 v/v; 20 kV, 12 bar. Reprinted from [46], with permission.

2 to 2.5-fold higher efficiency was observed in the CEC mode. The influence of the most common organic modifiers, methanol and acetonitrile, on the enantiomer separation of hexobarbital was investigated. With methanol, a higher theoretical plate number, chiral separation factor, and higher resolution were observed. Increasing the methanol concentration led to enhancement of the efficiency and a decrease in elution time, but also to a decrease in chiral separation factor and resolution. The negatively charged compounds ibuprofen and carprofen were transported by both the electrophoretic forces and the oppositely directed electroendosmotic flow. By varying the polarity of the applied voltage and the pressure support, the enantiomer separation can be optimized.

5 Comparison of o-CEC, p-CEC and rod-CEC

The most commonly used electrochromatographic method is p-CEC using packed capillaries. This method offers the advantages of high sample capacity, high resolution, and short analysis time. The selectivity is often greater than that found in o-CEC due to stronger analyte-bonded phase interactions. A high number of commercially available LC packing materials with different selectivities are applicable. An important drawback of this method, however, is the need for frits to retain the packing material. Frits are difficult to fabricate reproducibly with respect to permeability and mechanical stability. They are also fragile and often active, leading, *e.g.*, to peak tailing for basic analytes. Irreproducible column performance and bubble formation are consequences of retaining frits. Bubble formation inside the capillary leads to a noisy baseline, an unstable current, and, under unfavorable circumstances, to the breakdown of the current. To prevent these deleterious effects, pressurization of one or both buffer vials is required. However, pressurization complicates instrumentation and CEC operation. Other possibilities of reducing bubble formation are degassing the solvent, using a low buffer concentration, a high proportion of organic modifier, a zwitterionic buffer, or a surfactant as buffer additive. Uniformly packing very small particles into capillaries can be difficult and thermal effects can influence the chiral analysis. The problems associated with frits and uniform particle density can be avoided with the alternative technique of open tubular CEC (o-CEC). This method requires no frits and the column length can be easily reduced by simply cutting it. However, a significant disadvantage is the low sample capacity. This problem may be alleviated by etching the internal capillary wall to increase the surface area. Coating or derivatizing the capillary wall often leads to very low EOFs. Capillaries with monolithic stationary phases display the desirable features of robustness, stability, and high sample capacity, and are often

simply prepared. Retaining frits are not necessary and thus the risk of bubble formation is reduced. Monoliths with very narrow channels can be easily prepared by *in situ* polymerization or by a fusion process, and therefore these separation media are suitable for chip technology.

6 Future trends

CEC is still a growing area of research and the full potential of CEC will only be attained by using tailor-made CSPs. The use of particles with very small diameters ($< 3\ \mu m$) improves column efficiency. Very fast and efficient separations with achiral modified nonporous silica particles with a diameter of $1.5\ \mu m$ have been demonstrated and similar results can be expected for chiral analysis. Stationary phases with incorporated permanent pH-independent charges promote the EOF, and future research, also in the field of enantiomer separation, will expand in this area. Unfortunately, this specially designed chiral mixed-mode material is not commercially available to date. Recently, several papers have demonstrated the successful use of nonaqueous mobile phases in chiral CEC. This promising method allows the analysis of compounds with solubility and/or stability problems in aqueous media, but the present understanding of this approach is still limited. A rapidly growing column technology is the preparation of monolithic stationary phases possessing desirable features (*e.g.*, no frits) for CEC. These continuous separation media are an attractive alternative to the packed separation bed and surely the most promising development in the area of CEC.

The authors gratefully acknowledge Graeme J. Nicholson for helpful advices in preparing the manuscript.

Received October 18, 2000

7 Reference

[1] Wistuba, D., Schurig, V., *J. Chromatogr. A* 2000, *875*, 255–276.

[2] Gübitz, G., Schmid, M. G., *Enantiomer* 2000, *5*, 5–11.

[3] Schurig, V., Wistuba, D., *Electrophoresis* 1999, *20*, 2313–2328.

[4] Dermaux, A., Sandra, P., *Electrophoresis* 1999, *20*, 3027–3065.

[5] Altria, K. D., *J. Chromatogr. A* 1999, *856*, 443–463.

[6] Mayer, S., Schurig, V., *High Resolut. Chromatogr.* 1992, *15*, 129–131.

[7] Mayer, S., Schurig, V., *J. Liq. Chromatogr.* 1993, *16*, 915–931.

[8] Mayer, S., Schurig, V., *Electrophoresis* 1994, *15*, 835–841.

[9] Mayer, S., Schleimer, M., Schurig, V., *J. Microcol. Sep.* 1994, *6*, 43–48.

[10] Armstrong, D. W., Tang, Y., Ward, T., Nichols, M., *Anal. Chem.* 1993, *65*, 1114–1117.

[11] Sezemán, J., Ganzler, K., *J. Chromatogr. A* 1994, *668*, 509–517.

[12] Francotte, E., Jung, M., *Chromatographia* 1996, *42*, 521–527.

[13] Liu, Z., Zou, H., Ni, J. Y., Zhang, Y., *Anal. Chim. Acta* 1999, *378*, 73–76.

[14] Liu, Z., Zou, H., Ye, M., Ni, J. Y., Zhang, Y., *Chin. J. Chromatogr.* 1999, *17*, 245–248.

[15] Liu, Z., Zou, H. Ye, M., Ni, J. Y., Zhang, Y., *Electrophoresis* 1999, *20*, 2891–2897.

[16] Hofstetter, H., Hofstetter, O., Schurig, V., *J. Microcol. Sep.* 1998, *10*, 287–291.

[17] Pesek, J. J., Matyska, M. T., Menezes, S., *J. Chromatogr. A* 1999, *853*, 151–158.

[18] Sinibaldi, M., Vinci, M., Frederici, F., Fieger, M., *Biomed. Chromatogr.* 1997, *11*, 307–310.

[19] Brüggemann, O., Freitag, R., Whitcombe, M. J., Vulfson, E. N., *J. Chromatogr. A* 1997, *781*, 43–53.

[20] Tan, Z. J., Remcho, V. T., *Electrophoresis* 1998, *19*, 2055–2060.

[21] Schweitz, L., Andersson, L. I., Nilsson, S., *J. Chromatogr. A* 1998, *817*, 5–13.

[22] Li, S., Lloyd, D. K., *Anal. Chem.* 1993, *65*, 3684–3690.

[23] Li, S., Lloyd, D. K., *J. Chromatogr. A* 1994, 666, 321–335.

[24] Lloyd, D. K., Li, S., Ryan, P., *J. Chromatogr. A* 1995, *694*, 285–296.

[25] Lelièvre, F., Yang, C., Zare, R. N., Gareil, P., *J. Chromatogr. A* 1996, *723*, 145–156.

[26] Wolf, C., Spence, P. L., Pirkle, W. H., Derrico, E. M., Cavender, D. M., Rozing, G. P., *J. Chromatogr. A* 1997, *782*, 175–179.

[27] Wistuba, D., Czesla, H., Roeder, M., Schurig, V., *J. Chromatogr. A* 1998, *815*, 183–188.

[28] Lämmerhofer, M., Lindner, W., *J. Chromatogr. A* 1998, *829*, 115–125.

[29] Dermaux, A., Lynen, F., Sandra, P., *J. High Resolut. Chromatogr.* 1998, *21*, 575–576.

[30] Krause, K., Chankvetadze, B., Okamoto, Y., Blaschke, G., *Electrophoresis* 1999, *20*, 2772–2778.

[31] Wistuba, D., Schurig, V., *Electrophoresis* 1999, *20*, 2779–2785.

[32] Krause, K., Girod, M., Chankvetadze, B., Blaschke, G., *J. Chromatogr. A* 1999, *837*, 51–63.

[33] Carter-Finch, A. S., Smith, N. W., *J. Chromatogr. A* 1999, *848*, 375–385.

[34] Wikström, H., Svensson, L. A., Owens, P. K., *J. Chromatogr. A*, 2000, *869*, 395–409.

[35] Lin, J.-M., Nakagama, T., Wu, X.-Z., Uchiyama, K., Hobo, T., *Fresenius J. Anal. Chem.* 1997, *357*, 130–132.

[36] Schweitz, L., Andersson, L. I., Nilsson, S., *Chromatographia* 1999, *49*, S93–S94.

[37] Schweitz, L., Andersson, L. I., Nilsson, S., *J. Chromatogr. A* 1997, *792*, 401–409.

[38] Nilsson, S., Schweitz, L., Petersson, M., *Electrophoresis* 1997, *18*, 884–890.

[39] Schweitz, L., Andersson, L. I., Nilsson, S., *Anal. Chem.* 1997, *69*, 1179–1183.

[40] Vallano, P. T., Remcho, V. T., *J. Chromatogr. A* 2000, *887*, 125–135.

[41] Peters, E. C., Lewandowski, K., Petro, M., Svec, F., Fréchet, J. M. J., *Anal. Commun.* 1998, *35*, 83–86.

[42] Svec, F., Lämmerhofer, M., Fréchet, J. M. J., Lindner, W., *23rd International Symposium on Capillary Chromatography*, Riva del Garda, June 5–10, 2000.

[43] Schmid, M. G., Grobuschek, N., Tuscher, C., Gübitz, G., Végvári, Á., Machtejevas, E., Maruska, A., Hjertén, *Electrophoresis* 2000, *21*, 3141–3144.

[44] Koide, T., Ueno, K., *J. High Resolut. Chromatogr.* 2000, *23*, 59–66.

[45] Végvári, Á., Földesi, A., Hetényi, C., Kocnegarova, O., Schmid, M. G., Kudirkaite, V., Hjertén, S., *Electrophoresis* 2000, *21*, 3116–3125.

[46] Wistuba, D., Schurig, V., *Electrophoresis* 2000, *21*, 3152–3159.

[47] Chirica, G., Remcho, V. T., *Electrophoresis* 1999, *20*, 50–56.

[48] Kato, M., Dulay, M. T., Bennett, B., Chen, J., Zare, R. N., *Electrophoresis* 2000, *21*, 3145–3151.

[49] Jakubetz, H., Czesla, H., Schurig, V., *J. Microcol. Sep.* 1997, *9*, 421–431.

[50] Schurig, V., Jung, M., Mayer, S., Negura, S., Fluck, M., *Angew. Chem. Int. Ed. Engl.* 1994, *33*, 2222–2224.

[51] Schurig, V., Jung, M., Mayer, S., Fluck, M., Negura, S., Jakubetz, H., *J. Chromatogr. A* 1995, *694*, 119–128.

[52] Ishii, D., Niwa, T., Ohta, K., Takeuchi, T., *J. High Resolut. Chromatogr.* 1988, *11*, 800–801.

[53] Jakubetz, H., Juza, M., Schurig, V., *Electrophoresis* 1998, *19*, 738–744.

[54] Vallano, P. T., Remcho, V. T., *J. Chromatogr. A* 2000, *887*, 125–135.

[55] Steiner, F., Scherer, B., *J. Chromatogr. A* 2000, *887*, 55–83.

[56] Pursch, M., Sander, L. C., *J. Chromatogr. A* 2000, *887*, 313–326.

[57] Wang, F., Khaledi, M. G., *J. Chromatogr. A* 2000, *875*, 277–293.

[58] Zhu, J., Zou, H., Shi, W., Rui, J., Ni, J., Zhang, Y., *Sci. China* 1999, *42*, 511–519.

[59] Zhang, M., El Rassi, Z., *Electrophoresis* 2000, *21*, 3126–3134.

[60] Zhang, M., El Rassi, Z., *Electrophoresis* 2000, *21*, 3135–3140.

[61] Meyering, M., Chankvetadze, B., Blaschke, G., *J. Chromatogr. A* 2000, *876*, 157–167.

[62] Girod, M., Chankvetadze, B., Blaschke, G., *J. Chromatogr. A* 2000, *887*, 439–455.

[63] Mayer, S., Briand, X., Francotte, E., *J. Chromatogr. A* 2000, *875*, 331–339.

[64] Otsuka, K., Mikami, C., Terabe, S., *J. Chromatogr. A* 2000, *887*, 457–463.

[65] Wei, W., Luo, G., Xiang, R., Yan, C., *J. Microcol. Sep.* 1999, *11*, 263–269.

[66] Deng, Y., Zhang, J., Tsuda, T., Yu, P. H., Boulton, A. A., Cassidy, R. M., *Anal. Chem.* 1998, *70*, 4586–4593.

[67] Wolf, C., Spence, P. L., Pirkle, H. P., Cavender, D. M., Derrico, E. M., *Electrophoresis* 2000, *21*, 917–924.

[68] Lämmerhofer, M., Lindner, W., *J. Chromatogr. A* 1999, *839*, 167–182.

Electrophoresis 2000, *21*, 4136–4158

[69] Tobler, E., Lämmerhofer, M., Lindner, W., *J. Chromatogr. A* 2000, *875*, 341–352.

[70] Lämmerhofer, M., Tobler, E., Lindner, W., *J. Chromatogr. A* 2000, *887*, 421–437.

[71] Lin, J.-M., Uchiyama, K., Hobo, T., *Chromatographia* 1998, *47*, 625–629.

[72] Lin, J.-M., Nakagama, T., Uchiyama, K., Hobo, T., *J. Liq. Chromatogr. Rel. Technol.* 1997, *20*, 1489–1506.

[73] Lin, J.-M., Nakagama, T., Uchiyama, K., Hobo, T., *Biomed. Chromatogr.* 1997, *11*, 298–302.

[74] Gusev, I., Huang, X., Horvárth, C., *J. Chromatogr. A* 1999, *855*, 273–290.

[75] Svec, F., Peters, E. C., Sýkora, D., Yu, C., Fréchet, J. M. J., *J. High Resolut. Chromatogr.* 2000, *23*, 3–18.

[76] Svec, F., Peters, E. C., Sýkora, D., Fréchet, J. M. J., *J. Chromatogr. A* 2000, *887*, 3–29.

[77] Purch, M., Sander, L. C., *J. Chromatogr. A* 2000, *887*, 313–326.

[78] Minakuchi, H., Nakanishi, K., Soga, N., Ishizuka, N., Tanaka, N., *Anal. Chem.* 1996, *68*, 3498–3501.

[79] Fields, S. M., *Anal. Chem.* 1996, *68*, 2703–2712.

[80] Ishizuka, N., Minakuchi, H., Nakanishi, K., Soga, N., Hosoya, K., Tanaka, N., *J. High Resolut. Chromatogr.* 1998, *21*, 477–479.

[81] Constantin, S., Freitag, R., *J. Chromatogr. A* 2000, *887*, 253–263.

[82] Dulay, M. T., Kulkarni, R. P., Zare, R. N., *Anal.Chem.* 1998, *70*, 5103–5107.

[83] Tang, Q., Xin, B., Lee, M. L., *J. Chromatogr. A* 1999, *837*, 35–50.

[84] Tang, Q., Lee, M. L., *J. Chromatogr. A* 2000, *887*, 265–275.

[85] Dittmann, M. M., Rozing, G. P., Ross, G., Adam, T., Unger, K. K., *J. Capil. Electrophor.* 1997, *4*, 201–212.

[86] Asiaie, R., Huang, X., Farman, D., Horváth, C., *J. Chromatogr. A* 1998, *806*, 251–263.

[87] Li, D., Remcho, V. T., *J. Microcol. Sep.* 1997, *9*, 389–397.

[88] Adam, T., Unger, K. K., Dittmann, M. M., Rozing, G. P., *J. Chromatogr. A* 2000, *887*, 327–337.

[89] Koide, T., Ueno, K., *Anal. Sci.* 1998, *14*, 1021–1023.

[90] Koide, T., Ueno, K., *Anal. Sci.* 1999, *15*, 791–794.

[91] Guttman, A., Paulus, A., Cohen, A. S., Grinberg, N., Karger, B. L., *J. Chromatogr.* 1988, *448*, 41–53.

[92] Cruzado, I. D., Vigh, G., *J. Chromatogr.* 1992, *608*, 421–425.

[93] Lin, J.-M., Nakagama, T., Okazawa, H., Wu, X.-Z. Hobo, T., *Fresenius J. Anal. Chem.* 1996, *354*, 451–454.

Electrophoresis 2000, *21*, 4159–4178

Review

Bezhan Chankvetadze[*]
Gottfried Blaschke

Institute of Pharmaceutical
Chemistry, University of
Münster, Münster, Germany

Enantioseparations using capillary electromigration techniques in nonaqueous buffers

This review summarizes recent developments in the field of enantioseparations in capillary electromigration techniques using nonaqueous background electrolytes. The more established and rather intensively reviewed field of nonaqueous chiral capillary electrophoresis (NAQ-CE) is covered in less detail whereas more attention is paid to the relatively new field of nonaqueous capillary electrochromatography (NAQ-CEC).

Keywords: Enantioseparation / Capillary electromigration / Nonaqueous buffers / Review

EL 4227

Contents

Correspondence: Prof. Dr. Bezhan Chankvetadze, Institute of
Pharmaceutical Chemistry, University of Münster, Hittorfstr. 58–
62, D-48149 Münster, Germany
E-mail: chankve@uni-muenster.de
Fax:+49-251-8332144

Abbreviations: APEC, amylose tris(*S*-phenylethylcarbamate);
ADMPC, amylose tris(3,5-dimethylphenylcarbamate); **CDCPC**,
cellulose tris(3,5-dichlorophenylcarbamate); **CDMPC**, cellulose
tris(3,5-dimethylphenylcarbamate); **CMB**, cellulose tris(4-methyl-
benzoate); **CSP**, chiral stationary phase; **NAQ-CE**, nonaqueous
capillary electrophoresis; **PDPM**, poly(diphenyl-2-pyridylmethyl-
methacrylate)

1 Introduction

Capillary electromigration techniques such as capillary electrophoresis (CE) and capillary electrochromatography (CEC) are powerful tools for analytical-scale enantioseparations [1–4]. Moreover, the potential of these techniques

---

[*] Permanent address: Molecular Recognition and Separation
Science Laboratory, School of Chemistry, Tiblisi State University, Chavchavadze Ave 1, 380028 Tbilisi, Georgia

for micropreparative purposes has been demonstrated [5–8]. The medium for performing the electrophoretic experiment, the background electrolyte (BGE) in capillary zone electrophoresis (CZE), affects the electroosmotic flow (EOF) generated on the inner surface of a separation capillary as well as the charge, solubility, and solvation of the analytes. The effect of a BGE is even more significant in chiral CE and CEC because, in addition to the above-mentioned effects, a BGE has a major influence on selector-selectand interactions. This is an unavoidable part of any enantioseparation process independent of whether the experiment is performed by CE, CEC, or chromatographic techniques.

The complimentary potential of nonaqueous CE (NAQ-CE) to aqueous CE for chiral separations was realized several years ago and at present about 20 research papers [9–23] and one recent review paper [24] have been published on this subject. CEC enantioseparations in packed capillaries is a rather new technique [25–29]. This applies especially to CEC enantioseparations in nonaqueous buffers, which has only been developed for the last several years [30–42]. At present, only a few papers have been published on nonaqueous chiral CEC. However, some interesting trends can already be noted and summarizing them may generate further research interest in this promising field.

2 Enantioseparations in capillary electromigration techniques in nonaqueous *vs.* aqueous BGEs

Water as a medium for electrophoresis offers some unique advantages such as low viscosity, nonvolatility, safety, availability, and low costs. These characteristics in combination with the well-known acid-base chemistry of water, its compatibility with various detection schemes, high solvating capacity of electrolytes, *etc.*, make aqueous electrolytes the most commonly used BGEs in CE [43]. However, aqueous buffers may be undesirable in certain cases. Some analytes or buffer additives may be insoluble or unstable in aqueous solutions; the electric current, and, consequently, Joule heat generation is commonly higher in aqueous buffers; and aqueous solutions are not always ideally compatible with a mass spectrometric detector. In the case of chiral CE and CEC separations some special effects of the BGE on selector-selectand interactions have to be considered. Some intermolecular interactions such as hydrogen-bonding, electrostatic, dispersive, *etc.* may be selectively favored or disfavored depending on the separation medium. Thus, application of nonaqueous BGEs may significantly extend the potential of both chiral CE and CEC in chemical, biochemical, pharmaceutical, clinical, and environmental analyses. Nonaqueous electrolytes in classical electrophoresis

have been known since the early 1950s [44, 45]. The first applications of nonaqueous electrolytes in CE were reported by Walbroehl and Jorgenson [46, 47] and the first enantioseparation in NAQ-CE by Ye and Khaledi in 1994 [48]. The developments in chiral CE in nonaque-ous buffers in the last five years are briefly summarized below.

3 Enantioseparations in NAQ-CE

Enantioseparations in aqueous buffers in CE work extremely well. In addition, most of the biomedical, clinical and environmental samples (which are commonly aqueous solutions) are better compatible with CE in aqueous rather than nonaqueous buffers. Therefore, chiral CE in nonaqueous buffers must offer something significant from the viewpoint of unique separation mechanisms, extremely high efficiency, environmental and laboratory safety, *etc.* In order to meet these requirements, both the BGE and the chiral selector must be selected carefully.

3.1 Selection of nonaqueous electrolytes in chiral CE

There are several basic requirements for the solvents used as BGE in CE. Some organic solvents meet these requirements without any addition of electrolytes, whereas others need dissolved inorganic or organic electrolytes. The nature of the BGE may affect the separation factor, peak efficiency, analysis time, and detection sensitivity. One of the basic requirements of the BGE is to be conductive and to alllow the generation of a sufficiently strong EOF. According to the Smolukhovski equation

$$\mu_{EOF} = \frac{\varepsilon_0 \varepsilon \xi}{4\pi\eta} \qquad (1)$$

the mobility of the EOF (μ_{EOF}) depends on the dielectric constant (ε), the viscosity (η) of a medium, and the zeta potential (ζ) on the liquid-solid interface. Until recently, it has been assumed that the ability of a solvent to generate an EOF is proportional to the ε/η ratio. In several recent studies, however, it has been shown that this assumption may not always be valid because the ζ potential is still unpredictably affected by the solvent. The aspects of EOF generation in fused-silica capillaries filled with aqueous and nonaqueous buffers have been thoroughly discussed in the recent summary by Valko *et al.* [21]. The authors detected the EOF in the range of 3.4×10^{-9} -1.8×10^{-7} m^2/Vs without the addition of an ionic species in the following solvents: water, deuterium oxide, acetonitrile, acetone, 2-butanone, formamide (FA), *N*-methylformamide (NMF), *N,N*-dimethylformamide (DMF), methanol, ethanol, 1-propanol, dimethyl sulfoxide (DMSO), ethyl acetate, tetrahydrofuran, and morpholin, as well as in *N*-methylacetamide, which is solid at room temperature. Not all of these solvents will allow the generation of an EOF sufficient for performing a separation with acceptable characteristics in a reasonable period of time.

FA, NMF, DMF, and DMSO have been studied in chiral CE as nonaqueous BGEs without the addition of ionic species. One advantage of these solvents is the easier CE-MS coupling because most MS ion sources do not easily tolerate high amounts of ionic additives. The main problem when applying these solvents in chiral CE is the strongly diminished intermolecular selector-selectand interactions. For example, almost all of these solvents strongly inhibit the hydrogen-bonding interactions and unfavorably affect hydrophobic, electrostatic, *etc.*, forces. This leads to markedly higher concentrations of cyclodextrin-type chiral selectors used in combination with these solvents compared to aqueous solutions. When higher amounts of a chiral selector are required in nonaqueous BGEs, the advantage compared to aqueous BGEs becomes questionable providing that the analyte and the chiral selector are both soluble and also stable in aqueous solvents. One interesting option from the mechanistic point of view could be a comparison of the structure of selector-selectand complexes and the nature of the forces involved in the intermolecular interactions in aqueous and nonaqueous buffers. However, at least to the best of our knowledge, studies of this kind have not yet been published.

CE enantioseparations have been reported in alcoholic media containing ionic species as electrolytes. Nonvolatile ionic additives may, in principle, create problems for on-line CE-MS coupling. However, a good choice of volatile ionic species exists. A significant advantage of alcoholic solvents is that they do not inhibit specific intermolecular interactions as strongly as the above-mentioned solvents. Therefore, the concentrations of chiral selectors may be comparable and even lower than those used in aqueous buffers; in some cases even the enantiomers unresolvable in aqueous buffers can be resolved with the same chiral selector in a nonaqueous medium [12].

3.2 Chiral selectors for NAQ-CE

Native cyclodextrins (CDs) and their derivatives were the first chiral selectors used for enantioseparations in NAQ-CE [40]. Neutral CD derivatives have not been studied in NAQ-CE. However, randomly substituted anionic sulfobutyl-β-CD (SBE-β-CD) [9], sulfated β-CD (SU-β-CD) [14], and cationic trimethylammonium-β-CD (TMA-β-CD) [13] have been used for the enantioseparations in nonaqueous buffers. Recently, negatively charged single-isomer

β-CD sulfates such as heptakis-(2,3-diacetyl-6-sulfato)-β-CD (HSDAC-β-CD) [16] and heptakis-(2,3-dimethyl-6-sulfato)-β-CD (HSDM-β-CD) [20, 22] were shown to be very useful for the enantioseparation of a wide range of chiral analytes in CE under nonaqueous conditions. Of other chiral selectors used in NAQ-CE, quinine and camphorsulfonic acid are also well-known reagents for the enantioseparations using a classical diastereomeric crystallization; their application for enantioseparations in aqueous CE are either limited or impossible [12].

Bjørnsdottir *et al.* [12] employed the principle of diastereomeric ion-pair formation in acetonitrile for CE enantioseparations. The authors showed that optically pure (*R*)- and (*S*)-camphorsulfonates not allowing enantioseparations in aqueous buffers may effect a baseline enantioseparation for several racemates in acetonitrile, which was made conductive by the addition of ionic substances. No enantioseparation of a test racemic compound was observed in aqueous buffer when using 30 mM (+)-*S*-camphorsulfonate as buffer additive. However, under nonaqueous conditions a baseline enantioseparation was obtained at concentrations of (+)-*S*-camphorsulfonate above 5 mM [12]. In contrast to the CD-type chiral selector, chiral (+)- and (–)-camphorsulfonates are available in both configurations, easily allowing a designed reversal of enantiomer migration order, which is a certain advantage. Otsuka *et al.* [49] reported chiral separations in nonaqueous conditions with cellulose tris(3,5-dimethylphenylcarbamate) (CDMPC). This water-insoluble material has been used for the preparation of one of the most widely used chiral stationary phases (CSP) for enantioseparations in high-performance liquid chromatography (HPLC). Therefore, the adaptation of CDMPC or its analogs to chiral separations in CE may markedly increase the potential of this technique.

Quinine [10] and quinine derivatives [15, 19] represent another group of effective chiral selectors used in NAQ-CE. As noted in the pioneering study by Stalcup and Gahm [10], quinine is readily available, relatively inexpensive, and well-studied in other enantiorecognition and enantioseparation techniques such as NMR spectrometry [50], HPLC [51, 52], and TLC [53]. As proposed [10], the ion-pairing interaction between the positively charged quinine and anionic chiral analytes is the major intermolecular interaction responsible for the enantioseparation; π-π interactions between the selector and the selectand may also positively contribute to enantioseparations using quinine as a chiral selector in methanolic solution [10].

Piette *et al.* [15] investigated the potential of quinine and *tert.*-butylcarbamoylquinine as chiral selectors for the enantioseparation of *N*-protected amino acids in nonaqueous electrolytes consisting of 12.5 mM ammonium acetate in methanol or a 60:40 ethanol:methanol mixture. Recently, the same group also extended the studies to other derivatives and analogs of quinine such as quinidine, cinchonine, cinchonidine, *tert.*-butylcarbamoylquinine, dinitrophenylcarbamoylquinine and cyclohexycarbamoylquinine [19]. Cinchona alkaloids and their derivatives exhibit high molar absorbtivities at the detection wavelengths commonly used in CE. In order to avoid the problems of decreased sensitivity due to high background absorbance of the chiral selector, the technique of countercurrent selector-selectand migration [54] was used [15, 19].

As shown in [19], quinine and quinidine carbamates exhibit higher enantiomer resolving ability compared to native cinchona alkaloids. This effect was attributed to the hydrogen donor-acceptor capability of the carbamate moiety. The enantiomer migration order of DNB-Leu was opposite when native quinine and its *tert.*-butylcarbamoylated derivative were used as chiral selectors. Although very high resolution factors (R_s) in the range of 50–70 were observed for several amino acid derivatives, the applicability of this type of chiral selector to other groups of chiral analytes still needs to the proven.

3.3 Challenges and future trends in enantioseparations using NAQ-CE

Stalcup and Gahm [10] noted technical difficulties when working with methanolic solutions due to the greater volatility and lower boiling point of methanol. Hence, in methanolic BGE, it was necessary to maintain operating currents well below levels tolerated in aqueous CE. Failure to maintain lower currents resulted in a frequent interruption of current. Other solvents used in nonaqueous CE also do not offer any significant advantages compared to water from the viewpoints of ruggedness, robustness of a method, costs, and safety. Thus, providing that the analyte is soluble and stable in aqueous solutions, and that the chiral selectors work effectively in aqueous buffers, there are not many practical arguments for nonaqueous conditions. However NAQ-CE may be of significant theoretical interest. In addition, for the chiral selectors which are insoluble, unstable, or do not exhibit a desired chiral recognition ability in aqueous solutions, the application under nonaqueous conditions seems promising.

NAQ-CE is a suitable technique in order to study the chiral recognition properties of a large number of organic macrocyclic compounds, such as chiral calixarens, resorcarens, cryptands, coronands, *etc.* In certain cases, the chiral recognition ability may appear as a unique tool in order to demonstrate the chirality of topologically chiral molecules with a complex structure where the presence

of the elements of chirality are not obvious. Many of these highly hydrophobic compounds are insoluble in water, which opens the opportunities for NAQ-CE. In summary, chiral CE separations under nonaqueous conditions is a challenging field with a certain future. The potential of this technique for solving problems of practical importance in biomedical, pharmaceutical, and environmental analyses still needs to be explored.

4 Enantioseparations in NAQ-CEC

Two questions need to be answered before going into details of CEC enantioseparations in nonaqueous buffers: (i) why CEC, and (ii) why under nonaqueous conditions? The general arguments for developing NAQ-CEC are the same as for developing NAQ-CE. There are some special aspects caused by the presence of the stationary bed which are discussed below. CEC as a hybrid technique of CE and HPLC offers certain complementary challenges compared to CE. Thus, for example, a free, nonimmobilized chiral selector as an additive to the BGE is not only responsible for the advantages of CE, such as versatility, flexibility, *etc.*, but also causes severe problems for on-line CE-MS coupling and when the detector possesses a response to the chiral selector used. These problems become evident, for example, when glycopeptide antibiotics, proteins, or the above-mentioned phenylcarbamates of polysaccharides or carbamates of cinchona alkaloids are applied as chiral selectors. The same problem will appear for any type of chiral selector if one decides to use a polaryrmetric or circular dichroism detector in chiral CE.

One additional motivation for performing enantioseparations under NAQ-CEC conditions is that many effective CSPs for HPLC such as polysaccharide derivatives, chiral polymethacrylates, *etc.*, do not exhibit as high enantiomer resolving ability in aqueous buffers as they do in the nonaqueous ones. These are the advantages that NAQ-CEC could offer. These advantages as well as certain problems in this challenging field are discussed below.

4.1 Generation of the EOF in packed capillary columns under nonaqueous conditions

The EOF is the major driving force for uncharged analytes in CEC. Therefore, the aspects of EOF generation and measurements belong to the basic problems of CEC. It is known that in packed capillaries the surface and, under suitable conditions, also the intraparticle channels of the packing material are the major contributors to the generation of EOF. The inner wall of the capillary does not significantly contribute to the EOF in packed capillaries. Therefore, the geometry and morphology of a particle, its diameter and pore size, the nature and loading of a chiral selector immobilized on the surface, and characteristics of the BGE all contribute significantly to the EOF.

4.1.1 Effect of the nature of silica support on EOF generation

From various types of silica materials, porous native silica gel with unmodified silanol groups, aminopropylsilanized silica, and octadecylsilica modified with C-18 groups have been studied as supports of a chiral selector for enantioseparations under nonaqueous conditions. In several studies, these supports have been modified by covalent attachment of *tert.*-butylcarbamoylquinine *via* a 3-mercaptopropyltriethoxysilane spacer [31] or by coating with poly-(diphenyl-2-pyridylmethylmethacrylate) (PDPM) [30, 33] and polysaccharide derivatives such as cellulose and amylose tris(3,5-dimethylphenylcarbamates) (CDMPC and ADMPC, respectively) [32, 34, 41], cellulose tris(3,5-dichlorophenylcarbamate) (CDCPC) [42], cellulose tris (4-methylbenzoate) (CMB) [34], amylose tris(*S*-phenylethylcarbamate) [40], or a designed mixture of two different polysaccharide phenylcarbamates [32]. The application of macrocyclic antibiotics, such as vancomycin and teicoplanin chemically immobilized on silica [35–37], and organic methacrylate monoliths containing quinine derivatives [38, 39] under NAQ-CEC conditions were also reported mainly after submission of this manuscript for publication. A significant EOF could be generated in the capillaries packed with all above-mentioned modified silica materials in the nonaqueous solutions of electrolytes in methanol, ethanol, isopropanol, acetonitrile, and mixtures of some of these solvents. As expected, the EOF in the capillaries packed with aminopropylsilanized silica gel was directed to the anode (Fig. 1) [29, 30, 32–34, 40–42].

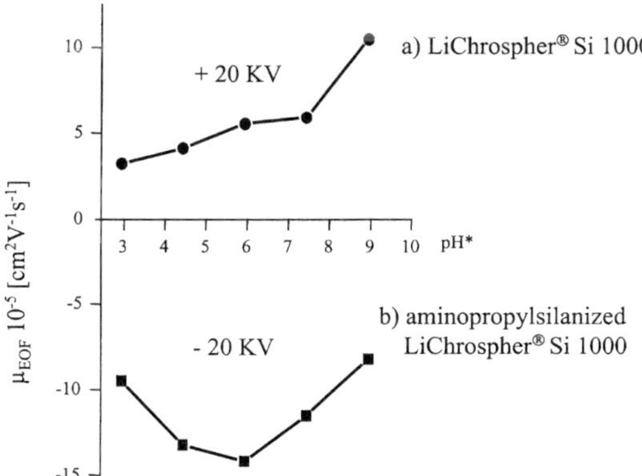

Figure 1. Dependence of (a) the cathodic and (b) anodic EOF generated in capillaries packed with (a) Lichrosper Si 1000 and (b) aminopropylsilanized Lichrosper Si 1000 on the apparent pH* of 2.5 mM ammonium acetate solution in methanol. Reprinted from [33], with permission.

4.1.2 Effect of silica pore size

The effect of the silica pore size on the generation of the EOF under nonaqueous conditions has been studied for aminopropylsilanized silica with a nominal particle size of 5 µm modified with 20% w/w CDMPC [42]. The study was performed in a pore size range of 60–2000 Å. Although the surface area of the packing material decreased from 464 to 15 m²/g with increasing pore size from 60 to 2000 Å, the EOF increased by a factor of 3.4 (Fig. 2). This effect in nonaqueous methanolic BGEs is in agreement with the results previously observed in packed capillary columns in aqueous BGEs [55, 56]. Thus, the EOF apparently collapses in narrow intraparticle channels due to double-layer overlapping. For this reason, wide-pore silica materials may offer some advantages from the viewpoint of a generation of a sufficiently strong EOF and achieving high separation efficiencies in CEC.

An increase of the EOF with increasing pore size has recently also been reported under nonaqueous conditions for a monolithic capillary column with quinidine functionality [38]. A similar effect was observed with a vancomycin-containing capillary column in aqueous BGE but could not be confirmed clearly under nonaqueous conditions [35].

4.1.3 Effect of nature of chiral selector

The chiral selectors applied in NAQ-CEC at present are either low-molecular-weight compounds containing cationic groups [31], polymethacrylate-type chiral polymers containing pyridyl groups (PDPM), [30, 33] neutral derivatives of polysaccharides such as CDMPC, ADMPC, CMB, CDCPC, amylose tris(*S*-phenylethylcarbamate) (APEC) [32, 34, 40–42], macrocyclic antibiotics such as vancomycin and teicoplanin [35–37], and chiral polymeric mono-

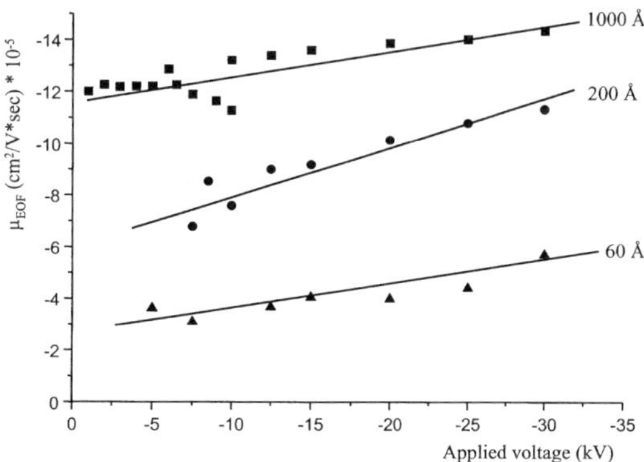

Figure 2. Dependence of the anodic EOF on the pore size of aminopropylsilanized silica gel (5 µm) coated with 20% w/w CDMPC. Reprinted from [41], with permission.

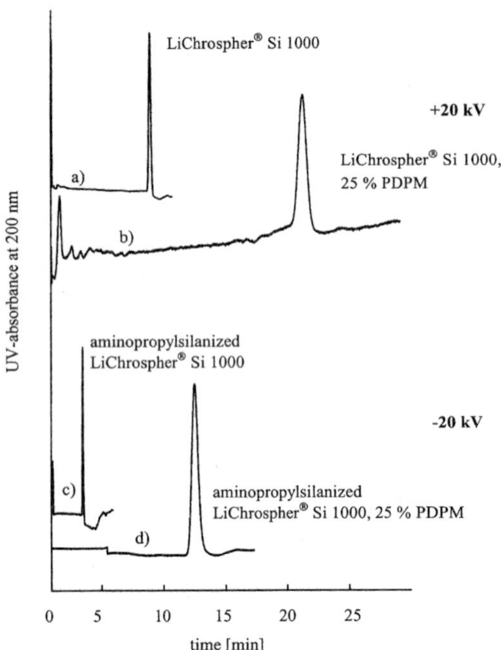

Figure 3. Effect of the coating with 25% w/w PDPM on (a, b) the cathodic and (c, d) the anodic EOF. Reprinted from [33], with permission.

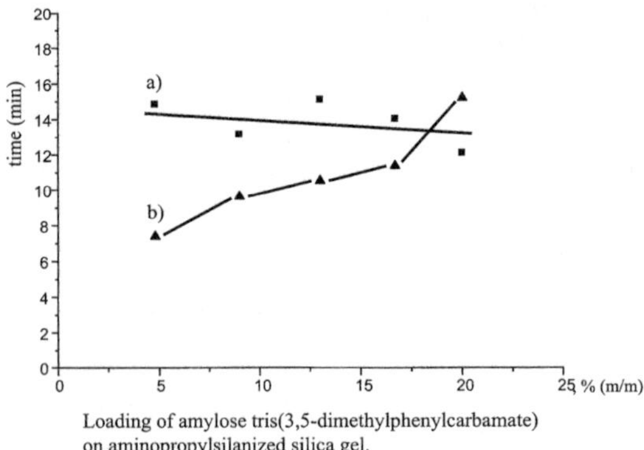

Loading of amylose tris(3,5-dimethylphenylcarbamate) on aminopropylsilanized silica gel.

Figure 4. Effect of the loading of ADMPC onto the aminopropylsilanized silica gel on (a) the retention and (b) migration time of thiourea in (a) chromatographic and (b) CEC runs. Reprinted from [34], with permission.

liths containing quinidine functionality [38, 39]. As shown, the coating of polymeric-type chiral selectors onto the surface of either native or aminopropylsilanized silica drastically diminishes the EOF (Fig. 3) [33]. It appears that the shielding of the surface charge is a major reason for this effect. As demonstrated previously [30], a chargeable pyridyl group does not significantly contribute to the generation of an anodic EOF, at least in methanolic ammonium acetate solution at an apparent pH of 4.5. A similar effect was reported for the antibiotic-type CSPs [35–37]. In contrast to these results, the tertiary amino group of the quinine moiety of *tert.*-butylcarbamoylquinine is responsible for the anodic EOF observed in capillaries packed with native silica which has been previously modified with the above-mentioned chiral selector [31].

4.1.4 Effect of chiral selector loading

The coating of an uncharged chiral selector onto the surface of silica diminishes the EOF. This effect may be caused basically by: (i) shielding of the surface charge, and (ii) narrowing the intraparticle channels and overlapping of the electrical double layers. The migration time of the nonretained compound thiourea increased 2.3-fold

when increasing the loading of ADMPC from 5 to 20% w/w onto the surface of aminopropylsilanized silica in the electrokinetically driven mode (CEC mode) but the elution time did not change significantly in the pressure-driven mode (HPLC mode) (Fig. 4) [34]. This result indicates that the decreased EOF is solely responsible for the migration time increase with the increasing coating of ADMPC. Thus, obtaining enantioseparations at the lowest possible loading of a neutral chiral selector onto the surface of silica clearly may be advantageous. However, this will require chiral selectors with extremely high chiral recognition abilities.

4.1.5 Effect of BGE pH

It does not seem trivial in NAQ-CEC to adjust the ionic strength and the apparent pH of the BGE independently. In addition, acid-base equilibria are shifted in organic solvents. The pK_{auto} value of methanol is 17.2, and of acetonitrile >33.3, instead of 14.0 as for water [43]. For this reason, the pH value measured with a common pH meter does not say much about the actual concentrations of $[H^+]$ and $[OH^-]$ ions in solution. Therefore, when working with nonaqueous BGEs, it is advised to indicate the actual amount of acid and basic electrolytes contained in the solution. The dependence of the anodic EOF on the apparent pH* of the methanolic ammonium acetate solution may appear somewhat unusual at first glance (Fig. 5) [34].

In particular, the increase of the anodic EOF with decreasing pH* in the range of 7.5 to 5.0 seems logical and may be explained by an increase of the effective

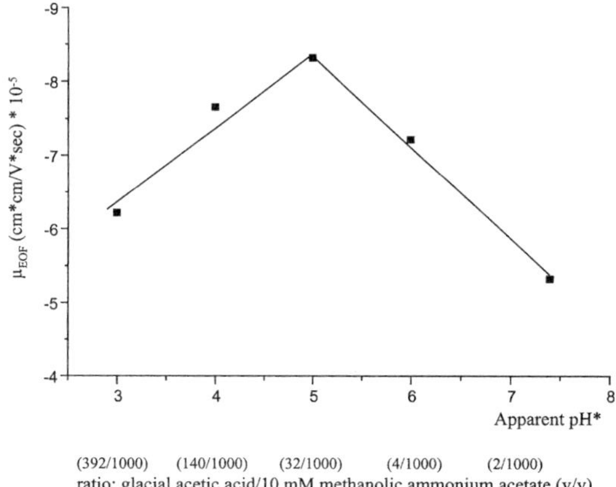

Figure 5. Dependence of EOF on the apparent pH* in a fused-silica capillary packed with aminopropylsilanized silica gel coated with 20% w/w CDMC in 10 mM methanolic ammonium acetate solution. Reprinted from [34], with permission.

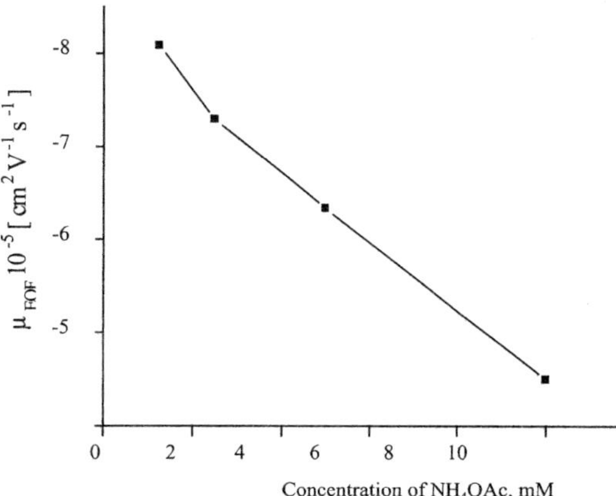

Figure 6. The effect of the ammonium acetate concentration on the EOF generated in a methanolic BGE. Reprinted from [33], with permission.

charge of the cationic aminopropyl groups while decreasing it at the residual silanol groups of the silica material. A decrease of the anodic EOF with a further decrease of pH seems unusual but may be explained by the increasing ionic strength of the BGE due to the significant amount of glacial acetic acid required for the pH* adjustment in this range [34]. Similar to the results shown in Fig. 7, the anodic EOF decreased with increasing amounts of triethylamine in acetonitrile-methanol (80:20 v/v) mixtures containing 200 mM acetic acid [31].

4.1.6 Effect of the ionic strength of the BGE

The influence of increasing amounts of glacial acetic acid on the EOF (Fig. 5) provides a preliminary indication of the decrease of the EOF with increasing ionic strength of the buffer. This effect, which is similar to that observed in CE and in CEC in aqueous buffers, has been confirmed in an independent study where the amount of ammonium acetate in the methanol varied between 2 and 10 mM (Fig. 6) [33]. Further studies of this effect depending on the pore size of silica are required because increasing ionic strength may have a dual effect on the EOF. On the one hand, increasing the ionic strength of the BGE will decrease the double-layer thickness and this also decreases the EOF generated on the surface of the silica particles and bigger interparticle channels. On the other hand, decreasing the double-layer thickness may allow avoiding the collapse of the EOF in the narrow intra- and interparticle channels.

4.1.7 Effect of water in the BGE

To date, all CSPs used for enantioseparations in nonaqueous buffers may also be applied for the same purpose in aqueous buffers [29, 35–37, 56–59]. Therefore, studies of the effect of the additives of water on the EOF generation in organic solvents represents a certain interest from the practical as well as mechanistic points of view. At first, it seems interesting to note that in capillaries packed with aminopropylsilanized silica a higher EOF was observed in methanolic ammonium acetate solution compared to ammonium acetate solution in acetonitrile (Fig. 7) [33]. This indicates that the ratio of the dielectric constant (ε)/viscosity of a solvent (η) can not always be applied as a characteristic for the ability of a solvent to generate an EOF.

The ε/η ratio is 61 for methanol and 110 for acetonitrile [43]. In the range of additives of 0–20% water added to methanol, the EOF decreased continuously, instead of the increase expected according to the rule because the ε/η ratio for water is 90. There are two possible explanations for this effect: (i) the known nonadditive change of the viscosity of a water-methanol mixture, and (ii) the effect of the liquid phase composition of the ζ potential on the liquid-solid interface. The ζ potential has been considered to be constant when deriving the ε/η rule. Again, in contrast to the expectations based on this rule, the EOF increased in acetonitrile-water mixtures with increasing amounts of water in the range of 0–20% (Fig. 7) [33]. The latter phenomenon seems to be of interest and may lead to unique effects when performing enantioseparations under the conditions mentioned here (see Section 4.2.8).

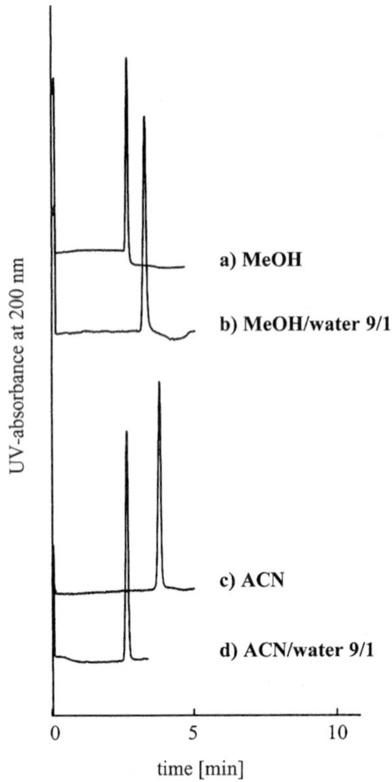

Figure 7. Migration time of thiourea at −20 kV in a capillary packed with aminopropylsilanized Licrospher Si 1000 using (a) 2.5 mM ammonium acetate in methanol (100%), (b) methanol/water 90/10 v/v, (c) acetonitrile, (d) acetonitrile/water 90/10. Reprinted from [33], with permission.

Similar trends in the range of low water content in methanol and acetonitrile were observed by Lämmerhofer and Lindner [57] but these authors did not extend their studies to pure nonaqueous solvents. Together with the addition of water, a mixing of nonaqueous solvents may allow a desirable EOF to be adjusted and it may significantly affect all parameters of an enantioseparation [31, 32, 35–39].

4.2 Adjustment of enantioseparations in NAQ-CEC in packed capillaries

4.2.1 CSPs used in NAQ-CEC

CSPs employed for enantioseparations under nonaqueous conditions are listed in Table 1. One paper has been published on the use of quinine carbamate type chiral

selectors as chiral buffer additives under nonaqueous conditions in a capillary packed with octadecyl silica [60]. It seems somewhat logical that most of these materials are polysaccharide derivatives. To date, polysaccharide derivatives represent the most widely used CSPs for practical problem solving in pharmaceutical, biomedical, clinical, chemical, and environmental analyses. In addition, although effective in aqueous mobile phases, these CSPs were originally developed and are more effective under nonaqueous conditions. The high enantiomer resolving ability of polysaccharide derivatives in polar organic solvents has been known for several years [61] and several interesting applications of this mode in HPLC appeared recently [62–66].

Although the CSPs based on polysaccharide derivatives belong to the same group, they may differ markedly from the viewpoint of their chiral recognition pattern and ability [59]. Thus, an opposite elution order of enantiomers has been described, depending on the structure of both polysaccharide backbone or substituents on the hydroxyl groups. The example depicted in Fig. 8 shows that the enantiomers of the chiral anesthetic veterinary drug metomidate can not be resolved on a capillary column packed with a CSP based on CDMPC but almost baseline enantioseparation was obtained when the amylose derivative with the same substituent was used for the preparation of CSP [34]. Extending the number of CSPs adapted for CEC enantioseparations in nonaqueous buffers will further strengthen this technique. In the rich arsenal of CSPs suitable for enantioseparations in HPLC there are without doubt many suitable candidates for the same purpose under the conditions of NAQ-CEC.

4.2.2 Effect of the nature of a silica support on enantioseparations

The surface chemistry of silica together with its geometry and morphology seems to play an important role not only for the generation of the EOF but also for obtaining highly efficient enantioseparations. Significant EOF generated on silica particles coated with a rather thick film (25% w/w) of a neutral polymeric material indicates that the functional groups on the silica surface are not absolutely masked and may also be involved in selector-selectand interactions. As mentioned above, a significant EOF can be generated in capillaries packed with native silica without any previous functionalization. This material becomes suitable for enantioseparation after coating with a chiral polymer [33]. The same applies for aminopropylsilanized silica gel as well as reversed-phase materials modified by covalent attachment of octadecyl groups [41].

  Capillary electromigration techniques

Table 1. CSPs studied in NAQ-CEC

Chiral selector	BGE	Chiral analytes	Ref.
PDPM (Chiralpak OP)[a]	Ammonium acetate in MeOH or MeCN	Benzoin and its derivatives, Tröger's base, *trans*-stilbene oxide, 1,1-binaphthyl-2,2'-diol, cyclobutyldianilide carbamate	[30, 33]
Weak anion exchanger (*tert*-butylcarbamoyl-quinine)	Triethylamine+acetic acid in MeOH+MeCN	Amino acid and arylpropionic acid derivatives	[31]
CMB (Chiralcel OJ)	Ammonium acetate in MeOH or ETOH	Piprozolin, glutethimide, etozolin, Tröger's base, indapamide, *trans*-stilbene oxide	[34]
CDMPC (Chiralcel OD)	Ammonium acetate in MeOH or ETOH	Econazole, *trans*-stilbene oxide, 2,2'-diamino-6,6'-dimethylbiphenyl, glutethimide	[34, 41]
CDCPC	Ammonium acetate in MeOH	Chiral sulfoxides, etozolin, piprozolin	[42]
ADMPC (Chiralpak AD)	Ammonium acetate in MeOH or EtOH	Aminoglutethimide, *trans*-stilbene oxide, metomidate	[34, 41]
APEC (Chiralpak AS)	Ammonium acetate in MeOH or EtOH	Aminoglutethimide, *trans*-stilbene oxide, metomidate	[40]
Mixed (OD+AD)	Ammonium acetate in MeOH, EtOH or mixed with MeCN	Thalidomide and its metabolites	[32]
Vancomycin[b] (Chirobiotic V)	Triethylamine (TEA)/OHAc+ MeOH/MeCN	Alprenolol, atenolol, benzoin, 1,1-binaphthyl-2,2'-diol, bupivacaine, dichloroprop, ephedrine, fenetorol, ibuprofen, isoprenaline, labetalol, ketamine, ketoprofen, metaqualone, methadone, methoxyfenandrin, metoprolol, ornidazole, pindolol, practolol, propranolol, phenyl-propanolamine, 5-(4-methylphenyl)-5-phenylhydantoin, bromacil, dopa, felodipine, sotalol, thalidomide, terbutaline, verapamil, warfarin	[35, 36]
Teicoplanin (Chirobiotic T)	TEA/OHAc+ MeOH/MeCN	Alprenolol, atenolol, benzoin, bupivacaine, fenetorol, ibuprofen, labetalol, ketoprofen, metoprolol, pindolol, propranolol, phenylpropanolamine, 5-(4-methylphenyl)-5-phenylhydantoin, dopa, felodipine, sotalol, thalidomide, terbutaline, verapamil, warfarin, coumachlor, tryptophan, *N*-CBZ-L-glutamic acid, β-hydroxyphenethylamine	[37]
Chiral monolith containing quinidine functionality	TEA/OHAc+ MeOH/MeCN	DNB-amino acids, other amino acid derivatives, mecoprop, fenoprop	[38, 39]

a) Commercially available columns for HPLC enantioseparations containing the same chiral selector
b) Not all analytes were resolved successfully

4.2.3 Effect of silica particle diameter on enantioseparations

One of the potential advantages of CEC compared to LC together with the plug-like profile of an electrokinetically driven flow is the possibility to use packing materials with a smaller particle diameter without any problems with high back pressure. The particle size determines the dimensions of intraparticle channels and this way may affect mass transfer in the mobile phase and Eddy diffusion. These contributions to band-broadening depend on the particle size of the packing material in the following ways:

$$H_{ed} = 2\lambda d_p \quad (2)$$

$$H_{mm} = \frac{\omega d_p^2 u}{D_m} \quad (3)$$

where H_{ed} and H_{mm} are the contributions of the Eddy diffusion and mass transfer due to the mobile phase to band-broadening, λ is a measure of flow inequality in the packed column, ω is a column packing parameter and d_p is the particle size. It seems worth mentioning that H_{mm} will be affected more significantly by the particle size as a

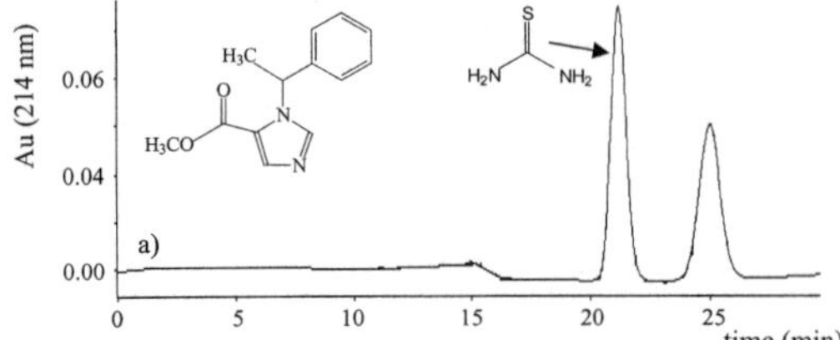

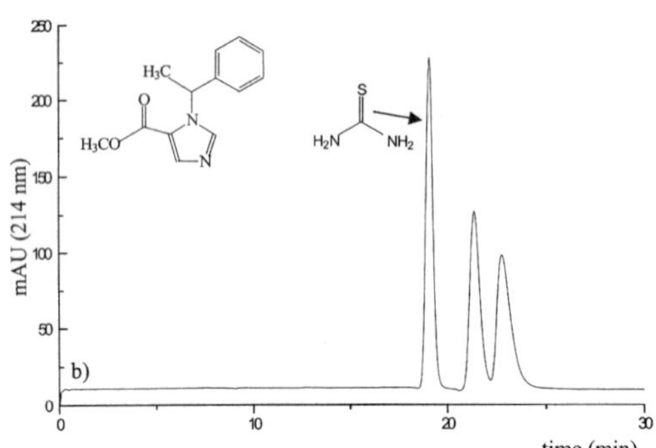

Figure 8. NAQ-CEC enantioseparations of metomidate in fused-silica capillaries packed with CSPs prepared based on (a) cellulose- and (b) amylose- tris(3,5-dimethyl-phenylcarbamate)s. Reprinted from [34], with permission.

quadratic function. In addition, this contribution may increase with increasing flow velocity. The results represented in Fig. 9 show that the contributions of H_{ed} and H_{mm} to band broadening may be minimized with decreasing particle diameter [41]. However, further studies are required in order to differentiate between these two contributions to band broadening and the particle diameter must be optimized in combination with pore size.

4.2.4 Effect of silica pore size on enantioseparations

As shown in several studies [29, 30, 33, 41, 42], mass transfer kinetics between a mobile and a stationary phase seems to be a major factor limiting plate height in chiral CEC with polymeric-type CSPs. Thus, increasing perfusive transport will reduce the portion of the analyte distributed in the so-called stagnant mobile phase pool and, thus, improve mass transfer kinetics between mobile and stationary phases. Therefore, wide-pore silica may provide higher plate numbers for separations where slow mass transfer kinetics is a limiting factor. On the other hand, it should be taken into account that interparticle and intraparticle "routes" may not be adequate from the viewpoint of analyte-chiral selector interactions. In particular, it is less probable that high-molecular-weight chiral selec-

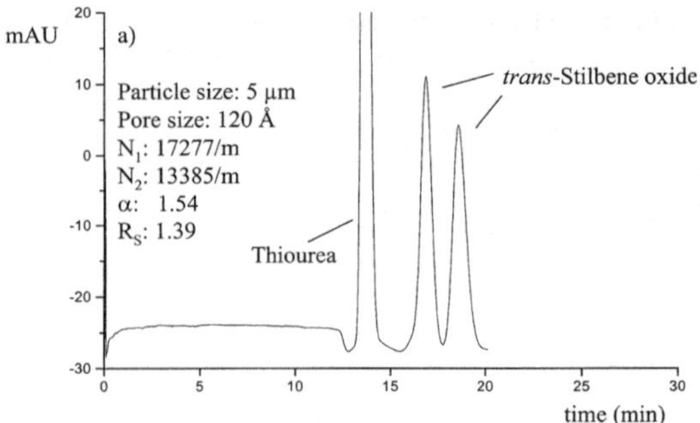

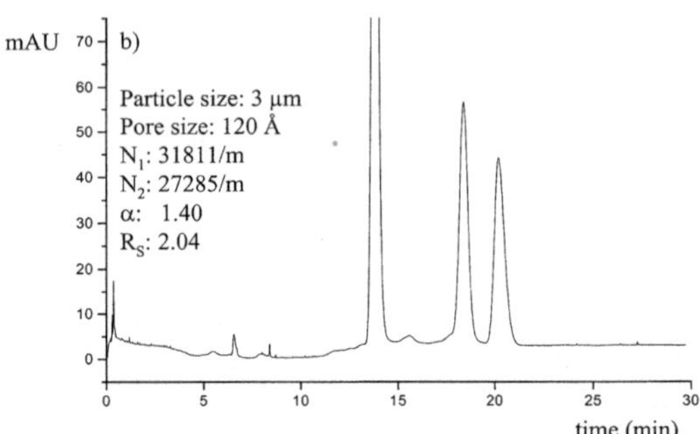

Figure 9. NAQ-CEC enantioseparation of *trans*-stilbene oxide in fused-silica capillaries packed with CSPs prepared based on CDMPC (20% w/w) on aminopropylsilanized silica with (a) 3 µm and (b) 5 µm particle diameter. Reprinted from [41], with permission.

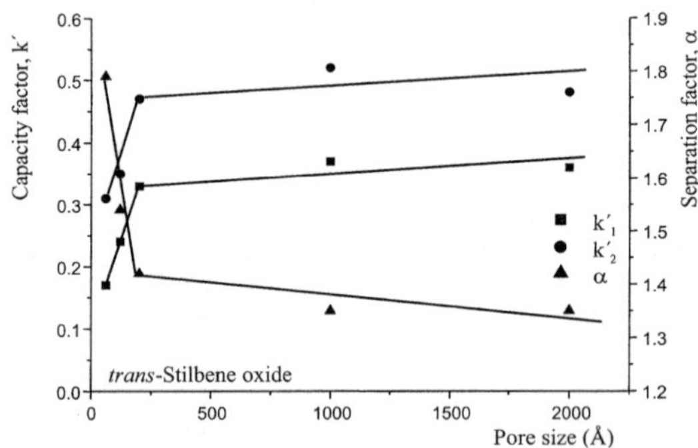

Figure 10. Dependence of capacity (k') and separation (α) factors of *trans*-stilbene oxide on the silica pore size used for preparation of CSP. The chiral selector was the same as in the experiment shown in Fig. 11. Reprinted from [41], with permission.

tors easily penetrate the pores of even wide-pore silica. Thus, with increasing contribution of the perfusive transport, a decrease of the separation selectivity is possible due to the analyte migration *via* the routes which contain

Table 2. Effect of silica pore size on bulk flow increment and the retention (k') and separation (α) factors of the enantiomers of *trans*-stilbene oxide in CEC mode

Pore size (Å)	$\dfrac{k'_{1\,(\times\,\text{Å})}}{k'_{1\,(60\,\text{Å})}}$	$\dfrac{k'_{2\,(\times\,\text{Å})}}{k'_{2\,(60\,\text{Å})}}$	$\dfrac{\mu_{EOF\,(\times\,\text{Å})}}{\mu_{EOF\,(60\,\text{Å})}}$	Separation factor (α)
60	1.00	1.00	1.00	1.83
120	1.29	1.10	2.70	1.54
1000	2.24	1.68	3.37	1.35

Reprinted from [41], with permission.

less chiral selector compared to the interparticle channels.

As shown for *trans*-stilbene oxide in Fig. 10, the capacity factor actually increases and the separation factor decreases in the pore size range of 60–200 Å. This result clearly indicates that the intraparticle route is at least less retentive and less enantioselective compared to interparticle channels. The 3.3- to 3.4-fold increase of the bulk flow with increasing pore size from 60 to 1000 Å corresponds only to a 1.6- to 2.3-fold increase of the capacity factors (Table 2). This means that with increasing pore size the flow is increasingly mobilized basically through the less retentive and less enantioselective channels. This effect also seems to be responsible for the decreasing separation factors of the enantiomers as shown in Table 2 and Fig. 10. The above-mentioned decrease of the capacity and enantioseparation factors may also in part be caused by the decreasing surface area of silica with increasing pore diameter.

The improved mass transfer characteristics basically affect the higher linear flow velocity range of the van Deemter plot. Therefore, this plot constructed for the materials with different pore size may be used as an important diagnostic tool in order to differentiate between various band-broadening contributions in chiral CEC. Plots of the plate height, *h, vs.* the mobile phase linear velocity for CSPs, prepared based on the silica materials of different pore size, are shown in Fig. 11 for thiourea and the first and second peak of *trans*-stilbene oxide, respectively. The curves become flat in the range of higher linear flow rates with increasing pore size. This is a clear indication that the plate height is actually controlled by mass transfer characteristics.

The mass transfer as a combined term may be divided into the following contributions: (i) mass transfer in the mobile phase; (ii) mass transfer in the stagnant mobile phase residing in the pores of a stationary phase, and (iii) mass transfer in the stationary phase. The latter contribution accounts for diffusion through the chiral seletor film attached onto the silica as well as for specific intermolecu-

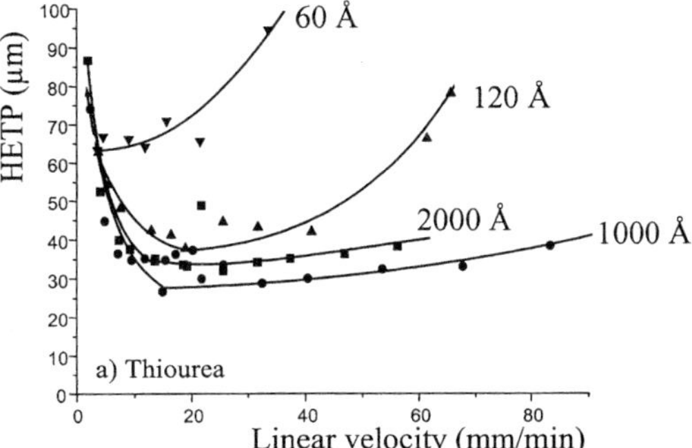

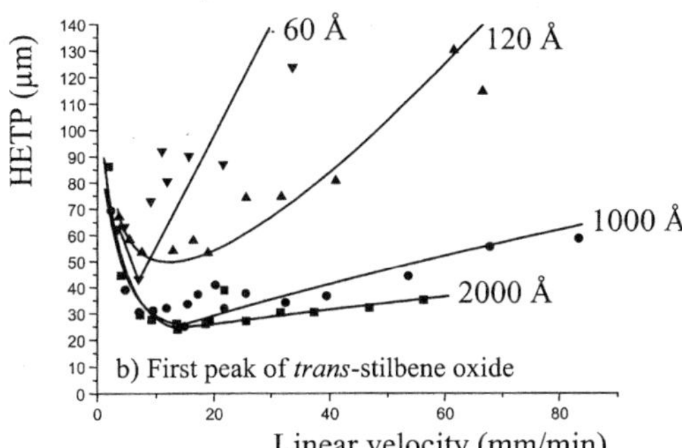

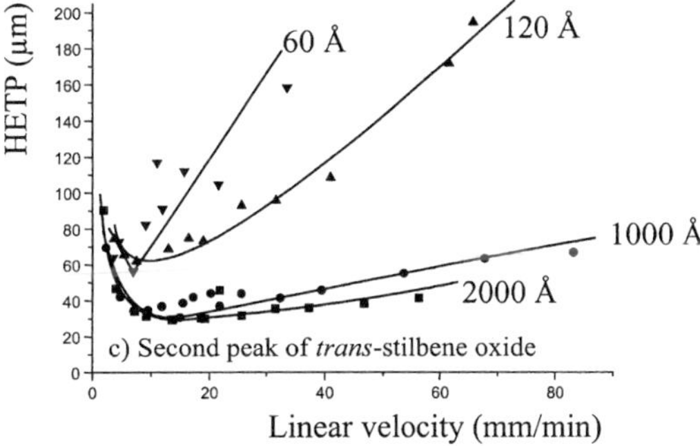

Figure 11. Van Deemter plots for (a) thiourea and (b) the first and (c) the second eluted enantiomers of *trans*-stilbene oxide for CSPs prepared by coating of CDMPC (20% w/w) on aminopropylsilanized silica of various pore size. Reprinted from [41], with permission.

lar interactions between an analyte and a chiral selector. The porosity of silica in the CEC mode may basically affect the latter two. Thus, perfusive transport will reduce the portion of stagnant mobile phase pools inside the pores of silica as well as improve the exchange of the analyte between stationary and mobile phases by continuous replacement of the mobile phase in the pores. Similar effects of the pore size on the van Deemter curves for thiourea (which does not significantly interact with the sta-

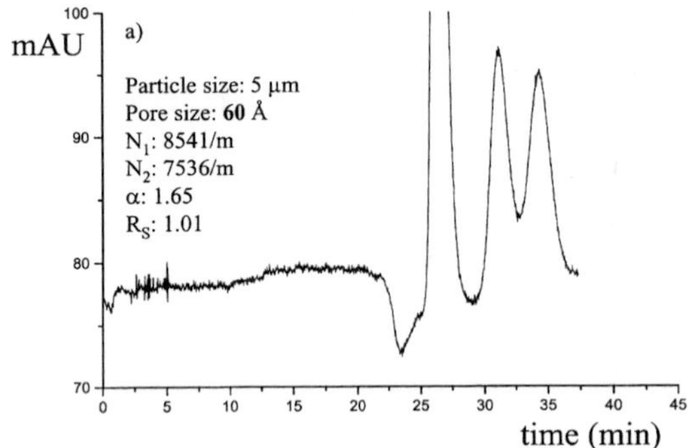

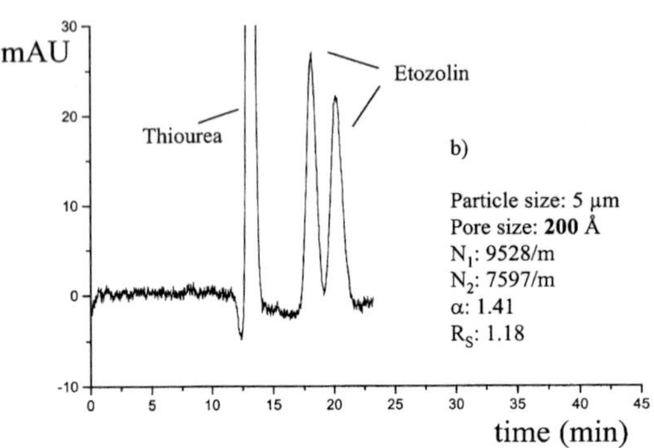

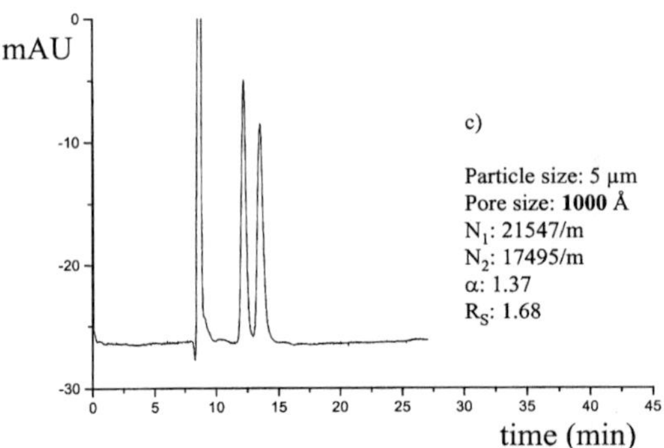

Figure 12. NAQ-CEC enantioseparations of etozolin in fused-silica capillaries packed with CSPs prepared based on CDMC (20% w/w) on aminopropylsilanized silica of (a) 60 Å, (b) 200 Å, and (c) 1000 Å pore size. Reprinted from [41], with permission.

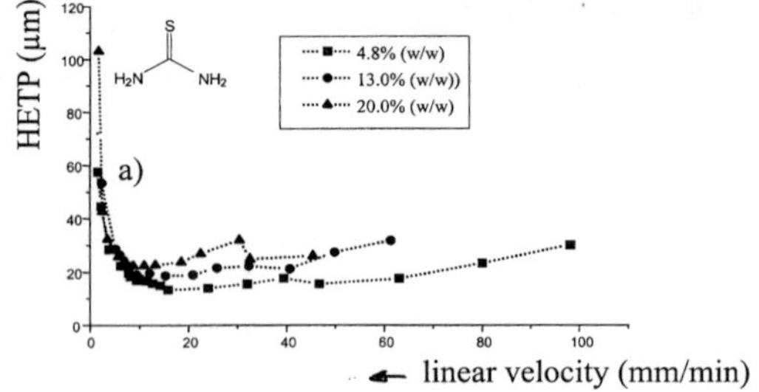

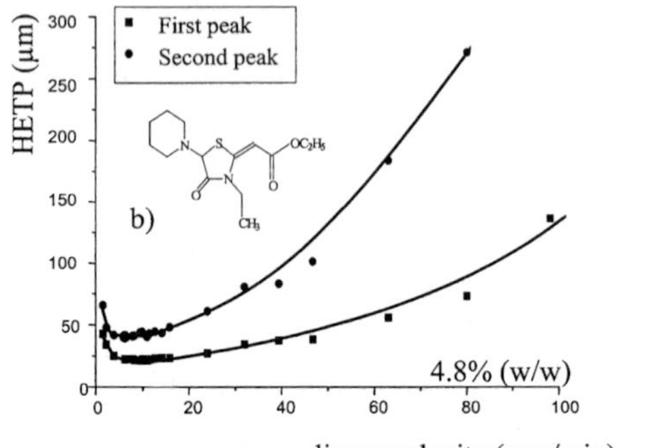

Figure 13. Van Deemter plots for (a) thiourea and (b) the first and (c) the second eluted enantiomers of piprozolin in fused-silica capillaries packed with CSPs prepared by coating of ADMC (a) 4.8% and 13.0% w/w on aminopropylsilanized silica. Reprinted from [34], with permission.

tionary phase) and both enantiomers of *trans*-stilbene oxide (see Fig. 11) indicate that the perfusive transport favorably affects mass transfer in the stagnant mobile phase and this effect is responsible for the improvement of the peak efficiency with increasing pore size. The favorable effect of increasing pore size of silica is also illustrated in the enantioseparations of etozolin in Fig. 12 [42]. The above given explanation for the dependence of

separation characteristics on the pore size of silica sounds reasonable and is also in agreement with observations in achiral CEC separations [55, 56].

4.2.5 Effect of chiral selector loading on the silica surface on enantioseparation

The switching from pressure-driven to electrokinetically driven migration mechanism makes it possible to flatten the parabolic flow profile caused by longitudinal diffusion. The flow profile mainly affects parameter *B* in the modified van Deemter equation for reduced plate heights [67]:

$$h = Au^{0.33} + B/u + Cu \tag{4}$$

where u is the linear flow rate of a mobile phase, A is the Eddy diffusion and flow distribution component, B is the longitudinal diffusion component, and C is the mass transfer component. However, as mentioned above, the mass transfer kinetics between the mobile phase and CSP may be significantly improved in the CEC mode due to the generation of a significant intraparticle flow. The effect of mass transfer kinetics on the separation efficiency is especially important for high-molecular-weight CSPs [68, 69].

In order to study the effect of the mass transfer kinetics on the plate heights, van Deemter curves were constructed for CSPs with various loadings of a chiral selector. In addition, this was studied for analytes with different retention characteristics. Coated-type CSPs offer certain advantages from this viewpoint compared to covalently bound ones because a continuous variation of a chiral selector loading onto the silica surface is possible. The minimal plate heights observed for the thiourea peak were in the range of 13.4–14.0 μm (Fig. 13a). In contrast to thiourea, the plate height for retained chiral analytes strongly decreased with increasing loading of the polysaccharide derivative on the silica gel (Fig. 13b, c). This effect was more pronounced for the more retained second enantiomer. Thus, the minimal plate height of 21.5 μm for the first peak of piprozolin increased to 58.3 μm with increasing loading of ADMPC from 4.8 to 13% w/w. The related numbers for the second eluted enantiomer were 40.3 and 102.7 μm. These data indicate that the mass transfer kinetics between the stationary and mobile phase significantly affect peak dispersion of the piprozolin enantiomers [34]. Interestingly, no significant effect of the loading was observed on the peak efficiency of the enantiomers of *trans*-stilbene oxide in the range of 13–20% w/w. This result agrees with expectations because the enantiomers of *trans*-stilbene oxide, especially the less retained enantiomer, possess a lower affinity to the CSP compared to the enantiomers of piprozolin. The significant effect of chiral selector loading on the enantioseparation of piprozolin is shown in Fig. 14.

4.2.6 Effect of the separation medium on enantioseparations

The separation medium together with the analyte characteristics determine its longitudinal molecular diffusion in the mobile phase (D_m). The contribution of the latter in peak variance (σ_{md}) is given by the Einstein equation:

$$\sigma_{md} = \sqrt{2\gamma_m D_m t} \tag{5}$$

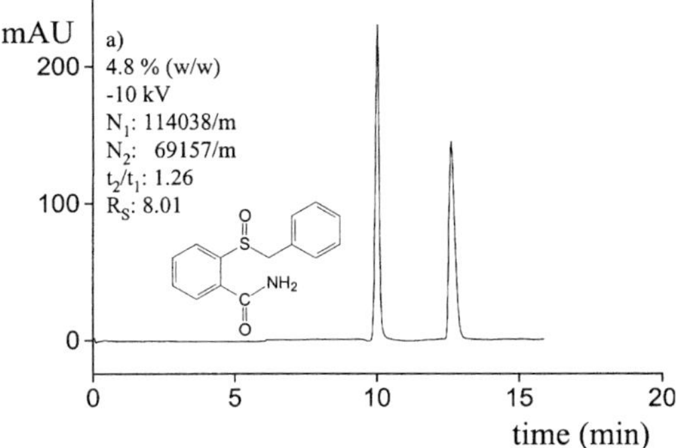

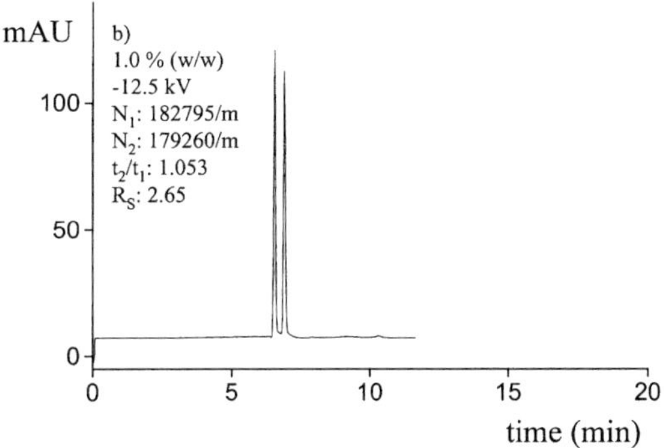

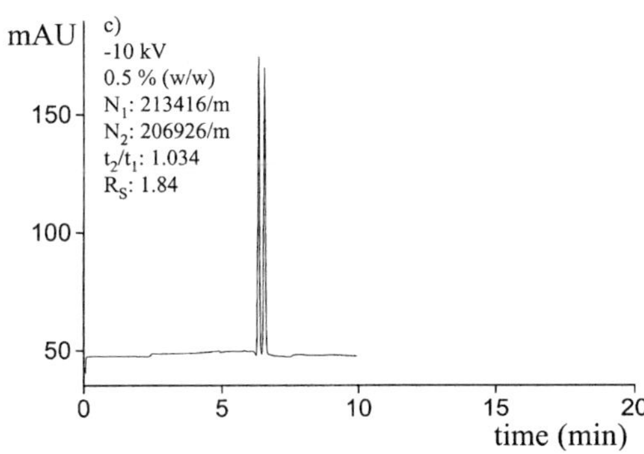

Figure 14. CEC enantioseparation of 2-(benzylsulfinyl)-benzamide in capillaries packed with CSPs containing (a) 4.8%, (b) 1.0% and (c) 0.5% w/w of CDCPC. Reprinted from [42], with permission.

where γ_m is an obstruction factor ($\approx$ 0.5–1.0) and t is the time spent by the analyte during its passage through the column. Equation (5) may be transformed to:

$$H_{md} = \frac{2\gamma_m D_m}{u} \tag{6}$$

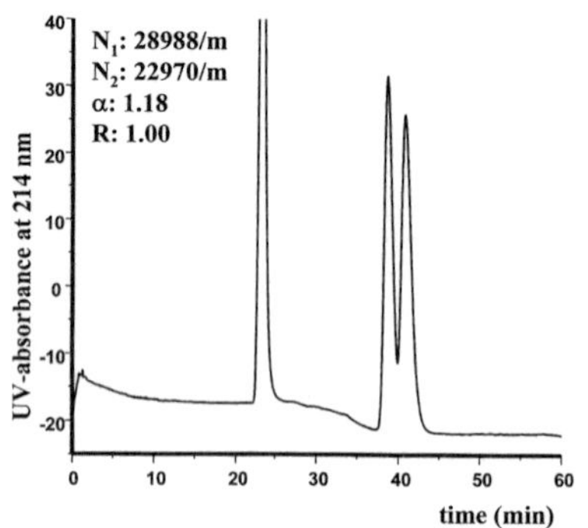

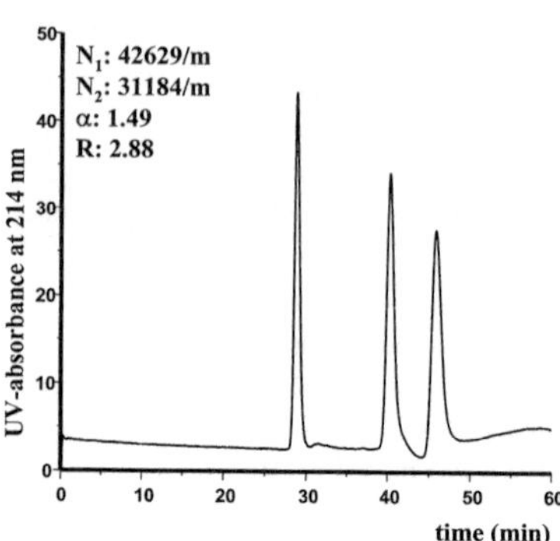

Figure 15. NAQ-CEC enantioseparations of *trans*-stilbene oxide in fused-silica capillaries packed with CSPs prepared based on ADMC (20% w/w) on Lichrospher RP-18 with 3 µm particle diameter. The separation was performed in 2.5 mM ammonium acetate solution in (a) methanol and (b) ethanol. Reprinted from [41], with permission.

where H_{md} is the plate height contribution due to diffusion in the mobile phase and u is the linear flow velocity.

The longitudinal molecular diffusion basically affects the low linear velocity range of the van Deemter plot. Thus, according to Eq. (5), mobile phases with lower diffusion coefficients may offer some advantages in the low-linear velocity range. This was actually observed when methanol and ethanol were compared as separation media [34, 41]. On the other hand, ethanol as a separation medium, which may also offer advantages from the viewpoint of the separation factor (Fig. 15), may not be advantageous at the higher linear flow rates. The reason for this is that in this range the van Deemter plot is favorably controlled by the mass transfer characteristics between the mobile and stationary phases that is inferior in ethanol compared to methanol due to the lower diffusion coefficient in the former. Thus, as shown in this example, the optimization of separation medium and the linear flow rate need to be performed in combination.

4.2.7 Effect of the apparent pH* of the BGE on enantioseparations

The effect of the pH* can be multivariate in CEC separations. First the EOF and consequently the linear flow rate of a mobile phase significantly changes depending on the pH*. Further, selector-selectand interactions may vary depending on the pH*. The more significant pH*-dependent effects may appear for those CSPs and analytes which change their effective charge in the pH* range studied. The enantioseparations of indapamide in methanolic ammonium acetate solutions with different apparent pH*

in capillaries packed with Chiralcel OD material is shown in Fig. 16 [34]. Both the CSP and the analyte are neutral and most likely their effective charge does not change in the pH* range studied. This seems to be a reason for the almost constant capacity factors for both enantiomers. The selectivity of enantioseparation decreases slightly in the apparent pH* range of 7.4-4.0 and drastically from pH* 4.0 to pH* 3.0. Thus, apparent pH* can significantly affect the enantioseparation in nonaqueous CEC and may be used for the optimization of separation.

4.2.8 Effect of water in the BGE on enantioseparations

In Section 4.1.7, the nontrivial effect of water additives to nonaqueous solvents on the EOF generation in the capillaries packed with aminopropylsilanized LiChrospher Si 1000 was described. After coating this material with PDPM (25% w/w) the enantioseparations were compared in ammonium acetate solutions in pure organic and aqueous organic solvents. The effect of water additives to methanolic BGE was similar to what would be expected in HPLC conditions. With increasing water content, the migration times and a separation factor of neutral chiral analyte ethylbenzoin increased and peak efficiency decreased [33].

In contrast to the methanolic solution, the EOF increased in acetonitrile-water mixtures with increasing amounts of water in the range of <20% v/v. This phenomenon allows us to observe interesting effects in chiral CEC. Thus, the addition of increased amounts of water to the ammonium

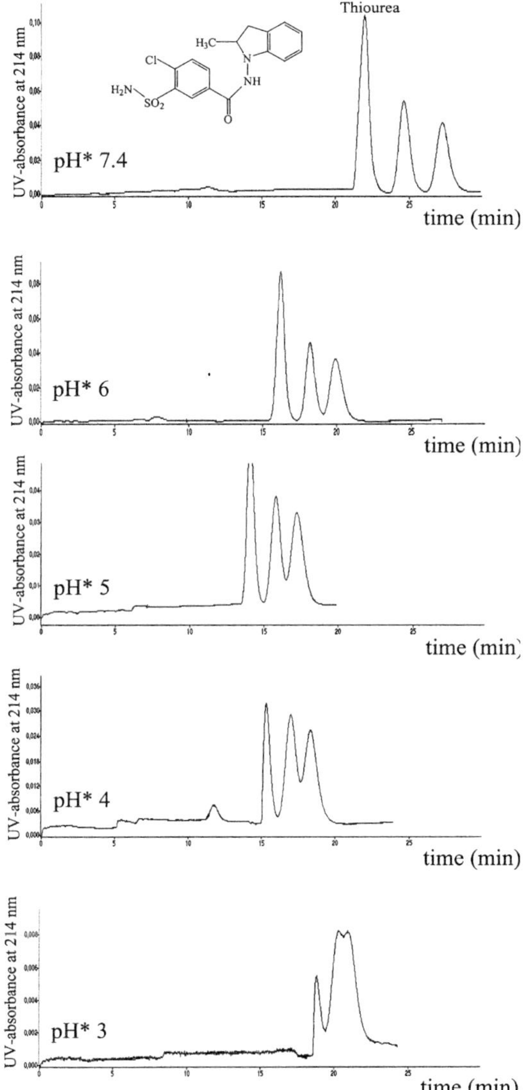

Figure 16. Enantioseparations of indapamide at different apparent pH* of 10 mM methanolic ammonium acetate solution in a capillary packed with CDMPC (20% w/w) on aminopropylsilanized silica (5 μm, 1000 Å). Reprinted from [34], with permission.

acetate solution in acetonitrile simultaneously allowed a shortening of the analysis time, and an improvement in peak efficiency and separation selectivity (Fig. 17) [33]. Although the mechanisms of this effect needs further investigation it seems interesting from the viewpoint of adjustment of enantioseparations in CEC.

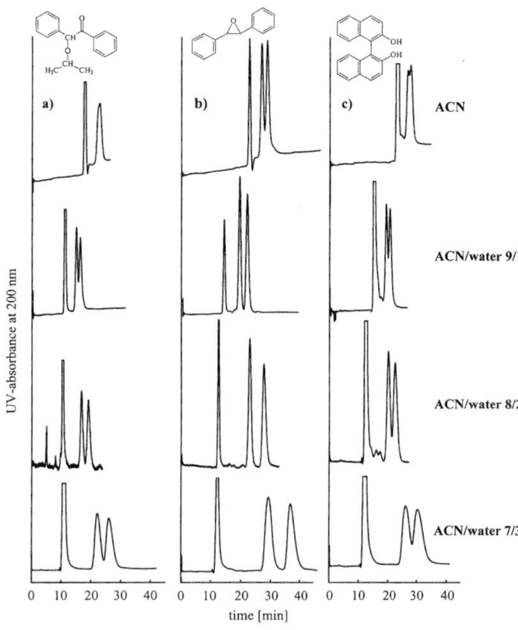

Figure 17. Effect of increasing amount of water additives on the enantioseparation of (a) isopropylbenzoin, (b) *trans*-stilbene oxide, and (c) 1,1'-binaphthyl-2,2'-diol in a solution of 2.5 mM ammonium acetate in acetonitrile. Reprinted from [33], with permission.

4.2.9 Effect of temperature on enantioseparations

Temperature represents an important variable for the optimization of separation as well as for better understanding various band-broadening contributions to peak variance. The influence of temperature on enantioseparation of *trans*-stilbene oxide in the capillary packed with Chiralcel-OD material is shown in Fig. 18 [41]. The proportional effect of the temperature on the capacity factors of both enantiomers resulted in constant separation factors in the temperature range of 10–60°C, which seems somewhat surprising. Interestingly, the plate numbers of thiourea, which is apparently controlled by the diffusion in the mobile phase, decreases with increasing temperature (Fig. 18).

However, the plate numbers for the enantiomers, which together with diffusion in the mobile phase may also be controlled by the diffusion and mass transfer in the stationary phase, do not change significantly in the temperature range studied. The most likely explanation for this effect seems to be counterbalancing of two favorable and unfavorable effects on the peak efficiency of retained compounds depending on the temperature. Thus, the dif-

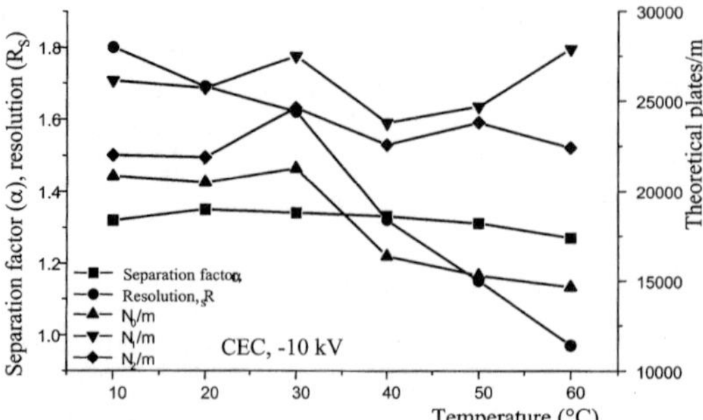

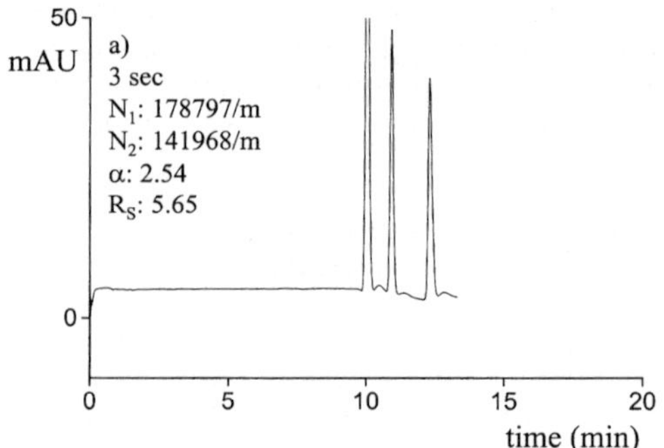

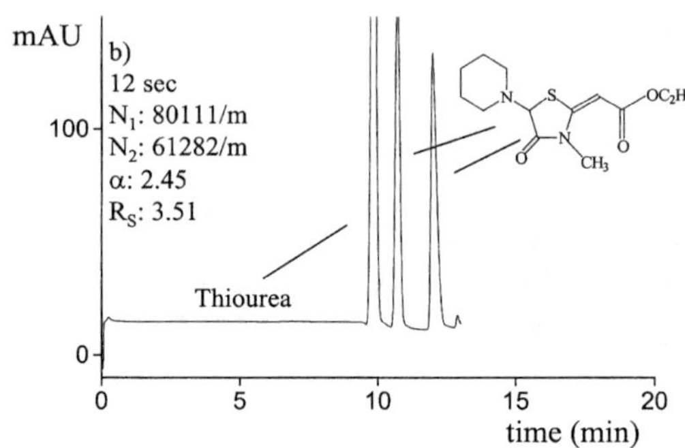

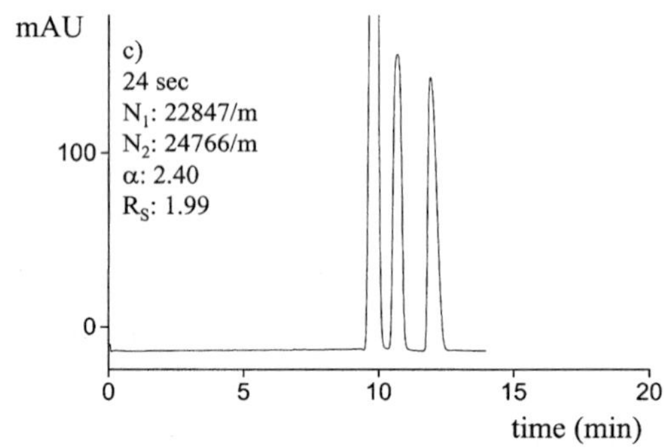

Figure 18. Temperature dependence of CEC enantio-separation parameters of *trans*-stilbene oxide. Reprinted from [41], with permission.

Figure 19. Effect of etozolin loading on the enantiosepa-ration in capillaries packed with CSP prepared by coating with CDCPC (4.8% w/w) on aminopropylsilanized silica with a particle diameter of 5 μm and a pore size of 2000 Å. Reprinted from [42], with permission.

fusion in the mobile phase will increase with increasing temperature and this may result in lower plate numbers for the compounds for which the diffusion in the mobile phase is the major contributor to the band broadening. This is perhaps the case with thiourea. A similar negative effect of the increasing temperature will also be present for the retained components such as the enantiomers of *trans*-stilbene oxide in this particular case. However, for the latter it might be possible that the negative effect of increasing diffusion in the mobile phase will be compen-sated by a positive effect of increasing temperature on mass transfer (diffusion through the solid film of a poly-saccharide and specific intermolecular interactions be-tween a stationary and a mobile phase). The multivariate effect of temperature on various band broadening contrib-utors does not allows us to clearly attribute its effect to one or another phenomenon. The effect of temperature on enantioseparation characteristics similar to that shown in Fig. 18 has also been reported for vancomycin-[36] and teicoplanin-[37] based CSPs.

4.2.10 Effect of chiral analyte loading on enantioseparations

A drastic decrease in peak efficiency due to overloading of capillary columns has been observed in several studies on enantioseparations in CEC using aqueous [29] as well as nonaqueous BGEs [41, 42]. This effect seems to be more significant for CSPs with a lower loading of a chiral selector (Fig. 19) [42].

4.3 Towards high peak efficiency

Step-by-step optimization of CEC enantioseparations in NAQ-CEC led to a better understanding of band broad-ening mechanisms and allowed us to achieve high peak

efficiencies in the range of 200 000 counts per meter [42]. The experience in preparation of CSP, packing capillary columns, injection of a sample, and performing CEC sep-aration in nonaqueous buffers are briefly summarized below. Thus, native, aminopropylsilanized or octadecyl-modified silica can be used as a "inert" support for prepa-ration of covalently immobilized or coated-type CSPs for NAQ-CEC enantioseparations. The pore size of a silica is extremely important for a generation of sufficiently strong EOF and performing highly efficient enantioseparations. Decreasing the particle diameter of silica is certainly another promising domain, which has not yet been studied in detail.

Although electrically neutral, the polysaccharide derivatives even at higher loadings onto the silica allow a generation of sufficiently strong EOF. However, it is recommended to perform a separation at the lowest possible loading of a chiral selector onto the silica [41]. This will allow the generation of a stronger EOF and avoids the limitations in peak efficiency caused by low mass transfer rates. The preference of a lower loading from the viewpoint of peak efficiency was also observed for monolithic-type CSPs in CEC mode [38, 39]. The amount of injected analyte becomes another important issue in capillary separation techniques and must be kept at a reasonable level – one that is possible to detect with required quantification parameters but still does not overload a capillary. Thus, considering the aforementioned, recently the peak efficiencies in the range of 207 000–213 000 counts per meter were reported in NAQ-CEC (Figs. 14 and 20) with

particulate modified silica [34]. The efficiency of the first peak as high as 242 000 counts per meter was recently reported with the monolithic-type CSP in one example [39]. The plate numbers on this level make a clear difference between a chiral CEC and capillary LC and evoke further research interest in this field.

4.4 Enantioseparations in CEC *vs.* CE

From chiral selectors studied in NAQ-CEC, polysaccharide derivatives could not be effectively used in CE mode due to their insolubility or low chiral recognition ability in buffers for aqueous and nonaqueous-CE [49]. Cinchona alkaloids have been studied in the NAQ-CE mode but exhibited a rather limited chiral recognition ability [15, 19] compared to their potential demonstrated in the CEC mode [31, 38, 39]. The enantioseparation properties of macrocyclic antibiotics in CE and CEC seem to be somewhat complementary. Thus, it has been shown in previous CE studies that macrocyclic glycopeptide antibiotics, such as vancomycin, teicoplanin, and ristocetin, are better suited chiral selectors for the enantioseparation of acidic compounds [70–72]. In contrast to this, the CSPs

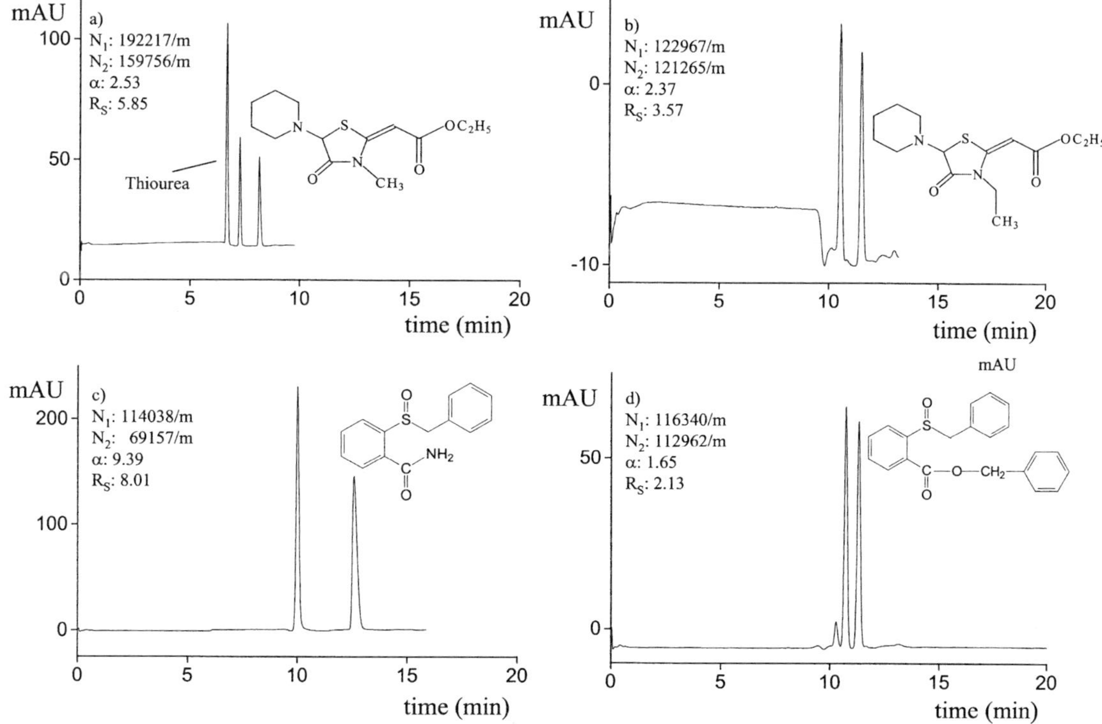

Figure 20. CEC enantioseparations of (a) etozolin, (b) piprozolin, (c) 2-(benzylsulfinyl)benzamide, and (d) 2-(benzylsulfinyl)benzoic acid benzylester. Capillary and separation conditions as in Fig. 19. Reprinted from [42], with permission.

Electrophoresis 2000, *21*, 4159–4178

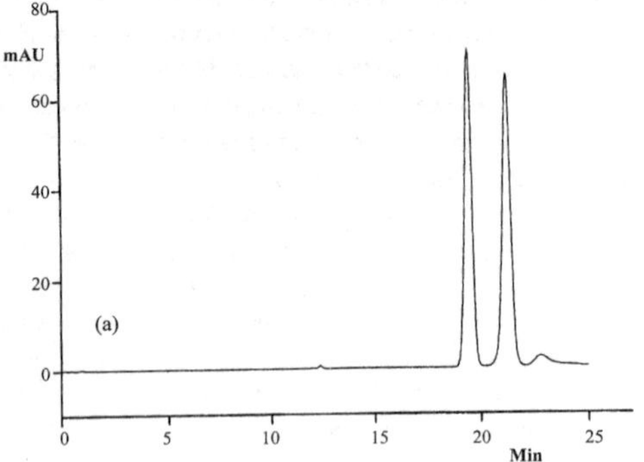

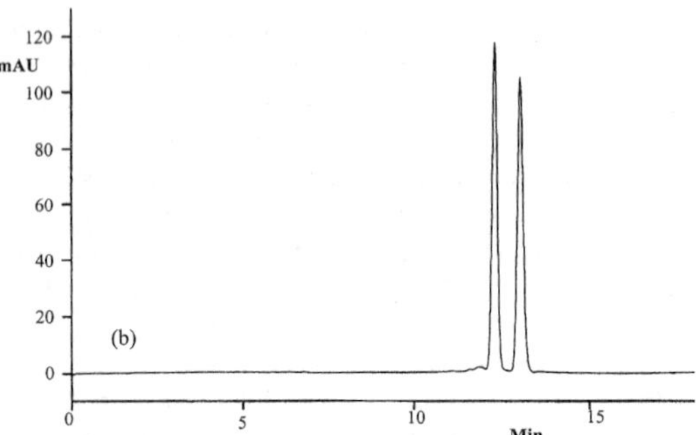

Figure 21. CEC enantioseparation of thalidomide on vancomycin-based CSP under (a) aqueous and (b) nonaqueous conditions. Reprinted from [35], with permission.

prepared based on vancomycin and teicoplanin were ineffective for CEC enantioseparation of acidic compounds and exhibited a significant enantioseparation ability towards neutral and basic chiral analytes [35–37].

4.5 Enantioseparations in nonaqueous *vs.* aqueous CEC

Almost all CSPs examined in NAQ-CEC have also been studied in the aqueous CEC mode. Although polysaccharide- and polymethacrylate-type CSPs are studied less from this point of view [29, 30, 32, 34, 40–42], the peak efficiencies reported in the nonaqueous mode [34, 41, 42] are somewhat higher than under aqueous conditions [29, 73, 74]. The particulate-type CSPs based on cinchona alkaloids seem to be equally effective for both aqueous [28, 57] and nonaqueous mode [31]. However, the highest peak efficiency for this type of monolithic CSP was reported under nonaqueous conditions [39]. More detailed studies in both aqueous and nonaqueous CEC have been performed for macrocyclic antibiotic-type CSPs [35–37]. It has been shown that more chiral analytes could be resolved with macrocyclic antibiotic-type

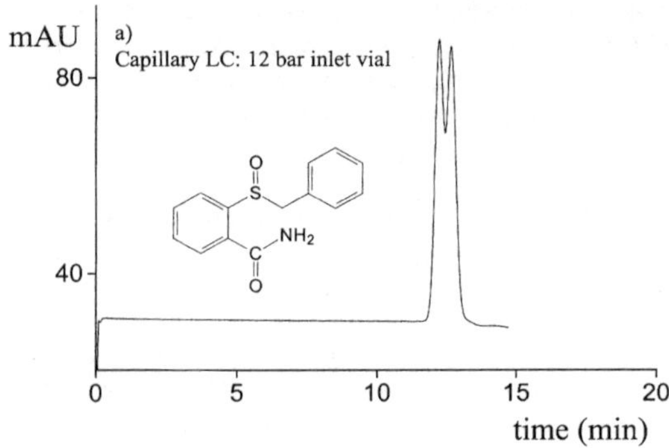

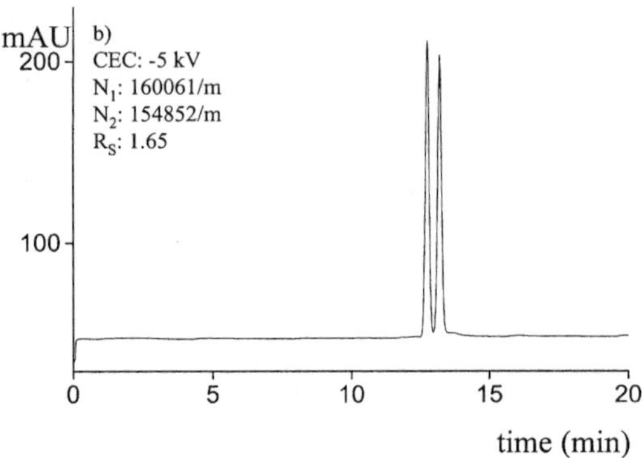

Figure 22. Enantioseparation of 2-(benzylsulfinyl)benzamide in (a) capillary LC and (b) CEC mode. Capillary and separation conditions as in Fig. 19. Reprinted from [42], with permission.

CSPs in polar organic mode compared to reversed-phase mode. Moreover, peak efficiencies were almost always higher under nonaqueous conditions than in aqueous BGEs (Fig. 21) [35].

4.6 CEC *vs.* capillary LC

CEC offers clear advantages compared to HPLC in common-size columns due to miniaturization. However, all advantages of miniaturization are also offered by capillary LC. Technical problems related to a generation of constant flow of a mobile phase in a low µL and nL range, as well as injection of minute samples without introducing additional band-broadening sources in the separation system, was a problem in capillary LC until the past few years. For these reasons, this technique has not found wide acceptance for solving analytical problems of practical importance. However, with the development of the newest software-assisted instruments and flow-splitting devices, these problems seem to be solved in capillary

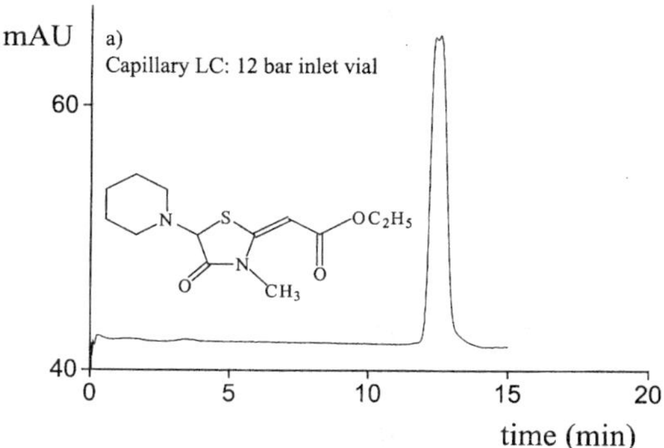

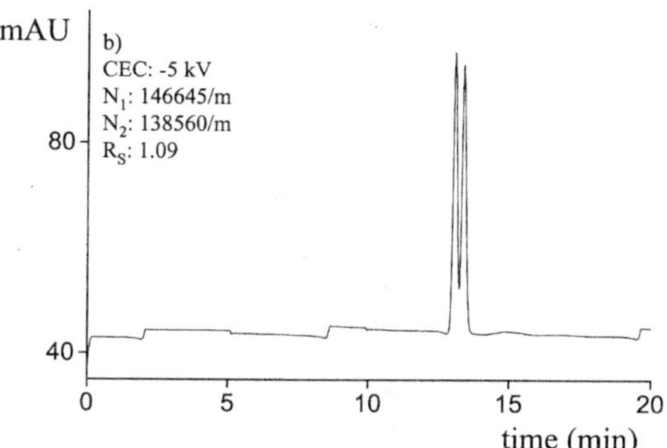

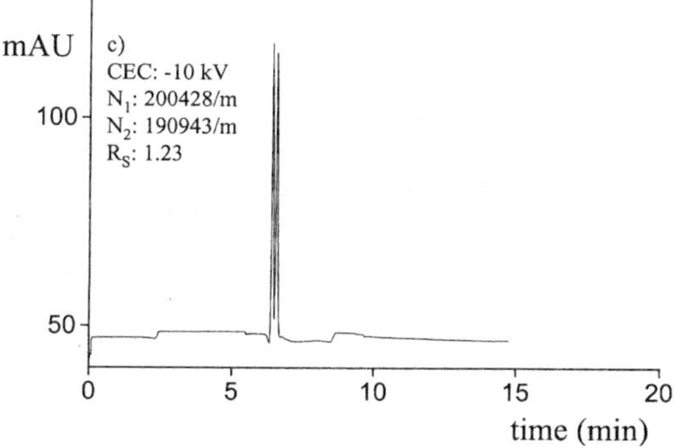

Figure 23. Enantioseparation of etozolin in (a) capillary LC and (b, c) CEC mode. Capillary and separation conditions as in Fig. 19. Reprinted from [42], with permission.

LC, which may appear to be experimentally a somewhat simpler technique than CEC because the latter implies additional requirements on the mobile phases, such as to be conductive and to allow a generation of sufficient EOF.

For the above-mentioned reasons, CEC has to offer a significant gain in separation characteristics in order to be accepted by practitioners in analytical laboratories. It is obvious that CEC is still not developed to a status where it can claim that it routinely meets the aforementioned strict requirements in day-to-day analyses. However, the first examples illustrating great potential of chiral CEC in packed capillaries were obtained under aqueous [26–28, 31] as well as nonaqueous conditions [35, 39, 42]. Two examples are shown below in (Figs. 22 and 23). In Fig. 22, the low separation and retention factors of the enantiomers of a chiral sulfoxide, 2-(benzylsulfinyl)benzamide, do not allow acceptable enantioseparation in capillary LC mode. In contrast to this, baseline enantioseparation was observed in the CEC mode with the same capacity and separation factors (Fig. 22b). For the antidiuretic drug etozolin almost no enantioseparation was observed in the capillary LC mode (Fig. 23a) but good enantioseparation was observed with the same capacity factor in the CEC mode (Fig. 23b), which could be improved further by optimizing the linear flow rate of the mobile phase (Fig. 23c).

Financial support by Deutsche Forschungsgemeinschaft (DFG) of our project on capillary electrochromatography is acknowledged.

Received September 18, 2000

5 References

[1] Nishi, H., Terabe, S., (Eds.), *J. Chromatogr. A* 2000, *875*, 1–492.

[2] Chankvetadze, B., *Capillary Electrophoresis in Chiral Analysis*, Wiley & Sons, Chichester, UK 1997.

[3] Fanali, S., *J. Chromatogr. A* 1997, *792*, 227–267.

[4] Vigh, G., Sokolowski, P., *Electrophoresis* 1997, *18*, 2305–2310.

[5] Kaniansky, D., Simunicova, E., Ölvecka, E., Ferancova, A., *Electrophoresis* 1999, *20*, 2786–2793.

[6] Chankvetadze, B., Burjanadze, N., Bergenthal, D., Blaschke, G., *Electrophoresis* 1999, *20*, 2680–2685.

[7] Glukhovskiy, P., Vigh, G., *Electrophoresis* 2000, *21*, 2010–2015.

[8] Lanz, M., Caslavska, J., Thormann, W., *Electrophoresis* 1998, *19*, 1081–1090.

[9] Wang, F., Khaledi, M., *Anal. Chem.* 1996, *68*, 3460–3467.

[10] Stalcup, A. M., Gahm, K. H., *J. Microcol. Sep.* 1996, *8*, 145–150.

[11] Valko, I. E., Siren, H., Riekkola, M.-L., *J. Chromatogr. A* 1996, *737*, 263–272.

[12] Bjørnsdottir, I., Hansen, S. H., Terabe, S., *J. Chromatogr. A* 1996, *745*, 37–44.

[13] Wang, F., Khaledi, M. G., *J. Chromatogr. A* 1999, *817*, 121–128.

[14] Wang, F., Khaledi, M. G., *J. Chromatogr. B* 1999, *731*, 187–197.

[15] Piette, V., Lämmerhofer, M., Lindner, W., Crommen, J., *Chirality* 1999, *11*, 622–630.

[16] Vincent, J. B., Vigh, G., *J. Chromatogr. A* 1998, *816*, 233–241.

Electrophoresis 2000, *21*, 4159–4178

[17] Valko, I. E., Siren, H., Riekkola, M. L., *Chromatographia* 1996, *43*, 242–246.

[18] Mori, J., Ueno, K., Umeda, T., *J. Chromatogr. A* 1997, *757*, 328–332.

[19] Piette, V., Fillet, M., Lindner, W., Crommen, J., *J. Chromatogr. A* 2000, *875*, 353–360.

[20] Tacker, M., Glukhovskiy, P., Cai, H., Vigh, G., *Electrophoresis* 1999, *20*, 2794–2798.

[21] Valko, I. E., Siren, H., Riekkola, M.-L., *J. Microcol. Sep.* 1999, *11*, 199–208.

[22] Cai, H., Vigh, G., *J. Pharm. Biomed. Anal.* 1998, *16*, 615–621.

[23] Sahota, R. S., Khaledi, M. G., *Anal. Chem.* 1994, *66*, 1141–1146.

[24] Wang, F., Khaledi, M. G., *J. Chromatogr. A* 2000, *875*, 277–293.

[25] Li, S., Lloyd, D. K., *Anal. Chem.* 1993, *65*, 3684–3690.

[26] Wolf, C., Spence, P. L., Pirkle, W. H., Derrico, E. M., Cavender, D. M., Rozing, G. P., *J. Chromatogr. A* 1997, *782*, 175–179.

[27] Wistuba, D., Schurig, V., *J. Chromatogr. A* 2000, *875*, 255–276.

[28] Lämmerhofer, M., Lindner, W., *J. Chromatogr. A* 1998, *829*, 115–125.

[29] Krause, K., Girod, M., Chankvetadze, B., Blaschke, G., *J. Chromatogr. A* 1999, *837*, 51–67.

[30] Krause, K., Chankvetadze, B., Okamoto, G., Blaschke, G., *Electrophoresis* 1999, *20*, 2772–2778.

[31] Tobler, E., Lämmerhofer, M., Lindner, W., *J. Chromatogr. A* 2000, *875*, 341–352.

[32] Meyring, M., Chankvetadze, B., Blaschke, G., *J. Chromatogr. A* 2000, *876, 157–167.*

[33] Krause, K., Chankvetadze, B., Okamoto, G., Blaschke, G., *J. Microcol. Sep.* 2000, *12*, 398–406.

[34] Girod, M., Chankvetadze, B., Blaschke, G., *J. Chromatogr. A* 2000, *887*, 439–455.

[35] Wikström, H., Svensson, L. A., Torstensson, A., Owens, P. K., *J. Chromatogr. A* 2000, *869*, 395–409.

[36] Karlsson, C., Karlsson, L., Armstrong, D. W., Owens, P. K., *Anal. Chem.* 2000, 4394–4401.

[37] Karlsson, C., Wikström, H., Armstrong, D. W., Owens, P. K., *J. Chromatogr. A* 2000, 4614–4622.

[38] Lämmerhofer, M., Peters, E. C., Yu, C., Svec, F., Frechet, J. M., Lindner, W., *Anal. Chem.* 2000, in press.

[39] Lämmerhofer, M., Svec, F., Frechet, J. M., Lindner, W., *Anal. Chem.* 2000, 4623–4633.

[40] Girod, M., Chankvetadze, B., Blaschke, G., unpublished results.

[41] Girod, M., Chankvetadze, B., Blaschke, G., *Electrophoresis* 2000, *21*, in press.

[42] Girod, M., Chankvetadze, B., Okamoto, G., Blaschke, G., *J. Sep. Sci.*, in press.

[43] Miller, J. L., Khaledi, M. G., in: Khaledi, M. G., (Ed.), *High Performance Capillary Electrophoresis*, John Wiley and Sons, New York 1998, pp. 525–555.

[44] Hayek, M., *J. Phys. Colloid Chem.* 1951, *55*, 1527–1533.

[45] Korchemnaya, E. K., Ermakov, A. N., Bochlova, L. P., *Anal. Khim. (J. Anal. Chem. USSR)* 1978, *33*, 816–821.

[46] Walbroehl, Y., Jorgenson, J. W., *J. Chromatogr.* 1984, *315*, 135–143.

[47] Walbroehl, Y., Jorgenson, J. W., *Anal. Chem.* 1986, *58*, 479–481.

[48] Ye, B., Khaledi, M. G., *6th International Symposium on High Performance Capillary Electrophoresis*, San Diego, CA 1994, abstract No. 133.

[49] Otsuka, K., Kamamoto, C., Terabe, S., *Chromatography (Japan)* 1996, *17*, 304–305.

[50] Rosini, C., Uccello-Barretta, G., Pini, D., Abete, C., Salvadori, P., *J. Org. Chem.* 1988, *58*, 4579–4581.

[51] Pettersson, C., *J. Chromatogr.* 1984, *316*, 553–567.

[52] Salvadori, P., Rosini, G., Pini, D., Bertucci, C., Uccello-Barretta, G., *Chirality* 1989, *1*, 161–166.

[53] Pettersson, C., Schill, G., *J. Liq. Chromatogr.* 1986, *9*, 269–290.

[54] Chankvetadze, B., Endresz, G., Blaschke, G., *Electrophoresis* 1994, *15*, 804–807.

[55] Li, D., Remcho, V. T., *J. Microcol. Sep.* 1997, *9*, 389–397.

[56] Venema, E., Kraak, J. C., Poppe, H., Tijssen, R., *J. Chromatogr. A* 1999, *837*, 3–15.

[57] Lämmerhofer, M., Lindner, W., *J. Chromatogr. A* 2000, *887*, 421–437.

[58] Okamoto, Y., Mohri, H., Hatada, K., *Chem. Lett.* 1988, 1879–1882.

[59] Okamoto, Y., Yashima, E., *Angew. Chemie* 1998, *37*, 1020–1043.

[60] Lämmerhofer, M., Lindner, W., *J. Chromatogr. A* 1998, *839*, 167–181.

[61] Krstulovic, A. M., Rossey, G., Porziemsky, J. P., Long, D., Chekreum, I., *J. Chromatogr.* 1987, *411*, 461–465.

[62] Aboul-Enein, H. Y., Serignese, V., Bojarski, J., *J. Liq. Chromatogr.* 1993, *16*, 2741–2749.

[63] Dingenen, J. in: Subramanian, G. (Ed.), *A Practical Approach to Chiral Separations by Liquid Chromatography*, VCH, New York 1994, pp. 115–181.

[64] Chankvetadze, B., Yamamoto, C., Okamoto, Y., *Chem. Lett.* 2000, 352–353.

[65] Miller, L., Orihuella, C., Fronek, R., Murphy, J., *J. Chromatogr. A* 1999, *865*, 211–226.

[66] Miller, L., Orihuella, C., Fronek, Honda, D., Depremert, O., *J. Chromatogr. A* 1999, *849*, 309–317.

[67] Meyer, V. R., *Practical High-Performance Liquid Chromatography*, John Wiley & Sons, Chichester, UK 1994.

[68] Guiochon, G., Golshan-Shirazi, S., Katti, A., *Fundamentals of Preparative and Nonlinear Chromatography*, Academic Press, New York 1994.

[69] Chen, Y., Kele, M., Sajonz, P., Sellergren, B., Guiochon, G., *Anal. Chem.* 1999, *71*, 928–938.

[70] Armstrong, D. W., Rundlett, K. L., Chen, J.-R., *Chirality* 1994, *6*, 496–509.

[71] Armstrong, D. W., Gasper, M. P., Rundlett, K. L., *J. Chromatogr. A* 1995, *689*, 285–304.

[72] Rundlett, K.-L., Gasper, M. R., Zhou, E. Y., Armstrong, D. W., *Chirality* 1996, *8*, 88–107.

[73] Mayer, S., Briand, X., Francotte, E., *J. Chromatogr. A* 2000, *875*, 331–339.

[74] Otsuka, K., Mikami, C., Terabe, S., *J. Chromatogr. A* 2000, *887*, 457–463.

Review

Mark R. Hadley[1]
Patrick Camilleri[2]
Andrew J. Hutt[3]

[1]Department of Analytical
Sciences, SmithKline
Beecham Pharmaceuticals,
Tonbridge, Kent, UK
[2]Department of Analytical
Sciences, SmithKline
Beecham
Pharmaceuticals,Harlow,
UK
[3]Department of Pharmacy,
King's College London,
Franklin-Wilkins Building,
London, UK

Enantiospecific analysis by capillary electrophoresis: Applications in drug metabolism and pharmacokinetics

Enantiospecific analysis has an important role in drug metabolism and pharmacokinetic investigations and its now no longer acceptable to determine total drug, or metabolite, concentrations following the administration of a racemate. Inspite of the fact that capillary electrophoresis (CE) has become an essential technique in pharmaceutical and enantiospecific analysis, the chromatographic methodologies remain the most commonly used approach for the determination of the enantiomeric composition of drugs in biological fluids. The application of CE to bioanalysis has been slow, which is in part associated with the complexity of biological matrices together with the relatively poor concentration limits of detection achievable. However, as a result of its verstility, high separation efficiency, minimal sample requirements, speed of analysis and low consumable expense CE is likely to play an increasingly significant role in the area. This review present an oberview of enantiospecific CE in bioanalysis in which the approaches to enantiomeric resolution and the problems associated with biological matrices are briefly discussed. The application of enantiospecific CE to samples of biological origin is illustrated using examples where the methodology has either solved an analytical problem, or provided a useful alternative to the currently available chromatographic methods. Such improvements in methodology are associated with either the high separation efficiency and/or microanalytical capabilities of the technique. Enantiospecific CE will not replace the chromatographic methodologies but does provide the bioanalyst with a useful addition to his armamentarium.

Keywords: Enantiospecific analysis / Capillary electrophoresis / Drug metabolism / Pharmacokinetics / Biological fluids / Review

EL 3874

Contents

Correspondence: Dr. Andrew J. Hutt, Department of Pharmacy, King's College London, Franklin-Wilkins Building, Stamford Street, London, SE1 8WA, UK
E-mail: pharmacy-dept@klc.ac.uk
Fax: + 44-20-7848-4800

Abbreviations: CSP, chiral stationary phase; **DMPK**, drug metabolism and pharmacokinetics; **EDDP**, 2-ethylidene-1,5-dimethyl-3,3-diphenylpyrrolidine; **FMO**, flavin-containing monooxygenase; **TM-β-CD**, heptakis-(2,3,6-tri-*O*-methyl)-β-cyclodextrin

1 Introduction

Drug metabolism and pharmacokinetic (DMPK) studies have made, and continue to make, fundamental contributions to the drug discovery and development process and the regulatory guidelines for drug safety evaluation all make reference to metabolic and pharmacokinetic data. The development of analytical methodology suitable for the accurate and reproducible determination of both drug and metabolite concentrations in biological media is obviously an essential prerequisite for the determination of

409

    Electrophoresis 2000, *21*, 1953–1976

the pharmacokinetic parameters of a drug. If a chiral compound is to be administered, either as a single enantiomer or a racemate, then such analytical methodology must be stereospecific for meaningful pharmacokinetic analyses to be performed. In recent years, as a result of the interest of the regulatory authorities and increased realisation of the potential significance of the pharmacodynamic and pharmacokinetic properties of the enantiomers of chiral drugs, it has become unacceptable to determine "total biofluid concentrations" of drugs administered as racemates.

While a number of analytical techniques provide stereochemical information, those employed by the bioanalyst are, in the main, based on the separation sciences, *i.e.*, the chromatographic techniques, particularly gas chromatography (GC) and high performance liquid chromatography (HPLC). With respect to enantiospecific methodology, the most significant advance in recent years has been the development of chiral stationary phases (CSPs) for both these techniques, and the availability of essentially off-the-shelf technology for enantiomeric resolution has greatly facilitated stereospecific bioanalysis.

In contrast to the chromatographic methods, there are relatively few reports concerning the bioanalytical applications of capillary electrophoresis (CE) and even fewer which deal with enantiospecific bioanalysis. For example, a recent review of bioanalytical applications of CE cited only 27 publications, from a database of over 3000, concerned with drug analysis in biological matrices [1]. Indeed, the author made the point that of the 27 reports cited, not all presented fully validated methods. A similar review, published in 1997, listed 74 publications involving a variety of CE techniques and detection systems, concerned with drug analysis in biological media including blood, plasma, serum, urine, tissue extracts, hepatic microsomal preparations and gastric fluid [2]. The various studies were described by the authors as being quantitative (validated), semiquantitative (partially validated) and qualitative [2]. However, these same authors were able to cite only six publications where CE had been used for pharmacokinetic investigations. Shihabi [3] recently reviewed the use of CE in therapeutic drug monitoring, an application where the minimal sample preparation, high efficiency, speed of analysis and low operating costs of CE offer distinct advantages over the chromatographic approaches. Thirty-eight publications were cited listing applications of CE for drug analysis in either serum or urine and examination of these indicated that in a number of instances the methodology was associated with the analysis of "artificially prepared" rather than "real" samples [3]. With respect to enantiospecific bioanalytical CE methods, the review by Perrett [1] lists eight publications

(up to early 1996) and that by Shihabi [3] cites 11, the majority of which employ either cyclodextrin (CD) or a modified CD as a chiral selector.

The aim of the present article is to illustrate the application of enantiospecific CE methodologies to DMPK investigations and to attempt to indicate some of the advantages and limitations of the methodologies employed. Additional reviews concerning the application of CE for the analysis of drugs and metabolites include Desiderio and Fanali [4], Bojarski and Aboul-Enein [5] and Boone *et al.* [6]. Xu [7] has recently (1999) summarised the clinical chemistry applications, including the analysis of drugs and metabolites, of CE using the *Chemical Abstract Selects for Capillary Electrophoresis*, through the period October 1996 to October 1998.

2 Enantiomeric resolution by CE

Enantiomeric resolution by CE has been extensively investigated and is the subject of a number of reviews [8–13], single-topic issues of specialised journals [14, 15], and monographs [16]. The most commonly used enantioselective CE techniques involve the direct, rather than the indirect, approach to enantiomeric analysis, using free-solution CE or micellar electrokinetic chromatography (MEKC), with resolution being achieved by the addition of a chiral selector to the run buffer. A number of chiral selectors have been employed including CDs, crown ethers, proteins, oligosaccharides and macrocyclic glycopeptide antibiotics [8–16], some of which show remarkable versatility and selectivity, *e.g.*, over 100 racemates have been resolved using ristocetin A and vancomycin as chiral selectors [17, 18]. In addition, the native and derivatised CDs have proved to be exceptionally useful chiral selectors and it has been estimated that more than 70% of all enantioseparations by CE have employed such modifiers [16]. However, the use of such selectors is not without problems and both source and between-batch variation in material from the same source, may result in differences in enantioselectivity and possible loss of analyte resolution [19]. This is obviously a potential complication when such materials are to be used for routine analysis. Also, the addition of multiple selectors to the run buffer, either chiral or achiral, may facilitate and enhance the resolution and separation of structurally similar analytes [20–22], a situation frequently encountered in metabolic investigations.

There is no doubt that the development and commercialisation of chromatographic CSPs, particularly for HPLC, greatly facilitated enantiospecific analysis. However, a number of problems, the majority of which are associated with the low efficiencies of the phases, remain. In contrast, the high efficiencies typically achievable with CE facilitate the baseline resolution of enantiomers with very low separation factors *e.g.*, 1.01, which would be either impossible, or extremely difficult using conventional HPLC or GC [6, 16]. In addition, while the enantioselectivity of a CSP may be satisfactory, the chemical selectivity may be inadequate, such that structurally similar metabolites coelute with the parent drug peak [23, 24]. The separation efficiency of CE, however, can enable the simultaneous separation and resolution of structurally similar metabolites [25].

CE also has additional advantages compared to chromatography in terms of the speed and relative ease of method development. For example, the run buffer composition may be changed rapidly in order to establish the optimal separation conditions required; small capillary volumes allow the use of relatively expensive chiral selectors as a result of the small quantities required; and the consumable costs of the technique are relatively low [8, 10, 26]. Enantiospecific chromatographic methodology has also been described as relatively "delicate" and interlaboratory method transfer may prove difficult, whereas data presented by Altria *et al.* [27] has indicated acceptable robustness in an intercompany cross-validation exercise concerned with the analysis of clenbuterol enantiomers by CE.

3 Bioanalysis by CE

3.1 General aspects

Analytical samples arising from DMPK investigations frequently contain a variety of structurally related analytes, present in relatively low concentrations in a complex matrix. To add to this the problem of distinguishing between stereoisomers presents the bioanalyst with a challenging task. An additional problem is associated with the number of samples which may be generated during a pharmacokinetic study and the speed of analysis; together with the associated costs this may be of considerable significance. CE therefore has a number of advantages compared to chromatography for analyses of this type, particularly with respect to the high separation efficiencies achievable for the resolution of structually similar compounds, speed, and hence cost of analysis and reduction in the use of organic solvents [8, 10, 26]. An additional advantage of CE is that relatively low UV detection wavelengths (typically 190–200 nm) may be employed, which, due to the absorptivity of typical mobile phases, are generally not feasible with HPLC. Use of such wavelengths does have the disadvantage that the number of potentially interfering components is generally increased, with the result that electropherograms become more complex, particularly if biological fluids are injected

on-column (see below). However, bioanalytical data derived from a number of CE methods have been found to compare favourably with those obtained using established chromatographic assays, examples of which are cited below.

A frequently cited problem associated with the CE analysis of drugs and metabolites in biological media are the relatively poor limits of detection achievable, which are associated with limited sample volume loading capacity and the short optical path length available when employing on-capillary detection [5, 28–31]. This factor has probably contributed to the relatively small number of CE applications to bioanalysis reported to date [28–31]. The problem has been addressed, in terms of the technology, in a number of ways including: modification of injection technique (*e.g.*, analyte stacking, field amplification, transient isotachophoresis), on-line preconcentration approaches (*e.g.*, solid-phase preconcentration, membrane preconcentration) and increasing the sensitivity of detector systems (*e.g.*, improved detector optics, specialised capillary detection windows, electrochemical detection, laser-induced fluorescence (LIF) detection, on-line mass spectrometry). These various approaches, together with their advantages and limitations, have been reviewed elsewhere [28–33].

Direct on-column injection of samples of biological origin with minimal sample pretreatment is an obvious advantage of CE compared to HPLC where such an approach could easily result in irreversible modification or damage to an expensive CSP. The lack of an extensive extraction process during the analytical procedure saves time, effort and removes a potential analytical variable [31, 32]. This approach, in combination with the low sample injection volume required, also has the advantage that a much smaller sample is necessary, which may be of significance for certain studies, *e.g.*, investigations involving paediatric patients and small experimental animals [28, 34–36]. It is obviously the case that for this approach to be useful, the limit of detection (LOD) and limit of quantification (LOQ) must not present a significant problem.

However, direct analysis of biological fluids by CE is not without problems. Samples containing protein, *e.g.*, plasma, may produce complex electropherograms due to the presence of broad protein peaks and also adhesion of protein to bare silica can result in changes to the electro-osmotic flow due to modification of the capillary surface. Nakagawa *et al.* [37] overcame such problems by the use of MEKC, employing sodium dodecyl sulphate (SDS) as a pseudostationary phase. SDS is known to bind proteins and denature them [38], thus preventing capillary wall adsorption and, in addition, releasing any protein-bound drugs. The additional selectivity available due to the presence of the pseudostationary phase also facilitated resolution between the analytes of interest and the peaks corresponding to the proteins present [37]. However, even with the addition of SDS, some modification of the capillary surface may occur, *e.g.*, the migration times of the diastereomers of L-buthionine-(*R,S*)-sulphoximine increased by approximately 2 min following 20 repeated injections of plasma samples [38].

Direct analysis of urine is easier than plasma as appreciable modification of the capillary by endogenous materials is not a significant problem. Urine does, however, contain a large number of components and the resulting electropherograms may be complex and problems may arise due to peak identification and resolution [39, 40]. In addition, both plasma and urine contain high electrolyte concentrations which may result in problems due to the high conductivity of the matrix, yielding a reduction in field strength in the sample zone and possible band broadening.

If direct analysis of biological media is not feasible, either as a result of the endogenous components present in the matrix or due to the limit of quantitation, then conventional drug/metabolite isolation procedures may be used, *i.e.*, liquid-liquid or solid-phase extraction. Such approaches have the advantage of removing proteins, other endogenous compounds and inorganic salts, and facilitate analyte concentration. Provided the residue is reconstituted in either a buffer of lower ionic strength than the background electrolyte or an aqueous-organic mixture, then analyte stacking will also occur. However, if such methods are adopted then one of the significant advantages of CE compared to chromatography, *i.e.*, its microanalytical ability, is not utilised and the sample manipulations are similar to those used in chromatographic analysis. However, CE still retains the advantage in terms of separation efficiency which is obviously of significance for both enantiomeric resolution and separation of structurally related metabolites.

A recent approach to analyte isolation prior to CE analysis is the supported liquid membrane (SLM) technique. This approach, which enables selective extraction of analytes from complex mixtures, has recently been applied to the isolation of the terbutaline prodrug bambuterol from plasma prior to determination of enantiomeric composition by CE [19]. Using this approach, together with a double stacking procedure to increase sensitivity, no loss in separation was observed on introduction of a 3 µL sample, obtained by 50% methanol dilution of a sample isolated by SLM, onto the capillary. Enantiomeric resolution was achieved using dimethyl-β-CD as a chiral selector and analytical repeatability was within 10% at a concen-

tration of 4 nM of each enantiomer. While the methodology was not employed for the analysis of samples of biological origin, the potential of the approach for bioanalysis in CE is considerable [19].

3.2 Coupled techniques

Coupled column or multidimensional techniques for the enantiomeric analysis of drugs in biological media have been used in chromatography for a number of years [41, 42] and the extension of this approach to electrophoretic separations has obvious potential. To date, coupled CE has not been used for the analysis of pharmaceuticals, but two recent publications have employed model analytes to illustrate the methodology [43, 44]. Krásenky *et al.* [43] reported the separation and enantiomeric resolution of mandelic and *m*-methoxymandelic acids using a coupled two capillary system. The background electrolytes for the two capillaries were selected to provide optimal stereochemical selectivity in the first capillary and optimal detection sensitivity in the second. Enantiomeric resolution was achieved in the first capillary using copper (II)-aspartame as the chiral selector in 20 mM acetate buffer, pH 4.4. This was followed by a second capillary containing acetate buffer of the same concentration but lower pH (3.1). The on-line capillary coupling resulted in an increase in sensitivity by a reduction in the LOD by six orders of magnitude in comparison to a single capillary system containing the chiral selector in the run buffer [43].

This approach has been extended by Danková *et al.* [44] using CE coupled with on-line capillary isotachophoresis (ITP). Using tryptophan as a model analyte it was possible to determine the enantiomeric composition of the amino acid in urine samples and in an artificially prepared, 90-component mixture of organic anions. Resolution of the tryptophan enantiomers was achieved in the CE stage using 80 mM α-CD as the chiral selector in a 50 mM borate buffer, pH 9.0, containing 0.2% w/v methylhydroxyethylcellulose 30 000. Various working modes of the coupled system were investigated but analysis of urine samples required a postcolumn on-line capillary ITP clean-up for the successful determination of the analyte. Using this methodology, greater than 99% of the anionic sample constituents migrating in the ITP stack were removed and the high sample load capacity of the on-line ITP stage allowed analysis of samples corresponding to 3–6 µL of undiluted urine with an LOD of 10 nM per tryptophan enantiomer [44].

An additional approach involving microcolumn liquid chromatography coupled on-line with CE has also been reported [45]. Using this methodology, sample pretreatment takes place on the microcolumn, the analyte elution volume being totally transferred to the CE capillary which may be partially, or completely, filled with sample. This is then followed by a double stacking procedure used for analyte focusing [45]. As a result, sample pretreatment, and hence selectivity, together with an increase in sensitivity may be achieved. The applicability of the technique was demonstrated by the analysis of terbutaline spiked into plasma at an enantiomeric concentration of 5 nM. Enantiomeric resolution was achieved in the electrophoretic step using dimethyl-β-CD as the chiral selector. While the application of the above methodologies to the analysis of samples of biological origin has yet to be reported, their potential in terms of both selectivity and sensitivity is considerable.

4 Applications of enantiospecific CE in drug metabolism and pharmacokinetics

The remainder of this article will address selected topics where the application of enantiospecific CE methodology has either solved a bioanalytical problem or provided a useful alternative to the chromatographic methods currently available. A number of studies have reported the use of "spiked" rather than "real" biological samples to illustrate the application, or potential application, of CE methodology to bioanalytical investigations. In addition, drug/metabolite mixtures have also been used to illustrate the selectivity and specificity of the methodology rather than its applicability to samples of biological origin and the nature/origin of the samples will be addressed in the text. Examples of enantiospecific CE methods not discussed in any detail are presented in Table 1.

4.1 Microdialysis sampling

As pointed out above, the microanalytical nature of CE is frequently discussed as a limitation of the technique with respect to the limits of quantitation in bioanalytical investigations. However, the limited sample volume requirement of CE does have distinct advantages in some instances, particularly for those which employ a microsampling technique, *e.g.*, ultrafiltration probes and microdialysis. Microdialysis perfusion rates are typically of the order of 1 µL min^{-1} and therefore analysis of such samples by chromatography requires collection of material over several minutes for a sufficient sample volume to be acquired for analysis [2]. In contrast, microdialysis, either on- or off-line, in combination with CE analysis offers a significant advantage [34]. Hadwiger *et al.* [35, 36] used this approach to examine the pharmacokinetic properties of the enantiomers of the catecholamine isoprenaline, following intravenous administration of the racemate to rats. The plasma half-life of isoprenaline is relatively short, approximately 10 min, and therefore frequent sampling is essential in order to define the pharmacokinetic profile of the drug.

Resolution of the enantiomers of isoprenaline was achieved using 100 mM lithium acetate pH 4.75, run buffer containing 0.5 mM EDTA and 0.1 g mL^{-1} methyl-β-CD as the chiral selector, with electrochemical detection [35, 36]. Under these conditions the enantiomeric elution order was (–)- before (+)-isoprenaline with an overall run time of less than 11 min. The achiral catecholamines 3,4-dihydroxybenzylamine and dopamine were used as an internal standard for the analytical method and as an indicator of the microdialysis probe performance, respectively [35, 36]. Using a pH-mediated stacking approach, the LOD was 0.63 ng mL^{-1} with a linear response up to 1.06 µg mL^{-1} [36]. Microdialysis, in combination with peak stacking and electrochemical detection, allowed sampling at 1 or 2 min intervals for over 1 h, with sample volumes of 2 or 4 µL, and it was possible to monitor the plasma profile of both enantiomers over six half-lives. Analysis of the data obtained indicated a biexponential decline in plasma concentration and a significant difference between the elimination half-lives of the enantiomers [36]. This investigation illustrates the potential of CE, in combination with microsampling techniques, for the analysis of rapidly eliminated compounds.

4.2 Determination of metabolic phenotype

Interindividual variation in drug response, both in terms of therapeutic and adverse effects, is a considerable problem. Such variability is frequently associated with variations in drug metabolism and over the last twenty years it has been appreciated that genetic polymorphisms of the drug metabolising enzyme systems are of considerable significance [46, 47]. Examples of particular interest include the acetylation polymorphism and the debrisoquine and mephenytoin oxidative polymorphisms, with the population falling into distinct subgroups (phenotypes) depending on their ability to carry out a particular biotransformation. In the case of the above examples, the population is divided into either slow or rapid acetylators and extensive or poor metabolisers. The determination of an individual's phenotype may be readily carried out by the administration of a test or probe drug, followed by the analysis of both drug and appropriate metabolite(s) excreted in the urine. As the analysis involves urine, or enzyme-treated urine samples (for the hydrolysis of conjugates), CE offers a useful, rapid alternative analytical approach to either GC or HPLC for the determination of metabolic phenotype.

The freely available antitussive agent dextromethorphan undergoes *O*-demethylation to dextrophan, the reaction being mediated by the polymorphically expressed enzyme cytochrome P450 2D6 (see below). Li *et al.* [48] developed an achiral CE method for the determination of

Table 1. Enantiospecific electrophoretic analysis of drugs and metabolites in biological fluids

Analyte	Matrix	Chiral selector (concentration)	Run buffer	Detection	Application	Comment	Ref.
Bupivacaine	Serum	2,6-Dimethyl-β-CD (10 mM)	Tris (18 mM; pH 2.9) adjusted with phosphate, MHEC 4000 (0.1% w/w), HTAB (0.03 mM)	UV 220 nm	Analaysis of artificially prepared samples	Migration order R before S, run time *ca.* 13 min, calibration range 0.2–3.8 µg mL^{-1}, mepivacaine used as internal standard	[103]
L-Buthionine-(R,S)-sulfoximine	Plasma	–	Phosphate buffer (20 mM, pH 6.8), SDS (170 mM)	UV 190 nm	Pharmacokinetic investigation following administration to patients	Migration order L, R- before L,S-diastereomer, internal standard acetaminophen, 2.5–500 µg mL^{-1} calibration range, LOD 3.9 µg mL^{-1}	[104]
Carvedilol	Serum	Hydroxypropyl-β-CD (10 mM)	Sodium phosphate (25 mM; pH 2.5)	UV 200 nm	Determination of drug enantiomers in serum	(–)-Propranolol used as internal standard, calibration linear over the range 1–5 µg mL^{-1}, intra- and interassay coefficient of variation ranged from 4 to 15%, serum extracted using diethylether and residues redissolved in ethanol, good correlations between CE and HPLC methods	[105]
Cicletanine	Plasma	γ-CD (25 mM)	Sodium borate (100 mM; pH 8.6), SDS (110 mM), CH$_3$CN (10% v/v)	UV 214 nm	Human pharmacokinetics	Migration order S before R, run time 12 min, linear range 10–500 ng mL^{-1}, LOD 10 ng mL^{-1}	[106]
Ciprofibrate	Urine	γ-CD (7.5 mM)	Phosphate buffer (100 mM; pH 6.0)	UV 230 nm	Urinary excretion of the drug and glucuronide conjugate	Migration order, glucuronide conjugate of (S)- before (R)-ciprofibrate (*ca.* 15 min), (S)- before (R)-ciprofibrate (*ca.* 20 min). Qualitative study, analytes isolated from urine by SPE	[107]
Deprenyl	–	2,6-Dimethyl-β-CD (6 mM)	Tris-phosphate (20 mM; pH 2.8), HPMC (0.1% w/v), methanol (10% v/v)	UV 190 nm	Separation and resolution of drug enantiomers and amphetamine derivative metabolites	Migration order of the drug enantiomers (–) before (+), drug separated from (–)-amphetamine, (+)- and (–)-methamphetamine and (–)-propargylamphetamine, run time 25 min	[108]
(–)-Deprenyl	Urine	2,6-Dimethyl-β-CD (12 mM)	Tris-phosphate (20 mM; pH 2.7), HPMC (0.5% w/v), methanol (15% v/v)	UV 190	Metabolite identification following administration to rats	Resolution of *rac*-propargylamphetamine not achieved	[109]

Table 1. continued

Analyte	Matrix	Chiral selector (concentration)	Run buffer	Detection	Application	Comment	Ref.
Dimethindene	–	Hydroxypropyl-β-CD (30 μg mL^{-1})	Phosphate buffer (50 mM; pH 3.3)	UV 205 nm	Separation and resolution of both drug and metabolite enantiomers: N-desmethyl, N-desmethyl-6-methoxy-, 6-methoxy and dimethindene N-oxide	Migration order of drug enantiomers S before R, run time ca. 8 min, elution order desmethyldimethindene, dimethindene, dimethindene N-oxide, desmethyl-6-methoxy-dimethindene, 6-methoxy-dimethindene	[110]
Flezelastine	Hepatic micro-somes	β-CD (16 mM)	Phosphate buffer (50 mM; pH 3.75)	UV 210 nm	Enantioselective metabolism	LOD ca. 500 μg mL^{-1}	[111]
Leucovorin, 5-methyltetra-hydrofolate	Plasma	γ-CD (200 mM)	Phosphate buffer (300 mM; pH 9) containing urea (6 M)	UV 289 nm	Analysis of artificially prepared samples	Migration order, (–)-6-methoxy–methyl-2-naphthalene-αacetate (internal standard), (6S)- before (6R)-5-methyltetrahydrofolate followed by (S)- before (R)-leucovorin and ascorbic acid (added to prevent oxidation of tetrahydrofolates); run time ca. 35 min. Isolation, addition of urea to denature plasma proteins followed by ultrafiltration. Authors concluded that improvements in sensitivity required before the methodology is suitable for routine clinical analysis.	[112]
Mepivacaine	Serum	2,6-Dimethyl-β-CD (20 mM)	Phosphate buffer (100 mM; pH 2.5), HTAB (0.03 mM)	UV 215 nm	Analysis of artificially prepared samples	Migration order, R (14.9 min) before S (15.4 min), (R)-prilocaine (internal standard) 17.9 min. Calibration range, 200 –2000 ng mL^{-1}; LOD, 150 ng mL^{-1}; LOQ, 200 ng mL^{-1}. Intra- and interday precision and accuracy <7.5% for both enantiomers. Isolation, liquid-liquid extraction using diethylether and ethylacetate (1:1 v/v), recovery 73%; good correlation between CE and HPLC methods	[113]

Table 1. continued

Analyte	Matrix	Chiral selector (concentration)	Run buffer	Detection	Application	Comment	Ref.
Mianserin	Plasma	Hydroxypropyl-β-CD (2 mM)	TEA phosphate (75 mM; pH 3.0)	UV 211 nm	Therapeutic drug monitoring	Desmethyl- and 8-hydroxy-mianserin also resolved, propyl-norclozapine used as internal standard, migration order (S)-desmethyl-, (S)-mianserin, (R)-desmethyl-, (R)-mianserin, (S)-8-hydroxy-, (R)-8-hydroxy-, run time ca. 30 min, calibration range 10–150 ng mL^{-1} for mianserin and desmethyl-mianserin, 20–300 ng mL^{-1} for 8-hydroxymianserin, LOQ for mianserin and desmethyl-mianserin 5 ng mL^{-1}, LOQ for 8-hydroxymianserin 15 ng mL^{-1}, good correlations between CE and HPLC methods	[114]
Ondansetron	Serum	2,6-Dimethyl-β-CD (15 mM)	Sodium phosphate (100 mM), pH 2.5), HTAB (0.03 mM)	UV 254 nm	Analysis of artificially prepared samples	Migration order S before R, 15–250 ng mL^{-1} calibration range, procainamide used as internal standard	[115]
Pentobarbitone	Serum	Hydroxypropyl-γ-CD (40 mM)	Sodium phosphate (50 mM; pH 9.0)	UV 254 nm	Analysis of artificially prepared samples	Migration order R before S, run time ca. 11 min, calibration range 1–60 µg mL^{-1}, LOQ 1 µg mL^{-1}, aprobarbitone used as internal standard	[116]
Prilocaine	Serum	2,6-Dimethyl-β-CD (15 mM)	Phosphate buffer (100 mM; pH 2.5), HTAB (0.03 mM)	UV 215 nm	Analysis of artificially prepared samples	Migration order, procainamide (internal standard) 8.9 min, followed by (R)- (16.0 min) then (S)- (16.4 min) prilocaine; run; time 17.0 min. Calibration range 45–750 ng mL^{-1}; LOD, 38 ng mL^{-1}; LOQ, 45 ng mL^{-1}. Intra- and interday precision and accuracy <8.5% for both enantiomers. Isolation, SPE; good correlation between CE and HPLC methods	[117]
Reduced haloperidol	–	Dimethyl-β-CD (10 mM)	Phosphate buffer (40 mM; pH 2.5)	UV 200 nm	–	Separation of achiral haloperidol and resolution of the reduction product in 11 min	[118]
Reduced haloperidol	–	Dimethyl-β-CD (10 mM)	Tris-phosphate (40 mM; pH 2.5): PEG 6000 (20 mg mL^{-1}) 10:1 v/v	UV 200 nm	–	Migration order of reduced haloperidol (–) before (+), 4-chlorohaloperidol used as an internal standard	[119]

Table 1. continued

Analyte	Matrix	Chiral selector (concentration)	Run buffer	Detection	Application	Comment	Ref.
Reduced haloperidol	Plasma	Dimethyl-β-CD (10 mM)	Tris-phosphate buffer (40 mM; pH 2.5): PEG 6000 (20 mg mL^{-1}) 1:0.1 v/v	UV 200 nm	Determination of the drug and the enantiomers of the metabolite in plasma	Migration order, haloperidol, (−)-before (+)-enantiomer of reduced haloperidol, chlorohaloperidol (internal standard); run time ca. 9.0 min. Calibration range, haloperidol, 40–400 nM, each enantiomer of the reduced metabolite, 50–500 nM; intra- and interday precision <20%. LOD for haloperidol, 15 ng mL^{-1} and 30 ng mL^{-1} for each metabolite enantiomer. Isolation, liquid-liquid extraction using heptane:isoamyl alcohol (98:2 v/v) involving alkaline extraction, acid back extraction followed by reextraction at alkaline pH	[120]
Secobarbitone	Serum	Hydroxypropyl-γ-CD (40 mM)	Phosphate buffer (50 mM; pH 9.0)	UV 254 nm	Analysis of artificially prepared samples	Migration order, R (10.8 min) before S (11.2 min), aprobarbitone (internal standard) 11.9 min. Calibration range, 1–60 µg mL^{-1}, intra- and interday precision and accuracy <9% for both enantiomers. Isolation, SPE. Metabolites (1'RS, 3'RS)-5-allyl-5-(3'-hydroxy-1'-methyl-butyl) barbituric acid and (1'RS, 3'SR)-5-allyl-5-(3'-hydroxy-1'-methylbutyl) barbituric acid elute at approximately 12.5 min	[121]
Thalidomide	Rat hepatic mirosomes	Carboxymethyl-β-CD (15 mM)	Phosphate buffer (50 mM; pH 6.0)	UV 225 nm	In vitro metabolism	Parent drug, 4-hydroxy and 5-hydroxy metabolite enantiomers resolved in 15 min. Enantiomers of N-hydroxy metabolite poorly resolved at pH 4.4. Achiral MEKC method also reported	[122]
Thalidomide	–	Carboxymethyl-β-CD, sulphobutyl-β-CD, 2-hydroxy-propyl-trimethyl-ammonium-β-CD and β-CD	–	–	Resolution of the drug enantiomers and the 5-hydroxy and one pair of the 5'-hydroxy metabolites in a single run	A variety of chiral selectors, capillaries (fused-silica and polyacrylamide-coated) and two separation modes (normal polarity and carrier mode) were evaluated to facilitate both the separation and enantiomeric resolution of the three analytes in a single electrophoretic run	[123]
Thiopentone, pentobarbitone	Plasma	Hydroxypropyl-γ-CD (5 mM)	Phosphate buffer (75 mM; pH 8.5)	UV, scanned between 195–300 nm or simultan-eous monitoring at 200, 245 and 300 nm	Determination of the stereochemical composition in patient plasma following administration of racemic thiopentone	Migration order, 3-isobutyl-1-methylxanthine (internal standard), (R)- before (S)-pentobarbitone followed by (R)- before (S)-thiopentone; run time ca. 17 min. Calibration range, 0.5–25 µg mL^{-1} and 0.5–12.5 µg mL^{-1} for the enantiomers of thiopentone and pentobarbitone, respectively. Intraday precision, <12% and <4% at enantiomeric concentrations of 12.5 µg mL^{-1} and 5 µg mL^{-1} for thiopentone and phenobarbitone, respectively. Isolation, liquid-liquid extraction under acidic conditions using dichloromethane	[124]

Table 1. continued

Analyte	Matrix	Chiral selector (concentration)	Run buffer	Detection	Application	Comment	Ref.
Tramadol	–	Methyl-β-CD (75 mM)	Phosphate buffer (50 mM; pH 2.5), urea (220 mM), HPMC (0.05% w/v)	UV 200 nm	Separation and resolution of the and N-desmethyl metabolites	Migration order O-desmethyl-tramadol enantiomers, (+)-tramadol, (–)-tramadol, N-desmethyltramadol enantiomers	[125]
Tramadol	Urine	Carboxymethyl-β-CD (5 mM)	Phosphate buffer (50 mM; pH 2.5)	UV 195 nm	Urinary excretion of the drug and O-demethylated, N-demethylated, N,O-didemethylated metabolites	Migration order, (+)-ephedrine (internal standard), O-demethylated metabolite enantiomers, (–) before (+)-tramadol, N,O-didemethylated metabolite enantiomers and N-demethylated metabolite enantiomers; run time ca. 20 min. Calibration range 0.5–10 µg mL^{-1} for the enantiomers of the N-demethylated metabolite and 0.5–20 µg mL^{-1} for all other analytes. Intra- and interday precision <10% for all analytes; LOD, 0.1 µg mL^{-1}; LOQ 0.3 µg mL^{-1} for each enantiomer of tramadol. Isolation, liquid-liquid extraction at alkaline pH using hexane:ethyl acetate (8:2 v/v), followed by a second extraction with t-butyl methyl ether, recoveries between 68 and 96% depending on the analyte. Under the conditions employed, resolution of the N,N-didemethylated and N,N,O-tridemethylated metabolites was also achieved	[126]
Tramadol	Urine	Carboxymethyl-β-CD (30 mg mL^{-1})	Borate buffer (50 mM; pH 10.1)	UV 214 nm	Urinary excretion of the drug and O-demethylated metabolite	Migration order, tramadol analogue enantiomers (internal standard), (–)-before (+)-demethylated metabolite, (–)-before (+)-tramadol; run time ca. 22 min. Calibration range 0.24–9.6 µg mL^{-1}; precision, <7.5% for both enantiomers of the drug and metabolite. Isolation, liquid-liquid extraction at alkaline pH using n-hexane:ethyl acetate (8:2 v/v), recovery ca. 90%. Reduction of the chiral selector concentration to 20 mg mL^{-1} resulted in the separation and resolution of the drug and O-desmethyl metabolite together with the N,O-didemethylated, N,N,O-tridemethylated and N-run time of 25 min	[127]
Verapamil	Plasma	Trimethyl-β-CD (60 mM)	Phosphate buffer (60 mM; pH 2.5)	UV 200 nm	Pharmacokinetic studies	Migration order, gallopamil, (R)-verapamil, (R)-norverapamil, (S)-verapamil, (S)-norverapamil, gallopamil used as internal standard, run time 10 min, isolation by liquid-liquid extraction; calibration range, 2.5–250 ng mL^{-1}; LOQ for each analyte, 2.5 ng mL^{-1}; within- and between-day variation, 7.9–13.5%	[128]

Table 1. continued

Analyte	Matrix	Chiral selector (concentration)	Run buffer	Detection	Application	Comment	Ref.
Zopiclone	Urine	β-CD (16 mM)	Phosphate buffer (100 mM; pH 2.75 adjusted with triethanolamine)	LIF λex 325 nm λem 450 nm	Urinary excretion of drug and metabolites	Run time 15 min, migration order (R)-desmethylzopiclone, (R)-zopiclone, (S)-desmethyl-zopiclone, (S)-zopiclone, (R)-zopiclone N-oxide, (S)-zopiclone N-oxide. Linear range, 0.6–30 µg mL^{-1}. Drug also detected in saliva	[129]

Abbreviations: *HPMC, hydroxypropylmethyl cellulose; HTAB, hexadecyltrimethylammonium bromide; MHEC, methylhydroxyethyl cellulose; TEA, triethylamine*

both compounds following hydrolysis of the conjugates present in urine samples with β-glucuronidase/arylsulphatase (see also Section 4.3). The CE methodology produced data comparable to that obtained by HPLC but with considerably easier sample preparation. In a more extensive investigation, Caslavska *et al.* [39] developed methodology based on MEKC for the determination of 4-hydroxymephenytoin (see below), dextromethorphan and three metabolites, together with the metabolites of caffeine for the determination of the acetylation phenotype. The main advantage of the methodology being that data for all three compounds could be determined simultaneously if required [39]. However, apart from a limited examination of the stereoselectivity of mephenytoin oxidation, the above methods involved achiral analysis. Two probe compounds where stereospecific methodology is of interest, however, are debrisoquine and mephenytoin.

Oxidation of the antihypertensive agent debrisoquine to yield 4-hydroxydebrisoquine is mediated by the enzyme cytochrome P450 2D6 (Fig. 1). Two phenotypes are seen within the population and are known as extensive (EMs) and poor (PMs) metabolisers. The phenotype is defined by the metabolic ratio of debrisoquine/4-hydroxydebrisoquine in an 8 h urine sample following administration of the drug, with EMs and PMs exhibiting metabolic ratios within the ranges of 0.01–12.5 and 12.5–200, respectively [49]. Within the Caucasion population, *ca.* 8–10% fall in the PM group. The formation of 4-hydroxydebrisoquine results in the introduction of a chiral centre into the molecule and within the EM group the formation of the *S*-enantiomer has been demonstrated with an enantiomeric excess (e.e.) of >97%. In contrast, in PMs not only is the formation of the metabolite decreased but there is a reduction in the stereoselectivity of oxidation, with the e.e. ranging from 28 to 90% with respect to the *S*-enantiomer [50].

Determination of the debrisoquine metabolic ratio is performed using either GC or HPLC. Recently, Lanz *et al.* [51] developed CE methodology for the determination of both debrisoquine and the enantiomeric composition of the 4-hydroxy metabolite using heptakis-(2,3,6-tri-*O*-methyl)-β-CD (TM-β-CD) as a chiral selector. The initial methodology employed a run buffer of 50 mM potassium

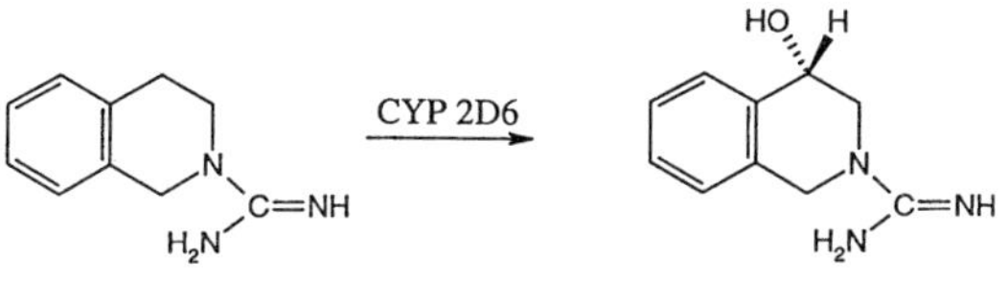

Figure 1. Metabolic oxidation of debrisoquine by cytochrome P450 2D6 to (*S*)-4-hydroxydebrisoquine.

phosphate pH 2.5, with UV detection at 195 nm. Under these conditions, debrisoquine eluted before the 4-hydroxy metabolite with an analysis time of less than 12 min. Addition of 50 mM TM-β-CD to the run buffer resulted in a reversal of the analyte elution order and complete enantiomeric resolution of 4-hydroxydebrisoquine with an overall run time of *ca.* 18 min. Under these conditions, the enantiomeric migration order was *S* before *R*. Baseline resolution of the metabolite enantiomers was also achieved using 2,6-di-*O*-methyl-β-CD at a lower selector concentration, but with a *ca.* twofold increase in total analysis time.

Using the above method, direct injection of urine samples following administration of debrisoquine was not possible because a number of peaks, presumably due to endogenous components, were detected in the region of interest in the electropherogram. Urine samples (2 mL) were subjected to solid-phase extraction (SPE) using a cross-linked polystyrene-based polymer. The conditioning, loading and washing of the SPE cartridges involved a fairly lengthy procedure, final analyte elution being achieved with 2 mL of a mixture of methanol:formic acid (4:1 v/v). The solvent was evaporated and the residue reconstituted in 100 µL of 4 mM aqueous HCl prior to injection on-capillary, to facilitate sample stacking and hence improve analytical sensitivity. The SPE cartridge wash procedure prior to analyte elution was found to be essential for enantiomeric resolution and peak shape [51]. Analysis of urine samples obtained from both EMs and PMs confirmed the reported stereoselectivity of metabolism within the EM group, the *R*-enantiomer of the metabolite being undetected. In the PMs however, the metabolite was typically not detected unless large volumes of urine were processed and the stereoselectivity of the oxidation could not be confirmed. This latter result is not surprising as the LOD of the CE assay was 150 ng mL^{-1} and the urinary concentration of 4-hydroxydebrisoquine in PMs is frequently less than 100 ng mL^{-1} [51].

Metabolic oxidation of the chiral antiepileptic agent mephenytoin is also under genetic control with 2–5% of the Caucasion population being deficient in this pathway [52, 53] (Fig. 2). In EMs, (*S*)-mephenytoin undergoes rapid oxidation, mediated by cytochrome P450 2C19 to yield (*S*)-4'-hydroxymephenytoin, whereas the *R*-enantiomer undergoes *N*-demethylation at a much slower rate, resulting in a 200-fold difference in the oral clearance of the mephenytoin enantiomers [54, 55]. In PMs, (*S*)-mephenytoin undergoes demethylation at a similar rate to the *R*-enantiomer.

The electrophoretic resolution of the enantiomers of both mephenytoin and 4-hydroxymephenytoin was reported originally by Okafo *et al.* [56] under MEKC conditions

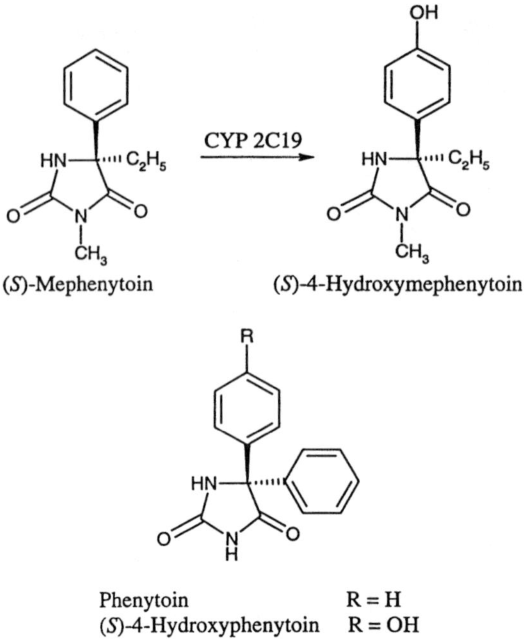

Figure 2. Metabolic oxidation of (*S*)-mephenytoin to (*S*)-4'-hydroxymephenytoin, together with the structures of phenytoin and (*S*)-4-hydroxyphenytoin.

using β-CD and taurodeoxycholate at a molar ratio of 2:5 in an electrolyte consisting of 30 mM sodium dihydrogen phosphate and 10 mM boric acid at pH 7.2. Resolution of the enantiomers of both the drug and metabolite was achieved within 15 min. The methodology was used to confirm the enantioselectivity of oxidation following incubation of racemic mephenytoin with human liver microsomes. Minor variations in analyte migration times were observed following analysis of the incubation extracts due to the presence of endogenous components affecting the electroosmotic flow [56] (Fig. 3).

Desiderio *et al.* [57] also developed an MEKC method for the determination of the enantiomers of both mephenytoin and the 4-hydroxy metabolite in urine. Baseline resolution of the enantiomers of 4-hydroxymephenytoin and partial resolution of the drug enantiomers was achieved using a 10 mM phosphate, 6 mM borate buffer, pH 9.1, containing 75 mM SDS and 30 mM β-CD with a total analysis time of 23 min. Increasing the SDS and β-CD concentrations to 100 mM and 50 mM, respectively, together with the addition of 10% v/v propan-2-ol to the run buffer resulted in the resolution of the enantiomers of both analytes within 34 min. The electrophoretic migration order was *S*- before

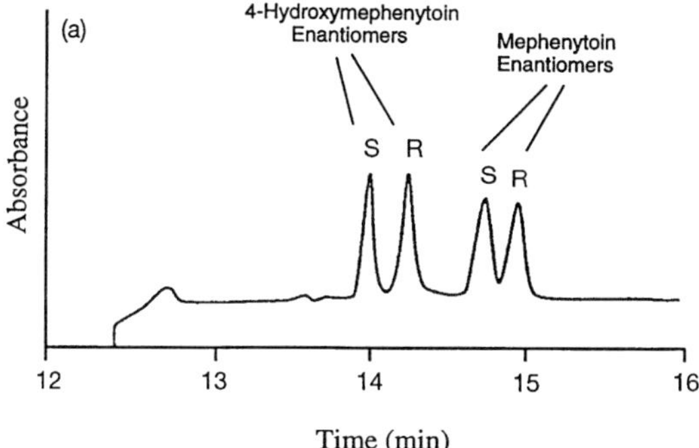

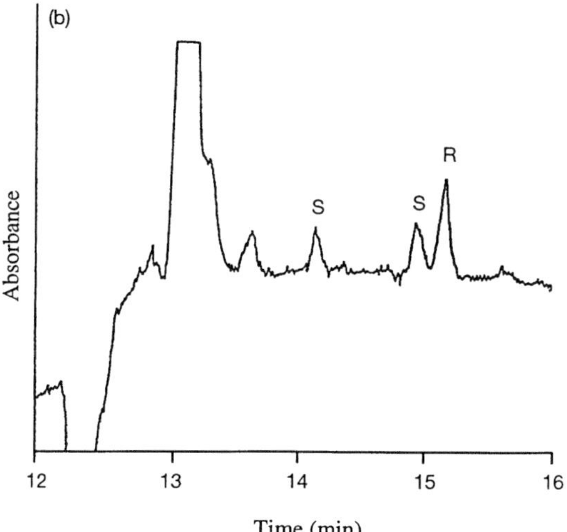

Figure 3. (a) Enantiomeric resolution of mephenytoin and 4'-hydroxymephenytoin and (b) analysis of an extract following incubation of the racemic drug with human hepatic microsomes. Conditions: instrument, Beckman P/ACE 2100; capillary, fused silica, 57 cm total length, 50cm effective length × 50 µm ID; buffer, 30 mM sodium dihydrogen phosphate / 10 mM boric acid containing 20mM β-CD and 50 mM taurodeoxycholate, pH 7.2; separation voltage, 9 kV; detection, UV at 214 nm; temperature, 25°C.

R- for both analytes. Examination of urine samples following administration of racemic mephenytoin to healthy volunteers, necessitated modification of the electrolyte due to the presence of endogenous compounds; neither mephenytoin nor the metabolite could be detected. The final conditions adopted were 8.4 mM sodium phosphate 5.6 mM sodium borate buffer, pH 9.1, containing 95 mM SDS, 40 mM β-CD and 8% v/v propan-2-ol with UV detection at 192 nm [57].

Analysis of urine samples required enzyme hydrolysis with β-glucuronidase/arylsulphatase, followed by direct injection for the determination of the metabolite. However, analysis of the drug required solvent extraction with dichloromethane, solvent removal and reconstitution of the residue in the run buffer (100 µL) due to the relatively high LOD (3 µg mL^{-1}) and less than 2% urinary recovery of the dose [57]. Following enzyme hydrolysis, the *S*-enantiomer of the 4-hydroxy metabolite could be readily detected in the urine of a volunteer of the EM phenotype but not in that of a PM phenotype. Analysis of nonenzyme treated urine extracts resulted in the detection of (*R*)-mephenytoin and (*S*)-4-hydroxymephenytoin in an EM sample but only (*R*)- and (*S*)-mephenytoin in that of a PM [57].

The above methodology was also applied to the analysis of the stereochemical composition of the 4-hydroxy metabolite of the related prochiral antiepileptic agent phenytoin (Fig. 2). In the case of phenytoin, metabolic oxidation of the enantiotopic phenyl rings is stereoselective for the formation of the *S*-enantiomer of the metabolite (e.e. > 90% in man) [58]. Resolution of the 4-hydroxyphenytoin enantiomers was achieved by increasing the concentration of SDS to 150 mM and following administration of the drug, only the *S*-enantiomer of the metabolite could be detected in urine postenzymatic hydrolysis and extraction [58]. The above investigations illustrate the utility of the CE methodologies for the determination of oxidation phenotype in terms of sample work-up procedures and ease of analysis.

4.3 Forensic applications

Stereospecific analytical methodology also has applications in forensic analysis [59]. For example, the stereochemical composition of a raw material may provide the forensic scientist with information associated with sample origin and the determination of the stereochemical composition of a drug and/or metabolites in a biological sample provides evidence of the nature of the material administered, *i.e.*, single enantiomer or racemate. Thus, enantiospecific analysis increases the information content of a sample which may have considerable legal significance. In the USA, Vicks Inhaler contains (*R*)-methamphetamine and following detection of the drug and/or demethylated metabolite, amphetamine, in urine the determination of the stereochemical composition of the analytes is essential to ascertain if an illicit substance has been consumed [60]. Dextromethorphan, a widely used antitussive agent, undergoes *O*-demethylation to yield dextrorphan, the enantiomer of which levorphanol ((−)-3-hydroxy-*N*-methylmorphinan) is a potent analgesic. The use of levorphanol by athletes is banned by the International Olympic Committee (IOC), whereas the use of preparations containing dextromethorphan is allowed. It is therefore essential that the forensic analyst can differentiate between the two stereoisomers dextrorphan and levorphanol to confirm that a banned substance has been taken [61] (see below).

The enantiomeric analysis of amphetamine derivatives by CE has been the subject of a number of reports. Lurie [59] used the indirect approach to enantiomeric analysis, using 2,3,4,6-tetra-*O*-acetyl-β-D-glucopyranosyl isothiocyanate as a homochiral derivatising agent. Resolution of the diastereomeric derivatives was achieved by MEKC using 100 mM SDS in a phosphate/borate buffer, pH 9.0, containing 20% v/v methanol. While the methodology was applied to samples of forensic interest, it was not used to examine samples of biological origin.

The majority of the CE methods for the analysis of amphetamines employ β-CD, or a derivative for enantiomeric resolution. Varesio and Veuthey [40] examined the resolution of amphetamine using methyl, dimethyl, 2-hydroxypropyl and native β-CD as chiral selectors. Satisfactory enantiomeric resolution was achieved using both native β-CD and the 2-hydroxypropyl derivative. The latter was selected for further investigation as a result of the slightly greater resolution, shorter analysis time and greater stability of the baseline with UV detection at 200 nm [40]. The final system adopted consisted of 200 mM phosphate buffer, pH 2.5, containing 20 mM 2-hydroxypropyl-β-CD using a fused-silica capillary equipped with an extended path-length detection window (150 μm ID). Using this system, enantiomeric resolution of amphetamine, methamphetamine, 3,4-methylenedioxyamphetamine (MDA), 3,4-methylenedioxymethamphetamine (MDMA; Ecstasy) and 3,4-methylenedioxyethylamphetamine (MDEA) was achieved in a single 30 min run [40]. In order to evaluate the methodology for the analysis of urine samples, the above compounds were dissolved in blank urine (drug-free) at a racemate concentration of 1 μg mL^{-1} and phenylethylamine was added as an internal standard. Direct injection of urine, both with and without ultrafiltration yielded complex electropherograms which could not be interpreted. Isolation of the analytes by either SPE or liquid-liquid extraction produced cleaner electropherograms, with those obtained following SPE yielding fewer potentially interfering peaks [40].

Ševčik *et al.* [62] carried out a similar study on methamphetamine. Using a 100 mM Tris-phosphate buffer, pH 2.5, containing 20 mM hydroxypropyl-β-CD and a detection wavelength of 190 nm, the enantiomers of methamphetamine, amphetamine, *p*-hydroxynorephedrine, *p*-hy-

droxymethamphetamine and *p*-hydroxyamphetamine could be resolved in a single 25 min run. The method was applied to the analysis of the enantiomeric composition of methamphetamine and amphetamine in urine following the administration of the racemic drug to a healthy volunteer. Samples were introduced on-capillary after passage through a 0.45 μm filter and in contrast to the previous study using artificially prepared samples [40], no interfering peaks due to endogenous constituents of urine were detected in the electropherograms presented [62]. However, no validation details were reported.

The resolution of the enantiomers of ephedrine, amphetamine and methamphetamine in a single run has also been achieved using 15 mM β-CD in 150 mM phosphate buffer, pH 2.5, with a detection wavelength of 200 nm [63]. Validation of the methodology was presented and detection of 1% enantiomeric impurity was achieved using amphetamine as a model analyte. The application of the methodology to the analysis of both "spiked" urine and amphetamine positive samples was presented and following drug isolation using Toxi-Tubes (full details not provided), the electropherograms presented were free from interfering endogenous components. The same method was also used to examine hair samples "spiked" with racemic amphetamine at a concentration of 10 ng mg^{-1} of hair. The samples were extracted overnight with 0.25 M HCl at 45°C, the mixture neutralised by the addition of NaOH and the drug isolated as for urine. Analysis of "blank" and "spiked" hair extracts yielded no interfering peaks in the electropherograms [63].

The recreational use of "designer amphetamines" MDA, MDMA and MDEA (see above) is currently the subject of considerable concern as a result of adverse reactions and neurotoxicity. As mentioned above, these agents may be resolved using 2-hydroxypropyl-β-CD as a chiral selector [40]. However, recently Lanz *et al.* [64] developed CE methodology for the determination of MDMA, MDA and 4-hydroxy-3-methoxymethamphetamine (HMMA) enantiomers in human urine. Resolution of the enantiomers of the three analytes was achieved in a single run (27 min total analysis time) using a run buffer consisting of 75 mM potassium dihydrogen phosphate, pH 2.5, containing 30 mM 2-hydroxypropyl-β-CD and a detection wavelength of 195 nm. Under these conditions, the analyte migration order was the enantiomers of HMMA, followed by those of MDA and then (*R*)- and (*S*)-MDMA. Analysis of "blank" drug-free urine by direct injection indicated the presence of potentially interfering peaks in the electropherograms due to endogenous components. Indeed, direct injection of urine obtained from a patient following MDMA administration allowed detection of both drug enantiomers but not those of either metabolite [64].

An analyte isolation procedure employing SPE was developed in which 5 mL of urine, to which the internal standard codeine had been added, was diluted with an equal volume of 100 mM phosphate buffer, pH 6.0, and applied to the cartridges. Elution of the analytes was achieved using, in turn, dichloromethane, methanol and dichloromethane:propan-2-ol (8:2, v/v), containing ammonia. The organic solvent was evaporated to dryness and the residue reconstituted in 100 µL of the 10-fold diluted run buffer to facilitate sample stacking. Urine samples treated with β-glucuronidase/arylsulphatase mixtures, for the hydrolysis of conjugates were treated in the same manner. Analysis of "spiked" urine samples indicated that the potentially interfering materials were removed using this procedure but that the hydrolysed samples produced additional peaks, one of which coeluted with the first eluting enantiomer of MDA. An additional problem was that the urinary concentrations of MDMA were high and that the majority of samples required considerable dilution prior to analysis. As a result of these complications, even though all three analytes could be resolved in a single run, three individual analyses were required, the first for MDMA, a second for both MDA and MDMA and a third for HMMA present in the hydrolysis samples [64]. Using the above isolation procedure, the analyte recovery ranged from 65 to 90%, with intraday and interday relative standard deviations ranging from 1.2 to 10.4% depending on both analyte and concentration. Analysis of urine samples collected over 72 h post-oral administration of 1.5 mg kg^{-1} racemic MDMA to two patients indicated stereoselective metabolism of the drug, significantly greater amounts of (*R*)-MDMA being recovered compared to the *S*-enantiomer. The recovery of the enantiomers of the two metabolites was also determined, but considerable intersubject variability was observed and the lack of stereochemically defined standards prevented data interpretation [64].

Methadone (6-dimethylamino-4,4-diphenylheptan-3-one; Fig. 4) is a potent analgesic and is also used in maintenance therapy of heroin addicts. The metabolism of methadone involves *N*-demethylation followed by cyclisation to yield the major urinary metabolite 2-ethylidene-1,5-dimethyl-3,3-diphenylpyrrolidine (EDDP). Two analytical methods concerned with the enantiospecific analysis of methadone and EDDP by CE have been reported. Lanz and Thormann [65] developed a method using 4.3 mM hydroxypropyl-β-CD as a chiral selector in a 100 mM potassium phosphate buffer, pH 3, with a UV detection wavelength of 195 nm. Under these conditions, resolution of the enantiomers of methadone was achieved (migration order *R* before *S*), but those of EDDP were not baseline resolved. Increasing the capillary length from 60 to 100 cm resulted in almost complete resolution but in-

creased the overall analysis time from 25 to 40 min. Direct injection of "blank" drug-free urine together with samples obtained from patients undergoing methadone therapy indicated a number of interfering peaks in the electropherograms. Extraction of the analytes from urine was performed using SPE. Samples (1 mL) were diluted to 3 mL with 100 mM phosphate buffer, pH 6, prior to application to the cartridges, the analytes were eluted using organic solvents which were then evaporated and the residues dissolved in 200–500 µL water. Using this approach, the majority of the patient samples analysed produced electropherograms free from interference in the regions of interest. Codeine was employed as an internal standard and calibration curves were constructed for each analyte enantiomer over the ranges 1.5–26.7 µg mL^{-1} and 1.1–21.1 µg mL^{-1} for methadone and EDDP, respectively, intraday variability being less than 10% for all four analytes [65]. Frost *et al.* [66] achieved resolution of both EDDP and methadone using heptakis-(2,6-di-*O*-methyl)-β-CD as a chiral selector in a 100 mM potassium phosphate buffer, pH 2.3, containing 10% v/v methanol and a detection wavelength of 200 nm. The use of the dimethyl-β-CD is of interest as Lanz and Thormann [65] reported no resolution of the two analytes using the same selector in a 50 mM sodium phosphate buffer, pH 2.5. Serum and urine samples were analysed following the addition of the internal standard, diphenhydramine, by extraction with hexane at alkaline pH. The organic extracts were evaporated to dryness and the residues reconstituted in methanol. Using this procedure, analyte recovery varied between 77 and 86% depending on both the matrix and analyte. Analysis of blank samples confirmed the absence of interfering peaks in the electropherograms and analytical precision was found to vary between 1–20% for both analytes over the ranges 5–250 ng mL^{-1} and 50–2500 ng mL^{-1} for serum and urine samples, respectively [66]. The methodology was also examined to ensure that additional compounds, *e.g.*, amphetamines, benzodiazepines, cocaine, frequently found in biological samples obtained from multidrugs users would not interfere. The methodology was applied

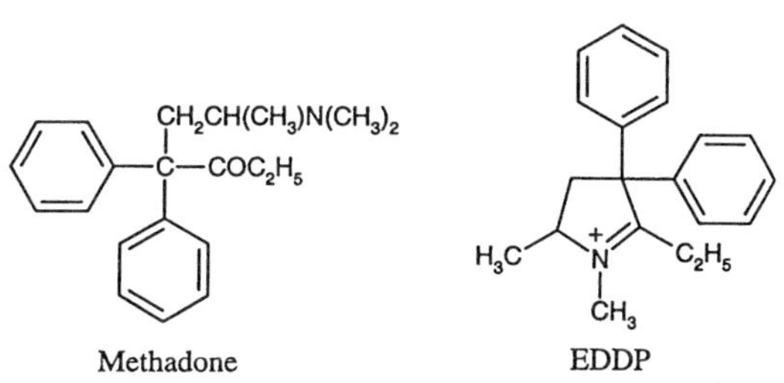

Figure 4. Structure of methadone and the cyclic metabolite EDDP.

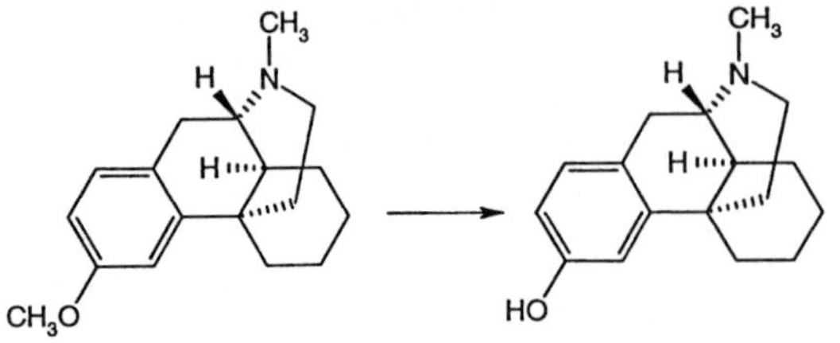

Electrophoresis 2000, *21*, 1953–1976

Figure 5. Metabolic *O*-demethylation of the antitussive agent dextromethorphan to dextrorphan, the enantiomer of which, levorphanol, is a potent analgesic.

to the analysis of both plasma and urine samples obtained following administration of both racemic and (*R*)-methadone to patients undergoing heroin withdrawal therapy. In addition, hair samples were analysed from a subject receiving (*R*)-methadone, the metabolite could not be detected but the drug, together with cocaine and nicotine, was readily detected [66].

The application of CE methods for the determination of oxidation phenotype using dextromethorphan as a probe drug have been addressed in Section 4.2 and the potential problems associated with the use of dextromethorphan and levorphanol by athletes have been outlined above (Fig. 5). The enantiospecific analysis of these agents in urine using CE has been reported by Aumatell and Wells [61]. Resolution of dextromethorphan and levomethorphan, together with their *O*-demethylated metabolites, dextrorphan and levorphanol, was achieved in a single 55 min electrophoretic run using an electrolyte consisting of 50 mM sodium tetraborate, pH 9.05, containing 50 mM SDS, 60 mM β-CD and 20% v/v propanol at 30°C with a separation voltage of 30 kV and UV detection at 200 nm [61]. Following the addition of internal standard, ethylmorphine, and acid hydrolysis of the conjugates, the analytes were isolated from urine by a combination of liquid-liquid extraction and SPE. The eluent from the SPE cartridges was evaporated and the residue reconstituted in 100 µL of run buffer, resulting in a 50-fold increase in analyte concentration and >70% recovery. The limit of detection of the method was reported to be between 0.02 and 0.04 ppm, depending on the analyte; the data obtained following the analysis of samples resulting from the oral administration of dextromethorphan to man were comparable to those obtained using GC-MS methodology [61].

4.4 β-Agonists and antagonists

A number of studies have examined the enantiomeric resolution of the arylethanolamine β-agonists and the aryloxypropanolamine β-antagonists using CE. However,

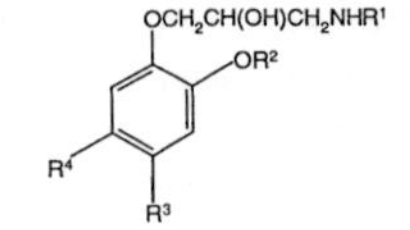

	R^1	R^2	R^3	R^4
Oxprenolol	CH(CH$_3$)$_2$	CH$_2$CH=CH$_2$	H	H
Desisopropyloxprenolol	H	CH$_2$CH=CH$_2$	H	H
Desallyloxprenolol	CH(CH$_3$)$_2$	H	H	H
4-Hydroxyoxprenolol	CH(CH$_3$)$_2$	CH$_2$CH=CH$_2$	OH	H
5-Hydroxyoxprenolol	CH(CH$_3$)$_2$	CH$_2$CH=CH$_2$	H	OH
Moprolol	CH(CH$_3$)$_2$	CH$_3$	H	H

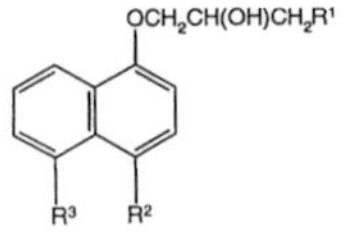

	R^1	R^2	R^3
Propranolol	NHCH(CH$_3$)$_2$	H	H
Desisopropylpropranolol	NH$_2$	H	H
Propranolol Glycol	OH	H	H
4'-Hydroxypropranolol	NHCH(CH$_3$)$_2$	OH	H
5'-Hydroxypropranolol	NHCH(CH$_3$)$_2$	H	OH

Figure 6. Structures of oxprenolol and propranolol and their metabolites.

there are relatively few investigations which examine the analysis of these agents in biological media or report data associated with the metabolites. Oxprenolol (Fig. 6) is a nonselective β-blocker, used as the racemate, which undergoes *N*-dealkylation, *O*-dealkylation and aromatic oxidation, to yield both the 4'- and 5'-hydroxy metabolites and conjugation with glucuronic acid [67]. Resolution of the enantiomers of oxprenolol (R_s = 1.73, migration (–) before (+)) has been achieved using a run buffer of 100 mM sodium phosphate, pH 2.5, containing 50 mM hydroxypropyl-β-CD, 0.05% w/v polyethylene glycol (PEG; M_r 5000–6000) and 0.03 mM tetrabutylammonium dihydrogen phosphate (TBA), with UV detection at 200 nm [67]. Under these conditions, in addition to the drug, complete resolution of the enantiomers of the two aromatic oxidation products and the *O*-desalkyl metabolite was also achieved, whereas only partial resolution of the enantiomers of the *N*-desisopropyl metabolite was obtained. In each case the migration order was (–) before (+). Quantitative analysis of 5 mL urine samples was carried out by liquid-liquid extraction (ethyl acetate), using the related compound (–)-moprolol as an internal standard, hydrolysis of conjugates being achieved using β-glucuronidase. Following evaporation of the solvent, the residue was reconstituted in 200 µL of 5 mM phosphate buffer prior to hydrodynamic injection onto the capillary. Calibration curves were constructed over the range of 0.4–16 µg mL^{-1} for each racemic analyte, an LOD of 0.2 µg mL^{-1} being achieved for each analyte enantiomer, at a signal to noise ratio of 3:1, with intra- and interassay precision of less than 6% [67].

Analysis of urine samples, following oral administration of the racemic drug to man, indicated the presence of two isomeric phenols together with the unchanged drug. Using this methodology it was possible to determine the urinary excretion profiles of the enantiomers of the free

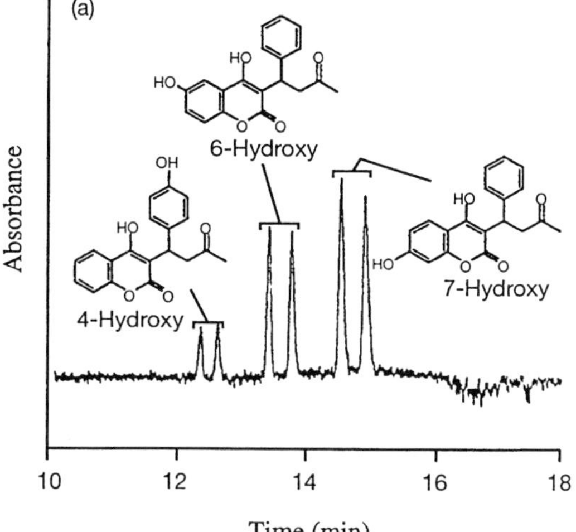

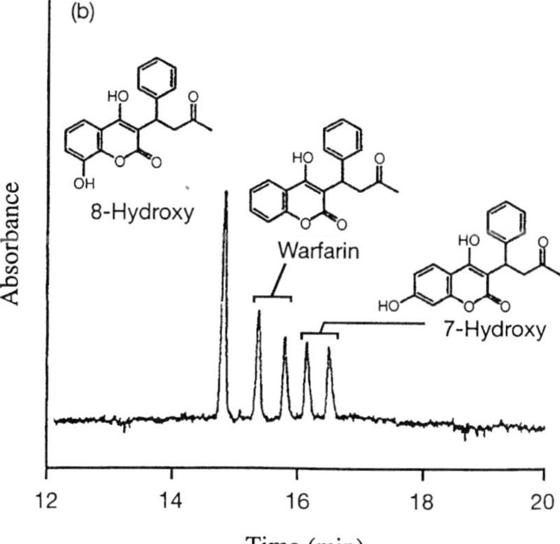

Figure 7. Resolution of the enantiomers of warfarin and some of its hydroxy metabolites. Conditions: instrument and capillary same as in Fig. 3; buffer, 10 mM potassium borate containing 3% v/v propan-2-ol and the novel surfactant, 40 mM *N*-lauroyl-6-aminopenicillanic acid, pH 7.0; separation voltage, 15 kV; detection, UV at 214 nm; temperature, 25°C. 7-Hydroxywarfarin was included in both electropherograms to allow comparison of migration times.

drug and both phenolic metablites up to 7 h post drug administration and the conjugated material for up to 14 h [67]. The enantiomeric resolution of the related compound propranolol has been achieved using a variety of derivatised β-CDs, cellulose, and the chiral surfactant *N*-dodecoxycarbonylvaline as chiral selectors [68, 69]. A recent report has addressed the problem of the separation and enantiomeric resolution of the drug and its 4'- and 5'-hydroxy, desisopropyl and glycol metabolites (Fig. 6) using hydroxypropyl-β-CD as a chiral selector [70]. The influence of various factors on the resolution of the drug and metabolite enantiomers were reported, including CD concentration (optimal 17.4 mM), buffer pH (optimised, 100 mM phosphoric acid-triethanolamine, pH 3–4), applied voltage, and methanol content. While separation of the drug and metabolites was readily achieved, as was enantiomeric resolution of the drug and the 4'- and 5'-hydroxy compounds, the desisopropyl metabolite was not resolved and the neutral glycol was not detected. Attempts to use MEKC, in the presence of hydroxypropyl-β-CD, were unsuccessful and the application of the methodology to samples of biological origin was not reported [70].

Clenbuterol, a potent β₂-agonist, is the subject of abuse in sport and is also used as a growth promoter in animals as a result of its anabolic effects. Gausepohl and Blaschke [71] recently reported a CE method for the determination of the enantiomeric composition of the drug in urine. Using a run buffer of 200 mM sodium phosphate, pH 3.3, containing 30 mg mL^{-1} hydroxyethyl-β-CD as a chiral selector, resolution of the enantiomers (elution order (–) before (+)) and separation from the bupranolol internal standard was achieved in less than 15 min. The analytes were extracted from 2 mL urine at alkaline pH with 10 mL of hexane:*t*-butylmethylether (99.5:0.5 v/v) and following evaporation of the solvent, the residue was reconstituted in 100 μL methanol followed by electrokinetic injection to facilitate sample stacking. Using this approach, analyte recovery was *ca.* 65% and the method was validated over the concentration range 0.5–10 ng mL^{-1} for each enantiomer. The LOQ was 0.5 ng mL^{-1}, interday precision being within 18% at this concentration but reducing to 6% at the upper limit of the concentration range. Using this methodology, it was possible to monitor the enantiomeric composition of clenbuterol in urine over 18 h following oral administration of 20 μg of the racemate to man [71].

4.5 Warfarin

The metabolism of the racemic anticoagulant warfarin involves both stereo- and regioselectivity in aromatic oxidation. For example, the major metabolite of (*S*)-warfarin in man is 7-hydroxywarfarin, the oxidation being mediated by cytochrome P450 2C9, whereas (*R*)-warfarin undergoes oxidation mainly at the 6-position. It has therefore been proposed that warfarin may be a useful probe compound for investigation of human cytochrome P450 profiles [72]. Using the novel surfactant, *N*-lauroyl-6-aminopenicillanic acid (40 mM) as a chiral selector, Bouzige *et al.* [73] resolved synthetic samples of racemic 4'-, 6- and 7-hydroxywarfarin, together with the drug enantiomers, using a 10 mM potassium borate buffer, pH 7, containing 3% v/v propan-2-ol within an 18 min electrophoretic run. The 8-hydroxy compound was also separated from the other metabolites but was not enantiomerically resolved (Fig. 7).

The enantiomeric resolution of warfarin, together with the related coumarin anticoagulants 4'-chlorowarfarin, phenprocoumon and 4'-chlorophenprocoumon, but not acenocoumarol, was reported by D'Hulst and Verbeke [74]. Resolution was achieved using the maltodextrin preparation Glucidex 2 as the chiral selector in a 10 mM Trisphosphate buffer, pH 7, with UV detection at 185 nm. The methodology was used to examine plasma samples "spiked" with 0.4–4.0 µg mL^{-1} of racemic warfarin using acenocoumarol as an internal standard. Following extraction of plasma (0.5 mL) with ether and evaporation of the solvent, the residue was reconstituted in 100 µL of 1 mM Tris-phosphate, pH 7.9, containing 25% v/v methanol. In order to optimise the resolution, the Glucidex 2 concentration in the run buffer was reduced to 2.5% and under these conditions the electrophoretic migration order was (*R*)- before (*S*)-warfarin with a run time of under 4.5 min [74].

Gareil *et al.* [75] reported a CE method for the determination of the enantiomeric composition of warfarin in human plasma, using 4'-chlorowarfarin as an internal standard. Using 8 mM methyl-β-CD as the chiral selector and a run buffer of 100 mM sodium phosphate, pH 8.35, containing 2% v/v methanol, enantiomeric resolution was achieved

within 12 min. Under these conditions 4'-chlorowarfarin was not resolved, but reduction in the chiral selector concentration (optimal 3 mM) resulted in partial resolution. The enantiomeric migration order under these conditions was determined to be (*S*)- before (*R*)-warfarin. The drug was isolated from 1 mL plasma by extraction with dichloromethane and following evaporation of the solvent the residue was reconstituted in 50 µL of the run buffer prior to hydrodynamic injection. Calibration and validation experiments were performed between 1–3 µg mL^{-1}, yielding relative standard deviations for relative peak height and area measurements of approximately 5% or less. The methodology was used to examine the enantiomeric composition of warfarin in the plasma of four patients who had been undergoing warfarin therapy at a dose of 10 mg day^{-1} for several months. Analysis of both drug-free and patient plasma indicated a lack of interfering peaks in the electropherograms and the data obtained confirmed the greater plasma concentrations of (*R*)- compared to (*S*)-warfarin [75].

4.6 2-Arylpropionic acids and ibuprofen metabolism

The 2-arylpropionic acids (2-APAs) are an importanmt class of nonsteroidal antiinflammatory drugs (NSAIDs) which exhibit stereoselectivity in both action and disposition [76–78]. The metabolism of these agents involves enzyme-mediated inversion of chirality of the weakly active, or inactive, *R*-enantiomers to their active *S*-antipodes [76–78]. As a result of the interest in the stereochemical aspects of the disposition of the 2-APAs, a number of chromatographic methods have been reported for the determination of their enantiomeric composition in biological fluids [79]. CE has also been employed for the analysis of the 2-APAs and enantiomeric resolution has been achieved using derivatised CDs either singularly [80–82] or as mixtures [20], maltooligosaccharides [83], the macrocyclic antibiotics vancomycin and ristocetin A [17, 18] and the basic protein avidin (using a polyacrylamide coated capillary) [84] as chiral selectors and also

Figure 8. Metabolic oxidation of ibuprofen to hydroxyibuprofen and the diastereomers of carboxyibuprofen.

by capillary electrochromatography using an open tubular capillary coated with permethylated β- or γ-CD (Chirasil-Dex) [85].

There have been relatively few reports concerning the application of CE methods for the analysis of the 2-APAs in biological fluids. Shihabi and collaborators [86, 87] have reported methods for the achiral analysis of both ibuprofen and ketoprofen in human serum. Acetonitrile was used to precipitate the plasma proteins and samples were then directly injected on capillary. However, in the case of ibuprofen, the detection limit at 214 nm was relatively high at 10 mg L^{-1}, the main advantages of the methodology being simplicity and speed. A more recent achiral assay for ibuprofen in rat plasma decreased the LOD to 0.3 mg L^{-1} using a detection wavelength of 200 nm [88]. Of the enantiospecific methods cited above, only that of Soini *et al.* [83] was examined with respect to "spiked" serum. Using a separation buffer consisting of 5% w/w maltrin M040 in 30 mM TAPS / 10 mM Tris (pH 7.8) and a UV detection wavelength of 220 nm the enantiomers of ibuprofen, extracted from serum, were resolved with an interference-free background using (*S*)-naproxen as an internal standard. The migration order was (*R*)- before (*S*)-ibuprofen followed by the internal standard, the total run time being less than 15 min.

In addition to chiral inversion, the metabolism of ibuprofen involves conjugation with glucuronic acid and oxidation to yield hydroxy- and carboxyibuprofen (Fig. 8). The urinary excretion of the two metabolites, both free and conjugated, in addition to the drug accounts for approximately 75% of the dose following oral administration of the racemate to man [89, 90]. The formation of carboxyibuprofen results in the introduction of a second chiral centre into the molecule and following administration of racemic ibuprofen, all four stereoisomers of the metabolite are detected in urine. The chromatographic resolution of the four stereoisomers has proved to be problematical, however, this was recently achieved using a derivatised amylose HPLC-CSP [91, 92].

Analysis of ibuprofen metabolites by CE has been examined by three groups. The initial investigation by Ashcroft *et al.* [93] involved SPE of urine prior to achiral analysis using 20 mM ammonium acetate, pH 9, run buffer and electrospray-tandem mass spectrometry (ES-MS/MS) for

analyte detection and identification. Using this methodology, one of the first applications of coupled CE-MS in drug metabolism, the glucuronide conjugates of both the drug and two major oxidation products could be readily characterized.

Fanali *et al.* [94] recently examined the CE resolution of ibuprofen and both metabolites, using vancomycin as a chiral selector and ES-MS for detection. The presence of the chiral selector in the ion source of the mass spectrometer may reduce analytical sensitivity due to increased noise and competition with the analyte for the available charge [95]. However, judicious choice of the chiral selector which migrates in the opposite direction to the analyte (countercurrent mode) can prevent source contamination. Using a polyacrylamide-coated capillary, an electrolyte of 50 mM ammonium acetate, pH 4.8, containing 5 mM vancomycin, which under these conditions is positively charged and migrates in the opposite direction to the acidic analytes, it was possible to detect and resolve all four stereoisomers of carboxyibuprofen together with the enantiomers of ibuprofen and the hydroxy metabolite in a single 15 min run. Peak overlap between the first eluting enantiomers of the hydroxy metabolite and the drug was observed, but this did not present a significant problem as detection was mass-selective. While the above approach illustrates the potential of highly efficient CE separations together with the specificity of MS detection for the analysis of structurally similar analytes present in complex mixtures, it has not yet been applied to the analysis of biological samples.

As mentioned above, the 2-APAs have been resolved by both addition of CDs and maltooligosaccharides to the background electrolyte in CE. Bjørnsdottir *et al.* [22] examined a range of both native and derivatised CDs to effect the resolution of ibuprofen and its metabolites in a single electrophoretic run. The method employed fused-silica capillaries dynamically coated with hexadimethrine bromide as a flow reversal agent such that the analytes migrated before the EOF. As observed by previous investigators [81], optimal chiral selectivity was found with 20 mM heptakis (2,3,6-tri-*O*-methyl)-β-CD at pH 5.0 using 100 mM 2-(*N*-morpholino)ethanesulphonic acid (MES). Under these conditions the stereoisomers of carboxyibuprofen and the enantiomers of ibuprofen were resolved, but only partial resolution of the enantiomers of hydroxyi-

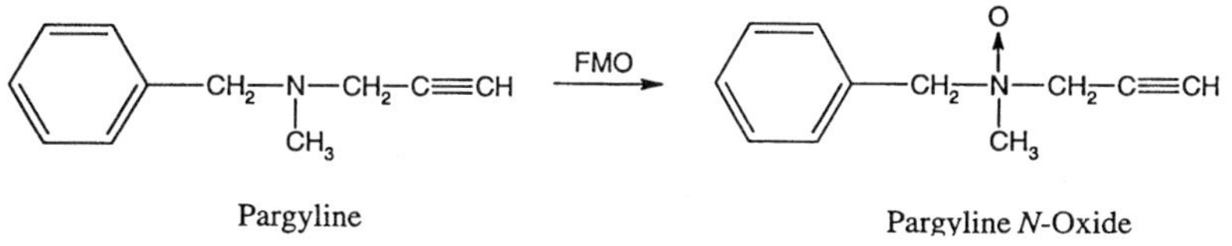

FMO

Pargyline

Pargyline *N*-Oxide

Figure 9. Structure of pargyline and pargyline *N*-oxide.

buprofen was achieved which also comigrated with the carboxy metabolite. Similarly, using 10% w/v dextrin 10 in ammonium acetate buffer, pH 9, the enantiomers of both the drug and hydroxy metabolite were resolved but carboxyibuprofen yielded three peaks. Mixtures of 10% w/v dextrin 10 and 20 mM TM-β-CD were investigated using 100 mM MES as the background electrolyte over the pH range 5–7. Analyte separation and stereoisomer resolution was found to be critically dependent on pH; at values greater than pH 5.5 the carboxy metabolite yielded three peaks and resolution of the enantiomers of both the drug and hydroxyibuprofen was achieved, whereas at pH 5.0

the four stereoisomers of the carboxy metabolite were resolved but the enantiomeric resolution of the hydroxy metabolite was lost. The final optimised system, allowing simultaneous separation and stereoisomeric resolution of all three analytes within a 30 min run, utilized a run buffer containing 10% w/v dextrin 10, 20 mM TM-β-CD in 100 mM MES containing 0.01% w/v hexadimethrine bromide at pH 5.26 [22]. Additional investigations also indicated that the resolution of the two metabolites improved with an increase in temperature to 35°C. The methodology was applied to the analysis of urine samples obtained from a volunteer following the administration of either 600 mg of racemic or (*S*)-ibuprofen. The samples were either subjected to SPE, dilution with acetonitrile:water (1:3 v/v) or treatment with β-glucuronidase prior to injection on capillary. Examination of the resulting electropherograms indicated the presence of the expected metabolite stereoisomers and, in addition, the glucuronide metabolites were also resolved.

4.7 *N*-Oxidation of prochiral tertiary amines

In order to examine the potential of tertiary amines as metabolic probes for the characterisation of the isoforms of the flavin-containing monooxygenase (FMO) enzyme system we have examined the stereochemistry of the *N*-

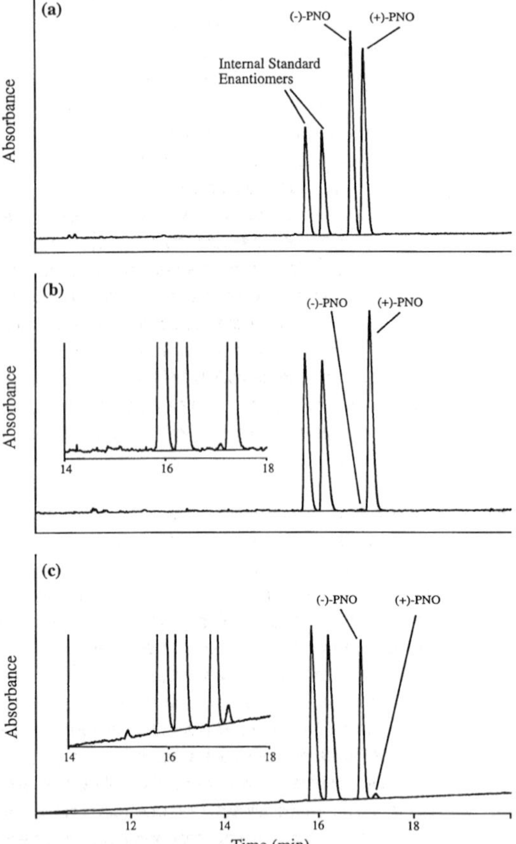

Figure 10. Enantiomeric composition of pargyline *N*-oxide following isolation of the racemic *N*-oxide (a) from inactivated enzyme and (b) following incubation of pargyline with human FMO1 and (c) rhesus macaque FMO2. Conditions: instrument, Beckman P/ACE 5010; capillary, same as in Fig. 3; run buffer, 150 mM lithium phosphate, pH 2.5, containing 50 mM hydroxyethyl-β-CD; separation voltage, 30 kV; detection, UV at 200 nm; temperature, 15°C.

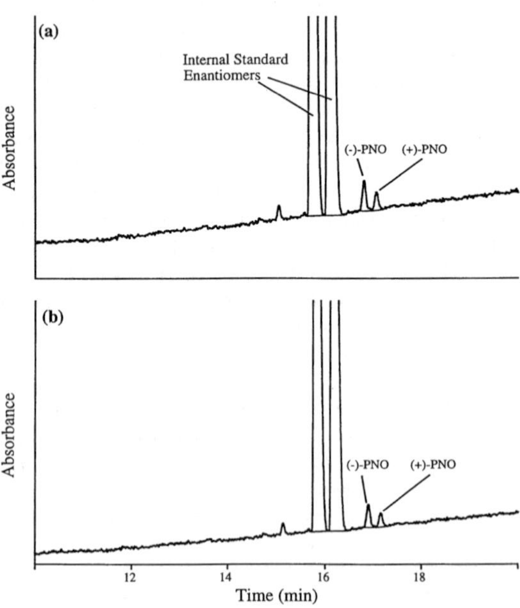

Figure 11. Enantiomeric composition of pargyline *N*-oxide following incubation of pargyline with (a) wild-type (E158) human FMO3 and (b) mutated (K158) human FMO3. Conditions as described in Fig. 10.

oxidation of the prochiral tertiary amines *N*-ethyl-*N*-methylaniline (EMA) and *N*-benzyl-*N*-methyl-2-propynylamine (pargyline) (Fig. 9) [96–98]. Enantiomeric resolution of the metabolically produced *N*-oxides examined was achieved using derivatised polysaccharide HPLC-CSPs [96–98]. Using this methodology, pargyline was shown to undergo stereoselective *N*-oxidation on incubation with hepatic microsomal preparations from various species [98], but when using purified porcine hepatic FMO and heterologously-expressed human FMO1, the oxidation appeared to show stereospecificity for the (+)-enantiomer [98, 99]. However, chromatographic analysis time was relatively long (*ca.* 80 min) and the enantiomeric resolution incomplete, such that minor quantities of the second eluting enantiomer (the (−)-isomer) could not be accurately determined [96, 98].

In order to improve our analytical methodology we recently examined the resolution of a series of racemic *N*-oxides by CE using both native and derivatised β-CDs as chiral selectors [100]. The enantiomers of pargyline *N*-oxide (PNO) and the selected internal standard (*N*-methyl-*N*-*n*-propylaniline *N*-oxide) were baseline-resolved in a single electrophoretic separation in less than 18 min using a 150 mM lithium phosphate buffer, pH 2.5, containing 50 mM hydroxyethyl-β-CD at 15°C with UV detection at 200 nm. Under these conditions, the electrophoretic migration order for the enantiomers of PNO was determined to be (−) before (+) by use of individual enantiomers obtained by semipreparative chromatography as previously described [98].

Examination of electropherograms obtained following incubation of pargyline in the presence of heterologously expressed human FMO1 and rhesus macaque FMO2 isoforms and isolation using SPE (Fig. 10), indicated a reversal of the stereoselectivity of PNO formation. *N*-Oxygenation was found to be highly stereoselective in the presence of both FMO1 and FMO2 isoforms (enantiomeric composition mean ± SD, *n* = 3; FMO1 (+):(−) = 99.4:0.6 ± 0.1; FMO2 (+):(−) = 2.8:97.2 ± 0.1), but not stereospecific [98, 100]. With the improved resolution and reversed migration order compared to the chromatographic method, levels of (−)-PNO present in the FMO1 incubations could be reliably determined at levels below 1% by area. In contrast, following human FMO3-mediated oxidation of pargyline (both wild type and K158 mutant), the stereoselectivity of formation of PNO was much less pronounced (Fig. 11) and similar for both enzymes (enantiomeric ratio (+):(−) = 36:64). The calculated activities of the enzyme preparations indicated that pargyline was a poorer substrate for expressed human FMO3 in comparison to human FMO1 and rhesus macaque FMO2 (enzyme activities, mean ± SD, *n* = 3 for FMO3 (wild

type), FMO3 (K158 mutant), FMO1 and FMO2 were determined to be 20 ± 0.5, 8 ± 0.5, 136 ± 9.3 and 165 ± 13 pmol of product formed per pmol FMO per min, respectively).

5 Conclusions

CE, as a result of its versatility, high efficiency, minimal sample requirements and low consumable expense has become an essential technique in pharmaceutical analysis [2, 101] and has considerable potential as a bioanalytical tool. In this article we have attempted to highlight the utility of CE methodology in enantiospecific bioanalysis. The relatively slow application of CE in bioanalysis, compared to applications in other analytical areas is associated with a number of factors including a lack of background knowledge of the technique and its potential applicability in a complex and demanding analytical area, together with a reluctance to invest the time and effort to the development of new technology in order to replace satisfactory existing methodology; in addition to which Bojarski and Aboul-Enein [5] have added a "psychological barrier" for bioanalysts to move away from the familiar and well-established chromatographic technologies, a sort of analytical conservatism. This view is supported to an extent by a recent survey amongst ten major pharmaceutical companies [102]. While the survey did not address bioanalytical applications specifically, it did examine the availability and use of enantiospecific CE in the drug discovery and development process. The majority of companies used enantiospecific CE during the discovery and development phases, but few cited CE as a method of choice and fewer still had either used enantiospecific CE data in regulatory submissions or envisaged doing so in the near future [102].

CE will obviously not replace HPLC in bioanalysis, in the same way that HPLC did not replace GC, but CE does provide a useful complementary technique. Despite the perceived limitations of CE compared to HPLC, the most frequently cited being sensitivity, there are a number of instances when the available sample volume is limited such that CE has obvious advantages. Imagine being able to determine the enantiomeric composition of a drug in cytoplasmic samples obtained from a single cell. It could be done!

Received January 31, 2000

6 References

[1] Perrett, D., in: Camilleri, P. (Ed.), *Capillary Electrophoresis: Theory and Practice*, CRC Press, Boca Raton, FL 1998, pp. 441–485.

[2] Levêque, D., Gallion-Renault, C., Monteil, H., Jehl, F., *J. Chromatogr. B* 1997, *697*, 67–75.

[3] Shihabi, Z. K., *J. Chromatogr. A* 1998, *807*, 27–36.

[4] Desiderio, C., Fanali, S., *Boll. Chim. Farm.* 1995, *134*, 541–546.

[5] Bojarski, J., Aboul-Enein, H. Y., *Electrophoresis* 1997, *18*, 965–969.

[6] Boone, C. M., Waterval, J. C. M., Lingeman, H., Ensing, K., Underberg, W. J. M., *J. Pharm. Biomed. Anal.* 1999, *20*, 831–863.

[7] Xu, Y., *Anal. Chem.* 1999, *71*, 309R–313R.

[8] Terabe, S., Otsuka, K., Nishi, H., *J. Chromatogr. A* 1994, *666*, 295–319.

[9] Bressolle, F., Audran, M., Pham, T.-N., Vallon, J.-J., *J. Chromatogr. B* 1996, *687*, 303–336.

[10] Fanali, S., *J. Chromatogr. A* 1996, *735*, 77–121.

[11] Aumatell, A., Wells, R. J., Wong, D. K. Y., *J. Chromatogr. A* 1994, *686*, 293–307.

[12] Desiderio, C., Fanali, S., *J. Chromatogr. A* 1998, *807*, 37–56.

[13] Arai, T., *J. Chromatogr. B* 1998, *717*, 295–311.

[14] Fanali, S. (Ed.), *Electrophoresis* 1994, *15*, 753–870.

[15] Fanali, S. (Ed.), *Electrophoresis* 1997, *18*, 841–1044.

[16] Chankvetadze, B., *Capillary Electrophoresis in Chiral Analysis*, Wiley, Chichester 1997, pp. 1–155.

[17] Armstrong, D. W., Rundlett, K. L., Chen, J.-R., *Chirality* 1994, *6*, 496–509.

[18] Armstrong, D. W., Gasper, M. P., Rundlett, K. L., *J. Chromatogr. A* 1995, *689*, 285–304.

[19] Pálmarsdottir, S., Mathiasson, L., Jönsson, J. Å., Edholm, L.-E., *J. Chromatogr. B* 1997, *688*, 127–134.

[20] Fillet, M., Bechet, I., Schomburg, G., Hubert, P., Crommen, J., *J. High Resol. Chromatogr.* 1996, *19*, 669–673.

[21] Armstrong, D. W., Chang, L. W., Chang, S. S. C., *J. Chromatogr. A* 1998, *793*, 115–134.

[22] Bjørnsdottir, I., Kepp, D. R., Tjørnelund, J., Hansen, S. H., *Electrophoresis* 1998, *19*, 455–460.

[23] Le Corre, P., Gibassier, D., Sado, P., LeVerge, R., *J. Chromatogr. Biomed. Appl.* 1988, *450*, 211–216.

[24] Sioufi, A., Colussi, D., Marfil, F., Dubois, J. P., *J. Chromatogr.* 1987, *414*, 131–137.

[25] Blaschke, G., in: Aboul-Enein, H. Y., Wainer, I. W. (Eds.), *The Impact of Stereochemistry on Drug Development and Use*, Wiley, New York 1997, pp. 107–123.

[26] Ward, T. J., Ward, K. D., in: Aboul-Enein, H. Y., Wainer, I. W. (Eds.), *The Impact of Stereochemistry on Drug Development and Use*, Wiley, New York 1997, pp. 317–344.

[27] Altria, K. D., Harlen, A. C., Hart, M., Hevizi, J., Hailey, P. A., Makwana, J., Portsmouth, M. J., *J. Chromatogr.* 1993, *641*, 147–153.

[28] Lloyd, D. K., in: Riley, C. M., Lough, W. J., Wainer, I. W. (Eds.), *Pharmaceutical and Biomedical Applications of Liquid Chromatography*, Pergamon/Elsevier, Oxford 1994, pp. 3–40.

[29] Naylor, S., Benson, L. M., Tomlinson, A. J., *J. Chromatogr. A* 1996, *735*, 415–438.

[30] Tomlinson, A. J., Benson, L. M., Guzman, N. A., Naylor, S., *J. Chromatogr. A* 1996, *744*, 3–15.

[31] Tomlinson, A. J., Guzman, N. A., Naylor, S., *J. Capil. Electrophor.* 1996, *2*, 247–266.

[32] Lloyd, D. K., *J. Chromatogr. A* 1996, *735*, 29–42.

[33] Guzman, N. A., *LC.GC Int.* 1999, *12*, 164–174.

[34] Hogan, B. L., Lunte, S. M., Stobaugh, J. F., Lunte, C. E., *Anal. Chem.* 1994, *66*, 596–602.

[35] Hadwiger, M. E., Torchia, S. R., Park, S., Biggin, M. E., Lunte, C. E., *J. Chromatogr. B* 1996, *681*, 241–249.

[36] Hadwiger, M. E., Park, S., Torchia, S. R., Lunte, C. E., *J. Pharm. Biomed. Anal.* 1997, *15*, 621–629.

[37] Nakagawa, T., Oda, Y., Shibukawa, A., Fukuda, H., Tanaka, H., *Chem. Pharm. Bull.* 1989, *37*, 707–711.

[38] Lloyd, D. K., in: Reid, E., Hill, H. M., Wilson, I. D. (Eds.), *Methodological Surveys in Bioanalysis of Drugs, Vol. 23, Biofluid and Tissue Analysis for Drugs Including Hypolipidaemics*, Royal Society of Chemistry, Cambridge 1994, pp. 41–50.

[39] Caslavska, J., Hufschmid, E., Theurillat, R., Desiderio, C., Wolfisberg, H., Thormann, W., *J. Chromatogr. B* 1994, *656*, 219–231.

[40] Varesio, E., Veuthey, J.-L., *J. Chromatogr. A* 1995, *717*, 219–228.

[41] Hutt, A. J., *Anal. Proc.* 1991, *28*, 185–186.

[42] Fried, K., Wainer, I. W., *J. Chromatogr. B* 1997, *689*, 91–104.

[43] Krásensky, S., Fanali, S., Křivánková, L., Boček, P., *Electrophoresis* 1995, *16*, 968–973.

[44] Danková M., Kaniansky, D., Fanali, S., Iványi, F., *J. Chromatogr. A* 1999, *838*, 31–43.

[45] Pálmarsdottir, S., Edholm, L.-E., *J. Chromatogr. A* 1995, *693*, 131–143.

[46] La Du, B. N., in: Kalow, W. (Ed.), *Pharmacogenetics of Drug Metabolism*, Pergamon Press, New York 1992, pp. 1–12.

[47] Meyer, U. A., *J. Pharm. Pharmacol. Suppl. 1* 1994, *46*, 409–415.

[48] Li, S., Fried, K., Wainer, I. W., Lloyd, D. K., *Chromatographia* 1993, *35*, 216–222.

[49] Idle, J. R., Smith, R. L., *Drug Metab. Rev.* 1979, *9*, 301–319.

[50] Eichelbaum, M., Bertilsson, L., Küpfer, A., Steiner, E., Meese, C. O., *Br. J. Clin. Pharmacol.* 1988, *25*, 505–508.

[51] Lanz, M., Theurillat, R., Thormann, W., *Electrophoresis* 1997, *18*, 1875–1881.

[52] Küpfer, A., Roberts, R. K., Schenker, S., Branch, R. A., *J. Pharmacol. Exp. Ther.* 1981, *218*, 193–199.

[53] Küpfer, A., Preisig, R., *Eur. J. Clin. Pharmacol.* 1984, *26*, 753–759.

[54] Wilkinson, G. R., Guengerich, F. P., Branch, R. A., in: Kalow, W. (Ed.), *Pharmacogenetics of Drug Metabolism*, Pergamon Press, New York 1992, pp. 657–685.

[55] Wedlund, P. J., Aslanian, W. S., Jacqz, E. McAllister, C. B., Branch, R. A., Wilkinson, G. R., *J. Pharmacol. Exp. Ther.* 1985, *234*, 662–669.

[56] Okafo, G. N., Bintz, C., Clarke, S. E., Camilleri, P., *J. Chem. Soc. Chem. Commun.* 1992, 1189–1192.

[57] Desiderio, C., Fanali, S., Küpfer, A., Thormann, W., *Electrophoresis* 1994, *15*, 87–93.

[58] Poupaert, J. H., Cavalier, R., Claesen, M. H., Dumont, P. A., *J. Med. Chem.* 1975, *18*, 1268–1271.

[59] Lurie, I. S., *J. Chromatogr.* 1992, *605*, 269–275.

[60] Fitzgerald, R. L., Ramos, J. M., Bogema, S. C., Poklis, A., *J. Anal. Toxicol.* 1988, *17*, 255–259.

[61] Aumatell, A., Wells, R. J., *J. Chromatogr. Sci.* 1993, *31*, 502–508.

[62] Ševčik, J., Lemr, K., Smysl, B., Jirovsky, D., Hradil, P., *J. Liq. Chromatogr. Rel. Technol.* 1998, *21*, 2473–2484.

[63] Scarcella, D., Tagliaro, F., Turrina, S., Manetto, G., Nakahara, Y., Smith, F. P., Marigo, M., *Forens. Sci. Int.* 1997, *89*, 33–46.

[64] Lanz, M., Brenneisen, R., Thormann, W., *Electrophoresis* 1997, *18*, 1035–1043.

[65] Lanz, M., Thormann, W., *Electrophoresis* 1996, *17*, 1945–1949.

[66] Frost, M., Köhler, H., Blaschke, G., *Electrophoresis* 1997, *18*, 1026–1034.

[67] Li, F., Cooper, S. F., Mikkelsen, S. R., *J. Chromatogr. B* 1995, *674*, 277–285.

[68] Ljungberg, H., Nilsson, S., *J. Liq. Chromatogr.* 1995, *18*, 3685–3698.

[69] Valtcheva, L., Mohammad, J., Pattersson, G., Hjertén, S., *J. Chromatogr.* 1993, *638*, 263–267.

[70] Pak, C., Marriott, P. J., Carpenter, P. D., Amiet, R. G., *J. Chromatogr. A* 1998, *793*, 357–364.

[71] Gausepohl, C., Blaschke, G., *J. Chromatogr. B* 1998, *713*, 443–446.

[72] Kaminsky, L. S., Dunbar, D. A., Wang, P. P., Beaune, P., Larrey, D., Guengerich, F. P., Schnellmann, R. G., Sipes, I. G., *Drug Metab. Disp.* 1984, *12*, 470–476.

[73] Bouzige, M., Okafo, G., Dhanak, D., Camilleri, P., *J. Chem. Soc. Chem. Commun.* 1996, 671–672.

[74] D'Hulst, A., Verbeke, N., *Chirality* 1994, *6*, 225–229.

[75] Gareil, P., Gramond, J. P., Guyon, F., *J. Chromatogr.* 1993, *615*, 317–325.

[76] Hutt, A. J., Caldwell, J., *J. Pharm. Pharmacol.* 1983, *35*, 693–704.

[77] Caldwell, J., Hutt, A. J., Fournel-Gigleux, S., *Biochem. Pharmacol.* 1988, *37*, 105–114.

[78] Evans, A. M., *Eur. J. Clin. Pharmacol.* 1992, *42*, 237–256.

[79] Davies, N. M., *J. Chromatogr. B* 1997, *691*, 229–261.

[80] Guttman, A., Cooke, N., *J. Chromatogr. A* 1994, *685*, 155–159.

[81] Fanali, S., Aturki, Z., *J. Chromatogr. A* 1995, *694*, 297–305.

[82] Blanco, M., Coello, J., Iturriaga, H., Maspoch, S., Pérez-Maseda, C., *J. Chromatogr. A* 1998, *793*, 165–175.

[83] Soini, H., Stefansson, M., Riekkola, M-L., Novotny, M. V., *Anal. Chem.* 1994, *66*, 3477–3484.

[84] Tanaka, Y., Matsubara, N., Terabe, S., *Electrophoresis* 1994, *15*, 848–853.

[85] Mayer, S., Schurig, V., *Electrophoresis* 1994, *15*, 835–841.

[86] Shihabi, Z. K., Hinsdale, M. E., *J. Chromatogr. B* 1996, *683*, 115–118.

[87] Friedberg, M., Shihabi, Z. K., *J. Chromatogr. B* 1997, *695*, 193–198.

[88] Kang, S. H., Chang, S.-Y., Do, K.-C., Chi, S.-C., Chung, D. S., *J. Chromatogr. B* 1998, *712*, 153–160.

[89] Lockwood, G. F., Albert, K. S., Gillespie, W. R., Cole, G. G., Harkcom, T. M., Szpunar, G. J., Wagner, J. G., *Clin. Pharmacol. Ther.* 1983, *34*, 97–103.

[90] Evans, A. M., Nation, R. L., Sansom, L. N., Bochner, F., Somogyi, A. A., *Biopharm. Drug Dispos.* 1990, *11*, 507–518.

[91] Tan, S. C., Baker, J. A., Stevens, N., deBiasi, V., Salter, C., Chalaux, M., Afarinkia, K., Hutt, A. J., *Chirality* 1997, *9*, 75–87.

[92] Tan, S. C., Jackson, S. H., D. Swift, C. G., Hutt, A. J., *J. Chromatogr. B* 1997, *701*, 53–63.

[93] Ashcroft, A. E., Major, H. J., Lowas, S., Wilson, I. D., *Anal. Proc.* 1995, *32*, 459–462.

[94] Fanali, S., Desiderio, C., Schulte, G., Heitmeier, S., Strickmann, D., Chankvetadze, B., Blaschke, G., *J. Chromatogr. A* 1998, *800*, 69–76.

[95] Sheppard, R. L., Tong, X., Cai, J., Henion, J. D., *Anal. Chem.* 1995, *67*, 2054–2058.

[96] Hutt, A. J., Hadley, M. R., Tan, S. C., *Eur. J. Drug Metab. Pharmacokin.* 1994, *19*, 241–251.

[97] Hadley, M. R., Oldham, H. G., Damani, L. A., Hutt, A. J., *Chirality* 1994, *6*, 98–104.

[98] Hadley, M. R., Švajdlenka, E., Damani, L. A., Oldham, H. G., Tribe, J., Camilleri, P., Hutt, A. J., *Chirality* 1994, *6*, 91–97.

[99] Phillips, I. R., Dolphin, C. T., Clair, P., Hadley, M. R., Hutt, A. J., McCombie, R. R., Smith, R. L., Shephard, E. A., *Chem. Biol. Interact.* 1995, *96*, 17–32.

[100] Hadley, M. R., Gabriac, S. D., Hutt, A. J., *Chirality* 1999, *11*, 409–415.

[101] Camilleri, P., *J. Chem. Soc. Chem. Commun.* 1996, 1851–1858.

[102] Soo, E. C., Salmon, A. B., Lough, W. J., *Chem. Ind.* 1999, *220–224*.

[103] Soini, H., Riekkola, M.-L., Novotny, M. V., *J. Chromatogr.* 1992, *608*, 265–274.

[104] Sandor, V., Flarakos, T., Batist, G., Wainer, I. W., Lloyd, D. K., *J. Chromatogr. B* 1995, *673*, 123–131.

[105] Clohs, L., McErlane, K. M., *Pharm. Res.* 1996, *13*, S3.

[106] Pruñonosa, J., Obach, R., Diez-Cason, A., Gouesclon, L., *J. Chromatogr.* 1992, *574*, 127–133.

[107] Hutteman, H., Blaschke, G., *J. Chromatogr. B* 1999, *729*, 33–41.

[108] Szökö, E., Magyar, K., *J. Chromatogr. A* 1995, *709*, 157–162.

[109] Szökö, E., Magyar, K., *Int. J. Pharm. Adv.* 1996, *1*, 320–328.5mm>

[110] Heuermann, M., Blaschke, G., *J. Chromatogr.* 1993, *648*, 267–274.

[111] Paris, S., Blaschke, G., Locher, M., Borbe, H. O., Engel, J., *J. Chromatogr. B* 1997, *691*, 463–471.

[112] Shibukawa, A., Lloyd, D. K., Wainer, I. W., *Chromatographia* 1993, *35*, 419–429.

[113] Siluveru, M., Stewart, J. T., *J. Pharm. Biomed. Anal.* 1997, *15*, 1751–1756.

[114] Eap, C. B., Powell, K., Baumann, P., *J. Chromatogr. Sci.* 1997, *35*, 315–320.

[115] Siluveru, M., Stewart, J. T., *J. Chromatogr. B* 1997, *691*, 217–222.

[116] Srinivason, K., Bartlett, M. G., *J. Chromatogr. B* 1997, *703*, 289–294.

[117] Siluveru, M., Stewart, J. T., *J. Chromatogr. B* 1997, *693*, 205–210.

[118] Wu, H-L., Otsuka, K., Terabe, S., *J. Liq. Chromatogr. Rel. Technol.* 1996, *19*, 1567–1577.

[119] Wu, H-L., Wu, S.-M., Chen, S.-H., Otsuka, K., Terabe, S., *J. Chin. Chem. Soc.* 1997, *44*, 141–144.

[120] Wu, S.-M., Ko, W.-K., Wu, H.-L., Chen, S.-H., *J. Chromatogr. A* 1999, *846*, 239–243.

[121] Srinivasan, K., Zhang, W., Bartlett, M. G., *J. Chromatogr. Sci.* 1998, *36*, 85–90.

[122] Weinz, C., Blaschke, G., *J. Chromatogr. B* 1995, *674*, 287–292.

[123] Meyring, M., Chankvetadze, B., Blaschke, G., *Electrophoresis* 1999, *20*, 2425–2431.

[124] Zaugg, S., Caslavska, J., Theurillat, R., Thormann, W., *J. Chromatogr. A* 1999, *838*, 237–249.

[125] Chan, E. C. Y., Ho, P. C., *J. Chromatogr. B* 1998, *707*, 287–294.

[126] Rudaz, S., Veuthey, J. L., Desiderio, C., Fanali, S., *J. Chromatogr. A* 1999, *846*, 227–237.

[127] Kurth, B., Blaschke, G., *Electrophoresis* 1999, *20*, 555–563.

[128] Dethy, J-M., DeBroux, S., Lesne, M., Longstreth, J., Gilbert, P., *J. Chromatogr. B* 1994, *654*, 121–127.

[129] Hempel, G., Blaschke, G., *J. Chromatogr. B* 1996, *675*, 139–146.

Review

Theo de Boer
Rokus A. de Zeeuw
Gerhardus J. de Jong
Kees Ensing

Department of Analytical
Chemistry and Toxicology,
University Center for
Pharmacy,
Groningen, The Netherlands

Recent innovations in the use of charged cyclodextrins in capillary electrophoresis for chiral separations in pharmaceutical analysis

A review is presented on the use of charged cyclodextrins (CDs) as chiral selectors in capillary electrophoresis (CE) for the separation of analytes in pharmaceutical analysis. An overview is given of theoretical models that have been developed for a better prediction of the enantiomeric resolution and for a better understanding of the separation mechanism. Several types of charged CDs have been used in chiral capillary electrophoretic separation (anionic, cationic, and amphoteric CDs). Especially the anionic CDs seem to be valuable due to the fact that many pharmaceutically interesting compounds can easily be protonated (*e.g.*, amine groups). For that reason several anionic CDs are now commercially available. Cationic and amphoteric CDs are less common in chiral analysis and only a few are commercially available. Attention is paid to the most common synthesis routes and the characterization of the CDs used in chiral capillary electrophoretic separations. The degree of substitution in the synthesized CDs may vary from one manufacturer to another or even from batch to batch, which may have a detrimental effect on the reproducibility and ruggedness of the separation system. In Sections 4, 5, and 6 the applications of anionic, cationic, and amphoteric CDs for the chiral separation in CE are described. Many interesting examples are shown and the influence of important parameters on the enantioselectivity is discussed.

Keywords: Anionic cyclodextrins / Amphoteric cyclodextrins / Capillary electrophoresis / Cationic cyclodextrins / Selectivity / Review
EL 4070

Contents

Correspondence: Dr. Theo de Boer, Department of Analytical Chemistry and Toxicology, University Center for Pharmacy, A. Deusinglaan 1, 9713 AV Groningen, The Netherlands
E-mail: t.de.boer@farm.rug.nl
Fax: +31-50-3637582

Abbreviations: AEA-β-CD, mono-(6-β-aminoethylamino-6-deoxyl)-β-cyclodextrin; **CM-β-CD**, carboxymethyl-β-CD; **CME-β-CD**, carboxymethylethyl-β-CD; **DAS-β-CD**, 2,3-di-*O*-acetyl-6-sulfato-β-CD; **DMA-β-CD**,-6,6-dimethylamino-6-deoxy-β-CD; **DAS-β-CD**, 2,3-(diacetyl)-6-sulfato-β-CD; **DMS-β-CD**, 2,3-dimethyl-6-sulfato-β-CD; **DS**, degree of substitution; **MA-β-CD**, methylamino-β-CD; **NSAID**, nonsteroidal anti-inflammatory drug; **QA-β-CD**, 2-hydroxypropyl-trimethyl-ammonium-β-CD; **S-β-CD**, sulfated-β-CD; **SBE-β-CD**, sulfobutyl ether-β-CD; **SEE-β-CD**, sulfoethyl ether-β-CD; **SPE-β-CD**, sulfopropyl ether-β-CD

1 Introduction

The field of enantiomer separation in pharmaceutical analysis and bioanalysis for the monitoring of drugs, drug impurities (*i.e.*, degradation products) synthetic precursors, side products, or metabolites has been extensively explored over the last decades. Indirect chiral separations, *i.e.*, *via* derivatization of the enantiomers with optically pure reagents, have been performed with several chromatographic separation techniques (HPLC, GC, TLC) [1–3]. Although GC usually offers a higher efficiency due to a larger number of plates available in the column

434

than HPLC, the obtained diastereomers are often less volatile. This limits the use of GC for these purposes. Furthermore, the indirect approach is rather time-consuming, needs very pure optical agents and, most important, for derivatization a functional group with sufficient reactivity towards the derivatization agent is necessary. In both GC and HPLC, direct separations with CDs as chiral selectors have also been described. The CDs are either coupled to the stationary phase in the column (GC, HPLC) or are added to the mobile phase (HPLC). Both principles are limited in their success. First, when chiral selectors are covalently bound to the stationary phase, an increased peak broadening is expected due to a relatively slow mass transfer (the *C*-term of the van Deemter equation). Second, inherent to HPLC, a large amount of chiral selector is needed when added to the mobile phase.

In the last two decades, the use of CE as a novel, fast, and efficient separation technique has proven to be suitable for chiral separations with CDs added to the run buffer (see reviews [4–8]). Some major advantages of chiral separations in CE in comparison with HPLC are the low consumption of the chiral selector (reduced costs) and the high plate numbers due to a reduced peak broadening as a consequence of the absence of Eddy diffusion and mass transfer between two phases (the *A*- and *C*-terms of the van Deemter equation, respectively). Also, the selectivity, which was extensively discussed as one of the key parameters in electrokinetic separation sciences [9], is often relatively high in CE. This implies that because of the combination of the high plate numbers and a high selectivity, baseline separations can be achieved at much lower CD concentrations. Besides the often used neutral CD derivatives, a relatively new class of CDs, the charged CDs, are gaining much interest because of their ability to perform fast chiral separations at low concentrations and because of the possible chiral separation of neutral racemates. Moreover, charged CDs are expected to give the best resolving power when the analytes are oppositely charged, because interactions of the CDs with the analytes are now not only based on inclusion complexation but also on strong electrostatic interactions. For example, for an anionic CD, the electrophoretic mobility of the CD is towards the anode, implying a larger difference between the effective mobilities of the analyte and the complexed analyte, resulting in an increased selectivity factor. The mobilities of an anionic CD, a cationic chiral analyte, and their inclusion complexation products are depicted in Fig. 1. The figure shows the chiral separation of the racemate under ideal conditions, *i.e.*, the anionic CD as well as the cationic analytes are fully charged and the EOF is eliminated. Obviously, the mobilities of the enantiomers are equal ($\mu A1 = \mu A2$), whereas the mobility of the CDA1 complex (top) is smaller than the mobility of the CDA2

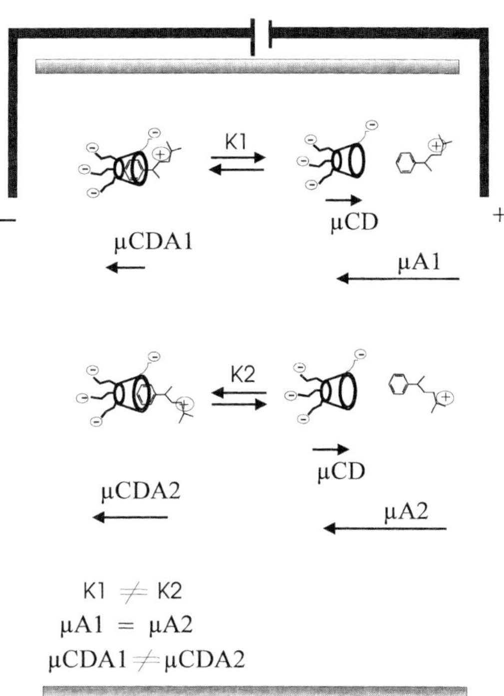

Figure 1. Illustration of the separation of a cationic analyte with an anionic CD, under the conditions that the EOF is eliminated and that both analyte and CD are fully charged.

complex (bottom). This is due to the fact that in the latter case interaction is only based on inclusion complexation, whereas in the former case interaction is based on inclusion complexation and strong ion-interaction (diastereomeric interaction). However, the most important factor in discrimination between the enantiomers is the difference in the binding constants (K1 and K2) between the CD and the analytes as will be explained in the next section. It is also possible that only diasteromeric ion-pair interactions occur, *i.e.*, without inclusion complexation. In such cases, the capability of enantioseparation is reduced because of the (much) lower chiral recognition for one of the enantiomers. In analogy with separations using neutral (substituted) CDs as chiral selectors, the migration order of the analytes may be altered when using different types of charged CDs, as depicted in Fig. 2, and sometimes even the migration order of the enantiomers may be altered. A reversal of migration order can have several advantages, for instance when dealing with impurity profiling where enantiomers are not completely baseline-resolved. It is likely that in this case integration errors occur, especially when the distomer (smaller peak) is located at the tail of the eutomer (larger peak) [10, 11].

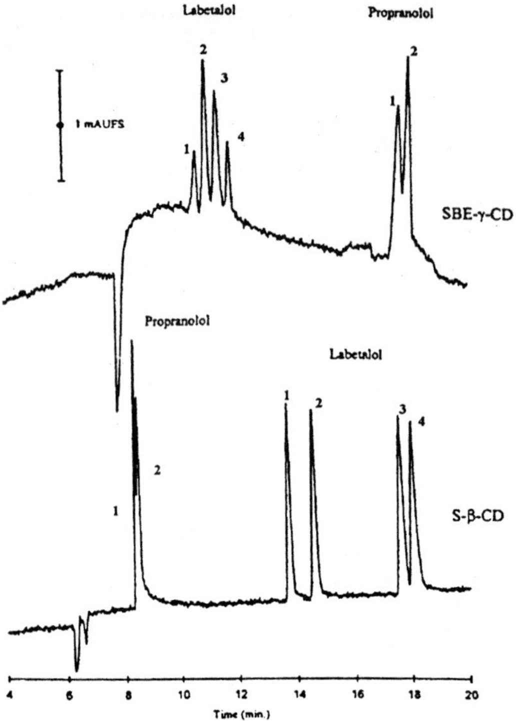

Figure 2. Electropherogram (modified from [97]) showing the reversal of the migration order of racemic labetalol (L) and propranolol (P) using 30 mM phosphate buffer containing either 20 g/L of SBE-γ-CD or 7.7 g/L of S-β-CD.

Charged CDs are chemically modified naturally occurring oligosaccharides with 6, 7, or 8 glucopyranose units, usually designated as α-, β-, and γ-CD, respectively. Particularly the β-CDs seems to have an excellent resolving power for chiral molecules containing a (substituted) aromatic ring and, for instance, an amino group or a carboxyl group [9]. Modification of the CDs occurs with either strongly charged functional groups (*e.g.*, sulfate or sulfoalkyl ether groups) or weakly charged functional groups (carboxy(m)ethyl, phosphate or amino groups). Sulfated CDs or the often used sulfobutyl ether-β-CD (SBE-β-CD) can be used over a wide pH range without affecting their net charge. Quaternary ammonium CDs have the advantage that they are always positively charged, irrespective of the pH of the background electrolyte (BGE). The objective of this paper is to review the use of charged CDs as chiral selectors for baseline separations in pharmaceutical analysis, starting with the pioneering work of Terabe and co-workers [12, 13]. Furthermore, it provides the most common synthesis routes and characterization techniques for the anionic and cationic CDs.

2 Theoretical aspects

The advantage of using charged CDs has been described in the theoretical model for CD-based separations, developed by Wren and Rowe [14, 15]. From their model it can be derived that chiral baseline separations are achieved at rather low CD concentrations in comparison with neutral CDs. After the model of Wren and Rowe, which relates mobility differences to the concentrations of the CD and the organic solvent, other mathematical models were developed. Vigh and co-workers [16–19] described a multi-equilibria-based model for both chiral weak acids and weak bases to account for the effects of the pH of the buffer and the CD concentration in the buffer. Surapaneni *et al.* [20] developed a theoretical model for the separation of enantiomers of neutral species by employing a combination of charged and neutral CDs. Their model is able to calculate resolution and selectivity based on model parameters, which was demonstrated by resolving the enantiomers of LY213829 and its sulfoxide metabolites with a run buffer containing anionic SBE-β-CD and the neutral β-CD. The model is based on simultaneous multiple interactions between the neutral analyte with the charged CD and the neutral CD, with the assumptions that the EOF is negligible and the charged CD is fully charged under the given pH conditions. As a consequence, the neutral analyte becomes mobile when it is complexed with the charged CD. The use of a dual CD system is based on the principle that the free fraction of the neutral analyte, which is able to complex with the charged CD, decreases when neutral CDs are added to the solution. The neutral CDs stereoselectively affect the free fraction of the chiral analyte available for complexation with charged CDs. This simplifies fine-tuning the separation system. The apparent electrophoretic mobility of the enantiomer (S) can be given as:

$$\mu_S = \frac{\mu_{S,ICD} K_{S,ICD}[ICD]}{1 + K_{S,ICD}[ICD] + K_{S,NCD}[ICD]} \tag{1}$$

where $\mu_{S,ICD}$ is the mobility of the analyte (S)/ionized-CD (ICD) complex $K_{S,ICD}$ is the equilibrium constant of the analyte-charged CD system, and $K_{S,NCD}$ is the equilibrium constant of the analyte-neutral CD system.

Although there are advantages in using charged CDs in the separation of oppositely charged analytes (and to a lesser extent for neutral analytes) some limitations have been mentioned [21]. For instance, anionic CDs can be used in CE systems with normal polarity (counter-EOF) and reversed polarity (co-EOF). When using reversed polarity, the anionic CD concentration should be increased to generate a negative net mobility for cationic analyte-anionic CD complexes. Yet, higher CD concentrations can lead to electrodispersion causing band broadening. Also, cationic CDs (containing an amine group) may adsorb to the wall of the capillary due to electrostatic interactions between the negatively charged wall and the positively charged amine groups, leading to a reduction or reversal of the EOF and possible peak broadening. Williams and Vigh [22] mentioned some additional difficulties with the use of charged CDs: (i) Most commercially available materials are complicated mixtures containing a large number of isomers; (ii) the ionic strength may change dramatically if the concentration of the charged CD in the BGE is changed; (iii) there is no method that can be used to measure EOF rates in the presence of charged CDs. The latter is important to monitor the experimental variables that affect the separation. In their paper, Williams and Vigh [22] derived a set of guidelines to optimize the separation of chiral compounds using charged CDs. This so-called CHARM model (characteristics of the charged resolving agent migration model), that focuses on method development for enantiomeric separations using a rational and predictable selection of the operating conditions, was experimentally verified in aqueous and nonaqueous solvents [23–26]. Charged CDs can be used for the separation of nonelectrolyte enantiomers, strong electrolyte enantiomers, and for the separation of weak electrolyte enantiomers. For the most common and, in pharmaceutical analysis, most favorable application of charged CDs, *i.e.*, the separation of positively charged analytes using strong anionic CDs, they suggest that at (or near) the CD concentration at which the enantiomers start to migrate in the same direction, the resolution can be increased, but at the expense of the ruggedness of the separation and the run time [22].

Finally, Wang and Khaledi [27] expressed the resolution between two enantiomers, based on the equation originally developed by Giddings [28] and Jorgenson and Lukacs [29],

$$R_s = \frac{\sqrt{N}}{4} \left(\frac{\Delta K[\mathrm{CD}]}{(1 + K_1[\mathrm{CD}])(1 + K_2[\mathrm{CD}])} \right) \left(\frac{\mu^f - \mu^c}{\mu_{avg} + \mu_{eof}} \right) \tag{2}$$

where N is the number of theoretical plates, K_1 and K_2 are the binding constants between the enantiomers and the CDs, μ^f and μ^c are the mobilities of the enantiomers in free and in fully complexed forms, μ_{avg} is the average electrophoretic mobility of the two enantiomers and μ_{eof} is the mobility of the EOF.

In the above equation μ^c is considered the same for both enantiomer-CD complexes, but in theory they are not. Although the differences in μ^c of the two enantiomer-CD complexes may be small, it should be described as originally done by Wren and Rowe [14]. They expressed the

difference in the apparent electrophoretic mobility between the two enantiomer-CD complexes as

$$\Delta\mu^c = \frac{[\mathrm{CD}](\mu_1 - \mu_2)\Delta K}{1 + [\mathrm{CD}](K_1 + K_2) + K_1 K_2[\mathrm{CD}]^2} \tag{3}$$

Wang and Khaledi [27] mentioned that for separations with charged CDs of oppositely charged analytes, a higher separation selectivity can be achieved due to a larger "separation window", *i.e.*, the $\mu^f - \mu^c$ term in Eq. (2), than when using neutral or equally charged CDs and analytes. Furthermore, they showed that in nonaqueous media the ion-pair effect between the charged CD and the oppositely charged analyte was even stronger, which resulted in chiral baseline separations at lower CD concentrations than in aqueous solutions [30]. Finally, according to Wren and Rowe [14], the optimal concentration of CD is inversely related to the binding constants of the enantiomers:

$$[\mathrm{CD}]_{opt} = 1\sqrt{K_1 K_2} \tag{4}$$

Several papers have been published describing the applicability of charged CDs in separation systems. A small number is directed to HPLC [31–35], but the majority of the publications concerns the use in aqueous and nonaqueous CE [4, 8, 36–40].

3 Synthesis and characterization of charged CDs

Croft and Bartsch [41] wrote a comprehensive review on the synthesis of chemically modified CDs in which they summarized the synthesis of acylated, alkylated, deuterated, mesylated, tosylated, and rigidly capped CDs. Furthermore, they described the synthesis of CDs containing amino, azido, halogen, nitrate, phosphorous, imidazole, pyridine, sulfur, alcohol, aldehyde, keto, oxime, carboxyl, carbonate, carbamate, silicon, boron and tin as functional groups. In this review, we will discuss the synthesis and characterization of those CDs that are in use for pharmaceutical analysis with CE. Several cationic and anionic CDs (*e.g.*, SBE-[I-VII]-CDs, carboxy(m)ethylated-CDs, sulfated-CDs, mono-(6-[2-hydroxy]propyl-trimethylamino-6-deoxy)-β-CD; degree of substitution (DS = 3.5) are nowadays commercially available from several companies (Supelco, Bellefonte, PA, USA; Wacker-Chemie, Munich, Germany and Cyclolab, Budapest, Hungary) in contrast to only one amphoteric derivative, *i.e.*, β-CD substituted at the 6-position with 2-hydroxypropyl-trimethylammonium chloride (DS = 2.0) or a sodium acetate group (DS = 2.0, Supelco).

3.1 Synthesis of anionic CDs

Vigh and co-workers described the synthesis of several anionic CDs, namely heptakis(2,3-diacetyl-6-sulfato)-β-CD (CD1) [42], hepta-6-sulfato-β-CD (CD2) [43] and heptakis (2,3-dimethyl-6-sulfato)-β-CD (CD3) [44]. The synthesis is based on principles described in the papers of Takeo *et al.* [45] and Newton *et al.* [46] and is basically similar for all three CDs. The principle is shown (without the experimental conditions and purification steps) in Fig. 3, scheme 1. Sulfopropyl-ether-β-CD (SPE-β-CD) was synthesized by Mayer and Schurig [47] (Fig. 3, scheme 2). The structural analog SBE-γ-CD was synthesized using 1,4-butanesulfone and γ-CD. It was shown that a lower concentration of 1,4-butanesulfone, a higher reaction temperature or a longer reaction time, resulted in a lower DS [48]. Terabe *et al.* [12] described the synthesis of 2-*O*-carboxymethyl-β-CD (CM-β-CD) *via* the reaction of β-CD with sodiumiodoacetate in DMSO according to a slightly modified synthesis route originally developed by Kitaura and Bender [49].

3.2 Synthesis of cationic CDs

6^A-Methylamino-β-CD was synthesized by Nardi *et al.* [50] (see Fig. 4, scheme 1). 6^A, 6^D-Dimethylamino-β-CD, was synthesized [50] through regioselective capping of β-CD with biphenyl-4,4'-disulfonylchloride as described previously [51]. The glucose-unit at the A-position was linked at the 6-methoxy group *via* the 6-methoxy group of the glucose-unit at the D-position with the disulfonate group. Substitution of this biphenyl-4,4'-disulfonyl group with methylamine was as described previously [52]. Another cationic CD, 2-hydroxypropyltrimethylammonio-β-CD (or sometimes referred to as 2-hydroxy-3-trimethyl-ammonio-propyl-β-CD) was first synthesized by Parmeter *et al.* [54] (Fig. 4, scheme 2). Haynes III *et al.* [55] described the synthesis of methoxyethylamine-β-CD (see Fig. 4,

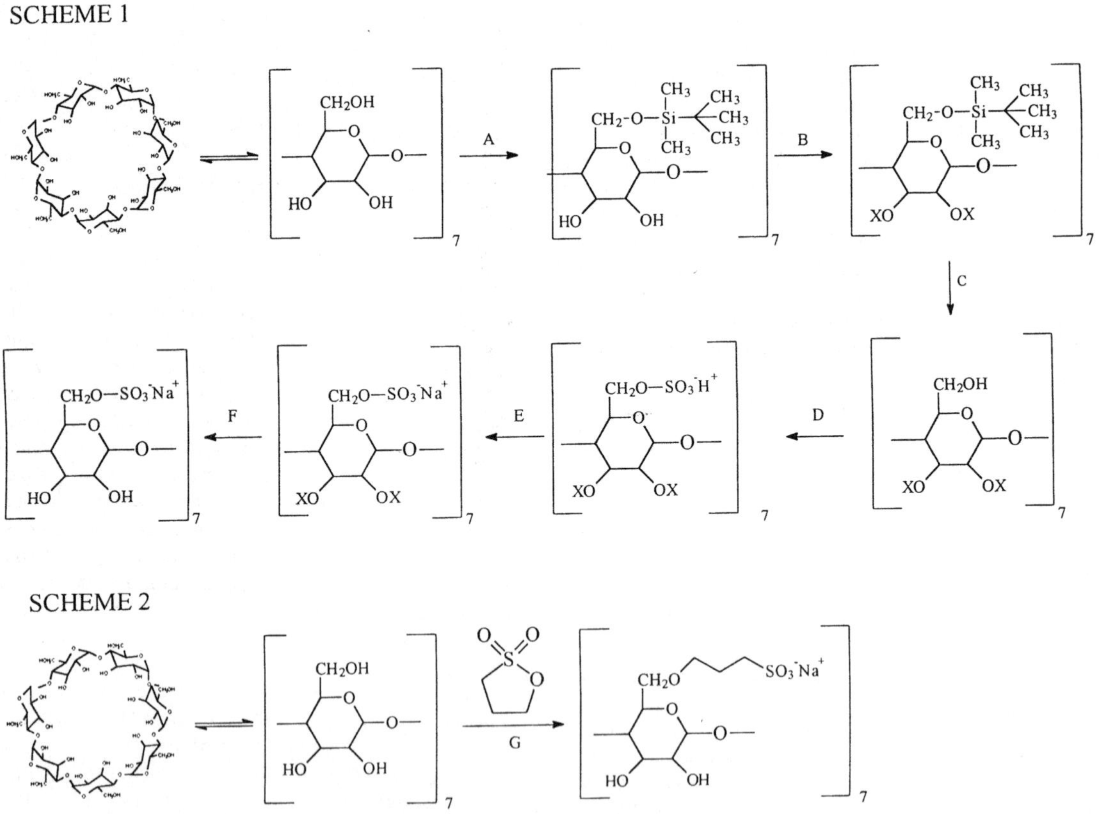

SCHEME 1

SCHEME 2

Figure 3. Synthesis of anionic CDS. Scheme 1: Synthesis of 2,3-DMS-β- or DAS-β-CDs or S-β-CD. The reaction is started by the addition of β-CD to dimethyl-*tert*-butylchlorosilane (step A). The intermediate is then peracetylated or methylated (step B). The dimethyl-*tert*-butylsilyl protecting group is removed (step C), followed by complete sulfation of the primary hydroxyl groups of the CD (step D). After neutralization with NaOH (step E) the S-β-CD is obtained by complete deacetylation (step F). Scheme 2: Synthesis of SPE-β-CD by reaction of β-CD with 1,3-propane sulfone.

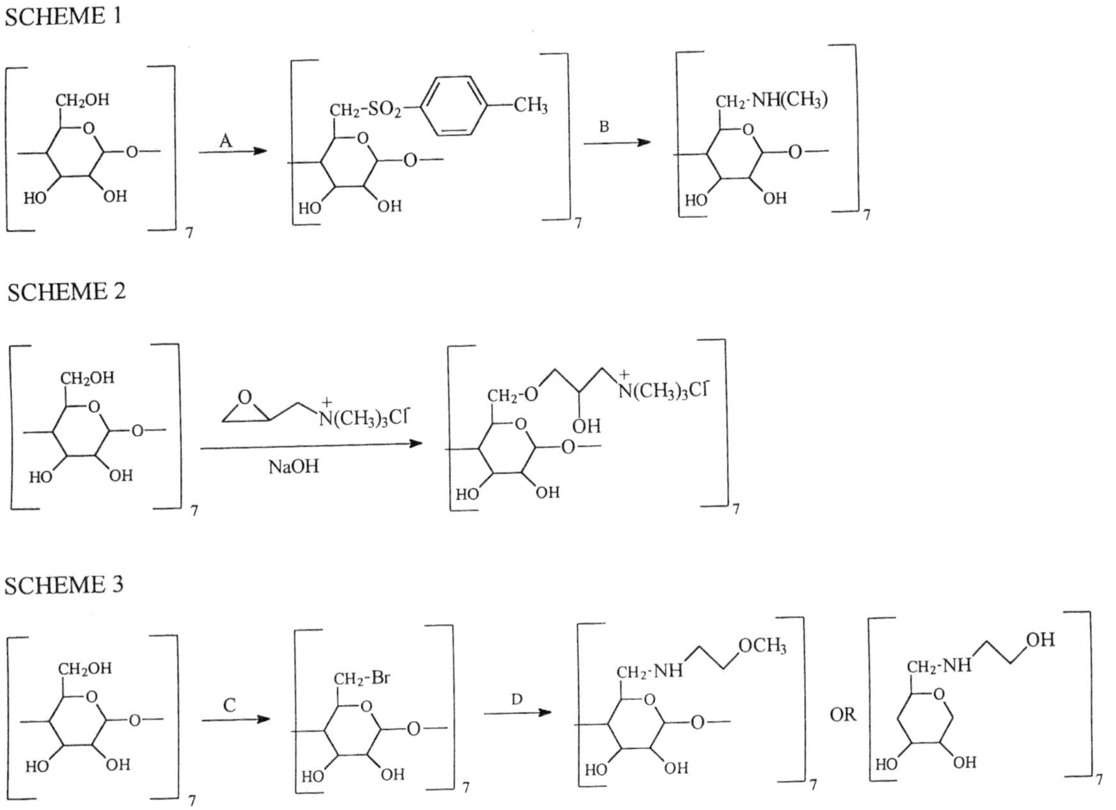

SCHEME 1

SCHEME 2

SCHEME 3

Figure 4. Synthesis of cationic CDs. Scheme 1: Synthesis of 6^A-methylamino-β-CD by reacting β-CD with *p*-toluenesulfo-nylchloride according to [155] (step A). The tosyl groups are substituted by methylamine according to [52] (step B). Scheme 2: Synthesis of QA-β-CD by reacting β-CD with 2,3-epoxypropyl-trimethylammonium chloride. Scheme 3: Synthesis of methoxyethylamine-β-CD. Plain β-CD was perbrominated *via* the procedure described in [156] (step C). After that, the bro-minated product was dissolved in methoxyethylamine or hydroxyethylamine (step D).

scheme 3). A similar derivative, hydroxyethylamine-β-CD where the methoxy group was substituted by a hydroxyl group, was synthesized by using ethanolamine instead of methoxyethylamine [56].

3.3 Characterization and inclusion behavior of charged CDs

3.3.1 Characterization

Characterization of SBE-β-CD with various DS was de-scribed by Luna *et al.* [57, 58]. They applied the fractiona-tion of the different substituted CDs by preparative anion exchange chromatography, after which characterization took place by ^{1}H-NMR spectroscopy and CE analysis with MS detection. Chankvetadze *et al.* [59] demonstrated the use of electrospray ionization-mass spectrometry (ESI-MS), matrix assisted laser desorption/ionization-time of

flight-mass spectrometry (MALDI-TOF-MS), and fast atomic bombardment-mass spectrometry (FAB-MS) for the determination of molecular weight, DS, and purity of several charged CD derivatives. Charged CD derivatives were dissolved in H_2O and introduced into the MS system at final concentrations of approximately 100 µg/mL (ESI), 1 µg/mL (MALDI-TOF) and 3–5 µg/mL (FAB), respective-ly. They concluded that ESI-MS and MALDI-TOF-MS were suitable methods for the rapid analysis of the DS and purity of the derivatized CDs, whereas the FAB-MS method was more suitable for the determination of the molecular weight of pure molecules. The determined DS was verified by CE separations.

Tanaka *et al.* [60] described the composition analysis of some charged CD derivatives by on-line CE-ionspray-MS (CE-ISP-MS), using a pneumatically assisted ESI inter-face. They dissolved the CD derivatives at final concentrations of 1–5 mg/mL in 0.05% formic acid in water:methanol 1:1 v/v. Samples were infused into the ISP interface directly at 5 µL/min with a syringe pump and the ISP voltage was maintained at 5 kV for the cationic CDs and at −4.5 kV for the anionic CDs. Typical mass spectra of charged CDs are shown in Fig. 5. The figure shows the heterogeneity of the synthesized CDs. Therefore, the peaks obtained by CE separation should be identified by CE-MS. In addition, CE-MS seems to be a good method to monitor the batch-to-batch differences in synthesized CDs. It should be kept in mind that due to the heterogeneity of the derivatized CDs, the reproducibility of the obtained selectivity factors and/or resolutions can be influenced severely. Furthermore, different synthesis routes may lead to different selectivities, and even differences in batch-to-batch productions of the same manufacturer might influence the outcome of the analysis.

3.3.2 Inclusion behavior

NMR spectroscopy can provide suitable information about the environment of individual atoms and intermolecular interactions and thus provide data about the structure and molecular dynamics of complexes. Furthermore, it is able

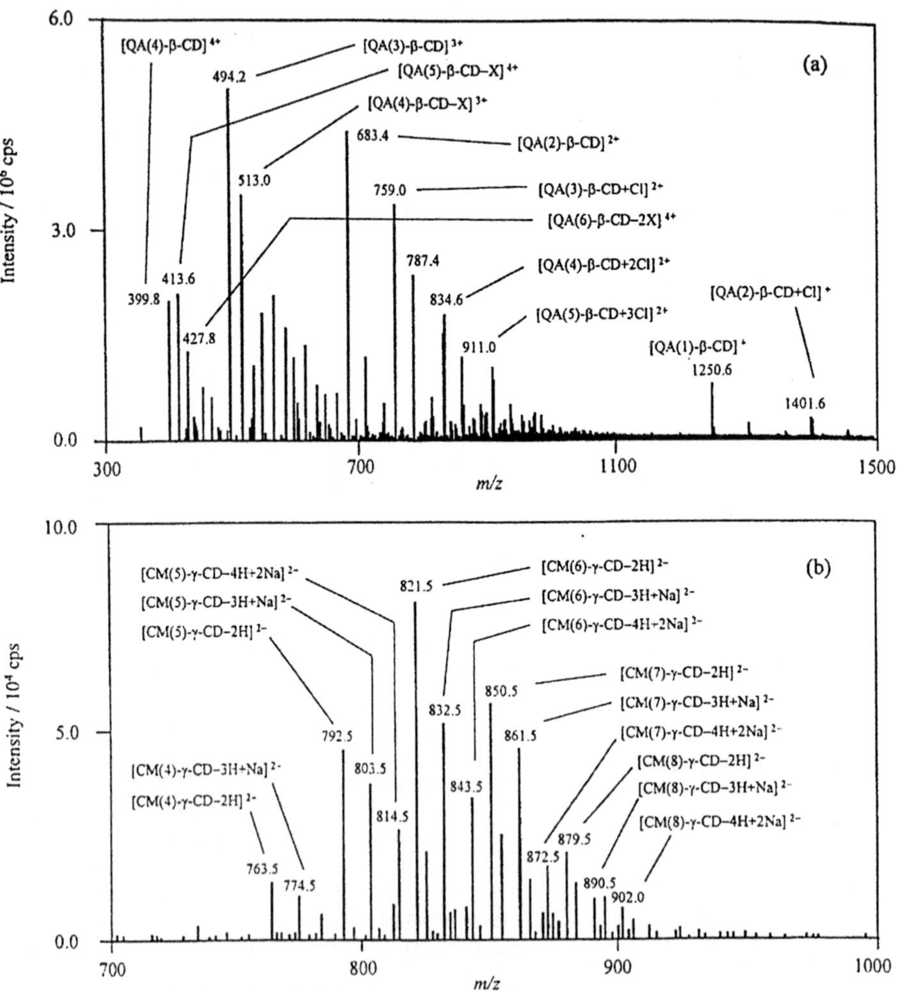

Figure 5. Mass spectra of charged CDs. (a) QA-β-CD in 0.05% formic acid in water:methanol 1:1 v/v (5 mg/mL); (b) CM-γ-CD in 20 mM ammonium acetate in 0.05% formic acid in water:methanol 1:1 v/v (1 mg/mL). X is the fragment ion $HN^+(CH_3)_3$ from QA-β-CD. Reprinted from [157], with permission.

to dynamically monitor inclusion complexation, *i.e.*, the chiral recognition mechanism. Still, although NMR spectroscopy can provide useful information about complexation patterns in the liquid phase, some restrictions should be taken into account [4]. For instance, the NMR signals observed in racemate/CD solutions are the time-averaged signals of both the complexed and the free substances. This implies that because there is a difference in binding constants between the two enantiomers and the CD, the more strongly complexed enantiomer will subsequently be more shifted, which leads to a difference in the intensity of the signal. If two enantiomers have the same binding constants but the obtained diastereomeric complexes have different NMR spectra, this will consequently lead to signal splitting in NMR.

In CE it is a common fact that the mobility of the analyte is strongly dependent on the characteristics of the buffer solution (pH and ionic strength). Furthermore, electrodispersion can be reduced by an increased ionic strength. Chankvetadze *et al.* [4] described the effect of the ionic strength and the pH of the run buffer on the chiral recognition of CM-β-CD with ^{1}H-NMR and CE. They showed that the resolution was substantially increased upon increasing the ionic strength of the phosphate buffer. Yet, it was shown that the ionic strength had no significant effect on the chiral recognition process monitored by ^{1}H-NMR. Hence, the resolution was increased mainly by a change of the effective mobility of the enantiomers and not by an improved chiral recognition of the CD derivative. Changing the pH of the run buffer will have an effect on the solute-CD interaction and therefore on the resolution. Chankvetadze *et al.* [4] showed that the upfield chemical shift of the 2-H signal of the imidazole moiety of (±)-metomidate and complexation-induced nonequivalence of the chemical shift between the enantiomers and the carboxymethylated CDs was enhanced by an increase of the pH. This increase in chiral recognition was also observed in CE, illustrated by an increased resolution. The latter observations show a good correlation between the data obtained by CE separation and ^{1}H-NMR spectroscopy. Kitae *et al.* [61] used 6-(2-thioglycolic acid)-6-deoxy-β-CD and mono-(6-amino)-β-CD to study the chiral recognition of α-amino acids by means ^{1}H-NMR spectroscopy. They verified that inclusion of the guest into the host cavity and intermolecular Coulomb interactions participate cooperatively in the complexation. This, and the observation that charged CDs possess a higher chiral recognition ability compared to native or neutral CDs, was also confirmed by other authors using ^{1}H-NMR or ^{13}C-NMR [62, 63]. A different approach to study the inclusion behavior of analytes with CDs was described by Shuang *et al.* [64]. They used steady-state fluorometry to study complexation interactions of DM-β-CD with procaine at different concentrations. Procaine exhibits at weak fluorescence emission in the absence of CDs. When the concentration of CM-β-CD increases, an enhanced fluorescence emission of procaine is observed. They concluded that the formation of a inclusion complex affected the ground state properties of the procaine molecule and protected it from quenching occurring in aqueous solutions. The formation constants of the solutes and the CDs were obtained from fluorescence data, evaluated at different pH values (assuming a 1:1 inclusion model), by the modified Benesi-Hildebrand equation [65].

4 Applications of anionic CDs in CE

Terabe and co-workers [12, 13] were the first to introduce an anionic CD, *i.e.*, CM-β-CD, for separation purposes, although in their first paper they did not use them for chiral separations. Since that time different types of negatively charged CDs have been described in the literature. Besides the carboxymethylated CDs, the most commonly used are the sulfated and the sulfoalkyl ether-substituted CDs. More than four years passed before another paper reported the use of CM-β-CD for the chiral separation of several pharmaceuticals at different pHs [66]. With this material, at low pH (< 4) all carboxylic functions were protonated and the CDs behaved like ordinary neutral (substituted) CDs, whereas at high pH (> 5) the carboxylic functions were deprotonated to give negatively charged CDs as a moving stationary phase that could separate neutral analytes. Later, CM-β-CDs were used for the chiral separation of anticoagulants [67], anti-epileptics [68], antitumor agents [69, 70], antihistaminics [71], anxiolytics [68], antidepressants [62, 72], sedatives (*e.g.*, barbiturates) [68, 73] calcium channel blockers [74], diuretics [68], corticosteroids [75], and several peptides [76–78]. In these cases either a fused-silica capillary, a polyacrylamide-coated capillary or, as once reported, a 3-(trimethoxysilyl)propyl methacrylate-coated capillary [72] were used.

Although most applications on the use of anionic CDs concern the separations of basic compounds, the theoretically less favorable separation of acidic compounds like barbiturates and nonsteroidal anti-inflammatory drugs (NSAIDS), at a pH where they are undissociated, have been described [79]. The latter approach is assumed to be less favorable due to the equal charge of the CD and the analyte. Table 1 lists the analyzed neutral or positively charged compounds, along with the used anionic chiral selector. The table mentions only those analytes that have been baseline-separated. Besides CM-β-CD, four other structurally related carboxylated CDs have been described in the literature, namely carboxyethylated-β-CD

Table 1. Applications of anionic CDs for chiral and achiral separations

Compound	CD	References
Acebutolol	S-β-CD, CM-β-CD, SBE-β-CD, Ph-β-CD	[83, 94, 98, 132]
Acenocoumarol	CM-β-CD, S-β-CD	[67]
Acetanilide	SBE-β-CD	[133]
Acetaminophen (paracetamol)	SBE-β-CD	[133]
Adrenaline	CM-β-CD, S-β-CD	[134, 135]
Alclofenac	CM-β-CD	[79]
Alprenolol	S-β-CD, CM-β-CD, SBE-β-CD	[94, 129, 132]
Amino acids and peptides (*e.g.*, dinitrobenzoyl amino acids and dansylated amino acids	CM-β-CD, CE-β-CD, SBE-β-CD, S-β-CD, TGA-β-CD, CML-β-CD, SBE-γ-CD	[48, 61, 76, 81, 83, 108, 136–138]
Aminoglutethimide	CM-β-CD, S-β-CD, Ph-β-CD, SBE-β-CD, CM-γ-CD	[69, 83, 94]
Amlodipine	CM-β-CD	[74, 129]
Amphetamine impurities	SBE-β-CD, SBE-γ-CD, S-α-CD, S-β-CD, S-γ-CD	[86, 89, 92]
Anisodamine	SBE-β-CD	[139]
1-(9-Anthryl)-2,2,2,-trifluoro-ethanol	SBE-γ-CD	[48]
Aromatic compounds (hydrophobic, achiral)	SBE-β-CD	[140]
Atenolol	SBE-β-CD, S-β-CD, CM-β-CD	[83, 89, 94, 129, 132, 136, 141]
Azelastine	Ph-β-CD, CM-γ-CD	[83]
Bencynonate	SBE-β-CD	[139]
Benzhexol	SBE-β-CD	[139]
Benzoin	CM-β-CD, CE-β-CD, SBE-β-CD, S-β-CD, DMS-β-CD, Ph-β-CD	[23, 81, 83, 94]
Bepridil	SBE-β-CD	[139]
Betaxolol	CM-β-CD, SBE-β-CD	[129]
Bevantolol	CM-β-CD, SBE-β-CD	[129]
Binaphthol	CM-β-CD, CE-β-CD, SUCC-β-CD, SBE-γ-CD	[48, 66]
R/S-1,1'-binaphthyl-2,2'-diyl hydrogen phosphate	CM-β-CD, SBE-β-CD + SEE-β-CD	[142]
Bisoprolol	SBE-β-CD, Ph-γ-CD	[83, 139]
Bromoidoles	SBE-β-CD	[143]
Brompheniramine	CM-β-CD, CE-β-CD, S-β-CD	[71, 80, 94]
Bunitrolol	CM-β-CD, SBE-β-CD, CE-β-CD, Ph-β-CD, Ph-γ-CD	[81, 83, 132]
Bupivacaine	SBE-β-CD, S-β-CD, Ph-β-CD, CM-γ-CD, Ph-γ-CD	[83, 94, 136]
Bupropion	S-β-CD	[94]
Butaclamol	SBE-γ-CD	[138]
Caffeine	SBE-β-CD	[133]
Canadine	S-β-CD	[94]
Carbamazepine	CM-β-CD	[72]
Carbinoxamine	S-β-CD	[94]
Carprofen	CM-β-CD	[79]
Carvedilol	SBE-β-CD	[139]
Catecholamines + related compounds	S-β-CD	[95]
Chlorcyclizine	S-β-CD	[144]
Chlormezanone	CM-β-CD, SBE-β-CD	[68, 83]
Chloroquine	S-β-CD, CE-β-CD	[80, 94]
Chlorpheniramine	CM-β-CD, CE-β-CD, S-β-CD, SBE-β-CD, DM-γ-CD, Ph-γ-CD	[71, 80, 83, 94, 135, 144, 145]
Chlorprenaline	Ph-β-CD, SBE-β-CD, CM-γ-CD, Ph-γ-CD	[83]
Chlorthalidone	CM-β-CD, SBE-β-CD	[68]
Cimaterol	SBE-β-CD	[89]
Clenbuterol	SBE-β-CD	[89]

Table 1. continued

Compound	CD	References
Clozapine	S-β-CD	[99]
Cocaine impurities	SBE-β-CD	[91]
Cyclopentolate	S-β-CD	[144]
Denopamine	S-β-CD, SBE-β-CD, Ph-γ-CD	[83, 135]
Desipramine	CM-β-CD	[72]
Diisopyramide	S-β-CD, SBE-β-CD	[94, 129, 141]
Dimetindene	CM-β-CD, CE-β-CD, SUCC-β-CD, S-β-CD	[66, 94, 145]
Doxylamin	CM-β-CD, CE-β-CD, SUCC-β-CD, S-β-CD	[66, 80, 94, 144]
Eperisone	SBE-β-CD, CM-γ-CD, Ph-γ-CD	[83]
Ephedrine	CM-β-CD, CE-β-CD, SUCC-β-CD	[66, 145]
Ephedrine + derivatives	SBE-β-CD, S-α-CD, S-β-CD, S-γ-CD	[87, 103]
Epinastine	SBE-β-CD, Ph-γ-CD, CM-γ-CD	[83]
Esmolol	SBE-β-CD	[139]
Estrogens	S-α-CD, S-β-CD, S-γ-CD	[100]
Etilefrin	SBE-β-CD, Ph-γ-CD	[83]
Fencamfamine isomers	SBE-β-CD	[146]
Fenfluramine	CM-β-CD	[145]
Fenoprofen	SBE-β-CD	[147]
Fenoterol	CM-β-CD, CE-β-CD, SBE-β-CD, Ph-γ-CD	[81, 83]
Flufenamic acid	CM-β-CD	[79]
Flurbiprofen	CM-β-CD	[79]
Glutamine	SBE-β-CD	[89]
Glutethimide	SBE-β-CD	[139]
Glycopyrrolate	SBE-β-CD	[139]
GR50360A and GR57732A	CME-β-CD	[84]
Guaifenesin	SBE-β-CD	[133]
Heroin impurities	SBE-β-CD	[90]
Hexobarbital	CM-β-CD, CE-β-CD, SBE-β-CD, SPE-β-CD, SUCC-β-CD	[66, 68, 118, 147]
Homochlorcyclizine	S-β-CD	[144]
Hydantoins	S-β-CD	[94]
Hydroxyxhloroquine	S-β-CD	[94]
Idazoxan	S-β-CD	[94]
Imazalil	CM-β-CD, CE-β-CD, Ph-β-CD, SBE-β-CD, CM-γ CD	[81, 83]
Indapamide	SBE-β-CD	[139]
Indomethacin	CM-β-CD	[79]
Indoprofen	CM-β-CD	[79]
Isoxuprine	S-β-CD	[94]
Ketamine	S-β-CD, Ph-β-CD	[83, 94]
Ketoprofen	CM-β-CD, SBE-β-CD	[79, 147]
Labetalol	CM-β-CD, SBE-β-CD, S-β-CD	[97, 98, 132]
Laudanosine	S-β-CD	[94]
Lobeline	SBE-β-CD	[139]
Loxapine	S-β-CD	[99]
Mandelic acid esters	CM-β-CD, CE-β-CD, Ph-β-CD	[81, 83]
Meclizine	SBE-β-CD, CM-γ-CD	[83]
Mepenzolate	S-β-CD	[94]
Mephenytoin	CM-β-CD, SBE-β-CD	[68]
Mephobarital	CM-β-CD, SBE-β-CD	[68, 141]
Mepivacaine	S-β-CD	[94]
Methanephrine	S-β-CD, Ph-β-CD, Ph-γ-CD	[83, 98]
Methoxamine	SBE-β-CD	[139]
Methoxyphenamine	S-β-CD	[94]
Methylphenidate	SBE-β-CD	[139]

Table 1. continued

Compound	CD	References
Metoprolol	SBE-β-CD, S-β-CD, CM-β-CD, Ph-γ-CD	[94, 129, 132, 136]
Mexiletine	S-β-CD, SBE-β-CD, Ph-γ-CD	[83, 94]
Mianserine	CM-β-CD, SBE-β-CD, SEE-β-CD, S-β-CD, SBE-γ CD	[62, 138, 144]
Midodrine	S-β-CD	[94]
Nadolol	SBE-β-CD, S-β-CD	[88, 98]
Nafronyl	S-β-CD	[98]
Naproxen	CM-β-CD, SBE-β-CD	[79, 148]
Nebracetam	Ph-γ-CD	[83]
Nefopam	S-β-CD	[94, 144]
Nicardipine	Ph-β-CD, SBE-β-CD, CM-γ-CD	[83]
Niflumic acid	CM-β-CD	[79]
Nitro aromatic explosives	SBE-β-CD, SUCC-β-CD	[107]
Noradrenaline	S-β-CD	[135]
Norephedrine	Ph-β-CD	[83]
Norlaudanosoline	S-β-CD	[135]
Normethanephrine	S-β-CD	[98]
Ofloxacin	S-β-CD	[101]
Oligosaccharides (2-ammino-benzamide derivatized)	SBE-β-CD, SBE-γ-CD	[149]
Ondansetron	SBE-β-CD	[139]
Opipramol	CM-β-CD	[72]
Orphenadine	S-β-CD	[94]
Oxazolidinon	CM-β-CD, CE-β-CD, SUCC-β-CD	[66]
Oxprenolol	CM-β-CD, SBE-β-CD, S-β-CD	[94, 132, 136]
Oxyphencyclimine	CM-β-CD, CE-β-CD, S-β-CD, SBE-β-CD	[81, 83, 94]
Pentobarbital	CM-β-CD, SBE-β-CD	[68, 150]
Phenacetin	SBE-β-CD	[133]
Pheniramine	CM-β-CD, CE-β-CD, S-β-CD, SBE-β-CD	[71, 80, 94, 141]
Phensuximide	S-β-CD	[94]
Phenylephrine	SBE-β-CD, Ph-β-CD	[83, 139]
Pinacidil	SBE-β-CD	[139]
Pindolol	SBE-β-CD, S-β-CD, CM-β-CD, CM-γ-CD	[83, 94, 129, 132, 136, 141]
Piperoxan	S-β-CD	[94]
Piroxicam	CM-β-CD	[79]
Praziquantel + metabolites	SBE-β-CD	[151]
Primaquine	S-β-CD, Ph-β-CD, SBE-β-CD, CM-γ-CD, Ph-γ-CD	[83, 94]
Procaine	CM-β-CD	[64]
Promethazine	SBE-β-CD, Ph-β-CD, CM-γ-CD, Ph-γ-CD	[83, 136]
Propranolol + metabolites	CM-β-CD, CE-β-CD, SBE-β-CD, SBE-γ-CD, SUCC-β-CD and S-β-CD	[66, 94, 97, 129, 132, 135, 136, 141, 152]
Protriptyline	CM-β-CD	[72]
Pyridoxine	SBE-β-CD	[133]
Salbutamol	SBE-β-CD, Ph-β-CD	[83, 89]
Secobarbital	CM-β-CD, SBE-β-CD, SBE-γ-CD	[48, 68, 138]
Sotalol	CM-β-CD, SBE-β-CD	[129, 132]
Substance P + metabolites	SBE-β-CD	[93]
Sulconazole	SBE-β-CD, CM-γ-CD	[83]
Sulfonium ions	S-β-CD	[96]
Sulindac	CM-β-CD, SBE-β-CD	[79, 147]
Sulpiride	Ph-β-CD, SBE-β-CD	[83]
Suprofen	CM-β-CD	[79]
Synephrine	SBE-β-CD	[139]
Terazosin	SBE-β-CD	[139]

Table 1. continued

Compound	CD	References
Terbutaline	SBE-β-CD, S-β-CD, CM-β-CD	[83, 89, 94, 136, 145, 153]
Tetrahydropapaveroline	S-β-CD	[94]
Tetramisole	S-β-CD	[94, 144]
Thalidomide + metabolites	CM-β-CD, SBE-β-CD	[73, 85, 89]
Thiopental	CM-β-CD, SBE-β-CD	[68]
Thioridazine	S-β-CD, Ph-β-CD, CM-γ-CD, Ph-γ-CD	[83, 144]
Tiaprofenic acid	CM-β-CD	[79]
Timepidium	S-β-CD	[135]
Toliprolol	CM-β-CD, SBE-β-CD	[132]
Tolperisone	S-β-CD, CM-γ-CD, Ph-γ-CD	[83, 94]
Tramadol	CM-β-CD, SBE-β-CD	[70, 139]
Tranylcypromine	S-β-CD	[94]
Triamcinolone acetonide	CM-β-CD, CE-β-CD, S-β-CD, Ph-β-CD	[75]
Trihexyphenidyl	S-β-CD, SBE-β-CD	[83, 94]
Trimetoquinol	S-β-CD, SBE-β-CD, CM-γ-CD, Ph-γ-CD	[83, 135]
Trimebutine	Ph-β-CD, SBE-β-CD, CM-γ-CD	[83]
Trimipramine	CM-β-CD, CE-β-CD, S-β-CD, SBE-β-CD, CM-γ-CD Ph-γ-CD	[81, 83, 94, 144]
Trimeprazine	S-β-CD	[144]
Troger's base	S-β-CD	[94]
Tropa-alkaloids	SBE-β-CD, S-β-CD, DAS-β-CD	[106]
Verapamil	S-β-CD, Ph-β-CD, SBE-β-CD, CM-γ-CD, Ph-γ-CD	[83, 94, 135, 141]
Voriconazole	SBE-β-CD	[154]
Warfarin	CM-β-CD, SBE-β-CD, S-β-CD, DMS-β-CD	[23, 67, 94, 141, 147]

Abbreviations: CE-β-CD, carboxyethyl-β-CD; CML-β-CD, carbamoylated-β-CD; Ph-β-CD, phosphated-β-CD; SUCC-β-CD, succinylated-; TGA-β-CD, 6-(2-thioglycolic acid)-6-deoxy-β-CD

(CE-β-CD) [66, 75, 80–82] a γ-CD derivative, CM-γ-CD [83], and carboxymethylethyl-β-CD (CME-β-CD) [84] for the development of GR50360A and GR57732A.

Nowadays the most widely used anionic CD derivative is probably sulfatobutylether-β-CD, ofter referred to (although incorrect according to chemical nomenclature) as SBE-β-CD. It was introduced in 1994 for the chiral separation of the neutral enantiomers of thalidomide [85]. Other early applications include cationic drugs of forensic interest [86], ephedrine and related compounds [87], nadolol [88], illicit drugs like cocaine and khat [86], and many other pharmacologically active substances (β-agonists, β-antagonists, phenylethylamines, stimulants, *etc.*) [89]. After these first experiments, SBE-β-CD was used for the analysis of acidic and neutral impurities in illicit heroin [90], cocaine [91], and methamphetamine [92], and for the separation of substance P, a cationic peptide (Arg-Pro-Lys-Pro-Gln-Gln-Phe-Phe-Gly-Leu-Met-NH$_2$) [93]. In the latter study, phytic acid was employed as a run buffer additive to eliminate the interaction of substance P and its cationic *N*-terminus metabolites with ionized silanol groups. Besides SBE-β-CD the use of the structurally related sulfoethyl ether-β-CD (SEE-β-CD), SPE-β-CD, and SBE-γ-CD has been described (see Table 1).

Sulfated CDs (S-β-CD), *i.e.*, without any alkyl chain located between the 6-O-position and the sulfate-group, are recently gaining much interest. Like the sulfoalkyl ether substituted CDs, S-β-CD (DS usually between 7–10) are negatively charged over the entire pH range that is common in CE separations (pH 2–12). This is in contrast to CM-β-CD, for instance. Stalcup and Gahm [94] and Gahm and Stalcup [95] described the use of mixtures of substituted S-β-CDs for the separation of 56 compounds of pharmaceutical interest, among them anesthetics, antidepressant, antihistamines, and catecholamines, as shown in Fig. 6. Also the separation of sulfonium ions [96], β-antagonists [97, 98] and antipsychotic drugs [99] has been described. Wang and Khaledi [98] mentioned the use of capillaries coated dynamically with poly(vinylalcohol) for the separation of some β-antagonists. Munro *et al.* [100] applied on-line charged CD-mediated sample stacking techniques in MEKC for the separation of neutral hydrophobic estrogens. They showed that the sensitivity was increased tremendously by replacing the organic modifier methanol by S-β-CDs.

We are now developing a rapid chiral separation system for ofloxacin, a synthetic antimicrobial agent. Separations are performed in urine samples obtained from patients

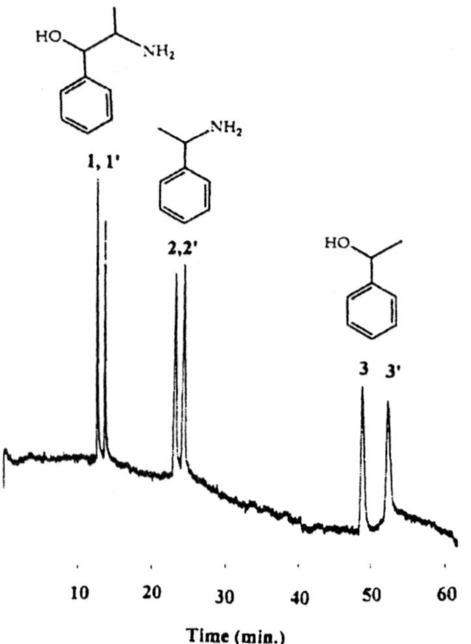

Figure 6. Separation of norephedrine (1,1'), α-methyl-benzylamine (2,2'), and sec-phenethyl alcohol (3,3') using S-β-CD as the chiral selector. Reprinted from [95], with permission.

suffering from hemophilic diseases. The objective of the study is to investigate the use of S-β-CD in combination with several neutral CD derivatives for dual chiral separation systems in order to monitor possible metabolic racemization of S(–)ofloxacin [101]. Only a few reports have been written on the use of sulfated α- and γ-CDs, among them a European patent application [92, 102]. Highly sulfated α-, β-, and γ-CDs, with an average number of sulfate groups per CD of 11, 12, and 13, respectively, were just recently described by Chen and Evangelista [103] and Verleysen *et al.* [104]. They used highly sulfated CDs developed by scientists at the Beckman Company, which are not commercially available at this time. The C-6 primary hydroxyl groups of these CDs are completely sulfated, and the C-2 secondary hydroxyl groups are 70% sulfated, but there is no sulfate substitution at the C-3 secondary hydroxyls. Verleysen *et al.* compared the use of the highly sulfated CDs (DS = 12) to the use of sulfated CDs with a DS of 4 for the chiral separation of isomeric α- and β-aspartyl containing di- and tripeptides. They experienced a higher selectivity for the highly sulfated CDs. A special type of sulfated CDs are the 2,3-subsituted CDs that are sulfated at the 6-position. So far, two types have been mentioned in the literature: the 2,3-dimethyl-6-sulfato-β-CD (DMS-β-CD) by Cai and Vigh [23, 105] and Cai

et al. [44] and the 2,3-di-*O*-acetyl-6-sulfato-β-CD (DAS-β-CD) by Wedig and Holzgrabe [106] and Vincent *et al.* [42]. These 2,3-substituted S-β-CD are expected to give even higher resolutions than plain S-β-CDs or sulfoalkyl ether-β-CD. Figure 7 shows the enantioseparation of 2-indanol using DMS-β-CD as the chiral selector.

Finally, some rare anionic derivatized CDs have been described. Among these are succinylated-β-CD [66, 107], 6-(2-thioglycolic acid)-6-deoxy-β-CD [61], phosphated-CDs (both γ- and β-CD) [75, 83], and carbamoylated β-CDs [108]. Wang and Khaledi [109] recently published an article in which for the first time an anionic β-CD (*i.e.*, sulfated β-CD) was used in nonaqueous solutions. They achieved chiral separations of the enantiomers of several pharmaceuticals in formamide. In nonaqueous systems, the CD concentration (Eq. 4) is not as critical as in aqueous solutions, due to the lower binding constants between solutes and CDs observed in less polar solvents [110].

Szeman *et al.* [111] found that the DS in CM-β-CD can have a significant effect on resolution. This is probably a consequence of a difference in electrostatic interaction with the analytes. Also the width of the CD "mouth" may be affected, so that the inclusion complexation may change. Francotte *et al.* [48] studies the influence of the DS of SBE-γ-CD derivatives (DS = 3.5–7.5) on the chiral separation of several solutes (binaphthol, trifluoroanthranyl ethanol, secobarbital and dansylated amino acids). Conceivably, the number of ionic groups will affect the electrophoretic mobility of the CD due to a difference in charge density. Figure 8 shows the influence of the DS of SBE-γ-CD (15 mM) on resolution of trifluoroanthranyl ethanol and binaphthol. Trifluoroanthranyl ethanol cannot

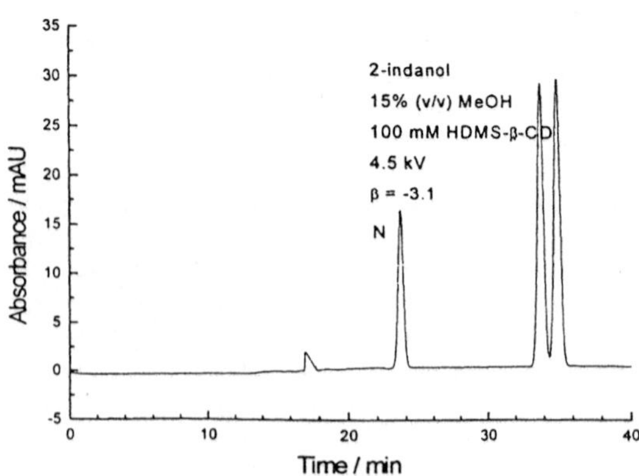

Figure 7. Electropherogram of the chiral separation of 2-indanol, migrating to the anode in the acidic DMS-β-CD BGE. N, nitromethane, Reprinted from [23], with permission.

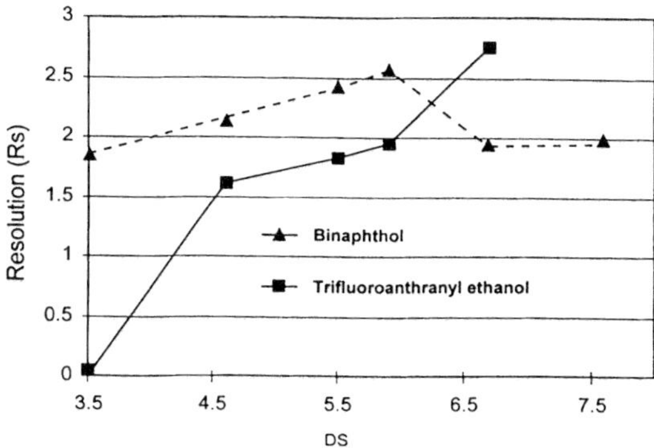

Figure 8. Influence of the DS of SBE-γ-CD on the resolution (R_s) of the enantiomers of binaphthol and of 1-(9-anthryl)-2,2,2-trifluoroethanol. Reprinted from [48], with permission.

be baseline separated with a DS below 4.5 while for binaphthol the influence is less significant.

5 Applications of cationic CDs in CE

The most common application is the separation of carboxylic acids or amino acids. A wide variety of mono-polysubstituted amino derivatives of CDs were described in a review by Croft and Bartsch [41]. As with anionic CDs, Terabe [13] was the first in 1989 to achieve separation of six dansyl amino acids enantiomers using mono-(6-β-aminoethylamino-6-deoxy)-β-CD (AEA-β-CD). After a period of four years in which no other publications appeared using cationic CDs in CE separations, Fanali and co-workers [50] mentioned the use of 6^A-methylamino-β-CD (6^A-MA-β-CD) and $6^A,6^D$-dimethylamino-β-CD ($6^A,6^D$-DMA-β-CD) for the enantiomeric resolution of some 2-hydroxy acids. Since then only a few other cationic derivatives have been mentioned in the literature. 6-Methylamino-β-CD (MA-β-CD), with several DSs at the 6-position, was used in the chiral separation of several acidic and basic compounds [112], as listed in Table 2. The latter showed that the DS at the 6-position affected the chiral resolution of fenoprofen: it was not baseline separated using mono-MA-β-CD, but was baseline-separated when using hepta-MA-β-CD.

Mono-MA-β-CD was described for the chiral separation of some propionic acid derivatives belonging to the NSAIDs [113], several organic acids [114], chlorthalidone (diuretic) and precursors [114], and binaphthyls [63]. The most widely used cationic CD is 2-hydroxypropyl-trimethylammonio-β-CD, which is the only commercially available cationic (quaternary ammonium substituted) β-CD. Its usual abbreviation is QA-β-CD. The first application of this cati-

onic derivative in CE was in 1996 for the chiral baseline separation of cyclodrine and cyclopentolate [115]. QA-β-CD was further used in the chiral separation of several barbiturates [116–119], hydantoins [116], binaphthyls [116], tricyclic amines (*e.g.*, antidepressants, antipsychotics, antihistaminics) [21], sympathomimetics [117, 120], sympatholytics [21], antimuscarinics [120], diuretics [117], antihistaminics [117], anticoagulants [117], NSAIDs [27], organic acids [120, 121] and several derivatized amino acids [27], peptides, and proteins [122]. In these cases, either a linear polyacrylamide-coated fused-silica capillary or additives like tetraalkylammonium ions and pyridinium ions were used to reduce the EOF and/or to prevent wall absorption.

Finally, some rare cationic CDs have been described for the chiral analysis of several acidic compounds. NSAIDs and phenoxypropionic herbicides (polyaromatic hydrocarbons, PAHs) were enantio-separated using methoxyethylamine-β-CD [55] or hydroxyethylamine-β-CD [56]. The use of hexylamino-β-CD was described for the separation of some amino acid derivatives [123] and monosubstituted ethylenediamine-β-CD for the separation of (among others) several dansylated amino acids [124]. α-Hydroxy acids and carboxylic acids were separated using either 6-*N*-histamino-6-deoxy-β-CD or [4-(2-aminoethyl)imidazolyl]-6-deoxy-β-CD [125, 126]. The former material was superior in these separations. Furthermore, the pH played an important role as shown in Fig. 9. This may allow us to modulate the number and position of the positive charges as a function of the pH, thus providing a model system for investigating the role of charge in the recognition process.

Wang and Khaledi [27] were the first to perform chiral separations using a cationic β-CD (*i.e.*, QA-β-CD) in non-aqueous solutions. They achieved chiral separations of several NSAIDs and several dansyl amino acids in formamide, *N*-methylformamide, methanol, and dimethylsulfoxide. Formamide was the best organic solvent in the latter application.

6 Applications of amphoteric CDs in CE

Amphoteric compounds (*e.g.*, carprofen, see Tables 1 and 2) can be separated with cationic and with anionic CDs, depending on the pH of the run buffer. Obviously, the most interesting approach in these cases is the use of amphoteric CDs. So far, only two papers were published using an amphoteric or zwitterionic CD. Tanaka and Terabe [121] used a commercially available amphoteric CD, *i.e.*, a mixture of 2-hydroxypropyl-trimethylamonium chloride (DS = 2.0) and a sodium acetate group (DS = 2.0) at the 6-position for the chiral separation of several solutes (see Table 3). Lelievre *et al.* [127] used a zwitterionic CD,

Table 2. Applications of cationic CDs for chiral and achiral separations

Compound	CD	References
Abscisic acid	QA-β-CD	[121]
Acenocoumarol	MA-β-CD	[112]
Amethopterin	ED-β-CD	[124]
α-Amino acids (dansylated)	NH$_2$-β-CD, HEA-β-CD, HA-β-CD, QA-β-CD, ED-β-CD, Im-β-CD, Hm-β-CD	[27, 56, 61, 123, 124, 126]
Atrolactic acid	NH$_2$-β-CD	[114]
Barbituric acid derivatives	QA-β-CD	[117]
Benzocyclobutene carbonxylic acid	ED-β-CD, QA-β-CD	[124]
Benzoin	NH$_2$-β-CD	[114]
1,1'-Binaphthyl-2,2'-diyl phosphate 1,1'-binaphthyl-2,2'-dicarboxylate	NH$_2$-β-CD, QA-β-CD, ED-β-CD	[27, 63, 124]
4-Bromomandelic acid	QA-β-CD	[120, 121]
Brompheniramine	QA-β-CD	[117]
Carprofen	NH$_2$-β-CD, MEA-β-CD, QA-β-CD	[27, 55, 113]
(Chlorophenoxy)propionic acid derivaties	HEA-β-CD, MEA-β-CD, ED-β-CD	[55, 56, 124]
p-Chlorowarfarin	QA-β-CD	[121]
Chlorthalidon	NH$_2$-β-CD, QA-β-CD	[114, 117]
Chrysanthemum monocarboxylic acid	QA-β-CD	[121]
Trans-4-cotinine carboxylic acid	ED-β-CD, QA-β-CD	[124]
Cyclodrine	QA-β-CD	[115, 120]
Cyclopentolate	QA-β-CD	[115, 120]
2-(2,4-Dichlorophenoxy)propionic acid	HEA-β-CD	[56]
Fenoprofen	MA-β-CD, HEA-β-CD, MEA-β-CD, QA-β-CD	[27, 55, 56, 112]
Fluriprofen	MA-β-CD, HEA-β-CD, MEA-β-CD, QA-β-CD	[27, 55, 56, 112]
Hexobarbital	QA-β-CD	[116–118]
Hydrobenzoin	NH$_2$-β-CD	[114]
2-(4-Hydroxyphenoxy)propionic acid	EC-β-CD, QA-β-CD	[124]
3-(4-Hydroxyphenyl) lactic acid	EC-β-CD, QA-β-CD, Im-β-CD, Hm-β-CD	[124, 125]
Hydroxymandelic acid derivatives	6^A-MA-β-CD, 6^A,6^D-DMA-β-CD, NH$_2$-β-CD, ED-β-CD, QA-β-CD, Im-β-CD, Hm-6-deoxy-β-CD	[50, 114, 124, 125]
Ibuprofen	HEA-β-CD, MEA-β-CD	[55, 56]
Indollactic acid	ED-β-CD, QA-β-CD	[124]
Indoprofen	MEA-β-CD, QA-β-CD, ED-β-CD	[27, 55, 124]
Ketoprofen	MA-β-CD, HEA-β-CD, MEA-β-CD, QA-β-CD	[27, 55, 56, 112]
Labetalol	QA-β-CD	[21]
Mandelic acid	6^A-MA-β-CD, 6^A,6^D-DMA-β-CD, NH$_2$-β-CD, QA-β-CD	[50, 114, 120]
Mendadione bisulfite	QA-β-CD	[121]
2-Methoxyphenylacetic acid	QA-β-CD	[120]
Methylether benzoin	NH$_2$-β-CD	[114]
Methylpheno barbital	QA-β-CD	[116]
5-Methyl-5-phenylhydantoin	QA-β-CD	[116]
Nafronyl	QA-β-CD	[21]
N-[1-(naphthyl)-ethyl]phthalamic acid	QA-β-CD	[27]
Naproxen	MA-β-CD	[112]
Nefopam	QA-β-CD	[21]
Norpseudoephedrine	QA-β-CD	[120]
Phendimetrazine	QA-β-CD	[120]
2-Phenoxypropionic acid	QA-β-CD, HEA-β-CD	[56, 121]
2- and 3-Phenyllactic acid	6^A-MA-β-CD, 6^A,6^D-DMA-β-CD, MA-β-CD, QA-β-CD, ED-β-CD	[50, 112, 114, 121, 124]
2- and 3-Phenylbutyric acid	NH$_2$-β-CD, QA-β-CD, Im-β-CD, Hm-β-CD	[114, 120, 121, 125]
Phenylpropionic acid	NH$_2$-β-CD, QA-β-CD, Im-β-CD, Hm-β-CD	[114, 120, 125]
Pholedrine	QA-β-CD	[120]
Promethazine	QA-β-CD	[21]

Table 2. continued

Compound	CD	References
Propiomazine	QA-β-CD	[21]
Salbutamol	QA-β-CD	[117]
Suprofen	MEA-β-CD, QA-β-CD	[27, 55]
Terbutaline	QA-β-CD	[117, 120]
Thalidomide	QA-β-CD	[116]
Thioridazine	QA-β-CD	[21]
Tiaprofenic acid	MA-β-CD	[112]
Trimeprazine	QA-β-CD	[21]
Trimipramine	QA-β-CD	[21]
Tropic acid	QA-β-CD	[120, 121]
Warfarin	MA-β-CD, QA-β-CD	[112, 117]

Abbreviations: ED-β-CD, 6-ethylenediamino-6-deoxy-β-CD; HA-β-CD, 6-hexylamino-6-deoxy-β-CD; HEA-β-CD, 6-hydroxy-ethylamino-6-deoxy-β-CD; Hm-β-CD, 6-*N*-histamino-6-deoxy-β-CD; Im-β-CD, [4-(2-aminoethyl) imidazolyl]-6-deoxy-β-CD; MEA-β-CD, 6-methoxyethylamino-6-deoxy-β-CD; NH$_2$-β-CD, 6-amino-6-deoxy-β-CD

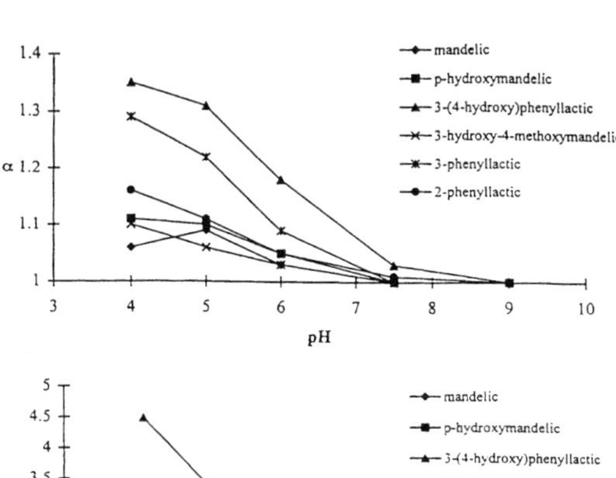
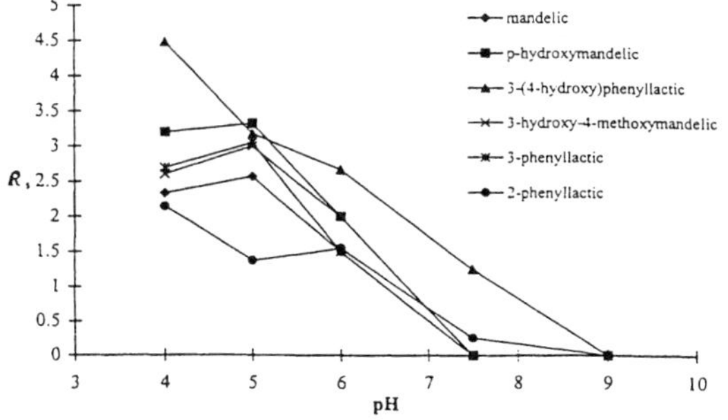

Figure 9. Effect of pH on the enantioselectivity (α) and on the resolution (R_s) using 1 mM Hm-β-CD. Reprinted from [125], with permission.

glutamylamino-6-deoxy-β-CD for the chiral separation of chlorthalidone at pH 2.3, where the zwitterionic CD is partially protonated. Comparison with the cationic mono-(6-amino-6-deoxy)-β-CD showed that larger differences in mobility between the free analyte and the analyte-CD

Table 3. Applications of amphoretic CDs for chiral separations

Compound	CD	References
4-Bromomandelic acid	HPTMA/SA-β-CD	[121]
Carprofen	GluA-β-CD	[127]
Chlorthalidone	GluA-β-CD	[127]
4-Chromomandelic acid	HPTMA/SA-β-CD	[121]
Chrysanthenum mono-carboxylic acid	HPTMA/SA-β-CD	[121]
Dansylated amino acids	HPTMA/SA-β-CD	[121]
Hydrobenzoin	GluA-β-CD	[127]
3-Phenylbutyric acid	HPTMA/SA-β-CD	[121]
2-Phenoxypropionic acid	HPTMA/SA-β-CD	[121]

Abbreviations: GluA-β-CD, glutamylamino-6-deoxy-β-CD; HPTMA/SA-β-CD, 2-hydroxypropyl-trimethylammonium chloride (DS = 2.0) or a sodium acetate group (DS = 2.0)

complex provide a better resolution. Hydrobenzoin enantiomers were separated at pH 11.2, a pH at which the zwitterionic CD is anionic. Under these conditions, the migration order was opposite to that observed in the presence of β-CD-NH2 at pH 2.3. When no separation was obtained directly with GluA-β-CD, a dual CD system was developed. Carprofen enantiomers, which are neutral at the given pH, were resolved at pH 2.3 in the presence of a GluA-β-CD/trimethyl-β-CD (TM-β-CD) system in which the charged CD confers a non-zero mobility to the analyte, while the neutral CD allows chiral recognition. In this case the tuning of the chiral separation is somewhat different from that of other more commonly used dual CD systems. Here the neutral CDs are used for the chiral separation of the neutral enantiomers whereas the charged CDs / analyte complexes provide the necessary mobilities.

7 Conclusions

The use of charged CDs has proven to be successful in the chiral analysis of a wide variety of pharmaceutically interesting racemic compounds. Their potential as chiral selectors are likely to be explored further in the next decade(s), not in the least because the US Food and Drug Administration (FDA), as well as regulatory authorities in Europe, China, and Japan have provided guidelines indicating that preferably only the active enantiomer of a chiral drug should be brought to market [128, 129]. Exceptions are that both enantiomers have a comparable pharmacological effect, or that racemization occurs. Furthermore, for each enantiomeric drug, racemization by metabolic conversion should be studied [130].

Chiral separations with CE can be used to monitor drug impurities, *i.e.*, degradation products, synthetic precursors, and side products (pharmaceutical analysis), or in metabolic profiling to study pharmacokinetics of active and/or inactive compounds (bioanalysis). Especially for charged analytes the use of oppositely charged CDs can provide a unique capability of chiral recognition at much lower concentration levels than commonly used in chiral separations with neutral (substituted) CDs. On the other hand, high concentrations of charged CDs, which can lead to an extreme increase in the current should be avoided. Also, commercially available CDs with a defined DS may lead to better modeling, optimization, reproducibility, and to a more rugged separation system. On the other hand, with the use of different types and numbers of cationic or anionic groups attached to the CDs, unique chiral separation systems can be exploited, as long as the synthesized CDs can be produced in reproducible fashion and can be well characterized. A special application of charged CDs seems to be the addition of the charged CDs as chiral selectors in CE-MS [131]. When the CDs enter the ion source of the MS, they can show a detrimental effect on sensitivity. However, due to the countermigration principle described in the introduction, charged CDs do not necessarily enter the ion-source and may therefore retain the unique information of MS.

Received April 4, 2000

8 References

[1] Witte, D. T., PhD Thesis, Groningen, The Netherlands, 1992.

[2] Duncan, J. D., *J. Liq. Chromatogr.* 2000, *13*, 2737–2755.

[3] Souter, R. W., *Chromatographic Separation of Stereoisomers*, CRC Press, Boca Raton, FL 1985.

[4] Chankvetadze, B., Endresz, G., Blaschke, G., *Chem. Soc. Rev.* 1996, *25*, 141–153.

[5] Lelievre, F., Gareil, P., Jardy, A., *Anal. Chem.* 1997, *69*, 385–392.

[6] Luong, J. H. T., Nguyen, A. L., *J. Chromatogr. A* 1997, *792*, 431–444.

[7] Lurie, I. S., *J. Chromatogr. A* 1997, *792*, 297–307.

[8] Verleysen, K., Sandra, P., *Electrophoresis* 1998, *19*, 2798–2833.

[9] De Boer, T., De Zeeuw, R. A., De Jong, G. J., Ensing, K., *Electrophoresis* 1999, *20*, 2989–3010.

[10] De Boer, T., Ensing, K., *J. Pharm. Biomed. Anal.* 1998, *17*, 1047–1056.

[11] De Boer, T., Bijma, R., Ensing, K., *J. Pharm. Biomed. Anal.* 1999, *19*, 529–537.

[12] Terabe, S., Ozaki, H., Otsuka, K., Ando, T., *J. Chromatogr.* 1985, *332*, 211–217.

[13] Terabe, S., *TRAC* 1989, *8*, 129–134.

[14] Wren, S. A. C., Rowe, R. C., *J. Chromatogr.* 1992, *603*, 235–241.

[15] Wren, S. A. C., Rowe, R. C., *J. Chromatogr.* 1993, *636*, 57–62.

[16] Biggin, M. E., Williams, R. L., Vigh, G., *J. Chromatogr. A* 1995, *692*, 319–325.

[17] Rawjee, Y. Y., Staerk, D. U., Vigh, G., *J. Chromatogr.* 1993, *635*, 291–306.

[18] Rawjee, Y. Y., Williams, R. L., Buckingham, L. A., Vigh, G., *J. Chromatogr. A* 1994, *688*, 273–282.

[19] Rawjee, Y. Y., Vigh, G., *Anal. Chem.* 1994, *66*, 619–627.

[20] Surapaneni, S., Ruterbories, K., Lindstrom, T., *J. Chromatogr. A* 1997, *761*, 249–257.

[21] Wang, F., Khaledi, M. G., *Electrophoresis* 1998, *19*, 2095–2100.

[22] Williams, B. A., Vigh, G., *J. Chromatogr. A* 1997, *777*, 295–309.

[23] Cai, H., Vigh, G., *J. Chromatogr. A* 1998, *827*, 121–132.

[24] Cai, H., Vigh, G., *J. Pharma Biomed. Anal.* 1998, *18*, 615–621.

[25] Vincent, J. B., Vigh, G., *J. Chromatogr. A* 1998, *817*, 105–111.

[26] Vincent, J. B., Vigh, G., *J. Chromatogr. A* 1998, *816*, 233–241.

[27] Wang, F., Khaledi, M. G., *J. Chromatogr. A* 1998, *817*, 121–128.

[28] Giddings, J. C., *Sep. Sci.* 1969, *4*, 181–189.

[29] Jorgenson, J. W., Lukacs, K. D., *Anal. Chem.* 1981, *53*, 1298–1302.

[30] Wang, F., Khaledi, M. G., *Anal. Chem.* 1999, *68*, 3460–3467.

[31] Owens, P. K., Fell, A. F., Coleman, M. W., Berridge, J. C., *Chirality* 1997, *9*, 184–190.

[32] Owens, P. K., Fell, A. F., Coleman, M.W., Berridge, J. C., *Chirality* 1996, *8*, 466–476.

[33] Roussel, C., Favrou, A., *J. Chromatogr. A* 1995, *704*, 67–74.

[34] Thuaud, N., Sebille, B., Deratani, A., Lelievre, G., *J. Chromatogr.* 1990, *503*, 453–458.

[35] Takahashi, K., Nakada, S., Mikami, M., Hattori, K., *Nippon Kagaku Kaishi* 1986, 1032–1039.

[36] Chankvetadze, B., *J. Chromatogr. A* 1997, *792*, 269–295.

[37] Fanali, S., Aturki, Z., Desiderio, C., *Enantiomer* 1999, *4*, 229–241.

[38] Janini, G. M., Issaq, H. J., Muschik, G. M., *J. Chromatogr. A* 1997, *792*, 125–141.

[39] Tanaka, Y., Yanagawa, M., Terabe, S., *J. High Resolut. Chromatogr.* 1996, *19*, 421–433.

[40] Vigh, G., Sokolowski, A. D, *Electrophoresis* 1997, *18*, 2305–2310.

[41] Croft, A. P., Bartsch, A. P., *Tetrahedron* 1983, *39*, 1417–1474.

[42] Vincent, J. B., Sokolowski, A. D., Nguyen, T. V., Vigh, G., *Anal. Chem.* 1997, *69*, 4220–4233.

[43] Vincent, J. B., Kirby, D. M., Nguyen, T. V., Vigh, G., *Anal. Chem.* 1997, *69*, 4419–4428.

[44] Cai, H., Nguyen, T. V., Vigh, G., *Anal. Chem.* 1998, *70*, 580–589.

[45] Takeo, K., Hisayoshi, M., Uemura, K., *Carbohydr. Res.* 1989, *187*, 203–208.

[46] Newton, R. F., Reynolds, D. P., Finch, M. A. W., Kelly, D. R., Roberts, S. M., *Tetrahedron Lett.* 1979, *41*, 3981–3989.

[47] Mayer, S., Schurig, V., *Electrophoresis* 1994, *15*, 835–841.

[48] Francotte, E., Brandel, L., Jung, M., *J. Chromatogr. A* 1997, *792*, 379–384.

[49] Kitaura, Y., Bender, M. L., *Bioorg. Chem.* 1975, *4*, 237–248.

[50] Nardi, A., Eliseev, A., Boček, P., Fanali, S., *J. Chromatogr.* 1993, *638*, 247–253.

[51] Tabushi, I., Yamamura, K., Nabeshima, T., *J. Am. Chem. Soc.* 1984, *106*, 5267–5270.

[52] Breslow, R., Czarniecki, M. F., Emert, J., Hamaguchi, H., *J. Am. Chem. Soc.* 1980, *102*, 762–770.

[53] Jakubetz, H., Juza, M., Schurig, V., *Electrophoresis* 1997, *18*, 897–904.

[54] Parmeter, S. M., Allen, E. E., Hull, G. A., *US Patent 3,426,001 A, Chem. Abstr.* 1969, *71*, 13331z.

[55] Haynes III, J. L., Shamsi, S. A., O'Keefe, F., Darcey, R., Warner, I. M., *J. Chromatogr. A* 1998, *803*, 261–271.

[56] O'Keeffe, F., Shamsi, S. A., Darcy, R., Schwinte, P., Warner, I. M., *Anal. Chem.* 1997, *69*, 4773–4782.

[57] Luna, E., Bornancini, E. R. N., Thompson, D. O., Rajewski, R. A., Stella, V. J., *Carbohydr. Res.* 1997, *299* 103–110.

[58] Luna, E., Van der Velde, D. G., Tait, R. J., Thompson, D. O., Rajewski, R. A., Stella, V. J., *Carbohydr. Res.* 1997, *229*, 111–118.

[59] Chankvetadze, B., Endresz, G., Blaschke, G., Juza, M., Jakubetz, H., Schurig, V., *Carbohydr. Res.* 1996, *287*, 139–155.

[60] Tanaka, Y., Kishimoto, Y., Terabe, S., *Anal. Sci.* 1998, *14*, 383–388.

[61] Kitae, T., Nakayama, T., Kano, K., *J. Chem. Soc. Perkin Trans. II* 1998, *2*, 207–212.

[62] Chankvetadze, B., Endresz, G., Bergenthal, D., Blaschke, G., *J. Chromatogr. A* 1995, *717*, 245–253.

[63] Kano, K., Kitae, T., Takashima, H., *J. Inclus. Phenom. Mol. Recognit. Chem.* 1996, *25*, 243–248.

[64] Shuang, S. M., Guo, S. Y., Pan, J. H., Li, L., Cai, M. Y., *Anal. Lett.* 1998, *31*, 879–889.

[65] Catena, G. C., Bright, F. V., *Anal. Chem.* 1989, *61*, 905.

[66] Schmitt, T., Engelhardt, H., *Chromatographia* 1993, *37*, 475–481.

[67] Chankvetadze, B., Endresz, G., Blaschke, G., *J. Capil. Electrophor.* 1995, *2*, 235–240.

[68] Fillet, M., Fotsing, L, Crommen, J. *J. Chromatogr. A* 1998, *817*, 113–119.

[69] Anigbogu, V. C., Copper, C. L., Sepaniak, M. J., *J. Chromatogr. A* 1995, *705*, 343–349.

[70] Rudaz, S., Veuthey, J. L., Desiderio, C., Fanali, S., *Electrophoresis* 1998, *19*, 2883–2889.

[71] Wu, H. L., Huang, C. H., Chen, S. H., Wu, S. M., *J. Chromatogr. Sci.* 1999, *37*, 24–30.

[72] Spencer, B. J., Zhang, W., Purdy, W. C., *Electrophoresis* 1997, *18*, 736–744.

[73] Weinz, C., Blaschke, G., *J. Chromatogr. B* 1995, *674*, 287–292.

[74] Owens, P. K., Fell, A. F., Coleman, M. W., Kinns, M., Berridge, J. C., *J. Pharm. Biomed. Anal.* 1997, *15*, 1610–1619.

[75] Izumoto, S., Nishi, H., *Bunseki Kagaku* 1998, *47*, 739–746.

[76] Sabah, S., Scriba, G. K. E., *J. Microcol. Sep.* 1998, *10*, 255–258.

[77] Sabah, S., Scriba, G. K. E., *J. Chromatogr. A* 1998, *822*, 137–145.

[78] Verleysen, K., Sandra, P., *J. High Resolut. Chromatogr.* 1999, *22*, 33–38.

[79] Fillet, M., Hubert, P., Crommen, J. *Electrophoresis* 1997, *18*, 1013–1018.

[80] Jin, L. J., Li, S. F. Y., *J. Chromatogr. B* 1998, *708*, 257–266.

[81] Tanaka, Y., Yanagawa, M., Terabe, S., *Int. Symp. Chromatogr., 35th Anniv. Res. Group Liq. Chromatogr. Jpn*, Nippon Boehringer Ingelheim Co., Kawanishi, 666–01, Japan, 1995 World Scientific; Abstract pp. 395–400.

[82] Szolar, O. H. J., Brown, R. S., Luong, J. H. T., *Anal. Chem.* 1995, *67*, 3004–3010.

[83] Tanaka, Y., Yanagawa, M., Terabe, S., *J. High Resolut. Chromatogr.* 1996, *19*, 421–433.

[84] Smith, N. W., *J. Chromatogr.* 1993, *652*, 259–262.

[85] Chankvetadze, B., Endresz, G., Blaschke, G., *Electrophoresis* 1994, *15*, 804–807.

[86] Lurie, I. S., Klein, R. F., X., Dal Cason, T. A., LeBelle, M. J., Brenneisen, R., Weinberger, R. E., *Anal. Chem* 1994, *66*, 4019–4026.

[87] Dette, C., Ebel, S., Terabe, S., *Electrophoresis* 1994, *15*, 799–803.

[88] Aumatell, A., Wells, R. J., *J. Chromatogr. A* 1994, *688*, 329–337.

[89] Aumatell, A., Wells, R. J., Wong, D. K. Y., *J. Chromatogr. A* 1994, *686*, 293–307.

[90] Lurie, I.S., Chan, K. C., Spratley, T. K., Casale, J. F., Issaq, H. J., *J. Chromatogr. A* 1995, *669*, 3–13.

[91] Lurie, I. S., Hays, P. A., Casale, J. F., Moore, J. M., Castell, D. M., Chan, K. C., Issaq, H. J., *Electrophoresis* 1998, *19*, 51–56.

[92] Lurie, I. S., Odeneal II, N. G., McKibben, T. D., Casale, J. F., *Electrophoresis* 1998, *19*, 2918–2925.

[93] Kostel, K. L., Freed, A. L., Lunte, S. M., *J. Chromatogr. A* 1996, *744*, 241–248.

[94] Stalcup, A. M., Gahm, K. H., *Anal. Chem.* 1996, *68*, 1360–1368.

[95] Gahm, K. H., Stalcup, A. M., *Chirality* 1996, *8*, 316–324.

[96] Valenzuela, F. A., Green, T. K., Dahl, D. B., *Anal. Chem.* 1998, *70*, 3612–3618.

[97] Le Potier, I., Tamisier-Karolak, S. L., Morin, P., Megel, F., Taverna, M., *J. Chromatogr. A* 1998, *829*, 341–349.

[98] Wang, F., Khaledi, M. G., *J. Microcol. Sep.* 1999, *11*, 11–21.

[99] Pucci, V., Raggi, M., Kenndler, E., *J. Chromatogr. A* 1999, *853*, 461–468.

[100] Munro, N. J., Palmer, J., Stalcup, A. M., Landers, J. P., *J. Chromatogr. B* 1999, *731*, 369–381.

[101] De Boer, T., Mol, R., De Zeeuw, R.A., De Jong, G. J., Ensing, K., *Electrophoresis* 2000, *21*, in press.

[102] Chen, F. T., Shen, G. G. Y., Evangelista, R. A., Oh, C. S., edited by Beckman Coulter, Eur. Pat. Appl., 19 pp. CODEN: EPXXDW, 1998.

[103] Chen, F. T. A., Evangelista, R. A., *J. Chin. Chem. Soc.* 1999, *46*, 847–860.

[104] Verleysen, K., Sabah, S., Scriba, G. K. E., Sandra, P., *J. Chromatogr. A* 1998, *824*, 91–97.

[105] Cai, H., Vigh, G., *J. Microcol. Sep.* 1998, *10*, 293–299.

[106] Wedig, M., Holzgrabe, U., *Electrophoresis* 1999, *20*, 1555–1560.

[107] Luong, J. H. T., Guo, Y., *J. Chromatogr. A* 1998, *811*, 225–232.

[108] Gahm, K. H., Stalcup, A. M., *Anal. Chem.* 1995, *67*, 19–25.

[109] Wang, F., Khaledi, M. G., *Anal. Chem.* 1996, *68*, 3460–3467.

[110] Smith, N. W., *J. Chromatogr.* 1993, *652*, 259–262.

[111] Szeman, J., Ganzler, K., Salgo, A., Szejtli, J., *J. Chromatogr. A* 1996, *728* 423–431.

[112] Fanali, S., Camera, E., *Chromatographia* 1996, *43*, 247–253.

[113] Lelievre, F., Gareil, P., Bahaddi, Y., Galons, H., *Anal. Chem.* 1997, *69*, 393–401.

[114] Lelievre, F., Gareil, P., Jardy, A., *Anal. Chem.* 1997, *69*, 385–392.

[115] Bunke, A., Jira, T., *Pharmazie* 1996, *51*, 672–673.

[116] Schulte, G., Chankvetadze, B., Blaschke, G., *J. Chromatogr. A* 1997, *771*, 259–266.

[117] Jakubetz, H., Juza, M., Schurig, V., *Electrophoresis* 1997, *18*, 897–904.

[118] Jakubetz, H., Juza, M., Schurig, V., *GIT Labor-Fachz.* 1998, *42*, 479–481.

[119] Jakubetz, H., Juza, M., Schurig, V., *Electrophoresis* 1998, *19*, 738–744.

[120] Bunke, A., Jira, T., *J. Chromatogr. A* 1998, *798*, 275–280.

[121] Tanaka, Y., Terabe, S., *J. Chromatogr. A* 1997, *781*, 151–160.

[122] Zhang, R., Zhang, H. X., Eaker, D., Hjertén, S., *J. Capil. Electrophor.* 1997, *4*, 105–112.

[123] Egashir, N., Mutoh, O., Kurauchi, Y., Ohga, K., *Anal. Sci.* 1996, *12*, 503–505.

[124] Nair, U. B., Armstrong, D. W., *Microchem. J.* 1997, *57*, 199–217.

[125] Galaverna, G., Corradini, R., Dossena, A., Marchelli, R., *Electrophoresis* 1999, *20*, 2619–2629.

[126] Galaverna, G., Corradini, R., Dossena, A., Marchelli, R., Vecchio, G., *Electrophoresis* 1997, *18*, 905–911.

[127] Lelievre, F., Gueit, C., Gareil, P., Bahaddi, Y., Galons, M., *Electrophoresis* 1997, *18*, 891–896.

[128] DeCamp, W. H., *J. Pharm. Biomed. Anal.* 1993, *11*, 1167–1172.

[129] Fillet, M., Hubert, P., Crommen, J., *Electrophoresis* 1998, *19*, 2834–2840.

[130] Witte, D. T., Ensing, K., Franke, J. P., De Zeeuw, R. A., *Pharm. World Sci.* 1993, *15*, 10–16.

[131] Schulte, G., Heitmeier, S., Chankvetadze, B., Blaschke, G., *J. Chromatogr. A* 1998, *800*, 77–82.

[132] Vargas, M. G., Van der Heyden, Y., Maftouh, M., Massart, D. L., *J. Chromatogr. A* 1999, *855*, 681–693.

[133] Tanaka, Y., Kishimoto, Y., Otsuka, K., Terabe, S., *J. Chromatogr. A* 1998, *817*, 49–57.

[134] Fanali, S., Furlanetto, S., Aturki, Z., Pinzauti, S., *Chromatographia* 1998, *48*, 395–401.

[135] Izumoto, S., Nishi, H., *Electrophoresis* 1999, *20*, 189–197.

[136] Desiderio, C., Fanali, S., *J. Chromatogr. A* 1995, *716*, 183–196.

[137] Janini, G. M., Muschik, G. M., Issaq, H. J., *Electrophoresis* 1996, *17*, 1575–1583.

[138] Jung, M., Francotte, E., *J. Chromatogr. A* 1996, *755*, 81–88.

[139] Ren, X. Q., Dong, Y. Y., Liu, J. Y., Huang, A. J., Liu, H. W., Sun, Y. L., Sun, Z. P., *Chromatographia* 1999, *50*, 363–368.

[140] Szücs, R., Beaman, J. V., Lipczynski, A. M., *J. Chromatogr. A* 1999, *836*, 53–58.

[141] Xie, G. H., Skanchy, D. J., Stobaugh, J. F., *Biomed. Chromatogr.* 1997, *11*, 193–199.

[142] Chankvetadze, B., Endresz, G., Blaschke, G., *J. Chromatogr. A* 1995, *704*, 234–237.

[143] Szücs, R., Usman, M., Beaman, J. V., Lipczynski, A. M., *J. High Resolut. Chromatogr.* 1999, *22*, 59–62.

[144] Wang, F., Khaledi, M. G., *J. Chromatogr. B* 1999, *731*, 187–197.

[145] Fillet, M., Bechet, I., Hubert, P., Crommen, J., *J. Pharm. Biomed. Anal.* 1996, *14*, 1107–1114.

[146] Thunhorst, M., Otte, Y., Jefferies, T. M., Branch, S. K., Holzgrabe, U., *J. Chromatogr. A* 1998, *818*, 239–249.

[147] Fillet, M., Bechet, I., Schomburg, G., Hubert, P., Crommen, J., *J. High Resolut. Chromatogr.* 1996, *19*, 669–673.

[148] Fillet, M., Fotsing, L., Bonnard, J., Crommen, J., *J. Pharm. Biomed. Anal.* 1998, *18*, 799–805.

[149] Tran, N. T., Taverna, M., Deschamps, F. S., Morin, P., Ferrier, D., *Electrophoresis* 1998, *19*, 2630–2638.

[150] Fillet, M., Fotsing, L., Schomburg, G., Hubert, P., Crommen, J., *Biomed. Chromatogr.* 1998, *12*, 131–132.

[151] Lerch, C., Blaschke, G., *J. Chromatogr. B* 1998, *708*, 267–275.

[152] Pak, C., Marriott, P. J., Carpenter, P. D., Amiet, R. G., *J. High Resol. Chromatogr.* 1998, *21*, 640–644.

[153] Gratz, S. R., Stalcup, A. M., *Anal. Chem.* 1998, *70*, 5166–5171.

[154] Owens, P. K., Fell, A. F., Colean, M. W., Berridge, J. C., *Enantiomer* 1999, *4*, 79–90.

[155] Matsui, Y., Okimoto, A., *Bull. Chem. Soc. Jpn.* 1978, *51*, 3030–3034.

[156] Ahern, C., Darcy, R., O'Keefe, F., Schwinte, P., *J. Incl. Phenom. Mol. Recogn.* 1996, *25*, 43–48.

[157] Tanaka, Y., Kishimoto, Y., Terabe, S., *Anal. Sci.* 1998, *14*, 383–388.

Electrophoresis 2000, *21*, 4029–4045

Review

Joop C. M. Waterval[1]
Henk Lingeman[2]
Auke Bult[1]
Willy J. M. Underberg[1]

[1]Universiteit Utrecht, Faculty of
 Pharmacy,
 Utrecht, The Netherlands
[2]Free University Amsterdam,
 Department of Analytical
 Chemistry and Applied
 Spectroscopy,
 Amsterdam, The Netherlands

Derivatization trends in capillary electrophoresis

This survey gives an overview of recent derivatization protocols, starting from 1996, in combination with capillary electrophoresis (CE). Derivatization is mainly used for enhancing the detection sensitivity of CE, especially in combination with laser-induced fluorescence. Derivatization procedures are classified in tables in pre-, on- and postcapillary arrangements and, more specifically, arranged into functional groups being derivatized. The amine and reducing ends of saccharides are reported most frequently, but examples are also given for derivatization of thiols, hydroxyl, carboxylic, and carbonyl groups, and inorganic ions. Other reasons for derivatization concern indirect chiral separations, enhancing electrospray characteristics, or incorporation of a suitable charge into the analytes. Special attention is paid to the increasing field of research using on-line precapillary derivatization with CE and microdialysis for *in vivo* monitoring of neurotransmitter concentrations. The on-capillary derivatization can be divided in several approaches, such as the at-inlet, zone-passing and throughout method. The postcapillary mode is represented by gap designs, and membrane reactors, but especially the combination of separation, derivatization and detection on a chip is a new emerging field of research. This review, which can be seen as a sequel to our earlier reported review covering the years 1991–1995, gives an impression of current derivatization applications and highlights new developments in this field.

Keywords: Derivatization / Capillary electrophoresis / Precapillary / Postcapillary / On-capillary / On-line derivatization / Laser-induced fluorescence / Microdialysis / Review   EL 4165

Contents

Correspondence: Dr. Willy J. M. Underberg, Universiteit Utrecht, Faculty of Pharmacy, Department of Pharmaceutical Analysis, PO BOX 80082, 3508 TB Utrecht, The Netherlands
E-mail: w.j.m.underberg@pharm.uu.nl
Fax: +31-30-2535180

Abbreviations: 9-AMAC, 9-aminoacridone; **ANDSA**, 7-amino-1,3-naphthalene disulfonic acid; **APTS**, 8-aminopyrene-1,3,6-trisulfonate; **CLOD**, concentration limit of detection; **FQ**, 3-(2-furoyl)quinoline-2-carboxaldehyde; **HQS**, 8-hydroxyquinoline-5-sulfonic acid; **NDA**, napthalene-2,3-dicarboxaldehyde; **NIR**, near-infrared; **PAR**, 4-(2-pyridylazo)resorcinol; **TNS**, 2-(*p*-toluidino)-naphthalene-6-sulfonic acid

1 Introduction

Derivatization is a modification intended to give the analyte(s) of interest more suitable characteristics. In most of the cases, derivatization is detection-oriented, associated with the incorporation of UV-absorbing or fluorescent groups into the analytes to obtain better sensitivity and selectivity. Apart from these reasons, derivatization is also used to give the analyte a more suitable mass-to-charge ratio, to convert enantiomers to diastereomers, which can be separated in a nonchiral environment, to increase the hydrophobicity necessary for MEKC separations or to provide saccharides with better properties for mass spectrometric detection. This review, which strives to be complete, reports the different approaches in derivatization over the last four years, categorized in tables. The derivatizations are divided into subclasses based on how derivatization takes place (pre-, on-, or postcapillary, or on-line), which reagent is used and which functional group is derivatized. This review can be seen as a sequel to a previous review published by our group, covering the years 1991–1995 [1]. Since not so many new reagents were introduced the past four years, the reader is referred to the previous review which explains the mechanism behind derivatizations using isothiocyanate, succinimidyl ester, chloroformate, or isoindol-based reagents along with several other reagents. Many reviews have been published over the years, handling derivatization, each with their special interests, *e.g.*, the derivatization of low protein concentrations and its associated problems [2], amino acids [3], peptides for LC and CE analysis [4], the derivatizing dyes used for LIF detection [5], the labeling of carbohydrates [6–12], the pre- and postcolumn labeling on a chip [13], and the derivatization of inorganic cations and anions [14–16]. This review gives an overview of derivatization strategies combined with CE and will focus on new developments.

2 Precapillary derivatization

Although other modes of derivatization become more important, the precapillary mode is still the most employed mode for derivatization (Table 1). There are several reasons for this phenomenon. All types of reagents, both fluorophoric and fluorogenic can be employed, including those that require long reaction times or elevated temperatures. Manipulations such as evaporation and removal of excess reagent seem to be restricted to the precapillary mode. Furthermore, the absence of commercially available solutions for postcapillary derivatization prevents a boost for this derivatization mode.

3 On-line derivatization

The on-line applications are summarized in Table 2. In the on-line applications, precapillary mixing of an analyte solution takes place just before the capillary with the reagent solution using a T-junction. This method is frequently employed in combination with microdialysis. The low flow rate and the wish to monitor the analyte concentration with short intervals necessitates the use of a technique with low sample volume requirements combined with good sensitivity and short analysis time. CE with LIF detection can fulfill these requirements.

The extra value of an on-line coupling of microdialysis with derivatization is that it allows the handling of submicroliter volumes of sample and femtomole amounts of compounds without any loss in sensitivity, as was demonstrated by the work of Bert *et al.* [17]. The manual derivatization could not be performed on sample volumes lower than 3 µL.

Robert *et al.* [18] used an electric valve and a liquid junction interface to couple the microdialysis probe and continuous flow derivatization setup to a CE-LIF system, allowing a 2 min sampling interval. Since no sample handling for derivatization or injection was required, continuous monitoring of analyte concentrations was performed in 6–7 h (*i.e.*, 200 analyses).

In an early report, Lada and Kennedy [19] used a flow-gated interface, which allows dialysate samples to be automatically injected onto the separation capillary. During a separation, a gating flow of electrophoresis buffer was pumped at 0.34 mL/min through the gap between the capillaries. To perform an injection, the gating flow was stopped and the injection voltage was applied. Once the injection was complete, the gating flow was resumed and the separation voltage applied. Unfortunately, the injected amount depended not only on the voltage and time but also on the space between reaction and separation capillary, which varied from day to day. In order to obtain consistent performance in terms of sensitivity, the exact injection voltage and time were varied. The system setup was able to determine the analytes every 90 s, frequently reported as the temporal resolution of the system.

In a later report [20], the temporal resolution was further improved for *in vivo* monitoring. Optimization of the reaction pH and dialysis flow rate for the derivatization process decreased the necessary reaction time dramatically. In order to minimize the chance that the separation would limit the temporal resolution, CE analysis was accelerated using a high field strength (4.2 kV/cm) in combination with a short separation capillary (4.5 cm). These conditions resulted in a complete separation of the glutamate and aspartate derivatives within 5 s combined with a high peak efficiency. In this way the temporal resolution was improved to 12 s, allowing the first simultaneous monitoring of glutamate and aspartate overflow during acute electrical stimulation in the rat brain.

The gate function, necessary for the introduction of fixed amounts from the sample stream, can also be accomplished using gating optics [21, 22]. *o*-Phthaldialdehyde (OPA) derivatives of aspartate were separated with high speed using a CE-LIF setup with an optical gating injection. A gating beam, focused on the capillary, photo-

Table 1. Precapillary derivatization

Reagent	CE mode	Compounds	Details	Ref.
Amine function				
ACCQ	MEKC-UV	Amines	Quantitative determination in food; CLOD: 1–40 μM	[59]
AEOC	MEKC-UV	Amino acids	Chiral separation	[43]
APOC	MEKC-LIF	Amino acids	LOD: 10^{-18} mol	[41]
BMMC	CE-LIF	Herbicides, dicamba 2,4-D	In water samples; CLOD 10 ng/L	[60]
CBQCA	CE-LIF	Amino acids	Microdialysate of periaqueductal grey matter; CLOD: (sub)nanomolar range	[61]
CDITC	CEC-DAD	Amino acids and chiral amines	Comparison between RP-LC and pseudo-CEC using PVP as pseudostationary phase	[62]
CFSE	CE-LIF	Amino acids	Derivatization of low concentrations	[54]
Cy5.29.OSu	MEKC-LIF	Amantadine	Far red label; LOD: 115 amol	[63]
Cyanine derivative	CGE-LIF	Amino acids	Chiral separation using cyclodextrins; CLOD: 60 nM	[64]
DAB-Cl	CE-Forward-scattering four-wave mixing	Amino acids	LOD: 13 amol; CLOD: 1.7 μM	[65]
DMABSCl	MEKC-thermal lens detector	Glycine	Emphasis on new detector; LOD: 180 nM	[66]
DNSH	CE-LIF	Herbicides, chlorimuron	In water samples; CLOD 10 ng/L	[60]
DNSH	CE-LIF	Amino acids	Nonaqueous chiral CE using N-methylformamide	[67]
DNSH, AEOC, FMOC	MEKC-UV	Dipeptides	Chiral separation using vancomycin Evaluation of labeling effect on resolution	[68]
Eosin Y, RBITC, TRITC and fluorescamine	CE-CL	Proteins	CLOD: nanomolar range	[69]
FITC	CE-LIF	Glutamate	From microdialysates in the amygdala and in the lateral hypothalamus	[70]
FITC	CE-LIF	Glutamate and glutamine	From cerebrospinal fluid (CSF) samples	[71]
FITC	CE-LIF on a chip	Amphetamine and analogs	Direct labeling of spiked urine sample compared with labeling after solid-phase extraction; CLOD 200 ng/mL, 10 μg/mL, respectively	[72]
FITC	CE-LIF	Amines	Environmental water samples; CLOD: 10 ng/mL	[73]
FITC	CE-LIF	Amines	Variables investigated; CLOD: 1 nM	[55]
FITC	MEKC-LIF	Amines	Quantitation of biogenic amines in soy sauce using ion-pair solid-phase extraction as sample preparation step; CLOD: 0.6–35 nM	[74]
FITC	MEKC-LIF on a chip	Amines	15-fold shorter separation time compared with fused-silica capillary setup at the expense of separation and sensitivity; CLOD: 3–7 μM	[75]
FITC	CE-LIF	Amino acids	Analysis of the gamma-carboxyglutamic acid content of protein, urine, and plasma; CLOD: 0.25 nM	[76]
FITC	CE-LIF	Amino acids	Demonstrating the utility of a gap flow cell	[77]
FITC	CE-LIF on a chip	Amino acids	Chiral separation of labeled amino acids, analysis time less than 4 min	[78]
FITC	CE-LIF	Amino acids	LOD: 0.052–1.01 fmol	[79]
FITC	MEKC-LIF on a chip	Amino acids	CLOD: 3 nM	[80]
FITC	CE-LIF	Amphetamine	Cortex microdialysates; LOD: 3 amol, CLOD: 3 nM	[81]
FITC	CE-LIF-CCD	Amino acids	CLOD: 1.2–17.2 nM	[82]
FITC	CE-LIF on a chip	Peptides/proteins	Chip-based separation	[83]
FITC	MEKC-LIF	Amines	Analysis in wine	[84]
FITC	CE-LIF	Amino acid	In combination with miniaturized protein sequencer; LOD: 10 zmol	[85]
FITC	CE-LIF	Glutamate, aspartate	in vivo monitoring in freely moving rats using microdialysis	[86]

Table 1. continued

Reagent	CE mode	Compounds	Details	Ref.
FITC	MEKC-LIF chiral CE-LIF	Amino acids	Used for determination of amino acids racemization of pharmaceutical products	[87]
FLEC	MEKC-UV	Dipeptides	Derivatization used for indirect chiral separation	[40]
FLEC	MEKC-UV	Dipeptides	Derivatization used for indirect chiral separation	[42]
Fluorescamine	CE-LIF	Arginine,citrulline and other nitric oxide synthase related metabolites	Determination in neurons; LOD: 50 amol–17 fmol; CLOD: 5 nм–17 μм	[88]
Fluorescamine	CE-CL	Proteins	On-line chemiluminescence detection	[89]
Fluorescamine	CE-Two-photon excited fluorescence	Neurotransmitters	LOD: 22–57 zmol, CLOD: 110–250 nм	[90]
FMOC	MEKC-UV	Amino acids	Chiral separation using cyclodextrins	[91]
FMOC	MEKC-UV	Dipeptides	Chiral separation using vancomycin	[92]
FMOC	CE-UV	Amino acids	Fast chiral separation using vancomycin and coosmotic flow	[93]
FMOC and AP	CE-F	Amino acids and oligosaccharides	In combination with new detection cell; CLOD: 100 nм, 4 μм, respectively	[94]
FMOC+FLEC	CE-UV	Carnitines	Chiral separation using cyclodextrins; LOD: 5 fmol, CLOD: 10 μм	[95]
FQ	CE-LIF	Protein	Evaluating band broadening effect	[29]
FQ	CIEF-LIF	Proteins	Effect labeling on p*I*	[96]
IITC	CE-CL	Amino acids, peptides	Using chemiluminescence detection cell; LOD: 209–930 amol, CLOD: 23–380 nм	[97]
MCIDC	CE-F	Amikacin	In human plasma; CLOD: 0.5 μg/mL	[98]
MOPDYA	CE-LIF	Amines	New reagent, good stability in both acidic and basic solutions LOD: 1 pmol, CLOD: 200 nм	[99]
NBD	CITP-F	Lipoproteins	Quantitation of lipoprotein subpopulations	[100]
NBD-Cl	MEKC-LIF	Amines	In beer samples, in combination with easy-to-use on-column fiber optics LOD: 1 pmol, CLOD: 9–130 μм	[101]
NBD-F	CE-LIF	Amino acids	LOD: 100 nм, chiral separation using cyclodextrins	[102]
NBD-F	MEKC-LIF	Methamphetamine and related compounds	Detected in urine; LOD: 2–690 amol, CLOD: 0.4–5.9 μм	[103]
NBD-PyNCS	MEKC-LIF	Amino acids and peptides	Separation after chiral derivatization using nonionic micelles; determination of D-proline and D-aspartate in rat serum; CLOD: 50 nм	[104]
NDA	CE-F	Amino acids	Evaluating fluorescence signal enhancement by cyclodextrins (factor 1.73 max)	[105]
NDA	CE-ECD	Amines	Analysis in neurons	[106]
NDA	CE-LIF	Serotonin, glutamate, aspartate, GABA	*in vivo* monitoring in freely moving rats using microdialysis	[107]
NN382	CE-NIR-LIF	Amino acids	LOD: low zmol range	[58]
NN382	CE-NIR-LIF, MEKC-NIR-LIF	Peptides	Angiotensin I variants; LOD: 100–300 zmol, CLOD: 50 nм	[56]
NQS	CE-LIF	Amino acids	Pharmaceutical samples	[108]
NS	CE-UV	Amino acids	Chiral separation using cyclodextrins	[44]
NS	CE-UV	Amino acids	Chiral separation with cyclodextrins	[45]
OPA	CE-LIF	Amino acids	Analysis in plant matrices	[109]
OPA	CE-ECD	Amino acids	Electrochemical detection; CLOD: 100–200 nм	[110]
OPA	MEKC-LIF	Amino acids	LOD: 360 amol, CLOD: 98 nм	[111]
OPA followed by PITC	MEKC-UV	3- and 4-Hydroxyproline	Primary amines eliminated using OPA derivatization and precipitation followed by derivatization of prolines using PITC; LOD: pmol level	[112]
PITC	CE-UV	Phosphoamino acids	Derived from hydrolysis of proteasome subunits	[113]
PTH	MEKC-TOAD	Amino acids	In combination with sequencer	[114]
PTH	MEKC-UV	Amino acids	In combination with protein sequencer	[115]

Table 1. continued

Reagent	CE mode	Compounds	Details	Ref.
PTH	MEKC-TOAD	Amino acids	In combination with protein sequencer	[116]
TRITC	MEKC-LIF	Phenylalanine	Separation from deuterated isotopomers	[117]
TRITC	CE-LIF	[D-pen^{2,5}]enkephalin	Determined in rat serum; LOD: amol level, CLOD: nM level	[118]
TRSE	CGE-LIF	rec. monoclonal antibodies	Replacement for SDS-PAGE; LOD: 9 ng/mL	[119]
Reducing end of saccharides				
2-AMAC	CE-LIF	Disaccharides	Derived from chondroitin/dermatan	[120]
2-AMAC	MEKC-LIF	Saccharides	Derived from dextran/glycan	[48]
2-AMAC	MEKC-LIF	Saccharides	Used to differentiate between core fucosylated and nonfucosylated glycans	[121]
2-AMAC	CE-LIF	Glycans	Used for profiling complex glycans released from fetuin and human antibodies	[122]
AA	CE-F	Monosaccharides	Composition of hydrolyzed thickeners	[123]
AB	CE-UV	Saccharides	CLOD: 3 μM, LOD: 33 fmol	[124]
AB	CE-LIF	Saccharides	Optimization of labeling parameters examined	[125]
ABN + ANTS	CE-LIF	Malto-oligosaccharides	Profiling molecular weight distribution	[51]
ANDSA	CE-F	Oligosaccharides	Mixtures of oligosaccharides obtained after treatment of hyaluronic acid with hyaluronate lyase	[38]
ANDSA	CE-UV	Saccharides	Compared with LC-pulsed amperometric detection	[49]
ANDSA	CE-LIF	Acarbose and metabolite	In human urine; CLOD: 24–30 ng/mL	[126]
ANDSA	CE-LIF	Disaccharides	CLOD: nanomolar level	[127]
ANDSA	CE-LIF	Saccharides	Effect of the ionic strength of the running electrolyte on selectivity and resolution evaluated	[128]
ANDSA	CE-LIF	Glucosinolates	Glucose released enzymatically and converted to gluconic acid before derivatization	[129]
ANTS	CE-MS	Oligosaccharides	Hydrolyzed dextran; LOD: 1.6 pmol	[46]
ANTS	CE-LIF	Saccharides	Used for characterization of substrate specificity of endopolygalacturonases	[130]
ANTS	CE-LIF	Dextrans	SEC used for evaluation labeling process	[50]
AP	CE-F	Oligosaccharides	Multidimensional mapping using three buffer systems	[47]
APTS	CE-LIF	Oligosaccharides	Derived from glycoproteins; LOD: pmol level	[131]
APTS	CE-LIF	Dextrins	Investigation of carbohydrate-drug interactions	[132]
APTS	CE-LIF	Oligosaccharides	Profiling glycoprotein N-linked oligosaccharide	[133]
APTS	CE-LIF	Oligosaccharides	Quantitative analysis of sugar constituents of glycoproteins; LOD: 50–100 pmol	[134]
APTS	CE-LIF	Monosaccharides	Influence of matrix on derivatization yield investigated	[135]
APTS	CE-LIF	Oligosaccharides	Labeling efficiency evaluated	[52]
APTS	CE-LIF	Polysaccharides	Unhydrolyzed carrageen samples	[136]
APTS	CE with off-line MALDI-TOF-MS	Oligosaccharides	MALDI-MS used from molecular weight determination of CE fractions; LOD: 30 fmol	[137]
APTS/9-AMAC	CE-LIF	Saccharides	Derived from glycoproteins	[39]
BH	CE-UV	Carbohydrates	LOD: 30 fmol	[138]
Cu(II)	CE-UV	Neutral sugars	Chelation of Cu(II) by sugars under alkaline conditions yields absorption in the UV range; CLOD: 50–100 μM	[139]
DNSH	CE-LIF	Carbohydrates	Used for determination of glucose in tears; LOD: 100 amol	[140]
EAB	CE-UV	Saccharides	Specificities of the beta-galactosidases evaluated	[141]
EAB and ABN	CE-UV	Carbohydrates	Fast separation using co-electroosmotic CE	[142]
PEA	CE-UV	Carbohydrates	In food samples	[143]
PMP	CE-UV	Saccharides	Chiral separation using an optically active N-dodecoxycarbonylvaline	[144]
TMR	CE-LIF	Oligosaccharides	LOD: 80 yoctomoles (50 molecules); CLOD: nanomolar range	[145]

Table 1. continued

Reagent	CE mode	Compounds	Details	Ref.
Thiol function				
ABD-F/SBD-F	CE-UV/CE-DAD	(Homo)cysteine, glutathione	Thiols in human plasma; CLOD: 0.5–2 μM	[146]
Ellman's reagent	CE-UV	Thiols, disulfides	Glutathione in erythrocytes	[147]
Ellman's reagent	CE-UV	Glutathione	Determination in lymphocytes; LOD: 0.5 μM	[148]
FITC	CE-LIF	Homocysteine	Determination in plasma; CLOD: 1 μM	[149]
IAF	CE-LIF	Homocysteine	Determination in plasma; CLOD: 0.25 μM	[150]
IAF-LRB	CIEF-LIF	Cysteine residue	Indirect way for quantitation of trypsin activity	[151]
Carboxylic functions				
ADAM	CE-LIF	Carnitines and several acylcarnitines	Using nonaqueous SDS solution; CLOD: 10 μM	[152]
ADAM	CE-LIF	Carnitines	Determination in human plasma samples; CLOD: 60 nM	[153]
AF	CE-LIF	Carboxylic acids	Includes a phase transfer and a reextraction procedure; CLOD: 3–150 nM	[154]
ANDSA	CE-LIF	Phenoxy acid herbicides	Chiral separation using cyclodextrins; CLOD: 0.2 ng/mL	[155]
ANDSA	CE-LIF	Phenoxy acid herbicides	Chiral separation using surfactant in combination with field-amplified stacking; CLOD: picomolar range	[156]
ANDSA	CE-LIF	Insecticides	After hydrolysis derivatized	[157]
BMF	CE-LIF	fatty acids	C_8 to C_{11} acids; CLOD: 10^{-10} M range	[158]
BMMC	CE-LIF	Resin acids	CLOD: 10–20 μg/mL	[159]
PMC	CE-NIR-LIF	Fatty acids	CLOD: 60 pM in methanolic separation medium	[57]
Carbonyl function				
DNSH	CE-UV	Carbonyl compounds	Determination in raindrops	[160]
DNSH	CE-UV	Carbonyl compounds, acetone, acetaldehyde, propionaldehyde, benzaldehyde, formaldehyde	Determination in raindrops; CLOD: 170–300 nM	[37]
DNSH	CE-LIF	Steroids	Derivatization steroids inside single cells using DMSO-containing buffer; LOD: attomole range	[161]
DNSH	CE-UV	Formaldehyde	In the presence of an excess of dihydroxyacetone; CLOD: 0.2 μg/mL	[162]
Girard P and T reagent	CE-UV	Steroids	Transformed into the corresponding hydrazones; LOD: 6 pmol	[163]
Hydroxy function				
AAC	MEKC-LIF	Brevetoxines	CLOD: 4 pg/g fish	[164]
BTA	CE-UV	Poly(ethylene glycol)s and ethoxylated surfactants	Derivatization used for improving detection and incorporation of charge	[165]
CAA	tITP-LIF	Polyhydroxyl species	Adenosine; CLOD: low nanomolar range	[166]
DBTA	CE-LIF	Anticoagulant drugs	Diastereomeric separation evaluating pH, use of organic modifiers and PVP	[167]
MFPTS	CE-DAD	Phytosteroids	Nonaqueous separation media containing cyclodextrins	[168]
PA	CE-UV	Fatty alcohol ethoxylates	Determination in technical products and household formulations	[169]
TBA HS	CE-UV	Aliphatic alcohols	Derivatized to dithiocarbamates; CLOD: 1.5–7.9 μg/mL	[170]
Noncovalent interactions				
BODIPY	CE-LIF	Nucleotides	Used for detection of DNA adducts	[171]
DBD-IMI	CE-LIF	Nucleotides	Phosphomonoester are specifically derivatized	[172]

Table 1. continued

Reagent	CE mode	Compounds	Details	Ref.
DTTCl, IR-125	CE-LIF	Proteins	Noncovalent binding of these near infrared dyes with proteins lenhances their fluorescence emission	[173]
Indocyanine g	CE with diode-LIF	Albumin	Noncovalently bound to protein; CLOD: 1.5 nM	[174]
Nucleic acid probes	CE-LIF	Specific nucleic acid sequences	Detection at two different wavelength channels simultaneously, coincident Idetection of both dyes provides the necessary specificity to detect an Iunamplified, single-copy target DNA molecule	[175]
Zn(II)	CE-LIF	Kynurenic acid	Becomes fluorescent on complex formation with Zn(II), Iin microdialysate samples; CLOD: 50 mM	[176]
Inorganic ions				
EDTA, CDTA land DTPA	CE-UV	Co(III), Bi(III), Fe(III), Cr(III), V(IV), Pb(II), Hg(II), Co(II), Cu(II) and Ni(II)	Separated as complexes; CLOD: 2–8 µM	[177]
NTA	CE-UV, MEKC-UV	(in)organic lead, mercury and selenium	Complex formation; CLOD: 10–310 ng/mL, Iin combination with stacking 0.2–5 ng/mL	[178]
PAR	CE-UV	Fe(II), Co(II), Ni(II), Cu(II) and Zn(II)	Complex formation; CLOD: 264–680 nM	[179]
Phenanthrolin	CE-UV	Fe(II), Co(II), Ni(II), Cu(II) and Zn(II)	CLOD: Cu(II) 30 ng/mL	[180]

Abbreviations: AA, 2-aminoanthracene; AAC, acyl azide coumarin; AB, 2-aminobenzoic acid; ABD-F, 4-aminosulfonyl-7-fluoro-2,1,3-bezoxadiazole; ABN, 4-aminobenzonitrile; ACCQ, 6-aminoquinolyl-*N*-hydroxysuccinimidyl carbamate; ADAM, 9-anthryldiazomethane; AEOC, 2-(9-anthryl)ethyl chloroformate; AF, 4-aminofluorescein; ANTS, 8-aminonaphthalene-1,3,6-trisulfonate; AP, 2-aminopyridine; APOC, 1-(9-anthryl)-2-propyl chloroformate; BH, benzoyl hydrazine; BMF, 5-bromomethylfluorescein; BMMC, 4-bromomethyl-7-methoxycoumarin; BODIPY, 4,4-difluoro-5,7-dimethyl-4-bora-3a,4a-diaza-s-indacene-3-propionic acid; BTA, 1,2,4-benzenetreicarboxylic anhydride; CAA, chloroacetaldehyde; CBQCA, 3-(4-carboxybenzoyl)-2-quinoline-carboxaldehyde; CDITC, 2-isothiocyanatocyclohexyl-6-methoxy-4-quinolinylamide; CDTA, cyclohexane-1,2-diaminetetraacetic acid; CFSE, 5-carbodyfluorescein suddinimidyl ester; CNOD, 7-chloro-4-nitrobenzo-2-oxa-1,3-diazole; DAB-Cl, dabsyl chloride; DBD-IMI, 4-/*N,N*-dimethylaminosulfonyl)-2,1,3-benzoxadiazole; DBTA, (+)-*O,O′*-dibenzoyl-L-tartaric anhydride; DMABSCl, 4-dimethylazobenzne-4-sulfonyl-chloride; DNSH, dansylhydrazine; DTPA, diethylenetriaminepentaacetic acid; DTTC, sodium diethyldithiocarbamate; EAB, ethyl 4-aminobenzoate; EB, ethidium bromide; EDTA, ethylenediaminetetraacetic acid; Ellman's reagent, 5,5′-dithio-bis-(2-nitrobenzoic acid); FLEC, 9-fluorenyl-ethyl chloroformate; FMOC, 9-fluorenyl-methyl chloroformate; GABA, γ-aminobutyric acid; HPAA, p-hydroxyphenylacetic acid; IAF, 6-iodoacetamidefluorescein; IDA, 1-methoxycarbonylindolizine-3,5-dicarbaldehyde; IITC, isoluminol isothiocyanate; LNA, L-leucine-*p*-nitroanilide; IAF-LRB, iodoacetyl derivative of Lissamine rhodamine B; MBBR, monobromobimane; MCIDC, 1-methoxycarbonylindolizine-3,5-dicarbaldehyde; MFPTS, 1-methyl-2-fluoropyridinium *p*-toluenesulphonate; MOPDYA, 2-methyl-3-oxo-4-phenyl-2,3-dihydrofuran-2-yl-acetate; NBD, 7-nitrobenz-2-oxa-1,3-diazole (NBD)-ceramide; NBD-Cl, 7-chloro-4-nitrobenzo-2-oxa-1,3-diazole; NBD-F, 4-fluoro-7-nitro-2,1,3-benzooxadiazole; NBD-PyNCS, 4-(3-isothiocyanatopyrrolidin-1-yl)-7-nitro-2,1,3-benzooxadiazole; NQS, 1,2-naphthoquinone-4-sulfonate; NS, α- and β-naphthalenesulfonyl chloride; NTA, nitrilotriacetic acid; OPA, *o*-phthaldialdehyde; PA, phthalic anhydride; PC, propyl chloroformate; PEA, phenylethylamine; PITC, phenylisothiocyanate; PMC, polymethine cyanine; PMP, 1-phenyl-3-methyl-5-pyrazolone; PTH, phenylthiohydantoin; RBITC, rhodamine B isothiocyanate; SBD-F, ammonium 7-fluoro-2,1,3-benzoxadiazole-4-sulfonate; SEC, size-exclusion chromatography; TBA HS, tetrabutylammonium hydrogen sulfate; TMR, tetramethylrhodamine; TRITC, tetramethylrhodamine isothiocyanate; TRSE, 5-carboxytetramethylrhodamine succinimidyl ester; XO, xylenol orange

Table 2. On-line derivatization

Reagent	CE mode	Compounds	Details	Ref.
MBBr	CE-LIF	Thiols	On-line microdialysis-CE; CLOD: 20–40 nM	[181]
NDA	CE-LIF	Catecholamines	On-line microdialysis – precolumn derivatization – fraction collection CLOD: 0.3–1.8 nM LOD: 0.2–0.9 fmol	[17]
NDA	CE-LIF	Catecholamines / excitatory amino acids	On-line microdialysis – precolumn derivatization – fraction collection CLOD: 2.5–32 nM LOD: 0.42–16 fmol	[182]
NDA	CE-LIF	Glutamate	On-line microdialysis – precolumn derivatization – fraction collection	[183]
NDA	CE-LIF	Glutamate / phosphoethanolamine	On-line microdialysis – precolumn derivatization – on-line coupled to CE-LIF	[18]
OPA	CE-LIF MEKC-LIF	Primary amines	On-line microdialysis – precolumn derivatization – CE; sampling time 45 s –3 min CLOD: 20–50 nM	[19]
OPA	CE-LIF	Primary amines	On-line microdialysis – precolumn derivatization – CE; sampling time reduced to 12 s CLOD: 200 nM	[20]
OPA	CE-LIF	Aspartate enantiomers	On-line microdialysis – precolumn derivatization – chiral CE separation using β-cyclodextrins; chiral separation in less than 3 s	[21]
OPA	CE-LIF	Amino acids	Optically gated CE to monitor amino acids with high temporal resolution (< 2 s); CLOD: low nanomolar region	[22]
OPA	CE-LIF	Glutamate / aspartate	*in vivo* monitoring, on-line microdialysis – precolumn derivatization – CE	[184]

For abbreviations see Table 1.

bleached all the fluorophores migrating through the capillary. In this way the derivatized analytes are undetectable down-column. When the bleaching beam was briefly blocked by a shutter, a plug of intact analytes was injected onto the column which can be separated by electrophoresis and detected using the detection beam focused into the capillary downstream. In this way, the length to detector can be minimized to several mm without limitations, yielding extremely high speed separations.

4 On-capillary derivatization

The on-capillary mode (Table 3) is a form of derivatization which can be used without changing the setup of the commercially available CE devices. Several options have been described to do so. In the at-inlet technique, a sample and a reagent solution are introduced to the inlet of a capillary either by tandem or sandwich mode. These reactants are mixed by diffusion and allowed to react for a specified time. The zone-passing technique is based on derivatization in the middle of the capillary by passing either a sample or a reagent zone through the other in an electric field. In the throughout-capillary technique an analyte introduced from the inlet of a capillary continues to travel through an electrophoretic solution containing the reagent.

Taga *et al.* [23] compared the peak areas and the S/N ratios of various techniques of on-capillary derivatization and the following order was found: zone-passing < throughout-capillary < at-inlet. The theoretical plate number was generally in the order of zone-passing < at-inlet < throughout-capillary < precapillary; however, it varied slightly depending on the operation conditions. Derivatization in the middle of the capillary was most liable to be affected by various factors, including sample/reagent introduction times, their relative concentration and the applied voltage. An increase of applied voltage decreased the passing period, and thereby reduced the relative peak response. Also, the reproducibility of peak response was inferior compared to derivatization at the inlet of the capillary [24].

The ability to use longer derivatization times (up to 30 min) and increased reaction temperature are advantages of the at-inlet technique, as was demonstrated by Lee *et al.* [25]. Electrophoretically mediated microanalysis, as it was called, was explored for trace-level analysis of proteins. Protein and a fluorogenic reagent 3-(2-furoyl)-quinoline-2-carboxaldehyde (FQ) were introduced into the capillary in order of increasing mobility. By performing the on-capillary reaction at 65°C rather than at room temperature, a 10-fold improvement in detection limit for con-

Table 3. On-capillary derivatization

Reagent	CE mode	Compounds	Details	Ref.
Cu(II)	CE-UV	Single plant cell vacuole components, sugars	On-column chelation with Cu(II); CLOD: micromolar range	[26]
FQ	CE-LIF	Proteins, conalbumin	Protein extracts are profiled originating from 2.5 cells (calculated) from human colon adenocarcinoma cell line; CLOD: 0.3 pM	[25]
Glucose oxidase / homovanillic acid / peroxidase	CE-LIF	Glucose	Two enzymatic reactions combined with CE are used to determine glucose CLOD: 50 nM precolumn, 800 nM on-column	[185]
HQS	CE-LIF	Magnesium	Combination of short separation capillary and laser-excitation and emission-collection fibers in a low-volume cross; CLOD: ng/mL range	[186]
IDA	CE-F	Alanine	Capillary throughout filled with reagent-separation buffer mixure compared with sandwich method; CLOD: 1 μM	[187]
LNA	CE-LIF	Leucine aminopeptidase	Kinetic determination of enzyme activity on-column	[188]
NDA	CE-LIF	Amines / peptides	On-column derivatization of single vesicle	[28]
NDA	CE-LIF	Components from secretory vesicles	Single cell analysis, on-capillary lysis and derivatization	[27]
Oligreen dye	CGE-LIF	Short single-stranded oligonucleotides (< 20 bases)	Based on noncovalent interactions, bioanalysis in plasma CLOD: 20 ng/mL	[189]
OPA	MEKC-F	Single plant cell vacuole components, amino acids	Sample is injected between two plugs of reagents (sandwich mode)	[26]
OPA	CE-F	Amino acids	Capillary throughout filled with reagent-separation buffer mixure; CLOD: 2.5–10 μM	[190]
OPA	MEKC-F	Amines	Capillary throughout filled with reagent-separation buffer mixure; CLOD: 1–5 μM	[191]
OPA	MEKC-F	Spermidine	Determined in ultrafiltrated plasma samples, capillary throughout filled with reagent-separation buffer mixure; CLOD: 0.2 μM	[192]
OPA	MEKC-F	Amino acids	Used for determination of membrane permeation experiments	[193]
OPA	CE-UV	Amino acids	Quantitative evaluation of derivatization at capillary inlet sample-reagent introduction methods evaluated sandwich mode *vs.* tandem mode	[194]
OPA	CE-UV	Amino acids	Comparison between several on-capillary derivatization (at-inlet, throughout, zone-passing or precapillary)	[23]
OPA	CE-UV	Amino acids	Investigation of factors affecting zone-passing on-capillary derivatization	[24]
OPA	MEKC-LIF	Amino acids	Several chiral thiols were compared; LOD: low amol level	[195]
TNS	CE-LIF	Bovine whey proteins	Quantum yield of these fluorescence probes is enhanced after noncovalent association with proteins; CLOD: 60–3000 nM, comparable to CE-UV detection	[196]

For abbreviations see Table 1.

albumin was obtained, accompanied by a reduction of the total analysis time from 35 min to 5 min. Picomolar concentrations and zeptomole amounts of protein were analyzed using LIF detection. FQ has a relatively low reaction efficiency for amino acids compared to that for proteins. The reagent reacts much more efficiently with the ε-amine group of lysine residues of proteins.

On-column derivatization of saccharides have not been reported yet, as the reactions involved are slow (reaction time > 1 h), often multistep, and generally require rather high temperatures to proceed. These are conditions which are difficult to realize on-column. On the other hand, dynamic labeling does not have these limitations. In dy-

namic labeling the reaction of analytes with a complexing agent leads to a change in UV absorbance or fluorescence. In the case of saccharides, chelation with Cu(II) provides *in situ* charged species that can be separated according to their different mobilities. In addition, a marked increase in UV absorption between 230 nm and 320 nm results from this complexation. As the background noise of the electrolyte was sufficiently low, the detection of micromolar concentrations of saccharides was possible [26].

In case of single cell analysis, the on-capillary mode is the most frequently reported mode of derivatization [25–28]. Lillard *et al.* [28] separated and characterized amines from individual atrial gland vesicles of *Aplysia californica*. The inlet of the tapered separation capillary was placed in a droplet of a dilute suspension of cells. Using an optical imaging system built in-house, the inlet was positioned in front of a trapped vesicle. A low voltage was applied to introduce the vesicle into the capillary. After vesicle introduction, the tapered inlet was placed in a naphthalene-2,3-dicarboxaldehyde (NDA) solution which was also introduced electrokinetically. During the next 5 min the vesicle lysed and its contents were allowed to react. Finally, the inlet was returned to its buffer vial, and electrophoresis was initiated to separate the labeled components.

5 Postcolumn derivatization

The postcolumn design has several advantages over previously mentioned derivatization modes. Derivatization normally leads to a significant change in the analyte's mobility, due to changes in the mass-to-charge ratio and hydrophobicity before and after the labeling process. In the case of neutral saccharides these changes can be exploited in a positive way to some extent, but in many cases these changes may also become detrimental for the separation. An example of another negative effect of precolumn derivatization is described by Craig and Dovichi [29]. A large protein was derivatized using FQ for as little as 1 min. Severe band broadening and several distinct components were visible after the labeling, caused by the labeling with FQ on several sites in the molecule. Additionally, in contrast to precolumn derivatization, side products do not cause interfering peaks. Another advantage of postcolumn labeling is demonstrated in a study evaluating the DNA-protein interaction. The postcapillary instead of on-column addition of ethidium bromide (EB) preserved the integrity of the native interaction [30].

Disadvantages of postcolumn labeling are the negative effects on peak efficiency, loss of analyte, incomplete reactions, and the higher baseline noise caused by added reagent solution. Although OPA is considered to be fluorogenic, its presence increased the background noise by a factor of three [31]. Band broadening seems be an inherent problem of postcapillary detection schemes, since some turbulence can be expected at the site where the reagent solution is added. Several reactor designs have been explored to minimize this band broadening effect. Most approaches to postcolumn derivatization introduce the derivatizing reagents through a gap junction, using hydrodynamic pressure, electroosmotic flow [32], or diffusion. Good separation efficiencies were obtained with a sheath-flow cell reactor [33]. The sheath liquid, containing OPA, entered from the top of the flow cell and surrounded the capillary eluent as it exited the capil-

lary. In this way the original flow profile of the CE buffer was less disturbed compared with other gap reactors.

The use of porous membranes, positioned as sleeves over a gap between capillaries, has improved the results of gap reactors. The membrane allows the reagents to diffuse through the semipermeable membrane and to mix with the analyte of interest. It also acts as a restrictive barrier to limit loss of analyte. The latter factor is especially important for large molecules, such as substance P. It was found that the sample loss using the membrane reactor was approximately a factor of five less than with the standard gap-reactor design [34]. In another porous membrane design, the reagent solution was introduced through the porous membrane into the reaction capillary by means of air pressure [35].

Microfabrication offers substantial advantage for the postcolumn derivatization method in particular, since mixing intersections with essentially zero dead volume can be integrated onto the CE chip. Since these separation chips can be produced reproducibly on a mass scale with low costs, a further acceptance of postcapillary derivatization becomes plausible. Globulin-type proteins were separated and then, still within the chip channel structure, mixed with 0.2 mM 2-(*p*-toluidino)naphthalene-6-sulfonic acid (TNS) labeling reagent, as prepared in borate buffer. Just downstream from the mixing point, labeled proteins were detected by means of LIF. In this manner, a mixture of four human serum proteins were resolved in less than 60 s. On-chip separations of true human serum samples, however, has not yet been achieved since TNS preferentially associates with albumin rather than with globulin-type proteins [36].

6 Derivatization reasons other than for LIF detection

6.1 Derivatization to incorporate charge

Several examples are described where derivatization had a beneficial effect on the separation. Neutral aldehydes were separated under CE conditions as anionic derivatives [37]. Most examples can be found in the field of carbohydrate research, since most analytes contain no charge. Labeling with one charged reagent yields charged derivatives, which can be separated well from other saccharide derivatives. In case of carbohydrate polymer mixtures with different degrees of polymerization, the beneficial effect becomes even more pronounced. For example, since the oligosaccharides that were generated from hyaluronic acid, using hyaluronate lysase, have a uniform charge density (*i.e.*, one carboxylate group per disaccharide repeat), poor separation between oligosaccharides of

Table 4. Postcapillary derivatization

Reagent	CE mode	Compounds	Details	Ref.
EB	CE-LIF	DNA fragments	CLOD: low nanogram per mL range	[30]
Fluorescamine / OPA	CE-F	Amino acids, peptides	CLOD: submicromolar range	[197]
HPAA	MEKC-LIF	Fatty acid hydroperoxides	Reaction catalyzed by microperoxidase-11 covalently immobilized in a reaction capillary; CLOD: 10–15 µM	[198]
HQS	CE-LIF on a chip	Metal cations, magnesium, calcium	In combination with sample stacking and using a gated injection scheme; CLOD: 0.5 and 18 ng/mL	[199]
HQS	CE-F	Al, Ca, Mg, Zn, Cd	Pressure-controlled postcolumn reaction chamber connecting separation and reaction capillary with a porous tube; CLOD: 58–340 nM	[200]
NDA	CE-LIF	Substance P and metabolites	Used for metabolism study in brain in combination with microdialysis, two reactor designs evaluated; CLOD: 75 nM	[34]
OPA	CE-LIF	Amino acids / proteins	Using sheath-flow cell as a postcolumn reactor; CLOD: 0.83–69 nM, LOD: 1.8–170 amol	[33]
OPA	CE-LIF on a chip	Amino acids	Evaluation of influence postcolumn reactor geometry on peak efficiency	[201]
OPA	CE-LIF on a chip	Amino acids	Postcolumn derivatization using voltage-driven addition of reagent solution	[202]
OPA	CE-UV-CID	Amino acids, peptides	In the same setup pre- and postcolumn derivatization can be employed; LOD: 110 amol, CLOD: 94 nM	[31]
OPA	CE-F	Amino acids	Based on gap design, addition of reagents by gravity; CLOD: 0.67 µM	[203]
OPA	CE-LIF	Amines, amino acids and proteins, single cell components	Based on a coaxial design consisting of two narrow-bore capillaries; LOD: low attomole range, CLOD: 4 nM–13 µM	[32]
Permanganate	CE-CL	Alkaloids, morphine, 6-monoacetylmorphine, heroin	On-line permanganate CL detection; LOD: 23–115 fmol	[204]
Ru(bpy)$_3^{3+}$	CE-CL	Amines / amino acids	CLOD: 68 nM–100 µM, LOD: 3–4500 fmol	[205]
Ru(bpy)$_3^{3+}$	CE-CL	Amines / amino acids	CLOD: micromolar range	[206]
Terbium	CE-CL	Salicylic acid, diflunisal	CLOD: micromolar range	[207]
Terbium	CE-CL	Catecholamines and	Pressure-controlled postcolumn reaction chamber connecting separation related compounds and reaction capillary with a porous tube; CLOD: 70–130 nM	[35]
TNS	CE-LIF on a chip	Human serum proteins	Based on noncovalent association with the protein	[36]
XO and PAR	CE-UV	Trace metals, Cu, Cd, Co, Ni, Zn and Mn	Small intercapillary gap used surrounded by a permeable membrane to introduce the colorimetric reagent. PAR gave the most sensitive reaction. CLOD: 0.1 µg/mL	[208]

For abbreviations see Table 1.

differing size can be anticipated. This problem is circumvented if a single charge group can be introduced into each oligosaccharide chain. Reductive amination with 7-amino-1,3-naphthalene disulfonic acid (ANDSA), labeled each oligosaccharide at its reducing end, and, in this way, separation was accomplished [38]. Depending on the charge of the analytes a proper reagent choice should be made. Guttman [39] decided that the neutral and amino-sugars in the hydrolyzates should be labeled with the charged fluorophore, 8-aminopyrene-1,3,6-trisulfonate (APTS), while sialic acids should be labeled with 9-aminoacridone (9-AMAC). This change in derivatization agent was necessary for sialic acid because the electrophoretic mobilities of the APTS-labeled sialic acids were very close to the electroosmotic flow at pH 10; thus the migration time was found to be inconveniently long.

6.2 Derivatization used for indirect or direct chiral separation

Derivatization with a chiral pure reagent followed by separation of the diastereomeric derivatives in an achiral environment is a frequently used strategy. Since no chiral selector is needed for the indirect chiral separation, no optimization is needed for the concentration of the chiral selector. Disadvantages of the indirect method are the requirement for high chiral purity of the reagents and the corresponding prices of the reagents; furthermore, the intermolecular distance between the chiral centers of the derivatization reagent and of the analyte should not be too long [40]. For determination of the enantiomeric purity in pharmaceutical research, the indirect method can give an overestimation of the impurity in case the chiral reagent itself contains trace amounts of the unwanted enantiomer. Therefore, the degree of enantiomeric purity of the reagent was tested first in order to determine the enantiomeric purity of the amino acids to be analyzed [41].

The formed diastereomers are usually separated using MEKC. The hydrophobicity of the derivatization reagent is of great importance for the optimal SDS concentration. Derivatives obtained with 9-fluorenyl-ethyl chloroformate (FLEC) are more hydrophobic than those originating from 2,3,4,6,-tetra-*O*-acetyl-β-D-glucopyranosylisothiocyanate (GITC), OPA or Marfey's reagent [42]. In this way only 10–20 mM of SDS was required for good separation instead of the usual 100–200 mM. This is an advantage since high SDS concentrations may result in excessive Joule heating. The importance of derivatization is also demonstrated for direct chiral separation, using a nonchiral derivatizing agent. For enantioselectivity, inclusion into the cyclodextrin-cavity required the presence of the label moiety [43–45].

6.3 Derivatization for better MEKC separation

Monosaccharides are typically very hydrophilic; labeling them by a hydrophobic tag enables their separation by MEKC [39].

6.4 Derivatization for mass spectrometric detection

The analytes determined by a mass spectrometer are required to be surface-active in a viscous matrix. Unfortunately, free oligosaccharides are exceedingly hydrophilic. Although MALDI-MS can be used to analyze underivatized oligosaccharides, the sensitivity is low. Derivatization of the oligosaccharides is highly recommended to enhance hydrophobicity. Reductive amination has improved the sensitivity 1000- to 5000-fold in ESI and MALDI experiments over free oligosaccharides [46].

7 Derivatization of saccharides

Structural elucidations of the carbohydrate part of glycoproteins together with the determination of the degree of polymerization of polysaccharides are reasons for the increase in papers describing the derivatization of saccharides. The interest in the carbohydrate structure

results from the fact that the biological activity of recombinant glycoproteins is influenced by their carbohydrate structure [9]. Zieske *et al.* [47] described a method for multidimensional mapping by CE to identify the glycoprotein-derived structure using the relative migration values of oligosaccharides. The method was found to be simple and reproducible. It was based on the use of a combination of orthogonal separations in which each compensates the limitations of the other, and their selectivities complement each other. Most derivatizations use AMAC, ANDSA, or APTS, which are attached to the reducing end of sugars *via* reductive amination, *i.e.*, the amino group of the reagent reacts with the carbonyl group of the reducing end to form a Schiff base, which is further reduced with sodiumcyanoborohydride to form a stable secondary amine. Unreacted reagent is retarded by the micellar phase due to its higher hydrophobic nature, so that its retention time is well behind that of the analytes of interest [48].

Since maltodextrin or dextran polymers are labeled with a single charged reagent label per polymer, a good separation is obtained for the derivatives, giving an impression of the molecular weight distribution. In a few articles it was found that the molecular weight distribution found by CE differed significantly from that measured with more conventional methods, such as anion-exchange chromatography with pulsed amperometric detection or size-exclusion chromatography combined with light-scattering detection [49–51]. In all cases, the proportion of polymers with lower molecular weight was higher in CE than with the conventional technique. Different hypotheses for this effect have been brought up, such as degradation of higher molecular weight oligomers [49] or a lesser accessibility of the larger polymer's reducing end to label. On the other hand, using APTS the labeling efficiency was found to be highly reproducible and independent of average chain length between 3 and 135 degrees of polymerization, with an average efficiency of 80% [52].

A large peak for unreacted reagent is a frequently observed problem for many derivatizations. Mort *et al.* [53] found a solution to this problem. Excess reagent reacted with "scavenger beads", carrying an appropriate functional group to remove it from the sample solution. Examples were presented for removal of aminonaphthalene mono-, di-, and trisulfonic acid from mixtures in which they were used to label mono- or oligosaccharides by reductive amination. Aldehyde-containing scavenger beads were made by oxidizing Sephadex G-50 beads with sodium periodate. These were added to the labeling reaction mixtures after the reductive amination of the sugars was complete. Almost complete elimination of the peak from the labeling agent could be achieved.

8 Derivatization at low analyte concentrations

The reported concentration limit of detection (CLOD) values are in many cases not the lowest derivatizable concentrations, but are frequently obtained by extrapolation of observed signal at much higher concentrations or after extensive dilution of labeled compounds. Presenting the CLOD values in this way disregards the poorer labeling efficiency at low analyte concentrations. Actually, derivatization at subnanomolar concentrations seems to be impossible for most of the reagents. Differences in the rate of the reaction, the degree of substitution, and derivative stability between reagents do exist and determine the overall efficiency. 5-Carboxyfluorescein succinimidyl ester (CFSE) was compared with FITC for the derivatization of amino acids. CFSE was able to label amino acids at nanomolar concentrations, which is a 1000-fold improvement compared to FITC [54]. Derivatization of 10^{-6} M amine concentrations with isothiocyanate became problematic, since a derivatization has to be carried out at high pH. At pH 9, the OH^- concentrations is 10^{-5} M, so hydrolysis of FITC became the dominant mechanism for FITC depletion. It is likely that the rate constant for amino acid derivatization by CFSE is significantly larger than that for hydrolysis [54]. Another study evaluated the parameters affecting derivatization of amines using FITC, including the chemical composition, concentration, and pH of the buffer used, the amount of FITC, addition of organic solvents, reaction time, and temperature [55]. After all these optimizations, 10^{-7} M concentrations of amines were detectable.

9 Near-infrared labels

It is remarkable that no breakthrough in the acceptance of new near-infrared (NIR) labels for derivatization can be reported, keeping in mind all the advantages [56–58]. A plausible reason for this phenomenon is the delay in new NIR label development. For example, the NN382 NIR dye resulted in superior mass detection limits (zmole range), but some critical comments were made. First, the slow labeling kinetics requires 12 h of reaction. Second, the labeling efficiency for basic and neutral amino acids was highly variable and never exceeded 90%. Since the dye is highly anionic, labeling of acidic amino acids was found to be impossible [58].

The advantage of detection in the NIR region (670–1000 nm) results from the low level of background interference since few naturally occurring molecules can undergo electronic transitions in this low energy region of the electromagnetic spectrum. This is especially beneficial when working with bioanalytical and other complex samples, where autofluorescence in the visible region is a source of interference. Scatter/Raman and Rayleigh can be a large contributor of noise at low analyte concentrations and is reduced at higher wavelengths due to its dependence on the wavelength of detection by $1/\lambda$ [56].

Furthermore, excitation in the NIR region can be accomplished using compact and inexpensive GaAlAs laser diodes. The laser diodes have stable and relatively high power outputs and, in contrast to other laser systems, have long operating lifetimes and low maintenance costs and are easy to use. Detection can be achieved using silicon avalanche photodiode detectors. These detectors offer quantum efficiencies around 80% in the NIR region with low noise characteristics. They are rugged and have long usable operating lifetimes exceeding 100 000 h [56].

Received June 23, 2000

10 References

[1] Bardelmeijer, H. A., Waterval, J. C. M., Lingeman, H., Van't Hof, R., Bult, A., Underberg, W. J. M., *Electrophoresis* 1997, *18*, 2214–2227.

[2] Banks, P. R., *Trends Anal. Chem.* 1998, *17*, 612–622.

[3] Smith, J. T., *Electrophoresis* 1997, *18*, 2377–2392.

[4] Deantonis, K. M., Brown, P. R., *Adv. Chromatogr.* 1997, *37*, 425–452.

[5] Couderc, F., Causse, E., Bayle, C., *Electrophoresis* 1998, *19*, 2777–2790.

[6] Davies, M. J., Hounsell, E. F., *Biomed. Chromatogr.* 1996, *10*, 285–289.

[7] El Rassi, Z., *Electrophoresis* 1997, *18*, 2400–2407.

[8] El Rassi, Z., Mechref, Y., *Electrophoresis* 1996, *17*, 275–301.

[9] Kakehi, K., Honda, S., *J. Chromatogr. A* 1996, *720*, 377–393.

[10] Paulus, A., Klockow, A., *J. Chromatogr. A* 1996, *720*, 353–376.

[11] Suzuki, S., Honda, S., *Electrophoresis* 1998, *19*, 2539–2560.

[12] Grimshaw, J., *Electrophoresis* 1997, *18*, 2408–2414.

[13] Effenhauser, C. S., Bruin, G. J. M., Paulus, A., *Electrophoresis* 1997, *18*, 2203–2213.

[14] Liu, B. F., Liu, L. B., Cheng, J. K., *J. Chromatogr. A* 1999, *834*, 277–308.

[15] Timerbaev, A. R., *J. Chromatogr. A* 1997, *792*, 495–518.

[16] Timerbaev, A. R., Buchberger, W., *J. Chromatogr. A* 1999, *834*, 117–132.

[17] Bert, L., Robert, F., Denoroy, L., Stoppini, L., Renaud, B., *J. Chromatogr. A* 1996, *755*, 99–111.

[18] Robert, F., Bert, L., Parrot, S., Denoroy, L., Stoppini, L., Renaud, B., *J. Chromatogr. A* 1998, *817*, 195–203.

[19] Lada, M. W., Kennedy, R. T., *Anal. Chem.* 1996, *68*, 2790–2797.

[20] Lada, M. W., Vickroy, T. W., Kennedy, R. T., *Anal. Chem.* 1997, *69*, 4560–4565.

[21] Thompson, J. E., Vickroy, T. W., Kennedy, R. T., *Anal. Chem.* 1999, *71*, 2379–2384.

[22] Tao, L., Thompson, J. E., Kennedy, R. T., *Anal. Chem.* 1998, *70*, 4015–4022.

[23] Taga, A., Nishino, A., Honda, S., *J. Chromatogr. A* 1998, *822*, 271–279.

[24] Taga, A., Sugimura, M., Honda, S., *J. Chromatogr. A* 1998, *802*, 243–248.

[25] Lee, I. H., Pinto, D., Arriaga, E. A., Zhang, Z. R., Dovichi, N. J., *Anal. Chem.* 1998, *70*, 4546–4548.

[26] Lochmann, H., Bazzanella, A., Bachmann, K., *J. Chromatogr. A* 1998, *817*, 337–343.

[27] Chiu, D. T., Lillard, S. J., Scheller, R. H., Zare, R. N., Rodriguezcruz, S. E., Williams, E. R., Orwar, O., Sandberg, M., Lundqvist, J. A., *Science* 1998, *279*, 1190–1193.

[28] Lillard, S. J., Chiu, D. T., Scheller, R. H., Zare, R. N., Rodriguezcruz, S. E., Williams, E. R., Orwar, O., Sandberg, M., Lundqvist, J. A., *Anal. Chem.* 1998, *70*, 3517–3524.

[29] Craig, D. B., Dovichi, N. J., *Anal. Chem.* 1998, *70*, 2493–2494.

[30] Nirode, W. F., Staller, T. D., Cole, R. O., Sepaniak, M. J., *Anal. Chem.* 1998, *70*, 182–186.

[31] Oldenburg, K. E., Xi, X. Y., Sweedler, J. V., *Analyst* 1997, *122*, 1581–1585.

[32] Zhang, L. L., Yeung, E. S., *J. Chromatogr. A* 1996, *734*, 331–337.

[33] Coble, P. G., Timperman, A. T., *J. Chromatogr. A* 1998, *829*, 309–315.

[34] Kostel, K. L., Lunte, S. M., *J. Chromatogr. B* 1997, *695*, 27–38.

[35] Zhu, R. H., Kok, W. T., *Anal. Chem.* 1997, *69*, 4010–4016.

[36] Colyer, C. L., Mangru, S. D., Harrison, D. J., *J. Chromatogr. A* 1997, *781*, 271–276.

[37] Mainka, A., Bachmann, K., *J. Chromatogr. A* 1997, *767*, 241–247.

[38] Park, Y., Cho, S. H., Linhardt, R. J., *Bba-Protein Struct. Mol. Enzym.* 1997, *1337*, 217–226.

[39] Guttman, A., *J. Chromatogr. A* 1997, *763*, 271–277.

[40] Wan, H., Blomberg, L. G., *J. Chromatogr. A* 1997, *758*, 303–311.

[41] Thorsen, G., Engstrom, A., Josefsson, B., *J. Chromatogr. A* 1997, *786*, 347–354.

[42] Wan, H., Blomberg, L. G., *J. Chromatogr. A* 1997, *792*, 393–400.

[43] Wan, H., Engstrom, A., Blomberg, L. G., *J. Chromatogr. A* 1996, *731*, 283–292.

[44] Miura, M., Funazo, K., Tanaka, M., *Anal. Chim. Acta* 1997, *357*, 177–185.

[45] Miura, M., Kawamoto, K., Funazo, K., Tanaka, M., *Anal. Chim. Acta* 1998, *373*, 47–56.

[46] Che, F. Y., Song, J. F., Zeng, R., Wang, K. Y., Xia, Q. C., *J. Chromatogr. A* 1999, *858*, 229–238.

[47] Zieske, L. R., Fu, D. T., Khan, S. H., Oneill, R. A., *J. Chromatogr. A* 1996, *720*, 395–407.

[48] Okafo, G., Burrow, L., Carr, S. A., Roberts, G. D., Johnson, W., Camilleri, P., *Anal. Chem.* 1996, *68*, 4424–4430.

[49] Pfaff, P., Weide, F., Kuhn, R., *Chromatographia* 1999, *49*, 666–670.

[50] Chmelik, J., Chmelikova, J., Novotny, M. V., *J. Chromatogr. A* 1997, *790*, 93–100.

[51] Kazmaier, T., Roth, S., Zapp, J., Harding, M., Kuhn, R., *Fresenius J. Anal. Chem.* 1998, *361*, 473–478.

[52] Oshea, M. G., Samuel, M. S., Konik, C. M., Morell, M. K., *Carbohydr. Res.* 1998, *307*, 1–12.

[53] Mort, A. J., Zhan, D. F., Rodriguez, V., *Electrophoresis* 1998, *19*, 2129–2132.

[54] Lau, S. K., Zaccardo, F., Little, M., Banks, P., *J. Chromatogr. A* 1998, *809*, 203–210.

[55] Dabek-Zlotorzynska, E., Maruszak, W., *J. Chromatogr. B* 1998, *714*, 77–85.

[56] Baars, M. J., Patonay, G., *Anal. Chem.* 1999, *71*, 667–671.

[57] Gallaher, D. L., Johnson, M. E., *Analyst* 1999, *124*, 1541–1546.

[58] Legendre, B. L., Moberg, D. L., Williams, D. C., Soper, S. A., *J. Chromatogr. A* 1997, *779*, 185–194.

[59] Kovacs, A., Simonsarkadi, L., Ganzler, K., *J. Chromatogr. A* 1999, *836*, 305–313.

[60] Penmetsa, K. V., Leidy, R. B., Shea, D., *J. Chromatogr. A* 1996, *745*, 201–208.

[61] Bergquist, J., Vona, M. J., Stiller, C. O., Oconnor, W. T., Falkenberg, T., Ekman, R., *J. Neurosci. Methods* 1996, *65*, 33–42.

[62] Kleidernigg, O. P., Lindner, W., *J. Chromatogr. A* 1998, *795*, 251–261.

[63] Nagaraj, S., Rahavendran, S. V., Karnes, H. T., *J. Pharm. Biomed. Anal.* 1998, *18*, 411–420.

[64] Kaneta, T., Shiba, H., Imasaka, T., *J. Chromatogr. A* 1998, *805*, 295–300.

[65] Wu, Z. Q., Tong, W. G., *J. Chromatogr. A* 1997, *773*, 291–298.

[66] Seidel, B. S., Faubel, W., *J. Chromatogr. A* 1998, *817*, 223–226.

[67] Valko, I. E., Siren, H., Riekkola, M. L., *J. Chromatogr. A* 1996, *737*, 263–272.

[68] Wan, H., Blomberg, L. G., *Electrophoresis* 1996, *17*, 1938–1944.

[69] Tsukagoshi, K., Fujimura, S., Nakajima, R., *Anal. Sci.* 1997, *13*, 279–281.

[70] Tucci, S., Rada, P., Hernandez, L., *Brain Res.* 1998, *813*, 44–49.

[71] Tucci, S., Pinto, C., Goyo, J., Rada, P., Hernandez, L., *Clin. Biochem.* 1998, *31*, 143–150.

[72] Ramseier, A., Vonheeren, F., Thormann, W., *Electrophoresis* 1998, *19*, 2967–2975.

[73] Brumley, W. C., Kelliher, V., *J. Liq. Chromatogr.* 1997, *20*, 2193–2205.

[74] Rodriguez, I., Lee, H. K., Li, S. F. Y., *Electrophoresis* 1999, *20*, 1862–1868.

[75] Rodriguez, I., Lee, H. K., Li, S. F. Y., *Electrophoresis* 1999, *20*, 118–126.

[76] Britzmckibbin, P., Vo, H. C., Macgillivray, R. T. A., Chen, D. D. Y., *Anal. Chem.* 1999, *71*, 1633–1637.

[77] Gallaher, D. L., Johnson, M. E., *Appl. Spectrosc.* 1998, *52*, 292–297.

[78] Hutt, L. D., Glavin, D. P., Bada, J. L., Mathies, R. A., *Anal. Chem.* 1999, *71*, 4000–4006.

[79] Takizawa, K., Nakamura, H., *Anal. Sci.* 1998, *14*, 925–928.

[80] Vonheeren, F., Verpoorte, E., Manz, A., Thormann, W., *Anal. Chem.* 1996, *68*, 2044–2053.

[81] Paez, X., Rada, P., Tucci, S., Rodriguez, N., Hernandez, L., *J. Chromatogr. A* 1996, *735*, 263–269.

[82] Zhang, L., Chen, H., Hu, S., Cheng, J., Li, Z. W., Shao, M., *J. Chromatogr. B* 1998, *707*, 59–67.

[83] Raymond, D. E., Manz, A., Widmer, H. M., *Anal. Chem.* 1996, *68*, 2515–2522.

[84] Nouadje, G., Simeon, N., Dedieu, F., Nertz, M., Puig, P., Couderc, F., *J. Chromatogr. A* 1997, *765*, 337–343.

[85] Ireland, I. D., Lewis, D. F., Li, X. F., Renborg, A., Kwong, S., Chen, M., Dovichi, N. J., *J. Prot. Chem.* 1997, *16*, 491–493.

[86] Sepulveda, M. J., Hernandez, L., Rada, P., Tucci, S., Contreras, E., *Pharmacol. Biochem. Behav.* 1998, *60*, 255–262.

[87] Liu, J. P., Dabrah, T. T., Matson, J. A., Klohr, S. E., Volk, K. J., Kerns, E. H., Lee, M. S., *J. Pharm. Biomed. Anal.* 1997, *16*, 207–214.

[88] Floyd, P. D., Moroz, L. L., Gillette, R., Sweedler, J. V., *Anal. Chem.* 1998, *70*, 2243–2247.

[89] Tsukagoshi, K., Tanaka, A., Nakajima, R., Hara, T., *Anal. Sci.* 1996, *12*, 525–528.

[90] Wei, J., Gostkowski, M. L., Gordon, M. J., Shear, J. B., *Anal. Chem.* 1998, *70*, 3470–3475.

[91] Wan, H., Blomberg, L. G., *J. Chromatogr. Sci.* 1996, *34*, 540–546.

[92] Wan, H., Blomberg, L. G., *J. Microcol. Sep.* 1996, *8*, 339–344.

[93] Kang, J. W., Yang, Y. T., You, J. M., Ou, Q. Y., *J. Chromatogr. A* 1998, *825*, 81–87.

[94] Kaizu, T., Kim, D., *Yakugaku Zasshi* 1999, *119*, 674–680.

[95] Vogt, C., Kiessig, S., *J. Chromatogr. A* 1996, *745*, 53–60.

[96] Richards, D. P., Stathakis, C., Polakowski, R., Ahmadzadeh, H., Dovichi, N. J., *J. Chromatogr. A* 1999, *853*, 21–25.

[97] Hashimoto, M., Tsukagoshi, K., Nakajima, R., Kondo, K., *J. Chromatogr. A* 1999, *832*, 191–202.

[98] Oguri, S., Miki, Y., *J. Chromatogr. B* 1996, *686*, 205–210.

[99] Chen, P., Novotny, M. V., *Anal. Chem.* 1997, *69*, 2806–2811.

[100] Schmitz, G., Mollers, C., Richter, V., *Electrophoresis* 1997, *18*, 1807–1813.

[101] Preston, L. M., Weber, M. L., Murray, G. M., *J. Chromatogr. B* 1997, *695*, 175–180.

[102] Tsunoda, M., Kato, M., Fukushima, T., Santa, T., Homma, H., Yanai, H., Soga, T., Imai, K., *Biomed. Chromatogr.* 1999, *13*, 335–339.

[103] Kuroda, N., Nomura, R., Aldirbashi, O., Akiyama, S., Nakashima, K., *J. Chromatogr. A* 1998, *798*, 325–334.

[104] Liu, Y. M., Schneider, M., Sticha, C. M., Toyooka, T., Sweedler, J. V., *J. Chromatogr. A* 1998, *800*, 345–354.

[105] Church, W. H., Lee, C. S., Dranchak, K. M., *J. Chromatogr. B* 1997, *700*, 67–75.

[106] Swanek, F. D., Anderson, B. B., Ewing, A. G., *J. Microcol. Sep.* 1998, *10*, 185–192.

[107] Rocher, C., Bert, L., Robert, F., Trouvin, J. H., Renaud, B., Jacquot, C., Gardier, A. M., *Brain Res.* 1996, *737*, 221–230.

[108] Latorre, R. M., Saurina, J., Hernandezcassou, S., *J. Chromatogr. Sci.* 1999, *37*, 353–359.

[109] Bazzanella, A., Lochmann, H., Mainka, A., Bachmann, K., *Chromatographia* 1997, *45*, 59–62.

[110] Durgbanshi, A., Kok, W. T., *J. Chromatogr. A* 1998, *798*, 289–296.

[111] Tivesten, A., Lundqvist, A., Folestad, S., *Chromatographia* 1997, *44*, 623–633.

[112] Chu, Q. Y., Evans, B. T., Zeece, M. G., *J. Chromatogr. B* 1997, *692*, 293–301.

[113] Wehren, A., Meyer, H. E., Sobek, A., Kloetzel, P. M., Dahlmann, B., *Biol. Chem.* 1996, *377*, 497–503.

[114] Waldron, K. C., Li, X. F., Chen, M., Ireland, I., Lewis, D., Carpenter, M., Dovichi, N. J., *Talanta* 1997, *44*, 383–399.

[115] Kurosu, Y., Murayama, K., Shindo, N., Shisa, Y., Ishioka, N., *J. Chromatogr. A* 1996, *752*, 279–286.

[116] Li, X. F., Waldron, K. C., Black, J., Lewis, D., Ireland, I., Dovichi, N. J., *Talanta* 1997, *44*, 401–411.

[117] Culbertson, C. T., Jorgenson, J. W., *J. Microcol. Sep.* 1999, *11*, 175–183.

[118] Chen, C. P., Jeffery, D., Jorgenson, J. W., Moseley, M. A., Pollack, G. M., *J. Chromatogr. B* 1997, *697*, 149–162.

[119] Hunt, G., Nashabeh, W., *Anal. Chem.* 1999, *71*, 2390–2397.

[120] Lamari, F., Theocharis, A., Hjerpe, A., Karamanos, N. K., *J. Chromatogr. B* 1999, *730*, 129–133.

[121] Okafo, G. N., Burrow, L. M., Neville, W., Truneh, A., Smith, R. A. G., Reff, M., Camilleri, P., *Anal. Biochem.* 1996, *240*, 68–74.

[122] Harland, G. B., Okafo, G., Matejtschuk, P., Sellick, I. C., Chapman, G. E., Camilleri, P., *Electrophoresis* 1996, *17*, 406–411.

[123] Schneider, P. J., Grosche, O., Engelhardt, H., *J. High Resol. Chromatogr.* 1999, *22*, 79–82.

[124] Sato, K., Sato, K., Okubo, A., Yamazaki, S., *Anal. Biochem.* 1997, *251*, 119–121.

[125] Sato, K., Sato, K., Okubo, A., Yamazaki, S., *Anal. Biochem.* 1998, *262*, 195–197.

[126] Rethfeld, I., Blaschke, G., *J. Chromatogr. B* 1997, *700*, 249–253.

[127] El Rassi, Z., Postlewait, J., Mechref, Y., Ostrander, G. K., *Anal. Biochem.* 1997, *244*, 283–290.

[128] Mechref, Y., Ostrander, G. K., El Rassi, Z., *J. Chromatogr. A* 1997, *792*, 75–82.

[129] Karcher, A., Melouk, H. A., El Rassi, Z., *Anal. Biochem.* 1999, *267*, 92–99.

[130] Mort, A. J., Chen, E. M. W., *Electrophoresis* 1996, *17*, 379–383.

[131] Guttman, A., Ulfelder, K. W., *J. Chromatogr. A* 1997, *781*, 547–554.

[132] Hong, M. F., Soini, H., Baker, A., Novotny, M. V., *Anal. Chem.* 1998, *70*, 3590–3597.

[133] Chen, F. T. A., Evangelista, R. A., *Electrophoresis* 1998, *19*, 2639–2644.

[134] Chen, F. T. A., Dobashi, T. S., Evangelista, R. A., *Glycobiology* 1998, *8*, 1045–1052.

[135] Evangelista, R. A., Guttman, A., Chen, F. T. A., *Electrophoresis* 1996, *17*, 347–351.

[136] Roberts, M. A., Zhong, H. J., Prodolliet, J., Goodall, D. M., *J. Chromatogr. A* 1998, *817*, 353–366.

[137] Suzuki, H., Müller, O., Guttman, A., Karger, B. L., *Anal. Chem.* 1997, *69*, 4554–4559.

[138] Lin, Q. S., Zhang, R. N., Liu, G. Q., *J. Liq. Chromatogr.* 1997, *20*, 1123–1137.

[139] Bazzanella, A., Bachmann, K., *J. Chromatogr. A* 1998, *799*, 283–288.

[140] Perez, S. A., Colon, L. A., *Electrophoresis* 1996, *17*, 352–358.

[141] Zeleny, R., Altmann, F., Praznik, W., *Anal. Biochem.* 1997, *246*, 96–101.

[142] Nguyen, D. T., Lerch, H., Zemann, A., Bonn, G., *Chromatographia* 1997, *46*, 113–121.

[143] Noe, C. R., Lachmann, B., Mollenbeck, S., Richter, P., *Z. Lebensm. Unters. Forsch.* 1999, *208*, 148–152.

[144] Honda, S., Taga, A., Kotani, M., Grover, E. R., *J. Chromatogr. A* 1997, *792*, 385–391.

[145] Le, X. C., Zhang, Y., Dovichi, N. J., Compston, C. A., Palcic, M. M., Beever, R. J., Hindsgaul, O., *J. Chromatogr. A* 1997, *781*, 515–522.

[146] Kang, S. H., Kim, J. W., Chung, D. S., *J. Pharm. Biomed. Anal.* 1997, *15*, 1435–1441.

[147] Russell, J., Rabenstein, D. L., *Anal. Biochem.* 1996, *242*, 136–144.

[148] Raggi, M. A., Mandrioli, R., Sabbioni, C., Mongiello, F., Marini, M., Fanali, S., *J. Microcol. Sep.* 1998, *10*, 503–509.

[149] Causse, E., Terrier, R., Champagne, S., Nertz, M., Valdiguie, P., Salvayre, R., Couderc, F., *J. Chromatogr. A* 1998, *817*, 181–185.

[150] Causse, E., Siri, N., Bellet, H., Champagne, S., Bayle, C., Valdiguie, P., Salvayre, R., Couderc, F., *Clin. Chem.* 1999, *45*, 412–414.

[151] Shimura, K., Matsumoto, H., Kasai, K., *Electrophoresis* 1998, *19*, 2296–2300.

[152] Kiessig, S., Vogt, C., Werner, G., *Fresenius J. Anal. Chem.* 1997, *357*, 539–542.

[153] Kiessig, S., Vogt, C., *J. Chromatogr. A* 1997, *781*, 475–479.

[154] Kibler, M., Bachmann, K., *J. Chromatogr. A* 1999, *836*, 325–331.

[155] Mechref, Y., El Rassi, Z., *Anal. Chem.* 1996, *68*, 1771–1777.

[156] Mechref, Y., El Rassi, Z., *Electrophoresis* 1997, *18*, 220–226.

[157] Karcher, A., El Rassi, Z., *Electrophoresis* 1997, *18*, 1173–1179.

[158] Zuriguel, V., Causse, E., Bounery, J. D., Nouadje, G., Simeon, N., Nertz, M., Salvayre, R., Couderc, F., *J. Chromatogr. A* 1997, *781*, 233–238.

[159] Luong, J. H. T., Rigby, T., Male, K. B., Bouvrette, P., *J. Chromatogr. A* 1999, *849*, 255–266.

[160] Mainka, A., Ebert, P., Kibler, M., Prokop, T., Tenberken, B., Bachmann, K., *Chromatographia* 1997, *45*, 158–162.

[161] Malek, A., Khaledi, M. G., *Anal. Biochem.* 1999, *270*, 50–58.

[162] Feige, K., Ried, T., Bachmann, K., *J. Chromatogr. A* 1996, *730*, 333–336.

[163] Gorog, S., Gazdag, M., Kemenesbakos, P., *J. Pharm. Biomed. Anal.* 1996, *14*, 1115–1124.

[164] Shea, D., *Electrophoresis* 1997, *18*, 277–283.

[165] Barry, J. P., Radtke, D. R., Carton, W. J., Anselmo, R. T., Evans, J. V., *J. Chromatogr. A* 1998, *800*, 13–19.

[166] Wang, C. C., Mccann, W. P., Beale, S. C., *J. Chromatogr. B* 1996, *676*, 19–28.

[167] Schutzner, W., Fanali, S., Rizzi, A., Kenndler, E., *J. Chromatogr. A* 1996, *719*, 411–420.

[168] Morin, P., Daguet, D., Coic, J. P., Dreux, M., *J. Chromatogr. A* 1999, *837*, 281–287.

[169] Heinig, K., Vogt, C., Werner, G., *Anal. Chem.* 1998, *70*, 1885–1892.

[170] Lee, W. M., Liu, L. L., Lau, D. S., Chan, R., *Anal. Lett.* 1999, *32*, 1235–1243.

[171] Lan, Z. H., Wang, P. G., Giese, R. W., *Rapid Commun. Mass Spectrom.* 1999, *13*, 1454–1457.

[172] Lan, Z. H., Qian, X. H., Giese, R. W., *J. Chromatogr. A* 1999, *831*, 325–330.

[173] Legendre, B. L., Soper, S. A., *Appl. Spectrosc.* 1996, *50*, 1196–1202.

[174] Moody, E. D., Viskari, P. J., Colyer, C. L., *J. Chromatogr. B* 1999, *729*, 55–64.

[175] Castro, A., Williams, J. G. K., *Anal. Chem.* 1997, *69*, 3915–3920.

[176] Hansen, D. K., Lunte, S. M., *J. Chromatogr. A* 1997, *781*, 81–89.

[177] Padarauskas, A., Schwedt, G., *J. Chromatogr. A* 1997, *773*, 351–360.

[178] Liu, W., Lee, H. K., *J. Chromatogr. A* 1998, *796*, 385–395.

[179] Liu, B. F., Liu, L. B., Cheng, J. K., *J. Chromatogr. A* 1999, *848*, 473–484.

[180] Yokoyama, T., Akamatsu, T., Ohji, K., Zenki, M., *Anal. Chim. Acta* 1998, *364*, 75–81.

[181] Lada, M. W., Kennedy, R. T., *J. Neurosci. Methods* 1997, *72*, 153–159.

[182] Robert, F., Bert, L., Lambassenas, L., Denoroy, L., Renaud, B., *J. Neurosci. Methods* 1996, *70*, 153–162.

[183] Robert, F., Parisi, L., Bert, L., Renaud, B., Stoppini, L., *J. Neurosci. Methods* 1997, *74*, 65–76.

[184] Lada, M. W., Vickroy, T. W., Kennedy, R. T., *J. Neurochem.* 1998, *70*, 617–625.

[185] Jin, Z., Chen, R., Colon, L. A., *Anal. Chem.* 1997, *69*, 1326–1331.

[186] Dickens, J. E., Sepaniak, M. J., *J. Microcol. Sep.* 1999, *11*, 45–51.

[187] Oguri, S., Fujiyoshi, T., Miki, Y., *Analyst* 1996, *121*, 1683–1688.

[188] Harmon, B. J., Leesong, I., Regnier, F. E., *J. Chromatogr. A* 1996, *726*, 193–204.

[189] Reyderman, L., Stavchansky, S., *Anal. Chem.* 1997, *69*, 3218–3222.

[190] Oguri, S., Yokoi, K., Motohase, Y., *J. Chromatogr. A* 1997, *787*, 253–260.

[191] Oguri, S., Watanabe, S., Abe, S., *J. Chromatogr. A* 1997, *790*, 177–183.

[192] Oguri, S., Tsukamoto, A., Yura, A., Miho, Y., *Electrophoresis* 1998, *19*, 2986–2990.

[193] Oguri, S., Kumazaki, M., Kitou, R., Nonoyama, H., Tooda, N., *Bba-Gen Subjects* 1999, *1472*, 107–114.

[194] Taga, A., Honda, S., *J. Chromatogr. A* 1996, *742*, 243–250.

[195] Tivesten, A., Folestad, S., *Electrophoresis* 1997, *18*, 970–977.

[196] Benito, I., Marina, M. L., Saz, J. M., Diezmasa, J. C., *J. Chromatogr. A* 1999, *841*, 105–114.

[197] Zhu, R., Kok, W. T., *J. Chromatogr. A* 1998, *814*, 213–221.

[198] Schmitz, O., Melchior, D., Schuhmann, W., Gab, S., *J. Chromatogr. A* 1998, *814*, 261–265.

[199] Kutter, J. P., Ramsey, R. S., Jacobson, S. C., Ramsey, J. M., *J. Microcol. Sep.* 1998, *10*, 313–319.

[200] Zhu, R. H., Kok, W. T., *Anal. Chim. Acta* 1998, *371*, 269–277.

[201] Fluri, K., Fitzpatrick, G., Chiem, N., Harrison, D. J., *Anal. Chem.* 1996, *68*, 4285–4290.

[202] Harrison, D. J., Fluri, K., Chiem, N., Tang, T., Fan, Z. H., *Sensor Actuator B-Chem.* 1996, *33*, 105–109.

[203] Wei, H. P., Li, S. F. Y., *Anal. Chem.* 1998, *70*, 5097–5102.

[204] Gong, Z. L., Zhang, Y., Zhang, H., Cheng, J. K., *J. Chromatogr. A* 1999, *855*, 329–335.

[205] Bobbitt, D. R., Jackson, W. A., Hendrickson, H. P., *Talanta* 1998, *46*, 565–572.

[206] Dickson, J. A., Ferris, M. M., Milofsky, R. E., *J. High Resol. Chromatogr.* 1997, *20*, 643–646.

[207] Milofsky, R., Bauer, E., *J. High Resol. Chromatogr.* 1997, *20*, 638–642.

[208] Hardy, S., Jones, P., Riviello, J. M., Avdalovic, N., *J. Chromatogr. A* 1999, *834*, 309–320.

Review

Jozef L. Beckers[1]
Petr Boček[2]

[1]Eindhoven University of
 Technology, Department of
 Chemistry (SPO),
 Eindhoven, The Netherlands
[2]Institute of Analytical
 Chemistry,
 Academy of Sciences
 of the Czech Republic,
 Brno, Czech Republic

Sample stacking in capillary zone electrophoresis: Principles, advantages and limitations

The principles of stacking procedures are described and their properties are discussed, including the fundamentals of the behavior of zone boundaries and the consequences of the self-correcting properties of boundaries in moving boundary electrophoresis, isotachophoresis, and zone electrophoresis. Further, the diverse possibilities of stacking procedures and the unavoidable destacking are described, and several examples of practically applied stacking procedures are given, besides many references to applications. Some limitations in the use of stacking procedures are discussed. The paper is arranged in such a way that it can serve both as an introduction into the field and as a reference overview.

Keywords: Sample stacking / Capillary zone electrophoresis / Review   EL 4030

Contents

1 Introduction

Capillary zone electrophoresis (CZE) is a separation technique [1–5] especially for the analysis of ionic species and there is growing interest and research regarding its applicability in diverse areas, especially in biological and biochemical fields. The sensitivity of the method is of key importance for practical use. Generally, spectrometric detectors such as UV detectors are applied in CZE whereby the UV signal is determined by Beer-Lambert law [6]:

$$A = \epsilon c l \tag{1}$$

In this formula A is the absorbance (in absorbance units, AU), ϵ is the molar absorptivity ($dm^3 mol^{-1} cm^{-1}$), c (M) is the concentration of the UV absorbing species, and l is the optical path length (cm). Because the optical path length, approximately the size of the diameter of the capillary tube, is very short, rather high detection limits are the result compared with HPLC techniques, for instance where much longer optical path lengths are generally applied. For a 50 μm capillary tube, a concentration of approximately 10^{-5} M of an analyte is required at the detector to obtain a UV signal of 5×10^{-4} AU if the molar absorptivity is 10^4 $dm^3 mol^{-1} cm^{-1}$. In chromatographic

Correspondence: Prof. Petr Boček, Institute of Analytical Chemistry, Academy of Sciences of the Czech Republic, Véveří 97, 611 42 Brno, Czech Republic
E-mail: bocek@iach.cz
Fax: +420-5-41212113

Abbreviations: KRF, Kohlrausch regulation function; **MBE**, moving boundary electrophoresis; **ZE**, zone electrophoresis

techniques, the concentration of sample analytes generally decreases continuously during the separation process due to diverse dispersion processes; on the other hand, in electrophoretic processes, fortunately, sample analytes are generally concentrated when the sample analytes are introduced at low ionic strength, the so-called "intrinsic sample stacking procedure", and many other stacking procedures can also be applied. Otherwise, the high concentration limit at the detector would result in a considerably higher concentration limit in the sample solution. For the determination of components present in samples at low concentrations it is therefore possible to apply large injection volumes in order to introduce a detectable amount of the analytes; in order to obtain a good resolution and detectable concentration, the sample analytes can be concentrated in narrow bands by a stacking procedure. For the components to ultimately migrate in a CZE manner, they must destack; generally, the stacking and destacking procedures start simultaneously during the electrophoretic process.

As many authors have already made contributions in the field of sample stacking and discussed several facets and phenomena thereof, it is difficult to give a complete survey of all the work done. However, in many studies where stacking and destacking are observed and utilized, the interpretation of the process is missing or is incomplete. In this article we describe some basic concepts of electrophoresis and sample stacking. We describe basic laws, regulating the electrophoretic processes and the migration behavior of the boundaries and their properties. After a short description of the main techniques in electrophoresis we discuss how these rechniques are reflected in sample stacking. Then we describe the principles of sample stacking and diverse stacking procedures. Moreover, we show that sample stacking has limits that are connected to the decrease in resolution and uncontrolled migrating pH shifts, which may strongly affect the separation process itself, followed by a related discussion.

2 Basic concepts in electrophoretic processes

Electrophoretic processes are regulated by some basic physicochemical laws and they can be described by a set of rather simple basic equations. Besides all equilibrium constants and the mass balances for all ionic species, the electroneutrality equation, modified Ohms law, and Kohlrausch regulation function (KRF) are important. Several mathematical models that solve these basic equations are described [7–14] and, based on these models, simulation programs are presented. Also, many phenomena occurring during electrophoretic processes and diverse aspects of electrophoretic separations can be explained

applying these models. For electrophoretic systems consisting of strong electrolytes in the steady state, the related equations are algebraic ones and we will use, among others, the modified Ohm's law and KRF to understand the electrophoretic phenomena.

2.1 Modified Ohm's law

The modified OhmM law prescribes that

$$E\sigma = j \tag{2}$$

whereby E represents the electric field strength (Vm^{-1}), σ is the specific zone conductivity ($\Omega^{-1}m^{-1}$) and j is the current density (Am^{-2}). For a separation capillary with a constant open hole cross section, the current density j is constant for the whole electrophoretic system at each moment and this law indicates that the local electric field strength is higher in a zone with a lower local zone conductivity. Differences in electric field strengths in the different sample zones are of practical significance and they are responsible for several phenomena in electrophoretic process such as the fronting/tailing character of peaks and zone broadening. For the specific zone conductivity, σ can be written:

$$\sigma = \sum_i c|m_i|F \tag{3}$$

whereby c and $|m|$ refer to the concentrations and absolute values of the mobilities of all ionic species and F is the Faraday constant.

As already stated the specific conductivity σ is a very important quantity. If we know the composition of a zone formed inside the capillary, the σ can be calculated with Eq. (3). This equation is also used in the computer simulations of electrophoretic patterns. Applying Eq. (3), calculating σ for a 0.01 M solution of KCl at 25°C, *i.e.*, a concentration of 10 molm^{-3}:

$$\sigma = 10 \ mol \cdot m^{-3} \times 96497 \ Cmol^{-1} \times (71.34 + 74.17) \times 10^{-9}$$
$$m^{-2}V^{-1}S^{-1} \tag{4}$$

$$= 0.14 \ \Omega^{-1}m^{-1}$$

The ionic mobilities are calculated from the ionic mobilities at infinite dilution according to Bennewitz, Wagner and Kuchler [15]:

$$m^c = m^o - (0.229 \ m^o + 31.2 \times 10^{-9}) \sqrt{C} \tag{5}$$

This relationship can be applied to diluted solutions and the resulting specific conductivity of the KCl solution is comparable with values at 25°C.

2.2 KRF

The KRF [16] prescribes that its numerical value ω is locally invariant in time and is defined as:

$$\omega = \sum_i \frac{c_i}{|m_i|} \tag{6}$$

assuming only the presence of fully ionized monovalent ionic constituents. The c_i and m_i refer to the concentrations and actual mobilities of all ionic species. This function says that electrophoretic processes aare regulated by the initial conditions. If the KRF has a certain value at a given point along the migration path (along the capillary tube) prior to the start of the electric current this value is constant there. That is, if the capillary is filled with 0.003 M KCl, then the value of ω is

$$\omega = \frac{c_K}{m_K} + \frac{c_{cl}}{m_{cl}} = \frac{3}{76.2 \times 10^{-9}} + \frac{3}{79.1 \times 10^{-9}}$$

$$= 7.73 \times 10^7 \, \text{molVsm}^{-5} \tag{7}$$

If during the electrophoretic process a pure sodium zone migrates into the capillary, substituting for K^+, the concentration of Na^+ adjusts automatically to the value of 0.00242 M NaCl corresponding to the same value of KRF (in the calculation the ionic mobilities at infinite dilution are used).

In order to demonstrate some consequences of KRF, simulations will be used that are calculated as described earlier [17], because in practice it is difficult or even impossible to measure single concentrations of the ionic species present in an electrophoretic system. The simulations must be seen as "drawings true to nature" and are carried out by applying a separation capillary of ID 200 μm and constant current of 20 μA. In all simulations, cationic concentration profiles are given as functions of their position in the separation capillary. All cations migrate from left to right in the presented figures with constant mobility. In Table 1, the pK values and mobilities at infinite dilution are given for all ionic species used in the experiments and simulations. In Fig. 1 the cationic concentration profiles are given at four different times for an electrophoretic process of a sample of 0.003 M KCl, applying a background electrolyte (BGE) consisting of 0.01 M NaCl. We arbitrarily applied the ionic mobilities at infinite dilution in the simulation. Figure 1a states the initial concentration profiles. The original K^+ sample solution is 0.003 M (solid line) and the NaCl concentration is 0.01 M (dashed line). After a short time (Fig. 1b) the K^+ is partially replaced by Na^+ and according to the KRF, the Na^+ concentration is adapted to the KRF at the place of the sam-

Table 1. Mobilities at infinite dilution, m ($10^{-9}\text{m}^2\text{V}^{-1}\text{s}^{-1}$), and pK values for ionic species used in the simulations and experiments

Ionic species	m	pK
Acetic acid	−42.4	4.76
Hydrochloric acid	−79.1	−2.0
Histidine	29.5	6.0
Imidazole	52.0	7.15
Lithium	40.1	14.0
Sodium	51.9	14.0
Potassium	76.2	14.0
Tris	29.5	8.1

pling zone and is somewhat lower (0.00242 M) than that of K^+ because its mobility is lower. The sample cation is partially concentrated beyond the original boundary between the sampling zone and BGE owing to the adaptation to the ω value of the BGE. In Fig. 1c all K^+ ions are concentrated and, according to the KRF, the concentration in the K^+ zone is higher than 0.01 M whereas in the original sample zone the BGE NaCl is present at the ω value of the original K^+ zone, regardless of some anomalies present at the boundaries by the simulation. In Fig. 1d we see that when the K^+ zone has passed, the original Na^+ zone is restored at the original concentration of 0.01 M. Consequences of the KRF are that (i) if a sample solution is introduced with an ω value different from that of the BGE, this deviating ω value remains valid at the spot of injection during the whole experiment, (ii) due to the adaptation to the KRF diluted sample zones will be concentrated and stacked, and (iii) if the ionic species of the BGE penetrate into the sample zone and begin to appear there, the destacking process starts, *i.e.*, the sample zone is going to migrate in a zone electrophoretic way. Further on in the simulations, moving and stationary (nonmoving) boundaries are visible (see Section 2.3). The BGE zone, which is present at the spot where the original sample was injected, shows stationary boundaries. Of course, when an EOF is present, the injection spot migrates with the velocity of the EOF through the capillary; for this reason the EOF can be observed as peaks or dips in the baseline. If this ω value is lower than that of the BGE a "waterdip" is visible; if the ω value is higher, a peak can be observed. The boundaries around the sample zones are moving and can either be electrophoretically stabilized or not. The migrating potassium zone is a moving zone with a stabilized rear boundary, whereas the boundary at the front has a nonstabilized fronting character, which will be discussed in Section 2.3. In Fig. 2, in analogy and in contrast with Fig. 1, the concentration profiles are given for a sample of 0.003 M NaCl (solid line) applying a BGE of 0.01 M KCl (dashed line). In Fig. 2a the initial concentration profiles are given. After some time (Fig. 2b)

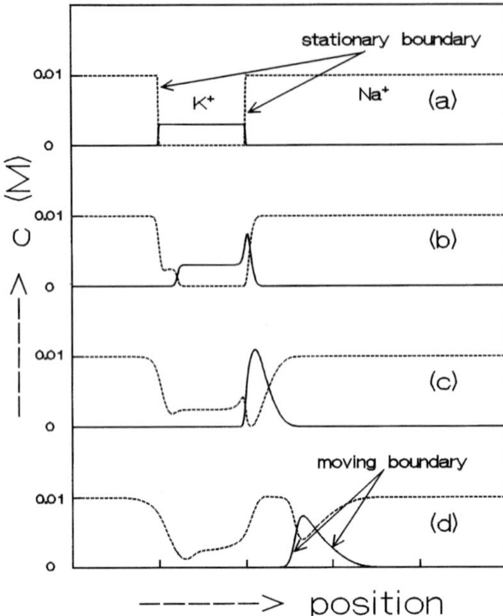

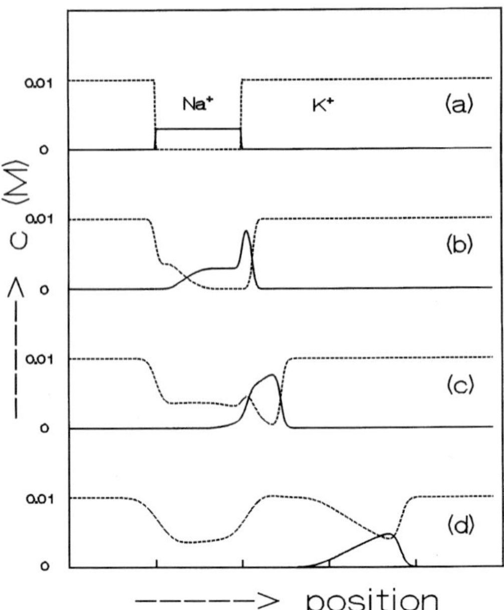

Figure 1. Cationic concentration profiles for an electrophoretic process of a sample of 0.003 M KCl (K⁺, solid line) applying a BGE of 0.01 M NaCl (Na⁺, dashed line). (a) Original concentration profile prior to the electrophoretic process. (b) After a short time the K⁺ zone is partially replaced by Na⁺ and K⁺ is concentrated beyond the sampling zone. (c) The K⁺ is concentrated completely. (d) K⁺ migrates in ZE mode. When the K⁺ zone has passed, the BGE zone of Na⁺ is restored.

Figure 2. Cationic concentration profiles for an electrophoretic process of a sample of 0.003 M NaCl (Na⁺, solid line) applying a BGE of 0.01 M KCl (K⁺, dashed line). (a) Original concentration profile prior to the electrophoretic process. (b) After a short time the Na⁺ zone is partially replaced by K⁺ and Na⁺ is concentrated beyond the sampling zone. (c) The Na⁺ is concentrated completely. (d) Na⁺ migrates in ZE mode. When the Na⁺ zone has passed, the BGE zone of K⁺ is restored.

the Na⁺ is replaced by K⁺ in the original sample zone, whereby the K⁺ concentration is higher than 0.003 M (Fig. 2c) and the maximal concentration of Na⁺ beyond the sampling zone is lower than 0.01 M according to the KRF. Figure 2d shows Na⁺ migrating in a zone electrophoretic (ZE) mode, whereby the front is stabilized and the rear side is diffuse and tailing. This is in contrast to Fig. 1. The behavior of boundaries will be discussed in Section 2.3.

2.3 Zone boundaries and their properties

In electrophoretic processes, several kinds of zone boundaries can be present, all with a different character and different properties. A main distinction can be made between stationary (nonmoving) boundaries and moving and stabilized between (steady-state) and nonstabilized (diffuse) moving electrophoretic boundaries. The latter can be fronting or tailing.

2.3.1 Stationary boundaries and moving boundaries

If the same ionic species are present on both sides of a boundary, although at different concentrations, and if the flow rates of mass across virtual planes before and behind the boundary are the same for all ionic species, the boundary does not migrate and we speak of a stationary (nonmoving) boundary. The flow rate of mass F, over a cross section of the separation capillary, is

$$F = vAc = mEAc \qquad [\text{mol·s}^{-1}] \qquad (8)$$

where A represents the cross sectional area.

The simplest case of such a stationary boundary is a binary BGE, consisting of a single cation and anion, present on both sides of the boundary. If the ionic species has constant mobilities on both sides of the boundary, the

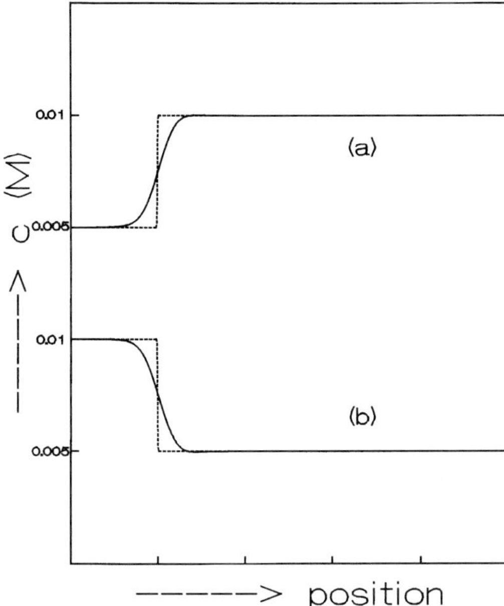

Figure 3. Cationic concentration profile for a concentration boundary between (a) a terminating electrolyte, 0.005 M KCl, and a leading electrolyte, 0.01 M KCl, and (b) a terminating electrolyte, 0.01 M KCl, and a leading electrolyte, 0.005 M KCl. The dashed lines represent the original concentration profiles and the solid line the concentration profiles after passage of an electric current.

product Ec is constant, combining Ohm's law and Eq. (3), and according to this the flow rates of mass transport on both sides of the boundaries are always equal, *i.e.*, the boundary is stationary. To demonstrate this, the calculated cationic concentration profiles are given (Fig. 3) for a boundary between (a) a terminating electrolyte, 0.005 M KCl, and a leading electrolyte, 0.01 M KCl, and (b) a terminating electrolyte, 0.01 M KCl, and a leading electrolyte, 0.005 M KCl. The dashed lines represent the original concentration profiles and the solid lines the concentration profiles after some time. Stationary boundaries are obtained, although boundary broadening occurs, due to numerical dispersion, mimicking the diffusion in real experiments. These boundaries are often called concentration boundaries or dilution boundaries.

As stated before, information of the initial conditions prior to the start of the electric current is kept in the form of the ω value at all points along the migration path during the whole electrophoretic process. That is, once a certain discontinuity of a given length is created in the composition of the electrolyte at a given location in the capillary, *e.g.*,

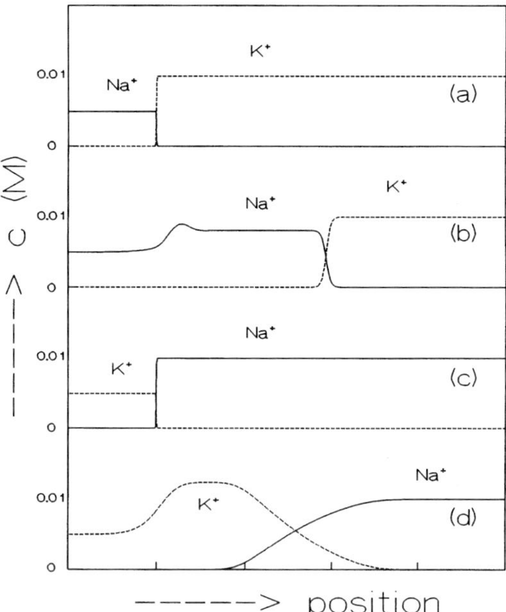

Figure 4. Cationic concentration profiles for a boundary between (a) a terminating electrolyte, 0.005 M NaCl, and a leading electrolyte, 0.01 M KCl, and (c) a terminating electrolyte, 0.005 M KCl, and a leading electrolyte, 0.01 M NaCl, and (b) and (d) the concentration profiles after passage of an electric current. Solid lines for Na$^+$ and dashed lines for K$^+$. For further information see Section 2.3.1.

at the spot of the original zone, this discontinuity survives at this location throughout the whole experiment and at this spot the BGE will be adapted to the original local ω value. A boundary is created whereby the same ionic species are present on both sides of this boundary, although at different concentration. As an example, Fig. 4 shows the cationic concentration profiles for a binary BGE present on both sides of a boundary, with different cations. Figure 4a shows the initial concentration profile for a boundary between a terminating electrolyte, 0.005 M NaCl, and a leading electrolyte, 0.01 M KCl, prior to the start of the electric current. Figure 4b shows the concentration profiles after some time of electrophoresis after the voltage is applied. Because all cations migrate through the capillary tube the Na/K boundary is a moving boundary. At the original spot a concentration boundary arose between 0.005 M NaCl and *ca.* 0.008 M NaCl. The latter concentration is the concentration of NaCl adapted to the original 0.01 M KCl solution according to the KRF. This concentration boundary is a stationary boundary. Figure 4c shows an initial boundary between a terminating electrolyte, 0.005 M KCl, and a leading electrolyte, 0.01 M

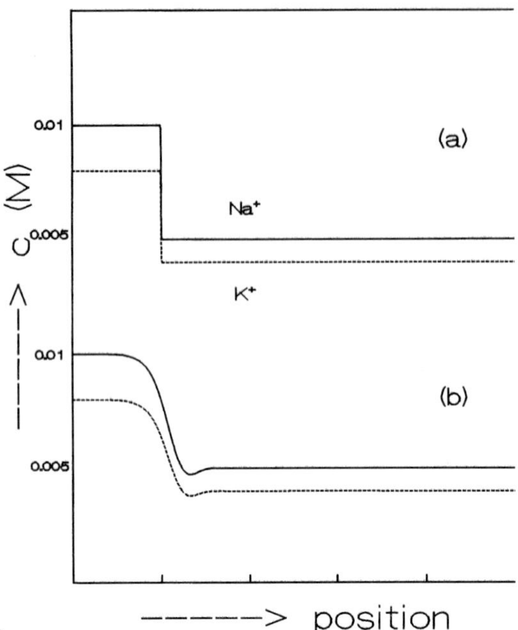

Figure 5. Cationic concentration profiles for a boundary between (a) a terminating mixture of 0.01 M NaCl and 0.008 M KCl and a leading mixture of 0.005 M NaCl and 0.004 M KCl, and (b) concentration profiles after some time. Solid lines for Na⁺ and dashed lines for K⁺. For further information see Section 2.3.1.

NaCl. When the voltage is applied, after some time, a concentration boundary between 0.005 M KCl and 0.012 M KCl is created, whereas the K/Na boundary is a moving boundary (Fig. 4d). Note the differences between the moving boundaries in Figs. 4b and d. The Na/K boundary is stabilized because the mobility of the leading ion K⁺ is larger than that of the terminating ion Na⁺. The K/Na boundary is diffuse and not stabilized because the leading ion Na⁺ has a smaller mobility than the terminating ion K⁺.

Matters become more complicated when several cationic and anionic species are present on both sides of a boundary. In that case several parameters decide whether stationary or moving boundaries are present. If the ratios of the concentrations on both sides of the boundary are equal for all ionic species, a stationary boundary is the result; otherwise, moving boundaries can also be present in the system. Different cases are considered (Figs. 5 and 6) to show that not only the mobilites of the ionic species but also their concentrations play an important part in the migration behavior. Figure 5 shows the cationic concentration profiles for (a) the original boundary between a ter-

minating mixture of 0.01 M NaCl and 0.008 M KCl and a leading mixture of 0.005 M NaCl and 0.004 M KCl, and (b) the profiles after passage of an electric current. It can be seen clearly that if more ionic species are present on both sides of the boundary and if the ratios of the concentrations on both sides of the boundary for all ionic species are equal, a stationary concentration boundary exists. If, however, the ratios are not constant, then different electrophoretic processes can be the result. In Fig. 6A the cationic concentration profiles after electric current passage are given for boundaries between an originally leading electrolyte, 0.01 M KCl, and terminating mixtures of (a) 0.01 M NaCl and 0.005 M KCl, (b) 0.01 M NaCl and 0.01 M KCl, and (c) 0.01 M NaCl and 0.02 M KCl. In all cases there is a stationary boundary at the place of the original boundary, and behind the leading zone there is a moving boundary of a mixture of Na⁺ and K⁺. The compositions of the moving zones and their velocities are different. All boundaries are stabilized. In Fig. 6B the cationic concentration profiles after the passage of electric current are given for boundaries between originally (a) a leading mixture of 0.01 M KCl and 0.002 M NaCl and a terminating mixture of 0.005 M KCl and 0.01 M NaCl, and (b) a leading mixture of 0.002 M KCl and 0.01 M NaCl and a terminating mixture 0.01 M KCl and 0.002 M NaCl. In both cases a stationary boundary exists at the location of the original boundary. In (a) there is a stabilized moving boundary behind the leading zone containing Na⁺ and K⁺ whereas in (b) there is a nonstabilized moving boundary behind the leading zone containing Na⁺ and K⁺. It is obvious that the concept of moving boundaries and stationary boundaries in electrophoretic processes is essential in any further theoretical and practical considerations. The theoretical and practical investigation of these phenomena was already a hot research topic a half century ago [18–26].

2.3.2 Stabilized and nonstabilized moving boundaries

As described before, moving boundaries are often present in electrophoretic processes. Sometimes the same ionic species are present on both sides of the moving boundaries and sometimes a specific ionic species is only present on one side of the boundary. Such a moving boundary can either be stabilized or not. The question whether a boundary is stabilized or not is generally regulated by Ohm's law. If an ionic species leaves its zone and passes the boundary in the direction of migration, and if it reaches a spot with higher electric conductivity, *i.e.*, a locally lower *E*, its velocity decreases and will be overtaken by its own zone, resulting in a stabilized zone boundary. If it reaches a spot with higher *E*, its velocity increases and a nonstabilized, diffuse boundary is the result. Boundaries will be stabilized if the leading zone has a lower electric field strength than the terminating

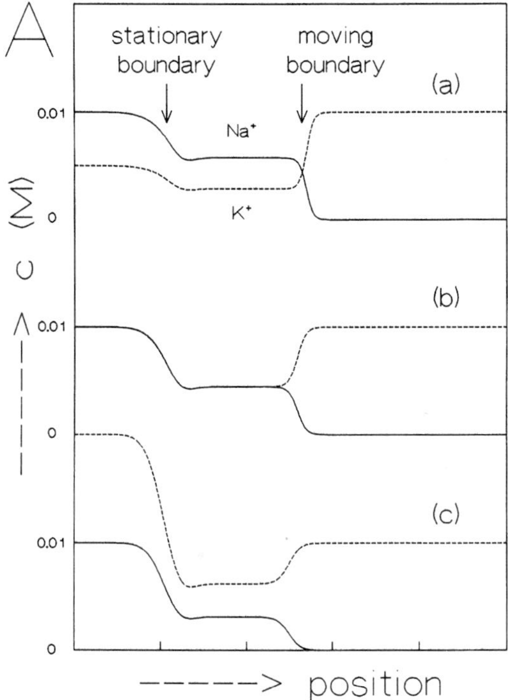

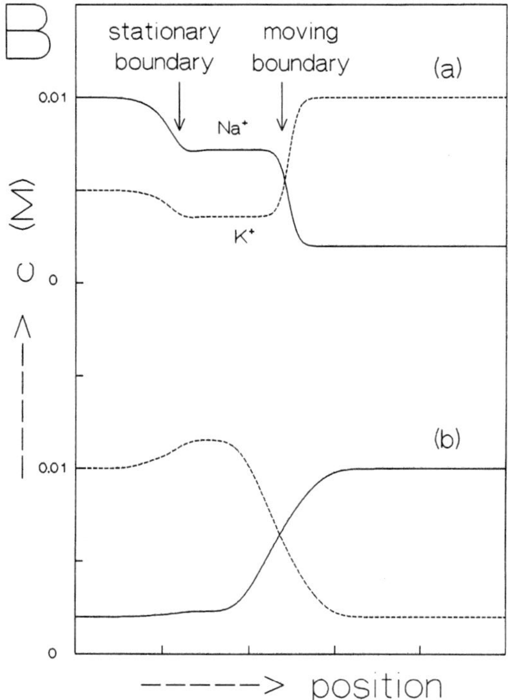

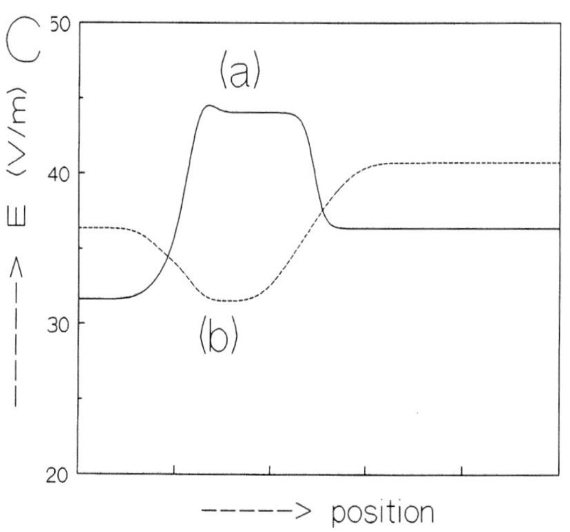

Figure 6. (A) Cationic concentration profiles after passage of an electric current across a boundary consisting of a leading electrolyte, 0.01 M KCl, and terminating mixtures of (a) 0.01 M NaCl and 0.005 M KCl, (b) 0.01 M NaCl and 0.01 KCl, and (c) 0.01 M NaCl and 0.02 M KCl. Solid lines for Na^+ and dashed lines for K^+. In all cases a concentration boundary is created at the place of the original boundary and a migrating moving boundary between a mixture of Na^+ and K^+ behind the leading electrolyte KCl. In the first part of (b) the dashed line of K^+ is covered by the solid line at Na^+. (B) (a) Terminating mixture of 0.01 M NaCl and 0.005 M KCl and leading mixture of 0.01 M KCl and 0.002 M NaCl; (b) terminating mixture of 0.01 M KCl and 0.002 M NaCl and a leading mixture of 0.002 M KCl and 0.01 M NaCl. Solid lines for Na^+ and dashed lines for K^+. (a) A stabilized moving boundary exists besides a stationary boundary at the original boundary, whereas in (b) a nonstabilized moving boundary exists. (C) Electric field profiles for the electrophoretic processes represented in (B). (a) The ratio $(c_{K+}/c_{Na+})_L$ in the leading electrolyte is larger than $(c_{K+}/c_{Na+})_T$ in the terminating electrolyte, resulting in a stabilized boundary. (b) The leading ratio is smaller than the terminating ratio, resulting in a nonstabilized diffuse boundary.

zone. To demonstrate this effect, Fig. 6C shows the electric field strength profiles for the cases of Fig. 6B: it is clear that a stabilized boundary is obtained for a decreasing E in the migration direction and a nonstabilized boundary for an increasing E. In the case of a stabilized boundary a self-restoring mechanism exists, often called the self-correcting or self-sharpening effect, which acts against diffusion [27–29]. Discussion of the constraint leading to stabilized electrophoretic boundaries, often called steady-state boundaries, are numerous. In 1980, for instance, Mikkers and Everaerts [30] described the concept of "in-stack" and "out-of-stack" in isotachophoretic systems and in 1985 Mosher and Thormann [31] gave a general description of conditions for steady-state boundaries. The theory clearly indicates that the general requirement for establishing steady-state migrating electrophoretic boundaries in strong electrolytic systems is a stepwise decrease in the intensity of the electric field strength across the boundary in the direction of migration. If the boundary exists between two different ionic species, this requirement is met if the mobility of the leading ion, *i.e.*, migrating in front of the boundary in the migration direction, is larger than that of the terminating ion. If both ionic species are present on both sides of the boundary, the concentration ratio between the ionic species with highest mobility *vs.* lowest mobility must be higher ahead of the moving boundary (Fig. 6Bb). In our case, where $m_K > m_{Na}$, it holds that

$$\left(\frac{c_K}{c_{Na}}\right)\text{ahead} > \left(\frac{c_K}{c_{Na}}\right)\text{behind} \tag{9}$$

If these requirements are not fulfilled, diffuse boundaries are the result.

The above discussions covered the simple cases where the mobilities of the ionic species are equal on both sides of the boundary. The cases where pH differences and changes in effective mobilities of weak electrolytes are taken into account are more complicated and it is not always possible to create a telling simple picture. In such cases even moving boundaries can be created if a BGE with different pH is introduced in a BGE. Hydrogen ions (H^+) act as second coion and instead of a concentration boundary, a special type of moving boundary migrates through the system with a specific mobility [32, 33]. Further, a specific pH change across a boundary can enforce the effective mobility in such as a way that this effect overrules the change in the electric field strength and creates a self-sharpening effect. Examples are the enforced migration in ITP [34, 35].

2.3.3 Electromigrational dispersion of the boundaries

As already described, stabilized boundaries represent situations in which electrophoretic self-sharpening effect acts against diffusion. When the boundaries are nonstabilized, it does not mean, however, that electrophoretic effects are absent. In this case, the electrophoretic migration contributes positively to the broadening of the migrating boundaries and brings additional dispersion. For example, an ionic species leaves its zone by diffusion across its front boundary in the direction of the migration, and if it reaches a spot with higher electric field strength, it is accelerated further and becomes more and more removed from its own zone. This is called electromigration dispersion. The theory herefore says that for simple strong electrolyte systems the magnitude of the electromigration dispersion is proportional to the difference in mobilities of an analyte and the coion of the BGE.

3 Electrophoretic modes

Electrophoresis is generally distinguished into four main techniques, *viz.*, moving boundary electrophoresis (MBE), isotachophoresis (ITP), capillary zone electrophoresis (CZE), and isoelectric focusing (IEF). The first three techniques will be discussed briefly, with special attention to the properties of the zone boundaries. IEF will not be discussed here; the interested reader is refered to Section 7 of [3].

3.1 MBE

As an example we consider MBE of cations (Fig. 7). The anode compartment is filled with the sample solution and the separation capillary and cathode compartment are filled with a leading electrolyte, the cation of which should have a mobility higher than that of any of the sample cations. Behind the leading zone, a series of mixed zones arises from the sample solution containing an ever-increasing number of cations with decreasing mobilities. From the KRF it can be concluded that for zones containing cations with lower mobilites, the concentrations decrease and the specific conductivity in these zones will be lower. According to modified Ohm's law this means that from the leading zone to the rear the electric field strengths increase and all zone boundaries are therefore stabilized moving boundaries and are sharp (see Section 2.3). Figure 7 shows an MBE procedure. In Fig. 7a the original situation prior to the electrophoretic process is given. The anode compartment is filled with the sample mixture consisting of 0.01 M TrisCl, LiCl, and NaCl. The cathode compartment and separation capillary are filled with the leading electrolyte, 0.01 M KCl. In MBE only the

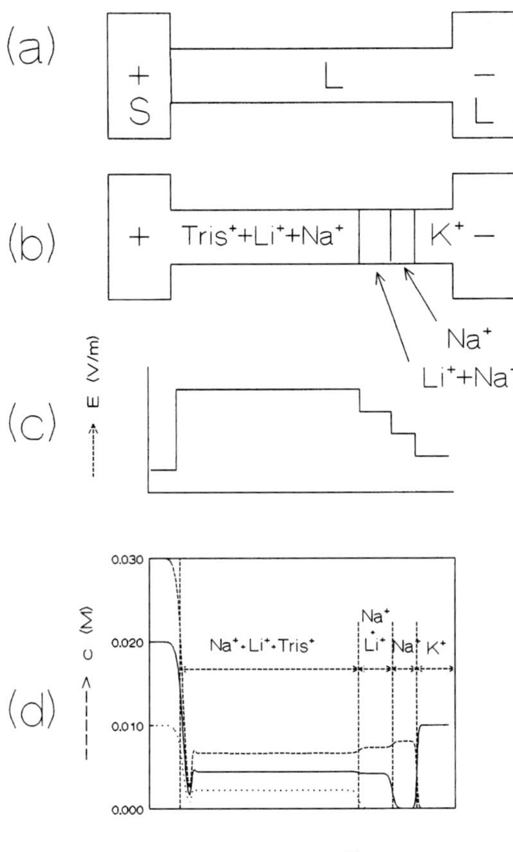

Figure 7. In MBE for a cationic mixture, the anode compartment is filled with the sample solution S whereas the separation and cathode compartment are filled with the leading electrolyte L. (a) Original situation. (b) After some time a series of mixed zones is formed for a sample mixture of Tris⁺, Li⁺, and Na⁺ applying a leading electrolyte of KCl. Only the sample cation Na⁺ with the highest mobility forms a pure zone. (c) Electric field strength profile over the separation capillary. (d) Simulated concentration profiles. For further information see Section 3.1.

sample ion with the highest mobility forms a pure zone behind the leading electrolyte; in Fig. 7b this is the Na⁺ zone. All other zones are mixed zones with an increasing number of sample ionic species. Behind the Na⁺ zone, the mixed zones contain the ions Na⁺ and Li⁺ as well as Tris⁺, Li⁺, and Na⁺, respectively. The original boundary between sample and leading electrolyte is a stationary boundary, and on both sides of the stationary boundary all sample ionic species are present although at different concentrations. According to Eq. (8) the ratio of the sam-

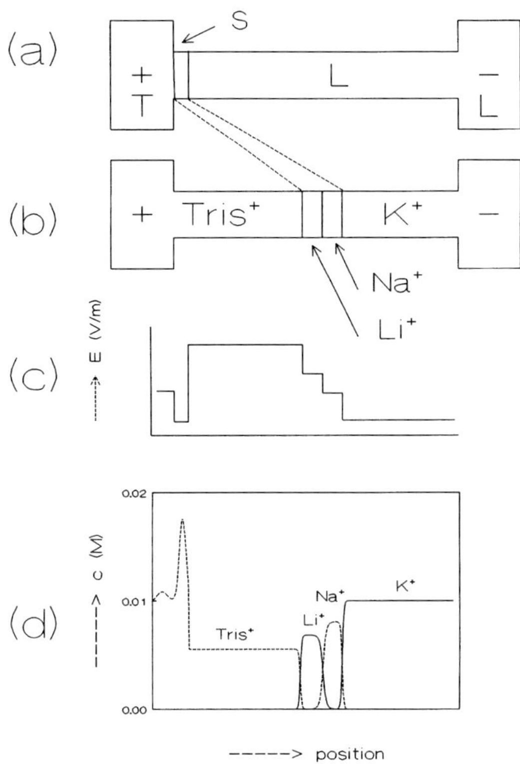

Figure 8. ITP of a cationic mixture. (a) The sample solution S is introduced between a terminating electrolyte T and a leading electrolyte L. (b) After some time the sample cationic species, here Li⁺ and Na⁺, are separated and migrate with equal speed between the leading ion K⁺ and terminating ion Tris⁺. (c) Electric field strength profile. (d) Simulated concentration profiles. For further information see Section 3.2.

ple concentrations on both sides of the stationary boundary is equal for all sample ionic species. The electric field strength profile is given in Fig. 7c and the simulation of the MBE process in Fig. 7d. The dotted line represents the Tris⁺ concentration, the solid line the sum of Tris⁺ and Li⁺ concentrations, and the dashed line the sum of Tris⁺, Li⁺, and Na⁺ concentrations. All moving boundaries are stabilized in MBE.

3.2 ITP

As an example of ITP [34,35], we consider the separation of cations again and we will use the same selection of cations (Fig. 8). The sample solution, S, is introduced between the so-called electrolyte L (placed in separation capillary and cathode compartment) and terminating elec-

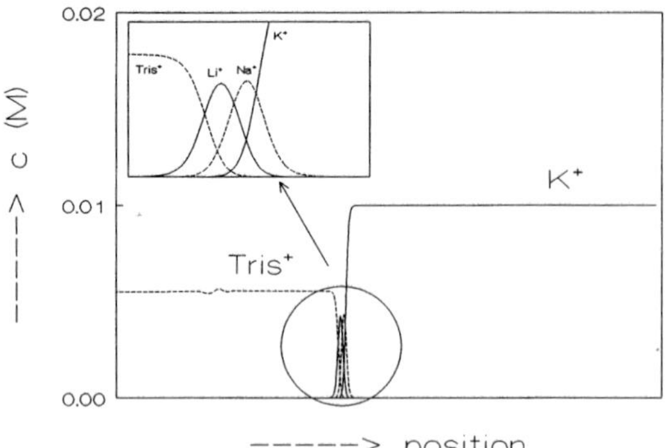

Figure 9. Simulated concentration profiles of an isotachophoretic separation of trace amounts of Li^+ and Na^+ applying a terminating ion, $Tris^+$, and a leading electrolyte, 0.01 M KCl. From the enlarged inset it can be seen that the sample cations migrate as very small zones. For further information see Section 3.2.

trolyte T (placed in anode compartment; Fig. 8a). The condition is that the mobility of the cation of the leading electrolyte is higher and the mobility of the cation of the terminating electrolyte is lower than those of any of the cations of the sample solution. Compared with MBE, a limited amount of sample is present through which a complete separation will occur after some time. In the so-called steady state all zones, containing only a single cation of the sample mixture, migrate with equal speed, close together through the capillary tube. Because all sample cations form separate zones, where they are present as the single cation, they are concentrated in order to fulfill the KRF. Because all cations are ordered according to decreasing mobilities, the electric field strengths increase from leading to terminating zone. If a cation stays behind, it comes into a higher electric field, migrates with a higher velocity and reaches its own zone. If a cation migrates into the foregoing zone, where the electric field strength is lower, its velocity decreases and it is overtaken by its proper zone. Therefore, all zone boundaries that are not only stabilized but also do not change with time are called steady-state boundaries.

In Fig. 8 an ITP separation is shown for a sample of 0.01 M Li^+ and Na^+ applying the leading electrolyte 0.01 M KCl and a terminator 0.01 M TrisCl. Figure 8a shows the original setup prior to the electrophoretic process. The anode compartment is filled with the terminator, and the separation capillary and cathode compartment are filled with the leading ion and the sample is introduced between leading and terminating electrolyte. In ITP all sample ions form pure zones behind the leading electrolyte (see Fig.

8b, the zones containing Na^+ and Li^+, respectively). The originall boundaries between sample and leading electrolyte and between sample and terminator are stationary boundaries, and on both sides of the stationay boundary all sample ionic species are present although at different concentration. All moving boundaries are steady-state boundaries. The electric field strength profile and the simulation of the ITP process are given in Fig. 8c and Fig. 8d, respectively. As can be seen from Fig. 8d, the concentrations of the analytes Na^+ and Li^+ in their ITP zone have fixed values and are adjusted to the KRF values of the leading electrolyte. Providing we inject double the amount of the sample, the adjusted concentrations are the same and therefore the length of the zones will be double. If we inject half the amount, the length will also be half. If we inject trace amounts which are not sufficient to reach the adjusted concentration, then we obtain narrow zones, sandwiched between the neighboring boundaries. It is important to know that even such zones migrate isotachophoretically and keep their shape like a real ITP zone, *i.e.*, boundaries are steady-state boundaries. As already mentioned, this is a very important practical isotachophoretic stacking of trace analytes. The simulated concentration profiles of an isotachophoretic separation of trace amounts of Li^+ and Na^+ applying a terminating ions, $Tris^+$, and a leading electrolyte, 0.01 M KCl, illustrate this phenomenon (see Fig. 9) and very small zones are the result. From the enlarged inset it can be seen that the sample cations migrate as very small zones. The simulation conditions for Figs. 8 and 9 were identical.

3.3 CZE

In CZE the whole system is filled with a BGE. Because the BGE carries the electric current, there is no need for the sample zones to migrate close together to provide the current transport. Therefore, all zones will be separated and will migrate with their specific velocities of $v = mE$, whereby the electric field strength E will be determined by the specific conductivity of the BGE. In all sample zones there is a small difference in composition, and therefore in specific conductivity, compared with that of the BGE. According to Ohm's law and the KRF the specific conductivity in zones containing a sample ion with a higher mobility than that of the coion of the BGE is higher than in the BGE and therefore the local E is lower in the zone and vice versa for zones with a sample ion with a lower mobility. This is shown in Fig. 10, where the specific conductivity profile and electric field profile are given for a sample zone in CZE (a) with the mobility higher sample ion than that of the coion of the BGE, and (b) with lower mobility. If the sample mobility is higher, also the specific conductivity in the sample zone is higher and, therefore, the electric

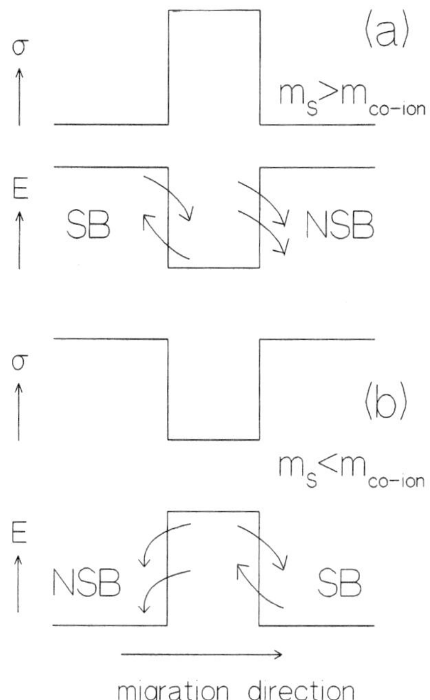

Figure 10. Stabilized (steady-state) boundaries (SB) show self-correcting properties; nonstabilized boundaries (NSB) are diffuse. In CZE, one boundary between the sample zones and the BGE is always stabilized and the other boundary is not stabilized, through which sample zones show electromigration dispersion. For further information see Section 3.3.

field strength is lower than that of the BGE. This means that if a sample ion penetrates into the BGE in front, it is under the influence of a higher E, attains a higher velocity, and will migrate in front of the sample zone. This zone boundary will be diffuse and forms a nonstabilized boundary (NSB). This sample peak will be fronting. If a sample ion stays behind, it is under the influence of a higher E, attains a higher velocity and returns to its own zone. The rear boundary is therefore a stabilized boundary (SB) and remains sharp. As a result, in zone electrophoresis one of the boundaries between sample zones and BGE is always stabilized and the other is nonstabilized, and therefore the peaks are triangled.

In Fig. 11 these properties of CZE are shown schematically. The original setup for CZE is given in Fig. 11a. The entire electrophoretic equipment is filled with the BGE and the sample solution is inserted in the BGE. After some time the sample ions are separated in a sequence of zones with decreasing mobilities (Fig. 11b). Figure 11c shows the electric field profile. If the mobility of the sam-

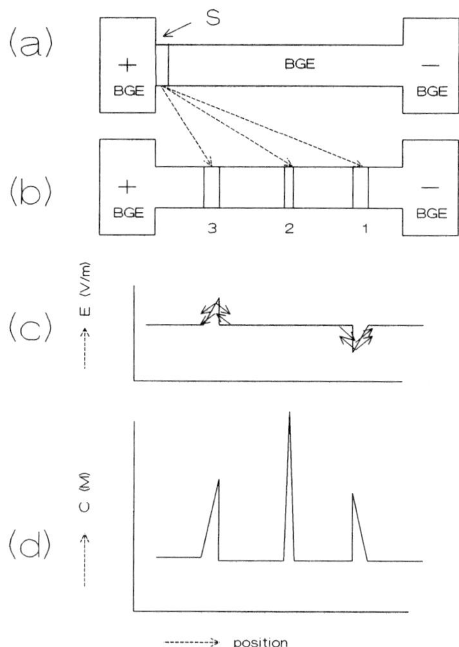

Figure 11. CZE of a cationic mixture. (a) Original setup. The whole equipment is filled with the BGE. The sample solution is introduced in this BGE. (b) After some time the sample species are separated. (c) Electric field strength profile. Sample zones with ionic species with higher mobilities than that of the coion of the BGE — component 1 — show a local lower E and are not stabilized at the front, resulting in (d) fronting peaks. Otherwise they are not stabilized at the rear side and are tailing — component 3.

ple ion is higher than that of the coion in the BGE — peak 1 — the local E is lower and the peak will be fronting with the stabilized rear side. If the sample mobility is lower — peak 3 — the local E is higher, resulting in a tailing peak with stabilized front. If the mobilities of sample ion and coion are equal there is no deviation in the E profile and there is no electrodispersion. Diffusion is the only peak broadening mechanism and the peak stays symmetric, narrow, and high. The difference between stabilized boundaries and steady-state boundaries can be summarized as follows. Both steady-state boundaries and stabilized boundaries show self-sharpening effects, *i.e.*, the step in electric field strengths acts against diffusion. The shape and the magnitude of the steady-state boundaries do not change with time as long as the voltage is applied, (see ITP). The magnitude of the stabilized boundaries do change with time, in such a way that the maximum concentration of the sample species becomes smaller and smaller.

4 Stacking and destacking

The preceding sections have shown that CZE zones containing sample ionic species with mobilities different from those of the coions of the BGE have stabilized boundaries on one side of the zone only, which will be extinguished in time. It is therefore important in CZE to concentrate the injected sample solution as much as possible at the beginning of the electrophoretic process, especially if large volumes of very diluted sample solutions are introduced. Procedures whereby sample ions are concentrated are called sample stacking. Many stacking procedures can be distinguished and some typical characteristics will be discussed.

4.1 Stacking procedures

Stacking of an analyte is only possible if the sample analyte to be stacked is present in the sample solution at a lower concentration than the BGE. If the sample analyte is present at a higher concentration it will always be diluted. Further, the velocity of that analyte must be smaller in the BGE in front of the sample zone (lower E). It will be clear that sample stacking will be simplest by introducing the sample at a lower ionic strength compared with the BGE. We call this procedure "intrinsic sample stacking" and this stacking will generally occur if $\sigma_{sample} <$ σ_{BGE}. If the mobility of the coion in the BGE is larger than that of the analyte, the front boundary is stabilized. This is called "front boundary stacking" or "stacking by the presence of a leader". If the mobility of the coion is smaller (Fig. 1), the rear side is stabilized. This procedure is called "rear boundary stacking" or "stacking by the presence of a terminator". The best stacking is to utilize both a stabilized front and rear boundary by addition of both a leader and terminator, which is called isotachophoretic stacking. In the isotachophoretic stacking even trace components of the sample can be stacked effectively between terminator and leader, regardless of whether the stacked trace component reaches the isotachophoretic concentration or not.

In the intrinsic sample stacking, we generally have either "front boundary stacking" or "rear boundary stacking". In order to optimize the stacking procedure, we can add a short zone in front of the sample, containing a leader, or a short zone behind the sample, containing a terminator, so that temporarily an extra "front" or "rear" boundary stacking is applied. This is called transient stacking. Of course, these leaders or terminators can also be added to the sample solution. The stacking capability of such a zone is limited. This procedure can be carried out in different ways, as will be discussed in Section 5.2.

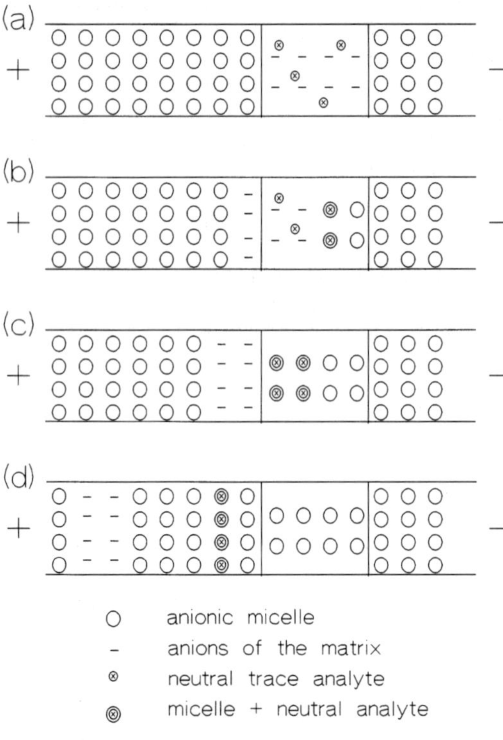

Figure 12. Sample stacking of neutral analytes in MEKC. (a) Original setup. (b) After some time the anionic micelles migrate to the anode, partially replace the anions of the sample solution and pick up the neutral analytes in the sample. (c) All anions of the sample are replaced and all neutral analytes are caught. (d) Finally the neutral analyte is concentrated depending on the distribution coefficient. For further information see Section 4.1.

Recently a totally different principle of stacking was developed in micellar electrokinetic capillary chromatography (MEKC), the so-called sweeping. The sweeping procedure [36–38] enables one to stack neutral analytes prior to MEKC analysis and is based on (i) formation of a zone of concentrated micellar agent by electrophoretic stacking and (ii) movement of the concentrated micellar zone across the zone of the originally injected sample, whereas the micellar zone accumulates the neutral analytes like a trap. In Fig. 12 the sweeping procedure is shown schematically. In Fig. 12a the original setup prior to the electrophoretic process is given. A sample containing a trace of a neutral analyte is placed between a BGE containing anionic micelles. After some time (Fig. 12b), the anionic micelles partially replace the anions of the sample solution, with a concentration according to the KRF, and pick up the neutral analytes in the sample. If all anions of the matrix are replaced they migrate in a ZE mode in the BGE, whereas all neutral particles are picked up by the micelles. For high distribution coefficients, the neutral analytes are concentrated and they move with the micelles to the anode. In this consideration the EOF is not taken into account, but if an EOF is present with a mobility higher than that of the micelles, the neutral analyte and the micelles are swept towards the cathode. A more favorable sample stacking of the neutral analyte seems to be obtained by adding an anion at high concentration to the sample [37]. In that case the concentration of the micelles in the sampling zone is adapted to the high concentration of the anions in the matrix, resulting in optimum stacking.

4.2 Destacking mechanism

Stacking of sample components is an important step in the analysis of analytes; however, it is only half of the whole story. In order to obtain a ZE separation process, the analytes must be destacked, *i.e.*, they must obtain mutually different velocities. Unless an ITP process is wanted by utilizing both a leader and terminator, the destacking process begins automatically during the stacking procedure, because the coions of the BGE will mix with the sample ions, starting at the non-stabilized boundary of a sample zone. This means that when stacking by the presence of a leader starts at the front of a zone (front boundary stacking), the destacking procedure starts simultaneously at the rear boundary where very mobile coions penetrate into the sample zone and mix with it. Electromigration dispersion of the rear boundary is thus strongly enhanced and, finally, the dispersion which had started at the rear boundary prevails in the whole sample zone – it is destacked (Fig. 2). In analogy, when stacking by the presence of a terminator starts at the rear boundary of a sample zone (rear boundary stacking), the destacking starts simultaneously at the front where the mobile sample ions penetrate into the BGE and they are more and more accelerated there by the higher electric field strength. The final result is similar again, the dispersion of this boundary prevailing in the whole zone – it is destacked (Fig. 1). After some time the analytes migrate in ZE mode. The destacking process of course needs a certain time and migration path in order to be finished and this depends on the way the stacking was performed and on the amount and concentration of the sample. In the case of transient stacking the properties and amount of the stacker also play a significant role. If the difference between the mobilities of the stacker and analyte is small and if a long stacker zone is applied, the time for destacking is larger. In that case the stacking time is also long. The extent of stacking procedures may not be

beyond a certain limit because the destacking must be finished before analyte zones reach the detector.

5 Practical procedures

5.1 Intrinsic sample stacking

5.1.1 General

This type of stacking is the simplest of all and it means that the sample solution is introduced at low ionic strength compared with that of the BGE. For a given sample a BGE can be chosen at higher ionic strength or for a given BGE the sample solution can be diluted before use. All sample components follow the situation as already described in Figs. 1 or 2, used for the explanation of the consequences of the KRF. The main advantage of this stacking procedure is its simplicity. There are, however, limitations in the application of the intrinsic sample stacking, which will be described; similar limitations can also play a role in other stacking procedures.

5.1.2 Limitations in sample stacking

If components to be analyzed are present at low concentrations in a sample matrix, large injection volumes must be introduced. To obtain a high resolution and adequate detection, the sample components have to be concentrated by a sample stacking procedure prior to the separation procedure. The resolution is given by

$$R_\mathrm{s} = \frac{\Delta t}{2(S_1 + S_2)} \tag{10}$$

where S_1 and S_2 represent the standard deviations of the peaks of the analytes to be separated and Δt is the difference between the migration times of the components. This formula indicates that for a good resolution a large difference between the migration times is favorable and that peak broadening works against it. Injecting large volumes at low ionic strength, some effects can have a bad influence on the resolution, due to the resulting nonuniform electric field strength distribution and an increased electroosmotic flow (EOF) [39–41]. Sometimes even pH shifts can occur, affecting the migration behavior [42].

5.1.3 Nonuniform electric field strength

To discuss the influence of these effects [39], a sample stacking procedure of cations is shown in Fig. 13. We assume that the analytes are dissolved in diluted BGE and that the ionic strengths of the analytes can be neglected compared with that of it. At the beginning (see Fig. 13a), the separation capillary is filled to a length *l* with the sample solution S consisting of the analytes and

Electrophoresis 2000, *21*, 2747–2767

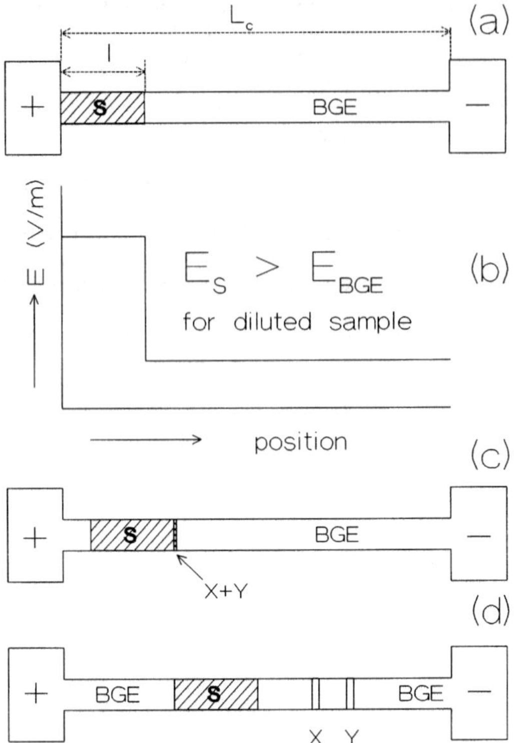

Figure 13. Effect of sample stacking. (a) Original situation with sampling zone S (sample components dissolved diluted BGE) behind BGE (L_C, capillary length; *l*, length-injected sampling zone (S)). (b) Electric field strength *E* distribution over the separation capillary. The E_S is much larger than E_{BGE} due to the higher local resistivity. (c) After a short stacking time, the cationic sample components are concentrated in a narrow zone between BGE and sampling zone S. The sampling zone S and BGE move by the velocity of the EOF. (d) The separation of the components X and Y is executed in BGE.

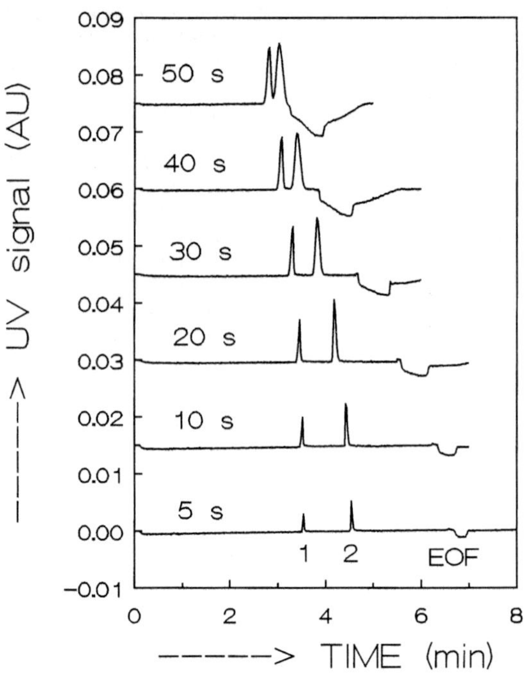

Figure 14. Measured electropherograms for the separation of a 0.0001 M solution of (1) imidazole and (2) histidine in water adjusted to pH 4.8 by adding acetic acid for different injection times. The BGE was 0.01 M Tris-acetate at pH 4.8. Decreasing resolution is obtained for longer injection times. (Beckman P/ACE System 2000; applied voltage, 10 kV; 214 nm; capillary length, 46.7 cm; distance between injection-detection, 40 cm; temperature 25°C).

diluted BGE at low ionic strength. The length of the capillary is L_C. When the voltage is applied, the electric field strength is distributed nonuniformly (Fig. 13b), where E_S, E_{BGE}, ρ_S and ρ_{BGE} are the electric field strengths [Vm^{-1}] and the resistivities [Ωm] of the diluted BGE in the sampling zone S and of the BGE in the capillary, respectively. If the capillary is filled for the fraction *x* (= l/L_C) during the sampling procedure, the electric field strengths in the zone S and BGE can be derived, respectively:

$$E_s = \frac{\rho_s V_{tot}}{[x\rho_s + (1-x)\rho_{BGE}]L_c} \qquad (0 < x < 1) \qquad (11)$$

$$E_{BGE} = \frac{\rho_{BGE} V_{tot}}{[x\rho_s + (1-x)\rho_{BGE}]L_c} \qquad (0 < x < 1) \qquad (12)$$

whereby V_{tot} it the total applied voltage [V] over the capillary with length L_C. The ratio of the concentrations of the BGE and diluted BGE S is known and denoted as the dilution factor *K*. The ratios E_S/E_{BGE} and ρ_S/ρ_{BGE} are approximately equal to the dilution factor *K*. From this formula it can be concluded that for large values of *x* and ρ_S, the electric field strength in the sampling zone E_S is considerably larger than the E_{BGE} as shown in Fig. 13b. Cationic sample components will migrate quickly out of the sampling zone and stack in front of sampling zone S, between this zone and BGE (see Fig. 13c). Then the components will be destacked and separated according to the zone

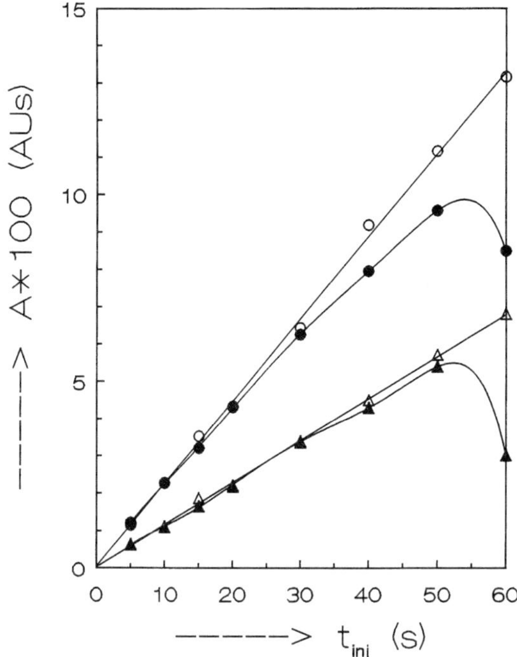

Figure 15. Measured relationships between peak area and pressure injection time for (▲) imidazole and (●) histidine dissolved in water (0.0001 M) and (△) imidazole and (○) histidine dissolved in the BGE Tris-acetate at pH 4.8. For apparatus see Fig. 14.

electrophoretic principle (Fig. 13d). What is very important and should be stressed here is that the separation process takes place for the largest part in the BGE at a low electric field strength, E_{BGE}.

EOF

If the capillary tube is filled with two different electrolyte solution, S and BGE, with mobilities of the EOF $[m^2V^{-1}s^{-1}]$ of $m_{EOF,S}$ and $m_{EOF,BGE}$, the overall velocity of the EOF, $v_{EOF,tot}$, can be derived [40]:

$$V_{EOF,tot} = X V_{EOF,S} + (1-x)\, v_{EOF,BGE} = x m_{EOF,S} E_S + (1-x)\, m_{EOF,BGE} E_{BGE} \tag{13}$$

Since the m_{EOF} is generally higher at lower ionic strength, the $m_{EOF,S}$ will be higher than the $m_{EOF,BGE}$. By applying Eqs. (11)-(13) it can be concluded that for increasing values of x and ρ_S, the $v_{EOF,tot}$ is determined to a large extent by the properties of the diluted BGE in the sampling zone S and is increased considerably. For large injection volumes, the effective separation length and the effective E_{BGE} decrease. Moreover, the increased overall EOF

results in smaller effective separation times and the resolution will decrease dramatically. In order to demonstrate these effects, the electropherograms are given (Fig. 14) for the separation of a solution of 0.0001 M histidine and imidazole adjusted to pH 4.8 by adding acetic acid, using several different pressure injection times. From Fig. 14 it is obvious that for larger injection times, a strong decrease in separation power occurs. The increase in velocity of the EOF due to injecting large volumes of diluted samples also has another effect which needs to be mentioned here. At large EOF, the sample zones are pushed along the detector with high speed, which results in smaller peak areas. Due to this effect the calibration graphs are no longer linear. This effect (Fig. 15) gives the measured relationships between peak area and pressure injection time for separations of 0.0001 M solutions of (▲) imidazole and (●) histidine dissolved in water, and (△) imidazole and (○) histidine dissolved in the BGE Tris-acetate at pH 4.8, applying a constant voltage of 10 kV. For all sample solutions dissolved in the BGE, all peaks were rectangular owing to the absence of any concentrating effect. Because the migration times were fairly constant, temporal peak areas are handled in the calibration graph and linear relationships are obtained. As can be seen for large sample volumes at low ionic strength, no linear relationships could be obtained for sample analytes dissolved in water, especially for histidine for which the shift in migration time, owing to an increased EOF, is largest (see Fig. 14).

5.2 On-line combination ITP + CZE

The application of ITP as stacking procedure prior to the CZE analysis is the most effective method. Even large volumes of sample can be preconcentrated on the ITP step and trace amounts of analytes can be stacked in a very sharp and short ITP zone between the leading and terminating electrolyte. Such a stack of sample zones represents an ideal sampling situation for zone electrophoresis. The full advantage of this combination may be utilized by using the instrumentation equipped with the column-coupling or column-switching system [43]. The procedure is in fact a two-stage analysis and its description enables one to show and explain nearly all important features of other stacking and destacking procedures also.

The instrumentation consists of two capillaries, connected *via* a T piece which enables one to carry out a first run ITP in the first capillary, which serves as an efficient preseparation and concentration (stacking) stage. As second stage, it follows the on-line transfer of the sample cut into the second capillary where analytical zone electrophoresis proceeds. The online transfer of the sample from the first capillary into the second is always accompanied by

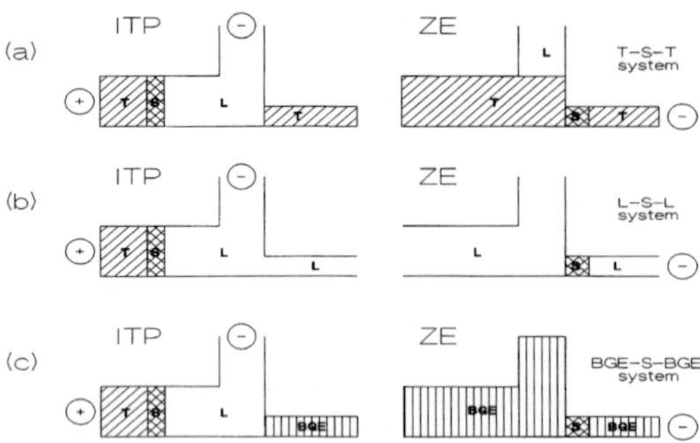

Figure 16. Possible online combinations of ITP and CZE. The left panels show the situation in the preseparation capillary at the end of the ITP separation. The right panels show the situation in the analytical capillary at the beginning of the zone electrophoretic separation for (a) the terminating electrolyte T, (b) the leading electrolyte L, and (c) another BGE being used as the electrolyte for the separation by zone electrophoresis. S, sample zone. The panels refer to cationic analyses; by changing the polarity they apply to anionic systems.

the segments of some additional amounts of the leader and/or terminator from the ITP step. This result in transient survivial of isotachophoretic migration during the second stage. Hence the second run can be characterized as the destacking of analytes followed by zone electrophoretic separation. The destacking process is, of course, dependent strongly both on the amount of the accompanying segments of the leader/terminator transferred and on the actual type of electrolyte system used. This is mainly due to the fact that the ITP process, after transferring into the second capillary, still survives for a certain time (transient ITP) and significantly affects the following CZE process. The sample components leave the transient ITP stack gradually, depending on the mobilities and electrolyte system used. Hence, each sample components starts its zone electrophoretic migration at a different time and at a different capillary position. This brings not only the variation of migration times but, when the segments of leader/terminator are very large, also the risk that the sample components may reach the detector while still stacked and thus not resolved. The situation may be advantageously explained by using the following figures depicting the electrolyte schemes. There are three general possible types of electrolyte systems [44] (Fig. 16), (i) the T-S-T system, where the terminating electrolyte T from the ITP stage serves as the BGE during CZE, (ii) the L-S-L system, where the leading electrolyte (L) from the ITP stage serves as the BGE during CZE of the sample S, and (iii) the BGE-S-BGE system, where L, T and the BGE are mutually different.

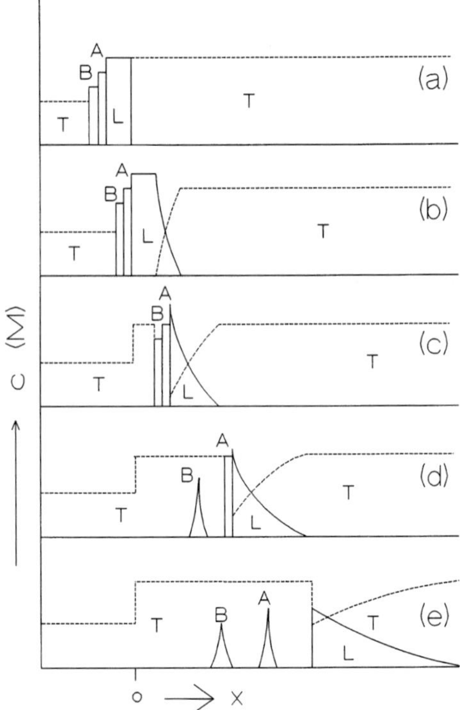

Figure 17. Scheme of transition of the stack of zones from isotachophoretic to zone electrophoretic migration in the T-S-T electrolyte system. The concentration profiles are given for the situations at the following time moments: (a) $t = 0$, when the current is switched to the analytical capillary; (b) when the entire zone of the leading electrolyte has just migrated into the analytical capillary; (c) when the ITP concentration plateau of the leading zone has just disappeared; (d) when zone B has destacked while A is still migrating in the stack; (e) when all sample zones migrate in a zone electrophoretic mode. L, leading electrolyte; T, terminating electrolyte; A, B, sample zones; x, longitudinal coordinate; 0, interface between the analytical and preseparation capillaries.

Then, the schemes of transition of the stack of analyte zones (A and B, with $m_A > m_B$) from isotachophoretic to ZE migration may be depicted as follows [45]. The scheme for the T-S-T electrolyte system is represented in Fig. 17. Figure 17a shows the stack of A and B together with a segment of L, placed between the zones of T serving as BGE. Figure 17b shows the situation after a certain time of electromigration has elapsed. The leading zone L penetrates into the T zone. Figure 17c shows the situation where the ITP concentration plateau of the leading zone has just disappeared. Figure 17d shows that the B zone

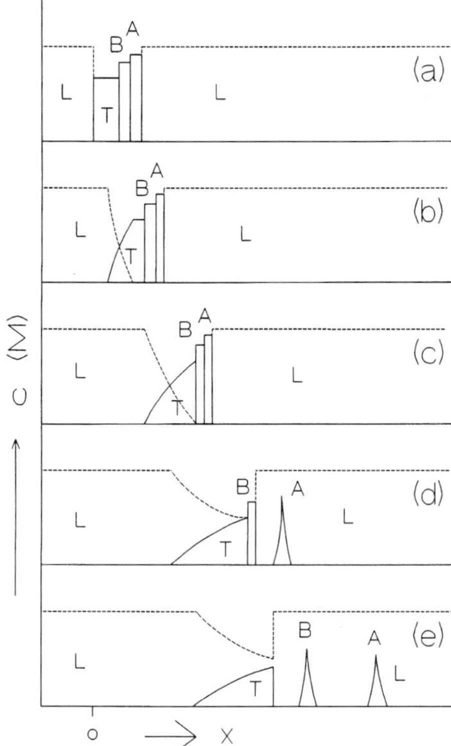

Figure 18. Scheme of transition of the stack of zones from isotachophoretic to zone electrophoretic migration in the L-S-L electrolyte system. The concentration profiles are given for the situations at the following time moments: (a) t = O, when the current is switched to the analytical capillary and the terminating electrolyte T in the preseparation capillary is replaced by the leading electrolyte L and the separation in the ZE mode starts; (b) when an ITP plateau of zone T still exists; (c) when the pure zone of T just disappeared and only the transition mixed zone of T and L is present; (d) when the zone A is already destacked and B is still migrating in the stack; (e) when all sample zones migrate in ZE mode. Abbreviations as in Fig. 17.

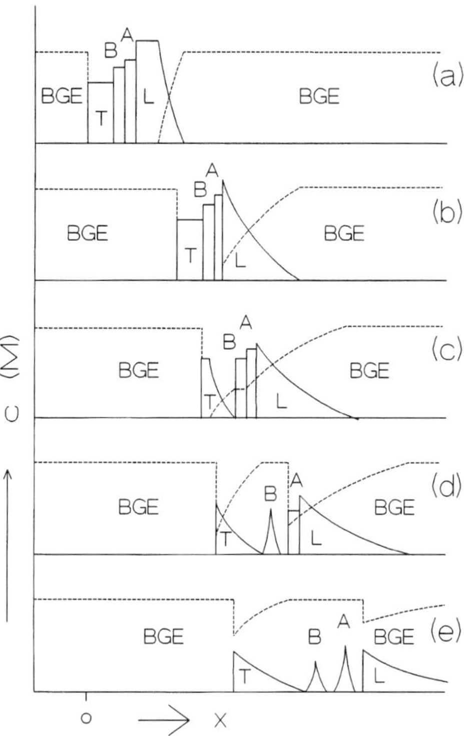

Figure 19. Scheme of transition of the stack of zones from isotachophoretic to zone electrophoretic migration in the BGE-S-BGE electrolyte system in case the coion of the BGE possesses the lowest mobility of the system, $m_{\text{coion}}<m_{\text{T}}<m_{\text{L}}$. The concentration profiles are given for the situations at the following time moments: (a) t = O, when the current is switched to the analytical capillary and the terminating electrolyte T in the preseparation capillary has been replaced by the BGE; the front boundary of zone L has already developed into a diffuse one; (b) when the ITP concentration plateau of zone L just disappeared; (c) the destacking of the T zone proceeds; (d) when zones T and B have destacked while zone A is still migrating in the stack; (e) when all sample zones migrate in ZE mode. Abbreviations as in Fig. 17.

is already destacked and migrates in a ZE mode. Zone A is still stacked. Figure 17e depicts the situation where both zones A and B migrate in the ZE mode. The scheme for an L-S-L electrolyte system is given in Fig. 18. The situation is in analogy to that in Fig. 17; however, in contrast, zone A is destacked first. The use of a BGE different from L and T of the ITP run brings about the situation where one must consider the effects of both L and T segments, which sandwich the stack of analytes A and B. The transient ITP and destacking is dependent first of all on the

relationship between the mobility of the coion of the BGE and those of L and T. One can distinguish two extreme cases. The scheme of transition of the stack of zones from the ITP to the ZE migration in the BGE-S-BGE electrolyte system, when the coion of the BGE possesses the lowest mobility of the system, *i.e.*, $m_{\text{coion}}<m_{\text{T}}<m_{\text{L}}$, is shown in Fig. 19. An analogous situation, when the mobility of this coion is the highest, *i.e.*, $m_{\text{coion}}>m_{\text{L}}>m_{\text{T}}$, is given in Fig. 20.

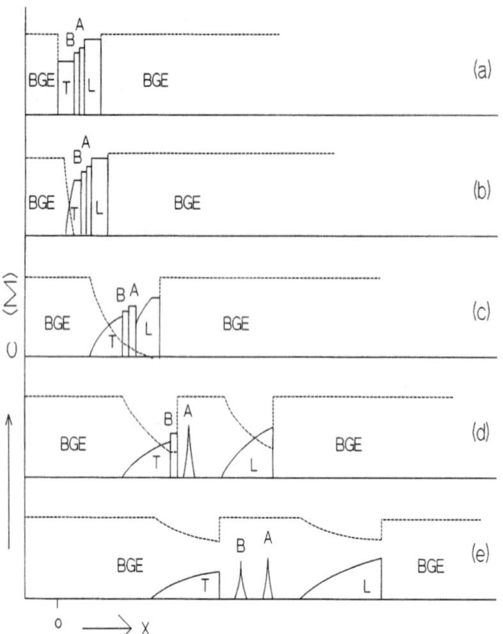

Figure 20. Scheme of transition of the stack of zones from isotachophoretic to zone electrophoretic migration in the BGE-S-BGE electrolyte system in case the coion of the BGE possesses the highest mobility of the system, $m_{coion} > m_L > m_T$. The concentration profiles are given for the situations at the following time moments: (a) $t = 0$, when the current is switched to the analytical capillary and the terminating electrolyte T in the preseparation capillary has been replaced by the BGE and zone electrophoretic separation starts; (b) when the ITP concentration plateau of zone T still exists; (c) the plateau of the T zone has disappeared; (d) when zones L and A have left the stack while zone B is still migrating in the stack; (e) when all sample zones migrate in a zone electrophoretic mode. Abbreviations as in Fig. 17.

6 Discussion

Although a complete survey of the literature on sample stacking in electrophoresis is beyond the scope of this article, in this discussion we consider some important contributions to sample stacking, which can be of service both to newcomers, as an introduction into the field, and to experts, as a reference literature survey. The history of stacking started already in the 1960s with comprehensive papers on methodology of disc electrophoresis by Davies [46] and Ornstein [47]. They derived the name from the use of discontinuous buffer systems and also from the shape of the migrating zones of proteins in their tubes.

What they did was in fact ITP, used in the first stage for concentrating the sample proteins into a narrow starting band, and then, in the second (online) stage this band was destacked and the proteins were separated by zone electrophoresis. Some possibilities to enhance the sensitivity in capillary electrophoresis are described in [48, 49], which also mentions improvements of the detection, the stacking procedures, and some related references. Moreover, extensive lists of references may be found in the following papers on selected stacking procedures [50–54] and on the combination ITP + CZE [44, 45, 55–57].

From a physicochemical point of view, stacking procedures are always based on suitable combinations of the sample composition, of the electrolytes placed in front of and behind the injected sample, and on the polarity of the voltage imposed. The variety of mutual combinations of the above-mentioned factors leads to complicated names for practical stacking procedures published, such as "head-column field-amplified sample stacking..." [54] and "field-amplified polarity switching sample injection..." [58]. Because we intend to give a description of the principles, which enables one to understand the physicochemical basis of all these stacking procedures, and to recognize all combinations of the effects involved, we would like to bring another characterization of the stacking procedures, which is based on their practical aspects. From a practical point of view, the stacking procedures may be classified mainly according to two aspects. The first aspect is connected with the instrumentation required, where the principal question to be answered is whether we use a single column system (single capillary) or a coupled column system (two interconnected capillaries) which can be filled independently with various solutions. The second aspect is connected with the answer to the question whether the stacking in induced by the sample or by the electrolyte system. If it is induced by the sample, it can be induced by the original sample, or by the diluted sample or by adding additives to the sample, such as a large amount of ionic species with high mobilitiy. If it is induced by the electrolyte system, important factors are the composition of the electrolyte systems, the arrangement of various operational electrolytes, and the use of modifying solutions in front of or behind the injected sample inside the capillary and/or in the electrode compartments. With respect to these two aspects we can distinguish between the following arrangements.

6.1 Single-column arrangement

This arrangement is the so-called classical one and is applied in most of the commercial instruments. Generally, a capillary is placed between two electrode chambers, and the sample, as well as all other electrolytes, are present in a series of vials in a caroussel. All solutions can be injected into this capillary by a suitable pneumatic system. The polarity of the voltage across the capillary may be selected as required and, moreover, it can be switched over during the run. Regardless of all the above possibilities, the single-column arrangement is suitable for the procedure based on sample-induced stacking, mostly of the intrinsic type, which means simple adjustment of the low concentrated sample to the higher concentrated BGE electrolyte, according to the KRF. In practice this procedure is frequently called field-amplified injection [50]. It is carried out in such a way that a diluted sample is hydrodynamically injected directly into the BGE system, and/or it is combined with an injection of a short zone (several millimeters) of water in front of the sample [58, 59]. Sometimes it is combined with EOF and changing of the polarity of the voltage, in order to remove the original sample spot from the column [50, 57, 58] to avoid subsequent problems with a low conductive plug inside the capillary (see Section 5.1).

Another practical procedure which belongs to this arrangement is simple electrokinetic injection of a diluted sample or its combination with a short plug of water, first injected into the capillary hydrodynamically [54]. The procedures of the concentration of neutral solutes in MEKC [60, 61] also belong to the above-mentioned simple procedure, since these procedures are based, in fact, upon the injection of a sample containing low concentrated charged micelles and the adjustment of this concentration to the KRF value of the BGE used. The neutral analytes are inside these micelles and are thus also concentrated.

An important group of stacking procedures can be characterized by transient isotachophoretic stacking induced by the leader coming from the sample matrix or added to the original sample. This situation often occurs, since often very large amounts of sodium chloride are present in, *e.g.*, many biological samples, and both sodium and chloride have much larger mobilities than many organic ionic compounds. Hence, sodium and/or chloride present in the sample or added to it may easily serve as the leaders for transient isotachophoretic stacking of cations and/or anions, respectively. On the other hand, when such a matrix is not present in the sample, it can easily be supplemented by a suitable leading ion. A proper selection of BGE then provides the suitable terminator. A comprehensive theoretical description of this stacking may be found elsewhere [51, 53]. A scheme of the destacking process related to these procedures can be found in Section 5.2, shown in Figs. 16–20. As practical experimental examples, the extensive studies of stacking of proteins [52, 56] is mentioned.

A similar type of stacking in a single capillary may be induced by the arrangement of a discontinuous electrolyte system in the capillary. A typical procedure may be described as follows. A fast coion of the BGE is used as leader. A suitable terminator is in the electrode chamber at the injection side of the capillary, *i.e.*, behind the injected sample. Then ITP is run for a certain time interval and next, the terminating electrode vial is substituted by another vial filled with the BGE suitable for destacking and zone electrophoretic analysis. An experimental analysis of shellfish-poisoning toxins [62] and peptide-like solutes [63, 64] is an example. Similar transient stacking may also be induced in a single capillary by creating the entire discontinuous electrolyte system inside this capillary. Stacking and separation of peptides serve as an example [65]. A somewhat different principle of stacking of peptides in a single capillary was published [66] and is based on the injection of a sample of peptides at a very high pH into the BGE having very low pH. The negatively charged peptides migrate first to the anode and are stopped at the boundary between the original sample and BGE, where they become neutral and are accumulated into a narrow band. Then they become positively charged, move to the cathode and are separated in ZE mode.

6.2 Coupled-column arrangement

This type of arrangement, which is used at present, consists of mostly laboratory-made [43, 55, 56] or laboratory-adapted commercial instrumentation for ITP [67] or for CZE [68, 69]. However, a commercial instrument with coupled columns is also available [70]. A sophisticated version of the related instrumenation has been described in detail previously [57]. The use of a coupled-column arrangement is practically unique and it is an online combination of ITP and CZE. An extensive description of this combination is given in Section 5 and the methodology may be found elsewhere [44, 55, 66, 67]. An extensive experimental study in the field of proteins [56] or ultratrace analysis of paraquat and diquat in water [57] may be mentioned here as examples. An exhaustive survey on the combination of ITP + CZE is given by Křivánková *et al.* [71, 72] where one can also find some example of practical analyses of real samples.

7 Conclusions

The stacking procedures represent an important tool for enhancement of the sensivity of capillary electrophoresis. They bring the accumulation of trace analytes from large volumes of a sample injected into a narrow band (stack) which then serves as an "ideal" sample for further electrophoretic separation. The fundamental principle of stacking is adjustment of concentration in accord with the KRF. In principle, the most efficient combination of the stacking and separation is the online ITP + CZE in coupled-column systems. In practice, however, only single-column systems are available in many cases. Therefore, procedures creating transient ITP in single-column system are also advantageously used. This article gives a survey of basic electrophoretic principles that are involved in stacking procedures. Hence, it can help newcomers, as well as capillary electrophoresis practitioners, to recognize the principles utilized by various authors regardless of the big variety of practical procedures described and their complicated names.

The co-author (P.B.) acknowledges the support of the Grant Agency of the Czech Republic, grant No. 203/99/ 0044, grant No. A4031703 of the Grant Agency of the Academy of Sciences of the Czech Republic, and grant No. VS-96021 and Nr. VS-97014 of the Ministry of Education (MSMT) of the Czech Republic.

Received January 3, 2000

8 References

[1] Li, S. F. H., *Capillary Electrophoresis, Principle, Practice and Applications*, Elsevier, Amsterdam 1992.

[2] Kuhn, R., Hoffstetter-Kuhn, S., *Capillary Electrophoresis: Principles and Practice*, Springer Verlag, Berlin, Heidelberg 1993.

[3] Mosher, R. A., Saville, D. A., Thormann, W., *The Dynamics of Electrophoresis*, VCH Verlagsgesellschaft mbH, Weinheim 1992.

[4] Foret, F., Křivánková L., Boček, P., *Capillary Zone Electrophoresis*, VCH Verlagsgesellschaft mbH, Weinheim 1993.

[5] El Rassi, (Ed.), *Electrophoresis* 1999, *20*, 2987–3328.

[6] Atkins, P. W., *Physical Chemistry*, Oxford University Press, Oxford 1982, p. 605.

[7] Thormann, W., *Electrophoresis* 1983, *4*, 383–390.

[8] Beckers, J. L., *J. Chromatogr. A* 1995, *696*, 285–294.

[9] Reijenga, J.C., Kenddler, E., *J. Chromatogr. A.* 1994, *659*, 403-415, 417–426.

[10] Gebauer, P., Deml, M., Póspíchal, J., Boček, P., *Electrophoresis* 1990, *11*, 724–731.

[11] Schafer-Nielsen, C., *Electrophoresis* 1995, *16*, 1369–1376.

[12] Dose, E. V., Guiochon, G. A., *Anal. Chem.* 1991, *63*, 1063–1072.

[13] Mikkers, F. E. P., *Anal. Chem.* 1999, *71*, 522–533.

[14] Ermakov, S.V., Righetti, P.G., *J. Chromatogr. A* 1994, *667*, 257–270.

[15] Falkenhagen, H., *Electrolyte*, Hirzel, Leipzig 1932, p. 325.

[16] Kohlrausch, F., *Ann. Phys. (Leipzig)* 1997, *62*, 209–239.

[17] Beckers, J. L., Verheggen, T.P.E.M., Everaerts, F. M., *J. Chromatogr.* 1988, *452*, 591–600.

[18] Longsworth, L. C., *J. Am. Chem. Soc.* 1945, *67*, 1109–1119.

[19] Dole, V. P., *J. Am. Chem. Soc.* 1945, *67*, 1119–1126.

[20] Longsworth, L. C., *J. Am. Chem. Soc.* 1944, *66*, 449–453.

[21] Alberty, R. A., Nichol, J. C., *J. Am. Chem. Soc.* 1948, *70*, 2297.

[22] Nichol, J. C., *J. Am. Chem. Soc.* 1950, *72*, 2367–2374.

[23] Alberty, R. A., *J. Am. Chem. Soc.* 1950, *72*, 2361–2367.

[24] Dismukes, E. B., Alberty, R. A., *J. Am. Chem. Soc.* 1954, *76*, 191–197.

[25] Nichol, J. C., Dismukes, E. B., Alberty, J. C., *J. Am. Chem. Soc.* 1958, *80*, 2610–2615.

[26] Nichol, J. C., Gosting, L. J., *J. Am. Chem. Soc.* 1958, *80*, 2601–2609.

[27] Martin, A.J.P., Everaerts, F. M., *Anal. Chim. Acta* 1967, *38*, 233–237.

[28] Martin, A.J.P., Everaerts, F. M., *Proc. Roy. Soc. Ser. A.* 1970, *316*, 493–514.

[29] Haglund, H., *Sci. Tools* 1970, *17*, 2–13.

[30] Mikkers, F., Everaerts, F. M., *Analytical Chemistry Symposia Series*, Vol. 6, Elsevier, Amsterdam 1981, pp. 1–17.

[31] Mosher, R. A., Thormann, W., *Electrophoresis* 1985, *6*, 477–482.

[32] Beckers, J. L., *J. Chromatogr. A* 1994, *662*, 153–166.

[33] Gebauer, P., Křivánková, L., Boček, P., *J. Chromatogr.* 1989, *470*, 3–20.

[34] Everaerts, F. M., Beckers, J. L., Verheggen, T.P.E.M., *Isotachophoresis - Theory, Instrumentation and Applications*, Elsevier, Amsterdam 1976, p. 264.

[35] Boček, P., Deml, M., Gebauer, P., Dolník, V., *Analytical Isotachophoresis*, VCH Verlagsgesellschaft mbH, Weinheim 1988, pp. 97–101.

[36] Quirino, J. P., Terabe, S., *Science* 1998, *282*, 465–468.

[37] Palmer, J. Munro, N. J., Landers, J. P., *Anal. Chem.* 1999, *71*, 1679–1687.

[38] Quirino, J. P., Terabe, S., Boček, P., *Anal. Chem.* 2000, *72*, 1934–1940.

[39] Huang, X., Ohms, J. I., *J. Chromatogr.* 1990, *516*, 233–240.

[40] Chien, R. L., Helmer, J. C., *Anal. Chem.* 1991, *63*, 1354–1361.

[41] Burgi, D. S., Chien, R. L., *Anal. Chem.* 1991, *63*, 2042–2047.

[42] Beckers, J. L., Ackermanns, M. T., *J. Chromatogr.* 1993, *629*, 371–378.

[43] Everaerts, F. M., Verheggen, T.P.E.M., Mikkers, F.E.P., *J. Chromatogr.* 1979, *169*, 21–38.

[44] Křivánková, L., Gebauer, P., Boček, P., *J. Chromatogr. A* 1995, *716*, 35–48.

[45] Křivánková, L., Gebauer, P., Thormann, W., Mosher, R. A., Boček, P., *J. Chromator.* 1993, *638*, 119–135.

[46] Davies, B. J., *Ann. N.Y. Acad. Sci.* 1964, *121*, 404–427.

[47] Ornstein, L, *Ann. N.Y. Acad. Sci.* 1964, *121*, 321–349.

[48] Nielen, M. W. F., *TRAC* 1993, *12*, 345–356.

[49] Albin, M., Grossman, P. D., Moring, S. E., *Anal. Chem.* 1993, *65*, 489A–497A.

[50] Chien, R.-L., Burgi, D. S., *Anal. Chem.* 1992, *64*, 489A–496A.

[51] Gebauer, P., Thormann, W., Boček, P., *Electrophoresis* 1995, *16*, 2039–2050.

[52] Thompson, T. J., Foret, F., Vorous, P., Karger, B. J., *Anal. Chem.* 1993, *65*, 900–906.

[53] Gebauer, P., Thormann, W., Boček, P., *J. Chromatogr.* 1992, *608*, 47–57.

[54] Zhang, C.-X., Thormann, W., *Anal. Chem.* 1996, *68*, 2523–2532.

[55] Foret, F., Sustáček, V., Boček, P., *J. Microcol. Sep.* 1990, *2*, 229–233.

[56] Foret, F., Szökö, E., Karger, B. L., *J. Chromatogr.* 1992, *608*, 3–12.

[57] Kaniansky, D., Ivanyi, F., Onuska, F. I., *Anal. Chem.* 1994, *66*, 1817–1824.

[58] Chien, R. L., Burgi, D. S., *J. Chromatogr.* 1991, *559*, 153–161.

[59] Chien, R. L., Burig, D. S., *J. Chromatogr.* 1991, *559*, 141–152.

[60] Liu, Z., Sam, P., Sirimanne, S. R., McClure, P. C., Grainger, J., Patterson Jr., D. G., *J. Chromatogr. A* 1994, *673*, 125–132.

[61] Nielsen, K. R., Foley, J. P., *J. Chromatogr. A* 1994, *686*, 283–291.

[62] Locke, S. J., Thibault, P., *Anal. Chem.* 1994, *66*, 3436–3446.

[63] Larsson, M., Nagard, S., *J. Microcol. Sep.* 1994, *6*, 107–113.

[64] Witte, D. T., Nagard, S., Larsson, M., *J. Chromatogr. A* 1994, *687*, 155–156.

[65] Schwer, C., Gaš, B., Lottspeich, F., Kenndler, E., *Anal. Chem.* 1993, *65*, 2108–2115.

[66] Aerbersold, R., Morisson, H. D., *J. Chromatogr.* 1990, *516*, 79–88.

[67] Stegenhuis, D. S., Irth, H., Tjaden, U.R., Van der Greef, J., *J. Chromatogr.* 1991, *538*, 393–402.

[68] Pálmarsdóttir, S., Edholm, L.-E., *J. Chromatogr. A* 1995, *693*, 131–143.

[69] Van der Vlis, E., Mazereeuw, M., Tjaden, U. R., Irth, H., Van der Greef, J., *J. Chromatogr. A* 1994, *687*, 333–341.

[70] CS Isotachophoretic Analyzer EA 100, Villa-Labeco, Spisska Nova Ves, Slovak Republik.

[71] Křivánková, L., Foret, F., Boček, P., *J. Chromatogr.* 1991, *545*, 307–313.

[72] Křivánková, L., Vrana, A., Gebauer, P., Boček, P., *J. Chromatogr. A* 1997, *772*, 283–295.

Electrophoresis 2000, *21*, 2768–2779

Review

Damon M. Osbourn
David J. Weiss
Craig E. Lunte

Department of Chemistry,
University of Kansas,
Lawrence, KS, USA

On-line preconcentration methods for capillary electrophoresis

The limits of detection (LOD) for capillary electrophoresis (CE) are constrained by the dimensions of the capillary. For example, the small volume of the capillary limits the total volume of sample that can be injected into the capillary. In addition, the reduced pathlength hinders common optical detection methods such as UV detection. Many different techniques have been developed to improve the LOD for CE. In general these techniques are designed to compress analyte bands within the capillary, thereby increasing the volume of sample that can be injected without loss of CE efficiency. This on-line sample preconcentration, generally referred to as stacking, is based on either the manipulation of differences in the electrophoretic mobility of analytes at the boundary of two buffers with differing resistivities or the partitioning of analytes into a stationary or pseudostationary phase. This article will discuss a number of different techniques, including field-amplified sample stacking, large-volume sample stacking, pH-mediated sample stacking, on-column isotachophoresis, chromatographic preconcentration, sample stacking for micellar electrokinetic chromatography, and sweeping.

Keywords: On-line preconcentration / Stacking / Capillary electrophoresis / Field-amplified stacking / pH-mediated stacking / Review   EL 4092

Contents

1 Introduction

With a separation based on physical phenomena different from those used in chromatography, capillary electrophoresis (CE) has been the focus of attention for developing new analytical methodology. The analysis of samples that cannot be separated by more common reversed-phase liquid chromatography (LC) are often resolved by CE. Unfortunately, the benefits from the high number of theoretical plates obtained with CE have been overshadowed by the poor detection limits achieved with UV detection. Because of the small dimensions of CE capillaries, typically 25–150 µm ID and 40–80 cm in length, only very small sample volumes may be loaded onto the column. Additionally, for the most common optical detection techniques, CE suffers from a drastically reduced pathlength as compared to LC. Since absorbance is directly proportional to pathlength and concentration, the concentration of the samples must be dramatically increased to obtain the same signal-to-noise ratio as would result from a typical LC experiment. Overcoming the poor sensitivity of CE has been the emphasis of many investigations. A number of techniques have been developed to preconcentrate samples and to increase the amount of sample that can be loaded onto the column without degrading the separation. The focus of this article is to explain the different approaches that have been developed over the last ten years. These approaches can be categorized into two groups based on the physical phenomena used to concentrate analytes. One method involves manipulating the electrophoretic velocity of the analyte and includes techniques such as field-amplified sample stacking, large-

Correspondence: Prof. Craig E. Lunte, University of Kansas, Department of Chemistry, Lawrence, KS, 66045 USA
E-mail: c-lunte@ukans.edu
Fax: +785-864-5396

Abbreviations: FASS, field-amplified sample stacking; **LVSS**, large-volume sample stacking

volume sample stacking, isotachophoresis, pH-mediated stacking, and matrix switching. The other group utilizes partitioning into a stationary or pseudostationary phase to affect the analyte preconcentration, including chromatographic preconcentration and sweeping.

Two effects, the electroosmotic flow EOF and the molecule's electrophoretic mobility determine the migration time of sample molecules in CE. The EOF is the bulk flow through the capillary due to the migration of cationic counterions, which are held closely to the anionic capillary wall by electrostatic forces. The electrophoretic mobility is a result of the force exerted on any charged particle in an electric field. This force is balanced by the frictional counterforce of the fluid surrounding the particle. With the exception of a very short initial time, these two forces are equal. This results in a steady-state velocity of the charged particle that can be calculated by

$$v_{ss} = qE/f \tag{1}$$

where q is the charge on the particle, E is the field strength, and f is the frictional coefficient. It is important to note that the electrophoretic mobility of an analyte is directly proportional to the local field strength inside the capillary. The field strength in a capillary filled with one buffer is defined by

$$E = V/L \tag{2}$$

where V is the voltage and L is the length of the capillary. In a capillary filled with two buffers of differing resistivities the field strength is given by

$$E_1 = \gamma E_0/[\gamma x + (1-x)] \tag{3}$$

$$E_2 = E_0/[\gamma x + (1-x)] \tag{4}$$

where E_1 and E_2 are the field strengths of the high and low resistance solutions, respectively. E_0 is the field strength in a system of only buffer 1 or 2, γ is the ratio of the resistivities of the low concentration buffer to that of the high, and x is the fraction of the capillary filled with low resistance buffer. It follows that the higher the resistivity of one buffer relative to the other, the more the field strength is amplified in that buffer. Since analyte velocity is directly proportional to field strength, the larger the difference in resistance of the two buffers inside the capillary, the faster the analytes will migrate through the high resistance buffer. This can be particularly detrimental when dealing with biological samples where the sample matrix consists of salts with a concentration on the order of 100 mM. In this instance, as sample is injected, a high field strength develops over the relatively low concentra-

tion background electrolyte (BGE) while a low field occurs over the portion of the capillary where low resistance, high conductivity sample matrix has been injected. The analytes migrate slowly through the sample matrix until they reach the BGE where they accelerate. This leads to band broadening and a decreased signal-to-noise ratio. The difficulty of making the BGE at a lower resistance than the sample matrix is the limitation created by Joule heating as a result of high currents with high ionic strength buffers.

Although the band broadening as a result of a relatively low field over the sample plug is detrimental, the reverse has proven to be important in understanding the techniques used for online preconcentration in CE. Almost all on-column concentration techniques manipulate the changes in electrophoretic mobility of analytes at the boundary between high-resistance and low-resistance buffers. Albeit this may seem insignificant when first investigating some of the more sophisticated techniques involving polarity switching, matrix switching, and the acid/base titration of a sample zone, the same phenomena is responsible for on-column sample concentration in each of these techniques.

Two factors affect the enhancement of analyte detection in stacking. First is the narrowing of analyte bands in the column. As the peak width of the analyte is decreased, the peak height is dramatically increased, resulting in a greater signal-to-noise ratio, improving the limits of detection. The second factor is the amount of sample that can be loaded onto the column. Because the width of the peaks is significantly reduced by the stacking procedure, much larger sample volumes may be injected without losing separation efficiency. This results in a greater mass of analyte in the capillary and therefore a greater response at the detector.

2 Field-amplified sample stacking

The simplest technique for sample stacking is field-amplified sample stacking (FASS). The effects of injecting samples in low-conductivity matrix were first explained by Mikkers *et al.* in 1979 [1]. In general, this method is based upon the idea that ions electrophoretically migrating through a low-conductivity solution into a high-conductivity solution slow down dramatically at the boundary of the two buffers. For example, consider a sample dissolved in a low-conductivity solution such as water. The electric field strength will be higher over the low-conductivity sample zone than the BGE. Therefore, the velocity of the analyte will be high in the sample zone until it reaches the buffer interface. As the sample ions reach the high-concentration BGE they slow down and stack into a narrow zone.

Electrophoresis 2000, *21*, 2768–2779

Burgi and Chien [2] described a model to optimize the buffer concentrations in the sample and capillary, and have discussed the limitations to stacking efficiency in FASS [3]. The stacking efficiency is limited by laminar flow created by differences in EOF arising from the discontinuous buffers. The flow profile created by laminar flow conditions can be convex or concave depending on whether the leading buffer has the higher or lower EOF. This hemispherical shape can cause band broadening in the narrow sample zone. Furthermore, Chien and Burgi [4] described enhanced stacking and sample loading by injection of a water plug into the capillary immediately prior to sample injection (Fig. 1). Further study has shown that injection of a high viscosity plug, such as ethylene glycol, prior to the preinjection plug of water, acts as a trap to slow the electrophoretic velocity of the analytes [5]. Stacking efficiencies were doubled using this procedure. Addition of organic solvent to the sample matrix has also been shown to provide an improved signal-to-noise ratio [6].

Sjogren and Dasgupta [7] describe a FASS technique in which there is no liquid junction between the anode and the capillary tip. After the sample is injected into the capillary, the capillary tip is positioned 1.5 cm above the anodic buffer reservoir and the potential is applied to the buffer reservoir. Since the field strength is significantly higher than the breakdown voltage of the intervening air, the circuit is closed and the analytes stack. Care must be taken to return the capillary to the anodic reservoir for separation, before air bubbles are introduced into the capillary tip. The goal of this approach is to increase reproducibility and efficiency. Significant improvements were not observed for small molecules while quantitative improvements were evident for larger molecules with slower diffusion coefficients.

Friedberg *et al.* [8] have reported the effects of pH, ionic strength and buffer composition on stacking with and without acetonitrile in the sample matrix. Different buffers were shown to amplify the stacking effects for different analytes. Additionally, increased efficiency was observed for the analytes studied by lowering the pH of the sample buffer relative to the BGE and by increasing the ionic strength of the sample matrix when using acetonitrile to dilute samples for stacking. Furthermore, Beckers and Ackermans [9] have discussed the effects of stacking on resolution and the pH in the capillary. The resolution was decreased as a result of large injection volumes used in stacking. This is due in part to the relatively low field strength in the BGE, and to the diminished effective separation length of the capillary when long stacking zones are employed. Stacking techniques can also create circumstances where a difference in field strength between sample and BGE zones can affect local pH inside the capillary. For instance, in a typical stacking experiment protons will migrate much more quickly through the sample zone than the BGE. Therefore, the number of protons leaving the sample zone will be greater than the number of protons entering and the pH will increase. This can slow the mobility of weak cations, countering the stacking effect.

Applications of FASS include analysis of DNA fragments [10], analysis of pharmaceuticals in serum [11] and drugs of abuse [12, 13]. Sensitivity enhancements up to 1000-fold have been reported [14]. Sample stacking for nonaqueous CE [15] and nonaqueous chiral separations [16] has also been performed. However, a limitation to FASS is that the ionic strength of the sample must be significantly lower than that of the BGE. This requirement may cause problems for analysis of some physiological solutions such as dialysate. FASS was discussed in detail by Chien and Burgi [17] as well as by Albin *et al.* [18].

3 Large-volume sample stacking

Large-volume sample stacking (LVSS) is a technique designed by Chien and Burgi [19] that is performed by dissolving the sample in water and hydrodynamically filling 1/3–1/2 of the capillary with the sample. Reverse polarity is applied with BGE at the detection end of the capillary. As a result, the EOF backs the sample plug out of the capillary while anionic analytes move toward the detection end and stack at the interface with the BGE. The electrophoretic current is monitored until it reaches approximately 95–99% of its original value. At this point the polarity is returned to normal, and the separation occurs in the usual fashion.

Under reverse polarity, the cations and neutrals should exit the capillary into the waste buffer reservoir before the polarity is returned to normal. However, if the electrophoretic current is not carefully monitored, some anionic analyte may be lost. Applications of this method include analysis of drugs [20, 21], dyes [21, 22], chelates [21, 22], metals in hair [23], chemicals of environmental concern [24], and phenols [25] with 2- to 100-fold enhancements reported. A variation of this technique has been used to improve the detection limits of cations [20, 21]. The EOF modifier cetyltrimethylammonium bromide (CTAB) has been used to reverse the EOF of the system and the separation performed with the CE system in reversed polarity mode. McGrath and Smyth [20] reported 10-fold concentration enhancement for analysis of cationic drugs in a urine sample after extraction and reconstitution in water.

LVSS is a demanding procedure since the current must be closely monitored by the analyst to obtain reproducible results. This technique will not separate anions and cations simultaneously, and is additionally limited to analytes with low mobilities [26]. However, variations on this technique do allow the analysis of fast moving ions. LVSS without polarity switching has been reported for the analysis of high mobility anions [27, 28]. In this technique an EOF modifier such as CTAB is present in the BGE. When the capillary is primarily filled with sample dissolved in water and a reversed polarity is applied to the inlet, the EOF pushes the sample plug out of the capillary. As this happens, BGE from the detector side of the capillary is pulled into the column. The CTAB present in the BGE coats the capillary and reverses the direction of the EOF, eliminating the need for polarity switching. This approach has also been reported using a BGE at low pH to suppress the EOF rather than reversing it [29]. Another variation on LVSS is termed double stacking [30]. In this procedure all potentials are applied in normal polarity and pressure at the detection side of the capillary is used to back the sample plug out of the capillary.

4 pH-Mediated stacking

FASS and LVSS are performed with the sample either dissolved in water or diluted with a low conductivity buffer. Unfortunately, this is not always an option, as is often the case when analyzing low concentrations in dialysate. One method for improving the detection limits with these samples is to neutralize the high-conductivity sample matrix with pH-mediated sample stacking. This is a technique in which FASS is triggered by titrating the injected sample zone to neutral, thus creating a low-conductivity region (Fig. 2). First, a sample in a high-ionic strength biological medium is electrokinetically injected. As the sample is injected the anions of the strong acid such as Cl⁻ are displaced by the anions of the weak acid in the BGE, such as acetate. Next, a plug of strong acid (HCl) is injected electrokinetically. The protons from the acid injection migrate quickly through the sample zone, titrating the acetate ions and creating a region of neutral charge and high resistivity. This allows the sample cations to migrate quickly through the titrated zone to the boundary with the BGE, where they stack into a narrow band [31, 32]. This method can also be used for the determination of anions by incorporating an EOF modifier such as CTAB into a basic BGE and running in reverse polarity [33] (Fig. 3). Applications of pH-mediated stacking have been reported for the analysis of pharmaceuticals [31] as well as for DNA sequencing [34].

Previous reports of pH-mediated stacking for the analysis of peptides have used different techniques for titrating the sample zone. Aebersold and Morrison [35] injected a

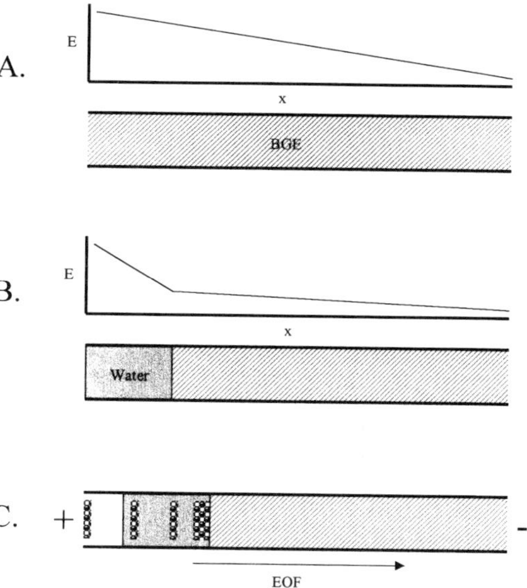

Figure 1. FASS with a preinjection plug of water. (A) Capillary filled with BGE prior to injection. The field strength across the capillary is shown qualitatively. (B) A short plug of water is injected hydrodynamically. The diagram of the field strength shows that high field strength is exhibited over the water plug, while the remainder of the capillary has a relatively low field strength. (C) As the sample is electrokinetically injected cations race across the water plug and stack at the boundary of water and BGE, while the EOF draws a small amount of the sample matrix into the capillary.

sample of dilute peptides in a highly basic matrix into a capillary containing an acidic buffer. In the basic sample solution the peptide has a net anionic charge, while in the acidic BGE the peptide is positively charged. Therefore, as the sample zone is titrated from basic to acidic during the CZE separation, the direction of the electrophoretic migration of the peptide is reversed. Schwer and Lottspeich [36] describe the hydrodynamic injection of strong base, followed immediately by injection of peptide sample, then strong acid. The OH⁻¹ and H⁺ migrate towards one another, temporarily creating a low-conductivity region to electrofocus the peptides.

If too large a volume of sample is injected, the separation efficiency will be dramatically diminished. This is caused by too large a portion of the capillary being used for stacking and an insufficient length left for the separation. Simply increasing the length of the capillary does not solve this problem, however, since the field strength is highly amplified in the sample zone and weakened in the BGE.

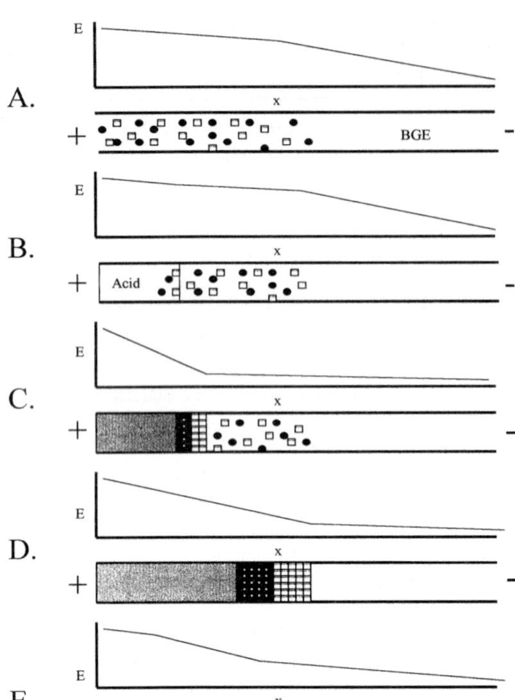

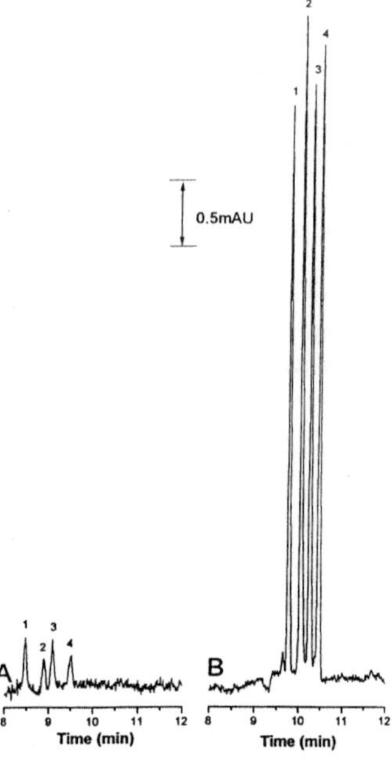

Figure 3. Example of pH-mediated stacking using base stacking. (A) Normal electrokinetic injection of 10 µM phenolic acids in 90% Ringer's solution. (B) Injection of 10 µM phenolic acids in 90% Ringer's solution using base stacking procedure.

Figure 2. Schematic of the acid stacking mechanism for pH-mediated stacking. (A) A sample in a high ionic strength sample matrix is injected electrokinetically. Because the resistivity of the sample matrix is lower than that of the BGE, most of the field strength is dropped across the BGE portion of the capillary. (B) Next a plug of strong acid is electrokinetically injected behind the sample plug. (C) As the separation voltage is applied the strong acid titrates the sample zone to neutral, creating a high resistance zone. Now mosts of the field strength is dropped across the sample zone. (D) The acid titrates the entire sample zone and the analytes are stacked into narrow bands at the boundary of the titrated zone and the BGE. (E) The electrophoretic separation proceeds through the remainder of the capillary.

To increase the sample loading capacity, further investigation into pH-mediated stacking incorporated a double-capillary system in which the capillaries are arranged in a "T". The analytes are stacked between one arm of the "T" and the injection end of the capillary. The separation potential is applied across the stacking arm of the "T" and the detection end of the capillary, allowing a separation without the sample zone affecting the field strength [33]. This increased the amount of sample that could be loaded

onto the column. A 300-fold enhancement in detection limits has been reported using pH-mediated stacking [33]. The most impressive aspect of this method is the complete simplicity with which it allows the user to analyze biological samples.

5 Isotachophoresis

Unlike CE, isotachophoresis (ITP) is performed with a discontinuous buffer. ITP is applicable to many compounds ranging from small charged analytes to proteins [37–39]. This method is particularly useful when samples are not simply dissolved in water, but have some conducting ions, such as biological samples. ITP is often performed with the sample zone between the BGE of higher (leading electrolyte) and lower electrophoretic mobilities (terminating electrolyte). Typically, either cations or anions can be separated in one run with some exceptions [40, 41]. For cation analysis the leading buffer is positioned on the cathodic end of the sample zone while the terminating buffer is injected after the sample zone. When high voltage is applied, a potential gradient develops and each of the analyte zones migrate with the same velocity. Where

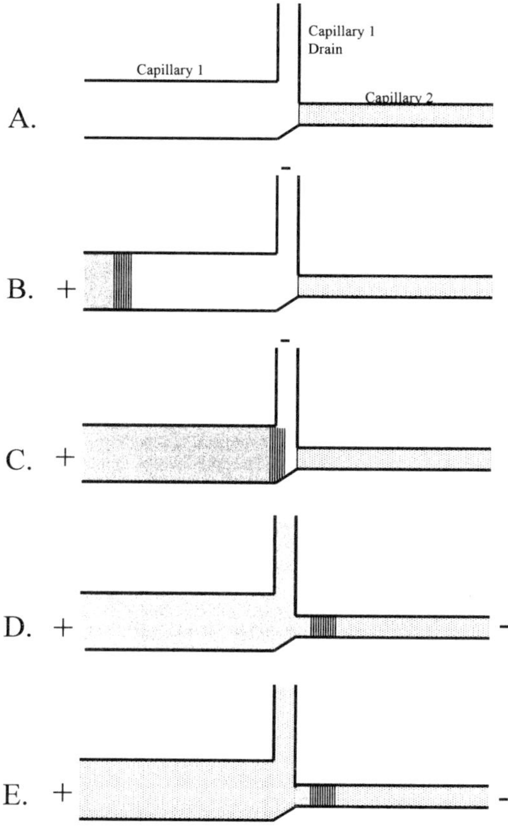

Figure 4. Double capillary system for ITP-CZE. (A) The second capillary is filled with BGE and the first capillary is filled with leading buffer. (B) The sample is injected and the anodic reservoir is filled with terminating buffer. (C) The ITP preconcentration is carried out in the first capillary. Once the boundary of the leading buffer and sample zone has reached the inlet of the second capillary the high voltage is turned off. (D) The high voltage is applied between the inlet of the first capillary and the outlet of the second capillary to transfer the analyte bands into the separation capillary. (E) The first capillary is flushed with BGE and the separation potential is applied.

cations of lower mobility are present, the electric field is stronger, making the velocity of the zone match the rest of the sample. If a solute moves too slowly and enters the band behind it, a region of higher field strength, the analyte will accelerate until it re-enters its own zone. Eventually a steady state is reached where each analyte moves as a discrete band according to its mobility. Bands with the highest mobility will elute before those of lowest mobility. Continuing the separation in the zone electrophoresis

mode can be performed by placing the capillary inlet in leading electrolyte. When the leading electrolyte catches up with the sample zone the field gradient will be lost and CE separation will begin. This is one of the main modes of transient ITP-CE, so named because the migration mode gradually changes from ITP to CE.

One of the advantages of ITP over CE is that with samples in a high ionic strength matrix, the matrix itself may be used as the leading electrolyte. This is another mode of transient-ITP, self-induced stacking [42], when the BGE has a lower mobility than the analytes. This was observed as one of the first examples of transient ITP-CE [43]. In this case the BGE serves as the terminating electrolyte. Reinhoud and co-authors [44] have developed a variation of transient ITP-CE referred to as counterflow ITP-CE. A large diameter capillary is used (100 µm ID) to load several microliters onto the capillary. To stack almost 100% of the capillary volume, hydrodynamic back pressure is used during the focusing step, to remove the terminating electrolyte before the CE separation. Using this method, 100-fold concentration of sample was reported, compared to CE alone. ITP followed by capillary gel electrophoresis has also been performed [45]. Using this method, half the capillary was filled with sieving gel and a 700 nL sample of DNA restriction fragments was focused prior to separation.

Coupled column ITP (Fig. 4) uses two capillaries for ITP followed by CE [38, 46–53]. The first capillary has an ID of approximately 500 µm while the second capillary has an ID of 50 µm with similar length [54]. ITP is used to focus analytes in the larger ID capillary. Once the sample bands have been injected into the second capillary, the terminating electrolyte is flushed from the first capillary with leading electrolyte. The CE separation then proceeds in the second capillary. In this way, volumes greater than the total volume of the CE capillary can be stacked into nanoliter volumes for the separation. Nanomolar limits of detection have been reported with this technique for separations of enantiomers in urine [49]. In terms of the effective on-column concentration, a 10 000-fold sensitivity enhancement has been reported using coupled column ITP preconcentration [52]. ITP can handle both complex and highly concentrated matrices [53, 55–57] including blood, plasma and urine [49, 58, 59]. Kaniansky *et al.* [57] demonstrated the determination of six anions in a matrix of 10^4 times higher concentration than the sample analytes. Applications of ITP are wide, ranging from peptides [38, 60] to arsenic speciation [61]. Since leading and terminating electrolytes must be chosen carefully, Křivánková and Boček have recommended several [39].

6 Chromatographic preconcentration

Sample pretreatment with solid-phase extraction (SPE) is commonly used off-line for separations. This is a useful technique that allows a large volume of low concentration sample to be loaded onto the solid phase and eluted in a small volume, providing concentrations that can be easily detected. Since this technique obviously consumes more analyst time, on-line methods have been investigated for CE. One method is to pack a short segment, about 2 mm, of the injection end of the capillary with LC stationary phase [62–73]. The sample is loaded onto the stationary phase by hydrodynamic injection, and then eluted by injection of a second solvent. Applications of this technique have included affinity chromatography stationary phases [71], analysis of pharmaceuticals isolated from urine [73] and the direct analysis of biological samples [72]. Alternatively, Cai and Rassi [74, 75] have developed an open-tubular preconcentrator for CE. In this approach the walls of a 20 cm capillary are modified with a C18 phase for herbicide analysis [74] or a metal chelate phase for protein analysis [75]. This preconcentration capillary is connected in series with a separation capillary. While these configurations improve sample stacking, a number of problems may arise, including tailing, loss of CE efficiency, and interference between the organic elution solvent and the CE electric field [76]. Attempts to alleviate these problems include development of a double-capillary system [72] and an on-line switching valve [77]. In the double-capillary design the sample matrix is pushed through the inlet and out a drain capillary while the analytical separation is performed between the inlet and a second capillary. With the on-line switching valve the analytes are retained on a stationary bed contained within the valve. To inject the analytes the valve is switched so that the packed bed lies within the path of the CE separation capillary.

Additionally, two-dimensional (2-D) LC-CE designs [78–80] have led to the investigation of on-line LC or SPE steps without an in-capillary solid-phase cartridge. The use of small-bore LC columns on-line with CE, coupled through electrokinetic injection manifolds, have been reported [81–85]. This allows the sample matrix, unretained analytes, and large volumes of LC mobile phase to be flushed through the stationary phase without traversing the capillary's volume. The difficulty in transferring analyte bands from the LC portion of the system to the CE portion is the major problem with this approach. While this is not an issue in 2-D separations, the coupling of these techniques for preconcentration makes it critical that as much of the analyte as possible is transferred to the CE. The LC portion of the separation must be rugged in order to have a reliable injection of analytes for CE.

Tomlinson, Naylor and co-workers [86–94] have reported a different technique for on-column partitioning-based preconcentration, termed membrane preconcentration (mPC), with the goal of solving some of the problems arising from the use of large packed beds in on-line SPE-CE. Specifically, mPC is designed to improve the CE efficiency by reducing the large volume of organic solvent often needed to elute compounds and to lessen the impact of the large packed bed on the EOF and backpressure [76]. In this technique a thin polymer membrane is placed between two short capillary segments. One end of this preconcentration cartridge serves as the capillary inlet, while the other end connects to the separation capillary. Polymeric phases such as styrene-divinyl benzene, C2 and C8 have been successfully used. The application of this preconcentration method has been shown to improve detection for on-line CE-MS [86, 88–91, 93, 94].

Precolumn and on-line concentration techniques using SPE to replace the high ionic strength sample matrix with a low ionic strength matrix for FASS have also been investigated. The substantial difference between this technique, matrix switching, and other on-line SPE techniques is the purposeful reliance on the elution of analytes in a low-conductivity plug to amplify field effect stacking rather than relying on the solid phase to provide the concentration enhancement. The instrumentation is similar to that described above.

Petersson *et al.* [95] used a 1–3 mm packed bed at the capillary inlet. The washing of buffer and sample was carefully controlled to prevent the problems previously encountered. All washing, sample loading, and electrolyte rinsing proceeded from the detection end of the capillary to the injection end. Organic solvent was then injected for a short time from the capillary inlet. The elution solvent was pushed further through the packed bed by a short injection of electrolyte, followed by application of the separation potential. The analytes are stacked at the boundary of the high resistance organic solvent and the BGE.

Zhao and co-workers [96] have made use of a four-capillary system arranged in a cross. Numbering the capillaries 1–4 in a clockwise manner, the LC mobile phase flows from a micro-LC column through capillary 1 and to waste through capillary 3. Once the retained analytes are transported into capillary 3 they may be electrokinetically injected into capillary 2. Stacking is observed due to the difference in conductivities between the LC mobile phase and the CE BGE. Once the analytes are stacked into narrow bands at the head of capillary 2, electrophoretic separation is performed between capillaries 4 and 2, thus preventing the low-conductivity mobile phase (present in capillaries 1 and 3) from interfering with the separation electric field. Using matrix switching techniques, detection enhancements of 400- to 500-fold [96, 97] and as high as 7000-fold [95] have been reported. These techniques are limited by their complexity, which can lower the reproducibility of the methods.

7 Stacking techniques in MEKC

Techniques for on-column sample stacking have also been applied to MEKC. These techniques fall into two categories; the first is a derivative of FASS and the second is termed sweeping. A number of techniques for enhancing the detection limits of neutral analytes by applying FASS in MEKC have been reported [5, 98–107]. Some of the methods reported are outlined below.

7.1 FASS

Liu *et al.* [98] reported stacking by hydrodynamically injecting a sample in a low-conductivity micellar solution into the capillary containing a high-conductivity micellar BGE. The analytes may be stacked in either normal or reverse polarity mode. In normal polarity mode the neutral analytes contained within the anionic micelles migrate quickly toward the inlet end of the capillary and stack at the boundary between the sample solution and the BGE that is being drawn into the capillary by the EOF. The net migration then switches toward the detector since the electroosmotic velocity is greater than the electrophoretic velocity of the micelles in the high conductivity buffer. In reverse polarity mode a negative high voltage is placed at the capillary inlet. The micelles stack towards the detector and the sample plug is backed out of the capillary. Once the current measured through the capillary returns to 90–99% of its preinjection value, the polarity is reversed and the separation proceeds in normal polarity mode. The two methods gave similar stacking efficiencies up to an injection volume of 80 nL, while with an injection volume of 80–160 nL normal polarity stacking yielded superior results. This is attributed to the decrease in stacking efficiency due to band broadening during polarity switching. With the instrumentation used in this study, polarity switching took about 1 min, allowing time for the analyte bands to diffuse. Above 160 nL, the reverse polarity stacking mode was found to be superior. This effect was due to the increase in sample injection volume, which disrupted the uniform field strength in the capillary, resulting in a decrease in efficiency. An 85-fold increase in response was reported using these methods.

Quirino and Terabe [100] have reported a series of studies investigating protocols for the stacking of neutral analytes in MEKC. The first method is termed normal stacking mode. In this technique a sample of neutral analytes,

in a low-conductivity sample solution such as water, is injected hydrodynamically into the capillary. Due to the high field strength when the separation voltage is applied, micelles race across the sample zone toward the capillary inlet incorporating the neutral analytes. Once the micelles reach the boundary between the sample zone and the BGE that has been drawn into the capillary by the EOF, they are stacked into narrow bands. The result was a 10-fold enhancement in detection limits. In reverse polarity stacking mode, analytes in a low-conductivity matrix are again injected hydrodynamically [101]. Stacking is then performed by applying a negative voltage to the capillary inlet. As the sample solution is being backed out of the capillary by the EOF, micelles from the inlet buffer reservoir migrate across the sample zone towards the detector, stacking the analytes. Once the current of the system has returned to 90–99% of its preinjection value, the polarity is reversed and the separation proceeds in normal polarity mode. Although some improvement in stacking efficiency is observed, as compared to normal stacking mode, the reproducibility of this method may be limited by the analyst's ability to determine when to switch the polarity from stacking to separation mode.

Stacking with reverse migrating micelles employs a highly acidic micellar BGE to reduce the EOF [102]. Samples in water are injected onto the capillary *via* pressure, and the

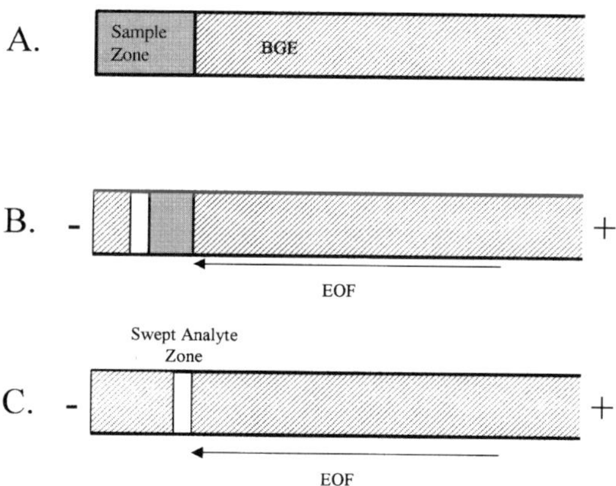

Figure 5. Schematic of sweeping for MEKC. (A) A sample in buffer of the same conductivity as the BGE but without micelles is injected into the capillary. (B) The inlet reservoir is replaced with BGE and a reverse polarity high voltage is applied. The anionic micelles from the BGE migrate across the sample zone, incorporating the neutral analytes into a narrow stacked band. (C) The analytes are stacked into a narrow band and the separation proceeds towards the detector. Although the EOF is in the direction of the capillary inlet, the low pH of the BGE suppresses the EOF so that the net migration of the micelles is towards the detector.

Electrophoresis 2000, *21*, 2768–2779

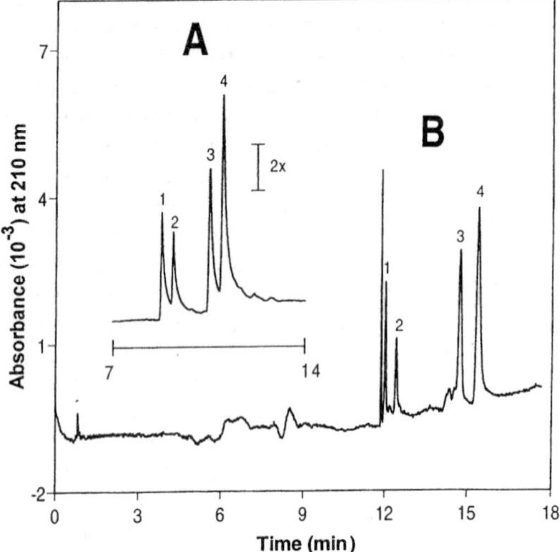

Figure 6. Example of sweeping for concentration of neutral analytes. (A) Injection of analytes between 190 and 265 ppm. (B) Injection of analytes between 19 and 26.5 ppb. Peak identities: 1, trimipramine; 2, nicardipine; 3, noscapine; 4, laudanosine.

Table 1. Comparison of stacking efficiencies and reproducibility for preconcentration methods

Technique	Limits of detection	Stacking efficiency (X-fold increase in peak height)	Reproducibility (RSD peak height, $n \geq 3$)
FASS	80 nM [11]	1000 [14]	3–6% [11]
			2.5–25% [14]
LVSS	2 ppb [29]	100 [29]	5–12% [29]
pH-mediated	35–50 µM [25]	8–20 [25]	1.4–8.8% [25]
	300 nM [33]	66 [33]	1.4–1.7% [33]
ITP - single capillary	50 nM [58]	100–200 [43, 58]	0.7–1.5% [43, 44]
ITP - double capillary	10 nM [49]	300–10 000 [49, 52]	0.3–2.6% [50–52]
Chromatographic preconcentration	5 nM [66]	100–200 [62, 66]	3.8–6.2% [62, 66]
Matrix switching	0.6–10 nM [95, 96]	500–7000 [95, 96]	10% [96]
FASS with MEKC	15.5–21.6 ppb [105]	28–102 [105]	1.2–4.6% [105]
Sweeping	1.7–9.6 ppb [110]	88–5044 [110]	3.9–13.8% [110]

separation is run in reverse polarity mode. Since the potential at the inlet is negative, the EOF slowly pushes the sample plug out of the capillary. However, because the high concentration of protons dramatically reduces the EOF, the micelle's net migration is towards the detector. This method was shown to achieve two orders of magnitude enhancement in detection. Quirino and Terabe have also reported stacking of neutral analytes in MEKC by field-enhanced injection [103], field-enhanced injection with reverse migrating micelles [104], and reverse migrating micelles with the injection of a water plug [105]. Enhancements in detection as determined from peak heights were found to be about 20-fold, 75-fold, and 100-fold with some compounds for each of the previous methods, respectively.

A 400-fold increase in peak heights was demonstrated utilizing a preinjection plug of high conductivity and high viscosity [5]. In this technique a 38 mm plug of 500 mM phosphate was first followed by injection of a 3.8 mm plug of water, then electrokinetic injection of low-conductivity sample solution. Next a negative voltage is applied while the capillary inlet is submersed in 500 mM surfactant for a short period of time. The separation is carried out in normal polarity mode with micellar BGE at the capillary inlet. Again, a major factor limiting the appeal of this method is the ability of the analyst to repeat the injection sequence in a precise manner.

7.2 Sweeping

Sweeping (Fig. 5) is a technique for on-column sample concentration of nonpolar molecules resulting in an 80- to 5000-fold enhancement (Fig. 6) based on the analytes' ability to partition into the pseudostationary phase in MEKC [108–110]. Samples are injected onto the column in a buffer solution with a similar conductivity as that of the BGE, but in the absence of a pseudostationary phase. The capillary inlet is then placed into an anionic micellar BGE solution and the separation is performed in reverse polarity mode. The BGE is kept at a low pH to suppress the EOF, allowing the anionic micelles to electrophoretically migrate towards the detector. As the micelles migrate towards the detector they "sweep" the neutral analytes along. This effect is dependent on a uniform electric field and the absence of micelles in the sample solution. The effectiveness of this sample concentration technique has been shown to be dependent on the analytes' affinity for the pseudostationary phase. The length of the sweep zone (l_{sweep}) is related to the length of the analyte zone ($l_{injected}$) by the equation

$$l_{sweep} = l_{injected}/(1+k) \tag{5}$$

where k is the retention factor.

8 Future trends

Sample stacking is often simple to perform and has a wide range of applications. Most analyses use water as the sample diluent with large injection volume or field amplification methods. With the availability of better commercial instruments, automation will provide increased reproducibility in the steps used to perform these stacking techniques. This makes it more likely that validated CE methods for routine analysis will take advantage of stacking procedures to enhance the detection of analytes in low concentrations or biological matrices.

A new direction in sample stacking involves analysis of samples in high ionic strength matrices. Hadwiger and co-workers [31] have illustrated the use of pH-mediated sample stacking for on-column concentration of isoproterenol in dialysate. Double-capillary pH-mediated stacking is also an important new area, with a 300-fold sample concentration being possible [33]. While the high ionic strength of the sample buffer is detrimental to normal CE separations, some sample stacking methods are actually improved when high ionic strength sample matrix is used instead of water. For example, Shihabi [6] has shown FASS to be enhanced when diluting samples with acetonitrile if the sample is initially in 1% saline. Further application and new methods of stacking of physiological solutions would be helpful in studying drug transport, metabolism, and pharmacokinetics and may help move CE out of the academic laboratory and further into the industrial setting.

Microfabrication is an area in which CE is now playing a major role. Jacobson and Ramsey [111] demonstrated field effect stacking of dansylated amino acids on a quartz microchip with a 31-fold concentration of sample. Detection limits of 70 nM were reported for dansyl-lysine. Microfabrication technology will provide the ability to create new stacking procedures, making it easier to design stacking systems with multiple channels and electrical circuits. In general, sample stacking applied to microfabricated CE systems would allow for on-column concentration with the advantages of the high efficiency and fast analysis of separations in the chip format.

Received February 15, 2000

9 References

[1] Mikkers, F. E. P., Everaerts, F. M., Verheggen, T. P. E. M., *J. Chromatogr.* 1979, *169*, 11–20.

[2] Burgi, D. S., Chien, R.-L., *Anal. Chem.* 1991, *63*, 2042–2047.

[3] Chien, R.-L., Helmer, J. C., *Anal. Chem.* 1991, *63*, 1354–1361.

[4] Chien, R.-L., Burgi, D. S., *J. Chromatogr.* 1991, *559*, 141–152.

[5] Zhang, C.-X., Thormann, W., *Anal. Chem.* 1998, *70*, 540–548.

[6] Shihabi, Z. K., *J. Chromatogr. A* 1999, *853*, 3–9.

[7] Sjogren, A., Dasgupta, P. K., *Anal. Chem.* 1996, *68*, 1933–1940.

[8] Friedberg, M. A., Hinsdale, M., Shihabi, Z. K., *J. Chromatogr. A* 1997, *781*, 35–42.

[9] Beckers, J. L., Ackermans, M. T., *J. Chromatogr.* 1993, *629*, 371–378.

[10] Tan, W. G., Tyrrell, D. L. J., Dovichi, N. J., *J. Chromatogr. A* 1999, *853*, 309–319.

[11] Zhang, C.-X., Aebi, Y., Thormann, W., *Clin. Chem.* 1996, *42*, 1805–1811.

[12] Tagliaro, F., Manetto, G., Crivellente, F., Scarcella, D., Marigo, M., *Forens. Sci. Int.* 1998, *92*, 201–211.

[13] Wey, A. B., Zhang, C.-X., Thormann, W., *J. Chromatogr. A* 1999, *853*, 95–106.

[14] Zhang, C.-X., Thormann, W., *Anal. Chem.* 1996, *68*, 2523–2532.

[15] Morales, S., Cela, R., *J. Chromatogr. A* 1999, *846*, 401–411.

[16] Wang, F., Khaledi, M. G., *J. Chromatogr. B* 1999, *731*, 187–197.

[17] Chien, R.-L., Burgi, D. S., *Anal. Chem.* 1992, *64*, 489A–496A.

[18] Albin, M., Grossman, P. D., Moring, S. E., *Anal. Chem.* 1993, *65*, 489A–497A.

[19] Chien, R.-L., Burgi, D. S., *Anal. Chem.* 1992, *64*, 1046–1050.

[20] McGrath, G., Smyth, W. F., *J. Chromatogr. B* 1996, *681*, 125–131.

[21] Harland, G. B., McGrath, G., McClean, S., Smyth, W. F., *Anal. Commun.* 1997, *34*, 9–11.

[22] Smyth, W. F., Harland, G. B., McClean, S., McGrath, G., Oxspring, D., *J. Chromatogr. A* 1997, *772*, 161–169.

[23] McClean, S., O'Kane, E., Coulter, D. J. M., McLean, S., Smyth, W. F., *Electrophoresis* 1998, *19*, 11–18.

[24] Hissner, F., Daus, B., Mattusch, J., Heinig, K., *J. Chromatogr. A* 1999, *853*, 497–502.

[25] Martinez, D., Borrull, F., Calull, M., *J. Chromatogr. A* 1997, *788*, 185–193.

[26] Albert, M., Debusschere, L., Demesmay, C., Rocca, J. L., *J. Chromatogr. A* 1997, *757*, 281–289.

[27] Burgi, D. S., *Anal. Chem.* 1993, *65*, 3726–3729.

[28] Albert, M., Debusschere, L., Demesmay, C., Rocca, J. L., *J. Chromatogr. A* 1997, *757*, 291–296.

[29] Quirino, J. P., Terabe, S., *J. Chromatogr. A* 1999, *850*, 339–344.

[30] Pálmarsdóttir, S., Mathiasson, L., Jonsson, J. A., Edholm, L.-E., *J. Chromatogr. B* 1997, *688*, 127–134.

[31] Hadwiger, M. E., Torchia, S. R., Park, S., Biggin, M. E., Lunte, C. E., *J. Chromatogr. B* 1996, *681*, 241–249.

[32] Park, S., Lunte, C. E., *J. Microcol. Sep.* 1998, *10*, 511–517.

[33] Zhao, Y., Lunte, C. E., *Anal. Chem.* 1999, *71*, 3985–3991.

[34] Xiong, Y., Park, S., Swerdlow, H., *Anal. Chem.* 1998, *70*, 3605–3611.

[35] Aebersold, R., Morrison, H. D., *J. Chromatogr.* 1990, *516*, 79–88.

[36] Schwer, C., Lottspeich, F., *J. Chromatogr.* 1992, *623*, 345–355.

[37] Mazereeuw, M., Tjaden, U. R., Reinhoud, N. J., *J. Chromatogr. Sci.* 1995, *33*, 686–697.

[38] Foret, F., Szökö, E., Karger, B. L., *J. Chromatogr.* 1992, *608*, 3–12.

[39] Křivánková, L., Boček, P., *J. Chromatogr. B* 1997, *689*, 13–34.

[40] Thormann, W., Dieter A., Ernst, S., *Electrophoresis* 1985, *6*, 10–18.

[41] Hirokawa, T., Watanabe, K., Yokota, Y., Kiso, Y., *J. Chromatogr.* 1993, *633*, 233–259.

[42] Gebauer, P., Thormann, W., Boček, P., *J. Chromatogr.* 1992, *608*, 47–57.

[43] Schoots, A. C., Verheggen, T. P. E. M., De Vries, P. M. E. M., Everaerts, F. M., *Clin. Chem.* 1990, *36*, 435–440.

[44] Reinhoud, N. J., Tjaden, U. R., van der Greef, J., *J. Chromatogr.* 1993, *641*, 155–162.

[45] van der Scans, M. J., Beckers, J. L., Molling, M. C., Everaerts, F. M., *J. Chromatogr. A* 1995, *717*, 139–147.

[46] Mikkers, F. E. P., Everaerts, F. M., Verheggen, T. P. E. M., *J. Chromatogr.* 1979, *169*, 21–38.

[47] Křivánková, L., Gebauer, P., Boček, P., *J. Chromatogr. A* 1995, *716*, 35–48.

[48] Křivánková, L., Pantuckova, P., Boček, P., *J. Chromatogr. A* 1999, *838*, 55–70.

[49] Dankova, M., Kaniansky, D., Fanali, S., Ivanyi, F., *J. Chromatogr. A* 1999, *838*, 31–43.

[50] Křivánková, L., Foret, F., Boček, P., *J. Chromatogr.* 1991, *545*, 307–313.

[51] Kaniansky, D., Marak, J., *J. Chromatogr.* 1990, *498*, 191–204.

[52] Hirokawa, T., Ohmori, A., Kiso, Y., *J. Chromatogr.* 1993, *634*, 101–106.

[53] Kaniansky, D., Marak, J., Lastinec, J., Reijenga, J. C., Onuska, F. I., *J. Microcol. Sep.* 1999, *11*, 141–153.

[54] Stegehuis, D. S., Irth, H., Tjaden, U. R., van der Greef, J., *J. Chromatogr.* 1991, *538*, 393–402.

[55] Boček, P., Deml, M., Kaplanova, B., Janak, J., *J. Chromatogr.* 1978, *160*, 1–9.

[56] Mikkers, F. E. P., Everaerts, F. M., Peek, J. A. F., *J. Chromatogr.* 1979, *168*, 317–332.

[57] Kaniansky, D., Zelensky, I., Hybenova, A., Onuska, F. I., *Anal. Chem.* 1994, *66*, 4258–4264.

[58] Larsson, M., Nagard, S., *J. Microcol. Sep.* 1994, *6*, 107–113.

[59] Foret, F., Sustáček, V., Boček, P., *J. Microcol. Sep.* 1990, *2*, 229–233.

[60] Schwer, C., Lottspeich, F., *J. Chromatogr.* 1992, *623*, 345–355.

[61] Michalke, B., Schramel, P., *Electrophoresis* 1998, *19*, 2220–2225.

[62] Hoyt Jr., A. M., Beale, S. C., Larmann Jr., J. P., Jorgenson, J. W., *J. Microcol. Sep.* 1993, *5*, 325–330.

[63] Strausbauch, M. A., Madden, B. J., Wettstein, P. J., Landers, J. P., *Electrophoresis* 1995, *16*, 541–548.

[64] Strausbauch, M. A., Xu, S. J., Ferguson, J. E., Nunez, M. E., Machacek, D., Lawson, G. M., Wettstein, P. J., Landers, J. P., *J. Chromatogr. A* 1995, *717*, 279–291.

[65] Beattie, J. H., Self, R., Richards, M. P., *Electrophoresis* 1995, *16*, 322–328.

[66] He, J., Shibukawa, A., Zeng, M., Amane, S., Sawada, T., Nakagawa, T., *Anal. Sci.* 1996, *12*, 177–181.

[67] Figeys, D., Ducret, A., Aebersold, R., *J. Chromatogr. A* 1997, *763*, 295–306.

[68] Knudsen, C. B., Beattie, J. H., *J. Chromatogr. A* 1997, *792*, 463–473.

[69] Li, J., Thibault, P., Martin, A., Richards, J. C., Wakarchuk, W. W., van der Wilp, W., *J. Chromatogr. A* 1998, *817*, 325–336.

[70] Tomlinson, A. J., Benson, L. M., Braddock, W. D., Oda, R. P., Naylor, S., *J. High Resol. Chromatogr.* 1994, *17*, 729–731.

[71] Guzman, N. A., Trebilcock, M. A., Advis, J. P., *J. Liq. Chromatogr.* 1991, *14*, 997–1015.

[72] Morita, I., Sawada, J., *J. Chromatogr.* 1993, *641*, 375–381.

[73] Swartz, M. E., Merion, M., *J. Chromatogr.* 1993, *632*, 209–213.

[74] Cai, J., El Rassi, Z., *J. Liq. Chromatogr.* 1992, *15*, 1179–1192.

[75] Cai, J., El Rassi, Z., *J. Liq. Chromatogr.* 1993, *16*, 2007–2024.

[76] Tomlinson, A. J., Benson, L. M., Guzman, N. A., Naylor, S., *J. Chromatogr. A* 1996, *744*, 3–15.

[77] Debets, A. J. J., Mazereeuw, M., Voogt, W. H., van Iperen, D. J., Lingeman, H., Hupe, K.-P., Brinkman, U. A. T., *J. Chromatogr.* 1992, *608*, 151–158.

[78] Bushey, M. M., Jorgenson, J. W., *Anal. Chem.* 1990, *62*, 978–984.

[79] Lemmo, A. V., Jorgenson, J. W., *Anal. Chem.* 1993, *65*, 1576–1581.

[80] Hooker, T. F., Jorgenson, J. W., *Anal. Chem.* 1997, *69*, 4134–4142.

[81] Pálmarsdóttir, S., Mathiasson, L., Jonsson, J. A., Edholm, L.-E., *J. Capil. Electrophor.* 1996, *3*, 255–260.

[82] Veraart, J. R., Gooijer, C., Lingeman, H., Velthorst, N. H., Brinkman, U. A. T., *Chromatographia* 1997, *44*, 581–588.

[83] Veraart, J. R., Gooijer, C., Lingeman, H., Velthorst, N. H., Brinkman, U. A. T., *J. Chromatogr. B* 1998, *719*, 199–208.

[84] Arce, L., Kuban, P., Rios, A., Valcarcel, M., Karlberg, B., *Anal. Chim. Acta* 1999, *390*, 39–44.

[85] Chen, H.-W., Fang, Z.-L., *Anal. Chim. Acta* 1997, *355*, 135–143.

[86] Tomlinson, A. J., Naylor, S., *J. High Resol. Chromatogr.* 1995, *18*, 384–386.

[87] Tomlinson, A. J., Benson, L. M., Braddock, W. D., Oda, R. P., Naylor, S., *J. High Resol. Chromatogr.* 1995, *18*, 381–383.

[88] Tomlinson, A. J., Naylor, S., *J. Liq. Chromatogr.* 1995, *18*, 3591–3615.

[89] Tomlinson, A. J., Naylor, S., *J. Capil Electrophor.* 1995, *2*, 225–233.

[90] Benson, L. M., Tomlinson, A. J., Mayeno, A. N., Gleich, G. J., Wells, D., Naylor, S., *J. High Resol. Chromatogr.* 1996, *19*, 291–294.

[91] Tomlinson, A. J., Benson, L. M., Jameson, S., Johnson, D. H., Naylor, S., *J. Am. Soc. Mass Spectrom.* 1997, *8*, 15–24.

[92] Rohde, E., Tomlinson, A. J., Johnson, D. H., Naylor, S., *J. Chromatogr. B* 1998, *713*, 301–311.

[93] Naylor, S., Tomlinson, A. J., *Talanta* 1998, *45*, 603–612.

[94] Yang, Q., Tomlinson, A. J., Naylor, S., *Anal. Chem.* 1999, *71*, 183A–189A.

[95] Petersson, M., Wahlund, K.-G., Nilsson, S., *J. Chromatogr. A* 1999, *841*, 249–261.

[96] Zhao, Y., Mclaughlin, K., Lunte, C. E., *Anal. Chem.* 1998, *70*, 4578–4585.

[97] Pálmarsdóttir, S., Edholm, L.-E., *J. Chromatogr. A* 1995, *693*, 131–143.

[98] Liu, Z., Sam, P., Sirimanne, S. R., McClure, P. C., Grainger, J., Patterson Jr., D. G., *J. Chromatogr. A* 1994, *673*, 125–132.

[99] Nielsen, K. R., Foley, J. P., *J. Chromatogr. A* 1994, *686*, 283–291.

[100] Quirino, J. P., Terabe, S., *J. Chromatogr. A* 1997, *781*, 119–128.

[101] Quirino, J. P., Terabe, S., *J. Chromatogr. A* 1997, *791*, 225–267.

[102] Quirino, J. P., Terabe, S., *Anal. Chem.* 1998, *70*, 149–157.

[103] Quirino, J. P., Terabe, S., *J. Chromatogr. A* 1998, *798*, 251–257.

[104] Quirino, J. P., Terabe, S., *Anal. Chem.* 1998, *70*, 1893–1901.

[105] Quirino, J. P., Otsuka, K., Terabe, S., *J. Chromatogr. B* 1998, *714*, 29–38.

[106] Wu, Y. S., Lee, H. K., Li, S. F. Y., *J. Microcol. Sep.* 1998, *10*, 529–535.

[107] Szucs, R., Vindevogel, J., Sandra, P., Verhagen, L. C., *Chromatographia* 1993, *36*, 323–329.

[108] Quirino, J. P., Terabe, S., Otsuka, K., Vincent, J. B., Vigh, G., *J. Chromatogr. A* 1999, *838*, 3–10.

[109] Quirino, J. P., Terabe, S., *Anal. Chem.* 1999, *71*, 1638–1644.

[110] Quirino, J. P., Terabe, S., *Science* 1998, *282*, 465–468.

[111] Jacobson, S. C., Ramsey, J. M., *Electrophoresis* 1995, *16*, 481–486.

Electrophoresis 2000, *21*, 691–698

Review

Georg Hempel

Institut für Pharmazeutische
Chemie der Universität
Münster, Münster, Germany

Strategies to improve the sensitivity in capillary electrophoresis for the analysis of drugs in biological fluids

Capillary electrophoresis (CE) is a useful method to quantify drugs in biological fluids. However, especially for blood or plasma samples, the sensitivity is not sufficient to quantify drugs and their metabolites as they often need to be quantified in the lower µg/L range. To overcome this limitation and to increase the sensitivity, two strategies are applied: first, to increase the amount of analyte added to the capillary and, second, to increase the sensitivity on the detector site. To improve the sensitivity on the detector site, alternative detection techniques to UV detection, *e.g.*, laser-induced fluorescence detection (LIF) or mass spectroscopy (MS), can be applied. However, LIF detection can only be used for fluorescent analytes and the current equipment for CE-MS coupling provides only small improvements in sensitivity compared to UV detection. The detection window for UV detection can be enhanced using capillaries with an extended light path (bubble cell) or Z-shaped capillaries. Sensitivity improvements up to a factor of 10 have been reported. Increasing the amount of analyte in the capillary can be done either by chromatographic or by electrokinetic methods. Chromatographic methods such as on-capillary membrane preconcentration have been used for several analytes. However, no validated application has been reported to date. In contrast, several validated examples can be found in which electrokinetic techniques like sample stacking have been applied to achieve limits of quantification in the lower µg/L range. In conclusion, to date, electrokinetic techniques such as field-amplified sample injection offer the most promising results in achieving a sufficient sensitivity to quantify drugs in biological fluids.

Keywords: Capillary electrophoresis / Drug analysis / Sensitivity enhancement / Review

EL 3810

Contents

1 Introduction

Capillary electrophoresis (CE) offers several advantages over chromatographic techniques for the analysis of drugs in biological fluids, especially when only small amounts of sample are available. The main advantages of CE over chromatographic methods are the higher separation efficacy, the smaller sample volume required, and the lower costs due to low, if any, organic solvent consumption and inexpensive capillaries instead of expensive HPLC or GC columns. In addition, under special conditions (with coated capillaries or with the addition of SDS) the method is insensitive to other sample constituents like proteins. Biomedical applications of CE have

Correspondence: Dr. Georg Hempel, Institut für Pharmazeutische Chemie der Universität Münster, Hittorfstr. 58–62, and Universitätsklinik, Abt. Hämatologie/Onkologie, 48149 Münster, Germany
E-mail: hempege@uni-muenster.de
Fax: +49-251-83-47828

Electrophoresis 2000, *21*, 691–698

been reviewed extensively [1–3]. In addition, several reviews can be found that list applications of CE for the analysis of drugs in biological fluids [4–8]. If the drug of interest is present in the higher mg/L range, as in urine, or when the compound is administered in gram amounts, direct injection of body fluids into the capillary can be possible [9, 10]. However, many drugs of interest are administered in low milligram amounts and many drugs show a high volume of distribution, resulting in low concentrations in the body fluids (plasma or cerebrospinal fluid). In comparison to HPLC, the sample volume applied in CE is about 1000-fold smaller and the detector cell path length is 100 times less than in HPLC. Therefore, CE with UV detection is not sensitive enough to quantify these drugs in biological fluids.

To increase the sensitivity, two main approaches are reported: increasing the amount of analyte added to the capillary and improving the sensitivity of the detector. This review gives an overview of methods to overcome problems concerning the sensitivity of CE for the determination of drugs in biological fluids and to emphasize their potential and limitations. It will focus on methods which can be applied in a clinical laboratory with commercially available equipment.

2 Improvements in detection

Due to the short detection pathlength, several problems arise. Some authors report an up to fivefold increase in sensitivity with new instrumentation, for example the Beckman MDQ instrument (Palo Alto, CA, USA) compared to older CE equipment [5, 11].

2.1 Bubble cells

An enhanced detection window with three- to fivefold the inner diameter of the remaining capillary can be etched in fused-silica capillaries using hydrofluoric acid [12]. These capillaries are also commercially available from Hewlett Packard (HP, Avondale, PA, USA). Although they are designed for the HP instrument they can also be used for other equipment by carefully burning off the plastic device designated to fix the capillary in the HP cartridge. Sensitivity improvement is in the range of a factor of 2–3 without substantial loss in separation efficacy. Several applications have been described for the analysis of drugs in biological fluids [13, 14]. Although bubble cells provide no substantial improvement in sensitivity, they can be useful if a slightly higher sensitivity is necessary. In addition, they can be combined with other techniques for sensitivity improvement (*e.g.*, sample stacking or LIF detection) [15]. The enhanced detection window is shown in Fig. 1.

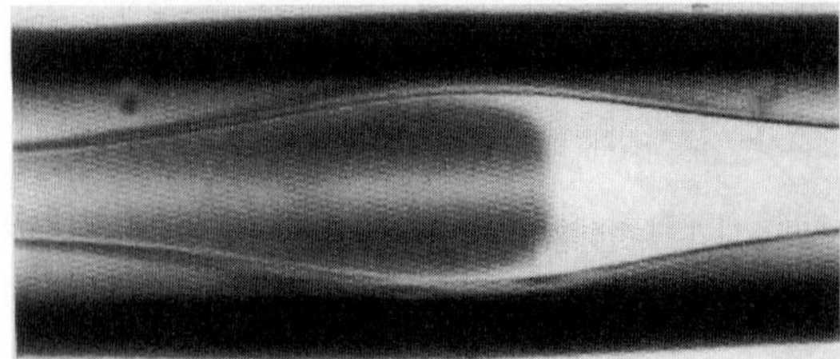

Figure 1. Detection window of a capillary with an extented light path (bubble cell). Courtesy of Hewlett-Packard.

2.2 Z-shaped cells

Bent capillaries, with a few millimeters of the capillary parallel to the light path, can provide a much longer optical path length. However, due to the longer path length the resolution is decreased, which might cause problems with peaks migrating closely together. It is obvious that the distance between the peaks must be at least the detection path length. In 1991, Chevret *et al.* [16] introduced a Z-shaped cell with a 3 mm path length (Fig. 2). The improvement in signal-to-noise ratio by a factor of 6 was lower than expected considering the increase in path length, possibly due to the presence of scattered light [16]. Recently, Hewlett-Packard introduced a so-called high sensitivity cell consisting of a Z-shaped capillary located in the detection window of a capillary cartridge [17]. In comparison to the earlier reports, several problems with scattered light were solved. This detection device is connected to the separation capillary by fittings with ferrules similar to connections in gas chromatography. However, in our experience these connections can cause problems due to disturbances in the fluid connec-

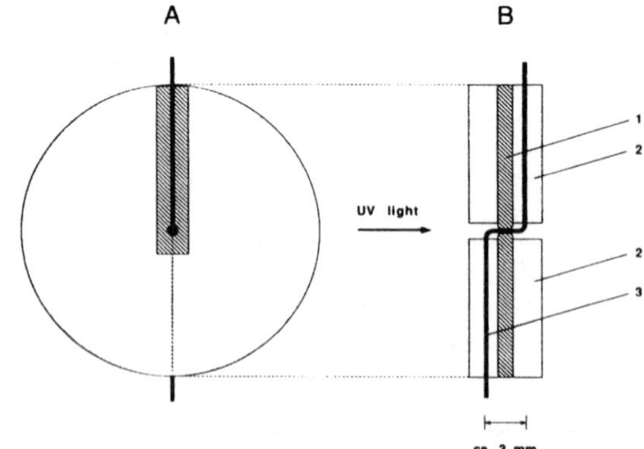

Figure 2. Diagram of the Z-shaped capillary flow cell. (A) Front view; (B) cross-sectional view. 1, shim (alumina); 2, platic disks; 3, capillary. Reproduced from[16], with permission.

tion between the detection cell and the remaining capillary. Thus, currently we are only aware of two applications reported in the literature for the high sensitivity cell, of which only one has been developed for a quantitative assay [18, 19]. In conclusion, this device needs to be improved before it can be recommended for use in analytical routines.

2.3 LIF detection

LIF detection provides extremely high mass sensitivity with single molecule detection being reported for some analytes [20]. Currently, however, direct LIF is only applicable for some analytes as lasers are only commercially available with wavelengths of 325 or 488 nm. An alternative to direct detection is derivatization of the analytes with fluorescent tags. However, derivatization means a loss in selectivity and is therefore not generally applicable. An example is the quantification of the antidiabetic acarbose and its main metabolite in human urine using 7-aminonaphthalene-1,3-disulfonic acid as a derivatization reagent and LIF detection [21]. Although reasonable limits of quantification were achieved, the method was not selective enough to quantify the pseudooligosaccharide acarbose in human plasma.

A LIF detector for CE equipped with an argon-ion laser emitting at 488 nm is available from several manufacturers. Applications are mainly in the field of DNA, peptide, and amino acid analysis where selective fluorescent markers are used. However, some drugs like anthracyclines fluoresce when excited at 488 nm and this can be used for the direct determination of these compounds by CE. Reinhoud *et al.* [22] were the first to show the applicability of CE-LIF for the analysis of anthracyclines. In our laboratory, we developed and validated methods for the quantification of idarubicin [23], doxorubicin [24], and their respective metabolites. Limits of quantification are 0.5 and 2 µg/L for idarubicin and doxorubicin, respectively, with a sample volume of 100 µL plasma. For the determination of the peak plasma levels of doxorubicin, a 10 µL plasma sample can simply be diluted with acetonitrile (Fig. 3).

Although no complete CE-LIF unit with an He-Cd laser has been commercialized, the laser can be connected with commercially available LIF detectors. With an He-Cd laser emitting at 325 nm, more examples can be found in the literature about drugs having been analyzed with CE-LIF. In this UV region, several drugs contain fluorophores which can be excited. Methotrexate was the first drug analyzed with CE-LIF [25], although it must be noted that this drug must be oxidized or photochemically decomposed before the fluorescence intensity is sufficient. A detection limit as low as 0.22 ng/L was reported. Methods

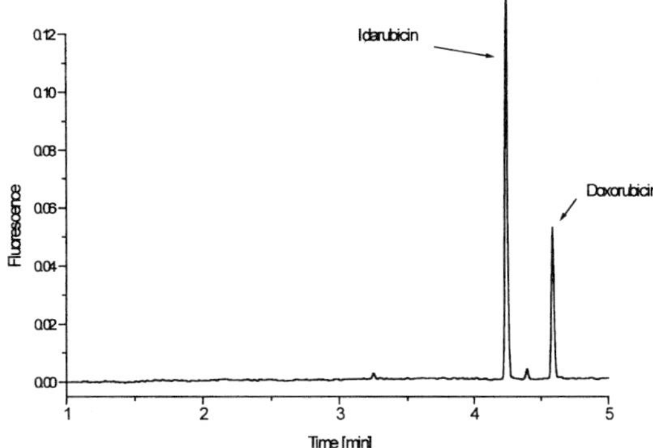

Figure 3. Quantification of doxorubicin in a 10 µL plasma sample after dilution with acetonitrile. Idarubicin is the internal standard. Conditions: Phosphate buffer pH 5 with 60 µM spermine and 70% acetonitrile. Doxorubicin concentration is 354 µg/L [24].

are also described for the determination of 6-mercaptopurine [26], naproxen [27] and the hypnotics zolpidem [28] and zopiclone [29]. In the latter application the drug and its main metabolites were simultaneously separated into the enantiomers.

The fluoroquinolones are a group of antibiotics showing native fluorescence when excited at 325 nm. Bannefeld *et al.* [29] first demonstrated the applicability of CE-LIF for the determination of fluoroquinolone ciprofloxacin. During sample preparation, plasma samples were diluted 50-fold. The method was validated down to a concentration of 20 µg/L. However, by avoiding the dilution, a lower LOQ would also be possible. Möller *et al.* [30] used CE-LIF for the determination of the fluoroquinolone moxifloxacine in a 20 µL plasma sample. The method has been cross-validated with an HPLC assay demonstrating the applicability of CE-LIF in a good laboratory practice (GLP) environment. An excellent review about the applications of CE-LIF in drug analysis was recently published by Couderc *et al.* [31]. With the availability of lasers emitting in the lower UV region, many applications for CE-LIF for the analysis of drugs would be possible. However, these lasers (*e.g.*, frequency doubled argon-ion lasers) are still too expensive for routine use. Thus, due to its limited applicability, CE-LIF cannot yet be regarded as a routine method for the determination of drugs in biological fluids.

2.4 MS detection

CE-MS has been shown to be useful in qualitative assay, for the identification of metabolites in biological fluids [32]. Although MS is generally far more sensitive than UV detection, the current CE/MS interfaces employ make-up liquids diluting the CE efflux to increase the flow up to

10 µL/min. This results in much lower concentration sensitivity compared to LC-MS or GC-MS, or even CE-UV. Attempts are ongoing to improve the interfaces between CE and MS to increase the sensitivity [33, 34]. Thus, only few quantitative methods can be found in the literature [35, 36] and, to our knowledge, only two reports describe attempts to validate CE-MS methods [37, 38] according to the general accepted criteria for bioanalytical methods [39]. Recently, Bach and Henion [38] reported a method for the determination of methylphenidate in urine with a limit of quantification of 1.5 µg/L from 4 mL of urine. A 20-fold preconcentration using liquid/liquid extraction and field-amplified sample injection (see Section 3.2.2) was applied to increase the sensitivity. Although these data are impressing, currently CE-MS cannot be regarded as a solution to overcome the lack of sensitivity in CE. CE-MS is still restricted to more specialized laboratories because MS detectors are still very expensive and the interface is difficult to handle.

3 Strategies to improve the sample amount in the capillary

3.1 Chromatographic techniques

Several groups have focused on developing techniques to concentrate the analytes in the capillary by chromatographic techniques. This can be done by introducing a small amount of reversed-phase silica material at the inlet of the capillary. The silica is held in place by two glass frits on both ends. Excellent linearity has been reported with a correlation coefficient of 1.00 at concentrations from 10 ppb to 10 ppm for doxepin and propranolol. In addition, with repeated injections, relative standard deviations as low as 1.74 and 1.9% for migration times and peak areas respectively, have been observed [40]. However, although the first paper utilizing this LC principle for sample enrichment in CE was published almost 10 years ago [41], to the best of our knowledge no quantitative application of this technique can be found in the literature. Obviously, problems arise in applying this technique for validated assays. One main problem seems to be increased back pressure disturbing the electroosmotic flow within the capillary due to the glass frits and the packing material [42]. In addition, sample enrichment of the packing material can take several minutes due to the high volume to be pumped through the capillary. This and purge steps to precondition the silica material can increase the overall analysis time to values comparable or even longer than with HPLC techniques. Thus, a major advantage of CE is lost. Another problem is the relatively large volume of organic solvents necessary to elute the sample from the silica materials. This may have a negative influence on the CE separation of the compounds.

To overcome some of the problems of this approach, another concept has been introduced using C_{18}-impregnated styrene-divinylbenzene membranes installed in a Teflon cartridge system [43] (Fig. 4). This technique can reduce the volume of organic solvent necessary to elute the analyte from the preconcentration device. In addition, the reduced backpressure in comparison to glass frits results in a more reproducible EOF and a better resolution in the CE separation [44]. In conjunction with transient isotachophoresis and MS, detection limits as low as 200 ng/L were reported [45]. Although this technique is attractive from a theoretical point of view, no reports can be found in the literature applying this approach to a quantitative assay of drugs or endogenous compounds in biological fluids. Recently, Barroso and de Jong [46] introduced a device consisting of a small frit and a 1 mm zone of 5 µm C_{18}-modified silica material placed in a 75 µm capillary. A very low backpressure and a low volume of organic solvent were necessary to elute the sample from the silica material. It remains to be seen if this approach can be applied to develop quantitative assays.

3.2 Electrophoretic techniques

In electrophoresis, analytes migrate in the electric field controlled by their mobility. Differences in the electric field can be achieved by addition or removal of ions or micelles from certain zones in the capillary. In addition, by using electrokinetic injection, analytes can be selectively applied to the capillary without injecting a large volume of sample that may lead to band broadening. Thus, electrophoretic techniques for sample enrichment utilize mechanisms typical for CE, whereas chromatographic techniques introduce other mechanisms into the system, making it more complicated to optimize. However, electrophoretic techniques are usually highly dependent on the sample matrix. Thus, biological samples have to be prepared by liquid/liquid extraction, solid-phase extraction or, ideally, they need only be diluted with acetonitrile [47]. Isotachophoresis is a technique in which samples are concentrated in zones using at least two different buffer systems [48].Although the same mechanism as for sample stacking is used, no applications have been reported for the determination of drugs in biological fluids. Therefore, this technique will not be discussed here in detail.

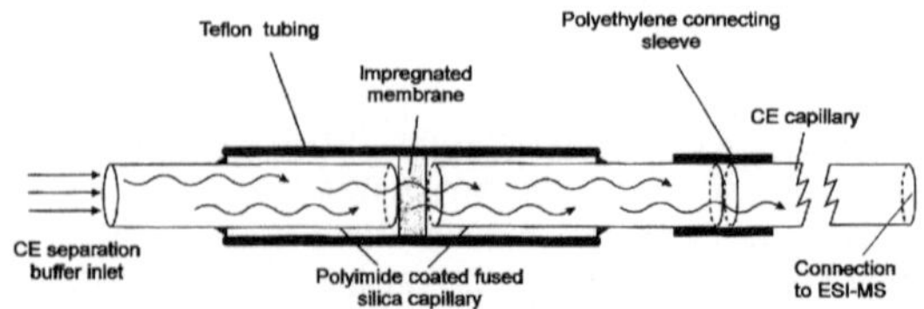

Figure 4. Membrane-preconcentration device with styrene-divinyl-benzene membrane to concentrate samples online in CE. Reproduced from [46], with permission.

3.2.1 Sample stacking

Sample stacking is a well known phenomenon in electrophoresis [49]. Burgi and Chien [50] investigated the factors and mechanism of this method for CE in detail. Injection of a long sample plug with a buffer concentration ten times less than that used for the separation resulted in an improvement in detectability of one order of magnitude [50]. Due to the low ion concentration in the sample plug, the resistivity is higher, causing a higher electric field in this region. Thus, the analytes migrate faster in the sample zone in comparison to their migration in the separation buffer and slow down when entering the separation buffer. Thus, the analytes stack at the boundaries of the sample zone with the separation buffer (anions at the anodic boundary, cations at the cathodic end of the sample zone). Peak broadening due to laminar flow resulting from a mismatch in the electroosmotic velocity of the different zones was identified as the limiting factor in sample stacking.

A higher enhancement in sensitivity can be achieved when the sample buffer is removed by applying a pressure from the outlet end of the capillary after the stacking process [51]. In an extreme case (and if the sample does not contain other ions interfering with the stacking process) the whole capillary can be filled with the sample. Compared to conventional injection, a hundredfold enhancement in sensitivity can be seen. Filling the whole capillary with a pretreated plasma sample and concentrating the analytes in a double stacking process has been reported for terbutaline [52] and bambuterol [53]. Although good reproducibility and limits of detection in the lower µg/L range are reported, to our knowledge no quantitative assays with the analysis of patient samples have been reported to date. In contrast, there are several reports about simple sample stacking having been applied for quantitative assays for drugs in biological fluids [14, 54, 55]. This underlines the applicability of this method in analytical practice for the quantification of drugs in real samples. One limitation of the method is the maximum volume of sample applied, which is limited by the capillary volume. With an injection volume of 0.5–2% in classical CE, the maximum enhancement factor in sample stacking is about 100.

3.2.2 Field-amplified sample injection / stacking

In CE, there are two methods for introducing sample into the capillary: hydrostatic injection (also termed hydrodynamic, pressure, or gravity injection) and electrokinetic injection, sometimes referred to as electromigration [56]. With the latter method, the sample can be introduced into the capillary through electroosmosis, *i.e.*, the electroosmotic flow acts as a pump sucking a certain volume of sample into the capillary. In addition, charged compounds can be selectively applied to the capillary due to their migration in the electric field without applying great sample volumes which might affect the separation. If the sample matrix has a lower ionic strength (and a lower conductivity) than the separation buffer, the electric field strength in the sample zone is higher than in the running buffer [57]. Charged samples migrate faster in the sample zone and can be stacked at the boundary to the separation buffer. However, during electroinjection the buffer boundary may be disturbed, possibly by ions migrating from the buffer into the sample zone. To overcome this, Chien and Burgi [58] showed that injecting a small water plug at the injector site before electroinjection establishes a high electric field at the injection point and, consequently, high stacking factors can be achieved. Minimum detectable concentrations in the order of 10^{-8} M are reported. By switching the polarity of the power source during injection, the method can be applied for both negatively and positively charged analytes [59].

Thormann and Zhang [60] investigated the applicability of field-amplified sample stacking for the analysis of drugs in biological fluids in detail. Under optimized conditions, they found an over 1000-fold sensitivity enhancement compared to hydrodynamic injection without sample stacking. A validated assay for the determination of the antiarrhythmic amiodaron and its main metabolites in human serum has been described [61]. Samples were prepared by liquid/liquid extraction with *n*-hexane at pH 6, thus providing a very clean sample. Electropherograms (Fig. 5B) showed a limit of detection of 50 µg/L with a plasma volume of 20 µL. Inter- and intraday precision and accuracy was found to be better than 7% for concentrations in the therapeutic range. Recently, the same group [62] presented an assay for the determination of dihydrocodeine in 20 µL of plasma with a limit of detection of 3 µg/L. Reproducibility data were comparable to those obtained with validated HPLC methods. Due to changes in the sample composition during electroinjection and partial depletion of the analyte, the samples can only be injected once.

The technique of field-amplified sample stacking has also been applied by other groups for the determination of drugs in biological samples [63, 64]. Thus, this approach provides a significant enhancement in sensitivity for CE in biological fluids while maintaining other features of CE, such as short analysis times and the low sample volume required. This can offer significant advantages over HPLC assays. Theoretically, the whole analyte present in a sample vial of 10–30 µL can be applied to the capillary, thus

Electrophoresis 2000, *21*, 691–698

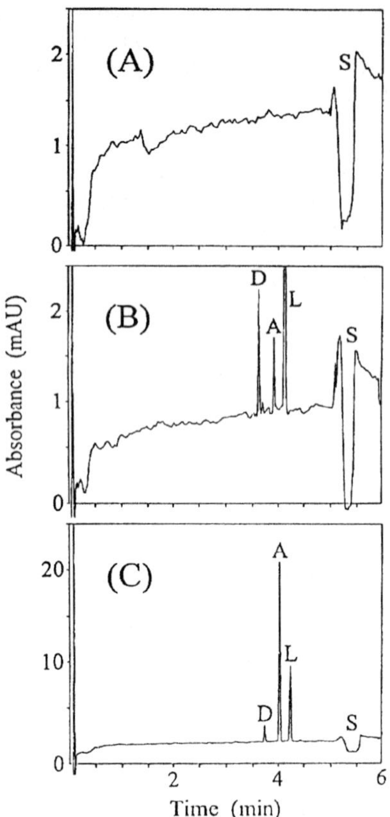

Figure 5. Electropherograms of (A) blank serum and (B, C) two patients' sera with desethylamiodaron (D) and amiodaron (A). L, internal standard. Reproduced from [62], with permission.

reaching concentration factors much higher than in sample stacking with hydrodynamic injection. However, one limitation of this technique is that the analyte must be chargeable under CE conditions; furthermore, other sample constituents, such as ions and proteins, have to be selectively removed before injection.

In a recent paper by Taylor *et al.* [65] addressing the question of sensitivity in CE for drug analysis, the authors pointed out that field-amplified sample injection provides dramatic improvements in sensitivity. However, it was difficult to apply this technique to plasma samples due to other plasma constituents disturbing the stacking process. This underlines the importance of sample pretreatment in sample stacking to produce a clean extract, which can be difficult for neutral and hydrophilic drugs.

3.2.3 Stacking in micellar electrokinetic chromatography

Micellar electrokinetic chromatography (MEKC) extends the applicability of CE to neutral compounds [66]. The separation is based on the affinity of the analytes to micelles added to the running buffer. The micelles act as a pseudostationary phase, thus providing conditions similar to reversed-phase HPLC. Sodium dodecyl sulfate (SDS) is the most commonly used micelle-forming molecule in MEKC. Until recently, only few reports were found in the literature dealing with stacking in MEKC [67, 68]. Lately, Quirino and Terabe [69–72] investigated the factors for stacking in MEKC in more detail. Different surfactants were investigated as well as different modes of stacking with both electrokinetic [73] and hydrodynamic [74] injections. The highest sensitivity improvement was achieved under acidic conditions (to decrease the EOF) with a method termed sweeping: the sample is dissolved in a matrix with the same conductivity as the running buffer but without micelles and is applied to the capillary by pressure injection. When a voltage is applied with the anode at the detector side, the SDS micelles collect the analytes from the sample zone, thus concentrating them in sharp zones. Very long sample plugs can be detected, resulting in an increase in sensitivity of three orders of magnitude [75]. This enhancement factor is higher than the factor theoretically achievable when filling the whole capillary with the sample (as discussed in Section 3.2.1). However, the authors state that a clean sample is necessary to apply the sweeping method because all impurities with affinity to the micelles are enriched as well. Therefore, it remains to be seen if this method can be applied for biological samples.

Another strategy introducing a sample with a higher ionic strength than the running buffer has been proposed by Landers *et al.* [76]. Using sodium cholate as the pseudostationary phase and simply adding sodium chloride (or other ions) to the sample matrix, a reasonable enhancement in sensitivity has been achieved for a series of corticosteroids. The stacking effect is dependent on the affinity of the analytes to the micelles because in the first step the micelles, and not the analytes, are stacked at the boundary between the sample and buffer zone. Subsequently, the analytes migrating with the electroosmotic flow are enriched in the zone with the high micelle concentration as the micelles are negatively charged and migrate in the opposite direction (Fig. 6). This approach is attractive due to its robustness towards other sample constituents. As this technique is relatively new, no quantitative applications have been reported to date.

3.2.4 Acetonitrile stacking

Shihabi [6] described a simple stacking method for plasma samples: the addition of two sample volumes of

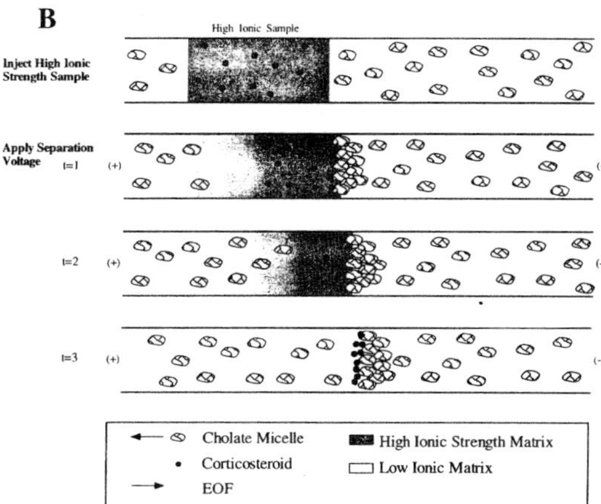

Figure 6. Mechanism for stacking in MEKC with a high ionic strength matrix. Reproduced from [77], with permission.

acetonitrile removes proteins from the sample and enables sample injections of up to 50% the capillary volume. Surprisingly, the addition of ions to the sample even increased the peak shape. A proposal for the mechanism of this phenomenon has not yet been given. Some applications for peptides and some drugs (*e.g.*, acetaminophen) are given [77–79]. Quantitative data, such as RSD values of repeated injections or calibration graphs, are reported for ketoprofen [79], ibuprofen [80] and theophilline [81]. This method is easy to use and well suited for the analysis of drugs in body fluids present in the mg/L range. However, the concentration with acetonitrile is not high enough to quantify analytes at lower concentrations.

4 Conclusions

Caution is necessary when interpreting the enhancement factors reported in the literature because, for comparison, "standard" conditions with very short injection times and not operating under optimized conditions are often used. If high sensitivity is necessary for a CE assay for drugs in biological fluids and LIF detection cannot be used, whenever possible field-amplified sample injection should be applied. This method provides the best results available with standard instruments and capillaries. The effort necessary for method development is limited as only a few factors have to be optimized. Problems may arise with more hydrophilic drugs as it might be difficult to obtain a clean extract necessary for electrokinetic injection. Bubble cells provide a slight improvement in sensitivity and can be used with all standard CE equipment. All other techniques are limited to experimental laboratories and cannot be used in clinical routine to date. Furthermore, no

general common method for concentrating all drugs exists. Some methods of stacking might work better with certain drugs while others may produce only a limited degree of concentration.

This work was supported by the Federal Department of Research and Technology (#1EC9401). The help of Rainer Lingg and Claudia Lanvers in correcting the manuscript is gratefully acknowledged.

Received August 6, 1999

5 References

[1] Lehmann, R., Voelter, W., Liebich, H. M., *J. Chromatogr. B* 1997, *697*, 3–35.

[2] Jenkins, M. A., Guerrin, M. D., *J. Chromatogr. B* 1996, *682*, 23–34.

[3] Deyl, Z., Tagliaro, F., Miksik, I., *J. Chromatogr. B* 1994, *656*, 3–27.

[4] Brunner, L. J., DiPiro, J. T., *Electrophoresis* 1998, *19*, 2848–2855.

[5] Wätzig, H., Degenhardt, M., Kunkel, A., *Electrophoresis* 1998, *19*, 2695–2752.

[6] Shibabi Z. K., *J. Chromatogr. A* 1998, *807*, 27–36.

[7] Guzman, N. A., *J. Chromatogr. B* 1997, *697*, 37–66.

[8] Brunner, L. J., Dipiro, J. T., *Electrophoresis* 1998, *19*, 2848–2855.

[9] Lloyd D. K., *J. Chromatogr. A* 1996, *735*, 29–42.

[10] Thormann, W., Lienhard, S., Wernly, P., *J. Chromatogr.* 1993, *636*, 137–148.

[11] Engelhardt, H., Faller, T., *12th International Symposium on High Performance Capillary Electrophoresis*, Palm Springs 1999, Book of Abstracts, p. 42

[12] Adam, S., Beck, W., Engelhardt, H., *Würzburger Kolloqium – Kapillarelektophorese – Fortschrittsberichte 94*, Bertsch Verlag, Straubing 1994, p. 97.

[13] Hempel, G., Lehmkuhl, D., Krümpelmann, S., Blaschke, G., Boos, J., *J. Chromatogr. A* 1996, *745*, 173–179.

[14] Olgemöller, J., Hempel, G., Boos, J., Blaschke, G., *J. Chromatogr B* 1999, *726*, 261–268.

[15] Frost, M., Köhler, H., Blaschke, G., *J. Chromatogr. B* 1997, *693*, 313–319

[16] Chevret, J. P., Van Soest, R. E. J., Ursem, M., *J. Chromatogr.* 1991, *543*, 439–449.

[17] Kaltenback, P., Ross, G., Heiger, D. N., *9th International Symposium on High Performance Capillary Electrophoresis*, Anaheim 1997, Book of Abstracts, p. 65.

[18] Mrestani, Y., Neubert, R., *Electrophoresis* 1998, *19*, 3022–3025.

[19] Havel, K., Wielgos, T., *12th International Symposium on High Performance Capillary Electrophoresis*, Palm Springs 1999, Book of Abstracts, p. 58.

[20] Pentoney Jr., S. L., Sweedler, J. V. in: Landers, J. P. (Ed.), *Handbook of Capillary Electrophoresis*, CRC Press, Boca Raton, FL 1992, pp. 379–423.

[21] Rethfeld, I., Blaschke, G., *J. Chromatogr. B* 1997, *700*, 249–253.

[22] Reinhoud, N. J., Tjaden, U. R., Irth, H., van der Greef, J., *J. Chromatogr.* 1992, *574*, 327–334.

[23] Hempel, G., Haberland, S., Schulze-Westhoff, P., Möhling, N., Blaschke, G., Boos, J., *J. Chromatogr. B* 1997, *698*, 287–292.

[24] Hempel, G., Schulze-Westhoff, P., Flege, S., Laubrock, N., Boos, J., *Electrophoresis* 1998, *19*, 2939–2943.

[25] Roach, M. C., Gozel, P., Zare, R. N., *J. Chromatogr.* 1988, *426*, 129–140.

[26] Rabel, S., Trueworthy, R., Stobaugh, J. F., *J. High Resol. Chromatogr.* 1993, *16*,326–327.

[27] Soini, H., Novotny, M., Riekkola, M. L., *J. Microcol. Sep.* 1992, *4*, 313–318.

[28] Hempel, G., Blaschke, G., *J. Chromatogr. B* 1996, *675*, 131–137.

[29] Bannefeld, K., Stass, H., Blaschke, G., *J. Chromatogr. B* 1997, *692*, 453–459.

[30] Möller, J. G., Stass, H., Heinig, R., Blaschke, G., *J. Chromatogr. B* 1998, *721*, 109–125.

[31] Couderc, F., Causse, E., Bayle, C., *Electrophoresis* 1998, *19*, 2777–2790.

[32] Heitmeier, S., Blaschke, G., *J. Chromatogr. B* 1999, *721*, 109–125.

[33] Wahl, J. H., Gale, D. C., Smith, R. D., *J. Chromatogr. A* 1994, *659*, 217–222.

[34] Wachs, T., Sheppard, R. L., Henion, J., *J. Chromatogr. B* 1996, *685*, 335–342.

[35] Henion, J. D., Mordehai, A. V., Cai, J., *Anal. Chem.* 1994, *66*, 2103–2109.

[36] Hsieh, F. Y. L., Cai, J., Henion, J., *J. Chromatogr. A* 1994, *679*, 206–211.

[37] Sheppard, R. L., Henion, J., *Anal. Chem.* 1997, *69*, 2901–2907.

[38] Bach, G. A., Henion, J., *J. Chromatogr. B* 1998, *707*, 275–285.

[39] Shah, V. P., Midha, K., Dighe, S., Mcgilveray, I. J., Skelly, J. P., Yacobi, A., Layloff, T., Viswanathan, C. T., Cook, C. E., Mcdowall, R. D., Pittman, K. A, Spector, S., *Eur. J. Drug Metab. Pharmacokinet.* 1991, *4*, 249–255.

[40] Schwartz, M. E., Merion M., *J. Chromatogr.* 1993, *632*, 209–213.

[41] Tsuda, T., Sweedler, J. V., Zare, R. N., *Anal. Chem.* 1990, *62*, 2149–2159.

[42] Tomlinson, A. J., Naylor, S., *J. Liq. Chromatogr.* 1995, *18*, 3591–3615.

[43] Tomlinson, A. J., Benson, L. M., Oda, R. P., Braddock, D., Riggs, B. L., Katzmann, J. A., Naylor, S., *J. Capil. Electrophor.* 1995, *2*, 97–104.

[44] Tomlinson, A. J., Benson, L. M., Braddock, D., Oda, R. P., Naylor, S., *J. High Resol. Chromatogr.* 1995, *18*, 381–383.

[45] Tomlinson, A. J., Guzman, N. A., Naylor, S., *J. Capil. Electrophor.* 1995, *2*, 247–266.

[46] Barroso, M. B., de Jong, A. P., *J. Capil. Electrophor.* 1998, *5*, 1–7.

[47] Shihabi, Z. K., *J. Capil. Electrophor.* 1995, *2*, 267–271.

[48] Gebauer, P., Boček, P., *Electrophoresis* 1997, *18*, 2154–2161.

[49] Mikkers, F. E. P., Everaets, F. M., Verheggen T., *J. Chromatogr.* 1979, *169*, 11–20.

[50] Burgi, D. S., Chien, R., *Anal. Chem.* 1991, *63*, 2042–2047.

[51] Chien, R., Burgi, D. S., *Anal. Chem.* 1992, *64*, 1046–1050.

[52] Palmasdottir, S., Edholm, L., *J. Chromatogr. A* 1995, *693*, 131–143.

[53] Palmasdottir, S., Mathiassn, L., Jönsson, J. A., Edholm, L., *J. Chromatogr. B* 1997, *688*, 131–143.

[54] Heitmeier, S., Blaschke, G., *J. Chromatogr. B* 1999, *721*, 93–108.

[55] Sczesny, F., Hempel, G., Boos, J., Blaschke, G., *J. Chromatogr. B* 1998, *718*, 177–185.

[56] Huang, X., Gordon, M. J., Zare, R. N., *Anal. Chem.* 1988, *60*, 375–377.

[57] Gross, L., Yeung, E. S., *J. Chromatogr.* 1989, *480*, 169–178.

[58] Chien, R., Burgi, D. S., *J. Chromatogr.* 1992, *559*, 141–152.

[59] Chien, R., Burgi, D. S., *J. Chromatogr.* 1992, *559*, 153–161.

[60] Zhang, C, Thormann, W., *Anal. Chem.* 1996, *68*, 2523–2532.

[61] Zhang, C., Aebi, Y., Thormann, W., *Clin. Chem.* 1996, *42*, 1805–1811.

[62] Zhang, C, Thormann, W., *Anal. Chem.* 1998, *70*, 540–548.

[63] Song, J. Z., Chen, H., Tian, S., Sun, Z., *J. Chromatogr. B* 1998, *708*, 277–283.

[64] Tagliaro, F., Manetto, G., Crivellente, F., Scacella, D., Marigio, M., *Forens. Sci. Int.* 1998, *92*, 201–211.

[65] Taylor, R., Toasaksiri, S., Reid, R. G., *Electrophoresis* 1998, *19*, 2791–2797.

[66] Nishi, N., Terabe, S., *J. Chromatogr. A* 1996, *735*, 3–27.

[67] Liu, Z., Sam, P., Sirimanne, S. R., McClure, P. C., Graingner, J., Patterson, D. G., *J. Chromatogr. A* 1994, *673*, 125–132.

[68] Nielsen, K. R., Foley, J. P., *J. Chromatogr. A* 1994, *686*, 283–291.

[69] Quirino, J. P., Terabe, S., *J. Chromatogr. A* 1997, *781*, 119–128.

[70] Quirino, J. P., Terabe, S., *J. Chromatogr. A* 1997, *791*, 255–267.

[71] Quirino, J. P., Terabe, S., *Anal. Chem.* 1998, *70*, 149–157.

[72] Quirino, J. P., Terabe, S., *J. Chromatogr. A* 1998, *798*, 251–257.

[73] Quirino, J. P., Terabe, S., *Anal. Chem.* 1998, *70*, 1893–1901.

[74] Quirino, J. P., Terabe, S., *J. Chromatogr. B* 1998, *714*, 29–38.

[75] Quirino, J. P., Terabe, S., *Science* 1998, *282*, 465–468.

[76] Palmer, J., Munro, N. J., Landers, J. P., *Anal. Chem.* 1999, *71*, 1679–1687

[77] Shihabi, Z. K., *J. Chromatogr. A* 1996, *744*, 231–240.

[78] Shihabi, Z. K., *J. Chromatogr. A* 1997, *781*, 35–42.

[79] Friedberg, M., Shihabi, Z. K., *J. Chromatogr. B* 1997, *695*, 193–198.

[80] Shihabi, Z. K., Hindsdale, M. E., *J. Chromatogr. B* 1996, *683*, 115–118.

[81] Shihabi, Z. K., *J. Chromatogr. A* 1993, *652*, 471–475.

Electrophoresis 2000, *21*, 1239–1250

1239

Review

Kelly Swinney
Darryl J. Bornhop

Detection in capillary electrophoresis

A review of the four major, on-line, capillary electrophoresis (CE) detection modalities is presented. It is shown that each detection method, fluorescence, absorbance (conventional and nonconventional), electrochemical and refractive index, have distinct advantages and limitations when applied to analysis in a CE format. Various aspects of CE detection are considered and a perspective regarding the applicability of the technique is provided. It is shown that because of widely varying detection limits (ranging from single molecule to 10^{-5} M) and detection scheme complexity, the particular application should dictate the selection of detection methodology in CE.

Keywords: Capillary electrophoresis / Detection techniques / Review   EL 3844

Contents

1 Introduction

This comprehensive description of the four major detection modalities (absorbance, fluorescence, electrochemical and refractive index), as well as brief discussion of the less used detection techniques for CE, focuses on direct, on-line detection techniques; therefore, detection *via* mass spectroscopy, which is an off-line analysis method, has not been included. The most sensitive on-line detection scheme for CE is laser-induced fluorescence detection, which is capable of single molecule detection but

Correspondence: Dr. Darryl Bornhop, Department of Chemistry and Biochemistry, Texas Tech University, Lubbock, TX, USA
E-mail: djbornhop@ttu.edu
Fax: +806-742-1289

Abbreviation: DABSYL, 4-dimethylaminoazobenzene-4′-sulfonyl; MIBD, microinterferometric backscatter detector; PMT, photomultiplier tube; RI, refractive index

most often requires chemical derivatization. For CE applications needing universal detection, refractive index methods or, to some degree, UV-Vis systems using the low UV (<190 nm) and conductivity detection are suitable. In general, the detection scheme of choice is application-specific and dictated by the level of sensitivity necessary (Table 1). If CE is to become a routine / robust and widely accepted analysis method, however, improvements in detection will still be needed.

2 Absorbance detectors

Absorbance detectors are the most commonly used detectors for microseparations, particularly for CE; they are often used in universally applicable detection techniques because many organics can be detected at 195–210 nm. In this region, molar extinction coefficients are often low. Essentially all organic molecules have appreciable absorption in the low UV (160–180 nm); nevertheless, access to this wavelength region is particularly problematic due to absorbance by optics and air gases. Furthermore, for compounds whose structures do not include pi bonds (*i.e.*, ions, carbohydrates, and inostol phosphates), absorption detection at wavelengths > 190 nm gives relatively poor signal-to-noise ratios. Therefore, while UV-Vis absorption techniques are applicable to an abundance of solutes, they are not truly universal. Moreover, in the small ID capillaries (<100 μm) in which CE detection is typically performed, the path length dependency of absorbance detection makes solutes with poor absorptivity difficult to detect.

2.1 Direct measurements

Since most commercial CE instruments are equipped with UV-Vis detectors, many applications have been published using absorbance detection. For example, CE with UV detection has been used to measure the binding constants between purified calf tyhmus DNA and a library of designed tetrapeptides [1]. CE was utilized to separate

Table 1. LODs for different detection techniques in CE

Detection technique	LOD (M)	References
Direct absorbance	10^{-5}–10^{-6} (standard pathlength)	[11]
	10^{-6} (extended pathlength)	[11]
Indirect absorbance	10^{-5}–10^{-6}	[19]
Photothermal refraction	10^{-7}–10^{-8}	[24, 26]
Direct on-column LIF	$< 10^{13}$ (chemical derivatization)	[11, 31, 32, 41,
	10^{-10}–10^{-11} (native)	42, 45–48]
Post-column LIF	10^{-16} (single molecule)	[34, 49, 50, 51]
Indirect LIF	10^{-5}–10^{-7}	[20, 40, 41]
Potentiometric	10^{-7}–10^{-8}	[56–58]
Conductivity	10^{-7}–10^{-8}	[59–64]
Amperometric	10^{-7}–10^{-8}	[71–74]
Refractive index	10^{-5}–10^{-6} (capillary)	[78–81, 88–90]
	10^{-5} (chip-scale)	
Raman	10^{-3}–10^{-6} (preconcentration needed)	[91–93]
Radiofrequency (NMR)	10^{-3}	[96, 97]
Radioisotope	10^{-10}	[98, 99]
Laser-induced capillary vibration	10^{-8} (chemical derivation)	[102, 103]
	10^{-5} (native)	[104]

unbound tetrapeptides from the DNA-peptide complex. The absorbance signal for the unbound tetrapeptide peaks were used to create a Scatchard plot in order to calculate the DNA-peptide binding constants, which ranged from 102 to 106 M^{-1} [1]. In another application, Hu *et al.* [2] described a procedure for the analysis of α-difluoromethylornithine (DFMO), an anticancer agent in plasma microdialysis samples, using CE with absorbance detection. In this procedure, a precolumn derivatization step is necessary to convert the non-UV active DFMO to *N*-substituted 1-cyanobenz[*f*]isoindole (CBI), which is UV active, by reacting it with napthalene-2,3-dicarboxalde-hyde-cyanide (NDA-CN) in pH 10.0 borate buffer. Using UV detection at 254 nm, the detection sensitivity of DFMO in plasma microdialyzate samples (LOD for DFMO in plasma dialyzate was reported as 5 μM) is adequate and has successfully been applied to monitoring the pharmacokinetics of DFMO by CE-UV.

Due to the pathlength sensitivity of absorbance measurements, several attempts have been made to improve absorption detection limits, without sacrificing resolution, by increasing the detection pathlength through modification of the capillary shape or diameter in the detection zone [3–6]. While improvements have been realized, theoretically predicted increases in absorbance due to longer pathlengths have not been achieved. For example, because of light attenuation, only a sixfold S/N ratio was observed despite a 60 times longer pathlength for the capillary modified Z-cell [7]. Yet, using a Z-cell, UV detection and chemical derivatization of the analytes, an improvement in the detection limit by a factor of 20 was achieved for the trace analysis of carbonyl compounds in rain water as reported by Bachmann and Mainka [8] (Fig. 1). Through chemical derivatization (derivatized with dansylhydrazine, DNSH), the carbonyl compounds were converted to compounds with higher molar extinction coefficients [8]. As a result, LODs in the range of 170–300 nmol/L were reached [8]. The improvement in the detection limit using a Z-cell compared to the results obtained by Chervet *et al.* [7] is due to a difference in detection volume and separation efficiency. However, Bachmann and Mainka [8] report that increasing the pathlength of the flow cell resulted in a reduction in resolution by 47%. More recently, Hewlett-Packard [9] introduced a "Z"-type flow cell which has a pathlength of 1.2 mm and a sample volume of 12 nL. This cell is comprised of black fused-silica and flat windows in order to minimize stray light that often leads to high detection limits in absorbance measurements. Hewlett-Packard reports that by minimizing stray light and increasing the pathlength of the cell, the S/N ratios have increased tenfold without sacrificing separation efficiency.

To date, the most promising attempt at improving detection limits by increasing the pathlength in absorbance detection with CE has been with the development of the multireflection nanoliter scale cell (Fig. 2) [10]. The multireflection cell is made directly onto the column by silver coating a short length in the detection region of the capillary. Two windows located on opposite sides of the capillary above and below the detection zone are separated by a distance of 1.5 mm and serve as the entrance and exit ports for the laser illumination. Using laser illumination and the multireflection cell a 40-fold improvement in absorption detection limits was obtained when compared to standard UV detection with a traditional illumination source directly probing a fused-silica capillary. Nonetheless, this detection cell geometry produces a larger probe volume (6.6 nL or a few milimeters of capillary) and as a result, separation efficiency is sacrificed. Furthermore, in order for the multireflection cell to be practical for CE, a broad band light source must be substituted for the laser allowing detection of many analytes. This incrases the dificulty in achieving total internal reflection with the sample cell [10] due to the optical constraints associated with tightly focusing a noncoherent light source and the wavelength dependency associated with refraction and reflection of light.

Even with increased pathlength cells, absorbance detectors are ultimately limited by Beer's Law and the constraints associated with measuring small signal changes

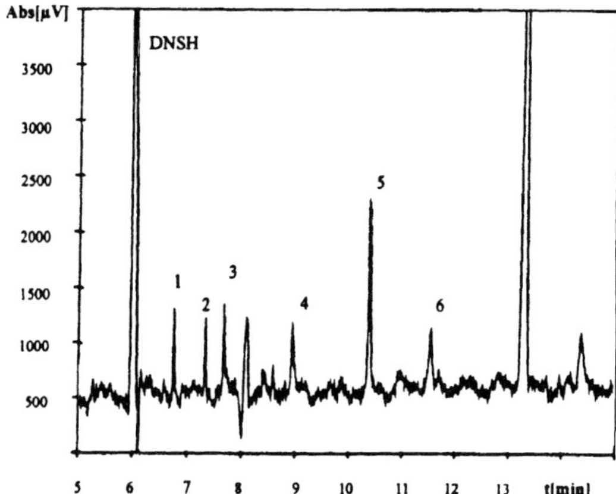

Figure 1. Electropherogram of the separation of derivatized carbonyl compounds found in rain water using a Z-cell flow cell and UV detection [6]. Capillary, 80 cm total length, 60 cm length to the detector; electrolyte, 5 mM Na_3PO_4, 10 mM $Na_2B_4O_7$, 20 acetonitrile, pH 8.0; detection, 218 nm; high voltage, 25 kV; injection, hydrostatic 1 min, 10 cm. 1 Acetone; 2, acetaldehyde; 3, propionaldehyde; 4, benzaldehyde; 5, formaldehyde; 6, methylglyoxal; each 40 μM.

in the presence of large backgrounds. As a result, the absorption techniques have relatively poor concentration detection limits, generally falling into the 10^{-6} M range [11]. Eventually, a compromise between loss in separation efficiency with increased pathlength will restrict performance and thus applicability of conventional absorbance monitors for CE detection. Furthermore, direct absorption measurements are difficult, if not nearly impossible, for solutes that possess miniscule absorption coefficients, thus eliminating detection of such solutes as ions or inorganic salts.

2.2 Indirect measurements

An alternative way to detect molecules that do not possess appreciable absorptivities is to use indirect absorbance methods. Indirect absorption measurements are basically vacancy techniques and in CE have primarily been used for the detection of small ions [12–14], carbohydrates [15], lanthanides [16], alkyl sulfates [17], and carboxylates [18]. In the indirect absorbance measurement [19] as shown in Fig. 3, an additive is introduced to the buffer system producing a relatively large background signal, which reduces the amount of light reaching the detector. When a solute band (having a lower absorbance than the additive) passes through the detection volume, the light throughput to the photodetector increases, providing a signal for the solute. The LODs for such measurements are comparable to "standard" UV-Vis absorbance and fall in the 10^{-6}–10^{-5} M range [20]. For the best performance, indirect detection requires a well-characterized displacement mechanism such as a large transfer ratio, S/A (the ratio of the number of signal-generating molecules, S, displaced in the detection zone by a single analyte molecule, A), a stable background, and careful selection of the signal-generating chromophore. Indirect detection is best suited for application-specific systems in which all parameters are optimized and remain unchanged from run to run.

2.3 Nontraditional absorbance measurements

Photothermal refraction (PTR) is a nontraditional absorbance technique that can provide impressive detection limits because the signal is shifted away from the background. For example, an absorbance event produces a thermal perturbation, which leads to an optical signal. PTR is pathlength-insensitive, can be used directly on capillary, and therefore is a particularly attractive detec-

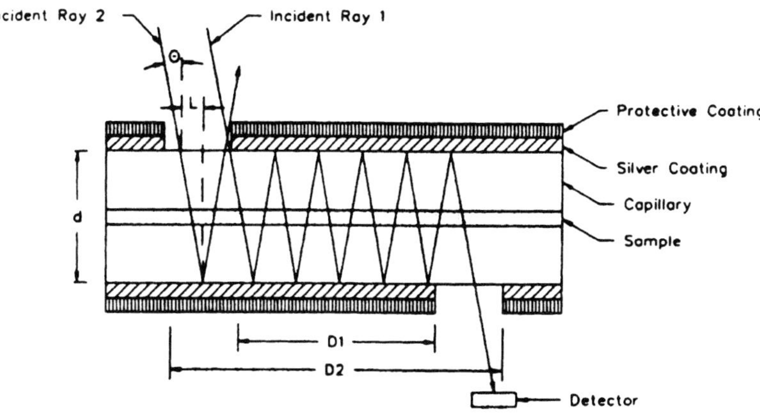

Figure 2. Diagram of the multireflection nanoliter scale cell [10]. The capillary is made of fused silica (75 μm ID × 364 μm OD). The silver layer was deposited by a redox reaction of $Ag(NH_3)_2$ and glucose on the section of the capillary where the polyimide coating had been removed. A protective coating of black paint was applied to the silver layer to preserve the integrity of the silver surface. The entrance and exit windows are separated by distance D_1 and D_2 (0.8 and 1.5 mm, respectively). The cell volume calculated from D_2 is 6.6 nL.

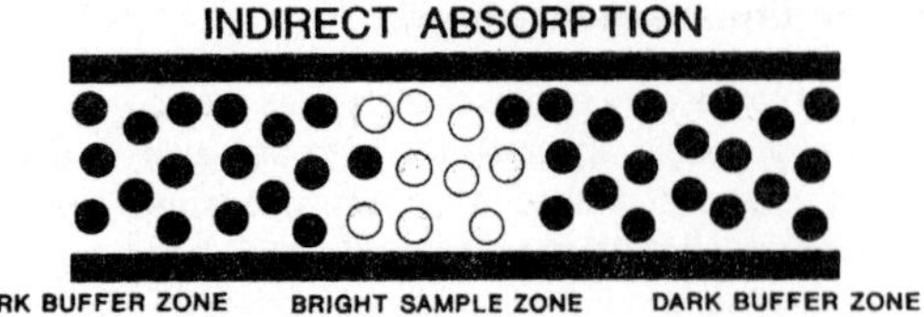

Figure 3. Cartoon of indirect absorption [19]

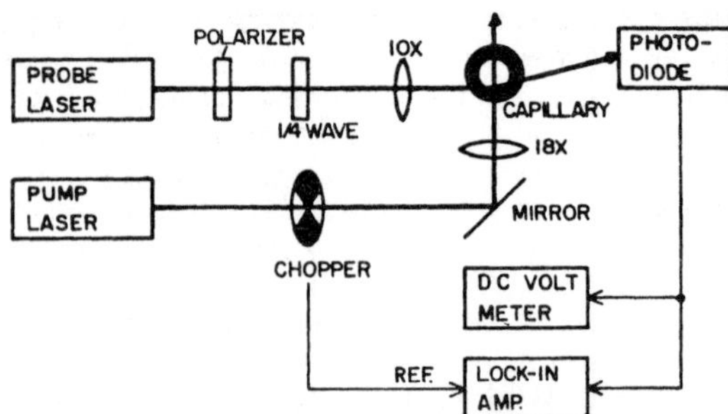

Figure 4. Block diagram of the cross beam thermal lens instrument developed by Bornhop and Dovichi [21].

tion method for CE. This technique requires that two lasers be employed, a pump laser and a probe laser. As shown in Fig. 4 for the cross beam thermal lens instrument developed by Bornhop and Dovichi [21], the pump laser and the probe laser are oriented such that they intersect at the capillary defining the detection region. When a molecule passes through the detection region, it absorbs light from the pump laser. Deactivation through a nonradiative decay pathway and the subsequent transfer of thermal energy to the surrounding buffer leads to a change in the refractive index of the buffer. As a result, the probe beam is refracted, and an analytical signal is produced. Thermooptical absorbance sensitivity is proportional to the pump laser power and, therefore, higher power lasers are useful for detecting weakly absorbing or low concentration analytes.

Thermooptical detection has been used for the detection of a variety of compounds separated by CE. Dovichi and colleagues [21–25] have described thermooptical absorbance systems for several different chromophore/laser combinations. For example, a 4mW HeCd laser was chosen as the pump source for the analysis of (DABSYL)-derivatized amino acids [23]. For this separation, a mass limit of detection of 200 amole for DABSYL-glysine was reported [23]. Yu and Dovichi [26] reported the detection of 4-dimethylaminoazobenzene-4'-sulfonyl DABSYL-methionine at a mass detection limit of 37 amole (5×10^{-8} M concentration detection limit) using a 130 mW argon-ion laser as the pump source. Amino acids derivatized with phenylthiohydantoin were detected by Waldron and Dovichi [24] using an excimer laser and reported detection limits of 9×10^{-7} M for glycine. The detection of 19 amino acids derivatized with phenylthiohydantoin produced in the Edman degradation reaction for protein sequencing was also accomplished with thermooptical absorption techniques by Chen *et al.* [27]. More recently, Dovichi *et al.* [28] have applied PTR successfully to CE and CEC for the detection of five tricyclic antidepressants. Others have been involved in using PTR systems for detection in CE. For example, Bruno *et al.* [29] have also reported the use of thermooptical absorbance detection for CE using a frequency doubled argon-ion laser as the pump beam. Danylated amino acids were analyzed in a 25 μm ID capillary [29] proving the pathlength insensitivity

of thermooptical techniques. PTR techniques have several advantages including very small probe volumes and high sensitivity. However, tedious alignment is required for PTR to function, the solute must have absorbance features that correspond to the wavelength of the excitation laser beam, $d\eta/dT$ for the mobile phase must be appreciable in order to produce a highly sensitive measurement, and the PTR techniques are relatively expensive to configure.

3 Laser-induced fluorescence

Laser-induced fluorescence (LIF) [30] is the most sensitive, small volume detection method developed to date. The LODs of the best CE/laser-based systems are well below 10^{-13} M [11, 31] producing mass detection limits of less than 10 molecules [32]. In fact, single molecule detection has been demonstrated in a capillary using LIF [33] and subsequently at the stochatic noise limit in CE [34]. Various arrangements or approaches to LIF have been implemented, including the first on-column LIF detection system developed by Zare *et al.* [35], the post-column sheath flow cuvette first demonstrated by Dovichi and co-workers [36–39], and indirect LIEF reported by Yeung and Kuhr [40, 41]. The advantages and limitations of each technique will be detailed below.

3.1 Direct on-column detection

The first on-column LIF detection system described for CE was reported by Zare *et al.* [35] and used fiber optics to illuminate the capillary and to collect the fluorescence signal. The optical configuration developed by Zare *et al.* [35] utilized a 325 nm HeCd laser whose output was focused into an optical fiber and was used to remotely illuminate a small region of the separation capillary. Using a collection fiber, the fluorescence signal was directed into a monochromator and detected with a photomultiplier tube (PMT). With this optical arrangement, LODs of approximately 200 amole (0.1 μM) for dansylated amino acids was possible. Hernandez and co-workers [42, 43] report zeptomole mass detection limits (10^{-13} M concentration LODs) for a series of derivatized amino acids using their collinear LIF detector. In this arrangement, the angle between the exciting laser beam and the collected fluorescence is $0°$, which reduces laser scattering by the walls of the capillary, leading to lower background signals and enhanced collection of the fluorescence signal. Novotny and co-workers [44] also reported the use of LIF detection for amino acids at the amole level. Here, a modulated (mechanical chopper) HeCd laser was chosen as the excitation source and focused directly onto the capillary. The fluorescence signal (50% peak transmission) was collected at $90°$ with respect to the incident beam using a fiber optic whose output was focused onto a PMT in communication with a lock-in amplifier. In addition, Novotny *et al.* [44] used precolumn derivatization of the analytes with 3-(4-carboxybenzoyl)-2-quinolinecarboxaldehyde that has the advantages of specifically labeling the analyte of interest with a fluorophore whose excitation maximum is coincident with the wavelength of the probe laser.

Although precolumn derivatization allows molecules that do not fluoresce at available laser lines [44] to be detected *via* LIF, the complicated chemistry and time-consuming procedures associated with derivatization make LIF less attractive than universal detection methods. Swaile and Sepaniak [45] were the first of demonstrate native fluorescence with on-column detection using UV excitation (frequency-doubled Ar-ion laser, 257 nm). They were capable of detection conalbumin at a detection limit of 1.4×10^{-8} M. Using the same laser, Nie *et al.* [46] demonstrated improved mass LODs (30–150 zmole mor 6 × 10^{-11} M) for the detection of native fluorescence for polycyclic aromatic hydrocarbons. Yeung and Chang [47] and Sweedler *et al.* [48] both have described the detection of tryptophan based on native fluorescence using a frequency-doubled Kr laser (284 nm). Both groups report subnanomolar concentration detection limits, 7×10^{-10} M and 2×10^{-10} M, respectively [47, 48].

The advantage of detecting native fluorescence is the elimination of precolumn derivatization steps simplifying the chemical workup and reducing the time of the total analysis. The disadvantage of using native fluorescence is that the signal is often weak and the background signal can be high, leading to poorer S/N ratios than those produced by fluorophore-tagged analytes. As a result, the sensitivity suffers. In addition, very few molecules exhibit appreciable native fluorescence (for example, peptides with tryptophan residues do have fluorescence only at λ_{ex} < 220 nm); therefore, the technique is limited to a relatively small class of molecules.

3.2 Post-column detection

Several LIF post-column detectors [34, 36–39, 49, 50, 55] have been reported but only one method has gained wide acceptance. In all cases, the goal has been to spatially or spectrally separate the fluorescence signal from the high background signal. To date, the best performing system, which was developed by Dovichi and co-workers [34, 36–39, 49, 50], involves the incorporation of a sheath flow cuvette at the end of the capillary and is comprised of a 200 μm fused-silica square flow chamber with 1 mm thick windows (Fig. 5). In general, sheath flow cuvettes use a flowing jacket of refractive index-matching buffer to hydrodynamically focus the sample stream as it exits the capillary. The excitation laser beam is focused, using a microscope objective, into the sample stream about 30 μm below the tip of the capillary. The fluorescence signal is collected on each side of the cuvette by two microscope objectives. The signal is filtered with a band pass interference filter and is imaged onto an iris (pinhole). Two PMTs connected by a summing circuit are used to detect the fluorescence.

The sheath flow cuvette eliminates many of the problems associated with on-column detection such as broad-band luminescence background and Rayleigh and Raman light scattering from the capillary. First, little light scattering occurs at the sheath flow/sample interface due to a lack of a refractive index interface. Second, a pinhole is incorporated into the optical configuration to further minimize light scattering, by assuring that only the fluorescence emission from the sample stream is detected. Third, the sheath flow cuvette is a hydrodynamic focusing technique which limits post-column band broadening and allows for high efficiency separations (1×10^{-6} theoretical plates). The sum of the advantages provided by the sheath flow

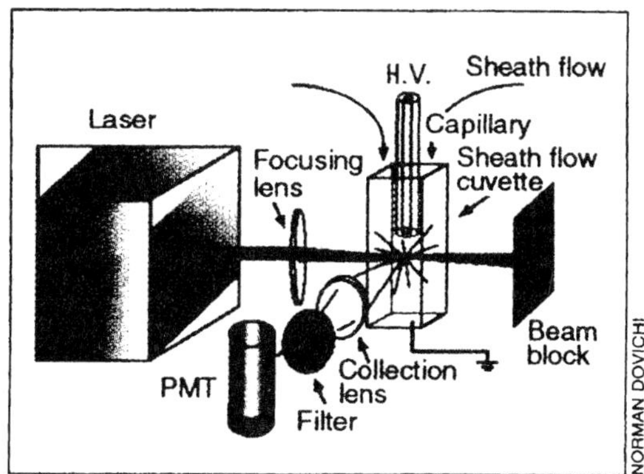

Figure 5. Block diagram of LIF using a sheath flow cuvette [39].

cuvette and LIF has led to single molecule detection [34] in CE. Note that Winefordner and co-workers [33] were the first group to demonstrate single molecule detection in a capillary; however, Chen and Dovichi [34] first showed that single molecule detection could be achieved under CE conditions. More recently, the ability to discern the individual behavior and activity of enzymes has been demonstrated by the detection of reaction products generated from enzyme action by several groups [49–51] (Fig. 6).

3.3 Indirect fluorescence detection

Indirect fluorescence detection using LIF was first reported by Yueng and Kuhr [40, 41]. The concept of indirect fluorescence measurements is essentially the same as indirect absorbance measurements, except the background (buffer) fluoresces instead of absorbing (Fig. 7). The LODs for indirect fluorescence techniques generally fall in the 10^{-6}–10^{-5} M range. Using the 325 nm line of an 8 mW HeCd laser for excitation and a capillary mounted at Brewster's angle to minimize the amount of stray light entering a PMT fitted with a spatial filter and a 405 nm interference filter, Kuhr and Yeung [40] reported LODs as low as 10^{-7} M for the separation of amino acids (Fig. 8). Compared to direct LIF, poor LODs for indirect fluorescence measurements are the result of the noisy fluorescence background, primarily caused by laser intensity instability (flicker noise). In addition, low fluorescent addi-

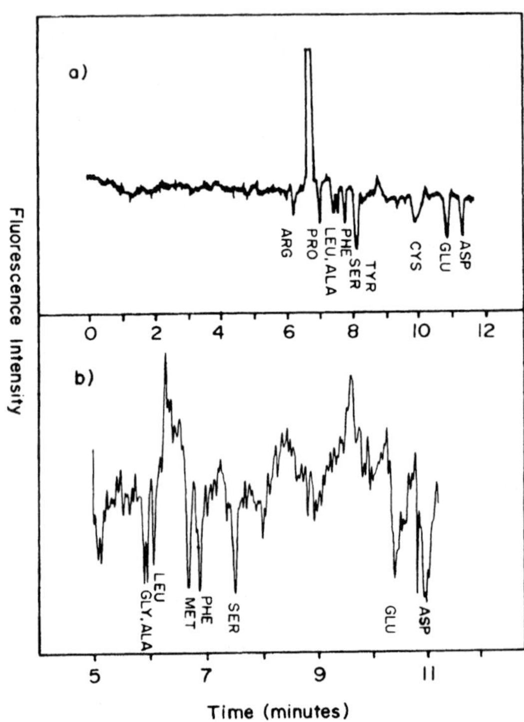

Figure 7. Cartoon of indirect fluorescence [19].

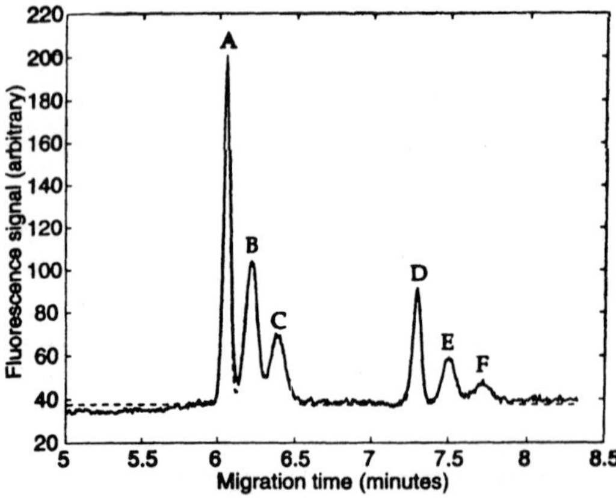

Figure 6. Electropherogram of alkaline phosphatase thermodynamic analysis using a sheath flow cuvette and LIF [49, 50]. A solution of 4.6×10^{-16} M alkaline phosphatase in 100 mM borate (pH 9.5) containing 1 mM AttoPhos was injected onto the capillary. Sample was incubated for three 15 min periods at 16, 24, and 30°C, interrupted by brief periods of high voltage. Following the third incubation, the capillary contents were electrokinetically swept through the detector. A set of three peaks is shown for each of two molecules of enzyme (peaks A, B, and C and peaks D, E, and F). Peaks A and D, incubation at 16°C; B and E, 24°C; C and F, 30°C.

Figure 8. Electropherograms of underivatized amino acids using indirect fluorescence detection [38]. (a) Electropherogram (raw data) of a 1 s (2 nL), 45 kV injection of 10^{-4} M (200 fmole) underivatized amino acids (arginine, proline, alanine, leucine, phenylalanine, serine, tyrosine, cysteine, glutamic acid and aspartic acid) in a 1.0 mM sodium salicylate and 200 μM sodium carbonate buffer, adjusted to pH 9.7 with 0.1 M sodium hydroxide. Electrophoresis was performed at 45 kV (2.3 μA). (b) Electropherogram (with digital filtering) of a 1 s (2 nL), 45 kV injection of 10^{-5} (20 fmole) underivatized amino acids in a buffer consisting of 200 μM sodium salicylate and 40 μM sodium carbonate buffer, adjusted to pH 9.7 with 0.1 M sodium hydroxide. Electrophoresis was performed at 45 kV (0.4 μA).

tive concentrations are required to maximize competition between the test compound and the additives. Because of these low additive concentrations, the ionic strength of the electrolyte solution (buffer) is low. Low ionic strength buffers are problematic in CE because peak asymmetry often occurs at the upper end of the linear dynamic range [20]. As a result, indirect fluorescence detection is not often employed in CE.

4 Electrochemical detection

Electrochemical detection for CE can be divided into three main categories: potentiometric, conductimetric and amperometric. In general, conductivity and potentiometric detectors provide a bulk property response with good sensitivity. Amperometric detection, on the other hand, is selective and can be "tuned" to the analyte of interest. While some successes have been realized, implementing electrochemical detection with CE can be extremely difficult and tedious since sensor preparation can involve complicated fabrication and handling procedures, and accurate spatial alignment of the electrode within the capillary is required.

4.1 Potentiometric detection

The first report of potentiometric detection in CE was by the group of Haber and Simon in 1990 [56, 57]. In general, potentiometric detectors are based on classical ion-selective microelectrodes and have the ability to detect extremely small quantities of inorganic and organic ions in small probe volumes ($\sim$0.5 $\times$ 10^{-15} L for calcium) [56]. The signal is produced when the ion of interest is transferred from the flowing sample stream into a lipophilic membrane phase of the detector. The presence of the analyte generates a change in the potential difference between the internal filling solution of the sensor and the sample stream. The potential difference is a measure of the ion's activity given by the Nernst equation and is directly related to the ion concentration. The potentiometric response is generally fast enough to prevent peak shape distortion, and therefore quantitative information about analyte concentration is possible. Yet, the disadvantages of potentiometric microsensors include tedious sensor preparation and handling procedures (*i.e.*, preparation of the ion-selective membrane phase of the microelectrode) and difficult micromanipulations necessary for alignment [56]. Another drawback of potentiometric detection is limited sensor lifetime, under normal CE conditions approximately 2–3 days. Also, recalibration of the sensor is recommended every 12 h due to changes in emf response of the electrode [57].

The selectivity and response of a potentiometric sensor is directly related to the CE electrolyte solution and the inter-

nal filling solution of the electrode. For the detection of cations, the running electrolyte solution normally consists of 20 m$_M$ MgAc$_2$/HCl at pH 5.14 and an internal filling solution of 20 m$_M$ MgCl$_2$ [57]. Under such conditions, the concentration LODs for the analysis of alkali and alkaline earth cations in a 25 µm ID capillary were reported by Haber and Simon's group [57] to be less than 10^{-8} M. Modifying the running electrolyte solution to be 20 m$_M$ MgAc$_2$ at pH 7.8, the detection of amphetamines at similar LODs ($<$ 10^{-8} M) are possible [56]. Anions can be detected using potentiometric techniques with an electrolyte solution that is anion selective. For example, an electrolyte solution consisting of 20 m$_M$ Tris-formate at pH 7 allows for the detection of inorganic anions such as perchlorate at an LOD of 5 $\times$ 10^{-8} M [11, 58].

4.2 Conductivity detection

Typically, ion analysis in CE has been accomplished with conductivity detection because ions have essentially no absorption and fluorescence features. In general, conductivity detectors consist of two electrodes, in contact with the electrolyte solution, across which an electrical potential has been applied. When an analyte passes through the electrode gap, the conductivity between the electrodes changes by a quantity directly related to the concentration of the ionic analyte. Conductivity LODs range between 10^{-6} and 10^{-7} for samples introduced by electrokinetic injection [11, 59–62]. An improvement in the LODs to 10^{-7}–10^{-8} M can be achieved by replacing standard conductivity detection with suppression conductivity detection for the analysis of anions [63, 64]. In suppression conductivity, the buffer counterion is replaced by ionic exchange or electrodialysis, producing a weakly dissociated acid (anion analysis) or base (cation analysis). Consequently, the total electrolyte (buffer) conductance is lowered and the conductivity difference between the sample ion and "background" is enhanced. Suppression conductivity, however, is problematic for cations because of its low pH buffer requirement since low pH buffers result in weak EOFs, which prevent efficient transport through the suppressor membrane to the detector. Furthermore, suppression conductivity detection is a post-column detection technique that induces extra column band broadening and limits separation efficiency by producing dispersion at the membrane/detector connection [11].

4.3 Amperometric detection

Amperometric CE detection was first reported as a detection technique for CE by Wallingford and Ewing in 1987 [65] for the quantitation of catechol and catecholamines (also see Wallingford and Ewing [66–68] and Ewing *et al.* [69]. Amperometric detection is based on electron transfer to or from the analyte of interest at an electrode surface that is under the influence on an applied DC voltage. The result of electron transfer is a redox reaction at the electrode that produces a current that is directly related to the analyte concentration. The major drawback of amperometric techniques is strong absorption to the electrode surface (carbon electrodes) of the intermediate reaction products of the analyte subsequently reducing the activity (electron transfer) of the electrode and interfering with detection. This problem can be reduced by using pulsed amperometric detection (PAD) which utilizes a continuous three-step potential waveform (step 1, detection; step 2, anodical cleaning of electrode surface; step 3, reactivation) [70]. PAD was first reported as a detection technique for CE for the detection of carbohydrates. Using gold electrodes in an end-column configuration and PAD, mass LODs ranging from 0.28 fmole (10^{-6} M) for inositol to 1.21 fmole (2×10^{-6} M) for maltose were achieved [71, 72].

Recently, Zare and co-workers [73] have shown that by replacing carbon electrodes with copper electrodes under standard amperometric conditions (no pulsing), there is no need to clean the electrode since intermediate absorption to the electrode does not occur. Therefore, a single waveform can be applied to the electrode. Using the copper electrode, 50 fmole mass LOD are possible for carbohydrates [73]. More recently, copper disk electrodes were employed by Lin *et al.* [74] to simultaneously detect ribonucleotides, ribonucleosides, and purine bases with reported mass LODs below 10 fmole (Fig. 9).

5 Refractive index detection

Refractive index (RI) detectors, in general, are bulk property, nondestructive sensors that are mass-sensitive, and as such they are potential candidates for use as universal detectors in CE. Yet implementation of RI detection schemes poses a special challenge when small-volume detection is required as in CE. First, for most conventional RI techniques (particularly those based on refraction-induced defection) physical constraints and an optical pathlength sensitivity make detection directly on capillaries essentially problematic. Second, the change in RI with temperature (dn/dT) is large (8×10^{-4} RIU/°C for water) [75] for most fluids, and thus small changes in temperature result in appreciable RI signals. Even so, many attempts have been made [75–87] to miniaturize these bulk property optical detectors to nanoliter volumes.

Because of the unique optical properties of the laser (*i.e.*, high spatial coherence and monochromaticity), they have become the "source of choice" in the effort to construct microvolume RI detectors [75, 77–85]. Many RI detection

Figure 9. Electropherogram of the separation of ribonucleotides, ribonucleosides, and purine bases using amperometric detection and copper disk electrodes [72].

methods have been investigated for CE, including a concentration gradient method, which probes the on-axis optical perturbation produced by a transient solute band [83] and the use of a tapered fiber optic [85], which probes the reduction in transmitted beam intensity due to refractive index interfacial beam intensity coupling. While powerful, these approaches to RI sensing have limitations. For example, the concentration gradient detector [83] is somewhat insensitive to thermal noise, but most suitable for detection schemes in capillary isoelectric focusing or for separation systems where post-column detection is acceptable.

Most state-of-the-art RI measurements involve some form of interferometry, a measurement technique that is critically dependent on the characteristics of laser light. In general, very small phase changes caused by optical path length differences, in response to the analyte, allow for high sensitivity [86, 87]. The use of interferometric techniques has produced some impressive results toward measuring RI changes in the small volumes [78, 79] characteristic of CE; however, sensitivity dependency on pathlength has persisted. Bruno and co-workers [78] further developed the forward scatter, off-axis technique originally developed by Bornhop and Dovichi [77] for HPLC detection. They showed enhanced performance is possible by using an RI-matching fluid to surround the capillary tube, a flow cell assembly with active thermal control and position-sensitive detection [78]. These improvements facilitated the use of the RI detector for CE for the detection of saccharose at a 2σ limit of detection of 10 μM [78], yet the device still requires off-axis alignment and is limited by the need to modify the capillary tube by removing the polymer coating to aid in index matching.

Another technique for detecting changes in *n* within capillary tubes uses a holographic grating to produce two-beam interference in a forward scatter configuration [79]. Krattiger and co-workers [79] were able to separate and detect metal ions by CE using a holographic grating, a capillary that is encapsulated in an index matching glue, and a photodiode array wired to produce position-sensitive detection. Manz *et al.* [84] recently reported the application of the holographic forward scatter RI detector for CE on a chip. The holographic technique is a significant advance toward developing a small-volume RI detector and eliminates the need for the capillary to serve as the optic, yet it is another arrangement of the forward-scattering refractive index technique with its limitations [77–79, 84]. The major drawback of holographic interferometric or grazing angle forward scatter optical configurations is an inherent pathlength dependency, which ultimately hinders detection limits in ultrasmall volumes.

Deng and Li [82], acknowledging the difficulties with forward scattering techniques, have recently shown that a retroreflected beam interference technique similar to the microinterferometric backscatter detector (MIBD) [75, 80, 81] can be used for RI detection in CE. In their scheme, a focused laser beam impinges on a bare capillary surface, causing interference of two retroreflected beams originating from the outer surface of the capillary. By observing the retroreflected interference pattern, RI measurements are possible. The main drawbacks of the retroreflected beam interference RI detector is poorer LODs than in previously reported RI detectors [75, 77–81] and enhanced complexity of the optical train [75, 80, 81].

Recently, Borhop and co-workers [75, 80, 81, 88–90] introduced an alternative approach for on-column refractive index measurements based on interferometric backscatter. Using a slightly tilted, side illuminated fluid-filled capillary tube, in a backscatter configuration, high contrast interference fringes are produced that move spatially in response to changes in optical pathlength allowing for refractive index detection at a level of two parts in 10^7 RIU [80, 81] (Fig. 10). The backscatter configuration exhibits essentially no pathlength sensitivity for capillaries ranging in tube size from 75–775 μm [81], requires no special optical alignment, no beam-conditioning optics (lens or microscope objectives), and no modification of the polymer-coated tube. Additionally, MIBD has been directly compared to a standard commercial UV-Vis

detector in CE. For the detection of a series of dyes with high absorption coefficients [89] (Fig. 11) in the red region of visible light spectrum, although it is possible for the solute molecules to absorb some of the HeNe laser light, due to the alignment of the MIBD system (fringe shifts due to RI changes result in an increase in measured intensity), it is highly unlikely that the observed signal is due to an absorption phenomenon. In fact, any light absorbed by the molecules would result in a decrease in intensity of the laser light increasing the LODs of the device and lessening the appeal of the detector. As is, the functional LODs of the MIBD-RI detector were shown to be a factor of 1.6–2.5 times better than those reported for the standard UV-Vis absorption detector for a 100 μm capillary [89]. Due to the pathlength sensitivity of absorption detectors (Beer's Law), upon reducing the capillary diameter to less than 50 μm ID, it was reported that the LOD for MIBD is expected to be at least five times better than for the standard UV detector. Moreover, MIBD is capable of picogram mass LODs in nanoliter volumes with minimal passive thermal control for the detection of carbohydrates [89], organic dyes [89], caffeine [90], and

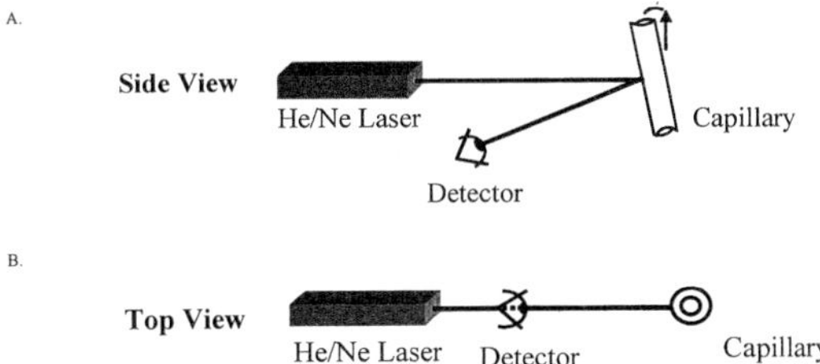

Figure 10. Block diagram of MIBD [75, 80, 81]. (A) Top view; (B) side view.

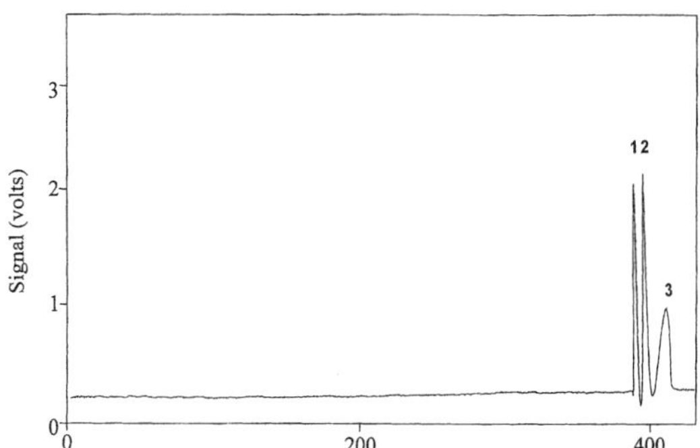

Figure 11. Electropherogram for a series of organic dyes detected using MIBD (RI) detection [89]. Peak 1, 30 μM bromothymol blue; 2, 60 μM thymol blue; 3, 30 μM bromocresol green.

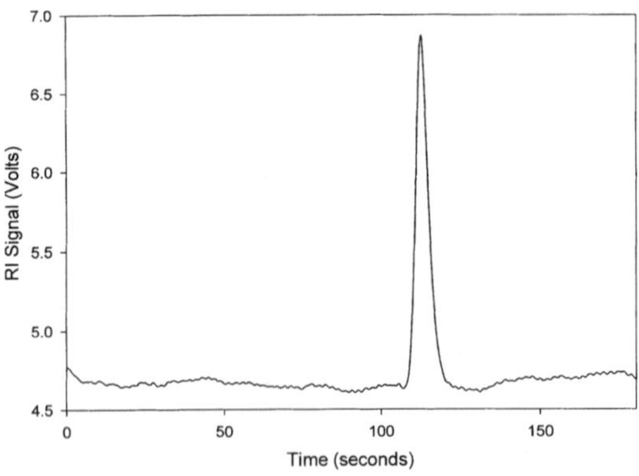

Figure 12. Chip-scale analysis. Electropherogram of a single solute (33 mM sucrose) detected using MIBD in a planar format with a 50 μm radius etched channel. The sample was injected for 3 s at 3 kV and electrokinetically eluted using 5 kV. A buffer consisting of 30 mM boric acid and 100 mM Tris was used [91].

cations [88]. MIBD has also been applied to on-chip detection for the analysis of sugars (Fig. 12). Ignoring dilution from the separation process, and even with an extremely wide elution event (significant band broadening), the 3σ LOD is about 875 μmole and more than an order of magnitude better than previously reported for RI detection on a chip [84].

Refractive index detectors are advantageous because they are universal in nature. As a result, they possess wide applicability. Furthermore, no chemical derivatization steps are necessary for detection as in LIF, thus no precolumn workup is involved. However, because of their universal nature, RI detectors are not very specific. Hence, in applications where specificity is required, RI detectors must be coupled with or replaced by other detection methods (LIF, absorbance, or amperometric detection).

6 Raman-based detectors

Raman-based detection is useful for obtaining qualitative information (*i.e.*, structural details) about the analytes being separated. However, in order for solutes to be detected, they must be Raman-active (polarizable). In general, the signal is obtained by monitoring changes in intensity and frequencies of scattered light induced by solutes passing through the detection zone. In the on-line Raman spectroscopy / CE system developed by Chen and Morris [91], a 40 mW HeCd laser is used for excitation and is focused by a microscope objective onto the capillary. The scattering signal is collected by an array of

10–200 μm quartz fiber optics that directs the scattered light into a monochromator with PMT detection. Using this optical configuration, LODs of 2.5×10^{-6} M were reported for methyl red and methyl orange under CE conditions. Such concentration LODs compare favorably with UV-Vis detection while providing structural information. More recently Walker and Morris [92] have reported the use of a fiber optic Raman sensor using a 2 W 532 nm Nd/YAG laser to obtain online normal Raman spectra of AMP, CMP, GMP, and UMP separated by capillary isotachophoresis. In addition, Walker *et al.* [93] have used a Raman detector based on an upright microscope equipped with a cyrogenically cooled CCD camera and a frequency-doubled Nd-YAG laser (532 nm, 400 and 700 mW) for the online capilary isotachophoresis analysis of ATP, ADP, and AMP in phosphate buffers. LODs are reported at 5×10^{-6} M for injected concentrations of the ribonucleotides onto the separation capillary [93]. However, on-capillary preconcentration of the injected sample to a concentration above 10^{-2} M is required to achieve micromolar LODs. Preconcentration of the sample was accomplished on-column using a 1×10^{-2} M phosphate buffer (pH 7.5), 0.1 M KCl, or Na_2SO_4 leading electrolyte, and a 0.1 M 4-morpholinepropane sulfonic acid terminating electrolyte [93].

7 Unconventional CE detection methods

Numerous other detection methods have been used with CE. For example, on-line NMR detection was first shown to work with CE by Wu *et al.* [94, 95] and has been used for the analysis of peptides [96]. More recently, Sweedler and co-workers [97] have shown that NMR detection can be used in stop-flow CE analysis and achieve nanogram sensitivity for arginine and triethylamine (TEA), 7 ng (330 pmole; 31 mM), and 9 ng (88 pmole; 11 mM), respectively. Although other methods are more sensitive (*i.e.*, LIF), using NMR detection structural information can be obtained. Another example of CE detection is radioisotope detection. Radioisotope detection is an extremely sensitive and highly selective technique that is based on scintillation counting of the CE eluent. Pentoney *et al.* [98, 99] first described the on-line semiconductor radioisotope detector which is positioned at the end of the capillary. Using this detector, ^{32}P-labeled solutes such as ATP and GTP can be detected at a limit of 1×10^{-10} M [98]. Pentoney *et al.* [99] have also described an on-column radioisotope detector for CE. Sensitivity of the on-column detector [99] is similar (10^{-10} M) to the post-column detector [98, 99] for the analysis of radiolabeled ATP and cytidine-5′-triphosphate (CTP). Although both radioisotope detectors are extremely sensitive, applications are limited to beta- and gamma-emitters. Furthermore, radioactive waste is produced which is dangerous to handle.

Laser-induced capillary vibration is yet another detection method that can be used with CE. Laser-induced capillary vibration detection was developed by Sawada and co-workers [100–104] and can be considered an unconventional absorbance measurement. In a detection volume of 100 pL it is possible to measure 6.0 fg (13 amole or 13×10^{-8} M) of sunset yellow dye, which corresponds to 1.5×10^{-5} absorbance units [100] using an argon-ion laser. Laser-induced capillary vibration detection has also been used in the analysis of DABSYL amino acids [102]. Under conditions optimized to reduce background signal generated from heat, 8 amole (8×10^{-8} M) mass LODs were achievable for the separation of derivatized amino acids [103]. More recently, Sawada and colleagues [104] successfully analyzed a series of nonderivatized amino acids at the fmole (10^{-5} M) level using an excimer laser operated at 248 nm.

8 Conclusions

In conclusion, there are a number of different detection methods that can be used with CE. In general, detector selection is dependent upon the properties of the solutes being analyzed and the sensitivity required for the analysis, and, as a result, is application-specific. While much reasearch has gone into detector development in the last decade, improvement in detector sensitivity is still needed if CE analysis is to become a widely accepted analysis technique, especially in the planar format.

Support for this paper was provided by the Welch Foundation (D1-1312) and the National Science Foundation (DBI-9876839).

Received September 16, 1999

9 References

[1] Li, C., Martin, L. M., *Anal. Biochem.* 1998, *263*, 72–78.

[2] Hu, T., Zuo, H., Riley, C. M., Stobaugh, J. F., Lunte, S. M., *J. Chromatogr. A* 1995, *716*, 381–388.

[3] Grant, I. H., Steuer, W., *J. Microcol. Sep.* 1990, *2*, 74–80.

[4] Tsudea, T., Sweedler, J. V., Zare, R. N., *Anal. Chem.* 1990, *62*, 2149–2152.

[5] Taylor, J. A., Yeung, E. S., *J. Chromatogr.* 1991, *550*, 831–837.

[6] Poppe, H., *Anal. Chim. Acta.* 1980, *114*, 59–70.

[7] Chervet, J. P., Van Soest, R. E. J., Ursem, M., *LC Packings*, Technical Communication, San Francisco, CA 1990.

[8] Mainka, A., Bachmann, K., *J. Chromatogr. A* 1997, *767*, 241–247.

[9] Kaltenbach, P., Ross, G., Heiger, D. N., *HPCE* 1997, Anaheim, CA 1997.

[10] Wang, T., Aiken, J. H., Huie, C. W., Hartwick, R. A., *Anal. Chem.* 1991, *63*, 1372–1376.

[11] Landers, J. P. (Ed.), *Handbook of Capillary Electrophoresis*, CRC Press, New York 1997.

[12] Weston, A., Brown, P. R., Jandik, P., Jones, W. R., Heckenberg, A. L., *J. Chromatogr.* 1992, *593*, 289–295.

[13] Jandik, P., Jones, W. R., *J. Chromatogr.* 1991, *546*, 431–443.

[14] Jandik, P., Jones, W. R., Weston, A., Brown, P. R., *LC.GC* 1991, *9*, 634–637.

[15] Vorddran, A., Oefner, P., Scherz, H., Bonn, G., *Chromatographia* 1992, *33*, 163–168.

[16] Foret, F., Fanali, S., Nardi, A., Boček, P., *Electrophoresis* 1990, *11*, 780–783.

[17] Nielen, M. W. W., *J. Chromatogr.* 1991, *588*, 321–326.

[18] Wang, T., Hartwick, R. A., *J. Chromatogr.* 1992, *589*, 307–313.

[19] Yeung, E. S., *Acc. Chem. Res.* 1989, *22*, 125–130.

[20] Weinberger, R., *Practical Capillary Electrophoresis*, Academic Press, Boston 1993.

[21] Bornhop, D. J., Dovichi, N. J., *Anal. Chem.* 1987, *59*, 1632–1636.

[22] Yu, M., Dovichi, N. J., *Anal. Chem.* 1989, *61*, 37–40.

[23] Yu, M., Dovichi, N. J., *Appl. Spectrosc.* 1989, *43*, 196–201.

[24] Waldron, K. C., Dovichi, N. J. *Anal. Chem.* 1992, *64*. 1396–1399.

[25] Earle, C. W., Dovichi, N. J., *J. Liq. Chromatogr.* 1989, *12*, 1989–1994.

[26] Yu, M., Dovichi, N. J., *Microchem. Acta* 1988, *27*, 25–35.

[27] Chen, M., Waldron, K. C. Zhao, Y., Dovichi, N. J., *Electrophoresis* 1994, *15*, 1290–1294.

[28] Li, X.-F., Liu, C.-S., Roos, P., Hansen Jr., E. B., Cerniglia, C. E., Dovichi, N. J., *Electrophoresis* 1998, *19*, 3178–3182.

[29] Bruno, A. E., Paulus, A., Bornhop, D. J., *Appl. Spectrosc.* 1991, *45*, 462–467.

[30] Gassmann, E., Kuo, J. E., Zare, R. N., *Science* 1985, *230*, 813–814.

[31] Timperman, A. T., Khatib, K., Sweedler, J. V., *Anal. Chem.* 1995, *67*, 139–144.

[32] Chen, D. Y., Aldelhelm, K., Cheng, X. L., Dovichi, N. J., *Analyst* 1994, *119*, 349–352.

[33] Lee, Y.-H., Maus, R. G., Smith, B. W., Winefordner, J. D., *Anal. Chem.* 1994, *66*, 4142–4149.

[34] Chen, D. Y., Dovichi, N. J., *Anal. Chem.* 1996, *68*, 690–696.

[35] Gassman, E., Kuo, J. E., Zare, R. N., *Science* 1985, *230*, 813–814.

[36] Chen, D. Y., Swerdlow, H. P., Harke, H. R., Zang, J. Z., Dovichi, N. J., *J. Chromatogr.* 1991, *559*, 237–246.

[37] Swerdlow, H., Wu, S., Harke, H., Dovichi, N. J., *J. Chromatogr.* 1990, *516*, 61–67.

[38] Wu, S., Dovichi, N. J., *J. Chromatogr.* 1989, *480*, 141–155.

[39] Cheng, Y. F., Dovichi, N. J., *Science* 1988, *242*, 562–564.

[40] Kuhr, W. G., Yeung, E. S., *Anal. Chem.* 1988, *60*, 1832–1834.

[41] Kuhr, W. G., Yeung, E. S., *Anal. Chem.* 1988, *60*, 2642–2646.

[42] Hernandez, L., Joshi, N., Excalon, J., Guzman, N., *J. Chromatogr.* 1990, *502*, 247–255.

[43] Hernandez, L., Joshi, N., Excalona, J., Guzman, N., *J. Chromatogr.* 1991, *559*, 183–196.

[44] Liu, J., Shirota, O., Novotny, M., *Anal. Chem.* 1991, *63*, 413–417.

[45] Swaile, D. F., Sepaniak, M. J., *J. Liq. Chromatogr.* 1991, *14*, 869–889.

[46] Nie, S., Dadoo, R., Zare, N., *Anal. Chem.* 1993, *65*, 3571–3575.

[47] Chang, H. T., Yeung, E. S., *Anal. Chem.* 1995, *67*, 1079–1083.

[48] Craig, D. B., Arriaga, E., Wong, J. C. Y., Lu, H., Dovichi, N. J., *Anal. Chem.* 1998, *70*, 39A–43A.

[49] Craig, D. B., Arriaga, E., Wong, J. C. Y., Lu, H., Dovichi, N. J., *J. Am. Chem. Soc.* 1998, *118*, 5254–5259.

[50] Xue, Q., Yeung, E. S., *Nature* 1995, *373*, 681–683.

[51] Timperman, A. T., Oldenburg, K. E., Sweedler, J. V., *Anal. Chem.* 1995, *67*, 3421–3426.

[52] McGregor, D. A., Yeung, E. S., *J. Chromatogr. A* 1994, *680*, 491–496.

[53] Eriksson, K. O., Palm, A., Hjertén, S., *Anal. Biochem.* 1992, *201*, 211–215.

[54] Cheng, Y. F., Fuchs, M., Andrews, D., Carson, W., *J. Chromatogr.* 1992, *608*, 109–116.

[55] Huang, X., Zare, R. N., *J. Chromatogr.* 1990, *516*, 185–187.

[56] Haber, C., Roosli, S., Tsuda, T., Scheidegger, D., Müller, S., Simon, W., *12ᵗʰ Int. Symp. Capillary Chromatography*, Kobe, Japan 1990.

[57] Haber, C., Silverstri, I., Roosli, S., Simon, W., *Chimia* 1991, *45*, 117–121.

[58] Nann, A., Pretsch, E., *J. Chromatogr. A* 1994, *676*, 437–442.

[59] Haung, X., Pang, T. K., Gordaon, M. J., Zare, R. N., *Anal. Chem.* 1987, *59*, 2747–2749.

[60] Huang, X., Gordon, M. J., Zare, R. N., *J. Chromatogr.* 1988, *425*, 385–390.

[61] Gordon, M. J., Huang, X., Pentoney, S. L., Zare, R. N., *Science* 1988, *63*, 224–228.

[62] Reay, J. R., Dadoo, R., Storment, C. W., Zare, R. N., Kovacs, G. T. A., *Proc. Solid-State Sensors and Actuator Workshop*, Hilton Head, SC 1994, p. 61.

[63] Dasgupta, P. K., Bao, L., *Anal. Chem.* 1993, *65*, 1003–1011.

[64] Kar, S., Dasgupta, P. K., Liu, H., Hwang, H., *Anal. Chem.* 1994, *66*, 2537–2543.

[65] Wallingford, R. A., Ewing, A. G., *Anal. Chem.* 1987, *59*, 1762–1766.

[66] Wallingford, R. A., Ewing, A. G., *Anal. Chem.* 1988, *60*, 1972–1975.

[67] Wallingford, R. A., Ewing, A. G., *J. Chromatogr.* 1988, *441*, 299–309.

[68] Wallingford, R. A., Ewing, A. G., *Anal. Chem.* 1988, *60*, 258–263.

[69] Ewing, A. G., Wallingford, R. A., Olefirowicz, R. A., *Anal. Chem.* 1989, *61*, 292A–294A.

[70] Ewing, A. G., Mesaros, J. M., Gavin, P. F., *Anal. Chem.* 1994, *66*, 527A–537A.

[71] O'Shea, T. J., Lunte, S. M., LaCourse, W. R., *Anal. Chem.* 1993, *65*, 948–951.

[72] Lu, W. Z., Cassidy, R. M., *Anal. Chem.* 1993, *65*, 2878–2881.

[73] Colon, L. A., Dadoo, R., Zare, R. N., *Anal. Chem.* 1993, *65*, 476–481.

[74] Lin, H., Xu, D. K., Chen, H. K., *J. Chromatogr. A* 1997, *760*, 227–233.

[75] Tarigan, H. J., Neill, P., Kenmore, C. K., Bornhop, D. J., *Anal. Chem.* 1996, *15*, 1763–1770.

[76] Jorgenson, J. W., Wit, J. D., in: Hill, H. H., McMinn, D. G. (Ed.), *Detectors for Capillary Chromatography*, Wiley, New York 1992, Chapter 15.

[77] Bornhop, D. J., Dovichi, N. J., *Anal. Chem.* 1986, *58*, 504–505.

[78] Bruno, A. E., Krattinger, B., Maystre, F., Widmer, H. M., *Anal. Chem.* 1991, *63*, 2689–2697.

[79] Krattiger, B., Bruno, A. E., Bruin, G. J., *Anal. Chem.* 1994, *66*, 1–8.

[80] Bornhop, D. J., *US Patent 5,235,170*, 1994.

[81] Bornhop, D. J., *Appl. Opt.* 1995, *34*, 3234–3239

[82] Deng, Y., Li, B., *Appl. Opt.* 1998, *37*, 998–1005.

[83] Wu, J., Pawliszyn, J., *Anal. Chem.* 1992, *64*, 224–227.

[84] Bruggraf, N., Krattiger, B., de Mello, A., de Rooij, N., Manz, A., *Analyst* 1998, *123*, 1443–1447.

[85] Buttry, D. A., Vogelmann, T. C., Chen, G., Goodwin, R., *US Patent 5,600,433*, 1997.

[86] Born, M., Wolf, E., *Principles of Optics*, Pergamon Press, New York 1975.

[87] Saleh, B. E. A., Teich, M. C., *Fundamentals of Photonics*, Wiley-Interscience, New York 1991.

[88] Swinney, K., Pennington, J., Bornhop, D. J., *Microchem. J.* 1999, *62*, 154–163.

[89] Swinney, K., Pennington, J., Bornhop, D. J., *Analyst* 1999, *124*, 221–225.

[90] Swinney, K., Bornhop, D. J., *J. Microcol. Sep.* 1999, *11*, 596–604.

[91] Chen, C. Y., Morris, M. D., *Appl. Spectrosc.* 1998, *42*, 515–518.

[92] Walker III, P. A., Morris, M. D., *J. Chromatogr. A* 1998, *805*, 269–275.

[93] Walker III, P. A., Kowalchyk, W. K., Morris, M. D., *Anal. Chem.* 1995, *67*, 4255–4260.

[94] Wu, N., Peck, T. L., Webb, A. G., Magin, R. L., Sweedler, J. V., *Anal. Chem.* 1994, *66*, 3849–3857.

[95] Wu, N. Peck, T. L., Webb, A. G., Sweedler, J. V., *J. Am. Chem. Soc.* 1994, *116*, 7929–7930.

[96] Olson, D. L., Peck, T. L., Webb, A. G., Sweedler, J. V., in: Kaumaya, P. T. P. and Hodges, R. S., (Eds.), *Peptides: Chemistry, Structure, and Biology*, ESCOM, Leiden, The Netherlands 1996, pp. 730.

[97] Olson, D. L., Lacey, M. E., Webb, A. G., Sweedler, J. V., *Anal. Chem.* 1999, *71*, 3070–3076.

[98] Pentoney Jr., S. L., Zare, R. N., Quint, J. F., *J. Chromatogr.* 1989, *480*, 259–270.

[99] Pentoney Jr., S. L., Zare, R. N., Quint, J. F., *Anal. Chem.* 1989, *61*, 1642–1647.

[100] Wu, J., Kitamori, T., Sawada, T., *Anal. Chem.* 1990, *62*, 1676–1678.

[101] Wu, J., Odake, T., Kitamori, T., Sawada, T., *Anal. Chem.* 1991, *63*, 2216–2218.

[102] Odake, T., Kitamori, T., Sawada, T., *Anal. Chem.* 1992, *64*, 2870–2871.

[103] Odake, T., Kitamori, T., Sawada, T., *Anal. Chem.* 1995, *67*, 145–148.

[104] Odake, T., Kitamori, T., Sawada, T., *Anal. Chem.* 1997, *69*, 2537–2540.

Review

Fluorescence line-narrowing detection in chromatography and electrophoresis

Ryszard Jankowiak
Kenneth P. Roberts
Gerald J. Small

A review of the basic aspects of fluorescence line-narrowing spectroscopy (FLNS) and its coupling with thin-layer chromatography (TLC) and polyacrylamide gel electrophoresis (PAGE) for off-line high-resolution low temperature spectral characterization is discussed. This is followed by a description of the on-line interfacing of capillary electrophoresis (CE) and capillary electrochromatography (CEC) with FLN detection. CE/CEC-FLNS instrumentation and its applications for spectral identification of closely related analytes are also presented. Future prospects of micro and capillary high performance liquid chromatography (HPLC) with on-line high-resolution low temperature spectroscopic identification are considered.

Keywords: Chromatography / Electrophoresis / Capillary electrophoresis – fluorescence line-narrowing spectroscopy / Analyte identification / On-line fluorescence detection / Review   EL 3808

Contents

Correspondence: Dr. Ryszard Jankowiak, Ames Laboratory USDOE, Iowa State University, Ames, IA 50011, USA
E-mail: jankowiak@ameslab.gov
Correspondence: Dr. Gerald J. Small, Department of Chemistry, Iowa State University, Ames, IA 50011, USA, **E-mail:** gsmall@ameslab.gov

Abbreviations: BA, benz[*a*]-anthracene; **B[*a*]P**, benzo[*a*]pyrene; **BPDE**, benzo[*a*]pyrene-7,8-dihydrodiol-9,10-epoxide or benzo; **DB[*a,l*]P**, dibenzo[*a,l*]pyrene; **DB[*a,l*]PDE**, dibenzo[*a,l*]-pyrene diol epoxide; **DB[*a,l*]PDE-14-N7Ade**, 14-(adenin-7-yl)-11,12,13-trihydroxy-11,12,13,14-tetrahydrodibenzo[*a,l*]-pyrene; **DB[*a,l*]PDE-14-N^{2}dG**, dibenzo[*a,l*]pyrene diol epoxide-N^2-deoxyguanosine; **DB[*a,l*]PDE-14-N7Gua**, 14-(guanin-7-yl)-11,12,13-trihydroxy-11,12,13,14-tetrahydrodibenzo[*a,l*]pyrene; **DB[*a,l*]P tetrol**, 11,12,13,14-tetrahydroxy-11,12,13,14-tetrahydrodibenzo[*a,l*]pyrene; ***anti*-DB[*a,l*]PDE-N^{6}dA**, *anti*-dibenzo[*a,l*]pyrene diol epoxide-14-N^{6}deoxyadenosine; ***syn*-DB[*a,l*]PDE-14-N^{6}dA**, *syn*-dibenzo[*a,l*]pyrene diol epoxide-14-N^{6}deoxyadenosine; **DOSS**, dioctyl sulfosuccinate; **FLNS**, fluorescence line-narrowing spectroscopy; **PAH**, polycyclic aromatic hydrocarbon(s); **PEI**, polyethylenimine; **ZPL**, zerophonon line

1 Introduction

Electrophoresis and chromatography are widely used analytical separation techniques. Both have found use in various chemical, biochemical, and biomedical sciences, and new applications continue to emerge [1–4]. Of increasing importance to research with electrophoresis and chromatography is their marriage with various "information-rich" detection methods (also known as hyphenated techniques) that are capable of providing unambiguous on-line identification of separated analytes [5]. Several information-rich methods are being explored, including Raman [6–12], nuclear magnetic resonance (NMR) [13–19], mass spectrometry (MS) [20–25], and high-resolution fluorescence line-narrowing spectroscopy (FLNS) [26–30]. In this review FLNS and its applications when combined with various separation techniques such as high performance liquid chromatography (HPLC), thin-layer chromatography (TLC), polyacrylamide gel electrophoresis (PAGE), capillary electrophoresis (CE), and capillary electrochromatography (CEC) are discussed. It is anticipated that these diverse combinations, with both off-line and on-line detection, will find applications in the environmental and bioanalytical sciences.

FLNS has proven to be a powerful, high-resolution technique for characterization and determination of molecular and biomolecular analytes. Its principles and attributes are discussed in several books [31–37] and reviews [38–42]. FLNS is a low temperature (typically 4.2 K) technique in which a tunable laser is used to excite a narrow region in the inhomogeneously broadened absorption band of an analyte imbedded in an amorphous host [31]. Since only excited analytes fluoresce, the inhomogeneous broadening contribution to the fluorescence vibronic bands is largely eliminated, *i.e.*, the fluorescence spectrum is line-narrowed. Typically, the inhomogeneous broadening is in the range of 100–300 cm^{-1}. The line-narrowed widths of vibronic transitions are usually lifetime-limited (≈ 5 cm^{-1}) due to vibrational relaxation.

FLNS has been used for identification and spectroscopic characterization of a large number of organic [31, 41, 43] and inorganic [44] molecules as well as a wide variety of biomolecules including photosynthetic pigments [45–50], antenna protein-pigment complexes [51], hemoglobin [52], and antibody-polycyclic aromatic hydrocarbon (PAH) complexes [53]. FLNS has also been applied to nucleosides/nucleotides, oligonucleotides, macromolecular DNA, and proteins adducted to metabolites of PAHs [28–30, 39, 41, 54]. In particular, FLNS has proven to be the most powerful optical spectroscopic technique for the detection and characterization of PAH-DNA adducts formed *in vitro* [41, 54, 55] and *in vivo* [56–60], where one is limited to picomole or sub-picomole quantities of bound metabolite. For example, one adduct in 10^8 base pairs for 100 µg of macromolecular DNA [54] can be detected, while at the nucleoside level 0.1 fmole of a moderately strong fluorophore such as benzo[*a*]pyrene can be detected with a spectral resolution of approximately 5 cm^{-1} [41, 54]. The distinction between adduction of a given metabolite to different bases and to different nucleophilic centers of a given base such as guanine (Gua) and adenine (Ade) has been demonstrated [55, 60]. Moreover, FLNS has been used to differentiate between external, base-stacked, and intercalated conformations of PAH-DNA adducts and stereoisomeric PAH-DNA adducts [57, 58, 61–64]. The success of FLNS is the result of a combination of high sensitivity, high selectivity, and versatility.

This review is organized as follows. In Section 2, the major information-rich detection modes in chromatography and electrophoresis are briefly discussed. The principles of FLNS are illustrated in Section 3, where it is shown that vibronic excitation within the $S_0 \rightarrow S_1$ absorption spectrum provides the highest selectivity for analyte identification [39, 41, 54]. The coupling of FLNS with TLC and PAGE is described in Sections 4 and 5. Section 6 is concerned with the recently developed CE-FLNS technique [26–28]. This section also includes the results of proof-of-principle experiments that establish that CEC can be coupled with FLNS for on-line detection. Applications of CE-

FLNS, which illustrate the attributes of CE-FLNS, are reviewed in Section 7. This article ends with concluding remarks and future prospects, including the possibility of interfacing HPLC with FLNS.

2 Information-rich detection in chromatography and electrophoresis

There is a multitude of detection techniques available for electrophoresis and chromatography and a detailed comparison is beyond the scope of this manuscript (for reviews see [5, 65, 66]). The most common detection modes include optical absorption, fluorescence, and radioactivity, where analyte identification is based on retention/migration times (or migration distances) that are determined using standards. Such identification requires the absence of coelution from other separating species, a high degree of reproducibility, and the availability of reliable standards. To circumvent these problems, a number of information-rich detection schemes have been developed to provide a higher level of confidence in the identification process. They include absorption and fluorescence spectroscopy at room temperature [5], MS [20–25], NMR [13–19], electrochemical (EC) detection [67, 68], circular dichroism (CD) [69], Fourier transform-infrared (FTIR) spectroscopy [70, 71], Raman spectroscopy [6–12], and FLNS [26–30]. A direct comparison of their attributes is difficult due to the inherent differences between the identification processes. On-line wavelength-resolved absorbance detection is the most frequently applied method for monitoring separations due to its low cost and universal applicability. Although this method offers a spectral signature of the separated components, the spectra of liquid solutions are often broad and quite structureless at room temperature. Moreover, the sensitivity of this approach is low due to shot-noise and the typically short optical paths. In contrast to absorbance, wavelength-resolved fluorescence emission/excitation detection provides higher sensitivity; this is especially true for laser-induced fluorescence (LIF) [72, 73]. Unfortunately, as with absorption, room temperature fluorescence spectra are often featureless, preventing identification of closely related analytes.

NMR detection in chromatography and electrophoresis can provide unequivocal structural and conformational information on separated species. However, due to field inhomogeneities from limited residence times and technical difficulties of probe/coil design, the sensitivity of this technique is severly compromised for on-line analysis [80, 81] with detection limits of about 0.1–1 M [5]. Also, the current spectral resolution of on-line NMR detection is rather low, excluding the possibility to differentiate between stereoisomers [5, 82, 83]. Furthermore, the need for solvent suppression methodologies or deuterated solvents imposes additional constraints for practical applications of on-line NMR analysis. Therefore, NMR is still mainly used as an off-line structural identification technique.

FTIR spectroscopy can also provide spectral information from which analyte identity may be obtained in both off- and on-line modes [70, 71]. However, since most solvents used in separation procedures absorb strongly in the IR region, flow cells with a small pathlength have to be used, severely decreasing the limit of detection (LOD) for on-line analysis. In the case of liquid chromatography (LC), the LOD can be improved by using off-line FTIR detection, where the LC effluent is deposited on a zinc selenide substrate followed by evaporation of the mobile phase prior to analysis [78].

Raman spectroscopy has also been used to provide on-line detection for LC [8, 9] and CE [6, 7]. The main advantage of Raman detection over room temperature absorption and fluorescence spectroscopies is that Raman spectra are highly structured. Resonant, near-resonant, surface-enhanced, and nonresonant excitation Raman spectroscopies have been employed [6–12]. As in the case of FTIR, on-line analysis by Raman spectroscopy suffers from inadequate sensitivity and limit of detection due to solvent interference.

On-line EC detection is limited to compounds that are selectively oxidized or reduced at the electrode surface [65]. Consequently, EC (potentiometric and/or amperometric) detection is suitable for analysis of select components in complex biological fluids [79, 80]. Detection limits for electroactive analytes in CE/LC with EC detection are typically in the low nanomolar range [65, 81].

Fluorescence-detected circular dichroism (FDCD) is another information-rich detection method that is applicable to chiral molecules whose CD signature varies with molecular conformation, thereby providing the analyst with important structural information [69]. Since many biomolecules are chiral, this type of detection has found applications in bioanalytical sciences [69, 82]. As an example of sensitivity, CE-FDCD detection allows for optical activity measurements of riboflavin in picoliter volumes with a detection limit in the subfemtomole range [69].

Detection by MS in CE and chromatography provides a high degree of structural information and, thus, is becoming widely used. Both off- and on-line MS detection modes have been successfully employed [20–25]. With MS detection, separated analytes are identified based on mass-to-charge ratios of parent ions or fragment ions pro-

duced from collision-induced dissociation. MS has very good sensitivity and excellent selectivity [5, 65]. However, in the case of structurally similar molecules (*e.g.*, geometric isomers), the highly discretionary, multipass process often used results in a severe decrease in throughput. Nonetheless, MS has a better LOD than absorbance-, EC-, NMR-, FTIR-, Raman-, and FDCD-based detection.

Finally, FLNS has been coupled both off- and on-line with chromatography and electrophoresis [26–30, 83–88]. FLNS is very sensitive and highly selective in both frequency and time domains and can provide structural/conformational information that lead to analyte identification. Low attomole LOD for on-line FLNS has been demonstrated [30]. The applicability of FLNS can be expanded to nonfluorescent compounds derivatized with a fluorescent marker which adopts specific interactions with the analytes of interest [89]. The principles and applications of FLNS detection in chromatography and electrophoresis are discussed in the remaining sections of this review.

3 Principles of FLNS

FLNS is a laser-based technique that eliminates the static inhomogeneous broadening (Γ_{inh}) contribution to the widths of vibronic transitions of chromophores imbedded in glasses, polymers, proteins, and DNA [31, 37, 54] at low temperatures. Γ_{inh} is the result of the chromophore adopting a large number of energetically inequivalent sites [31, 36, 90]. In this methodology only those sites whose (0,0) transitions overlap the laser profile are excited (often referred to as selecting an "isochromat"), resulting in an FLN spectrum. Thus, a key requirement of FLNS is the use of a spectrally narrow and tunable laser source for excitation. Moreover, to observe line narrowing the chromophore concentration has to be sufficiently low ($\leq 10^{-4}$ M) to prevent intermolecular energy transfer.

Two types of excitation can be employed in FLNS: the chromophore can be excited *via* its inhomogeneously broadened origin absorption profile (origin band excitation) or *via* vibronic absorption bands (vibronic excitation). Figure 1 illustrates origin band excitation where thermal broadening of the zero-phonon lines (ZPLs) is negligible (≤ 0.1 cm^{-1} at 4.2 K). The narrow bandwidth laser with frequency ω_L selectively excites chromophores within the band of origin. The resulting FLN spectrum shown at the bottom of the top frame is comprised of the highest energy ZPL which corresponds to the transition between the zero-point vibrational levels of the fluorescent (S_1) and ground electronic (S_0) states, and a series of vibronic ZPLs corresponding to transitions terminating at vibrational levels of S_0. This type of FLN spectrum provides

Electrophoresis 2000, *21*, 1251–1266

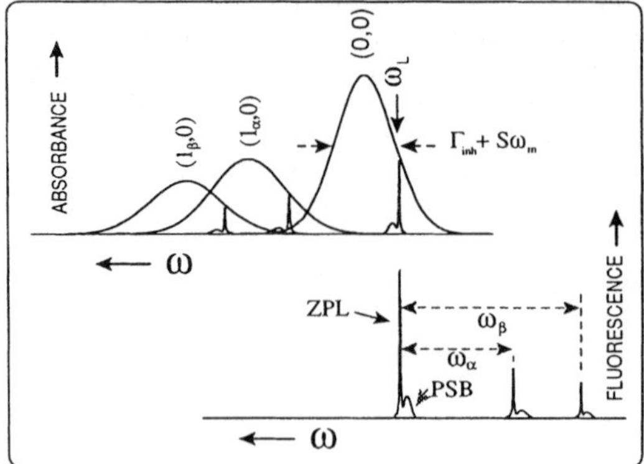

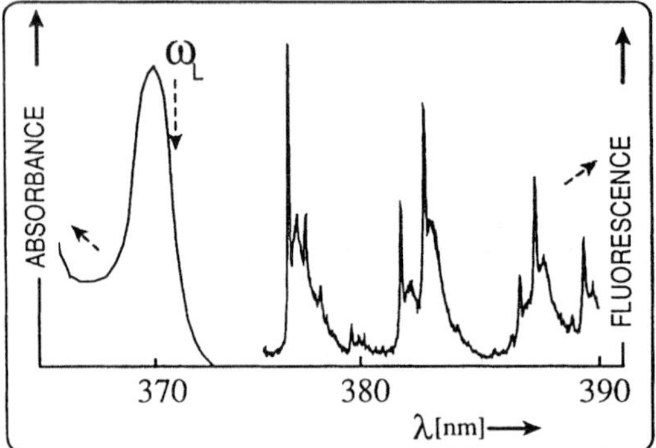

Figure 1. Origin band excitation in FLNS. Top frame: Γ_{inh} denotes the inhomogeneous broadening of the (0,0) band; S is the Huang-Rhys factor, and ω_m corresponds to the mean phonon frequency. Laser frequency, ω_L, excites only those chromophores whose (0,0) transition overlap the laser profile. ω_α and ω_β correspond the ground state vibrational frequencies, in cm^{-1}. Bottom frame: (0,0) absorption band (left) and FLN spectrum (right) obtained for pyrene in ethanol glass at T = 4.2 K, λ_{ex} = 371.5 nm. The sharp peaks (ZPLs) in the fluorescence spectrum correspond to pyrene ground-state vibrational frequencies.

ground state vibrational frequencies (ω_α and ω_β) which, when combined with the relative vibronic intensities, serve as a "fingerprint" of the chromophore.

The weak, broad (approximately 30–40 cm^{-1}) bands that accompany the ZPL in the FLN spectrum are the phonon-sidebands (PSBs) whose relative intensity depends on the strength of electron-phonon coupling. The PSB builds on each ZPL and is a consequence of the change in the chromophore-host matrix equilibrium geometry upon $S_0 \rightarrow S_1$ excitation. Since the PSBs contribute to the absorption and/or fluorescence bands of origin, the width of the Gaussian ((0,0) absorption band) shown in Fig. 1 (top) is

approximately given by $\Gamma_{inh} + S\omega_m$, where ω_m is the mean phonon frequency (for organic glasses and polymers typically about 20–30 cm^{-1}) and S is the Huang-Rhys factor [44, 90, 91]. In the low temperature limit, the Franck-Condon (FC)-factor for the ZPL and PSB is given by $\exp(-S)$ and $(1-\exp(-S))$, respectively. The physical significance of S is, perhaps, most easily appreciated by noting that $2S\omega_m$ corresponds to the Stokes shift for relaxed fluorescence. Strong and weak electron-phonon coupling is defined by $S > 1$ and $S < 1$, respectively [31, 32, 90]. Therefore, within the limit of strong coupling, the selectivity of FLN is lost because the fluorescence spectrum is dominated by PSB emission with very weak ZPLs. The bottom frame of Fig. 1 shows an example of an origin band excited FLN spectrum (right), along with the inhomogeneously broadened absorption spectrum (left) for pyrene embedded in ethanol glass at 4.2 K. For clarity, only part of the FLN spectrum is shown; the narrow peaks are ZPLS, and their frequencies (not shown) correspond to the ground-state vibrations. The prominence of the ZPLs indicates weak electron-phonon coupling. As the laser is tuned across the band of origin the fluorescence spectrum tracks the laser while maintaining the same vibronic structure.

Vibronically excited FLNS probes the S_1-state vibronic levels as shown in the top frame of Fig. 2. In that figure the laser (ω_L) excites only two overlapping vibronic transitions, $(1_\alpha,0)$ and $(1_\beta,0)$, which are not resolved in the absorption spectrum. The two selected isochromats undergo vibrational relaxation to two different points (A and B) in the zero-point site energy distribution of the S_1-state. This is followed by fluorescence from the energetically distinct isochromats (A and B). The result is an FLN spectrum that consists of two strong ZPLs – $(0,0)_A$ and $(0,0)_B$ – often referred to as a "multiplet origin structure". The displacements between ω_L and the doublet components of the origin transition – $(0,0)_A$ and $(0,0)_B$ – yield the excited-state vibrational frequencies ω_α' and ω_β'. The vibronic bands that build on $(0,0)_A$ and $(0,0)_B$ to lower energy are not shown for simplicity. In actual experiments it is not uncommon to observe 6–8 ZPLs that contribute to the multiplet origin structure for each excitation frequency.

Due to its superior selectivity (relative to origin-excited FLNS), vibronically exited FLNS has been used exclusively for identification and conformational analysis of closely related analytes [41, 54]. It has been established that the vibrational frequencies and intensities of the S_1-state are the most sensitive to subtle changes in the structure of the chromophore and its environment [39, 41]. This sensitivity has a firm theoretical understanding based on the Duschinsky effect (a mixing of the normal

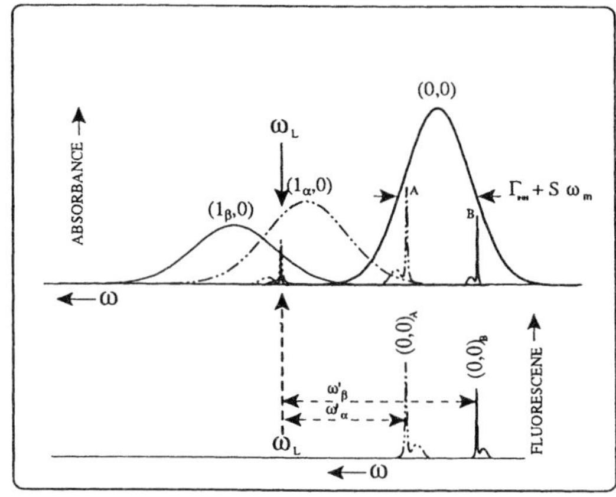

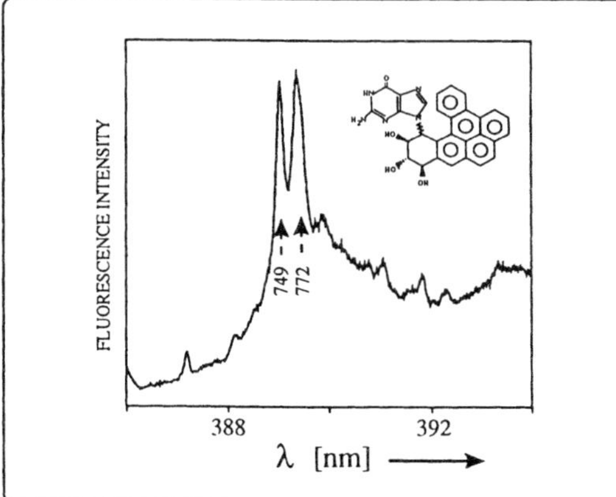

Figure 2. Principles of vibronic excitation in FLNS. Top frame: selective laser excitation (ω_L) of two subsets of molecules (A and B) within the vibronic region. Schematic of the resulting fluorescence spectrum with two $(0,0)_A$ and $(0,0)_B$ transitions is shown at the bottom of the top frame. ω_α' and ω_β' are the vibrational frequencies in the excited state. Bottom frame: an example of vibronically excited FLN spectrum is shown for the *syn*-DB[*a,l*]PDE-14-N7Gua adduct; λ_{ex}, 376.0 nm; *T*, 4.2 K. The sharp ZPLs at 749 and 772 cm^{-1} correspond to excited-state vibrational frequencies.

coordinates in the excited state and vibronically induced anharmonicity) [92]. An example of a vibronically excited FLN spectrum obtained at 4.2 K for the *syn*-dibenzo[*a,l*]-pyrene-diolepoxide-14-N7Gua adducts is shown in the lower frame of Fig. 2. Laser excitation at 378 nm selectively excites several modes, with the two strongest bands corresponding to 749 and 772 cm^{-1} excited-state vibrations. The prominence of the vibronic ZPLs indicates weak electron-phonon coupling. By tuning ω_L across the $S_1 \leftarrow S_0$ absorption spectrum one can determine all

active excited-state vibrations [31, 32, 34, 54]. For each excitation frequency ω_L, a distinct fingerprint of the analyte is obtained. We note that FLN is generally only observed with excitation frequencies located within the $S_0 \rightarrow S_1$ absorption spectrum. Excitation of the higher-energy states (S_n, $n \geq 2$) results in S_1 fluorescence spectra that exhibit, at best, only a slight degree of line narrowing. This is a consequence of the site excitation energies of different electronic states being largely uncorrelated.

4 TLC and FLNS

TLC is a well-established and inexpensive separation technique. However, visualization of separated components in TLC is often difficult and insufficient for analyte identification. This led to the coupling of TLC with FLNS for off-line analysis. The first experiments were performed with PAHs and PAH-DNA adducts deposited on filter paper [93] and silica gel TLC plates [84, 94]. For example, Cooper *et al.* [84] studied benzo[*a*]pyrene (B[*a*]P)-derived deoxyguanosine (dG) adducts (B[*a*]P-C8dG) deposited on an octadecylsilate TLC plate. A detection level of ~ 10 fmole was demonstrated. The vibronically excited FLN spectrum obtained with λ_{ex} = 395.7 nm for B[*a*]P-C8dG on the TLC plate is shown in Fig. 3. The numbers above the peaks reflect the displacement of the ZPLs from the laser frequency in cm^{-1} and correspond to excited state vibrational frequencies. These frequencies and the relative intensities of the ZPLs allowed for positive adduct identification.

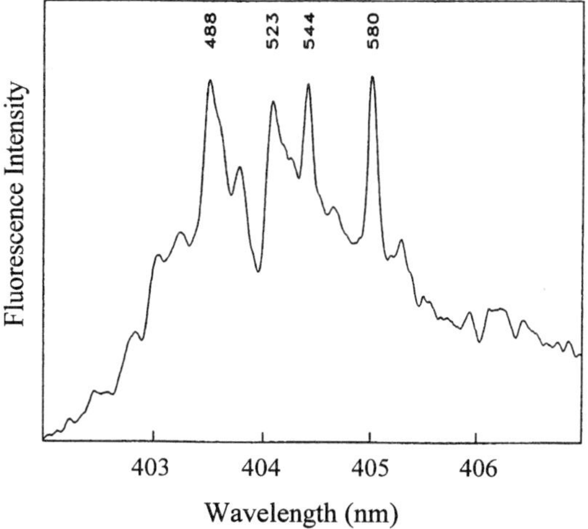

Figure 3. FLN spectrum of the B[*a*]P-C8dG nucleoside adduct on an octadecylsilate-TLC plate. λ_{ex} = 395.7 nm (vibronic excitation); *T*, 4.2 K. The numbers above the peaks correspond to excited-state vibrations (in cm^{-1}). Reprinted from [84], with permission.

Velthorst and co-workers [87, 88, 95–97] established that by coupling FLNS with HPLC-TLC for off-line detection, pyrene and halogenated pyrene derivatives could be identified on the chromatographic media (*e.g.*, silica gel, C18, and cellulose) of a TLC plate. The estimated limit of detection was in the nanogram range. The same group [83, 97] demonstrated that a second separation (perpendicular to the HPLC deposition trace) on a TLC plate, prior to FLNS analysis, further improved the quality of chromatographic resolution and the FLN spectra. The HPLC-TLC system is shown in Fig. 4. A fused-silica capillary is connected to the outlet of the HPLC detector flow cell which is coupled to a heated spray jet assembly. Heating the spray jet assembly with hot nitrogen gas up to 175°C allowed for the effluent to be gently deposited on the linearly translating TLC plate. Evaporation of the mobile phase limited the amount of diffusion of the deposited traces. Upon final drying the TLC plate was placed in a closed-cycle refrigeration system (10–20 K) for FLNS analysis [83]. An example of FLN spectra from this methodology for stereoisomeric B[*a*]P tetrol standards is shown in Fig. 5. Detection limits were at the fmole level [98]. With selective excitation, *e.g.*, 364 nm, the FLN spectra of the *trans-* and *cis*-stereoisomers of the *anti-* and *syn*-BP tetrols are clearly distinguishable at 12 K. The numbers in Fig. 5 correspond to excited-state vibrational frequencies. These results demonstrate that B[*a*]P-derived tetrols (originating from PAH-DNA adducts) can be identified using HPLC-TLC in combination with FLNS. However, application of this methodology to real-world samples has yet to be demonstrated.

It is well known that DNA adducts can be enzymatically digested, labeled with ^{32}P, and separated on a polyethylenimine (PEI)-cellulose TLC plate. In this ^{32}P-postlabeling

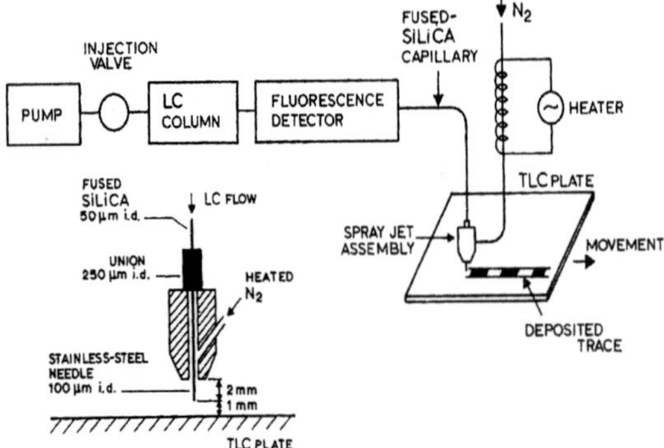

Figure 4. Schematic of the HPLC-TLC system and the coupling interface. The lower left corner shows the spray jet assembly in more detail (see Section 4). Reprinted from [97], with permission.

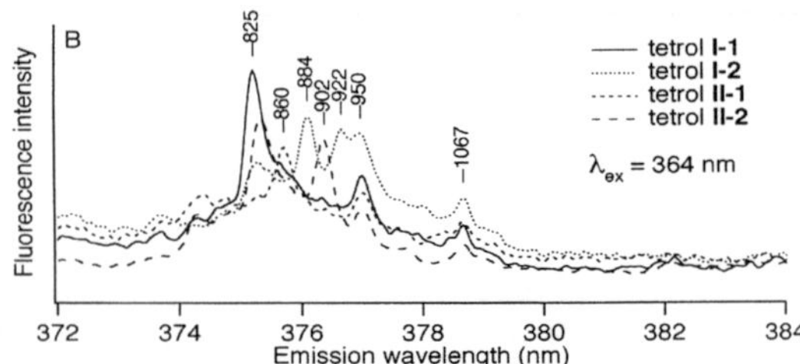

Figure 5. FLN spectra of *trans-* (1) and *cis-* (2) stereoisomers derived from the *anti-* (I) and *syn*-B[*a*]P tetrols (II) obtained with HPLC-TLC-FLNS. FLNS of the analytes were performed on the silica TLC substrate. λ_{ex}, 364 nm; *T*, 12 K. The numbers correspond to excited-state vibrational frequencies (in cm^{-1}). Reprinted from [98], with permission.

(Randerath) procedure [99], identification is typically accomplished by measuring the relative positions of the separated components autoradiographically. This methodology has recently been combined with FLNS [100]. The autoradiogram of benzo[*a*]pyrene-7,8-dihydrodiol-9,10-epoxide (BPDE)-adducted (1 adduct in 10^2 base pairs) and ^{32}P-postlabeled DNA digest on a PEI plate is shown in Fig. 6A. Direct on-line FLNS analysis of spot A was not possible due to strong luminescence background of the employed PEI plates. The background was several orders of magnitude stronger than that observed for the octadecylsilate TLC plates discussed above [84, 94]. Therefore, the analyte from spot A was extracted with an aqueous pyridinium formate solution, filtered, evaporated to dryness, and resuspended in the glass-forming matrix water/glycerol/ethanol for off-line FLNS analysis. An example of FLN spectrum obtained with an excitation wavelength of 361 nm is shown in Fig. 6B (top). Since this FLN spectrum is similar to the FLN spectrum obtained for the *anti*-BPDE-dAMP standard (Fig. 6B, bottom), it was suggested that the BPDE-dAMP adduct was formed [100].

Similar experiments were performed in this laboratory where the B[*a*]P-DNA digest at the adduction level of ~ 1:10^4 base pairs was analyzed by on-line TLC-FLNS. Extensive washing with ethanol decreased the luminescent background of the PEI plates. Based on well-established migration times/distances for the TLC separation of *anti*-BPDE-dGMP [57], the same major spot as shown in Fig. 6A was observed (data not shown). The on-line TLC-FLNS analysis revealed that the spot on the PEI-TLC plate corresponded to the *anti-trans*-BPDE-dGMP adduct. This indicates, in agreement with a large number of ^{32}P-postlabeling studies [57, 101, 102], that the top FLN spectrum of Fig. 6B most probably does not correspond

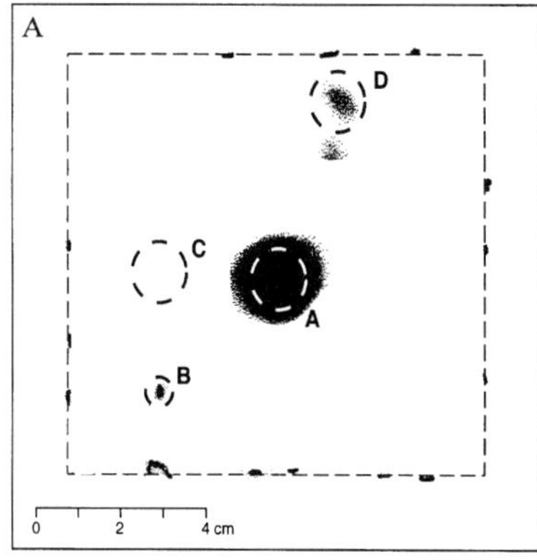

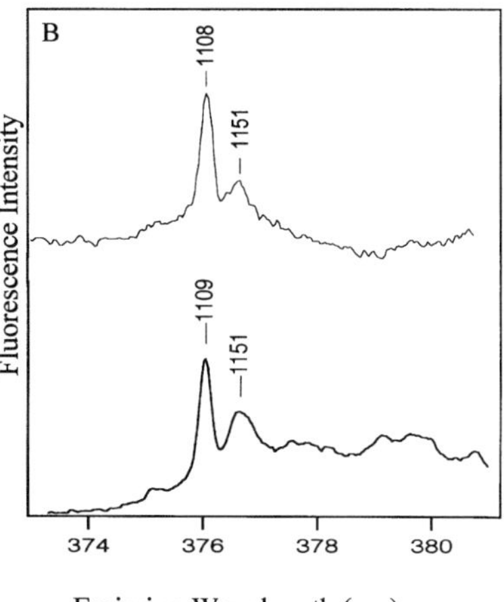

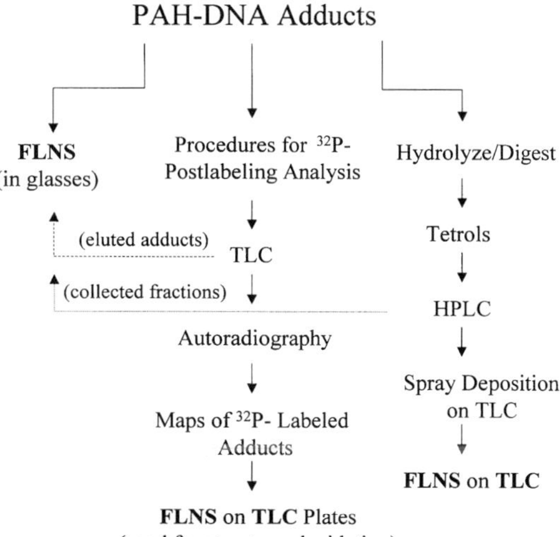

Figure 6. (A) Autoradiogram of BPDE adducted ($\sim 1:10^2$) and ^{32}P-postlabeled DNA after two-dimensional development on a PEI cellulose plate. (B) Top: FLN spectrum of spot A extracted from the PEI plate of (A) and resuspended in a water/glycerol/ethanol matrix. The bottom spectrum corresponds to the *anti*-BPDE-dAMP standard on the PEI-TLC plate. T, 12 K; λ_{ex}, 361.0 nm. Reprinted from [100], with permission.

to *anti-trans*-BPDE-dAMP but to *anti-trans*-BPDE-dGMP. Moreover, the presence of the -dGMP adduct was anticipated since 90–95% of total *anti*-BPDE-derived DNA adducts are formed with dGMP [103].

These results show that FLNS can be combined with ^{32}P-postlabeling, providing additional selectivity in the analysis of TLC spots of structurally similar analytes. We note that TLC-FLNS with PEI plates would be more practical if the luminescence background of PEI plates could be eliminated. Moreover, the four-dimensional chromatographic separation used in the Randerath procedure leads to large analyte zones, which limits the sensitivity of this approach to about 1 adduct in 10^4–10^5 base pairs (unpublished results). However, on-line FLNS analysis of TLC plates could be beneficial when an unresolved (diffuse) multicomponent zone is spatially probed with a focused laser beam. In summary, the ways in which FLNS can be utilized for the identification of DNA adducts and/or their hydrolysis products on TLC plates are illustrated in Fig. 7.

5 PAGE and FLNS detection

PAGE is a well-established technique for separating intact DNA adducts. Stable *anti*-B[*a*]P diolepoxide (*anti*-BPDE) adducts of various oligodeoxynucleotides were characterized and identified with PAGE and off-line FLNS detection [85, 86]. Since fluorescence analysis of PAGE-separated bands directly on gel was impractical due to background luminescence, the adducts were extracted and redissolved in glycerol/water for FLNS analysis. For the sequences studied, it was shown that the spectral features of the N^2-dG adducts of single-stranded (ss) and double-stranded (ds) oligomers were similar, limiting

Figure 7. Outline of a methodology for the investigation of PAH-DNA adducts utilizing FLNS in combination with ^{32}P-postlabeling and TLC (center), HPLC and TLC (right), and standard detection in glasses (left).

spectral differentiation between ss and ds oligomers. However, conformational information could be obtained for both ss and ds oligomers, as described below.

An example of conformational analysis by PAGE-FLNS of the major BPDE-induced lesions in ss oligomers is shown in Fig. 8. The autoradiogram (left) shows the modification of d(TTAAGGAATT) by *anti*-BPDE enantiomers (see figure caption for details). The ds oligomer denatured during electrophoresis. Figure 8A shows FLN spectra (λ_{ex} = 356.9 nm) of DNA ss oligomers adducted with (+)-*anti*-

Electrophoresis 2000, *21*, 1251–1266

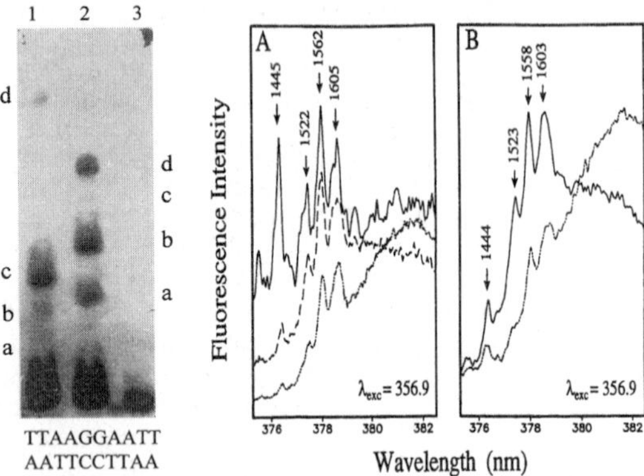

Figure 8. Left: autoradiogram of d(TTAAGGAATT) modified by (±)-*anti*-BPDE enantiomers. Lanes (1)–(3) correspond to (–)-*anti*-BPDE, (+)-*anti*-BPDE, and unreacted DNA oligomer under denaturing electrophoretic conditions, respectively. The anode is at the bottom of the figure; unmodified oligomer corresponds to the highly intense band exhibiting the fastest electrophoretic mobility. Bands of modified oligomers are labeled by lower-case letters. (A) FLN spectra of (+)-*anti*-adducts corresponding to bands d (solid line), b (dashed line), and a (alternating dotted and dashed line) of lane 2. (B) FLN spectra of (–)-*anti*-BPDE adducts corresponding to bands a (dotted-dashed line) and band c (solid line) of lane (1). FLN spectra from top to bottom correspond to external (top), partially base-stacked, and highly base-stacked adducts. λ_{ex}, 356.9 nm; glycerol/water matrix; *T*, 4.2 K. Reprinted from [86], with permission.

BPDE that were extracted from the polyacrylamide gel (lane 2 of the autoradiogram), bands a, b, and d. The slowly migrating modified DNA oligomers (assigned as external type adducts, solid line of Fig. 8A) showed prominent ZPLs, whereas the oligomers with the fastest electrophoretic mobilities exhibited weaker ZPLs (alternating dotted and dashed line) and the origin band shifted to ~381.5 nm, characteristic of highly base-stacked adducts [85, 86]. Adducted oligomers with a fluorescence origin at 379.5 nm and intermediate electrophoretic mobilities possessed weaker ZPLs (dashed line) compared to the external adduct type, but stronger than those of highly base-stacked adducts. Therefore, this type of adduct was listed as partially base-stacked [85]. Figure 8B shows spectra from two different (–)-*anti*-BPDE-adducted oligomers (lane 1; bands a and c). The solid spectrum is from the major adduct (extract from band c), and the alternating dotted and dashed spectrum is of the extract from the fastest migrating band (a). Based on FLN and nonline narrowed (NLN) spectra (data not shown), the conformations were assigned as partially and highly base-stacked, respectively. The FLN spectrum of the extract from band

d (line 1, data not shown) was similar to the FLN spectrum of adduct extracted from spot d (lane 2), and therefore it was assigned as an external (–)-*anti*-BPDE-dGMP adduct.

PAGE-FLNS was also used to study the binding of (–)-*anti*- and (+)-*anti*-BPDE to several sequence-defined ds oligomers [85]. Two of the oligomers contained central 5′-RAGGAR-3′ sequences (R, purine), which appear to be frequently mutated by racemic (±)-*anti*-BPDE [104–106]. Two other oligomers contained central 5′CCGG-3′ or 5′-TGGT-3′ sequences, which are strongly preferred for covalent binding but are less frequently mutated. Binding of the two enantiomers to the latter two sequences (data not shown) yielded a distribution of BPDE-N^2-dG adduct conformations similar to those formed in random sequence DNA *in vitro*, which means that the external conformation of the N^2-dG adduct was dominant. PAGE-FLNS studies established that binding of (–)-*anti*-BPDE to the 5′-RAGGAR-3′ sequences yielded more of the partially base-stacked and less of the quasi-intercalated conformers than observed for random sequence DNA [85, 86]. Importantly, the (+)-*anti*-BPDE bound to the more mutagenically inclined 5′-RAGGAR-3′ sequences yielded little external-type adduct in comparison to the other two sequences or random-sequence DNA. However, the (+)-*anti*-BPDE adducts formed with the 5′-RAGGAR-3′ sequences yielded an unusually high proportion of partially base-stacked adducts. Since the (+)-*anti*-BPDE appears to be more mutagenic, this result suggests a possible role of partially base-stacked adduct conformations in mutagenesis. Therefore, FLNS coupled with PAGE is capable of providing identification and quantitation of DNA adducts in different conformations.

6 On-line coupling of CE and CEC with FLNS

CE has been interfaced with low temperature on-line FLNS detection. A schematic of the CE-FLNS system [26, 29] is shown in Fig. 9. The apparatus consists of a modular CE apparatus, FLNS instrumentation (see figure caption) and a capillary cryostat (CC) mounted to a translational stage. The CC consists of a double-walled quartz cell with inlet and return lines for introducing liquid nitrogen or liquid helium. The outer portion of the CC is evacuated. The capillary, positioned in the central region of the CC, is cooled by a continuous flow of liquid nitrogen [77 K] or helium (4.2 K). The low thermal capacity of the capillary, and the small dimensions of the CC (inner portion, 4 mm ID × 22 cm length) allow rapid cooling to 4.2 K (in less than 1 min) and the ability to form disordered matrices for FLN to be operative in typical CE buffers. It was found that typical CE buffers consistently formed glassy

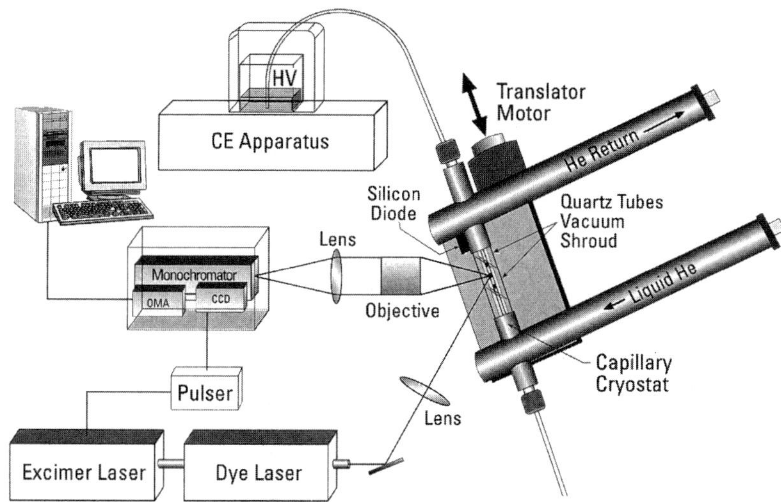

Figure 9. Schematic diagram of the CE-FLNS system. The instrumentation consists of a CE apparatus, capillary cryostat, translation stage, laser excitation source, a monochromator, and either an intensified photodiode array or charge-coupled device for detection. Pulse generators are employed for time-resolved spectroscopy. Reprinted from [29], with permission.

matrices. The latter was attributed to the presence of buffer salts, which are known to assist in the formation of the glassy state. Moreover, FLN spectra can be obtained in a variety of solvent and/or polymer matrices frozen to 4.2 K. An example of the cooling rate is shown in Fig. 10, where fluorescence spectra of pyrene, acquired as a function of time after opening the helium transfer line valve, are shown. Spectrum (a) was obtained at room temperature and spectra (b)–(e) were obtained 30, 35, 40, and 50 s,

respectively, after opening the valve. The five spectra, plotted using the same *y*-axis scale expansion (offset for clarity), illustrate a fast cooling rate (minimizing analyte zone dispersion) and a significant increase in fluorescence with decreasing temperature. Spectrum (e), obtained at 50 s, is identical to the FLN spectrum of pyrene obtained in a regular helium immersion cryostat.

In CE-FLNS, when the analytes enter the CC optical window, the capillary is rapidly frozen in the CC to 4.2 K with liquid helium. Once frozen, arbitrary detection times can be used to completely characterize the analyte zones *via* FLNS. A precision stage provides translation of the CC along the capillary axis by ± 4 cm, allowing the separated analytes to be sequentially characterized by FLNS. An intensified charge-coupled device and a specially designed objective were incorporated into the system to improve the fluorescence collection efficiency. With these changes the LOD of the modified CE-FLNS system is at the low-attomole level for moderately fluorescent chromophores [30]. When the fluorescence analysis is complete, efficient warming of the CC and capillary can be achieved by closing the valve of the helium transfer line and introducing (warm) helium gas into the CC [26, 27]. We emphasize that CE-FLNS, in addition to being highly selective and sensitive, is a practical technique. All apparatus components are commercially available, including the capillary cryostat.* Moreover, standards are not required each time a qualitative determination is to be made; rather, this methodology relies on an established

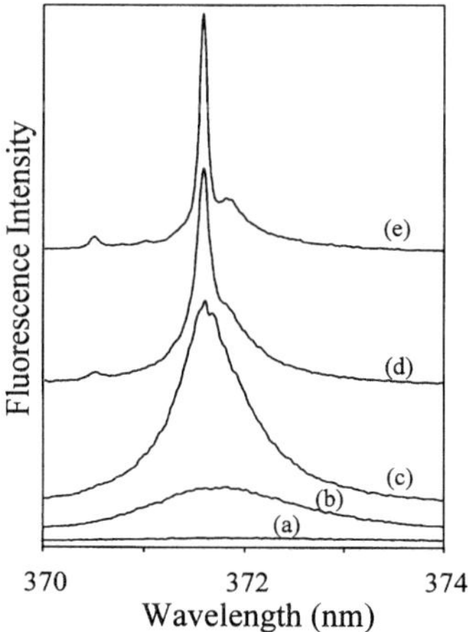

Figure 10. FLN spectra of pyrene in a CE tube (75 μm ID, 365 OD) as a function of time (temperature) after opening the valve of the helium transfer line to the capillary cryostat; matrix = ethanol; λ_{ex}, 364.0 nm. Spectrum (a) was obtained at room temperature and spectra (b)–(e) were obtained after 30, 35, 40, and 50 s, respectively. Reprinted from [28], with permission.

---

* The capillary cryostat (Patent 5,898,493) is available from Janis Research Company, Inc., Two Jewel Drive, P.O. Box 696, Wilmington, MA 01887-0696, USA

reference library of spectra. This attribute is particularly important when standards (*e.g.*, DNA⁻ and protein adducts/metabolites) are difficult or expensive to synthesize, or are unstable.

FLNS can also be interfaced with CEC. The same instrumentation is used; the only difference is that the open tubular capillary used for CE is replaced with a capillary packed with various stationary-phase particles, *e.g.*, C18, phenol, silica, *etc.* A preliminary result from a CEC-FLNS experiment is shown in Fig. 11 for B[*a*]P in a C18 capillary (excitation wavelength of 395.7 nm; unpublished data). Well-resolved ZPLs clearly indicate that the commercially available coated CEC capillaries from Unimicro Technologies (Pleasanton, CA), packed with small particles (3 μm

octadecylsilate), are suitable for interfacing with FLNS. In the above experiment the polyimide capillary coating was removed to create the optically transparent excitation and detection window. Currently, we are testing capillaries with UV-transmitting coatings. The advantages of CEC over standard CE include shorter analysis time, better resolution, and good stability for hundreds of injections. The disadvantages are limited packing materials, fragile capillaries, and rather high cost. Nevertheless, CEC is rapidly gaining acceptance as a high performance analytical technique. Therefore, arbitrary detection times, selectivity based on both the frequency and time domains, as well as increased fluorescence signal and/or decreased photodegradation (due to the low temperatures used), make CE/CEC-FLNS a practical methodology for identification and quantitation of closely related analytes [26–30].

7 Applications of CE-FLNS

7.1 PAHs

Analysis of a mixture of five PAHs – (B[*a*]P, benzo[*e*]pyrene (B[*e*]P), pyrene, 1-hydroxypyrene, and benz[*a*]anthracene (BA) – by CE-FLNS is shown in Fig. 12 [26]. The CE electropherogram is shown in Fig. 12A, While the on-line FLN spectra of the five separated PAHs are shown in Fig. 12B and C, respectively. The mode of CE used for separation was micellar electrokinetic chromatography (MEKC). Identification was based on comparison of the vibronically excited FLN spectra of the CE-separated peaks with those of off-line generated reference spectra of the five compounds. Another example of separation and spectral identification by CE-FLNS is shown in Fig. 13 for a mixture of protonated and deuterated benzo[*a*]pyrene (B[*a*]P-d$_{12}$). Figure 13A shows a portion of the MEKC, room-temperature fluorescence electrophero-

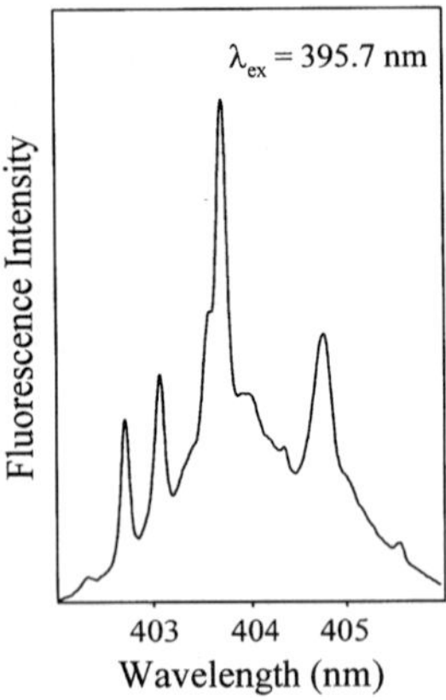

Figure 11. FLN spectra obtained for B[*a*]P in a CEC capillary column packed with 3 μm octadecylsilate particles; ethanol mobile phase. The numbers correspond to excited-state vibrational frequencies in cm⁻¹. λ$_{ex}$, 396 nm; *T*, 4.2 K.

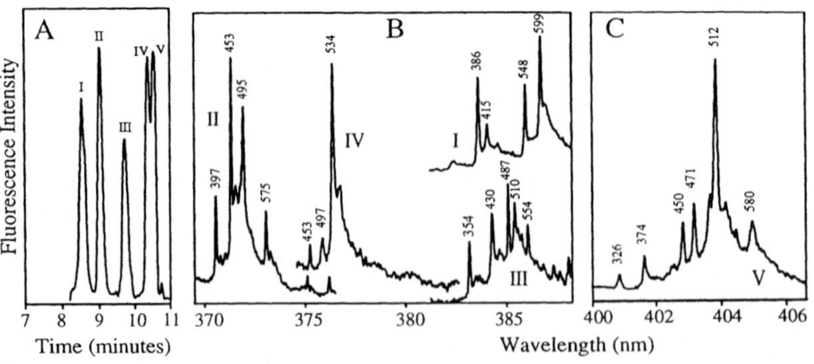

Figure 12. (A) Fluorescence-based CE electropherogram of the PAH mixture; (B) FLN spectra of the CE-separated PAHs; see [26] for separation conditions. Based on FLNS analysis the labeled peaks correspond to 1-hydroxypyrene (I), pyrene (II), BA (III), and B[*e*]P (IV), and B[*a*]P (V). Laser excitation wavelengths, 378.0 nm, 365.2 nm, 378.0 nm, 369.0 nm, and 395.7 nm, respectively; *T*, 4.2 K. Separation buffer, 40 mM DOSS, 8 mM sodium tetraborate, pH 9, ACN-water (30% v/v). Peaks are labeled with their excited-state vibrational frequencies, in cm⁻¹. Reprinted from [26], with permission.

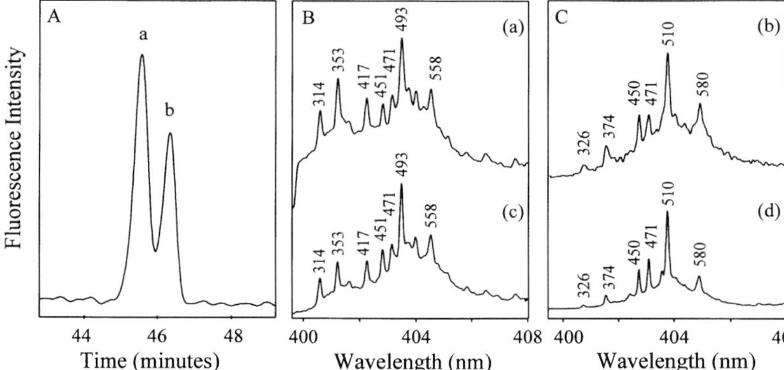

Figure 13. (A) Room-temperature fluorescence electropherogram for a mixture of B[*a*]P-d$_{12}$ (peak a) and B[*a*]P (peak b). Separation buffer, 40 mM DOSS, 8 mM sodium tetraborate, pH 9, ACN-water (30% v/v). (B) FLN spectra obtained for the CE-separated peak a (spectrum a), and the FLN spectrum from the reference library (spectrum c). (C) Identification of the CE-separated peak b (see Section 7.1 for details). Reprinted from [29], with permission.

gram for a B[*a*]P-d$_{12}$/B[*a*]P mixture (10^{-5} M each) [28]. The 4.2 K FLN spectra for the CE-separated peaks (a and b) obtained using selective laser excitation at 395.7 nm are shown in Fig. 13B (spectrum a) and C (spectrum b). The FLN peaks are labeled with their S_1 vibrational frequencies in cm^{-1}. Comparison of the CE-FLN spectra in Fig. 13 shows that spectrum (a) is virtually indistinguishable from spectrum (c) of the B[*a*]P-d$_{12}$ standard and that spectrum (b) is identical to spectrum (d) of the B[*a*]P standard. Obvious differences in vibrational frequencies and intensities between spectra (a) and (b) are observed. For example, there are strong modes at 353, 493, and 558 cm^{-1} for B[*a*]P-d$_{12}$, while B[*a*]P has strong modes at 510 and 580 cm^{-1}. Therefore, the peaks (a and b) of the electropherogram were definitively identified as being due to deuterated and protonated B[*a*]P, respectively.

7.2 Stable PAH-DNA adducts

CE-FLNS was also used to identify eight diastereomeric deoxyadenosine (dA) adducts derived from dibenzo[*a,l*]pyrene diolepoxide (DB[*a,l*]PDE) [29]. Electrophoretic separation of stereoisomers was accomplished using a mixed surfactant buffer – dioctyl sulfosuccinate (DOSS) and Brij-S – in which the surfactant was below the critical micelle concentration due to the high concentration (~25%) of organic solvent. The separation of the eight diastereometric-dA adducts is shown in Fig. 14A. Since eleven peaks are nearly baseline-resolved, three peaks of this electropherogram (see below) must be from decomposition products or impurities.

As an example, Fig. 14B and C demonstrates on-line FLNS identification of peaks 7 and 11 of the electropherogram, respectively [29]. All spectra were obtained at 4.2 K with an excitation wavelength of 372.0 nm. Figure 14B shows the FLN spectra of the CE-separated peak 7 (spectrum a) and the reference spectrum of the (+)-*trans*-

syn-dA adduct standard (spectrum b). The comparison reveals that the vibronic modes at 720, 757, and 796 cm^{-1} in spectra (a) and (b) are identical, proving that peak 7 corresponds to the (+)-*trans-syn*-dA. Figure 14C compares the on-line FLN spectra of the CE-separated peak 11 (spectrum a) with that of the off-line (+)-*cis-syn*-dA adduct standard (b). Again, identical vibrational modes at 736, 796, 848, and 930 cm^{-1}, are observed, proving that peak 11 corresponds to the (+)-*cis-syn*-dA adduct. The identification of the remaining peaks in the electropherogram is described elsewhere [29]; here we only note that peaks 3, 6, and 10 correspond to impurities. In addition, the high-resolution fluorescence spectra provided conformational information on the spatial relationship of the carcinogen and dA moiety [29, 62, 64].

7.3 Depurinating PAH-DNA adducts

Efficient formation of depurinating DNA adducts formed by the one-electron oxidation metabolic pathway of PAHs has been observed in many *in vitro* and *in vivo* experiments [54, 55, 59, 60], demonstrating the importance of the one-electron oxidation pathway in chemical carcinogenesis [54, 107, 108]. In general, PAH-DNA adducts are structurally similar, since the PAH moiety can bind at different nucleophilic sites of both Gua and Ade [54]. For example, dibenzo[*a,l*]pyrene (DB[*a,l*]P) forms three structurally similar adducts with Ade, *i.e.*, DB[*a,l*]P-10-N7Ade, DB[*a,l*]P-10-N1Ade, and DB[*a,l*]P-10-N3Ade [109]. FLNS experiments revealed that a mixture of such adducts cannot be resolved by FLNS alone. However, with CE-FLNS the above adduct-mixture can first be electrophoretically separated in the MEKC mode of CE and subsequently characterized on-line as shown in Fig. 15. The room temperature fluorescence electropherogram of the adduct mixture is shown in Fig. 15. Figure 15B shows FLN spectra obtained for CE-separated peaks II, III, and IV obtained with a laser excitation of 416.0 nm. Obvious differences in the vibrational frequencies were observed.

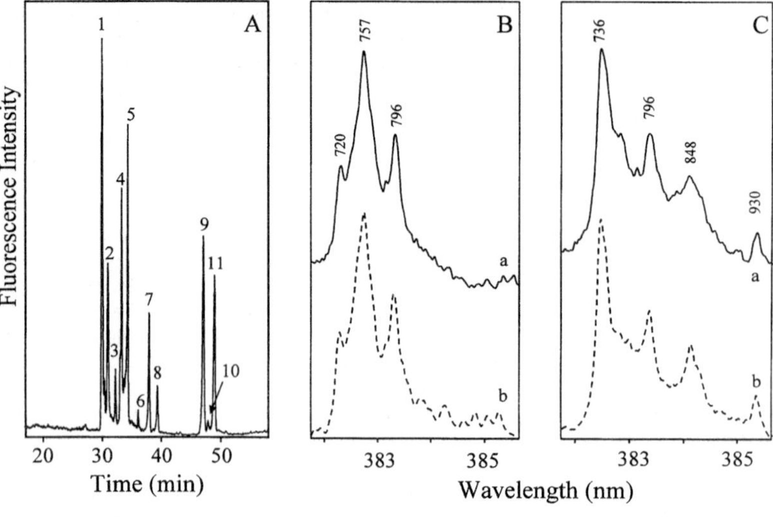

Figure 14. (A) Room-temperature fluorescence electropherogram acquired during CE separation of the eight HPLC purified DB[*a,l*]PDE-derived adduct standards. Separation buffer, 34.4 mM DOSS, 7.4 mM Brij-S, 6.8 mM sodium tetraborate, pH 8.5, and 25.5% ACN v/v. (B) Spectrum a was obtained on-line for peak 7 of (A), with an excitation wavelength of 372.0 nm at T = 4.2 K. Spectrum b corresponds to FLN spectra of the (+)-*trans-syn*-14-N^6-dA standard. Spectrum a of (C) was obtained on-line for peak 11; T 4.2 K; λ_{ex}, 372.0 nm. Spectrum b corresponds to the (+)-*cis-syn*-DB[*a,l*]PDE-14-N^5dA adduct standard. Reprinted from [29], with permission.

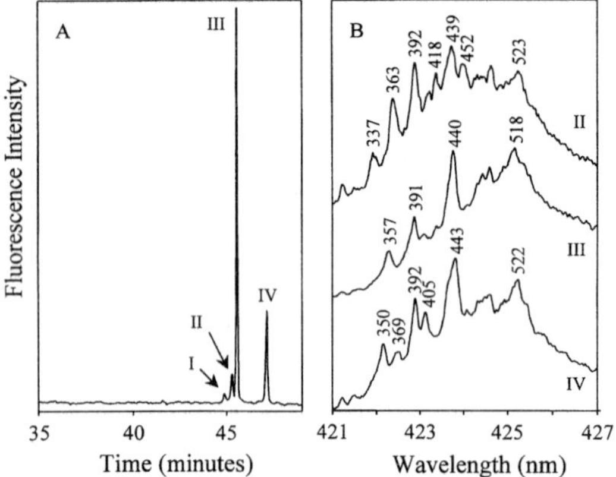

Figure 15. (A) Room-temperature fluorescence electropherogram acquired during CE separation of a mixture of (II) DB[*a,l*]P-10-N7Ade, (III) dB[*a,l*]P-10-N1Ade, and (IV) DB[*a,l*]P-10-N3Ade. Separation buffer, 40 mM DOSS, 8 mM sodium tetraborate, pH 9, ACN-water (30% v/v). Peak I marks an unidentified analyte. (B) FLN spectra for the three CE-separated adducts, obtained at 4.2 K using selective laser excitation at 416.0 nm. Reprinted from [28], with permission.

Based on comparison with the reference library of FLN spectra for DB[*a,l*]P-derived Ade adduct standards, spectra II, III, and IV correspond to DB[*a,l*]P-10-N7Ade, DB[*a,l*]P-10-N1Ade, and DB[*a,l*]P-10-N3Ade, respectively. Peak I was determined to be due to an impurity [109]. As in all cases discussed above, analyte identification was obtained using several different excitation wavelengths.

7.4 PAH-DNA adducts in human urine

It is thought that various chemicals (or their products with DNA and proteins) found in human fluids (*e.g.*, urine and serum) could serve as effective biomarkers for molecular dosimetry. Therefore, methods for screening human fluids to detect mutagenic metabolites and/or elevated levels of various biomolecules are of great interest. For example, CE coupled with sensitive fluorescence detection has been successfully used to study urinary indole derivatives, catecholamine metabolites, naproxen, and other metabolites in human serum and urine [110]. Is was also shown that HPLC coupled with fluorescence detection can be used to monitor B[*a*]P-tetrols (hydrolysis products of the BPDE derived DNA adducts) in urine [111] of psoriasis patients medicinally treated with coal tar. Additionally, B[*a*]P-derived depurinating adducts (*e.g.* B[*a*]P-6-N7Gua) in rat urine were identified using HPLC with off-line FLNS detection [112].

Recently, it was shown that the depurinating B[*a*]P-6-N7Gua adduct is formed in humans exposed to coal smoke (a known source of B[*a*]P). The adduct was found in urine using a combination of solid-phase extraction, HPLC, and CE-FLNS [30]. Identification of this adduct by CE-FLNS is shown in Fig. 16. Figure 16A shows a room-temperature electropherogram (left) of the "urine-extract" of a coal smoke-exposed individual along with room-temperature fluorescence spectra (right) of the CE-separated peaks. Based on the migration times, the intense peak at ~ 23 min (labeled with an asterisk) corresponds to the B[*a*]P-6-N7Gua standard. More definitive proof of structural identification was provided by on-line FLNS analysis, as shown in Fig. 16B, where the FLN spectrum obtained

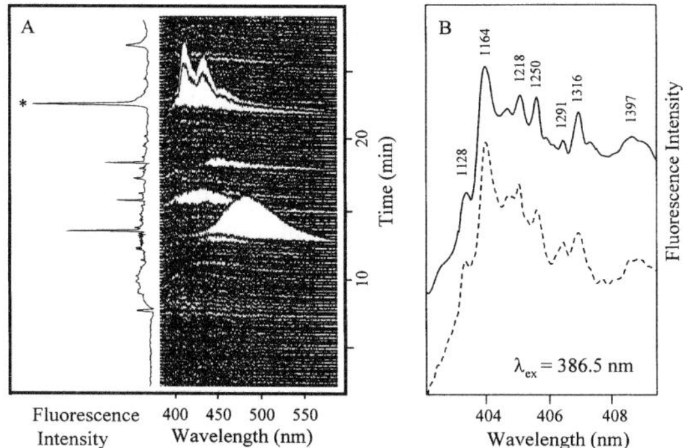

Figure 16. (A) CE electropherogram (left) and room temperature fluorescence spectra (right) of a "urine-extract" from an individual exposed to coal smoke. (B) The solid curve is the FLN spectrum obtained on-line for the peak labeled with an asterisk, whereas the dashed curve corresponds the library FLN spectrum of the B[*a*]P-6-N7Gua standard. Separation buffer, 20 mM DOSS, 4 mM sodium tetraborate, pH 9, ACN-water (15% v/v). *T*, 4.2 K; λ_{ex}, 386.5 nm. Reprinted from [30], with permission.

with an excitation wavelength of 386.5 nm for the ~23 min peak (solid line) is compared with the off-line FLN spectrum (dashed line) of the B[*a*]P-6-N7Gua standard. The spectra are nearly indistinguishable as revealed by the identical frequencies (and intensities) of the excited state vibronic modes at 1128, 1164, 1218, 1250, and 1316 cm^{-1}, proving that B[*a*]P-6-N7Gua is formed in humans exposed to B[*a*]P ([30], Casale *et al.*, in preparation). Quantitative standard addition studies revealed that the above adduct was excreted at a level of ~ 200 pmole per day, *i.e.*, ~ 226 pmole per µmole of creatinine. Since its recovery through the multistep preparative procedure was only 65–85%, the daily excretion of B[*a*]P-6-N7Gua could be 18–53% higher (G. P. Casate *et al.*, submitted). The enhanced fluorescence intensity (by a factor of ~ 10) of B[*a*]P-6-N7Gua at 4.2 K relative to room temperature resulted in an attomole detection limit [30].

8 Concluding remarks and future prospects

This review focused on the coupling of FLNS with various separation techniques, and provided several examples of how TLC-, PAGE-, CE/CEC-FLNS can be applied to the identification and characterization of closely related analytes, including DNA adducts. So far, the off- and on-line TLC-FLNS and off-line PAGE-FLNS approaches have shown limited applications due to poor detection limits arising from a high luminescence background originating from the chromatographic/electrophoretic separation media. However, in cases where extremely low detection

levels are not required, these methodologies can be utilized. In particular, the combination of HPLC-TLC-FLNS, where the coupling strategies have been extensively developed and tested, could be applicable to the analysis of complex mixtures.

The on-line hybrid of CE-FLNS has proven to be successful in the analysis of closely related analytes formed at low levels. This is the result of the superior separation afforded by CE and the high sensitivity/selectivity of FLNS detection. With on-line analysis, losses associated with fraction collection and sample handling are eliminated. For example, in the case of PAH-DNA adducts, low attomole detection levels have been obtained. Therefore, it is anticipated that CE/CEC-FLNS could play a role in identification of closely related compounds formed at low levels.

Furthermore, with the increasing complexity of analytical work there is a need for on-line high-resolution detection capabilities in micro- and capillary HPLC, well-established chromatographic techniques in sample-constrained applications (*e.g.*, bioanalysis). Based on preliminary results we are confident that interfacing micro-HPLC with FLNS is feasible, and will provide a new method for on-line, low temperature high-resolution spectroscopic identification and characterization. We predict that the main advantages of adapting HPLC to FLNS will be (i) versatile separation protocols (*e.g.*, gradient elution, varying stationary phases, temperature-programmed separation, *etc.*), and (ii) availability of fraction collection for postcolumn analysis/reaction. It is anticipated that on-line CE/CEC-FLNS and HPLC-FLNS will significantly expand the applicability of FLNS in electrophoresis and chromatography.

Ames Laboratory is operated for the U.S. Department of Energy by Iowa State University under contract No. W-7405-Eng-82. This work was supported by the Office of Health and Environmental Research, Office of Energy Research, and partly by NCI grant POI CA49210-05. We would like to acknowledge former group members S. Cooper, P. Lu, G. Marsh, M. Suh, F. Ariese, D. Zamzow, and C.-H. Lin for their valuable contributions to the research discussed herein.

Received September 13, 1999

9 References

[1] Brown, R. S., Luong, J. H. T., Szolar, O. H. J., Halasz, A., Hawari, J., *Anal. Chem.* 1996, *68*, 287–292.

[2] Kientz, C. E., Hulst, A. G., De Jong, A. L., Wils, E. R. J., *Anal. Chem.* 1996, *68*, 675–681.

[3] Lee, T. T., Yeung, E. S., *J. Chromatogr.* 1992, *595*, 319–325.

[4] Lurie, I. S., in: Adamovics, J. A. (Ed.), *Analysis of Addictive and Misused Drugs*, Marcel Dekker, New York 1995, pp. 151–219.

[5] Kok, S. J., Velthorst, N. H., Gooijer, C., Brinkman, U. A. T., *Electrophoresis* 1998, *19*, 2753–2776.

[6] Walker, P. A., Morris, M. D., *J. Chromatogr. A* 1998, *805*, 269–275.

[7] Walker, P. A., Kowalchyk, W. K., Morris, M. D., *Anal. Chem.* 1995, *67*, 4255–4260.

[8] Kennedy, B. J., Milofsky, R., Carron, K. T., *Anal. Chem.* 1997, *69*, 4708–4715.

[9] Cooper, S. D., Robson, M. M., Batchelder, D. N., Bartle, K. D., *Chromatographia* 1997, *44*, 257–262.

[10] Chen, C.-H., Morris, M., *J. Chromatogr.* 1991, *540*, 355–363.

[11] Kowalchyk, W. K., Walker, P. A., Morris, M. D., *Appl. Spectrosc.* 1995, *49*, 1183–1188.

[12] Somsen, G. W., ter Riet, G. J. H. P., Gooijer, C., Velthorst, N. H., Brinkman, U. A. T., *J. Planar Chrom.* 1997, *10*, 10–17.

[13] Olson, D. L., Lacey, M. E., Sweedler, J. V., *Anal. Chem.* 1998, *70*, 645–650.

[14] Webb, A. G., *Progr. Nucl. Magn. Reson. Spectr.* 1997, *31*, 1–42.

[15] Olson, D. L., Peck, T. L., Webb, A. G., Margin, R. L., Sweedler, J. V., *Science* 1995, *270*, 1967–1970.

[16] Albert, K., *J. Chromatogr. A* 1995, *703*, 123–147.

[17] Korhammer, S. A., Bernreuther, A., *Fresenius J. Anal. Chem.* 1996, *354*, 131–135.

[18] Albert, K., *Analysis* 1996, *24*, M17–M18.

[19] Olson, D. L., Lacey, M. E., Sweedler, J. V., *Anal. Chem.* 1998, *70*, 257A–264A.

[20] Cai, J. Y., Henion, J., *J. Chromatogr. A* 1995, *703*, 667–692.

[21] Banks, J. F., *Electrophoresis* 1997, *18*, 2255–2266.

[22] Ding, J., Vouros, P., *Am. Lab.* 1998, *30*, 15–29.

[23] Severs, J. C., Smith, R. D., in: Landers, J. (Ed.), *Handbook of Capillary Electrophoresis*, CRC Press, New York 1996, pp. 791–826.

[24] Kelly, J. F., Ramaley, L., Thibault, P., *Anal. Chem.* 1997, *69*, 51–60.

[25] Unger, M., Stöckigt, D., *J. Chromatogr. A* 1996, *752*, 271–277.

[26] Jankowiak, R., Zamzow, D., Ding, W., Small, G. J., *Anal. Chem.* 1996, *68*, 2549–2553.

[27] Zamzow, D., Small, G. J., Jankowiak R., *Mol. Cryst. Liq. Cryst.* 1996, *291*, 155–162.

[28] Zamzow, D., Lin, C.-H., Small, G. J., Jankowiak, R., *J. Chromatogr. A* 1997, *781*, 73–80.

[29] Roberts, K. P., Lin, C.-H., Jankowiak, R., Small, G. J., *J. Chromatogr. A* 1999, *853*, 159–170.

[30] Roberts, K. P., Lin, C.-H., Singhal, M., Casale, G. P., Small, G. J., Jankowiak, R., *Electrophoresis* 2000, *21*, in press.

[31] Personov, R. I., in: Agranovich, V. M., Hochstrasser, R. M. (Eds.), *Spectroscopy and Excitation Dynamics of Condensed Molecular Systems*, Vol. 4, North-Holland, Amsterdam, 1983, pp. 555–619.

[32] Osadko, I. S., in: Dusek, K. (Ed.), *Advances in Polymer Science*, Springer-Verlag, Berlin, Heidelberg 1994, pp. 123–186.

[33] Kohler, B. E., in Moore, C. B. (Ed.), *Chemical and Biochemical Applications of Lasers*, Academic Press, New York 1979, pp. 31–51.

[34] Hofstaat, J. W., Gooijer, C., Velthorst, N. H., in: Schulman, S. G. (Ed.), *Molecular Luminescence Spectroscopy: Methods and Applications*, John Wiley and Sons, New York 1988, pp. 383–459.

[35] Selzer, P. M., Yen, W. M., Selzer, P. M. (Eds.), *Topics in Applied Physics*, Vol. 49, Springer-Verlag, New York, Heidelberg, Berlin 1986, pp. 113–146.

[36] Weber, M. J., in: Yen, W. M., Selzer, P. M. (Eds.), *Topics in Applied Physics*, vol. 49, Springer-Verlag, New York, Heidelberg, Berlin 1986, pp. 1189–1240.

[37] Fidy, J., Vaanderkooi, J. M., in: Douglas, R. H., Moan, J., Rontó, G. (Eds.), *Light in Biology and Medicine* vol. 21, Plenum Press, New York 1991, pp. 367–374.

[38] Weber, J. M., *J. Lum.* 1987, *36*, 179–181.

[39] Jankowiak, R., Small, G. J., *Anal. Chem.* 1989, *61*, 1023A–1032A.

[40] Price, B. P., Wright, J. C., *Anal. Chem.* 1990, *62*, 1989–1994.

[41] Jankowiak, R., Small, G. J., *Chem. Res. Toxicol.* 1991, *4*, 256–269.

[42] Riesen, H., *Comments Inorg. Chem.* 1993, *14*, 323–328.

[43] Nakhimovski, L., Lamotte, M., Joussot-Dubien, J., *Handbook of Low Temperature Electronic Spectra of Polycyclic Aromatic Hydrocarbons*, Elsevier, Amsterdam, New York, Oxford, Tokyo 1989.

[44] Riseberg, L. A., *Phys. Rev.* 1973, *A7*, 67–70.

[45] Renge, I., Mauring, K., Sarv, P., Avarmaa, J., *Chem. Phys.* 1986, *90*, 6611–6616.

[46] Fünfschilling, J., Williams, D. F., *Photochem. Photobiol.* 1977, *26*, 109–118.

[47] Avarma, R., Rebane, K. K., *Spectrochim. Acta* 1985, *41A*, 1365–1380.

[48] Hala, J., Pelant, I., Ambroz, M., Pancoska, P., Vacek, K., *Photochem. Photobiol.* 1985, *41*, 643–648.

[49] Rebane, K. K., Avarmaa, R., *Chem. Phys.* 1982, *68*, 191–200.

[50] Fünfschilling, J., Zschokke-Gränacher, I., *Chem. Phys. Lett.* 1982, *91*, 122–125.

[51] Avarmaa, R., Renge, I., Mauring, K., *FEBS Lett.* 1984, *167*, 186–190.

[52] Jankowiak, R., Day, B. W., Lu, P., Doxtader, M. M., Skipper, P. L., Tannenbaum, S. R., Small, G. J., *J. Am. Chem. Soc.* 1990, *112*, 5866–5869.

[53] Singh, K., Skipper, P. L., Tannenbaum, S. R., Dasari, R. R., *Photochem. Photobiol.* 1993, *58*, 637–642.

[54] Jankowiak, R., Small, G. J., in: Neilson, A. H. (Ed.), *The Handbook of Environmental Chemistry: PAHs and Related Compounds*, Springer-Verlag, Berlin, Heidelberg, New York 1998, pp. 119–145.

[55] Li, K.-M., Todorovic, R., Rogan, E. G., Cavalieri, E. L., Ariese, F., Jankowiak, R., Small, G. J., *Biochemistry* 1995, *34*, 8043–8049.

[56] Marsch, G. A., Jankowiak, R., Small, G. J., Hughes, N. C., Phillips, D. H., *Chem. Res. Toxicol.* 1992, *5*, 765–772.

[57] Suh, M., Ariese, F., Small, G. J., Jankowiak, R., Hewer, A., Phillips, D. H., *Carcinogenesis* 1995, *16*, 2561–2569.

[58] Jankowiak, R., Ariese, F., Hewer, A., Luch, A., Zamzow, D., Hughes, N. C., Phillips, D. H., Seidel, A., Platt, K.-L., Oesch, F., Small, G. J., *Chem. Res. Toxicol.* 1998, *11*, 674–685.

[59] Devanesan, P. D., RamaKrishna, N. V. S., Padmavathi, N. S., Higginbotham, S., Rogan, E. G., Cavalieri, E. L., Marsch, G. A., Jankowiak, R., Small, G. J., *Chem. Res. Toxicol.* 1993, *6*, 364–371.

[60] Rogan, E. G., Devanesan, P. D., RamaKrishna, N. V. S., Higginbotham, S., Padmavathi, N. S., Chapman, K., Cavalieri, E. L., Jeong, H., Jankowiak, R., Small, G. J., *Chem. Res. Toxicol.* 1993, *6*, 356–363.

[61] Suh, M., Ariese, F., Small, G. J., Jankowiak, R., Liu, T.-M., Geacintov, N. E., *Biophys. Chem.* 1995, *56*, 281–296.

[62] Ariese, F., Small, G. J., Jankowiak, R., *Carcinogenesis* 1996, *17*, 829–837.

[63] Jankowiak, R., Ariese, F., Zamzow, D., Luch, A., Kroth, H., Seidel, A., Small, G. J., *Chem. Res. Toxicol.* 1997, *10*, 677–686.

[64] Jankowiak, R. Lin, C.-H., Zamzow, D., Roberts, K. P., Li, K.-M., Small, G. J., *Chem. Res. Toxicol.* 1999, *12*, 768–777.

[65] Holland, L. A., Chetwyn, N. P., Perkins, M. D., Lunte, S. M., *Pharm. Res.* 1997, *14*, 372–387.

[66] Huber, L., George, S. A., *Diode Array Detection in HPLC*, Marcel Dekker, New York 1993.

[67] Wallington, R. A., Ewing, A. G., *Anal. Chem.* 1987, *59*, 1762–1766.

[68] Ewing, A. G., Mesaros, J. M., Gavin, P. F., *Anal. Chem.* 1994, *66*, 527A–536A.

[69] Christensen, P. L., Yeung, E. S., *Anal. Chem.* 1989, *61*, 1344–1347.

[70] Somsen, G. W., van Stee, L. P. P., Gooijer, C., Brinkman, U. A. T., Velthorst, N. H., Visser, T., *Anal. Chim. Acta* 1994, *290*, 269–276.

[71] Somsen, G. W., Hooijschuur, E. W. J., Gooijer, C., Brinkman, U. A. T., Velthorst, N. H., *Anal. Chem.* 1996, *68*, 746–752.

[72] Kok, S. J., Kristenson, E. M., Gooijer, C., Velthorst, N. H., Brinkman, U. A. T., *J. Chromatogr. A* 1997, *771*, 331–341.

[73] Pentoney, S. L., Sweedler, J. V., in: Landers, J. (Ed.), *Handbook of Capillary Electrophoresis*, Marcel Dekker, New York 1997, pp. 379–423.

[74] Albert, K., Schlotterbeck, G., Tseng, L.-H., Braumann, U., *J. Chromatogr. A* 1996, *750*, 303–309.

[75] Sweedler, J. V., Olson, D., Lacey, M., Webb, A. G., *Proceedings of the 19th International Symposium on Capillary Chromatography and Electrophoresis*, Wintergreen, May 18–22, 1997, pp. 52–53.

[76] Wu, N., Peck, T. L., Webb, A. G., Magin, R. L., Sweedler, J. V., *Anal. Chem.* 1994, *66*, 3849–3857.

[77] Wu, N., Peck, T. L., Webb, A. G., Magin, R. L., Sweedler, J. V., *J. Am. Chem. Soc.* 1994, *116*, 7929–7930.

[78] Somsen, G. W., van de Nesse, R. J., Gooijer, C., Brinkman, U. A. T., Velthorst, N. H., Visser, T., Kootstra, P. R., de Jong, A. P. J. M., *J. Chromatogr.* 1991, *552*, 635–647.

[79] Zhou, W. H., Liu, J., Wang, E., *J. Chromatogr. A* 1995, *715*, 355–360.

[80] Ye, J., Baldwin, R. P., *Anal. Chem.* 1994, *66*, 2669–2674.

[81] Zhou, J., Lunte, S. M., *Electrophoresis* 1995, *16*, 498–503.

[82] Synovec, R. E., Yeung, E. S., *J. Chromatogr.* 1986, *368*, 85–93.

[83] Kok, S. J. Bakker, I., Gooijer, C., Brinkman, U. A. T., Velthorst, N. H., *Anal. Chim. Acta* 1999, *389*, 77–83.

[84] Cooper, R. S., Jankowiak, R., Hayes, J. M., Pei-qui, L., Small, G. J., *Anal. Chem.* 1988, *60*, 2692–2694.

[85] Marsch, G. A., Jankowiak, R., Suh, M., Small, G. J., *Chem. Res. Toxicol.* 1994, *7*, 98–109.

[86] Marsch, G. A., Jankowiak, R., Farhat, J. H., Small, G. J., *Anal. Chem.* 1992, *64*, 3038–3044.

[87] Hofstraat, J. W., Jansen, H. J. M., Hoornweg, G. P., Gooijer, C., Velthorst, *Anal. Chim. Acta* 1985, *170*, 61–71.

[88] van de Nesse, R. J., Vinkenburg, I. H., Jonker, R. H. J., Hoornweg, G. P., Gooijer, C., Brinkman, U. A. T., Velthorst, N. H., *Appl. Spectrom.* 1994, *48*, 788–795.

[89] Jankowiak, R., Zamzow, D., Stack, D. E., Todorovic, R., Cavalieri, E. L., Small, G. J., *Chem. Res. Toxicol.* 1998, *11*, 1339–1345.

[90] Jankowiak, R., Hayes, J. M., Small, G. J., *Chem. Rev.* 1993, *93*, 1471–1502.

[91] Jankowiak, R., in: *Shpol'skii Spectroscopy and Other Site Selection Methods: Applications in Environmental Analysis, Bioanalytical Chemistry and Chemical Physics*, John Wiley and Sons, New York 2000.

[92] Dushinsky, F., *Acta Physiochem.* 1937, *7*, 551–553.

[93] Vo-Dinh, T., Suter, G. W., Kallir, A. J., Wild, U. P., *Anal. Chem.* 1986, *58*, 3135–3139.

[94] Hofstraat, J. W., Engelsma, M., Gooijer, C., Velthorst, N. H., *Spectrochim. Acta* 1989, *45A*, 491–495.

[95] Hofstraat, J. W., Engelsma, M., van de Nesse, R. J., Gooijer, C., Velthorst, N. H., Brinkman, U. A. T., *Anal. Chim. Acta* 1986, *186*, 247–259.

[96] Hofstaat, J. W., Gooijer, C., Velthorst, N. H., *Appl. Sepctrom.* 1988, *42*, 614–619.

[97] van de Nesse, R. J., Hoogland, G. J. M., De Moel, J. J. M., Gooijer, C., Brinkman, U. A. T., Velthorst, N. H., *J. Chromatogr.* 1991, *552*, 613–623.

[98] Kok, S. J., Posthumus, R., Bakker, I., Gooijer, C., Brinkman, U. A. T., Velthorst, N. H., *Anal. Chim. Acta* 1995, *303*, 3–10.

[99] Randerath, K., Reddy, M. V., Gupta, R. C., *Proc. Natl. Acad. Sci. USA* 1981, *78*, 6126–6129.

[100] Kok, S. J., *PhD Thesis*, Vrije Universiteit, de Boelelaan, Amsterdam, The Netherlands 1999.

[101] Meehan, T., Straub, K., *Nature* 1979, *277*, 410–412.

[102] Cheng, S. C., Hilton, B. D., Roman, J. M., Dipple, A., *Chem. Res. Toxicol.* 1989, *2*, 334–340.

[103] Geacintov, N. E., Cosman, M., Hingerty, B. E., Amin, S., Broyde, S., Patel, D. J., *Chem. Res. Toxicol.* 1997, *10*, 111–146.

[104] Yang, J.-L., Maher, V. M., McCormick, J. J., *Proc. Natl. Acad. Sci. USA* 1987, *84*, 3787–3791.

[105] Maher, V. M., Yang, J.-L., McCormick, J. J., *Mol. Cell Biol.* 1987, *7*, 1267–1270.

[106] Marsch, G. A., *PhD Thesis*, Florida State University, Tallahassee, FL 1990.

[107] Cavalieri, E., Rogan, E., in: Neilson, A. H. (Ed.), *The Handbook of Environmental Chemistry: PAHs and Related Compounds*, Springer-Verlag, Berlin, Heidelberg, New York 1998, pp. 81–117.

[108] Cavalieri, E. L., Rogan, E. G., *Xenobiotica* 1995, *25*, 677–688.

[109] Li, K.-M., Byun, J., Gross, M. L., Zamzow, D., Jankowiak, R., Rogan, E. G., Cavalieri, E. L., *Chem. Res. Toxicol.* 1999, in press.

[110] Landers, J. P., *Clin. Chem.* 1995, *41*, 495–509.

[111] Bowman, E. D., Rothman, N., Hackl, C., Santella, R. M., Weston, A., *Biomarkers* 1997, *2*, 321–327.

[112] Rogan, E. G., RamaKrishna, N. V. S., Higginbotham, S., Cavalieri, E. L., Jeong, H., Jankowiak, R., Small, G. J., *Chem. Res. Toxicol.* 1990, *3*, 441–444.

Electrophoresis 2000, *21*, 1267–1280

Review

Scott McWhorter
Steven A. Soper

Near-infrared laser-induced fluorescence detection in capillary electrophoresis

As capillary electrophoresis continues to focus on miniaturization, either through reducing column dimensions or situating entire electrophoresis systems on planar chips, advances in detection become necessary to meet the challenges posed by these electrophoresis platforms. The challenges result from the fact that miniaturization requires smaller load volumes, demanding highly sensitive detection. In addition, many times multiple targets must be analyzed simultaneously (multiplexed applications), further complicating detection. Near-infrared (NIR) fluorescence offers an attractive alternative to visible fluorescence for critical applications in capillary electrophoresis due to the impressive limits of detection that can be generated, in part resulting from the low background levels that are observed in the NIR. Advances in instrumentation and fluorogenic labels appropriate for NIR monitoring have led to a growing number of examples of the use of NIR fluorescence in capillary electrophoresis. In this review, we will cover instrumental components used to construct ultrasensitive NIR fluorescence detectors, including light sources and photon transducers. In addition, we will discuss various types of labeling dyes appropriate for NIR fluorescence and finally, we will present several applications that have used NIR fluorescence in capillary electrophoresis, especially for DNA sequencing and fragment analysis.

Keywords: Near-infrared fluorescence / Capillary electrophoresis / Laser-induced fluorescence / Review
EL 3910

Contents

Correspondence: Dr. Steven A. Soper, Department of Chemistry, 232 Choppin Hall, Louisiana State University, Baton Rouge, LA 70803-1804, USA
E-mail: steve.soper@chem.lsu.edu
Fax: +225-388-3458

Abbreviations: Ang-I, angiotensin I; **CGE**, capillary gel electrophoresis; **HPMC**, hydroxypropylmethylcellulose; **LPA**, linear polyacrylamide; **NIR**, near-infrared; **PMMA**, poly(methyl methacrylate); **PMT**, photomultiplier tube; **SEM**, scanning electron micrograph; **S/N**, signal-to-noise ratio; **SPAD**, single photon avalanche diode; **TAG**, thiazole green

1 Introduction

Capillary electrophoresis (CE) has become a valuable tool for the-high speed and efficient separation of such analytes as pharmaceuticals [1, 2], environmental pollutants [3, 4], oligonucleotides [5–10] and proteins [3, 11]. The popularity of CE results primarily from the high speed in which separations can be carried out as well as the high plate numbers that can be generated. The speed of the separation is produced by using high electric fields and, in some cases, short column lengths. In order to further increase the speed of the separation, miniaturization of the column, both in length and internal diameter, will be required. As a result, injection volumes will need to be reduced as well in order to minimize zone variance introduced by the finite injection volumes placing severe demands on detection. In order to meet the challenges placed on detection by the small injection volumes required for CE, fluorescence, in particular laser-induced fluorescence (LIF), has become the detection protocol of choice.

The pioneering work of LIF in CE was carried out by Zare and Gassman [12] who separated and detected femtomole quantities of racemic mixtures of amino acids labeled with a fluorescent dansyl group. In this work, an He-Cd laser (442 nm excitation) was used to excite the fluorescence of the labeled amino acids. Since then, many researchers have demonstrated the utility of LIF detection in CE. The common LIF strategy is to use labeling dyes that absorb radiation in the visible region of the spectrum (350–650 nm), due primarily to the readily available instrumentation, such as excitation sources like He-Cd, Ar ion, He-Ne and Kr ion lasers. In addition, most photon transducers (photomultiplier tubes, PMTs) possess high single photon detection efficiencies in the visible region. And finally, fluorogenic labels that absorb radiation in the visible range for a variety of different functional groups (primary amines, alcohols, and thiols) are readily available from commercial sources.

Some of the disadvantages associated with visible LIF detection include the expensive and sophisticated instrumentation required to carry out detection. For example, ion lasers, such as Ar or Kr, are large lasers that can place heavy demands on utilities and have relatively short operational lifetimes, requiring the replacement of expensive ion tubes. Another issue is the signal-to-noise ratio (S/N) in the measurement. Many times, the use of visible excitation can produce large backgrounds in the form of impurity fluorescence (autofluorescence) or scattering from the sample matrix. This is particularly true in capillary gel electrophoresis, where the sieving gel can produce a large scattering background. Indeed, formats using off-column detection such as the sheath flow cell [13, 14] have been implemented to reduce this background problem.

Recently, researchers have been interested in the near-infrared (NIR) region for LIF detection because the NIR possesses many advantages over visible fluorescence, primarily due to photoprocesses of absorption and emission occurring above 700 nm [15–17]. In the NIR, background interference from impurity molecules can be reduced or even eliminated because few molecules demonstrate intrinsic fluorescence in this region. Also, since the amplitude of Raman or Rayleigh scattering is inverse-

ly proportional to the 4th power of the excitation wavelength ($1/\lambda^4$) [18], enhanced sensitivity can be achieved because of significant reductions in scattering effects. Moreover, with a major push towards miniaturization of CE instrumentation, the NIR is a viable detection strategy because of the emergence of semiconductor lasers and detectors, which have a compact size and lower cost compared to gas ion lasers and PMTs. Additionally, semiconductor lasers provide sufficient light power for ultrasensitive measurements. This review will focus specifically on the advances in NIR-LIF detection for CE applications, namely advances in instrumentation and fluorescent labeling dyes appropriate for ultrasensitive NIR fluorescence detection in CE. We will restrict our discussion to those applications which use fluorescence monitoring above 680 nm for CE detection.

2 Developments in NIR-LIF detection

2.1 Instrumentation

2.1.1 Excitation sources

Excitation of analytes for NIR fluorescence in CE initially involved the use of large, expensive, high-powered, water-cooled systems, such as the Ar ion pumped Ti:sapphire laser [15]. While the Ti:sapphire laser is basically a solid-state laser, it requires a pumping source to stimulate fluorescence of the Ti ions doped in the sapphire crystal, which in this case is the all-lines output of an Ar ion laser. The advantages associated with the Ti:sapphire laser include its wide tunability range, typically from 690–1000 nm, which allows matching the excitation wavelength to the absorption maximum of the labeling dye. In addition, it can provide sufficient laser power for ultrasensitive applications. And finally, it can be operated in a pulsed mode to allow performing time-resolved fluorescence measurements as well.

One of the major advancements in NIR fluorescence detection in terms of excitation sources was the use of semiconductor diode lasers. Semiconductor diode lasers are attractive excitation sources because they are very compact and inexpensive while supplying ample power (~ 100 mW) for ultrasensitive detection and also can be operated in a pulsed mode for time-resolved fluorescence [5]. In addition, diode lasers show an operational lifetime that is much longer than the lifetime of most plasma tubes in ion lasers. Another attractive feature of diode lasers is that they can be operated off a simple battery, making them amenable to implementation in miniaturized and portable systems. Diode lasers can be purchased in a variety of different lasing wavelengths, for example, 635, 680, 750, 780, 810 and 830 nm. In addition, they can be slightly frequency-tuned by regulating the temperature to the diode head (0.05 nm/°C). Diode lasers are semiconductors (p-n junction diodes) that are forward-biased, causing a splitting of the Fermi level (measure of electron population distribution in the crystal lattice). Electrons flow into the p-type region, while the holes flow into the n-type region. Recombination of the generated electron/hole pairs results in the spontaneous emission of light with the wavelength of the emitted light determined by the band gap, which is defined by the energy difference between the valence and conduction bands of the semiconductor material. A diode operated in this fashion is called a light emitting diode.

For the diode to operate as a laser, it must possess the characteristics typically encountered in lasers, namely, a population inversion in an excited state to allow sufficient light gain *via* stimulated emission. For diodes, this condition is met by sufficiently doping the crystal lattice with p- and n-type materials to create a population inversion of electrons and holes in a confined region near the junction of the p- and n-type material. In order to enhance the confinement of the generated electron/hole pairs in the junction region, double-heterojunction diodes are typically used, in which three layers are employed in the semiconductor. One layer is a thin active layer (thickness ~ 0.1 μm) which is surrounded by two different cladding layers that consist of a material with a higher bandgap. By enclosing the junction region in an optical cavity, lasing action can be achieved. The optical cavity can be created through large differences in the refractive index of the semiconductor material. Parallel facets (where reflection occurs) can be produced by cleaving the semiconductor material in a particular crystallographic direction.

In contrast to ion lasers, the spectral output of the diode laser is typically much broader, 20–100 MHz, whereas in ion lasers, the spectral output is on the order of 2×10^{-4} MHz. In addition, the output of an ion laser is a collinear beam of light, whereas the diode laser produces a divergent beam with the degree of divergence related to the lateral dimensions of the semiconductor active region. As such, for diode lasers, an external optic is required to collimate the beam or couple it into a fiber optic. However, in both the diode and ion laser, the beams exhibit a Gaussian intensity profile. An additional distinction between diode lasers and ion lasers is in terms of the beam shape, which is elliptical for diodes; for most ion lasers, the beam is circular. The elliptical shape of the beam arises from the asymmetry in the active region of the diode, which is typically 0.1 μm thick and 4 μm wide. One final note on diode lasers is that they produce a polarized beam, as do ion lasers, with the polarization ratio as high as 100:1.

2.1.2 Detectors

The common photon transducers used in most LIF applications are PMTs. While PMTs are an excellent choice due to their high single-photon detection efficiencies and low dark-count rates, the photoactive material typically shows poor quantum efficiencies in the NIR. For example, a common extended-red PMT will possess a quantum efficiency of only 10% around 800 nm and drop to 1% at 900 nm. Therefore, PMTs have been difficult to implement in ultrasensitive detection applications using NIR fluorescence.

Semiconductor materials, on the other hand, can be used effectively in the NIR. For example, Si-based semiconductor materials possess quantum efficiencies that approach 70% at 800 nm; even at 900 nm, the quantum efficiency is near 50%. Most semiconductor detectors, for example photodiodes, are p-n junction diodes that are reverse-biased and when light shines on the device it generates an electron/hole pair by shuttling the electron into the conduction band. These electron/hole pairs are swept out of the junction region (depletion region) and through an external circuit. The flow of electrons creates a current, which can be related to the amount of photons striking the photodiode. Unfortunately, conventional photodiodes do not possess gain. That is, every photon striking the photodiode will generate only a single electron-hole pair, and as such, the photodiode is not appropriate for single-photon counting. In most PMTs, the gain can be on the order of 10^6. In addition, most photodiodes have high dark noise characteristics.

To circumvent this limitation associated with conventional photodiodes, single photon avalanche diodes (SPADs) have been developed. A SPAD consists of a semiconductor material, which is reverse-biased above its breakdown voltage. When a photon strikes the diode, it creates an electron/hole pair, which generates a cascading effect of electrons due to the high energy imparted to the primary electron or photogenerated carrier [19]. The major advantage of this device is that the detector has gain, which can be on the order of 10^6–10^8. This results in a large signal response (current) for a single photon. As with PMTs, the photocurrent can be amplified using a low-noise amplifier and then conditioned using a leading edge discriminator to reduce dark counts.

The SPAD can be operated in one of two modes, actively or passively quenched, which basically determines the dynamic range of the device. In SPADs, once the avalanche process commences, the current that is generated causes the voltage to increase rapidly across a large current limiting resistor (typically 100 kΩ). The bias voltage momentarily drops below breakdown, causing the current to drop below the latch current, and the avalanche is quenched (passive quenching). The voltage is then restored to its initial value (above breakdown). However, during a time period in which the bias voltage drops below breakdown (determined by the value of the current limiting resistor and the stray capacitance in the device), it is impossible to produce an avalanche for a photon striking the detector during this period. The SPAD can recover from the avalanching process by designing an electronic circuit to stop the avalanche and reset the bias voltage once the pulse has been initiated, which is called active quenching. Typical dead times for passively quenched SPADs are ~ 2 µs (maximum count rate, 500 000 cps) and for actively quenched SPADs, the dead time is 0.5 µs (maximum count rate, 2 million cps). One drawback to the use of SPADs as detectors is their small photoactive area, which can be on the order of 150–200 µm in diameter. The small photoactive area is required to minimize the dark noise associated with the detector. However, this limitation can be alleviated in most detector configurations by using the proper focusing optics to effectively image the signal onto the photoactive area of the SPAD.

For timing applications, such as in time-resolved fluorescence, SPADs can produce instrument response functions which range from 150–300 ps, comparable to that observed for conventional PMTs [20]. While multichannel plates (MCP) produce timing responses typically less than 100 ps, they show poor single-photon detection efficiency in the NIR [20]. When operating a SPAD for fast timing, the bias voltage is adjusted to ~ 20 V or greater above breakdown and the radiation is tightly focused onto the photoactive area.

2.2 NIR fluorophores

The major type of fluorophores that exhibit NIR fluorescence belong to the cyanine-class of dyes [21]. Two of the more commonly used cyanine dyes are the tricarbocyanines and the naphthalocyanines. These dyes have found numerous applications in optical recording media [22, 23], dye lasers [24–27], and in photodynamic therapy [28–30]. However, the use of these dyes for fluorescence labeling applications has only arisen recently due to the lack of NIR dyes with functional groups to allow facile conjugation to target molecules.

The basic structures of the tricarbocyanine and naphthalocyanine dyes are shown in Fig. 1. The tricarbocyanines (Fig. 1A) possess a conjugated polymethine chain, which links together two heteroaromatic fragments. The characteristic feature of these dyes is a bathochromic shift in the absorbance and emission maxima that results from an increase in the length of the polymethine chain. For example, dicarbocyanines show absorption maxima near 630 nm, while the tricarbocyanines (heptamethine dyes) show absorption maxima near 780 nm. A naphthalocyanine dye (shown in Fig. 1B) possesses a conjugated ring structure, which is linked together by four naphthalenedicarbonitrile fragments. The characteristic feature of these dyes is the significant alteration in the fluorescence properties of the naphthalocyanine following coordination of a metal to the core of the molecule. The ability to tailor the spectroscopic properties with subtle changes in dye structure makes both the tricarbocyanine and naphthalocyanine dyes excellent candidates for bioanalytical applications.

Some of the common spectroscopic properties associated with these NIR dyes are shown in Table 1. The tricarbocyanine and naphthalocyanine dyes typically exhibit large extinction coefficients. The naphthalocyanine dyes typically have larger fluorescence quantum yields and longer fluorescence lifetimes compared to the tricarbocyanines. This stems from the fact that the ring structure of the naphthalocyanine dyes is much more rigid than the polymethine chain in the tricarbocyanines. The flexible nature of the structure results in high rates of internal conversion, significantly reducing the fluorescence quantum yield and lifetime [17, 31, 32].

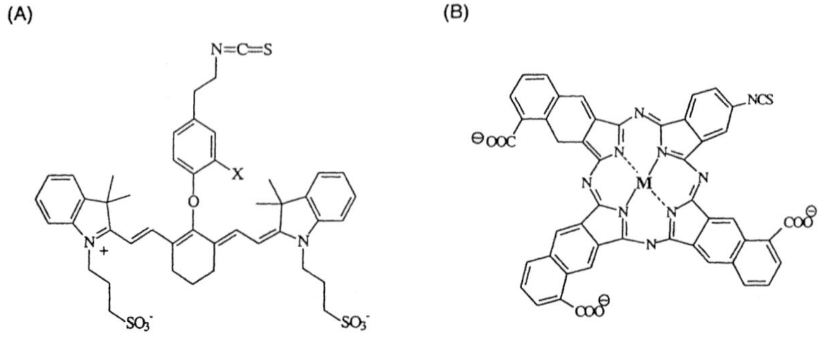

(A)

(B)

Figure 1. (A) Heavy-atom modified tricarbocyanine dyes which contain an isothiocyanate labeling group where X = I, Br, Cl, or F. Taken from [39], with permission. (B) Naphthalocyanine dye with an isothiocyanate labeling group, where M = Zn, Al, Co, Cu, or Sn. Taken from Owens *et al.*, submitted, with permission.

Table 1. Spectroscopic properties of typical NIR dyes

	Tricarbocyanine[a]	Naphthalocyanine[b]
Absorbance and emission range (nm)	750–900	700–1000
ε ($M^{-1}cm^{1}$)	~ 200 000	~ 250 000
Quantum yield	0.07	0.30–0.50
Fluorescence lifetime, τ_f (ns)	< 1.0	1.0–5.0

a) Taken from [40].
b) Taken from Owens *et al.* (submitted).

Several research groups have focused on the development of fluorogenic labels built from the tricarbocyanine molecule framework. Mujumdar and co-workers [33] reported isothiocyanate-derivatized cyanine dyes for flow cytometry applications. Proteins were labeled in human T-cells with a series of NIR dyes possessing unique emission properties, which facilitated a multicolor analysis. Strekowski *et al.* [34–36] reported the synthesis of several heptamethine cyanine derivatives that contained an isothiocyanate labeling group appropriate for primary amine-containing targets. In a subsequent paper, they reported the use of the heptamethine derivatives as labels for DNA sequencing applications [37]. Waggoner *et al.* [38] also discussed the synthesis of several carboxymethylindocyanine dyes, which contained a succinimidyl ester reactive group for labeling proteins or other amino-containing functional groups. Flanagan *et al.* [39, 40] synthesized a series of heavy-atom-modified tricarbocyanine dyes, which possessed succinimidyl esters or isothiocyanate groups for labeling primary amine-containing targets. The interesting features of these dyes was that they contained an intramolecular heavy atom to perturb the fluorescence lifetime of the base chromophore. Therefore, the dyes possessed distinct lifetimes, the value of which depended upon the identity of the heavy atom modification, but also showed similar absorption and emission wavelengths. Therefore, in multiplexed applications, the lifetimes could be used to identify the target, while processing all the fluorescence on a single channel.

While the naphthalocyanine dyes typically show better photophysical properties compared to the tricarbocyanines, the ability to make functional analogs has been somewhat difficult due to the need to assemble an asymmetric ring structure. For example, to improve water solubility, several charged moieties must be added to the ring structure and a functional group must also be inserted within the molecule. Because these dyes are usually assembled by reacting a dicarbonitrile with the appropriate metal halide in nonaqueous conditions, a mixture of both structural and positional isomers results ([41];

Owens *et al.*, submitted). The challenge is that these isomers are difficult to purify ([41]; Owens *et al.*, submitted). Owens and co-workers have described a method for the synthesis of asymmetrical naphthalocyanine derivatives with isothiocyanate functional moieties for labeling primary amine-containing analytes (Owens *et al.*, submitted). The authors reported the absorbance and emission maxima of the dye to be 755 and 765 nm, respectively, in an aqueous media. Also, once the dye was conjugated to an oligonucleotide, the absorbance and emission properties shifted to 735 and 780 nm, respectively.

3 Applications of CE with NIR-LIF detection

Applications using NIR-LIF detection in CE have grown over the past decade due to the increased availability of chromophores and semiconductor devices appropriate for NIR fluorescence. While many commercial sources currently exist from which laser diodes and diode detectors can be purchased to allow construction of NIR detectors appropriate for CE, work continues toward increasing the availability of labeling dyes that target particular functional groups of the analyte. In addition, modifications in the CE operating conditions to accommodate the photophysical and chemical properties of the NIR labeling dyes has been and continues to be a focus area. In the following sections, we will discuss CE applications using NIR fluorescence detection for the analysis of amino acids, peptides, and proteins, as well as both single-stranded and double-stranded DNAs.

3.1 Analysis of amino acids

Several groups have used CE coupled to NIR-LIF detection for the analysis of amino acids. Imasaka *et al.* [42] synthesized a thiazine chromophore, Azur B, which was fluorescent in the deep-red region and contained a succinimidyl ester for labeling amino acids. However, due to the poor labeling efficiency and photophysics of the dye in aqueous solvents, detection limits in the pmol range were demonstrated. In a subsequent report by the same group, detection limits in the sub-amol range (800 zmol) for amino acids labeled with a pyronin succinimidyl ester (λ_{abs} = 663 nm) separated by CE using NIR fluorescence detection were reported [43]. Detection was accomplished off-column in a sheath flow cuvette, which reduced the scattering contribution to the background originating from the cell wall. The authors attributed the improved LODs to higher labeling efficiency, higher fluorescence collection efficiency, and background reduction due to sheath flow detection.

Yeung and Mank [44] reported the separation and detection of amino acids using a diode laser-based system coupled to micellar electrokinetic chromatography. The system consisted of a diode laser (λ_{ex} = 670 nm) equipped with a 10 × microscope objective to focus the laser into the capillary. The fluorescence was then collected at a right angle to the excitation using a 20 × microscope objective with the collected emission subsequently focused onto a PMT. The results are shown in Fig. 2. Eighteen amino acids were labeled with a succinimidyl ester functionalized dicarbocyanine (λ_{abs} = 667/λ_{em} = 689 nm). The amino acids were separated by optimization of the running buffer conditions causing the elution order to be reversed compared to that of conventional CE, which was attributed to the fact that the higher charged species spent less time in the micelles compared to the less charged species. An LOD of 0.1 amol was reported for labeled glycine with similar detection limits for the other amino acids.

Further improvements in the LOD for amino acids analyzed by NIR were reported by Legendre *et al.* [45]. A block diagram of their diode-based NIR-LIF detection system for CE is shown in Fig. 3. The excitation source consisted of a GaAlAs diode laser which was tuned to 785 nm (T = 25°C). The fluorescence emission was collected at 90° with respect to the excitation using a 60 × microscope objective. The emission was then spatially and spectrally filtered and focused onto the photodetector by a 10 × microscope objective. The photodetector was a SPAD which was mounted on a thermoelectric cooler to

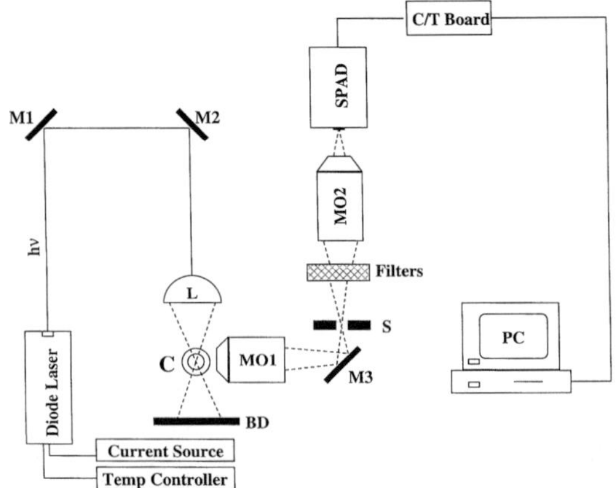

Figure 3. NIR-LIF system for CE analysis. M1, M2, and M3, mirrors; L, laser focusing lens; C, capillary tube; BD, beam dump; MO1 and MO2, microscope objectives; S, spatial filter; F, optical filters; SPAD, single-photon avalanche diode; C/T, counter/timer; PC, computer. Reprinted from [45], with permission.

reduce the dark count rate. Separation and detection of a mixture of six labeled amino acids after addition of a surfactant, cetyltrimethylammonium bromide, used below its critical micelle concentration to reverse the electroosmotic flow, is shown in Fig. 4. The amino acids were labeled with an anionic NIR chromophore, NN382 (Li-COR, Lincoln, NE, USA), which contained an isothiocyanate functional group for covalent attachment of the dye

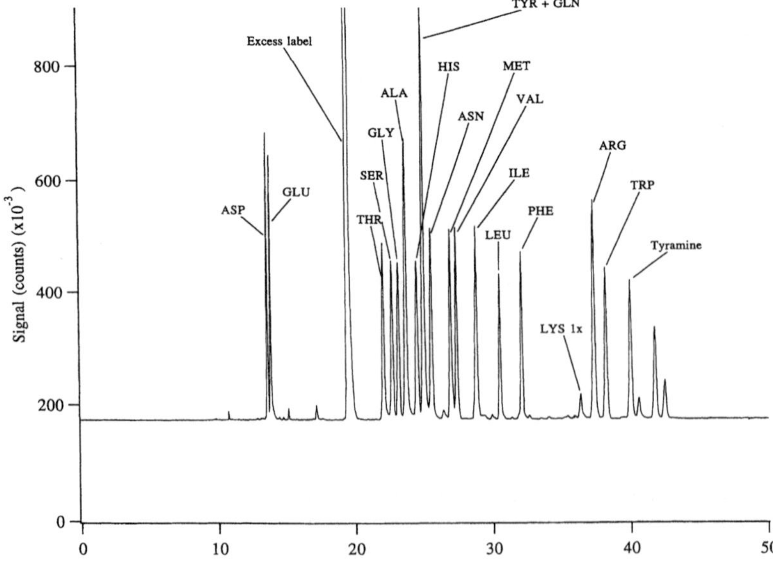

Figure 2. Electropherogram of a mixture of 18 amino acids plus tyramine that were derivatized with a NIR labeling dye (a succinimidyl ester functionalized dicarbocyanine; λ_{abs} = 667/λ_{em} = 689 nm). CE conditions: capillary length, 75 cm; 33 µm ID; applied voltage, 30 kV; run buffer, 20 mM borate (pH 9.0):methanol (72.5:27.5 v/v) and 17.5 mM sodium dodecyl sulfate. Reprinted from [44], with permission.

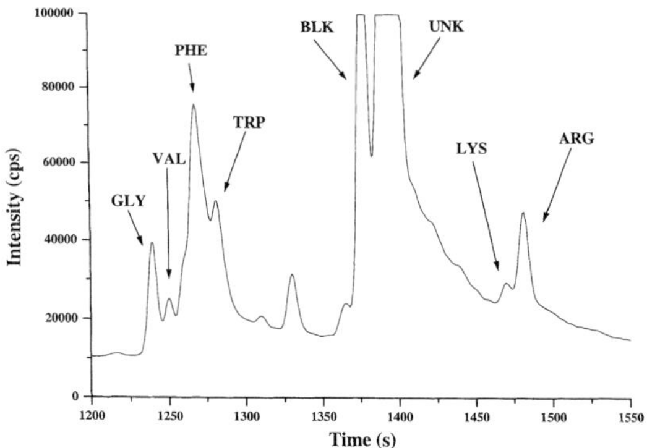

Figure 4. Electropherogram of a mixture of six amino acids labeled with NN382 and detected using a diode-based, NIR-LIF detector. The running buffer consisted of 50:50 methanol:water, 0.55 mM CTAB, 40 mM borate, pH 9.4, with the system run in reverse polarity. The electrophoretic conditions were as follows: [amino acid] = 1.3 $\times$ 10^{-9} M; [NN382] = 4 $\times$ 10^{-8} M; V_{inj} = 10 kV; t_{inj} = 5 s; E_{sep} = 270 V/cm. Reprinted from [45], with permission.

to the primary amine. This dye also contains several alkyl sulfonate groups to improve its water solubility. Detection limits on the order of zeptomoles were obtained (see Table 2) for all labeled amino acids.

3.2 Analysis of peptides

Baars and Patonay [46] evaluated NN382 (Li-COR) as a sensitive peptide labeling reagent for use in CE. The experimental setup consisted of a proprietary microscope and laser assembly (Li-COR) which was interfaced to a P/

ACE 5000 CE instrument (Beckman Instruments, Fullerton, CA, USA). NN382 was used to label six angiotensin I (Ang-I) variants whose charge-to-mass ratios were similar. The peptides were derivatized using optimized buffer conditions (150 mM borate buffer, pH 8; 2 M NaCl), which also consisted of a 15-fold excess of dye with respect to the highest concentration of peptide evaluated. The authors noted that this was a significant improvement in derivatization concentration levels compared to Legendre *et al.* [45]. The peptides were separated after hydrodynamic injection onto an uncoated fused-silica capillary in free solution (CZE). The results of a typical CZE separation are shown in Fig. 5. Four peptides (goosefish, elasmobranch, human and bullfrog Ang-I) were resolved using a 50 mM phosphate run buffer at pH 7.2. Two pairs of coeluting peptides were also separated using micellar electrokinetic chromatography with a nonionic surfactant, Triton X-100. The authors reported detection limits ranging from 100 to 300 zmol (S/N = 3) for the Ang-I variants.

3.3 Analysis of proteins

Legendre and Soper [47] used NIR-LIF detection coupled to free-solution CE to study the noncovalent labeling of proteins with NIR dyes. The experimental setup was similar to that shown in Fig. 3. Two NIR dyes, DTTCI (cationic) and IR-125 (anionic) were used to noncovalently label three model proteins, β-casein, β-lactoglobulin, and chicken egg albumin (CEA) in an aqueous buffer solution. In these experiments, the protein was mixed with the appropriate NIR dye and allowed to incubate at room temperature to form a complex. It was found that the fluorescence of the cationic NIR dye, DTTCI, showed significant enhancements in its fluorescence yield when complexed to these proteins, while IR-125 did not show dramatic

Table 2. Labeling efficiency, mobilities and detection limits for various NN382-labeled amino acids using CE with NIR-LIF detection

Amino acid	Migration time (s)	Conjugation efficiency (%)	Mobility (cm^2/Vs, $\times$ 10^4)	Plates (m^{-1}, $\times$ 10^{-5})	LOD (zmol; S/N = 3)
SER	960	30.1	1.43	2.42	65.3
GLY	963	64.8	1.43	2.21	26.2
VAL	968	43.0	1.42	2.62	37.8
GLN	975	44.4	1.41	2.66	33.2
MET	976	58.4	1.41	4.53	21.2
LEU	981	48.7	1.40	5.50	27.5
PHE	982	81.0	1.40	3.59	16.5
TRP	990	62.3	1.39	4.47	20.2
LYS	1137	30.7	1.21	3.87	64.3
ARG	1148	78.9	1.20	2.87	27.1
HIS	1156	20.0	1.19	0.77	401

CE was performed at a field strength of 400 V/cm and operated in a reverse polarity mode. The running buffer consisted of 50:50 methanol:water with 0.55 mM CTAB and buffered with borate (40 mM), pH 9.4. Reprinted from [45], with permission.

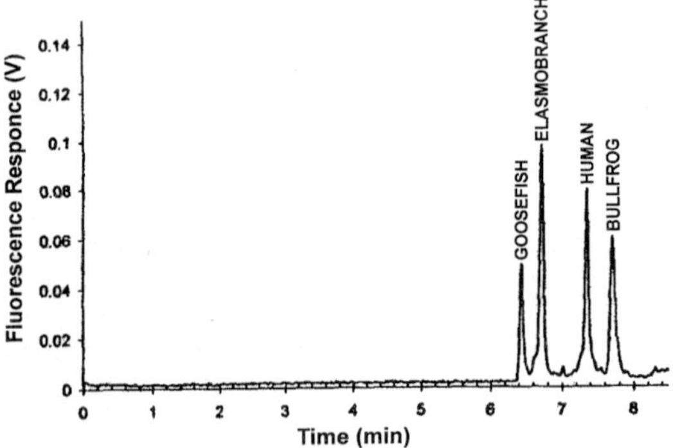

Figure 5. CZE separation of 13 amols of goosefish, elasmobranch, human, and bullfrog Ang-I peptides injected on column. A standard peptide mixture was prepared in water containing each peptide at 5×10^{-7} M. A 50 µL sample of the mixture was derivatized with the NIR dye NN382 and diluted 310-fold, to a derivative concentration of 5.4×10^{-10} M prior to injection. Approximately 23.5 nL of the sample was injected on-column during a 5 s hydrodynamic injection at 0.5 psi. Separation was conducted at 21 kV and 24°C on a bare fused-silica capillary (50 cm effective column length, 75 µm ID) with a 50 mM phosphate run buffer, pH 7.2. Reprinted from [46], with permission.

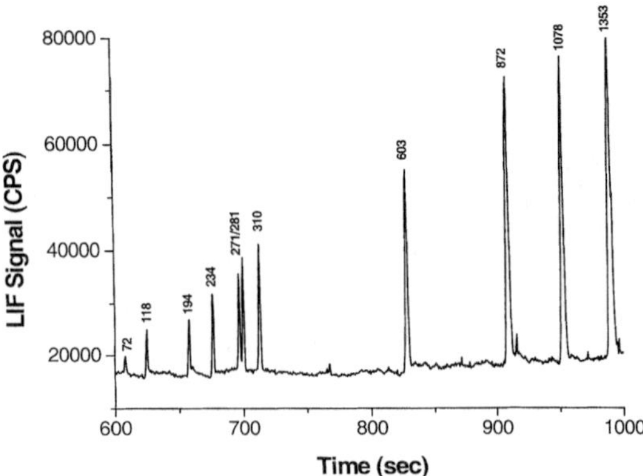

Figure 6. Electropherogram of ΦX174 DNA restriction fragments with LIF detection. The staining dye (TAG) concentration added to the running buffer was 1.0 µM, and the running buffer was composed of 40 mM TRIS, 20 mM sodium acetate, 2.0 mM EDTA, pH 7.6, and 0.25% HPMC, which served as the sieving matrix. The sample was electrokinetically injected onto the column for 6 s at −2.0 kV. The electric field used for the separation was 175 V/cm. The laser power was set at approximately 8.5 mW (λ_{ex} = 750 nm). Reprinted from [50], with permission.

alterations in its fluorescence when placed in a solution of these proteins. This mixture was then directly injected into the CE column, which consisted of a run buffer of 20 mM borate, pH 9.2. The mass detection limits for the proteins stained with DTTCI were found to range from 1.3 to 3.7 amol, which is approximately three orders of magnitude lower than that reported for UV absorption detection [48,49]. However, it was found that the plate numbers for the separation were low, which was attributed to dissociation of the dye/protein complex during the separation, resulting in a change in the mobility of the complex during migration.

3.4 DNA restriction fragment analysis

Owens and co-workers [50] recently reported on the high resolution separation of DNA restriction fragments using capillary gel electrophoresis (CGE) with NIR fluorescence detection. In this study, an NIR monomeric nuclear staining dye, thiazole green (TAG) was used. TAG is a fluorescent mono-intercalating dye that exhibits a fluorescent enhancement ratio of approximately 102 when bound to DNAs and shows a binding constant of 6.1×10^{6} M^{-1}. To investigate the CE separation and detection of the TAG-stained DNA in CE, the researchers used a diode-based laser-induced fluorescence detector, which consisted of a

GaAlAs diode laser with a principal lasing line at 750 nm and an avalanche photodiode for photon transduction. Using a sieving matrix of hydroxypropylmethylcellulose (HPMC), the 271/281 bp fragments were nearly baseline-resolved with plate numbers exceeding 1×10^{6} plates/m (Fig. 6), significantly better than the plate numbers obtained for ethidium-bromide-stained DNA fragments. A mass detection limit of 20 fg of DNA per electrophoretic band was reported [50], which the authors stated was comparable to those obtained using visible LIF detection [51].

3.5 NIR-LIF detection in DNA sequencing

LIF detection coupled to CGE has become an attractive detection strategy for many DNA sequencing applications due to its ultrahigh sensitivity [6, 52, 53]. Detection limits ranging from 6000–60 000 molecules have been reported using visible LIF detection [9, 54]. NIR fluorescence offers a viable alternative to visible fluorescence detection in CGE due to the intrinsically lower background levels that are observed in the NIR when compared to the visible. A direct comparison of the limits of detection for an NIR-labeled oligonucleotide to that of a visible-labeled oligonucleotide (fluorescein tag) was reported [55]. The oligos were electrophoresed in a capillary column filled with a

Table 3. S/N of NIR- and FITC-labeled M13 universal sequencing primers with on-column LIF detection

	NIR primer	FITC primer
Concentration (nM)	1.3	5.0
Migration time (s)	2555	660
Mobility (cm²/Vs)	5.6×10^{-5}	2.1×10^{-4}
Injection volume (nL)	2.9	11.0
Moles injected (amols)	0.38	58.0
Net signal (cps)	33 500	15 920
Background (cps)	10 000	19 490
S/N	335	114
Mass detection limit (moles, S/N = 3)	3.4×10^{-20}	1.5×10^{-18}
Quantum yield of dye	0.07	0.90

FITC-primer was excited at 488 nm (argon ion laser) and the NIR-primer excited at 795 nm (Ti:sapphire laser). Capillary gel column was 6%T, 5%C. Samples were electrokinetically injected onto the gel column for 10 s at 5 kV and electrophoresed at a field strength of 250 V/cm. Reprinted from [55], with permission.

gel matrix and excited with lasers operating at 488 or 780 nm. A summary of the data is presented in Table 3. Comparison of the detection limits (S/N = 3) indicated a significant improvement in the NIR case, with a mass detection limit of 3.4×10^{-20} moles compared to 1.5×10^{-18} moles for the visible labeled oligonucleotide. Even though the NIR-labeled oligo demonstrated a lower quantum yield, the significant reduction in the background observed in the NIR resulted in the improved LOD.

3.5.1 Single lane/single color NIR-DNA sequencing

To demonstrate the utility of NIR fluorescence in actual DNA sequencing runs using CGE, Williams and Soper [10] performed CGE using an M13mp18 plasmid and a single-lane, single-fluor peak height identification protocol with ddNTP concentration ratios of 4:2:1:0 (A:C:G:T) used during DNA polymerization. In this study, an NIR dye-labeled primer was used with excitation at 780 nm. In Fig. 7 a typical sequencing run is shown. The sequencing results were compared to the known sequence of the M13mp18 template using the Genetics Computer Group (GCG, Madison, WI, USA) program Pileup, which yielded a base-calling accuracy of 84% when reading up to 250 bases from the primer annealing site.

3.5.2 Fluorescence lifetime measurements in DNA sequencing

In many CE applications, one would potentially like to detect several types of labeling dyes, with each dye targeted for a specific analyte (multiplexed assays). For example, in DNA sequencing, one generates a series of fragments, which differ in base numbers, and are terminated in a common nucleotide base. These are then fractionated in a capillary gel column according to size. Since the DNA is composed of four bases, it becomes advantageous to identify the four bases in a single CE run to improve throughput. This is commonly done using spectral discrimination, in which each dye has a unique emission maximum and the color is sorted using filtering systems onto the appropriate read-out channel.

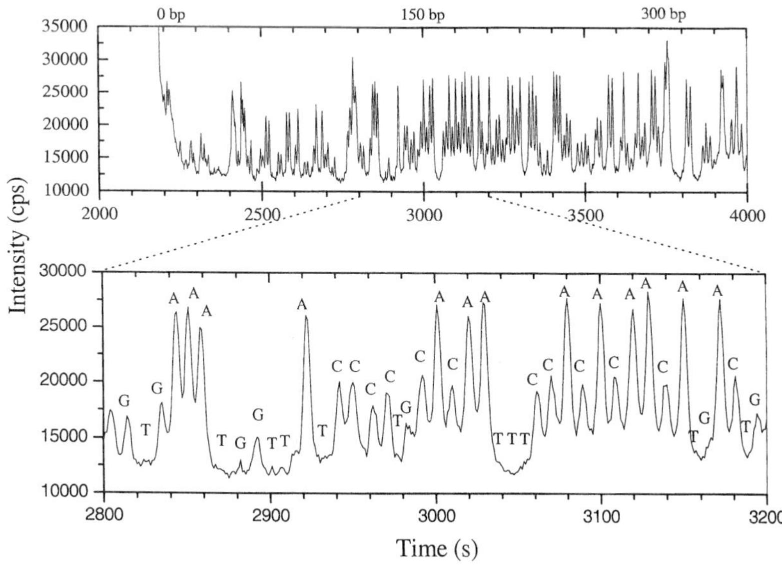

Figure 7. Capillary gel sequencing run of an M13 sequencing template. The molar ratios of the ddNTPs used were 4:2:1:0 (A:C:G:T). The sequencing ladders were electrokinetically injected onto the capillary column for 120 s at 250 V/cm. The separation was performed at a field strength of 250 V/cm. The gel column consisted of 3%T, 3%C polyacrylamide gel matrix. Laser excitation was accomplished with approximately 5 mW at 780 nm. Reprinted from [10], with permission.

As an alternative to spectral discrimination, various groups have suggested that fluorescence lifetimes can potentially serve as a viable method for identifying components in multiplexed CE applications [5, 18, 47, 56–66]. The principal advantages associated with lifetime discrimination are (i) because the calculated lifetime is immune to concentration differences, the identification can be accomplished with high accuracy, irrespective of the intensity of the CE band, (ii) lifetime values can be determined with higher precision than fluorescence intensities under appropriate conditions, (iii) lifetime determinations do not suffer from broad emission profiles associated with spectral discrimination, and (iv) the fluorescence can potentially be processed on a single detection channel without the need for spectral sorting to multiple detection channels.

While there are several ways to measure fluorescence lifetimes, one method involves a time-resolved approach. In this technique, a laser that operates in a pulsed mode is used to excite the fluorescence of the dye molecules traveling through the excitation beam. Typically, the laser is on for a short time and then off for an extended period of time. During this off state, the excited dye molecule can relax, emitting a fluorescent photon. In time-resolved fluorescence, the time-difference between excitation and fluorescence emission is measured. These time-differences are histogrammed and when photons are accumulated over many excitation/emission cycles, an exponential function is produced from which the lifetime can be extracted.

NIR fluorescence can be an appealing choice when attempting to perform lifetime measurements in CE separations. The advantages of NIR fluorescence monitoring include smaller Raman cross sections (lower scattering contribution to the background) and fewer fluorescence interferences [21, 61–75], which can improve the accuracy and precision in a measurement of lifetime. However, a major challenge associated with NIR fluorescence is the intrinsically short excited-state lifetimes associated with many NIR chromophores and their poor photophysical properties (*i.e.*, low quantum yields and poor photochemical stabilities) [17]. Due to the subnanosecond lifetimes of many NIR chromophores, the instrumentation should possess certain criteria, such as subnanosecond pulse widths associated with the excitation source and a detector with a small transit time spread, which are the main components to temporal broadening of the instrumental response function. On-line fluorescence lifetime determinations in capillaries presents many technical challenges as well because of the short residence times of the analytes in the detection zone

and the low loading masses, which limit the number of photocounts that can be accumulated into the decay profile producing poor photon statistics.

Soper and co-workers [58] have developed a system for on-line lifetime measurements in capillary gel DNA sequencing using NIR fluorescence. The instrumental setup is shown in Fig. 8. The instrument consisted of a solid-state, actively pulsed GaAlAs diode laser with a variable repetition rate of single shot to 80 MHz, and an average power of 5.0 mW at a lasing wavelength of 780 nm. A SPAD operated in a Geiger mode was used as photodetector. The electronics for making the time-resolved measurement were situated directly on a PC-board while plugged into the bus of the PC and powered by a Windows-based graphical user interface (GUI) panel for easy operation. A CGE electropherogram of NIR-labeled A- and *C*-terminated fragments generated from an M13mp18 plasmid is shown in Fig. 9A. In this example, the A-terminated fragments were labeled with the tricarbocyanine dye, IRD40 (Li-COR), while the C-terminated fragments were labeled with IRD41. In both cases, the 5′-end of the sequencing primers contained the dye label. Figure 9B shows an expanded view of a section of the electropherogram along with the integration periods that were used to construct the decay profiles. In Fig. 9C, the decay profiles for gel-only (prompt peak) and two dye-labeled oligonucleotides are shown. From the decay profiles of the dye-labeled oligonucleotides, the fluorescence lifetimes were determined to be 669 ps (IRD40) and 528 ps (IRD41).

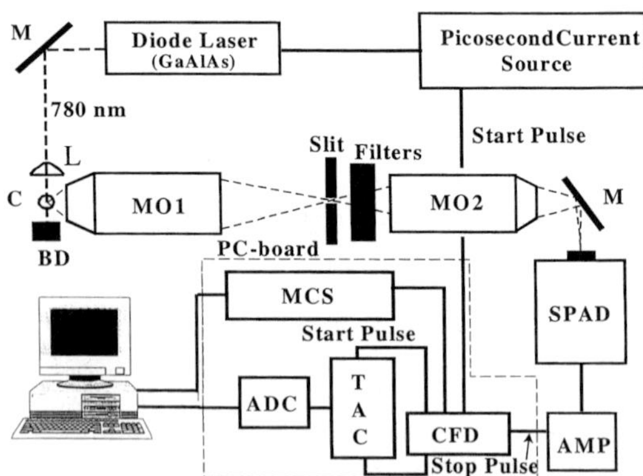

Figure 8. Block diagram of an NIR-TCSPC instrument. L, laser singlet focusing lens; C, capillary; BD, beam dump; MO1, collecting microscope objective; MO2, focusing microscope objective; SPAD, single photon avalanche diode; AMP, amplifier; CFD, constant fraction discriminator; TAC, time-to-amplitude converter; ADC, analog-to-digital converter; MCS, multichannel scaler. Reprinted from [5], with permission.

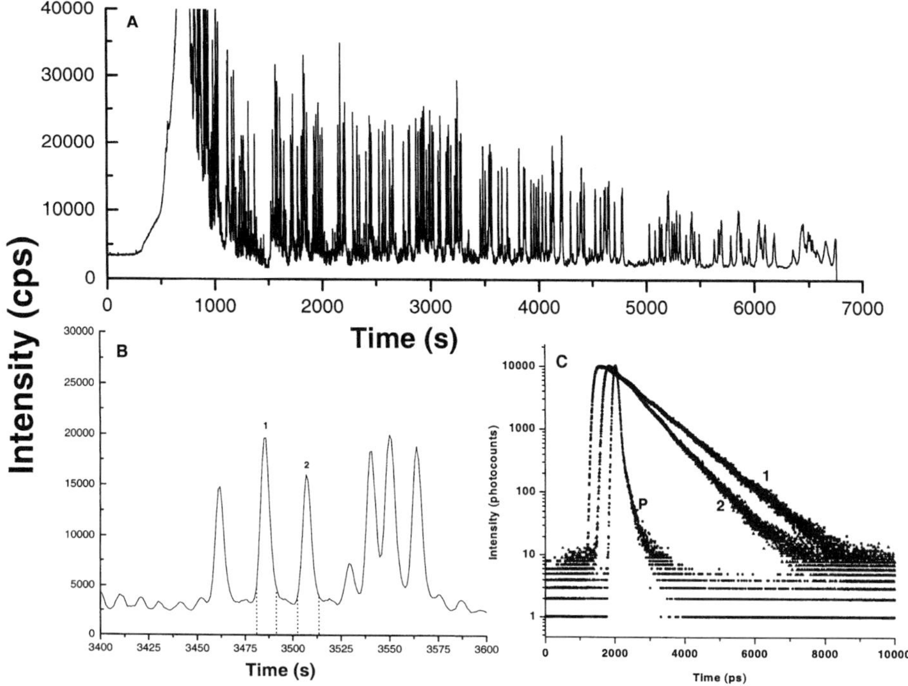

Figure 9. (A) Capillary gel electropherogram of A- and C-terminated fragments from an M13mp18 plasmid labeled on the 5′-end of the sequencing primer with the NIR fluorescent dyes, IRD40 (A-terminated fragments) and IRD41 (C-terminated fragments). (B) An expanded view of this electropherogram showing the time periods (dashed lines) which were used for accumulating photocounts into the decay profile. (C) Decay profiles for the dye-labeled oligonucleotides resident in the detection zone (1) and (2) and of the gel only (P). The laser wavelength was set at 780 nm with an operating power of 5 mW at the capillary. The sequencing ladder was electrokinetically injected onto the column for 1 min at 200 V/cm with the separation carried out using the same field strength. Reprinted from [5], with permission.

4 NIR detection on a CE microchip

Although CE is a valuable tool for many applications because of its high resolving power and short analysis times, there is a need for micrototal analysis systems (microchips) where sample injection, manipulation, reactions, and detection all occur on one small platform to minimize sample handling of submicroliter volumes. Manz *et al.* [76] reported the first use of microchips in 1992, and since that time their use has grown tremendously. NIR fluorescence can be a particularly attractive detection strategy for microchip applications, since the basic building blocks of the detectors are conducive to miniaturization. In addition, many groups have initiated the fabrication of microchips from materials other than glass, for example plastics [71–83]. The difficulty in using plastics for micro-

fluidic devices is that higher background levels can be generated from the substrate (compared to glass) when excitation is accomplished using a visible laser. However, NIR fluorescence alleviates this problem to a large extent, making it possible to perform ultrasensitive fluorescence measurements in plastic devices.

Recently, Ford *et al.* [82] fabricated a microchip device from poly(methyl methacrylate) (PMMA) using X-ray lithography techniques and injection molding with NIR fluorescence detection. The device topography is shown in Fig. 10A. The microdevice consisted of a fixed volume injector (V_{inj} = 100 pL), a separation channel and two channels to accommodate optical fibers for NIR-LIF excitation and detection. Figure 10B shows a scanning electron micrograph (SEM) of the dual fiber detector fabri-

A

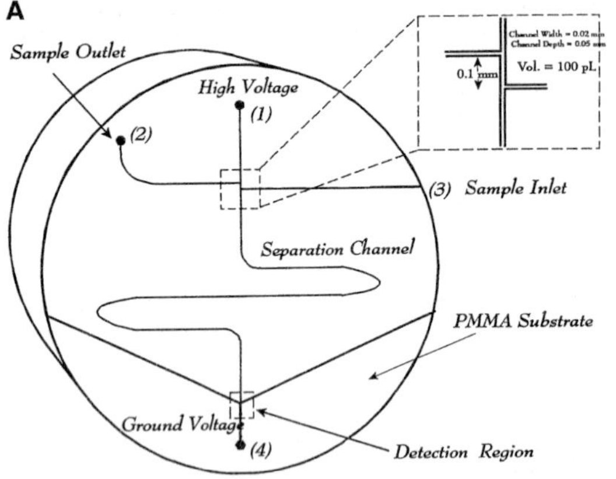

B

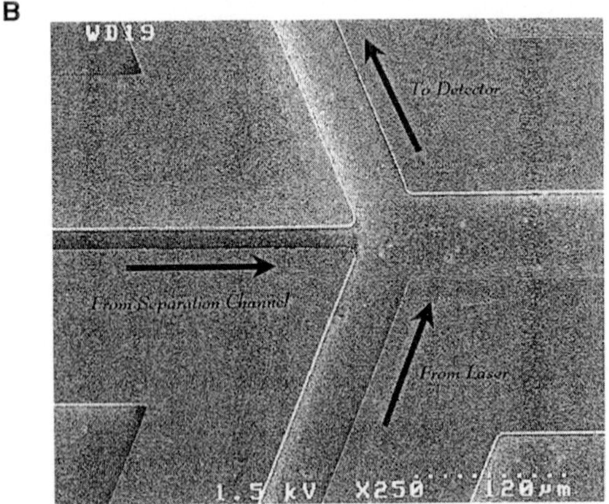

Figure 10. (A) Topographical layout of a microelectro-phoresis chip fabricated in PMMA. The channels were 50 µm in depth and varied in width (20 µm or 50 µm). The injector contained a volume of 100 pL. (B) SEM image of dual fiber-optic component micromachined in PMMA. Taken from [83], with permission.

cated in PMMA. In this device, LIF detection in the NIR was used with a battery-operated diode laser serving as the excitation source and a SPAD as the photon trans-ducer.

In order to show the ability to use PMMA-based micro-chips for oligonucleotide separations with NIR detection, the authors electrophoresed an NIR-labeled oligonucleo-tide using a 4% linear polyacrylamide (LPA) sieving matrix in both a PMMA-based microdevice and a conven-tional fused-silica capillary, which was modified by coating the wall with a 1% LPA polymer [83]. The results (see Fig. 11) show that the primer migrated through the fused-silica capillary in approximately 1720 s, while for the PMMA-based device, the migration time was only 194 s,

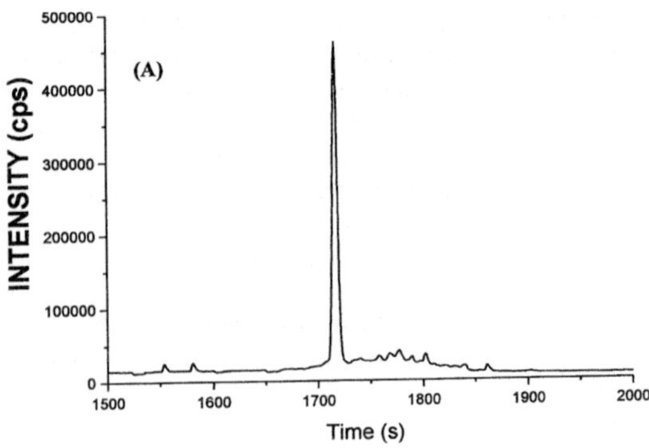

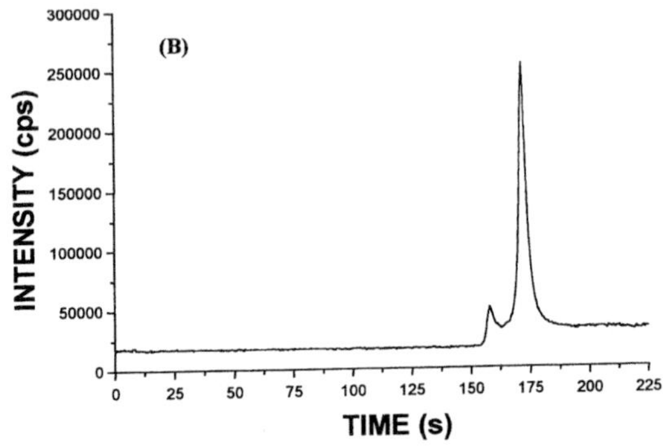

Figure 11. Electrophoretic separations of NIR dye-labeled sequencing primers (A) in a conventional capillary tube and (B) PMMA-based microchip. The labeling dye was NN382 (Li-COR), and the primer was an M13 (–21) oligonucleotide. The sieving matrix was a 4% LPA solu-tion containing 7 M urea as the denaturant. For the capil-lary gel separation, the wall was coated with a 1% LPA solution; for the microchip separation, the device walls were not modified. The shift in the background seen for the microelectrophoresis run after detection of the oligo resulted from leakage of sample from the waste channel into the separation channel. Taken from [83], with permis-sion.

an approximate 10-fold decrease in migration time. Detection limits for the PMMA-device using NIR fluores-cence were reported to be 1.2 zmol (S/N = 3), which, according to the authors, compared favorably to the limits of detection determined for a capillary-based system using NIR [45].

5 Conclusions

Over the past decade, NIR fluorescence has become an increasingly popular detection method for CE. One advantage associated with NIR fluorescence includes the lower background level typically observed, improving the

S/N in the measurement. Recent improvements in excitation and detection hardware plus the continued development of novel labeling chromophores should expand the potential applications of NIR-LIF detection in CE and microchip formats as well. As CE systems continue to shrink in size, NIR fluorescence will play a dominant role in detection due to the small size of the components required to carry out detection and the high-sensitivity NIR fluorescence displays when monitoring fluorescence in less than ideal conditions. Another particularly attractive aspect of NIR fluorescence is that traditional instrument-intensive measurements, such as time-resolved fluorescence, can be accomplished to a much simpler degree, making it feasible to develop detection with lifetime identification capabilities for multiplexed applications appropriate for CE.

The authors would like to thank the National Institutes of Health, National Human Genome Research Institute (NHGRI), for partial financial support of this work.

Received January 9, 2000

6 References

[1] Swartz, M. E., *J. Chromatogr.* 1993, *640*, 441–444.

[2] Tsai, E. W., Singh, M. M., Lu, H. H., Ip, D. P., Brooks, M. A., *J. Chromatogr.* 1992, *626*, 245–250.

[3] Schwartz, H. E., Ulfelder, K. J., Chen, F. A., Pentoney, S. L., *J. Capil. Electrophor.* 1994, *1*, 36–54.

[4] Crosby, D., El Rassi, Z., *J. Chromatogr.* 1993, *16*, 2161–2181.

[5] Legendre Jr., B. L., Williams, D. C., Soper, S. A., Erdmann, R., Ortmann, U., Enderlein, J., *Rev. Sci. Instrum.* 1996, *67*, 3984–3989.

[6] Luckey, J. A., Drossman, H., Kostichka, A. J., Mead, D. A., D'Cuhna, J., Swerdlow, H. P., Gesteland, R., *Nucleic Acids Res.* 1990, *18*, 4417–4421.

[7] Karger, A. E., Harris, J. M., Gestland, R. F., *Nucleic Acids Res.* 1991, *19*, 4955–4962.

[8] Huang, X. C., Quesada, M. A., Mathies, R. A., *Anal. Chem.* 1992, *64*, 2149–2154.

[9] Swerdlow, H. P., Ahang, J. A., Chen, D. Y., Harke, H. R., Grey, R., Wu, S., Dovichi, N. J., *Anal. Chem.* 1991, *63*, 2835–2841.

[10] Williams, D. C., Soper, S. A., *Anal. Chem.* 1995, *67*, 3427–3432.

[11] Mattusch, J., Dittrich, K. J., *J. Chromatogr. A* 1994, *680*, 279–285.

[12] Gassman, E., Kuo, J. E., Zare, R. N., *Science* 1985, *230*, 813–814.

[13] McGregor, D. A., Yeung, E. S., *J. Chromatogr. A* 1994, *680*, 491–496.

[14] Chen, D. Y., Dovichi, N. J., *J. Chromatogr. A* 1994, *657*, 265–269.

[15] Flanagan Jr., J. H., Legendre Jr., B. L., Hammer, R. P., Soper, S. A., *Anal. Chem.* 1995, *67*, 341–347.

[16] Shealy, D. B., Lipowska, M., Lipowski, J., Narayanan, N., Sutter, S., Strekowski, L., Patonay, G., *Anal. Chem.* 1995, *67*, 247–251.

[17] Soper, S. A., Mattingly, Q. L., *J. Am. Chem. Soc.* 1994, *116*, 3744–3752.

[18] Gilson, T. R., Hendra, P. J., *Laser Raman Spectroscopy*, Wiley-Interscience, New York 1970.

[19] Nordlund, T. M., *Streak Cameras for Time-Domain Fluorescence*, Plenum Press, New York 1991.

[20] Li, L., Davis, L. M., *Rev. Sci. Instrum.* 1993, *64*, 1524–1529.

[21] Patonay, G., Miquel, D., *Anal. Chem.* 1991, *63*, 321A–326A.

[22] Przhonska, O. V., *NATO ASI Ser., Ser. 3* 1998, *52*, 265–267.

[23] Umehara, H. Y., Masatoshi; T. Y., Hirose, S. M., Tsutayoshi, T., Hirosuke, T., *Jpn. Kokai Tokkyo Koho*, Mitsui Toatsu Chemicals, Japan 1994.

[24] Babenko, V. A. S., Ishchenko, A. A., *Kvantovaya Elektron. (Moscow)* 1995, *22*, 765–768.

[25] Babenko, V. A. S., Ishchenko, A.A., *J. Opt. Res.* 1996, *4*, 1.

[26] Dai, Q. C., Zhusheng, X. C., *Huadong Ligong Daxue Xuebao* 1996, *22*, 47–51.

[27] Ishchenko, A. A., *Kvantovaya Elektron. (Moscow)* 1994, *21*, 513–534.

[28] Kogan, B. Y., Butenin, A. V., Torshina, N. L., Kogan, E. A., Kaliya, O. L. L., Luzhkov, E. A., Yu, M., Derkacheva, V. M., Pankratov, A. A., Vorozhtzov, G. N., Volkova, A. I., *Proc. SPIE-Int. Soc. Opt. Eng.* 1999, *3563*, 148–151.

[29] Tolmachev, A. I. S., Yu, L., Kudinova, M. A., *Zh. Nauchn. Prikl. Fotogr.* 1997, *42*, 54–58.

[30] Yates, N. C., Moan, J., Western, A., *J. Photochem. Photobiol. B* 1990, *4*, 379–390.

[31] Mialocq, J. C., Jaradias, J., Goujon, P., *Chem. Phys. Lett.* 1977, *47*, 123–126.

[32] Hofer, L. J., Grabenstetter, R. J., Wiig, E. O., *J. Am. Chem. Soc.* 1950, *72*, 203–209.

[33] Mujumdar, R. B., Ernst, L. A., Mujumdar, S. R., Waggoner, A. S., *Cytometry* 1989, *10*, 11–19.

[34] Lipowska, M., Patonay, G., Strekowski, L., *Synth. Commun.* 1993, *23* 3081–3094.

[35] Strekowski, L., Lipowska, M., Patonay, G., *J. Org. Chem.* 1992, *57*, 4578–4580.

[36] Strekowski, L., Lipowska, M., Patonay, G., *Synth. Commun.* 1992, *22*, 2593–2598.

[37] Shealy, D. B., Lipowska, M., Lipowska, J., Narayanan, N., Sutter, S., Strekowski, L., Patonay, G., *Anal. Chem.* 1995, *67*, 247–251.

[38] Soutwick, P. L., Ernst, L. A., Tauriello, E. W., Parker, S. R., Mujumdar, R. B., Mujumdar, S. R., Clever, H. A., Waggoner, A. S., *Cytometry* 1990, *11*, 418–430.

[39] Flanagan Jr., J. H., Kahn, S. H., Menchen, S., Soper, S. A., Hammer, R. P., *Bioconjug. Chem.* 1997, *8*, 751–756.

[40] Flanagan Jr., J. H., Owens, C. V., Romero, S. E., Waddell, E., Kahn, S. H., Hammer, R. P., Soper, S. A., *Anal. Chem.* 1998, *70*, 2676–2684.

[41] Hall, T. W., Greenberg, S., McArthur, C. R., Khouw, B., Leznoff, C. C., *Nouv. J. Chim.* 1982, *6*, 653–658.

[42] Higashijima, T., Fuchigami, T., Imasaka, T., Ishibashi, N., *Anal. Chem.* 1992, *64*, 711–714.

[43] Fuchigami, T., Imasaka, T., Shiga, M., *Anal. Chim. Acta* 1993, *282*, 209–213.

[44] Mank, A. J. G., Yeung, E. S., *J. Chromatogr. A* 1995, *708*, 309–321.

[45] Legendre Jr., B. L., Moberg, D. L., Williams, D. C., Soper, S. A., *J. Chromatogr. A* 1997, *779*, 185–194.

[46] Baars, M. J., Patonay, G., *Anal. Chem.* 1999, *71*, 667–671.

[47] Legendre Jr., B. L., Soper, S. A., *Appl. Spectrosc.* 1996, *50*, 1196–1202.

[48] Lauer, H. H., McManigill, P., *Anal. Chem.* 1986, *58*, 166–170.

[49] Walbroehl, Y., Jorgenson, J. W., *J. Chromatogr.* 1984, *315*, 135–143.

[50] Owens, C. V., Davidson, Y. Y., Kar, S., Soper, S. A., *Anal. Chem.* 1997, *69*, 1256–1261.

[51] Zhu, H., Clark, S. M., Benson, S. C., Rye, H. S., Glazer, A. N., Mathies, R. A., *Anal. Chem.* 1994, *66*, 1941–1948.

[52] Drossman, H., Luckey, J. A., Kostichka, A., D'Cuhna, J., Smith, L. M., *Anal. Chem.* 1990, *62*, 900–903.

[53] Chen, D., Swerdlow, H. P., Harke, H. R., Zhang, J. Z., Dovichi, N. J., *J. Chromatogr.* 1991, *559*, 237–246.

[54] Swerdlow, H. P., Wu, S., Karke, H. R., Dovichi, N. J., *J. Chromatogr.* 1990, *516*, 61–67.

[55] Soper, S. A., Flanagan Jr., J. H., Legendre Jr., B. L., Williams, D. C., Hammer, R. P., *IEEE J. Select. Topics Quantum Electron.* 1996, *2*, 1–11.

[56] Zander, C., Sauer, M., Drexhage, K. H., Ko, D. S., Schulz, A., Wolfrum, J., Brand, L., Eggeling, C., Seidel, C. A. M., *Appl. Phys. B* 1996, *63*, 517–523.

[57] Warner, I. M., Soper, S. A., Mcgown, L. B., *Anal. Chem.* 1996, *68*, R73–R91.

[58] Soper, S. A., Legendre Jr., B. L., Williams, D. C., *Anal. Chem.* 1995, *67*, 4358–4365.

[59] Soper, S. A., Legendre Jr., B. L., *Appl. Spectrosc.* 1994, *48*, 400–405.

[60] Sauer, M., Arden-Jacob, J., Drexhage, K. H., Marx, N. J., Karger, A. E., Lieberwirth, U., Müller, M., Neumann, M., Nord, S., Paulus, A., Schulz, A., Seeger, S., Zander, C., Wolfrum, J., *Biomed. Chromatogr.* 1997, *11*, 81–82.

[61] Müller, R., Herten, D. P., Lieberwirth, U., Neumann, M., Sauer, M., Schulz, A., Siebert, S., Drexhage, K. H., Wolfrum, J., *Chem. Phys. Lett.* 1997, *279*, 282–288.

[62] Müller, R., Zander, C., Sauer, M., Deimel, M., Ko, D. S., Siebert, S., Arden-Jacob, J., Deltau, G., Marx, N. J., Drexhage, K. H., Wolfrum, J., *Chem. Phys. Lett.* 1996, *262*, 716–722.

[63] Lieberwirth, U., Arden-Jacob, J., Drexhage, K. H., Herten, D. P., Müller, R., Neumann, M., Schulz, A., Siebert, S., Sagner, G., Klingel, S., Sauer, M., Wolfrum, J., *Anal. Chem.* 1998, *70*, 4771–4779.

[64] Li, L. C., He, H., Nunnally, B. K., Mcgown, L. B., *J. Chromatogr. B* 1997, *695*, 85–92.

[65] Dreizler, A., Tadday, R., Monkhouse, P., Wolfrum, J., *Appl. Phys. B* 1993, *57*, 85–87.

[66] Bachteler, G., Drexhage, K. H., Arden-Jacob, J., Han, K. T., Kollner, M., Müller, R., Sauer, M., Seeger, S., Wolfrum, J., *J. Lumin.* 1994, *62*, 101–108.

[67] Williams, R. J., Narayanan, N., Casay, G. A., Lipowska, M., Strekowski, L., Patonay, G., Peralta, J. M., Tsang, V. C. W., *Anal. Chem.* 1994, *66*, 3102–3107.

[68] Williams, R. J., Lipowska, M., Patonay, G., Strekowsi, L., *Anal. Chem.* 1993, *65*, 601–605.

[69 Soper, S. A., Mattingly, Q. L., Vegunta, P., *Anal. Chem.* 1993, *65*, 740–747.

[70] Zen, J. M., Patonay, G., *Anal. Chem.* 1991, *63*, 2934–2938.

[71] Wilberforce, D. A., Patonay, G., *Spectrochim. Acta* 1990, *46A*, 1153–1162.

[72] Imasaka, T., Yoshitake, A., Ishibashi, N., *Anal. Chem.* 1984, *56*, 1077–1079.

[73] Imasaka, T., Ishibashi, N., *Anal. Chem.* 1990, *62*, 363A.

[74] Johnson, P. A., Barber, T. E., Smith, B. D., Winefordner, J. D., *Anal. Chem.* 1989, *61*, 861–863.

[75] Sauda, K., Imasaka, T., Ishibashi, N., *Anal. Chem.* 1986, *58*, 2649–2653.

[76] Manz, A., Harrison, D. J., Verporte, E. M. J., Fettinger, J. C., Ludi, H., Widmer, H. M., *J. Chromatogr.* 1992, *593*, 253–258.

[77] Martynova, L., Locascio, L. E., Gaitan, M., Kramer, G. W., Christensen, R. G., MacCrehan, W. A., *Anal. Chem.* 1997, *69*, 4783–4789.

[78] Duffy, D. C., McDonald, J. C., Schueller, O. J. A., Whitesides, G. M., *Anal. Chem.* 1998, *70*, 4974–4984.

[79] Effenhauser, C. S., Bruin, G. J. M., Paulus, A., Ehrat, M., *Anal. Chem.* 1997, *69*, 3451–3457.

[80] Ford, S. M., Davies, J., Kar, B., Qi, S. D., McWhorter, S., Soper, S. A., Malek, C. K., *J. Biomechan. Engineer.-Trans. Asme* 1999, *121*, 13–21.

[81] Qin, D., Xia, Y. N., Rogers, J. A., Jackman, R. J., Zhao, X. M., Whitesides, G. M., *Microsystem Technology in Chemistry and Life Science*, 1998, Vol. 194, 1–20.

[82] Ford, S. M., Kar, B., McWhorter, S., Davies, J., Soper, S. A., Klopf, M., Calderon, G., Saile, V., *J. Microcol. Sep.* 1998, *10*, 413–422.

[83] Soper, S. A., Ford, S. M., Xu, Y. C., Qi, S. Z., McWhorter, S., Lassiter, S., Patterson, D., Bruch, R. C., *J. Chromatogr. A* 1999, *853*, 107–120.

Electrophoresis 2000, *21*, 4017–4028

4017

Review

Richard P. Baldwin

Department of Chemistry,
University of Louisville,
Louisville, KY, USA

Recent advances in electrochemical detection in capillary electrophoresis

Recent advances in the design and application of electrochemical (EC) detection systems in capillary electrophoresis (CE) are reviewed, with the objective of providing the nonelectrochemist with a state-of-the-art picture of CEEC instrumentation and an overview of the principal analytes for which CEEC is best suited. The detection schemes considered here include those based both on amperometry and on potentiometry as both kinds of EC systems are being actively developed in CE and have the potential for broad application in analysis. Over the three-year period covered by this review, an important direction that CEEC has taken is the construction of more complex electrode systems beginning with the use of multiple EC electrodes and culminating with the adaptation of EC detection to microfabricated "lab-on-a-chip" analysis devices. In addition, CEEC applications have now grown to include a broad variety of inorganic, organic, and biochemical analytes and samples.

Keywords: Electrochemical detection / Amperometry / Potentiometry / Review   EL 4200

Contents

1 Introduction

Three years ago, in the 1997 review issue of this journal, we provided a summary of the developments that had

---

Correspondence: Professor Richard P. Baldwin, Department of Chemistry, University of Louisville, Louisville, KY 40292, USA
E-mail: rick.baldwin@louisville.edu
Fax: +502-852-8149

Abbreviations: EC, electrochemical; **ISE**, ion-selective electrode

taken place with respect to the use of electrochemical (EC) detection in capillary electrophoresis (CE) instrumentation [1]. This work described developments in both CEEC instrumentation and applications that had occurred from the initial report of CEEC by Ewing and Wallingford in 1987 [2] to research reported as recently as 1996. It is the intent of the present review to provide an update of our earlier work so as to include comparable CEEC developments at least through 1999.

A wide variety of different detection technologies, most commonly optically based, continue to be actively explored for use in CE [3]. As was true earlier, the principal driving force for EC-based techniques in CE remains the need for novel detection methodologies that are adaptable to the 10–50 μm size of the capillaries typically employed in CE and also offer the sensitivity suitable for the femtomole (and lower) analyte levels typically involved. At present, laser-induced fluorescence (LIF) continues to be the instrumental approach that is most widely accepted to meet these needs and certainly plays a dominant role in the high-profile CE applications related to DNA sequencing and analysis. However, many important analytes do not possess native functionalities that are strongly fluorescent and thus can be considered for LIF only after derivatization with suitable fluorophores. More recently, mass spectrometric (MS) detection with electrospray ionization has seen initial application to CE; and, because of the high information content of MS data, this detection technique will undoubtedly play an increasingly important role in the future development of CE.

As an alternative detection strategy, EC offers several unique and attractive features for CE. First and foremost, EC is ideally suited for the small physical dimensions required for CE sytems. Microelectrodes of a few µm in size are already commonly used in numerous EC applications. Not only is the construction of such electrodes rather easily carried out, but also their analytical performance – unlike optical detection methods – is not compromised by miniaturization since the small currents (nano- and picoamps) associated with microelectrodes can be easily and accurately measured and the noise generated at the electrode surface actually decreases with electrode area as fast as or faster than does the EC signal. Second, the associated EC instrumentation is probably less expensive than that for any other form of detection – certainly compared to LIF or MS systems. Finally, because of decades of earlier work with EC sensors (especially in HPLC applications), there already exists a wide variety of well-characterized electrode/analyte systems available for direct adaptation to CE. As will be discussed in detail subsequently, one of the major trends in CE instrumentation is miniaturization down to the "lab-on-a-chip" scale. It seems clear, even at this early stage, that the strengths of EC detection, mentioned above, will be even more attractive for use with these down-sized, microfabricated CE devices.

In general, there are two principal EC detection modes which can be applied to CE systems: amperometric and potentiometric.* Our earlier review focused on the first of these – amperometric – in which a potential is applied to the sensing or working electrode and the faradaic current flowing as a result of analyte oxidation or reduction at this potential is measured. Significant applications of potentiometric EC detection – in which changes in the potential of the sensing electrode are measured under zero current conditions – have also begun to appear and will be included in the discussion that follows. We will begin with a brief synopsis of the specific issues and related terminology in CEEC, will then review the important developments in CEEC instrumentation that have occurred since the earlier review, and will end by providing an overview of current CEEC applications.

2 CEEC instrumentation

From its start, the implementation of EC detection in CE has involved the solution of two unique instrumental prob-

lems: electrical decoupling of the CE and EC electronics and physical alignment of the EC electrode with the capillary exit. The need to isolate or "decouple" the CE and EC systems arises from the fact that the separation voltages used in CE typically fall in the 5–30 kV range and generate µA-level background electrophoresis currents. This is in contrast to typical EC detection potentials of 1 V or less and detection currents on the order of pA's (for amperometric detection). Therefore, for optimum CEEC performance, it is essential that electrical overlap of the two systems is controlled and minimized. The need to achieve close capillary/electrode alignment arises from the fact that the inner diameter (ID) of CE capillaries ranges from 5 to 100 µm while the EC electrodes have typically consisted of carbon fibers or metal wires only 10–50 µm in thickness. Therefore, optimum sensitivity dictates that the sensing electrode is positioned very near the capillary outlet while acceptable reproducibility requires that this positioning, once established, is maintained throughout an entire set of calibration and sample measurements.

2.1 Background

Our earlier review, along with other related reviews on CEEC that have appeared subsequently [4, 5], described in detail the instrumental procedures that have been developed to handle decoupling and alignment difficulties. Accordingly, these procedures do not need to be discussed at length here. Nevertheless, it is important to have a basic understanding of the various decoupling and alignment approaches in use, and especially of the terminology that has evolved to designate each of these. A brief description follows below.

There are two fundamental approaches to CEEC decoupling. In the first, which is termed "off-column" detection, an electrically conducting fracture is created in the capillary a short distance before the exit in order to allow the CE voltage and current to be dropped before reaching the EC electrodes which are placed at or near the physical end of the capillary. Alternatively, in place of such active decoupling, the sensing electrode is simply placed in the CE buffer reservoir at the end of a relatively narrow-bore capillary. The idea is that, as long as the capillary ID is small (*e.g.*, 25 µm or less), its ohmic resistance is high enough that the CE current is negligibly small and nearly all of the CE high voltage is dropped across the capillary. This latter approach has been termed "end-column" detection.

As far as alignment is concerned, several procedures have been developed, with varying degrees of effectiveness, to position the EC electrode and the CE exit in close proximity. These include "in-capillary", "wall-jet", and "on-

---

* Note that conductivity-based detection may be considered to be "electrochemical" in some circles. However, this is true only in the most permissive sense that electrodes are indeed required to measure electrical conductivity. Other than this, there is little electrochemical about the technique as oxidation-reduction plays no real role in the measurements. Accordingly, conductivity detection will not be considered in this work.

capillary" designs. Note that each of these can be used successfully to obtain high-quality analytical results. The most important functional difference is the relative ease with which acceptable alignment is initially achieved and then the constancy of this positioning over time. With the "in-capillary" approaches, the EC electrode is physically inserted into the exit opening of the capillary *via* three-dimensional micropositioners. Clearly, the electrode here must have smaller dimensions than the capillary ID (except for the "optimized" scheme where the capillary is enlarged somewhat *via* chemical etching). With the "wall-jet" approach, the EC electrode, which is usually disk-shaped and can be much larger physically than the usual microfibers or -wires, is simply placed, again with micropositioners, across from the capillary outlet. However, because of the electrode's greater size, good placement is easier to achieve; and slight drifting in its position over time is less critical. In addition, wall-jet procedures can be simplified further by use of very simple support devices to help guide the electrode and capillary together initially and then hold them in place for the necessary sequence of CE experiments. Finally, with the "on-capillary" approach, the capillary and electrode are physically integrated into a single fixed unit. This has been accomplished either by manually gluing the electrode onto the capillary tip or by sputter-coating the capillary tip with a thin film of the electrode material. Clearly, this approach, once achieved, renders alignment issues irrelevant as the integrated CEEC unit must simply be inserted into the terminating electrophoresis buffer reservoir at the start of the CE experiment.

2.2 New developments

Over the past two to three years, no fundamentally new decoupling and alignment strategies have been introduced; and with few exceptions, nearly all of the CEEC work reported has employed the general decoupling and alignment methods outlined above. Some interesting studies have served to clarify the capabilities and limitations of the nondecoupled end-column detection format. For example, Wallenborg *et al.* [6] investigated in detail the effect of CE separation conditions on the observed half-wave potentials with and without decoupler use. In addition, Matysik [7] demonstrated that, contrary to common wisdom, high-quality end-column detection is possible with capillaries as large as 75 µM ID as long as attention is paid to relevant experimental parameters such as proper positioning of the EC electrode. Other useful but very straightforward adaptations of conventional decoupling and alignment approaches will not be considered in detail here. Rather, our discussion will focus on those developments perceived to represent fundamentally new

directions in CEEC. For our purposes, such advances are grouped into four categories: potentiometric detection, multiple electrode detection, use of EC detection in microfabricated CE devices, and miscellaneous developments.

2.2.1 Potentiometric detection

Although amperometry was the EC detection mode initially applied to CE separations and remains the one that has been most frequently employed to date, the alternative potentiometric EC mode has long been the subject of limited study which has increased markedly in the recent past. The first report of potentiometric detection was made in the early 1990s by Simon's group, who demonstrated the detection of alkali and alkaline earth ions [8] and organic amines [9] at liquid membrane ion-selective electrodes (ISEs) operated in the end-column configuration. The ISEs used in the earlier investigations usually consisted of microelectrodes made from drawn-out capillaries or micropipets whose tips were filled with a lipophilic membrane solution containing the active ionophore and whose interiors contained an aqueous filling solution and internal reference electrode. In the current generation of potentiometric CEEC devices, the sensing electrodes are typically miniaturized "coated-wire" electrodes which consist of a metal wire whose exposed surface is covered with a thin polymer film – often poly(vinyl chloride) – containing the ionophore of interest. The obvious advantage of the coated-wire approach is their simpler (no internal reference electrode) and more rugged construction.

To date, potentiometric CEEC systems have been developed for a variety of inorganic and organic species. In addition to simple inorganic cations and anions, these include neurotransmitters [9, 10], alkylamines [11], amino acids [12], fatty acids [13–15], and alkylsulfonates [11, 16]. In all cases, the electrodes used have been based on specific ionophore systems or ISEs that had previously been developed and characterized in more routine potentiometry applications involving bulk solution, flow injection, or liquid chromatography. Thus, there remains a rather wide range of already well-characterized ISE systems for numerous interesting analytes that could easily be adapted for use in CEEC. The operational concentration range for potentiometric detection depends, of course, on the specific ISE system but typically is on the order of 10^{-4}–10^{-6} M. It is important to note that, in CE aplications, the selectivity of such electrodes can be widely variable, determined principally by the selectivity of the ionophore employed. Thus, the response can be nearly universal (*e.g.*, for all cations or anions) if a general cation- or anion-exchanger is used or highly selective if a more selective ionophore is employed. This represents a very different selectivity approach from that of amperometric detection systems where selectivity is determined largely by the value of the applied potential.

2.2.2 Amperometric detection

As the amperometric CEEC detection approach has matured, the most dominant instrumental trend over the past few years has been the incorporation of increasingly complex EC electrode systems. This trend can be seen in two distinct developments: the use of dual electrode detection formats and the incorporation of EC detection into microfabricated CE devices. Each of these developments will be considered in turn.

2.2.2.1 Dual electrode CEEC

The idea of simultaneously employing two (or more) independent sensing electrodes in CEEC is directly related to the dual electrode detection systems that were developed and characterized in the 1980s for EC detection in HPLC [17]. For both separation techniques, the strategies of increasing detection sensitivity and selectivity through the use of multiple electrodes are identical. Of course, the much smaller size of CEEC systems and the resulting problems associated with capillary/electrode alignment make the implementation of such schemes somewhat more difficult in CE than in HPLC. Nevertheless, there has been a significant increase in dual electrode CEEC work; and several clever dual electrode designs have been devised. Examples demonstrating both the "dual-parallel" (*i.e.*, two electrodes operating independently in a side-by-side arrangement) and the "dual-series" (*i.e.*, two electrodes operating sequentially in a one-after-the-other arrangement) configurations have been realized in the CEEC work to date.

The initial reports of dual electrode CEEC, covered in our earlier review, included systems for the detection of thiols and disulfides at dual-series Au/Hg amalgam electrodes [18], NADH and NAD$^+$ at dual-parallel carbon fibers [19], and catechols in the presence of electrochemically irreversible interferents at dual-series carbon fibers [19]. Although these systems were able to carry out their designed functions successfully and certainly demonstrated the viability of the dual electrode format in CEEC, the specific designs employed were relatively complicated to construct and use in practice. Subsequently, recent work carried out by the Susan Lunte group [20] has extended the dual electrode approach with a variety of new on-capillary designs which have improved both their experimental convenience and the analytical performance. In one application, a dual-series system where the

two electrodes consisted of an on-capillary Au tube epoxied to the capillary tip and an Au wire inserted into the tube was constructed for simultaneous thiol/disulfide determination. This arrangement was shown to provide a 200-fold improvement in detection limit for cystine over earlier designs because of the large surface area afforded by the Au tube used to convert cystine to the cysteine for detection at the downstream wire electrode. Another dual-series configuration involving two on-capillary Pt wire electrodes was developed for the detection of phenolic acids [21]. The first Pt wire was epoxied onto the capillary tip, inserted a short distance into the capillary outlet, and maintained at an oxidizing potential; and the second Pt wire, also glued to the capillary tip, was bent across the capillary exit and held at a reducing potential. In this manner, catechols and other sample constituents that undergo chemically reversible oxidations were detected at both electrodes while ascorbate and other irreversibly oxidized species were detected only at the upstream electrode.

In yet another application, Holland and Lunte [22] examined three different dual-series Pt electrode systems (tube/wire, wire/ring, and wire/wire) for use in a postcolumn reaction system. In this scheme, bromine was generated by oxidation of bromide at the upstream electrode and then reduced back to bromide at the downstream electrode. The analytical signal consisted of the decrease in reduction current at the second electrode due to the presence of analytes – in this case, sulfur-containing compounds – which react with the electrogenerated bromine and thereby decrease its concentration from its steady-state level. Finally, Voegel and Baldwin [23] described a very simple and versatile approach to dual-parallel CEEC in which two identical capillaries were each equipped with its own on-capillary electrode by sputter-coating the capillary exit tips with a thin metallic film. When the sample was injected simultaneously into the two capillaries, separation and detection were performed simultaneously at both electrodes. Different responses were obtained by poising the electrodes at different detection potentials or by making the electrodes from two different materials. Although limited to dual-parallel detection only, the flexibility of this approach and the absence of complicated electrode manipulations involved are extremely attractive.

Finally, two dual electrode studies involving microfabricated electrode assemblies have also been reported. Matysik *et al.* [24] used a commercial microband array electrode in a wall-jet configuration to carry out both dual-series and dual-parallel EC detection of model ferrocene compounds. The 1 × 1 mm array, which was obtained from a commercial source, contained four Au bands, each of which was 10 μm wide, 1 mm long, and separated by a 10 μm spacing. Because the bands were individually addressable electronically, all dual electrode detection modes could be demonstrated. These included dual-parallel detection showing enhanced selectivity by operation at two different oxidizing potentials and dual-series detection showing increased sensitivity due to both redox cycling and generation-collection mechanisms at adjacent electrodes maintained at alternate oxidizing and reducing potentials. Niwa *et al.* [25] employed conventional photolithographic techniques to prepare a 64×66 μm microarray consisting of eight pairs of 2 μm wide bands separated by a 2 μm gap. Alternating bands were connected electrically to form two dual-series interdigitated electrodes. When installed in a micromachined cell holder which served to orient the capillary and the microarray in a wall-jet mode, enhanced detection sensitivity was obtained for dopamine by redox cycling between the two sets of finger electrodes.

2.2.2.2 Microfabricated CEEC devices

Certainly the most dramatic development in CE instrumentation in the 1990s has been the use of fabrication technologies common to the microelectronics industry to create miniaturized "lab-on-a-chip" CE instruments [26–28]. Despite their greatly reduced size, these devices are able to provide highly efficient separations comparable to bench-scale CE instruments and at the same time integrate the CE process with a full range of sample handling and analysis operations to produce great advantages in terms of speed, cost, portability, and reagent/sample consumption. Not surprisingly, functional on-chip EC detection systems have started to appear in the past few years.

Although EC is not yet ready to rival LIF as the detection method of choice for microchip devices, it does offer several unique advantages. First and foremost, EC, especially in the amperometric mode, is among the most sensitive of the CE detection techniques. Second, because EC electrodes can be incorporated directly onto the chip along with the CE separation/injection channels without substantial ancillary hardware, this form of detection represents what is certainly the closest yet to a truly microsized, on-chip detection scheme. Third, EC detection is extremely compatible with mainstream microfabrication technologies as electrodes can be patterned onto a chip device by essentially the same photolithography/sputter-coating methods used to construct the chips themselves [29]. Thus, inclusion of EC electrodes can be viewed as simply an additional step in the microfabrication process, and the resulting CEEC packages should be well-suited for mass production. Finally, with conventional microfabri-

cation procedures, the resulting microelectrodes can be prepared in virtually any shape, size, number, or chip location needed to best suit the particular CE application.

EC detection on the microchip scale was first reported in 1996 by Mathies' group [30] who constructed a microchip device containing a 10-μm-wide Pt working electrode patterned within 30 μm of the end of the CE channel and a larger Pt counter electrode located farther out in the buffer reservoir. The electrodes, along with the CE channels, were photolitographically patterned onto a glass microscope slide; and the Pt electrodes were formed by plasma sputtering onto a Ti adhesion layer. The device was used to detect catechol neurotransmitters at the 10 μM level by direct oxidation and DNA digestion fragments and PCR products indirectly. Very recently, Wang and co-workers [31–34] have described two different lab-on-a-chip CEEC devices. In the first, a microfabricated CE chip was constructed so that the CE separation channel extended all the way to the edge of the chip; and a removable 33×10 mm strip containing a small screen-printed carbon line working electrode was mounted onto the chip in a wall-jet arrangement. Consequently, the CE effluent flowed off the chip and impinged directly onto the carbon electrode. In addition to the usual catechols, nitroaromatic explosives [31], glucose [32], and phenolic pollutants [33] were determined *via* this approach. In a slightly different approach, the EC working electrode was prepared by sputter-coating an Au film onto the outside of the chip where the CE channel exited [34]. In this manner, the sensing electrode was incorporated directly onto the chip in a format analogous to the on-capillary system previously developed by Voegel *et al.* [23, 35] for CEEC with conventional capillaries. Although reference and counterelectrodes were still located separately off-chip, this approach was certainly very simple experimentally and potentially very versatile for different electrode materials. Finally several microchip CEEC papers have been presented at recent conferences but have not yet reached the printed literature. These indicate that developments involving the use of on-chip Cu electrodes for DNA detection [36] and carbon paste electrodes for catechols [37] will soon be reported.

In addition to the above examples illustrating amperometric CEEC on a microchip scale, one example involving potentiometric detection has also appeared. Tantra and Manz [38] recently described the construction of a chip-based flow cell containing a microfabricated sensor for Ba^{2+} using a polymeric membrane-ionophore system adapted from a conventional scale Ba^{2+} ISE. Response down to the μM level was observed. However, the response time of the microfabricated electrode was rather slow; and no actual CE separations were attempted.

2.2.3 Miscellaneous developments

Three final developments that are of interest but do not easily fit into the above categories are reported here. In view of the emphasis on miniaturized CEEC analysis platforms highlighted above, it is appropriate that some work has been directed toward the reduction in size of the rest of the CE instrument. Most notably, Hauser's group [39] has reported the construction and characterization of a portable CEEC instrument suitable for possible field applications. In this design, the bench-top high voltage power supply usually associated with CE was replaced by commercial high-voltage modules powered by two rechargeable 12 V lead-acid batteries. In addition, custom control circuitry to manage both amperometric and potentiometric detection was also included to bring the total instrument to a weight of only 7.5 kg. The use of the device was illustrated by the potentiometric detection of alkali and alkaline earth cations at a coated wire ISE, the amperometric detection of nitrite and thiocyanate at an Au electrode, and the joint amperometric/potentiometric detection of amino acids at a Cu electrode.

Finally, Gerhard *et al.* [40, 41] have described two interesting innovations with respect to amperometric detection strategy. In the first, EC detection was performed with potential scanning so as to provide voltammetric (*i.e.*, current *vs.* potential) characterization of analyte species in addition to the usual electrophoretic (*i.e.*, current *vs.* time) information afforded by CE [40]. The technique was applied to the model catechols dopamine and epinephrine, and high-frequency square-wave voltammetry was developed in order to insure adequate time and potential resolution. An added benefit of the approach was that very attractive sensitivities were obtained for these analytes in the end-column configuration without the use of a decoupling device. In addition, the square-wave methodology was also used as the basis of a new EC detection method targeting traditionally nonelectroactive organic analytes [41]. In this approach, currents associated with non-faradaic interfacial processes due to analyte adsorption at a Pt microelectrode were used to provide the analytical signal. Detection sensitivities obtained by this mechanism for amino acids, sulfonamide antibiotics, and other compounds of forensic interest varied with the nature of the particular analyte – with compounds possessing significant π-electron density yielding the best response (down to approximately the 10^{-8} M level in the most favorable cases). However, in nearly all instances examined, the resulting sensitivities were up to an order of magnitude improved over what could be obtained by UV absorption. As one of the primary limitations of amperometric EC detection is that it works well only for analytes that are easily oxidized or reduced, the possibil-

ity of making EC more universal in applicability certainly merits attention.

3 CEEC applications

In the past reviews of CEEC, applications of the technique have tended to focus on the same analyte groups that have been previously targeted by HPLC-EC – namely, catechols, thiols, carbohydrates, and amino acids [1, 4]. In general, these analytes are not amenable to direct detection by other approaches (because they do not possess chromophores that absorb strongly at visible or near-visible wavelengths) but can be made to undergo oxidation at some electrode at relatively modest applied potential. The current review period has seen a considerable broadening of this focus although these analytes do still account for a major share of the CEEC applications reported.

The discussion below is organized into three sections: potentiometric applications, amperometric applications, and microfabricated device applications. With the specific intention of emphasizing and encouraging novel EC applications, we have not tried to be all-inclusive, especially in the more established amperometric area. Rather, we have tried to identify reported applications that are new and different or seem particularly challenging in nature. As always with EC detection, a critical issue is the selection of electrode material and configuration, which are specified as succinctly as possible in the accompanying tables. It is to be expected that the development of new and more effective electrodes will continue to form the basis of new CEEC applications.

3.1 Potentiometric applications

Potentiometric CEEC applications that have occurred during the review period are listed in Table 1. All applications shown employed either coated wires or metallic Cu as the sensing electrode, largely due to the simple structure and ruggedness of these devices. Of course, for the coated wire ISEs, the potentiometric response is determined by the properties of the particular ionophore incorporated into the coating, which in the studies reported to date has consisted of general-purpose, nonspecific ion-exchangers. Consequently, the detection provided was relatively nonspecific as well, generally allowing broad cation or anion detection. Similarly, the Cu electrode also is rather broad in its response selectivity, giving signals for any analyte that can complex ionic Cu and thereby change the Nernstian potential of the Cu surface. Examples of such analytes include not only numerous common anions [16] but also organic ligands such as amines and amino acids [12].

Table 1. Potentiometric CEEC applications

Analyte	Ref.	Electrode	Alignment	Decoupling	Notes
Amino acids	[12]	Cu disk	Wall-jet	End-column	Analytes included nine underivatized amino acids
Alkali and alkaline earth cations	[41]	Coated wire ISE	Wall-jet with cell holder	End-column	Analytes included nine Group 1 and 2 cations
Amines and sulfonic and carboxylic acids	[11]	Coated wire ISEs	Microelectrode with micropositioner	End-column	Analytes included anionic analgesics in pharmaceutical preparations and artificial sweeteners in soft drinks
Alkali and alkaline earth cations, NO_2^-, SCN^-, amino acids	[39]	Coated wire ISE, Au, Cu	Wall-jet with cell holder	End-column	CEEC was carried out with a field-portable instrument powered by two Pb-acid batteries
Fatty acids	[14]	Coated wire ISE	Wall-jet	End-column	Analytes included C_1 to C_{10} n-alkyl-carboxylates
Carboxylic acids and inorganic anions	[15]	Coated wire ISE	Wall-jet with cell holder	End-column	Analytes included C_1-C_6 n-alkyl-carboxylates, malic, lactic, and gluconic acids, and SO_3^{2-}, PO_4^{3-}, and CO_3^{2-}
Inorganic anions and alkylsulfonates	[16]	Cu disk	Microelectrode with cell holder	End-column	Analytes included 14 common anions and C_1-C_8 n-alkylsulfonates
Acetylcholine	[10]	Coated wire ISE, graphite	Wall-jet with cell holder	End-column	Compared potentiometric and amperometric CEEC of neuro-transmitters

To date, the potentiometric electrodes used in CE applications represent only a relatively small sampling of the great variety of ISEs that have been developed for stand-alone applications. As befitting a detection technique used in conjunction with a powerful separation technique such as CE, the relatively nonselective ISEs that have been used so far probably constitute good choices with which to start. However, it is clear that opportunities exist for inclusion of a much wider variety of potentiometric electrodes in CEEC systems and their application to a host of important analytical problems.

3.2 Amperometric applications

Amperometric CEEC applications that have occurred during the review period are listed in Table 2. Comparison of this table to the analogous one in our earlier review makes it apparent that the range of analytes addressed by CEEC has increased substantially over the past 2–3 years. Although catechols, thiols, and carbohydrates continue to be important CEEC targets, also featured prominently in the compilation are nucleic acid components, peptides and proteins, and a wide variety of specific organic and inorganic species. No attempt will be made here to describe each and every one of these which has been cited. Rather, we will mention a few examples which

we believe may represent larger CEEC trends that may have particular significance in the near future.

Certainly the application of amperometric CEEC systems for the determination of purine and pyrimidine compounds merits attention in view of potential implications for DNA sequencing and analysis. Thus far, work has been reported in this area mostly for the nucleic acid bases adenine and guanine (which can be oxidized at modest potentials) and their nucleotides and nucleosides [52–54]. A particularly impressive example is the assay developed by Weiss and Lunte [54] for 8-hydroxydeoxyguanosine, a urinary biomarker for oxidative DNA damage associated with cancer and other age-related diseases. This assay coupled the separation capability of CE and the selectivity of EC detection with a relatively simple solid-phase extraction for sample clean-up and preconcentration in order to enable the determination of this compound in urine samples at the nM level. Whether CEEC methods can be successfully scaled up to permit the determination of the DNA lengths required for identification, screening, and sequencing applications remains to be determined. However, a variety of EC approaches such as the use of Cu and other electrode materials and the electrochemical labeling of DNA are under active investigation and hold promise for the future.

Table 2. Amperometric CEEC applications

Analyte	Ref.	Electrode	Alignment	Decoupling	Notes
Catechols	[42]	C fiber	Optimized in-capillary	End-column	Cell extracts were analyzed for catecholamine neurotransmitters and other easily oxidized species; detection at the zeptomole level allowed study of related processes for single cells.
	[40]	Au disk	Microelectrode with cell holder	End-column	Detection of dopamine and epinephrine was performed by high-frequency square-wave voltammetry; this provided both high sensitivity and added current-voltage information.
	[21]	Dual-series Pt wires	On-capillary	Off-column	Dual-series detection was demonstrated for catecholamines and phenolic acids; ferulic acid was detected in beer.
	[43]	C disk	Wall-jet	End-column	Catecholamines and other monoamine neurotransmitters and metabolites were detected in caudate nucleus of rat; samples were obtained by microdialysis.
	[25]	Dual-series interdigitated Au microarray	Wall-jet with cell holder	Off-column	Enhancement of dopamine response was demonstrated by redox cycling at a dual-series microarray electrode.
Carbohydrates	[44]	Pt disk coated with glucose oxidase and amyloglucosidase	Wall-jet with cell holder	Off-column	Bienzyme layer served to hydrolyze α-glucans to glucose and then oxidize glucose to produce H_2O_2 for detection at the Pt electrode.
	[45]	Au tube	On-capillary	Off-column	Pulsed amperometric detection was used to determine glucose in human serum; the larger area of the tubular electrode produced improved sensitivity.
	[23]	Pt with adsorbed glucose oxidase	On-capillary	End-column	Glucose was detected at one of two dual capillary/dual-parallel electrodes.
	[46]	Cu disk	Wall-jet	End-column	α-, β, and γ-cyclodextrins were separated and detected.
	[47]	Cu disk	Wall-jet	End-column	Simple sugars, and tartaric and ascorbic acids were separated and determined in grapes and grape juice.
Thiols and disulfides	[48, 49]	Au/Hg-wire	Wall-jet	End-column	Cysteine [48] and glutathione [49] were determined in human plasma, blood, and urine.
	[20]	Dual-series Au/Hg tube and wire	On-capillary (tube)/in-capillary (wire)	Off-column	Cysteine, glutathione, and reduced glutathione were detected simultaneously; a tryptic digest of ribonuclease A was analyzed for disulfides.
	[22]	Dual-series Au wires	On-capillary	Off-column	Dual-series electrodes were used for postcolumn generation of Br_2 and indirect detection of Br_2-reactive analytes; glutathione, cysteine, and methionine were detected here.
	[50]	C disk	Wall-jet	End-column	Cysteine, glutathione, 6-thiopurine, and methimazole were detected at an unmodified C electrode.
	[51]	Au/Hg wire	Wall-jet	End-column	Single erythrocyte cells were injected and lysed in the capillary; the resulting glutathione was determined by EC.

Table 2. continued

Analyte	Ref.	Electrode	Alignment	Decoupling	Notes
Purines and pyrimidines	[52]	Cu disk	Wall-jet	End-column	Adenine and guanine, along with their nucleotides and nucleosides, were separated and detected in human plasma.
	[53]	C fiber microdisk array	Wall-jet	End-column	Adenine and guanine were detected after hydrolysis of yeast RNA and calf thymus DNA.
	[54]	C fiber	In-capillary	Off-column	8-Hydroxydeoxyguanosine, a biomarker for DNA damage, was determined in human urine after solid-phase extraction.
Petides and proteins	[55]	C fiber microdisk array	Wall-jet	End-column	Bovine serum albumin was detected in a serum sample.
	[56]	Au microdisk array coated with a layer of L-cysteine	Wall-jet	End-column	Cytochrome *c* was detected in a pig heart extract.
	[57]	C fiber	In-capillary	Off-column	Biuret complexes of vasopressin, neurotensin and other neuropeptides were characterized by CEEC prior to analysis in rat hypothalamus dialysates by capillary LCEC.
	[58]	C fiber microdisk array	Wall-jet	End-column	Myoglobin was detected in human urine.
Miscellaneous organics	[59]	C fiber	In-capillary	Off-column	Transdermal nicotine delivery to a rat was monitored by coupling *in vivo* microdialysis sampling and CEEC.
	[60]	Pt disk	Wall-jet with cell holder	Off-column	Nicotine in tobacco reference standards was determined by nonaqueous CE.
	[61]	C fiber	Wall-jet	End-column	Aesculin and aesculetin, phenolic Chinese medicinal constituents, were detected in spiked urine and ash brak.
	[62, 63]	Au or Au/Ag disk	Wall-jet	End-column	13 Nitroaromatic and nitramine explosives were separated by micellar electrokinetic chromatography [62] and capillary electrochromatography [63] and detected by electroreduction; samples included contaminated soil and groundwater.
	[64]	Au wire	Wall-jet	End-column	Metronidazole, a nitroimidazole, was determined in a spiked urine sample by electroreduction.
	[65]	C fiber microdisk array	Wall-jet with cell holder	End-column	Clozapine, a diazepine used for treatment of psychotics, was determined in spiked human blood.
	[66]	C fiber microdisk array	Wall-jet with cell holder	End-column	Promethazine, a phenothiazine used as an antihistamine, was determined in spiked human blood serum.
	[67]	C fiber microdisk array	Wall-jet	End-column	Reserpine was determined in spiked human serum.
Inorganics	[68]	Au disk	Wall-jet	End-column	Hg^{2+} and CH_3Hg^+ were detected at the ng/mL level in sediment samples.
	[69]	C fiber	Wall-jet	End-column	ClO_2^- was determined in drinking water by electro-oxidation.

Table 2. continued

Analyte	Ref.	Electrode	Alignment	Decoupling	Notes
	[70]	C coated with a mixed valence RuFeCN film	Wall-jet	End-column	Cs^+ (and also Li^+ and Na^+) were detected indirectly in spiked leukocyte culture media.
	[71]	Pt disk coated with a self-assembled 4-pyridyl hydro-quinone film	Wall-jet	End-column	N_2H_4, $N_2H_3(CH_3)$, and isoniazid were detected electrocatalytically at the sub-μM level.
	[72]	Au/Hg wire	Wall-jet	End-column	Tl^+ was detected by electroreduction; with deoxygenation, detection was possible at the μM level.

Many of the other novel CEEC applications also have targeted biochemical analysis problems. These include direct CEEC-based assays of drugs of varying chemical structure – *e.g.*, methimazole [50], nicotine [59, 60], clozapine [65], promethazine [66], reserpine [67], and isoniazid [71]. In addition, CEEC procedures have been paired in increasingly complex assay schemes that provide unique solutions to practical bioanalysis problems. Particularly nice examples here include the determination of catecholamines at the zeptomole levels required for single cell analysis [42], the integration of CEEC with microdialysis sampling to study bioprocesses such as neurotransmitter release [43] and transdermal nicotine delivery [59], and the development of a straightforward on-capillary reaction procedure to enhance the EC activity of peptides [57].

The above discussion is not intended to imply that all of the interesting CEEC activity has been directed toward things biological. Rather, some impressive applications of CEEC have taken place in the areas of organic and inorganic analysis as well. Hilmi *et al.* [62, 63] have demonstrated that reductive EC detection can be successfully coupled with micellar electrokinetic chromatography and capillary electrochromatography for the analysis of nitroaromatic and nitramine explosives in contaminated soil and groundwater. And Lai *et al.* [68] have applied CEEC to the simultaneous determination of Hg^{2+} and CH_3Hg^+ in sediment samples. In view of the great number of electroactive heavy metals, the demonstration of such speciation capabilities *via* CEEC certainly holds promise for environmental analysis.

3.3 CEEC applications in microfabricated analysis devices

Applications of CEEC in microfabricated lab-on-a-chip devices that have occurred during the review period are listed in Table 3. Both the potential strengths of EC detection for this instrumental regime and each of the specific applications that are included have already been discussed in an earlier section. Therefore, it is not necessary to comment in detail on these here. Clearly, the compatibility of electrode design and construction with standard microfabrication methods and the degree to which the EC detection elements may be integrated into the microchip itself are unique among the various CE detection options currently available. Furthermore, as microfabricated devices become more complex in operation, it is important to appreciate that the complexity of an associated EC detection system can be increased as needed with relatively little additional difficulty. For example, with photolithography, there is little difference in the time, cost, or fabrication effort required for construction of a "simple" EC scheme with a single sensing electrode compared to that needed for a "complex" EC setup containing multiple electrodes distributed at various locations on the chip surface. Furthermore, it seems likely that the difficulties experienced in conventional bench-scale CEEC with capillary/electrode alignment and CE/EC decoupling may well be rendered inconsequential with microfabricated systems where electrode placement is fixed precisely and many as yet unexplored approaches to performing effective on-chip decoupling can easily be envisioned.

4 Conclusions

From the above discussion, it seems obvious that EC detection techniques have a great deal to offer CE. Important analytical applications for which EC detection is highly suited abound, and CEEC instrumentation has evolved to the point that the experimenter no longer needs to be specifically trained in electrochemistry. Nevertheless, the actual use of the technique, while growing steadily, must still be considered to be rather modest. One of the reasons for this is undoubtedly that an intact CEEC instrument is still not commercially available.

Table 3. CEEC applications in microfabricated analysis systems

Analyte	Ref.	EC mode	Electrode	Fabrication	Notes
Neutrotransmitters, DNA	[30]	Amperometric	Pt strip, 10 µm wide and 30 µm from CE channel	Photolithography for patterning CE channels and working and counterelectrodes, plasma sputtering of Pt onto CE chip	Catechols used as model system for direct EC detection; restriction digest and PCR products separated by CE and detected by indirect EC
Neurotransmitters	[34]	Amperometric	Au film	Sputtering onto side of chip surrounding CE exit	Catechols used as model system
Neurotransmitters, explosives	[31]	Amperometric	C line, 300 µm wide	Screen-printing of working electrode which was positioned at side of chip across from CE exit	Catechols and nitrobenzenes used as model systems
Glucose, uric acid, ascorbic acid, and acetaminophen	[32]	Amperometric	Au film on 300 µm wide C-line	Screen printing of C electrode with electrodeposition of Au; electrode was positioned at side of chip across from CE exit	Glucose detected as H_2O_2 after addition of glucose oxidase to CE run buffer
Ba^{2+}	[38]	Potentiometric	PVC polymer membrane containing Vogtle ionophore	Photolithography and etching used to construct CE and ISE channels	On-chip Ba ISE functioned similarly to conventional ISE; no CE carried out
Phenolic compounds	[33]	Amperometric	Au film on 300 µm wide C line	Screen printing of C electrode with electrodeposition of Au; electrode was positioned at side of chip across from CE exit	Phenol and chlorophenols determined in river water

Rather, CEEC equipment is nearly always assembled in-house from free-standing high-voltage power supplies, potentiostats, and other components. While this should certainly not be considered to be a daunting task, it does serve to make CEEC much harder to access than most of the competing CE detection methods.

It is expected that, in the short term, continued incremental development of CEEC instrumentation and applications should continue to occur. This growth will include the adaptation of new electrode materials and electrode configurations to CEEC systems and the further increase in the number and variety of analytes and samples studied. As in recent years, interest in CEEC will continue to be driven predominantly by biotechnological needs and advances although other interesting applications certainly exist. Without question, incorporation of EC detection schemes onto lab-on-a-chip platforms is an area that should experience intense research and development activity over the next few years. Success in this endeavor would provide the most likely solution to the unavailability of off-the-shelf CEEC instruments noted above and, at the same time, highlight to full advantage the special strengths of EC detection.

This work was supported by a grant from the U.S. National Science Foundation XYZ-on-a-chip program.

Received September 6, 2000

5 References

[1] Voegel, P. D., Baldwin, R. P., *Electrophoresis* 1997, *18*, 2267–2278.

[2] Wallingford, R. A., Ewing, A. G., *Anal. Chem.* 1987, *59*, 1762–1766.

[3] Swinney, K., Bornhop, D. J., *Electrophoresis* 2000, *21*, 1239–1250.

[4] Holland, L. A., Lunte, S. M., *Anal. Commun.* 1998, *35*, 1H–4H.

[5] You, T., Yang, X, Wang, E., *Electroanalysis* 1999, *11*, 459–464.

[6] Wallenborg, S. R., Nyholm, L., Lunte, C. E., *Anal. Chem.* 1999, *71*, 544–549.

[7] Matysik, F.-M., *Anal. Chem.* 2000, *72*, 2581–2586.

[8] Haber, C., Silvestri, I., Roosli, S., Simon, W., *Chimia* 1991, *45*, 117–121.

[9] Nann, A., Silvestri, I., Simon, W., *Anal. Chem.* 1993, *65*, 1662–1667.

[10] Kappes, T., Schnierle, P., Hauser, P. C., *Electrophoresis* 2000, *21*, 1390–1394.

[11] Schnierle, P., Kappes, T., Hauser, P. C., *Anal. Chem.* 1998, *70*, 3585–3589.

[12] Kappes, T., Hauser, P. C., *Anal. Chim. Acta* 1997, *354*, 129–134.

[13] De Becker, B. L., Nagels, L. J., *Anal. Chem.* 1996, *68*, 4441–4445.

[14] Poels, I., Nagels, L. J., *Anal. Chim. Acta* 1999, *385*, 417–422.

[15] Poels, I., Nagels, L. J., *Anal. Chim. Acta* 1999, *401*, 21–27.

[16] Macka, M., Gerhardt, G., Andersson, P., Bogan, D., Cassidy, R. M, Haddad, P. R., *Electrophoresis* 1999, *20*, 2539–2546.

[17] Roston, D. A., Shoup, R. E., Kissinger, P. T., *Anal. Chem.* 1982, *54*, 1417A–1434A.

[18] Lin, B. L., Colon, L. A., Zare, R. M., *J. Chromatogr. A* 1994, *680*, 263–270.

[19] Zhong, M., Zhou, J., Lunte, S. M., Zhao, G., Giolando, D. M., Kirchoff, J. R., *Anal. Chem.* 1996, *68*, 203–207.

[20] Zhong, M., Lunte, S. M., *Anal. Chem.* 1999, *71*, 251–255.

[21] Holland, L. A., Harmony, N. M., Lunte, S. M., *Electroanalysis* 1999, *11*, 327–330.

[22] Holland, L. A., Lunte, S. M., *Anal. Chem.* 1999, *71*, 407–412.

[23] Voegel, P. D., Baldwin, R. P., *Electrophoresis* 1998, *19*, 2226–2232.

[24] Matysik, F.-M., Bjorefors, F., Nyholm, L., *Anal. Chim. Acta* 1999, *385*, 409–415.

[25] Niwa, O., Kurita, R., Liu, Z., Horiuchi, T., Torimitsu, K., *Anal. Chem.* 2000, *72*, 949–955.

[26] Manz, A., Harrison, D. J., Verpoorte, E. M. J., Fettinger, J. C., Paulus, A., Ludi, H., Widmer, H. M., *J. Chromatogr.* 1992, *593*, 253–258.

[27] Harrison, D. J., Fluri, K., Seiler, K., Fan, Z. H., Effenhauser, C. S., Manz, A., *Science* 1993, *261*, 895–897.

[28] Figeys, D., *Anal. Chem.* 2000, *72*, 330A–335A.

[29] Lindner, E., Buck, R. P., *Anal. Chem.* 2000, *72*, 336A–345A.

[30] Wooley, A. T., Lao, K., Glazer, A. N., Mathies, R. A., *Anal. Chem.* 1996, *70*, 684–688.

[31] Wang, J., Tian, B., Sahlin, E., *Anal. Chem.* 1999, *71*, 5436–5440.

[32] Wang, J., Chatrathi, M. P., Tian, B., Polsky, R., *Anal. Chem.* 2000, *72*, 2514–2518.

[33] Wang, J., Chatrathi, M. P., Tian, B., *Anal. Chim. Acta* 2000, *416*, 9–14.

[34] Wang, J., Tian, B., Sahlin, E., *Anal. Chem.* 1999, *71*, 3901–3904.

[35] Voegel, P. D., Zhou, W., Baldwin, R. P., *Anal. Chem.* 1997, *69*, 951–957.

[36] Brazill, S., Roy, B., Kuhr, W. G., *Pittsburgh Conference on Analytical Chemistry and Applied Spectroscopy*, New Orleans, LA, March, 2000, Abstract #866.

[37] Gawron, A. J., Martin, R. S., Henry, C., Lunte, S. M., *Pittsburgh Conference on Analytical Chemistry and Applied Spectroscopy*, New Orleans, LA, March, 2000, Abstract #869.

[38] Tantra, R., Manz, A., *Anal. Chem.* 2000, *72*, 2875–2878.

[39] Kappes, T., Schnierle, P., Hauser, P. C., *Anal. Chim. Acta* 1999, *393*, 77–82.

[40] Gerhardt, G. C., Cassidy, R. M., Baranski, A. S., *Anal. Chem.* 1998, *70*, 2167–2173.

[41] Gerhardt, G. C., Cassidy, R. M., Baranski, A. S., *Anal. Chem.* 2000, *72*, 908–915.

[42] Bergquist, J., Josefsson, E., Tarkowski, A., Ekman, R., Ewing, A., *Electrophoresis* 1997, *18*, 1760–1766.

[43] Jin, W., Jin, L., Shi, G., Ye, J., *Anal. Chim. Acta* 1999, *382*, 33–37.

[44] Wei, H., Wang, T., Li, S. F. Y., *Electrophoresis* 1997, *18*, 2024–2029.

[45] Zhong, M., Lunte, S. M., *Anal. Commun.* 1998, *35*, 209–212.

[46] Fang, Y., Gong, F., Fang, X., Fu, C., *Anal. Chim. Acta* 1998, *369*, 39–45.

[47] Fu, C., Song, L., Fang, Y., *Anal. Chim. Acta* 1998, *371*, 81–87.

[48] Jin, W., Wang, Y., *J. Chromatogr. A* 1997, *769*, 307–314.

[49] Jin, W., Wang, Y., *Anal. Chim. Acta* 1997, *343*, 231–239.

[50] Wang, A., Zhang, L., Zhang, S., Fang, Y., *J. Pharm. Biomed. Anal.* 2000, *23*, 429–436.

[51] Jin, W., Li, W., Xu, Q., *Electrophoresis* 2000, *21*, 774–779.

[52] Lin, H., Xu, D.-K., Chen, H.-Y., *J. Chromatogr. A* 1997, *760*, 227–233.

[53] Jin, W., Wei, H., Zhao, X., *Electroanalysis* 1997, *9*, 770–774.

[54] Weiss, D. J., Lunte, C. E., *Electrophoresis* 2000, *21*, 2080–2085.

[55] Jin, W., Weng, Q., Wu, J., *Anal. Chim. Acta* 1997, *342*, 67–74.

[56] Jin, W., Weng, Q., Wu, J., *Anal. Lett.* 1997, *30*, 753–769.

[57] Shen, H., Witowski, S. R., Boyd, B. W., Kennedy, R. T., *Anal. Chem.* 1999, *71*, 987–994.

[58] Jin, W., Dong, Q., Yu, D., Ye, X., *Electrophoresis* 2000, *21*, 1535–1539.

[59] Zhou, J., Heckert, D. M., Zuo, H., Lunte, C. E., Lunte, S. M., *Anal. Chim. Acta* 1999, *379*, 307–317.

[60] Matysik, F.-M., *J. Chromatogr. A* 1999, *853*, 27–34.

[61] You, T., Yang, X., Wang, E., *Anal. Chim. Acta* 1999, *401*, 29–34.

[62] Hilmi, A., Luong, J. H. T., Nguyen, A.-L., *Anal. Chem.* 1999, *71*, 873–878.

[63] Hilmi, A., Luong, J. H. T., *Electrophoresis* 2000, *21*, 1395–1404.

[64] Jin, W., Li, W., Xu, Q., Dong, Q., *Electrophoresis* 2000, *21*, 1409–1414.

[65] Jin, W., Xu, Q., Li, W., *Electrophoresis* 2000, *21*, 1415–1420.

[66] Jin, W., Xu, Q., Li, W., *Electrophoresis* 2000, *21*, 1527–1534.

[67] Jin, W., Ye, X., Yu, D., Dong, Q., *Anal. Chim. Acta* 2000, *408*, 257–262.

[68] Lai, E. P. C., Zhang, W., Trier, X., Georgi, A., Kowalski, S., Kennedy, S., MdMuslim, T., Dabek-Zlotorzynska, E., *Anal. Chim. Acta* 1998, *364*, 63–74.

[69] Wallenborg, S. R., Dorholt, S. M., Faibushevich, A., Lunte, C. E., *Electroanalysis* 1999, *11*, 362–366.

[70] Fu, C., Wang, L., Fang, Y., *Anal. Chim. Acta* 1999, *391*, 29–34.

[71] You, T., Niu, L., Gui, J. Y., Dong, S., Wang, E., *J. Pharm. Biomed. Anal.* 1999, *19*, 231–237.

[72] Jin, W., Dong, Q., Yu, D., Ye, X., Li, W., *Electrophoresis* 2000, *21*, 1540–1544.

Electrophoresis 2000, *21*, 1281–1290

Review

Anbao Wang
Yuzhi Fang

Department of Chemistry,
East China Normal
University, Shanghai, China

Applications of capillary electrophoresis with electrochemical detection in pharmaceutical and biomedical analyses

As a high efficiency separation technique, capillary electrophoresis has been widely used in various fields of analytical science. This review discusses the applications of electrochemical detection systems combined with capillary electrophoresis in pharmaceutical and biomedical analysis. These detection methods mainly involve amperometric detection but also include conductivity detection and potentiometric detection. Its applications in the field are divided into six parts, including catechol compounds, thiols, amino acids and peptides, carbohydrates, general pharmaceuticals, and other related compounds. A relatively detailed discussion is described for each compound under the current studied. On this basis, we have suggested several conceivable directions for capillary electrophoresis with electrochemical detection in the future.

Keywords: Capillary electrophoresis / Electrochemical detection / Pharmaceutical analysis / Biomedical analysis / Review
EL 3884

Contents

1 Introduction

1.1 Background

With the increase in longevity of human beings and the demands on medical treatment and health protection, more and more emphasis is being placed on analytical methods. These revolve around chemical properties for many compounds, *i.e.*, purity, quantitative assays, separation of chiral compound, inorganic ion content, election of new pharmaceutical candidate, and determination of the active component, whereby the analysis and control of the active ingredients and impurities are of great importance. Impurities may result from either synthetic or degradative sources and may therefore have widely differing structures and/or polarities. Impurity levels may be less than 1% of the main ingredient, which necessitates a detection system with a suitable linear range. In the development of new drugs, the analytical objectives are to collect information, to design methods for checking for the presence of contaminants, and to draw up guidelines for quality maintenance and control. Obviously, the total amount of impurities, as well as the quantity of each individual impurity in the starting material, is limited. Therefore, the development of methods to indicate the purity and stability is a necessary and often difficult task. Certainly, it is also important to resolve other problems such as separation of enantiomers and determination of some endogenous compounds. In addition, the need to analyze some basic compounds biochemically brings analytical techniques into broader application areas.

In the past decades, gas chromatography (GC), thin-layer chromatography (TLC), and high performance liquid chromatography (HPLC) were used in the majority of cases to analyze pharmaceutically and biochemically active components. Capillary electrophoresis (CE) has advanced tremendously since Jorgenson and Lukacs' pioneering

Correspondence: Dr. Yuzhi Fang, Department of Chemistry, East China Normal University, Shanghai, 200062, China
E-mail: yuzhi@online.sh.cn
Fax: +86-21-62451921

Abbreviations: ED, electrochemical detection; **PAD**, pulsed amperometric detection

work [1] in 1981; it has gradually been developed to perform these analytical tests and has already been shown to be an attractive alternative to the established techniques [2–7]. In CE, the separation of mixtures is driven by the combination of two forces, the electrophoretic velocity of the analyte and the electroosmotic flow generated by the ionized silanol groups present in the fused-silica capillary. The former is the movement of analyte under the electric field and is a combination of frictional and electrical forces. The latter is due to the bulk flow of charge in the background electrolyte, which results from the double layer produced on the inner surface of the capillary in the presence of the electric field. CE is primarily capable of providing extremely high separation efficiencies in short times and to do so with relatively simple instrumentation. Together with various separation modes, *e.g.*, capillary zone electrophoresis, micellar electrokinetic capillary chromatography, gel capillary electrophoresis *etc.*, CE can be used for the separation of anions, cations, neutral molecules [8, 9] or even optical isomers [10, 11]. Additionally, capillary electrophoresis is particularly suited to low-volume sampling because the total volume of the separation capillary is of the order of a microliter. This advantage made it successful in the use for analysis of single cells and microdialysis samples [12]. The small capillary volume and low flow rates in CE are highly desirable from economic and environmental standpoints as exotic or expensive electrophoresis solutions can be employed at lower cost and with fewer disposal problems than in HPLC.

1.2 Electrochemical detection

Although CE has shown many merits since being introduced into the analytical field, some limitations in the area of detection systems still exist. Currently, several detection techniques, such as UV-visible absorption, laser-induced fluorescence, mass spectroscopy, and electrochemical detection, have been used in combination with CE. Among these detection methods, the most widely used detection systems are based on UV-visible absorption [8, 10]. However, the detection approach is restricted in its sensitivity in CE by the extemely short optical path-lengths afforded by the small inside diameter of the capillaries. In addition, its disadvantages still consist of the absence of chromophoric groups for some compounds such as carbohydrates. A laser-induced fluorescence detector [13] provides the required sensitivity for narrow-bore CE; however, it usually requires extra procedures of pre- or postcolumn derivatization of the analytes. A mass spectroscopy detector [14] provides the highest sensitivity and also the structural information, but a cost problem arises compared with other modes of detection. The electrochemical detection (ED) methods, which possess high

sensitivity, low expense, and practicality, can make up the deficiency to a certain extent and could thus be turned into an important detection technique used in CE.

In CE-ED, the off-column [15, 16] and end-column [17, 18] ED modes have been developed. For the off-column mode, a conductive junction between the separation and detection capillary is used to isolate the high voltage applied to the separation capillary from the ED system. For the end-column detection, a microelectrode is placed directly at the end of the separation capillary, without a conductive joint. Although satisfactory results can be obtained from both modes of ED, these modes have seldom been applicable to routine analysis. This is primarily due to the difficulty in finding an appropriate material to make the conductive junction, and the fairly elaborate work that is missing for the construction of a reliable and sophisticated electrochemical cell. For end-column detection, in order to improve the sensitivity and to eliminate the noise from mechanical vibrations, precise alignment and stabilization of the working electrode are required. ED employed in CE include methods based on conductivity detection (CD) [19], potentiometric detection (PD) [20], and amperometric detection (AD) [21]. The main challenge in the development of electrochemical detectors for CE lies in the isolation of the high voltage electric field across the separation capillary from the detection system. With suitable design and improvement, the detection technique will become another popular and more sensitive detection method. To our knowledge, the majority of applications utilize amperometric detection in ED.

Conductivity detection is based on the measurement of the conductivity between indicator electrodes when a small alternating constant current is applied to them. It gains its popularity from its universality and is particularly useful for species not readily detected by UV absorption. Potentiometric detection was based on the measurement of the Nernst potential either across an ion selective barrier or at electrode surface with respect to a reference electrode. Amperometric detection with a microelectrode is potentially one of the most sensitive detection techniques for CE separation. A few reviews [22–26] based on applications and methodology of CE coupled with electrochemical detection have been published in recent years. The use of ED system with CE has grown rapidly since the report of such a system by Wallingford and Ewing in 1987 [27]. Amperometric detection in CE has been accomplished by either off-column or end-column detection. End-column detection, where the electrochemical sensor is placed directly at the end of the separation capillary, has been successfully employed in CE. However, in this method, because the detector is not isolated from the separation voltage used in CE, the actual detection potential is influenced by the electric field present at the end of the separation capillary. The magnitude of this electrical field is proportional to capillary diameters of less than 10 μm [28]. In addition, small fluctuation in the voltage of the power supply used for separation can translate into noise with end-column detection, and the electrode must reequilibrate each time the separation voltage is turned on. Due primarily to these factors, concentration detection limits using end-column detection have been reported to be consistently higher than those obtained by off-column detection.

In the case of off-column detection, the detector must be electrically isolated from the applied electric field associated with the separation system by an off-column interface joint. Thus, if CE-EF were developed and improved, it would be widely applied to the separation and determination of a variety of samples including pharmaceuticals and biomedicine. To date, with CE-ED being rapidly advanced and perfected, it is increasingly being applied to analyze drugs and some related compounds. But there is only one review [29] about this subject to our knowledge. In that review, Lunte and O'Shea's attention mainly focused on the use of CE with electrochemistry for the analysis of microdialysis samples obtained for pharmacokinetic and neurochemical studies; the compounds were dopamine, thiols, glucose, amino acids, and peptides. Furthermore, Lunte and O'Shea did not deal with the general pharmaceuticals. Now, we introduce applications of CE-ED, including conductivity detection, potentiometric detection, amperometric detection, pulsed amperometric detection, *etc.*, in pharmaceutical and biomedical analyses. These applications required an understanding of the various application areas of CE-ED in the pharmaceutical and biomedical analysis for the researchers focusing their attention on these specific aspects.

2 Applications

Many pharmaceutically or biomedically active compounds can be converted to an ionized state and, there, analyzed in principle by means of CE. As a separation of neutral molecules can only be accomplished by using micellar solutions, the majority of the selected compounds is more or less charged so that free zone electrophoresis can be applied. Currently, the potentiometric detection and conductivity detection in CE is mainly used to determine inorganic ions and some small organic species. The application of the amperometric detection, including constant potential and pulsed amperometric detection (PAD) *etc.*, is more widely used than that stated above. It can be directly applied to analyze both small and large electroactive compounds. For nonelectroactive species, they can be determined by using indirect procedures. In this sec-

tion, we intend to review some of the more interesting applications of pharmaceutical and biomedical analyses.

2.1 Catecholamine assay

First, analyzed compounds were mainly of a catecholamine family of the neurotransmitters dopamine, epinephrine, norepinephrine, and catechol. These analytes have been most frequently studied by HPLC-ED over the past decades. Therefore, it is understandable that they were the first ones on which CE-ED efforts were focused more than 10 years ago [27, 30–33] and that they remain one of the groups most frequently studied by CE-ED today [28, 34–64]. The reasons for this are related both to the important physiological role of these compounds in the transmission of nerve impulses and to the fact that, by virtue of their facile oxidation at low potential, they exhibit nearly ideal electrochemical properties for detection purposes. In these investigations most reports describing the separation and determination of the catecholamines by CE-ED were used to optimize the design of the detection modes [28, 34, 37, 45, 57], the betterment of the joint [28, 40, 56, 57], the improvement of the detection methods [39, 48, 49, 52, 55, 57, 64], the selection of the working electrode material [37, 38, 46, 49, 55, 58], the comparison of the separation modes [30], and the applicability of the detection cell [36, 45, 51, 59, 61], *etc.* All these studies obtained good results for their various purposes. Of course, in order to be practical, some of these procedures were also simultaneously applied to analysis of drugs and biomedicine.

The reports stated above show that the CE-ED, mainly CE with amperometric detection, separation and determination of the catechols, are generally not difficult to achieve. Common separation buffers such as pH 6 phosphate solutions or morphilinoethanesulfonic acid have been demonstrated to provide excellent separation of most catechols and catecholamines. To achieve the separation of many neutral catechol compounds, micellar electrokinetic chromatography may be used. Among these reports, a few studies have focused on *in vivo* analysis situations by using CE-ED combined with microdialysis samplings. Unlike most reports, using EC-ED methods, Nann *et al.* [63] determined the dopamine by applying capillary zone electrophoresis with potentiometric detection, where imidazole was used as internal standard. Hadwiger *et al.* [65] used a carbon fiber microelectrode setup to obtain pharmacokinetic data on the catecholamine isoproterenol after intravenous administration of the drug to rats. Hu *et al.* [60] have applied the CE amperometric detection system to analyze a single rat sympathetic nerve cell. The catecholamines, as neurotransmitter in a single cell, were separated and determined. Jin *et al.* [62] employed CE amperometric detection at a normal-sized carbon disk electrode to separate both monoamine transmitters and their metabolites in phosphate buffer. The monoamine transmitters from the caudate nucleus of an anesthetized male rat were determined by microdialysis sampling.

Except for the analyses of catecholamines stated above, Ewing *et al.* [66] applied CE-ED to the analysis of dopamine in a fully developed neuron in planorbis corneous. One method was by proceeding with scanning electrochemical detection. By manipulating the lyse time of the neuron after it was injected onto the capillary, the two vesicular compartments were observed as two individually resolved peaks. Scanning electrochemical detection allowed voltammetry to be obtained as the peaks eluted, confirming their identity. Dopamine concentrations were determined. The other method was directly applied to identify amino acids and amine from a fully developed neuron in planorbis corneous as well as cultured pheochromocytoma cells by CE with naphthalene-2,3-dicarboxaldehyde derivatization and ED [67]. Several compounds, including dopamine were identified. The results were tested and verified by scanning electrochemical detection.

2.2 Thiol compound assay

Because of the important role of thiols in the clinical, biological, and pharmaceutical fields, they have been continuously researched for many years by using various techniques. Just as HPLC-ED, CE-ED has also been used to analyze these species. The thiol compounds investigated are mainly the amino acids cystine and cysteine and the peptides glutathione and glutathione disulfide, whose levels in blood are of physiological interest and which have no strongly absorbing or fluorescing functional groups to make optical detection an attractive option. All these problems cause a change in the electrochemical detection. But, as we know, the thiols undergo oxidation at carbon electrode only at relatively higher positive potentials and disulfides generally are not oxidized at all in carbon. Accordingly, much of the effort in devising effective CE-ED methods for these compounds has been directed toward the development of alternative electrode systems.

The detection of thiols by CE UV-visible or fluorimetric detection often involves derivatization of the analyte prior to separation. Alternatively, these compounds can be determined by CE-ED using chemically modified electrodes. O'Shea and Lunte [68] were the first to introduce this method of analyzing the thiols by employing a gold/mercury microelectrode. This electrode was constructed by dipping a 50 µm gold microelectrode in mercury for

15 s. This detection method is based on the catalytic oxidation of mercury in the presence of thiols. Detection limits for glutathione are 21 nmol. It is also applied to the separation and detection of a number of biological important thiols, including glutathione and cysteine as well as the thiol-containing pharmaceuticals penacillamine, captopril, and 6-mercaptopurine. The electrode can easily be generated and fabricated. The other example is the use of the carbon electrode modified with cobalt phthalocyanine by fabricating in two modes (carbon paste or a conductive carbon cement) for the determination of the thiols, including cysteine, homocysteine, reduced glutathione, and dithiothreitol [69–71]. Detection limits for these methods were in the micromolar range, which was adequate for the determination of cysteine in urine. The third chemically modified microelectrode involved mixed-valent ruthenium cyanide for the determination of disulfide cystine and oxidized glutathione [72]. The detection limits and detector stability were improved over those reported previously for unoptimized systems [73]. An alternative approach for the detection of thiols is the use of PAD. Advantages of PAD over chemically modified electrodes include the fact that PAD can be operated over a wider pH range and will generally have a longer lifetime. It has been used for the determination of a host of sulfur compounds, including thiols and disulfides [74, 75]. Detection limits using optimized PAD were typically 0.5 µM for cysteine, glutathione, and glutathione disulfide.

2.3 General pharmaceutical assay

CE-ED continues to grow rapidly as an analytical technique in a wide range of application areas. It is also of great advantage in the determination of the pharmaceuticals and drug-related impurities. Malone *et al.* [76] developed a reductive electrochemical detection with CE to determine mitomycin C directly in human serum without extraction procedures. The method possesses major advantages over HPLC-based methods in terms of a less complicated deoxygeneration system and shorter deoxygeneration time.

The local anesthetics procaine and lidocaine were determined by amperometric detection at a carbon fiber microelectrode following their separation using CE. Under the optimized conditions, procaine and lidocaine were separated satisfactorily using a short capillary (33 cm) and eluted within 5 min [77]. Xin *et al.* [78] used CE-ED to detect the promethazine with the end-column method. The detection limit of promethazine was 3.5×10^{-8} mol/L (S/N = 3). 1×10^{-7} mol/L of promethazine in urine was separated and determined. Jin *et al.* [79] applied CE-amperometric detection to the analysis of chlorpromazine with a carbon fiber electrode. The detection limit was $1 \times$

10^{-7} mol/L. The method was applied to the determination of chlorpromazine in synthetic human urine.

Fang *et al.* [80] employed CE-amperometric detection to analyze impurities, including glycol, glycerol, erythritol, and mannitol, in the raw material of the sorbitol which acted as diuretic, additive, *etc.* The detection limits were 1 $\times 10^{-5}$ mol/L to 7×10^{-7} mol/L. Glycerol and sorbitol were separated and determined by CE-ED with a copper electrode. This method was used for the detection of glycerol and sorbitol in tooth paste [81]. The separation and determination of polyhydroxy antibiotics were proceeded by employing CE-ED at various metal disk electrodes, copper and nickel [82–84]. These antibiotics include doublemycin, streptomycin, micronomycin, brulamycin, lincomycin, lincomycin B, kanamycin, amikacin, and gentamycin. The analytes have demonstrated good responses at the two electrodes. The detection limit was 5×10^{-7} mol/L. These methods were applied to the detection of some antibiotics in real samples. In another field of application CE-ED was employed in the examination of active ingredients in composite pharmaceuticals. Fang *et al.* [85] used this method to analyze active components, including sulfadiazine, sulfamethazine, and trimethoprim, in three composite sulfonamide tablets. It was successfully used in the analysis of actual pharmaceutical samples. And it was applied to inspect the hydrolysate of chloramphenicol in ophthalmic solutions [86]. The results obtained from the experiment show that it is a useful method for testing and controlling the quality of commercial eyedrops. The two reports stated above were achieved under the relatively higher working potentials, with the carbon disk electrode resulting from the high-resolution efficiency of the CE. Adenosine monophosphate and adenosine diphosphate, as the impurities of adenosine triphosphate, always exist in the production and storage of adenosine triphosphate. The determination of adenosine triphosphate has great clinical and productive implications. One method [87], described for the determination of adenosine triphosphate, adenosine diphosphate, and adenosine monophosphate by using CE-ED at a carbon paste electrode, was employed for the analysis of preparations of adenosine triphosphate. The results obtained were satisfactory.

Currently there is great interest in the separation of the enantiomers of biologically important compounds in chemistry and pharmacology, especially regarding the different bioactivities of the enantiomers in the living world. Chiral separation is of importance not only when monitoring the enantiomeric purity of the drugs but also when studying the relationship between chirality and a drug's therpeutic efficacy. Today, CE with UV-visible detection employing a chiral additive is the dominant technique for chiral separation of enantiomers in CE. Because of the advantages of ED, it has also been applied to this area. Hadwiger *et al.* [65] first employed this method in the isoproterenol samples obtained by microdialysis. The samples were analyzed for both isoproterenol enantiomers which were able to be separated in a cyclodextrin-containing CE buffer. Fang *et al.* [88, 89] applied this method to directly separate the enantiomers of two amine derivatives, threo-2-amino-1-(4-nitrophenyl)-1,3-propanediol and threo-2-(dimethylamino)-1-(4-nitrophenyl)-1,3-propanediol by CE, employing β-cyclodextrin as a chiral additive in strongly alkaline solutions. Both the free enantiomers and the enantiomer-cyclodextrin inclusion complexes could be detected by amperometric detection, using a copper disk electrode. Under these optimum conditions, baseline separation of the enantiomers can be accomplished in less than 18 min. In addition, successful application of the method to the enantiomeric purity determination confirmed its validity and practicability.

Nann *et al.* [63] determined the histamine and inorganic cations by applying capillary zone electrophoresis with potentiometric detection, where imidazole and potassium were used as internal standard. Hu *et al.* [90] used CE-ED with copper microelectrodes to determine the histamine as neurotransmitter. The mass detection limits of 490 amol were achieved by using this method. Thioguanine was studied by using CE-ED at the electrocatalyst cobalt phthalocyanine [69]. A novel and sensitive method for the simultaneous determination of purine bases using CE with vertical wall jet amperometric detection at copper disk electrodes was developed by Chen *et al.* [91, 92]. The current response of high densitivity and stability was obtained in strong basic solutions, which were suitable for satisfactory CE separations for adenine, guanine, hypoxanthine, xanthine, and uric acid. This method was used for the separation and detection of these compounds present in human plasma samples. The mass detection limits were below 9 fmol. Under the reported separation conditions, Zhong *et al.* [46] applied CE-ED at dual electrodes to determine vitamin B_2. The detection limit was 2 $\times 10^{-7}$ mol/L. Yik *et al.* [93] determined the three vitamins B in vitamin tablets by employing CE with amperometric detection at carbon fiber microelectrode. O'Shea and Lunte [69] simultaneously separated ascorbic acid from other analytes to determine the thiol compounds.

2.4 Amino acid and peptide assay

As the smallest units of existing life, amino acids and lower peptides have continued to be a popular research field in CD-ED for more than ten years. Amino acids are not readily oxidized or reduced at carbon electrodes and are often classified as nonelectroactive; exceptions to this classification are tyrosine and tryptophan, which do undergo electrooxidation at accessible potentials at carbon and other electrodes. This characteristic is not the sole advantage of CE-ED; it can also be accomplished by the following approaches: indirect detection or direct detection of electroactive derivatives at carbon electrodes and direct detection by catalytic oxidation at metallic electrodes. Certainly, it can also be determined by conductivity or by potentiometry. To our knowledge, Ewing and Engstrom-Silverman [38] first reported that the detection of nonelectroactive native amino acids and dipeptides was accomplished by using CE-amperometric detection at a copper wire electrode inserted in the end of capillary. An anodic current is produced by a change in the copper oxide film solvency, resulting from complexation of copper ions with certain analytes at the electrode surface. Sub-femtomole detection limits in picoliter injection volumes were obtained without solute derivatization. A Cu(II)-coated capillary was developed for the determination of peptides by CE-ED [94]. Under alkaline conditions, peptides complex with Cu(II) present on the walls of the capillary to form Cu(II)-peptide complexes which can be detected oxidatively at a carbon fiber electrode. Di-, tri-, tetra-, and pentaglycine were determined with a detection limit of 7×10^{-7} mol/L for triglycine. A native amino acid mixture was separated and detected by CE-ED with a copper disk electrode under alkaline conditions. This procedure was applied to analyze the amino acid products from cytochrome *c* [95]. Baldwin and Voegel [96] applied CE-PAD to the analysis of underivatized amino acids at various metal electrodes under a strongly basic medium. Some peptides and proteins were detected. LaCourse and Owens [74] used CE-PAD for the determination of amino acids in mildly alkaline conditions. Fermier and Colon [97] determined the amino acids by CE-AD at a nickel microelectrode under alkaline conditions. Exceptions to the above are some directly analytical methods related to the separation and determination of amino acids and peptides [21, 39, 70, 72, 74, 75, 90, 98–104]. The most promising detection of amino acids and peptides was achieved by using CE-AD resulting from the oxidation of the metal electrode surface species in mildly alkaline medium by the amino acid complexes, or the amino acids appear to be oxidized electrocatalytically. In these cases, all amino acids and peptides can be detected at or below the 10^{-6} mol/L. Metal oxide-modified carbon paste electrodes and microelectrodes for the detection of amino acids and their application to CE were investigated in the strongly alkaline medium by Labuda *et al.* [105]. CuO and NiO have been found to be the most effective electrocatalysts.

Amino acids and peptides can also be detected by employing a derivatizing reagent to derivatize these analytes for electrochemical detection. These reagents are electrogenics, that is, the reagents themselves are not electroactive but the resulting cyano[f]benzoisoindole [43, 106] or dinitrophenyl [76] derivatives can be detected at moderate oxidation potentials. This methodology was used to analyze samples collected from rat brains for alanine, glutamate, and aspartate [43, 106]. Amino acid compounds from a fully developed neuron in planorbis corneus as well as cultured pheochromocytoma cells have been identified by CE-ED using their derivatives with naphthalene-2,3-dicarboxaldehyde [66].

In addition, CE with conductivity detection or potentiometric detection can also be used to separate and determine the amino acids. Wu *et al.* [107] employed a self-assembled CE with a conductivity detection system to determine the histidine and glutamic acid in 5×10^{-3} mol/L NaAc-HAc solutions, carried out within 8 min. Liu and Liu [108] obtained satisfactory results when using this method to detect the amino acids in human hair. Kappes and Hauser [109] separated a mixture of nine underivatized amino acids by using CE with potentiometric detection in basic carbonate and borate buffers as running electrolyte. The detector response is based on the complexation of copper ions at the electrode surface. The detection limits reached a magnitude of 10^{-6} mol/L.

2.5 Carbohydrate assay

The analysis of carbohydrates is another challenge in the area of bioanalysis. Carbohydrates, as an important group of biological compounds, do not possess distinct chromophores or fluorophores and are impossible to detect with any selectivity by UV-visible or fluorescence detection. But for these compounds, the development and use of CE-AD at metallic electrodes are optimal as a result of their direct oxidation at low potentials. Because carbohydrates are not electroactive with carbon, it is essential to employ alternative electrode materials such as Au, Cu, Pt, and Ni. A feature shared by all of these metal-based electrode systems is that they function well for carbohydrates only in strongly alkaline solutions.

For most methods based on CE-ED, the detectors of carbohydrates were Cu and Au electrodes. When copper acts as the detection electrode, the detection potentials were generally constant at about +0.6 V (*vs.* Ag/AgCl). The detection limits of these methods can reach 1×10^{-6} mol/L or below [50, 55, 96, 99, 100, 110–119]. In these studies, some were used to detect simple sugars and to analyze these sugars in real samples, including human urine and plasma [50, 110, 111, 116, 118]. Some were applied to determine simple sugars and sugar derivatives [55, 96, 99, 112] or oligosaccharides and polysaccharides [113, 115]. Others were employed to study their physicochemical properties [114, 117, 118]. These studies differed from others in that Fang *et al.* [119] used a disk-shaped copper electrode as a detector to simultaneously determine sugars and organic acids by coelectroosmotic CE-AD, whereby satisfactory results were obtained. When the working electrodes are gold materials in CE-ED, a PAD mode is required to obtain stable, reproducible results. This avoids the fouling of the electrode surface by carbohydrate oxidation products that occurs with constant potential operation [16, 49, 120–123]. The results obtained from these investigations demonstrated that the detection limits are lower than that obtained at copper electrodes. The same feature of the two methods, taking copper or gold as electrode materials, respectively, is that they can be applied to a variety of different carbohydrates in various samples. It is obvious that not only simple carbohydrates are detected well by ED at these metal electrodes but also oligo- and polysaccharides, carbohydrate derivatives such as alditols, sugar acids, *etc.* For these procedures, derivatization is not necessary.

Not only can copper and gold acting as detectors, but nickel and platinum can also be prepared as detection electrodes for the determination of carbohydrates in CE-ED [50, 55, 58, 97, 124]. Fermier and Colon [97] used CE-AD at nickel microelectrodes in alkaline medium to detect sugars. This method was applied to the analysis of sugars in two real urine and beverage samples. Chen and Huang [124] also used nickel electrodes as detectors to determine the alditols and alcohols in two alcohol beverages. Using chemically modified electrodes as detectors is another means of analyzing carbohydrates CE-AD. The modified platinum electrodes with glucose oxidase were employed to determine the glucose following separation by using CE [55, 58]. The other electrode, a carbon paste electrode modified with glucose oxidase, was prepared for the detection of glucose [69]. In addition, metal oxide-modified microelectrodes were also applied for underivatized sugar determination [125]. A mixture of sugars can be separated baseline in 0.10 mol/L sodium hydroxide solutions; detection limits of 1–2 μmol/L were found for various carbohydrates.

2.6 Other applications

We have classified the analytes analyzed by CE-ED according to their properties or purposes, whereas a few of the described substances still can not be included. Huang *et al.* [126] used CE with conductivity detection in direct determinations of cations in samples of biological interest such as blood serum. This method can analyze nanoliter samples. As a versatile method, conductivity detection can also be used for the analysis of counterions in pharmaceutical substances [127]. The counterion and inorganic impurities can be separated and detected in the same way. Additionally, CE-ED was also applied in the separations and determinations of several cationic biogenic amines and anionic carboxylic acid metabolites in samples taken from single nerve cells [31], ribonucleosides at a CoPC electrode [72], serotonin and homovanillic acid [47], the plant hormone indole-3-acetic acid and one of its primary metabolites indole-3-acetylaspartic acid [128], creatinine and urine acid [99], procatechuric aldehyde [59], and DNA restriction fragment and polymerase chain reaction product [129].

3 Future developments

CE-ED is now firmly established as a viable option for the analysis of pharmaceuticals and biomedicals. With this survey of a few primary applications of the separation and detection techniques, we have tried to demonstrate the potential of CE-ED as an analytical technique for the separation and determination of very similar structures, both qualitatively and quantitatively. Current advantages of CE-ED are good selectivity, high sensitivity, and low cost. Simultaneously, its disadvantages of poor stability and difficult manipulation made it difficult to become a routine analytical method. To date, it is still primarily the domain of specialists in electrochemistry.

In order to aid future pharmaceutical and biomedical analyses, it is undoubtedly necessary to maintain its advantages, overcome its disadvantages, and broaden its new applications in technological developments and progression in methodology. To increase sensitivity, techniques such as sinusoidal voltammetry, used to achieve picomolar detection limits, have been applied to CE. The development of new electrode materials and chemically or biologically modified electrodes and miniaturization of cells and electrodes would extend the range of analytes separated and detected using CE-ED. The coupling of channel CE and array microelectrode detection is another new and powerful tool for monitoring constantly changing microenvironments. Presently, except for amperometric detection, the applications of CE with conductivity and potentiometric detection in pharmaceutical and biological compounds, especially in impurities, are of great potential. With the development of new general and chiral pharmaceuticals, another important direction is also to establish routine analytical procedures for these compounds by using this method.

The coupling of CE-ED to sampling and separation techniques for continuous monitoring of analytes is also an important future direction. For example, the direct coupling of microdialysis to CE-ED will make it possible to observe the pharmacokinetics of drugs and the release of

Electrophoresis 2000, *21*, 1281–1290

neurotransmitters in the near future. The small volume requirements of CE and its fast separation times make it possible to achieve good temporal resolution with microdialysis sampling.

4 Conclusions

In conclusion, electrochemical detection has emerged as one of the more powerful detection methods for CE because of its selectivity, sensitivity, and ability to detect low levels of biologically and pharmaceutically relevant molecules without prior derivatization. It is clear that the future of CE-ED will be bright in the pharmaceutical and biomedicals applications. This was evidenced by recent developments in the construction of new electrodes and the implementation of different detection schemes. Although CE-ED is not yet a routine detection technique in pharmaceutical controlling and industry productivity or in clinical medicine, ongoing research is expected to improve the practicality of this analytical method, so that it might become the alternative method in the near future.

Received August 31, 1999

5 References

[1] Jorgenson, J. W., Lukacs, K. D., *Anal. Chem.* 1981, *53*, 1298–1302.

[2] Smith, N. W., Evans, M. B., *J. Pharm. Biomed. Anal.* 1994, *12*, 579–585.

[3] Holland, L. A., Chetwyn, N. P., Perkins, M. D., Lunte, S. M., *Pharm. Res.* 1997, *14*, 372–387.

[4] Pluym, A., Ael, W. V., Smet, M. D., *Trends Anal. Chem.* 1992, *11*, 27–32.

[5] Altria, K. D., *J. Chromatogr.* 1993, *646*, 245–257.

[6] Altria, K. D., Kelly, M. A., Clark, B. J., *Trends Anal. Chem.* 1998, *17*, 204–214.

[7] Altria, K. D., Kelly, M. A., Clark, B. J., *Trends Anal. Chem.* 1998, *17*, 214–226.

[8] Ludi, H., Gassmann, E., Grossenbacher, H., Marki, W., *Anal. Chim. Acta* 1988, *213*, 215–219.

[9] Wallingford, R. A., Ewing, A. G., *J. Chromatogr. A* 1988, *441*, 299–309.

[10] Gozel, P., Gassmann, E., Michelsen, H., Zare, R. N., *Anal. Chem.* 1987, *59*, 44.

[11] Honda, S., Iwase, S., Makino, A., Fujiwara, S., *Anal. Biochem.* 1989, *176*, 72–77.

[12] Ewing, A. G., Wallingford, R. A., Olefirowicz, T. M., *Anal. Chem.* 1989, *61*, 297A–303A.

[13] Cheng, Y., Dovichi, N. J., *Science* 1988, *242*, 562–565.

[14] Smith, R. D., Wahl, J. H., Goodlett, D. R., Hofstadler, S. A., *Anal. Chem.* 1993, *65*, 574A–584A.

[15] Wallingford, R. A., Ewing, A. G., *Anal. Chem.* 1989, *61*, 98–100.

[16] O'Shea, T. J., Lunte, S. M., LaCourse, W. R., *Anal. Chem.* 1993, *65*, 948–951.

[17] Colon, L. A., Dandoo, R., Zare, R. N., *Anal. Chem.* 1993, *65*, 476–481.

[18] Ye, J., Baldwin, R. P., *Anal. Chem.* 1993, *65*, 3525–3527.

[19] Huang, X., Zare, R. N., *Anal. Chem.* 1991, *63*, 2193–2196.

[20] Harber, C., Silvestri, I., Roosil, S., Simon, W., *Chimia* 1991, *45*, 117–126.

[21] Olefirowicz, T. M., Ewing, A. G., *J. Chromatogr. A* 1990, *499*, 713–719.

[22] Yik, Y. F., Li, S. F. Y., *Trends Anal. Chem.* 1992, *11*, 325–332.

[23] Buchberger, W., *Fresenius J. Anal. Chem.* 1996, *354*, 797–802.

[24] Zhou, W., Wu, M., Wang, E., *Chin. J. Anal. Chem.* 1995, *23*, 343–348.

[25] Voegel, P. D., Baldwin, R. P., *Electrophoresis* 1997, *18*, 2267–2278.

[26] Holland, L. A., Lunte, S. M., *Anal. Commun.* 1998, *35*, 1H–4H.

[27] Wallingford, R. A., Ewing, A. G., *Anal. Chem.* 1987, *59*, 1762–1766.

[28] Kok, W. T., Sahin, Y., *Anal. Chem.* 1993, *65*, 2497–2501.

[29] Lunte, S. M., O'Shea, T. J., *Electrophoresis* 1994, *15*, 79–86.

[30] Wallingford, R. A., Ewing, A. G., *Anal. Chem.* 1988, *60*, 258–263.

[31] Wallingford, R. A., Ewing, A. G., *Anal. Chem.* 1988, *60*, 1972–1975.

[32] Wallingford, R. A., Ewing, A. G., *J. Chromatogr. A* 1988, *441*, 299–309.

[33] Ewing, A. G., Wallingford, R. A., *Anal. Chem.* 1989, *61*, 98–100.

[34] Sloss, S., Ewing, A. G., *Anal. Chem.* 1993, *65*, 577–581.

[35] Chen, I., Whang, C., *J. Chromatogr.* 1993, *644*, 208–212.

[36] Tudos, A. J., Van Dyck, M. M. C., Poppe, H., Kok, W. H., *Chromatographia* 1993, *37*, 79–85.

[37] Lu, W., Cassidy, R. M., Baranski, A. S., *J. Chromatogr. A* 1993, *640*, 433–440.

[38] Engstrom-Silverman, C. E., Ewing, A. G., *J. Microcol. Sep.* 1991, *3*, 141–145.

[39] Ferris, S. S., Luo, G., Ewing, A. G., *J. Microcol. Sep.* 1994, *6*, 263–268.

[40] Park, S., Lunte, S. M., Lunte, C. E., *Anal. Chem.* 1995, *67*, 911–918.

[41] Park, S., Lunte, C. E., *Anal. Chem.* 1995, *67*, 4366–4370.

[42] Zhou, J., Lunte, S. M., *Anal. Chem.* 1995, *67*, 13–18.

[43] O'Shea, T. J., Greenhagen, R. D., Lunte, S. M., Lunte, C. E., Smyth, M. R., Radzik, D. M., Watanabe, N., *J. Chromatogr.* 1992, *593*, 305–312.

[44] Huang, X., Zare, R. N., Sloss, S., Ewing, A. G., *Anal. Chem.* 1991, *63*, 189–192.

[45] Chen, M., Huang, H., *Anal. Chem.* 1995, *67*, 4010–4014.

[46] Zhong, M., Zhou, J., Lunte, S. M., Zhao, G., Giolando, D. M., Kirchhoff, J. R., *Anal. Chem.* 1996, *68*, 203–207.

[47] Matysik, F., *J. Chromatogr. A* 1996, *742*, 229–234.

[48] Gavin, P. F., Ewing, A. G., *J. Am. Chem. Soc.* 1996, *118*, 8932–8936.

[49] Zhong, M., Lunte, S. M., *Anal. Chem.* 1996, *68*, 2488–2493.

[50] Fermier, A. M., Gostkowski, M. L., Colon, L. A., *Anal. Chem.* 1996, *68*, 1661–1664.

[51] Wu, M., Xu, L., *Chin. J. Anal. Chem.* 1995, *23*, 604–607.

[52] Xu, D., Hua, L., Chen, H., Zhu, S., *Chem. J. Chin. Univ.* 1996, *17*, 707–709.

[53] Hu, S., Hu, Y., Li, P., Cheng, J., *Chin. J. Anal. Chem.* 1996, *24*, 1028–1031.

[54] Hu, S., Li, P., Cheng, J., *Chem. J. Chin. Univ.* 1996, *17*, 710–712.

[55] Voegel, P. D., Zhou, W., Baldwin, R. P., *Anal. Chem.* 1997, *69*, 951–957.

[56] Hu, S., Wang, Z., Li, P., Cheng, J., *Anal. Chem.* 1997, *69*, 264–267.

[57] Chen, D., Hsieh, R. R., Chen, C., *J. Chin. Chem. Soc.* 1998, *45*, 257–262.

[58] Voegel, P. D., Baldwin, R. P., *Electrophoresis* 1998, *19*, 2226–2232.

[59] Liu, Z., You, T., Wang, E., *Chin. J. Anal. Chem.* 1998, *26*, 786–791.

[60] Hu, S., Pang, D., Wang, Z., Cheng, J., Li, Z., Fan, Y., Hu, H., *Chin. J. Anal. Chem.* 1998, *26*, 752–756.

[61] Durgbanshi, A., Kok, W. T., *J. Chromatogr. A* 1998, *798*, 289–295.

[62] Jin, W., Jin, L., Shi, G., Ye, J., *Anal. Chim. Acta* 1999, *382*, 33–37.

[63] Nann, A. Silvestri, I., Simon, W., *Anal. Chem.* 1993, *65*, 1662–1667.

[64] Gerhardt, G. C., Cassidy, R. M., Baranski, A. S., *Anal. Chem.* 1998, *70*, 2167–2173.

[65] Hadwiger, M. E., Torchia, S. R., Park, S., Biggin, M. E., Lunte, C. E., *J. Chromatogr. B* 1996, *681*, 241–249.

[66] Swanek, F. D., Chen, G., Ewing, A. G., *Anal. Chem.* 1996, *68*, 3912–3916.

[67] Swanek, F. D., Anderson, B. B., Ewing, A. G., *J. Microcol. Sep.* 1998, *10*, 185–192.

[68] O'Shea, T. J., Lunte, S. M., *Anal. Chem.* 1993, *65*, 247–250.

[69] O'Shea, T. J., Lunte, S. M., *Anal. Chem.* 1994, *66*, 307–311.

[70] Huang, X., Kok, W. T., *J. Chromatogr. A* 1995, *716*, 347–353.

[71] Ye, J., Zhao, X., Fang, Y., *Chin. J. Anal. Chem.* 1997, *25*, 322–325.

[72] Zhou, J., Lunte, S. M., *Anal. Chem.* 1995, *67*, 13–18.

[73] Zhou, J., O'Shea, T. J., Lunte, S. M., *J. Chromatogr. A* *1994, 680, 271–277.*

[74] LaCourse, W. R., Owens, G. S., *Electrophoresis* 1996, *17*, 310–318.

[75] Owens, G. S., LaCourse, W. R., *J. Chromatogr. B* 1997, *695*, 15–25.

[76] Malone, M. W., Weber, P. L., Smyth, M. R., Lunte, S. M., *Anal. Chem.* 1994, *66*, 3782–3787.

[77] Zhou, W., Liu, J., Ding, J., Wang, E., *Chin. J. Anal. Chem.* 1995, *23*, 880–884.

[78] Xin, H., Liu, Z., Wu, M., *Chin. J. Anal. Chem.* 1997, *25*, 555–558.

[79] Jin, W., Zhang, J., *Chin. J. Anal. Chem.* 1998, *26*, 671–673.

[80] Fang, X., Xie, Z., Ye, J., Fang, Y., *Chin. J. Anal. Chem.* 1996, *24*, 861.

[81] Fang, X., Xie, Z., Ye, J., Fang, Y., *Chin. J. Chromatogr.* 1996, *14*, 467–469.

[82] Fang, Y., Ye, J., Fang, X., *Chem. J. Chin. Univ.* 1995, *16*, 1514–1518.

[83] Fang, X., Ye, J., Fang, Y., *Anal. Chim. Acta* 1996, *329*, 49–55.

[84] Fang, X., Liu, X., Ye, J., Fang, Y., *Anal. Lett.* 1996, *29*, 1975–1984.

[85] Wang, A., Gong, F.. Li, H., Fang, Y., *Anal. Chim. Acta* 1999, *386*, 265–269.

[86] Wang, A., Zhang, L., Fang, Y., *Anal. Chim. Acta* 1999, *394*, 309–316.

[87] Fu, C., Song, L., Fang, Y., *Anal. Chim. Acta* 1999, *399*, 259–263.

[88] Fang, X., Liu, X., Ye, J., Fang, Y., *Chem. J. Chin. Univ.* 1997, *18*, 1944–1948.

[89] Fang, X., Gong, F., Fang, Y., *Anal. Chem.* 1998, *70*, 4030–4035.

[90] Hu, S., Huang, W., Pang, D., Wang, Z., Cheng, J., *Wuhan Univ. J. Nat. Sci.* 1998, *3*, 341–344.

[91] Xu, D., Lin, H., Chen, H., *Anal. Chim. Acta* 1996, *335*, 95–101.

[92] Lin, H., Xu, D., Chen, H., *J. Chromatogr. A* 1997, *760*, 227–233.

[93] Yik, Y. F., Lee, H. K., Li, S. F., Khoo, S. B., *J. Chromatogr.* 1991, *585*, 139–144.

[94] Deacon, M., O'Shea, T. J., Lunte, S. M., Smyth, M. R., *J. Chromatogr.* 1993, *652*, 377–384.

[95] Gluo, Y., Colon, L. A., Dadoo, R., Zare, R. N., *Electrophoresis* 1995, *16*, 493–497.

[96] Voegel, P. D., Baldwin, R. P., *Am. Lab.* 1996, *28*, 39–45.

[97] Fermier, A. M., Colon, L. A., *J. High Resolut. Chromatogr.* 1996, *19*, 613–616.

[98] Ye, J., Baldwin, R. P., *Anal. Chem.* 1994, *66*, 2669–2674.

[99] Hong, J., Baldwin, R. P., *J. Capil. Electrophor.* 1997, *4*, 65–71.

[100] Ye, J., Zhao, X., Fang, Y., Baldwin, R. P., *Chin. J. Chem.* 1998, *16*, 226–233.

[101] Malone, M. A., Zuo, H., Lunte, S. M., Smyth, M. R., *J. Chromatogr. A* 1995, *700*, 73–80.

[102] Zhou, J., Lunte, S. M., *Electrophoresis* 1995, *16*, 498–503.

[103] Weber, P. L., Lunte, S. M., *Electrophoresis* 1996, *17*, 302–309.

[104] Matysik, F. M., Backofen, U., *Fresenius J. Anal. Chem.* 1996, *356*, 169–172.

[105] Labuda, J., Meister, A., Glaser, P., Werner, G., *Fresenius J. Anal. Chem.* 1998, *360*, 654–658.

[106] O'Shea, T. J., Weber, P. L., Bammel, B. P., Lunte, C. E., Lunte, S. M., Smyth, M. R., *J. Chromatogr. A* 1992, *608*, 189–195.

[107] Wu, J., Zou, H., Ma, W., Lei, K., Huang, S., Deng, M., *Chin. J. Anal. Instrum.* 1996, *15*, 37–40.

[108] Liu, H., Liu, W., *Chin. J. Anal. Chem.* 1996, *24*, 1354–1359.

[109] Kappes, T., Hauser, P. C., *Anal. Chim. Acta* 1997, *354*, 129–134.

[110] Colon, L. A., Dadoo, R., Zare, R. N., *Anal. Chem.* 1993, *65*, 476–481.

[111] Ye, J., Baldwin, R. P., *Anal. Chem.* 1993, *65*, 3525–3527.

[112] Ye, J., Baldwin, R. P., *J. Chromatogr. A* 1994, *687*, 141–148.

[113] Zhou, W., Baldwin, R. P., *Electrophoresis* 1996, *17*, 319–324.

[114] Fang, X., Gong, F., Ye, J., Fang, Y., *Chromatographia* 1997, *46*, 137–140.

[115] Xu, D., Chen, H., Zhu, S., *Chem. J. Chin. Univ.* 1997, *18*, 720–722.

[116] Zhu, S., Xiao, Q., Chen, H., Xu, D., Li, Z., Hua, L., *Chin. J. Anal. Instrum.* 1998, *17*, 9–12.

[117] Ye, J., Zhao, X., Sun, Q., Fang, Y., *Mikrochim. Acta* 1998, *128*, 119–123.

[118] Ye, J., Zhao, X., Jin, W., Fang, Y., *Chin. J. Anal. Instrum.* 1998, *17*, 34–36.

[119] Fu, C., Song, L., Fang, Y., *Anal. Chim. Acta* 1998, *371*, 81–87.

[120] Lu, W., Cassidy, R. M., *Anal. Chem.* 1993, *65*, 2878–2881.

[121] Roberts, R. E., Johnson, D. C., *Electroanalysis* 1994, *6*, 269–273.

[122] Roberts, R. E., Johnson, D. C., *Electroanalysis* 1995, *7*, 1015–1019.

[123] Lunte, S. M., Zhong, M., Zhong, M., *Anal. Commun.* 1998, *35*, 209–212.

[124] Chen, M., Huang, H., *Anal. Chim. Acta* 1997, *341*, 83–90.

[125] Huang, X., Kok, W. T., *J. Chromatogr. A* 1995, *707*, 335–342.

[126] Huang, X., Pang, T. K. J., Gordon, M. J., Zare, R. N., *Anal. Chem.* 1987, *59*, 2747–2749.

[127] Williams, R. C., Boucher, R., Brown, J., Scull, J. R., Walker, J., Paolini, C., *J. Pharm. Biomed. Anal.* 1997, *16*, 469–479.

[128] Olsson, J. C., Andersson, P. E., Karlberg, B., Nordstroem, A. C., *J. Chromatogr. A* 1996, *755*, 289–298.

[129] Woolley, A. T., Lao, K., Glazer, A. N., Mathies, R. A., *Anal. Chem.* 1998, *70*, 684–688.

Review

Lars G. Blomberg
Hong Wan

Department of Chemistry,
Karlstad University,
Karlstad, Sweden

Determination of enantiomeric excess by capillary electrophoresis

Capillary electrophoresis (CE) is becoming an established method for the determination of chiral trace impurities. This paper provides an overview of the state of the art of CE for such determinations. Detection limits of 0.1% impurity is widely accepted as a minimum requirement for chiral trace impurity determinations. This can be relatively easily achieved with CE. However, determination of lower concentrations requires careful optimization of the separation system. Four factors that are of particular significance for trace enantiomeric determinations: resolution, limit of detection, linear range and type of detection, are discussed. Further, the advantages and disadvantages of derivatization in this context are treated as well as the separation approach, *i.e.*, direct chiral separation or separation after the formation of diastereomers. It is concluded that the limit of impurity detection can be about 0.05% when UV detection is employed. Using laser-induced fluorescence detection, a quantitative determination at the 0.005% level is often possible.

Keywords: Capillary electrophoresis / Enantiomeric excess / Enantiomeric impurity / Pharmaceuticals / Chiral separation / Review
EL 3927

Contents

Correspondence: Dr. Lars G. Blomberg, Department of Chemistry, Karlstad University, SE-651 88 Karlstad, Sweden
E-mail: lars.blomberg@kau.se
Fax: +46-54-7001424

Abbreviations: ANP, *threo*-2-amino-1-(4-nitrophenyl)-1,3-propanediol; **DANP**, *threo*-2-(dimethylamino)-1-(4-nitrophenyl)-1,3-propanediol; **ee**, enantiomeric excess; **FLEC**, (+) or (–)-1-(9-fluorenyl)ethyl chloroformate; **HP-β-CD**, 2-hydroxypropyl-β-CD; **IPA**, 2-propanol

1 Introduction

The importance of stereochemical specificities in biological systems is widely recognized nowadays. Chiral stereoisomers often differ greatly in bioactivity, and the application of a chiral active agent, for example a drug, may be accompanied by an equal or even larger dose of an inactive agent, the inactive isomer. The latter does not contribute to the desired effect, but brings about a toxicological risk, which is to be avoided as much as possible [1]. Differences in bioactivity of enantiomers are of importance in sciences like pharmacology/toxicology and agricultural/environmental. The current awareness of the differences in biological activity of enantiomers, for example in drugs, has created an urgent need for stereoselective analytical methods.

Determination of enantiomeric purity, or enantiomeric excess (ee), is of special importance in a number of contexts, for example for control of the purity of chiral synthetic building blocks and of chiral pharmaceuticals. Such impurities may come from the synthesis or they may be a consequence of poor configurational stability. In many cases, determination of chiral impurities at concentration levels below 0.1% is required, and this puts heavy demands on the analytical methods. Quite good results have been obtained with HPLC [2, 3]. To date, however, the chiral HPLC technique has some drawbacks. (i) Chiral columns for HPLC lack general applicability, in which a number of columns are required to cover a reasonably wide application range. (ii) Column enantioselectivity is still unpredictable; users have to resort to a trial and error approach. (iii) Column lifetime tends to be relatively short. (iv) Chiral HPLC columns are relatively expensive. In practice, these difficulties make chiral HPLC less robust. This is mainly because column properties change with time and the column-to-column reproducibility is poor. Further, the low separation efficiencies make the analytical systems more sensitive to disturbances. Moreover, it may be difficult to reverse chiral elution order in chiral HPLC. These problems have led to an increasing interest in the application of capillary electrophoresis (CE) for the determination of enantiomeric excess [4–6]. The major advantage of CE over HPLC is its simple, inexpensive routine operation. Capillary and reagent costs are minimal compared to those for HPLC.

However, CE and HPLC are complementary. The main reason for this is that the methods function according to widely different mechanisms; thus the two methods are suitable for confirmatory purposes. For example, good agreement between HPLC and CE results strongly supports a comprehensive evaluation of the purity of a sample. During recent years, CE has been successfully employed for the determination of enantiomeric excess in a number of cases, and the topic has been discussed in reviews on chiral CE separations [7–16]. The differences between HPLC and CE for ee determinations were clearly demonstrated by Noroski *et al.* [17]. The examined analytes were a cholesterol-lowering drug and its impurities (Fig. 1). Figures 2 and 3 show separation on HPLC, and separation on CE of the same substances is presented in Fig. 4. For this application, two runs on different HPLC columns were required. Only partial separations were obtained leading to uncertainties in peak area integration. Separation on CE, on the other hand, was efficient and relatively rapid. The resolution achieved allows further optimization to shorten the separation time. The determination of impurities in a product, *e.g.* a drug, is an important issue. Often, such impurities are nonchiral and their determination has been described (*cf.* [7, 18–23]. In the pharmaceutical industry, a major part of the products are

chiral and it is frequently important to be able to monitor the chiral purity of such products. Special methods are required for that purpose, and this paper presents an overview and a discussion of CE methods for determination of chiral trace impurities.

2 Optical purity, enantiomeric excess and enantiomeric impurity

This section describes some basic concepts and definitions in stereochemistry; for a thorough discussion of the different terms see [2, 3, 24–26]. Stereochemistry is thus the study of the three-dimensional structure of chemical compounds, and stereoisomers are compounds that only differ in the spatial arrangement of their atoms. A pair of stereoisomers whose molecules are nonsuperimposable mirror images of one another are enantiomers. A chiral molecule is a molecule that has the potential to exist as two nonsuperimposable structures that are mirror images. Optical activity is the rotation of the plane of monochromatic polarized light. Optical isomer is a synonym for enantiomer but the term is less frequently used since most enantiomers lack optical activity at some wavelength of light. Diastereoisomers (diastereomers) are stereoisomers with two or more centers of dissymmetry and whose molecules are not mirror images of one another. Diastereomers are characterized by differences in physical properties, and by some differences in chemical behavior toward achiral as well as chiral reagents [25]. The enantiomeric content of a mixture of isomers can be expressed in different ways.

Optical purity, *P* (%), is the percent of one enantiomer in a mixture of the two according to

$$P(\%) = \left[\frac{\alpha}{\alpha_{max}}\right]100 \qquad (1)$$

where $[\alpha]_{max}$ is the specific rotation of an optically pure compounds and $[\alpha]$ is the optical rotation of a given sample.

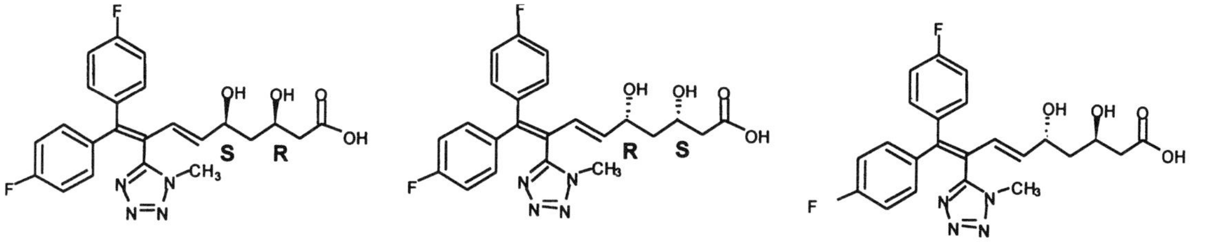

Figure 1. Structures of BMS-180431-09 and two stereoisomers. Reprinted from [17] with permission.

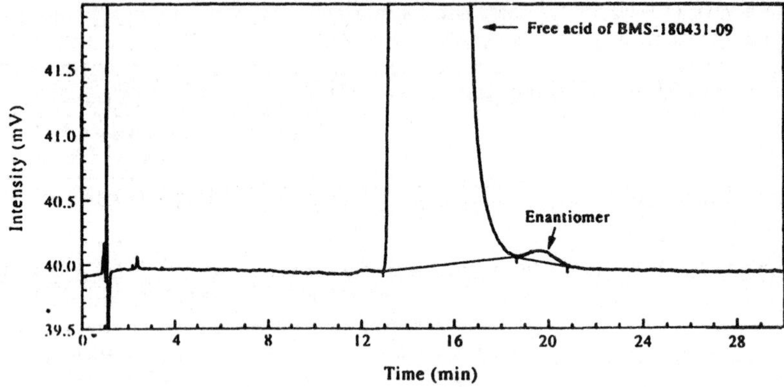

Figure 2. HPLC of BMS-180431-09 and added opposite enantiomer (0.1% w/w). Conditions: chiral α-acidic glycoprotein (AGP) column; 96% 10 mM potassium phosphate buffer (pH 4.0), 4% acetonitrile; column temperature 40°C, 1.0 mL min^{-1}, UV absorption detection at 296 nm; sample concentration, was 0.1 mg mL^{-1} BMS-180431-09 with 0.1% w/w enantiomer added. Reprinted from [17], with permission.

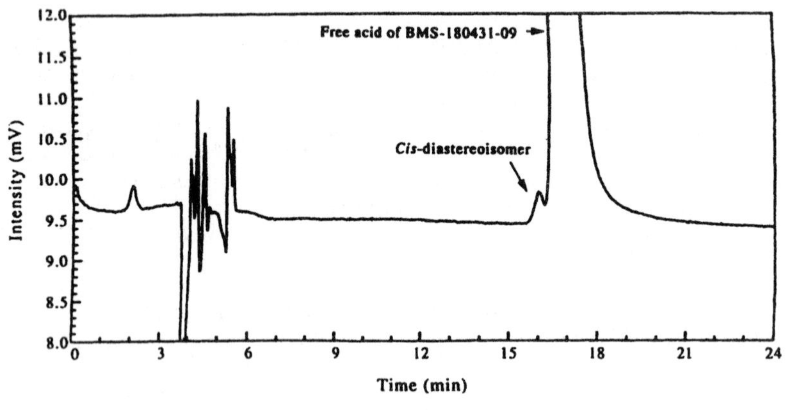

Figure 3. HPLC of BMS-180431-09 and added *cis*-diastereomer (0.1% w/w). Conditions: β-CD 3,5-dimethylphenyl carbonate (cyclobond I DMP) column; 99.95% methanol, 0.05% acetic acid, 1.0 mL min^{-1}, UV absorption detection, at 296 nm; sample concentration, 0.1 mg mL^{-1} BMS-180431-09 with 0.1% w/w *cis*-diastereomer added. Reprinted from [17], with permission.

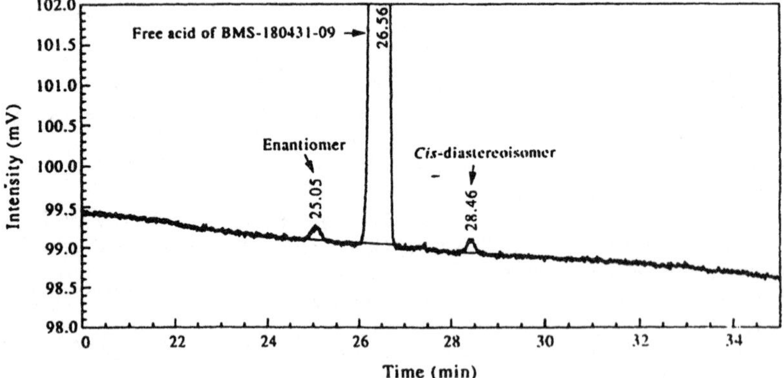

Figure 4. CD-MEKC of BMS-180431-09 with 0.1% w/w enantiomer and *cis*-diastereoisomer added. Conditions: 0.01 M (1.5 g/100 mL) HP-β-CD 0.1 M sodium borate, 0.03 M SDS, pH 9.3; 20 kV; current, *ca.* 65 μA; capillary, 50 cm effective length, 50 μm ID; UV absorption detection, at 200 nm; sample, 0.3 mg mL^{-1} in water. Reprinted from [17], with permission.

Enantiomeric excess, ee (%) is given by

$$ee(\%) = \left[\frac{(R-S)}{(R+S)}\right]100 \qquad (2)$$

where R and S represent the proportion of the respective isomers in the mixture.

Enantiomeric impurity, ei (%) is given by

$$ei(\%) = \left[\frac{(S)}{(R+S)}\right]100 = \left[\frac{A_S}{A_R + A_S}\right]100 \approx \left[\frac{A_S}{A_R}\right]100 \qquad (3)$$

Here $R \gg S$, and A_R and A_S are the areas of the large and small peak, respectively. Enantiomeric impurity and enantiomeric excess are the most commonly used expressions in analytical chemistry. The magnitude of ee extends from ee = 0 for a racemate to ee = 1 for the pure enantiomer.

3 Quantitation of minor peaks

As is common in HPLC, calculations may be based on response factors obtained from calibration solutions. Often, pure standards are not readily available and enantiomeric excess or enantiomeric impurity is then reported simply as % area/area of the electropherogram [27]. Peak areas are generally used in quantitative analysis since the peak height increases in a nonlinear manner at high sample concentrations [7, 28]. The nonlinear part of the peak height plot may be attributed to peak broadening effects. Internal standards can be incorporated to decrease injection based errors [29]. In general, injection errors are somewhat higher in CE than in HPLC [30]. This is because fixed volume injectors are not available for CE. Further, matrix effects can have more influence in CE compared to HPLC [31].

In CE, two separated enantiomers have different mobilities and therefore different residence times in the detection zone leading to different response factors for the enantiomers. This requires normalization of the peak area and such a normalization is simply achieved by dividing the area of each peak with its corresponding migration time. This function is included in the software of most commercial CE instruments. Another source of differences in detector response to the enantiomers may be the differences of complexation of the enantiomers to the chiral selectors. In this case, the transient diastereomeric complexes of the enantiomers have different detector responses. This effect was shown by Hempel and Blaschke [32] for the fluorescence detection of some drugs using β-CD as chiral selector.

Before choosing a method for the analysis, the goal has to be determined. Do we want a quantitative determination or is the goal merely to test whether the concentration of the impurity enantiomer is below a certain level? What precision and accuracy are needed? A large number of enantiomeric impurity determinations of chiral drugs have been reported (Table 1). The major part of the applications, numbers 1–35, were performed according to the direct method without derivatization. Derivatization was employed for numbers 36–43. The indirect approach was reported in two cases, numbers 42 and 43. Detection was mainly by UV (39 cases). LIF detection was used in three cases, numbers 38, 39 and 43. Electrochemical detection was used in one case, number 24, and mass spectro-

scopy in three cases, numbers 25, 28, 34. Finally, the majority of the entries in Table 1 were qualitative rather than quantitative. In general, the methods were not validated regarding basic factors such as specificity, linear range, limit of detection, accuracy, precision, and robustness. Typically, data were reported as "it was possible to detect 0.2%". Altria [6] has published a table summarizing applications of chiral trace determinations up to 1997.

4 Separation approaches

The separation can be performed with two principally different approaches, the indirect and the direct. In the indirect method, enantiomers are derivatized by complete chemical reaction with an enantiomerically pure reagent to form diastereomers. These are subsequently separated on a nonchiral system. In the direct method, chiral separation is performed on a chiral system. Here, derivatization is, in principle, not necessary but derivatization may still be advantageous. This is mainly in cases when detection is a problem but derivatization can also result in improved chiral selectivities of chiral selectors, for example, in connection with CDs [72].

A derivatization reagent must fulfill a number of requirements. (i) It should be stable and give rapid reactions in high yields at low temperatures and the reaction products should be sufficiently stable. (ii) Excess reagent or by-products from the reaction should not disturb the separation. (iii) The reagent should be selective for the target analytes. In those cases where diastereomers are formed, the reagent must have a very high degree of chiral purity. A small impurity can lead to large errors, especially in the determination of enantiomeric excess [3, 69]. (iv) The reagent should contain or produce a strong chromophore or fluorophore. The reagent should be commercially available and, for chiral reagents, it is desirable that they are available in the L- as well as in the D-form, so that elution orders can be reversed when required. (v) The reagents should be inexpensive, although chiral reagents in high purity are inherently costly. The drawbacks of the derivatization approach are well known [4, 73]. A general drawback is the low yield obtained on derivatization of analytes occurring in very low concentrations, and matrix effects could also be disadvantageous. Further, the rate of diastereomer formation can be different. In such cases, the final ratio of diastereomers would not reflect the exact ratio of the starting enantiomers. Moreover, derivatization takes some time and, as mentioned above, costs may be relatively high for the indirect method. However, there are a great number of applications for which these drawbacks are irrelevant. Finally, when derivatization is needed for the detection, the indirect approach becomes attractive.

Table 1. Some reported ee determnations and separation conditions applied

No.	Enantiomer determined	Enantiomeric impurity (%)	Separation conditions	Detection	Detection limit[a] (linear range)[b]	Ref.
Direct method, without derivatization						
1	D-Loxiglumide	0.2	50 mM phosphate (pH 6.0), 1 mM vancomycin	UV 255 nm	0.5µg/mL[a] (0.4–17.5 µg/mL)[b]	[32]
2	(*S*)-Compound I	0.4	40 mM borate (pH 8.5), 15 mM SBE-β-CD, 25% MeOH, 25 kV	UV 215 nm	-	[33]
	(*R*)-Compound I	0.9			-	
3	(–)-Isoproterenol	0.1	100 mM phosphate-2-propanol (30:70 v/v; pH 7.0), 25 mM rifamycin, 8 kV	Indirect UV 350 mm	- -	[34]
4	LY248686	0.1	50 mM Tris-phosphate, pH 2.35, 4 mM HP-β-CD	UV 214 nm	-	[35]
5	BMS-180431-09	0.1 0.06 (estimated)	100 mM borate (pH 9.3), 30 mM SDS, 10 mM HP-β-CD	UV 200 nm	- (0.15–0.40 mg/mL)	[17]
6	(+)-Fluparoxan	1.0	10 mM borate-10 mM Tris, 150 mM β-CD, 6 M urea-2-propanol (80:20 v/v; pH 2.5)	UV 214 nm	- (1.5–125%)	[36]
7	Ephedrine Norephedrine	0.5	30 mM Tris-phosphate (pH 2.5), 40 g/L of alkylated β-CD	UV 200 nm	10 µM (2.5-83×10^{-5}M)	[37]
8	2*S*,3*S*-AMP	0.88	50 mM Tris-phosphate (pH 2.5), 50 HP-β-CD	UV 220 nm	5 µM (5×10^{-6}-8×10^{-4}M)	[39]
9	(*R*)-PrPPX	0.1	100 mM phosphate-Tris (pH 3.0), 10 mM 2,6-DM-β-CD	UV 206 nm	3 µM) (3–30 µM)	[40]
10	L-Tryptophan	0.1	25 mM triethanolamine-phosphate (pH 2.5), 75 mM α-CD	UV 200 nm	- (0.4m–2.5 mg/mL))	[41]
11	(*R*)-Ropivacaine	0.1	100 mM phosphate-triethanolamine (pH 3.0), 20 mM 2,6-DM-β-CD	UV 206 nm	- -	[42]
12	Dihydroxy-2-2 -aminotetralin	0.3	10 mM Tris-citric acid (pH 2.2), 30 mM 18C6H$_4$	UV 280 nm	- (0.5–5.5%)	[43]
13	(*S*)-Propanolol	0.1	100 mM phosphate-triethanolamine (pH 3.0), 10 mM CM-β-CD	UV 206 nm	(0.1–10%)	[44]
	(*R*)-Propranolol	0.5			(0.5–10%)	
14	(*R*)-Propranolol	0.1	25 mM phosphate-Tris (pH 3.0), 10 mM CM-β-CD + 8 mM DM-β-CD	UV 214 nm	- (0.025–0.07 mg/mL)	[13]
15	D-Tryptophan	0.05	40 mM phosphate-Tris (pH 3.0), 75 mM α-CD	UV 214 nm	- (0.25–0.7 mg/(mL)	[12]
16	L-Carbidopa	0.08	22 mM phosphate (pH 2.5), 20 mM HDAS-β-CD, 8% PEG-900, 16°C	UV 280 nm	- (0.077 µg–2.49 mg/mL)	[45]
17	(*S*)-MK-0677	0.35	40 mM L-tartartic acid-20 mM phosphate (pH 4.2), 30 mM β-CD, 25% ethanol (v/v)	UV 200 nm	0.96 µg (0.32 µg–0.54 mg/mL)	[46]
18	(*S*)-1,1'-Binaphthyl-2,2'-diamine	0.2 <0.1 (possible)	10 mM 18C6H$_4$, 20 mM Tris--phosphate (pH 2.06), 15°C	UV 234 nm	- -	[47]
19	(+) or (–)-Ephedrine	0.5–1	25 mM phosphate (pH 8.0), 50 mM (*R*)-or (*S*)-DDCV, 30°C	UV 214 nm	- -	[48]
20	(+)-Terbutaline	0.1	100 mM phosphate (pH 2.5), 10 mM HE-β-CD, 10% PEG-2000, 15°C, 30 kV	UV 210 nm	(0.1–1.3%)	[49]
21	Fenfluramine	2%	100 mM phosphate-Tris (pH 3.0), 10 mM DM-β-CD, 20°C, 25 kV	UV 214 nm	- 8–32 µg/mL	[50]
22	(*S*)-Ibuprofen	1	10 mM phosphate (pH 7.0), 2.5% Glucidex 2	UV 185 nm	- (1–100 µg/mL)	[51]

Table 1. continued

No.	Enantiomer determined	Enantiomeric impurity (%)	Separation conditions	Detection	Detection limit[a] (linear range)[b]	Ref.
23	(–)-Galanthamine	0.1	50 mM phosphate (pH 8.0), 50 mM tetrabutylammoniumdihydrogen, 30 mM DM-β-CD, 26 kV	UV 216 nm	100 µg/mL (0.4–1.0 ng)	[52]
24	L-ANP L-DANP	1 1	100 mM NaOH, 12 mM β-CD, 30°C	ECD: 675 mV Ag/AgCl	- -	[53]
25	(*R*)-Ropivacaine	0.25	50 mM acetic acid, 100 mg/mL Met-β-CD (partial filling β-CD), 30 kV	CE-MS	90 nM -	[54]
26	(*S*)-Ropivacaine	0.1	30 mM phosphate (pH 3.0), 76 mM Met-β-CD, 20°C	UV 214 nm	1.6 µg/mL (1.6–62.6 µg/mL)	[55]
27	(*R*)-Ropivacaine	0.1	100 mM phosphate-Tris (pH 3.0), 10 mM 2,6-DM-β-CD	UV 206 nm	- (0–3%)	[40]
28	(–)-Ephedrine	1	(i) 5 mM phosphate (pH 2.5), 10 mM 2,6-DM-β-CD (ii) Tris/formic acid (pH 2.5–3.0)	UV 200 nm CE-MS	MS offering 1000-fold higher sensitivity than UV	[81]
29	(+)-Clenbuterol	1	200 mM phosphate (pH 3.3), 30 mg/mL HE-β-CD	UV 210 nm	0.5 ng/mL (with sample pretreatment)	[82]
30	(–)-Terbutaline	0.1	100 mM phosphate (pH 2.5), 10 mM HE-β-CD, 10% PEG-2000, 15°C	UV 210 nm	- (0.1–1.3%)	[83]
31	(*RR*)-Diltiazem hydrochloride	0.2	(i) 20 mM phosphate-borate (pH 2.4), 3% chondroitin sulfate C, 23°C	UV 214 nm	- -	[84]
	(*R*)-Trimetoquinol hydrochloride	0.2	(ii) pH 2.8; other conditions as in (i)	UV 235 nm	- -	
32	(*R*)-Naproxen	0.11	200 mM MES/TBAH, 10 mM HP-β-CD, 0.4% polymeric additive, 20°C	UV 230 nm	300 nM (1 µg-10 mg/mL)	[85]
33	(+)-ENX792	0.2	100 mM phosphate (pH 2.5), 20 mM β-CD	UV 200 nm	- -	[87]
34	(*R*-(–)-Camphor-sulfonic acid	0.9	40 mM ammonium formate (pH 4.0), 50 mM DM-β-CD (partial filling DM-β-CD)	CE-MS	- -	[88]
35	(*R*)-Ropivacaine	0.08	100 mM phosphate-triethanolamine (pH 3.0), 20 mM 2,6-DM-β-CD	UV 206 nm	- (0–3%)	[56]

Direct method; with derivatization

No.	Enantiomer determined	Enantiomeric impurity (%)	Separation conditions	Detection	Detection limit[a] (linear range)[b]	Ref.
36	(+), (–)-Cizolitine (I) (DTT)	>99 (ee)	100 mM phosphate (pH 2.5), 25 mM HP-β-CD, 20°C, 30 kV	UV 205–220 nm	- -	[38]
37	D-Phe (DNS)	0.064	20 mM phosphate (pH 8.0), 5 mM HP-β-CD	UV 214 nm	- -	[57]
38	(+)- and (–)-FLEC (–)-FLEC (Glycine)	0.1 (UV) 0.005 (LIF)	40 mM HEPES-Tris, 40 mM SDS, 10 mM γ-CD, 15% 2-propanol (v/v)	UV 256 nm LIF: 244/330 nm	(2–2000 µM), UV (0.04–2000 µM), LIF	[58]
39	L-Phe (NBD-F)	0.05	100 mM Tris-borate (pH 8.3), 2 mM DM-β-CD	LIF: 488/520 nm	140 ppm >10³)	[59]
40	(+)-EAA494 (DNS)	0.1	100 mM phosphate-borate (pH 11), 20 mM γ-CD, 15 kV	UV 214 nm	- (0.1–1.5%	[86]
41	(+)-FLEC (Glycine)	0.1	(i) 50 mM phosphate (pH 6.0), 14 mM β-CD, 18% IPA(v/v), 30 kV,	UV 256 nm	0.1 µM (2.3 cm sample plug	[60]

Electrophoresis 2000, *21*, 1940–1952

Table 1. continued

No.	Enantiomer determined	Enantiomeric impurity (%)	Separation conditions	Detection	Detection limit[a] (linear range)[b]	Ref.
	L-Arg-Gly (FLEC)	0.024	(ii) 20 mM borate, 15 mM phosphate (pH 9.2), 10 mM SDS, 18% 2-propanol, 30 kV	UV 256 nm	injected) -	-
Indirect method						
42	D-Carnitine (FLEC)	0.2	50 mM phosphate (pH 3.4), 14 kV	UV 214 nm	- (0.2–10%)	[61]
43	L-Peptides (APOC-)	0.12–0.28	10 mM borate, 7.5 mM phosphate (pH 9.2), 15 mM SDS	LIF: 351/412 nm	- -	[62]

Abbreviations: APOC, (+) or (−)-1-(9-anthryl)-2-propyl chloroformate; CM-β-CD, carboxymethyl-β-CD; DDCV, (*S*)-and /*R*-*N*-dodecoxycarbonylvaline; DM-β-CD, heptakis(2,6-di-*O*-methyl)-β-CD; DNS, 5-dimethylaminonaphthalene-1-sulfonyl chloride; DTT, di-*p*-toluoyltartaricacid; ECD, electrochemical detection; HDAS, heptakis-(2,3-diacetyl-6-sulfato)-β-CD; HE-β-CD, 2-hydroxyethyl-β-CD; Met-β-CD, methyl-β-CD; MK-0677, {(*R*)-2-amino-*N*-[2-(1,2-dihydro-1-(methylsulfonyl) [3H-indol-3,4'piperidin-1'yl]-2-oxo-1-[(phenylmethoxy)-methyl]ethyl-2-methylpropanamide-monomethane sulfonate}; NBD-F, 4-fluoro-7-nitrobenz-2,1,3-oxadiazole; SBE-β-CD, sulfobutyl ether β-CD; TBAH, tetrabutylammonium hydroxide.
a) indicates detection limit
b) indicates linear range

In such cases, derivatization may lead not only to improved detection but also to improved separation efficiency. Depending on the conditions, kinetics may be more rapid in nonchiral than in chiral systems, and rapid kinetics may lead to high plate numbers. As mentioned above, derivatization may lead to enhanced selectivity; for example, the aromatic part of a derivatization reagent may fit in the cavity of a CD molecule.

In the indirect method, separation is often performed in micellar electrokinetic chromatography (MEKC) mode. The micelles contribute to the BGE conductivity and in order to avoid excessive Joule heating, the micellar concentration should be kept at a low level. The hydrophobicity of the derivatization reagent is of great importance for the optimal concentration of surfactant, *e.g.*, sodium dodecyl sulfate (SDS). Most reagents give relatively polar derivatives that require high SDS concentrations for their separation. For example, the necessary SDS concentration for the separation of amino acids strongly depends on the nature of the derivatization reagent. Thus, the required SDS concentrations were: 200 mM for amino acids (AA) derivatized with 2, 3, 4, 6-tetra-*O*-acetyl-β-D-glucopyranosyl-isothiocyanate (GITC) [74], 150 mM for AA derivatized with *o*-phthalaldehyde [75] and about 100 mM for AA derivatized with Marfey's reagent [76]. Derivatives obtained with (+) or (−)-1-(9-fluorenyl)-ethyl chloroformate (FLEC) are more hydrophobic and only 10–20 mM SDS was required here [77]. Similarly, the nature of the derivatization reagent is of crucial importance for the selectivity (Fig. 5) [72].

5 CE for ee determinations

The determination of enantiomeric excess by means of CE depends on a number of factors, *e.g.* separation efficiency, resolution, limit of detection and dynamic linear range. These factors will be discussed in this section.

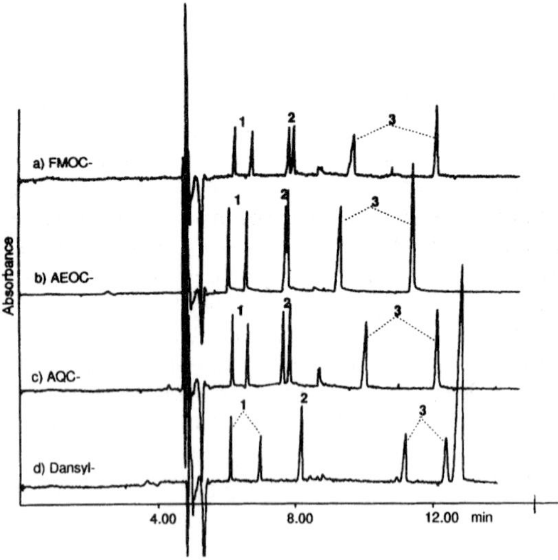

Figure 5. Electropherograms of differently derivatized peptides. Separation capillary, 67 cm total length (46.5 cm to detector window), 25 µm ID; buffer, 25 mM phosphate/Tris (pH 6.25), 1.2 mM teicoplanin, 40% v/v ACN. Peaks: 1, Leu-Gly-Gly; 2, Leu-Gly; 3, Gly-Asp. Reprinted from [72], with permission.

5.1 Separation efficiency

For a given selectivity (α) resolution increases as the number of theoretical plates increases. This is in favor of the indirect method where plate numbers tend to be higher than in the direct method. Depending on conditions, the kinetics of analyte interactions with chiral selectors, *e.g.* CDs, may be slower than with nonchiral selectors, *e.g.*, micelles. Further, the analyte band width is dependent on a number of factors such as focusing of analytes on injection, Joule heating and electrophoretic band broadening. To obtain a narrow starting band it is essential to obtain a good stacking condition. This is typically achieved when the ionic strength in the sample solution is about 10 times lower than in the running buffer. However, to be able to detect trace concentrations when the detection system has relatively poor concentration sensitivity (e.g., UV), it is desirable to inject a relatively concentrated sample solution. To achieve stacking also of more concentrated sample solutions, the buffer concentration in BGE should be increased accordingly. But with increasing BGE conductivity, the current will increase and thereby the Joule heating will increase, leading to broadening of the analyte bands. Thus, the Joule heating has to be decreased, and that can be accomplished by increasing the resistance in the capillary; for that purpose, a low conductivity buffer can be used, *e.g.*, a biological buffer. Another way to decrease the current is to use a narrower capillary. However, there is a trade-off between capillary diameter and optical path length for on-column detection [66]. A further drawback of highly concentrated sample solutions is that electrophoretic band broadening will result. This happens when the buffer ion mobility is not matched to the mobility of the analytes. In summary, if a relatively high sample concentration is employed to improve the limit of detection, the large peak then becomes broadened. The degree of peak broadening that can be accepted, of course, depends on the selectivity. These problems are most prominent when using less sensitive detection, *e.g.*, UV. When a highly sensitive detector, *e.g.*, laser-induced fluorescence (LIF), is used, a wide linear range can be achieved without overloading the system (*cf.* Section 5.4). However, the upper limit of the linear range may depend on the detection as well as on the overloading of the separation system. The nature of the analytes determines which of these two factors will be limiting.

5.2 Resolution

In order to achieve high precision in peak area determination the peaks should be fully resolved. For UV detection of compounds occurring in low concentrations, a high sample loading is required, but in CE the resolution decreases when the separation system becomes overloaded. Thus, an electrophoretic system giving merely baseline resolution of a racemate will not fully be able to resolve a minor and a major peak; thus, relatively high resolution is required. Further, it is always advantageous to elute the minor peak before the major peak [44, 78, 79]. Due to column overloading, the major peak tends to tail and there is a big risk that the minor peak will appear on such a tail. Therefore, it is important to be able to elute the small peak first. For example, Fillet *et al.* [44] demonstrated that a separation system giving a resolution of 4.4 facilitated the determination of 0.1% of an *S*-isomer, the first eluting peak, while only 0.5% of the *R*-isomer, the last eluting peak, could be determined. In the indirect method, elution order can be controlled by the selection of derivatization reagent; the (+) or (−) reagent gives a different elution order of a pair of diastereomers. In the direct method, chiral elution order can be altered by means of a change of chiral selector, for example, by changing the selector configuration from (+) to (−). CDs occur only in one enantiomeric form, but a change from an uncharged to a charged CD will often facilitate alteration of the elution order. Enantiomer migration order in chiral CE has been extensively discussed [80].

5.3 Limit of detection and linear range

UV detection has to be performed on-column; the optical path length will therefore be quite short and, in addition, the walls of the "detection cell" are not rectangular but curved. This results in relatively poor concentration sensitivity, typically in the μM range. The linear range will only be about two, or, at most, three orders of magnitude, thus corresponding to an impurity of 0.1%. These properties, of course, strongly limit the possibilities to perform ee determinations by means of one injection. (If the concentration in the large peak is above the linear range, the small peak will be overestimated).

Some different approaches can be taken to circumvent the problem of the limited linear range. First, to be able to quantify both peaks in an enantiomeric excess determination, a variable sample loading (high-low sample loading) can be employed [81]. This technique comes from HPLC [82]. In this technique, an initial run is performed to obtain a main component just within the operating range of the detector. Sample loading is then increased, *e.g.* ten times, to determine the small peak. The increase in sample loading can be obtained either by the application of a prolonged injection time or by the use of a more concentrated sample solution [81]. An LOD of 0.01% area/area was reported for the nonchiral impurities of ranitidine and fluparoxan when using the high-low sample loading tech-

    Electrophoresis 2000, *21*, 1940–1952

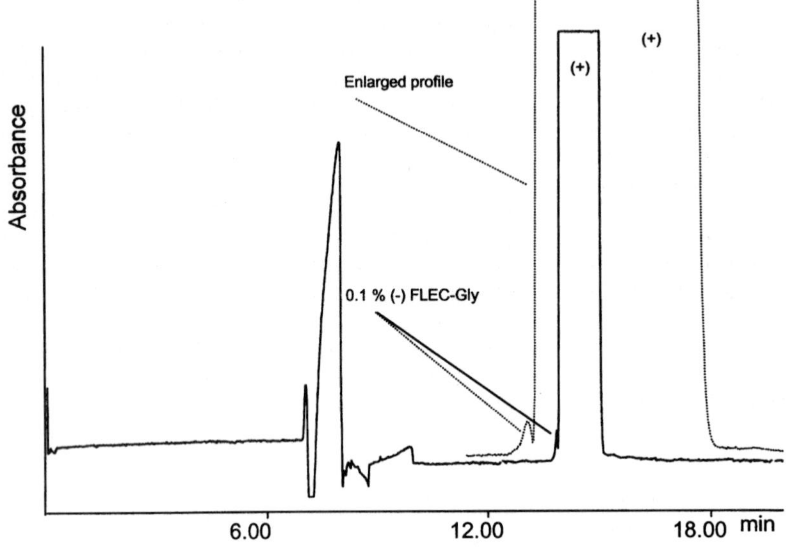

Figure 6. Determination of added 0.1% (–)-FLEC-Gly using a 50 µm ID capillary. Conditions: 40 m_M_ HEPES (pH 7.50), 40 m_M_ SDS, 10 m_M_ γ-CD, 15% v/v 2-propanol (IPA); 30 kV; 32 µA; injection, 1.6 min (50 mbar). Reprinted from [66], with permission.

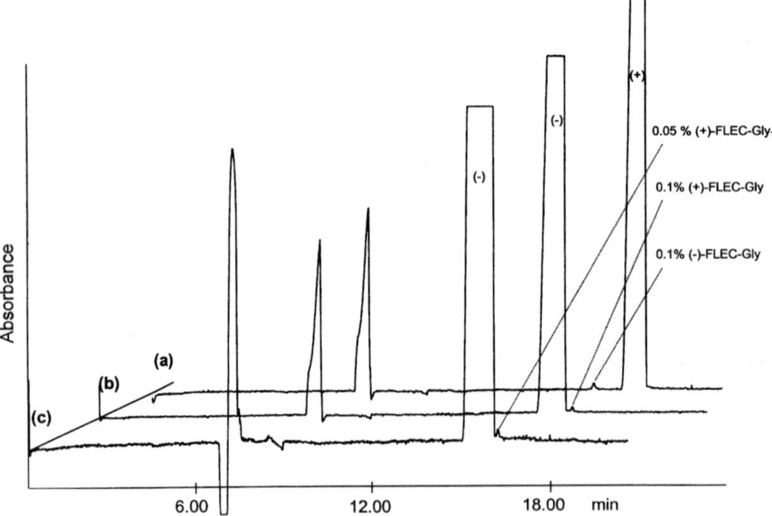

Figure 7. Determination of added 0.05%–0.1% (+) and (–)-FLEC-Gly using a 25 µm ID capillary. (a) 0.1% (–)-FLEC; (b) 0.1% (+)-FLEC; (c) 0.05% (–)-FLEC. Injection, 2 min (50 mbar). Other separation conditions as in Fig. 6; I = 6.6 µA. Reprinted from [66], with permission.

nique [9, 81]. However, the linear range was not given. Second, when the chemical purity of the sample is known, only the minor peak has to be measured [83]. For quantitative purposes, external standard calibration and standard addition approaches have been compared, and found to be in good agreement [84]. In general, injection reproducibilities are poorer in CE than in HPLC, typical values being about 1–2% and 0.5–1% RSD, respectively. Precision can, as mentioned above, be improved by applying the internal standard technique [28]. In a recent paper [66] we evaluated different approaches to ee determination using CE/UV. It was found that UV detection, in general, is problematic for the determination, on the basis of one injection, of 0.1% or lower impurity due to the relatively high LOD and the limited linear range of this detector. Using the high-low injection technique, or the sample dilution approach, chiral impurity determinations could be extended to about 0.05% (Figs. 6 and 7) [66]. In our work, higher accuracy was obtained with the sample dilution technique than with high-low injection [66].

5.4 LIF detection

This technique provides several properties that are advantageous for ee determinations. (i) The LOD is several orders of magnitude lower than for UV detection. Thereby the detection can be shifted to lower concentrations, and electrophoretic band broadening of the large peak will be virtually eliminated. However, if the intention is to determine extremely low levels of impurity, the way to do this is to increase the amount of sample injected. It seems that in such a case, the electrophoretic band broadening will become one of the limiting factors concerning the level of impurity that it will be possible to determine. (ii). The linear range will be increased when using LIF detection; depending on conditions, it can be four orders of magnitude or more [85–87]. (iii). An increase in plate numbers can be expected as a consequence of the short width of capillary illumination in fluorescence detection mode, about 10 µm; in UV mode this section is usually about 0.5 mm. LIF detection is presently not being used frequently, because a laser is typically involved, which makes the technique somewhat cost prohibitive. Further, LIF detection is offered only by few manufacturers of CE instruments. The performance of CE/LIF for ee determinations was recently investigated [66]. Using this technique, enantiomeric impurities as low as 0.005% could readily be determined (Fig. 8) [66]. In this case, a normal sample concentration and standard injection were utilized, thus minimizing band broadening and enabling the direct measurement of trace impurities with good accuracy within the dynamic linear range.

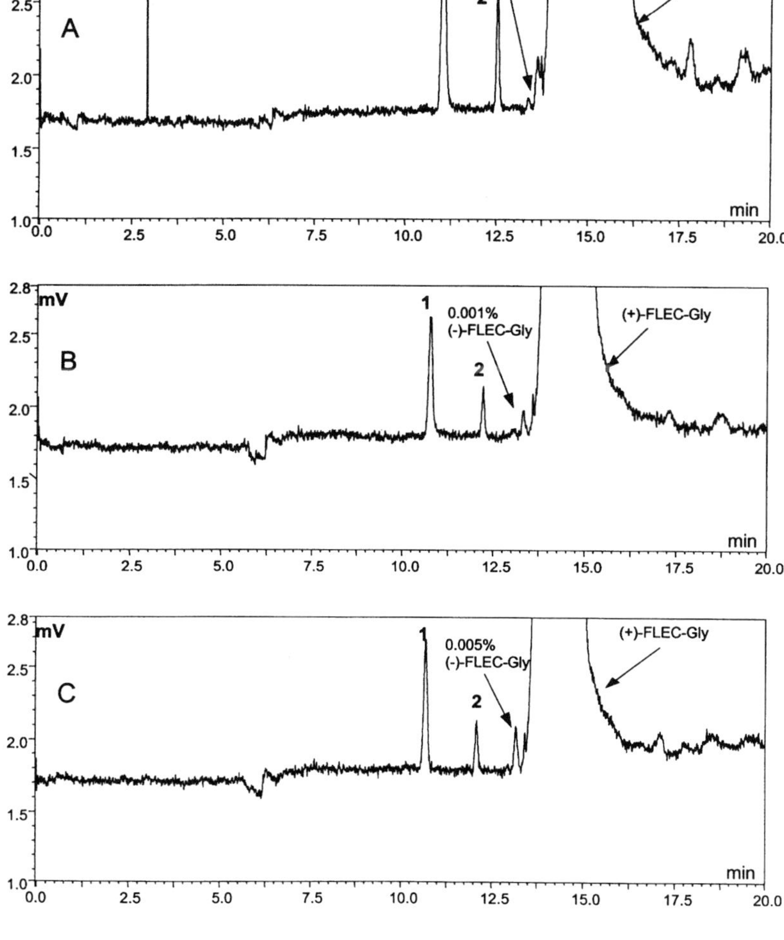

Figure 8. Quantitative determination of trace FLEC-Gly impurity with LIF detection. Separation capillary, 50 µm ID × 55.5 cm (40 cm to LIF detector and 47 cm to UV detector); buffer, 40 mM HEPES (pH 7.50), 40 mM SDS, 10 mM γ-CD, 15% v/v IPA; applied voltage, 26 kV; *I* = 18.6 µA; temperature, 25°C; injection, 50 mbar × 20 s. (A) 4 mM (+)-FLEC-GLy; (B) 0.001% (–)-FLEC-GLy spiked, a sample mixed with 0.04 µM (–)-FLEC-Gly and 4 mM (+)-FLEC-Gly (1:1 v/v); (C) 0.005% (–)-FLEC-Gly spiked, a sample mixed with 0.2 µM (–)-FLEC-Gly and 4 mM (+)-FLEC-Gly (1:1 v/v); 1, 2, unknown impurities. Reprinted from [66], with permission.

5.5 Mass spectrometric detection

This technique is under rapid development. Presently, the separation system has to be adopted to the mass spectrometer because nonvolatile buffers and micelles still detract from detector performance [88]. Thus, for direct chiral separations, the chiral selector should not be allowed to enter the MS. This can be avoided by the partial filling technique [89]. Two papers have been published dealing with MS detection in combination with the partial filling technique [54, 62].

5.6 Electrochemical detection

Electrochemical detection was successfully employed for trace enantiomeric determination of *threo*-2-amino-1-(4-nitrophenyl)-1,3-propanediol (ANP) and *threo*-2-(dimethylamino)-1-(4-nitrophenyl)-1,3-propanediol (DNAP) [53].

6 Conclusions

Determination of trace enantiomeric impurity is not a simple task. The success of direct determination of trace enantiomeric impurities depends on a number of interrelated factors. The main factors are: resolution, LOD, dynamic linear range, and type of detection. First, when UV detection is being used, the resolution has to be relatively high. When injecting a racemate, for example, a system giving merely baseline resolution is not enough for trace enantiomeric determination. A resolution of 10 for equally large peaks may lead to a minimum in impurity determination of 0.05%. Second, the trace impurity determination is limited by the LOD. Application of UV detection is, in general, problematic for the direct determination of 0.1% or lower impurity due to the limited dynamic linear range of this detector. Using the high-low injection approach, the limit of enantiomer impurity determination can be extended to about 0.05%, provided the resolution is good enough. On determination of low % impurities on the basis of one injection, the large peak becomes overloaded and results in tailing; for reliable results it is then necessary to elute the small peak first. In CE the chiral elution order can, in general, be easily shifted, which is a definitive advantage of this technique over HPLC.

In our opinion, LIF detection is ideally suited for determination of trace enantiomeric impurities. First, LIF greatly decreases the LOD, and there is no immediate need to inject a large sample to be able to detect low concentrations. Thereby, electrophoretic band broadening can be minimized. Second, the linear range is extended to > four orders of magnitude (depending on the analyte), and this makes quantitative determination from one injection possible down to the 0.005% level. Finally, it should be

mentioned that only relatively few analytes have natural fluorescence and that a derivatization can be applied in many, but not all, cases.

A literature survey was performed to investigate the state of the art. We found that direct determination, without any derivatization, is the most commonly used technique for determination of trace enantiomeric impurities. Derivatization was employed in eight cases and the indirect separation approach was used in two cases. Detection was by UV in thirty-nine cases. LIF detection was used in three cases, electrochemical detection in one case, and MS in three cases. The major part of the published methods was qualitative rather than quantitative. Obviously, the method of choice is presently direct chiral separation followed by UV detection. However, as shown in this paper, there are other approaches and techniques that are powerful for the determination of ee. These include derivatization, indirect chiral separation, and detection by LIF or MS. In our opinion, these methods should receive increased attention.

This work was supported by the Swedish Natural Science Research Council.

Received February 8, 2000

7 References

[1] Ariëns, E. J., in: Ariëns, E. J., Soudijn, W., Timmermans, P. B. M. W. M. (Eds.), *Stereochemistry and Biological Activity of Drugs,* Blackwell, Oxford 1983, pp. 11–32.

[2] Beesley, T. E., Scott, R. P. W., *Chiral Chromatography,* Wiley, Chichester 1998.

[3] Allenmark, S. G., *Chromatographic Enantioseparation,* Wiley, New York 1988.

[4] Chankvetadze, B., *Capillary Electrophoresis in Chiral Analysis,* Wiley, Chichester 1997.

[5] Altria, K. D., *Capillary Electrophoresis Guidebook,* Humana Press, Totowa 1996.

[6] Altria, K. D., *Analysis of Pharmaceuticals by Capillary Electrophoresis,* Chromatographia CE Series, Vol. 2, Vieweg, Braunschweig 1998.

[7] Altria, K. D., *J. Chromatogr.* 1993, *646*, 245–257.

[8] Altria, K. D., *LC.GC Int.* 1993, *6*, 164–172.

[9] Altria, K. D., *J. Chromatogr. A* 1996, *735*, 43–56.

[10] Altria, K. D., Kelly, M. A., Clark, B. J., *Trends Anal. Chem.* 1998, *17*, 204–214.

[11] Altria, K. D., Kelly, M. A., Clark, B. J., *Trends Anal. Chem.* 1998, *17*, 214–226.

[12] Assi, K. H., Abushoffa, A. M., Altria, K. D., Clark, B. J., *J. Chromatogr. A* 1998, *817*, 83–90.

[13] Assi, K. A., Clark, B. J., Altria, K. D., *Electrophoresis* 1999, *20*, 2723–2725.

[14] Vespalec, R., Boček, P., *Electrophoresis* 1999, *20*, 2579–2591.

[15] Altria, K. D., *J. Chromatogr. A* 1999, *856*, 443–463.

[16] Altria, K. D., Rogan, M. M., *Introduction to Quantitative Applications of Capillary Electrophoresis in Pharmaceutical Analysis*, Beckman Instruments, Fullerton, CA 1994.

[17] Noroski, J. E., Mayo, D. J., Moran, M., *J. Pharm. Biomed. Anal.* 1995, *13*, 45–52.

[18] Altria, K. D., Chanter, Y. L., *J. Chromatogr.* 1993, *652*, 459–463.

[19] Williams, R. C., Edwards, J. F., Ainsworth, C. R., *Chromatographia* 1994, *38*, 441–446.

[20] Lurie, I. S., Chan, K. C., Spratley, T. K., Casale, J. F., Issaq, H. J., *J. Chromatogr. B* 1995, *669*, 3–13.

[21] Stålberg, O., Westerlund, D., Rodby, U.-B., Schmidt, S., *Chromatographia* 1995, *41*, 287–294.

[22] Van Schepdael, A., Van den Bergh, I., Roets, E., Hoogmartens, J., *J. Chromatogr. A* 1996, *730*, 305–311.

[23] Degenhardt, M., Wätzig, H., *J. Chromatogr. A* 1997, *768*, 113–123.

[24] Stinson, S. C., *Chem. Eng. News* 1992, *70*, 46–79.

[25] Müller, P., *Pure Appl. Chem.* 1994, *66*, 1077–1184.

[26] Reist, M., Testa, B., Carrupt, P.-A., Jung, M., Schurig, V., *Chirality* 1995, *7*, 396–400.

[27] Altria, K. D., *Chromatographia* 1993, *35*, 177–182.

[28] Wätzig, H., Degenhardt, M., Kunkel, A., *Electrophoresis* 1998, *19*, 2695–2752.

[29] Kunkel, A., Degenhardt, M., Schirm, B., Wätzig, H., *J. Chromatogr. A* 1997, *768*, 17–27.

[30] Altria, K. D., Fabre, H., *Chromatographia* 1995, *40*, 313–320.

[31] Altria, K. D., Bestford, J., *J. Capil. Electrophor.* 1996, *3*, 13–23.

[32] Hempel, G., Blaschke, G., *J. Chromatogr. B* 1996, *675*, 139–146.

[33] Fanali, S., Desiderio, C., *J. High Resolut. Chromatogr.* 1996, *19*, 322–326.

[34] Liu, L., Nussbaum, M. A., *J. Pharm. Biomed. Anal.* 1995, *14*, 65–72.

[35] Ward, T. J., Dann III, C., Blaylock, A., *J. Chromatogr. A* 1995, *715*, 337–344.

[36] Rickard, E. C., Bopp, R. J., *J. Chromatogr. A* 1994, *680*, 609–621.

[37] Altria, K. D., Walsh, A. R., Smith, N. W., *J. Chromatogr.* 1993, *645*, 193–196.

[38] Nielen, M. W. F., *Anal. Chem.* 1993, *65*, 885–893.

[39] Kang, J.-W., Zhang, X.-M., Zhang, S.-M., Ou, Q.-Y., *Chromatographia* 1999, *50*, 317–320.

[40] Sänger-van de Griend, C. E., Gröningsson, K., *J. Pharm. Biomed. Anal.* 1996, *14*, 295–304.

[41] Altria, K. D., Harkin, P., Hindson, M. G., *J. Chromatogr. B* 1996, *686*, 103–110.

[42] Sänger-van de Griend, C. E., Gröningsson, K., Westerlund, D., *Chromatographia* 1996, *42*, 263–268.

[43] Castelnovo, P., Albanesi, C., *J. Chromatogr. A* 1995, *715*, 143–149.

[44] Fillet, M., Bechet, I., Chiap, P., Hubert, P., Crommen, J., *J. Chromatogr. A* 1995, *717*, 203–209.

[45] Vincent, J. B., Vigh, G., *J. Chromatogr. A* 1998, *817*, 105–111.

[46] Zhou, L., Trubig, J., Dovletoglou, A., Locke, D. C., *J. Chromatogr. A* 1997, *773*, 311–320.

[47] Nishi, H., Nakamura, K., Nakai, H., Sato, T., *J. Chromatogr. A* 1997, *757*, 225–235.

[48] Swartz, M., Mazzeo, J. R., Grover, E. R., Brown, P. R., *J. Chromatogr. A* 1996, *735*, 303–310.

[49] Nishi, H., Izumoto, S., Nakamura, K., Nakai, H., Sato, T., *Chromatographia* 1996, *42*, 617–630.

[50] Boonkerd, S., Detaevernier, M. R., Vander-Heyden, Y., Vindelvogel, J., Michotte, Y., *J. Chromatogr. A* 1996, *736*, 281–289.

[51] D'Hulst, A., Verbeke, N., *Electrophoresis* 1994, *15*, 854–863.

[52] Rizzi, A., Schuh, R., Brückner, A., Cvitkovich, B., Kremser, L., Jordis, U., Fröhlich, J., Hüenburg, B., Czollner, L., *J. Chromatogr. B* 1999, *730*, 167–175.

[53] Fang, X., Gong, F., Fang, Y., *Anal. Chem.* 1998, *70*, 4030–4035.

[54] Jäverfalk, E. M., Amini, A., Westerlund, D., Andrén, P. E., *J. Mass Spectrom.* 1998, *33*, 183–186.

[55] Amini, A., Wiersma, B., Westerlund, D., Paulsen-Sörman, U., *Eur. J. Pharm. Sci.* 1999, *9*, 17–24.

[56] Sheppard, R. L., Tong, X., Cai, J., Henion, J. D., *Anal. Chem.* 1995, *67*, 2054–2058.

[57] Gausepohl, C., Blaschke, G., *J. Chromatogr. B* 1998, *713*, 443–446.

[58] De Boer, T., Ensing, K., *J. Pharm. Biomed. Anal.* 1998, *17*, 1047–1056.

[59] Nishi, H., Nakamura, K., Nakai, H., Sato, T., *Anal. Chem.* 1995, *67*, 2334–2341.

[60] Guttman, A., Cooke, N., *J. Chromatogr. A* 1994, *685*, 155–159.

[61] Kuhn, R., Stoecklin, F., Erni, F., *Chromatographia* 1992, *33*, 32–36.

[62] Tanaka, Y., Kishimoto, Y., Terabe, S., *J. Chromatogr. A* 1998, *802*, 83–88.

[63] Sänger-van de Griend, C. E., Wahlström, H., Gröningsson, K., Widahl-Näsman, M., *J. Pharm. Biomed. Anal.* 1997, *15*, 1051–1061.

[64] Torrens, A., Castrillo, J. A., Frigola, J., Salgado, L., Redondo, J., *Chirality* 1999, *11*, 63–69.

[65] Guttman, A., Brunet, S., Cooke, N., *LC.GC Int.* 1996, *9*, 88–100.

[66] Wan, H., Schmidt, S., Carlsson, L., Blomberg, L. G., *Electrophoresis* 1999, *20*, 2705–2714.

[67] Ruyters, H., van der Wal, S., *J. Liq. Chromatogr.* 1994, *17*, 1883–1897.

[68] Werner, A., Nassauer, T., Kiechle, P., Erni, F., *J. Chromatogr. A* 1994, *666*, 375–379.

[69] Engstaröm, A., Wan, H., Andersson, P. E., Josefsson, B., *J. Chromatogr. A* 1995, *715*, 151–158.

[70] De Witt, P., Deias, R., Muck, S., Galletti, B., Meloni, D., Celletti, P., Marzo, M., *J. Chromatogr. B* 1994, *657*, 67–73.

[71] Thorsén, G., Engström, A., Josefsson, B., *J. Chromatogr. A* 1997, *786*, 347–354.

[72] Wan, H., Blomberg, L. G., *Electrophoresis* 1997, *18*, 943–949.

[73] Vespalec, R., Böcek, P., *Electrophoresis* 1994, *15*, 755–762.

[74] Nishi, H., Fukyama, T., Matuso, M., *J. Microcol. Sep.* 1990, *2*, 234–240.

[75] Kang, L., Buck, R. H., *Amino Acids* 1992, *2*, 103–109.

[76] Tran, A. D., Blanc, T., Leopold, E. J., *J. Chromatogr.* 1990, *516*, 241–249.

[77] Wan, H., Blomberg, L. G., *J. Chromatogr. A* 1997, *792*, 393–400.

[78] Meyer, V. R., *Chromatographia* 1995, *40*, 15–22.

[79] Meyer, V. R., *J. Chromatogr. Sci.* 1995, *33*, 26–33.

[80] Chankvetadze, B., *Capillary Electrophoresis in Chiral Analysis*, Wiley, Chichester 1997, Chapter 12, pp. 428–457.

[81] Altria, K. D., *Chromatographia* 1993, *35*, 493–496.

[82] Inman, E. L., Tenbarge, H. J., *J. Chromatogr. Sci.* 1988, *26*, 89–94.

[83] Houben, R. J. H., Gielen, H., van der Wal, S., *J. Chromatogr.* 1993, *634*, 317–322.

[84] Altria, K. D., *J. Chromatogr.* 1993, *634*, 323–328.

[85] Albin, M., Weinberger, R., Sapp, E., Moring, S., *Anal. Chem.* 1991, *63*, 417–422.

[86] Swart, R., Elgersma, J. W., Kraak, J. C., Poppe, H., *J. Microcol. Sep.* 1997, *9*, 591–600.

[87] Beckman Coulter, Instrument Specifications, www. Beckmancoulter.com January 2000.

[88] Ding, J., Vouros, P., *Anal. Chem.* 1999, *71*, 378A–385A.

[89] Amini, A., Paulsen-Sörman, U., Westerlund, D., *Chromatographia* 1999, *50*, 497–506.

Author Index

Keywords Index